T0343168

Basics in Human Evolution

Basics in Human Evolution

Edited by

Michael P. Muehlenbein
Department of Anthropology, University of Texas at San Antonio, USA

AMSTERDAM • BOSTON • HEIDELBERG • LONDON • NEW YORK • OXFORD • PARIS
SAN DIEGO • SAN FRANCISCO • SINGAPORE • SYDNEY • TOKYO

Academic Press is an imprint of Elsevier

Academic Press is an imprint of Elsevier
125 London Wall, London EC2Y 5AS, UK
525 B Street, Suite 1800, San Diego, CA 92101-4495, USA
225 Wyman Street, Waltham, MA 02451, USA
The Boulevard, Langford Lane, Kidlington, Oxford OX5 1GB, UK

British Library Cataloguing-in-Publication Data
A catalogue record for this book is available from the British Library

Library of Congress Cataloging-in-Publication Data
A catalog record for this book is available from the Library of Congress

ISBN: 978-0-12-802652-6

For information on all Academic Press publications
visit our website at http://store.elsevier.com/

Acquisition Editor: Kristi Gomez
Editorial Project Manager: Pat Gonzalez
Production Project Managers: Karen East and Kirsty Halterman
Designer: Alan Studholme

Typeset by TNQ Books and Journals
www.tnq.co.in

Dedication

This book is dedicated to our students and mentors.

Contents

Contributors xv
Preface xvii

Part I
Positioning Human Evolution

1. Basic Evolutionary Theory

Douglas J. Futuyma

Synopsis 3
Introduction 4
The Origin of Genetic Variation 4
Variation within Populations 4
Genetic Drift 5
The Neutral Theory of Molecular Evolution 6
Natural Selection 6
Modes of Selection 7
Components of Fitness 8
Modeling Adaptation 9
Levels of Selection 9
Speciation 10
From Microevolution to Macroevolution 12
Evolutionary Theory Today 13
References 13

2. Evolution, Creationism, and Intelligent Design

Andrew J. Petto

Synopsis 15
Introduction 16
A Scientific Creationism 16
Pillar 1: Theory in Crisis 17
Pillar 2: Atheism 18
Pillar 3: Fairness 19
Creation Science Winds Down 19
Intelligent Design 20
Pillar 1 20
Pillar 2 23
Pillar 3 24
Intelligent Design Sputters Along 24
Conclusions 25
References 26

Part II
Primates

3. Primate Evolution

Robert D. Martin

Synopsis 31
Introduction 32
Extant Groups of Primates 32
Higher-Level Relationships 33
Primates on an Ascending Scale? 34
Defining Features of Primates 34
Fossil Primates 37
Overall Evolutionary Relationships 38
References 40

4. Comparative Anatomy of Primates

Rui Diogo, Magdalena N. Muchlinski and Adam Hartstone-Rose

Synopsis 43
Osteology 44
Introduction 44
The Strepsirrhine Skeleton 44
The Tarsiiform Skeleton 45
The Anthropoid Skeleton 45
The Platyrrhine Skeleton 45
The Catarrhine Skeleton 45
The Hominoid Skeleton 45
Myology 46
Introduction 46
General Notes on the Evolution of the Primate Muscles 46
Head and Neck Muscles of Primates 48
Pectoral and Forelimb Muscles of Primates 48
Variation and Evolutionary History of Primate Muscles, and Comments on the "Scala Naturae" 49
External Features and Internal Organs 49
Introduction 49
The Nervous and Sensory Systems 50
The Digestive System 51
The Reproductive System 52
References 53

5. Primate Behavior

David P. Watts

Synopsis 57
Social Organization, Predation, and Group Living 58
Social Organization 58
Sex Differences in Optimal Social Dispersion, Diets, and Feeding Competition 58
Social Organization and Social Relationships 61
Dominance, Leverage, and Power 61
The "Socioecological Model" 62
Sexual Conflict and "Intersexual Mutualism" 64
Cooperation 65
Kin Selection and Primate Social Evolution 65
Reciprocity and Mutualism 65
Sociality and Fitness 66
References 67

6. Primate Models for Human Evolution

Robert W. Sussman and Donna Hart

Synopsis 73
Models of Human Evolution 74
Dentition and Diet 75
Locomotion 76
Habitat of Our Earliest Ancestors 77
The Macaque Model 78
Fossils and Living Primates 79
Man the Hunted 80
References 80

Part III
Hominins

7. Early Hominin Ecology

Jeanne Sept

Synopsis 85
The Early Hominin Record 86
Macro Paleoenvironmental Context 86
Paleogeography 86
Paleoclimate 87
Reconstructing Terrestrial Habitats 89
Geological Evidence 89
Site Taphonomy 89
Paleolandscapes 90
Fossils 91
Inferring Hominin Habitats and Adaptations 94
The Habitats of the Earliest Bipeds 94
Late Pliocene Adaptive Radiations 96
Later "Robust" Hominins: Chewing On a Problem 96
Early Homo: Ecological Entrepreneurs 96
Summary: The Limits and Potential of Our Paleoecological Knowledge 98
References 98

8. Bipedalism

Kevin D. Hunt

Synopsis 103
How Do Humans Walk? 104
Anatomical Features Associated with Bipedalism 105
When Did Bipedalism Evolve, and What Evolutionary Stages of Bipedalism Did We Pass Through? 107
Why Did Bipedalism Evolve? 108
References 112

9. Early Hominins

Kevin D. Hunt

Synopsis 113
Taxonomy 114
Purported Early Hominin Species 114
Context: Chimpanzee Morphology 116
Context: Middle Miocene Apes 118
Context: Late Miocene Apes 120
The Morphology of Purported Early Hominins 120
Early Possible Hominins' Adaptations 124
Early Possible Hominins' Place in Nature 125
References 127

10. Australopithecines

Carol V. Ward

Synopsis 129
Australopithecus 130
"Robust" Australopithecines 131
Australopithecus afarensis 133
Australopithecus bahrelghazali 135
Australopithecus anamensis 135
Australopithecus garhi 136
Kenyanthropus (Australopithecus) platyops 136
Australopithecus sediba 137
Australopithecine Adaptations 137
References 138

11. Early Pleistocene *Homo*

Scott W. Simpson

Synopsis 143
Defining the Genus *Homo* 144
History of Discovery of Early *Homo* Fossils 144
Taxonomic Diversity in the Genus *Homo* 145
Homo habilis 146
Homo rudolfensis 148
Homo erectus 148
Homo ergaster 150
Taxonomic and Phylogenetic Issues: *Homo habilis* and *Homo rudolfensis* 151
Phyletic Origins and the Dating of Early *Homo* 153
Context of Origins and Existence 154
Musculoskeletal Adaptations in Early *Homo* 155
Association with Stone Tools 155
Future Directions 156
References 157

12. Archaic *Homo*

James C.M. Ahern

Synopsis 163
Geographical Distribution 164
Chronological Variation 164
Regional Variation 168
Europe 169
Africa 169
West Asia 171
South Asia 171
East Asia 171
Australasia 171
Growth, Development, and Energetics 172
Language, Behavior, and Culture 172
Evolutionary Relationships and Taxonomy 173
Summary 173
References 174

13. Anatomically Modern *Homo sapiens*

Brigitte M. Holt

Synopsis 177
History of Discovery 178
Earliest Fossil Evidence of Modern Humans 178
Climatic Conditions for the Spread of Modern Humans 180
Origins of Modern Humans 181
Relationships with Archaic *Homo* 181
Paleobiology of Anatomically Modern *Homo sapiens* 181
Cranial Morphology 181
Robusticity 182
Stature 183
Hand Morphology 183
Life History 183
Childbirth 184
Health and Diseases 184
Technology 185
Diet 186
Mobility 186
Symbolic Behavior 186
Funerary Behavior 187
Language 189
References 189

14. Evolution of Tool Use

Nicholas Toth and Kathy Schick

Synopsis 193
Evolving Hominin Forms and Technologies: A Review 194
Tools in the Animal World 196
Chimpanzee Tool Use as a Model for Early Hominins 196
The Earliest Known Stone Tools 197
The Oldowan (Earlier Lower Palaeolithic/Early Stone Age) 197
The Acheulean and Contemporaneous Industries (Later Lower Palaeolithic) 201
The Middle Palaeolithic/Middle Stone Age 202
The Upper Palaeolithic/Late Stone Age 204
Later Developments 205
References 206

Part IV
Genetics and Biology

15. Contemporary Human Genetic Variation

John H. Relethford

Synopsis 211
Introduction 212
A Brief History of Human Evolution 212
Our Species Has a Relatively Low Level of Genetic Variation 213
There Is Higher Genetic Variation in Sub-Saharan African Populations than in Other Geographic Regions 213
Genetic Diversity Declines with Distance Out of Africa 213
There Are Low Levels of Genetic Differentiation between Geographic Regions 213
Genetic Distance Is Correlated with Geography 214

The Human Genome Shows Evidence of Admixture from Neandertals ... and Others 215
Natural Selection Has Further Affected the Pattern of Contemporary Genetic Variation 216
Summary 217
References 217

16. Human Population Movements: A Genetic Perspective

Oscar Lao and Manfred Kayser

Synopsis 219
Introduction 220
Out of Africa Migration of Archaic Humans 220
Out of Africa Migration of Anatomically Modern Humans 220
Genetic Admixture between Archaic and Modern Humans 222
Technological Advances and Migration: The Case of the European Farmers 225
Historic Migrations into a Previously Occupied Territory: The Case of the Romani 227
Most Recent Migrations into a Newfound Land: The Case of the Polynesians 227
Conclusions and Future Prospects 230
References 230

17. Brain Evolution

Ralph L. Holloway

Synopsis 235
Introduction 236
Lines of Evidence 236
Direct Evidence 236
Indirect Evidence 237
Characteristics of the Human Brain 237
Brain Size, Absolute and Relative 237
Encephalization 238
Brain Organization and Reorganization 241
Human Brain Asymmetry 246
Synthesis: Putting Together Size, Organization, and Asymmetry during Human Evolution 247
And to the Future? 249
References 249

18. Physiological Adaptations to Environmental Stressors

William R. Leonard

Synopsis 251
Introduction 252
Principles of Thermoregulation 252
Climate and Body Morphology 253
Physiological Adaptations to Cold Stress 256
Acclimation to Cold Stress 256
Responses among Indigenous Arctic Populations 258
Physiological Adaptations to Heat Stress 261
Acclimation to Heat Stress 261
Responses among Indigenous Populations of the Tropics 262
Physiological Adaptations to High Altitude Hypoxia 263
Acclimation to Hypoxic Stress 264
Responses among Indigenous High Altitude Populations 265
Summary 269
References 269

19. Evolution of Skin Color

Peter M. Elias and Mary L. Williams

Synopsis 273
When and Why Epidermal Pigmentation Evolved 274
Protection Against Skin Cancer and Eccrine Gland Destruction 274
Protection Against Folate Degradation and Vitamin D Toxicity 274
Barrier Requirements Likely Stimulated the Development of Epidermal Pigmentation 276
Sweating was Critical for Heat Dissipation, but Placed Additional Stress on the Barrier 276
While Erythemogenic UV-B is Toxic, Pigmentation Shifts the Dose–Response Curve toward the Beneficial Effects of Suberythemogenic UV-B 277
Basis for Pigment Dilution in Modern Humans 277
Cultural Theories for Pigment Dilution 277
Most Pigment Dilution Did Not Occur from a Greater Need for Vitamin D 278
Non-pigment-based Mechanisms That Likely Enhance Vitamin D Bioavailability 279
Conservation of Metabolic Energy 279
Summary 280
References 280

20. Human Growth and Development

Barry Bogin

Synopsis 285
Growth and Evolution 286
Human versus Chimpanzee Growth 286
Adolescence 289

Evolution of Human Adolescence 289
Girls and Boys: Separate Paths through Adolescence 290
Adolescent Contributions to the Reproductive Success of Adults 292
Risks of Childhood and Adolescence 292
Conclusions 292
References 292

21. Human Reproductive Ecology

Claudia R. Valeggia and Alejandra Núñez-de la Mora

Synopsis 295
Introduction 296
Human Life History 296
Human Reproductive Physiology: The Basics 297
Female Reproductive Ecology 299
  Menarche 299
  Variation in Adult Female Reproductive Function 300
  Pregnancy 300
  Lactation 301
  Menopause 301
  Male Reproductive Ecology 302
Testicular Function during Early Development 302
  Puberty 302
  Variation in Adult Male Reproductive Function 302
  Male Reproductive Senescence 303
Challenges and Future Directions of the Field of Human Reproductive Ecology 304
References 304

22. Human Senescence

Lynnette L. Sievert

Synopsis 309
Introduction 310
  Definition of Senescence 311
The Evolution of Senescence 313
  Disposable Soma Theory 313
  Mutation Accumulation and Late-Acting Genes 314
  Antagonistic Pleiotropy 315
  In Summary 315
Reproductive Senescence 316
  Male Reproductive Senescence 316
  Female Reproductive Senescence 316
Future Directions 319
References 319

Part V
Lifeways

23. Hunter-Gatherers

Michael A. Little and Mark A. Blumler

Synopsis 323
Early Hunting and Gathering Subsistence 324
Hunter-Gatherers during the Historical Era 325
Migration, Biogeography, and Contemporary Populations 326
Hunter-Gatherers and Evolution 327
  Demographic Characteristics 327
  Diet and Nutrition 328
  Child Growth, Body Size, and Life History 329
  Disease and Morbidity 331
Hunter-Gatherers in Evolutionary Perspective: Summary 332
References 332

24. Pastoralism

Michael A. Little

Synopsis 337
Pastoralism as Subsistence 338
Prehistory of Pastoralism 338
Biogeography of Pastoralism 339
Food, Diet, and Cuisine 340
Milk and the Evolutionary Basis for Lactose Tolerance 342
Health, Disease, and Pastoralism 343
Coevolution of Livestock and Their Human Hosts 345
References 345

25. Agriculturalism

Mark A. Blumler

Synopsis 349
Introduction 350
Agricultural Origins 351
Competing Hypotheses 351
The Spread of Agriculture 354
The Columbian Exchange 356
"Scientific Breeding," and the Industrial Revolution 357
Crop Evolution 357
  Crop Choice 357
  Crop Mimics 358
  Unconscious (Automatic) versus Conscious Evolution 358

Loss of the Competitive, Protective, and Dispersal Functions 358
Speciation 359
Coevolution 359
Palatability 360
Evaluating the Paleo Diet Hypothesis 360
Conclusions 361
References 362

Part VI Health

26. Evolutionary and Developmental Origins of Chronic Disease

Felicia M. Low, Peter D. Gluckman and Mark A. Hanson

Synopsis 369
Evolutionary Perspective on Human Disease 370
Developmental Origins of Health and Disease 370
Developmental Plasticity and Predictive Adaptive Responses 371
Empirical Support for the PAR Hypothesis 372
Epigenetics as an Underpinning Mechanism 372
Nonhuman Animal Evidence 373
Human Evidence 373
Evolutionary and Developmental Mismatch: The Case of Obesity and Related Chronic Diseases 374
Evolutionary Mismatch 374
Developmental Mismatch 375
Maternal Constraint 375
Excessive Nutrition and Gestational Diabetes Mellitus 376
Transgenerational Inheritance 377
Concluding Remarks 377
References 377

27. Modernization and Disease

William W. Dressler

Synopsis 383
Introduction 384
Modernization and Disease: Basic Findings 384
Critiques of Studies of Modernization 386
New Approaches to the Study of Modernization and Disease 387
Modernization and Stress 387
Cultural Consonance 388
Political Economy and Health 389
Discussion 390
References 391

28. Modern Human Diet

Andrea S. Wiley

Synopsis 393
Introduction 394
The Role of Diet in Hominin Evolution 394
Bipedalism 394
Brain and Body Size 394
Agricultural Transition: Dietary and Evolutionary Consequences 396
Starch Digestion 396
Variation in Adult Milk Digestion 397
Industrialization of the Diet and Consequences for Human Biology 398
Paleolithic Prescriptions 399
Conclusion 402
References 402

29. Diversity and Origins of Human Infectious Diseases

Serge Morand

Synopsis 405
The Diversity of Infectious Diseases in Space 406
Origins of Infectious Diseases in Nonhuman Primates 406
The First Epidemiological Transition: Out of Africa 406
The Second Epidemiological Transition: Animal Domestication 409
The Third Epidemiological Transition: First Globalizations 410
The Fourth Epidemiological Transition: Recent Emergences and the Homogenization of Infectious Diseases 411
Concluding Remarks 412
References 412

30. Coevolution of Humans and Pathogens

Lisa Sattenspiel

Synopsis 415
The Importance of Coevolution between Hosts and Pathogens 416
Ensuring Pathogen Persistence Over Time: Modes of Transmission 416
Mechanisms of Host–Pathogen Coevolution 417
Examples of Pathogens That Have Coevolved with Humans 418
ABO Blood Groups and Infectious Diseases 418
Genetic Adaptations to Malaria 419

Interacting Species and the Evolution of Influenza Viruses 421
Why Does the CCR5-Δ32 Allele Reach Such High Frequencies in European Populations? 422
Conclusions 423
References 424

31. Paleopathology

Della C. Cook

Synopsis 427
Fossil Humans and Disease 428
Diseases in the Past 429
Bioarchaeology 432
Osteobiography 433
Interdisciplinarity and Difficulties in Communication 434
Questions for the Future 434
References 434

Part VII
Behavior and Culture

32. Evolutionary Biology of Human Stress

Carol M. Worthman

Synopsis 441
Evolving Concepts of Stress and Adaptation 442
Life History 442
Endocrine Architecture of Life History 443
How Stress Works 443
Stressors 443
Detection–Assessment 444
Response 445
The Brain Talks to the Body 445
And the Body Talks Back 447
Stress Moderators and Buffers 447
Developmental and Intergenerational Processes 448
Developmental Processes 448
Transgenerational Effects 449
Gender Differences 450
Stress and Health 450
Stress and Life History 450
References 451

33. Aggression, Affiliation, and Parenting

Mark V. Flinn

Synopsis 455
Brain, Childhood, and Parenting 456
The Human Family 457
Fathers 458
Grandparents 459
Extended Kinship and Control of Mating Relationships 459
Neurological and Physiological Mechanisms 460
Hormonal Basis for Attachment and Family Love 460
Conclusions 462
References 463

34. Human Mating Systems

Steven W. Gangestad and Nicholas M. Grebe

Synopsis 467
Mating Systems: Basic Concepts and Understandings 468
A Basic Classification of Mating Systems 468
A Traditional View 468
Variants with Stable and Unstable Unions 468
Mixed Systems 468
Sexual Selection 468
What Selection Pressures Give Rise to Mating Systems? 469
The Polyandry "Revolution" within Behavioral Biology 469
Extra-Pair Mating 469
New Models of Sexual Selection 470
Human Mating Systems 470
Mating in Hominoids and the Ancestral State 470
Chimpanzee Mating 470
Bonobo Mating 471
Human Marital Systems 471
The Human Case: Do Males Possess Adaptations for Care? 471
Hunting-As-Paternal-Effort Views 471
Hunting-As-Mating-Effort Views 472
A Blended View 472
Extra-Pair Paternity: Sexual or Social Monogamy? 472
Adaptations for Mating 473
Male Adaptations for Pair-Bonding and Parenting 473
Mutual Mate Choice 473
Discriminative Care 473
Female Estrus and Extended Sexuality 474
Conclusions 475
References 475

35. Evolution of Cognition

Jennifer Vonk and Chinmay Aradhye

Synopsis 479
Evolution of Cognition 480
Consciousness 482

Language 483
Abstraction 484
Theory of Mind 485
Causal Reasoning 485
An Economic Mind 486
Cooperation 487
Conclusions 487
References 488

36. Evolution of Language

Philip Lieberman

Synopsis 493
Evidence from Comparative Studies 494
Human Speech 494
The Supralaryngeal Vocal Tract and Speech Encoding 495
The Unique Human Tongue and Supralaryngeal Vocal Tract 496
The Speech Capacities of Other Species 497
Neural Circuits and the Evolution of Human Language 497
Broca's Area 497
Circuits Linking Cortex and the Basal Ganglia 497
Neuroimaging 498
The Implausibility of Universal Grammar 498
Fully Human Linguistic and Cognitive Capability 499
Brain Size 499
Transcriptional Genes 500
When Did Fully Human Language Appear? 500
References 501

37. Evolution of Moral Systems

Douglas Allchin

Synopsis 505
Introduction 506
Moral Behavior 506
Moral Systems 507
Network Reciprocity 507
Strong Reciprocity 508
Indirect Reciprocity 509
Morality and Social Organization 509
Moral Sentiments 510
Morality and Neurophysiology 510
Morality and Development 511
The Evolution of Moral Psychology 511
Morality and Levels of Selection 511
References 512

38. Race and Ethnicity

Catherine Bliss

Synopsis 515
Introduction 516
A History of Racial Science 516
Postwar Debates 517
The Social Construction of Race 518
Race and Health 519
New Genetic Sciences 519
Postgenomic Developments 520
Commercial Ancestry Estimation 521
References 522

39. Evolution of Culture

Robert S. Walker

Synopsis 525
Human Expansions 526
Comparative Cultural Data 526
Phylogenetic Comparative Methods 526
The Problem of Borrowing 528
Cultural Transition Rates 528
Future Work 530
References 530

Glossary 533
Index 549
Color Plates

Contributors

James C.M. Ahern Department of Anthropology, University of Wyoming, Laramie, WY, USA

Douglas Allchin Minnesota Center for the Philosophy of Science University of Minnesota, St. Paul, MN, USA

Chinmay Aradhye Department of Psychology, Oakland University, Rochester, MI, USA

Catherine Bliss Department of Social and Behavioral Sciences, University of California, San Francisco, CA, USA

Mark A. Blumler Department of Geography, State University of New York, Binghamton, NY, USA

Barry Bogin Centre for Global Health & Human Development, School of Sport, Exercise & Health Sciences, Loughborough University, Loughborough, UK

Della C. Cook Department of Anthropology, Indiana University, Bloomington, IN, USA

Rui Diogo Department of Anatomy, Howard University College of Medicine, Washington, DC, USA

William W. Dressler Department of Anthropology, University of Alabama, Tuscaloosa, AL, USA

Peter M. Elias Dermatology Service, Department of Veterans Affairs Medical Center and Department of Dermatology, University of California, San Francisco, CA, USA

Mark V. Flinn Department of Anthropology, University of Missouri, Columbia, MO, USA

Douglas J. Futuyma Department of Ecology and Evolution, Stony Brook University, Stony Brook, NY, USA

Steven W. Gangestad Department of Psychology, University of New Mexico, Albuquerque, NM, USA

Peter D. Gluckman Liggins Institute, The University of Auckland, Auckland, New Zealand

Nicholas M. Grebe Department of Psychology, University of New Mexico, Albuquerque, NM, USA

Mark A. Hanson Institute of Developmental Sciences, University of Southampton, Southampton General Hospital, Southampton, UK

Donna Hart Department of Anthropology, University of Missouri–St. Louis, MO, USA

Adam Hartstone-Rose Department of Cell Biology and Anatomy, University of South Carolina School of Medicine, Columbia, SC, USA

Ralph L. Holloway Department of Anthropology, Columbia University, New York, NY, USA

Brigitte M. Holt Department of Anthropology, University of Massachusetts, Amherst, MA, USA

Kevin D. Hunt Human Evolution and Animal Behavior, Indiana University, Bloomington, IN, USA

Manfred Kayser Department of Forensic Molecular Biology, Erasmus MC University Medical Center Rotterdam, Rotterdam, The Netherlands

Oscar Lao Department of Forensic Molecular Biology, Erasmus MC University Medical Center Rotterdam, Rotterdam, The Netherlands

William R. Leonard Department of Anthropology, Northwestern University, IL, USA

Philip Lieberman Department of Cognitive, Linguistic, and Psychological Sciences, Brown University, Providence, RI, USA

Michael A. Little Department of Anthropology, Binghamton University, State University of New York, Binghamton, NY, USA

Felicia M. Low Liggins Institute, The University of Auckland, Auckland, New Zealand

Robert D. Martin Department of Anthropology, The Field Museum, Chicago, IL, USA

Serge Morand CNRS ISEM–CIRAD AGIRs, Centre d'Infectiologie Christophe Mérieux du Laos, Vientiane, Lao PDR, France

Magdalena N. Muchlinski Department of Anatomy and Neurobiology, University of Kentucky, College of Medicine, Lexington, KY, USA

Alejandra Núñez-de la Mora Instituto de Investigaciones Psicológicas, Universidad Veracruzana, Xalapa, Veracruz, Mexico

Andrew J. Petto Department of Biological Sciences, University of Wisconsin–Milwaukee, Milwaukee, WI, USA

John H. Relethford Department of Anthropology, State University of New York College at Oneonta, Oneonta, NY, USA

Lisa Sattenspiel Department of Anthropology, University of Missouri-Columbia, Columbia, MO, USA

Kathy Schick The Stone Age Institute, Gosport, IN, USA; Indiana University, Bloomington, IN, USA

Jeanne Sept Department of Anthropology, Indiana University, Bloomington, IN, USA

Lynnette L. Sievert Department of Anthropology, Machmer Hall, UMass Amherst, Amherst, MA, USA

Scott W. Simpson Department of Anatomy, Case Western Reserve University School of Medicine, Cleveland, OH, USA

Robert W. Sussman Department of Anthropology, Washington University, St. Louis, MO, USA

Nicholas Toth The Stone Age Institute, Gosport, IN, USA; Indiana University, Bloomington, IN, USA

Claudia R. Valeggia Department of Anthropology, Yale University, New Haven, CT, USA

Jennifer Vonk Department of Psychology, Oakland University, Rochester, MI, USA

Robert S. Walker Department of Anthropology, University of Missouri, Columbia, MO, USA

Carol V. Ward Department of Pathology and Anatomical Sciences, University of Missouri, Columbia, MO, USA

David P. Watts Department of Anthropology, Yale University, New Haven, CT, USA

Andrea S. Wiley Department of Anthropology, Indiana University, Bloomington, IN, USA

Mary L. Williams Departments of Dermatology and Pediatrics, University of California, San Francisco, CA, USA

Carol M. Worthman Department of Anthropology, Emory University, Atlanta, GA, USA

Preface

I don't really remember anyone first telling me not to edit a book. However, shortly after agreeing to edit my first volume (*Human Evolutionary Biology*, 2010, Cambridge University Press), colleagues from all over began asking me why I agreed to do something like this; a commitment that would take years when I could have spent the time writing a book myself, or a dozen articles. Yet *Human Evolutionary Biology* was produced to fill a gap: a comprehensive reference text written by authorities on subjects varying from theory and methods and phenotypic/genotypic variation to reproduction, growth/development, and evolutionary medicine. There still is no text as broad on the general topic of human evolutionary biology.

I distinctly remember telling myself never to edit another book. That plan clearly didn't work. I was originally invited to edit the "evolution" section for the third edition of Elsevier's *Encyclopedia of Human Biology*, and I recruited authors for about 30 appropriate chapters. A year into the process, we were informed that the encyclopedia would be put on hold. A new plan was formulated to expand our section into a stand-alone text (thank you to Pat Gonzalez, Janice Audet, Graham Nisbet, Kristi Gomez, Kristy Halterman, Karen East, and several anonymous reviewers for making this happen; and thank you to Maentis for providing thought-provoking illustrations). I subsequently tried to fill in missing topics with an additional 10 chapters, and the result is what you have before you.

Now, rather than repeat the obvious information that the table of contents provides, allow me to act briefly as my own reviewer:

> *Muehlenbein has brought together an impressive list of 48 authors to produce 39 chapters organized into seven different sections (positioning human evolution, primates, hominins, genetics and biology, lifeways, health, and behavior and culture). The purpose of this book is clearly to provide a general reference to a wide audience of students and professionals interested in anthropology, modern human biology, human evolution, and health and behavior. Given the long length of the text and the careful level of detail provided within the chapters, this is not the type of book designed for use in the classroom, except as a supplemental reference. There are already great textbooks on human evolution and biological anthropology, and the current book is not designed to replace these.*
>
> *What the current book offers (that introductory textbooks do not) is much more detail on specific topics, written by a diversity of experts on those specific topics (rather than a single author trying a hand at everything). The people who authored these chapters are known authorities; there are few better people in the world to learn this information from. Furthermore, introductory texts do not typically offer dedicated chapters on, for example, primate comparative anatomy, primate models of human evolution, the evolution of tool use, human population movements, brain evolution, human reproductive ecology, human senescence, developmental origins of disease, modernization and disease, paleopathology, evolutionary biology of human stress, human mating systems, evolution of moral systems, and cultural evolution. The downside is that this volume is inappropriate for use as a textbook, except under special circumstances such as a graduate seminar. We hope that the title will not be misleading; "basics" is meant to imply the general topics that advanced students should be exposed to and experts should know. It does not imply that learning this information is going to be easy, or that this is an introductory book.*
>
> *No text can be perfect, and this is especially the case for edited volumes. For one, there is variation in chapter length and writing style. As the editor, Muehlenbein worked to recruit experts, review and correct content, make suggestions for including specific information, unite the chapters through cross-referencing, and compile a (fantastic) glossary, while also allowing authors natural flexibility in their products. It is true that some chapters could have used more figures. It is true that some chapters provide relatively more detail than others, and this represents the preferred approach of individual authors. Sometimes, authors disagree with one another, and this is to be expected given the topic of human evolution. It is not a flaw that authors in this book may disagree about the classification of, for example, early hominins, paranthropines, and Homo habilis. Again, this is to be expected.*
>
> *Other editors could have organized this volume differently. For example, the current section on hominins is organized by taxonomy; this did not have to be the case. Additionally, other topics could have been included. Like most edited projects involving so many authors, several authors probably agreed to participate, but then dropped out; fellow editors can empathize. This may be a reason why the current book lacks some chapters on specific subjects, particularly genetics.*

Despite the deficiencies, my fictitious reviewer and I are both pleased with the outcome of this project. We hope that readers will value the depth of knowledge that individual authors bring to the table and the diversity of topics reviewed. We also value the ability to search and access specific chapters, and take advantage of the extensive cross-referencing among chapters, through Elsevier's online ScienceDirect system. Online or offline, no matter what, just keep reading.

… and a mind needs books as a sword needs a whetstone, if it is to keep its edge.

Tyrion Lannister in George R. R. Martin's (1996) A Game of Thrones, p. 124.

Michael P. Muehlenbein
Department of Anthropology
University of Texas at San Antonio

Part I

Positioning Human Evolution

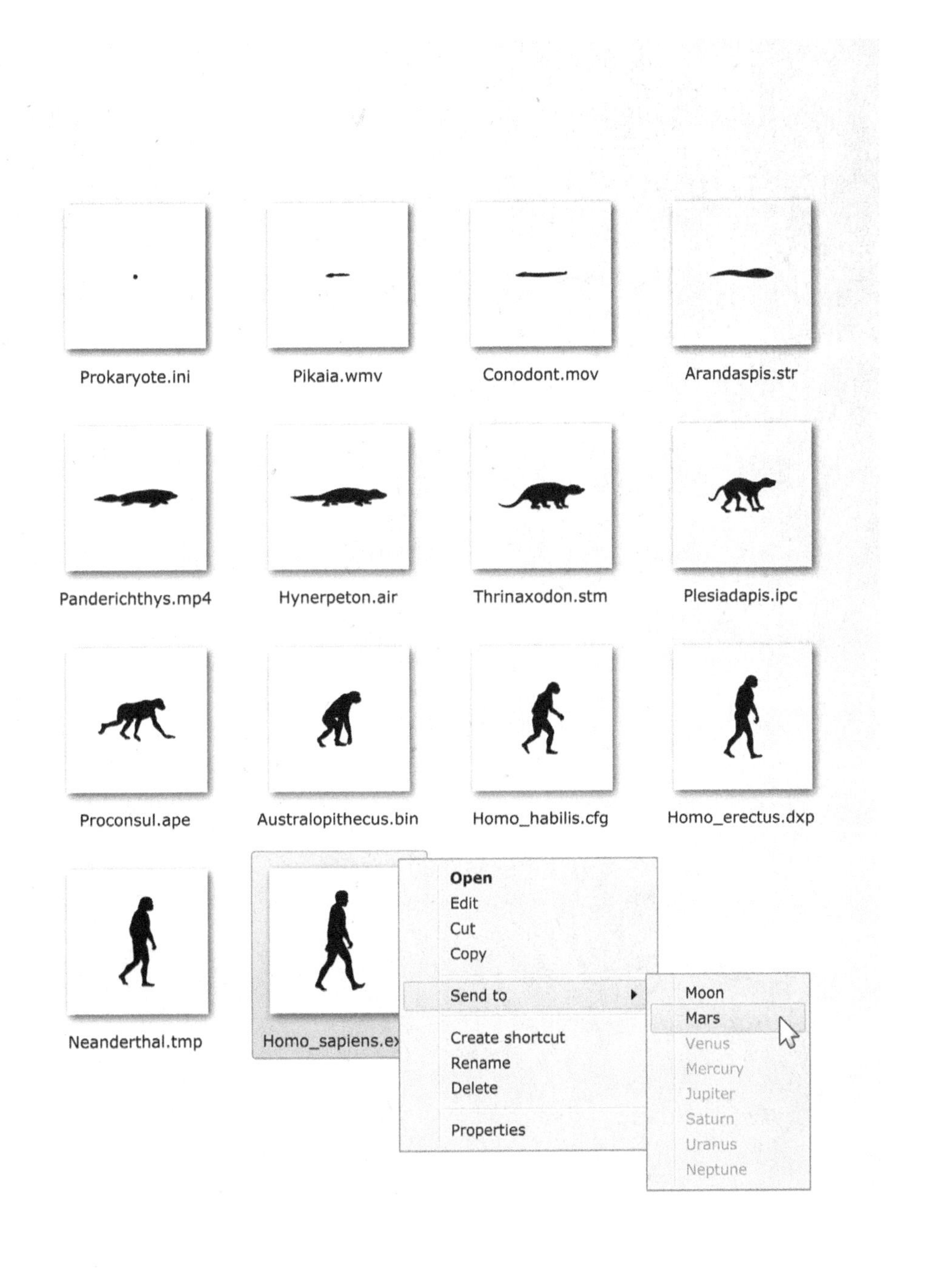

Chapter 1

Basic Evolutionary Theory

Douglas J. Futuyma
Department of Ecology and Evolution, Stony Brook University, Stony Brook, NY, USA

SYNOPSIS

This chapter provides a cursory view of the basic principles of modern evolutionary theory. The major topics treated include the origin and nature of genetic variation, effects of genetic drift and various forms of natural selection on phenotypic traits and the genetic constitution of populations, fitness and its components, models of adaptation, levels of selection, the origin of species by speciation, and some aspects of macroevolution, chiefly phylogenies, evolutionary trends and diversification, gradualism, and the role of development in phenotypic evolution.

INTRODUCTION

The modern theory of evolutionary change has grown out of the "Evolutionary Synthesis," from 1930 to 1950, in which researchers in genetics, zoology, botany, and paleontology united Darwin's theory of evolution by natural selection with Mendelian genetics. They showed that this formulation accurately described the variation and diversity of animals and plants, both living and extinct, and explained why competing theories, such as neo-Lamarckism and simple mutationism, were inadequate or simply false. Reduced to its simplest elements, the modern theory may be summarized as follows. Elementary evolutionary change consists of changes in the genetic constitution of a population of organisms, or in a group of populations of a species. These genetic changes may be reflected as changes in one or more phenotypic characteristics. Genetic change is based on variation that has originated by mutation and/or recombination of DNA sequences. The most elementary evolutionary process is an increase in the frequency of a mutation, or a set of mutations, within a population, and the corresponding decrease in the frequency of previously common alleles. The major causes of such frequency changes are random genetic drift and diverse forms of natural selection. Successive such changes in one or more characteristics cumulate over time, so that potentially indefinite divergence of a lineage from the ancestral state may result. Different populations of a species may remain similar due to gene flow and perhaps uniform selection, but they can diverge (become different from one another) due to differences in mutation, drift, and/or selection. Some of the genetic differences between them can generate biological barriers to gene exchange, resulting in speciation: the formation of different biological species from their common ancestor. The particulars of these processes in any specific population depend on many aspects of the physical and biological environment, and on the existing features of the population, resulting from its previous evolutionary history (see Futuyma, 2013, for elaboration).

THE ORIGIN OF GENETIC VARIATION

Mutational changes in DNA sequences range from single base-pair alterations to insertions, deletions, and rearrangements of genetic material, and even changes in ploidy (the number of sets of chromosomes). Mutations that have no effect on "fitness" (i.e., survival and/or reproduction) are said to be "selectively neutral." These may include synonymous mutations in protein-coding regions (those that do not alter amino acid sequence), and mutations in pseudogenes and other apparently nonfunctional regions. Nonsynonymous mutations in coding regions and mutations in regulatory sequences are more likely to affect fitness. The rate of mutation (usually on the order of 10^{-9} per base pair per gamete) is usually too low to appreciably drive allele frequency change within a population, but it can determine the rate of DNA sequence change in the long term, and can influence the level of genetic variation within a population. Whether or not the supply of suitable mutations often constrains rates and directions of phenotypic evolution is uncertain (Blows and Hoffmann, 2005; Futuyma, 2010). There is no known mechanism by which the environment can direct the mutational process in advantageous directions; in that sense, mutation is random with respect to utility.

VARIATION WITHIN POPULATIONS

Populations of most species carry substantial sequence variation in many gene loci, and most quantitative traits exhibit some heritable variation. The presence of two or more fairly common alleles or genotypes within a population is referred to as "polymorphism." Such variation has arisen by mutation. Variation is enhanced by mutation, recombination (often), gene flow from other populations, and some forms of natural selection (see below). Variation is eroded by genetic drift and by most forms of natural selection. Analysis of genetic variation is based on the frequencies (proportions, in a population) of the alleles and genotypes at individual genetic loci (see Hartl and Clark, 2007). For sexually reproducing populations, the Hardy–Weinberg (H–W) theorem states that the frequency of each allele (p_i for allele i) will remain constant from generation to generation unless perturbed by mutation, gene flow, sampling error (genetic drift), or natural selection, and that the frequencies of the several genotypes will likewise remain constant, at values given by the binomial theorem (p_i^2 for homozygote A_iA_i, and $2p_ip_j$ for heterozygote A_iA_j) if mating occurs at random. Alleles at two or more polymorphic loci are eventually randomized with respect to each other by the process of recombination (a state of linkage equilibrium) so that different alleles at one locus are not associated with those at the other locus. (If they are associated to any degree, the loci are in linkage disequilibrium.) These principles have important consequences; for example, at H–W equilibrium, a rare allele exists mostly in a heterozygous state, and so is concealed if it is recessive. Closely linked mutations can remain in linkage disequilibrium for many generations, enabling geneticists to use detectable mutations as genetic markers for nearby mutations of interest, such as those that cause inherited disease.

Phenotypic variation in most quantitative traits is polygenic, based on segregating alleles at several or many loci, and also includes environmental (e.g., dietary) effects on the development or expression of a character (Falconer and Mackay, 1996). Thus, the variance in phenotype (V_P) includes a genetic component (genetic variance, V_G) and an environmental component (V_E), and often an interaction effect ($V_{G.E}$) as well. Although the individual phenotypic effects of alleles at various segregating loci are difficult to measure, most of the many loci appear to have small effects, and a few have fairly large effects, relative to the range of variation. One component of V_G, the additive genetic variance (V_A), is important for evolution by natural selection because it expresses the correlation between the phenotype of parents and their offspring. This component is attributable to the "additive" effects of alleles, that is, the phenotypic effect of each allelic substitution, averaged over all the genetic backgrounds in which it occurs. V_A depends on the number of loci contributing to the character, on the evenness of allele frequencies at each locus, and on the average magnitude of the phenotypic effect of different alleles. The ratio V_A/V_P is termed the "heritability" of a trait (in the narrow sense); it is valid only for the particular population and the particular environment in which it was estimated, since other populations might differ in allele frequencies or in environment (which affects V_E). All else being equal, the higher the V_A (or V_A/V_P), the greater the potential rate of evolution of a character, in response to natural selection.

A gene commonly affects two or more characters (pleiotropy), and so can contribute to a genetic correlation (r_G) between them. Another possible cause of genetic correlation is linkage disequilibrium, nonrandom association of certain alleles at two or more loci within a population (e.g., an excess of *AB* and *ab* combinations and a deficiency of *Ab* and *aB*). Genetic correlations are important because if the population mean of one character is altered, perhaps by natural selection, the other character will also be changed.

GENETIC DRIFT

Random genetic drift is simply random change in the frequency of alleles (and consequently, of genotypes) over the course of generations. Let us, for the moment, consider selectively neutral alleles, those that are not affected by natural selection, in a diploid population of *N* individuals (and therefore with 2*N* copies of the gene locus). The genes carried by a generation of newly formed zygotes in a population are a sample of the genes carried by the previous generation, to which the parents belong. The frequency (p) of an allele, say A_i, among the zygotes is unlikely to be exactly the same as in the previous generation because of random sampling error, owing to random mortality and random variation in female reproduction (fecundity) and male reproduction (number of mates) among individuals in the previous generation. Although the allele frequency in a new generation of zygotes is p on average (the same as in the previous generation), the frequency distribution of possible allele frequencies has a variance, given by the binomial expression $\mathrm{Var}\,(p)=p(1-p)/(2N)$. The greater the Var ($p$) is, the greater the random change in allele frequency is likely to be, from generation to generation, and thus the faster the process of evolutionary change by genetic drift. The expression for Var (p) tells us that this happens faster, the smaller the population size *N*. (*N* in this theory refers to the effective size of the population, which is smaller than the "census size" if individuals vary in reproductive rate, if the sex ratio among breeding individuals departs from 1:1, or if the population fluctuates in size.)

In this process, p fluctuates at random from generation to generation with no corrective tendency to return to its starting point, in a "random walk" to a boundary from which no return is possible: either loss of the allele A_i from the population or *fixation* of the allele A_i, that is, attainment of $p=1$. (Movement away from this boundary is possible, however, if new variation enters the population by mutation or by gene flow from other populations.) Hence, genetic drift results in the loss of genetic variation within a population.

If several separate populations of the species all began with the same initial p, different populations would have different random paths, and A_i may become fixed in some and lost in others; thus, genetic drift results in variation (divergence) among populations. An allele is more likely to be lost than to be fixed if its frequency is near zero, and conversely if its frequency is near 1.0; in fact, the probability, at any time, t, that an allele will eventually become fixed is p_t, its frequency at that time. A new mutation often exists, at first, as a single gene copy among the 2*N* genes in a population, so its initial frequency is $1/(2N)$, and this is the probability that it will eventually be fixed (if it is selectively neutral).

Another theoretical approach to studying the dynamics of genetic variation, coalescent theory, is often used for analyzing DNA sequence data (Hein et al., 2005). Looking back in time from the present, the gene copies (at a particular gene locus) in the population today are descended from only some of the genes carried by the previous generation's zygotes, due to sampling error; those zygotes in turn carried genes descended from only some of those in their parents' generation; and so on. Pursuing this logic, we realize that, inevitably, all the gene copies in the population today are descended from one single ancestral gene copy (one DNA molecule) at some time in the past. The descendants of that gene form lineages of genes, replicating down through the generations to the present time, the set of lineages forming a gene tree that, like a phylogenetic tree of species, portrays their ancestry back to ("coalesces to") the common ancestral gene, which existed t_{CA} generations ago. That ancestor was one of the (2*N*) genes in the population at that time, any of which might have been the ancestor of the gene copies in the present population. The speed of genetic drift is inversely related to population size: for a population of constant effective population size *N* (2*N* genes at a diploid locus), the average time back to the common ancestor of all contemporary genes, t_{CA}, is 4*N* generations (e.g., four million if the effective population size is one million individuals).

Over the course of those t_{CA} generations, nucleotide mutations occur in various descendant gene copies, and are copied down through the later descendants from those mutated genes. Gene copies (DNA sequences) that share recent mutations are descended from more recent common ancestral sequences than are sequences that share fewer, older, mutations, just as closely related species are marked by shared characteristics that originated in a recent common ancestor and distinguish them from all other species. Thus, the methods used to infer phylogenetic relationships among species can also estimate a gene tree, using the nucleotide substitutions that have accrued among the gene lineages during their descent from their common ancestor.

The Neutral Theory of Molecular Evolution

Using the theory of genetic drift, Motoo Kimura developed a neutral theory of molecular evolution that is the basis for analyzing DNA sequence variation within and among species, and is often considered the "null hypothesis" that should be rejected if alternative hypotheses, such as natural selection, are to be invoked (Kimura, 1983; Nei and Kumar, 2000). Mutational changes occur at many sites in a DNA sequence, at a total rate of, say, u_T per gene per generation. If some fraction f is selectively neutral, the neutral mutation rate is $u = fu_T$. (The fraction f is likely to be higher for nonfunctional sequences such as many pseudogenes, and for synonymous mutations in functional genes, than for nonsynonymous mutations in a gene with a critical function.) Since $2N$ genes are carried by (diploid) zygotes in each generation, the total number of new neutral mutations in the population each generation is $2Nu$, on average. The probability of fixation of a new neutral mutation is $1/(2N)$, so $2Nu \times 1/(2N) = u$ new mutations occur each generation that will eventually be fixed (after $4N$ generations, on average). Thus, population-wide substitutions of nucleotides in a DNA sequence occur at a roughly constant rate, accumulating ut substitutions over the course of t generations and theoretically providing a molecular clock. If two populations (or species) are derived from a common ancestor and do not exchange genes for t generations, and if mutations at different sites in the DNA sequence are fixed in each population, the expected difference D between sequences taken from the two populations will be $D = 2ut$. If u (the neutral mutation rate, which can vary among genes because of functional differences or DNA repair processes) can be calibrated, then the time since the two populations separated can be estimated from the observed difference D, as $t = D/2u$. (Calibration is usually based on geologically dated events, such as fossils of related lineages, or separation of two land masses on which related taxa reside.) Eventually, the increase in D slows down and levels off, because mutational substitutions occur repeatedly at the same nucleotide sites within the sequence and erase evidence of previous substitutions. This must be taken into account when DNA sequence divergence is used to estimate the time that has elapsed since species diverged from their common ancestor.

According to the neutral theory, evolutionary change is more rapid if mutations do not affect organismal function, since mutations that affect protein function are more likely to be deleterious and eliminated by natural selection. Consequently, evolution is predicted, and found, to be more rapid in nonfunctional sequences, such as pseudogenes, than in functional sequences, and more rapid at third-base than second-base positions in codons, because third-base mutations are more often synonymous.

Genetic variation is lost from a population by genetic drift, but is regenerated by mutations at many sites in a DNA sequence. At equilibrium, when the rate of input by neutral mutation balances the rate of loss by genetic drift, the level of variation, measured as the average proportion (π) of base pairs that differ between two gene copies taken at random from a population, equals $4Nu$; that is, it is proportional to the population size and the mutation rate. Given the knowledge of the mutation rate u, the effective population size can be estimated from $\pi/4u$.

Thus, DNA sequence variation, interpreted under the theory of genetic drift, provides a basis for many important inferences about effective population size, time since the separation of populations (or since speciation), historical relationships among populations, and whether or not natural selection has affected DNA sequence divergence and polymorphism. All of these have been applied to human population history.

NATURAL SELECTION

The concept of natural selection is the most important of Darwin's ideas, and one of the most important ideas that anyone has ever had. Nevertheless, many definitions of natural selection have been proposed (see the chapters by M.J.S. Hodge and J.A. Endler in Keller and Lloyd, 1992). In *The Origin of Species*, Darwin defined natural selection somewhat indirectly, writing, "this preservation of favourable variations and the rejection of injurious variations, I call Natural Selection;" but this leaves open the questions of what makes a variation "favourable" or "injurious," and what preserves or rejects. A concise, comprehensive definition is this: Natural selection is consistent (nonrandom, or biased) differences in the rate of survival or reproduction among classes of entities that differ in inherited characteristics. The term "reproductive success" (or fitness) is often used for "survival and reproduction," since survival to reproductive age is a prerequisite for reproduction. "Entities" is deliberately vague, because selection can (in principle) act among various kinds of biological "individuals," such as genes or larger sections of genetic material, individual organisms, groups of conspecific organisms, species, or clades (Williams, 1992). We speak of "classes" of genes, individuals, etc. because we cannot tell if a difference in reproductive success is nonrandom if we have information about only a single individual of each kind; we require samples of similar genes or individuals in order to see if there is a consistent difference between different types of alleles or phenotypically different organisms. Natural selection, in distinction from genetic drift, is marked by a consistent difference in mean reproductive success within a given environment, not a random, unpredictable difference; thus, natural selection is the antithesis of chance.

A special form of natural selection, termed spatial sorting, occurs when the rate or direction of dispersal differs among individuals with different inherited traits (e.g., longer wings or greater stamina). This process may reduce the frequency of high-dispersal types within the source population and increase it in newly colonized areas.

Modes of Selection

Most analyses of evolution by natural selection are concerned with individual selection: differences in fitness, owing to a genetically variable phenotypic character, among individual organisms within a population. In the simplest models, the character is affected by variation at a single locus, and fitness is defined as a genotype's relative rate of increase, that is, growth in numbers from one generation to the next, which depends on several life-history parameters. The rate of increase is a complex function of the probability of survival at each age from birth to the oldest reproductive age class, and on the values, at each age, of female reproduction (fecundity) and male reproduction (based on mating success and sometimes sperm competition).

Consider selection among individual organisms in a sexually reproducing, diploid population that differs in genotype at a single locus with two alleles, *A* and *a*. In the simplest case, the fittest of the three genotypes *AA*, *Aa*, and *aa* is a homozygote. If *aa* is the fittest genotype, we speak of "directional selection" for *aa*. This genotype might be rare because the environment previously favored *AA* and has only recently changed. Directional selection for *aa* increases the frequency of the *a* allele at a rate proportional to spq, that is, the fitness advantage (s) of *aa* (i.e., the strength of natural selection) and the frequencies ($p.q$) of both alleles. (Thus, the rate of evolution is greatest when both the favored and the disfavored alleles have high frequency.) Once *aa* becomes the prevalent genotype, allele *A*, and any other disadvantageous alleles that may arise by mutation, is reduced in frequency, and selection is often termed "purifying." Directional selection and purifying selection are two faces of the same coin. The frequency q of the advantageous allele *a* attains the deterministic equilibrium $q=1$ if only selection is operating, but if other alleles repeatedly arise by mutation, the equilibrium frequency will be set by the mutation rate (u) relative to the strength of purifying selection (s). Similarly, if a locally disadvantageous allele (perhaps *A*) that is advantageous in a different geographic population enters the population by gene flow, the genetic equilibrium is determined by the relative strength of gene flow and purifying selection. Gene flow, owing to interbreeding with immigrants from other populations, can sometimes severely diminish the degree of adaptation of populations to their local environment.

The H–W frequencies of the two genotypes that contain the advantageous *a* allele, Aa and *aa*, are $2pq$ and q^2. If allele *a* is very rare, the vast majority of the *a* genes are carried by heterozygotes. (For example, if $q=0.01$, $2pq=0.0198$, $q^2=0.0001$, and the ratio of heterozygotes to homozygotes is 198:1.) Whether or not the *a* allele can increase (or "invade" the population) depends almost entirely on the fitness of *Aa* relative to the prevalent homozygous genotype (*AA*); at this stage the high fitness of *aa* is almost irrelevant because it is so rare. So, even if *aa* is the fittest genotype, the *a* allele will not increase if it reduces the fitness of the heterozygote. This example illustrates that natural selection acts only in the present, and cannot look forward toward the best possible outcome. It also shows the importance of the H–W principle.

Although directional selection eliminates genetic variation, several other modes of selection (balancing selection) may maintain genetic polymorphism. The simplest model is "heterozygous advantage," in which the fitness of *Aa* is greater than that of either *AA* or *aa*, and all three genotypes segregate each generation due to random mating. Several hemoglobin polymorphisms in human populations, including sickle-cell hemoglobin, are the best known of the few well-documented examples of this mode of selection. Unquestionably more important is frequency-dependent selection, in which each genotype is more and more advantageous, the rarer it is. Many biological phenomena, including competition for resources, social interactions, and resistance to different genotypes of parasites, can give rise to such frequency-dependent effects. For example, pathogens may become adapted to the most common genotype of their host, and less adapted to rare genotypes, which therefore suffer less from disease. Variable selection, in which different homozygotes are advantageous at different times or in different microhabitats within the area occupied by a breeding population, can also maintain polymorphism, although mathematical models show that this occurs only within a rather narrow range of combinations of selection intensities and environmental frequencies.

Models of natural selection on a quantitative, polygenic phenotypic trait are based on models of selection at a single locus. In the simplest ("additive") model (which fits real data surprisingly well), variation in a character, measured in, for example, millimeters, is based on "+" and "–" alleles, at each of k loci, that add or subtract the same amount. The mean and variance of the character are determined by the frequency of the alleles at all of the loci; the mean will clearly be higher (and the variance lower) if most of the + alleles have high frequency. Directional selection occurs when there is a monotonic relationship between phenotype and fitness. For example, selection may favor larger phenotypes, namely those with more +alleles in their genetic makeup, so + alleles rise in frequency. As the frequency of + alleles increases at many loci, some genotypes are formed by recombinations that have so many + alleles that they exhibit new phenotypes, above the original range of variation. Selection may ultimately fix + alleles at all loci; then genetic variation is eliminated, and evolution ceases unless new + alleles, perhaps at other loci, arise by mutation.

If the relationship between fitness and phenotype is not monotonic, but instead has an intermediate maximum ("optimum"), the character will evolve to the optimum, and then is subject to *stabilizing selection*: deviations in either direction from the mean are disadvantageous. If, instead, two or more phenotypes have higher fitness than the intermediates between them, selection is said to be disruptive, or diversifying. Disruptive selection at a single locus generally implies that the heterozygote for two alleles *A* and *a* has lower fitness than both homozygotes. Such a polymorphism is unstable, however, and the population will become fixed for the initially more common allele. In simple models of both stabilizing and disruptive selection, genetic variation is eroded, unless a factor such as frequency-dependent selection occurs. Under diversifying selection, for example, genetically different phenotypes may be adapted for different food or other limiting resources, so that competition among individuals of a particular phenotype becomes more intense, and fitness declines, as that phenotype becomes more abundant and depletes its resource.

In studies of natural populations, directional selection on many characteristics has been documented much more often than stabilizing or disruptive selection (Kingsolver et al., 2012). This is surprising and puzzling, because biologists often assume that populations are so well adapted to their current environment that most characteristics should be nearly optimal, and therefore subject to stabilizing selection. This is one of the many aspects of evolution that require more research.

Components of Fitness

Fitness has several components (life-history features: Stearns, 1992) that contribute to the rate of increase (numbers/time) of a genotype, relative to others. All else being equal, a difference (in, for example, survival probability or fecundity) expressed at an earlier age generally has a bigger impact on growth in numbers (fitness) than a similar difference expressed at a later age. Suppose individuals reproduce from age 3 until 10, and then die. A mutation that increases the chance of survival from age 8 to 9 has a smaller selective advantage than the one that provides a similar survival advantage from age 2 to 3, because survival enhancement in the older age classes will have much less effect on the number of offspring they might yet produce (and the number of genes passed on). Similarly, a mutation that increases reproductive output at age 3 has a greater impact on the increase of the mutation's frequency than that affects reproduction at age 9, because (1) fewer individuals survive to age 9, so they do not receive the benefit of the mutation; and (2) the mutation expressed at the younger age effectively shortens the generation time, so that more descendants (grandchildren, great-grandchildren, etc.) are produced per time unit than are produced by the genotype whose reproductive capacity is enhanced only at an older age.

Consequently, mutations that enhance survival or the number of offspring (e.g., number of eggs or young) are expected to increase fitness, but the magnitude of increase depends on the age at which these effects come into play. Moreover, there often exist trade-offs between different fitness components or between a given component expressed at different ages, because an organism must partition energy or nutrients among different functions (the principle of allocation). For example, if reproduction reduces growth, it may be advantageous to delay reproduction until the individual is larger, which may ensure a longer life and higher fecundity that together more than make up for the reproduction foregone at an earlier age. All these theoretical conclusions have been supported by experiments with fruit flies, guppies, plants, and many other organisms.

Darwin coined the term "sexual selection" to describe variation in reproductive success achieved through success in mating (Andersson, 1994). In many species of animals, males vary more than females in the number of their mates, and therefore in reproductive success, so that the intensity of sexual selection is greater in males than in females. The two most commonly discussed modes of sexual selection are conflict between males, with winners gaining access to more females, and female "choice" of some males over others, based on one or more characteristics that usually are actively displayed to females. (In many species, the same trait seems to play a role in both male–male and male–female interactions.) There is considerable evidence that conflict between males selects for larger size, greater weaponry, and many other kinds of traits that are used to establish dominance. Male investment in features that enhance mating success, such as mating activity, weaponry, or display features, is known to reduce investment in maintenance (e.g., immune system) and survival.

Female choice is known to impose sexual selection in birds, insects, and other animals, but there is considerable uncertainty about why females choose particular male phenotypes, such as males with more vigorous displays or more highly elaborated ornaments or vocalizations. According to the "good genes" hypothesis, exaggerated male features indicate high physiological vigor that may stem from superior genetic constitution, and females that choose such males will have fitter offspring (which inherit alleles that influence female choice). There is some evidence both for this hypothesis and for several contenders. In models of runaway sexual selection, a nonrandom association (linkage disequilibrium) develops between genes that affect a male ornament and genes that affect the degree of female preference for this character. Females that prefer more highly ornamented males have daughters that inherit this preference (as well as unexpressed genes for large male ornamentation) and sons that inherit larger ornaments (as well as unexpressed genes for heightened female preference). (Note that most features expressed by a single sex are encoded in the genome of both sexes.) Therefore, any increase in the average male ornament in the population will cause a correlated increase in the average female preference, and vice versa, ratcheting both toward more extreme values until the process is halted either by counteracting selection or by running out of genetic variation.

In a twist on the sexual selection theory, females and males are engaged in sexual conflict: males reduce females' fitness in various ways (e.g., incessantly attempting to mate), females are selected to resist, and selection favors males with ever more stimulating characteristics that can overcome female resistance (Arnqvist and Rowe, 2005). Moreover, females of many species mate with multiple males, even in species that form a supposedly monogamous pair bond. As a result, sperm from different males often compete for fertilization within the female reproductive tract. Probably because of the strong, long-continued selection exerted by sperm competition and sexual selection, reproductive characteristics of many groups of animals, including male display features, genital morphology, proteins from accessory reproductive glands, sperm morphology, and cell-surface proteins of gametes, evolve more rapidly than most other characteristics, and are often the major differences among closely related species.

Modeling Adaptation

The evolution of some features is best analyzed by genetic models. For example, linkage disequilibrium is an essential component of models of sexual selection by female choice, so this topic requires an explicit genetic approach. The major alternative is optimization, an approach that attempts to specify what the optimal character state ought to be, given some assumptions about benefits, costs, and constraints. This approach assumes that there has been enough time and enough genetic variation for natural selection to bring the characteristic nearly to its optimum value, and that the genetic details do not matter very much.

Optimization is a common approach in the fields of functional morphology and physiology, in which it is assumed that fitness is enhanced by maximizing some function, subject to constraints such as costs in energy or materials, or compromises with other functions. For example, aerodynamic models have been used to model flight and optimal wing morphology in birds, in which compromises among speed, maneuverability, and energy expenditure are taken into account. Optimal models have been extensively used to study the evolution of social interactions, in which genetic modeling is difficult because the optimal behavior of an individual often depends on the behavior of other individuals. Among the most widely used approaches is game theory (Maynard Smith, 1982). Suppose, for example, that the problem is whether or not parental care, by either or both mated partners, will evolve by individual selection. One might postulate two "strategies," "Stay and provide **C**are to offspring" and "**D**efect and attempt to reproduce again." "**C**are" (**C**) has the benefit of enhancing offspring survival, but sacrifices immediate additional reproductive opportunities; "**D**efect" (**D**) has the opposite benefit and cost. For each parent, **D** might have greater benefit than cost (i.e., greater "payoff") if the other parent provides sufficient care. So for each possible pair (**C♀/C♂**, **C♀/D♂**, **D♀/C♂**, **D♀/D♂**), one postulates for each partner the expected reproductive payoff. The average fitness of each strategy, for each sex, is then its payoff averaged over the possible pairings, and weighted by their frequency in a random-pairing population. The best strategy, within the set of strategies considered (here, **C** and **D**), is the one that, if fixed in the population, will remain fixed even if individuals with alternative strategies attempt to invade. This is the "evolutionarily stable strategy," or ESS.

LEVELS OF SELECTION

Natural selection was defined above as "consistent (nonrandom) differences in the rate of survival or reproduction among classes of entities that differ in inheritable characteristics." These "entities" may be at different, nested levels, and the effects of selection at different levels may be opposite (Okasha, 2006). Consider, for example, the levels "individual organism" and "somatic cell lineage" within multicellular organisms. If a cell lineage experiences a mutation that causes rapid, unrestricted cell division, that lineage has a "selective advantage" relative to other cells, and will constitute an increasing proportion of cells within the domain of the single organism. This proliferation—cancer—is clearly disadvantageous to the higher-level entity (the organism), if it occurs before or during the organism's reproductive ages. Selection among genetically variable individual organisms may then favor genotypes that can suppress cancerous tumors—but such individual selection will be weaker if cancers develop later in life than if earlier.

Likewise, conflicts can arise in sexually reproducing organisms between selection at the level of the individual gene (locus) and selection among individual organisms. A famous example is "segregation distortion" at the "*t* locus" during meiosis in male house mice. More than 90% of the sperm of males heterozygous for a normal allele (*T*) and a recessive allele (*t*) carry the *t* allele. Thus, there is "genic selection" for *t* alleles. However, various *t* alleles cause either embryonic death or male sterility under homozygous conditions, so there is "individual selection" against *t* alleles. Genic selection accounts for many phenomena, such as the proliferation of transposable elements ("selfish genetic elements"): DNA sequences that replicate more frequently than most of the genome.

Genic selection provides one way of viewing the evolution of cooperation, which stands in contrast to the selfish individualism that generally characterizes individual selection (Dawkins, 1982). Cells in multicellular organisms cooperate because they are (generally) genetically identical: a gene in a liver cell is replicated by virtue of the replication of identical copies in the germ cell line. Likewise, the rate of increase of a parent's gene over generations depends on the survival and replication of identical copies of that gene in the parent's offspring—and so alleles that program parental care may increase in frequency. This is an example of "kin selection": selection in which alleles differ in fitness by influencing the effect of their bearers on the reproductive success of individuals (kin) who carry the same allele due to common descent. Genes that enhance their bearers' propensity to help more distant relatives may also increase in frequency, but the benefit to the relative must be greater, since more distant relatives have a lower probability of sharing the "helping allele." (Kin selection is one of several explanations of the evolution of cooperation among genes, cells, or conspecific organisms. For example, reciprocity [reciprocal altruism] may evolve if individuals remain associated or can recognize one another and can repeatedly benefit one another.)

Although cooperation among family members can evolve by kin selection, each individual is more closely related to itself than to any other, so intrafamilial interactions are riddled with conflict. For example, conflict can exist between parents and offspring, because a parent maximizes her (his) fitness by allocating care among a number of offspring, present and future, so the optimal investment of care in any one offspring is lower from the parent's perspective than from the offspring's (Trivers, 1974). Genes expressed in human and mouse fetuses that enhance uptake of sugar and other nutrients from the mother are counteracted by maternally expressed genes that prevent the fetus from extracting too much (Haig, 1993).

If an individual with an "altruism allele" dispenses benefits indiscriminately to both related and unrelated individuals, the survival of copies of both that allele and its nonaltruistic ("selfish") alternative allele is equally enhanced, but the donor suffers a cost (*c*), so the selfish allele increases. As a rulc, individual selfishness increases within populations, even if the population as a whole suffers. In principle, extinction of whole populations of selfish individuals, and survival of populations of cooperative individuals, could cause evolution of cooperation in the species as a whole. This would be "group selection," in opposition to individual selection. Some authors hold that group selection can play a role in evolution, especially if the groups are very small and temporary (Eldakar and Wilson, 2011). For example, groups of nestling birds that include cooperators may yield more survivors than groups that do not. Other authors, however, argue that because most such groups consist of relatives, this form of group selection does not differ from kin selection (West et al., 2007). In any case, selection among large, long-lasting populations is unlikely to prevail against individual selection, because populations of cooperators are likely to be invaded by immigrant selfish genotypes ("cheaters"), which rapidly increase within these populations. Most evolutionary theorists therefore conclude that group selection, evolution by differential extinction or reproductive productivity of whole populations, is unlikely to play a major role in evolution. Consequently, characteristics that benefit the population or species, but which are disadvantageous to the individual or its kin, are unlikely to evolve.

Species differ in their probability of extinction or of speciation per unit time, giving rise to differences in the number of species among "clades" (i.e., branches of a phylogenetic tree). Characteristics of individual organisms (e.g., physiological tolerances) or of entire species (e.g., breadth of geographic distribution) that affect extinction and speciation rates may therefore become more or less frequent among all species taken together. This process is "species selection," in which the species is the unit of selection. For example, herbivory in insects, sexual dichromatism in passerine birds, and low body mass in several orders of mammals seem to be associated with increased diversification, compared to equally old-related groups that have the opposite characteristics (Coyne and Orr, 2004).

SPECIATION

There are several contending concepts of "species" because this word serves several purposes. As a term in classification, it may simply label phenotypically distinguishable populations; this is the basis of the phylogenetic species concept (PSC) that many systematists use. Most evolutionary biologists concerned with evolutionary processes use one or another version of the biological species concept (BSC), articulated by Ernst Mayr (1942). A biological species is a group of actually or potentially interbreeding populations that are reproductively isolated (by biological differences) from other such groups. The reproductive isolating barriers (RIBs), which are usually genetically based, include "prezygotic" barriers that reduce gene exchange before zygote formation (e.g., differences in habitat association, timing of reproduction, behavior, pollination [for plants], and failure of gametes to unite) and "postzygotic" barriers that are expressed as diminished survival or reproduction of hybrid genotypes, usually because of "incompatibility" between genes from the two parent populations. The BSC applies only to sexual organisms and may be more difficult to apply in practice than the PSC, since the potential ability of spatially separated (allopatric) populations to interbreed may be difficult to evaluate. However, reproductive isolation (RI) plays a critical role in long-term evolution, since it enables populations, even if they eventually meet and overlap, to retain their distinct characteristics and to generate clades of species that subsequently elaborate those differences. Speciation might facilitate sustained evolution of morphological and other characteristics, by preventing the "slippage" that interbreeding with other populations would cause (Futuyma, 1987). Speciation is the process of cladogenesis ("branching" in evolution), the basis of biodiversity.

Many species of animals and plants form by genetic divergence of spatially disjunct (allopatric) or neighboring (parapatric) populations of an ancestral species, since spatial separation reduces gene flow enough to allow divergent changes, by natural selection or genetic drift, in the frequencies of alleles that underlie RI. Whether or not species are often formed sympatrically, that is, by evolution of a single randomly mating population into two reproductively isolated forms, is controversial. Except for speciation by chromosome doubling (polyploidy), which is common in plants but rare in animals, the evolution of RI between populations, like the evolution of most phenotypic differences, is gradual: arrays of populations can be found that display all degrees of pre- or postzygotic isolation from none to complete. Thus, it is common to find partially reproductively isolated populations (semispecies) that are not readily classified as the same or as different species. Hybridization between such forms in nature can result in some parts of the genome readily "introgressing," or penetrating from one form into the other, while other parts of the genome remain much more differentiated, because stronger divergent natural selection on these regions counteracts gene flow.

Because RIBs are usually polygenic, full RI requires that two distinct clusters of different alleles be formed at loci that affect a RIB (*AABBCC*... and *aabbcc*...), whereas recombination generates allelic mixtures (*AaBBCc*..., etc.) that are only partially reproductively isolated and so form a "bridge" for gene exchange between diverging populations. An extrinsic barrier (e.g., topography or unsuitable habitat) between diverging populations, if it is seldom surmounted by dispersing individuals, reduces or eliminates this problem, which is why allopatric speciation is easy. After allopatric populations have diverged sufficiently, they may expand their ranges and overlap without interbreeding. Speciation with gene flow can occur if selection for some trait differences between initially interbreeding parapatric populations is strong enough to counteract gene flow and recombination. Fully sympatric speciation is generally considered difficult because divergent selection, unaided by an extrinsic barrier, must be very strong to overcome recombination (Gavrilets, 2004; Coyne and Orr, 2004).

What causes the evolution of RI between spatially separated populations? If some genetic incompatibility has evolved between allopatric populations, and if they subsequently expand their range, mate, and form hybrids with low fitness (i.e., low survival or reproduction), individuals that mate conspecifically will have more successful offspring that those that hybridize, and alleles that enhance (reinforce) mating discrimination may be transmitted and increase in frequency. Such reinforcement of prezygotic isolation is the major context in which there is direct selection for RI. As long as the populations are allopatric, however, there cannot be selection to avoid hybridization, so RI must evolve as a by-product of genetic divergence that transpires for other reasons. Genetic drift, ecological selection, and sexual selection are the postulated processes.

Recent research has pointed to ecological speciation as an important process, that is, the evolution of RI due to genetic differences that underlie divergent ecological adaptation of different populations (Nosil, 2012). There is also considerable indirect evidence that divergent sexual selection, resulting in different female preferences and male display traits, can result in behavioral isolation. The high rate of evolutionary divergence of genitalia, sperm, and other features associated with reproduction is consistent with this hypothesis.

Genetic drift plays a role in Ernst Mayr's influential hypothesis of founder-effect speciation (also called peripatric speciation). Mayr believed that the selective advantage of one allele over another depends strongly on which alleles it interacts with at other loci (epistasis for fitness). He postulated that a population founded by few individuals will undergo genetic drift at some loci, so that some previously rare alleles become common by chance. Because of this change in the "genetic environment," previously disadvantageous alleles at some interacting loci become advantageous. Fixation of previously rare alleles at many such loci so alters the genetic constitution of the newly founded population that it may become reproductively isolated from the parent population. Mayr's hypothesis is considered implausible by some theoreticians, but others feel that it still warrants consideration, albeit in the modified form (Gavrilets, 2004).

The divergence of populations into distinct species often includes ecological differentiation, often initiated in allopatry due to environmental differences among regions, and sometimes enhanced via character displacement, the evolution of accentuated ecological difference between sympatric populations of two species, to reduce competition. Bursts of speciation, accompanied by ecological divergence, are referred to as adaptive radiations, macroevolutionary episodes that account for much of the extraordinary adaptive variety of organisms (Schluter, 2000).

FROM MICROEVOLUTION TO MACROEVOLUTION

Macroevolution refers to evolutionary phenomena above the species level, such as evolutionary trends, the evolution of the highly divergent and sometimes novel features of "higher taxa," and patterns of change in diversity, owing to the origination and extinction of taxa. Most such changes are generally thought to result from the processes of evolution already outlined in this chapter, repeated and accumulated over long periods of geological time. Explaining the patterns, however, requires information such as phylogenetic, geological, and climate history.

Most questions about macroevolution (and many about evolution within species) are best considered within the framework of phylogenies or evolutionary trees. Most phylogenetic trees are branching diagrams, with an explicit or implicit axis of time, from the tree's "root" in the more remote past, toward the more recent "tips," which often refer to named species (or higher taxa). Each branch is an evolving lineage; each bifurcation represents the splitting of one lineage into two descendant lineages; and all the lineages (which together may be called a clade) are descended from the root, or ancestral lineage. A tree represents only the temporal order in which the lineages diverged from common ancestors. Rotating any pair of lineages around their branch point does not alter the temporal sequence of divergence, and so does not change the information conveyed by the tree (Figure 1.1). All the lineages, whether they are shown as more "basal" branches (e.g., frogs, in a phylogeny of tetrapod vertebrates) or more "recent" branches (e.g., human),

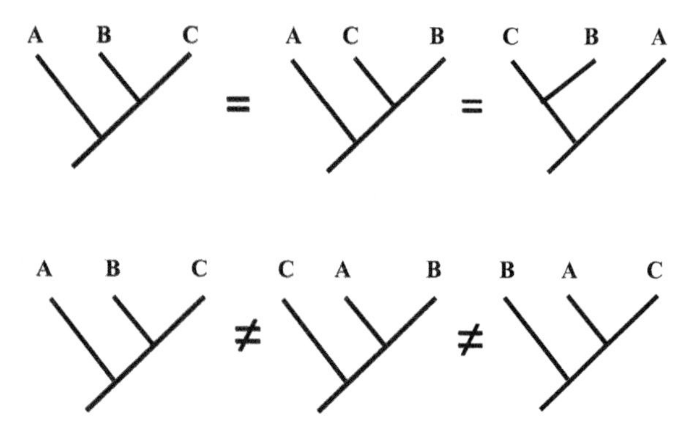

FIGURE 1.1 A tree represents the temporal order in which lineages diverged from common ancestors. Rotating any pair of lineages around their branch point does not alter the temporal sequence of divergence, and so does not change the information conveyed by the tree.

can have a mixture of relatively ancestral ("primitive") and relatively derived ("advanced") characteristics. All lineages, whenever they separated from other lineages, can continue to evolve new features.

The most common pattern in macroevolution is diversification, not a trend in any one direction. Nevertheless, trends do occur. In evolutionary biology, "trend" often refers to similar changes that have occurred, by convergent evolution, in many clades, such as reduction of the number of digits in many tetrapod vertebrates. The word can also refer to the sequence of changes, through intermediate steps, that lead from an ancestral to a derived character state, such as the stepwise reduction in toe number and the increase in tooth height in the lineage leading from the "dawn horse" *Hyracotherium* to *Equus*, the most familiar lineage of horses. Although the idea of a trend often summons connotations of "progress," evolutionary biologists generally reject the notion of progress in evolution, except in the limited sense that features may be steadily "improved" to serve specific functions in certain taxa—as exemplified by the successive changes in foot morphology that enabled greater speed in the horse lineage. Any such improvement or progress is entirely context dependent, just as I can claim to have progressed in knowledge of biology but not of automobile repair, while my car mechanic can claim the opposite. In particular, evolutionary biologists cannot identify any uniform standard of progress, and certainly do not single out humans as a criterion of evolutionary progress. And any notion that evolution has foresight, purpose, or goals (frequently connoted by the word progress) is thoroughly rejected by almost all evolutionary scientists.

During the 1930s and 1940s, the major authors of the Evolutionary Synthesis, affirming Darwin's hypothesis that evolution is a gradual process, argued that the distinctive features of higher taxa evolved more or less independently at different rates (mosaic evolution), and by incremental change in each character (Simpson, 1953). Such gradual evolution has been paleontologically documented for many skeletal features of mammals, and for many other lineages. Often, evolution of a character is accelerated, and takes surprising directions, because a feature may serve a new and different function (e.g., the hands of bats, serving for flight). True novelty is relatively uncommon, but does occur. For example, the multicusp teeth that evolved in the synapsid ancestors of mammals were a departure from the simple, single-pointed teeth of most other vertebrates.

A recurring challenge to gradualism is the possibility that mutations of large effect, which radically alter developmental pathways, have contributed to evolution, a possibility that is not denied by the many examples of evolution based on mutations with modest effects. Mutations in key regulatory genes, such as the *Hox* genes that are important in establishing the fundamental body plans of animals, can have drastic effects, sometimes switching development of one body region into another; but whether or not comparable mutations in these genes were the origin of major evolutionary changes in body plan is not yet known. The complex developmental pathways that these genes now control may have evolved incrementally. A major challenge in modern biology is to understand how differences in genes are translated into major evolutionary changes in phenotype. This is a task of evolutionary developmental biology (or "evo–devo") (Carroll, 2007).

Another major theme in macroevolutionary studies is how differences in the rate of origin and extinction of species and higher taxa have shaped the numbers of species and their phenotypic and ecological diversity over evolutionary time. For example, the rate of origin of marine animal taxa in the fossil record has been lower during times of higher diversity, as expected if ecological "niches" become increasingly filled, leaving less opportunity for new taxa to fit in (Foote, 2010). Another pattern is adaptive radiation, the rapid proliferation of ecologically diverse species within a clade. Time-calibrated phylogenies of some taxa show that, again consistent with the niche-filling hypothesis, the rate of origin of ecologically divergent forms is often greater early in an adaptive radiation (Gavrilets and Losos, 2009).

EVOLUTIONARY THEORY TODAY

When Darwin referred to "my theory," he was speaking of a little-tested hypothesis. Since then, all the major elements of his theory—the common ancestry of all organisms, the bifurcation of lineages in the origin of species, the evolution of adaptations by the action of natural selection on hereditary variation, the diversification of species by adaptation to different ecological niches, or places in the "economy of nature"—have been abundantly supported, elaborated, and extended. When biologists speak today of "evolutionary theory," they refer not to a speculation or hypothesis, but to a mature scientific theory, an accepted complex of general principles that explain a wide variety of natural phenomena, as quantum theory and atomic theory do in the physical sciences.

No biologist today would think of publishing "new evidence for evolution"—it has not been a scientific issue for a century.

Nevertheless, many questions remain unanswered and substantial controversy persists, as it does in all active sciences, about some important questions. Because of increasingly powerful and affordable molecular and computing technologies, discoveries of new phenomena (e.g., in genomics) that call for evolutionary explanation, and the development of new theory addressing both new and old questions, evolutionary research is probably more active now than ever before, occupied with analyzing the phylogeny of all major groups of organisms, discerning and explaining patterns of evolution of genes and genomes, explaining puzzling behaviors and other phenotypic traits, using advances in developmental biology to understand the evolution of phenotypes, and many other concerns. New approaches are being developed to answer old but difficult questions, such as the causes of speciation, the evolution of functionally integrated characteristics, and the conditions under which populations adapt to environmental change or become extinct—one of the most important questions now, when humans are changing the Earth at a frightening pace.

REFERENCES

Andersson, M.B., 1994. Sexual Selection. Princeton University Press, Princeton.

Arnqvist, G., Rowe, L., 2005. Sexual Conflict. Princeton University Press, Princeton.

Blows, M.W., Hoffmann, A.A., 2005. A reassessment of genetic limits to evolutionary change. Ecology 86, 1371–1384.

Carroll, S.B., 2007. The Making of the Fittest: DNA and the Ultimate Forensic Record of Evolution. W.W. Norton, New York.

Coyne, J.A., Orr, H.A., 2004. Speciation. Sinauer, Sunderland, MA.

Dawkins, R., 1982. The Selfish Gene. Oxford University Press, Oxford.

Eldakar, O.T., Wilson, D.S., 2011. Eight criticisms not to make about group selection. Evolution 65, 1523–1526.

Falconer, D.S., Mackay, T.F.C., 1996. Introduction to Quantitative Genetics. Longmans, Harlow.

Foote, M., 2010. The geological history of biodiversity. In: Bell, M.A., Futuyma, D.J., Eanes, W.E., Levinton, J.S. (Eds.), Evolution since Darwin: The First 150 Years. Sinauer, Sunderland, MA, pp. 479–510.

Futuyma, D.J., 1987. On the role of species in anagenesis. American Naturalist 130, 465–473.

Futuyma, D.J., 2010. Evolutionary constraint and ecological consequences. Evolution 64, 1865–1884.

Futuyma, D.J., 2013. Evolution. Sinauer, Sunderland, MA.

Gavrilets, S., 2004. Fitness Landscapes and the Origin of Species. Princeton University Press, Princeton.

Gavrilets, S., Losos, J.B., 2009. Adaptive radiation: contrasting theory with data. Science 323, 732–737.

Haig, D., 1993. Genetic conflicts in human pregnancy. Quarterly Review of Biology 68, 495–532.

Hartl, D.L., Clark, A.G., 2007. Principles of Population Genetics. Sinauer, Sunderland, MA.

Hein, J., Schierup, M.H., Wulff, C., 2005. Gene Genealogies, Variation and Evolution: A Primer in Coalescent Theory. Oxford University Press, Oxford.

Keller, E.F., Lloyd, E.A. (Eds.), 1992. Keywords in Evolutionary Biology. Harvard University Press, Cambridge, MA.

Kimura, M., 1983. The Neutral Theory of Molecular Evolution. Cambridge University Press, Cambridge.

Kingsolver, J.G., Diamond, S.E., Siepelski, A.M., Adam, M., Carlson, S.M., 2012. Synthetic analyses of phenotypic selection in natural populations: lessons, limitations, and future directions. Evolutionary Ecology 26, 1101–1118.

Maynard Smith, J., 1982. Evolution and the Theory of Games. Cambridge University Press, Cambridge.

Mayr, E., 1942. Systematics and the Origin of Species. Columbia University Press, New York.

Nei, M., Kumar, S., 2000. Molecular Evolution and Phylogenetics. Oxford University Press, New York.

Nosil, P., 2012. Ecological Speciation. Oxford University Press, Oxford.

Okasha, S., 2006. Evolution and the Levels of Selection. Oxford University Press, Oxford.

Schluter, D., 2000. The Ecology of Adaptive Radiation. Oxford University Press, Oxford.

Simpson, G.G., 1953. The Major Features of Evolution. Columbia University Press, New York.

Stearns, S.C., 1992. The Evolution of Life Histories. Oxford University Press, Oxford.

Trivers, R., 1974. Parent-offspring conflict. American Zoologist 11, 249–264.

West, S.A., Griffin, A.S., Gardner, A., 2007. Social semantics: altruism, cooperation, mutualism, strong reciprocity and group selection. Journal of Evolutionary Biology 20, 415–432.

Williams, G.C., 1992. Natural Selection: Domains, Levels, and Challenges. Oxford University Press, New York.

Chapter 2

Evolution, Creationism, and Intelligent Design

Andrew J. Petto
Department of Biological Sciences, University of Wisconsin–Milwaukee, Milwaukee, WI, USA

SYNOPSIS

The last half of the twentieth century saw an increased attention to science education, including the study of evolution, in public schools. As a result, a powerful creationist movement sought first to prevent teaching evolution, then to require teaching creationism in the public schools. After several defeats in federal courts, creationism, with its overt emphasis on conservative Christian views of the Bible, was replaced by intelligent design (ID). ID still held that life, or at least certain aspects of it, was the result of a purposeful action of an intelligent agent, and that this action could be detected by scientific study. In the early twenty-first century, ID still follows the trail of opposition to evolution blazed by the creationists: pointing out unanswered questions and disagreements among scientists about details of their studies and methods, arguing that things we do not know now are unknowable, calling up laws of thermodynamics and probability to prove the impossibility of evolution, drawing out credentialed dissenters and skeptics, focusing on sociopolitical organization rather than scientific research, and calling for fairness and openness in science education. In the end, their concerns echo the cultural understanding that life should have a purpose and meaning, and they object that natural scientists do not seem to concern themselves with this important issue.

INTRODUCTION

One of the hallmarks of human success as a species is the ability to recognize regularities in the world around us and to anticipate predictable events based on these observations. The obvious benefit is being able to anticipate changes associated with cycles of tides for coastal populations or seasons for both agricultural and hunter-gatherer populations; knowing "how the world works" is essential to survival and for well-being. Traditional knowledge, often passed from generation to generation as folklore, formalizes this understanding and also admonishes that there are exceptions from well-established patterns, which sometimes appear to contradict the cultural experience of past generations.

Perhaps to improve our predictions, perhaps to give us greater control or influence over natural phenomena, humans typically construct explanatory frameworks that guide our understanding—a search for the underlying causes of the events and phenomena that we experience in the world. If we understand how and why something happens—or particularly when its occurrence does not match our expectations—then we might, at least, be prepared better for when it does happen.

For most of human history, these underlying causes were mysterious, so the observations—whether regular and predictable or rare and unpredictable—were attributed to entities with supernatural or extranatural powers to influence natural phenomena. Often these entities were capricious in their behaviors, but for practical purposes, it mattered little whether the observations of the Sun in the daytime sky were due to Apollo's chariot's racing across the sky or whether our star's apparent motion was due to the movement of our planet.

Of course, in the Christian era, which dominated the Western tradition, the contradictory actions of competing gods were replaced by the purposeful creative actions of the God of the Bible; and the purpose was to prepare a place for humans at the crown of God's Creation. The view of Nature as God's palette, which would reveal the plan for His creation, dominated philosophic and what would be recognized later as early scientific scholarship well into the eighteenth century. This tradition culminated most notably in William Paley's *Natural Theology; or Evidences of the Existence and Attributes of the Deity* (1802)—a highly regarded and influential publication arguing that observations in the natural world provided direct evidence for the existence and actions of God.

Though many others used a similar line of reasoning, Paley is perhaps best known for using the argument from design, as summarized by the watch-on-the-heath analogy: an object known to require a maker (such as a watch lying on the ground) is in itself evidence of the existence of that maker. This formal argument reinforced the ethnoscientific view that complex natural events and phenomena required an intentional action by a purposeful agent, and it seemed to provide for biology a scientific basis that matched the advances in astronomy, mathematics, physics, and chemistry in the previous centuries.

As Daniel Dennett (1995), among others, has pointed out, these other scientific fields were uncovering and codifying laws of Nature that did not require any purposeful agency to account for the interactions of matter and energy that produced natural phenomena. It did not take long for those interested in biological sciences to recognize that their field of study also needed a similar foundation in the natural world. Gruber (1974, pp. 417–418) noted Charles Darwin's musings on this very problem as shown in comments he wrote in the margins of another work of natural philosophy, McCullough's (1843) *Proofs and Illustrations of the Attributes of God*:

> *The explanation of types of structures in classes—as resulting from the* will *of the deity, to create animals on certain plans, is no explanation*—it has not the character of a physical law /& *is therefore utterly useless.—it foretells nothing / because we know nothing of the will of the Deity, how it acts & whether constant or inconstant like that of man. — the cause given we know not the effect.*

Of course, it was Darwin (1859) who would later provide the foundation for the naturalistic explanations in the biological sciences: the formulation of descent with modification that would develop into an evolutionary theory for the history and diversity of life on earth. It was the acceptance of this evolutionary theory into the center of biology—and of biology education—that replaced the biblical narrative of life's history and indirectly spurred the movement to construct a scientific justification for the creationist explanatory model.

A SCIENTIFIC CREATIONISM

The Western scientific tradition conceded a created universe at the outset, but this was a starting assumption, not a foundation for systematic scientific research. As scientists in other disciplines discovered laws of Nature that operated without any apparent intervention from a purposeful agent, biologists and natural historians looked more deeply into Nature for the explanations of natural phenomena and events that they studied. It was, as noted by J.P. Moreland (2008)—a defender of creationism—a shift in what scholars meant by *scientific:* explanations that relied on Divine agency were replaced by those that were based on only natural laws and processes.

The challenge for creationists—and for their intellectual heirs, proponents of ID—was either to surrender the claim to scientific validity or to develop an alternative to the naturalistic methodology of modern science that could reconcile the facts uncovered in the natural world with revelations about the nature and actions of a supernatural entity. Unwilling to concede the cultural high ground occupied by "science," creationists both claimed scientific

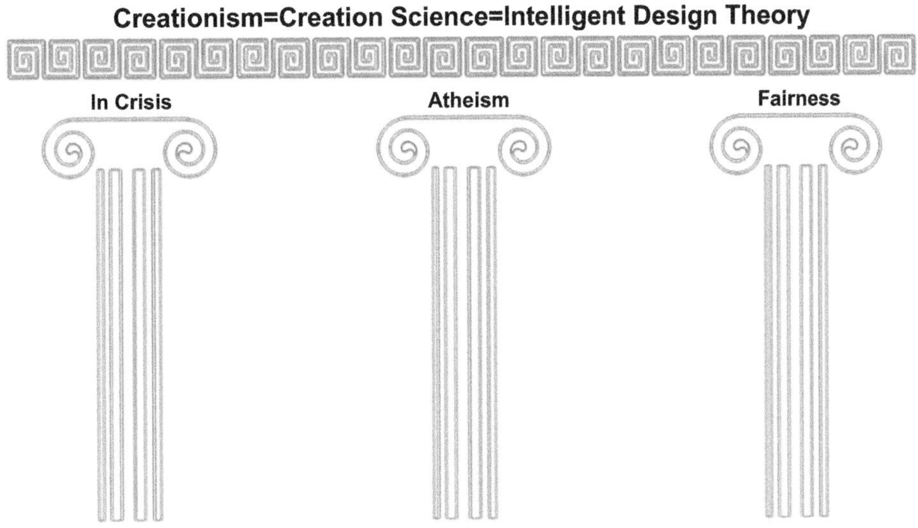

FIGURE 1 The three pillars of creationism.

validity for their positions and attempted to undermine contemporary scientific research. There are several excellent resources that describe these responses to the challenge (see, for example, Bowler, 2007, 2010; Numbers, 1998, 2006; Scott, 2009).

To resist this movement toward naturalism in the sciences—but especially in public-school science education—creationists rallied around three major themes to impede public acceptance of evolution. Eugenie Scott (2008) dubbed these themes the "three pillars" of creationism (Figure 1): (1) that evolution is a theory in crisis; (2) that evolution is equivalent to denying the existence of God; and (3) that fairness required that "both sides" (or both "theories") be taught. Almost all the antievolutionary activism in the last century has been based on one of the three pillars; the emphasis changes as the audience does. So, at a forum at a university, one might hear about the shortcomings of evolution as a scientific theory, while at a church gathering or evangelical rally, the emphasis might be more on evolution as a part of the threat of atheism and secularism in society.

Twentieth century creationism used a three-pronged strategy to attack evolution in science research and education. Almost all the arguments are related to one of three main themes shown here: that evolution is a theory in crisis; that evolution is atheistic; that evolution should not be the only "theory" taught in public schools (Scott, 2008).

Pillar 1: Theory in Crisis

Creationists read scientific journals relentlessly, looking for contradictions among them, areas of disagreement, or unanswered or incompletely answered questions. A good review of typical creationist arguments against the science of evolution can be found in McIver (1992) with a less comprehensive but more recent review by Alston (2003).

In an interesting reflection of creationists' own view of historic documents, their attacks on evolution showed a tendency to treat scientific publications as doctrinal texts: If Charles Darwin wrote about difficulties in explaining the evolution of complex structures, such as the eye, in 1859, then this must be an enduring problem for evolutionary science (but see Petto and Mead (2008)). If there were any errors in Darwin's original framework and its dependent hypotheses, then the whole theory must unravel.

With each new scientific discovery, creationists leaped to point out how this invalidated previous evolutionary thinking—another example to prove that evolution was wrong. This was especially true when scientists described "assumptions" or "adjustments" and corrections in their data; this was *prima facie* evidence that the scientific models themselves were unreliable and should be questioned. Seldom was this more aggressive than in the case of new fossils, with media—and unfortunately, sometimes scientists themselves—trumpeting that a certain fossil changes everything we know about some evolutionary field (but see Meilke and Petto, 2010 or Tarver et al., 2011).

Creationists have failed to publish primary peer-reviewed research journals. Thus they have published their own, and in these journals and in press releases and public communications, they promote their position that scientific findings are unreliable. The most common approach is to claim that inconsistencies or unexplained data are "evidences" against evolution. One of the favorite "evidences" is the out-of-place artifact (OOPART; for example, Heinrich, 2008).

Creationists also attempted to redefine science itself. One strategy was to divide the sciences into "historical" and "operational" sciences (Patterson, 2006). This strategy suggests that data produced by direct observation and experimentation are superior to those that are produced by analyzing the outcomes of natural processes in the past; that these latter are somehow less reliable or less valid because there was no one to witness them, and so they are essentially leaps of faith. Among many others, Cleland (2001) refutes this claim with focus especially on the geosciences. Earlier, Shea (1983) had pointed out that many creationist critiques of geologic sciences relied on an outdated concept of uniformitarianism—the working assumption that the natural laws and processes that produced the features that we can observe in the geologic record are the same ones that govern those processes today.

Another strategy was to use the neologism "origin(s) science" (Geisler and Anderson, 1987). In this view, origin events are unique and therefore cannot be studied scientifically, because scientific study requires the appearance of patterns that arise from the occurrence of many events (for example, Alvarez (1997)). Miller and Totten (2009) demonstrate that engaging appropriate scholarship from many disciplines can help us to study these unique past events fruitfully.

This strategy consistently failed to gain traction in scientific circles—among researchers and educators—but they did serve to convince the general public that science in general, and evolutionary science in particular, could not provide reliable and consistent answers about the history and diversity of life on earth, the age of the earth and the universe, or how organisms that were obviously quite different in their structure and function could in any way be related by descent from a common ancestor.

Pillar 2: Atheism

The second pillar was more successful, as its target was a general public that expressed a high degree of religiosity (Miller et al., 2006). Miller et al. (2006) noted that religiosity was a key predictor of the general acceptance of evolution. Barone et al. (2014) explored the role of religiosity further in acceptance of evolution among visitors to a natural history museum and found that religiosity could offset education as an influence on the acceptance of evolution.

With a general population so committed to religious belief, one way to resist the acceptance of evolution is to convince the believing public that evolution is equivalent—or a slippery slope—to atheism (for example, Colson and Pearcey, 2001). The stakes were high: take a chance on an uncertain and changeable science (Pillar 1) and risk eternal damnation, or stay faithful to God and reject evolution in favor of one's faith tradition. This was—and remains—a powerful argument for people of faith; and there is nothing as compelling in the scientific alternative.

The other reason for the success of Pillar 2 is that United States culture tends to look for "two sides" to every cultural conflict, as we see in politics and media. So, if one side is staking its claim as the "religious" position, then the other side must surely be atheistic. Of course, there are activists in both the religious and the scientific camps that agree with and reinforce this dichotomy, but there is not an insignificant number of citizens who form what philosopher Stachan Donnelley (1989) has called "the troubled middle"—recognizing the benefits that accrue from both positions, and unwilling to give up their stake in either one.

Furthermore, creationists often cast themselves as an embattled minority. Colson and Pearcey (2001) argue that "the dominant view in our culture today" is the "radically one-dimensional" view that "*this life is all there is, and nature is all we need to explain everything that exists*" (p. 18, emphasis in the original). This take on the state of religiosity in the shared culture in the United States, at least, is completely at odds with almost every survey of American attitudes toward naturalistic explanations of life (Miller et al., 2006; Gallup poll: http://www.gallup.com/poll/170822/believe-creationist-view-human-origins.aspx), which consistently show that fewer than 20% of Americans take this distinctly atheistic position.

Some creationists, like Ken Ham of Answers in Genesis, argue that the slide toward secularism is abetted by those believers who give into the teaching of evolution in the public schools (Ham, 2014). Even though these people profess a faith in God, writes Ham, they still are undermining the faithful when they admit naturalism into their world view.

In contrast with the dichotomous view, Scott (2000) has organized the attitudes of the general public on matters of science and faith into a continuum of creationist positions (Figure 2). This continuum reflects a variety of doctrinal positions that relate to the authority of the Bible, the extent to which the biblical text should be interpreted (and how and by whom), the nature of human understanding, and the degree of acceptance of modern scientific knowledge.

Scott (2000) observed that "creationism" was an umbrella term that masked a range of views among religious believers about the role of extranatural and natural causes in the history and diversity of life. She argued that the camps in the creation–evolution dispute were not best described by a simple dichotomy, but by a continuum.

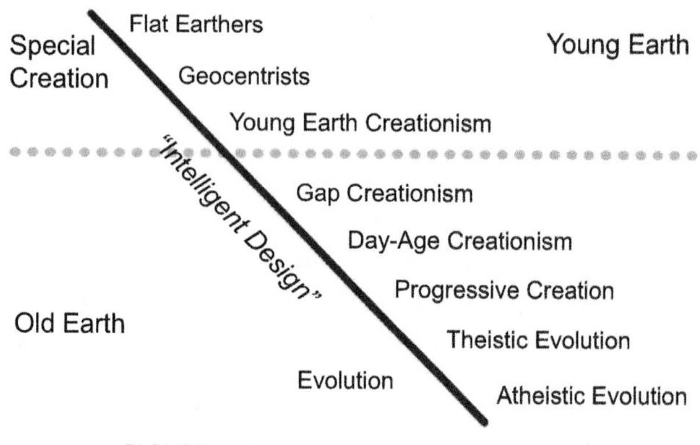

FIGURE 2 The creation–evolution continuum.

Some creationists are content with the authority of the Bible and require no material or earthly confirmation of the scriptural details of the history and diversity of life. Others argue that true science will always corroborate biblical texts ... if only the data were collected and analyzed properly (Plavcan, 2008). This branch of creationism focused on using scientific research methods to prove that the events revealed in the Judeo–Christian Bible correctly recounted the early history of the universe, of the planet, and of life on earth—which is the foundation for Pillar 3.

Pillar 3: Fairness

The third pillar has had notable successes and failures. Its basic appeal is that it is "unfair" to exclude creationism in the public schools for one or more of these reasons: (1) there is compelling evidence to support its scientific status (Pillar 1); (2) there are First Amendment grounds for not forcing children to accept ideas that they (or their parents) do not believe in (Pillar 2); and (3) there is a violation of local control or community standards when schools teach evolution "as fact" in districts where at least a vocal minority opposes it and accepts creationism (Pillars 1 and 2). The successes were in capturing the enthusiasm of those opposing evolution and their representatives in legislatures, school boards, and in some governors' mansions. The failures were in the courts, where federal and state courts repeatedly struck down these laws (Matsumura and Mead, 2007).

When the last state law prohibiting teaching evolution was struck down in 1968 as an unconstitutional endorsement of a sectarian religious restriction that was forced on the general public (Matsumura and Mead, 2007), evolution opponents raised a number of state laws using terms like "Balanced Treatment" or "Fair Treatment" to require that creationism be taught in schools where evolution was also taught. Versions of these laws were struck down in 1982 (*McLean v. Arkansas Board of Education*) and in 1987 (*Edwards v. Aguillard*).

Though *Edwards* was the turning point in the transition from "creation science" to "intelligent design" as the alternative to evolution, there were still some successes to be wrung from the "fairness" argument—at least in the short term. The next strategy—which was also a main approach used in ID—was to allow evolution alone in the curriculum, but to require "disclaimers" to be pasted into textbooks or read aloud in class (Matsumura and Mead, 2007) because doing so would promote "critical thinking" rather that uncritical acceptance of evolution. These policies consistently lost in the courts both as having no merit relative to their intended aims of promoting critical thinking and as being thinly veiled attempts to oppose evolution on religious grounds.

In the end, creationism as science quickly collapsed after this series of losses in court, which all had in common a rejection of the claims that creationism represented an alternative scientific theory (Pillar 1), and that it was the sectarian religious position of the creationists, and not the teaching of evolution in the public schools, that threatened First Amendment principles. With those two pillars in ruins, there was no foundation for the third.

Creation Science Winds Down

There were two key strategies for promoting scientific creationism (also known as creation science): one scientific and one political (see Scott (2009), for details and examples). The scientific strategy was based on research that borrowed the methods and materials used by natural scientists. The goal was to show that the facts could support either view, and the embrace of a naturalistic explanation was a worldview bias of atheistic scientists. The political strategy was based on an appeal to cultural values, calling for equal time for the two theories (arguing that creationism was as much a scientific theory as evolution was; or conversely that evolution was as much built on faith as creationism), fairness, balanced treatment, and rejection of bias. The goal was to pressure legislatures and school boards to require (or at least allow) creationism to be taught as a scientific theory, either on its own or whenever evolution was also taught.

The scientific strategy was largely fruitless. Creationists failed to convince mainstream scientific societies and institutions of their arguments, which focused on the age of the earth, the geologic record and gaps in the fossil record, the evidences [sic] against evolution, supposed deficiencies in genetics and natural selection as making evolution impossible, the inability to explain the emergence and success of complex anatomic structures, the violation(s) of the laws of thermodynamics and probability, and the rare hoax or persistent misinterpretation of fossil evidence. Creationists touted excavations that showed dinosaur trackways associated with what the discoverers determined were human footprints, "proving" that humans and dinosaurs lived at the same time rather than separated by several dozens of millions of years, as evolutionary science (biology and geology) argued.

The political strategy was generally quite successful, at least in the executive and legislative bodies of government. Several states either passed laws or implemented policies that required teachers to include creationism as science in the classroom. State laws had terms like "Balanced Treatment" or "Equal Time" in their titles, showing that legislators accepted the premise that creationism was scientifically valid and only being suppressed by mainstream scientists simply because it acknowledged a Creator's actions in the world.

On the other hand, the legal environment has been a lot less kind to creationism's claims to scientific status (Matsumura and Mead, 2007). Courts have persistently overturned creationism on First Amendment grounds; that is, they see that creationism in all these cases is grounded in a particular sectarian view of the Judeo–Christian Bible. It is not shared by all believers in that Bible, and therefore presenting this view in public school classrooms constitutes a government preference for one religious expression over all others.

In addition, two federal court cases also took extensive evidence on the claims to the scientific status of creationism (*McLean v. Arkansas Board of Education*) and ID (*Kitzmiller v. Dover, 2005*)—scholarship. In both cases, these proposed "alternatives" to evolutionary science failed to convince the courts of their scientific *bona fides*. In fact, the historical evidence points to the defeat of creationism in the *McLean* case as the impetus for what later emerged as "intelligent design theory" in the middle 1980s.

Creationism still persists, of course. Although it still argues that the "historical sciences" are unreliable and that there are "alternative" interpretations of the data, they tend to be more frank about the sectarian underpinnings than creation science was. That is, they see themselves as an evangelical mission, rather than as a scientific one. That role has been almost entirely ceded to ID proponents.

INTELLIGENT DESIGN

The antievolution movement known as "intelligent design theory" (ID) emerged in the middle 1980s, and gained momentum after a series of federal court decisions striking down laws supporting teaching "creation science" in the public schools. Readers should be aware that the term "intelligent design" is commonly used by engineers and industrial designers to develop materials and work spaces that are suited for efficient fulfillment of their intended functions by the humans who will be using them. This long-established meaning is, of course, different from the use of the term by evolution deniers. Instead, ID, as used by antievolutionists is a vague idea that natural processes are simply "not enough" to explain the history and diversity of life on earth. Like creation science, ID rests on the same three pillars.

Pillar 1

The first publication generally agreed on as leading to what would be known as ID theory came shortly after the decision in McLean (1982) in which creationism was rebuffed as science; Judge William Overton ruled that the creation science that was the subject of Arkansas' Act 590 was a thinly disguised religious position. When Thaxton et al. (1984) published *The Mystery of Life's Origins*, the book was almost entirely devoid of the words creation and creationism and based its arguments on contemporary scientific models, rather than on biblical texts. The authors also matter of factly accepted a 4.6 billion year age of the earth. In the tradition of twentieth-century creationism, however, the book recounted the many difficulties of producing the complex biochemical constituents that are considered fundamental to life on earth. They also introduced the concept of the "information" contained in a biological molecule.

These two concepts—complexity and information—would form the basis of the two new lines of argumentation that would mark the scientific foundations of the transition from creation science to ID; these, and not biblical narratives, would be the focus of the evidence against evolution that many ID proponents would present to their audiences. Ultimately, many traditional creationists gave up promoting a strict biblical creationism in the science class in favor of a version of ID, and by the mid-1990s, the Mere Creationism movement held its first conference. Of this event, Schaefer (1996) wrote that "virtually none of the conference participants were creationists of the sort one frequently reads about in the popular press ...a very large majority of the participants had no stake in treating Genesis as a scientific text" (p. 9).

The arguments in Thaxton et al. (1984) also illustrate another two themes in the attempts to validate the opposition to evolution as scientific. The first theme is a shift in the scale and locus of the argument. In matters of scale, the focus shifts from gross anatomy to the cellular and molecular aspects of living things and the aspects of their form and function that make them well suited to their biological roles. Although the argument is essentially the same as it was when creationists objected that natural selection could not allow, for example, half an eye in an evolving lineage (Petto and Mead, 2008), the evidence now takes aim at biochemical structures and processes, which are invisible and difficult for many people to imagine.

In matters of the locus of the argument, there is a shift from descent with modification within evolving lineages to origins: the origin of novelties in the history of life or the first emergence of life itself. This is essentially the strategy used by creationists to change the terms of the discussion by introducing new terms and redefining science itself.

The second theme was to bolster the appearance of scientific *bona fides* by making the exposition of the evidence more technical and mathematical. Although the main conclusion was the same as that of creation science—contemporary mainstream science has failed to provide a plausible explanation for some event or condition that defines the essential qualities of living organisms—the arguments could not be refuted without reference to the same technical and sometimes esoteric disciplinary research. After all, it was the mainstream scientists' own research that was on the basis of the ID formulations for why evolutionary science failed to account for at least some aspects of the history and diversity of life on earth (though many, like Yockey (2005) would argue that these relied on misinterpretations and misapplications of the scientific fields they claimed to embrace, and that a proper understanding of these fields would lead to the opposite conclusions).

The backbone of ID's claims to scientific legitimacy was built on complexity. Two seminal works laid the groundwork for much of what would come later. The earlier of these was Michael Behe's *Darwin's Black Box* (1996). Behe's main argument has to do with complexity: irreducible complexity.

> By irreducibly complex *I mean a single system composed of several well-matched, interacting parts that contribute to the basic function, wherein the removal of any one of the parts causes the system to effectively cease functioning. An irreducibly complex system cannot be produced directly (that is, by continuously improving the initial function, which continues to work by the same mechanism) by slight, successive modifications of a precursor system, because any precursor to an irreducibly complex system that is missing a part is by definition nonfunctional (p. 39).*

This argument echoes a concern of Darwin (1859): "if it could be demonstrated by any complex organ existed which could not possibly have been formed by numerous, successive, slight modifications, my theory would absolutely break down" (p. 189) and is also one of the rhetorical threads that connect creationist and ID attacks on evolution (Pillar 1): tying the evolution of complexity to the gradualistic model of biological change that Darwin and his colleagues assumed must be at the basis of descent with modification, despite the fact that there are several other scientifically well-studied processes of organismic change that can provide the raw material for evolution (Petto and Godfrey, 2008).

Behe's (1996) case rested on so-called molecular machines, such as flagella, and complex biochemical pathways, such as those in blood clotting and the immune system. In each case, biological researchers have demonstrated plausible naturalistic pathways for producing these features without the intervention of an intentional agent. The genes that produce flagella and the proteins that make them up in many bacteria suggest that they were produced by gene duplication (Blocker et al., 2003) in which the first gene retains its original function, but the new duplicate can be modified and develop a new function without harming the cell.

Long before Behe wrote *Darwin's Black Box,* however, Robinson et al. (1969) had demonstrated that, among other vertebrates, the cetacea (whales and dolphins) lacked one of the clotting factors in the presumed "irreducibly complex" biochemical sequence that produces blood clots in mammals. In the 40 years since, researchers have confirmed these findings and understood how clotting can happen in these systems.

Behe's argument presumes that Darwin's (1859) conception of biological change is doctrinal and has been unaltered by over 150 years of subsequent research. Darwin did not know about gene duplication (or genes for that matter), which is a non-Darwinian, but evolutionary, explanation for the processes that Behe (1996) argues cannot be produced in a Darwinian model. Strictly speaking, Behe's argument is correct: Darwin's model did not include this sort of biological change in organisms. On the other hand, the phylogenetic predictions of descent with modification—the basis of evolutionary theory—could not posit any other pattern than the unique biochemical pathway for blood clotting pioneered in an ancestor of modern whales and dolphins would be shared by all its descendants, marking them as members of the same evolving lineage.

William Dembski's first book, *The Design Inference*, laid out a case that what he called "specified complexity" could be evidence for purposeful design in Nature, and he proposed an Explanatory Filter (Dembski, 1998, p. 37) that could separate purposeful design from other causes for complex features. Regularity he defined as an "event that will (almost) always happen" and chance as based on laws of probability (p. 36).

> *To attribute an event to design is to say that it cannot reasonably be referred to either regularity or chance. Defining design as the set-theoretic complement of the disjunction regularity-or-chance guarantees that the three modes of explanation are mutually exclusive and exhaustive (p. 36).*

Dembski (1998) lays out an elaborate derivation to support the value of his explanatory filter, but the most powerful critique of this approach is the same as that for the creationist approach to probabilistic events: our inability to know in advance what the true probabilities are for these events and the lack of an accepted way of estimating those probabilities in the light of this ignorance (Van Till, 2011). Elsberry (2008) summarizes the difficulties in actually applying this explanatory filter in biological research—a judgment apparently confirmed by scientific consensus in the failure of researchers to adopt and apply this approach.

Dembski's (2002, 1998) books are densely written, drawing on his background in philosophy and mathematics. They are not the first theoretical constructs that could not be applied except in a much simplified form. One of his most successful arguments in terms of public acceptance of his critique of evolution relates to information theory. Information theory focuses on the complexity of a message and the uncertainty in our ability to predict the components of that message. Information relates to the level of uncertainty: less uncertainty means more information (see Stenger (2008) for an explanation and critique of Dembski's use of information). It is important to note that the "information" in information theory is not related to meaning or content of a message, but to the probability of getting the message right. Obviously, the more complex the message, the more likely that we will get some part of it wrong.

Information-based arguments in Dembski's work resonate with antievolutionists and are quite common in arguments against evolution. The general thrust of these arguments is that the process of mutation degrades genetic information, and so the message of the DNA in our genes is negatively affected. Meyer (2009) makes this sort of argument in *Signature in the Cell* in which he argues that Dembski's "specified complexity" is equivalent to the content of the DNA. However, Meyer's argument falters on the specific critique that Stenger (2008) made of Dembski's work: uncertainty is always reduced by knowing the outcome. The content and structure of DNA in a cell are not uncertain once they are in place regardless of how complex the molecule is. The uncertainty in information theory is in predicting and transmitting the information (such as the sequences of our DNA), not in describing the molecule once these sequences are known.

It was during the early 1980s that the first ID textbook was developed (Kenyon and Davis, 1989). Originally proposed as a supplement to standard biology textbooks, *Pandas* (as it came to be known) focused on the origins of life and the unresolved state of contemporary science. Indeed, at the end of the book they tell the reader

> *Throughout the six previous chapters we have sampled representative data related to origins from different areas of biology. We saw mounting problems in the assumptions underlying chemical evolution accounts of the origin of life. We observed how biological organisms exhibited the characteristics of manufactured things. We became aware of the enormous difficulties for the change production of functional information in the message text of DNA.*
>
> Davis and Kenyon (1993), p. 147.

This text is the essential message of ID, and the subtext is that if mainstream science has failed to answer these research questions, then the only alternative is that there was an intentional intervention in the emergence of life on earth. This line of argumentation is another legacy from creation science.

Although the first edition was published in 1989, historian Barbara Forrest (2006), in preparing her testimony for *Kitzmiler v. Dover,* discovered that the original manuscripts contained the terms and phrases relating to creationism which had been replaced by those relating to ID after the 1987 *Edwards v. Aguillard* case.

> *Beginning with the 1986 draft, "creation" was defined using the classic creationist concept of "abrupt appearance": "Creation means that the various forms of life began abruptly through the agency of an intelligent creator with their distinctive features already intact—fish with fins and scales, birds with feathers, beaks, and wings, etc." The 1989 and 1993 published versions preserve this definition verbatim, except that "intelligent design" and "agency" are substituted for "creation" and "creator", respectively.*
>
> Forrest (2006), p. 48.

Despite the objections of ID proponents that their proposition was scientific and distinct from creationism, the supporting text from *Pandas* apparently required no substantive changes to make its case for ID instead of creationism. Figure 3 shows both the extent and the "abrupt appearance" of the substitutions of the ID-related terms in place of the creation-related terms.

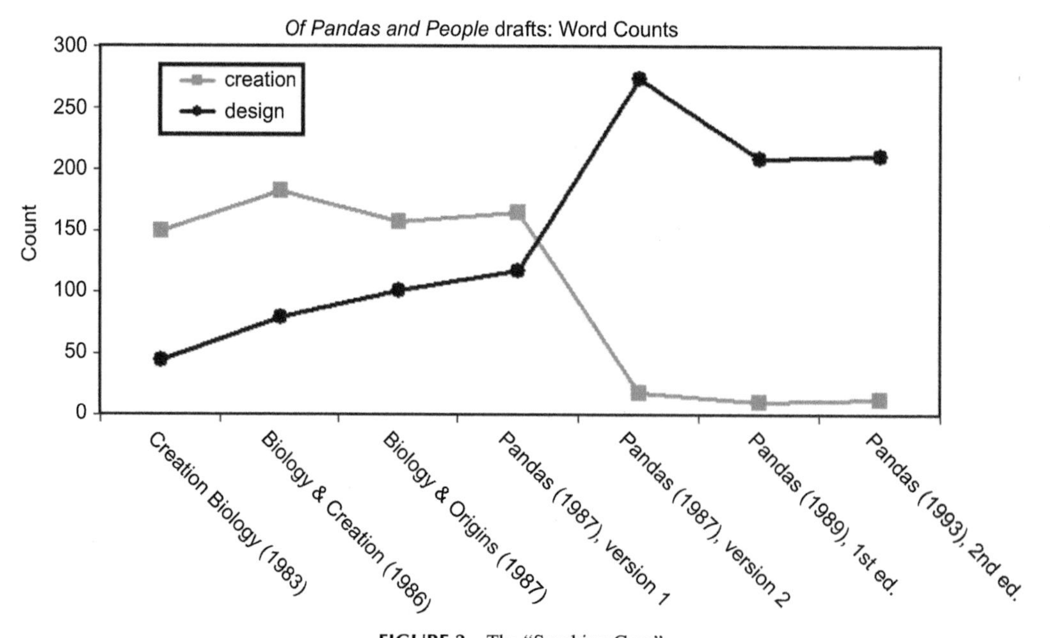

FIGURE 3 The "Smoking Gun."

Forrest (2006) shows the creationist intellectual history of the premier ID textbook supplement. The United States Supreme Court definitely ruled "creation science" as a religious, not scientific, construct in 1987. Almost immediately terms in the text with the root "creation" in them were replaced globally with "intelligent design." For example, "creationists" was replaced by "intelligent design proponents" with almost no other changes in the text.

It was a donation of *Pandas* to the Dover (PA) Area School Board that was the catalyst in the *Kitzmiller v. Dover* case (Matzke, 2004). Matzke (2004) points out that almost all the common arguments used in ID in the past few decades are found first in the pages of that textbook supplement.

Perhaps the most significant of the efforts to undermine the validity of evolutionary science is a publicity campaign that began as a newspaper advertisement: "A Scientific Dissent From Darwinism" (http://www.dissentfromdarwin.org/). The original list had 100 signatures, though the term "scientist" included historians and philosophers of science (Evans, 2002b). Those dissenters signed this statement:

> *We are skeptical of claims for the ability of random mutation and natural selection to account for the complexity of life. Careful examination of the evidence for Darwinian theory should be encouraged.*

Like Behe's (1996) argument, "Darwinian theory" in this dissent is a straw man—an incomplete description of evolutionary science. Random mutation and natural selection are surely important aspects of evolutionary change; however, they do not complete the toolbox of biological change that contributes to evolution (see, for example, Petto and Godfrey, 2008).

This list has grown to several 100 dissenters since its inception in 2001 (Evans, 2002b). To emphasize the vanishingly small percentage of scientists that have signed on, the National Center for Science Education initiated "Project Steve" in honor of Stephen J. Gould (http://ncse.com/taking-action/project-steve), showing that even limiting the list to scientists named Steven (or Stephen or Stephanie, etc.), the support for evolution among scientists dwarfed the dissenters' ranks. However, the campaign is effective because the 800 or so "scientists" who have signed the dissent seems like a large number. In fact, it represents something less than one-half of 1% of the 2.2 million scientists and engineers awarded PhDs in the decade between 1998 and 2008 http://www.nsf.gov/statistics/indusworkforce/)—most of whom we would expect are still working.

Pillar 2

The major works of ID have focused mainly on the perceived shortcomings of natural science—a legacy of the "theory in crisis" pillar from the creationist movement in earlier decades. However, like their predecessors, ID creationists had a grander vision. This vision was first glimpsed in 1998 in a working document called "The Wedge" (http://www.antievolution.org/features/wedge.pdf) produced at the Center for the Renewal of Science and Culture at the Discovery Institute (http://www.discovery.org/csc/). This document laid out a strategic plan for the Center to achieve its broad goals:

> *Discovery Institute's Center for the Renewal of Science and Culture seeks nothing less than the overthrow of materialism and its cultural legacies. Bringing together leading scholars from the natural sciences and those from the humanities and social sciences, the Center explores how new developments in biology, physics and cognitive science raise serious doubts about scientific materialism and have re-opened the case for a broadly theistic understanding of nature (p. 2).*

The document was reportedly retrieved from a copy center in Seattle and distributed by those opposed to the programs of the CRSC (which later changed its name to Center for Science and Culture). The Center for Science and Culture finally acknowledged the document in 2003, defending the chief statements in the document (http://www.discovery.org/a/2101). In addition to its goals for scientific progress (Phase I), the "Wedge" strategy included projects in publicity and opinion-making (Phase II), and "cultural confrontation and renewal" (Phase III). Phase II focused on publicity, media productions, legislative contacts, opinion pieces in major news outlets, and various seminars and educational programs.

Phillip Johnson, one of ID's "founding fathers," was promoting the wedge strategy in print pretty much right out of the gate (Johnson, 1999). The call then was not for a specific supernatural being, but some vague intelligent agency. Though trying hard to avoid invoking a *specific* intelligent agent, the actor in question must have the power to intervene in nature in a way that produces effects that are not possible under natural laws or to prevent effects that natural laws should produce; there are not many qualified candidates for that position other than the Creator God of the Middle-Eastern monotheistic religions that we see now in contemporary Christianity, Judaism, and Islam—all of which have strong creationist traditions.

As Johnson's note in *Touchstone* (Johnson, 1999) shows, ID proponents use carefully crafted language to couch the religious argument. This part of the strategic plan, however, makes it abundantly clear that they set themselves in opposition to a secular—and in their view an atheistic—natural science.

Johnson also told a reporter in an interview a few years earlier:

> *This isn't really, and never has been, a debate about science, ... it's about religion and philosophy.*
>
> Belz (1996); p. 18.

Another leading ID proponent, William Dembski, writing in the same issue of *Touchstone* described ID as "the *Logos* theology of John's Gospel restated in the idiom of information theory" (Dembski, 1999, p. 84).

Despite the public disclaimers that ID is about improving the practice of scientific inquiry, both these leading proponents make clear that ID is also supported by the second pillar of creationism: evolution equals atheism.

Pillar 3

ID proponents have used a variety of tactics to move ID into the public schools. Two examples are works by Wells (2002) and Meyer et al. (2007), which followed the creationist playbook. Starting on Pillar 1, they take aim squarely at evolution education by focusing on the shortcomings of evolutionary science—condensed into the phrase "neo-Darwinism"—and perpetuating the notion that evolution should be regarded as scientifically controversial. Wells (2002) focuses on errors and misrepresentations (often gross oversimplifications) in public school textbooks making the logical leap to presume that the science behind these ideas is also wrong.

Meyer et al. (2007) supposedly encourage critical examination of the strengths and weaknesses of evolutionary science, but Mead (2008) found that evolutionary content was either weakened or presented incorrectly—a charge that has been leveled at Meyer's other works (for example, Cook, 2013). What these texts do show is that ID proponents no longer appear to be attempting to insert ID into United States public schools per se. Rather, the strategy is to sow doubt about evolution, focusing on its weaknesses and the questions still unanswered. If the science behind evolution is weak, then it is "only fair" to allow other views into the classroom (Pillar 3).

The "teach the controversy" and "critical thinking" slogans of ID proponents give the impression that we are being unfair to students if we withhold ID materials from them—either because there are "a lot" of scientists who disagree with so-called neo-Darwinism or because they need to learn how to be critical thinkers. As with creation science supporters, ID proponents have proposed a number of model bills and policies for legislators and school districts, urging schools to examine evolutionary science as though it were scientifically controversial. Model legislation introduced in several states includes evolution on a short list, along with human cloning, climate change, and chemical origins of life (Anonymous, 2013a) as "controversial" topics. These campaigns obviously confuse the public into thinking that the social controversy over evolution reflects a similar conflict among scientists, and since a significant proportion of United States citizens finds evolution objectionable, these statements seem to be eminently reasonable.

In addition, to support offering students a full range of views on controversial issues, model legislation calling for "academic freedom" (Anonymous, 2009) proposed that students had the right to hear and teachers had the right to present in their science classes a wide variety of opinions on matters related to the curriculum. In the second decade of the twenty-first century, model legislation that uses one or both of these ideas is regularly introduced to state legislatures (Anonymous, 2013b).

This tactic was quite successful in promoting ID as a viable scientific program and one that ought to be included in public school curricula. The most successful of these efforts was in Kansas, where the State Board of Education officially sanctioned ID as a part of the science curriculum, at least temporarily (Cunningham, 1999; Anonymous, 2001). Shortly thereafter Ohio was the focus of an intense, but short-lived, challenge to teaching evolution in the public schools (Evans, 2002a; Miller, 2002).

A significant feather in the ID cap came in the form of what became known as "the Santorum Amendment" to the Elementary and Secondary Education Act Authorization in 2001 (Branch, 2002; Discovery Institute, 2012). The amendment inserted carefully framed language alluding to evolution as controversial and referring back to its potential negative influence on cultural and religious values (Pillar 2).

Despite their success in the forum of public opinion (and legislative action), ID proponents and their propositions fared much less well in the legal sphere. For example, Behe's arguments about the irreducible complexity of the immune system were a feature of his testimony in *Kitzmiller v. Dover*, echoing his judgment that the evolution by natural selection of the human immune system was impossible. Bottaro et al. (2006) discuss research in the field that contradicts Behe's testimony, but Presiding Judge John E. Jones evaluated the testimony this way:

> *He was presented with fifty-eight peer-reviewed publications, nine books, and several immunology textbook chapters about the evolution of the immune system; however, he simply insisted that this was still not sufficient evidence of evolution, and that it was not "good enough".*
>
> Kitzmiller v. Dover (2005)

It is clear that ID shares the same track record as creationism on the third pillar: it holds up a significant portion of public opinion, while there is simply no scientific support to keep it upright.

Intelligent Design Sputters Along

Of all three phases of the wedge strategy, the least successful has been to change the acceptance of ID by the scientific community. ID remains a scientific backwater because it is not widely adopted by scientific researchers, even as it remains quite popular with the general public (and with some legislators). It is true that evolution is a controversial subject with the general United States citizenry, but the success of ID has been due, in large part, to convincing the public that evolution is also controversial in the scientific community.

Like its previous incarnation, creationism, ID's scientific program is still fruitless after almost 30 years. Its success is in its appeal to religious believers who cannot or will not accept a universe without direct Divine agency, so they accept the claims that evolution is controversial and is about ready to fall apart under its own weight for failure to answer critical biological questions.

ID is also successful politically, having its concepts and ideas (both about evolution and about ID) incorporated into legislation or into school-board policies. However, when put to a legal test, it inevitably fails, as it does in the sciences.

The success of ID in the arena of public opinion and legislative bodies contrasts with its failure to gain traction in the sciences. This is because the price of admission to the arena of scientific ideas is successfully conducting original research. In the case of ID, Petto and Godfrey (2008) contrast ID's lack of success in that realm compared to 14 other non-Darwinian ideas that are now a standard part of the evolutionary mechanisms discussed in biology textbooks. These ideas were included in textbooks after years of research and testing, but there is no similar record of research to support the claims for a place in the science curriculum or in scientific policy of so-called ID theory.

CONCLUSIONS

Opposition to evolution in the United States has been transformed through the late twentieth century. Emerging first as an overtly biblical objection to the scientific models for the history and diversity of life, creationism first presented itself as a scientific alternative to evolution. Early legislation outlawed teaching evolution at all, and later laws called for equality or balance or fairness in presenting "both models," but the United States courts repeatedly struck down these laws as being frankly religious and having no secular purpose. Despite subtle changes in language in a series of these laws, at least two federal courts have now ruled that the claims of creationists, and later ID proponents, to scientific status were unsupportable.

While frankly religious organizations still promote a literal biblical perspective on the history of life on earth, it is ID that has taken center stage since the 1990s. Although many in ID acknowledge the obvious links between their science and Christian theology, they deny that ID is a religious model, and they eschew specific references to the Bible, particularly Genesis. Still, their public presentations and publications often argue that the naturalistic, or materialistic, bent in modern science is responsible for a variety of social ills.

The intellectual legacy of creationism, however, is writ large in ID. The objections to the shortcomings of evolutionary science (Pillar 1) are essentially the same: arguments about entropy (once about energy and disorganization, now about information), complexity, randomness and purpose, emergence of new structures, and modifications of old ones. The basic complaint seems to be (1) that evolutionary scientists content themselves with searching for naturalistic cause-and-effect relationships in their study of the history and diversity of life on earth; and (2) evolutionary scientists do not seek (or seem to find) in their studies any grand plan for life—especially human life.

In their presentations to the general public, ID proponents complain that natural scientists are artificially excluding useful lines of investigation because they will not use "all the tools in the tool kit" to study life on earth. By excluding design and by rejecting the analytical tools that could suggest purposeful design, they say, evolutionary scientists are missing out on powerful explanatory models that could solve a raft of problems that are as yet unsolved using naturalistic methods (Pillar 2). The implication is that using all the available tools, as it were, would give us better, more robust, and more useful answers, but the metaphor leaves one wondering how we would react if a roofer that we had hired showed up with plumbing tools using a similar explanation of the usefulness of wrenches, faucet clamps, and pipe cutters.

ID proponents also argue that we can solve crimes from evidence of crime scenes because we can infer purposeful action on the part of alleged perpetrators, and so, it is also possible to infer evidence of purposeful action in Nature. Of course, our criminals are ordinary humans who (1) are bound by natural laws and (2) act in ways that we know from experience. Neither of these is true of the unnamed intelligent agent behind ID.

The larger stakes in the antievolution movements of the last several decades are about shared values: traditional knowledge that gives authority to social practices and institutions. Kehoe (2008) argues that resistance to the authority of science is a symptom, or perhaps a manifestation, of an increasing secularization of society. A secular society confers authority based on evidence that can be evaluated empirically, while traditional knowledge is received as true. This authority influences the values and ideals that are shared in civil society and passed among generations—something Eve (2009) referred to as "a struggle for the means of cultural reproduction" (p. 31). In other words, opposition to evolution—and other modern scientific disciplines—is an effort to preserve the role of traditional narratives in defining and guiding future generations (Pillar 3).

The issue at the heart of the social controversy over evolution is just that: the purpose and meaning of life—the proper subject matter of philosophy and religion. To borrow an ID metaphor, science does not have the tools for this job, nor should it. At the heart of the matter, of course, is the place of humans in the order of things in the universe.

Both creationism and ID preserve a special role for humans over other living things on the planet. What drives many of the objections to evolutionary science relate to the total reevaluation of the place of humans in relation to the rest of Nature. The evolutionary perspective places humans at the end of one branch of the tree of life not as the angel at the top of the Christmas tree.

This is perhaps the most objectionable bit of evolutionary science from the point of view of many religious believers. It is certainly the one that bothers most of the creationists that have discussed this with me over the years—that if evolution is true, then we humans are no longer special.

As other chapters in this work will demonstrate, evolutionary science considers humans to be a remarkable species in many ways, but still a part of Nature, not apart from it. All the natural laws and processes that apply to the rest of life on earth also apply to us. And the result of operation of all those laws and processes is that, of more than 20 species of hominins that have dwelled on the earth at one time, only a single species remains today. Evolution helps us explain that history—*our* history (a bumper sticker from the National Center for Science Education reads: "*Australopithecus* ends in US").

On the other hand, we know of over 300,000 species of beetles (Haldane, 1949, p. 248), and so, if life on earth represents the intentional planning of an intelligent agent, then the response attributed to Haldane—that the Creator must have an inordinate fondness for beetles—must be seriously entertained.

REFERENCES

Anonymous, 2001. Kansas Board of Education Reinstates Evolution. National Center for Science Education. Available from: http://ncse.com/news/2001/02/kansas-board-education-reinstates-evolution-00224 (last accessed 20.09.13).

Anonymous, 2009. Discovery Institute's "Model Academic Freedom Statute on Evolution". National Center for Science Education. Available from: http://ncse.com/creationism/legal/discovery-institutes-model-academic-freedom-statute-evolution (last accessed 20.09.13).

Anonymous, 2013a. Antiscience Bill Dies in Oklahoma. National Center for Science Education. Available from: http://ncse.com/news/2013/02/antiscience-bill-dies-oklahoma-0014724 (last accessed 20.09.13).

Anonymous, 2013b. Chronology of "Academic Freedom" Bills. National Center for Science Education. Available from: http://ncse.com/creationism/general/chronology-academic-freedom-bills (last accessed 20.09.13).

Alston, J.P., 2003. The Scientific Case Against Scientific Creationism. iUniverse, New York.

Alvarez, W., 1997. T Rex and the Crater of Doom. Princeton University Press, Princeton, NJ.

Barone, L.M., Petto, A.J., Campbell, B.C., 2014. Predictors of evolution acceptance in a museum population. Evolution Education and Outreach 7 (23). Available online at: http://www.evolution-outreach.com/content/7/1/23.

Behe, M.J., 1996. Darwin's Black Box: The Biochemical Challenge to Evolution. The Free Press, New York.

Belz, J., 1996. Witnesses for the prosecution. World 11 (28), 18. Available from: http://www.worldoncampus.com/1996/11/witnesses_for_the_prosecution (last accessed 20.09.13).

Blocker, A., Komoriya, K., Aizawa, S.-I., 2003. Type III secretion systems and bacterial flagella: insights into their function from structural similarities. Proceedings of the National Academy of Sciences of the United States of America 100 (6), 3027–3030. Available from: http://www.pnas.org/content/100/6/3027.full (last accessed 20.09.13).

Bottaro, A., Inlay, M.A., Matzke, N.J., 2006. Immunology in the spotlight at Dover 'intelligent design' trial. Nature Immunology 7 (5), 433–435.

Bowler, P.J., 2007. Monkey Trials and Gorilla Sermons: Evolution and Christianity from Darwin to Intelligent Design. Harvard University Press, Cambridge, MA.

Bowler, P.J., 2010. Reconciling Science and Religion the Debate in Early Twentieth-century Britain. University of Chicago Press, Chicago.

Branch, G., 2002. Farewell to the Santorum amendment. Reports of the National Center for Science Education 22 (1–2), 12–14.

Cleland, C.E., 2001. Historical science, experimental science, and the scientific method. Geology 28 (11), 987–990.

Colson, C., Pearcey, N., 2001. Developing a Christian Worldview of Science and Evolution. Tyndale House Publishers, Inc., Wheaton, IL.

Cook, G., 2013. Doubting "Darwin's Doubt". New Yorker Online. Available from: http://www.newyorker.com/online/blogs/elements/2013/07/doubting-stephen-meyers-darwins-doubt.html (last accessed 20.09.13).

Cunningham, D.L., 1999. Creationist tornado rips evolution out of Kansas science standards. Reports of the National Center for Science Education 19 (4), 10–15.

Darwin, C.R., 1859. On the Origin of Species by Means of Natural Selection, or the Preservation of Favoured Races in the Struggle for Life. John Murray, London. Available from: http://darwin-online.org.uk/contents.html#origin/ (last accessed 20.09.13).

Davis, P., Kenyon, D.C., 1993. Of Pandas and People: The Central Question of Biological Origins, second ed., Haughton Publishing Co., Dallas.

Dembski, W.A., 2002. Intelligent Design: The Bridge between Science and Theology. InterVarsity Press, Downer's Grove, IL.

Dembski, W.A., 1999. Signs of intelligence: a primer on the discernment of intelligent design. Touchstone 12 (4), 76–84.

Dembski, W.A., 1998. The Design Inference: Eliminating Chance through Small Probabilities. Cambridge, New York.

Dennett, D., 1995. Darwin's Dangerous Idea: Evolution and the Meaning of Life. Simon and Schuster, New York.

Discovery Institute, 2012. Rick Santorum, the Santorum Amendment, and Intelligent Design. Available from: http://www.discovery.org/a/18071 (last accessed 20.09.13).

Donnelley, S., 1989. Speculative philosophy, the troubled middle, and the ethics of animal experimentation. Hastings Center Report 2, 15–21.

Edwards v. Aguillard, 1987. 482 U.S. 578. Available from: http://www.law.cornell.edu/supct/html/historics/USSC_CR_0482_0578_ZS.html (last accessed 20.09.13).

Elsberry, W.R., 2008. Logic and math turn to smoke and mirrors: William Dembski's "design inference". In: Petto, A.J., Godfrey, L.R. (Eds.), Scientists Confront Creationism: Intelligent Design and Beyond. W. W. Norton, New York, pp. 250–271.

Evans, S., 2002a. Ohio: the next Kansas? Reports of the National Center for Science Education 22 (1–2), 4–6.

Evans, S., 2002b. Doubting Darwin through Creative License. National Center for Science Educationhttp://ncse.com/creationism/general/doubting-darwinism-creative-license (last accessed 20.09.13).

Eve, R.A., 2009. Reflections on a visit to the Creation Museum. Reports of the National Center for Science Education 29 (5), 31–33.

Forrest, B., 2006. My role in *Kitzmiller v Dover*. Reports of the National Center for Science Education 26 (1–2), 47–48.

Geisler, N.L., Anderson, J.K., 1987. Origin Science. Baker Books, Grand Rapids, MI.

Gruber, H.E., 1974. Darwin on Man: A Psychological Study of Scientific Creativity. EP Dutton, New York.

Haldane, H.B.S., 1949. What Is Life? A Layman's View of Nature. L. Drummond, London.

Ham, K., 2014. One Nation Under ...? Answers Magazine. 21 November, p. 39.

Heinrich, P.V., 2008. The mysterious spheres of Ottosdal, South Africa. Reports of the National Center for Science Education 29 (1), 28–33.

Johnson, P.E., 1999. The wedge: breaking the modernist monopoly on science. Touchstone 12 (4), 18–24.

Kehoe, A.B., 2008. Why target evolution? The problem of authority. In: Petto, A.J., Godfrey, L.R. (Eds.), Scientists Confront Creationism: Intelligent Design and Beyond. W. W. Norton, New York, pp. 381–404.

Kenyon, D.C., Davis, P., 1989. Of Pandas and People: The Central Question of Biological Origins, second ed., Foundation for Thought and Ethics, Richardson, TX.

Kitzmiller v. Dover (Tammy Kitzmiller, et al. v. Dover Area School District, et al.), 2005. Case No. 04cv2688. Available from: http://ncse.com/files/pub/legal/kitzmiller/highlights/2005-12-20_Kitzmiller_decision.pdf (last accessed 20.09.13).

McCullough, J., 1843. Proofs and Illustrations of the Attributes of God: From the Facts and Laws of the Physical Universe: Being the Foundation of Natural and Revealed Religion. J Duncan, London.

McIver, T., 1992. Anti-evolution: A Reader's Guide to Writings before and after Darwin, second ed., Johns Hopkins, Baltimore.

McLean v. Arkansas Board of Education, 1982. 529 F. Supp. 1255, 50 U.S. Law Week 2412. Available from: http://supreme.justia.com/cases/federal/us/211/539/ (last accessed 20.09.13).

Matsumura, M., Mead, L.S., 2007. Ten Major Court Cases about Evolution and Creationism. [Internet] Available from: http://ncse.com/taking-action/ten-major-court-cases-evolution-creationism (accessed 26.08.13).

Matzke, N.J., 2004. Design on trial in Dover, Pennsylvania. Reports of the National Center for Science Education 24 (5), 4–9.

Mead, L.S., 2008. Explore evolution: notes from the field. Reports of the National Center for Science Education 28 (1), 11–12.

Mielke, W.E., Petto, A.J., 2010. Fossils that change everything we know about human evolution (... or not). Evolution Education and Outreach 3, 477–480.

Meyer, S.C., 2009. Signature in the Cell: DNA and the Evidence for Intelligent Design. HarerOne, New York.

Meyer, S.C., Minnich, S., Moneymaker, J., Nelson, P.A., Seeke, P., 2007. Exploring Evolution: The Arguments for and against Neo-Darwinism. Hill House Publishing, Melbourne.

Miller, J.D., Scott, E.C., Okamoto, S., 2006. Public acceptance of evolution. Science 313, 765–766.

Miller, K.B., Totten, I., 2009. Developing and implementing an interdisciplinary origins course at a state university. Journal of College Science Teaching 38 (4), 24–29.

Miller, K.R., 2002. Goodbye, Columbus. Reports of the National Center for Science Education 22 (1–2), 6–7.

Moreland, J.P., 2008. Intelligent design and the nature of science. In: House, H.W. (Ed.), Intelligent Design 101: Leading Experts Explain the Key Issues. Kregel, Grand Rapids, MI, pp. 43–65.

Numbers, R.L., 1998. Darwinism Comes to America. Harvard University Press, Cambridge, MA.

Numbers, R.L., 2006. The Creationists: From Scientific Creationism to Intelligent Design. Harvard University Press, Cambridge, MA.

Patterson, R., 2006. Evolution Exposed: Biology. Evolution Answer Book for the Classroom. Answers in Genesis, Peterborough, KY.

Paley, W., 1802. Natural Theology: or, Evidences of the Existence and Attributes of the Deity, Collected from the Appearances of Nature. Gould, Kendall, and Lincoln, Boston.

Petto, A.J., Godfrey, L.R., 2008. Why teach evolution? In: Petto, A.J., Godfrey, L.R. (Eds.), Scientists Confront Creationism: Intelligent Design and Beyond. W. W. Norton, New York, pp. 405–441.

Petto, A.J., Mead, L.S., 2008. Misconceptions about the evolution of complexity. Evolution Education and Outreach 1 (4), 505–508.

Plavcan, J.M., 2008. The invisible Bible: the logic of creation science. In: Petto, A.J., Godfrey, L.R. (Eds.), Scientists Confront Creationism: Intelligent Design and Beyond. W. W. Norton, New York, pp. 361–380.

Robinson, A.J., Kropatkin, M., Aggeler, P.M., 1969. Hageman Factor (Factor XII) Deficiency in Marine Mammals. Science 166, 1420–1422.

Schefer III, H.F., 1996. Foreword. In: Dembski, W.A. (Ed.), Mere Creationism. InterVarsity Press, Downer's Grove, IL, pp. 9–12.

Scott, E.C., 2000. The Creation/evolution Continuum. [Internet] National Center for Science Education. Available from: http://ncse.com/creationism/general/creationevolution-continuum (last accessed on 26.08.13).

Scott, E.C., 2008. The Pillars of Creationism. National Center for Science Education. Available from: http://ncse.com/taking-action/pillars-creationism (last accessed on 29.11.14).

Scott, E.C., 2009. Evolution vs Creationism: An Introduction. Greenwood Press, Westport, CT.

Shea, J.H., 1983. Creationism, uniformitarianism, geology and science. Journal of Geological Education 31 (2), 105–110.

Stenger, V.J., 2008. Physics, cosmology, and the new creationism. In: Petto, A.J., Godfrey, L.R. (Eds.), Scientists Confront Creationism: Intelligent Design and Beyond. W. W. Norton, New York, pp. 131–149.

Tarver, J.E., Donoghue, P.C.J., Benton, M.J., 2011. Is evolutionary history repeatedly rewritten in light of new fossil discoveries? Proceedings of the Royal Society B 278 (1705), 599–604.

Thaxton, C.B., Olsen, R.L., 1984. The Mystery of Life's Origin: Reassessing Current Theories. Philosophical Library, New York.

Van Till, H.A., 2011. *E coli* at the no free lunch room. Metanexus Institute, New York. Available from: http://www.metanexus.net/essay/e-coli-no-free-lunchroomwww.aaas.org/spp/dser/evolution/perspectives/vantillecoli.pdf (last accessed 20.09.13). Originally published as *E coli* at the no free lunchroom: bacterial flagella and William Dembski's case for intelligent design.

Wells, J.A., 2002. Icons of Evolution: Science or Myth? Why Much of What We Teach about Evolution Is Wrong. Regnery Publishing, Washington, DC.

Yockey, H.P., 2005. Information Theory, Evolution, and the Origin of Life. Cambridge, New York.

Part II

Primates

Chapter 3

Primate Evolution

Robert D. Martin
Department of Anthropology, The Field Museum, Chicago, IL, USA

SYNOPSIS

The order Primates, containing some 400 extant species, is clearly defined by a set of shared derived features. There is a basal divergence between strepsirrhines (lemurs and lorisiforms) and haplorhines (tarsiers, monkeys, and apes). Primates are typically tree-living inhabitants of low-latitude forests, with grasping hands and feet, flat nails, enhanced tactile sensitivity, and hind-limb domination. Uniquely among mammals, the petrosal bone forms the auditory bulla. Vision is especially important. Enlarged, forward-facing eyes scan a wide binocular field and special brain organization permits refined stereoscopic vision. Despite extensive overlap with other mammals, primates have relatively larger brains on average because fetal brain development is prioritized. Dentitions are relatively unspecialized, but anterior teeth are reduced while molar cusps are low and rounded. Various features contribute to a slow reproductive turnover. A patchy fossil record extends back 55 million years, but fossils documenting earliest primate evolution remain elusive. The little-studied colugos (Dermoptera) may be the closest relatives of primates.

INTRODUCTION

In his widely influential 1945 classification of mammals, George Gaylord Simpson's pivotal synopsis of primates began with a now-famous sarcastic comment: "The primates are inevitably the most interesting of mammals to an egocentric species that belongs to this order" (p. 180). Hubris easily leads us to forget that *Homo sapiens*, the name originally applied to humans by Carl Linnaeus in the mid-eighteenth century, is only one of over 400 extant species allocated to the order Primates and just one of over 6000 species in the class Mammalia. Nonetheless, because of interest in our own origins, inclusion of humans in Primates has sparked intensive research and emergence of primate biology as a discipline. An original focus on anatomy has been gradually expanded to include physiology, behavior, and ecology, and eventually studies of chromosomes and molecular genetics. The advent of chromosomal and molecular studies has been a particular boon in several ways. For instance, in 1967 Napier and Napier—in a survey that helped to launch the subject of primate biology—recognized 180 primate species. In the meantime, that number has been more than doubled (Groves, 2001; Petter, 2013). Although recognition of additional species is partly due to revisions following anatomical reinvestigation and new information supplied by burgeoning field studies, studies of chromosomes and molecular biology have also made a major contribution. For all subgroups of primates, comparisons of chromosomes and both mitochondrial and nuclear DNA have permitted identification of numerous "cryptic" species. In many cases, notably for species with nocturnal habits, differences between distinct populations are not easily recognizable from general appearance or even from detailed studies of morphology.

EXTANT GROUPS OF PRIMATES

Extant (currently living) primates are readily divisible into five "natural groups" that have provided a basic framework for all classifications and evolutionary trees (Martin, 1990; Rowe, 1996; Groves, 2001). These groups (classified here as infraorders) are to some extent indicated by geographical distribution alone (Figure 1) but also indicated by many morphological features of the skull, dentition, and postcranial skeleton. The first group contains the lemurs (infraorder Lemuriformes). These are the only primates on Madagascar—more a small continent than a large island—where they underwent adaptive radiation in isolation to yield a spectacular array of more than 80 extant species. In fact, around 20 additional lemur species, mostly larger-bodied

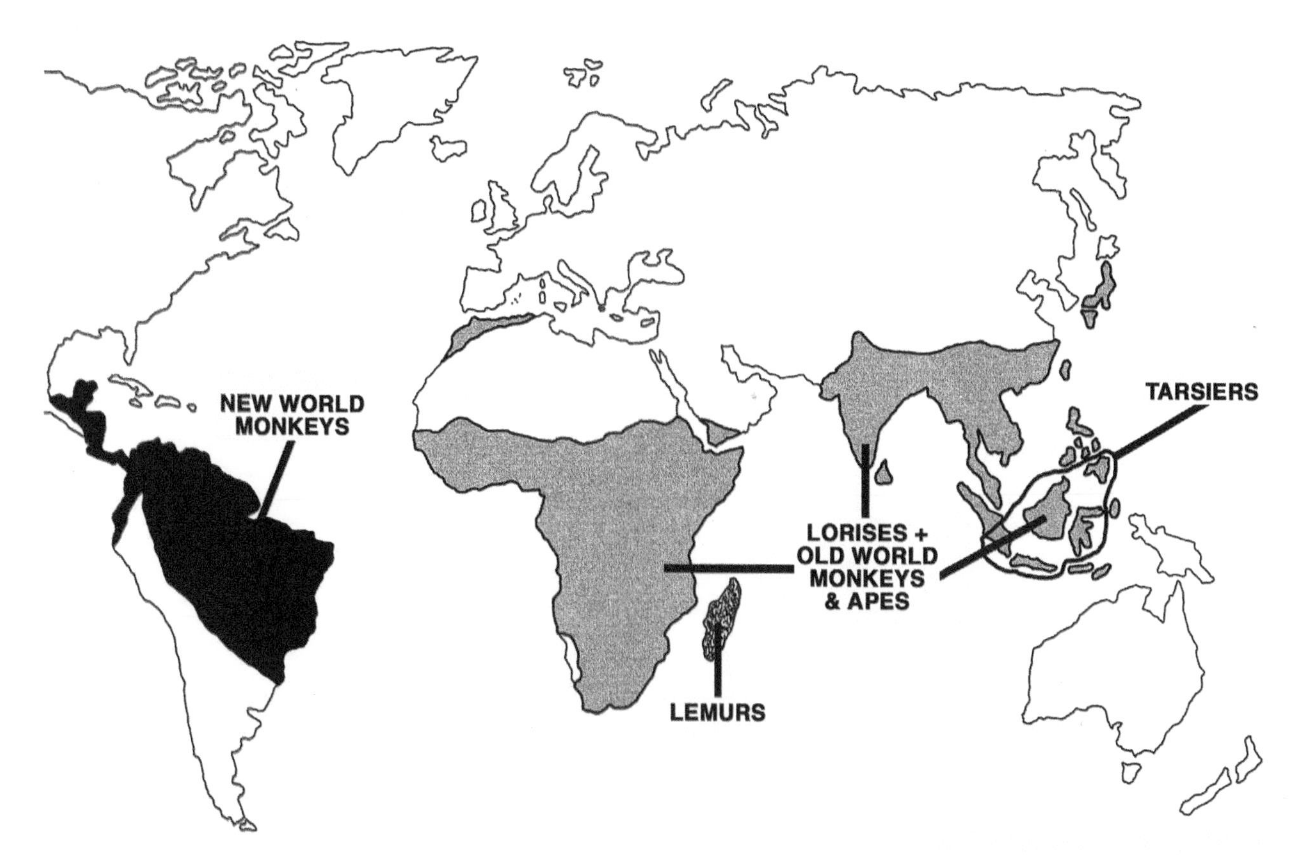

FIGURE 1 Geographical distribution of extant nonhuman primates (excluding tree shrews). Shaded areas indicate inclusive ranges of the five natural groups: (1) lemurs, (2) lorisiforms, (3) tarsiers, (4) New World monkeys, (5) Old World monkeys and apes. (Note that the geographical range of lorisiforms largely coincides with that of Old World monkeys and apes.)

than their extant relatives, died out during the last 2000 years since human colonizers first arrived on Madagascar. Those extinct species, some of which reached the size of a female gorilla, are documented only by subfossil (barely mineralized) remains. So over 100 lemur species were living on Madagascar when humans first settled on the island.

The second natural group of primates, including lorises, pottos, and bushbabies (infraorder Lorisiformes), is much smaller and comprises only 30 species or so. Their adaptive array is notably more restricted in both morphology and behavior. The primary distinction is between slow-moving lorises and pottos (family Lorisidae) and actively leaping bushbabies (family Galagidae). Despite the smaller number of species in the lorisiform group, its members have an extensive geographical distribution, covering much of Africa and South and Southeast Asia. But the third natural group, composed of perhaps a dozen broadly similar tarsier species (infraorder Tarsiiformes), constitutes an even smaller adaptive array, confined to islands in the Southeast Asian archipelago. Tarsiers provide a very good example of increasing recognition of cryptic species, as only three species (on Borneo, in the Philippines, and on Sulawesi) were recognized by Napier and Napier in 1967.

The fourth natural group of extant primates contains the New World monkeys (infraorder Platyrrhini). They are the only extant primates outside the Old World. They have diversified extensively in South and Central America to form another impressive adaptive array with around 130 known species. Two fairly distinct subgroups of New World monkeys are immediately evident. Relatively small-bodied, claw-bearing marmosets and tamarins are quite distinct from the larger-bodied "true" monkeys.

The fifth natural group of primates, which is also the largest, consists of the Old World monkeys and apes (infraorder Catarrhini). The group contains over 150 species, which also have a wide geographical distribution throughout most of Africa, South Asia, and Southeast Asia. Here too, there are two clearly recognizable sister groups with extensively overlapping distributions: monkeys and apes. Old World monkeys, in turn, fall into two distinct subgroups, one containing specialized leaf-eating species that are unique among primates in having a complex stomach like that of a cow, and the other including guenons, macaques, and baboons. Apes can similarly be subdivided into two subgroups: small-bodied lesser apes (gibbons), the only primates to show true brachiation, and the larger-bodied great apes and humans. Although humans and their fossil relatives belong to the great ape subgroup, the geographical range has expanded far beyond the original African homeland over the past two million years.

HIGHER-LEVEL RELATIONSHIPS

Members of the first three natural groups of primates (lemurs, lorisiforms, tarsiers) have generally remained closer to the ancestral condition for primates. For this reason, they have often been collectively labeled "prosimians" or "lower primates," contrasting them with monkeys and apes ("simians" or "higher primates"). To take one example, the volume of the brain relative to body size is typically smaller in prosimians. Average relative brain size in monkeys and apes is about twice as large as that in lemurs, lorisiforms, and tarsiers, although values overlap to some degree. One conspicuous outlier among Madagascar lemurs is the aye-aye (*Daubentonia*), which has a monkey-sized brain. Prosimians are generally relatively small, with a modal body weight close to 500 g. Another important characteristic is that most prosimian species are active at night (nocturnal). Daytime activity (diurnality) is shown only by certain lemurs, and most species are nocturnal even among lemurs. As a rule, primates are either clearly nocturnal or clearly diurnal. But a few species—brown lemurs (*Eulemur* species) and gentle lemurs (*Hapalemur* species)—show an unusual combination of nocturnal and diurnal activity labeled "cathemerality" by Tattersall in 1988. In fact, several lines of evidence indicate that ancestral mammals were nocturnal and that ancestral primates retained this primitive activity pattern. One indication that ancestral mammals were nocturnal is that the original complement of four different types of cone (receptors for color vision) in the retina of early vertebrates was evidently reduced to only two. Most primates still have only two cone types, although true trichromacy (with three different cone types) emerged as a special development in the common ancestor of Old World monkeys and apes.

In addition to their relatively large brains, simians share a suite of advanced features in their teeth, jaws, and reproductive system that distinguish them from prosimians. Moreover, simians are typically large bodied, with a modal body weight of about 5 kg, 10 times larger than the modal value for prosimians. Additionally, the vast majority of simians are diurnal, the only exception being the nocturnal owl monkeys (*Aotus* species) of the New World. This fits with a clear association between diurnal habits and comparatively large body size among mammals generally. This is also seen among Madagascar lemurs, as nocturnal species are typically quite small bodied, whereas diurnal species all exceed 1 kg in body weight. For this reason alone, the large-bodied subfossil lemurs were most likely diurnal as well, as is confirmed by other features such as relatively small eye sockets. Among extant primates, there is an extensive overall range of body sizes, encompassing a 3000-fold difference between the smallest, the pygmy mouse lemur (body weight 40 g), and the largest, an adult male gorilla (body weight 120 kg).

But the traditional distinction between prosimians and simians is not the only higher-level division that can be recognized among extant primates. Like other prosimians, tarsiers have retained numerous primitive features such as a relatively small brain, yet they also show an impressive number of special similarities to simians in the visual system, the olfactory apparatus, and certain aspects of reproduction. Moreover, similarities in visual and olfactory senses are reflected in specific shared features of brain morphology. As a result, for over a century various authors have inferred that tarsiers and simians probably shared a specific common ancestor in the primate evolutionary tree. Numerous lines of molecular evidence have recently confirmed this inference. Accordingly, there has been increasing support for an alternative higher-level subdivision among primates, distinguishing lemurs and lorisiforms in one group from tarsiers and simians in the other. Because lemurs and lorisiforms have retained the primitive feature of a naked, moist area of skin (rhinarium) surrounding the nostrils, they are collectively labeled strepsirrhines. On the other hand—uniquely among mammals—the rhinarium has been completely suppressed as a secondary development in tarsiers and simians. For this reason, they are collectively known as haplorhines.

PRIMATES ON AN ASCENDING SCALE?

It has often been stated that primates have no clear defining characteristics comparable to those seen in several other orders of mammals. Rodents, for example, have continuously growing incisor teeth adapted for gnawing, while even-toed hoofed mammals (artiodactyls) have a unique double-pulley arrangement of their ankle bones, carnivores have special carnassial cheek teeth to slice through animal tissue, and bats (chiropterans) have wings. Instead, a traditional view has been that extant primates can be arranged on an ascending ladder (*Scala naturae*), extending from primitive lemurs on the lowest rung up to humans on the highest. The concept of an ascending scale among extant primates is directly allied to the notion of a smooth transition between insectivores (generally seen as representative primitive mammals that occupy an even lower rung of the ladder) and lemurs. This opens the way to the idea of an "insectivore–primate boundary," which was at the heart of a long-running debate about the relationships among tree shrews, insectivores, and primates.

Tree shrews are squirrel-like inhabitants of rainforests in South and Southeast Asia. Early classifications of mammals commonly included them in the order Insectivora. Starting in the 1920s, however, a succession of authors—notably including Wilfred Le Gros Clark—emphasized morphological similarities between tree shrews and primates (see Le Gros Clark, 1959). Eventually, this persuaded Simpson to include tree shrews in the order Primates in his influential 1945 classification of mammals. For some 25 years thereafter, this allocation was widely accepted. Authors who adopted the notion of an evolutionary ladder with extant forms occupying successive rungs saw tree shrews as ideal intermediates between a typical insectivore such as a hedgehog and a relatively primitive primate, for instance, a ringtail lemur. But if extant species are taken as direct models for evolutionary stages they are effectively treated as "frozen ancestors." Of course, hedgehogs did not evolve into tree shrews, and tree shrews are not precursors of lemurs. A virtual revolution in the way in which individual characters are evaluated to infer evolutionary relationships was needed before it was clearly appreciated that extant species are all distinct tips on branches of the tree of life. Appropriate reconstructions revealed that tree shrews are at most only distantly related to primates.

DEFINING FEATURES OF PRIMATES

Once tree shrews have been excluded from consideration, it is entirely possible to identify a substantial array of features shared by all extant primates. These features, in combination, set them apart from other placental mammals. A composite definition (Martin, 1990) can be broken down into the following components for clarification:

1. Primates are typically arboreal inhabitants of tropical and subtropical forest ecosystems. Their extremities are essentially adapted for prehension, rather than grappling of arboreal supports. A widely divergent hallux (big toe) provides the basis for a powerful grasping action of the foot in all genera except *Homo*, while the hand usually exhibits at least some prehensile capacity. The digits typically bear flat nails rather than bilaterally compressed claws; the hallux always bears a nail. The ventral surfaces of the extremities bear tactile pads with cutaneous ridges (dermatoglyphs) that reduce slippage on arboreal supports and provide for enhanced tactile sensitivity in association with dermal Meissner's corpuscles.

Tree-living (arboreal) habits are quite rare among placental mammals, and primates differ from other arboreal species in their special commitment to that lifestyle and resulting adaptations. Both hands and feet are clearly suited for grasping twigs and branches, rather than serving as hooks for negotiating broad tree trunks. Contrasting with the pointed claws of other tree-living mammals, the blunt, flat nails that are typically present on primate fingers and toes provide a stiff backing for ventral tactile pads. In addition to ensuring traction on branch surfaces, the ridged pads have enhanced tactile sensitivity because of underlying Meissner's corpuscles, which are absent from most mammals, including tree shrews. In some primates—the aye-aye among lemurs and marmosets and tamarins among New World monkeys—most nails have been secondarily converted into claw-like structures, but the big toe always bears a nail.

2. Locomotion is hind-limb dominated, with the center of gravity of the body located closer to the hind limbs, such that the typical walking gait follows a diagonal sequence (forefoot precedes hind foot on each side).

The dominant role of the hind limbs in primate locomotion is associated with a typical posterior location of the body's center of gravity, closer to the hind limbs than the forelimbs. This, in turn, is linked to a characteristic walking gait in which the hand precedes the foot on each side (diagonal sequence), rather than the foot preceding the hand (lateral sequence) as in most other placental mammals. The unique striding bipedal pattern that emerged during human evolution owes its origin to the typical primate feature of hind-limb domination.

3. The foot is typically adapted for tarsi-fulcrimation, with at least some degree of relative elongation of the distal segment of the calcaneus, commonly resulting in reverse alternation of the tarsus (calcaneo-navicular articulation).

The heel bone (calcaneus) of primates is unusual in that the distal segment is relatively elongated, particularly in small-bodied, actively leaping species. This is connected with the fact that the fulcrum in foot movement is located further back in the foot (involving tarsal bones rather than metatarsals) because the big toe diverges from the other toes as a grasping adaptation.

4. The visual sense is greatly emphasized. The eyes are relatively large and the orbits possess (at least) a postorbital bar. Forward rotation of the eyes ensures a large degree of binocular overlap. Ipsilateral and contralateral retinofugal fibers are approximately balanced in numbers on each side of the brain and organized in such a way that the contralateral half of the visual field is represented. Enlargement and medial approximation of the orbits are typically associated with ethmoid exposure in the medial orbital wall (though there are several exceptions among the lemurs).

As suggested by their relatively large, forward-facing eyes, vision is particularly important for primates. The emphasis on vision has often been attributed to their arboreal habits, notably in connection with feeding on fruits (Sussman, 2013); however, it has also been explained as an adaptation for nocturnal hunting of animal prey in early primates (Cartmill, 1992). Perhaps both things are true, connected through a fundamental adaptation for moving and feeding in a network of fine branches (Martin, 1990). It is certainly the case that the eyes are generally even larger in nocturnal primates than in diurnal species. Yet other nocturnal mammals typically have quite small eyes. Other senses—touch, hearing, and smell—are brought into play during nocturnal activity. So the strikingly large eyes of nocturnal primates bear eloquent witness to their heavy reliance on vision. The bony bar around the eye socket (orbit) that all primates have is uncommon among mammals; but it is not confined to primates. However, tarsiers and simians (haplorhine primates) are truly unique compared to all other mammals. A bony partition extends from the postorbital bar to the side of the skull, essentially isolating the eye in a cup. But it is forward rotation of the eyes that is particularly significant in primates. Forward orientation of the orbits, which is in fact more complete in simians than prosimians, ensures extensive overlap between left and right visual fields. This permits binocular viewing of objects, which, in turn, provides the basis for refined stereoscopic vision enabled by enhanced organization of visual centers in the brain. In the primitive arrangement for vertebrates, inputs from each eye predominantly pass to the opposite side of the brain because the optic nerves cross over almost completely. Tree shrews, which have a very limited binocular field because their eyes typically face more-or-less sideways, have remained close to this primitive condition. In primates, inputs from the optic nerves have been fundamentally reorganized. Only half of the nerve inputs cross over to the opposite side of the brain (contralateral fibers), while the remainder pass to the brain on the side where the eye is located (ipsilateral fibers). This arrangement ensures that approximately balanced inputs from both eyes can be directly matched up and processed in each side of the brain.

5. The ventral floor of the well-developed auditory bulla is formed predominantly by the petrosal.

Compared to reptiles, mammals have particularly sensitive hearing, with three small ear ossicles rather than just one large bone to transmit sound waves from the eardrum to the inner ear. As the ear evolved, a ventral floor developed to protect the chain of ossicles in the middle ear chamber. But this was a parallel development in most orders of mammals, and a ventral floor evolved in different ways, forming an auditory bulla that may be inflated, as it is in many primates. Formation of an auditory bulla as an outgrowth of the petrosal bone (which houses the inner ear and the semicircular canals) is unique to primates. The bulla of tree shrews is formed from a separate entotympanic element.

6. The olfactory system is unspecialized in most nocturnal forms and reduced in diurnal forms.

It is often stated that the olfactory system has been reduced in all primates, but the basic apparatus and relative size of the olfactory bulbs have in fact remained close to the primitive mammalian condition in several nocturnal strepsirrhines (lemurs and lorisiforms). But reductions in the olfactory apparatus and in olfactory bulb size are clearly evident in haplorhines (tarsiers and simians). Indeed, reduction of the vomeronasal system, involving an accessory olfactory structure known as Jacobson's organ, is directly connected with characteristic loss of the rhinarium in haplorhine primates. In parallel with monkeys and apes, anatomical elements of the olfactory apparatus have also been reduced in some diurnal lemurs. Overall, the evidence suggests that reduction of the olfactory system in primates is connected with diurnal habits.

7. Partly because of the increased emphasis on vision, the brain is typically moderately enlarged, relative to body size, in comparison to other living mammals. The brain of living primates always possesses a true Sylvian sulcus (confluent with the rhinal sulcus) and a triradiate calcarine sulcus. Primates are unique among living mammals in that the brain constitutes a significantly larger proportion of fetal body weight at all stages of gestation.

Primates typically have large brains, at least in part because of increased emphasis on vision. Unfortunately, brain enlargement is often exaggerated with unqualified claims that primates have larger brains than other mammals. This certainly does not apply to actual brain size, as a sperm whale (the record holder among mammals) has a brain almost six times larger than a human brain. If appropriate allowance is made for the influence of body size, humans do prove to have the largest relative brain size among mammals. But other primates are not clearly separated from nonprimates, and some dolphins have relative brain sizes almost as large as those in humans. Relative brain sizes vary widely within individual orders of mammals, and values for some primate species in fact lie below the mammalian average. The most that can be said is that overall average relative brain size is greater in primates than in any other order of mammals. This reflects the fact that primates do differ from other mammals in that the developing brain is relatively larger throughout fetal development. Relative to neonatal body size, a primate infant's brain is about twice as big as that of any other newborn mammal. With respect to the two unique features of brain anatomy in primates, possession of a Sylvian sulcus reflects the increased importance of the temporal lobe (important for vision and memory), while the three-branched calcarine sulcus on the inner side of the occipital lobe marks boundaries between visual areas.

8. Male primates are characterized by permanent precocial descent of the testes into a postpenial scrotum; female primates are characterized by the absence of a urogenital sinus (i.e., urinary and reproductive tracts entirely separate). In all primates, involvement of the yolk sac in placentation is suppressed, at least during the latter half of gestation. Primates have long gestation periods, relative to maternal body size, and produce small litters of precocial neonates. Fetal growth and postnatal growth are characteristically slow in relation to maternal size. Sexual maturity is attained late and life spans are correspondingly long relative to body size. Primates are, in short, adapted for slow reproductive turnover.

Permanent descent of the testes into scrotal sacs is widespread among both placental mammals and marsupials. Unusually, however, testicular descent in primates typically occurs close to the time of birth, despite the fact that sexual maturity is achieved relatively late. A spike in testosterone production occurs soon after birth in male primates, so perhaps early descent of the testes permits socially important recognition of an offspring's sex. Complete separation of the urinary and reproductive tracts in female primates is also unusual among mammals. In many nonprimates, including tree shrews, the lower ends of the tracts are combined to form a common passage (urogenital sinus). Charnov and Berrigan (1993) appropriately summarized the primate pattern of slow reproductive turnover as "life in the slow lane." On the one hand, primates have low daily investment of resources in reproduction; on the other, this adds up to substantial long-term investment in individual offspring. Regarding placentation, all primates share certain advanced features, such as suppression of the yolk sac at least by the second half of pregnancy. But conformation of the placenta and its associated membranes is of special interest because of an unusual dichotomy among primates. In strepsirrhines, placentation is of the least invasive (epitheliochorial) kind, whereas haplorhines have a highly invasive (hemochorial) type of placenta. The traditional interpretation has been that a noninvasive placenta is primitive and inefficient and that increasing invasiveness is an advanced feature associated with greater efficiency in the transfer of resources from mother to fetus. However, analyses of the distribution of placenta types among mammals—using phylogenetic trees based on DNA sequences (e.g., Bininda-Emonds et al., 2007)—revealed that it is more parsimonious to infer that the placenta was invasive in ancestral placental mammals (Carter and Enders, 2004). Indeed the ancestral condition was probably of the moderately invasive, endotheliochorial kind (Martin, 2008a). So noninvasive placentation in strepsirrhines and highly invasive placentation in haplorhines probably arose by divergence from an intermediate condition in ancestral primates.

9. The dental formula exhibits a maximum of 2.1.3.3/2.1.3.3. The size of the premaxilla is very limited, in association with the reduced number of incisors, which are arranged more transversely than longitudinally. The cheek teeth are typically relatively unspecialized, though cusps are generally low and rounded and the lower molars possess raised, enlarged talonids.

Because fossil mammals are predominantly documented by isolated teeth (which are most resistant to degradation) and tooth-bearing jaw fragments, reconstruction of their evolutionary relationships has traditionally relied heavily on dental features. From front to back in each jaw, teeth are defined as incisors (I), canines (C), premolars (P), and molars (M) and their numbers are represented in a dental formula (I. C. P. M) for upper and lower jaws. The maximal dental formula of extant placental mammals is three incisors, one canine, four premolars, and three molars on either side of both upper and lower jaws. This formula (3.1.4.3/3.1.4.3) is commonly taken as the likely ancestral condition, from which evolution has typically occurred through loss of teeth. Extant primates have a maximal formula of 2.1.3.3/2.1.3.3, reflecting the loss of one incisor and one premolar in each tooth row. Relationships among mammals are often inferred primarily from cusp patterns on molars. But this is problematic with primates because their molars have remained comparatively primitive and lack the distinctive features found, for example, in rodents, carnivores, or hoofed mammals. In the common ancestor of marsupials and placental mammals, molars were tritubercular: There was a triangular array of main cusps in each upper molar and an equivalent triangle in each lower molar with additional cusps borne on a low posterior heel (talonid). Among extant primates, there is a general tendency for a fourth cusp to be present on each upper molar, while the original triangle on the lower molar has typically been reduced to two cusps. The low, rounded cusps that generally characterize primate molars are thought to reflect to a dietary shift from insects and other arthropods toward fruits. In tree shrews, by contrast, molar cusps are sharp and form a distinctive array.

The classic definition outlined above is derived primarily from Martin (1990). At that time, one key feature of extant primates was omitted: All species live in quite elaborate social networks in which there is enhanced communication between individuals. Many prosimian primates are nocturnal and have often described as "solitary," implying the opposite of "social." But burgeoning field studies revealed that in nocturnal species there are well-established social interactions based on shared ranging areas, nighttime encounters, and communal nesting. Individuals also have long-lasting associations. So all primates are social, but "solitary" nocturnal primates simply do not move around in groups, whereas diurnal primates live in recognizable social groups. So the opposite of "solitary" is "gregarious" not "social." The pronounced social tendency that is now evident for all primates is surely linked to heavy investment in individual offspring, including intensive parental care, and drawn-out life histories.

FOSSIL PRIMATES

Clearly recognizable, direct fossil relatives of extant primates are known as "euprimates" or "primates of modern aspect." Known fossils date back to the beginning of the Eocene epoch, about 55 million years ago (mya). Most classifications of the order Primates include an array of "archaic primates" (infraorder Plesiadapiformes), which are mainly documented from the Palaeocene epoch (65–55 mya), although some representatives survived into the Eocene. But plesiadapiforms share few of the features that define extant primates and their affinities remain uncertain. Although whether or not plesiadapiforms had some remote ancestral connection with primates, it is widely accepted that they branched away prior to the common ancestor that gave rise to euprimates (extant primates and their direct fossil relatives). The edited volume by Hartwig (2002) provides a comprehensive review of fossil euprimates. Within the adaptive radiation of euprimates, members of most modern groups are clearly identifiable in the fossil record by the earliest Miocene (about 25 mya).

There are two main families of early euprimates that are documented mainly from the Eocene (35–55 mya), although a few representatives survived into the late Miocene. Members of both families are clearly adapted for arboreal life and possess many of the defining features in the skull, dentition, and postcranial skeleton recognized for modern primates. Species belonging to both families are widely distributed across the northern continents, occurring in North America, Europe, and Asia. Species in the family Adapidae (for example, *Adapis* and *Europolemur* in Europe, *Notharctus* and *Smilodectes* in North America, *Adapoides* in Asia) are commonly medium sized and generally resemble lemurs. Species in the family Omomyidae (for instance, *Microchoerus* and *Necrolemur* in Europe, *Shoshosius* and *Tetonius* in North America, *Teilhardina* in Asia) are generally smaller and show various similarities to tarsiers. Adapids have often been linked to extant strepsirrhines and omomyids to modern haplorhines. However, a reasonable alternative interpretation is that adapids and omomyids belong to independent radiations from ancestral primates.

The early fossil record of early euprimates was originally limited to adapids and omomyids, but early relatives of simians (notably *Eosimias*) are now known from middle Eocene fossil deposits in China. Prior to their discovery, a single site in Egypt, the Fayum Depression, had yielded most of the earliest known fossils reliably identifiable as simians. Initially, Fayum simians were documented only from the early Oligocene (for instance, *Aegyptopithecus* and *Apidium*), but older representatives were subsequently discovered in deposits from the latest Eocene (notably *Catopithecus*). Adapids and omomyids have also been reported from the productive Fayum site, thus expanding their known range to the southern continents. Early fossils allied to simians have also been increasingly documented from late Eocene sites in Southeast Asia, including Myanmar (for example, *Amphipithecus* and *Pondaungia*) and Thailand (such as *Siamopithecus*). But these generally seem to be early offshoots rather than direct relatives of extant simians.

New fossil finds have increasingly illuminated the evolution of four of the five natural groups of extant primates, but Madagascar lemurs remain frustratingly undocumented. Apart from the recently extinct subfossils, which are really part of the extant fauna, the fossil record has remained entirely silent regarding the adaptive radiation of lemurs. However, early lorisiforms, the sister group of lemurs, are documented by a few fossil fragments. For some time, Miocene representatives from Kenya dating back about 20 mya (for instance, *Mioeuoticus* and *Progalago*) were the earliest known lorisiforms. But the known age of this natural group of primates was dramatically doubled by new discoveries at the Fayum site in Egypt from deposits dating back around 40 mya (late Eocene). In fact, those fragmentary fossils included an early relative of lorises and pottos (*Karanisia*) along with a bushbaby (*Saharagalago*), so divergence between those two main branches of lorisiforms had already occurred. Even in the complete absence of a fossil record for lemurs from Madagascar, it can therefore be concluded that they diverged from lorisiforms at least 40 mya.

As far as direct relatives of haplorhines are concerned, Middle Eocene fossil deposits in China have yielded a species that is very similar in its known features to modern tarsiers. Indeed, it is so similar that it has been included as a species in the same genus—*Tarsius eocaenus*. This has led some authors to proclaim *Tarsius* as a "living fossil," particularly because a skull fragment indicates that the eye sockets were already strikingly large in *T. eocaenus*. However, apart from that fragment, the fossil evidence is essentially limited to teeth and jaws, so it is simply unknown, for example, whether *T. eocaenus* already had the specialized adaptations for active leaping that characterize modern tarsiers.

Fossil relatives of extant groups of simians are documented back to about 30 mya, to late Oligocene times. The known early fossil record for New World monkeys begins in South America with late Oligocene *Branisella* and continues through the Miocene. Early Miocene representatives from Argentina, such as *Homunculus* and *Tremacebus*, appear to lack any direct connection with extant lineages of New World monkeys. By the late Miocene, however, members of individual modern families are recognizable, including a fossil owl monkey allocated to the same genus (*Aotus*) and a relative of saki monkeys (*Cebupithecia*). A skull fragment indicates that the eye sockets were enlarged in the Miocene *Aotus* species.

The earliest known representatives of the fifth natural group of primates, Old World monkeys and apes, come from early Miocene deposits of Africa dating back about 20 mya. East African sites have yielded fossil relatives of both monkeys (notably *Victoriapithecus*) and apes (particularly *Proconsul*). The early fossil record of Old World monkeys, up to about 10 mya, is in fact quite poorly documented, whereas fossil apes are relatively well known. It seems that in Miocene times apes were far more diverse and abundant than monkeys, whereas the reverse is the case today. The split between leaf monkeys and the guenon subgroup of Old World Monkeys is not well established by the fossil record, but it seems to have occurred quite early. Extant apes are clearly divided into relatively small-bodied lesser apes (gibbons) and the radiation of great apes, to which humans belong. No direct fossil relatives of gibbons have yet been found. Although small-bodied Miocene apes are known, they share no distinctive characters with modern gibbons. The fossil history of great apes is somewhat clearer, but it is noteworthy that no direct fossil relatives of orangutans, gorillas, or chimpanzees are known, apart from a few chimpanzee teeth dating back only half a million years. The widely accepted inference that humans are more closely related to chimpanzees than to other great apes is based essentially on molecular data rather than on comparative anatomy or fossil evidence.

Early fossil evidence for human evolution is confined to Africa. The earliest potential member of our own evolutionary lineage is *Sahelanthropus* from Chad, dated at about 7 mya (but see the chapter by Hunt in the present volume). Unfortunately, only a single skull is known, but it does indicate initial adaptation for upright bipedal locomotion. Adaptation for bipedalism is also suggested by fragmentary thigh bones from the somewhat younger Kenyan fossil *Orrorin*, dated at about 6 mya. Unfortunately, apart from a few teeth, nothing is known of the skull. Indisputable evidence for early hominid evolution dates back only to australopithecines, starting about 4 mya (see chapter by Ward). Known species belong only to the genus *Australopithecus*, while more robust (previously classified as *Paranthropus*) specimens emerged later, eventually disappearing about 1 mya. Unusually, the earliest convincing evidence for striding bipedal locomotion is provided not by skeletal remains but by a trail of footprints in Tanzania that are some 200,000 years older than the "Lucy" skeleton of *Australopithecus afarensis* (about 3.2 million years old) from Ethiopia. Species belonging to the more advanced hominid genus *Homo* first appear in the fossil record about 2.5 mya in Africa (see the chapter by Simpson). Reliably identified stone tools first appear in the record at about the same time (see chapter by Toth and Schick). *Homo* had expanded out of Africa by 2 mya, as is clearly indicated by 1.7 million year old fossil remains of *Homo georgicus* from Dmanisi (Republic of Georgia) (see the chapter by Ahern). From that point onward, *Homo* eventually expanded around the world (see the chapter by Holt).

OVERALL EVOLUTIONARY RELATIONSHIPS

Geographical isolation influences patterns of evolution, so it is notable that the five "natural groups" of primates, with their distinct distributions, correspond to major subdivisions within the primate evolutionary tree (Figure 2). Findings from chromosomal and molecular comparisons have broadly confirmed earlier conclusions derived from morphological evidence alone, and a clear consensus interpretation is now emerging. Each of the five natural groups of primates—lemurs, lorisiforms, tarsiers, New World monkeys, Old World monkeys, and apes—evidently descended from a distinct common ancestor. That is, all are

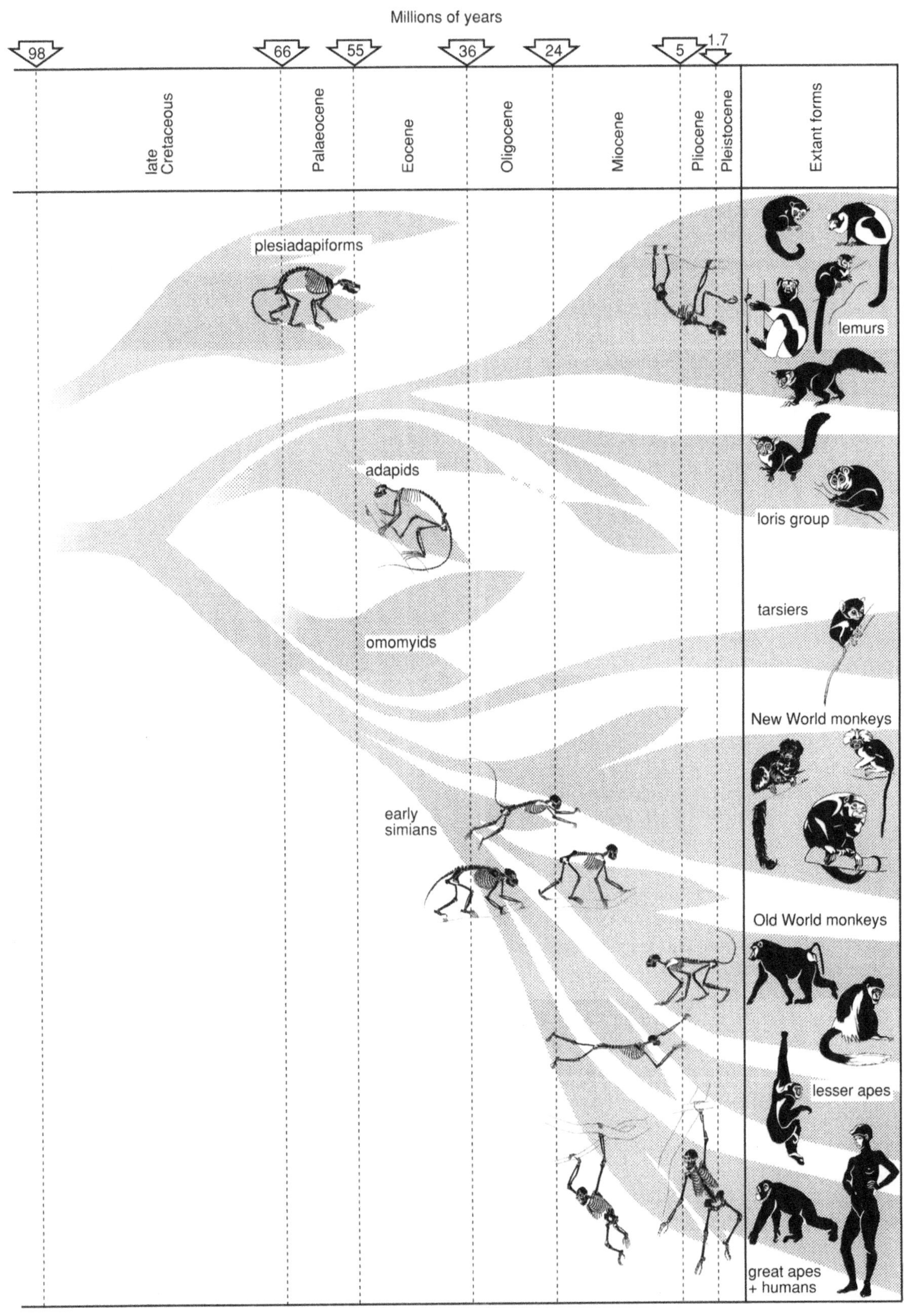

FIGURE 2 Evolutionary relationships among primates. Tree showing inferred phylogenetic connections among the five natural groups of extant primates. It includes well-documented fossil groups of "primates of modern aspect" (euprimates) and "archaic primates" (plesiadapiforms), whose relationships are uncertain. Note that the initial date of divergence for primates of modern aspect is set at about 80 mya (Wilkinson et al., 2011). *Original illustration by Lukrezia Bieler-Beerli, adapted from Martin, R.D., 1993. Nature 363, 223–234.*

monophyletic. At a higher level in the evolutionary tree, strepsirrhines (lemurs and lorisiforms) and haplorhines (tarsiers and simians) can also be recognized as monophyletic sister groups. Monkeys, apes, and humans also constitute a monophyletic group (simians), traceable to a specific common ancestor.

Understanding the origin of primates requires exploration of their relationships to other mammals. Primates are one of 18 orders of placental mammals with extant representatives. Among other things, to sharpen the focus on distinctive features of early primates, we must reliably identify the sister group among the other 17 orders. In broad outlines, our understanding of evolutionary relationships within placental orders such as Primates has been established for some time. However, higher-level affinities among orders long remained unsettled. Because of the ubiquitous problem of convergent adaptation, morphological evidence alone does not yield sufficient resolving power to identify very early divergences.

Although it is widely accepted that euprimates are monophyletic, relationships with other mammals, notably tree shrews, have remained uncertain. By general agreement, tree shrews are now allocated to the separate order Scandentia; yet many authors still see them as the sister group of primates among extant mammals. Although it has often been inferred that tree shrews and primates share several derived features, those similarities may be interpreted instead either as primitive retentions or as convergent adaptation for arboreal life (Martin, 1990). The affinities of tree shrews hence remained controversial.

The advent of broad-based molecular trees for placental mammals has now ushered in a fairly stable, and in some respects unexpected, consensus interpretation of higher-level relationships. One notable development has been subdivision of the mammalian tree into four monophyletic clusters (superorders). A surprising feature was clear emergence of the new superorder Afrotheria. This is a cluster of endemic African mammals, including aardvarks, certain insectivores (Afrosoricida: golden moles, otter shrews, tenrecs), elephant shrews, elephants, sirenians, and hyraxes. But Afrotheria does not include primates, whose origin had hitherto been firmly placed in Africa. Primates are in fact allocated to a second cluster containing colugos and tree shrews along with a subgroup containing lagomorphs and rodents (Glires). This superorder has been endowed with the ungainly label "Euarchontoglires" (Figure 3). The third superorder, named Laurasiatheria, is a large cluster containing bats, carnivores, all remaining insectivores (Eulipotyphla: hedgehogs, moles, shrews, solenodons), pangolins, artiodactyls, cetaceans, and perissodactyls. By contrast, the fourth superorder is very small and confined to the single order Xenarthra, which includes armadillos, anteaters, and sloths.

Within the superorder Euarchontoglires, a distinction is generally recognized between Euarchonta (colugos, primates,

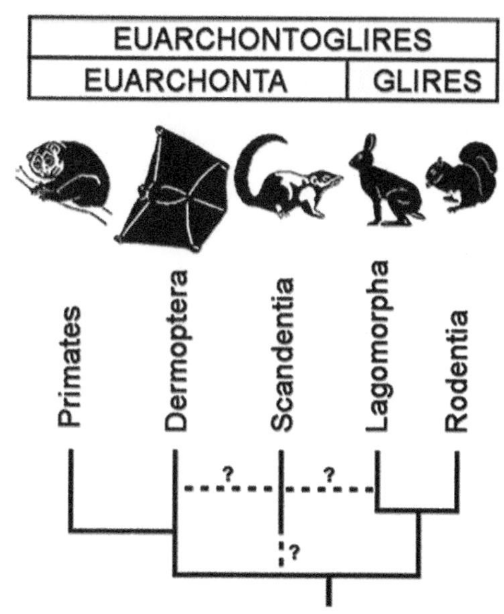

FIGURE 3 Inferred relationships within the superorder Euarchontoglires, indicated by shared similarities in chromosomal features and DNA sequences (Martin, 2008b). A sister-group relationship between lagomorphs and rodents (Glires) is widely accepted, and the consensus of available evidence indicates that primates and colugos are sister groups. Dashed lines with question marks indicate the uncertain relationships of tree shrews, which may be related either to primates and colugos (Euarchonta) or to lagomorphs, or may be an independent lineage. DNA sequences have generally provided little evidence for a specific relationship between tree shrews and primates.

tree shrews) and Glires (lagomorphs, rodents). The inference that colugos might be close relatives of primates was another unexpected finding, as this had not been indicated by morphological evidence (other than an unconvincing suggestion that colugos might be linked to plesiadapiforms). Whereas molecular evidence has consistently indicated monophyly for Euarchontoglires, there is residual uncertainty about relationships among the five orders included. Relationships among colugos, primates, and tree shrews remain undecided. Several studies have identified colugos and primates as sister groups, while others have linked colugos to tree shrews instead. But molecular comparisons have hardly ever identified tree shrews as the sister group of primates. In fact, a link between tree shrews and lagomorphs (usually regarded as the sister group of rodents) has emerged in several trees based on molecular evidence. So the relationships of tree shrews remain as enigmatic as ever.

REFERENCES

Bininda-Emonds, O.R.P., Cardillo, M., Jones, K.E., MacPhee, R.D.E., Beck, R.M.D., Grenyer, R., Price, S.A., Vos, R.A., Gittleman, J.L., Purvis, A., 2007. The delayed rise of present-day mammals. Nature 446, 507–512.

Carter, A.M., Enders, A.C., 2004. Comparative aspects of trophoblast development and placentation. Reproductive Biology and Endocrinology 2, 1–15.

Cartmill, M., 1992. New views on primate origins. Evolutionary Anthropology 1, 105–111.

Charnov, E.L., Berrigan, D., 1993. Why do female primates have such long lifespans and so few babies? or life in the slow lane. Evolutionary Anthropology 1, 191–194.

Groves, C.P., 2001. Primate Taxonomy. Smithsonian Institution Press, Washington, D.C.

Hartwig, W.C. (Ed.), 2002. The Primate Fossil Record. Cambridge University Press, Cambridge, UK.

Le Gros Clark, W.E., 1959. The Antecedents of Man. Edinburgh University Press, Edinburgh.

Martin, R.D., 1990. Primate Origins and Evolution: A Phylogenetic Reconstruction. Princeton University Press, Princeton, NJ.

Martin, R.D., 2008a. Evolution of placentation in primates: implications of mammalian phylogeny. Evolutionary Biology 35, 125–145.

Martin, R.D., 2008b. Colugos: obscure mammals glide into the evolutionary limelight. Journal of Biology 7 (Art 13), 1–5.

Napier, J.R., Napier, P.H., 1967. A Handbook of Living Primates. Academic Press, London.

Petter, J.-J., 2013. Primates (translated by Martin, R.D.). Princeton University Press, Princeton, NJ.

Rowe, N., 1996. The Pictorial Guide to the Living Primates. Pogonias Press, East Hampton, NY.

Sussman, R.W., Rasmussen, D.T., Raven, P.H., 2013. Rethinking primate origins again. American Journal of Primatology 75, 95–106.

Tattersall, I., 1988. Cathemeral activity in primates: a definition. Folia Primatologica 49, 200–202.

Wilkinson, R.D., Steiper, M.E., Soligo, C., Martin, R.D., Yang, Z., Tavaré, S., 2011. Dating primate divergences through an integrated analysis of palaeontological and molecular data. Systematic Biology 60, 16–31.

Chapter 4

Comparative Anatomy of Primates

Rui Diogo,[1] Magdalena N. Muchlinski[2] and Adam Hartstone-Rose[3]

[1]Department of Anatomy, Howard University College of Medicine, Washington, DC, USA; [2]Department of Anatomy and Neurobiology, University of Kentucky, College of Medicine, Lexington, KY, USA; [3]Department of Cell Biology and Anatomy, University of South Carolina School of Medicine, Columbia, SC, USA

SYNOPSIS

Primate comparative anatomy is a field of research that has profoundly interested scientists since several centuries ago, particularly due to its implications to the knowledge of the origin, biology, and/or evolution of our own species, *Homo sapiens*. In this chapter we provide a short summary of the comparative anatomy of humans and other primates that is based on a collaborative work of three experts in the field. A different author has written each of the sections. The first section focuses on osteology (bones and cartilages), the second on myology (muscles and tendons), and the third on external and internal organs. Within each section the information will be presented within an evolutionary context and often includes brief historical and/or functional considerations.

OSTEOLOGY

Introduction

The skeletal configuration of primates has often been related to arboreal and perhaps "visual predation" adaptations (Cartmill, 1972, 1974; see the chapter by Martin in this volume). For instance, the unguiculate (the state of possessing flattened nails) and opposable digits relate to a clear (though not unique or ubiquitous) grasping ability and the presence of a well-defined clavicle in primates that adds solidity to the shoulder in ways that are advantageous for arboreal stability all seem to argue for a life in the trees that more or less defines the order. Likewise, the orbital modifications—not only the bony encircling, but also a general reorganization to move the eyes forward for improved binocular vision—could have evolved to allow visual hunting, perhaps on terminal branches in trees. In short, though there is great diversity in the skeletal adaptations within the order, the primate skull and skeleton typically seem to have evolved for arboreality (Hill, 1972).

The Strepsirrhine Skeleton

There is almost as much skeletal diversity among the extant galagos, lemurs, and lorises (modern members of the suborder Strepsirrhini; see Figure 1) as there is in the whole order. If you include a consideration of the recently extinct lemurs, then this suborder certainly has had more variation in size, shape, posture, and locomotion than all of the living anthropoids combined. This group ranges in size from the diminutive mouse lemurs (*Microcebus*)—as small as 30 g, the smallest of all living primates—to the ~6 kg indri (*Indri*). While this largest of living lemurs is not particularly impressive in size relative to hominoids and some monkeys,

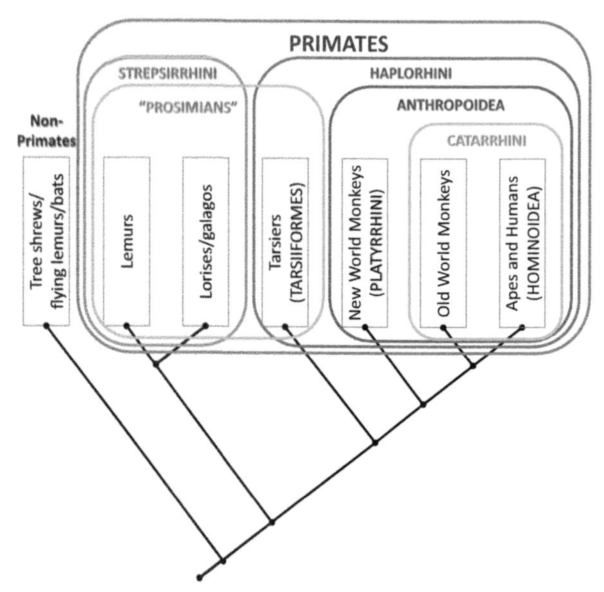

FIGURE 1 Scheme showing the main primate clades and their phylogenetic relationships. (See color plate section).

all of the recently extinct lemurs were larger than their surviving relatives. The largest of these, *Archaeoindris*, may have been nearly 200 kg—larger than most of the living hominoids. As impressive as they are in size, most of the recently extinct lemurs were even more interesting in their locomotor adaptations, with several species (*Archaeolemur* and *Hadropithecus*) converging on a more terrestrial monkey-like form, other giant forms displaying a bizarre form of arboreal morphology that included exceptionally long and curved digits and other skeletal elements that have given them the moniker the "koala lemurs" (*Megaladipis*). As strange as this taxon was, another extinct lineage (including *Babakotia* and *Paleopropithecus*) had distinct adaptations for underbranch suspension that has given them the name "sloth lemurs" (Mittermeier et al., 2006).

The living strepsirrhines mostly have more typical body forms that can generally be divided into four categories: (1) relatively small slow bodies (e.g., those of the slow loris, *Nycticebus*, and dwarf lemurs, *Cheirogaleus*) that may represent the body plan of the most primitive ancestral primate; (2) small bodies built for quick movement (e.g., the bush babies, *Galago*, and mouse lemurs, *Microcebus*); (3) arboreal quadrupeds (e.g., ringtail lemurs, *Lemur*, and the "true" lemurs, *Eulemur*)—what most may think of as a typical lemur form; and (4) the "vertical clinging and leaping" lemurs (e.g., the indri, *Indri*, and sifakas, *Propithecus*). Almost all of the living strepsirrhines fall more or less within one of these categories with one amazing exception: the aye-aye (*Daubentonia*). This truly unique animal has a suite of morphology unlike any other primate: as the largest of all nocturnal primates, it has huge ears that it uses in combination with exceptionally long fingers to echolocate wood-boring insects in a feeding method known as "tap foraging." Once it locates an airspace beneath the surface of the wood, it uses its ever-growing incisors (also unique among primates and almost all other mammalian orders) to gouge into those cavities, and then it inserts its long, thin, and highly flexible middle finger into the hole to probe and fish out grubs. This, along with its impressive long and bushy tail, vestigial premolars and molars, inguinal mammaries (breast located in the groin) and highly cryptic behavior, clearly makes this sole member of its own family a truly mysterious primate (Mittermeier et al., 2006). It is arguably one of the strangest of all mammals.

Exceptions aside, strepsirrhines have a fairly typical primate skeleton: substantial clavicles and opposable, powerful big toes. All have fairly large eyes surrounded by "postorbital bars"—bony struts connecting the frontal bone to the zygomatic arch to either support or protect the relatively convergent eyes. Most strepsirrhines have long tails and typical primate molars and premolars (albeit relatively primitive). Most also have "tooth combs"—a reorganization of the lower anterior dentition (incisors and canines) in which these teeth are long and thin and aligned as an apparatus used for grooming. They also have a "grooming" or "toilet" claw—a long sharp claw generally found on the second toe. The truly defining feature of the group—the wet "rhinarium," essentially a continuity between the upper lip and nose that allows improved use of the vomeronasal organ—is a predominantly soft-tissue feature. However, the strepsirrhine emphasis on olfaction does highly influence the shape of the skull: more than any of the other primates, lemurs have fairly elongated rostra. This anatomy allows for greater olfaction—a more important sense in this group than primates such as hominoids (as exemplified by the nearly ubiquitous scent glands used by these animals)—and makes the skull longer than almost all other primates (Hill, 1953; Fleagle, 1999).

The Tarsiiform Skeleton

The small number of species of this group that is found in the islands off of Southeast Asia are unique among primates in several features. For one thing, they are strictly carnivorous, feeding exclusively on insects and small vertebrates that they capture through fast visual predation. This feeding strategy has a strong effect on their bony anatomy: the tarsiers have very long ankle bones, giving them enormous leverage for leaping. They also have very sharp teeth for consuming soft animals. The most distinctive osteological features of the tarsiers are their orbits and huge eyes. Each is bigger, in fact, than the animal's brain. This is a consequence of both the animal's nocturnal visual predation strategy and its lack of a tapetum lucidum—the reflective layer that is found on the back of the eye of strepsirrhines and many other mammals, but not of primates such as hominoids. Aside from their size, the orbits of tarsiers are notable because they are much more enclosed in the back than those of strepsirrhines. There is a debate over whether this bony septum represents a true "postorbital wall"—a defining feature separating strepsirrhines from monkeys and hominoids—or whether this is an example of convergence, but suffice it to say, this is one piece of anatomy that seems to affirm the place of this unique genus somewhere as an offshoot between strepsirrhines and anthropoids (Hill, 1955; Fleagle, 1999).

The Anthropoid Skeleton

In addition to the full postorbital wall, unlike strepsirrhines, anthropoids generally have fused frontal bones (tarsiers have variability in the fusion of their frontal) and fused mandibular symphyses, that is, a single, unpaired lower jaw, while strepsirrhines and tarsiers have right and left mandibles that they can move, with varying degrees depending on genus, independently. Anthropoids also differ from "prosimians" in their possession of generally larger brains, lack of claws (with the exception of one South American monkey group), generally broad incisors, and marked sexual dimorphism in most species (most commonly seen as larger canines in males than in females) (Fleagle, 1999).

The Platyrrhine Skeleton

Osteologically, New World monkeys can be distinguished from other anthropoids in their retention of three premolars in each quadrant of the jaw (i.e., twelve total while catarrhines have only eight). There are also substantial (though fairly technical) details that define this group in terms of the morphology if the bones of their ear region and specific bones of the braincase come in contact with each other. The smallest monkeys, the pygmy marmosets (*Cebuella*) who weigh ~100 g, are members of this group. Some platyrrhines are also the only primates with truly prehensile tails. In some species (e.g., the spider monkeys, *Ateles*) this remarkable limb can support the full weight of the animal and is truly as agile as a fifth hand/foot. Some platyrrhines have completely nonprehensile tails (e.g., tamarins, *Saguinus*) while other species (e.g., the capuchin monkeys, *Cebus*) have some prehensile abilities that are not as complete as the most derived members of the group. Individual South American genera have osteological features that are unique. For instance, owl monkeys (*Ateles*) have exceptionally large orbits because these nocturnal animals (the only nocturnal haplorhines) lack tapeda lucida and thus have to collect as much light as possible by having very large eyes. Another orbital feature, a hole connecting the right and left orbits in squirrel monkeys (*Saimiri*), has less obvious adaptive significance. Possibly the most dramatic unique osteological adaptation is the hyoid (the bone at the top of the voice box) of howling monkeys (*Alouatta*), the largest of the Platyrrhines (~7–10 kg). In these monkeys, particularly in the males, this normally small bone at the top of the throat is expanded into a massive resonance chamber nearly as large as their skulls. This functions to amplify their impressively loud vocalizations (Hill, 1957, 1960, 1962; Fleagle, 1999).

The Catarrhine Skeleton

Skeletally catarrhines have more reduced dentition than the other primates, having lost one of the premolars. Members of this clade can be distinguished from other primates in details of the ear region, but beyond this, there are not many defining osteological features of the group. These are generally the largest primates, with the smallest extant monkey, the talapoin (*Miopithecus*), at ~1 kg, which is fairly large compared to many of the smallest strepsirrhines and platyrrhines, and the largest living members of the Catarrhini, the gorillas (*Gorilla*), are clearly the largest living taxon in the order at up to ~230 kg in the wild. One extinct catarrhine, the hominoid *Gigantopithecus*, was the largest primate to ever live at ~300 kg (Hill, 1966, 1970, 1974; Fleagle, 1999).

The Hominoid Skeleton

One obvious feature often used to define this group is the lack of an external tail. While some other catarrhines have also lost their tails (e.g., the Barbary macaque, *Macaca*

Sylvanus, is often called the Barbary "ape" because this species has no tail) (Hill, 1974), none of the hominoids have external tails nor did any of their common ancestors. In general, hominoids have a more orthigrade or upright posture. The hylobatids, that is, gibbons and siamangs, are highly specialized for a form of below-branch suspension that is so fast and explosive that it is often referred to as "ricochet" brachiation. Gorillas (*Gorilla*) and chimpanzees (*Pan*) exhibit a form of quadrupedal locomotion known as "knuckle walking" in which the weight of the front of the body is supported by curled fingers. The adaptations found in humans can generally be broken into two major categories: (1) changes related to our bipedal locomotion (e.g., changes in our spines, hips, angles, knees) and (2) changes related to advances in our cognition (not only our large brain cases, but also modifications to our teeth and hands in response to technological advances in feeding and tool use). Human adaptations are of such interest that whole courses and indeed whole academic disciplines are devoted to the subject, and thus it is clearly beyond the scope of this chapter to discuss those adaptations in detail (Fleagle, 1999).

MYOLOGY

Introduction

The comparative myology of primates has interested researchers for centuries. This fascination can be observed in Tyson's (1699) illustration that shows the muscles of a chimpanzee (probably a bonobo, *Pan paniscus*) standing in a biped position, as if it were a modern human. Extant primates are characterized by various myological synapomorphies, which are shared, derived features (Diogo and Abdala, 2010; Diogo and Wood, 2012b), including the presence of certain muscles in the hand (i.e., opponens pollicis and opponens digiti minimi) that increase dexterity and the fact that a specific arm muscle (the biceps brachii) does not insert onto one of the forearm bones (the ulna). Such features concern the movements of the forearm and hand and are probably related to the arboreal behavior of the first primates, although some of these features (e.g., biceps brachii not inserting onto the ulna and the differentiated opponens digiti minimi) are found in a few other mammals, including nonarboreal taxa such as rats. Below we will list the various groups of head and neck and appendicular muscles found in primates and briefly discuss their evolution within primates.

General Notes on the Evolution of the Primate Muscles

A list of the number of head, neck, pectoral, and forelimb muscles present in representative genera from all major primate groups is shown in Table 1. Diogo and Wood (2012a) have pointed out that the taxa that appear in more derived positions within the primate tree, for instance, humans, are also generally more derived morphologically in the sense that there is a higher number of unambiguous evolutionary steps (concerning these groups of muscles) from the base of the tree to those taxa. For example, ringtail lemurs (*Lemur*) are one of the least morphologically derived primates in the tree and have accumulated only 25 myological changes since the base of the primate tree, while *Homo* has 75 accumulated steps ("number of steps," or NS = 75). Previous studies have shown that the transition rate of mitochondrial DNA has also been extremely low in *Lemur* (e.g., Hasegawa et al., 1990). However, there are exceptions to this generalization. For example, there are more accumulated steps leading to *Saimiri* (NS = 46) than to the Old World monkeys *Cercopithecus* and *Colobus* (44); to *Callithrix* (51) than to *Cercopithecus* and *Colobus* (44) and to *Macaca* and *Papio* (49); and to *Hylobates* (72) than to *Pongo* (59), *Gorilla* (64), and *Pan* (70) (see Glossary). The unique evolutionary history of hylobatids (see Glossary) is reflected in a peculiar mix of plesiomorphic ("ancestral") and derived anatomical structures. Hylobatids are an example of mosaic evolution for although their head and neck muscles (partial NS = 26) are anatomically less derived than those of hominids (partial NS *Pongo* = 28, *Gorilla* = 28, *Pan* = 32, *Homo* = 33) and various monkeys (partial NS *Saimiri* = 27, *Callithrix* = 30, *Macaca* = 28, *Papio* = 27) their pectoral and upper limb muscles are more derived (partial NS = 46) than those of any other taxon, including modern humans (partial NS = 42). The highly specialized pectoral and upper limb musculature of hylobatids is most likely related to their peculiar brachiating lifestyle (e.g., Andrews and Groves, 1976; Michilsens et al., 2009).

Bakewell et al. (2007, p. 7492) suggest that their molecular studies show that "in sharp contrast to common belief, there were more adaptive genetic changes during chimp evolution than during human evolution," that is, their finding "suggests more unidentified phenotypic adaptations in chimps than in humans." However, the results of Diogo and Wood's (2012b) parsimony and Bayesian analyses indicate that, at least regarding the gross morphology of the head and neck and pectoral and upper limb muscles, since the *Pan*/*Homo* split humans have evolved faster than chimpanzees (c. 2.3 times faster according to the lengths of the branches leading to modern humans and to chimpanzees in their most parsimonious tree, that is, the primate evolutionary tree obtained from their cladistic–phylogenetic analysis). In turn, since the *Gorilla*/Hominini split, gorillas have only accumulated two unambiguous muscular apomorphies (derived features), while there are respectively 8 (4 + 4) and 13 (4 + 9) unambiguous apomorphies leading to extant chimpanzees and to modern humans.

It is, however, important to stress that in this case having a higher number of character state changes in a branch does not mean having a higher number of muscles. For example,

TABLE 1 Table Summarizing the Total Number of Mandibular, Hyoid (Not Including the Small Facial, Extrinsic Muscles of the Ear), Branchial, Hypobranchial, Pectoral, Arm, Forearm, and Hand Muscles in Adults of Primate Genera

	Lemur	*Propithecus*	*Loris*	*Nycticebus*	*Tarsius*	*Pithecia*	*Aotus*	*Saimiri*	*Callithrix*	*Colobus*	*Cercopithecus*	*Papio*	*Macaca*	*Hylobates*	*Pongo*	*Gorilla*	*Pan*	*Homo*
Mandibular muscles	8	8	8	8	8	8	8	8	8	7–8	7–8	8	8	8	7	8	8	8
Hyoid muscles (not extrinsic ear)	25	24	24–26	26	24	22	23	21	22	24–25	26–27	25–26	26	26	26	26	26	27
Branchial muscles	14–16	14–16	15–17	14–17	16–17	14–16	14–16	15–16	14–16	13–14	16	14–15	16	17	14–15	15–16	15	16
Hypo-branchial muscles	12	12	12–15	12–15	12	12–13	11–12	12	13	12	12	13	13	13	12–13	13	13	13
Pectoral muscles	17	15–16	16	16	17	15	16	16	17	16	17	17	17	14	15	14	14	14
Arm muscles	5	5	5	5	5	5	5	5	5	5	5	5	5	5	5	5	5	4
Forearm muscles	19	19	18	18	19	19	19	19	19	19	19	19	19	19	18	18	19	20
Hand muscles	30	30	30	34	32–36	22	22	22	21	27	27	27	27	27	20	20	26	21
Total number of muscles	130–132	127–130	128–135	133–139	133–138	117–120	118–121	118–119	119–121	123–126	129–131	128–130	131	129	117–119	119–120	126	123

The taxonomic groups in which the genera seen in this table are included are given in the Glossary.
Modified from Diogo and Wood (2012a).

some of the nine apomorphies of modern humans listed by Diogo and Wood (2012a) involve the loss of muscles (e.g., absence of levator claviculae; absence of dorsoepitrochlearis). In fact, as shown in Table 1, chimpanzees, along with most other extant primates, have more muscles than modern humans. The data provided in that table therefore contradict the *scala naturae* idea that there is a "progress" toward a greater "complexity" of modern humans, at least with regard to the number of muscles and muscle bundles. In the next paragraphs, we briefly summarize and discuss the information provided in Table 1 and compiled from Diogo and Wood's (2012a) studies. We focus on the muscles of the head and neck and pectoral and upper limb muscles and follow the myological nomenclature of Diogo and Abdala (2010). The pectoral and upper limb muscles are divided into five subgroups: pectoral, arm, ventral forearm, dorsal forearm, and hand. Regarding the head and neck muscles, we focus on four main subgroups: (1) Mandibular—these are generally innervated by the Vth cranial nerve, the trigeminal (e.g., the muscles of mastication and one of the middle ear muscles, the tensor tympani); (2) Hyoid—these are usually innervated by the VIIth cranial nerve, the facial (e.g., muscles of facial expression and the other middle ear muscle, the stapedius); (3) Branchial—these are usually innervated by the IXth (glossopharyngeal) and Xth (vagus) cranial nerves and include the majority of the intrinsic laryngeal muscles, although the trapezius and sternocleidomastoideus are mainly innervated by the XIth (accessory) cranial nerve; and (4) Hypobranchial—according to Edgeworth (1935) the muscles in this group are developed from the anterior myotomes of the body and thus they have all migrated into the head (e.g., infrahyoid muscles). Although they retain their main innervation from spinal nerves, they may also be innervated by the XIth and XIIth (hypoglossal) cranial nerves, but they usually do not receive any branches from the Vth, VIIth, IXth, and Xth cranial nerves. We therefore do not discuss the internal and external ocular muscles.

Head and Neck Muscles of Primates

The number of mandibular muscles is essentially constant ($N=8$) within extant primates, including modern humans; an exception is *Pongo*, which lacks the muscle digastricus anterior. With respect to the hyoid muscles, the predicted plesiomorphic condition for the primate clade is four nonfacial hyoid muscles, that is, more than most extant primates, including modern humans ($N=3$). This is because the jugulohyoideus is usually lacking in *Tarsius* and is almost always, if not always, absent in anthropoids. Some phylogenetically derived primates also have more nonfacial hyoid muscles than modern humans (e.g., *Pongo* has four muscles because it has a peculiar muscle stylolaryngeus). However, with respect to the facial muscles, despite lacking three muscles that were plesiomorphically present in primates (platysma cervicale, sphincter colli profundus, and mandibulo-auricularis), modern humans do have more facial muscles than any other primate studied by us ($N=24$, not including the small extrinsic muscles of the ear). This illustrates the role played by facial muscles in primate and particularly human evolution. Examples of facial muscles that are present in modern humans and missing in most other primates are the risorius and the temporoparietalis.

It is predicted that plesiomorphically ("ancestrally"), primates had four, or even five (if both the spinotrapezius and the acromiotrapezius were present as distinct muscles, as is the case in extant taxa such as *Tarsius*), true branchial muscles. Modern humans usually only have three muscles (trapezius, sternocleidomastoideus, and stylopharyngeus) for, as in other hominoids, modern humans usually lack a ceratohyoideus. The number of pharyngeal muscles is relatively constant within extant primates; modern humans have no autapomorphic, unique pharyngeal muscles nor do they lack any muscle that was predicted to be plesiomorphically present in primates. However, modern humans, as well as gorillas, do have more laryngeal muscles ($N=5$) than most other primates (which usually do not have a distinct arytenoideus obliquus). This stresses the importance of vocal communication in primate and human evolution. With a few exceptions, the number of hypobranchial muscles is essentially constant ($N=11$–13) within extant primates, including modern humans. The exceptions mainly concern the presence/absence of the palatoglossus and/or of the omohyoideus, although in some rare cases small muscles derived from the genioglossus may be recognized as distinct structures (e.g., genio-epiglotticus and hyo-epigloticcus in at least some specimens of *Loris*).

Pectoral and Forelimb Muscles of Primates

It is predicted that plesiomorphically, primates probably had 17 pectoral muscles (including a deltoideus acromialis et clavicularis and a deltoideus scapularis, as well as a single rhomboideus—not differentiated into a rhomboideus major and a rhomboideus minor). However, modern humans usually only have 14 pectoral muscles (the rhomboideus occipitalis, levator claviculae, and panniculus carnosus are missing and the deltoideus is not differentiated into a deltoideus acromialis et clavicularis and a deltoideus scapularis, but the rhomboideus minor and the rhomboideus major are differentiated). Extant primates generally have the same number of arm muscles ($N=5$), but because modern humans lack a dorsoepitrochlearis they have only four muscles. Forearm muscles in primates usually number between 18 and 19; two of the 19 muscles that are predicted to be plesiomorphically present in primates may be missing in some groups (e.g., the epitrochleoanconeus is usually absent in hominoids except *Pan* and the anconeus is usually absent in *Hylobates*). Modern humans, because they usually lack the epitrochleoanconeus but have a flexor pollicis longus and an extensor pollicis brevis, have more forearm muscles than any other primate studied by us (N.B., hylobatids also have a flexor pollicis longus and an extensor pollicis brevis, but they normally lack an anconeus). With respect to the hand muscles, phylogenetically plesiomorphic primates such as strepsirrhines and *Tarsius* usually have more than 30 muscles, but modern humans usually have only 21 muscles. The muscles that are conserved as separate structures in modern humans are those that insert onto the thumb; modern humans also have an additional pollical muscle "volaris primus of Henle" (TDAS-AD *sensu* the present work). The muscles that modern humans have lost relative to phylogenetically plesiomorphic primates (e.g., contrahentes, intermetacarpales, and interossei accessorii) attach to digits 2, 3, 4, and 5. This stresses the role played by thumb movements (produced by hand muscles, but also, importantly, by peculiar muscles such as the flexor pollicis longus and extensor pollicis brevis) in human evolution, probably related to tool manipulation (Diogo et al., 2012).

Variation and Evolutionary History of Primate Muscles, and Comments on the "Scala Naturae"

The muscle groups that show the least variation in number within the primate clade are the mandibular muscles (range 7–8), the arm muscles (range 4–5), and the forearm muscles (range 18–20); then come the pectoral muscles (range 14–17), the branchial and hypobranchial muscles (range 13–17 and 11–15, respectively); the hyoid muscles (range 21–27) and in particular the hand muscles (range 20–36) show the greatest variation in number. With respect to the overall number of muscles in the primate subclades, the New World monkeys are the least variable clade (range 117–121); this is remarkable because the Platyrrhini is the taxonomically more diverse group of extant primates. Then come the Old World monkeys (range 123–131), the hominoids (range 117–126), and the strepsirrhines (range 127–139). Interestingly, the phylogenetically more plesiomorphic groups of living primates (i.e., the strepsirrhines (range 127–139) and tarsiiforms (range 133–138)) are the ones with the greatest number of muscles in total, followed by the Cercopithecinae (range 128–131) and *Hylobates* (129), then by *Pan* (126) and *Colobus* (range 123–126), then by modern humans (123), and lastly by *Gorilla* (range 119–120), *Pongo* (range 117–119), and the New World monkeys (range 117–121 m).

In summary, regarding the evolution of these muscles listed in Table 1 within the order primates, it can be said that there is no simple relationship between muscle number and relative position on the primate evolutionary tree (cladogram). Modern humans have fewer head, neck, pectoral, and upper limb muscles than most extant primates, including phylogenetically plesiomorphic primates such as strepsirrhines and tarsiiforms, as well as Old World monkeys and chimpanzees. It is remarkable that *Nycticebus* and *Tarsius*, two nonanthropoid genera that include some of the smallest living primates and that are often considered to be anatomically plesiomorphic in comparison to anthropoid primates, have more head, neck, pectoral, and upper limb muscles than any other primate taxon included in Table 1. These data clearly refute the contention (*scala naturae*) that modern humans are the most anatomically complex primate. The only muscle groups for which modern humans have more muscles than most other primates are the muscles of the face, larynx, and forearm. In modern humans the latter include two peculiar muscles that are related to the movements of the thumb; among nonhuman primates extensor pollicis brevis and flexor pollicis longus are only present in hylobatids and as explained above modern humans usually have an additional muscle inserting onto the thumb, the volaris primus of Henle. Therefore, the summary provided here shows that modern humans have fewer muscles than most other living primates, but it is also consistent with the proposal that facial and vocal communication and specialized thumb movements have probably played an important role in human evolution.

EXTERNAL FEATURES AND INTERNAL ORGANS

Introduction

The constructional morphology framework presents three factors that interact to determine the biological form. These are a *phylogenetic factor*, which is guided by an individual's evolutionary history, a *morphogenetic factor*, which includes both ontogeny and biomechanics, and finally an *adaptive factor*, which acknowledges the role natural selection has on an animal's body plan (Thomas, 1979). As you have read above, there is an incredible variation in the musculoskeletal anatomy among primates. Discussed below is the comparative anatomy of the major organ systems in these mammals. It is important to recognize that although function is often attributed to many of the features described below, there are other factors beyond natural selection that shape the anatomy of a primate.

The internal organs of primates can be relatively conservative in appearance and size. This means that some organs do not deviate from what is expected for an individual's size. Examples of conservative organs are the heart, spleen, liver, and kidneys (Leonard and Robertson, 1992; Aiello and Wheeler, 1995). Some organs do differ in morphology, but on closer examination, these differences most likely are influenced more by phylogenetic or developmental factors, rather than adaptive factors. The lungs are an example of these differences. Humans typically have three lobes on the right and two lobes on the left (Saladin, 2012). However, despite lobe differences, the segmental architecture of the lung on each side remains the same, and thus functionally equivalent. Other primates, for example, the squirrel monkey (*Saimiri*), the orangutan (*Pongo*), and the gorilla (*Gorilla*) have fewer lobes but the segmental architecture remains the same (Ankel-Simons, 2007). Small differences in morphology, in many instances, do not result in differences in physiology or function.

Below we will highlight the major structural and functional differences of homologous organs and accessory structures across primates, including the nervous system, which includes a discussion on brain size; the visual, auditory, and olfactory systems; the digestive system, which includes comparative lingual anatomy; and lastly the reproductive system.

The Nervous and Sensory Systems

Primates are unique when compared to most other mammals because we have relatively larger brains than expected for our body size (Radinsky, 1972; Fleagle, 2013). Among primates, there is a notable increase in relative brain size between strepsirrhines and haplorhines, and then again between monkeys and hominoids. Measures of relative brain size are hypothesized to be a rough estimate of the intelligence of an animal but, although there is some truth to this hypothesis, the brain cannot be regarded as a single evolutionary or functionally homogeneous entity (Fleagle, 2013). The brain has several different regions, each specialized for a particular function (e.g., occipital lobe = vision). There are gross differences across primates in the relative size of these functional regions that might be masked if brain size alone is examined. For example, if a tarsier (*Tarisus*) and a mouse lemur (*Microcebus*) are compared, both have similar brain sizes and both are similar in size. However, on a closer examination of the functional regions of these animals, the tarsier devotes a larger percentage of the brain to visual processing (Fleagle, 2013).

In primates, the visual system is referenced as one of the most important sensory modalities relative to the other special senses (Le Gros Clark, 1959; Martin, 1990). Although this is true to some degree across primates, this generalization is most fitting for haplorhine primates (Kirk, 2004; Kirk and Kay, 2004). There is remarkable diversity in the visual abilities within haplorhines, but broadly anthropoids are adapted for extremely acute diurnal vision. Visual acuity is the ability to distinguish between two closely spaced visual stimuli (Kirk and Kay, 2004). Comparative studies have pinpointed several features that enhance visual acuity. These are relatively large eyes, an absence of a tapetum lucidum, the presence of short-wavelength filters in the lens and retina that limit chromatic aberration, and the presence of a central retina with a high photoreceptor density. There are several hypotheses on why these features have evolved in primates, one hypothesis suggesting that this combination of features is well developed in vertebrates that rely on visually directed predators (Ross, 1996).

The ability of primates to discriminate color is unique among mammals (Ahnelt and Kolb, 2000; Arrese et al., 2002; Dominy et al., 2003). Some primates possess three types of retinal cones (Dominy et al., 2003). All anthropoids exhibit at least dichromatic, and in catarrhines trichromatic, color vision (Dominy et al., 2004). Trichromacy in catarrhines is regarded as routine or uniform, meaning both males and females possess three opsins tuned to three key wavelengths: ~430, ~530, and ~560 nm (Jacobs, 1994). Several hypotheses have been proposed for why trichromacy evolved. The visual system of primates may be tuned to stimuli of vital importance, such as food. Some researchers do suggest that routine trichromachy in primates evolved to detect color differences in mature leaves, which are typically green, and immature leaves, which are typically red in color. Immature leaves have fewer toxins and are easier to digest (Dominy and Lucas, 2004). The ability to detect and select red leaves over green may be a foraging adaptation in primates. Other researchers, however, believe that color vision and enhanced visual acuity in catarrhine primates may have evolved to facilitate long-range visual communication and facial recognition. For example, in females of many catarrhine taxa, signal receptive sexual behavior changes with the size, shape, and color of their perianal "sexual" skin (Rowell, 1972). Signaling of this sort can communicate subtle differences in the reproductive state on an individual. Again, the visual system in primates is complex and has long been a topic of interest among scientists. Teasing out differences in the visual system can shed light on the evolution of both the feeding and the social ecology of our order.

Olfaction (smell), in contrast to vision, has been greatly reduced in primates. There are two main olfactory systems, the main olfactory system and the vomeronasal system. Primates vary greatly in the size and complexity of these systems. The vomeronasal system is finely tuned to detect sociosexual signals (pheromones) and the main olfactory system is more broadly tuned to detect both ecological and social signals; however, there is evidence for a wide degree of overlap of the two (Garrett and Steiper, 2013). The main olfactory system appears to be relatively well developed across all primates, while the development of the vomeronasal system is variable across the major primate clades. Strepsirrhines maintain a functional vomeronasal system comparable to many nonprimate mammals, while haplorhines exhibit a reorganization or loss of its peripheral sensory organ, the vomeronasal organ (Smith et al., 2007). Platyrrhines and tarsiers have functional vomeronasal organs that have a different distribution of sensory epithelium compared to other mammals (although research on olfactory marker protein activity indicates that *Saguinus tamarins* has nonfunctional vomeronasal organs), and catarrhine primates have either lost the vomeronasal organ entirely or exhibit nonfunctional vestiges (Kappeler and van Schaik, 2002; Smith et al., 2004, 2005, 2011). The overall reorganization/reduction of the vomeronasal organ in haplorhine primates may have been a consequence of acquiring a complex visual system that was more energetically expensive, and reducing a more finely tuned olfactory system for which the main olfactory system could compensate (Garrett and Steiper, 2013).

The auditory system of primates does not deviate significantly from most other mammals. Primates have "average" hearing. The auditory system is composed of six receptor endorgans that help us detect sound and facilitate movement. Together, these sensory endorgans comprise the "membranous labyrinth" of the inner ear (Purves et al., 2001; Smith et al., 2004). The six receptor endorgans include three semicircular canals, the utricle, the sacculus, and the cochlea. The semicircular canals help with detecting angular acceleration of the head and assist in stabilizing the eyes and head. The utricle and sacculus help detecting linear acceleration of the body and head tilt relative to the plane of gravity. The cochlea detects sounds that are transmitted through the fluid filled canal by the external and middle ear (Kirk and Gosselin-Ildari, 2010). These endorgans are suspended within the petrous temporal bone, and they are difficult to examine without the assistance of high-resolution imagining, such as computed tomography. Primates that are agile and have fast jerky locomotion have significantly larger canals relative to body size than those that are more cautious (Spoor et al., 2007). For example, the gibbon has large semicircular canals for its body size, while slower moving apes like the chimpanzee and gorilla have smaller semicircular canals (Spoor, 2003). In a recent study of 33 primate species Kirk and Gosselin-Ildari (2010) found differences in the length of the cochlea in primates. The dimensions of the cochlea influence hearing abilities in mammals. The basilar membrane length (cochlea) is correlated with both the high- and the low-frequency limits of hearing in mammals with unspecialized cochleae. In primates, as choclear size increases, the range of audible frequencies shifts downward (Kirk and Gosselin-Ildari, 2010). Establishing correlations among skeletal structures (inner ear), soft tissues (cochlear membrane), and function (agility and hearing) has important implications for interpretations of the fossil record (Spoor, 2003).

The Digestive System

The digestive system includes the oral cavity (including the tongue), esophagus, stomach, intestines, and rectum. The tongue is an important structure and the centerpiece of the gustatory system. The function of the gustatory system (taste) is to help determine the chemical contents of foods. Failure to recognize food with a high enough caloric content could mean death, albeit slow, from malnutrition. Obtaining adequate nutrients from food is critical for both males and females. However, food selection (and rejection) may be more important to the reproductive success of females because their reproductive costs are higher than male reproductive costs. As taste is a very important sense involved with food selection and rejection, females may benefit more from greater taste sensitivity than males. In a comparative analysis of primates, taxa with higher reproductive cost (e.g., large brained primates such as *Cebus*, *Gorilla*, *Pan*, and *Homo*) do show a sex-based difference in gustatory anatomy (Alport, 2009; Muchlinsk et al., 2011). The dorsum of the tongue is carpeted with four types of papillae: filiform, fungiform, foliate, and circumvallate. Filiform papillae can be found across the entire tongue and are nongustatory, meaning that they are not capable of transmitting taste information to the brain. Circumvallate, foliate, and fungiform papillae are gustatory and distinct in their location and anatomy (Purves et al., 1997). Circumvallate are located on the posterior aspect of the tongue, whereas foliate are located on the posterolateral aspect of the tongue. Fungiform papillae are located on the anterior two-thirds of the tongue. Fungiform papillae are the first gustatory structure that comes in contact with an ingested chemical (Purves et al., 1997; Buck, 2000). Accordingly, fungiform papillae are critical in food selection and show the most sexual dimorphism. Each papilla contains taste buds that include taste cells. Comparative gustatory research in primates has found that females in some species have a greater number of fungiform papillae, larger fungiform papillae, and a higher density of fungiform papillae than males (Muchlinski et al., 2011). In addition to documented sex-based differences, there is evidence for interspecific variation in fungiform papillae densities among primates who consume sweeter food, like fruit, and those that focus on less sweet food like leaves. Fruit eaters tend to have a higher density of fungiform papillae compared with fruit and insect eaters (Alport, 2009).

The digestive system is mainly a hollow muscular tube that can broadly be divided into three: the foregut, midgut, and hindgut. The foregut is the anterior part of the alimentary canal, which includes the pharynx, esophagus, stomach, liver, pancreas, and half of the duodenum. The midgut is the middle section of the alimentary canal, which includes a portion of the duodenum, the ileum, the jejunum, and the first two-thirds of the colon. The hindgut includes the last one-third of the transverse colon, the descending colon, sigmoid colon, rectum, and the upper two-thirds of the anal canal. Each region serves a different function in digestion (Chivers and Hladik, 1980; Hladik et al., 1980a; Hladik, 1981). Many animals are referred to as either foregut or hindgut fermenters. Beyond enzymes, the gut houses and nurtures a complex microorganism ecosystem. These microorganisms help animals digest foods, particularly cellulose. A symbiotic relationship exists between almost all herbivorous mammals and microbes. The animals provide the food and space for the microbes, and the animals receive, in return, the by-product of fermentation, which includes fatty acids and sugars (Ankel-Simons, 2007). Fermentation is a slow process, and it requires space for it to occur. Animals that rely on fermentation have expanded alimentary canal regions to facilitate digestion.

In primates, foregut fermenters like the Colobidae (e.g., *Colobus* spp., *Presbytis* spp., *Trachypithecus* spp., *Pygathrix nemaeus*, *Nasalis larvatus*) have enlarged saculated stomachs. The advantage of foregut fermentation is that the animals can absorb the by-products of fermentation because digestion of the cellulose occurs prior to entering into the small intestines, which is a region where the majority of nutrient absorption occurs (Fleagle, 2013). There are hindgut fermenters as well; these include members of the genus *Alouatta* and *Indridae*. Hindgut fermentation is facilitated by enlargement of the cecum and colon (Ankel-Simons, 2007). Hindgut fermenting primates tend to be more variable in their diets and focus more on young leaves, flowers, and fruit, than mature leaves (Milton, 1980, 1981). Mountain gorillas are different from other gorillas because they depend more on herbaceous vegetation. Most other gorillas focus primarily on fruit, and then fall back on herbaceous food during periods of food scarcity (Tutin et al., 1991; Nishihara, 1995; Remis, 1997; Rogers et al., 2004). Mountain gorillas are neither foregut nor hindgut fermenters, despite relying primarily on leaves and terrestrial herbaceous vegetation (Rogers et al., 2004). They have a generalized hominoid gut adapted for frugivory (Kay, 1975; Chivers and Hlakik, 1980; Martin et al., 1985), but do have a high number of cellulose-digesting protozoans in their colon (Collet et al., 1984). Thus, although primates may not be specialized for foregut or hindgut fermentation morphologically, they may possess wildly divergent microorganisms that can facilitate cellulose digestion. Faunivores (insect and small vertebrate feeders) are not classified as hindgut fermenters. Many faunivores do have notable enlargement of the cecum relative to other primates. The insect exoskeleton shares structural similarities to that of mature leaves (Chivers and Hladik, 1980). Cecal enlargement in many faunivores (like the tarsiers) slows down digestion by increasing the distance food must travel, and as a result, increases absorption. Beyond an enlarged cecum, all other digestive structures remain relatively simple (Chivers and Hladik, 1980; Ankel-Simons, 2007).

The Reproductive System

All primates have a basic mammalian reproductive system: sperm fertilizes a primary occycte (egg) housed within the female, and the embryo then develops in utero for several months after fertilization. The reproductive organs of both the male and the female are similar to those found among most nonprimate mammals. The male possesses testes, a spermatic cord, seminal vesicles, a prostate, and a urethra. The female has ovaries, fallopian tubes, and a uterus. Humans are no different from most other primates. However, the differences seen in both the external and the internal anatomy of primates is great and is a result of differences in socioecology, a field of biology that examines how social structure (including reproductive strategies) and organization are influenced by an organism's environment. The way females have organized themselves around food (a limiting resource for females) and how males have organized themselves around the females (females are the limiting resource for males) have had a huge influence on the mating behaviors and as a result the reproductive anatomy of primates.

Female primates typically possess two ovaries. These ovaries are connected to the uterus via the fallopian tubes, which are associated with the ovaries via one fimbria. There is considerable variation in the size and length of the fallopian tubes. In most strepsirrhines the fallopian tubes are large relative to the body of the uterus. Additionally, most strepsirrhines and the tarsiers have a uteruses that are bicornuate, or "two horned" (Ankel-Simons, 2007; Fleagle, 2013). This anatomy is most commonly among animals that have twins or have litters (Ankel-Simons, 2007). In contrast, haplorhines tend to have fallopian tubes that are relatively slender and the body of the uterus is not split and is relatively large compared to those observed in strepsirrhines (Fleagle, 2013). This anatomy is most common among animals that generally give birth to one offspring at a time.

One of the most fascinating displays of diversity among female primates is clitoral anatomy. The clitoris is located just anterior to the external urethral orifice and the vagina. In many primates, including humans, the clitoris is hidden under a "hood," which is a double fold of skin and is also the most anterior extension of the labia minora. In contrast, some primates like some Malagasy primates (*Strepsirrhini*: Lemuriformes) and atelines the clitoris is pendulous and larger than their male counterparts penises (Ostner et al., 2002; Drea and Weil, 2008; Fleagle, 2013). Adult males tend to dominate females because of their size and androgen-mediated aggressiveness (Ostner et al., 2002). However, in some primates, particularly the Malagasy strepsirrhines, females clearly dominate males in dyadic agonistic interactions (Tecot et al., 2013). Interestingly, adult female Malagasy strepsirrhines do not exhibit higher androgens than males. The reason behind female masculinization is not completely understood, but is mostly a result of prenatal androgen levels (Ostner et al., 2002). While most lemurs and some New World monkeys present with an enlargement of the clitoris, most Old World monkeys and hominoids do not; however, most Old World monkeys and hominoids do have enlarged sexual swellings that signal sexual receptivity. As noted above (sensory system section), all primates have a functional main olfactory system, but most catarrhine primates have lost the vomeronasal organ that is responsible for detecting pheromones. Pheromones are excreted hormonal signals. In animals that possess a vomeronasal organ, pheromones can communicate subtle differences in physiology. Researchers have suggested that the loss of the vomeronasal organ in catarrhines is closely correlated with the evolution of the sexual swelling. However, there is a long-standing dispute over whether there is a vomeronasal organ present in adult humans. Embryonic studies do show that developing humans do present with a vomeronasal organ. Recent microscopic studies show that some adult humans, if not most, have retained the vomeronasal organ on at lease one side (Meredith, 2001).

Males match females in genital diversity. Most males, including the apes, have androgen-dependent, keratinized "spines" (some small, and some not so small) along the shaft of the penis. These spines can be simple, like a single-pointed spine observed in the macaque, or more complex. Many strepsirrhines have complicated multispiked spines. Dixson (2012) attributes this variation to mating behavior. Simplified penile morphology tends to be associated with monogamous reproductive strategies in primates, while the more complex morphologies are associated with the more promiscuous strategies. These spines may increase tactile sensitivity of the glans because each spine is associated with tactile mechanoreceptors and is highly innervated. The spine may also help maximize reproductive success (Dixson, 2012). Testes size is also closely linked to androgen levels. In *Mirza coquereli*, and some other Malagasy strepsirrhines, the testes of the male enlarge during the mating season (Fleagle, 2013). The mating system of *M. coquereli* is described as scramble competition polygyny (Kappeler, 1999). Primates that show little mating competition, for example, monogamous (e.g., titi monkey) and single-male multifemale polygynous species (e.g., gorillas), have relatively small testes. Conversely, large testes are observed in primates that are multimale polyandrous where there is a significant amount of male–male competition (e.g., chimpanzees) (Wrangham, 1979; Van Schaik and Van Hooff, 1985; Kappeler and van Schaik, 2002). Interestingly, human male testicular anatomy falls somewhere between polygynous and polyandrous species, but closer to the polyandrous side of the spectrum (Wright, 1994). Interestingly, in many cultures, the correlations between mating systems and anatomy are not as closely linked as in many other primate groups. Based on testicular weight, humans should be polyandrous. Although polyandry is common in Africa (e.g., Northern Nigeria—cohusbands), Europe (e.g., Celtics), Asia (Lahaul-Spiti of the Himalayas), and in areas of the New World, today, monogamy or polygyny are more often the established culturally imposed behavioral norm. However, based on anatomical clues, one could hypothesize that monogamy and polygyny are a more recent phenomenon among humans than many people would like to think (Wright, 1994; see the chapter by Gangestad and Gebe).

REFERENCES

Ahnelt, P., Kolb, H., 2000. The mammalian photoreceptor mosaic-adaptive design. Progress in Retinal and Eye Research 19, 711–777.

Aiello, L.C., Wheeler, P., 1995. The expensive tissue hypothesis: the brain and the digestive system in human and primate evolution. Current Anthropology 36, 199–221.

Alport, L.J., 2009. Lingual Fungiform Papillae and the Evolution of the Primate Gustatory System (unpublished Ph.D. thesis). University of Texas, Austin.

Andrews, P., Groves, C.P., 1976. Gibbons and brachiation. In: Rumbaugh, D.M. (Ed.), Gibbon and Siamang, vol. 4. Karger, Basel, pp. 167–218.

Ankel-Simons, F., 2007. An Introduction: Primate Anatomy, third ed. Academic Press, Burlington.

Arrese, C.A., Hart, N.S., Thomas, N., Beazley, L.D., Shand, J., 2002. Trichromacy in Australian marsupials. Current Biology 12, 657–660.

Bakewell, M.A., Shi, P., Zhang, J., 2007. More genes underwent positive selection in chimpanzee evolution than in human evolution. PNAS 104, 7489–7494.

Buck, L.M., 2000. Smell and taste: the chemical senses. In: Kandel, E.R., Schwartz, J.H., Jessell, T.M. (Eds.), Principles of Neural Science. McGraw-Hill, New York, pp. 625–647.

Cartmill, M., 1972. Arboreal adaptations and the origin of the order primates. In: Tuttle, R.L. (Ed.), The Functional and Evolutionary Biology of Primates. Aldine-Atherton, Chicago, pp. 97–122.

Cartmill, M., 1974. Rethinking primate origins. Science 184 (4135), 436–443.

Chivers, D., Hladik, C., 1980. Morphology of the gastrointestinal tract in primates: comparisons with other mammals in relation to diet. Journal of Morphology 166, 337–386.

Clark, W.E.L.G., 1959. The Antecedents of Man. Harpers, New York.

Collet, J., Bourreau, E., Cooper, R., Tutin, C., Fernandez, M., 1984. Experimental demonstration of cellulose digestion by *Troglodytella gorillae*, an intestinal ciliate of lowland gorillas. International Journal of Primatology 5, 328.

Diogo, R., Abdala, V., 2010. Muscles of Vertebrates – Comparative Anatomy, Evolution, Homologies and Development. Taylor and Francis, Oxford.

Diogo, R., Wood, B., 2012a. Soft-tissue anatomy of the primates: phylogenetic analyses based on the muscles of the head, neck, pectoral region and upper limb, with notes on the evolution of these muscles. Journal of Anatomy 219, 273–359.

Diogo, R., Wood, B., 2012b. Comparative Anatomy and Phylogeny of Primate Muscles and Human Evolution. Taylor and Francis, Oxford.

Diogo, R., Richmond, B.G., Wood, B., 2012. Evolution and homologies of primate and modern human hand and forearm muscles, with notes on thumb movements and tool use. Journal of Human Evolution 63, 64–78.

Dixson, A.F., 2012. Primate Sexuality: Comparative Studies of the Prosimians, Monkeys, Apes and Human Beings, second ed. Oxford University Press, Oxford.

Dominy, N.J., Lucas, P.W., 2004. Significance of color, calories, and climate to the visual ecology of catarrhines. American Journal of Primatology 62, 189–207.

Dominy, N.J., Ross, C.F., Smith, T.D., 2004. The evolution of the special senses in primates: past, present, and future. Anatomical Record 281A, 1078–1082.

Dominy, N.J., Svenning, J.-C., Li, W.-H., 2003. Historical contingency in the evolution of primate color vision. Journal of Human Evolution 44, 25–45.

Drea, C., Weil, A., 2008. External genital morphology of the Ringtailed lemur (*Lemur catta*): females are naturally 'masculinized'. Journal of Morphology 269, 451–463.

Edgeworth, F.H., 1935. The Cranial Muscles of Vertebrates. Cambridge University Press, London.

Fleagle, J.G., 1999. Primate Adaptation and Evolution. Academic Press, New York.

Fleagle, J.G., 2013. Primate Adaptation and Evolution. Academic Press, San Diego.

Garrett, E.C., Steiper, M.E., 2013. Testing the Color Vision Priority Hypothesis in Primates: Acquisition of Trichromatic Color Vision Affects the Vomeronasal System and Not the Main Olfactory System. PaleoAnthropology Meeting [abstract].

Hasegawa, M., Kishino, H., Hayasaka, K., Horai, S., 1990. Mitochondrial DNA evolution in primates: transition rate has been extremely low in lemur. Journal of Molecular Evolution 31, 113–121.

Hladik, C.-M., 1981. Diet and the evolution of feeding strategies among forest primates. In: Harding, R.S., Teleki, G. (Eds.), Ominivorous Primates: Gathering and Hunting in Human Evolution. Columbia University Press, New York, pp. 215–254.

Hladik, C., Charles-Dominique, P., Petter, J., 1980a. Feeding strategies of five nocturnal prosimians in the dry forest of the west coast of Madagascar. In: Charles-Dominique, P., Cooper, H., Hladik, A., Hladik, C., Pages, E., Pariente, G., Petter-Rousseaux, A., Schilling, A., Petter, J. (Eds.), Nocturnal Malagasy Primates: Ecology, Physiology, and Behavior. Academic Press, London, pp. 41–73.

Hladik, C.M., Charles-Dominique, P., Petter, J.J., 1980b. Nocturnal Malagasy Primates: Ecology, Physiology, and Behavior. Academic Press New York.

Hill, W.C.O., 1953. Primates: Comparative Anatomy and Taxonomy; I Strepsirhini. Edinburgh Edinburgh University Press.

Hill, W.C.O., 1955. Primates: Comparative Anatomy and Taxonomy; II Haplorhini: Tarsioidea. Edinburgh Edinburgh University Press.

Hill, W.C.O., 1957. Primates: Comparative Anatomy and Taxonomy; III Pithecoidea Platyrrhini. Edinburgh Edinburgh University Press.

Hill, W.C.O., 1960. Primates: Comparative Anatomy and Taxonomy; IV Cebidae, Part a. Edinburgh Edinburgh University Press.

Hill, W.C.O., 1962. Primates: Comparative Anatomy and Taxonomy; V Cebidae, Part B. Edinburgh Edinburgh University Press.

Hill, W.C.O., 1966. Primates: Comparative Anatomy and Taxonomy; VI Catarrhini Cercopithecoidea: Cercopithecinae. Edinburgh Edinburgh University Press.

Hill, W.C.O., 1970. Primates: Comparative Anatomy and Taxonomy; VIII Cynopithecinae (Papio, Mandrillus, Theropithecus). Edinburgh Edinburgh University Press.

Hill, W.C.O., 1972. Evolutionary Biology of the Primates. Academic Press, London.

Hill, W.C.O., 1974. Primates: Comparative Anatomy and Taxonomy; VII Cynopithecinae (Cercocebus, Macaca, Cynopithecus). Edinburgh Edinburgh University Press.

Jacobs, G.H., 1994. Variations in primate color vision: mechanisms and utility. Evolutionary Anthropology: Issues, News, and Reviews 3, 196–205.

Kappeler, P.M., 1999. Lemur social structure and convergence in primate socioecology. In: Lee, P.C. (Ed.), Primate Socioecology. Cambridge University Press, Cambridge, pp. 273–299.

Kappeler, P.M., van Schaik, C.P., 2002. Evolution of primate social systems. International Journal of Primatology 23, 707–740.

Kay, R.F., 1975. The functional adaptations of primate molar teeth. American Journal of Physical Anthropology 43, 195.

Kirk, E.C., 2004. Comparative morphology of the eye in primates. The Anatomical Record 281A, 1095–1103.

Kirk, E.C., Gosselin-Ildari, A.D., 2010. Cochlear labyrinth volume and hearing abilities in primates. Anatomical Record 29, 765–776.

Kirk, E.C., Kay, R.F., 2004. The evolution of hight visual acuity in the anthropoidea. In: Ross, C., Kay, R.F. (Eds.), Anthropoid Origins: New Visions. Kluwer Academic/Plenum Publishing, New York, pp. 539–602.

Leonard, W.R., Robertson, M.L., 1992. Nutritional requirements and human evolution: a bioenergetics model. American Journal of Human Biology 4, 179–195.

Martin, R., 1990. Primate Origins and Evolution: A Phylogentic Reconstruction. Princeton University Press, Princeton.

Martin, R.D., Chivers, D.J., Maclarnon, A.M., Hladik, C.M., 1985. Gastrointestinal allometry in primates and other mammals. In: Jungers, W.L. (Ed.), Size and Scaling in Primate Biology. Plenum Press, New York, pp. 61–89.

Meredith, M., 2001. Human vomeronasal organ function: a critical review of best and worst cases. Chem Senses 26, 433–445.

Michilsens, F., Vereecke, E.E., D'Août, K., Aerts, P., 2009. Functional anatomy of the gibbon forelimb: adaptations to a brachiating lifestyle. Journal of Anatomy 215, 335–354.

Milton, K., 1980. Foraging Strategy of Howler Monkeys: A Study of Primate Economics. Columbia University Press, New York.

Milton, K., 1981. Food choice and digestive strategies of two sympatric primate species. American Naturalist 117, 496–505.

Mittermeier, R.W.R., Konstant, F., Hawkins, E.E., Louis, O., Langrand, J., Ratsimbazafy, R., Rasoloarison, J.U., Ganzhorn, S., Rajaobelina, I., Tattersall, Meyers, D.M., 2006. Lemurs of Madagascar. Conservation International.

Muchlinski, M.N., Paesani, S.M., Burrows, A.M., Smith, T.D., Alport, L.J., 2011. Behavioral and ecological consequences of sex based differences in taste bud densities in *Cebus apella*. Anatomical Record 294, 2179–2192.

Nishihara, T., 1995. Feeding ecology of western lowland gorillas in the Nouabale-Ndoki Park, northern Congo. Primates 36, 151–168.

Ostner, J., Kappeler, P.M., Heistermann, M., 2002. Seasonal variation and social correlates of androgen excretion in male redfronted lemurs (*Eulemur fulvus rufus*). Behavioral Ecology and Sociobiology 19, 1150–1158.

Purves, D., Augustine, G.J., Fitzpatrick, D., Katz, L.C., LaMantia, A.S., McNamara, J.O., Williams, S.M., 2001. The auditory system. Neuroscience 278–310.

Purves, D., Fitzpatric, D., Katz, L.C., LaMantia, A.S., McNamara, J.O., 1997. Neuroscience. Sinaur Associates, Inc, Sunderland.

Radinsky, L.B., 1972. Endocasts and studies of primate brain evolution. In: Tuttle, R. (Ed.), The Functional and Evolutionary Biology of Primates. Aldine, Chicago, pp. 175–184.

Remis, M., 1997. Western lowland gorillas (*Gorilla gorilla gorilla*) as mixed folivore/frugivores: use of variable resources. American Journal of Primatology 43, 87–109.

Rogers, M.E., Abernethy, K., Bermejo, M., Cipolletta, C., Doran, D., Mcfarland, K., Nishihara, T., Remis, M., Tutin, C.E.G., 2004. Western gorilla diet: a synthesis from six sites. American Journal of Primatology 64, 173–192.

Ross, C., 1996. Adaptive explanation for the origins of the Anthropoidea (primates). American Journal of Primatology 40, 205–230.

Rowell, T.E., 1972. Female reproductive cycles and social behavior in primates. Advances in the Study of Behavior 4, 69–105.

Saladin, K.S., 2012. Anatomy and Physiology: The Unity of Form and Function, sixth ed. McGraw Hill, New York.

Smith, T., Garrett, E., Bhatnagar, K., Bonar, C., Bruening, A., Dennis, J., Kinzinger, J., Johnson, E., Morrison, E., 2011. The vomeronasal organ of new world monkeys (Platyrrhini). Anatomical Record 2158–2178.

Smith, T.D., Bhatnagar, A.M., Tuladhar, P., Burrows, A.M., 2004. Distribution of olfactory epithelium in the primate nasal cavity: are microsmia and macrosmia valid morphological concepts? Anatomical Record 281A, 1173–1181.

Smith, T.D., Bhatnagar, K.P., Burrows, A.M., Shimp, K.L., Dennis, J.C., Smith, M.A., Maico-Tan, L., Morrison, E.E., 2005. The vomeronasal organ of greater bushbabies (*Otolemur* spp.): species, sex, and age differences. Journal of Neurocytology 34, 135–147.

Smith, T.D., Rossie, J.B., Bhatnagar, K.P., 2007. Evolution of the nose and nasal skeleton in primates. Evolutionary Anthropology 16, 132–146.

Spoor, F., 2003. The semicircular canal system and locomotor behaviour with special reference to hominin evolution. Courier Forschungsinstitut Senckenberg 243, 93–104.

Spoor, F., Garland Jr., T., Krovitz, G., Ryan, T.M., Silcox, M.T., Walker, A., 2007. The primate semicircular canal system and locomotion. PNAS 104, 10808–10812.

Tecot, S., Baden, A., Romine, N., Kamilar, J., 2013. Reproductive strategies in Malagasy strepsirhines. In: Rutherford, J., Hinde, K., Clancy, K. (Eds.), Building Babies: Primate Development in Proximate and Ultimate Perspective. Springer, New York, pp. 321–359.

Thomas, R., 1979. Constructional morphology. In: Fairbridge, R., Jablonski, D. (Eds.), Encyclopedia of Paleontology. Jutchinson and Ross, Dowden, pp. 482–487.

Tutin, C.E.G., Fernandez, M., Rogers, M.E., Williamson, E.A., McGrew, W.C., Altmann, S.A., Southgate, D.A.T., Crowe, I., Whiten, A., 1991. Foraging profiles of sympatric lowland gorillas and chimpanzees in the Lope Reserve, Gabon [and discussion]. Philosophical Transactions of the Royal Society of London Series B: Biological Sciences 334, 179.

Tyson, E., 1699. Orang-outang Sive Homo Sylvestris, or the Anatomy of a Pygmie Compared to that of a Monkey, an Ape and a Man. T. Bennet, London.

Van Schaik, C., Van Hooff, J., 1985. On the ultimate causes of primate social systems. Behavior 198, 91–117.

Wrangham, R., 1979. On the evolution of ape social systems. Social Science Information 18, 336.

Wright, R., 1994. The moral animal: the new science of evolutionary psychology. Pantheon Books, New York.

Chapter 5

Primate Behavior

David P. Watts
Department of Anthropology, Yale University, New Haven, CT, USA

SYNOPSIS

The order Primates comprises hundreds of species that vary widely in size, life history, ecology, and behavior. Most diurnal species and some nocturnal ones form stable social groups characterized by permanent association between males and females. Predation has been the main factor selecting for the evolution of group living. Typical group size and composition vary along with variation in predation risk, diet, food distribution, feeding competition, and mating strategies. Even in species that are mostly solitary while foraging (e.g., many nocturnal strepsirhines), individuals may associate nonrandomly and factors such as kinship can influence social structure. Social relationships within primate groups reflect complex mixes of competition and cooperation and are influenced by many factors, including variation in food distribution and resulting regimes of feeding competition, in relatedness, and in the availability and value of potential social partners; conflicts and convergences of reproductive interest between males and females; and correlated evolution of suites of behavioral traits in particular lineages. Kin biases in social behavior are common and much evidence indicates that kin selection has been important in primate evolution. Cooperation among nonkin is also common; much of this apparently is mutualism, in which cooperating individuals obtain immediate net benefits from their actions. Factors that account for permanent male–female association, whether as pairs or in groups that contain multiple females and, sometimes, multiple males, include the anti-predation benefits of group living; mate guarding by males; paternal care, including protection of offspring (notably against potentially infanticidal conspecific males); and defense of food supplies on which females and their offspring rely. Social bonds among the members of primate groups vary in strength and flexibility. In many cases (e.g., female baboons, male chimpanzees), some pairs of individuals maintain long-lasting, strong bonds, and increasing evidence indicates that how successful individuals are at maintaining strong social bonds and at negotiating networks of cooperation and competition influences their fitness.

The order Primates includes hundreds of extant species that vary widely in size, morphology, physiology, habitat tolerance, and life histories (see chapters by Hunt and by Martin). Nonhuman primates occur, or recently occurred, in most of the nonarid Old and New World tropics, including much of island Southeast Asia. Several species of *Macaca*, several colobines, and Hoolock gibbons in the Old World live in temperate habitats, as do howler monkeys, spider monkeys, and capuchins in the New World. Madagascar is home to the highly diverse lemur radiation. Dietary variation among and, often, within species is extensive, and variation in social systems and social behavior is correspondingly broad. Many primates form stable groups, and their social worlds often include complex social networks and durable and/or highly flexible social bonds, complex mixes of cooperation and competition, and variable mating tactics. This diversity and complexity, combined with phylogenetic proximity to humans, the fact that many primates are easily observable (once habituated), the availability of behavioral and life history data from long-term studies and of comparative data from multiple species and multiple populations of single species, and with methodological advances such as noninvasive collection of genetic and hormonal data, have led to an enormous literature on social behavior, ecology, and behavioral evolution.

No short review can summarize this literature. Instead, this chapter provides selective overviews of variation in social structure, social organization, and social relationships; the evolution of group living; cooperation and competition; sexual conflict; and how sociality affects fitness. For exhaustive reviews of these and related topics, see Mitani et al. (2012) and Campbell et al. (2007).

SOCIAL ORGANIZATION, PREDATION, AND GROUP LIVING

Social Organization

Social organization refers to how members of a population or species are dispersed in space and time: whether males and/or females are clustered in space or solitary; whether stable groups form and, if so, what their typical size and composition are; and whether groups are cohesive or exhibit fission–fusion dynamics (divide into temporary subgroups that vary in size and composition; Aureli et al., 2008). Diversity in nonhuman primate social organization is remarkable (Table 1). Most diurnal primates and a few nocturnal ones form stable groups, with memberships that change only with births, deaths, immigration, and emigration or expulsion. Typical group size and composition vary from single male/female pairs with associated dependent offspring to large multimale, multifemale groups. Pairs in pair-living species can be cohesive, as in gibbons (Brockelman, 2009), or dispersed with paired males and females usually not in visual contact while active (e.g., fork-marked lemurs, *Phaner furcifer*: Schülke and Kappeler, 2003) and not necessarily regularly sharing sleeping sites (e.g., sportive lemurs, *Lepilemur leucopus*: Droscher and Kappeler, 2013). A few species have multilevel societies, with stable mixed sex groups that usually have single adult males and multiple adult females at the base and multiple one-male units clustering more or less consistently at one or more higher levels of association (e.g., Hamadryas baboons, *Papio hamadryas*; geladas, *Theropithecus gelada*; Table 1). In contrast, orangutans (*Pongo* spp.) and most nocturnal species are solitary: individuals are mostly alone while active, although daytime sleeping groups with relatively stable memberships form in some solitarily foraging nocturnal species (e.g., gray mouse lemurs, *Microcebus murinus*; Table 1).

Sex Differences in Optimal Social Dispersion, Diets, and Feeding Competition

Social Dispersion and Feeding Competition

Social dispersion is essentially synonymous with social structure and highlights the fact that even solitary species have social systems (Kappeler, 2012). Emlen and Oring (1977) developed a conceptual scheme for analyzing variation in mammalian social dispersion that greatly influenced socioecological theory in primatology. They argued that food distribution determines the optimal social dispersion pattern for females because the high energetic and nutritional demands of reproduction place a premium on foraging efficiency. The optimal social dispersion pattern for males then depends on that of females, particularly potentially fertile females. Male responses to female distribution can impose costs on females (e.g., those associated with sexual conflict; below), which changes the female optimum; adjustments by females change the male optimum; and so on. The realized social system eventually results from confluences and compromises of male and female interest.

Wrangham (1979) imported this framework into primatology and applied it to the remarkably varied social systems of extant apes other than bonobos (*Pan paniscus*), about which nothing was then known. He proposed that the ideal female strategy in all apes was solitary foraging, which would minimize feeding competition. Such competition occurs when foraging by one individual or group reduces another's net energy (or nutrient) gain, either because it finds and harvests food first (indirect, or "scramble," competition) or because of agonistic interactions over access to food items or feeding sites that give winners access to food at losers' expense (direct, or "contest," competition). Scramble competition inevitably ensues if too many individuals simultaneously use the

TABLE 1 Forms of Nonhuman Primate Social Organization

Modal Form of Grouping	Examples
1. Females solitary; no stable or long-term associations	*Perodicticus potto* (Charles-Dominique, 1977)
2. Females solitary; consistent associations while inactive	*Microcebus ravelobensis* (stable MM/FF[a] sleeping groups) (Braune et al., 2005) *Microcebus murinus* (FF kin cosleeping groups) (Radespiel et al., 2001) *Galagoides demidovii* (FF cosleeping groups) (Nekaris and Bearder, 2007)
3. Females mostly solitary; varying association while active, dispersed community	*Pongo* spp. (Rodman and Mitani, 1987; Sugardjito et al., 1987; van Schaik, 1999; Watts, 2012)
4. Pair living Dispersed pairs Cohesive pairs	*Phaner furcifer* (M/F pairs may sleep apart) (Schülke and Kappeler, 2003) *Cheirogeleus medius* (M/F pairs cosleep) (Fietz, 1999) *Lepilemur leucopus* (M/F pairs may sleep apart) (Droscher and Kappeler, 2013) Gibbons (*Bunopithecus* spp. *Hylobates* spp.; *Symphalangus syndactylus*; *Nomascus* spp.) (Brockelman, 2009; Reichard et al., 2012) *Aotus* spp.; (Fernandez-Duque, 2007)
5. Stable M/FF groups, cohesive or low fission–fusion dynamics Female philopatry, male dispersal Both sexes disperse, but male bias	Most guenons (*Cercopithecus* spp.) (Cords, 2012) *Gorilla* spp. (Watts, 2012)
6. Stable MM/FF groups, cohesive or low fission–fusion dynamics male dispersal/female philopatry Same, but male philopatry and female dispersal	*Saimiri sciureus* (Boinski, 1999) *Macaca* spp. (Cords, 2012; Thierry, 2007a) *Cebus* spp. (Jack, 2007) Most *Papio* spp. (Cords, 2012) *Saimiri oerstedi* (Boinski, 1999)
7. MM/FF communities, stable membership, high fission–fusion dynamics	*Pan troglodytes* and *Pan paniscus* (Hohmann and Fruth, 2002; Watts, 2012) *Ateles* spp. (Di Fiore and Campbell, 2007)
8. Multilevel societies Stable M/FF units at base, multiple units cluster at several higher levels with moderate consistency; fission–fusion dynamics; male dispersal/female philopatry Stable M/FF groups at base, consistent clustering among multiple units at several higher levels; high fission–fusion dynamics; male philopatry; limited female dispersal	*Theropithecus gelada* (Dunbar and Dunbar, 1975; Snyder-Mackler et al., 2012) *Papio hamdryas* (Kummer, 1968; Swedell et al., 2011)

Categories are based on whether females forage in stable groups. The list is not exhaustive, and selected examples illustrate typical patterns for a given taxon and ignore within-genus and within-species variation and flexibility. Fission–fusion dynamics, rate and extent of group division into subgroups and of subgroups merging.
[a]*M, single male; MM, multimale; F, single female; FF, multifemale.*

same foraging area; thus its intensity is positively related to foraging group size. The intensity of contest competition depends on how clumped food is in space (Harcourt, 1987) and on overall food abundance (Isbell, 1991); it can also increase with group size.

Female orangutans come close to the solitary ideal because their large size, arboreality, and heavy reliance on fruit lead to an extremely high potential for feeding competition. Female gregariousness varies among populations and habitats along with spatial and temporal variation in fruit availability, but nowhere do females form stable groups, and in all known populations they spend the majority of time alone or accompanied only by dependent offspring (Sugardjito et al., 1987; van Schaik, 1999). Males—particularly fully grown, "flanged" males—are highly solitary and mutually intolerant (Sugardjito et al., 1987; van Schaik, 1999; Rijksen, 1978; Rodman and Mitani, 1987); they associate with females mostly in temporary feeding aggregations and when trying to mate. Despite low rates of association and of social interaction, though, individuals whose foraging ranges overlap may have long-term social relationships and form dispersed social communities (van Schaik, 1999). Gibbons also are predominantly frugivorous, and females typically forage independently of each other and defend feeding territories against neighboring females. Unlike orangutans, however, females associate permanently with males, usually in stable, cohesive pairs, although both multifemale and multimale groups occur occasionally (e.g., *H. lar*: Brockelman, 2009; Reichard, 2009; Reichard et al., 2012; siamangs, *Symphalangus syndactylus*: Lappan, 2007; Table 1). The degree of frugivory varies considerably among gorilla species and populations, but large body size allows females to reduce feeding competition by using perennially available herbaceous leaves, pith, and stems as staple foods or as fallbacks when fruit is scarce (Watts, 1996, 2012; Doran and McNeilage, 1998; Rogers et al., 2004; Harcourt and Stewart, 2007). Females live in stable, cohesive groups, usually with other females and with single males. Both sexes disperse; some proportion of males is always solitary, and new groups form when these females join these males. Chimpanzees (*Pan troglodytes*) are also highly frugivorous; they form stable social groups ("communities") with fixed memberships, but with high fission–fusion dynamics: individuals form subgroups ("parties") that vary in size, composition, and duration in response to variation in fruit abundance and spatial dispersion, the availability of sexually receptive females, and other factors. Females are less gregarious than males and more often in small parties or alone, as expected if foraging efficiency constrains their reproductive success more (Wrangham, 1979; Hasegawa, 1990; Lehmann and Boesch, 2008; Wakefield, 2008). Bonobos also form stable multimale, multifemale communities and exhibit high fission–fusion dynamics associated with heavy reliance on fruit, but sex differences in gregariousness are less pronounced than in chimpanzees (Hohmann and Fruth, 2002). Dispersal is female-biased in both *Pan* species: males are philopatric (stay in their natal groups as adults), while most females emigrate and join neighboring communities.

Diets and Food Distribution

Wrangham's (1979) logic should apply to all nonhuman primates: females would maximize net energy intake by foraging solitarily. Indeed, reliance on nonshareable, small and/or relatively sparsely distributed food items and feeding locations, like fruit from small trees and shrubs or sources of gum, may help explain why most nocturnal strepsirhines (e.g., mouse and dwarf lemurs, lorisids; Table 1) do not forage in groups (Dammhann and Kappeler, 2009). But even if feeding competition imposes only loose constraints, the question of why females in most species forage in stable, cohesive groups that contain males and often contain other females remains. Any answer must consider diets and food distribution. Most primates are omnivores: they consume food from two or more trophic levels. Many have morphological and/or physiological adaptations to processing particular kinds of food. Body size also influences the range of acceptable food items because of its negative allometric relationship to basal metabolic rate (reviewed in Lambert, 2007; Chapman et al., 2012). Tarsiers are strictly faunivorous (Gurski, 2007), and some other small-bodied species (e.g., squirrel monkeys, *Saimiri* spp.: Terborgh, 1983) rely heavily on insects and other invertebrates, but plant parts compose most of the diet for most primates. For many, this predominantly means relatively easily digested reproductive parts—especially fruit, also flowers and nectar—that are good energy sources. Seeds, potentially good energy and/or protein sources but often heavily defended chemically, are sometimes important. The importance of nonreproductive plant parts (leaves, stems, pith, bark, cambium, roots, corms, etc.) varies as a function of body size and of morphological and physiological specializations for dealing with plant structural carbohydrates. For example, colobine monkeys have complex, multichambered stomachs; one chamber has a neutral pH and harbors an enormous, diverse population of bacteria and protozoa that ferment the cellulose and hemicellulose in plant cell walls, thereby producing volatile fatty acids that the monkeys can use for energy, exposing cell contents to the monkeys' digestive enzymes, and otherwise facilitating nutrient extraction from diets higher in structural carbohydrates than expected given the animals' body sizes (Chivers, 1984; Kay and Davies, 1984).

All primate habitats are seasonal to some extent. In tropical habitats, seasonality is mostly in rainfall. Depending on the length and intensity of dry seasons and on variation in plant reproductive strategies, the availability of plant reproductive parts can vary considerably and more or less predictably over time. Community-wide dry season fruiting troughs occur even in some lowland tropical rainforests (e.g., Manu, Peru, where the primate community includes 13 species; Terborgh, 1983). Also, the spatial distribution of sources of fruit, flowers, nectar, and seeds varies greatly among habitats and within habitats over time. Nonreproductive plant parts are generally more evenly distributed in space and time than reproductive parts. The potential for scramble competition, hence the cost of foraging in groups, decreases as the density and evenness of food patches and of items within patches increase; it is also lower for food available more or less perennially than for ephemeral food sources. The potential for contest competition is directly related to the proportion of food that occurs in locally dense concentrations that one or a few individuals can monopolize or that one group can monopolize at the expense of another (Harcourt, 1987; Isbell, 1991).

Then Why Live in Groups?

The main counterweights to feeding competition, and the main benefits of group living, are improved protection against predators via dilution effects (safety in numbers), increased vigilance, preferential occupation of safer places within a group's perimeter, and active defense (van Schaik, 1983, 1989; Fichtel, 2012). Considerable evidence for predation as a major cause of mortality exists. For example, predation by fossa (*Cryptoprocta ferox*) is the main source of mortality for adult Verraux's sifaka (*Propithecus verrauxi*) at Kirindy (Kappeler and Fichtel, 2012), predation is the main source of adult mortality for chacma baboons (*Papio ursinus*) at Moremi (Cheney et al., 2004), and chimpanzees prey heavily on red colobus monkeys (*Procolobus* spp.) and apparently limit some red colobus populations (Stanford, 1998; Lwanga et al., 2011; Teelen, 2008; Watts and Amsler, 2013). Much indirect evidence (reviewed in Fichtel, 2012; also see chapters in Miller, 2002) indicates that predation has been a major selective force. This includes existence of alarm calls, increased vigilance in larger groups, avoidance of risky areas during foraging, and formation of mixed-species associations that reduce vulnerability for some or all participating species. For example, red colobus (*Procolobus pennantii*) and Diana monkey (*Cercopithecus dianae*) groups are better protected against raptors when associating with each other (Bshary and Noë, 1997).

Optimal group size for a species that gains such antipredator benefits should be that at which the costs of within-group feeding competition balance these benefits (Terborgh and Janson, 1986). However, forming large groups may not always reduce predation, depending on how predators detect prey. For example, predation by predators that locate prey auditorily (e.g., owls preying on mouse and dwarf lemurs of the family *Cheirogaleidae*) may help explain why many nocturnal species are solitary while active (Table 1).

SOCIAL ORGANIZATION AND SOCIAL RELATIONSHIPS

Primates that form stable groups and even some that forage solitarily regularly engage in a variety of social interactions with conspecifics other than dependent offspring. That is, they have social relationships. In Hinde's (1983) influential framework, social relationships are "abstractions" (Silk, 2012, p. 555) derived from the histories of interactions between individuals, while social organization emerges from the patterning of social relationships and can assume species-typical forms. Characterizing social relationships involves recording the frequencies, types, quality, and, when relevant, durations of affiliative and agonistic interactions (agonism includes both aggression and submissive behavior). It can also involve recording how much time individuals spend in close proximity and whether either member of a dyad is more responsible for initiating and maintaining proximity. Merely tolerating close proximity, especially physical contact, can reflect relationship quality. For example, the proportion of approaches between two females that are met with aggression or tolerance and the amount of time that female dyads rest in physical contact are important components of female–female relationships in mountain gorillas (Watts, 1994). Choice of association partners is also an aspect of social relationships in species whose groups often fission into subgroups (e.g., chimpanzees: Newton-Fisher, 1999).

From a complementary perspective, relationships result from strategies in which the "goal" is to maximize inclusive fitness (Dunbar, 1988). Thus aggression rates may be relatively high and/or most interactions being agonistic between serious competitors (e.g., males in multimale groups), while individuals may direct potentially beneficial acts mostly to others whose interests they most share and/or from whom they can benefit most in return. Strategic concerns help explain why social networks usually are differentiated and kin biases in behavior are common, as is typically true for social grooming, the most common affiliative behavior in many primates. In species that form multifemale groups and in which females are philopatric, female–female grooming is most common between close relatives (e.g., baboons: Silk et al., 2006a; macaques: Silk, 2002; Thierry, 2007a,b). Allogrooming is also common between maternal brothers in chimpanzees, in which males remain in their natal communities as adults (Langergraber et al., 2007). Allogrooming provides health benefits via ectoparasite removal, cleaning of wounds, and stress reduction in both groomers and recipients of grooming (Dunbar, 2010b). Close relatives share genetic interests and can gain indirect fitness benefits by grooming. More broadly, its potential health benefits make allogrooming a valuable "service" exchangeable for other benefits, such as return grooming or coalitionary support. For example, male chimpanzees show reciprocity in grooming and interchange of grooming with coalitionary support against third parties and with meat sharing (Watts, 2002; Mitani, 2009a). Effective agonistic support, especially when it occurs regularly between "allies," can help males to increase and maintain their dominance ranks (de Waal, 1982; Nishida and Hosaka, 1996). Male Assamese macaques (*Macaca assamensis*) also use grooming to help maintain alliances that in turn influence dominance ranks (Schülke et al., 2010).

DOMINANCE, LEVERAGE, AND POWER

"Dominance" is a much misunderstood concept. When two individuals have repeated agonistic interactions and one consistently wins, the winner is by definition dominant to the loser, who is subordinate to the winner. Dominance is thus a property of social relationship, not individuals, as is clear from the fact that the same individual can be dominant to some fellow group members and subordinate to others. When most or all members of a given class of individuals have decided agonistic relationships with each other and most or all of these are transitive (A dominant to B and B dominant to C→A dominant to C, etc.), linear dominance hierarchies exist. This is often the case for females in species that form multifemale groups (e.g., yellow baboons, *Papio cynocephalus*: Walters, 1980;

Barbary macaques, *Macaca sylvanus*: Paul and Kuester, 1987; white-faced capuchins: Perry, 1996), although considerable variation in the consistency of contest outcomes and the directionality of aggression between higher and lower ranking females, hence in the linearity and strength of hierarchies, exists (Koenig et al., 2013). For example, females in all macaque species form dominance hierarchies, but in some (e.g., rhesus macaques, *M. mulatta*), aggression is unidirectional and females are intolerant of subordinates, whereas in others (e.g., Tonkean macaques, *M. tonkeana*), tolerance is higher, serious aggression less common, and bidirectional aggression within dyads more common (Thierry, 2007a,b; Thierry et al., 2008). Moreover, female dominance hierarchies are absent or are poorly defined and/or nonlinear in some species groups (e.g., chimpanzees: Wakefield, 2008; mountain gorillas: Watts, 1994; ringtailed and brown lemurs: Pereira, 1993; Pereira and Kappeler, 1997).

Nonlinearity, nontransitivity, and hierarchy instability are likely when female status depends on individual fighting ability (e.g., hanuman langurs, *Semnopithecus entellus*: Borries, 1993; Koenig, 2000; ringtailed lemurs, *Lemur catta*: Pereira, 1993). In contrast, females form stable, transitive dominance hierarchies in which the rank of each individual depends crucially on who supports her, against whom, in macaques (de Waal, 1989; Chapais, 1992), most savanna baboons (Walters, 1980; Smuts, 1985; Engh et al., 2006), vervet monkeys (*Chlorocebus aethiops*; Walters and Seyfarth, 1987), and white-faced capuchins (*Cebus capucinus*; Perry, 1996). In these taxa, females are philopatric and matrilines form. As they mature, females usually acquire ranks immediately below their mothers' (although how consistently this happens varies; Thierry, 2000, 2007a), so matrilines also are ranked in relation to each other. This pattern, initially described for Japanese macaques (*M. fuscata*) by Kawai (1958), is often misleadingly called "inheritance" of maternal ranks; in reality, rank acquisition is a process that involves challenges by an immature female to peers and then to older females, the success of which depends on whether the challenger's female relatives support her, on which females from other matrilines also support her, and on the ranks of her supporters relative to those of her targets (Walters, 1980; Walters and Seyfarth, 1987; Chapais, 1992). Maintaining dominance after rank reversals also depends importantly on continued kin and nonkin support. Consistency and long-term stability ensue because females usually follow the "support the high-born" rule when they intervene in conflicts between other females, both kin and nonkin: they support the female whose mother outranks her opponent and other members of the opponent's matriline (Chapais, 1992). Females also support their younger daughters against older daughters.

Males in many species that form multimale groups also form dominance hierarchies (e.g., chimpanzees: Watts, 1998; yellow baboons: Hausfater, 1975; Alberts et al., 2003a,b). Male status typically depends on individual fighting ability, with some partial exceptions like chimpanzees and Assamese macaques (see above). Male contests occur in many contexts, but ultimately are most important regarding access to mating opportunities. Male dominance hierarchies are mating queues (Alberts et al., 2003b), and abundant evidence shows that high rank confers priority of access to sexually receptive females and is positively correlated with both mating and reproductive success (e.g., yellow baboons: Alberts et al., 2003b; chimpanzees: Langergraber et al., 2013). The strength of the rank–paternity success correlation depends on variation in male competitive ability and in the number of males per group (Alberts et al., 2003b). Also, correlations can be weakened by the use of alternative mating tactics (reviewed in Alberts, 2012), such as coalitions that middle-to-low ranking male yellow baboons use to disrupt consortships between higher ranking males and estrous females (Noë, 1992), and by female choice of males (e.g., rhesus macaques: Manson, 1994) and male choice of females (e.g., yellow baboons: Alberts et al., 2003a,b). Dependence by alpha males on lower ranking allies to acquire and maintain their status can lead to tolerance of the allies' mating activities (chimpanzees: de Waal, 1982; Duffy et al., 2007; Assamese macaques: Schülke et al., 2010).

The potential for some males to gain mating opportunities by giving agonistic support to higher ranking males illustrates the fact that dominance is one component (the agonistic component) of the broader category of "power" (Chapais, 1991; Lewis, 2002). Power also includes leverage, which exists when one individual controls a resource or service that others cannot forcibly appropriate (Chapais, 1991; Lewis, 2002). Leverage effects can open the door to competition for social partners best able to provide valuable services like grooming and coalitionary support (Noë, 1992; Barrett and Henzi, 2005).

THE "SOCIOECOLOGICAL MODEL"

Social relationships among females vary greatly with regard to the existence, strength, and stability of dominance hierarchies; the relative frequency of severe aggression and how much aggression is directed up a hierarchy; tolerance for close proximity; kin biases in behavior, including coalitionary support; and the tendency to reconcile—that is, to use affiliative behavior to resolve conflicts (Wrangham, 1980; de Waal, 1989; Cords and Aureli, 2000; Thierry, 2007a,b; Thierry et al., 2008; Silk, 2012). van Schaik (1989), building on work by Wrangham (1980), developed a conceptual scheme to map this variation onto ecological variation. The resulting "socioecological model," subsequently modified by others (Isbell, 1991, 2004; Sterck et al., 1997; Cords and Aureli, 2000; Koenig, 2002; Schülke and Ostner, 2012; Koenig et al., 2013), starts with the premise that predation is the main cause of group living and that food distribution and competitive regimes group size and composition and the nature of female–female relationships (Figure 1). When females form multifemale groups and food distribution generates within-group contest competition strong enough to influence their fitness, they are likely to form dominance hierarchies because high rank can confer priority of access to food at the expense of subordinates. Resulting energetic and nutritional advantages can lead to positive relationships between rank and reproductive success. Related females can benefit by residing together and forming alliances that influence the outcomes of within-group feeding contests; hence female philopatry and male dispersal are expected. When the potential nutritional advantages of high rank are large, food-related aggression should be relatively common and hierarchies steep, with aggression mostly directed down the hierarchy and females rarely winning contests against dominant opponents. Strict (or "despotic"; Sterck et al., 1997) dominance relationships make it risky for subordinates to approach unrelated high-ranking opponents after aggressive interactions; thus the tendency for opponents to reconcile should be relatively low and reconciliation should occur mainly between kin. The effects of between-group contest competition are likely to be minor compared to those of within-group contest competition. However, when between-group contest effects influence fitness, low-ranking females have leverage over those who outrank them because they can withhold participation in these contests. Dominance relationships should thus be more relaxed: food-related aggression should be less common and less often escalated, kin biases in behavior less pronounced (although alliances between kin occur), and nonkin more likely to reconcile after conflicts, and more aggression should go up the hierarchy. If within-group contest competition is minor, but

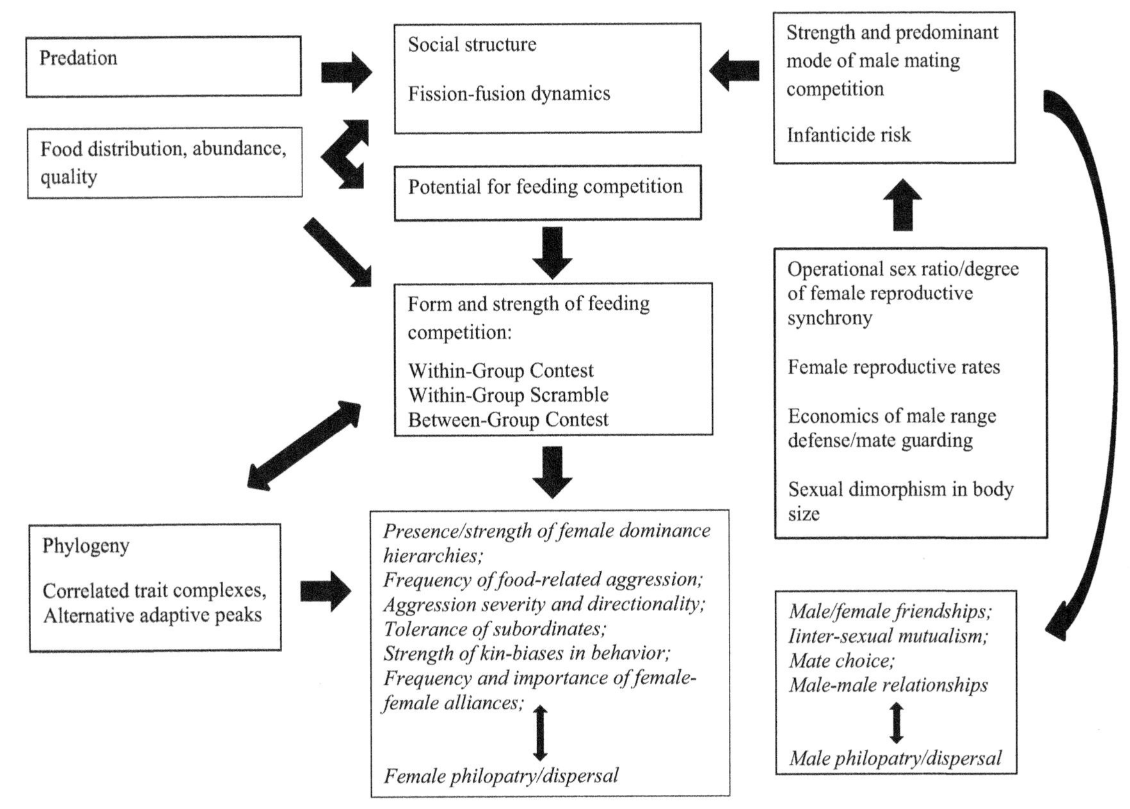

FIGURE 1 Schematic representation of the "socioecological" model, modified to incorporate the potential role of correlated evolution among traits (e.g., aggressiveness, tolerance) subject to the same neurobiological and hormonal influences (Thierry, 2007a,b) and feedback among female social dispersion, male mating and reproductive strategies, and social relationships. Predation is the main factor selecting for group living. When females do not forage in groups with each other, the distribution of fertile females in space and time combined with predation risk and food distribution determines whether males range widely and try to encounter multiple females, or share ranges with single females and form dispersed or cohesive pairs. When females form multifemale groups, food distribution and the strength of contest feeding competition within and between groups and of within-group scramble competition influence whether dominance hierarchies form and how steep these are, the extent of kin biases in behavior, and other aspects of social relationships. However, correlated trait evolution can lead to different social outcomes in similar ecological contexts. Group living species are characterized by permanent male–female association. Males can also gain protection against predation by grouping, and their presence may also represent mate-guarding, be tied to paternal care, and/or be tied to protection of females and offspring (especially against infanticide) and defense of food supplies. Whether female groups are typically single male or multimale depends largely on the effects of female group size on male mating competition.

between-group contest competition affects female fitness, female philopatry is still expected because related females can gain both direct and indirect fitness benefits by cooperating in between-group aggression, but dominance hierarchies are likely to be weak or absent. Finally, when food is dispersed evenly enough that neither within- nor between-group contest competition significantly influences fitness, female dominance hierarchies are also likely to be weak or absent, and agonistic relationships should be individualistic and not influenced by alliance formation. Also, female dispersal should be common: with little to gain by cooperating with kin in competition for access to food, females can make residence decisions that minimize scramble competition and/or reflect mate choice.

Many individual cases fit the predictions of the modified socioecological model (reviewed in Koenig, 2002; Schülke and Ostner, 2012). Broadening the focus from female–female relationships to include mating strategies and sexual conflict (Sterck et al., 1997; see below) helps explain why stable female groups always have males associated with them and, in species for which infanticide by males is an important threat to female fitness, to account for both voluntary female dispersal (e.g., mountain gorillas: Watts, 2002; Robbins et al., 2009) and aggressive eviction of some females by others (e.g., red howlers, *Alouatta seniculus*: Janson and Crockett, 2000). However, the model has also received much criticism. It does not adequately account for large differences in rates of agonism among major primate clades (cercopithecines, colobines, and platyrrhines; Koenig et al., 2013), and the predicted associations of female philopatry with high within-group contest competition and of low within-group contest competition with female dispersal do not consistently hold (Isbell, 2004; Schülke and Ostner, 2012). Notably, wide variation in social styles among macaques does not map onto ecology as predicted. Instead, variation in aggressiveness and tolerance seems to show phylogenetic signals that may derive from coevolution of a set of correlated traits, all subject to the same hormonal and neurobiological influences; such coevolution could have led to multiple adaptive peaks (Thierry, 2000, 2007a; Thierry et al., 2008; Balasubramaniam et al., 2012). Phylogenetic/epigenetic hypotheses and ecological hypotheses are not mutually exclusive (Koenig, 2002; Koenig and Borries, 2009; Figure 1). For example, considerable within-species variation in hierarchy steepness exists in macaques (Balasubramaniam et al., 2012); this may reflect variation among habitats in the intensity of feeding competition. In fact, with a few exceptions (e.g., squirrel monkeys, *Saimiri* spp.: Boinski, 1999; brown capuchins, *Cebus apella*: Janson, 1985), detailed quantitative data on food distribution and on the energetic consequences of feeding competition are sparse (Schülke and Ostner, 2012; Koenig et al., 2013). Future progress in explaining socioecological variation depends on collection of more data for more taxa and on integrating models focused on the ecology of social relationships with those focused on the ecology of mating systems (Schülke and Ostner, 2012; Koenig et al., 2013).

SEXUAL CONFLICT AND "INTERSEXUAL MUTUALISM"

Focusing on conflict and convergence in reproductive interests between males and females offers rich opportunities for such integration. The possibility of sexual conflict exists on many levels, from epigenetics to social interactions (Stumpf et al., 2011). Sexual coercion—male behavior that constrains or negates female mate choice, at some cost to the females and to the benefit of the males (Smuts and Smuts, 1993)—occurs in many primates (reviewed in Wrangham and Muller, 2009; Stumpf et al., 2011). In particular, much evidence supports the hypothesis that sexually selected infanticide is a threat to female fitness in many primates due to slow female reproductive rates and strong limits on male reproductive opportunities (reviewed in van Schaik, 2000; Palombit, 2012). Dispersal is male biased in most primates, and male dispersal strategies are driven largely by reproductive competition. For example, dispersing male yellow baboons tend to go from groups where they have few current mating opportunities to others where operational sex ratios are more favorable (Altmann and Alberts, 1995), and males in species that typically form one-male groups may gain mating opportunities mostly or only by aggressively invading groups and winning challenges against current resident males (e.g., most guenons, *Cercopithecus* spp.: Cords, 2012; geladas, *Theropithecus gelada*: le Roux and Bergman, 2012). Immigrant males who have not mated with the females whose groups they enter often pose the threat of infanticide.

Wrangham (1979) argued that permanent male–female association in gibbons, chimpanzees, and gorillas resulted from male mate guarding, combined with male protection against aggression from conspecific males in the case of gorillas (cf. Sterck et al., 1997). Also, males benefit females by defending food sources or feeding areas against neighboring groups (Wrangham, 1980). Mate guarding is a conflict of interest if it constrains female choice, but protection of females against conspecific aggression and of infants against infanticide represents a convergence of interests if it increases male mating success and if the infants are the males offspring, as does male defense of food on behalf of mates and offspring. Males might also protect females against predators, something mutually compatible with protection against infanticide (Harcourt and Stewart, 2007). Behavior resulting from such convergences of interest qualifies as intersexual mutualism (Wrangham, 1982).

Much evidence for intersexual mutualism exists. For example, female anti-infanticide tactics include extended periods of sexual receptivity and willingness to copulate with multiple males; this can mask paternity and induce males at least not to harm infants with whose mothers they have copulated. Tactics also include formation of strong social bonds with particular males whose probability of paternity is high and who protect infants against potentially infanticidal males (Palombit, 2012). In baboons, such "friendships" are not necessarily durable and may end with infant death, but at least in chacma baboons (*Papio ursinus*) females with male friends are less likely to lose infants to newly immigrated, infanticidal males than are females without friends (Palombit et al., 2000). Female mountain gorillas depend on males in their groups for protection against outside males, and those in multimale groups are less likely to lose infants to infanticide than are those in single-male groups (Watts, 1989, 2000; Robbins et al., 2007, 2009). Female transfer between groups or from groups to solitary males is common in mountain gorillas and apparently universal in western gorillas (Harcourt and Stewart, 2007; Caillaud et al., 2008); transfer represents mate choice, which presumably is largely based on assessment of male ability to provide protection against both infanticide and predation (Watts, 1990; Harcourt and Stewart, 2007; Caillaud et al., 2008; Robbins et al., 2009). Likewise, males often participate in intergroup aggression over food, either with female participation (e.g., gray-cheeked mangabeys, *Lophocebus albigena*, and redtail monkeys, *Cercopithecus ascanius*: Brown, 2013) or without it (e.g., guerezas, *Colobus guereza*: Harris, 2006). Maintaining and increasing food availability for females are probably the ultimate functions of cooperative male intergroup aggression in chimpanzees (Williams et al., 2004; Mitani et al., 2010).

COOPERATION

Kin Selection and Primate Social Evolution

Cooperation is broadly defined as behavior other than parental care by one individual that provides benefits to one or more others (West et al., 2006). Cooperative behavior may or may not have direct fitness costs. When costs exist, individuals can compensate for them by directing help to close relatives and thereby gaining indirect fitness benefits, that is, via kin selection. The widespread existence of kin biases in behavior indicates that kin selection has been important in social evolution in many primates (Chapais, 2001; Silk, 2002; Chapais and Berman, 2004; Langergraber, 2012). These include solitary species in which sociogenetic population structuring results in relatively frequent association among relatives (Kappeler, 2012). For example, close female kin form stable sleeping groups in gray mouse lemurs (*M. murinus*); these can provide protection against predation, thermoregulatory benefits, and benefits due to alloparental care (Radespiel et al., 2001).

The extent and benefits of alloparental care are particularly striking in callitrichids (marmosets and tamarins), which are cooperative breeders. Allocaretaking includes infant carrying, guarding infants while mothers forage, and sharing food with infants (reviewed in Garber, 1997; Digby et al., 2007). It reduces the energetic costs of reproduction for mothers and is associated with fast infant growth and high reproductive rates for breeding females (Mitani and Watts, 1997), and female reproductive success increases with the number of helpers per group (Garber, 1997). However, it imposes energetic costs on caretakers and perhaps increases their vulnerability to predators, and it is associated with reproductive suppression of subordinate females in multifemale groups. Much of the care is provided to younger siblings by natal individuals who are not yet reproductively mature or have not yet dispersed and lack personal opportunities to reproduce; caretakers can reap indirect fitness gains and acquire skills needed to provide competent care to any subsequent infants of their own (Goldizen, 1987; Garber, 1997). Kin selection might likewise have favored allocaretaking in many other primates, including humans (Hrdy, 2009).

Kinship can influence relationships among males when related males reside together as adults. For example, male chimpanzees are generally more likely to maintain close proximity, groom, share meat, and form coalitions with maternal brothers than with other social partners, although social bonds between unrelated males can be as strong as or stronger than those between maternal sibs (Langergraber et al., 2007; Mitani, 2009a,b).

Reciprocity and Mutualism

Reciprocal altruism can explain costly cooperation between nonrelatives. Besides the chimpanzee examples noted above, reciprocity in grooming and agonistic support is widespread in nonhuman primates (Schino, 2007) and interchange of grooming and agonistic support occurs in some other species, although apparent interchange can result spuriously from correlations of these behaviors with a third variable such as dominance rank (Schino, 2007; Hemelrijk, 1990). Exchange of grooming has been called reciprocal altruism, but whether grooming someone is costly is debatable. Its energetic costs are minimal, and groomers may benefit from the stress-reduction effects of the physical contact and tactile stimulation (Dunbar, 2010b). Opportunity costs exist—grooming one individual precludes simultaneous grooming of others, and grooming is incompatible with foraging—but these are probably slight and the attendant health benefits and the benefits of effective partner choice may outweigh them. Providing agonistic support to a nonrelative entails energetic costs and some risk of injury, which makes coalition formation a more likely candidate for reciprocal altruism. Revolutionary coalitions, in which coalition partners are subordinate to their opponents (Chapais, 1992, 2001), may sometimes qualify. However, many coalitions are conservative (both or all partners outrank their opponent) or bridging (one partner outranks the opponent, but the second is subordinate); in these, coalition partners have high probabilities of winning and low risk of injury, and they stand to gain by reinforcing their dominance over opponents or reversing rank with them (Chapais, 1992, 2001). Such coalitions—including most between females in cercopithecines in which female dominance ranks are maternally "inherited"—are acts of mutualism (or are "mutually beneficial," a term that distinguishes within-species cooperation from between-species mutualism; West et al., 2007): both or all participants achieve immediate net fitness gains (Chapais, 1992, 2001). Also, differences in fighting ability can curtail turn-taking in sequences of coalitions and enable some participants to benefit disproportionately, as Noë (1990) documented for alliances among unrelated yellow baboon males; this does not satisfy criteria for reciprocal altruism, although such variation in partner quality leads to competition for allies (Noë, 1990, 1992).

In some primates that form multimale groups, males cooperate in intergroup aggression. Kinship influences such cooperation in chimpanzees, in which maternal brothers participate jointly in patrols of territory boundaries more often than expected by chance (Langergraber et al., 2007), but apparently not in black howler monkeys (Van Belle et al., 2014). Van Belle et al. (2014) proposed instead that male black howlers gain mutual benefits through such defense. Males may also gain mutual benefits in tufted capuchins (Scarry, 2013) and in chimpanzees (Williams et al., 2004; Mitani et al., 2010), given that cooperative territorial defense can maintain access to food available to their communities and potentially lead to territory expansion and thus increase the amount of food available. Also, by making lethal coalitionary attacks on members of neighboring communities, they can reduce the strength of rival groups (Wrangham, 1999; Wilson and Wrangham, 2003). Territorial defense can pose collective action problems, particularly if benefits are unequally distributed. Evidence of collective action problems exists (e.g., white-faced capuchins: Crofoot and Gilby, 2012), and whether and how they are resolved can depend on multiple factors, including the number of potential defenders and the location of between-group interactions (Crofoot et al., 2008; Wilson et al., 2001, 2012).

Chimpanzees hunt a variety of vertebrate prey and mostly hunt red colobus monkeys wherever the two species are sympatric (see chapter by Sussman and Hart). Disagreement exists about whether group hunts of monkeys by chimpanzees and transfers of meat following successful hunts qualify as cooperation. Boesch and Boesch (1989; cf. Boesch and Boesch-Achermann, 2000; Boesch, 2002) reported that many hunts of red colobus at Taï involved collaboration—that is, individuals took different, complementary roles during pursuits and thereby increased the probability of capturing prey. Similar collaboration has not been reported from Gombe (Boesch, 1994, 2002) or Mahale (Uehara et al., 1992), and how often it happens at Ngogo is uncertain because of constraints on visibility (Watts and Mitani, 2002), but its occurrence in chimpanzees is unsurprising given that lions in some populations hunt collaboratively (Stander, 1992). Boesch (1994) argued that males at Taï maintain cooperation by withholding meat from others who are present at hunts but do not participate. Based on his estimates of energy expended during hunting and per capita meat intake, he concluded that discrimination against nonparticipants made group hunting energetically profitable, with net energy gain maximized for groups of three to five hunters; this meets an ecologically defined criterion for cooperative hunting (Creel and Creel, 1995). In contrast, estimated net energy intake was higher for solitary hunts than for group hunts at Gombe (Boesch, 1994). In red colobus hunts at Ngogo, the probability of capturing at least one monkey, the mean number of prey captured per hunt, mean overall meat availability per hunt, and the number of males who received some meat per hunt all increased significantly with the number of male hunters present (Watts and Mitani, 2002). However, per capita meat availability did not increase significantly, implying that hunting there may not satisfy the ecological criterion for cooperation. Based on data from Gombe, Gilby (2006) challenged the claim that meat sharing is a form of cooperation and argued instead that meat possessors allow others to have shares in response to harassment that decreases their energy intake rates. However, meat is probably primarily important as a source of amino acids, vitamins, and other macro- and micronutrients uncommon in plant foods, not of calories (Tennie et al., 2008). Moreover, considerable evidence that males use meat sharing to develop and maintain social bonds and facilitate other forms of cooperation exists (Nishida et al., 1992; Nishida and Hosaka, 1996; Mitani and Watts, 2002), and meat transfers at Ngogo are often active and voluntary, not passive responses to harassment (Watts and Mitani, unpublished data).

SOCIALITY AND FITNESS

The idea that relative success at negotiating the complex social worlds of many nonhuman primates influences fitness has a long history (reviewed in Dunbar, 2010a; Silk, 2014). Detailed observational data and field experiments have produced compelling evidence that wild monkeys and apes possess sophisticated social knowledge (e.g., chimpanzees: Mitani et al., 2010; baboons: Cheney and Seyfarth, 2007), but the slow life histories of most primates make testing the fitness effects of variation in social relationships and social strategies difficult. However, long-term data on females in two baboon populations have revealed significant links between components of fitness and the strength of social bonds, where strength was assayed by an index of general gregariousness that combined information on grooming and on the amount of time females had neighbors in close proximity. Female yellow baboons at Amboseli interact with many partners but maintain a few strong bonds, principally with maternal kin; their strongest bonds are likely to be with their mothers, and they invest more in bonds with maternal sisters following maternal deaths (Silk et al., 2006a,b). Bonds strength is positively associated with infant survival at Amboseli, independently of the effects of maternal rank, kinship, or environmental fluctuations (Silk et al., 2003). A similar relationship holds for female chacma baboons at Moremi, where the affect of maternal sociality on offspring survival also applies to juveniles (Silk et al., 2009a,b). Moreover, the strength and consistency of social bonds are positively associated with female life span, and variation in longevity accounts for a substantial proportion of variation in lifetime reproductive success (Silk et al., 2010).

Effects of variation in gregariousness may be widespread among females. Strong social bonds also lead to higher reproductive success in male Assamese macaques (Schülke et al., 2010), and such effects may occur in other species in which males establish strong bonds with each other (e.g., Guinea baboons, *Papio*: Patzelt et al., 2014). A probable proximate method underlying these effects is that strong bonds can buffer individuals against stress. For example, data on fecal glucocorticoid levels in female chacma baboons at Moremi show that having several closely bonded social partners (including male friends) can alleviate stress associated with instability in female dominance hierarchies, challenges for alpha male status, and the threat of infanticide posed by new male immigrants (Beehner et al., 2005; Engh et al., 2006; Crockford et al., 2008; Wittig et al., 2008). Also, baseline glucocorticoid levels were negatively related to dominance rank among the male Assamese macaques studied by Ostner et al. (2008); high-ranking males maintained stronger bonds than low-ranking males, which could have explained this effect. Several other factors could also contribute to the long-term positive effects of general sociality in mutually compatible ways (Silk, 2014). Individuals with strong social bonds may be better protected against predation than those with weaker bonds because they have neighbors close to them more of the time, and they may be better at avoiding or resolving social conflicts, which could allow them to feed more efficiently and also help buffer against stress. Continued investigation of such topics holds great promise for understanding the diversity, complexity, and flexibility of primate social worlds.

REFERENCES

Alberts, S.C., 2012. Magnitude and sources of variation in male reproductive performance. In: Mitani, J.C., Call, J., Kappeler, P.M., Palombit, R.A., Silk, J.B. (Eds.), The Evolution of Primate Societies. University of Chicago Press, Chicago, pp. 412–431.

Albert, S.C., Altmann, J., 1995. Balancing costs and opportunities: dispersal in male baboons. American Naturalist 145, 279–306.

Alberts, S.C., Watts, H.E., Altmann, J., 2003a. Queuing and queue-jumping: long-term patterns of reproductive skew in male savanna baboons, *Papio cynocephalus*. Animal Behaviour 65, 821–840.

Alberts, S.C., Buchan, J.C., Altmann, J., 2003b. Sexual selection in wild baboons: from mating opportunities to paternity success. Animal Behaviour 72, 1177–1196.

Aureli, F., Schaffner, C.M., Boesch, C., Bearder, S.K., Call, J., Chapman, C.A., Connor, R.A., Di Fiore, A., Dunbar, R.I.M., Henzi, S.P., Holekap, K., Korstjens, A.H., Layton, R., Lee, P., Lehmann, J., Manson, J.H., Ramos-Fernandez, G., Strier, K.B., van Schaik, C.P., 2008. Fission-fusion dynamics: new research frameworks. Current Anthropology 49, 627–654.

Balasubramaniam, K.N., Dittmar, K., Berman, C.M., Butovskaya, M., Cooper, M.A., Majolo, B., Ogawa, H., Schino, G., Thierry, B., de Waal, F.B.M., 2012. Hierarchical steepness and phylogenetic models: phylogenetic signals in Macaca. Animal Behaviour 83, 1207–1218.

Barrett, L., Henzi, P., 2005. Monkeys, markets, and minds: biological markets and primate sociality. In: Kappeler, P.M., van Schaik, C.P.M. (Eds.), Cooperation in Primates and Humans. Springer, Berlin, pp. 209–232.

Beehner, J.C., Bergman, T.J., Cheney, D.L., Seyfarth, R.M., Whitten, P.L., 2005. The effect of new alpha males on stress in free-ranging baboons. Animal Behaviour 69, 1211–1221.

Boesch, C., 1994. Cooperative hunting in wild chimpanzees. Animal Behaviour 48, 653–667.

Boesch, C., 2002. Cooperative hunting roles among Taï chimpanzees. Human Nature 13, 27–46.

Boesch, C., Boesch, H., 1989. Hunting behavior of wild chimpanzees in the Taï National Park. American Journal of Physical Anthropology 78, 547–573.

Boesch, C., Boesch-Achermann, H., 2000. The Chimpanzees of the Taï Forest. Oxford University Press, Oxford.

Boinski, S., 1999. The social organizations of squirrel monkeys: implications for ecological models of social evolution. Evolutionary Anthropology 8, 101–112.

Borries, C., 1993. Ecology of female social relationships: hanuman langurs (*Presbytis entellus*) and the van Schaik model. Folia Primatologica 61, 21–30.

Braune, P., Schmidt, S., Zimmermann, E., 2005. Spacing and group coordination in a nocturnal primate, the golden brown mouse lemur (*Microcebus ravelobensis*): the role of olfactory and acoustic signals. Behavioral Ecology and Sociobiology 58, 587–596.

Brockelman, W., 2009. Ecology and social system of gibbons. In: Whittaker, D.J., Lappan, S. (Eds.), Gibbons: New Perspectives on Small Ape Socioecology and Population Biology. Springer-Verlag, New York, pp. 211–239.

Brown, M., 2013. Food and range defense in group-living primates. Animal Behaviour 85, 807–816.

Bshary, R., Noë, R., 1997. Red colobus and Diana monkeys provide mutual protection against predators. Behavioral Ecology and Sociobiology 54, 1461–1474.

Caillaud, D., Levrero, F., Gatti, S., Menard, N., Raymond, M., 2008. Influence of male morphology on male mating status and behavior during interunit encounters in western lowland gorillas. American Journal of Physical Anthropology 135, 379–388.

Campbell, C.J., Fuentes, A., MacKinnon, K.C., Panger, M., Bearder, S.K. (Eds.), 2007. Primates in Perspective. Oxford University Press, Oxford.

Chapman, C.A., Rothman, J.M., Lambert, J., 2012. Food as a selective force in primates. In: Mitani, J.C., Call, J., Kappeler, P.M., Palombit, R.A., Silk, J.B. (Eds.), The Evolution of Primate Societies, pp. 149–168.

Chapais, B., 1991. Primates and the origins of aggression, politics, and power among humans. In: Loy, J.D., Peters, C.B. (Eds.), Understanding Behavior: What Primate Studies Tell Us about Human Behavior. Oxford University Press, Oxford, pp. 190–218.

Chapais, B., 1992. The role of alliances in social inheritance of rank among female primates. In: Harcourt, A.H., de Waal, F.B.M. (Eds.), Coalitions and Alliances in Humans and Other Animals. Oxford University Press, Oxford, pp. 285–321.

Chapais, B., 2001. Primate nepotism: what is the explanatory value of kin selection? International Journal of Primatology 22, 203–229.

Chapais, B., Berman, C. (Eds.), 2004. Kinship and Behavior in Primates. Oxford University Press, Oxford.

Charles-Dominique, P., 1977. Ecology and Behaviour of Nocturnal Primates. Duckworth, London.

Cheney, D.M., Seyfarth, R.M., Fischer, J.A., Beehner, J.C., Bergman, T.J., Johnson, S.E., Kitchen, D.M., Palombit, R.A., Rendall, D., Silk, J.B., 2004. Factors affecting reproduction and mortality among baboons in the Okavango Delta, Botswana. International Journal of Primatology 25, 401–428.

Cheney, D.L., Seyfarth, R.M., 2007. Baboon Metaphysics. University of Chicago Press, Chicago.

Chivers, D.J., 1984. Functional anatomy of the gastrointestinal tract. In: Davies, A.G., Oates, J.F. (Eds.), Colobine Monkeys: Their Ecology, Behaviour, and Evolution. Cambridge University Press, Cambridge, pp. 205–228.

Cords, M., 2012. The behavior, ecology, and social evolution of cercopithecine monkeys. In: Mitani, J.C., Call, J., Kappeler, P.M., Palombit, R.A., Silk, J.B. (Eds.), The Evolution of Primate Societies. University of Chicago Press, Chicago, pp. 90–112.

Cords, M., Aureli, F., 2000. Reconciliation and relationship qualities. In: Aureli, F., de Waal, F.B.M. (Eds.), Natural Conflict Resolution. University of California Press, Berkeley, pp. 177–198.

Creel, S., Creel, N.M., 1995. Communal hunting and pack size in African wild dogs, *Lycaon pictus*. Animal Behaviour 50, 1325–1339.

Crockford, C., Wittig, R.M., Whitten, P.L., Seyfarth, R.M., Cheney, D.L., 2008. Social stress and coping mechanisms in wild female baboons. Hormones and Behavior 53, 254–265.

Crofoot, M.C., Gilby, I.C., Wilkelski, M.C., Kays, R.W., 2008. Interaction location outweighs the competitive advantage of numerical superiority in *Cebus capucinus* intergroup contests. Proceedings of the National Academy of Sciences 105, 577–581.

Crofoot, M.C., Gilby, I.C., 2012. Cheating monkeys undermine group strength in enemy territory. Proceedings of the National Academy of Sciences 109, 101–105.

Dammhann, M., Kappeler, P.M., 2009. Females go where the food is: does the socioecological model explain variation in social organization in solitary foragers? Behavioral Ecology and Sociobiology 63, 939–952.

Digby, L., Ferrari, S.F., Saltzman, W., 2007. Callitrichines. In: Campbell, C.J., Fuentes, A., MacKinnon, K.C., Panger, M., Bearder, S.K. (Eds.), Primates in Perspective. Oxford University Press, Oxford, pp. 85–106.

Doran, D., McNeilage, A., 1998. Gorilla ecology and behavior. Evolutionary Anthropology 6, 120–131.

Droscher, I., Kappeler, P.M., 2013. Defining the low end of primate social complexity: the social organization of the nocturnal white-footed sportive lemur (*Lepilemur leucopus*). International Journal of Primatology 34, 1225–1243.

Duffy, K.G., Wrangham, R.W., Silk, J.B., 2007. Male chimpanzees exchange political support for mating opportunities. Current Biology 17, R586–R587.

Dunbar, R.I.M., 1988. Primate Social Systems. Cornell University Press, Ithaca.

Dunbar, R.I.M., 2010a. Brains and behavior in social evolution. In: Kappeler, P.M., Silk, J.B. (Eds.), Mind the Gap. Springer, Heidelberg, pp. 315–330.

Dunbar, R.I.M., 2010b. The social role of touch in humans and primates: behavioural function and neurobiological mechanisms. Neuroscience and Biobehavioral Review 34, 260–268.

Dunbar, R.I.M., Dunbar, E.P., 1975. Social Dynamics of Gelada Baboons. Columbia University Press, New York.

Di Fiore, A., Campbell, C.J., 2007. The atelines: variation in ecology and social organization. In: Campbell, C.J., Fuentes, A., MacKinnon, K.C., Panger, M., Bearder, S.K. (Eds.), Primates in Perspective. Oxford University Press, Oxford, pp. 155–185.

Emlen, S.T., Oring, L.W., 1977. Ecology, sexual selection, and the evolution of mating systems. Science 197, 215–223.

Engh, A.L., Beehner, J.C., Bergman, T.J., Whitten, P.L., Hoffmeier, R.R., Seyfarth, R.M., Cheney, D.M., 2006. Behavioral and hormonal responses to predation in female chacma baboons (*Papio hamadryas ursinus*). Proceedings of the Royal Society, London, B 273, 707–712.

Fernandez-Duque, E., 2007. Aotinae: social monogamy in the only nocturnal haplorhines. In: Campbell, C.J., Fuentes, A., MacKinnon, K.C., Panger, M., Bearder, S.K. (Eds.), Primates in Perspective. Oxford University Press, Oxford, pp. 139–154.

Fichtel, C., 2012. Predation. In: Mitani, J.C., Call, J., Kappeler, P.M., Palombit, R.A., Silk, J.B. (Eds.), The Evolution of Primate Societies. University of Chicago Press, Chicago, pp. 169–194.

Fietz, J., 1999. Monogamy as a rule rather than exception in nocturnal lemurs: the case of the fat-tailed dwarf lemur, *Cheirogaleus medius*. Ethology 105, 259–272.

Garber, P., 1997. One for all and breeding for one: cooperation and competition as a tamarin reproductive strategy. Evolutionary Anthropology 5, 187–222.

Gilby, I., 2006. Meat sharing among the Gombe chimpanzees: harassment and reciprocal exchange. Animal Behaviour 71, 953–963.

Goldizen, A.M., 1987. Tamarins and marmosets: communal care of offspring. In: Smuts, B.B., Cheney, D.L., Seyfarth, R.M., Wrangham, R.W., Struhsaker, T.T. (Eds.), Primate Societies. University of Chicago Press, Chicago, pp. 34–43.

Gurski, S., 2007. Tarsiiformes. In: Campbell, C.J., Fuentes, A., Mackinnon, K.C., Panger, M., Bearder, S.K. (Eds.), Primates in Perspective. Oxford University Press, Oxford, pp. 73–84.

Harcourt, A.H., 1987. Dominance and fertility among female primates. Journal of Zoology, London 213, 417–487.

Harcourt, A.H., Stewart, K.J., 2007. Gorilla society: Conflict, Compromise, and Cooperation between the Sexes. University of Chicago Press, Chicago.

Harris, T.R., 2006. Between-group contest competition for food in a highly folivorous population of black and white colobus monkeys (*Colobus guereza*). Behavioral Ecology and Sociobiology 61, 321–329.

Hasegawa, T., 1990. Sex differences in ranging patterns. In: Nishida, T. (Ed.), The Chimpanzees of the Mahale Mountains: Sexual and Life History Strategies. University of Tokyo Press, Tokyo, pp. 99–114.

Hausfater, G., 1975. Dominance and Reproduction in Baboons (*Papio cynocephalus*). S. Karger, Basel.

Hemelrijk, C.K., 1990. A matrix partial correlation test used in investigations of reciprocity and other social interaction patterns at a group level. Journal of Theoretical Biology 143, 405–420.

Hinde, R., 1983. Primate Social Relationships: An Integrated Approach. Blackwell Scientific, Oxford.

Hohmann, G., Fruth, B., 2002. Dynamics in social organization of bonobos (*Pan paniscus*). In: Boesch, C., Hohmann, G., Marchant, L.F. (Eds.), Behavioral Diversity in Chimpanzees and Bonobos. Cambridge University Press, Cambridge, pp. 138–155.

Hrdy, S.B., 2009. Mothers and Others: The Evolutionary Origins of Mutual Understanding. Harvard University Press, Cambridge.

Isbell, L.A., 1991. Contest and scramble competition: patterns of female aggression and ranging behavior among primates. Behavioral Ecology 2, 143–155.

Isbell, L.A., 2004. Is there no place like home? ecological bases of female dispersal and philopatry and their consequences for the formation of kin groups. In: Chapais, B., Berman, C. (Eds.), Kinship and Behavior in Primates. Oxford University Press, Oxford, pp. 71–108.

Jack, K.M., 2007. The cebines: towards an explanation of variable social structure. In: Campbell, C.J., Fuentes, A., MacKinnon, K.C., Panger, M., Bearder, S.K. (Eds.), Primates in Perspective. Oxford University Press, Oxford, pp. 107–122.

Janson, C., 1985. Aggressive competition and individual food consumption in wild brown capuchin monkeys (*Cebus apella*). Behavioral Ecology and Sociobiology 18, 125–138.

Janson, C.H., Crockett, C., 2000. Infanticide in red howlers: female group size, male membership, and a possible link to folivory. In: Janson, C.H., van Schaik, C.P. (Eds.), Infanticide by Males and its Implications. Cambridge University Press, Cambridge, pp. 75–98.

Kawai, M., 1958. On the system of social ranks in a natural group of Japanese monkeys. Primates 1, 11–48.

Kay, R.N.B., Davies, A.G., 1984. Digestive physiology. In: Davies, A.G., Oates, J.F. (Eds.), Colobine Monkeys: Their Ecology, Behaviour, and Evolution. Cambridge University Press, Cambridge, pp. 229–250.

Kappeler, P., 2012. The behavioral ecology of strepsirhines and tarsiers. In: Mitani, J.C., Call, J., Kappeler, P.M., Palombit, R.A., Silk, J.B. (Eds.), The Evolution of Primate Societies. University of Chicago Press, Chicago, pp. 17–42.

Kappeler, P.M., Fichtel, C., 2012. A 15-year perspective on the social organization and life history of sifaka in Kirindy Forest. In: Kappeler, P.M., Watts, D.P. (Eds.), Long-term Field Studies of Primates. Springer, Heidelberg, pp. 101–124.

Koenig, A., 2000. Competitive regimes in forest dwelling hanuman langurs (*Semnopithecus entellus*). Behavioral Ecology and Sociobiology 48, 93–109.

Koenig, A., 2002. Competition for resources and its behavioral consequences among female primates. International Journal of Primatology 23, 759–783.

Koenig, A., Borries, C., 2009. The lost dream of ecological determinism: time to say goodbye?…or a white queen's proposal? Evolutionary Anthropology 18, 176–184.

Koenig, A., Scarry, C.J., Wheeler, B.C., Borries, C., 2013. Variations in grouping patterns, mating systems, and social structure: what socio-ecological models attempt to explain. Philosophical Transactions of the Royal Society, B 368, 20120348.

Kummer, H., 1968. Social Organization of Hamadryas Baboons: A Field Study. University of Chicago Press, Chicago.

Lambert, J., 2007. Primate nutritional ecology: feeding biology and diet at ecological and evolutionary scales. In: Campbell, C.J., Fuentes, A., Mackinnon, K.C., Panger, M., Bearder, S.K. (Eds.), Primates in Perspective. Oxford University Press, Oxford, pp. 482–495.

Langergraber, K.G., 2012. Cooperation among kin. In: Mitani, J.C., Call, J., Kappeler, P.M., Palombit, R.A., Silk, J.B. (Eds.), The Evolution of Primate Societies. University of Chicago Press, Chicago, pp. 491–513.

Langergraber, K.G., Mitani, J.C., Vigilant, L., 2007. The limited impact of kinship on cooperation in wild chimpanzees. Proceedings of the National Academy of Sciences 104, 7786–7790.

Langergraber, K.E., Mitani, J.C., Watts, D.P., Vigilant, L., 2013. Male–female socio-spatial relationships and reproduction in wild chimpanzees. Behavioral Ecology and Sociobiology 67, 861–873.

Lappan, S., 2007. Social relationships among males in multi-male siamang groups. International Journal of Primatology 28, 369–387.

Lehman, J., Boesch, C., 2008. Sex differences in chimpanzee sociability. International Journal of Primatology 29, 65–81.

Lewis, R.J., 2002. Beyond dominance: the importance of leverage. Quarterly Review of Biology 77, 149–164.

Lwanga, J.S., Struhsaker, T.T., Struhsaker, P.J., Butynski, T., Mitani, J.C., 2011. Primate population dynamics over 32.9 years at ngogo, Kibale National Park, Uganda. American Journal of Primatology 73, 1–15.

Manson, J., 1994. Mating patterns, mate choice, and birth-season heterosexual relationships in free-ranging rhesus macaques. Primates 45, 417–433.

Miller, L.E. (Ed.), 2002. Eat or Be Eaten: Predator Sensitive Foraging Among Primates. Cambridge University Press, Cambridge.

Mitani, J.C., 2009a. Male chimpanzees form enduring and equitable social bonds. Animal Behaviour 77, 633–640.

Mitani, J.C., 2009b. Cooperation and competition in chimpanzees: current understanding and future challenges. Evolutionary Anthropology 18, 215–227.

Mitani, J.C., Amsler, S.J., Sobelewski, M.E., 2010. Chimpanzee minds in nature. In: Lonsdorf, E.V., Ross, S.R., Matsuzawa, T. (Eds.), The Mind of the Chimpanzee: Ecological and Experimental Perspectives. University of Chicago Press, Chicago, pp. 181–191.

Mitani, J.C., Call, J., Kappeler, P.M., Palombit, R.A., Silk, J.B., 2012. The Evolution of Primate Societies. University of Chicago Press, Chicago.

Mitani, J.C., Watts, D.P., 1997. The evolution of non-maternal caretaking in primates: do helpers help? Behavioral Ecology and Sociobiology 40, 213–220.

Mitani, J.C., Watts, D.P., 2002. Why do chimpanzees hunt and share meat? Animal Behaviour 61, 915–924.

Mitani, J.C., Watts, D.P., Amsler, S.J., 2010. Lethal intergroup aggression leads to territorial expansion in wild chimpanzees. Current Biology 20, R507–R508.

Nekaris, A., Bearder, S.K., 2007. The lorisiform primates of Asia and mainland Africa: diversity shrouded in darkness. In: Campbell, C.J., Fuentes, A., MacKinnon, K.C., Panger, M., Bearder, S.K. (Eds.), Primates in Perspective. Oxford University Press, Oxford, pp. 24–45.

Nishida, T., Hasegawa, T., Hayaki, H., Takahata, Y., Uehara, S., 1992. Meat-sharing as a coalition strategy by an alpha male chimpanzee? In: Nishida, T., McGrew, W.C., Marler, P., Pickford, M., de Waal, F.B.M. (Eds.), Topics in Primatology. Human Origins, vol. 1. University of Tokyo Press, Tokyo, pp. 159–174.

Nishida, T., Hosaka, K., 1996. Coalition strategies among adult male chimpanzees of the Mahale Mountains, Tanzania. In: McGrew, W.C., Marchant, L.F., Nishida, T. (Eds.), Great Ape Societies Cambridge University Press, Cambridge, pp. 114–134.

Noë, R., 1990. A veto game played by baboons: a challenge to the prisoner's dilemma as an paradigm for reciprocity and cooperation. Animal Behaviour 39, 78–90.

Noë, R., 1992. Alliance formation among male baboons: shopping for profitable partners. In: Harcourt, A.H., de Waal, F.B.M. (Eds.), Coalitions and Alliances in Humans and Other Animals. Oxford University Press, Oxford, pp. 285–321.

Newton-Fisher, N., 1999. Association by male chimpanzees: a social tactic? Behaviour 136, 705–730.

Ostner, J., Heistermann, M., Schülke, O., 2008. Dominance, aggression, and physiological stress in wild male Assamese macaques (*Macaca assamensis*). Hormones and Behavior 54, 613–619.

Palombit, R., 2012. Infanticide: male strategies and female counterstrategies. In: Mitani, J.C., Call, J., Kappeler, P.M., Palombit, R.A., Silk, J.B. (Eds.), The Evolution of Primate Societies. University of Chicago Press, Chicago, pp. 432–468.

Palombit, R., Cheney, D.L., Fischer, J., Johnson, S., Rendall, D., Seyfarth, R.M., Silk, J.B., 2000. In: Janson, C.H., van Schaik, C.P. (Eds.), Infanticide by Males and its Implications. Cambridge University Press, Cambridge, pp. 123–152.

Patzelt, A., Kopp, G.H., Ndao, I., Kalbitzer, U., Zinner, D., Fischer, J., 2014. Male tolerance and male-male bonds in a multilevel primate societies. Proceedings of the National Academy of Sciences 111, 14740–14745.

Pereira, M.E., 1993. Agonistic interactions, dominance relation, and ontogenetic trajectories in ringtailed lemurs. In: Pereira, M.E., Fairbanks, L.A. (Eds.), Juvenile Primates. University of Chicago Press, Chicago, pp. 285–305.

Paul, A., Kuester, J., 1987. Dominance, kinship, and reproductive value in female barbary macaques (*Macaca sylvanus*) at Affenberg, Salem. Behavioral Ecology and Sociobiology 21, 323–331.

Pereira, M.E., Kappeler, P.M., 1997. Divergent systems of agonistic relationships in lemurid primates. Behavior 134, 225–274.

Perry, S., 1996. Female-female social relationships in wild white-faced capuchin monkeys (*Cebus capucinus*). American Journal of Primatology 40, 167–182.

le Roux, A., Bergman, T.J., 2012. Indirect rival assessment in a social primate, *Theropithecus gelada*. Animal Behaviour 83, 249–255.

Radespiel, U., Sarikaya, Z., Zimmermann, E., Bruford, M.W., 2001. Sociogenetic structure in a free-living nocturnal primate population: sex-specific differences in the grey mouse-lemur (*Microcebus murinus*). Behavioral Ecology and Sociobiology 50, 493–502.

Reichard, U., 2009. Social organization and mating system of Khao Yai gibbons, 1992–2006. In: Whittaker, D.J., Lappan, S. (Eds.), Gibbons: New Perspectives on Small Ape Socioecology and Population Biology. Springer-Verlag, New York, pp. 347–384.

Reichard, U., Gapanakngam, M., Barelli, C., 2012. White-handed gibbons of Khao Yai: social flexibility, complex reproductive strategies, and a slow life history. In: Kappeler, P.M., Watts, D.P. (Eds.), Long-term Field Studies of Primates. Springer, Heidelberg, pp. 237–260.

Rijksen, H., 1978. A Field Study on Sumatran Orangutans (Pongo pygmaeus Abelii, Lessen 1827): Ecology, Behavior, and Conservation. H. Veemen and Zonen BV, Wageningen.

Robbins, A.M., Stoinski, T.M., Fawcett, K.A., Robbins, M.M., 2009. Socioecological influences on the dispersal of wild female mountain gorillas: evidence of a second folivore paradox. Behavioral Ecology and Sociobiology 63, 477–489.

Robbins, M.M., Robbins, A.M., Gerald-Steklis, N., Steklis, H.D., 2007. Socioecological influences on the reproductive success of female mountain gorillas (*Gorilla gorilla beringei*). Behavioral Ecology and Sociobiology 61, 919–931.

Rodman, P.S., Mitani, J.C., 1987. Orangutans: sexual dimorphism in a solitary species. In: Smuts, B.B., Cheney, D.L., Seyfarth, R.M., Wrangham, R.W., Struhsaker, T.T. (Eds.), Primate Societies. University of Chicago Press, Chicago, pp. 146–152.

Rogers, E., Abernethy, K., Bermejo, M., Cipoletta, C., Doran, D., McFarland, K., Nishihara, T., Remis, M., Tutin, C.E.G., 2004. Western gorilla diet: a synthesis from six sites. American Journal of Primatology 64, 173–192.

Scarry, C.J., 2013. Between-group contest competition among tufted capuchin monkeys, *Sapajus nigritus*, and the role of male resource defence. Animal Behaviour 85, 931–939.

Schino, G., 2007. Grooming and agonistic support: a meta-analysis of primate reciprocal altruism. Behavioral Ecology 18, 115–120.

Schülke, O., Bhagavatula, J., Vigilant, L., Ostner, J., 2010. Social bonds enhance reproductive success in male macaques. Current Biology 20, 2207–2210.

Schülke, O., Kappeler, P.M., 2003. So near and yet so far: territorial pairs but low cohesion between pair partners in a nocturnal lemur, *Phaner furcifer*. Animal Behaviour 65, 335–343.

Schülke, O., Ostner, J., 2012. Ecological and social influences on sociality. In: Mitani, J.C., Call, J., Kappeler, P.M., Palombit, R.A., Silk, J.B. (Eds.), The Evolution of Primate Societies. University of Chicago Press, Chicago, pp. 432–468.

Silk, J.B., 2002. The importance of kin selection in primates. International Journal of Primatology 23, 849–875.

Silk, J.B., 2012. The adaptive value of sociality. In: Mitani, J.C., Call, J., Kappeler, P.M., Palombit, R.A., Silk, J.B. (Eds.), The Evolution of Primate Societies. University of Chicago Press, Chicago, pp. 552–564.

Silk, J.B., Alberts, S.C., Altmann, J., 2003. Social bonds of female baboons enhance infant survival. Science 302, 1231–1234.

Silk, J.B., Altmann, J., Alberts, S.C., 2006a. Social relationships among adult female baboons. I: variation in the strength of social bonds. Behavioral Ecology and Sociobiology 61, 183–195.

Silk, J.B., Altmann, J., Alberts, S.C., 2006b. Social relationships among adult female baboons. II: variation in the quality and stability of social bonds. Behavioral Ecology and Sociobiology 61, 197–204.

Silk, J.B., Beehner, J.C., Bergman, T.J., Crockford, C., Engh, A.L., Muscovice, L.R., Wittig, R.M., Seyfarth, R.M., Cheney, D.L., 2009a. The benefits of social capital: close social bonds among female baboons enhance offspring survival. Proceedings of the Royal Society, B 276, 3099–3104.

Silk, J.B., 2014. Looking inside the black box: mechanisms linking social behavior to health and fitness. Evolutionary Anthropology 23, 83–84.

Silk, J.B., Beehner, J.C., Bergman, T.J., Crockford, C., Engh, A.L., Muscovice, L.R., Wittig, R.M., Seyfarth, R.M., Cheney, D.L., 2009b. Strong and consistent social bonds enhance the longevity of female baboons. Current Biology 20, 1359–1361.

Silk, J.B., Beehner, J.C., Bergman, T.J., Crockford, C., Engh, A.L., Moscovice, L.R., Wittig, R.M., Seyfarth, R.M., Cheney, D.L., 2010. Strong and consistent social bonds enhace the longevity of female baboons. Current Biology 20, 1359–1361.

Smuts, B.B., 1985. Sex and Friendship in Baboons. Aldine, Chicago.

Smuts, B.B., Smuts, R., 1993. Sexual coercion in primates and other mammals. Advances in the Study of Behavior 22, 1–63.

Snyder-Mackler, N., Beehner, J.C., Bergman, T.J., 2012. Defining higher levels in the multi-level societies of geladas (*Theropithecus gelada*). International Journal of Primatology 33, 1054–1068.

Stander, P.E., 1992. Cooperative hunting in lions: the role of the individual. Behavioral Ecology and Sociobiology 29, 445–454.

Stanford, C., 1998. Chimpanzees and Red Colobus: The Ecology of Predator and Prey. Harvard University Press, Cambridge.

Sterck, E.M., Watts, D.P., van Schaik, C.P., 1997. The evolution of social relationships in female primates. Behavioral Ecology and Sociobiology 41, 291–309.

Stumpf, R., Martinez-Mota, R., Milich, K.M., Rhigini, M., Shattuck, M.R., 2011. Sexual conflict in primates. Evolutionary Anthropology 20, 62–75.

Sugardjito, J., te Boekhoerst, I.J.A., van Hooff, J.A.R.A.M., 1987. Ecological constraints on the grouping of wild orangutans in (*Pongo pygmaeus*) in the Gunung Leuser National Park, Sumatra, Indonesia. International Journal of Primatology 8, 17–41.

Swedell, L., Saunders, J., Schrier, A., David, B., Tesfaye, T., Pines, M., 2011. Female "dispersal" in hamadryas baboons: transfer among social units in a multi-level society. American Journal of Physical Anthropology 145, 360–370.

van Schaik, C.P., 1983. Why are diurnal primate living in groups? Behavior 87, 120–144.

van Schaik, C.P., 1989. The ecology of social relationships amongst female primates. In: Standen, V., Foley, R.A. (Eds.), Comparative Socioecology: The Behavioral Ecology of Humans and Other Mammals. Blackwell, Oxford, pp. 195–218.

van Schaik, C.P., 1999. The socioecology of fission-fusion sociality in orangutans. Primates 40, 69–87.

van Schaik, C.P., 2000. Infanticide by male primates: the sexual selection hypothesis revisited. In: Janson, C.H., van Schaik, C.P. (Eds.), Infanticide by Males and its Implications. Cambridge University Press, Cambridge, pp. 27–60.

Teelen, S., 2008. Influence of chimpanzee predation on the red colobus population at Ngogo, Kibale National Park, Uganda. Primates 49, 41–49.

Tennie, C., Gilby, I., Mundry, R., 2008. The meat scrap hypothesis. Behavioral Ecology and Sociobiology 63, 421–431.

Terborgh, J., 1983. Five New World Primates: A Study in Comparative Ecology. Princeton University Press, Princeton.

Terborgh, J., Janson, C., 1986. The socioecology of primate groups. Annual Review of Ecology and Systematics 17, 111–135.

Thierry, B., 2000. Covariation of conflict management patterns in macaque societies. In: Aureli, F., de Waal, F.B.M. (Eds.), Natural Conflict Resolution. University of California Press, Berkeley, pp. 106–128.

Thierry, B., 2007a. The macaques: a double-layered social organization. In: Campbell, C.J., Fuentes, A., MacKinnon, K.C., Panger, M., Bearder, S.K. (Eds.), Primates in Perspective. Oxford University Press, Oxford, pp. 224–239.

Thierry, B., 2007b. Unity in diversity: lessons from macaque societies. Evolutionary Anthropology 16, 224–238.

Thierry, B., Aureli, F., Nunn, C.L., Petit, O., Abegg, C., de Waal, F.B.M., 2008. A comparative study of conflict resolution in macaques: insights into the nature of trait covariation. Animal Behaviour 75, 847–860.

Uehara, S., Nishida, T., Hamai, M., Hasegawa, T., Hayaki, H., Huffman, M., Kawanaka, K., Kobayashi, S., Mitani, J., Takahata, Y., Takasaki, H., Tsukahara, T., 1992. Characteristics of predation by the chimpanzees in the Mahale Mountains National Park, Tanzania. In: Nishida, T., McGrew, W.C., Marler, P., Pickford, M., de Waal, F.B.M. (Eds.), Topics in Primatology. Human Origins, vol. 1. University of Tokyo Press, Tokyo, pp. 143–158.

Van Belle, S., Garber, P.A., Estrada, A., De Fiore, A., 2014. Social and genetic factors mediating male participation in collective group defence in black howler monkeys. Animal Behaviour 98, 7–17.

de Waal, F.B.M., 1982. Chimpanzee Politics. Johns Hopkins University Press, Baltimore.

de Waal, F.B.M., 1989. Dominance "style" and primate social organization. In: Standen, V., Foley, R.A. (Eds.), Comparative Socioecology: The Behavioral Ecology of Humans and Other Mammals. Blackwell, Oxford, pp. 243–264.

Wakefield, M.L., 2008. Grouping patterns and competition among female chimpanzees (Pan troglodytes schweinfurthii) at Ngogo, Kibale National Park. International Journal of Primatology 29, 907–929.

Walters, J.R., 1980. Interventions and the development of dominance relationships in female baboons. Folia Primatologica 34, 61–89.

Walters, J.R., Seyfarth, R.M., 1987. Conflict and cooperation. In: Smuts, B.B., Cheney, D.L., Seyfarth, R.M., Wrangham, R.W., Struhsaker, T.T. (Eds.), Primate Societies. University of Chicago Press, Chicago, pp. 306–317.

Watts, D.P., 1989. Infanticide in mountain gorillas: new cases and a reconsideration of the evidence. Ethology 81, 1–18.

Watts, D.P., 1990. Ecology of gorillas and its relationship to female transfer in mountain gorillas. International Journal of Primatology 11, 21–45.

Watts, D.P., 1994. Social relationships of immigrant and resident female mountain gorillas, II: relatedness, residence, and relationships between females. American Journal of Primatology 32, 13–30.

Watts, D.P., 1996. Comparative socioecology of gorillas. In: McGrew, W.C., Nishida, T., Marchant, L.A. (Eds.), Great Ape Societies. Cambridge University Press, Cambridge, pp. 16–28.

Watts, D.P., 1998. Coalitionary mate guarding by male chimpanzees at Ngogo, Kibale National Park, Uganda. Behavioral Ecology and Sociobiology 4, 43–55.

Watts, D.P., 2000. Causes and consequences of variation in the number of males in mountain gorilla groups. In: Kappeler, P.M. (Ed.), The Socioecology of Primate Males. Cambridge University Press, Cambridge, pp. 169–180.

Watts, D.P., 2002. Reciprocity and interchange in the social relationships of wild male chimpanzees. Behaviour 139, 343–370.

Watts, D.P., 2012. The apes: taxonomy, biogeography, life histories, and behavioral ecology. In: Mitani, J.C., Call, J., Kappeler, P.M., Palombit, R.A., Silk, J.B. (Eds.), The Evolution of Primate Societies. University of Chicago Press, Chicago, pp. 113–143.

Watts, D.P., Amsler, S.J., 2013. Chimpanzee-red colobus encounter rates show a red colobus population decline associated with predation by chimpanzees at Ngogo. American Journal of Primatology 75, 927–937.

Watts, D.P., Mitani, J.C., 2002. Hunting behavior of chimpanzees at Ngogo, Kibale National Park, Uganda. International Journal of Primatology 23, 1–28.

West, S.A., Griffin, A.S., Gardner, A., 2006. Social semantics: altruism, cooperation, mutualism, strong reciprocity, and group selection. Evolutionary Biology 20, 415–432.

West, S.A., Griffin, A.S., Gardner, A., 2007. Social semantics: altruism, cooperation, mutualism, strong reciprocity, and group selection. Journal of Evolutionary Biology 21, 415–432.

Williams, J.M., Oehlert, G.W., Carlis, J.V., Pusey, A.E., 2004. Why do male chimpanzees defend a group range? Animal Behaviour 68, 523–532.

Wilson, M.L., Hauser, M.D., Wrangham, R.W., 2001. Does participation in intergroup conflict depend on numerical assessment, range location, or rank for wild chimpanzees? Animal Behaviour 61, 1203–1216.

Wilson, M.L., Kahlenberg, S.M., Wells, M., Wrangham, R.W., 2012. Ecological and social factors affect the occurrence and outcomes of intergroup encounters in chimpanzees. Animal Behaviour 83, 277–291.

Wilson, M.L., Wrangham, R.W., 2003. Intergroup relations in chimpanzees. Annual Review of Anthropology 32, 363–392.

Wittig, R.M., Crockford, C., Lehmann, J., Whitten, P.L., Seyfarth, R.M., Cheney, D.L., 2008. Focused grooming networks and stress alleviation in wild baboons. Hormones and Behavior 54, 170–177.

Wrangham, R.W., 1979. On the evolution of ape social systems. Social Science Information 18, 335–368.

Wrangham, R.W., 1980. An ecological model of female-bonded primate groups. Behavior 75, 262–300.

Wrangham, R.W., 1982. Mutualism, kinship, and social evolution. In: Wrangham, R.W., Rubinstein, D.I., Dunbar, R.I.M., Bertram, B.C.R., Clutton-Brock, T.H. (Eds.), Current Problems in Sociobiology. Cambridge University Press, Cambridge, pp. 269–290.

Wrangham, R.W., 1999. Evolution of coalitionary killing. Yearbook of Physical Anthropology 42, 1–30.

Wrangham, R.W., Muller, M.M., 2009. Sexual Coercion in Primates and Humans. Harvard University Press, Cambridge.

Chapter 6

Primate Models for Human Evolution

Robert W. Sussman[1] and Donna Hart[2]

[1]*Department of Anthropology, Washington University, St. Louis, MO, USA;* [2]*Department of Anthropology, University of Missouri–St. Louis, MO, USA*

SYNOPSIS

While many models have been developed to depict the behavior and ecology of our earliest relatives, the Man the Hunter model has been the most widely accepted view of human evolution. Many human traits (e.g., bipedalism, tools, and fire) are often linked to this perspective. Theories of human aggressive hunters abound but are rarely based on evidentiary approaches. Here is outlined a methodology using the fossil record and extant primate ecology and behavior. Data on fossil humans, modern primates, and rates of predation indicate that Man the Hunted may be the most accurate descriptor of our earliest relatives.

MODELS OF HUMAN EVOLUTION

What kinds of evidence should be used in attempts to reconstruct the behavior of our earliest ancestors? The best evidence of early hominin behavior must include careful examination and understanding of the actual skeletal remains of the creatures. However, evidence of behavior also includes other substantiation (such as tools or footprints) left by the creatures, and other fossil materials that give us clues about the environment in which they lived (such as fauna, flora, or water sources). These fossils provide the most important data for an accurate reconstruction. Interestingly, some past reconstructions of early human behavior lacked a critical examination of the fossil evidence. We might say they were virtually fossil-free.

Besides fossils, any other types of secondary evidence used in reconstructions are less reliable, but nonetheless offer insights. These are ranked in the following order as far as applicability to developing models of early hominin lifestyles: (1) The behavior of nonhuman primates living under ecological conditions similar to those of our earliest ancestors (see also Elton, 2006). However, these conditions are different for different ancestors. Hominins likely began as edge species but moved onto the savanna about two million years ago (Conroy and Pontzer, 2012), and there were often several species living simultaneously. (2) The behavior of our genetically closest primate relatives, such as chimpanzees, bonobos, and gorillas (McGrew, 2010). However, the behavior of the great apes is very diverse and each species is unique, so simple analogs are not very useful. Yet, some behavioral characteristics may remain conservative within a taxonomic group. For example, monogamous pair bonds among the lesser apes (gibbons including siamangs), or upright posture among the great apes, might be considered phylogenetically conservative traits shared by all species within a taxon. (3) Characteristics shared by certain (or all) modern humans that also might be similar in our earliest ancestors. Modern foragers (see the chapter by Little and Blumler in the present volume), however, are just as advanced and evolved within their own culture and environment as any Western urban dwellers. Our least confident recommendation is (4) the behavior of other animal species that might be living under similar conditions or share some aspects of the lifestyle of early humans, such as certain carnivore or prey species. A cat is still a carnivore even if it eats some grass; early hominins opportunistically included a few vertebrates in their diet, but they cannot legitimately be compared to obligate meat eaters (Hladik et al., 1999; McDougall, 2003).

In using any of these types of secondary evidence, if we are not extremely careful (because in many cases similar-looking behaviors are not the same), we can end up comparing apples with oranges, lions with hominins, or even strangler figs with purse snatchers! Obviously, words with loaded meaning for humans—war, rape, murder, infanticide, and genocide to name a few—must be used with extreme caution when referring to the activities of nonhuman species. In this regard, Jonathan Marks (2002, p. 104) warns against "a science of metaphorical, not of biological, connection."

We cannot necessarily impute correlation, therefore, between human ancestors and data based on extant carnivores, modern human foragers, or great apes. For example, even the concept of hunting in chimps and humans is quite different. Present-day human hunters purposely search for prey, but chimpanzees do not: "Instead, they forage for plant foods and eat prey animals opportunistically in the course of looking for fruits and leaves" (Stanford, 1999, p. 48). Even though they do hunt for prey occasionally, they are basically frugivorous (Hladik, 1977; Milton and Demment, 1989; Boesch and Boesch-Achermann, 2000). Furthermore, reconstructions must always be compatible with the actual fossil data—the fossils are real but the models we construct are hypothetical and must constantly be tested and reconfirmed. Lastly, when attempting to construct models of our early ancestors' behavior, it is necessary to be precise about timing (Tattersall, 2010). If we say our earliest human ancestors (those who lived seven million years ago; see the chapter by Hunt) behaved in a certain way, we cannot use fossil evidence from two million years ago.

For example, could hunting by early hominins have occurred without tools? The first evidence of stone tools comes from around 2.6 million years ago (mya) (Semaw, 2000) (see chapter by Toth and Schick). The earliest hominin fossils, however, date from almost 7 mya, at least 4 million years before the first stone tools. In fact, when we look at the fossil evidence, hunting may have come quite late to our human family. Interpretations of hominin behavior, therefore, should be conservative and cautious, as stated by Jurmain et al. (2003) (Figure 1):

> *The mere presence of animal bones at archaeological sites does not prove that hominids were killing animals or even necessarily exploiting meat. Indeed, as was the case in the earliest South African sites, the hominid remains themselves may have been the meal refuse of large carnivores.*

FIGURE 1 Predation by leopards on our earliest ancestors.

In examining one of the most accepted models of human evolution, the "Man the Hunter" model, we can see how fossil evidence may have been misused. The transition to hunting as a dominant way of life does not appear to have started until after the appearance of our own genus, *Homo*, and may not have even begun with the earliest members of our genus. *Homo erectus* has been given credit in the past for existing as a large animal hunter, with dates as far back as 1.75 mya hypothesized for such a lifestyle (Klein, 1999). But if a conservative approach to this subject is taken, looking only at facts and fossils, the first indications of hunting are amazingly recent. In fact, according to some paleontologists, the first unequivocal evidence of large-scale systematic hunting by humans is available from only 60,000–80,000 years ago (Binford, 1992; Klein, 1999). The earliest hominin fossils, therefore, existed almost 7 million years before the first factual evidence of systematic hunting by humans.

No actual fossil evidence of tools designed for hunting exists earlier than approximately 400,000 years ago—that evidence is a finely shaped wooden spear excavated at Schöningen, Germany (Dennell, 1997; Theime, 1997). Well-known sites in Spain, dated at 500,000 years ago, contain a huge number of large mammal bones and were thought to represent unquestionable evidence of megafauna killed by Pleistocene hunters. Now these sites are being reconsidered in light of better archaeological analysis. Elephant bones in situ could just as likely represent natural deaths or carnivore kills as the remains of human hunting (Klein, 1999; Klein et al., 2007). Further, no hominins were large-scale hunters before they had the use of fire (because of their dentition and alimentary tract, points we elucidate below), although insects, small vertebrates, lizards, and birds likely were eaten opportunistically. The best evidence for the controlled use of fire appears around 800,000 years ago in Israel (Goren-Inbar et al., 2004). Klein (1999, p. 160) states: "The assumption of consistent hunting has been challenged, especially by archaeologists who argue that the evidence does not prove the hunting hypothesis … it is crucial to remember (although not as exciting) that probably the majority of calories [came] from gathering plant foods."

DENTITION AND DIET

Whether *H. erectus* or any other hominin before 800,000 years ago hunted or scavenged may be a moot question. Hunting would only be an activity undertaken if early hunters could eat what they killed, and to eat raw meat it is necessary to have teeth capable of processing it.

Obviously, Man the Hunter models of human evolution assume that a significant portion of our earliest ancestors' diets must have come from killing and eating meat from relatively large mammals. By comparing the characteristics of the dental and jaw morphology of various living primates with those of fossils, we can make inferences about the diets of early hominins. Teaford and Ungar (2000), Ungar (2004) carried out such a comparison. Using the features of tooth size and shape, enamel structure, dental microwear, jaw biomechanics, occlusal slope, and relief of lower molars, they found that the earliest humans had a unique combination of dental characteristics and a diet different from those of modern apes and modern humans.

Australopithecus afarensis, who lived between 3.6 and 2.9 mya (and possibly as far back as 5 mya) (see the chapter by Ward), is characterized by thick jawbones, with comparatively small incisors and canines in relation to molars (Teaford and Unger, 2000). The molars, by comparison with other primates, are huge, flat, and blunt, show less slope and relief, and lack the long shearing crests necessary to mince flesh. *Australopithecus afarensis* also had larger front than back molars. The dental enamel is thick, and microwear on the teeth is a mosaic of gorilla-like fine wear striations (indicating leaf eating) and baboon-like pits and microflakes (indicating fruits, seeds, and tubers in the diet). This evidence all points away from meat eating.

In studies of mid- to large-sized primates such as macaques, baboons, chimpanzees, and modern human foragers for which the amount of time spent obtaining animal protein has been quantified, the total time is very low, usually making up less than five percent of time spent feeding (Garber, 1987; Sussman, 1999). Given these facts, we hypothesize that early humans were able to exploit a wide range of dietary resources, including both hard, brittle foods (e.g., tough fruits, nuts, seeds, and pods) and soft, weak foods (e.g., ripe fruits, young leaves and herbs, and flowers and buds). They may also have been able to eat abrasive objects, including gritty plant parts such as grass seeds, roots, rhizomes, and underground tubers. As stated by Teaford and Unger (2000, pp. 13508–13509), "this ability to eat both hard and soft foods, plus abrasive and nonabrasive foods, would have left early hominids particularly well suited for life in a variety of habitats, ranging from gallery forest to open savanna." Dental morphology indicates that the earliest hominins would have had difficulty breaking down tough pliant plant foods such as seed coats and mature leaves. Another tough pliant food that our early ancestors would have had difficulty processing was meat. Teaford and Unger state (2000, p. 13509): "The early hominids were not dentally preadapted to eat meat—they simply did not have the sharp, reciprocally concave shearing blades necessary to retain and cut such foods."

Both modern chimpanzees and humans have alimentary tracts that are specialized for the eating of neither leaves nor animal protein, but instead are more generalized and similar to those of the majority of primates who are omnivorous and eat a mixture of food types (Martin et al., 1985; Martin, 1990). Modern humans, especially in Western cultures, think of themselves as meat eaters. For Americans and many other cultures, meat defines that ephemeral status of wealth for which we strive. Because they themselves were rooted in these cultural stereotypes, anthropologists egregiously misnamed the modern forager cultures as hunter-gatherers and initially emphasized the contributions of male hunters. Nevertheless, more than two-thirds of modern-day foragers' food comes from gathering of plant foods by women, who in the process opportunistically capture small mammals and reptiles. Less than one-third of the diet (the meat portion brought in by male hunters) serves to supplement their foraged nutritional intake, except in cold climates or where fishing is prevalent (Marlowe, 2005).

Lastly, we believe that hominins did not hunt on a large scale before the advent of controlled fire. Again, we have neither the dentition nor the digestive tract of a carnivore. Our anatomy and physiology did not particularly suit us for digesting meat until the mastery of cooking solved the problem. Our intestinal tract is short, and predigestion by cooking with fire had to precede any major meat eating. As stated above, the oldest known hearths with good evidence for controlled use of fire are around 800,000 years old.

LOCOMOTION

By far the best known australopith is *A. afarensis*. Collections from Hadar, Ethiopia alone have yielded 250 specimens, representing at least 35 individuals, and there are a number of other East African sites with remains of this species (Kimbel and Delezene, 2009). Specimens include the famous Lucy (dated 3.2 mya), the most complete adult skeleton from this time period, and fossil footprints from Laetoli, Tanzania (dated 3.6 mya). Most hypotheses concerning human evolution position *A. afarensis* as a pivotal species from which all other later hominins, including *Homo*, evolved (Fleagle, 1999; Tattersall, 2010; Conroy and Pontzer, 2012). Given the above facts, we see *A. afarensis* as a good species to examine when attempting to develop models of early human behavior (Figure 2).

FIGURE 2 A reconstruction of two *Australopithecus afarensis* (used by permission of American Museum of Natural History AMNH Library 4936[2]). (See color plate section).

Terrestrial bipedalism is a hallmark of the fossil hominin family. This mode of locomotion can be inferred from fossil specimens that are nearly seven million years old (Galik et al., 2004—but see the chapter by Hunt for full explanation). It appears long before the vast growth of open grasslands in Africa and before the expansion of human brain size and recognizable stone toolmaking. Besides the fossilized bones, direct evidence of early bipedalism comes from the fossilized footprints at Laetoli (White, 1980; Tuttle, 1985). Looking at the skeletal evidence, however, the locomotion of these early hominins was not exactly identical to ours.

Australopithecus afarensis seems to have been a primate equally at home in the trees and on the ground, as indicated by several factors. First, the limb proportions are different from those of modern humans. The arms are similar in proportion to those of modern humans, but the legs are relatively shorter and more apelike, implying the use of suspensory locomotion in trees (Kimbel et al., 1994). Other aspects of the upper limbs retain features indicating an ability to move easily in the trees. The wrist and hand bones are quite chimpanzee-like, and the finger and toe bones are slender and curved as in apes, giving *A. afarensis* grasping capabilities compatible with suspensory behaviors. The toe bones are relatively longer and more curved than in *Homo sapiens*, and the joints of the hands and feet, as well as the overall proportions of the foot bones, reinforce evidence for climbing adaptations and arboreal activity. Nevertheless, the relative thumb length of these hominins is closer to that of modern humans than it is to chimpanzees (Susman et al., 1984; Conroy and Pontzer, 2012). The pelvis and lower limbs of *A. afarensis* are a mixture of humanlike and apelike features. Overall, Rak (1991, p. 283) summarizes: "Although clearly bipedal and highly terrestrial, Lucy evidently achieved this mode of locomotion through a solution of her own."

It appears that the combination of skeletal characteristics found in *A. afarensis* enabled these hominins to be quite versatile. They were able to use the ground and the trees equally and successfully for a very long time. We believe these early hominins were well adapted to their environment and not in the least inhibited by switching back and forth from bipedalism on the ground to quadrupedalism in the trees.

There have been several models proposed to account for the evolution of human bipedalism, each using specific behavioral or morphophysiological traits (e.g., tool using, carrying, vigilance, heat dissipation, and energy efficiency). Each model has some merit, but none of the theories seem to catch the significance of switching a method of locomotion. Furthermore, there are many other primates who spend most of their time on the ground, yet none has developed bipedalism, even though each of the theorized advantages presumably also could have accrued to them.

It is difficult to separate consequence from causation. We cannot conclude that any of the suggested models above caused hominins to become bipedal (again, see the chapter by Hunt). None of the theories may be causative; instead, all the theoretical causes may be results of a preadaptation to being bipedal. All the great apes are preadapted to bipedalism. When our ancestors came down from the trees, bipedalism was possible because of body proportions and suspensory adaptations—longer arms and shorter legs that allow gibbons, orangutans, and chimpanzees to hang from trees and forage for fruit. All apes have varying capacities for erect posture and are able to walk upright; bonobos, especially, will stride upright with humanlike posture. Further, we propose that when the earliest hominins began using the ground for a major portion of their activities, their body proportions were more suited for bipedalism than for other forms of locomotion. As stated by Fleagle (1999, p. 528):

> *Although it is important to see early hominids in the context of hominoid evolution, it is equally important to realize that in the same way that they were not little people, they also were not just bipedal chimps, but the beginning of a new radiation of very different hominoids. It is this uniqueness that makes reconstructing hominid origins so difficult. Thus although early hominids and their bipedal adaptations are certainly derived from an African apelike ancestry, human bipedalism is morphologically and physiologically different from the occasional facultative bipedal behaviors occasionally seen in other primates. The morphological and behavioral commitment to bipedalism that characterized early hominids suggests unique ecological and historical circumstances as well.*

Some primate species are intrinsically adapted to edge habitats and therefore able to take advantage of changing environments. We hypothesize that the earliest hominins were edge species (see below), and that they exploited a terrestrial habitat due to a developing mosaic environment that included climate change. Rather than seeking the factors that *caused* early human ancestors to become bipedal, we propose that it was a preadaptation that already existed, and it was efficient in a new habitat; the successes or added advantages were simply by-products.

Besides bipedalism and limb use, there are solid conjectures of what our earliest ancestors were like as far as body build, height, weight, and brain capacity.

From various *A. afarensis* specimens and by examining the skeleton of Lucy, it seems that there was a considerable size difference between males and females. Although the canines of both sexes were relatively small and not at all daggerlike, they were larger and longer in the males than in the females. The range of body size for *A. afarensis* individuals is estimated to be 30–45 kg (Fleagle, 1999). The height of the adults has been estimated at 1.0–1.7 m (Klein, 1999). Lucy stood slightly over 1 m tall and weighed around 30 kg (she was definitely on the small side) (Conroy and Pontzer, 2012). If these weights are accurate, we can extrapolate that female *A. afarensis* were the size of male baboons, and males were the size of female chimpanzees.

The cranial capacity of these hominins is estimated at 400–500 cubic centimeters—about the size of a modern chimpanzee, but twice as large as early, Miocene fossil apes. On average, australopiths and modern chimpanzees have brains that are two to three times larger than those of similar-sized mammals, whereas the modern human brain is six to seven times larger than that of other mammals.

Looking at brain size relative to body size, using the encephalization quotient as a measurement, the brain of *A. afarensis* was slightly larger in relation to its body than that of modern chimpanzees (EQ = 2.4 for *A. afarensis* vs. 2.0 for chimpanzees) (Boaz, 1997). Thus, our ancestors were midsized to relatively large-sized primates with brains that were slightly larger than that of any nonhuman primate, although only a fraction bigger than those of modern chimpanzees.

HABITAT OF OUR EARLIEST ANCESTORS

Although many early theories of human evolution stress the importance of arid, savanna environments, these seem to have become primary habitats beginning only about 2 mya (see chapter by Sept for complete discussion). The African climate was becoming more arid in the time between 12 and 5 mya, and equatorial forests were undoubtedly shrinking (Conroy and Pontzer, 2012). However, the process that led to this climatic phenomenon also greatly enlarged transitional zones between forest and adjacent savanna. Closed woodland forests were still widespread in East Africa 3.5 mya, whereas the proportion of dry shrub to grassland habitats began to increase around 1.8 mya (Schekleton, 1995).

It is in these transitional zones that the behavioral and anatomical changes were initiated in early hominin evolution. The flora and fauna remains found in association with fossil hominins of this time period indicate a mixed, mosaic environment—ecologically diverse and subject to seasonal and yearly changes (Potts, 1996; Wolpoff, 1998; Conroy and Pontzer, 2012). These environments were wetter than those in which later fossil hominins are found, and most fossil sites of this time period contained water sources such as rivers and lakes (WoldeGrabriel et al., 2001; Wrangham et al., 2009).

Thus, the earliest hominins appear to be associated with variegated fringe environments or edges between forest and grassland (Sept, 2013). These habitats usually contain animal and plant species of both the forest and the grassland, as well as species unique to the borders between the two, often referred to as *edge* species. During these earliest times, it appears that hominins began to take advantage of the growing fringe environments, lessening competition with their sibling ape species that were better adapted to exploit the dense forest, thus partitioning the niche occupied by the parent species of both apes and hominins into two narrower and less overlapping adaptive zones (Klein, 1999; Lee-Thorp et al., 2003; Reed, 2008; Conroy and Pontzer, 2012).

From available evidence, we speculate that our early ancestors were able to exploit a great variety of food resources but were mainly fruit eaters, probably supplementing this diet with some young leaves and other plant parts, social insects, and a small number of opportunistically captured small vertebrate prey—lizards, small snakes, birds, and mammals. They also likely exploited some freshwater and marine resources (Cunnane and Stewart, 2010).

Several other primate species are intrinsically adapted to edge habitats and thus can also take advantage of changing environments. Madagascan ring-tailed lemurs, African baboons and vervet monkeys, and some Asian macaques and langurs are nonhuman primate examples. Not coincidentally, these are some of the most common and numerous of all living nonhuman primates. The macaque genus (*Macaca*), for example, has the widest geographical distribution of any nonhuman primate in Asia. Many macaque species in Asia are endangered, but the ones that have the healthiest populations (e.g., long-tailed macaques, *Macaca fascicularis*, and rhesus macaques, *M. mulatta*) are edge-adapted.

Certain ecological niches may breed certain behavioral repertoires. Many argue that the closer the DNA comparison, the more similar the behaviors between two related species (McGrew, 2010). In that case, chimps and bonobos would be the best prototypes for early human ancestors. However, if ecology is paramount, chimps and bonobos may be less suitable prototypes (although some traits between these close relatives may still be important and phylogenetically conservative), and the best models for early humans may be edge species.

Nearly 40 years ago, Robin Fox (1967, p. 419) declared:

> *But the problem of taking the great apes as models lies in the fact of their forest ecologies. Most modern students of primate evolution agree that we should pay close attention to ecology in order to understand the selection pressures at work on the evolving primate lines. This has been shown to be crucial in understanding ... evolution.*

Even if one were to learn everything about the hominin-ape common ancestor, many of the most crucial questions about distinctively hominin evolution would remain unanswered. Although there is a fairly impressive record of human fossils during the period of 7 to 2 mya, there is a lack of great ape fossils at these early sites. Therefore, it seems likely that chimpanzee and gorilla ancestors did not inhabit these fringe environments and were instead restricted to closed forest ecosystems—areas where fossils are less likely to be preserved (Stewart, 2010). Some populations of chimpanzees moved into more mosaic, open habitats relatively recently, long after humans had moved into these environments. Furthermore, modern chimpanzees do not live in habitats in which modern humans lived in the past or are found today. The historical geographic range of chimpanzees is quite restricted, probably more restricted than even that of early humans before leaving Africa.

In our opinion, the best primate models to use as a basis for extrapolation about behavioral characteristics of our earliest ancestors are modern primate species living in edge habitats. Macaques can be extremely good colonizers of edge habitats. The macaque genus spread throughout Asia before humans reached that continent (Delson, 1980). By the time *Homo erectus* arrived in Asia 1.8 mya, most hominins were no longer edge species—our more recent ancestors were exploiting more open habitats by this date (Tattersall, 2010)—so hominins likely did not displace macaques. We propose that the macaques, true "weed" species, are excellent models for reconstructing how our earliest ancestors lived. Many of the features of the behavior and ecology of macaques are very similar to those of other primates living in similar habitats. In our opinion, it is these shared characteristics that make this such a strong model. After all, it is the environment in which species find themselves that determines many of their evolutionary adaptations.

THE MACAQUE MODEL

Long-tailed macaques (*M. fascicularis*) are small edge-dwelling species that spend a good proportion of time both in the trees and on the ground. They are omnivorous and very versatile in their locomotion, although mainly quadrupedal. The most widespread of any Southeast Asian monkey, they occur from Burma through Malaysia and from Thailand to Vietnam, while offshore populations are found on Java, Borneo, and numerous smaller islands as far east as the Philippines and Timor. Throughout this area, broadleaf evergreen and other forest types are interspersed with secondary and disturbed habitats, and it is the latter that long-tailed macaques prefer. They are most commonly found in secondary forest habitat, preferably near water (Sussman et al., 2011). The success of the long-tailed macaque throughout its Asian distribution is widely credited to its being an "adaptable opportunist" (MacKinnon and MacKinnon, 1980, p. 187). Researchers emphasize that these monkeys are extremely adaptable and able to flourish in highly disturbed land (Figure 3).

These are slender, active monkeys; average weight is 4–5 kg for females and 6–7.75 kg for males. Long-tailed macaque society is organized around matrilineal hierarchies. There are always one or two dominant males within the group, as well as some lower-ranking adult males, plus the adolescent and subadult male offspring of the females. At sexual maturity, males migrate to a new group. Female offspring are philopatric, but mate with unrelated males who join their group (Jamieson, 1998).

In most primates adapted to edge environments, it is the males who migrate. However, in the closest genetic relatives of humans, the African great apes, females normally change

FIGURE 3 Long-tailed macaques (*Macaca fascicularis*). (See color plate section).

groups when they mature. This appears to be a phylogenetically conservative characteristic among hominoids, which makes it possible that among our earliest ancestors, females rather than males migrated between groups. However, among most modern human foragers, individuals reside with their maternal relatives at times and with their paternal relatives at other times, or sometimes with neither (Marlowe, 2005).

The ability of edge species to exploit a wide variety of environments is accompanied by substantial behavioral flexibility. Long-tailed macaques appear to be primarily arboreal where suitable vegetation exists, but they come to the ground along riverbanks, seashores, and in open areas—and in some portions of their recently colonized range, such as Mauritius, they are highly terrestrial (Sussman and Tattersall, 1981). They are eclectic omnivores with a distinct preference for fruit. But the variety of habitat they exploit is reflected in a wide selection of food items—leaves, grasses, seeds, flowers, buds, shoots, mushrooms, water plants, gum, sap, bark, insects, snails, shellfish, bird eggs, and small vertebrates (Sussman et al., 2011). Human-disturbed habitat or proximity to human settlements is not avoided; rather, they tend to live in proximity to humans throughout their range, which results in crop raiding of sugar cane, rice, cassava, and taro fields.

Long-tailed macaques live in large multimale, multifemale groups of up to 80 individuals, although in some areas groups are much smaller. They show distinct flexibility in structure; the large basic social unit tends to split into smaller subgroups for daytime foraging activities. Subgroups may be all males, but most often consist of adult males accompanying females and their young. The number and size of subgroups vary with season and resource availability (Jamieson, 1998). The entire troop reforms each evening and returns to the same sleeping site each night, usually on the edge of a water source. Because of their unique behavior of returning to a home base each night, long-tailed macaques are referred to as "refuging" species.

FOSSILS AND LIVING PRIMATES

Looking at the fossil evidence, it is apparent that human ancestors living from 7 to 2.5 mya were intermediate-sized primates not smaller than male baboons or larger than female chimpanzees. Given their relative brain size, they were at least as clever as the great apes of today. They had diverse locomotor abilities, exploiting both terrestrial and arboreal habitats. They used climbing and suspensory postures when traveling in the trees and were bipedal when on the ground. We believe that their bipedalism was a preadaptation, but walking on two feet freed the arms and hands and proved to be advantageous in many ways.

Given their relatively small size and small canines, there is no reason to think that our early ancestors were any less vulnerable to predation than are modern monkeys—some of which have yearly predation rates generally comparable to those of gazelles, antelopes, and deer (Hart, 2000; Hart and Sussman, 2009). Indeed, edge species can be highly vulnerable to predation, and because of this usually live in relatively large social groups with many adult males and adult females; adult males often serve as sentinels and provide protection against predators. Because a primate group with only one male and 10 females can have the same reproductive output as a group with 10 males and 10 females, often the male role in primate groups is to act as a first line of defense; if he gets eaten there are other males to take his place. If a sexually mature female gets eaten, then she and all her potential offspring are lost.

We propose that, like long-tailed macaques, our hominin ancestors may have lived in multimale, multifemale groups of variable size that were able to split up depending on the availability of food and re-form each evening at home base refuges. However, certain facts such as the exact size of the groups and subgroups, whether males or females migrated from the group when they reached sexual maturity, the internal structure of the group (whether matrilineal or formed along male kinship lines), are impossible to determine accurately. Indications of these social parameters cannot be found in the fossil record and are quite variable even in closely related living primates.

In sum, the best archetype of early hominins may be multimale, group-living, midsized, omnivorous, quite vulnerable creatures living in an edge habitat near a large water source. These primates may well have been a refuging species returning to the same well-protected sleeping site each night. Most modern foragers are considered central-place foragers focusing their activities around a principal location, as are many birds, social carnivores, and primates (Marlowe, 2005). These creatures were adept at using both the trees and the ground, but when they exploited the terrestrial niche, they had upright posture and were bipedal. They depended mainly on fruit, including both soft fruits and some that were quite brittle or hard, but also ate herbs, grasses, seeds, and gritty foods such as roots, rhizomes, and tubers. A very small proportion of their diet was made up of animal protein, mainly social insects (ants and termites) and occasionally small vertebrates captured opportunistically. These early hominins did not regularly hunt for meat and could process it neither dentally nor in their digestive tracts.

Like all other primates, especially ground-living and edge species, these early hominins were very vulnerable to predators and this trait did not diminish greatly over time. Fossil evidence to this effect exists from South Africa, the Zhoukoudian cave in China (Boaz et al., 2004), skulls uncovered at Dmanisi in the Republic of Georgia, and a fossil found at Olorgesailie, Kenya (Hart and Sussman, 2009).

MAN THE HUNTED

Given that the earliest hominin ancestors were medium-sized primates without any inherent weapons to fight off the many predators that lived then—and given that they lived in edge environments that incorporated open areas and wooded forests near rivers—they were vulnerable to predation like other primates were. Because of this, we hypothesize that rates of predation were just as high in our early ancestors as they are in modern species of primates, and our origins are those of a hunted species (Hart and Sussman, 2009).

Protection from predation is one of the most important aspects of group living, and we believe that this was true of our earliest ancestors. Based on the long-tailed macaque model, social groups of early hominins may have been organized in a way that allowed efficient exploitation of a highly variable and changing environment and also protected its members from predators. Thus for early hominins, we propose several strategies for protection from predators based on the behavior and social organization we observe in long-tailed macaques.

- Relatively large groups of 25–75 individuals: Since safety lies in numbers, a main reason that all diurnal primates live in groups is for predator protection. In his research on modern human foragers, Marlowe (2005) found that the median group size is 30 individuals.
- Versatile locomotion that exploits both arboreal and terrestrial milieus: The major advantage of agility in the use of diverse habitats is safety in trees and dense underbrush. An added advantage of upright posture is the ability to scan for predators.
- Flexible social organization: Gathering scarce resources in small groups, but reuniting as a larger group when predation requires strength in numbers, allows small groups to quickly disperse and hide while large groups can mob and intimidate predators. Again, modern human foragers fit this pattern of flexibility (Marlowe, 2005).
- Multimale social structure: This provides more male protection when traveling through open areas and when the group settles in the evening or at midday. When large groups break into subgroups, females and young are accompanied by one or more larger males.
- Males as sentinels: Males are usually larger in these species. Upright posture adds to the appearance of large size and also allows for better vigilance and improved ability to wave arms, brandish sticks, and throw stones. Males mob or attack predators, since they are the more expendable sex.
- Careful selection of sleeping sites: Refuging species bring the whole group together at night in a safe area. During daytime rest periods, staying in very dense vegetation is essential. Males stay on high alert during these inactive periods and when the group is on the move.
- Remain one step ahead of predators: Intelligence endows primates with the ability to monitor the environment, communicate with other group members, and implement effective antipredator defenses (Hart and Sussman, 2009).

Our reconstruction of the behavior and ecology of our earliest hominin ancestors reflects the pervasive influence of large ferocious predatory animals throughout human evolution. Many circumstances have been proposed as a catalyst for the evolution of the human species—competition for resources, intellectual capacity, male–male conflicts, and hunting. But looking at our primate relatives and the fossil record, predation pressure was probably one of the most critical components in shaping human evolution.

REFERENCES

Binford, L., 1992. Subsistence—a key to the past. In: Jones, S., Martin, R., Pilbeam, D. (Eds.), Cambridge Encyclopedia of Human Evolution. Cambridge University Press, Cambridge, pp. 365–368.

Boaz, N., 1997. Eco Homo. Basic Books, New York.

Boaz, N., Ciochon, R., Xu, Q., Liu, J., 2004. Mapping and taphonomic analysis of the *Homo erectus* loci at locality 1 Zhoukoudian, China. Journal of Human Evolution 46, 519–549.

Boesch, C., Boesch-Achermann, H., 2000. The Chimpanzees of Tai Forest: Behavioural Ecology and Evolution. Oxford University Press, Oxford.

Conroy, G.C., Pontzer, H., 2012. Reconstructing Human Origins: A Modern Synthesis, third ed. W. W. Norton & Company, New York.
Cunnane, S.C., Stewart, K.M. (Eds.), 2010. Human Brain Evolution: The Influence of Freshwater Fish and Marine Resources. John Wiley & Sons, New York.
Delson, E., 1980. Fossil macaques, phyletic relationships and a scenario of deployment. In: Lindburg, D.G. (Ed.), The Macaques: Studies in Ecology, Behavior and Evolution. Van Nostrand-Reinhold, New York, pp. 10–30.
Dennell, R., 1997. The worlds oldest spears. Nature 385, 787–788.
Elton, S., 2006. Forty years on and still going strong: the use of hominin-cercopithecid comparisons in paleoanthropology. Journal of the Royal Anthropological Institute 12, 19–38.
Fleagle, J., 1999. Primate Adaptation and Evolution, second ed. Academic Press, New York.
Fox, R., 1967. In the beginning: aspects of hominid behavioural evolution. Man 2, 415–433.
Galik, K., Senut, B., Pickford, M., Gommery, D., Treil, J., Kuperavage, A.J., Eckhardt, R.B., 2004. External and internal morphology of the BAR 1002'00 *Orrorin tugenensis* femur. Science 305, 1450–1453.
Garber, P.A., 1987. Foraging strategies among living primates. Annual Review of Anthropology 16, 339–364.
Goren-Inbar, N., Alperson, N., Kislev, M., Simchoni, O., Melamed, Y., Ben-Nun, A., Werker, E., 2004. Evidence of hominin control of fire at Gesher Benot Ya'aqov, Israel. Science 304, 725–727.
Hart, D., 2000. Primates as Prey: Ecological, Morphological, and Behavioral Relationships between Primate Species and Their Predators (Ph.D. thesis). Washington University, St. Louis.
Hart, D., Sussman, R.W., 2009. Man the Hunted: Primates, Predators, and Human Evolution, expanded ed. Westview Press, New York.
Hladik, C., 1977. Chimpanzees of Gabon and chimpanzees of Gombe: some comparative data on the diet. In: Clutton-Brock, T. (Ed.), Primate Ecology: Studies of the Feeding and Ranging Behaviour in Lemurs, Monkey and Apes. Academic Press, New York, pp. 481–501.
Hladik, C., Chivers, D., Pasquet, P., 1999. On diet and gut size in nonhuman primates and humans. Is there a relationship to brain size. Current Anthropology 40, 695–697.
Jamieson, R.W., 1998. The Effects of Seasonal Variation in Fruit Availability on Social and Foraging Behavior in *Macaca fascicularis* in Mauritius (Ph.D. thesis). Washington University, St. Louis.
Jurmain, R., Kilgor, L., Trevathan, W., Nelson, H., 2003. Introduction to Physical Anthropology, ninth ed. Wadsworth/Thomson Learning, California.
Kimbel, W., Delezene, L.K., 2009. "Lucy" redux: a review of research on *Australopithecus afarensis*. Yearbook of Physical Anthropology 52, 2–48.
Kimbel, W., Johanson, D., Rak, Y., 1994. The first skull and other new discoveries of *Australopithecus afarensis* at Hadar, Ethiopia. Nature 368, 449–451.
Klein, R.G., 1999. The Human Career: Human Biological and Cultural Origins, second ed. University of Chicago Press, Chicago.
Klein, R.G., Avery, G., Cruz-Uribe, K., Steele, T.E., 2007. The mammalian fauna associated with an archaic hominin skull cap and later *Acheulean artifacts* at Elandsfontain, Western Cape Province, South Africa. Journal of Human Evolution 52, 164–186.
Lee-Thorp, J.A., Sponheimer, M., van der Merwe, N.J., 2003. What do stable isotopes tell us about hominid dietary and ecological niches in the Pliocene? International Journal of Osteoarchaeology 13, 104–113.
MacKinnon, J.R., MacKinnon, K.S., 1980. Niche differentiation in a primate community. In: Chivers, D.J. (Ed.), Malayan Forest Primates. Plenum Press, New York, pp. 167–190.
Marks, J., 2002. What it Means to Be 98% Chimpanzee: Apes, People, and Their Genes. University of California Press, Berkeley.
Marlowe, F.W., 2005. Hunter-gatherers and human evolution. Evolutionary Anthropology 14, 54–67.
Martin, R.D., 1990. Primate Origins and Evolution: A Phylogenetic Reconstruction. Princeton University Press, Princeton.
Martin, R.D., Chivers, D.J., MacLarnon, A.M., Hladik, C.M., 1985. Gastrointestinal allometry in primates and other mammals. In: Jungars, W.L. (Ed.), Size and Scaling in Primate Biology. Plenum Press, New York, pp. 61–90.
McDougall, J., 2003. Meat in the human diets. The McDougall Newsletter. 2 (7). www.drmcdougall.com.
McGrew, W.C., 2010. In search of the last common ancestor: new findings on wild chimpanzees. Philosophical Transactions of the Royal Society B 365, 3267–3276.
Milton, K., Demment, M., 1989. Features of meat digestion by captive chimpanzees *(Pan Troglodytes)*. American Journal of Physical Anthropology 18, 45–52.
Potts, R., 1996. Evolution and climate variability. Science 273, 922–923.
Rak, Y., 1991. Lucy's pelvic anatomy: its role in bidpedal gait. Journal of Human Evolution 20, 283–290.
Reed, K.E., 2008. Paleoecological patterns at the Hadar hominin site, Afar Regional State, Ethiopia. Journal of Human Evolution 54, 743–768.
Schekleton, N.J., 1995. New data on the evolution of Pliocene climatic variability. In: Vrba, E.S., Denton, G.H., Partridge, T.C., Burckle, L.H. (Eds.), Paleoclimate and Evolution with Emphasis on Human Origins. Yale University Press, New Haven, pp. 242–248.
Semaw, S., 2000. The world's oldest stone artifacts from Gona, Ethiopia: their implication for understanding stone technology and pattern of human evolution between 2.6–1.5 million years ago. Journal of Archaeological Science 27, 1197–1214.
Sept, J., 2013. Plants and proto-people: paleobotanical reconstruction and modeling early hominin ecology. In: Sponheimer, M., Lee-Thorp, J.A., Reed, K.E., Unger, P. (Eds.), Early Hominin Paleoecology. Boulder: University Press of Colorado, pp. 355–396.
Stanford, C., 1999. The Hunting Ape: Meat Eating and the Origins of Human Behavior. Princeton University Press, Princeton.
Stewart, K.M., 2010. The case for exploitation of wetlands environments and foods by presapiens hominins. In: Cunnane, S.C., Stewart, K.M. (Eds.), Human Brain Evolution: The Influence of Freshwater Fish and Marine Resources. John Wiley & Sons, New York, pp. 137–171.
Sussman, R.W., 1999. The species-specific dietary patters in primates, and human dietary patterns. In: Sussman, R.W. (Ed.), The Biological Basis of Human Behavior: A Critical Review. Prentice Hall, New Jersey, pp. 143–157.
Susman, R.L., Stern Jr., J.T., Ungars, W.L., 1984. Arboreality and bipedality in the Hadar hominids. Folia Primatologica 43, 113–156.
Sussman, R.W., Tattersall, I., 1981. Behavior and ecology of *Macaca fascicularis* in Mauritius: a preliminary study. Primates 22, 192–205.
Sussman, R.W., Shaffer, C.W., Guidi, L.M., 2011. *Macaca fascicularis* in Mauritius: implications for macaque-human interactions and for future research on long-tailed macaques. In: Gumert, M.D., Fuentes, A., Jones-Engel, L. (Eds.), Monkeys on the Edge. Cambridge University Press, New York, pp. 207–235.

Tattersall, I., 2010. Macroevolutionary patterns, exaptation, and emergence in the evolution of the human brain and cognition. In: Cunnane, S.C., Stewart, K.M. (Eds.), Human Brain Evolution: The Influence of Freshwater Fish and Marine Resources. John Wiley & Sons, New York.

Teaford, M., Ungar, P., 2000. Diet and the evolution of the earliest human ancestors. Proceedings of the National Academy of Sciences 97, 13506–13511.

Theime, H., 1997. Lower Paleolithic hunting spears from Germany. Nature 385, 807–810.

Tuttle, R.H., 1985. Ape footprints and Laetoli impressions: a response to the SUNY claims. In: Tobias, P.V. (Ed.), Hominid Evolution: Past, Present and Future. Alan R. Liss, New York, pp. 129–130.

Ungar, P., 2004. Dental topography and diets of *Australopithecus afarensis* and early *Homo*. Journal of Human Evolution 46, 605–622.

White, T.D., 1980. Evolutionary implications of Pliocene hominid footprints. Science 208, 175–176.

WoldeGrabriel, G., Haile-Selassie, Y., Renne, P.R., Hart, W.K., Ambrose, S.H., Asfaw, B., Heiken, G., White, T., 2001. Geology and palaeontology of the Late Miocene Middle Awash valley, Afar rift, Ethiopia. Nature 412, 175–181.

Wolpoff, M., 1998. Paleoanthropology, second ed. McGraw Hill, New York.

Wrangham, R., Cheney, D., Seyfarth, R., Sarmiento, E., 2009. Shallow-water habitats as sources of fallback foods for hominins. American Journal of Physical Anthropology 140, 630–642.

Part III

Hominins

Chapter 7

Early Hominin Ecology

Jeanne Sept
Department of Anthropology, Indiana University, Bloomington, IN, USA

SYNOPSIS

Patterns of human evolution can best be understood in their paleoecological context. Understanding the ecology of the earliest hominins requires analyzing evidence for the paleobiology and cultures of the extinct species in the context of reconstructions of the environments in which they lived, seeking evidence of interactions between the hominins and their habitats to infer ecological relationships. Since the fossil record is incomplete and preservation is biased, knowledge of the variation and overlap in populations of different hominin species and the animal and plant communities in which they lived is very limited. Nonetheless, the fossil and archaeological record allows us to discern ecological differences between sympatric species of early hominin, and potentially to explain longer-term, macroevolutionary-scale patterns of speciation, lineage diversification, and extinction in relation to broad climatic and paleoenvironmental trends. While any causal effects of paleoenvironmental conditions on the origins of our lineage and the emergence of bipedalism remain unresolved, it does seem clear that early hominins were ecologically fairly generalized and able to survive in a wide range of habitats. Through time, as climate change caused habitats to become patchier and less predictable, hominins diversified, either becoming adapted to more specialized diets or developing more versatile, omnivorous diets founded on tool use.

Patterns of human evolution can best be understood in their paleoecological context. To understand the ecology of the earliest hominins, one must take what is known about the paleobiology and cultures of these extinct species, reconstruct the environments in which they lived, and infer ecological relationships by looking for evidence of interactions between the hominins and their habitats. These tasks are complicated by an incomplete fossil record and biased preservation, so our knowledge of the variation and overlap in populations of different hominin species and the animal and plant communities in which they lived is very limited, compared with what one can learn about living species. Nonetheless, the fossil and archaeological record allows us not only to discern ecological differences between sympatric species of early hominin, but also to "connect the dots" and try to explain longer-term, macroevolutionary-scale patterns of speciation, lineage diversification, and extinction in relation to broad climatic and paleoenvironmental trends.

THE EARLY HOMININ RECORD

Fossil and genetic evidence indicates that our human ancestors diverged from a common lineage with apes sometime between 5 and 7 million years ago (mya) in Africa (see chapters by Hunt and Ward in this volume). From that initial split, the new lineage continued to evolve and diversify. While researchers can differ in their classifications of individual fossil hominin taxa, there is no question that, overall, many of the big patterns of early hominin evolution developed through a series of adaptive radiations—nature's paleoecological experiments in survival and extinction.

Our own genus, *Homo* (see the chapter by Simpson), first emerged from a diverse array of African hominin taxa (Anton and Snodgrass, 2012). Early *Homo* was initially not that different from other early hominin species, but just after 2 mya *Homo* populations became the first hominins to colonize Eurasia—expanding out of Africa into an ecological range unprecedented in any other mammal. Other hominin taxa survived for a time in Africa, but by 1 mya *Homo erectus* was the only surviving hominin, the founder of our own lineage. With *H. erectus* we can infer that a novel balance had coevolved between biological adaptations and sociocultural solutions to ecological problems (Potts, 2012a), a behavioral transition that ultimately gave the human lineage the entrepreneurial skills to populate the planet.

To understand the role that environment may have played in shaping the evolution of early hominin biology and culture, it helps to review how large-scale patterns of geography and climate framed the environmental context of hominin evolution, before illustrating the challenges of reconstructing the palaeoecology of specific early hominin species.

MACRO PALEOENVIRONMENTAL CONTEXT

Global patterns of geographical and climatic change shaped the terrestrial environments in Africa in which early hominins evolved. Many researchers have tried to correlate macropatterns of climate change with major events in human evolution (deMenocal, 2004; Maslin et al., 2013; Potts, 2012b), but evaluating hypotheses of such large-scale cause–effect relationships can be difficult because our planet is a complex, dynamic system.

Paleogeography

The configuration and shape of the continents have shifted significantly through time in a process known as continental drift. While continental plates move slowly/imperceptibly within any individual's lifetime, the cumulative impact through time can be dramatic. For example, Africa is mostly one large continental plate. The African plate was a free-floating "island" for millions of years during primate evolution; consequently, populations of early apes evolved in Africa and were isolated on the continent until the middle Miocene, around 17 mya, when the African plate moved into the Eurasian plate. This tectonic process had several major effects on the environmental context of human evolution (Kingston, 2007; Maslin et al., 2013), two of which are highlighted here.

First, the Tethys Sea, which had been an open seaway to the north of Africa, pinched shut and created the closed Mediterranean Basin as well as land bridges between Africa and Eurasia, thus disrupting global circulation patterns in the oceans and atmosphere and causing global cooling and aridification. It was at this time that terrestrial mammals first began to mix between Africa and Eurasia, and apes first expanded out of Africa. Later, beginning approximately 6 mya, continental shifting in combination with lower sea levels actually cut the Mediterranean off from the Atlantic, closing the Straits of Gibraltar for over half a million years; this "Messinian Salinity Crisis" created an extensive dry, inland basin with fluctuating brackish lakes and salt flats, sequestering significant portions of global salts (and thus affecting the chemistry of oceans, and the climate, around the globe) until sea water inundated the area again (Maslin et al., 2013).

Second, the eastern edge of the African plate was subducted under the Eurasian and Arabian plates, creating a geologically unstable, tectonically active area. The resulting geological uplift created a complex chain of volcanic highlands along the eastern side of the continent, and associated faulting caused a system of deep valleys to form running north–south—the Great Rift Valley—that filled with runoff and sediments from the flanking highlands, creating a distinctive mosaic of volcanoes, lakes, and river valleys. During the early Miocene, apes were restricted to Africa and ranged through tropical lowlands across the continent (see the chapter by Martin). However, the Rift Valley development created a heterogeneous topography in the east and influenced climate patterns in a way that created an array of new habitats during the period when Late Miocene apes and early hominins were evolving (Elton, 2008). The Rift Valley also created a perfect set of conditions for the preservation of fossils, and many early hominin fossils have been recovered from Rift Valley sediments. However, early hominin fossils of comparable age are also abundant in cave deposits in South Africa, and as new fossil localities continue to be discovered in other regions of the continent, it seems clear that the biogeography of early hominin origins is still poorly understood (Bailey et al., 2011; Behrensmeyer and Reed, 2013).

Paleoclimate

Astronomical Parameters

Climate patterns are driven by global energy balance. Because earth's energy budget is based on solar radiation, the geometry of the earth's orbit drives climatic patterns (Maslin et al., 2013). The familiar cycle of the seasons, for example, is a consequence of the earth's tilt on its axis in relation to its position around the annual orbit, but the angle of tilt has varied cyclically through time. Indeed, the angle of tilt, in relation to and in combination with other cyclical variables such as orbital shape, has altered the intensity and distribution of solar radiation on earth's surface, and thus systematically changed climate patterns through time. Orbital forcing of climatic change was first formally developed into a theory by the mathematician Milutin Milankovitch early in the twentieth century, and has since been broadly supported and refined by climate research. For example, the "Ice Ages"—well-documented 100,000-year glacial–interglacial cycles of the last 700,000 years—are patterned primarily by the forcing effects of orbital parameters enhanced by regional surface conditions (Imbrie and Imbrie, 1979).

Geographical Parameters

While broad, global patterns of climate change can be predicted by a model of earth's general energy budget, such as Milankovitch's model, global climate is also affected by regional geographical patterns (Anderson et al., 2007; Bradley, 1999). It is hotter on the equator than on the poles because higher latitudes receive relatively less radiation per unit of surface area due to the curve of the earth's surface. Temperature differentials around the globe tend to equalize in the atmosphere, producing flows of hot air from equatorial regions toward the cooler poles. These airflows form convection currents in the atmosphere, and since the earth rotates on its axis, these currents are deflected longitudinally, creating westerly and easterly winds and seasonal weather patterns. Atmospheric circulation is also strongly influenced by the distribution of land cover, water, and ice. Water and land conduct heat differently, so large bodies of water have more stable temperatures than land surfaces, which heat and cool more quickly. Air temperatures are even more volatile. In addition, lighter-colored areas (like ice or open sand) reflect more energy back up into the atmosphere than darker areas like oceans or forest cover. This can create feedback effects. For example, during very cold periods in earth's history when ice sheets were expanding, increasing surface areas of ice reflected more solar energy back into space, decreasing surface temperatures and promoting a subsequent expansion of the ice sheets—a chilling feedback loop that influenced the longevity and stability of glacial periods (Imbrie et al., 1993).

Since the closing of the Tethys Sea during the Miocene, Africa has straddled the equator, stretching into temperate latitudes in both the northern and the southern hemispheres.

- *Equatorial Climate*: In the summers, moist equatorial air is drawn from the high pressure areas of the Atlantic to the low-pressure zone of the hot continent, resulting in high monsoon rainfall along the West African coast. As the storms move eastward across the continent, they gradually lose their moisture and rainfall levels gradually drop. Any remaining moisture is precipitated when the winds meet the western flanks of the Rift Valley Highlands—as air gains altitude it cools and dries—creating a rain shadow in the Rift Valley itself. Thus fossil sites preserved in the Rift Valley have been in a relatively dry, highly seasonal area since the highlands formed in the Miocene, a much drier climate than might be otherwise expected for a region of the humid tropics. It is also clear from growing evidence that even though local conditions in Rift Valley basins varied based on their geological configurations—broadly they were all strongly influenced by astronomical forcing patterns (Campisano and Feibel, 2007; Joordens et al., 2011; Kingston et al., 2007).
- *Midlatitude Climate*: Moving north or south from the equatorial regions in Africa, the climate becomes progressively drier, grading today into deserts in the midlatitudes: the Sahara in the northern hemisphere and the Kalahari in the south. However, in the past, climatic conditions fluctuated between low-to-moderate net moisture in these regions, creating the so-called "Green Sahara" for example (Drake et al., 2011; Larrasoana et al., 2003), filled with shallow lakes and large river channels; these geological features are now poorly exposed on the surface, but can be mapped by a combination of both traditional stratigraphic methods and remote sensing techniques. While very few early hominin fossils have been discovered in these regions, that is largely due to circumstances of fossil preservation and discovery, and does not mean that these

regions were not occupied by early hominins at different points in the past. In fact, one of the proposed earliest hominins, *Sahelanthropus* (see the chapter by Hunt), was found in the desert of modern Chad—an area that was relatively lush when *Sahelanthropus* lived there 6–7 mya (Lee-Thorp et al., 2012).

- *Temperate Climate*: The northern and southern tips of the continent extend into the temperate zones, with freezing winters and more precipitation. In the north, this Mediterranean zone includes a few archaeological sites associated with early *Homo*, but little evidence of earlier occupation (Sahnouni et al., 2010, 2011). To the south, fossil cave sites in the so-called "Cradle of Humankind" in South Africa are located in an interior highland plateau, a zone of temperate climate and varied, wooded hominin habitats (Reed, 1997).

Some researchers have suggested that the unique geographical conditions present in the Rift Valley or around the South African fossil sites created distinctive selection pressures that led to the origins of the hominin lineage in these specific regions (Coppens, 1994). However, while some of the best sedimentary circumstances for fossil hominin preservation occurred in semiarid regions, the extent to which early hominins preferred such habitats is an open question.

Proxy Evidence of Climate Change

As illustrated in Figure 1, the global climate changed significantly during the time period in which early hominins evolved, shifting from relatively warm and humid conditions with moderate fluctuations to a prolonged period when cold, dry conditions recurred with greater frequency and amplitude, culminating in the "Ice Age" cycles of the last 700,000 years (Bonnefille, 2010; deMenocal, 2004; Maslin et al., 2013). Evidence for these climate trends comes from a number of sources (Bradley, 1999).

- *Sediment Cores*: Samples of deep-sea cores include preserved long continuous samples of sediments, some dating back millions of years, that can contain a variety of proxy indicators of climate, including for example fossils of ocean microorganisms such as foraminifera, whose chemical composition (e.g., carbon and oxygen isotopes) and morphology can provide evidence of salinity levels and water temperatures at the time they lived;

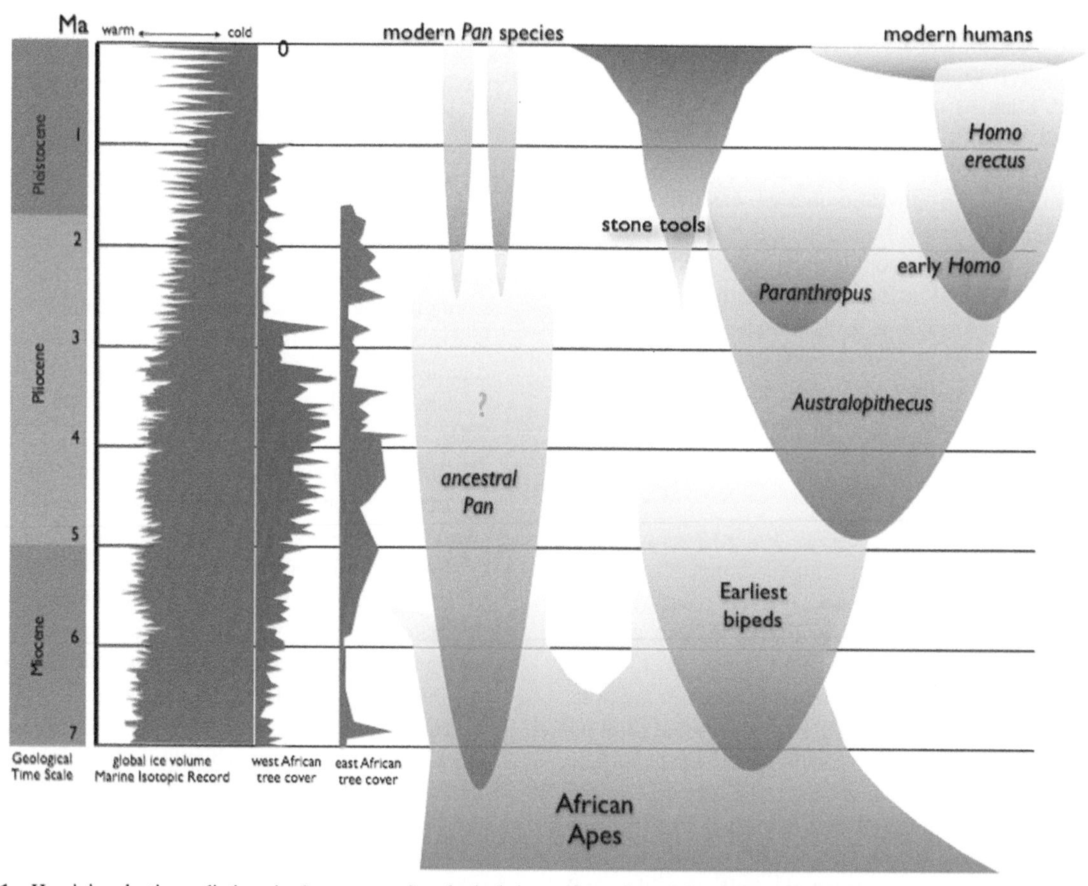

FIGURE 1 Hominin adaptive radiations in the context of geological timescale, and records of climate change generalized from marine isotope and paleobotanical evidence, after Bonnefille (2010), Kingston (2007), and Potts (2013). (See color plate section).

plant microfossils (e.g., pollen, spores, and phytoliths) that can provide evidence of the vegetation growing on adjacent continents; windblown sediments that can indicate the strength and direction of ancient winds, and the relative amount of vegetation cover on adjacent continents. Sediment cores from lakes can also provide analogous sequences of freshwater organisms, plant microfossils, terrestrial sediments, and organic compounds, but these lacustrine cores are generally shorter and more discontinuous than deep-sea sediment cores, and more likely to reflect regional rather than global climatic conditions.

- *Ice Cores*: Cores taken from undisturbed polar ice caps include preserved layers of ice that formed at different times, sometimes containing ancient air bubbles. When analyzed chemically, these samples provide evidence of levels of different gases in the paleoatmosphere when the ice formed as well as the isotopic composition of the ice itself, which is another proxy for ocean salinity and climate.

By dating and correlating such evidence of climatic fluctuations sampled from around the globe, paleoclimatologists have been able to construct a master yardstick of global climate change.

Several hypotheses have been proposed that link the impact of these large-scale climatic fluctuations in Africa to major phylogenetic events in hominin evolution (Bonnefille, 2010; deMenocal, 2004; Foley, 1994; Kingston, 2007; Potts, 2012b, 2013). Broadly, researchers have argued that:

- The Late Miocene climatic trends reduced forest distributions in Africa, favoring terrestrial adaptations among apes and the emergence of bipedal hominins (Cerling et al., 2011; Foley, 1994; Isbell and Young, 1996; Lovejoy, 1981; Rodman and McHenry, 2005);
- Increasing aridity, seasonality, instability, and patchiness of habitats in the Late Pliocene drove many species extinct, and favored dietary niche diversification and speciation among many taxa including early hominins (Lee-Thorp et al., 2011; Potts, 2013; Ungar and Sponheimer, 2011; Vrba, 1985, 1988);
- Increasingly frequent and intense climatic fluctuations attuned to orbital precession timescales (~20,000 year cycles) during the Pleistocene favored early hominins who had the behavioral flexibility and skills to cope with ecological variability, rather than hominins with specialized biological adaptations (Maslin and Trauth, 2009; Potts, 1998, 2007, 2012a; Potts and Teague, 2010; Ungar et al., 2006).

To evaluate these hypotheses and assess the impact of climatic shifts on fossil hominin taxa, researchers must first assess the effects that changing climate had on local vegetation and terrestrial communities.

RECONSTRUCTING TERRESTRIAL HABITATS

The settings in which hominin species lived can be reconstructed from a number of different kinds of evidence as proxies for various environmental conditions.

GEOLOGICAL EVIDENCE

Understanding the geological context of a fossil find or archaeological site is the necessary first step in being able to link an ancient species to its paleoecological context. Geological evidence provides vital clues to where and how a hominin species died, and the parts of ancient landscapes where different activities took place. Studies of the structure and distribution of sediments at a site can identify the local depositional circumstances that preserved fossils or artifacts. For example, microsedimentological work on a concentration of bones of a number of *Australopithecus afarensis* (see the chapter by Ward) individuals discovered at Hadar, Ethiopia, dubbed the "First Family," has revealed no evidence that these individuals all died in a catastrophic event like flood (Behrensmeyer, 2008); the bones were buried by gentle, seasonal floods in a shallow channel crossing an open, grassy plain. Early hominin fossils and archaeological sites in Africa have been preserved in sedimentary environments that range from extensive deposits formed along lake margins, river floodplains, swamps, and open plains, to localized pockets of accumulation in caves and rock shelters (Bailey et al., 2011). It is rare for fossils or early archaeological sites to be preserved in uplands or dry plateaus, where high levels of erosion and weathering create poor circumstances for burial and fossil preservation. One of the few such early hominin sites is Laetoli, a dry, open plain near volcanic highlands in Tanzania, where successive eruptions of volcanic ash blanketed the landscape and buried and preserved footprints and other fossils (Harrison, 2011).

Site Taphonomy

The science of analyzing the fossilization process is called "taphonomy" (Behrensmeyer, 1984). In particular, for paleoecological interpretation it is important to determine whether evidence was buried close to where a hominin lived or died (in a primary context) or transported before burial or disturbed and reburied (in a secondary context). For example, the location of a near-complete *H. erectus* (see the chapter by Simpson) skeleton like the "Turkana Boy" (discovered at the Nariokotome site near Lake Turkana, Kenya) can be examined to learn the circumstances of death; as interpreted by the research team of Alan Walker, over 1.5 mya this adolescent boy died in the seasonally marshy floodplain of a large river, and his body then floated and decomposed gently in the wetlands, where it became naturally buried in fine-grained sediments (Walker and Leakey, 1993). This type of primary geological context at Nariokotome contrasts markedly with nearby examples of hominin fossils in the secondary context. In the paleochannel deposits of the same large river, a number of individual fossil teeth of contemporary robust australopithecines have been recovered; these isolated fossil teeth had been transported like pebbles downstream in the swiftly moving current. But unlike the Turkana Boy, who had died in the flat floodplains near the river, these australopithecines had died somewhere upstream, perhaps as much as 50 km away in a very different habitat, and their skeletons had become broken up and disarticulated with only the most durable remnants coming to rest in sandy gravel bars downstream (Behrensmeyer and Reed, 2013).

Information about site formation processes can also be gleaned from the fossils and artifacts themselves. For example, the first australopithecine fossil discovered, the "Taung Child," was recognized almost a century ago in South Africa by Raymond Dart. Dart chiseled the fossil out of an encasing block of breccia—calcium carbonate deposits that had cemented the little skull in cave sediments millions of years ago. Dart thought that the child had probably lived and died in the cave in which it had become buried, and assumed that the australopithecine's environment was much like the open grasslands that surround the site today. Later, examining the bones of more australopithecines and other savanna animals found in other South African caves, Dart even suggested that early hominins could only have survived in such demanding environments if they had been both tool users and meat eaters; he argued that the broken bones of other large animals found in the South African caves had been the prey of the australopithecines. However, more recent studies of bone surface damage patterns to the Taung Child (Berger and McGraw, 2007) and dental microwear of other primate fossils from the site (L'Engle Williams and Patterson, 2010) lead to different paleoecological conclusions; skull damage suggests that the child had been captured and eaten by a large predatory bird, and its remains dribbled onto the ground, where they ultimately got washed into an underground cavern. In subsequent taphonomic studies of other South African cave deposits (Brain, 1981, 1993), C.K. Brain has demonstrated that most of the bones in these caves, including those of the australopithecines, were the remains of killings by carnivores such as leopards and hyenas—these hominins were the most likely prey rather than predators, raising interesting questions about the adaptations and ecology of different australopithecine species.

Paleolandscapes

To provide a regional context for the ecology of early hominins, local sedimentary sequences can sometimes be correlated to reconstruct areas of paleolandscape. For example, if a paleolake was large and its ancient mud layers are exposed along extensive geological outcrops, they can be traced laterally to identify the location of a lake basin and its margins.

- *Tephrochronology*: Volcanic sediments can be valuable for reconstructing paleolandscapes. Lava flows can form extensive deposits, and eruptions can blanket a region with volcanic ash that becomes consolidated as a "tuff." Because each volcanic eruption is a unique event, individual tuffs can be identified chemically and morphologically and traced across sedimentary outcrops. They can also be dated, using radiometric techniques, providing another valuable tool for comparing different geological samples. Some widespread tuffs have even been identified in ocean cores, allowing precise correlations between local sequences and the global sequence of climate change. These methodologies were particularly critical for correlating geographically separated outcrops in the ancient Turkana Basin in northern Kenya, where a number of important fossil hominin specimens and archaeological sites have been found (McDougall and Brown, 2006).
- *Paleosols and Geochemistry*: When a stable land surface is not subject to erosion it can weather, allowing the development of soil horizons. Even after burial by new sediments, some ancient soils retain their integrity and can be traced laterally across a paleolandscape as "paleosols." Such paleosols can preserve chemical evidence of the climate and land cover at the time they formed. For instance, pedogenic carbonates form in soils in drier climates, and the oxygen and carbon isotope composition of these carbonates can provide evidence of paleoenvironmental conditions (Quade and Levin, 2013). Oxygen isotopes are an indicator of local climate, tracking the balance of humidity, temperature, and evaporation. Carbon isotopes can indicate the type of plant cover that grew on the soil—specifically the abundance of tropical grasses and some other plants that have a specialized C_4 photosynthetic pathway adapted to hot/dry conditions. (These are ecologically distinct from trees, shrubs, and most herbaceous plants that have a more generalized C_3 photosynthesis.) In East Africa, carbon isotope signatures in paleosols reveal that the C_4 grasses that dominate the African savannas today began to displace C_3 species in existing grasslands 8–5 mya (Late Miocene), associated with a climatic shift to drier and more seasonal conditions in the region (Cerling et al., 2011).

For early hominin sites, this type of paleolandscape work was pioneered successfully by the geologist Richard Hay, who worked with Mary and Louis Leakey at the famous site of Olduvai Gorge in Tanzania (Hay, 1976). Correlating sedimentary layers across 25 km^2, Hay documented that several species of early hominin discovered at different sites at Olduvai had shared the environments along the margins of the fluctuating shoreline of an ancient, shallow lake for hundreds of thousands of years. His research laid the foundations for very detailed "landscape archaeology" studies, which have subsequently confirmed and refined his paleoecological hypotheses. For example, the carbon isotopes in Olduvai paleosols show that the soils near the shallow, fluctuating lakes frequented by early hominins were fairly open, covered with sparse grasslands or wetlands, while low ridges and hills nearby were covered by more hospitable but discontinuous patches of woodland—a savanna mosaic. Several of the most famous archaeological sites at Olduvai accumulated on these wooded spots; early hominin toolmakers repeatedly carried tools (see the chapter by Toth and Schick) and food resources up to these shady refuges overlooking the more exposed and dangerous lake margins (Blumenschine et al., 2012). It has also become evident that local conditions at Olduvai fluctuated dramatically (Magill et al., 2013).

Fossils

While geological evidence can provide indicators of ancient terrain, more direct and detailed evidence of the extinct plant and animal communities that lived on those landscapes comes from fossilized botanical and faunal remains. Not only can fossils provide evidence of the diversity of species that lived in a locale or region, but also they can help establish the kinds of resource competition and predation risks that early hominins would have faced in those communities (Reed et al., 2013b).

Plants

Vegetation was a core component of the early hominin landscape, whether providing hominins with staple plant foods, shade, and arboreal refuge, or creating habitats and hiding places for their predators. Trying to reconstruct and understand the paleoecological relationships between hominin species and the plant communities in which they lived is a challenge because direct fossil evidence of ancient plants and vegetation patterns is relatively rare, compared with the frequency of vertebrate or invertebrate fossil records, so many of our reconstructions of ancient vegetation patterns are based on indirect analogies to modern contexts (Sept, 2007, 2013).

The intersection of topographic and climatic patterns across Africa produces distinctive regional vegetation belts (Figure 2) (White, 1983). The vegetation covering the

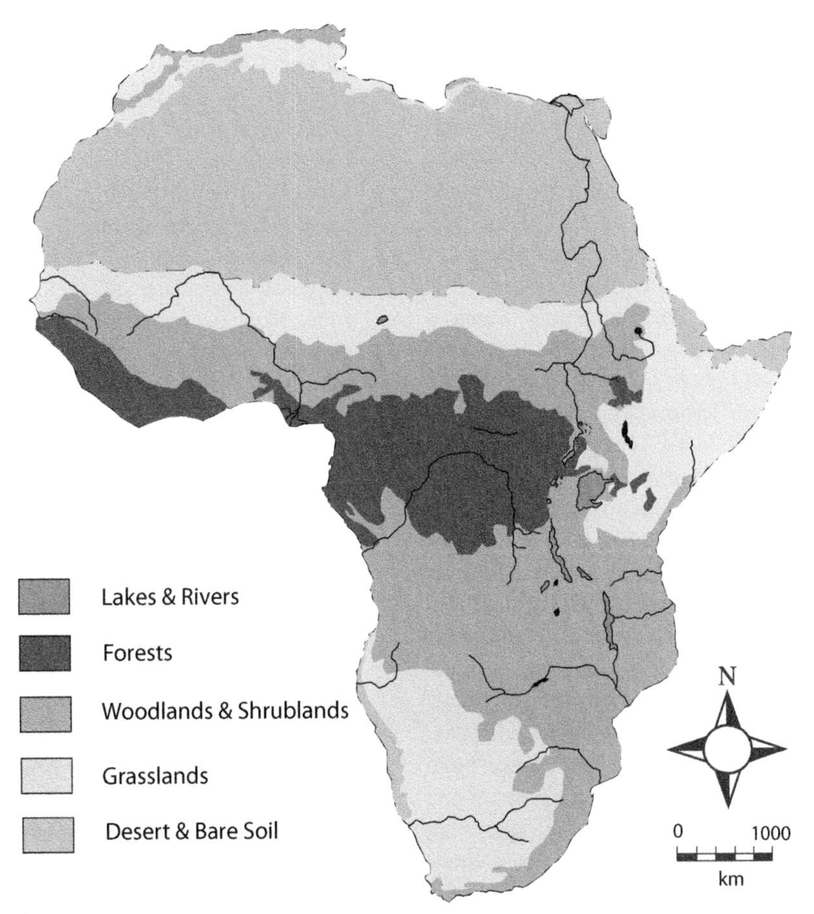

FIGURE 2 African vegetation map, after White (1983). (See color plate section).

low-altitude gentle topography across most of the continent is largely determined by rainfall patterns. Humid equatorial forests are flanked in both the north and the south by belts of drier, deciduous woodlands fringed by a zone of savanna grasslands, then transitioning into arid scrub and deserts. In the east, the highland rain-shadow effect creates a zone of savanna grasslands and open woodlands in the Rift Valley (Lind and Morrison, 1974). The temperate zones on northern and southern edges of the continent have shrubby Mediterranean vegetation. And poking up out of the lowland grasslands, the mountain ranges in the north, east, and south are all characterized by strong vertical zonation in montane vegetation types, particularly in the Rift highlands. These altitudinal vegetation zones range from moist evergreen and deciduous forests on the flanks of the mountains, through distinctive belts of montane forest including bamboo to moorland above the tree line on the highest peaks. These distinctive montane habitats stand out as isolated biogeographical zones, like "islands" in a "sea" of more open lowland habitats today, creating discontinuous habitats for many mammals, notably a number of primates including the endangered mountain gorilla.

What impact did fluctuating climate have on African vegetation patterns? To a great extent, it appears that belts of forest, woodland, grassland, and desert expanded and contracted in response to climate change, although regional patterns were complex (Bonnefille, 2010). During drier times in the past, the forested zones were smaller and patchier while open grasslands and scrub expanded. During more humid times both montane and lowland tropical forests were more extensive, grasslands more wooded, and open deserts significantly reduced.

In addition to the soil chemistry studies described earlier, evidence for these vegetation changes is preserved as both microbotanical and macrobotanical remains (Bonnefille, 2010).

- *Microbotanical Fossils*: Microscopic pollen grains and spores are the most ubiquitous types of plant fossil because they are produced in abundance by most plant species and their exines are extremely durable in a wide variety of sedimentary conditions. Their morphology can commonly be identified at the family level and sometimes even at the generic or specific level of plants, so they offer wide potential for reconstructing the relative abundance and distribution of plant communities in the past. The interpretation of fossil pollen, however, is complicated by differences in plant ecology: wind-pollinated plants such as grasses and evergreen trees make large quantities of pollen that blow over very large areas, while insect-pollinated plants make relatively little pollen that is distributed only in local areas. Thus pollen from wind-pollinated plants can be very abundant in a sediment sample even if the plants were not growing immediately in the area. Typically, pollen samples recovered from early hominin sites have been dominated by grass pollen, but that does not necessarily mean that the hominins lived in grassland—pollen provides a consistent signature of regional vegetation patterns, not local conditions (Sept, 2007, 2013). Still, it is clear from the pollen record in marine cores that a major grassland expansion took place in East Africa 10 mya and in West Africa 7 mya, predating the divergence date for early hominins (Elton, 2008). Other very useful microscopic plant fossils are called phytoliths: biogenic minerals that form within plant tissues and can be preserved in sediments like tiny endocasts. They are rarely taxonomically distinctive, but can be good indicators of alternative structures in plant tissue. They can be used to identify woody plants and monocots, for example, and thus provide fossil evidence of woody vegetation cover versus more open cover. Taxonomically, they can sometimes be used to distinguish the presence of various monocot families and subfamilies in fossil assemblages such as types of grasses, sedges, and palms. While they can become incorporated in windblown dust, most phytoliths accumulate in the location where the plants grew. At several sites at Olduvai, for example, phytolith samples offer a finer-grained spatial and temporal resolution to vegetation reconstruction that complements the regional record of the pollen assemblages (Barboni et al., 2010); phytoliths document the presence of palms, and fluctuating communities of sedges, various grasses, and bushlands along the eastern margins of the large, shallow, saline–alkaline paleolake, as well as dense patches of groundwater palm forest close to freshwater springs.
- *Macrobotanical Fossils*: Plant macrofossils such as impressions or casts of wood, fruits, seeds, leaves, or roots have been occasionally preserved, but are less common in African sediments than either pollen or phytoliths. Macrobotanical remains provide a valuable glimpse of the plant communities that grew near ancient sites, complementing the microbotanical record (Sept, 2013). An example comes from the Laetoli Beds in Tanzania, a site where fossils and footprints of *A. afarensis* were found in tuffaceous sediments. While pollen and soil analyses had suggested that this site was a semi-arid inland plain covered by sparsely wooded grassland, preliminary identifications of silicified wood specimens include a liana typical of Central African gallery forests today, and other plants typical of deciduous and riverine woodlands (Bamford, 2011a, b). Although the footprints preserved at the site document that the australopithecines walked across the open plains, clearly there were also patches of forest and woodland in the area, and it is unclear from the site context to what extent the australopithecines were dependent upon resources in different regional microhabitats.

Animals

The fossil record preserves samples of the animal communities in which early hominins lived. In addition to providing evidence for taphonomic analysis as discussed above, fossil animal remains can provide many types of evidence useful for reconstructing hominin paleoecology (Reed et al., 2013b). If the fossil assemblage is collected from the surface of an entire bed or stratigraphic horizon, it represents a depositional time span of hundreds or even thousands of years. Such assemblages can give only a time-averaged view of the range of animal species that lived in the general environment. On the other hand, if a fossil assemblage has been excavated from a tightly defined stratigraphic context, it can be a source of more refined paleohabitat reconstructions. For example, in the South African australopithecine cave sites, taphonomic analyses can provide evidence of whether carnivores likely accumulated the bone assemblages, thus providing evidence of the likely feeding preferences and paleohabitat range of those extinct carnivores, or whether the animals accidentally died in the cave, therefore providing a different type of paleoecological signal (Pickering, 2004).

- *Taxonomic Representation*: When recovered in the primary context, a collection of fossil animal remains can be analyzed to reconstruct the structure and composition of paleoenvironments. Such analyses often compare the taxonomic representation and diversity in a fossil assemblage with the abundance of their closest living relatives in analogous habitats today (Reed, 1997). An influential study of this type was done by Elisabeth Vrba, who noted that in the antelope family, two tribes (the Alcelaphini and the Antilopini) dominate open plains and grassland habitats in Africa today. She argued, therefore, that the South African australopithecine cave assemblages that had high proportions of those antelopes were also likely to have formed in open habitats (Vrba, 1985, 1988). While this type of study is useful as a starting point for paleoenvironmental research, it can be biased and misleading if the taxonomic groups being compared have evolved through time, and many of them have. Similarly, because of the heterogeneous nature of many paleohabitats, coupled with taphonomic biases in the record, documenting the first and last appearance of different taxa in the fossil record of a region can be problematic and perhaps more indicative of sampling bias than macroevolutionary patterns (Bobe and Leakey, 2009). On the other hand, multivariate changes in the diversity of assemblages within a region can indicate broad patterns of extinction, speciation, relative abundance, and species turnover—key biogeographical evidence to compare with other indicators of paleoenvironmental change (Vrba, 1995a, b). The key is to carefully contextualize multiple lines of evidence and search for consistent trends in carefully sampled regional fauna over time.
- *Ecomorphology and Community Structure*: To avoid the potential biases inherent in taxonomic comparisons, a variety of "taxon-free" approaches have been developed that focus on using morphological traits of individual species to reconstruct their likely ecological adaptations (Reed et al., 2013b)—distinguishing grazing antelope from browsing antelope or mixed feeders, for example, on the basis of a suite of craniomandibular traits. Such analyses are time-consuming, but contribute valuable details to the analysis of fossil assemblage composition. Other approaches analyze the overall locomotor habits or dietary adaptations of a range of different species in an assemblage, and use these as indices to reconstruct the likely vegetation structure of a site (e.g., arboreal frugivores are common in forests, but become infrequent when trees are more widely spaced in open woodlands and savannas). For example, at the site of Aramis in Ethiopia, fossils of the early hominin *Ardipithecus ramidus* (see chapter by Hunt) were associated with the bones of a number of large and small animals including monkeys, closed-habitat antelope, as well as a variety of small mammals, birds, and invertebrates (White et al., 2009a; WoldeGabriel et al., 2009). The assemblages that include early hominins are considered to be both taxonomically and morphologically characteristic of woodland zones, including a number of frugivorous species and herbivores that ate a mix of leaves rather than grazing on grasses. It is particularly interesting at the Aramis site that more open paleohabitats have also been documented, but early hominin fossils have not yet been found in those localities.
- *Biogeochemistry*: You are what you eat. Just as the isotopic composition of soils provides evidence of the climate and vegetation cover at a site, so the isotopic composition of fossil teeth reflects the chemistry of the foods the animal ate, with oxygen isotopes potentially indicating sources of drinking water (or moisture from plant foods) and carbon isotopes indicating the relative contributions of different foods derived from C_3 and C_4 plants. At the Aramis site, for example, fossil giraffes have chemical signatures that suggest that they were water-independent and browsed on the leaves of trees and shrubs, while the chemistry of fossil hippo teeth fit the water-dependent grazing lifestyle typical of hippos today; overall, the isotopic composition of the Aramis mammals was consistent with the habitat range suggested by their ecomorphology and diversity (White et al., 2009a). The biogeochemistry of the fauna at a site can also be used comparatively for interpreting the chemical signatures and diet of early hominins themselves. For comparative purposes, however, because the geochemical contexts of sites vary, calibrating the spectrum of chemical signatures between taxa within a single fossil assemblage is necessary before trying to compare values for early hominin fossils across different sites (Lee-Thorp et al., 2011; Ungar and Sponheimer, 2011).

INFERRING HOMININ HABITATS AND ADAPTATIONS

To evaluate the paleoecology of a particular early hominin species, it is important to focus on habitat parameters that would have been particularly important to that species, such as the terrestrial substrate for locomotion; access to food, water, and shelter; and risks from competing species and predators. Did fossil species have preferred habitats, or what might have limited their ability to survive in different habitats? The paleobiological challenge is to situate early hominin taxa in particular paleohabitats, not just in generalized paleoenvironmental settings (Behrensmeyer and Reed, 2013). If different types of paleoenvironmental proxy data can be recovered from a carefully sampled, primary context site, these can be compared and integrated to develop hypotheses about local habitat.

The Habitats of the Earliest Bipeds

We know that bipedalism, the founding adaptation of the hominin lineage, first emerged during the Late Miocene (see chapter by Hunt). Globally, the Late Miocene was characterized by cooler climates than in previous periods, and just before 6 mya very arid conditions caused a significant drop in tree cover across the African tropics, as evidenced in pollen from deep-sea cores off both the western and the eastern coasts. The Messinian Salinity Crisis from 6 to 5.3 mya probably compounded these climatic effects in Africa through a complex feedback loop (Imbrie et al., 1993), both through the effects of increased local albedo and through salt sequestration in the basin that impacted ocean chemistry and affected the strength of circulation patterns.

These environmental shifts would have shaped selection pressures for Late Miocene ape populations: splintered wooded habitats decreased the abundance and increased the patchiness of fruits, sleeping trees, and other forest resources; amplified seasonality increased the unpredictability of plant foods and drinking water; and trophic level shifts included diversified large terrestrial carnivores adapted to preying on greater numbers and varieties of open-country antelopes. Indeed, it seems that Eurasian and African apes either evolved ecological specializations or went extinct during this period when the hominin lineage split from that of chimpanzees. Thus many scholars have wondered if hominin bipedalism was also an adaptation to the spreading savannas (see chapter by Hunt for complete discussion). Certainly an upright posture and efficient stride could have been advantageous to early hominins in open habitats, whether traveling between patchy resources or avoiding predation risks. However, such basic adaptive scenarios are oversimplifications. As new paleoenvironmental evidence is synthesized, it has become clear that climatic fluctuations created complex patterns of vegetation change across Africa, both continentally and regionally (Bonnefille, 2010; Cerling et al., 2011; Kingston, 2007). As a result, such a generalized "savanna hypothesis" is not easy to test.

Savannas or Woodlands?

The earliest suggested hominin is currently *Sahelanthropus tchadensis* (see the chapter by Hunt), recovered from Chad in a region currently covered by sand dunes, but the site formed during a humid phase 7 mya that saw significant forest development in West Africa, and the locality was situated close to a lake probably surrounded by groundwater forest (Lebatard et al., 2008; Vignaud et al., 2002) but also close to sandy desert. After 7 mya, a dramatic increase in aridity significantly reduced the extent of tropical forest across Africa, and it is clear from the fossil faunal, floral, and isotopic records between 7 and 6 mya in the Tugen Hills that at least parts of the Rift Valley continued to support patches of lowland rainforest as well as open habitats (Kingston, 2002). Two suggested early bipeds that lived around 6 mya, *Orrorin tugenensis* in Kenya (Bamford et al., 2013) and *Ardipithecus kadabba* in Ethiopia (Hailie-Selassie and WoldeGabriel, 2009), were recovered with fossils that suggest they inhabited humid wooded locales. While these wooded early biped localities were likely riparian death sites rather than living sites, they still offer no support for the savanna biped hypothesis.

Of the earliest hominin taxa, *A. ramidus* is associated with the most extensive paleoecological evidence, but the interpretation of these data remains controversial. The Aramis site, at 4.4 mya, falls within a period between 4.5 and 4.0 mya when pollen evidence documents an expansion of forested habitats in both western and eastern Africa (Bonnefille, 2010), potentially allowing arboreally adapted animal communities to spread. The Aramis research team has integrated evidence from a variety of sources including topography and paleosol geochemistry, paleobotanical remains, and a wide range of carefully contextualized animal fossils, and concluded that these early hominins lived and died in habitats that ranged from patches of closed forest to woodlands, but did not exploit more open grassy habitats that were also present in the region (Louchart et al., 2009; White et al., 2009b; WoldeGabriel et al., 2009). They support this interpretation with paleobiological evidence from the hominin fossils themselves, such as the apparent climbing ability of these hominins, and isotopic values suggesting a diet dependent on woodland foods close to the frugivorous diets of living chimpanzees. Thus the Aramis researchers rejected the savanna hypothesis as an explanation for the origins of bipedalism. However, other scientists have argued that because of sampling biases and taphonomic issues, the same data (particularly paleosol isotopic data) are consistent with an interpretation that the hominins lived in a wooded savanna mosaic habitat ranging from riparian forests to open wooded grasslands (Cerling et al., 2010). Indeed, even chimpanzees that live in semiarid savanna mosaic environments today still obtain most of their foods from C_3 plants—they must simply range more widely to get them than forest chimps do (Sponheimer et al., 2006a). The small cheek teeth, thin enamel, and gracile masticatory cranial architecture of *A. ramidus* are consistent with isotopic values of the tooth enamel that suggest a diet of fruits and other C_3 plant resources from trees and shrubs (Suwa et al., 2009a, b; White et al., 2009a). Considering the entirety of the evidence (White et al., 2010), it seems reasonable to conclude that these early hominins probably preferred foraging in wooded habitats for food, shade, and predator avoidance, whatever the density of trees in grasslands they may have sometimes crossed.

Early Australopithecines in Varied and Variable Habitats

By 4 mya early australopithecines had become established in different parts of the continent, including *Australopithecus africanus* in South Africa and *Australopithecus anamensis* followed by *A. afarensis* in Rift Valley sites, and these lineages persisted during a million years of fluctuating environments (Behrensmeyer and Reed, 2013; Reed et al., 2013a). This early Pliocene period was characterized by climatic amelioration, and both paleobotanical evidence and isotopic data from paleosols complement fossil vertebrate and invertebrate studies to suggest several of the tectonically controlled basins in the Rift Valley were filling up with sediments, supporting a mosaic of well-watered wooded and wooded grassland habitats including extensive riverine forest.

Australopithecus anamensis is an interesting species that lived roughly 4 mya in such a Rift Valley basin, with an extensive woodland corridor along a major river flanked by more open environments. However, because the site evidence accumulated across significant periods of time, it is difficult to determine from the site context—evidence from site geology or associated fossil assemblages—whether this hominin preferred the woodlands, the grasslands, or frequented both (Behrensmeyer and Reed, 2013; Walker and Leakey, 1993). In contrast to earlier hominins, *A. anamensis* was a dedicated biped and had larger cheek teeth with thicker enamel, and was thus capable of processing a more heavily chewed diet. However, it lacked the robust face and jaws adapted to eating hard or tough foods; the wear patterns on its teeth suggest a diet of fairly soft foods (Ungar et al., 2010); and the tooth isotopes of this species suggest it was heavily dependent on a diet of C_3 plant resources. So perhaps as it foraged within riparian forests and between patches of trees and shrubs in adjacent grasslands, *A. anamensis* had evolved teeth capable of chewing tough foods, but these adaptations were only needed during difficult seasons when soft preferred foods were not available and hominins had to resort to tougher "fallback foods" to survive.

By 3.5–3.0 mya early australopithecines had begun to exhibit both masticatory adaptations (e.g., craniofacial buttressing and thicker, larger, and more heavily worn molar crowns), and isotopic signatures and wear patterns on their tooth enamel that suggest their diets had begun to incorporate more tough or abrasive foods (such as legume pods, seeds, or corms) and foods derived from C_4 plants (such as grasses, sedges, or the flesh of grass-eating animals) (Ungar and Sponheimer, 2011). An interesting example is *Australopithecus bahrelghazali*, recovered from this period in Chad. Although the sample size is small, the *A. bahrelghazali* fossil teeth all have an isotopic signature suggesting a heavily C_4 diet, and were recovered in association with aquatic fauna and grazing animals that suggest local environments were a patchwork of fairly open and moist grasslands, marshes, and streams near the margins of a large, shallow lake (Lee-Thorp et al., 2012). Thus it is likely that these hominins had foraged heavily on the shoots and shallow roots of tropical grasses and C_4 marsh plants such as sedges—a diet very different from that of any earlier hominin (Lee-Thorp et al., 2011). Several Rift Valley australopithecine sites from this period were also situated in quite open environments. For example, *A. afarensis* fossils were preserved in a wide variety of situations, ranging from lake and river margin habitats that seemed to be fluctuating ecotones between humid grassy floodplains and fringes of gallery forest along water margins (Bedaso et al., 2013), to dry plains like the Laetoli *A. afarensis* site (Andrews and Bamford, 2008; Aronson et al., 2008; Campisano and Feibel, 2007; Kingston and Harrison, 2007). However, just because the fossils were preserved in these situations does not mean that these hominin species preferred to live in such places. Because of the biases inherent in the sedimentary record of the Rift Valley, almost any locality where fossils were preserved would be close to a lake or stream and within walking distance of the margins of the sedimentary basin, possibly even highlands.

A bit later in time, South African australopithecine sites also included a mix of different local habitats in their temperate climate zone, ranging from open areas to densely forested patches in some valleys—habitats that were spatially patchy as well as shifting through time. The isotopic signatures of the South African australopithecines show a high degree of dietary variability among individuals. Thus *A. africanus* could have been a generalist, either feeding in diverse habitats in similar ways, or using similar habitats in diverse ways. Bone damage patterns suggest that they faced significant risk of predation from both open-country and closed-habitat carnivores. Overall, it is impossible to assign these early australopithecines to any one preferred habitat; they seem to have had the generalized ecological flexibility to cope with varying conditions in a range of geographical settings and diverse plant communities (Behrensmeyer and Reed, 2013; Potts, 2013).

LATE PLIOCENE ADAPTIVE RADIATIONS

By the Late Pliocene, hominin taxa had diversified. While there is still no consensus about the precise patterns of the phylogenetic affiliations of these different taxa, it is clear that individual species had evolved divergent ecological niches, suggesting adaptive radiations linked to diet. These species lived in South Africa and East Africa during a period of global climatic deterioration after 2.5 mya, when African woodlands were shrinking and fossil species adapted to exploiting dry, open grasslands (such as antelope grazing on C_4 grasses) were becoming extremely common.

Later "Robust" Hominins: Chewing On a Problem

A fascinating story has emerged in recent years, comparing two species of hominin that share many adaptations to a heavily chewed diet including very robust skulls, large chewing muscles, and very large, heavily worn cheek teeth. Many researchers assign these species to the genus *Paranthropus*, suggesting that they represent one robust lineage that spread and diversified between southern (*Paranthropus robustus*) and eastern (*Paranthropus boisei*) Africa (both Hunt and Ward in the present volume retain these robust specimens within the genus *Australopithecus*). Other researchers would argue, however, that the robust masticatory adaptations of these two taxa developed independently: parallel East African and South African lineages responding to Plio-Pleistocene environmental changes. In either case, researchers assumed for many years that the diets of these "robust" taxa were similar, and debated whether they lived in open or closed habitats around the continent (Wood and Constantino, 2007). As discussed earlier, a major difficulty in determining the preferred habitat of such species is that their fossils are rarely preserved in primary context living sites, and more often found in locations where their remains were transported after death, whether by carnivores or by river waters (Behrensmeyer and Reed, 2013). Measuring the isotopic signatures and dental microwear patterns of these fossil taxa has made this taphonomic issue less problematic, however, because both types of dental evidence document aspects of the diets of these individuals while they were alive. These behavioral signatures provide a valuable comparison with interpretations of the masticatory morphology of the specimens, in particular since the anatomical adaptations a hominin was born with may both have been affected by developmental processes and also provide evidence of capability, while not necessarily indicating a biological role (Daegling et al., 2013).

Despite their anatomical similarities, the robust taxa were apparently adapted to different diets. In South Africa, *P. (Au.) robustus* had tooth wear patterns and isotopic signatures that suggested it commonly ate a wide range of fairly soft C_3 foods, as had earlier australopithecines in the region, but supplemented these seasonally with significant amounts of small, hard, or tough foods such as C_4 grass seeds, roots, nuts, or even grass-eating termites or small vertebrates (Lee-Thorp et al., 2011; Sponheimer et al., 2007; Sponheimer et al., 2005; Sponheimer et al., 2006b). There are also suggestions, based on strontium isotopes for ranging behavior of *P. (Au.) robustus* individuals, that some individuals had traveled across geological/ecological zones during their lifetimes, perhaps providing evidence of relatively small territory sizes as well as potential patterns of social dispersion (Copeland et al., 2011). Finger bones assigned to *P. (Au.) robustus* show considerable dexterity, and they likely could have used tools to acquire different types of foods; for example, sharp long-bone fragments found at some of the cave sites with *P. (Au.) robustus* fossils seem to have been used as digging tools either to dig up roots or to break open termite nests (Brain, 1993; d'Errico and Blackwell, 2009). In East Africa, hominins such as *P. (Au.) boisei* were hyperrobust, suggesting a heavily masticated diet (Wood and Constantino, 2007). But what were they chewing for long periods? Their isotopic signatures suggest a diet heavily dominated by C_4 foods, which could theoretically include anything from grass blades, to the shallow roots of grasses and sedges, to the meat of antelopes. Detailed analyses of the complexity of their tooth microwear patterns suggest that they were not eating grasses, but were likely eating soft foods with fine gritty textures—such as shallow roots (Ungar et al., 2008). These types of plant foods would have been most abundant near rivers and shallow lakes, so perhaps in contrast to the generalized diets of southern African robusts, the East African robusts survived the increasingly challenging Rift Valley environments by becoming highly specialized foragers in perennial wetland habitats (Lee-Thorp et al., 2011).

Early Homo: Ecological Entrepreneurs

The precise phylogenetic origins of our genus, *Homo*, are still unresolved (Anton and Snodgrass, 2012). A number of fossils of relatively gracile hominins have been found in both the Rift Valley and South African caves. In East Africa, one or perhaps two early species of *Homo* (*Homo habilis*, *Homo rudolfensis*) (see the chapter by Simpson) have less robust faces and smaller cheek teeth, and share relatively larger brains, than found in contemporary "robust" hominins. Larger brain sizes would have placed developmental and metabolic demands on early *Homo* not faced by their contemporary hominins (Aiello and Wheeler, 1995; Foley and Lee, 1991; Leonard et al., 2007a, b; Leonard et al., 2003; Leonard et al., 2010). Fossils of their postcranial skeletons are poorly sampled, but their walking and climbing abilities seem to have been comparable to those of australopithecines. These fossil taxa appear first in Africa around 2.3 mya, and some survive in Africa until at least 1.5 mya. One of these species is the likely direct ancestor of *H. erectus* and our own lineage.

Fossils of *H. erectus* have been found both in Africa and outside of Africa in Eurasia, at sites close to 1.85 million years old (Wood, 2013). This species seems to have been highly variable in body size and cranial capacity, though larger than earlier *Homo*, with body proportions that may have been better adapted to long-distance walking or running (Aiello and Wells, 2002). *H. erectus* populations seem to have shared ancient landscapes for hundreds of thousands of years with both robust hominins and surviving populations of gracile species, and must have had different patterns of resource use (Wood and Strait, 2004). One of these, a distinctive gracile australopithecine species (*Australopithecus sediba*), has been recently discovered at a two-million-year-old South African site; although *A. sediba* had australopithecine-sized brains, some researchers have found their small teeth, gracile skulls, and derived manual dexterity and locomotor patterns to be reminiscent of early *Homo*, their contemporary (Berger et al., 2010). So a variety of gracile hominin populations, perhaps including several species, coexisted between 2.5 and 1.5 mya, a time of significant environmental change associated with increased aridity and fluctuating global temperatures. But ultimately only *H. erectus* endured and continued to spread around the globe. What ecological strategies led to the success of early *Homo*?

Fortunately for paleoecologists, in addition to fossilized skeletal evidence at death sites, evidence of activity sites has been preserved beginning at 2.6 mya. Buried in primary contexts near lakes and on river floodplains, archaeological sites have preserved stone tools often associated with fossil animal remains (Isaac, 1984; Plummer, 2004). These sites document that some early hominins had learned how to select and carry stones to use as hammers to pound materials and break apart brittle rocks; the resulting sharp slivers of stone were used as primitive knives and chopping tools, and fossil bones associated with a number of these early stone tools had been cut or smashed with stones, providing evidence of carcass butchery and the eating of meat and marrow (Braun, 2012; Bunn, 2007; Pickering and Bunn, 2012) (see the chapter by Toth and Schick). While no direct evidence excludes the possibility that robust australopithecines could have made and used such tools, most archaeologists assume that *Homo* had the dexterity and cognitive ability for toolmaking as well as strong incentives to use tools to access calorie-rich foods, given their small teeth and metabolically costly brains (see chapter by Wiley). Animal remains butchered at the early sites include everything from catfish and turtles at sites near wetlands, to haunches of small antelope, warthog, zebra, and other larger savanna and bushland animals repeatedly carried to sites on floodplains or near gallery forests of rivers and lakes (Braun et al., 2010). There is no direct evidence of plant foods preserved at such sites (Sept, 2013), and considerable debate about whether hominins lived or camped at these sites (Braun, 2013). However, the fact that many sites were visited repeatedly, over decades, suggests that early *Homo* was developing strategies for acquiring foods and other resources that were scattered widely across landscapes, and that their omnivorous diets, locomotor and planning abilities, and perhaps social cooperation, allowed them to exploit larger territories, and perhaps more diverse, unpredictable habitats, than was possible for their contemporary hominin species (Potts, 2012a; Sept, 2007).

Isotopic evidence from fossil teeth suggests that early *Homo* maintained a mixed diet of C_3/C_4 foods. C_3 foods could include everything from fish and small game to nuts, deep tubers, or even honey, while meat or marrow from grazing mammals would add to the C_4 balance (Ungar et al., 2006). Tooth microwear seems to be highly variable with no clear patterns, but some early *H. erectus* individuals had highly worn teeth, suggesting significant grit in their diet, perhaps from digging up roots. Overall, they seem to have been capable of an adaptively versatile diet (Ungar et al., 2006). One suggestion, promoted in recent years by primatologist Richard Wrangham (2003, 2009), is that significant shifts in *H. erectus* anatomy compared with those of earlier hominins (increased stature and brain size, with smaller teeth and a less capacious torso/gut size) could be explained if *H. erectus* had controlled the use of fire and begun to cook foods, which would decrease the potential toxicity and increase the energetic value and palatability of those foods (see the chapter by Wiley). Fire also could have protected *H. erectus* from predators in ways not possible for earlier hominins (Clark and Harris, 1985). While this cooking hypothesis is plausible, and there are tantalizing suggestions that some early stone tools were associated with patches of scorched earth that had been burnt in antiquity, the oldest widely accepted archaeological evidence of cooking is closer to one million years old, rather than two million years ago when *H. erectus* first evolved (Alperson-Afil, 2008; Alperson-Afil and Goren-Inbar, 2010; Berna et al., 2012; Brain and Sillen, 1988; Gowlett and Wrangham, 2013; Roebroeks and Villa, 2011).

Finally, early *H. erectus* fossils and archaeological sites have been found in more varied and disparate ecological settings than those of any early hominin (Potts and Teague, 2010). In addition to a variety of savanna woodland habitats in both open-air Rift Valley sites and South African caves, early members of the genus have been recovered from dry coastal plateau habitats in North Africa and as far north as central Eurasia at Dmanisi, a site in the inland plains of the Republic of Georgia near the Caucasus mountains. During the Early Pleistocene, *H. erectus* also swiftly moved eastwards across the Eurasian continent to found populations in Indonesia and China. It is clear, from the other fauna found at many of the earliest archaeological sites outside of Africa, that *H. erectus* was exploiting a variety of novel habitats in both temperate woodlands and humid tropics (Belmaker, 2010).

That *H. erectus* apparently had the social, cultural, and biological abilities to survive in such varied habitats also suggests that they would have been able to cope effectively with environmental change (Potts and Teague, 2010). Indeed, Richard Potts and others have argued that the success of the Homo lineage was a consequence of what he calls "variability selection"—effectively, that selection pressures favored the behavioral flexibility and creativity to adapt to varied and unpredictable environments (Potts, 2012a, b). Certainly the records of both global climate change and its impact on local early hominin habitats suggest that as Pleistocene climates began to oscillate strongly, selection pressures would have been very strong (Maslin and Trauth, 2009). Hominin species that had evolved biological specializations, such as wetland-focused plant-food diets, would have been more vulnerable than more generalized species to the geographical shifts and habitat degradation associated with many of these climatic changes.

SUMMARY: THE LIMITS AND POTENTIAL OF OUR PALEOECOLOGICAL KNOWLEDGE

While it remains unresolved whether paleoenvironmental conditions had any causal effects on the origins of our lineage and the emergence of bipedalism, it does seem clear that early hominins were ecologically fairly generalized and able to survive in a wide range of habitats. Through time, as climate change caused habitats to become patchier and less predictable, hominins diversified. During the Early Pleistocene, adaptive radiation species either became adapted to more specialized diets, like the later "robust" australopithecines, or developed more versatile, omnivorous diets founded on tool use, like early *Homo*. Ultimately, while all other hominin species went extinct, Early *Homo* seems to have expanded its ecological range and spread geographically throughout Africa, Europe, and Asia.

Because our collection of early hominin fossils is contingent on the distribution of sedimentary environments that preserved fossils that are also exposed as outcrops today, our samples of landscapes in which early hominins lived and died is geographically restricted and highly biased. So we should be cautious in using the localities of our current sample of early hominin fossils to infer ecological preferences of different early hominin species.

However, given recent advances on several fronts, the potential to reconstruct aspects of the fine-grained paleoecology of our ancestors has never been greater. Multiple sources of data can now be compared to develop more rigorous hypotheses about early hominin paleoecology than have previously been possible, including the analysis of evidence of the ecological life histories of individual fossils, such as isotopic clues to diet; taphonomic signatures of the circumstances of their deaths; and primary context archaeological evidence of the habitats in which they were active and specific evidence of tool resources and food remains. Once individual hominin fossils—ideally multiple individuals from each taxon—can be situated in the paleohabitats they frequented, we can step back with greater confidence to reevaluate the large-scale question: how were the macroevolutionary patterns of human evolution influenced by global patterns of climatic change?

REFERENCES

Aiello, L.C., Wells, J.C.K., 2002. Energetics and the evolution of the genus homo. Annual Review of Anthropology 31, 323–338.

Aiello, L.C., Wheeler, P., 1995. The expensive-tissue hypothesis. Current Anthropology 36, 199–222.

Alperson-Afil, N., 2008. Continual fire-making by hominins at gesher benot ya'aqov, Israel. Quaternary Science Reviews 27, 1733–1739.

Alperson-Afil, N., Goren-Inbar, N., 2010. The Acheulian Site of Gesher Benot Ya'aqov: Ancient Flames and Controlled Use of Fire. Springer, Dordrecht.

Anderson, D.E., Goudie, A.S., Parker, A.G., 2007. Global Environments through the Quaternary: Exploring Environmental Change. Oxford University Press, Oxford.

Andrews, P., Bamford, M., 2008. Past and present vegetation ecology of Laetoli, Tanzania. Journal of Human Evolution 54, 78–98.

Anton, S.C., Snodgrass, J.J., 2012. Origins and evolution of genus homo: new perspectives. Current Anthropology 53, S479–S496.

Aronson, J.L., Hailemichael, M., Savin, S.M., 2008. Hominid environments at Hadar from paleosol studies in a framework of ethiopian climate change. Journal of Human Evolution 55, 532–550.

Bailey, G.N., Reynolds, S.C., King, G.C.P., 2011. Landscapes of human evolution: models and methods of tectonic geomorphology and the reconstruction of hominin landscapes. Journal of Human Evolution 60 (3), 257–280.

Bamford, M.K., 2011a. Fossil leaves, fruits and seeds. In: Harrison, T. (Ed.), Paleontology and Geology of Laetoli: Human Evolution in Context. Geology, Geochronology, Paleocology and Paleoenvironment, vol. 1. Springer, Dordrecht, pp. 235–252.

Bamford, M.K., 2011b. Fossil woods. In: Harrison, T. (Ed.), Paleontology and Geology of Laetoli: Human Evolution in Context. Geology, Geochronology, Paleocology and Paleoenvironment, vol. 1. Springer, Dordrecht, pp. 217–233.

Bamford, M.K., Senut, B., Pickford, M., 2013. Fossil leaves from Lukeino, a 6-million-year-old formation in the Baringo Basin, Kenya. Geobios 46, 253–272.

Barboni, D., Ashley, G.M., Domínguez-Rodrigo, M., Bunn, H., Mabulla, A.Z.P., Baquedano, 2010. Phytoliths infer locally dense and heterogeneous paleovegetation at Flk North and surrounding localities during upper bed 1 time, Olduvai Gorge, Tanzania. Quaternary Research 74, 344–354.

Bedaso, Z.K., Wynn, J.G., Alemseged, Z., Geraads, D., 2013. Dietary and paleoenvironmental reconstruction using stable isotopes of herbivore tooth enamel from middle Pliocene Dikika, Ethiopia: implication for *Australopithecus afarensis* habitat and food resources. Journal of Human Evolution 64, 21–38.

Behrensmeyer, A.K., 1984. Taphonomy and the fossil record: the complex processes that preserve organic remains in rocks also leave their own traces, adding another dimension of information to fossil samples. American Scientist 72, 558–566.

Behrensmeyer, A.K., 2008. Paleoenvironmental context of the Pliocene A.L. 333 'first family' hominin locality, Hadar Formation, Ethiopia. In: Quade, J., Wynn, J.G. (Eds.), The Geology of Early Humans in the Horn of Africa: Geological Society of America Special Paper, vol. 446. Geological Society of America, Inc., Boulder CO, pp. 203–214.

Behrensmeyer, A.K., Reed, K.E., 2013. Reconstructing the habitats of australopithecus: paleoenvironments, site taphonomy, and faunas. In: Reed, K.E., Fleagle, J.G., Leakey, R.E. (Eds.), The Paleobiology of Australopithecus. Springer, Dordrecht, pp. 41–62.

Belmaker, M., 2010. On the road to China: the environmental landscape of the early Pleistocene in Western Eurasia and its implications for the dispersal of homo. In: Norton, C.J., Braun, D.R. (Eds.), Asian Paleoanthropology: from Africa to China and beyond. Springer, Dordrecht, pp. 31–40.

Berger, L.R., de Ruiter, D.J., Churchill, S.E., Schmid, P., Carlson, C.J., Dirks, P.H.G.M., Kibii, J.M., 2010. *Australopithecus sediba*: a new species of homo-like australopith from South Africa. Science 328, 195–204.

Berger, L.R., McGraw, W.S., 2007. Further evidence for eagle predation of, and feeding damage on, the Taung Child. South African Journal of Science 103, 496–498.

Berna, F., Goldberg, P., Horwitz, L.K., Brink, J., Holt, S., Bambford, M., Chazan, M., 2012. Microstratigraphic evidence of in situ fire in the Acheulian strata of Wonderwerk Cave, Northern Cape province, South Africa. Proceedings of the National Academy of Sciences 109, E1215–E1220.

Blumenschine, R.J., Masao, F.T., Stanistreet, I.G., Swisher, C.C. (Eds.), 2012. Five Decades after Zinjanthropus and *Homo habilis*: Landscape Paleoanthropology of Plio-Pleistocene Olduvai Gorge, Tanzania.

Bobe, R., Leakey, M.G., 2009. Ecology of Plio-Pleistocene mammals in the Omo-Turkana Basin and the emergence of *homo*. In: Grine, F.E. (Ed.), The First Humans: Origin and Early Evolution of the Genus Homo. Springer, Berlin, pp. 173–184.

Bonnefille, R., 2010. Cenozoic vegetation, climate changes and hominid evolution in tropical Africa. Global and Planetary Change 72, 390–411.

Bradley, R.S., 1999. Paleoclimatology: Reconstructing Climates of the Quaternary. Academic Press, San Diego, Calif.

Brain, C.K., 1981. The Hunters or the Hunted? An Introduction to African Cave Taphonomy. The University of Chicago Press, Chicago.

Brain, C.K. (Ed.), 1993. Swartkrans: A Cave's Chronicle of Early Man. Transvaal Museum, Pretoria.

Brain, C.K., Sillen, A., 1988. Evidence from the Swartkrans cave for the earliest use of fire. Nature 336, 464–466.

Braun, D.R., 2012. What does Oldowan technology represent in terms of hominin behavior? In: Dominguez-Rodrigo, M. (Ed.), Stone Tools and Fossil Bones. Debates in the Archaeology of Human Origins. Cambridge University Press, Cambridge, pp. 222–244.

Braun, D.R., 2013. The behavior of Plio-Pleistocene hominins: archaeological perspectives. In: Sponheimer, M., Lee-Thorp, J.A., Reed, K.E., Ungar, P.S. (Eds.), Early Hominin Paleoecology. The University Press of Colorado, Boulder, pp. 325–351.

Braun, D.R., Harris, J.W.K., Levin, N., McCoy, J.T., Herries, A.I.R., Bamford, M.K., Biship, L.C., Richmond, B.G., Kibunjia, M., 2010. Early hominin diet included diverse terrestrial and aquatic animals 1.95 ma in East Turkana, Kenya. Proceedings of the National Academy of Sciences 107, 10002–10007.

Bunn, H.T., 2007. Meat made us human. In: Ungar, P.S. (Ed.), Evolution of Human Diet. The Known, the Unknown, and the Unknowable. Oxford University Press, Oxford, UK, pp. 191–211.

Campisano, C.J., Feibel, C.S., 2007. Connecting local environmental sequences to global climate patterns: evidence from the hominin-bearing Hadar Formation, Ethiopia. Journal of Human Evolution 53, 515–527.

Cerling, T., Wynn, J.G., Andanje, S.A., Bird, M.I., Korir, D.K., Levin, N.E., Mace, W., Macharia, A.N., Quade, J., Remien, C.H., 2011. Woody cover and hominin environments in the past 6 million years. Nature 476, 51–56.

Cerling, T.E., Levin, N.E., Quade, J., Wynn, J.G., Fox, D.L., Kingston, J.D., Klein, R.G., Brown, F.H., 2010. Comment on the paleoenvironment of *Ardipithecus ramidus*. Science 328, 1105.

Clark, J.D., Harris, J.W.K., 1985. Fire and its roles in early hominid lifeways. The African Archaeological Review 3, 3–28.

Copeland, S.R., Sponheimer, M., de Ruiter, D., Lee-Thorp, J.A., Codron, D., le Roux, P., Grimes, V., Richards, M.P., 2011. Strontium isotope evidence for landscape use by early hominins. Nature 474, 76–79.

Coppens, Y., May 1994. East side story: the origin of humankind. Scientific American 88–95.

d'Errico, F., Blackwell, L., 2009. Assessing the function of early hominin bone tools. Journal of Archaeological Science 36, 1764–1773.

Daegling, D.J., Judex, S., Ozcivici, E., Ravosa, M.J., Taylor, A.B., Grine, F.E., Teaford, M.F., Ungar, P.S., 2013. Viewpoints: feeding mechanics, diet, and dietary adaptations in early hominins. American Journal of Physical Anthropology 151, 356–371.

deMenocal, P.B., 2004. African climate change and faunal evolution during the Pliocene-Pleistocene. Earth and Planetary Science Letters 220, 3–24.

Drake, N.A., Blench, R.M., Armitage, S.J., Bristow, C.,S., White, K.H., 2011. Ancient watercourses and biogeography of the Sahara explain peopling of the desert. Proceedings of the National Academy of Sciences 108, 458–462.

Elton, S., 2008. The environmental context of human evolutionary history in Eurasia and Africa. Journal of Anatomy 212, 377–393.

Foley, R.A., 1994. Speciation, extinction and climatic change in hominid evolution. Journal of Human Evolution 26, 275–289.

Foley, R.A., Lee, P.C., 1991. Ecology and energetics of encephalization in hominid evolution. Philosophical Transactions of the Royal Society of London, B 63–73.

Gowlett, J.A.J., Wrangham, R., 2013. Earliest fire in Africa: towards the convergence of archaeological evidence and the cooking hypothesis. Azania: Archaeological Research in Africa 48, 5–30.

Hailie-Selassie, Y., WoldeGabriel, G. (Eds.), 2009. *Ardipithecus kadabba*. Lake Miocene Evidence from the Middle Awash, Ethiopia. University of California Press, Berkeley, CA.

Harrison, T. (Ed.), 2011. Paleontology and Geology of Laetoli: Human Evolution in Context. Geology, Geochronology, Paleoecology and Paleoenvironment, vol. 1. Springer.

Hay, R.L., 1976. The Geology of Olduvai Gorge. University of California Press, Berkeley and Los Angeles.

Imbrie, J., Berger, A., Boyle, E.A., Clemens, S.C., Duffy, A., Howard, W.R., Kukla, G., Kutzbach, J., Martinson, D.G., McIntyre, A., Mix, A.C., Molfino, B., Morley, J.J., Peterson, L.C., Pisias, N.G., Prell, W.L., Raymo, M.E., Shackleton, N.J., Toggweiler, J.R., 1993. On the structure and origin of major glaciation cycles 2. The 100,000-year cycle. Paleoceanography 8, 699–735.

Imbrie, J., Imbrie, K.P., 1979. Ice Ages. Solving the Mystery. Harvard University Press, Cambridge MA.

Isaac, G.L., 1984. The archaeology of human origins: studies of the lower Pleistocene in East Africa 1971–1981. In: Wendorf, F. (Ed.), Advances in World Archaeology. Academic Press, New York, pp. 1–87.

Isbell, L.A., Young, T.P., 1996. The evolution of bipedalism in hominids and reduced group size in chimpanzees: alternative responses to decreasing resource availability. Journal of Human Evolution 30, 389–397.

Joordens, J.C.A., Vonhof, H.B., Feibel, C.S., Lourens, L.J., Dupont-Nivet, G., van der Lubbe, J.H.J.L., Sier, M.J., Davies, G.R., Kroon, D., 2011. An astronomically-tuned climate framework for hominins in the Turkana Basin. Earth and Planetary Science Letters 307, 1–8.

Kingston, J., 2002. Stratigraphy, age and environments of the late Miocene Mpesida Beds, Tugen Hills, Kenya. Journal of Human Evolution 42, 95–116.

Kingston, J., Harrison, T., 2007. Isotopic dietary reconstructions of Pliocene herbivores at laetoli: implications for early hominin paleoecology. Palaeogeography, Palaeoclimatology, Palaeoecology 243, 272–306.

Kingston, J.D., 2007. Shifting adaptive landscapes: progress and challenges in reconstructing early hominid environments. Yearbook of Physical Anthropology 50, 20–58.

Kingston, J.D., Deino, A.L., Edgar, R.K., Hill, A., 2007. Astronomically forced climate change in the Kenyan Rift Valley 2.7–2.55 ma: implications for the evolution of early hominin ecosystems. Journal of Human Evolution 53, 487–503.

L'Engle Williams, F., Patterson, J.W., 2010. Reconstructing the paleoecology of Taung, South Africa, from low magnification of dental microwear features in fossil primates. Palaios 25, 439–448.

Larrasoana, J.C., Roberts, A.P., Rohling, E.J., Winklhofer, M., Wehausen, R., 2003. Three million years of monsoon variability over the Northern Sahara. Climate Dynamics 21, 689–698.

Lebatard, A.E., Bourles, D.L., Duringer, P., Jolivet, M., Braucher, R., Carcaillet, J., Schuster, M., Arnaud, N., Monie, P., Lihoreau, F., Likius, A., Mackaye, H.T., Vignaud, P., Brunet, M., 2008. Cosmogenic nuclide dating of *Sahelanthropus tchadensis* and *Australopithecus bahrelghazali*: Mio-Pliocene hominids from Chad. Proceedings of the National Academy of Sciences 105, 3226–3231.

Lee-Thorp, J.A., Likius, A., Mackay, H.T., Vignaud, P., Sponheimer, M., Brunet, M., 2012. Isotopic evidence for an early shift to C_4 resources by Pliocene hominins in Chad. Proceedings of the National Academy of Sciences 109, 20369–20372.

Lee-Thorp, J.A., Sponheimer, M., Passey, B., de Ruiter, D., Cerling, T.E., 2011. Stable isotopes in fossil hominin tooth enamel suggest a fundamental dietary shift in the Pliocene. Philosophical Transactions of the Royal Society B: Biological Sciences 365, 3389–3396.

Leonard, W.R., Robertson, M.L., Snodgrass, J.J., 2007a. Energetic models of human nutritional evolution. In: Ungar, P.S. (Ed.), Evolution of Human Diet. The Known, the Unknown, and the Unknowable. Oxford University Press, Oxford, U.K, pp. 344–359.

Leonard, W.R., Robertson, M.L., Snodgrass, J.J., Kuzawa, C.W., 2003. Metabolic correlates of hominid brain evolution. Comparative Biochemistry and Physiology, Part A 136, 5–15.

Leonard, W.R., Snodgrass, J.J., Robertson, M.L., 2007b. Effects of brain evolution on human nutrition and metabolism. Annual Review of Nutrition 27, 311–327.

Leonard, W.R., Snodgrass, J.J., Robertson, M.L., 2010. Evolutionary perspectives on fat ingestion and metabolism in humans. In: Montmayeur, J.-P., le Coutre, J. (Eds.), Fat Detection: Taste, Texture and Post Ingenstive Effects. CRC Press, Taylor & Francis Group, Boca Raton FL, pp. 3–18.

Lind, E.M., Morrison, M.E.S., 1974. East African Vegetation. Longman Group Ltd., Bristol.

Louchart, A., Wesselman, H., Blumenschine, R.J., Hlusko, L.J., Njau, J.K., Black, M.T., Asnake, M., White, T.D., 2009. Taphonomic, avian, and small-vertebrate indicators of *Ardipithecus ramidus* habitat. Science 326, 66e1–66e4.

Lovejoy, C.O., 1981. The origin of man. Science 211, 341–350.

Magill, C., Ashley, G.M., Freeman, K.H., 2013. Ecosystem variability and early human habitats in Eastern Africa. Proceedings of the National Academy of Sciences 110, 1167–1174.

Maslin, M.A., Christensen, B., Wilson, K.E., 2013. Tectonics, orbital forcing, global climate change, and human evolution in Africa. In: Sponheimer, M., Lee-Thorp, J.A., Reed, K.E., Ungar, P.S. (Eds.), Early Hominin Paleoecology. University Press of Colorado, Boulder, pp. 103–162.

Maslin, M.A., Trauth, M.H., 2009. Plio-Pleistocene East African pulsed climate variability and its influence on early human evolution. In: Grine, F.E. (Ed.), The First Humans: Origin and Early Evolution of the Genus Homo. Springer, Dordrecht, pp. 151–158.

McDougall, I., Brown, F.H., 2006. Precise 40ar/39ar geochronology for the upper Koobi Fora Formation, Turkana Basin, Northern Kenya. Journal of the Geological Society London 163, 205–220.

Pickering, T., 2004. The context of Stw 573, an early hominid skull and skeleton from Sterkfontein member 2: taphonomy and paleoenvironment. Journal of Human Evolution 46, 277–295.

Pickering, T.R., Bunn, H.T., 2012. Meat foraging by Pleistocene African hominins: tracking behavioral evolution beyond baseline inferences of early access to carcasses. In: Dominguez-Rodrigo, M. (Ed.), Stone Tools and Fossil Bones. Debates in the Archaeology of Human Origins. Cambridge University Press, Cambridge, pp. 152–173.

Plummer, T., 2004. Flaked stones and old bones: biological and cultural evolution at the dawn of technology. Yearbook of Physical Anthropology 47.

Potts, R., 1998. Variability selection in hominid evolution. Evolutionary Anthropology 7, 81–96.

Potts, R., 2007. Environmental hypotheses of Pliocene human evolution. In: Bobe, R., Alemseged, Z., Behrensmeyer, A.K. (Eds.), Hominin Environments in the East African Pliocene: An Assessment of the Faunal Evidence. Springer, Dordrecht, The Netherlands, pp. 25–49.

Potts, R., 2012a. Environmental and behavioral evidence pertaining to the evolution of early homo. Current Anthropology 53, S299–S317.

Potts, R., 2012b. Evolution and environmental change in early human prehistory. Annual Review of Anthropology 41, 151–167.

Potts, R., 2013. Hominin evolution in settings of strong environmental variability. Quaternary Science Reviews 73, 1–13.

Potts, R., Teague, R., 2010. Behavioural and environmental background to 'Out of Africa I' and the arrival of *Homo erectus* in East Asia. In: Fleagle, J.G. (Ed.), Out of Africa I: The First Hominin Colonization of Eurasia. Springer, Dordrecht.

Quade, J., Levin, N.E., 2013. East African hominin paleoecology: isotopic evidence from paleosols. In: Sponheimer, M., Lee-Thorpe, J.A., Reed, K.E., Ungar, P.S. (Eds.), Early Hominin Paleoecology. University Press of Colorado, Boulder, pp. 59–102.

Reed, K.E., 1997. Early hominid evolution and ecological change through the African Plio-Pleistocene. Journal of Human Evolution 32, 289–322.

Reed, K.E., Fleagle, J.G., Leakey, R.E. (Eds.), 2013a. The Paleobiology of Australopithecus. Springer, Dordrecht.

Reed, K.E., Spencer, L.M., Rector, A.L., 2013b. Faunal approaches in early hominin paleoecology. In: Sponheimer, M., Lee-Thorp, J.A., Reed, K.E., Ungar, P.S. (Eds.), Early Hominin Paleoecology. University of Colorado Press, Boulder, pp. 3–34.

Rodman, P.S., McHenry, H.M., 2005. Bioenergetics and the origins of hominid bipedalism. American Journal of Physical Anthropology 52, 103–106.

Roebroeks, W., Villa, P., 2011. On the earliest evidence for habitual use of fire in Europe. Proceedings of the National Academy of Sciences 108, 5209–5214.

Sahnouni, M., Van der Made, J., Everett, M., 2010. Early North Africa: chronology, ecology and hominin behavior: insights from Ain Hanech and El-Kherba, northeastern Algeria. Quaternary International 223–224, 436–438.

Sahnouni, M., Van der Made, J., Everett, M., 2011. Ecological background to Plio-Pleistocene hominin occupation in North Africa: the vertebrate faunas from Ain Boucherit, Ain Hanech and El-Kherba, and paleosol stable-carbon-isotope studies from El-Kherba, Algeria. Quaternary Science Reviews 30, 1303–1317.

Sept, J., 2013. Plants and protopeople: paleobotanical reconstructions and early hominin ecology. In: Sponheimer, M., Lee-Thorp, J.A., Reed, K.E., Ungar, P.S. (Eds.), Early Hominin Paleoecology. University Press of Colorado, Boulder, pp. 355–396.

Sept, J.M., 2007. Modeling the significance of paleoenvironmental context for early hominin diets. In: Ungar, P.S. (Ed.), Evolution of Human Diet. The Known, the Unknown, and the Unknowable. Oxford University Press, Oxford, England, pp. 289–307.

Sponheimer, M., Lee-Thorp, J., De Ruiter, D., 2007. Icarus, isotopes, and australopith diets. In: Ungar, P.S. (Ed.), Evolution of the Human Diet. The Known, the Unknown, and the Unknowable. Oxford University Press, Oxford, U.K, pp. 132–149.

Sponheimer, M., Lee-Thorp, J., de Ruiter, D., Codron, D., Codron, J., Baugh, A.T., Thackeray, F., 2005. Hominins, sedges, and termites: new carbon isotope data from the Sterkfontein Valley and Kruger National Park. Journal of Human Evolution 48, 301–312.

Sponheimer, M., Loudon, J., Codron, D., Howells, M., Pruetz, J., Codron, J., Deruiter, D., Leethorp, J., 2006a. Do "savanna" chimpanzees consume C_4 resources? Journal of Human Evolution 51, 128–133.

Sponheimer, M., Passey, B.H., de Ruiter, D.J., Guatelli-Steinberg, D., Cerling, T.E., Lee-Thorp, J.A., 2006b. Isotopic evidence for dietary variability in the early hominin *Paranthropus robustus*. Science 314, 980–982.

Suwa, G., Asfaw, B., Kono, R.T., Kubo, D., Lovejoy, C.O., White, T.D., 2009a. The *Ardipithecus ramidus* skull and its implications for hominid origins. Science 326, 68e1–68e7.

Suwa, G., Kono, R.T., Simpson, S.W., Asfaw, B., Lovejoy, C.O., White, T.D., 2009b. Paleobiological implications of the *Ardipithecus ramidus* dentition. Science 326, 94–99.

Ungar, P.S., Grine, F.E., Teaford, M.F., 2006. Diet in early homo: a review of the evidence and a new model of adaptive versatility. Annual Review of Anthropology 35, 209–228.

Ungar, P.S., Grine, F.E., Teaford, M.F., Petraglia, M., 2008. Dental microwear and diet of the Plio-Pleistocene hominin *Paranthropus boisei*. PLoS One 3, e2044.

Ungar, P.S., Scott, R.S., Grine, F.E., Teaford, M.F., 2010. Molar microwear textures and the diets of *Australopithecus anamensis* and *Australopithecus afarensis*. Philosophical Transactions of the Royal Society B: Biological Sciences 365, 3345–3354.

Ungar, P.S., Sponheimer, M., 2011. The diets of early hominins. Science 334, 190–193.

Vignaud, P., Duringer, P., Mackaye, H.T., Likius, A., Blondel, C., Boisserie, J.-R., de Bonis, L., Eisenmann, V., Etienne, M.-E., Geraads, D., Guy, F., Lehmann, T., Lihoreau, F., Lopez-Martinez, N., Mourer-Chauvire, C., Otero, O., Rage, J.-C., Schuster, M., Viriot, L., Zazzo, A., Brunet, M., 2002. Geology and palaeontology of the Upper Miocene Toros-Menalla hominid locality, Chad. Nature 418, 152–155.

Vrba, E., 1988. Late Pliocene climatic events and hominid evolution. In: Grine, F.E. (Ed.), Evolutionary History of the Australopithecines. Aldine de Gruyter, New York, pp. 405–426.

Vrba, E.S., 1985. Ecological and adaptive changes associated with early hominid evolution. In: Delson, E. (Ed.), Ancestors: The Hard Evidence. Alan R. Liss, Inc., New York, pp. 63–71.

Vrba, E.S., 1995a. On the connections between paleoclimate and evolution. In: Vrba, E.S., Denton, G.H., Partridge, T.C., Burckle, L.H. (Eds.), Paleoclimate and Evolution, with Emphasis on Human Origins. Yale University Press, New Haven, pp. 24–48.

Vrba, E.S., 1995b. The fossil record of African antelopes (mammalia, bovidae) in relation to human evolution and paleoclimate. In: Vrba, E.S., Denton, G.H., Partridge, T.C., Burckle, L.H. (Eds.), Paleoclimate and Evolution, with Emphasis on Human Origins. Yale University Press, New Haven, pp. 385–424.

Walker, A., Leakey, R.E.F. (Eds.), 1993. The Nariokotome *Homo erectus* Skeleton. Harvard University Press, Cambridge MA.

White, F., 1983. The Vegetation of Africa, a Descriptive Memoir to Accompany the Unesco/Aetfat/Unso Vegetation Map of Africa, La Chaux-de-Fonds. UNESCO.

White, T.D., Ambrose, S.H., Suwa, G., Su, D.F., DeGusta, D., Bernor, R.L., Boisserie, J.R., Brunet, M., Delson, E., Frost, S., Garcia, N., Giaourtsakis, I.X., Haile-Selassie, Y., Howell, F.C., Lehmann, T., Likius, A., Pehlevan, C., Saegusa, H., Semprebon, G., Teaford, M., Vrba, E., 2009a. Macrovertebrate paleontology and the Pliocene habitat of *Ardipithecus ramidus*. Science 326, 87–93.

White, T.D., Ambrose, S.H., Suwa, G., WoldeGabriel, G., 2010. Response to comment on the paleoenvironment of *Ardipithecus ramidus*. Science 328, 1105-e.

White, T.D., Asfaw, B., Beyene, Y., Haile-Selassie, Y., Lovejoy, C.O., Suwa, G., WoldeGabriel, G., 2009b. *Ardipithecus ramidus* and the paleobiology of early hominids. Science 326, 75–86.

WoldeGabriel, G., Ambrose, S.H., Barboni, D., Bonnefille, R., Bremond, L., Currie, B., DeGusta, D., Hart, W.K., Murray, A.M., Renne, P.R., Jolly-Saad, M.C., Stewart, K.M., White, T.D., 2009. The geological, isotopic, botanical, invertebrate, and lower vertebrate surroundings of *Ardipithecus ramidus*. Science 326, 65e1–65e5.

Wood, B., 2013. Did early homo migrate "out of" or "in to" Africa? Proceedings of the National Academy of Sciences 108, 10375–10376.

Wood, B., Constantino, P., 2007. *Paranthropus boisei*: fifty years of evidence and analysis. Yearbook of Physical Anthropology 50, 106–132.

Wood, B., Strait, D., 2004. Patterns of resource use in early *homo* and *paranthropus*. Journal of Human Evolution 46, 119–162.

Wrangham, R., 2003. Cooking as a biological trait. Comparative Biochemistry and Physiology – Part A: Molecular & Integrative Physiology 136, 35–46.

Wrangham, R., 2009. Catching Fire: How Cooking Made Us Human. Basic Books, New York.

Chapter 8

Bipedalism

Kevin D. Hunt
Human Evolution and Animal Behavior, Indiana University, Bloomington, IN, USA

SYNOPSIS

Bipedalism evolved in a stagewise fashion. There is some evidence that prehominins were more bipedal than living African apes, perhaps similar to orangutans or gibbons. Australopiths had cone-shaped rib cages, gorilla-like scapulae, short but chimpanzee-like fingers, and long, curved toes, all consistent with adaptation to arm-hanging. Australopiths lack chimpanzee vertical-climbing adaptations such as gripping great toes and stiff backs. Chimpanzees are bipedal when feeding from the short trees that are common in dry habitats, where the feed by reaching up from the ground and with an arm-hanging, bipedal posture in the trees. The small-diameter branches in short trees discourage sitting and encourage arm-hanging; lateral toes grip the flexible branches, the body is balanced bipedally over the gripping feet, and short fingers are adequate for the small branches. Such trees can be entered without hand-over-hand climbing and opposable toes that allow it. Two million years ago hominins evolved longer legs and narrower hips that allowed a more modern human locomotion.

Alternate-foot or striding bipedalism is a locomotor mode unique to humans (Figure 1). Human positional behavior research has focused on five interrelated issues of bipedalism: (1) the muscle activity and movements involved in walking or running; (2) the morphology uniquely adapted to bipedalism; (3) the timing of its evolution; (4) the stages through which our ancestors evolved as modern bipedalism emerged; and (5) the selective pressures that caused bipedalism to evolve.

HOW DO HUMANS WALK?

As a neophyte might expect, bipedal walking requires substantial muscle activity in support of body weight, but a surprisingly large proportion of muscle activity also goes to acceleration and deceleration of the hind limbs and other body parts (Boakes and Rab, 2006). The bipedal step cycle begins with heel strike or initial contact (Figure 1(A)), the moment the heel of the leading leg touches the ground and the *stance* or *support phase* for that leg begins (the left leg in Figure 1). As the heel touches down, the weight of the body exerts rotational forces on the ankle and foot such that the forefoot and toes would flop onto the ground if unchecked. To prevent this, muscles in the anterior compartment of the lower leg (e.g., anterior tibialis) fire to slow the rotation of the ankle, allowing the sole to roll into the ground at a moderate speed and shifting weight onto the outer sole of the foot. At heel strike, muscles in the posterior compartment of the thigh (hamstrings) pull the weight-bearing leg backward, extending the hip. As the hip extends and the leg is pulled backward, muscles in the anterior compartment of the thigh (a group collectively referred to as the vasti; see Figures 2 and 3) fire to prevent the knee from collapsing or overflexing, thus supporting body weight during the swing phase (Figure 1(B) and (C)). The hip abductors, gluteus medius and less actively the gluteus maximus and minimus, fire to prevent the hip from collapsing under the combined weight of the upper body and non-weight-bearing leg (Figure 1(B–D)).

Focusing now on the non-weight-bearing leg, just after heel strike the trailing leg begins to swing forward into its *swing phase* (Figure 1(B)). During the swing phase, rectus femoris and muscles internal to the body, the iliopsoas muscles, are most active in pulling the trailing leg forward. Muscles in the anterior compartment of the lower leg (anterior tibialis) lift the toes toward the shin, allowing the forefoot to clear the ground. As the swing phase nears its end and the leg approaches heel strike, the hamstrings and gluteus maximus fire to slow the limb.

Returning to the weight-bearing limb, as body weight passes over the supporting foot, muscles in the posterior compartment of the lower leg (gastrocnemius and soleus) fire to rotate the ankle, raising the heel. The toes hyperextend against the midfoot at first, but as the support phase nears it end, the big toe in particular is flexed or pointed, driven largely by the peroneus longus and peroneus brevis, ending the step cycle with toe-off. Now the opposite leg begins its support cycle with a heel strike.

Other muscles are responsible for the finer details of hip support, toe movement, foot orientation, and torso stabilization.

Viewed from above, the hips rotate against the rib cage first one way and then the other around the long axis on the body. As the swing phase leg reaches forward, anticipating heel strike, the hip is rotated to allow the reaching leg to reach farther than it would if motion were just flexing (raising the knee) and extending the hip (lowering the knee). This rotation lengthens each stride slightly.

During walking, the upper body remains relatively stable with the exception of the arms. As the left hind limb reaches forward toward heel strike, the right forelimb swings forward, and as the right hind limb moves forward, the left arm swings forward. This alternate swinging helps to stabilize the upper torso against the rotating hips.

An important aspect of bipedal walking, and indeed of any locomotor activity, is the spring-like nature of ligaments (the fibrous tissue linking bones), tendons (the fibrous tissue linking muscle to bone), bones, and muscles. The elastic

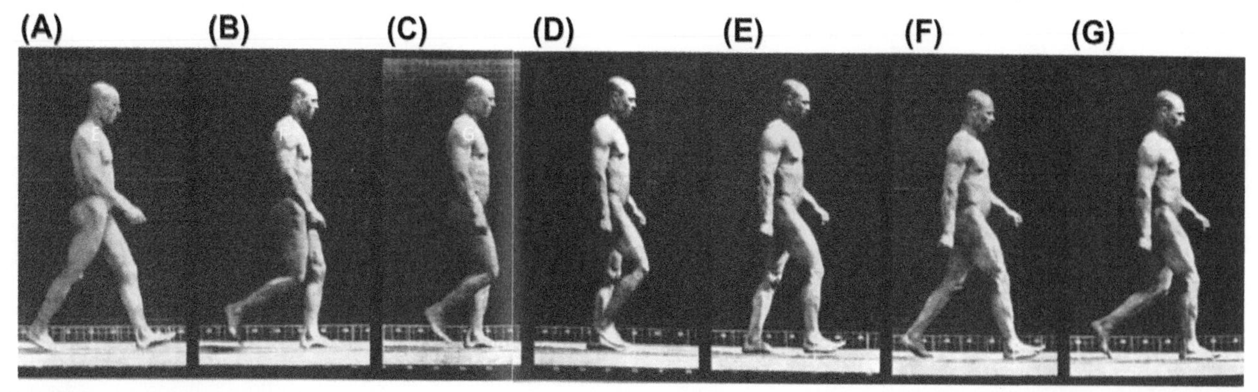

FIGURE 1 (A) Left heel strike and toe-off, (B, C, D, and E) right swing phase, left stance phase, (F) right heel strike, left toe-off, (G) right stance phase. *from Muybridge (1887).*

nature of these tissues allows them to first store and then release energy, helping to drive the cycle of limb movement much like a spring or slinky toy walking down stairs. As the trailing stance-phase limb has reached its hindmost extension, tension in the iliofemoral ligaments, ligaments binding the vertebrae together, and muscles in the anterior compartment of the thigh, help to pull the hind limb forward to begin the swing phase. Gravity plays a role as well, first decelerating the leading limb as the swing phase ends and accelerating the trailing limb forward as the swing phase begins. This dynamic cycle shares physical properties with a pendulum.

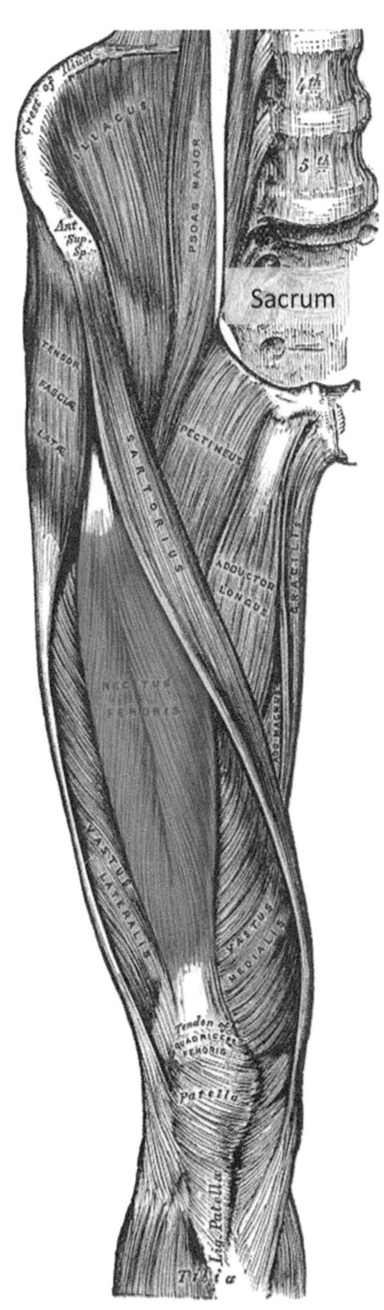

FIGURE 2 Muscles of the thigh. (See color plate section). *From Gray and Carter (1858).*

When weight is borne by bones, tendons, and ligaments rather than muscles, or when movement is driven by the passive elastic properties of muscles, little energy expenditure is required, and efficiency is high compared with activities where muscle action is primary. Because the knees are nearly completely extended in the mid stance phase, a large proportion of body weight is supported by bone and ligament rather than muscle activity. Humans are among the most efficient animals as walkers, while as runners they are merely average in efficiency.

Moderate-speed running in particular involves many of the same movements and muscles as walking; rapid running involves more-dramatic joint excursions and greater muscle activity. Slower running speeds rely more on elastic tension and pendular movement. Gluteus maximus is much more active during running than walking (Bramble and Lieberman, 2004), and most active during lunging (Marzke et al., 1988) or other rapid acceleration (Bartlett et al., 2014). Even though humans are not particularly energy-efficient runners, we are nevertheless exceptional endurance runners, capable of sustaining moderate running speeds for extended periods of time, probably as an adaptation to hunting (Carrier, 1984). Perhaps unexpectedly, it is elevated body core temperature that leads to fatigue; humans have a highly evolved temperature-regulation mechanism that relies on sweating and evaporative perspiration, aided by naked skin (Carrier, 1984; Wheeler, 1991).

ANATOMICAL FEATURES ASSOCIATED WITH BIPEDALISM

Anatomists and paleontologists isolate morphology that is functionally related to bipedalism by comparing human anatomy with that of closely related nonbipeds (see the chapter by Diogo et al. in the present volume). As our closest genetic relatives, chimpanzees and bonobos are likely to be most similar to our prehuman ancestors, and the similarity of the 4.4-million-year-old *Ardipithecus* fossils to living chimpanzees and bonobos reinforces this presumption.

A critical selective force on hominin anatomy is that bipeds bear weight on their hind limbs alone, whereas quadrupeds distribute weight approximately equally between the forelimbs and the hind limbs. The evolution of bipedalism required reorienting the upper body so that it was directly over the hips, which required bending the spine back to reorient it from horizontal to vertical and create a concave "small" of the back (technically, "lordosis"). The pelvis was reshaped as the spine was rotated to a vertical rather than horizontal position. The sacrum (Figures 3 and 4) along with the spine was pulled backward, leaving the anterior superior iliac spine in front of the hip joint, retaining the lever arm for the muscles of the anterior thigh. The internal organs of bipeds hang less from the spine and rest in a more bowl-shaped than blade-shaped pelvis (Figures 3 and 4).

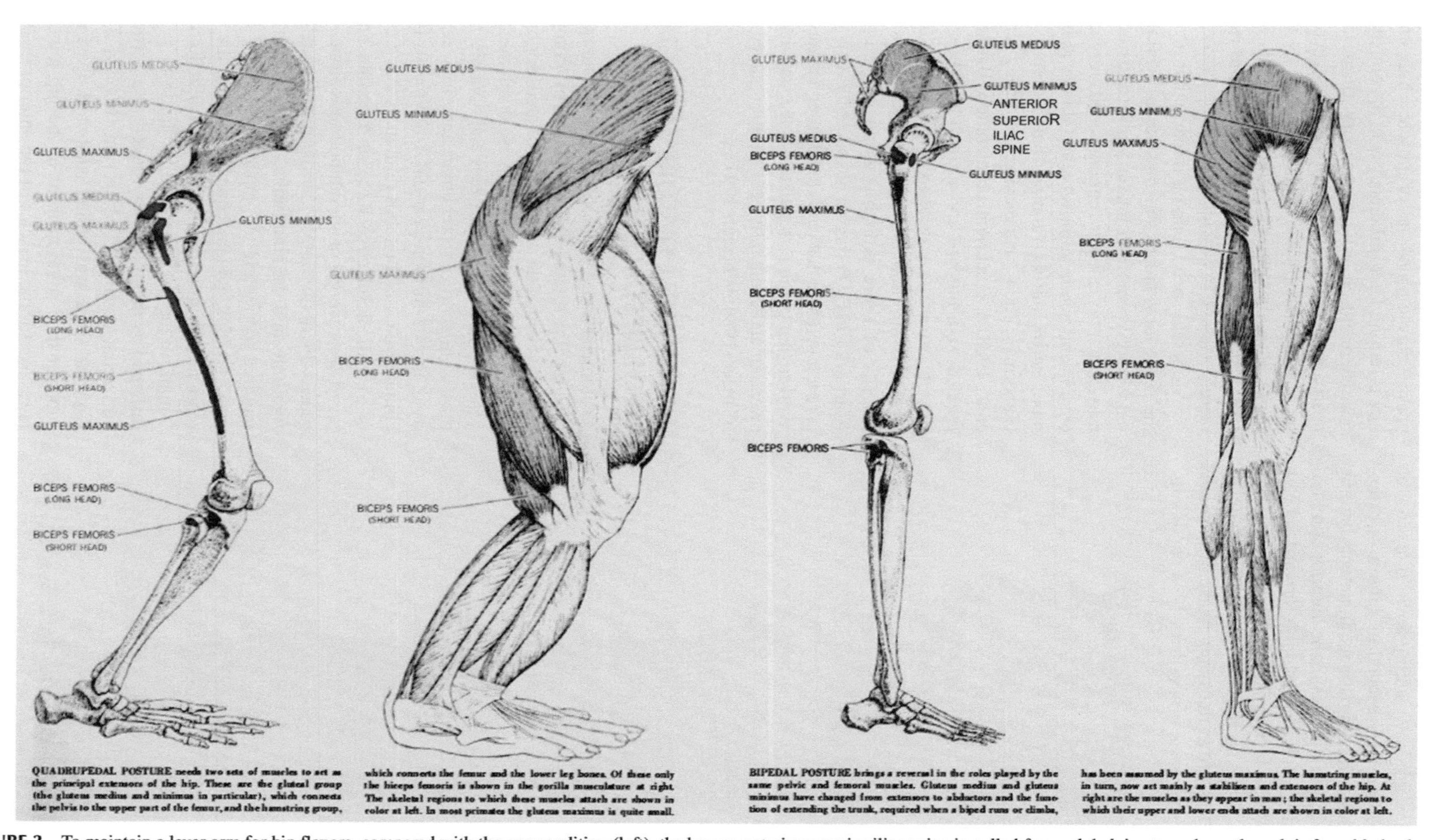

FIGURE 3 To maintain a lever arm for hip flexors, compared with the ape condition (left), the human anterior superior iliac spine is pulled forward, helping to reshape the pelvis from blade-shaped to curved and bowl-shaped. (See color plate section). *From Napier (1967).*

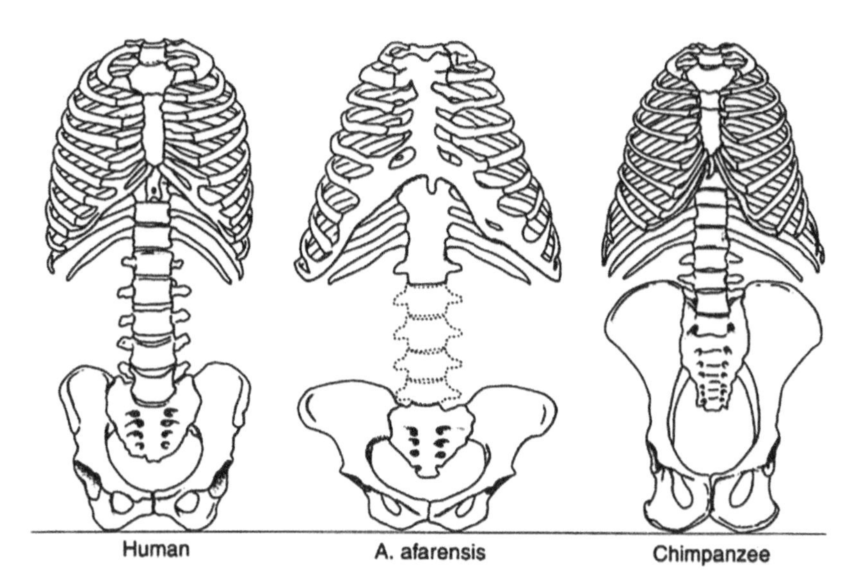

FIGURE 4 The rib cage of *Australopithecus afarensis* is cone-shaped like that of a chimpanzee, whereas the pelvis is short and bowl-shaped like that of a modern human.

Along with the reorganized pelvis, the human foot is dramatically reorganized compared with the ancestral condition. The ape foot is a grasping organ that resembles a human hand. The toes are long and curved, and preshaped to the form of the cylindrical branches the foot is adapted to grasping. The first or great toe is rotated away from the other toes, as with the human thumb, so that the foot can function as a clamp, allowing apes to grip and push off when climbing pole-like vertical supports. The gripping great toe allows apes to climb a tree bole using limb movements similar to those a human would use climbing a ladder. In contrast, the human foot has lost its gripping function, and instead the toes are in line so that all the toes are roughly parallel. The foot has been elongated into a strut; the plantar aponeurosis functions like the string of a bow, strung between the calcaneus (heel bone) and the toes so that the foot is a springy but sturdy lever. The calcaneus is broadened, thus reducing pressure during heel strike on tissue covering it. The internal structure of the heel is spongy rather than solid to allow shock absorption.

Because all body weight is borne on first one and then the other leg, the head of the femur, or hip joint, is large compared with that of apes, to accommodate the additional stress. During walking, the feet are placed close to the centerline of the body to reduce sway. The lower leg bone, the tibia, bears on its upper surface what is essentially a platform for the femur to rest on during the stance phase. This platform is roughly perpendicular to the ground; otherwise, the femur would tend to slide off the tibia, dislocating the knee laterally. The placement of the lower leg along the centerline requires that the upper leg bone, the femur, angles inward from its lateral position at the hip to articulate with the tibia near the centerline. The articular surface of the femur is angled to allow it to meet the horizontal tibial platform evenly, giving humans a knock-kneed appearance. This angle results in a *valgus* femur, a very distinctive morphology, compared with the bowlegged or *varus* chimpanzee knee. The same articular surface is elongated in humans from front to back, stabilizing the knee during the stance phase.

Some few features above the pelvis are directly or indirectly related to bipedalism. Most primates have highly evolved grasping and manipulative hands, but the hands must also support body weight during walking. The human forelimb is released from weight-bearing duties, allowing further refinement of its manipulative and tool-using functions. The semicircular canals, organs that help to maintain balance and stabilize the visual field during locomotion, are reshaped in bipeds. Australopiths have chimpanzee-like semicircular canals (Spoor et al., 1994). Humans have a specialized ligament, the nuchal ligament, that reduces the need for muscle action during running (Bramble and Lieberman, 2004). Because the head sits atop a vertical spine, the opening for the spinal cord, the foramen magnum, is closer to the very bottom of the skull rather than opening partly to the rear as it does in apes.

WHEN DID BIPEDALISM EVOLVE, AND WHAT EVOLUTIONARY STAGES OF BIPEDALISM DID WE PASS THROUGH?

Bipedalism may have evolved in roughly three stages. Monkeys and ancestral apes are quadrupeds, evolved as walkers and leapers. During typical locomotor activities, their spines are relatively horizontal; bipedal posture is rare. Apes engage in suspensory behavior, hanging from an arm or both arms and orienting the body so that the spine is vertical and the legs are extended as they are during bipedal walking (Keith, 1891, 1923). This posture, Keith speculated, preadapts gibbons for a clumsy kind of bipedalism. Gibbons brachiate, swinging under branches by hanging from first one arm and then the other; when a branch presents itself at their feet, they support their body weight with the hind limbs as well, sometimes even taking several steps bipedally. The most arboreal of the apes, orangutans and gibbons, are also the most bipedal (Hunt, 1991a). The more terrestrial chimpanzees and gorillas are less bipedal, perhaps having secondarily readapted for quadrupedal locomotion. Paleontologists have not yet positively identified postcranial elements of the common ancestor of humans and chimpanzees, so the locomotor habits of prebipeds are speculative. However, the fossil *Ardipithecus ramidus* is dated to near the branching node of humans and African apes, and may well resemble that common ancestor. *Ardipithecus ramidus* lacks adaptations for terrestrial locomotion, specifically dorsal ridges on their knuckles, making them more like gibbons or orangutans. The first stage in the evolution of bipedalism was a gibbon- or orangutan-like *facultative* or part-time bipedalism.

Habitual bipedalism may have evolved just over 4 million years ago (mya). The 4.2-million-year-old *Australopithecus anamensis* fossils have a tibial platform typical of bipeds, though fossils of the two most diagnostic skeletal elements, the pelvis and the distal (i.e., lower) femur, are as yet undiscovered. The Laetoli footprint track, dated to 3.6 million years, was clearly made by bipeds, even though the prints suggest a less-than-modern foot. Humans have a well-developed rounded ball, or swelling, at the base of the great toe, and a well-developed arch. These tracks suggest that these features were not yet evolved, and some argue that the prints reveal a somewhat divergent great toe. The well-known 2.9-million-year-old *A. afarensis* Lucy fossils are the earliest of a human ancestor to display clear skeletal hallmarks of bipedalism. The bowl-shaped *A. afarensis* pelvis is remarkably humanlike (Figure 4), though it did have attachments for the gluteals facing slightly more posteriorly. The *A. afarensis* pelvis is wide (Figure 4), giving Lucy a pear-shaped body; wide hips create more stress at the hip joint. *Australopithecus afarensis* had short legs and a knee joint that was less elongated front to back and therefore more apelike. Rather than a pair of convex femoral articular surfaces fitting into matching concave sockets on the lower leg bone, australopith tibia have rather convex surfaces both top and bottom, with round surface meeting roundish surface, and morphology that yields a flexible rather than stable joint. In contrast to this somewhat ape morphology, when viewed from the front the femur is angled in or valgus as it is in modern humans. The toes are long and curved, and the joint at the base of the big toe, deep in the foot near the ankle, is shaped to function like a hinge. As with a hinge, it allows little movement up or down and is therefore a well-developed rigid strut for pushing off during walking or running, but the toe can swing side to side to allow gripping of small branches. These traits suggest a less refined, less efficient bipedalism compared with that of modern humans, an anatomy that retained some adaptations to arboreal food collecting. Thus, by 3.6 mya, human ancestors had reached stage 2; they were striding bipeds differing from living humans only in retaining some climbing adaptations and less-than-fully-evolved energetic efficiency.

The skeletal morphology of the 1.5-million-year-old Nariokotome fossil (Walker and Leakey, 1993), often assigned to the taxon *Homo erectus*, closely resembles a modern human, though a robust one. While bone wall thickness was greater than that found in modern humans, leg length in relation to arm length, foot shape, joint size, and joint morphology, including the size of the hip joint, is virtually indistinguishable from that of modern humans, suggesting that modern bipedalism arose in this species. There is one important difference—the diameter of the vertebral column is small. This narrow spine is a feature also found in apes and earlier hominins. A broader vertebral body increases spine stability, preventing buckling, and is likely an adaptation to carrying. *Homo erectus* might then be said to be quite modern in locomotion, though less able to carry heavy objects than modern humans are. The earliest uncontested *H. erectus* fossils (e.g., ER 3733) are dated to 1.8 mya, and the 2.0-million-year-old ER 1472 femur is the first hominin femur similar in length to that of modern humans, differing from humans in only minor respects (McHenry and Corruccini, 1978). These fossils suggest that a gait virtually indistinguishable from that of modern humans had evolved in our ancestors by ~2.0 mya.

WHY DID BIPEDALISM EVOLVE?

Theorists who set out to explain the circumstances or forces that led to the evolution of bipedalism face a challenge that entangles few other human adaptations. Bipedalism is not only slow and unstable compared with quadrupedalism—unlike most adaptations (say, adaptations to chewing harder foods), it is clumsy and inefficient in its earliest evolutionary stages. This transition barrier would push most animals under evolutionary pressure to improve their locomotion toward refining quadrupedalism rather than adopting a new and inefficient locomotor mode. Either the anatomy of the nonbipedal ancestor was such that bipedalism was not significantly more inefficient or ineffective than the locomotion it typically engaged in, or the advantages of bipedalism were great enough to overcome its initial inefficiency.

The first bipedalism origin hypothesis was that of Darwin (1871). At the time, most scholars took for granted that bipedalism was superior to the locomotion of other animals, and that these inferior beings simply had not reached the advanced condition of humans. Darwin hypothesized that two of the most striking human characteristics, enormous brains and upright locomotion, were linked in a "tool-use hypothesis." Quite the opposite of our current understanding, Darwin suggested that humans evolved large brains while they were still apelike tree dwellers, and that large brains in turn led to the invention of tools, which required a freeing of the hands to carry and manipulate the tools, thus selecting for bipedalism. Canines were reduced in humans because they had been replaced by handheld weapons. This first true evolutionary hypothesis for the origin of bipedalism is now disproven by the fossil record: human-sized brains evolved only in the last million years, whereas the first stone tools appeared over a million years earlier. Bipedalism appeared ~4 mya, separated by millions of years from humanlike brains.

Sir Arthur Keith (1891, 1923) observed that gibbons held their legs extended during their distinctive hand-over-hand suspensory locomotion ("brachiation") and suggested that human bipedalism was the end product of a trend for progressively more erect posture from quadrupedalism in monkeys, to a more erect quadrupedalism in apes, to a quite erect posture during brachiation, finally culminating in bipedalism among humans. Keith noted more than 80 traits shared by apes and humans that are not shared with other primates. As this hypothesis was elaborated on through the mid-twentieth century, suspensory primates came to be seen as advanced and superior because suspensory behavior allowed foragers to move more directly among feeding sites in the periphery of tree crowns. Keith's "suspensory hypothesis" relied on evolutionary inertia to propel humans from the gibbon-like stage of bipedalism to a fully human one. Contemporary evolutionary theory discounts evolutionary inertia, leaving the suspensory hypothesis without a selective force necessary to push prebipeds over the bipedalism transition barrier. In the mid-twentieth century when the timing of brain size, stone toolmaking, and bipedalism was still unknown, Darwin's tool-use hypothesis was fused onto Keith's suspensory hypothesis to model bipedalism as having evolved in a brachiating ancestor as it began to include increased tool use in its daily behavior.

The hybrid tool-use/suspensory hypothesis had a solid internal consistency, but by midcentury the discovery of the *Proconsul* fossils, with their apelike face and teeth but monkey-like body, were argued by Louis S.B. Leakey and colleagues as evidence against a gibbon-like prehuman ancestor. Absolute dating and new fossil discoveries have now placed *Proconsul* early enough, and offered other more suspensory adapted fossils closer to the chimpanzee–human split, to give parts of the suspensory hypothesis new strength.

The "killer-ape hypothesis" (Dart, 1959) adopted parts of the tool-use hypothesis. Erect posture evolved to increase sight lines by raising the eyes above tall grasses endemic to savanna habitats and had the further advantage of freeing the hands to better dispatch prey with bone, horn and stone tools, rather than sharp teeth. As lethal open-habitat-adapted predators, early hominins were seen as relying on their increased bipedal stature to spot prey or avoid predators. Bipedalism had a further advantage—in the 1960s the tool-use hypothesis was embellished to suggest that throwing in particular was a vital hominin adaptation; other variants saw freed hands as valuable for grappling with prey. The killer-ape hypothesis is in conflict with evidence that bipedalism evolved well before stone tool use; australopith teeth are more characteristic of frugivores; their diet, as indicated by tooth wear, consisted of fruits, seeds, and leaves (Hunt, 1998); minute traces of food constituents left in fossils suggest plant diets; robust mandibles and faces are not expected in carnivores that used tools to butcher animals or crack open bones to get at marrow; no other predators have evolved a similar hunting adaptation; and australopith habitats likely contained diverse ground cover, only a small part of which would be tall grasses.

The "social display hypothesis" (Livingstone, 1962) was that bipedalism evolved because it increased the apparent size of the bipedal displaying individual, thereby increasing social status and access to food or mates, or intimidating predators. Among primates where bipedalism is part of social display, upright posture is used selectively and infrequently; there would seem to be no advantage to remaining bipedal outside of immediate dominance interactions. If bipedalism were engaged in indiscriminately and persistently, it would be disadvantageous for any but the most dominant animals, since among living primates aggressive signals by subordinate individuals challenge dominants and draw reprisals. Infants, juveniles, and young adults, low-ranking due to immaturity and small body size, would be at a particular disadvantage.

Some saw bipedalism as freeing the hands to carry either infants or food items. The "infant-carriage hypothesis" (Tanner, 1981; Iwamoto, 1985) recognized that brain expansion came late in human evolution, and instead identified the need for females to support infants with their arms as selecting for bipedalism. Counterevidence for the "infant-carriage hypothesis" is the fact that no other primate has evolved bipedalism to carry infants, and the small cranial capacity of early hominins suggests that infants would be no more helpless than chimpanzee infants. Young infant chimpanzees cling to their mother's belly, seemingly requiring more assistance than would be necessary to cling to a vertical torso.

The "food-carriage hypothesis," the idea that bipedalism freed the hands to carry food to mates or other relatives or group members, emerged midcentury (Bartholomew and Birdsell, 1953; Etkin, 1954) and gained momentum during the 1970s and 1980s (Isaac, 1978; Lovejoy, 1981). Lovejoy emphasized the disadvantages of upright locomotion, focusing attention on the extraordinary circumstances that must have been present to select for this slow and unstable positional behavior. Elaborating on the killer-ape hypothesis, early hominins were seen as tool-using hunters, where males gathered food and shared it with monogamous female mates and offspring; such monogamy accounts for the lack of visible estrus or sexual swellings in humans (Etkin, 1954; Lovejoy, 1981). Variants viewed early hominins as carrying scavenged carcasses back to a central gathering place. The hypothesis is not entirely consistent with primate and hominin morphology. The narrow vertebral body diameters of early hominins are contrary to the hypothesis that heavy loads were carried in the arms. Short legs, wide pelves, backward-facing pelvic muscle attachment areas, short front-to-back knee dimensions, flexible rather than stable knees, and long toes among early hominins are at odds with carrying, since bearing a load should increase selective pressure for locomotor efficiency. Carrying occupies the arms and makes further gathering difficult, and still more difficult if the food items are small or can only be collected in trees. Carrying implements might seem to solve these inconsistencies, but carrying devices would preclude the need for the hands in carrying. Multiple animal lineages, including birds and carnivores, have solved the carriage challenge by ingesting and regurgitating foods rather than adopting bipedalism. Australopith male body weights are reconstructed as twice that of females, whereas among monogamous primates, body weight differs little between the sexes (gibbons, for example). Furthermore, with males foraging away from the social group, lone females would be at risk for infanticide and forced copulation; gorilla females, for instance, follow their mates closely for protection against infanticidal males.

The "aquatic ape hypothesis" (Hardy, 1960) holds that bipedalism evolved as a wading posture. Its most popular permutation was diffuse in that it speculated that unique human traits evolved at different times for different aquatic activities, including diving, swimming and wading (Morgan, 1972). Robust faces, large molars, dental microwear, and reconstructed habitats with abundant aquatic resources are consistent with a wading/underground storage organ collection (Wrangham et al., 2009). Against the hypothesis is the fact that natural buoyancy reduces the need to evolve bipedal anatomical adaptations, since facultative bipedalism would be adequate for keeping the head above water. Much greater stresses during terrestrial behavior would seem to act to retain quadrupedal anatomy and are against selection to reorganize anatomy for bipedalism. Other wading mammals, including primates that forage in shallow water, have not evolved bipedalism.

The "locomotor efficiency hypothesis" (Rodman and McHenry, 1980) considers the efficiency of bipedalism over quadrupedalism as critical. Bipedalism evolved in response to a drying trend in the Miocene, which increased travel distances during foraging. Some carrying hypotheses coopt aspects of efficiency arguments. Chimpanzees, the best model for prehominins, are inefficient walkers, adapted for vertical climbing and arboreal quadrumanous climbing as much as terrestrial quadrupedalism. The locomotor efficiency hypothesis views prehominins as inefficient enough quadrupedally that bipedalism was not distinctly less efficient; or some individuals may have been more efficient as bipeds than quadrupeds (Pontzer et al., 2009). If efficiency selected for bipedalism in this manner, chimpanzees and other great apes might be expected to have evolved bipedalism as well. If other selection pressures are invoked to overcome the transition barrier—feeding for example—the hypothesis is more properly seen as a feeding hypothesis, wherein bipedalism was selected for improved efficiency after evolving for other reasons. Otherwise, evolutionary pressure to evolve greater efficiency seems more likely to select for increasingly efficient quadrupedalism rather than engaging in inefficient facultative bipedalism. Some quadrupeds are even more efficient walkers than humans. The initial inefficiency of a quadruped adopting facultative bipedalism is a seemingly impassable transition barrier.

The "thermal radiation avoidance hypothesis" (Wheeler, 1991) suggested that bipedalism evolved to reduce exposure to solar radiation. A bipedal hominin exposes only 7% of its surface to sunlight, whereas quadrupeds expose 20% of their surface to the sun, and wind speed is greater and temperatures lower only a meter above ground level (Wheeler, 1991). Against the hypothesis is the fact that living species that adapted to high heat environments are slender with long limbs to better dissipate heat, whereas the earliest bipeds have the short legs, wide hips, and robust upper bodies of australopiths that suggest higher energy generation and retention than in modern humans; inefficient locomotion is heat-generating and thus inconsistent with adaptation to heat stress. Further, reduction in thermal radiation absorption accrues only in direct sun and when the sun is directly overhead. Few animals are active at noon, and primates avoid direct sun when the weather is hot. Hominin habitats were likely partly wooded, further decreasing solar exposure. Wheeler's heat-stress hypothesis better accounts for the long-legged, slender body proportions of the early *Homo*.

Other hypothesis speculated that bipedalism freed the hands for gathering food (Du Brul, 1962). The "seed eating hypothesis" (Jolly, 1970) proposed that early hominins stood erect to gather grass seeds, and that the small size of seeds demanded a bipedal, alternate-hand plucking action to achieve a reasonable rate of ingestion. Robust jaws and teeth among early hominins and dexterous hands in living humans were hypothesized to have evolved for seed gathering. Drawbacks to the hypothesis are that grass seeds need not be gathered bipedally, since grass stems are easily pulled over to allow gathering from a sitting or squatting posture. Hominin dental microwear suggests a diet of fruits, leaves, and piths more than seeds. A completely terrestrial adaptation is at odds with arboreal australopith adaptations such as cone-shaped torsos (Figure 4), curved robust fingers, apelike scapulae, long curved toes, and short hind limbs.

The "small-object feeding hypothesis" (Rose, 1976) substitutes small-diameter fruits in short trees and bushes

FIGURE 5 Bipedal foraging in early hominins is used to harvest small fruits found in small trees. Some individuals reach up from the ground. Others stand bipedally, using the lateral toes to grip, but not the great toe. The smaller branches in such small trees are difficult to balance on, requiring the gripping of overhead branches using a forelimb as in arm-hanging. *From Hunt (1994).*

for grass seeds. Trees are not easily pulled over. When trees are distributed in semimonospecific stands, bipedal shuffling to move between trees hypothetically uses less energy than repeatedly lowering and raising the torso to walk quadrupedally (Wrangham, 1980). Against the hypothesis is morphology indicative of arboreal adaptations and inefficient bipedalism among early hominins.

The "postural feeding hypothesis" (Hunt, 1996) adds an arboreal component to the small-object feeding hypothesis. When feeding on fruits from short trees, chimpanzees collect fruits bipedally by reaching up to feed from the ground, but also when harvesting fruits arboreally (Hunt, 1994). Small trees elicited arboreal bipedalism because their small branches are too flexible to allow sitting posture or unassisted bipedalism. Instead, chimpanzees stand bipedally, gripping small branches with their lateral toes, and stabilize this bipedal posture by arm-hanging from an overhead branch (Figure 5). Small trees are more common in the driest parts of chimpanzee ranges and in dry habitats such as those in which early hominins are found. If prehominins were confined to habitats drier and more open than those in which chimpanzees are found, bipedalism might evolve first as a posture effective in terrestrial and arboreal gathering (assisted with a semi-arm-hanging posture arboreally) (Hunt, 1994).

If chimpanzees are appropriate stand-ins for prebipedal hominin ancestors, bipedalism is rare except in feeding contexts. Most behaviors offered as selective advantages to bipedalism are rare among chimpanzees: social display, tool use, carrying, throwing, and peering over obstacles constitute only 1%–2% of bipedal behavior each, whereas feeding makes up 85% of bipedal episodes (Hunt, 1994).

At Hadar, pollen from 15 tree species was recovered from strata containing *A. afarensis* fossils (Bonnefille et al., 2004). Two of those species, *Garcinia* and *Grewia*, were the first and ninth most common species gathered bipedally by chimpanzees at Gombe and Mahale (Hunt, 1994).

A postural feeding hypothesis is more consistent with early australopith morphology than other hypotheses. Robust, curved fingers, chimpanzee-like in morphology but more humanlike in length, are consistent with a partly arm-hanging adaptation in short trees with flexible small-diameter branches. Australopith torsos are cone-shaped (Schmid, 1983; Figure 4) and scapulae are intermediate between chimpanzees and humans, traits that reduce stress and fatigue during arm-hanging (Hunt, 1991b). Short hind limbs and a broad pelvis lower the center of gravity (Hunt, 1998), providing stability on flexible, small-diameter branches. Australopith semicircular canals are apelike and suggest arboreal adaptations (Spoor et al., 1994). While australopiths share arm-hanging traits with chimpanzees, they do not share climbing adaptations such as gripping great toes. Because small trees can be ascended with a single agile leap, harvesting fruits from small trees requires none of the vertical-climbing anatomy forest-adapted chimpanzees need to access fruits in tall trees (Hunt, 1998; Figure 5).

Arboreal adaptations found in australopiths are absent in *H. erectus*, suggesting that arboreal traits in early australopiths are reliable indicators of behavior since they disappear quickly, presumably once the trees are abandoned.

The postural feeding hypothesis views bipedalism as having evolved when an apelike prehominin, more arboreal than extant African apes and therefore more bipedal, was isolated in a dry habitat with abundant small fruits from short-statured trees. Such a semibipedal ape evolved bipedal posture for harvesting fruits by reaching up from the ground and with a semi-arm-hanging posture in trees. A discontinuous tree canopy and the opportunity for harvesting fruits from the ground required terrestrial movement, which selected for a more refined bipedalism than that of gibbons or orangutans. *Ardipithecus ramidus* lacks dorsal ridges, an adaptation to terrestrial knuckle-walking, but possesses chimpanzee-like upper body morphology, suggesting a gibbon-like adaptation. Faced with awkward bipedalism or awkward quadrupedalism, chimpanzees and bonobos, forced to the ground by more dispersed food resources due to competition from monkeys (Wrangham et al., 1998), were restricted to taller forests where no fruit could be harvested from the ground and took the alternate path, evolving into more effective quadrupeds. Faced with the same choice between an initially awkward bipedalism and awkward quadrupedalism, the advantages of bipedalism for fruit gathering might tip the balance to bipedalism among prehominins, leading first to postural and finally to striding bipedalism.

REFERENCES

Bartholomew Jr., G.A., Birdsell, J.B., 1953. Ecology and the protohominds. American Anthropologist 55, 481–498.

Bartlett, J.L., Sumner, B., Ellis, R.G., et al., 2014. Activity and functions of the human gluteal muscles in walking, running, sprinting, and climbing. American Journal of Physical Anthropology 153, 124–131.

Boakes, J.L., Rab, G.T., 2006. Muscle activity during walking. In: Human Walking. Lippincott Williams and Wilkins, Baltimore, pp. 103–118.

Bonnefille, R., Potts, R., Chalié, F., Jolly, D., Peyron, O., 2004. High-resolution vegetation and climate change associated with Pliocene *Australopithecus afarensis*. Proceedings of the National Academy of Sciences of the United States of America 101, 12125–12129.

Bramble, D.M., Lieberman, D.E., 2004. Endurance running and the evolution of *Homo*. Nature 432.7015, 345–352.

Carrier, D.R., 1984. The energetic paradox of human running and hominid evolution. Current Anthropology 25, 483–495.

Dart, R.A., 1959. Adventures with the Missing Link. Harper, New York.

Darwin, C., 1871. The Descent of Man and Selection in Relation to Sex. J. Murray, London.

Du Brul, E.L., 1962. The general phenomenon of bipedalism. American Zoologist 2, 205–208.

Etkin, W., 1954. Social behavior and evolution of man's mental faculties. The American Naturalist 88, 129–143.

Gray, H., Carter, H.V., 1858. Anatomy Descriptive and Surgical. John W. Parker and Son, London.

Hardy, A., 1960. Was man more aquatic in the past? The New Scientist 7, 642–645.

Hunt, K.D., 1991a. Positional behavior in the Hominoidea. International Journal of Primatology 12, 95–118.

Hunt, K.D., 1991b. Mechanical implications of chimpanzee positional behavior. American Journal of Physical Anthropology 86, 521–536.

Hunt, K.D., 1994. The evolution of human bipedality: ecology and functional morphology. Journal of Human Evolution 26, 183–202.

Hunt, K.D., 1996. The postural feeding hypothesis: an ecological model for the evolution of bipedalism. South African Journal of Science 92, 77–90.

Hunt, K.D., 1998. Ecological morphology of *Australopithecus afarensis*: traveling terrestrially, eating arboreally. In: Strasser, E., Fleagle, J.G., McHenry, H.M., Rosenberger, A. (Eds.), Primate Locomotion: Recent Advances. Plenum, New York, pp. 397–418.

Isaac, G. Ll., 1978. The food sharing behavior of protohuman hominids. Scientific American 238 (4), 90–108.

Iwamoto, M., 1985. Bipedalism of Japanese monkeys and carrying models of hominization. In: Kondo, S. (Ed.), Primate Morphophysiology, Locomotor Analysis and Human Bipedalism. Tokyo University Press, Tokyo, pp. 251–260.

Jolly, C.J., 1970. The seed-eaters: a new model of hominid differentiation based on a baboon analogy. Man 5, 1–26.

Keith, A., 1891. Anatomical notes on Malay apes. Journal of the Straits Branch of the Royal Asiatic Society 23, 77–94.

Keith, A., 1923. Man's posture: its evolution and disorder. British Medical Journal 1, 451–454 499–502.

Livingstone, F.B., 1962. Reconstructing man's Pliocene ancestor. American Anthropologist 64, 301–305.

Lovejoy, C.O., 1981. The origin of man. Science 211, 341–350.

Marzke, M.W., Longhill, J.M., Rasmussen, S.A., 1988. Gluteus maximus muscle function and the origin of hominid bipedality. American Journal of Physical Anthropology 77, 519–528.

Morgan, E., 1972. The Descent of Woman. Bantam Books, New York.

Muybridge, E., 1887. Complete Human and Animal Locomotion. Dover, New York.

Napier, J.R., 1967. The antiquity of human walking. Scientific American 56–67.

Pontzer, H., Raichlen, D.A., Sockol, M.D., 2009. The metabolic cost of walking in humans, chimpanzees, and early hominins. Journal of Human Evolution 56, 43–54.

Rodman, P.S., McHenry, H.M., 1980. Bioenergetics and the origin of hominoid bipedalism. American Journal of Physical Anthropology 52 (1), 103–106.

Rose, M.D., 1976. Bipedal behavior of olive baboons (*Papio anubis*) and its relevance to an understanding of the evolution of human bipedalism. American Journal of Physical Anthropology 44, 247–261.

Schmid, P., 1983. Eine Reconstrucktion des skelettes von A.L. 288-1 (Hadar) und deren konsequenzen. Folia Primatologica 40, 283–306.

Spoor, F., Wood, B., Zonneveld, F., 1994. Implications of early hominid labyrinthine morphology for evolution of human bipedal locomotion. Nature 369, 645–648.

Tanner, N.M., 1981. On Becoming Human. Cambridge University Press, Cambridge.

Walker, A.C., Leakey, R.E.F. (Eds.), 1993. The Nariokotome Homo Erectus Skeleton. Harvard University Press, Cambridge.

Wheeler, P.E., 1991. The thermoregulatory advantages of hominid bipedalism in open equatorial environments: the contribution of increased convective heat loss and cutaneous evaporative cooling. Journal of Human Evolution 21, 107–115.

Wrangham, R.W., Cheney, D., Seyfarth, R., Sarmiento, E., 2009. Shallow-water habitats as sources of fallback foods for hominins. American Journal of Physical Anthropology 140, 630–642.

Wrangham, R.W., 1980. Bipedal locomotion as a feeding adaptation in gelada baboons, and its complications for hominid evolution. Journal of Human Evolution 9, 329–331.

Wrangham, R.W., Conklin-Brittain, N.L., Hunt, K.D., 1998. Dietary response of chimpanzees and cercopithecines to seasonal variation in fruit abundance. I. Antifeedants. International Journal of Primatology 19, 949–970.

Chapter 9

Early Hominins

Kevin D. Hunt
Human Evolution and Animal Behavior, Indiana University, Bloomington, IN, USA

SYNOPSIS

Fossils of three genera of possible human ancestors (*Sahelanthropus*, *Orrorin*, and *Ardipithecus*) have been dated to 7–4.4 mya. Where the same skeletal elements have been recovered the morphology of the three genera is similar. When compared to chimpanzees and bonobos all the three exhibit reduced incisors, smaller canines, and less sectorial lower first premolars. A single fragmentary pelvis displays a mixture of chimpanzee- and hominin-like features. Interpretation of these fossils is confounded by the many australopith-like features in *Ouranopithecus*, the purported common ancestor of African apes and humans. *Ouranopithecus* shares with australopiths (but not these three possible hominins) a robust, prognathic face, tall zygomatics, large incisors, large cheek teeth, and thick molar enamel. *Ouranopithecus* shares with australopiths and the three possible hominins small canines and a shallow supratoral sulcus. The robusticity of *Ouranopithecus* makes it less likely that *Sahelanthropus*, *Orrorin*, and *Ardipithecus* are in the hominin lineage, since such a lineage would require an evolutionary progression that starts robust, becomes nonrobust, and then reacquires robusticity. Some paleontologists, however, are skeptical of the place of *Ouranopithecus* in the human lineage. They find compelling indirect evidence that the common ancestor of African apes and humans was quite chimpanzee-like, which would exclude robust species such as *Ouranopithecus* from the human lineage. If the common ancestor was chimpanzee-like, then the argument that *Sahelanthropus*, *Orrorin*, and *Ardipithecus* are part of the human ancestry is stronger.

TAXONOMY

Until recently, humans and human ancestors after the ape–human divergence were placed in their own exclusive family, the Hominidae. Members of this taxon were referred to informally as "hominids" (see the chapter by Martin). DNA sequencing and comparative anatomical studies, however, have shown that African apes and humans are more closely related than previously supposed, which means that they form a natural grouping to the exclusion of orangutans and the lesser apes. To reflect the newly recognized close relationship between humans and African apes, they are all placed together in the taxonomic family Hominidae. Because chimpanzees, bonobos, and humans are more closely related to one another than to the gorilla (Patterson et al., 2006; see Table 1), the family Hominidae is split into two subfamilies, the Gorillinae (encompassing the living genus *Gorilla* and its exclusive ancestors) and the Homininae, which includes humans, bonobos, chimpanzees, and their exclusive ancestors.

While there is considerable controversy over the exact divergence dates of the apes (see Rogers and Commuzzie, 1995; Langergraber et al., 2012), many would date the split of gorillas from the lineage that led to humans and *Pan* at ~10 mya (Scally et al., 2012). Humans are thought to have branched off from panins around 6–7 mya (Scally et al., 2012; Patterson et al., 2006). Genetics suggests that humans and paleochimpanzees did not make a clean break of it but rather underwent a period of interbreeding that lasted over a million years (Patterson et al., 2006). After the divergence of humans, the lineage that led to bonobos and chimpanzees evolved on its own until the chimpanzee–bonobo split at a little less than 2 mya (Stone et al., 2010). In the taxonomy used here each of the three living genera, *Homo*, *Pan*, and *Gorilla*, is placed in its own tribe: Hominini, Panini, and Gorillini. The chimpanzee/bonobo tribe Panini is referred to informally as "panins" and fossil humans and their immediate ancestors in the tribe Hominini are "hominins." The shift continues into lower taxonomic levels. The term "australopithecine" for the now defunct subfamily Australopithecinae has been replaced by "australopith," the colloquial name for the subtribe Australopithecina. Robust australopiths are left in the genus *Australopithecus* here, though some advocate placing these fossils in their own genus, *Paranthropus*. The validity of *Paranthropus* has been questioned in part as the close genetic similarity of chimpanzees and humans: if chimpanzees and humans are sister species, as closely related as animals usually placed in the same genus, then all of the hominins must be very closely related indeed, meaning that they should all be in the same genus. Members of the genus *Homo* are in the subtribe Hominina (informally, "homins"). Hominoidea (informally, "hominoids") refers to apes, including the gibbon and siamang and humans (Table 1).

PURPORTED EARLY HOMININ SPECIES

Four species in three genera dating between 7 and 4.4 mya have been proposed by their discoverers as hominins: *Sahelanthropus tchadensis* (6.5 mya), *Orrorin tugenensis* (~6 mya), *Ardipithecus kadabba* (5.8–5.2 mya), and *Ardipithecus ramidus* (4.4–4.2 mya). None of these species has been accepted unequivocally by the paleontological community as a human ancestor; rather, most experts agree that the first indisputable hominin is the 4.2 mya *Australopithecus anamensis* (Harrison, 2010) and they take a wait-and-see position on *Sahelanthropus*, *Orrorin*, and the two *Ardipithecus* species. We will refer to fossils placed in these three genera as Early Possible Hominins (EPHs).

The determination of where these four EPH species fall in the human phylogenetic tree would be considerably easier if we could be certain we have correctly identified the last common ancestor (LCA) of the hominids (i.e., the African apes and hominins). It is tempting to make the simplifying assumption that our closest genetic relatives—chimpanzees and bonobos—are good representatives of our common ancestor. Proponents of this perspective see panins as having remained in the traditional ape habitat, closed canopy forest, in contrast to hominins, the apes that moved into a drier, more open savanna/woodland habitat. In this view, the savanna habitat drove the evolution of human-like traits, traits the panins did not evolve because they remained in the forest. The chimpanzee-as-LCA argument is not merely a simplistic, first assumption; some scholars who are intimately familiar with apes, EPHs, and australopiths offer cogent arguments that the LCA was nearly indistinguishable from the living chimpanzee (Pilbeam and Young, 2004). If this turns out to be true, the morphology of the EPHs is reassuringly intermediate between chimpanzees and australopiths, and the case for EPHs as true hominins is strong.

Inconveniently, this straightforward state of affairs does not exist. Rather than being chimpanzee-like, the most often recognized candidate for the hominine LCA is the large-molared, robust-faced, australopith-like (at least in the face) *Ouranopithecus macedonensis* (Andrews, 1992; see Figure 1). If *Ouranopithecus* or some close relative with similar morphology is the LCA, the simplest evolutionary progression would leave EPHs as a better fit in the chimpanzee/bonobo lineage than the hominin one. However, while *Ouranopithecus* is on a very short list of most-likely LCAs, few ape fossils have been preserved from the late Miocene (Pilbeam and Young, 2004; Harrison, 2010) and the LCA may well have escaped fossilization or eluded discovery. Miocene (23–5.5 mya) apes are both diverse and fragmentary, leaving the puzzle of ape evolution in general and the identity of the common ancestor of the Homininae (chimpanzees and humans) in particular unresolved, thus complicating the interpretation of the EPHs.

TABLE 1 Early Hominin Taxonomy

Taxon	Dates
Order: Primates	
Suborder: Haplorhini	
Superfamily Hominoidea (apes and humans; informally "hominoids")	
Family Hylobatidae (informally "hylobatids")	
Genus *Hylobates* (gibbons and siamangs)	
Family Pongidae (informally "pongids")	
Subfamily Ponginae	
Pongo pygmaeus (orangutan)	
Family Hominidae (African apes and humans; informally "hominids")	
Subfamily Gorillinae	
Tribe Gorillini	
Gorilla gorilla (gorilla)	
Subfamily Homininae (chimpanzees and humans, "hominines")	
Tribe Panini (chimpanzees and bonobos; informally "panins")	
Pan paniscus	
Pan troglodytes	
Tribe *incertae sedis*	
Genus and species	Dates
Sahelanthropus tchadensis	(7–6 mya)
Orrorin tugenensis	(5.9–5.7 mya)
Ardipithecus kadabba	(5.8–5.2 mya)
Ardipithecus ramidus	(4.4–4.2 mya)
Tribe Hominini (informally "hominins," replacing hominids)	
Subtribe Australopithecina (informally "australopiths")	
Genus and species	Dates
Australopithecus anamensis	(4.2–3.9 mya)
Australopithecus afarensis	(3.6–2.9 mya)
synonym: Kenyanthropus platyops	
synonym: Australopithecus bahrelghazali	
Australopithecus aethiopicus	(2.7–2.3 mya)
Australopithecus boisei	(2.3–1.4 mya)
Australopithecus africanus	(3–2 mya)
Australopithecus garhi	(2.5 mya)
Australopithecus robustus	(1.9–1.5 mya)
Australopithecus "habilis"	(1.9–1.4 mya)
Subtribe Hominina (informally "homins")	
Homo habilis (e.g., OH 7 and ER 1470)	
Homo erectus (including *ergaster* and *antecessor*)	
Homo sapiens (including *heidelbergensis* and *neanderthalensis*)	

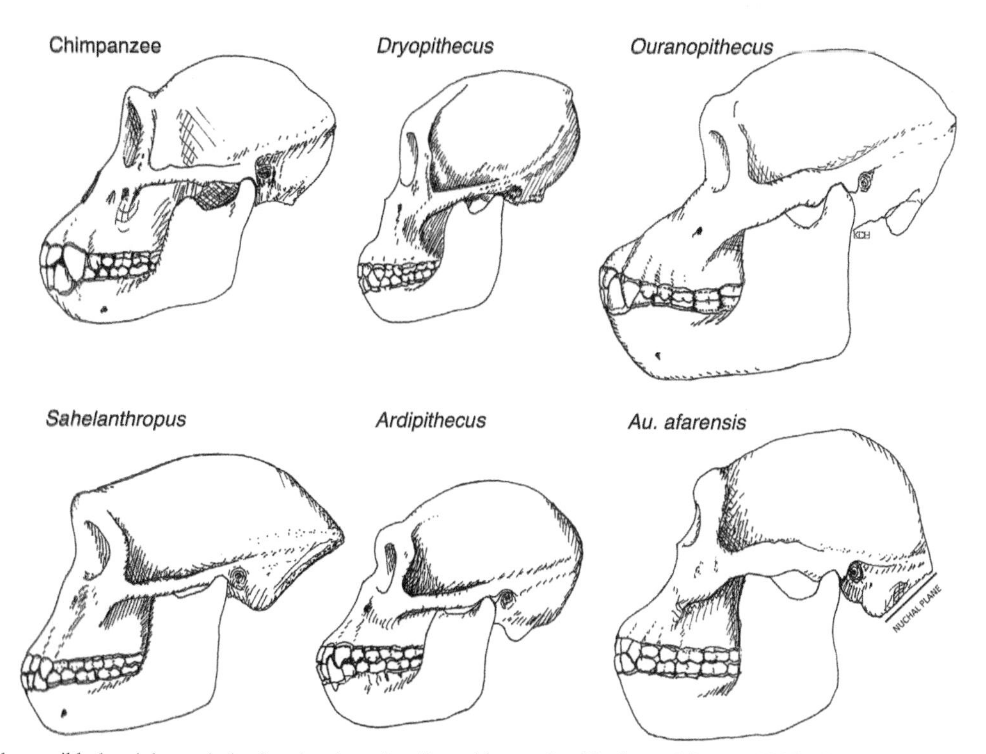

FIGURE 1 Early possible hominins and closely related species. *Dryopithecus* after Kordos and Begun (1997). *Ouranopithecus* after de Bonis and Koufos (1994). *Sahelanthropus* after Zollikofer et al. (2005) (mandible conjectural). *Ardipithecus* after Suwa et al. (2009); ARA-VP-6/500 cranium and ARA-VP-1/401 ARA-VP-6/500 mandible, with some reconstruction. *Australopithecus afarensis* after Kimbel and Rak (2005).

CONTEXT: CHIMPANZEE MORPHOLOGY

Chimpanzees and bonobos have prognathic faces, projecting out from between the eyes at a 45° angle in the side view (see Figure 1). Above the eyes, they have a large bar of bone called the supraorbital torus, set off from the braincase behind by a distinct depression or concavity called the supratoral sulcus. Panins have cranial capacities (brain sizes) of about 350 cubic centimeters (cc), or to be precise 367 cc for chimpanzees (males 381 cc, females 352) and 343 for bonobos (males 356, females 329; Schultz, 1969). Chimpanzees have a highly angled nuchal plane (see Figure 1, bottom right) that orients their neck (i.e., nuchal) muscles backward; the exit of the spinal cord—the foramen magnum—is placed relatively rearward in the skull. These features are often interpreted as accommodations to the horizontal orientation of the spine during quadrupedal locomotion, and conversely a downward facing nuchal plane and foramen magnum are often viewed as indicating bipedal posture. This connection is at the very least unreliable, and perhaps we should simply conclude that it is wrong. Both features vary wildly among quadrupedal primates. The subnasal (also called *alveolar* or *premaxillary*) area of the panin face is convex as viewed either from above (superiorly) or from the side (laterally). The chin is receding, retreating from the teeth at 45° angle (Figure 1). Viewed from the front, the inferior edge of the chimpanzee zygomatic melds into the upper jaw well above the tooth row, resulting in a high zygomatic origin. Again viewed from the front, the top to bottom dimension of the zygomatic is small, compared to *Ouranopithecus* or the australopiths. This is often referred to as a *short* or *gracile* zygomatic. Viewed from above (superiorly), the zygomatics slope backward (posteriorly) from the midline to either side, each forming a 45° angle to the coronal plane, a plane that would divide the body from head to toe into a front half and back half. This is in contrast to australopiths (particularly robust australopiths) where the zygomatics are flat or squared off as viewed from above (see the chapter by Ward for further details).

The interior of the panin mandible behind the chin has two horizontal (or transverse) ridges, called a superior (upper) transverse torus and an inferior (lower) transverse torus. The inferior torus is often called a *simian shelf* when it extends farther backward than the superior torus. Although the panin mandible is thick and dense compared to the human jaw, it is thin compared to australopiths and *Ouranopithecus,* not much thicker than the molars it holds.

Panin incisors are broad and have large roots compared to australopiths, humans, and monkeys. Male canines are tall (protruding well past the chewing plane of the cheek teeth), broad, and cone shaped (see Figure 1), with the broadest part of the crown near the gumline. There is a distinct diastema (space) between the upper canine and the upper lateral incisor, and between the lower canine and the first premolar. It is this space that allows the canines to interlock (Figure 1) when the mouth is closed; the lower

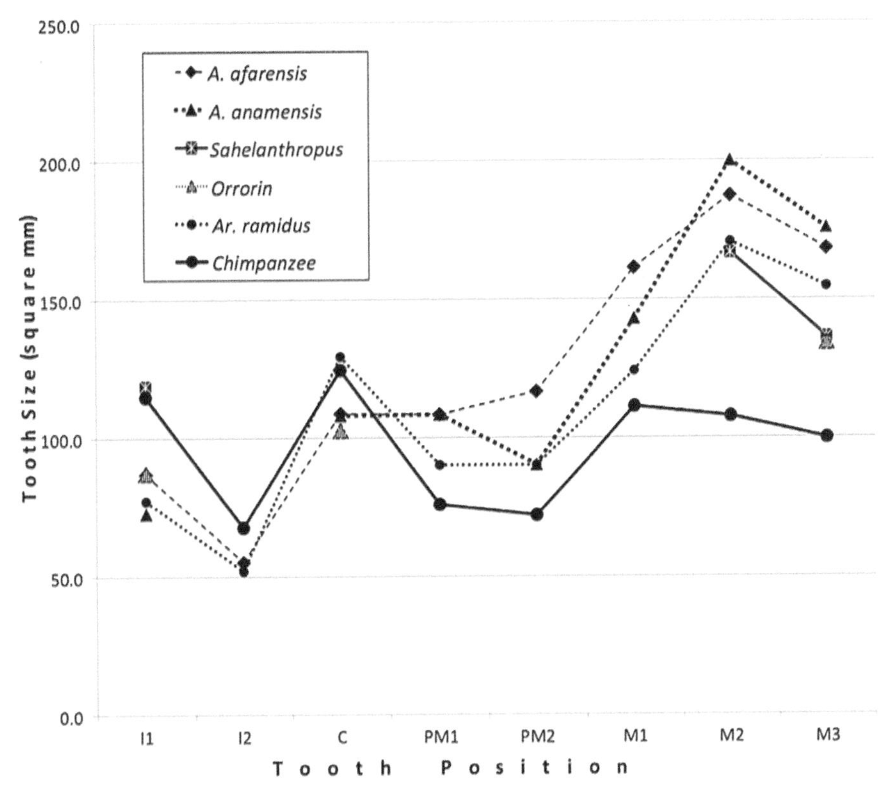

FIGURE 2 Tooth areas in relevant species, length × breadth. I1, central incisor; I2, lateral incisor; C, canine; PM1, first premolar; PM2, second premolar; M1, first molar; M2, second molar; M3, third molar. Chimpanzees (unbroken line) have large incisors, large canines but small premolars and molars. Australopiths have smaller incisors, smaller canines, and larger PMs and Ms. EPHs are intermediate.

canines fit just in front of the upper canines. Opening and closing the jaw cause the back edge of the upper canine to hone against the front edge of a specialized, elongated lower first premolar. This *sectorial* first lower premolar usually has a single large cusp. The long sloping honing-face makes the tooth asymmetrical, compared to the rather symmetrical, elliptical, bicuspid first premolar in humans. In juveniles the deciduous first molar is in this tooth position, and it is sectorial, too. In contrast to this honing mechanism, most hominins (and some female great apes) have canines that are small enough that there is no honing function and they wear on the tip rather than the trailing edge. A reduced canine is sometimes inappropriately offered as definitive evidence that a particular fossil is a hominin, but the feature is not diagnostic because a number of fossil apes (see *Dryopithecus* and *Ouranopithecus,* Figure 1) also have reduced canines.

While the incisors and canines of panins are larger than those of australopiths (Figure 2), the molars and premolars are smaller. The combination of large incisors and cheek teeth (premolars and molars) that are aligned one directly behind the other in chimpanzees and bonobos causes the teeth to be arranged in two approximately parallel left and right rows. This gives apes a U-shaped dental arcade. Because australopith incisors and canines are narrower the tooth rows are either V shaped or curved (parabolic).

Below the neck, chimpanzees have long, robust fingers, relatively small thumbs, and long upper limbs in relation to lower limbs giving them an intermembral index (IMI) above 100. Chimpanzees have IMIs of 110 whereas bonobos have an IMI of 102; the human IMI is about 67 (Schultz, 1937; Ankel-Simons, 2007). The panin ribcage is somewhat cone shaped; their scapulas are narrow. Long clavicles (collar bones) cause the scapulas to be positioned on the back of the torso rather than the side. Ape thoraxes are wide, compared to monkeys, giving apes shoulders. The shoulder joint is mobile in panins and other apes, similar to the human condition, allowing them to raise their arm completely above their heads. They can orient the forelimb at many other angles, as well, an ability denied most other animals, including monkeys, which have restricted shoulder mobility. Both apes and humans have the ability to rotate their wrists, a condition that contrasts to the stiff and nonrotatory wrists of monkeys.

Whereas humans have triangular scapulas, panin scapulas are narrow and elongated, an adaptation to arm-hanging that is also present in gibbons. Panins (and gorillas) have four lumbar vertebrae, in contrast to the human five (Schultz, 1969) and australopith six (Lovejoy et al., 2009c). Among the great apes, the last or fourth lumbar vertebra (and not infrequently the third as well) is nested between the iliac blades of the pelvis (i.e., "invaginated," see Figure 3). This

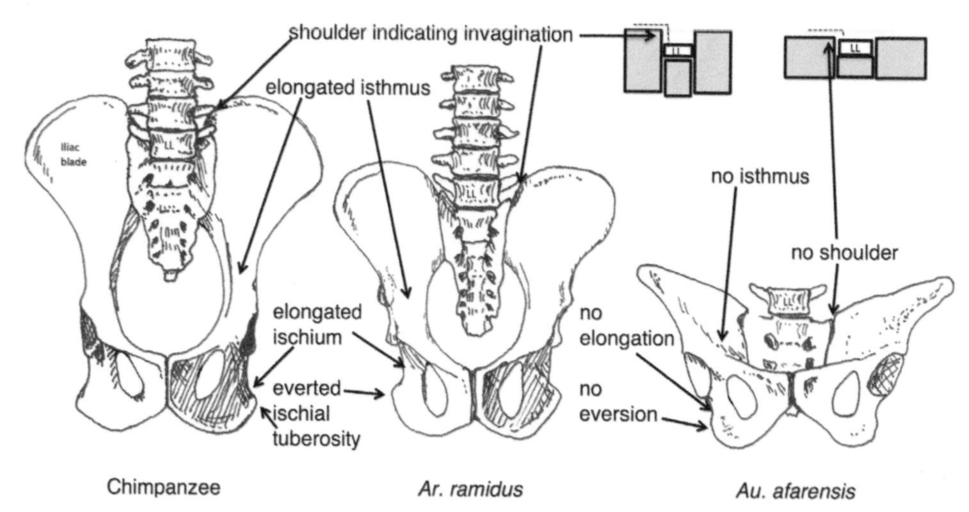

FIGURE 3 Pelves of the chimpanzee (left, after A.H. Schultz and lab specimens), *Ardipithecus* (after Lovejoy et al., 2009c, with chimpanzee rather than *Au. afarensis* sacrum), A. L. 288 ("Lucy") specimen. Note that both chimpanzees and *Ar. ramidus* have distinct shoulders where the sacrum articulates with the iliac blade. Both species have a last lumbar vertebra (LL) that is depressed downward between the iliac blades. *Australopithecus afarensis* has no shoulder. Both chimpanzees and *Ar. ramidus* have an elongated section of the pelvis, the "isthmus" that increases the height of the pelvis. Both chimpanzees and *Ar. ramidus* have an enlarged and everted ischial tuberosity. These features suggest that *Ar. ramidus* like chimpanzees has a short stiff back, likely an adaptation to vertical climbing, and engaged in long periods of sitting.

leaves very little space between the lowest rib and the top of the pelvis (Figure 3), giving great apes a short, stiff back and limited side-to-side movement (Tuttle, 1975). Not only lumbar length but also back musculature is reduced, compared to monkeys. The two main pelvic bones, the os coxae, are elongated top to bottom so that the middle portion of the os coxa is stretched into a narrow isthmus (Figure 3), as if a twisted disc were elongated into a twisted bow tie. The panin foot is hand-like with a large gap between the big toe and the other toes, giving them powerful, clamp-like gripping ability. The foot can grip a vertical bole powerfully enough to propel the body upward quickly enough that it can be called a vertical run; they can climb a pole much like humans climb a ladder.

While there is still controversy in linking specific panin morphology to specific behaviors, a well-accepted perspective is that narrow scapulas, a cone-shaped thorax, long arms, and long curved and robust fingers are adaptations to arm-hanging, a posture used most often when gathering fruit among small branches in the tree periphery (Hunt, 1991b). The gripping great toe, short lumbar segment, elongated os coxae (pelvic bones), and long arms are widely viewed as adaptations to vertical climbing. Dorsal ridges on the metacarpals function to prevent hyperextension of the fingers against the palm bones during knuckle-walking (Tuttle, 1967, 1975). Other traits in the wrist make it a better weight-bearing joint by limiting its range of motion, especially by limiting the action of cocking the wrist (dorsiflexion).

CONTEXT: MIDDLE MIOCENE APES

In the middle Miocene (~12 mya) a group of apes, the dryopiths, had evolved from relatively monkey-like forebears so that they resembled chimpanzees in many ways (Table 2). They inhabited closed canopy forests in what is now Europe where they were common, more abundant (Merceron et al., 2007) even than in Africa. *Dryopithecus* molars and premolars were chimpanzee-sized in relation to their body size and the enamel outer layer of the teeth was thin, again like panins (Alba et al., 2010). Their faces and mandibles were unreinforced and the zygomatics were short, resembling living chimpanzees (Figure 1). They had convex premaxillas, sloping chins, simian shelves, and backward facing nuchal areas. However, not all of their morphology was chimpanzee-like. Most significantly their faces were less prognathic, reduced, and retracted enough that as viewed from above their zygomatics were less angled back than those of chimpanzees and thus a little like australopiths. Their supraorbital torus and the sulcus behind it were small compared to panins (Figure 1). A particularly interesting species, *Anoiapithecus bevirostris* (12 mya), considered by some researchers to be *Dryopithecus,* has an extremely retracted face, leading some researchers to suggest that a retraced, australopith-like face might well have been the common condition among middle Miocene apes. *Dryopithecus* canines were smaller than those of chimpanzees, but chimpanzee-like in their conical shape and the tendency to wear along the trailing edge. Their incisors were smaller than those of chimpanzees.

Other dryopith fossils from this time period such as *Pierolapithecus* and *Oreopithecus* shared with chimpanzees broad torsos and long arms. Fingers are intermediate in *Pierolapithecus* and long in *Oreopithecus*. Unlike chimpanzees, however, *Oreopithecus* is described as having a more human-like short, broad pelvis, somewhat like that of *Ar. ramidus,* and *Pierolapithecus* has a rather gibbon-like

TABLE 2 Features of Critical Species[b]

Feature	*Dryopithecus*	*Ouranopithecus*	EPHs	Chimpanzees	Australopiths[a]
Date (mya)	12	9.6	6.5–4.2	0.00	4.2–2.5
Nuchal plane angle	Backward	?	Intermediate	Backward	Downward
Facial prognathism	Moderate	Great	Moderate	Great	Moderate
Premaxilla shape	Convex	Convex	Convex	Convex	Convex or flat
Chin shape	Very retracted	Very retracted	Moderately retracted	Very retracted	Moderately retracted
Simian shelf	Prominent	Moderate	None	Prominent	None
Supratoral sulcus	None	Small	Small	Large	Intermediate
Zygomatic origin/shape	High/Gracile	Low/Robust	High/Gracile	High/Gracile	Intermediate/Intermediate
Zygomatic angle	Moderately angled	Moderately angled	Highly angled	Highly angled	Not angled/flat
Mandibular robusticity[c]	Gracile	Robust	Intermediate	Gracile	Robust
Central incisor size	Moderate	Large	Moderate	Large	Moderate
Molar enamel	Thin	Thick	Intermediate	Thin	Thick
Molar dimensions	Small	Large	Intermediate	Small	Large
Canine size/shape	Conical/Large	Con-Diam/Moderate	Con-Diam/Moderate	Conical/Large	Con-Diam/Smaller
Lower 1st premolar shape	Sectorial	Sectorial/Bicuspid	Sectorial/Bicuspid	Sectorial	Sectorial/Bicuspid
Deciduous lower M1	Narrow/Sectorial	Narrow/Sectorial	Narrow/Sectorial	Narrow/Sectorial	Molarized or narrow
Tooth row shape	U shaped	U shaped	U shaped	U shaped	V shaped
Finger length	Long	As with *Dryopithecus*?	Long	Long	Short
Finger curvature	Great	As with *Dryopithecus*?	Great	Great	Great
Thumb length	Short	As with *Dryopithecus*?	Short	Short	Medium
Arm length	Very long	As with *Dryopithecus*?	Very long	Very long	Long
Gripping great toe	Yes	As with *Dryopithecus*?	Yes	Yes	No/Vestigial
Wrist mobility	Great	As with *Dryopithecus*?	Great	Great	Great
Rotatory wrist	Great	As with *Dryopithecus*?	Great	Great	Great
Thorax shape	Cone shaped	As with *Dryopithecus*?	?	Cone shaped	Cone shaped
Scapula shape	?	?	?	Narrow	Intermediate
Lumbar vertebra number	?	?	?	4	5–6
Inferred lumbar muscles	Medium	As with *Dryopithecus*?	?	Small	Small
Os coxa shape[d]	Elongated	As with *Dryopithecus*?	Somewhat elongated	Very elongated	Not elongated
Leg length	Short	As with *Dryopithecus*?	Short	Short	Medium
Foot shape	Gripping	As with *Dryopithecus*?	Gripping	Gripping	Human-like

[a]*Not including robust australopiths.*
[b]*Kordos and Begun (1997).*
[c]*Koufos and de Bonis (2005).*
[d]*Hammond et al. (2013).*

pelvis (Hammond et al., 2013). Where there are shared skeletal elements, other late Miocene apes are similar to these species; for example, *Hispanopithecus and Nakalipithecus* have been reconstructed with broad torsos, long arms, and long, chimpanzee-like fingers (Alba, 2012; Begun, 2010; Kunimatsu et al., 2007 Nakatsukasa and Kunimatsu, 2009).

Most paleontologists agree that these differences mean that dryopiths were more arboreal than chimpanzees, rarely coming to the ground. Chimpanzees spend nearly half their day on the ground, requiring adaptations for walking, among them dorsal ridges on the back of the palm bones that prevent hyperextension of the fingers during knuckle-walking and highly evolved adaptations to vertical climbing, necessary because they must reascend trees after each terrestrial travel bout (Hunt, 1991a). Gibbons, siamangs, and orangutans are nearly completely arboreal. Perhaps surprisingly, these highly arboreal species engage in more bipedalism than more terrestrial apes such as chimpanzees, bonobos, and gorillas (Hunt, 1991b). It may be that late Miocene apes have short pelves as do hominins because they are arboreal, and arboreal apes are more bipedal.

If we did not have fossils from the late Miocene, the dryopiths, with their gracile faces, largish canines, and small, thin-enameled molars would be taken as positive confirmation that the LCA of hominins and panins was more or less a chimpanzee.

CONTEXT: LATE MIOCENE APES

The late Miocene species *Ouranopithecus macedoniensis* from Greece is dated to 9.6 mya. Its morphology makes the chimpanzee-as-LCA scenario less likely. *Ouranopithecus* is a large ape with a body mass somewhat greater than that of chimpanzees, estimated at 72 kg (160 lbs) for males; the females are much smaller, perhaps half that of the males (de Bonis and Koufos, 1994). The *Ouranopithecus* face resembles that of gorillas and to a lesser extent chimpanzees but is more robust with thick facial bones, a low zygomatic origin, tall zygomatics, and an australopith-like flat face (e.g., the XIR-1 specimen in Figure 1). Like the African apes it has great prognathism, a receding chin, and a supraorbital torus and sulcus (though one less pronounced than in chimpanzees, Table 2, Figure 1). This small browridge resembles that of EPHs and australopiths more than chimpanzees or gorillas. *Ouranopithecus* mandible has a moderate simian shelf, resembling EPHs and australopiths rather than chimpanzees or dryopiths. *Ouranopithecus* molar cusps cluster toward the center of the tooth, reminiscent of the teeth of *Au. africanus.* Incisors are large and thus chimpanzee-sized, compared to the smaller incisors of EPHs and australopiths (Guleç et al., 2007).

The great *sexual dimorphism* of *Ouranopithecus* is typical of species with a gorilla- or orangutan-like social system, one where males battle one another singly. Chimpanzee males engage in group conflicts, and they have less sexual dimorphism. While it is not clear exactly how closely they are related, *Ouranopithecus* and its African relatives *Nakalipithecus* and *Samburupithecus* all differ from *Dryopithecus* in having greater jaw robusticity, large (larger than any of the EPHs or even *Au. afarensis*), thick-enameled molars, large premolars, and small canines (Figure 2). They may be similar because they had similar habitats and diets or they may resemble one another because they shared a recent ancestry. Either way there is a trend toward robust faces late in the Miocene.

The morphology of *Ouranopithecus* canines is important because not only are they relatively smaller than those of chimpanzees, but also some females have diamond-shaped canines—pointy but narrowing near the gumline—that are quite similar to those of australopiths (see the chapter by Ward). Some lower first premolars are bicuspid rather than sectorial (de Bonis and Koufos, 1994), again a condition seen in australopiths. *Oreopithecus,* a gibbon-like Miocene ape, also has bicuspid first premolars (Harrison, 1986; Sarmiento, 1987), suggesting that this is a common late Miocene ape condition. The postcranials of *Ouranopithecus* are poorly known but often assumed to be similar to those of other European Miocene apes, somewhat chimpanzee-like but without such long fingers or knucklewalking adaptations (Begun, 2010).

Ouranopithecus thus resembles australopiths as much as chimpanzees. They had smaller canines, larger, more thickly enameled molars and more robust faces than chimpanzees. If their pelvis is similar to that of other late Miocene apes like *Pierolapithecus* or *Oreopithecus* we begin to see surprisingly australopith-like apes.

Ouranopithecus shares more than just morphology with australopiths; their habitats were drier or more seasonal (de Bonis and Koufos, 1994) than that typical of panins and therefore more like australopith habitats. Many paleontologists see a connection between robust teeth and faces and dry habitats, since these habitats offer fewer fleshy fruits and more seeds and other tough foods compared to rainforests.

THE MORPHOLOGY OF PURPORTED EARLY HOMININS

Sahelanthropus tchadensis (roughly dated to 6.5 mya) was uncovered at Toumai, Chad, ~2500 km west and 1000 km or more farther from the equator than other early hominin sites, significantly extending the range of possible hominins (Brunet et al., 1995). The habitat has been reconstructed as relatively open, with woodland, savanna, and riverine forest (Vignaud et al., 2002), suggesting a dry-habitat ecology similar to that of *Ouranopithecus* and australopiths (see the chapter by Sept). The collection includes a cranium, mandible, and several isolated teeth.

The mandible is said to be robust (Brunet et al., 2002) but measurements on the image yield a corpus thickness of ~1 cm, which is well within the range of chimpanzees. The skull was crushed and otherwise deformed during preservation but much of this distortion was corrected with a digital reconstruction that "glues" together 3D images of the individual fragments to yield a relatively undistorted finished product (Zollikofer et al., 2005). It has a chimpanzee-sized brain (365 cc) and a proportionately massive supraorbital torus, thicker than typical in living African apes or even *Homo erectus*; despite the large torus, the sulcus or depression behind it is not as deep as in African apes. It is difficult to assess this feature since it does not resemble any other living or fossil species; it may be simply an unusual feature of this particular individual. The specimen has greater postorbital constriction than chimpanzees; that is, as viewed from above there is a distinct narrowing just behind the face and in front of the braincase (Brunet et al., 2002). *Sahelanthropus* is not terribly prognathic; in the side view the face does not jut out as much as chimpanzees, inspiring some to compare it to *H. erectus*, but this retracted condition is also similar to that of Miocene apes such as *Dryopithecus* (Figure 1) and *Anoiapithecus*. The subnasal area is convex both in lateral and in superior view. The zygomatics are receding as in chimpanzees, unlike the squared-off australopith zygomatics. The nuchal area is large, indicating large chimpanzee-like neck muscles in life. The angle of the nuchal plane has been a point of contention. The discoverers describe it as having a shallow slope, suggesting downward facing neck muscles and—they would contend—bipedalism. However, even in the digital reconstruction the slope of the nuchal plane is nearly as steep as that of a typical chimpanzee (Figure 1) and in any case as discussed above there is only the most tenuous relationship between nuchal plane angle and posture. The foramen magnum is said to be moved forward compared to chimpanzees, so that the opening faces directly downward, rather than posteriorly (rearward). A downward oriented foramen magnum is argued to be necessary to articulate with a vertical (or erect) spinal column, whereas a rearward facing opening would fit with a horizontal spine. This downward orientation is argued to be an indicator of bipedalism, but others point out that many quadrupedal primates, the squirrel monkey, for example, have downward oriented openings. The point may be moot, in any case, for two reasons. First, the foramen position is not outside the range for chimpanzees (Ahern, 2005); second, bonobos and gibbons have more human-like anteriorly placed foramen magnums than chimpanzees (Cramer, 1977).

In most measures *Sahelanthropus* tooth size is intermediate between chimpanzees and australopiths (Figure 2). Enamel thickness is intermediate as well. The upper first incisor is chimpanzee-sized, a distinct contrast to other EPHs and australopiths, which have smaller incisors than chimpanzees. As might be expected given the large incisor, the palate is U shaped, long and narrow, like that of living apes (Guy et al., 2005). The canine is chimpanzee-sized for the one measure we have; the root is large and the crown is conical. The discoverers emphasize that it has apical (australopith-like) wear, that is, wear on the tip, but there is also a thin strip of wear on the back margin of the tooth, suggesting some chimpanzee-like contact with the lower premolar. There is only a small space between the canine and the adjacent incisors; the canines did not interlock as completely as they do in living apes (Figure 1). This morphology is ambiguous. Some *Ouranopithecus* specimens have similar wear and the large male specimen (XIR-1) has a bicuspid lower first premolar, not the honing or sectorial morphology of living apes. Several other Miocene apes have this canine/PM morphology, as well (Sarmiento, 2010; Wood and Harrison, 2011). The discoverers compared the specimen to later hominins and demonstrated a close affinity to *Homo habilis* (Guy et al., 2005) but the comparative sample lacked Miocene apes. While the fossil resembles *H. habilis* when the choice is among *H. habilis*, *Australopithecus*, and chimpanzees, it may well have fallen closer to Miocene apes if they had been included.

Twenty fossils from four sites in the Tugen Hills of Kenya have been attributed to the species *O. tugenensis*. They include mandible fragments, the upper 2/3 of a femur including a femoral head, two femoral fragments, part of a humerus, a finger bone, a distal radius, and a number of teeth including a canine and molars (Figure 1; Senut et al., 2001; Senut and Pickford, 2004). The fossils are dated to ~5.8 mya. The equatorial latitude, the presence duikers and other forest animals, and the presence of tree trunks in contemporaneous deposits nearby suggest that the habitat was a mosaic of bushland, woodland, and closed canopy forest (Sawyer et al., 2007).

The *O. tugenensis* upper canine is small relative to the other teeth, more *Ouranopithecus*- or *Australopithecus*-sized than chimpanzee sized. Despite the small size it is conical and has an ape-like groove on the front where it contacts the lower canine as the jaws close, suggesting that the canines were interlocking (Senut et al., 2001). Incisors are small compared to chimpanzees (Figure 2) and are instead similar in size to *Au. afarensis*. As with *Sahelanthropus, Orrorin* molars are intermediate in size between chimpanzees and australopiths (Figure 2). The discoverers made the case for *O. tugenensis* as a human ancestor by noting that it shares with australopiths thick enamel, whereas later EPHs *Ar. kadabba* and *Ar. ramidus* have thinner enamel, possibly linking them with apes rather than hominins. However, this assertion was immediately questioned (Haile-Selassie, 2001). It is noteworthy that Miocene apes such as *Ouranopithecus* also have thick enamel, leaving it unclear whether *Orrorin* retains a primitive feature or has evolved toward australopiths.

The *Orrorin* humerus is ape-like in that it is robust and has a large supracondylar crest for the attachment of brachioradialis, an indication of powerful elbow-flexing capacity as would be used during vertical climbing. The finger bone is robust, curved, and therefore ape-like, but not as long as that of a chimpanzee. The humerus and fingers features suggest that *Orrorin* engaged in chimpanzee-like arboreal climbing and arm-hanging.

In contrast to the clearly ape-like morphology of the upper limb, one analysis of eight features in the lower limb found a clear resemblance to robust australopiths (Richmond and Jungers, 2008). The analysis, however, was limited to humans and living apes and did not include Miocene apes. Another analysis did include Miocene apes such as *Dryopithecus* and *Hispanopithecus,* and *Orrorin* falls near them (Almécija et al., 2013). The resemblance to Miocene apes suggests they may have engaged in a gibbon-like bipedal walking in trees rather than human-like striding bipedalism. As noted above, apes that live almost exclusively in trees are more bipedal than the partly terrestrial African apes (Hunt, 1991a). *Orrorin* has an intertrochanteric line, a feature that runs between two bony projections on the top of the femur, the greater and lesser trochanter. This feature might be argued to indicate bipedality, since it is found in *Au. afarensis* and modern humans, but it is actually of little help here since it is also found in orangutans (Tuttle, 1981). *Orrorin* shares with modern humans (but not australopiths) a medially projecting lesser trochanter: when you look at the femur straight on from the front, the smaller of these bony projections, the lesser trochanter, can be seen projecting medially toward the body's centerline; however, in australopiths the lesser trochanter is not visible because it projects backward (see the chapter by Ward). Yet again, this feature is less definitive than it first appears since the human condition is observed in Miocene apes (Almécija et al., 2013) and even in female chimps.

The *Orrorin* femoral head is large in comparison to the shaft, even larger than in *Au. afarensis,* a feature that links it to modern humans; frustratingly, orangutans also have a large femoral head. The femoral neck, the bony link between the shaft and the ball portion of the ball-and-socket hip joint, is angled upward as it is in humans; australopiths, in contrast, have femoral necks less strongly angled upward. This feature does not necessarily indicate bipedalism because chimpanzees have the human-like condition. The angled-upward femoral neck (human-like) and long femoral neck (australopith-like) are interesting but less definitive than they might be when function is considered. These features move the greater trochanter away from the hip joint, allowing greater hip flexibility; the orangutan hip is by far the most flexible of the living apes.

Crude CT scans have been interpreted as showing that bone in the femoral neck is thicker along the lower edge, a similarity to humans, australopiths, and (perhaps surprisingly) langurs and colobus monkeys (Galik et al., 2004, Rafferty, 1998), but not living apes; however, other studies find the pattern to be chimpanzee-like (Almécija et al., 2013).

In short, in feature after feature *Orrorin* resembles humans and australopiths rather than chimpanzees when only those three are compared, but when Miocene apes are introduced into the comparison it is unclear whether *Orrorin* most closely resembles living apes, Miocene apes, or australopiths. Its closest relatives are most likely other EPHs.

Ardipithecus kadabba fossils from the Middle Awash and Asa Koma of Ethiopia (and perhaps Lothagam in Kenya (Schwartz and Tattersall, 2005)) include mandibular fragments, a number of loose lower teeth, finger bone fragments, humeral fragments, an ulnar fragment, a clavicle fragment, a toe bone (Haile-Selassie, 2001), an upper canine, and a number of loose lower teeth (Haile-Selassie et al., 2004); all of these fossils are dated to ~5.5 mya.

The canine is conical and therefore ape-like, but as is typical of EPHs it is small compared to chimpanzees. It is worn both apically and distally, typical of other EPHs. Enamel thickness is intermediate between chimpanzees and australopiths, as is also the case with *Sahelanthropus* and *Ar. ramidus*—but perhaps not *Orrorin.* The lower first premolar is asymmetrical and therefore somewhat ape-like (Haile-Selassie et al., 2004). Wear on the canine and premolar suggest more of an interlocking, honing adaptation than is seen in *Ouranopithecus, Sahelanthropus, Orrorin,* and the later australopiths; the fossil is said to be female chimpanzee-like rather than Miocene ape- or australopith-like (Haile-Selassie et al., 2004). The molars and premolars are similar in size to other EPHs, that is, intermediate between chimpanzees and australopiths.

The finger bones are robust and curved and therefore similar to chimpanzees in shape but shorter in length. The ulna is curved, as is common among apes. The fragmentary humerus is robust and similar to that of chimpanzees, typical of EPHs and australopiths. The toe bone is curved and therefore chimpanzee-like, but the joint is tilted upward as in humans, possibly suggesting bipedal adaptations. The clavicle is robust. These features would be expected if *Ar. kadabba* is related to *Sahelanthropus* and *Orrorin,* so interpreting its evolutionary relationship to australopiths is fraught with the same difficulties.

Fossils attributed to *Ar. ramidus* were found in the Middle Awash, Ethiopia, and are dated to ~4.4 mya. Some slightly younger fossils attributed to *Ar. ramidus* from Gona, Ethiopia, may be as young as 4.15 Ma (Semaw et al., 2005) and their close geographic and temporal proximity argues that they may well be part of the same population. If so, *Ar. ramidus* is contemporaneous with the more australopith-like *Au. anamensis.* Although the specimens are fragmentary and lack critical elements such as the sacrum and scapula, *Ar. ramidus* fossils are abundant compared to other EPHs. There are 145 teeth (not including matching teeth on the opposite side) including 14 canines (Suwa et al., 2009). Important skeletal elements include a nearly complete cranium and mandible, parts of all long bones, a poorly preserved pelvis, various finger and toe bones, metacarpals (palm bones), wrist bones including a capitate, ankle bones including tarsals, a talus, and metatarsals (Lovejoy et al., 2009b). *Ardipithecus ramidus* is reconstructed as having a body weight of 50 kg (110 lbs), or the size of a small male chimpanzee, though 40 kg may be a better estimate. Low levels of variation in fossil size suggested to the discoverers that sexual dimorphism was low, which would make *Ar. ramidus* quite unlike the sexually dimorphic australopiths and *Ouranopithecus*.

Ardipithecus ramidus premolar and molar dimensions are similar to other EPHs, intermediate in absolute size between chimpanzees and *Au. afarensis* (Figure 3); dental enamel likewise is intermediate (Suwa et al., 2009). Incisors are small compared to chimpanzees and *Sahelanthropus* and very similar to *Orrorin*. The lower canine is conical and female chimpanzee- or bonobo-like in shape. Upper canines, though large in cross-sectional area (Figure 3), are variable in shape; some canines are diamond shaped and comparable to later australopiths and others (e.g., holotype ARA-VP-6/1; White et al., 1994) are more conical and thus chimpanzee-like. The low variation in canine size is offered as further evidence that males and females were relatively similar in body size.

As with other EPHs, many of the features the discoverers cite to show that the fossils are in the human lineage are also found in Miocene apes. In fact, of the 26 characters purported to indicate hominin status for *Ar. ramidus*, 14 are part of the canine/premolar complex (Sarmiento, 2010), a weakness in the analysis, since *Ouranopithecus* is rather australopith-like. Of the remaining characters in the skull and postcranial skeleton, all are also found in Miocene apes (Sarmiento, 2010).

Considering that *Sahelanthropus* and *Ar. ramidus* are separated by a little more than 2 million years, the skulls are quite similar (Figure 3). *Sahelanthropus* has an immense browridge, but both have relatively short faces, small supratoral sulcuses, relatively forward-placed foramen magnums, and nuchal area orientations that are intermediate between chimpanzees and later australopiths (as in Figure 3). Zygomatics are short: when viewed from the front there is a distinct space between the tooth row and the bottom edge of the zygomatics, a condition like that of chimpanzees and different from that in australopiths and *Ouranopithecus*. Zygomatics are closer to chimpanzees than to australopiths not only in size but also in orientation; they are angled back as viewed from the top, rather than squared-off as in the flat-faced australopiths. Both *Sahelanthropus* and *Ar. ramidus* have great postorbital constriction, the narrowing just behind the face and in front of the braincase and similar cranial capacities, 300–350 cc for *Ar. ramidus*. The tooth row is not quite U shaped, but slightly narrower at the front, clearly a consequence of their smaller incisors.

In most ways the arms and hands of *Ar. ramidus* resemble chimpanzees and bonobos more than australopiths. Finger bones are long, robust, and curved. While the hands have been described as short by the discoverers and others (Sarmiento and Meldrum, 2011), when finger length is compared to the length of the best preserved *Ar. ramidus* arm bone, the radius, the proportions are indistinguishable from bonobos (Figure 4). The thumb is short, as it is in living great apes, but more robust than that of panins, as in australopiths. The bones that make up the palm (the metacarpals, not the finger bones) are shorter than those of chimpanzees and they lack *dorsal ridges*, a feature found in apes that walk on their knuckles on the ground, as do all African apes. The discoverers reconstruct a wrist joint with greater flexibility than is seen in chimpanzees, particularly in the ability to cock (or dorsiflex) the wrist (Lovejoy et al., 2009a). Robust humeri and articular surfaces of the elbow joint are typical of the robust arms of EPHs and panins. Forearm bones are robust and curved, as in chimpanzees, indicating great power during rotatory wrist movement.

The discoverers reconstruct the *Ar. ramidus* lower limbs as slightly longer than the forelimbs, making them human-like. The reported proportions, however, are still in the range of bonobos. Moreover, the proportions cannot be determined with certainty since the lengths for both the partial humerus (upper arm bone) and the femur (upper leg bone) are only estimates. Allowing for some play with the reconstruction, *Ar. ramidus* forelimb/hind-limb proportions are not just within the range of bonobos, but possibly near the bonobo average.

Toe bones are long, curved, and robust. The big toe is divergent and similar in function to that of chimpanzees and bonobos. While the foot differs in some details from panins, it is ape-like.

The pelvis was deformed during preservation and broken into tens of fragments requiring some interpretation and guesswork during reconstruction. The sacrum was missing; the describers (Lovejoy et al., 2009c) substituted the sacrum from *Au. afarensis*, the Lucy specimen (A.L. 288-1). This is a problem because chimpanzees have narrow sacrums and humans have wide sacrums. Lucy's sacrum is broader than the human sacrum, and its substitution makes the *Ar. ramidus* pelvis even broader than the already extraordinarily broad Lucy pelvis (Lovejoy et al., 2009c). The insertion of the broad Lucy sacrum into the pelvis rotates the lateral iliac crest portion of the os coxae forward, making them appear more human-like than even the Lucy pelvis. This seems unlikely given

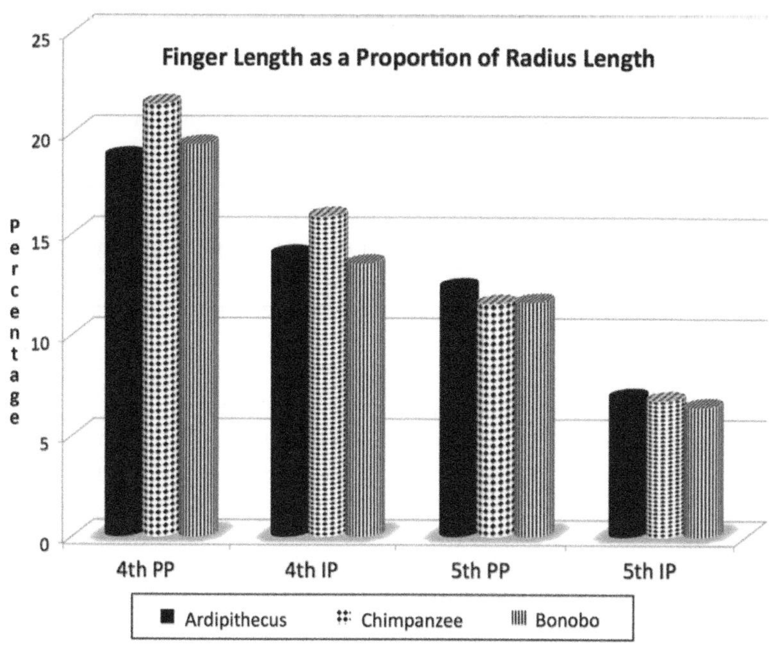

FIGURE 4 Relative finger lengths in *Ar. ramidus*, chimpanzees, and bonobos expressed as a proportion of radius length; PP proximal phalanx, IP intermediate phalanx. Radii are 25 cm for *Ar. ramidus* and 28 cm for the panins. Relative finger lengths are almost identical in bonobos and *Ar. ramidus*.

the ape-like character of nearly every other skeletal element. If the pelvis is reconstructed with a narrower, panin-like sacrum (Figure 3), a more believable profile as viewed from the top is obtained. The bottom half of the *Ar. ramidus* pelvis is wholly chimpanzee-like (Figure 3). It has a flared ischium, the area of the pelvis that bears weight during sitting. The os coxa has a distinct isthmus, or narrowed and elongated area between the top and the bottom of the bone (Figure 3). There is a pronounced shoulder just above the articulation of the sacrum (Figure 3), an ape-like trait; *Au. afarensis* and humans lack this shoulder. This steep slope suggests that the sacrum of *Ar. ramidus* was depressed downward or "invaginated" in a manner similar to that of chimpanzees or bonobos, yielding a shorter, stiffer back than in humans. *Australopithecus afarensis* has six lumbar vertebrae, compared to five for humans; panins have four. The invaginated sacrum of *Ar. ramidus* suggests four lumbar vertebrae.

Viewed from the side, the *Ar. ramidus* pelvis is more compressed from top to bottom than the chimpanzee and thus similar to the condition seen in *Oreopithecus*. Along the rear edge of the os coxa there is a curved area that in the still more compressed human pelvis makes a tight curve termed the sciatic notch. The *Ar. ramidus* pelvis also has a distinct bump in front of the hip joint called an anterior inferior iliac spine. This spine is rare in living apes, but present in humans. The compressed top-to-bottom proportions, pseudo-sciatic notch, and anterior inferior iliac spine are indeed human-like traits, but as with many other EPH traits these human-like features are less definitive than they appear at first glance, since they are also found in the one relatively complete Miocene ape pelvis, that of *Oreopithecus* (Kohler and Moya-Sola, 1997). The elongated os coxa, invaginated sacrum, long isthmus, *Oreopithecus*-like upper half of the pelvis, and distinctly chimpanzee-like lower halfare quite unlike those of australopiths and most resembles that of Miocene apes.

EARLY POSSIBLE HOMININS' ADAPTATIONS

The discoverers argue that *Ar. ramidus* was a careful climber in trees and that they combined upright, walking on branches with quadrupedal walking with their palms rather than knuckles contacting supports. While the discoverers seem to see this as a kind of locomotion unknown in living apes, it is in fact a locomotor mode chimpanzees regularly engage in on branches smaller than 5 cm (Hunt, 1991a). The divergent big toe of *Ar. ramidus* was said to be useful for gripping branches during slow, cautious movement in the trees, at the same time giving *Ar. ramidus* a stable, tripod-like foot when on the ground. It is a settled issue among functional morphologists that a gripping great toe is an adaptation to vertical climbing, not to other cautious arboreal movements (Rose, 1988; Hunt, 1991a). The discoverers argue that bipedalism was certainly the most common kind of terrestrial locomotion and may have been *Ar. ramidus*' only means of progression on the ground.

This behavioral reconstruction has not been greeted with enthusiasm by other experts. Some suggest that bonobo-like body proportions functioned in *Ar. ramidus* as they supposedly do in bonobos, to facilitate leaping

(Sarmiento and Meldrum, 2011). These scholars accept the body weight estimate of the discoverers, which gives *Ar. ramidus* short hands in relation to body weight, comparable to the proportions of gorillas but short compared to other apes. In this view, such short hands mean that arm-hanging suspensory behavior, a vital fruit-collecting posture for chimpanzees, must have been unusual for *Ar. ramidus*. These researchers suggest that *Ar. ramidus* walked quadrupedally on its palms both in the trees and on the ground (Sarmiento and Meldrum, 2011). They find little support for the contention that *Ar. ramidus* was a human rather than African ape ancestor.

A different interpretation is offered here. It is a long-accepted tenet among functional morphologists that curved, robust, and long fingers are linked to arm-hanging (Rose, 1988; Hunt, 1991b) and that a gripping foot is related to vertical climbing. *Ardipithecus ramidus* lacked dorsal ridges and may have had greater dorsiflexion (wrist-cocking) ability than extant African apes, together suggesting that *Ar. ramidus* did not often knuckle-walk (Lovejoy et al., 2009c). Likewise, there is little doubt that the intermediate nature of the *Ar. ramidus* pelvis—resembling chimpanzees, gibbons, and Miocene apes such as *Oreopithecus*—suggests that *Ar. ramidus* engaged in a higher frequency of bipedalism than is found among living great apes. The discoverers suggest that this bipedalism extends to frequent terrestrial bipedalism. The intermediate nature of the pelvis, long forelimbs, and chimpanzee-like fingers instead argue for a more gibbon-like *arboreal* bipedalism. Although the link is less secure, long forelimbs in relation to hind limbs and short and stiff backs have been linked to vertical climbing (Sarmiento, 1995). *Ardipithecus ramidus* has short metacarpals, a slightly shorter pelvis than panins, and a more flexible back than panins; it therefore lacks a full expression of vertical climbing adaptations, suggesting a lower frequency of vertical climbing compared to panins. Since living apes that confine their behavior to the trees are more bipedal than African apes, *Ar. ramidus* are best viewed as more arboreally adapted than chimpanzees. The lack of knuckle-walking features suggests the same. If *Ar. ramidus* came to the ground rarely, they would have had less need to reascend into trees, which would be consistent with their less-expressed vertical climbing adaptations. This interpretation views *Ar. ramidus* as an arm-hanger that moved with a chimpanzee-like quadrupedalism when in trees but rarely traveled on the ground and did not often engage in African ape-like knuckle-walking, making them a less specialized version of living gibbons or siamangs.

This locomotor interpretation is consistent with what we know of the *Ar. ramidus* diet. An examination of traces of radioactive isotopes that indicate diet places *Ar. ramidus* close to chimpanzees, suggesting that they subsisted on ripe fruits and other arboreal foods, not on resources found in the savanna (Sponheimer et al., 2013).

There is a diversity of opinions. *Ardipithecus ramidus* has been described as a terrestrial biped and arboreal careful climber (Lovejoy et al., 2009a) and a bonobo-like leaper and terrestrial quadruped (Sarmiento and Meldrum, 2011). Its anatomy is most consistent with the view that it is an arboreal arm-hanger and gibbon-like biped (Hunt, 1991a,b, 1994).

It is impossible to say without further evidence whether EPHs are members of a single evolving lineage, but their similarity makes this an attractive first interpretation. Their brains were chimpanzee-sized, and their molar dimensions, enamel thickness (with the possible exception of *Orrorin*) and incisors were intermediate in size between chimpanzees and australopiths. Of the living great apes, their faces are most like bonobos, and somewhat less robust than chimpanzees. Their canines are similar in shape to those of panins, though shorter and less interlocking. Their hands and arms were heavily muscled as in panins. Feet and toes were curved and adapted for gripping, which in turn suggests a locomotor repertoire with frequent arm-hanging and cautious arboreal walking and some underbranch suspensory locomotion. The ladder-climbing-like vertical climbing locomotion that is important for panins may have been less important for EPHs. There is only one pelvis for the EPHs (from *Ar. ramidus*) and it is shorter than that of living great apes, but resembles chimpanzees more than australopiths. There are enough *Ar. ramidus* fossils to suggest that there is little body size dimorphism, but this issue is far from resolved. If dimorphism were low, *Ar. ramidus* was unlike australopiths and *Ouranopithecus* and more like gibbons and siamangs. All in all, there are many intriguing gibbon-like features among the EPHs.

EARLY POSSIBLE HOMININS' PLACE IN NATURE

The determination of whether EPHs are in the human lineage or instead were just another of many different apes that existed in the Miocene would be simpler if we could identify the LCA of gorillas, chimpanzees, bonobos, and humans.

One particularly important issue is that reduced canines, small spaces for the canines to interlock, bicuspid rather than sectorial first premolars, short retracted faces, and short sideways facing pelvic bones are each found in at least one ancient ape, and most of these traits are found among a number of Miocene apes (Wood and Harrison, 2011). Far from being unique, the short, orthognathic face of the EPHs is found in *Dryopithecus* (Figure 1) and the 12 mya Spanish specimen *Anoiapithecus* (Wood and Harrison, 2011; Moya-Sola et al., 2009). Paleontologists encounter this dilemma often: retained primitive features in a lineage are often

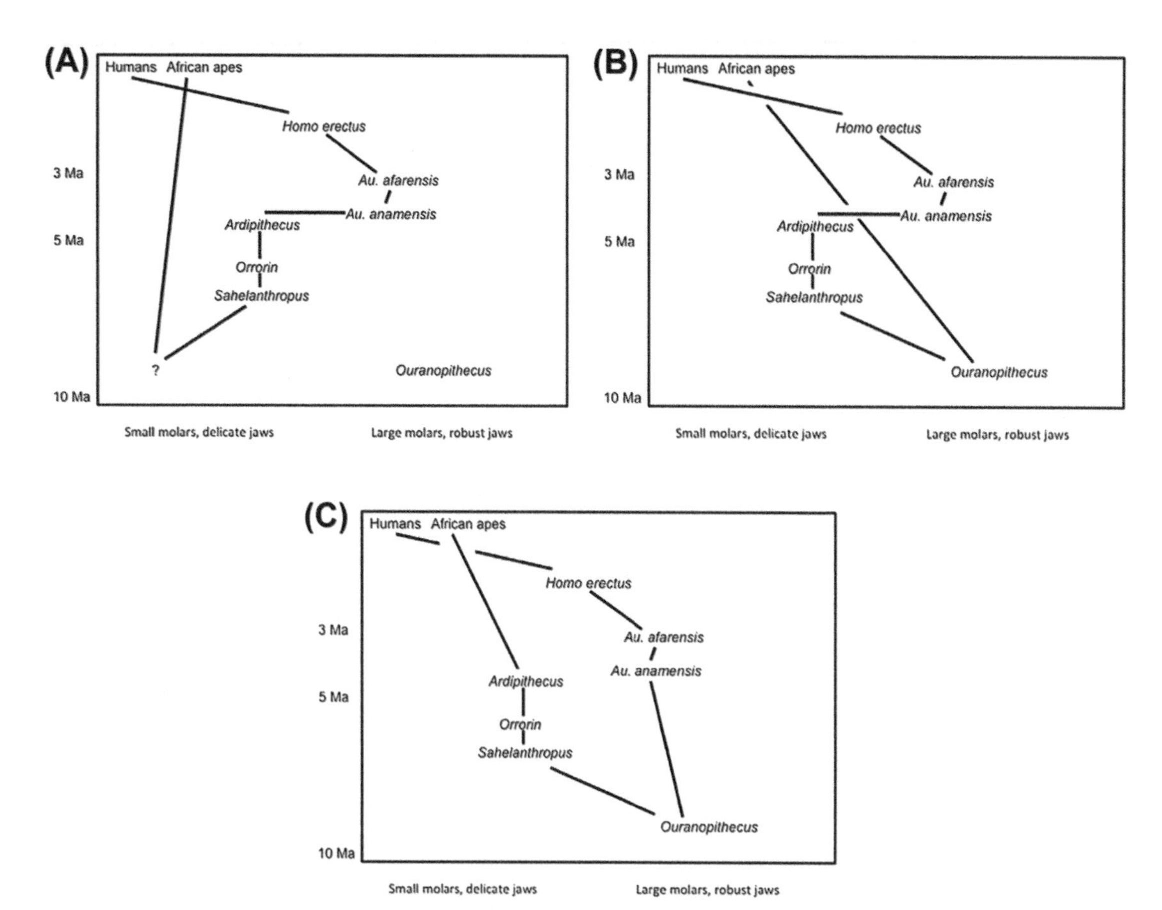

FIGURE 5 (A) EPHs are morphologically intermediate between a *Dryopithecus*-like common ancestor and australopiths; (B) including EPHs in an *Ouranopithecus*–EPH–australopith evolutionary lineage requires that faces and molars increase in size and robusticity, and then decrease in a relatively brief amount of time, around 4.4 Ma; (C) if EPHs are in the African ape lineage no reversals are required for either of the ape and hominin lineages.

deceptive. *Sahelanthropus*, *Orrorin*, and *Ardipithecus* are, in short, as at least as much Miocene ape-like as hominin-like.

If the common ancestor of panins and humans were a late surviving *Dryopithecus* or something like it (the "?" in Figure 5(A)), then with their small molars, thin enamel, and less robust faces the EPHs are good intermediates between *Dryopithecus* and australopiths. The difficulty here is that in the late Miocene robust faces became common among apes, perhaps related to decreased rainfall and less-forested habitats.

The most widely cited LCA of African apes and humans is *Ouranopithecus* (Harrison, 2010; Begun, 2010). It is very australopith-like with its robust face, large molars, thick enamel, and great sexual dimorphism. If *Ouranopithecus* is the LCA of humans and African apes, then the EPHs are not intermediate between the australopiths and the LCA (Figure 5(B)). Let us consider the course of human evolution, if *Ouranopithcus* is the LCA. An *Ouranopithecus*–EPH–australopith lineage would start with robust faces and teeth and great sexual dimorphism, it would have evolved to have a more delicate face and reduced sexual dimorphism (characteristics of the EPHs), and then it would have regained robust faces and great sexual dimorphism in the australopiths (Figure 5(B)).

There is also a time component. *Ardipithecus ramidus* dates to 4.4 mya and *Au. anamensis* (widely considered the earliest australopith) dates to 4.2 mya. Some possible *Ar. ramidus* fossils date to 4.15 mya, overlapping with *Au. anamensis*. The latter has fully australopith dentition (Figure 3) and evidence of bipedal locomotion. An EPH ancestry for the australopiths requires very rapid evolution between the EPHs and the *Au. anamensis* (Figure 5(B)), a scenario that is certainly possible, but not likely.

If EPHs are placed in the chimpanzee lineage, from Miocene apes to EPHs to chimpanzees there is a single evolutionary trend of facial and molar reduction, a more straightforward evolutionary progression requiring no reversals (Figure 5(C)).

For the present, *S. tchadensis* (crudely estimated at 6.5 mya), *O. tugenensis* (5.65–6.2 mya), *Ar. kadabba* (5.77–5.18 mya), and *Ar. ramidus* (4.4–4.15 mya) are an intriguing cluster of species of uncertain status. As we discover more Miocene ape fossils and more EPH fossils, we can look forward to clarifying their places in the natural world.

REFERENCES

Ahern, J.C.M., 2005. Foramen magnum position variation in *Pan troglodytes*, Plio-Pleistocene hominids, and recent *Homo sapiens*: implications for recognizing the earliest hominids. American Journal of Physical Anthropology 127, 267–276.

Alba, D.M., Almécija, C.-V.I., Méndez, J.M., Moyå-Solá, S., 2012. A partial skeleton of the fossil great ape *Hispanopithecus laietanus* from Can Feu and the mosaic evolution of crown-hominoid positional behaviors. PLoS ONE 7, e39617.

Alba, D.M., Fortuny, J., Moya-Sola, S., 2010. Enamel thickness in the middle Miocene great apes *Anoiapithecus*, *Pierolapithecus* and *Dryopithecus*. Proceedings of the Royal Society B (Biological Sciences) 277, 2237–2245.

Almécija, S., Tallman, M., Alba, D.M., Pina, J., Moyá-Solá, S., Jungers, W.L., 2013. The femur of *Orrorin tugenensis* exhibits morphometric affinities with both Miocene apes and later hominins. Nature Communications. http://dx.doi.org/10.1038/ncomms3888, 1–12.

Andrews, P.J., 1992. Evolution and environment in the Hominoidea. Nature 360, 641–646.

Ankel-Simons, F., 2007. Primate Anatomy, third ed. Academic Press, London.

Begun, D.R., 2010. Miocene hominids and the origins of the African apes and humans. Annual Review of Anthropology 39, 67–84.

de Bonis, L., Koufos, G.D., 1994. Our ancestors' ancestor: *Ouranopithecus* is a Greek link in human ancestry. Evolutionary Anthropology 3, 75–83.

Brunet, M., Beauvilain, A., Coppens, Y., Heintz, E., Moutaye, A.H.E., Pilbeam, D., 1995. The first australopithecine 2,500 kilometres west of the rift valley (Chad). Nature 378, 273–274.

Brunet, M., Guy, F., Pilbeam, D., Mackaye, H.T., Likius, A., Ahounta, D., Beauvilain, A., Blondel, C., Bocherens, H., Boisserie, J.R., De Bonis, L., Coppens, Y., Dejax, J., Denys, C., Duringer, P., Eisenmann, V.R., Fanone, G., Fronty, P., Geraads, D., Lehmann, T., Lihoreau, F., Louchart, A., Mahamat, A., Merceron, G., Mouchelin, G., Otero, O., Campomanes, P.P., De Leon, M.P., Rage, J.C., Sapanet, M., Schuster, M., Sudre, J., Tassy, P., Valentin, X., Vignaud, P., Viriot, L., Zazzo, A., Zollikofer, C., 2002. A new hominid from the Upper Miocene of Chad, central Africa. Nature 418, 145–151.

Cramer, D.L., 1977. Craniofacial morphology of *Pan paniscus*. A morphometric and evolutionary appraisal. Contributions to Primatology 10, 1–64.

Galik, K., Senut, B., Pickford, M., Gommery, D., Treil, J., Kuperavage, A.J., Eckhardt, R.B., 2004. External and internal morphology of the BAR 1002'00 *Orrorin tugenensis* femur. Science 305, 1450–1453.

Güleç, E.S., Sevim, A., Pehlevan, C., Kaya, F., 2007. A new great ape from the late Miocene of Turkey. Anthropological Science 115, 153–158.

Guy, F., Lieberman, D.E., Pilbeam, D., Ponce de León, M., Likius, A., Hassane, T.M., Vignaud, P., Zollikofer, C., Brunet, M., 2005. Morphological affinities of the *Sahelanthropus tchadensis* (late Miocene hominid from Chad) cranium. Proceedings of the National Academy Science 102, 18836–18841.

Haile-Selassie, Y., 2001. Late Miocene hominids from the Middle Awash, Ethiopia. Nature 412, 178–181.

Haile-Selassie, Y., Suwa, G., White, T.D., 2004. Late Miocene teeth from middle Awash, Ethiopia, and early hominid dental evolution. Science 303, 1503–1505.

Hammond, A.S., Alba, D.M., Almecija, S., Moya-Sola, S., 2013. Middle Miocene *Pierolapithecus* provides a first glimpse into early hominid pelvic morphology. Journal of Human Evolution 64, 658–666.

Harrison, T., 1986. A reassessment of the phylogenetic relationships of *Oreopithecus bambolii* Gervais. Journal of Human Evolution 15, 541–583.

Harrison, T., 2010. Apes among the tangled branches of human origins. Science 327, 532–534.

Hunt, K.D., 1991a. Mechanical implications of chimpanzee positional behavior. American Journal of Physical Anthropology 86, 521–536.

Hunt, K.D., 1991b. Positional behavior in the Hominoidea. International Journal of Primatology 12, 95–118.

Hunt, K.D., 1994. The evolution of human bipedality: ecology and functional morphology. Journal of Human Evolution 26, 183–202.

Kimbel, W.H., Rak, Y., 2010. The cranial base of Australopithecus afarensis: new insights from the female skull. Philosophical Transactions of the Royal Society (B) 365, 3365–3376.

Kohler, M., Moya-Sola, S., 1997. Ape-like or hominid-Like? The positional behavior of *Oreopithecus bambolii* reconsidered. Proceedings of the National Academy of Science 94, 11747–11750.

Kordos, L., Begun, D.R., 1997. A new reconstruction of RUD 77, a partial cranium of Dryopithecus brancoi from Rubabanya, Hungary. American Journal of Physical Anthropology 103, 277–294.

Kunimatsu, Y., Nakatsukasa, M., Sawada, Y., Sakai, T., Hyodo, M., Hironobu, H., Itaya, T., Nakaya, H., Saegusa, H., Mazurier, A., Saneyoshi, M., Tsujikawa, H., Yamamoto, A., Mbual, E., 2007. A new late Miocene great ape from Kenya and its implications for the origins of African great apes and humans. Proceedings of the National Academy of Sciences 104, 19220–19225.

Langergraber, K.E., Prüfer, K., Rowney, C., Boesch, C., Crockford, C., Fawcett, K., Inouef, E., Inoue-Muruyama, M., Mitani, J.C., Muller, M.N., Robbins, M.M., Schubert, G., Stoinski, T.S., Viola, B., Watts, D., Wittig, R.M., Wrangham, R.W., Zuberbühler, K., Pääbo, S., Vigilant, L., 2012. Generation times in wild chimpanzees and gorillas suggest earlier divergence times in great ape and human evolution. Proceedings of the National Academy of Sciences 109, 15716–15721.

Lovejoy, C.O., Simpson, S.W., White, T.D., Asfaw, B., Suwa, G., 2009a. Careful climbing in the Miocene: the forelimbs of *Ardipithecus ramidus* and humans are primitive, with supporting online material. Science 326, 70. http://dx.doi.org/10.1126/science.1175827.

Lovejoy, C.O., Latimer, B., Suwa, G., Asfaw, B., White, T.D., 2009b. Combining prehension and propulsion: the foot of *Ardipithecus ramidus*, with supporting online material. Science 326, 72. http://dx.doi.org/10.1126/science.1175832.

Lovejoy, C.O., Suwa, G., Simpson, S.W., Matternes, J.H., White, T.D., 2009c. The great divide: *Ardipithecus ramidus* reveals the postcrania of our last common ancestors with African apes, with supporting online material. Science 326, 73. http://dx.doi.org/10.1126/science.1175833.

Merceron, G., Schulz, E., Kordos, L., Kaiser, T.M., 2007. Paleoenvironment of *Dryopithecus brancoi* at Rudabánya, Hungary: evidence from dental meso- and micro-wear analyses of large vegetarian mammals. Journal of Human Evolution 53, 332–349.

Moya-Sola, S., Alba, D.M., Almecija, S., Casanovas-Vilar, I., Köhler, M., de Esteban-Trivigno, S., Robles, J.M., Galindo, J., Fortuny, J., 2009. A unique middle Miocene European hominoid and the origins of the great ape and human clade. Proceedings of the National Academy of Science 106, 9601–9606.

Nakatsukasa, M., Kunimatsu, Y., 2009. *Nacholapithecus* and its importance for understanding hominoid evolution. Evolutionary Anthropology 18, 103–119.

Patterson, N., Richter, D.J., Gnerre, S., et al., 2006. Genetic evidence for complex speciation of humans and chimpanzees. Nature 441, 1103–1108.

Pilbeam, D., Young, N., 2004. Hominoid evolution: synthesizing disparate data. Comptes Rendus Palevol 3, 305–321.

Rafferty, K.L., 1998. Structural design of the femoral neck in primates. Journal of Human Evolution 34, 361–383.

Richmond, B.G., Jungers, W.L., 2008. *Orrorin tugenensis* femoral morphology and the evolution of hominin bipedalism. Science 319, 1662–1665.

Rogers, J., Commuzzie, A.G., 1995. When is ancient polymorphism a potential problem for molecular phylogenetic. American Journal of Physical Anthropology 98, 216–218.

Rose, M.D., 1988. Functional anatomy of the cheiridia. In: Schwartz, J.H. (Ed.), Orangutan Biology. Oxford University Press, New York, pp. 299–310.

Sarmiento, E.E., 1987. The phylogenetic position of *Oreopithecus* and its significance in the origin of the Hominoidea. American Museum Novitates 2881, 1–44.

Sarmiento, E.E., 1995. Cautious climbing and folivory: a model of hominoid differentiation. Human Evolution 10, 289–321.

Sarmiento, E.E., 2010. Comment on the paleobiology and classification of *Ardipithecus ramidus*. Science 328, 1105b.

Sarmiento, E.E., Meldrum, D.J., 2011. Behavioral and phylogenetic implications of a narrow allometric study of *Ardipithecus ramidus*. Homo: Journal of Comparative Human Biology 62, 75–108.

Sawyer, G.J., Deak, V., Sarmiento, E.E., Milner, R., Tattersall, I., Leakey, M., Johanson, D.C., 2007. The Last Human. Yale University Press, New Haven.

Schultz, A.H., 1937. Proportions, variability and asymmetries of the long bones of the limbs and clavicle in man and the apes. Human Biology 9, 281–328.

Schultz, A.H., 1969. The skeleton of the chimpanzee. In: Bourne, G.H. (Ed.), The Chimpanzee. Karger, Basel, pp. 50–103.

Schwartz, J.H., Tattersall, I., 2005. The Human Fossil Record. Craniodental Morphology of Early Hominids (*Genera Australopithecus, Paranthropus, Orrorin*) and Overview, vol. 4Wiley, New York.

Scally, A., Dutheil, J.Y., Hillier, L.W., Jordan, G.E., Goodhead, I., Herrero, J., Hobolth, A., Lappalainen, T., Mailund, T., Marques-Bonet, T., McCarthy, S., Montgomery, S.H., Schwalie, P.C., Tang, Y.A., Ward, M.C., Xue, Y., Yngvadottir, B., Alkan, C., Andersen, L.N., Ayub, Q., Ball, E.V., Beal, K., Bradley, B.J., Chen, Y., Clee, C.M., Fitzgerald, S., Graves, T.A., Gu, Y., Heath, P., Heger, A., Karakoc, E., Kolb-Kokocinski, A.K., Laird, G.K., Lunter, G., Meader, S., Mort, M., Mullikin, J.C., Munch, K., O'Connor, T.D., Phillips, A.D., Prado-Martinez, J., Rogers, A.S., Sajjadian, S., Schmidt, D., Shaw, K., Simpson, J.T., Stenson, P.D., Turner, D.J., Vigilant, L., Vilella, A.J., Whitener, W., Zhu, B., Cooper, D.N., de Jong, P., Dermitzakis, E.T., Eichler, E.E., Flicek, P., Goldman, N., Mundy, N.I., Ning, Z., Odom, D.T., Ponting, C.P., Quail, M.A., Ryder, O.A., Searle, S.M., Warren, W.C., Wilson, R.K., Schierup, M.H., Rogers, J., Tyler-Smith, C., Durbin, R., 2012. Insights into hominid evolution from the gorilla genome sequence. Nature 483, 169–175.

Senut, B., Pickford, M., Gommery, D., Mein, P., Cheboi, K., Coppens, Y., 2001. First hominid from the Miocene (Lukeino Formation, Kenya). Comptes Rendus Palevol Paris 332, 137–144.

Semaw, S., Simpson, S.W., Quade, J., Renne, P.R., Butler, R.F., McIntosh, W.C., Levin, N., Dominguez-Rodrigo, M., Rogers, M.J., 2005. Early Pliocene hominids from Gona. Ethiopia Nature 433, 301–305.

Senut, B., Pickford, M., 2004. La dichotomie grands singes-homme revisitée. Comptes Rendus Palevol 3, 265–276.

Sponheimer, M., Alemesged, Z., Cerling, T.E., Grine, F.E., Kimbel, W.H., Leakey, M.G., Lee-Thorp, J.A., Kyalo Manthi, F., Reed, K.E., Wood, B.A., Wynn, J.G., 2013. Isotopic evidence of early hominin diets. Proceedings of the National Academy of Science 110, 10513–10518.

Stone, A.C., Battistuzzi, F.U., Kubatko, L.S., Perry, G.H., Trudeau, E., Lin, H.M., Kumar, S., 2010. More reliable estimates of divergence times in Pan using complete mtDNA sequences and accounting for population structure. Philosophical Transactions of the Royal Society (B) 365, 3277–3288.

Suwa, G., Asfaw, B., Kono, R.T., Kubo, D., Lovejoy, C.O., White, T.D., 2009. The *Ardipithecus ramidus* skull and its implications for hominid origins. Science 326 68–68e6.

Tuttle, R.H., 1967. Knuckle walking and the evolution of hominoid hands. American Journal of Physical Anthropology 26, 171–206.

Tuttle, R.H., 1975. Primate Functional Morphology and Evolution. Mouton, The Hague.

Tuttle, R.H., 1981. Evolution of hominid bipedalism and prehensile capabilities. Philosophical Transactions of the Royal Society 292, 89–94.

Vignaud, P., Duringer, P., Mackaye, H.T., Likius, A., Blondel, C., Boisserie, J.R., de Bonis, L., Eisenmann, V., Etienne, M.E., Geraads, D., Guy, F., Lehmann, T., Lihoreau, F., Lopez-Martınez, N., Mourer-Chauviré, C., Otero, O., Rage, J.C., Schuster, M., Viriot, L., Zazzo, A., Brunet, M., 2002. Geology and palaeontology of the Upper Miocene Toros-Menalla hominid locality, Chad. Nature 418, 152–155.

White, T.D., Suwa, G., Asfaw, B., 1994. *Australopithecus ramidus*, a new species of early hominid from Aramis, Ethiopia. Nature 371, 306–312.

Wood, B., Harrison, T., 2011. The evolutionary context of the first hominins. Nature 470, 347–352.

Zollikofer, C.P.E., Marcia, S., Ponce de Leon, M.S., Lieberman, D.E., Guy, F., Pilbeam, D., Likius, A., Mackaye, H.T., Vignaud, P., Brunet, M., 2005. Virtual cranial reconstruction of *Sahelanthropus tchadensis*. Nature 434, 755–759.

Chapter 10

Australopithecines

Carol V. Ward
Department of Pathology and Anatomical Sciences, University of Missouri, Columbia, MO, USA

SYNOPSIS

Australopithecines are an adaptive radiation of early hominins, all of which to some extent were bipedal, had brains only slightly larger than those of apes, and developed adaptations to a diet that involved at least occasionally difficult-to-chew foods. They are known from perhaps 10 species that existed between about 4.2 million and 1.0 million years ago and inhabited at least central, eastern, and southern Africa. It is out of this radiation that our genus *Homo* likely evolved, although we are not certain from which species, so understanding australopithecines is key to interpreting not only the diversity of early hominins but also the origins of *Homo*.

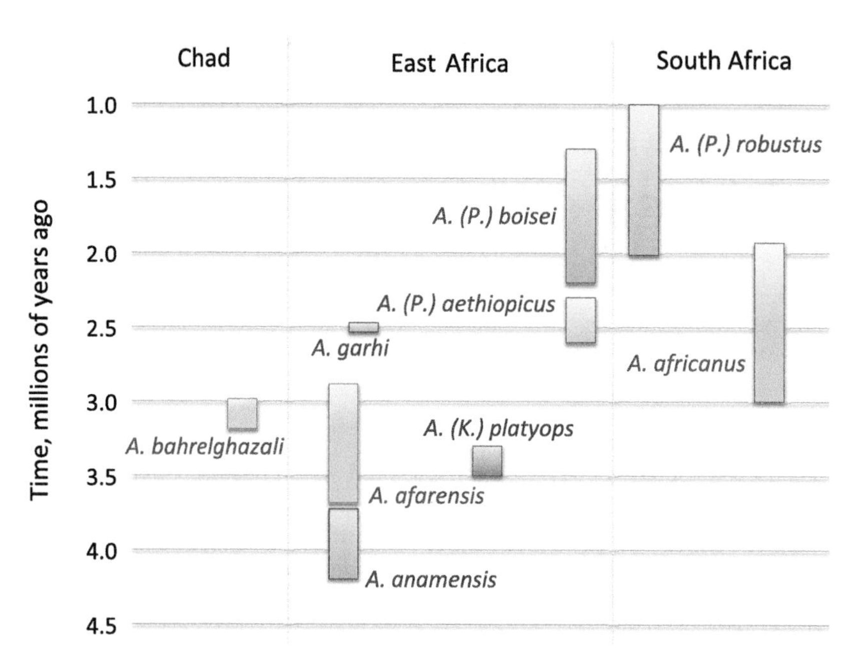

FIGURE 1 Timeline of australopithecine species. Bars represent the range of dates for which each species is known. (See color plate section). *Dates summarized from Brown et al. (2013) and Herries et al. (2013).*

The term "australopithecines" refers to a collection of perhaps as many as 10 Plio-Pleistocene hominin species from eastern, central, and southern Africa from about 4.2 million to 1.0 million years ago (mya) (Figure 1). They are not members of the genus *Homo*, but are hominins that share a suite of morphologies with *Homo* including terrestrial bipedal locomotion, reduced canine and incisor sizes as compared with earlier hominins (e.g., *Ardipithecus, Sahelanthropus*, and *Orrorin*), smaller brains than in *Homo*, thick dental enamel, and to varying extents, craniodental adaptations to processing a diet consisting of hard and/or tough foods. Thus australopithecines represent a grade rather than a clade of early hominin. Because of these features, we know that australopithecines are more closely related to *Homo* than they are to earlier hominins or apes.

The specific evolutionary relationships among these species are not completely clear, but together they represent an adaptive radiation of small-brained, thick-enameled terrestrial bipeds out of which our genus *Homo* likely evolved. Although the genus *Australopithecus* is used to characterize some or all of these species, technically it is not a valid taxon, as it would not include a single ancestor and all of its descendants, since *Homo* likely evolved from one of the species in this radiation (see chapters by Simpson and Holt in the present volume). Still, until evolutionary relationships among the australopithecine species and *Homo* are clear, the broad genus name is retained for pragmatic purposes. Over the years, all have been attributed at one time or another to the genus *Australopithecus* itself, but the recognition of considerable morphological variability among species has resulted in some species being attributed to other genera, such as *Paranthropus* or *Kenyanthropus*. This chapter reviews australopithecine species, and what we know about their adaptations, biology, and behavior.

AUSTRALOPITHECUS

Australopithecus was first discovered in South Africa. The genus *Australopithecus* was identified by Raymond Dart in 1925 with the Taung skull, which he named *Australopithecus africanus* ("southern ape from Africa;" Dart, 1925) (Figure 1). This child's skull has a foramen magnum, the hole through which the spinal cord exits the skull, tucked underneath the cranium, demonstrating that this animal stood and moved upright on two feet. It also has expanded molar teeth, reduced anterior teeth, and a shorter face compared with those of apes. Together, these features identified the Taung skull as being from a hominin. However, it differs from our genus *Homo* in having a much smaller brain and larger postcanine teeth, which is why Dart assigned it to a new genus.

The Taung specimen was found by workers at a limestone quarry, the type of cave deposit that has yielded all South African hominins. Over subsequent decades, more specimens were recovered from South Africa at other quarry sites, first from a cave site called Sterkfontein. The Sterkfontein fossils were initially referred to as *Australopithecus transvaalensis* (Broom, 1936) and then *Plesianthropus transvaalensis* (Broom, 1938), but are now generally recognized as belonging to *A. africanus* as well. These initial specimens included an adult cranium as well as some postcranial elements. They confirmed the morphology seen by

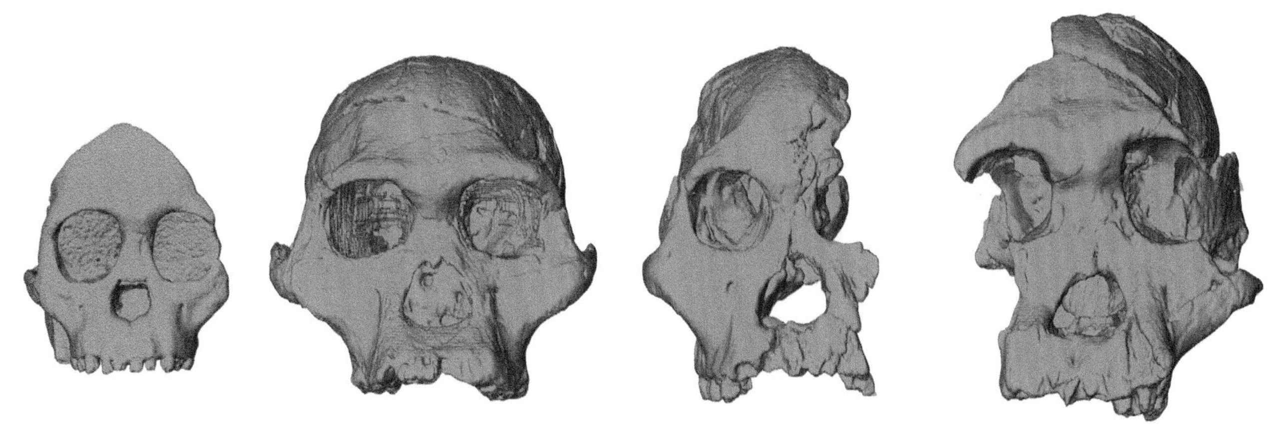

FIGURE 2 Four *Australopithecus africanus* specimens, from left to right: Taung 1, Sts 5, Sts 71, and StW 505. *Reconstructed CT scan images courtesy of F. Spoor.*

the Taung skull (Broom and Schepers, 1946). *Australopithecus africanus* is now known from Makapansgat (Dart, 1948; White et al., 1981; Schmid, 2002) and Gladysvale (Berger et al., 1993) as well. More than 20 skulls have been recovered, along with at least two partial skeletons and many isolated postcranial bones (Robinson, 1972; Partridge et al., 2003; Haüsler and Berger, 2001; Lockwood and Tobias, 2002; Toussaint, 2003; Clarke, 2013), making *A. africanus* the most well represented australopithecine species in the fossil record in terms of skulls, and one of the best in terms of postcranial remains (Figure 2). A partial skeleton from lower in the Sterkfontein site, in a deposit known as Jacovec Cavern (Clarke, 1998, 1999, 2008), may also represent *A. africanus*, but most of it remains unpublished in detail aside from some information about the foot (Clarke and Tobias, 1995), so little more can be said about it at present.

Morphologically, *A. africanus* is distinguished from other australopithecines by pronounced anterior pillars of bone flanking either side of the nasal aperture (Rak, 1985; Kimbel et al., 2004) and a concave face when viewed laterally (White et al., 1981; Kimbel and Delezene, 2009) (Figure 3). It also had a strong anterior temporalis chewing muscle, as evidenced by visible attachment lines on the frontal and parietal bones near the anterior (front) portion of the braincase. They probably weighed between 30 and 45 kg and stood about 105–142 cm tall (McHenry, 1986). They had brains ranging from 440 to 515 cc (Falk et al., 2000), slightly larger than those of apes for their body size, with evidence of some level of neural reorganization (White, 2002; Holloway, 2008; Schoenemann, 2013) (also see chapter by Holloway).

Dating of the South African sites is notoriously difficult due to the complex depositional environment of these caves, but comparative biostratigraphy, paleomagnetic, and uranium dating all yield dates between 3 and 2 mya for *A. africanus*, though they may extend in a range as wide as 3.3 to 1.5 mya (Berger et al., 2002; Walker et al., 2006; Pickering et al., 2010; Herries et al., 2013). The paleoenvironment in South Africa in the region where *A. africanus* fossils have been found included open forests and scrublands, and was wetter than the region is today (Rayner et al., 1993; Pickering et al., 2004).

"Robust" Australopithecines

In 1938, a second kind of hominin (Robinson, 1956, 1968) was found at Kromdraai, also in South Africa, and referred to a new genus, *Paranthropus* (Broom, 1938). The Kromdraai cranium had a shorter face than that of the Sterkfontein and Taung fossils, smaller anterior teeth, and flatter, squarer premolars and molars. Still, it differs from *Homo* in retaining a smaller brain coupled with larger posterior teeth. In the 1940s more *Paranthropus* fossils were recovered from the nearby site of Swartkrans and originally assigned to *Paranthropus crassidens* (Broom, 1949), but are now recognized as sufficiently similar to *Paranthropus robustus* to be attributed to that species as well. Since that time, more specimens have been found at Gondolin (Menter et al., 1999; Grine et al., 2012), Drimolen (Keyser, 2000; Keyser et al., 2000), and Cooper's Cave (Berger et al., 2003) (reviews in Wood and Constantino, 2007; Wood and Schroer, 2013). All of these fossils display adaptations to heavier chewing—even though the molar teeth are not necessarily larger than those of *A. africanus* (Figure 3)—with large cheekbones for attachment of the masseter muscle, large temporal fossae for passage of a massive temporalis muscle, and large sagittal and nuchal crests for attachment of these muscles. Also present is a very shallow glenoid fossa—the temporal half of the temporomandibular joint—that would allow for considerable transverse movement during chewing. The mandible is also large and robust. Molar microwear studies (Scott et al., 2005) and analysis of carbon isotopes preserved in the teeth (Sponheimer et al., 2006) suggest an eclectic diet that included hard objects as well as softer foods. Hence, the *P. robustus* fossils were termed "robust," in contrast to the *A. africanus* fossils that were termed "gracile" by Robinson (1954); general descriptions still often used today. Despite

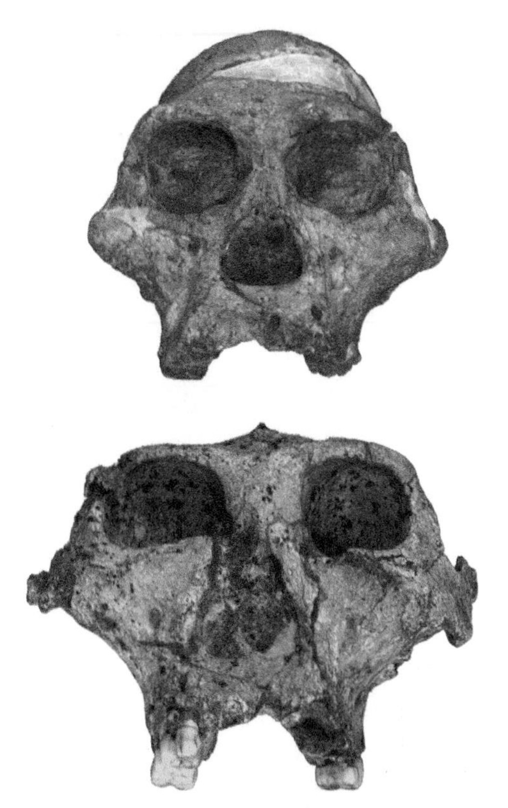

FIGURE 3 Comparison photographs of original fossils: *Australopithecus africanus* cranium (Sts 5) above, and *Australopithecus robustus* (SK 46) below. The *A. robustus* specimen displays the large zygomatics and sagittal crest typical of "robust" australopiths. *Photographs courtesy of M.H. Wolpoff.*

these differences, because of its clearly hominin status combined with the shared retention of primitive features not seen in later *Homo*, even then *Paranthropus* was referred to informally as one of the "australopithecines."

Only later were australopithecines discovered in East Africa, with some teeth in 1955 and a skull in 1959 (Leakey, 1959, 1960; Leakey and Leakey, 1964). These specimens had even smaller anterior and larger posterior teeth than *P. robustus*, and the skull showed an even more robust face and skull than South African specimens, so they were attributed to a new taxon, *Zinjanthropus boisei* (Leakey, 1960), now generally referred to *Australopithecus (or Paranthropus) boisei* (review in Wood and Constantino, 2007; Wood and Schroer, 2013) (Figure 4). Since this time, numerous crania and mandibles have been found at Koobi Fora, Kenya (Leakey and Walker, 1988; Wood, 1991; Brown et al., 1993; Prat et al., 2003); Chesowanja, Kenya (Szalay, 1971; Howell, 1978; Gowlett et al., 1981; Wood, 1991); Omo, Ethiopia (Coppens and Sakka, 1983; Alemseged et al., 2002); Konso Gardula, Ethiopia (Suwa et al., 1997) and perhaps Malema, Malawi (Kullmer et al., 1999); that are attributable to *A. (P.) boisei*. These fossils all date to between 2.2 and 1.3 mya (summary in Wood and Schroer, 2013). All show extremely small anterior dentition, very large molars, and quite large molarized premolars with very little occlusal relief (very flat chewing surfaces). The mandibles are especially large and robust, the zygomatic bones where the masseter chewing muscle would attach is huge, and the skulls have large flaring zygomatic arches with strong crests on the braincase reflecting a hypertrophied temporalis muscle. Thus, they show even more extreme adaptations to heavy mastication than does *P. robustus* (Wood and Constantino, 2007). Interestingly, microwear on the molars and evidence from carbon isotopes trapped in the tooth enamel suggest that they were not eating hard foods (van der Merwe et al., 2008; Cerling et al., 2011; Ungar et al., 2008, 2012), despite the extreme morphological adaptations in their faces, teeth, and jaws.

A third species of "robust" australopithecine is also known from East Africa, dating back to discoveries in the Omo River valley in Ethiopia during the 1960s. These fossils were assigned to *Paraustralopithecus aethiopicus* (Arambourg and Coppens, 1968), now recognized as *Australopithecus* or *Paranthropus aethiopicus* (Figure 5). It is known best from West Turkana, Kenya (Walker et al., 1986), and is also found at Laetoli, Tanzania (Harrison, 2011). This species has a similar dentition to *A. (P.) boisei*, but a more prognathic face, perhaps a smaller brain, and better-developed posterior attachments for the temporalis muscle (Arambourg and Coppens, 1968; Rak and Kimbel, 1991, 1993; Suwa et al., 1994; Walker et al., 1993, White and Falk, 1999; Wood and Constantino, 2007; Wood and Schroer, 2013). It dates to an earlier period, however—2.6–2.35 mya (summary in Wood and Schroer, 2013)—and therefore could at least hypothetically be an ancestor of *A. (P.) boisei*.

As the reader will have noted by now, nomenclature of the "robust" australopithecines is far from settled. The uncertainty stems from debates regarding the phylogenetic relationships among these species. According to proper zoological nomenclature, a genus must include an ancestor and all of its descendants, termed a monophyletic clade. Most formal phylogenetic analyses do in fact interpret the "robust" species *Australopithecus (P.) robustus, boisei*, and *aethiopicus* as a monophyletic clade supported by their numerous similarities in dental and craniofacial features. As such, they should most correctly be placed in their own genus, *Paranthropus* (discussions in Wood and Constantino, 2007; Wood and Schroer, 2013) (also see Hunt in the present volume).

Significant regional differences exist between the East and South African taxa (Tobias, 1967; see also Delson, 1997), raising doubts about their monophyly. The East African species have even smaller anterior and larger posterior teeth than the South African one (but there is one similarly sized molar from Gondolin, South Africa; Grine et al., 2012), with even more heavily built skulls. The South African species tends to have molar teeth that look "puffier," and the remnants of anterior pillars flanking the nasal aperture, somewhat like *A. africanus*. All of the features uniting a "robust" clade are related to adaptations for heavy mastication and appear

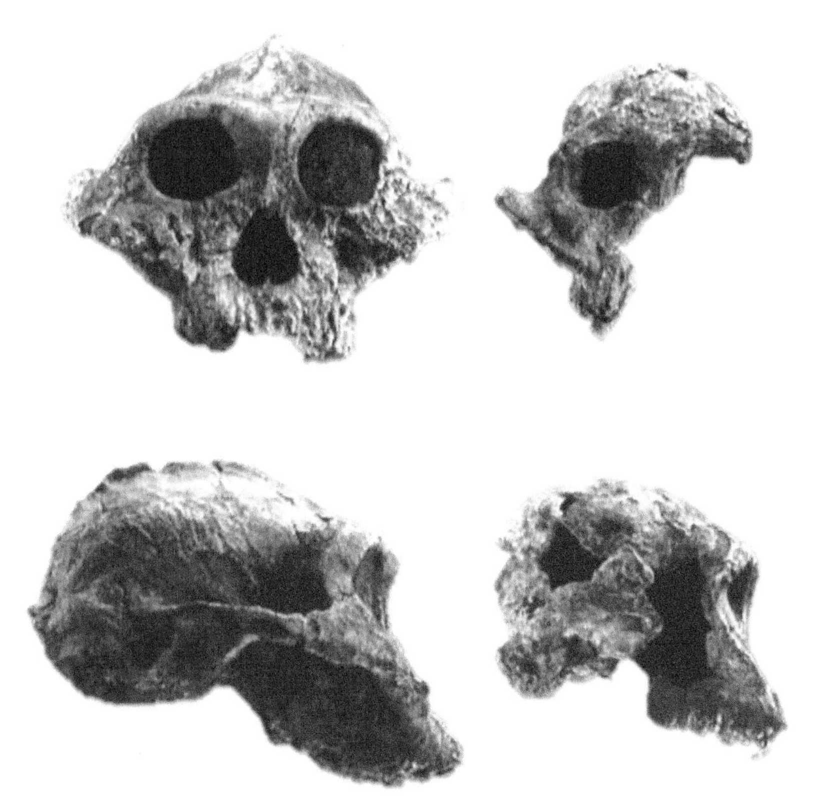

FIGURE 4 Australopithecus boisei crania KNM-ER 406 (left) and KNM-ER 732 (right) in anterior (above) and lateral views (below). KNM-ER 406 is widely regarded as belonging to a male individual and KNM-ER 732 to a female, suggesting strong sexual dimorphism in this species. *Photographs courtesy of B.A. Wood.*

FIGURE 5 *Australopithecus aethiopicus* cranium KNM-WT 17000. *Photograph courtesy of B.A. Wood.*

to be functionally integrated. Therefore, it is also likely that there may be homoplasy (independently evolved but superficially similar) traits that evolved separately in East and South Africa in response to a similar selective pressure, the need to chew harder and/or tougher food items. If the "robust" taxa are not monophyletic, it is inappropriate to refer to all of them as *Paranthropus*, but rather we should continue to use the traditional genus name *Australopithecus*.

That said, the genus *Australopithecus* itself is certainly not monophyletic and so itself is an invalid taxon. It is the opinion of this author that although it is technically invalid to assign any more than *A. africanus* to the genus *Australopithecus*, it is also confusing to adjust an uncertain phylogeny until we can be more certain about the relationship among the members of the australopithecine grade of hominin. Assigning some taxa to a particular genus name makes a phylogenetic statement that may not be supported in the future, which would add further confusion to an already difficult situation. Therefore, the genus name *Australopithecus* is retained here as a pragmatic approach to a currently thorny problem.

AUSTRALOPITHECUS AFARENSIS

In 1974, the type mandible of a second species of "gracile" australopithecine was found in East Africa at Laetoli, Tanzania by Mary Leakey (Leakey, 1987). The species *A. afarensis* was formally described by Johanson and colleagues upon the discovery of additional similar fossils from Hadar, a site in the Afar region of Ethiopia that gives the species its name (Johanson et al., 1978). Further fieldwork at Hadar in the 1970s yielded the famous "Lucy" skeleton, the "first family," over 200 fossils from between five and nine individuals ranging in age from juvenile to adult, along with numerous other dental, mandibular, and postcranial remains of *A. afarensis*.

Fieldwork throughout the 1990s and the first part of the twenty-first century yielded the first nearly complete cranium of this species (Kimbel et al., 1994; Figure 6), other dental and mandibular remains (review in Kimbel and Delezene, 2009), as well as 59 more postcranial fossils (Ward et al., 2012). *Australopithecus afarensis* is also known from craniodental and postcranial remains from Maka, Ethiopia (White et al., 1993, 2000), a cranium and partial skeleton of a child from Dikika, Ethiopia (Alemseged et al., 2006), another partial adult skeleton from Korsi Dora, Ethiopia (Haile-Selassie et al., 2010a), partial calvaria and teeth from East (Kimbel, 1988) and West (Brown et al., 2001) Turkana, Kenya, and more fossils from Laetoli (Puech, 1986). Some enigmatic hand bones from South Turkwel, Kenya were provisionally attributed to *A. afarensis*, but there is no real way to know to which taxon they properly belong (Ward et al., 1999a). There is a set of hominin footprints likely made by *A. afarensis* at Laetoli as well (Leakey and Hay, 1979; Leakey, 1981; recent review in Raichlen et al., 2010). Across all these sites, *A. afarensis* is known from 3.6 to 2.9 mya (Kimbel and Delezene, 2009; Brown et al., 2013).

Most of the skeletal elements are known for *A. afarensis*, and there are multiple examples of most bones, making *A. afarensis* by far the most well represented australopithecine species in the fossil record for everything except skulls. Because of this, much of our understanding of the biology and adaptations of this radiation derive from studies of *A. afarensis*. *Australopithecus afarensis* also forms the basis of our understanding of australopithecine origins, because it predates *A. africanus*, and until 1994 with the discovery of the earlier *Australopithecus anamensis* (below), was the earliest australopithecine species known.

The *A. afarensis* brain was about the same size as that of other australopithecines, ranging from about 400 to 550 cc (Holloway and Yuan, 2004) (see chapter by Holloway). Like *A. africanus*, *A. afarensis* was specialized for heavy chewing, compared with modern apes and humans, but not as much as the "robust" australopithecine species. It has robust mandibles and expanded postcanine teeth similar to those of *A. africanus* overall (Kimbel and Delezene, 2009). The braincase displays morphologies consistent with masticating hard and/or tough foods, although not to the extent seen in "robust" species, including large crests for attachment of the temporalis muscle, a shallow glenoid fossa, and a forward-positioned masseter muscle resulting in greater leverage for producing powerful bite forces. The lower face is prognathic (projecting) as in *A. africanus*, and more so than in *A. robustus* or *A. boisei*. Microwear data suggest differential usage of the anterior and posterior teeth for processing food, specifically preparation of leaves, seeds, and rhizomes on the incisors (Ryan and Johanson, 1989), and tough or hard food on the molars (Teaford and Ungar, 2000); carbon isotopic analysis suggests a range of diets (Sponheimer et al., 2013). *Australopithecus afarensis* inhabited a variety of habitats, from wooded at some sites to more open at others, although all sites show a mix of closed and open settings (Bonnefille et al., 2004; Wynn et al., 2006; Campisano and Feibel, 2007).

FIGURE 6 *Australopithecus afarensis* cranium A.L. 444-2. *Photograph courtesy of W.H. Kimbel.*

Body size estimated for this species ranges from about 28 to 50 kg, and stature from about 105 to 151 cm (McHenry, 1992). The postcranial skeleton of *A. afarensis* is also highly dimorphic in size, with males almost twice the body weight of females (McHenry, 1986, 1991, 1992; Kimbel et al., 1988; Richmond and Jungers, 1995; Lockwood et al., 1988; Plavcan et al., 2005; Gordon et al., 2008; see also Haile-Selassie et al., 2010a; but see Reno et al., 2003). High levels of body size dimorphism are only found in extant primates that have high levels of male–male competition over access to females, but not in species with a monogamous social system (as in Lovejoy, 1981; Plavcan, 1997) (see chapter by Watts).

The *A. afarensis* postcranial skeleton displays numerous derived characters related to committed terrestrial bipedal locomotion (reviews in Stern, 2000; Ward, 2002, 2013a,b; Richmond and Hatala, 2013; see also Ward et al., 2011, 2012) (see chapter by Hunt). This evidence includes an anteriorly situated foramen magnum showing that the head sat atop a vertical vertebral column. The spine exhibits pronounced humanlike sinusoidal curvatures and the thoracic (rib-bearing) vertebral column is invaginated into the rib cage, providing increased leverage for the spinal musculature to hold the body upright. The pelvis has laterally facing and widely flaring ilia, positioning the hip muscles to balance side to side over one leg at a time, which occurs during human bipedal gait. The femur (thigh bone) is angled in at the knee so that the knee is positioned directly above the foot and directly under the body's center of gravity when an individual is standing on one foot, and the shank is vertical rather than angled to the side as it is in climbing apes. The femoral condyles, or joints where the femur articulates at the knee, are elliptical in contour, providing a flattened bottom surface to dissipate high loads during heel strike in gait. They had arches in their feet, and the feet were stiff like ours, providing a firm platform for pushing off when walking. Perhaps most importantly, they had a large hallux (big toe) in line with the other digits, not divergent and opposable as in all other primates.

Australopithecus afarensis was not built exactly like humans, though. The pelvis, though unambiguously indicative of habitual bipedalism for this taxon, is nevertheless distinct from that of modern humans, who underwent further selection for obstetric sufficiency in giving birth to larger-brained offspring (Tague and Lovejoy, 1986). Compared with modern humans, *A. afarensis* also had longer arms relative to its thighs, longer and more-curved finger and toes bones, a more cranially oriented (upward-facing) shoulder joint, and smaller hind limb joints and vertebral bodies (reviews in Stern, 2000; Ward, 2002; Ward et al., 2012; Kimbel and Delezene, 2009; see also Alemseged et al., 2006; Green and Alemseged, 2012; Larson, 2013; Richmond and Hatala, 2013). These features are considered primitive for *Australopithecus*, as they also characterize apes and *Ardipithecus* (see chapter by Hunt).

There has been debate in the literature about what the retention of these primitive features of *Australopithecus* means for its biology and adaptation, with some suggesting that the presence of some ancient apelike traits must mean at least partial retention of tree climbing, the locomotor behavior that distinguishes apes (reviewed in Stern, 2000). Certainly *A. afarensis* would have been a more capable climber than are humans, but without key morphologies such as a grasping big toe, their ability to move about easily in the trees, particularly for females holding infants, would have been severely compromised compared with apes (see Ward, 2002, 2013a,b and references therein). This demonstrates that selection for terrestrial bipedality was far more significant than that favoring arboreal climbing. It has also been argued that certain aspects of their anatomy that differ from those of humans, including widely splayed iliac blades, large navicular tuberosity in the foot, and long pedal phalanges, would have made bipedal walking less efficient than and kinematically distinct from that of humans. However, the presence of a well-developed lumbar lordosis, femoral bicondylar angle, and pedal arches demonstrates that *A. afarensis* was adapted for walking with fully extended lower limbs. If *A. afarensis* was a less efficient bipedal walker than are humans, it would have been simply that they had shorter lower limbs, bulkier upper bodies, and a smaller body size, but not different limb or joint postures during walking or running (see also Hammond and Ward, 2013).

AUSTRALOPITHECUS BAHRELGHAZALI

In 1995, the first australopithecine fossils west of the Rift Valley was recovered from Chad in central Africa. *Australopithecus bahrelghazali* is a species named for a mandible from this site (Brunet et al., 1995). It is contemporaneous with *A. afarensis* 3.0–3.4 mya (Brunet et al., 1995; Brown et al., 2013). The specimens known for this species do not display any derived characteristics distinct from those of *A. afarensis*, and so are generally included within *A. afarensis* (see White et al., 2000; Alemseged et al., 2006; Kimbel and Delezene, 2009; MacLatchy et al., 2010). It is possible, though, that recovery of further fossils in the future from these localities will provide more information confirming the separate status of *A. bahrelghazali*.

The major significance of *A. bahrelghazali* is that it confirmed what had been long suspected—australopithecines inhabited a much greater part of Africa than just the areas for which we have fossil records. They even may have been found across the entire continent.

AUSTRALOPITHECUS ANAMENSIS

For many years, *A. afarensis* was the oldest known australopithecine and indeed one of the earliest hominins, and formed the basis of our understanding of hominin origins. Today, even earlier hominin genera are known (*Sahelanthropus, Orrorin,* and *Ardipithecus*; summary in MacLatchy et al., 2010; also see chapter by Hunt), and recently an earlier species of *Australopithecus* was identified from Kanapoi as well as Allia Bay, Kenya, *A. anamensis* (Figure 7) (Leakey et al.,

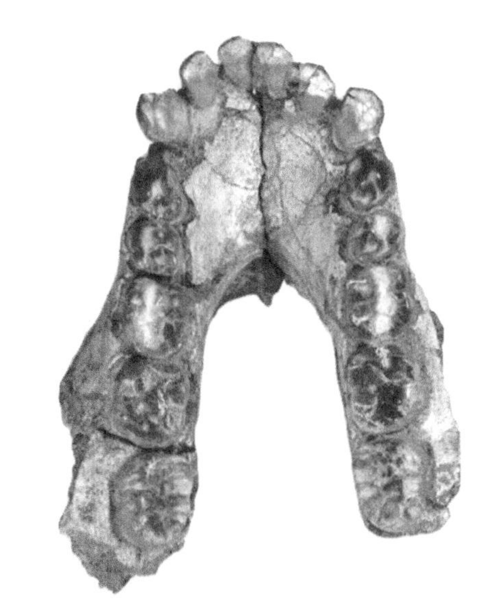

FIGURE 7 *Australopithecus anamensis* holotype mandible KNM-KP 29281 in occlusal and right lateral views. *Photo credit: C.V. Ward.*

1995, 1998). It is now known from these Kenyan sites (Coffing et al., 1994; Heinrich et al., 1993; Ward et al., 1999b, 2001; Manthi et al., 2012; Ward et al., 2013), plus Asa Issie, Fejej, and Woranso Mille in Ethiopia (Fleagle et al., 1991; White, 2002; White et al., 2006; Haile-Selassie, 2010; Haile-Selassie et al., 2010b; MacLatchy et al., 2010). This species ranges in age from 4.2 to 3.8 mya (McDougall and Brown, 2008; Haile-Selassie et al., 2010b; Brown et al., 2013). Thus, *A. anamensis* is the earliest *Australopithecus* and the earliest member of the *Australopithecus-Homo* clade, so its morphology and adaptations are key evidence of those that characterized the origins of this genus.

Australopithecus anamensis shares, with later *Australopithecus*, molars that are larger and more thickly enameled than seen in African apes and *Ardipithecus* (White et al., 1994; White, 2002; Suwa et al., 2009), signaling an adaptation for the ability to process tough and/or abrasive foods, possibly opening up greater ecological niches than were available to African apes, perhaps even being key to the origins of the genus (White et al., 2000, 2006). Still, *A. anamensis* retained narrower jaws, larger canine tooth roots, more primitive canine and premolar crown shapes, lower crowned molars, and other morphologies not only distinguishing it from later *A. afarensis*, but also suggesting that the dietary adaptations were not the same in these species (Ward et al., 2010; Manthi et al., 2012). Differences in the chemical stable carbon isotopes in dental enamel suggest a dietary difference as well, with *A. afarensis* eating a greater range of foods than consumed by *A. anamensis* (Sponheimer et al., 2013).

In most postcranial regions published so far, *A. anamensis* appears to be similar to *A. afarensis*, although only a handful of postcranial fossils are known for this species. *Australopithecus anamensis* appears to have been bipedal (see both chapters by Hunt), as its tibia was oriented vertically relative to the ankle joint as in all later hominins, indicating knee placement directly over the ankle during single-limb stance during bipedal walking (Latimer et al., 1987; Ward et al., 1999a; DeSilva, 2009). The upper limb of *A. anamensis* is similar to that of *A. afarensis* (Patterson and Howells, 1967; Lague and Jungers, 1996; Ward et al., 2001). However, a middle phalanx from Asa Issie is described as longer than those from Hadar (White et al., 2006), and the capitate, a bone from the wrist, may be more African apelike than that of *A. afarensis* in the joints between the wrist and metacarpals (Leakey et al., 1998). These small features may indicate some differences in locomotor or manipulatory function, but until more fossils are recovered our ability to infer postcranial variation between species is limited.

As one examines the samples of *A. anamensis* and *A. afarensis* through time, the distinctive derived features of *A. afarensis* appear to change throughout the period known for each species, and *A. anamensis* appears to grade into *A. afarensis* morphologically. For this reason, it is hypothesized that these species represent two ends of a single evolving lineage. Thus, some scholars could argue that they be combined into a single species, but because at least the early and later parts of this lineage are morphologically distinguishable, most scholars maintain their status as separate species for pragmatic reasons (Kimbel et al., 2006; Haile-Selassie, 2010; Ward et al., 2010, 2013; Manthi et al., 2012).

AUSTRALOPITHECUS GARHI

In 1999, *Australopithecus garhi* was announced from Bouri Hata, Ethiopia, dating to 2.5 mya (Asfaw et al., 1999; Brown et al., 2013) based on a cranium and dentition. This specimen differs from other australopithecines by having very large postcanine teeth like the "robust" species, but with relatively large canine teeth and a facial skeleton that is less heavily built and more prognathic than *A. (P.) robustus*, *A. (P.) boisei*, and *A. (P.) aethiopicus*. Thus it does not fall easily into the general "robust" versus "gracile" dichotomy. Its cranial capacity is 450 cc, similar to that of other australopithecines. Postcranial bones with *Homo*-like proportions, having relatively shorter upper limbs and longer hind limbs than typical for australopithecines, were found at the site, but are not necessarily associated with the skull and so cannot be attributed to *A. garhi*.

KENYANTHROPUS (AUSTRALOPITHECUS) PLATYOPS

Another fossil was discovered in 2001 that could be considered an australopithecine (Leakey et al., 2001). A nearly complete cranium and a maxilla were found at Lomekwi, Kenya (Figure 8), and this fossil is sufficiently different

FIGURE 8 *Kenyanthropus (Australopithecus) platyops* holotype mandible KNM-WT 40000 in anterolateral view. *Photo courtesy of M.G. Leakey.*

from all other *Australopithecus* and *Homo* fossils that its discoverers attributed it to a new genus and species, *Kenyanthropus platyops*. They note, however, that the genus name "*Kenyanthropus*" is only appropriate if one places the "robust" species into the genus *Paranthropus*. If not, as in this paper, these Lomekwi fossils should most appropriately be referred to as *A. platyops* (Leakey et al., 2001). *A. (K.) platyops* differs from other australopithecine taxa in having much smaller postcanine teeth, deeper mandibular fossa (the cranial portion of the temporomandibular joint), and anteriorly placed zygomatic bones flanking a fairly vertical, flattened lower face, from which its species name derives. In the latter feature it seems to resemble *Homo rudolfensis*, raising the possibility of a close phylogenetic relationship between the two (but see Strait and Grine, 2004). It dates to 3.5–3.3 mya, and so is one of the earliest members of the australopithecine radiation (Leakey et al., 2001; Brown et al., 2013). The environment of *A. (K.) platyops* was a partly wooded landscape similar to that at the *A. afarensis* sites, although perhaps wetter, as suggested by the presence of more water-dependent fauna (Leakey et al., 2001).

AUSTRALOPITHECUS SEDIBA

Australopithecus sediba (Figure 9) is the most recently discovered australopithecine species, with two remarkably complete partial skeletons and other bones known from the site of Malapa, near Sterkfontein in South Africa (Berger et al., 2010; de Ruiter et al., 2013a). Malapa appears to date to 1.98 mya (Pickering et al., 2011; Dirks et al., 2010), making *A. sediba* the youngest known gracile australopithecine.

Australopithecus sediba is similar to other gracile australopithecines, especially *A. africanus*, in a number of ways. Its brain size was about 420 cc (Carlson et al., 2011), and its body size was roughly similar as well. Its teeth were morphologically similar to those of other *Australopithecus* species but perhaps smaller on average, falling within the lower range of *A. africanus*.

FIGURE 9 *Australopithecus sediba* cranium MH 1. *Photograph courtesy of L.R. Berger.*

Australopithecus sediba is noted for sharing some apparently derived features with *Homo*. It seems to have had a less strongly developed masticatory system than other australopiths (Berger et al., 2010; Pickering et al., 2011; de Ruiter et al., 2013b). The sagittal crest for attachment of the temporalis is smaller, and the mandible is more gracile with a relatively narrow and vertical symphysis. The anterior pillars that characterize the *A. africanus* face are lacking. The cranium also shares similarities in shape with *Homo*, with a transversely wide braincase and widely spaced temporal lines, as well as minimal postorbital constriction. The phylogenetic position of *A. sediba* relative to *Homo* is unclear, but it is unlikely to have been ancestral because it postdates other *Homo* fossils in East Africa (Pickering et al., 2011; Dirks et al., 2010).

Postcranially, *A. sediba* was similar to other australopiths, with relatively long forearms, large upper limb joint surfaces, cranially oriented shoulder joints, and long and curved hand phalanges (finger bones). The pelvis is wide and has small joints, but may have an expanded posterior region like that of *Homo* (Berger et al., 2010; Kibii et al., 2011). The foot appears more primitive than that of *A. afarensis*, with perhaps a more mobile midfoot and calcaneus with less inflated tuberosity (Zipfel et al., 2011). These features and other details of foot anatomy have been cited as indicating a unique kinematic pattern of locomotion (DeSilva et al., 2013), although there is evidence of unusual soft tissue pathologies near the hip and knee, raising questions of whether the species itself would have differed in locomotor mode. The hand and wrist bones of *A. sediba* have *Homo*-like proportions and were likely capable of precision grasping (Kivell et al., 2011). As there has not been a stone tool industry associated with Malapa hominins, it is unclear whether these features relate to stone tool use and production or something else entirely.

AUSTRALOPITHECINE ADAPTATIONS

Although there is a great deal of morphological variability among species, overall the samples constituting the "grade" of australopithecines share a basic pattern of features. Notably, all samples display adaptations throughout the postcranial skeleton for upright terrestrial bipedal locomotion. That said, not all taxa are known from postcranial elements, though (*Australopithecus/Kenyanthropus platyops, A. aethiopicus*, and perhaps *A. garhi*) are known for these elements, so there is possibly more postcranial diversity within this adaptive radiation than currently appreciated. Only the recovery of more associated fossils will answer this question.

Australopithecines include a variety of species, each experimenting in being a bipedal capable of exploiting a range of dietary resources including tough or hard-to-chew foods when necessary, yet without having become encephalized to the extent seen in *Homo* (Hammond and Ward, 2013). They clearly represent the sister group to *Homo*, as they share numerous traits, including committed terrestrial bipedal locomotion, thick tooth enamel, and reduced facial projection, compared with apes. So, *Australopithecus* and *Homo* represent a single clade. This means that understanding this early adaptive radiation is key to understanding the origins of the genus *Homo* (Hammond and Ward, 2013). Often the differences between australopithecines and *Homo* are used to describe this radiation and discuss the adaptations of these species. However, interpreting the adaptations of australopithecines should not be done with reference to *Homo*, but rather to consider how they differ morphologically and behaviorally from earlier hominins and apes. These latter comparisons will reveal the magnitude and direction of change associated with the origins of australopith features, and thus the type and strength of selection led to their appearance and diversification (Hammond and Ward, 2013).

The causes influencing the origin and radiation of australopithecines remain unclear. Between about 4.0 and 2.0 mya, East Africa experienced an increase in aridity and a shift from more wooded to more fragmented environments (recent review in Behrensmeyer and Reed, 2013). This may be at least partially responsible for the increased hominid craniodental diversity related to exploitation of new dietary niches during the Middle Pliocene. However, there are no clear environmental triggers. Being able to process difficult-to-chew foods may have been associated with their success, giving them the ability to exploit a variety of adaptations across the landscape, rather than specializing on fruits that would tie them to forested habitats. It may be possible that being terrestrial bipeds, capable of traveling on the ground efficiently and over distance, was also related and opened up new evolutionary options to these hominins.

REFERENCES

Alemseged, Z., Coppens, Y., Geraads, D., 2002. Hominid cranium from Omo: description and taxonomy of Omo-323-1976-896. American Journal of Physical Anthropology 117, 103–112.

Alemseged, Z., Spoor, F., Kimbel, W.H., Bobe, R., Geraads, D., Reed, K., Wynn, J.G., 2006. A juvenile early hominin skeleton from Dikika, Ethiopia. Nature 443, 296–301.

Arambourg, C., Coppens, Y., 1968. Decouverte d'un australopithecien nouveau dans les gisements de l'Omo (Ethiopie). South African Journal of Science 64, 58–59.

Asfaw, B., White, T.D., Lovejoy, C.O., Latimer, B., Simpson, S., Suwa, G., 1999. *Australopithecus garhi*: a new species of early hominid from Ethiopia. Science 284, 629–635.

Behrensmeyer, A.K., Reed, K.E., 2013. Reconstructing the habitats of *Australopithecus*: paleoenvironments, site taphonomy and faunas. In: Reed, K.E., Fleagle, J.G., Leakey, R.E. (Eds.), The Paleobiology of Australopithecus. Springer, New York, pp. 41–60.

Berger, L.R., de Ruiter, D.J., Steininger, C.M., Hancox, J., 2003. Preliminary results of excavations at the newly investigated Coopers D Deposit, Gauteng, South Africa. South African Journal of Science 99, 276–278.

Berger, L.R., Keyser, A.W., Tobias, P.V., 1993. Brief communication: Gladysvale: first early hominid site discovered in South Africa since 1948. American Journal of Physical Anthropology 92, 107–111.

Berger, L.R., de Ruiter, D.J., Churchill, S.E., Schmid, P., Carlson, K.J., Dirks, G.M., Kibii, J.M., 2010. *Australopithecus sediba*: a new species of *Homo*-like australopith from South Africa. Science 328, 195–204.

Berger, L.R., Lacruz, R., de Ruiter, D.J., 2002. Revised age estimates of *Australopithecus*-bearing deposits at Sterkfontein, South Africa. American Journal of Physical Anthropology 119, 192–197.

Bonnefille, R., Potts, R., Chalie, F., Jolly, D., Peyron, O., 2004. High resolution vegetation and climate change associated with Pliocene Australopithecus afarensis. Proceedings of the National Academy of Sciences 101, 12125–12129.

Broom, R., 1936. A new fossil anthropoid skull from Sterkfontein, near Krugersdorp, South Africa. Nature 138, 486–488.

Broom, R., 1938. Pleistocene anthropoid apes of South Africa. Nature 142, 377–379.

Broom, R., 1949. Another new kind of fossil ape-man (*Paranthropus crassidens*). Nature 163, 57.

Broom, R., Schepers, G.W.H., 1946. The South African fossil ape-men: the Australopithecinae. Transvaal Museum Memoir 2, 1–272.

Brown, B., Brown, F.H., Walker, A., 2001. New hominids from the Lake Turkana Basin, Kenya. Journal of Human Evolution 41, 29–44.

Brown, B., Walker, A., Ward, C.V., Leakey, R.E., 1993. New *Australopithecus boisei* calvaria from east Lake Turkana, Kenya. American Journal of Physical Anthropology 91, 137–159.

Brown, F.H., McDougall, I., Gathogo, P.N., 2013. Age ranges of *australopithecus* species, Kenya, Ethiopia and Tanzania. In: Reed, K.E., Fleagle, J.G., Leakey, R.E. (Eds.), The Paleobiology of Australopithecus. Springer, New York, pp. 7–20.

Brunet, M., Beauvilain, A., Coppens, Y., Heintz, E., Moutaye, A.H.E., Pilbeam, D., 1995. The first australopithecine 2,500 kilometres west of the Rift Valley (Chad). Nature 378, 273–275.

Campisano, C.J., Feibel, C.S., 2007. Connecting local environmental sequences to global climate patterns: evidence from the hominin-bearing Hadar Formation, Ethiopia. Journal of Human Evolution 53, 515–527.

Carlson, K.J., Stout, D., Jashashvili, T., De Ruiter, D.J., Tafforeau, P., Carlson, K., Berger, L.R., 2011. The endocast of MH1, *Australopithecus sediba*. Science 333, 1402–1407.

Cerling, T.E., Mbua, E., Kirera, F.M., Manthi, F.K., Grine, F.E., Leakey, M.G., Sponheimer, M., Uno, K.T., 2011. Diet of *Paranthropus boisei* in the early Pleistocene of East Africa. Proceedings of the National Academy of Sciences 108, 9337–9341.

Clarke, R.J., 1998. First ever discovery of a well-preserved skull and associated skeleton of *Australopithecus*. South African Journal of Science 94, 460–463.

Clarke, R.J., 1999. Discovery of complete arm and hand of the 3.3 million-year-old *Australopithecus* skeleton from Sterkfontein. South African Journal of Science 95, 477–481.

Clarke, R.J., 2008. Latest information on Sterkfontein's *Australopithecus* skeleton and a new look at *Australopithecus*. South African Journal of Science 104, 443–450.

Clarke, R.J., 2013. *Australopithecus* form Sterkfontein Caves, South Africa. In: Reed, K.E., Fleagle, J.G., Leakey, R.E. (Eds.), The Paleobiology of *Australopithecus*. Springer, New York, pp. 105–123.

Clarke, R.J., Tobias, P.V., 1995. Sterkfontein member 2 foot bones of the oldest South African hominid. Science 269, 521–524.

Coffing, K., Feibel, C.S., Leakey, M.G., Walker, A., 1994. Four-million year old hominids from East Lake Turkana, Kenya. American Journal of Physical Anthropology 93, 55–65.

Coppens, Y., Sakka, M., 1983. Un nouveau crane d'australopitheque: evolutive morphogenese du cranie et anthropogenese. In: Sakka, M. (Ed.), Morphologie evolutive: Morphogenese du crane et origine d'homme. Pre-congress symposium du VIII Congres Internationational de Primatologie. CNRS, Paris, pp. 185–194.

Dart, R.A., 1925. *Australopithecus africanus*: the man-ape of South Africa. Nature 115, 195–199.

Dart, R.A., 1948. The Makapansgat protohuman *Australopithecus prometheus*. American Journal of Physical Anthropology 6, 259–283.

Delson, E., 1997. One skull does not a species make. Nature 389, 445–446.

DeSilva, J.M., 2009. Functional morphology of the ankle and the likelihood of climbing in early hominins. Proceedings of the National Academy of Sciences 106, 6567–6572.

DeSilva, J.M., Holt, K.G., Churchill, S.E., Carlson, K.J., Walker, C.S., Zipfel, B., Berger, L.R., 2013. Lower limb and mechanics of walking in *Australopithecus sediba*. Science 340 (6129). http://dx.doi.org/10.1126/science.1232999.

Dirks, G.M., Kibii, J.M., Kuhn, B.F., Steininger, C.M., Churchill, S.E., Kramers, J.D., Pickering, R., Farber, D.L., Meriaux, A.-S., Herries, A.I.R., King, G.C.P., Berger, L.R., 2010. Geological setting and age of *Australopithecus sediba* from Southern Africa. Science 328, 205–208.

Falk, D., Redmond, J.C., Guyer, J., Conroy, C., Recheis, W., Weber, G.W., Seidler, H., 2000. Early hominid brain evolution: a new look at old endocasts. Journal of Human Evolution 28, 695–717.

Fleagle, J., Rasmussen, D., Yirga, S., Brown, T., Grine, F., 1991. New hominid fossils from Fejej, Southern Ethiopia. Journal of Human Evolution 21, 145–152.

Gordon, A.D., Green, D.J., Richmond, B.G., 2008. Strong postcranial size dimorphism in *Australopithecus afarensis*: results from two new resampling methods for multivariate data sets with missing data. American Journal of Physical Anthropology 135, 311–328.

Gowlett, J.A.J., Harris, J.W.K., Walton, D., Wood, B.A., 1981. Early archaeological sites, hominid remains and traces of fire from Chesowanja, Kenya. Nature 294, 125–129.

Green, D.J., Alemseged, Z., 2012. *Australopithecus afarensis* scapular ontogeny, function and the role of climbing in human evolution. Science 338, 514–517.

Grine, F.E., Jacobs, R.L., Reed, K.E., Plavcan, J.M., 2012. The enigmatic molar from Gondolin, South Africa: implications for *paranthropus* paleobiology. Journal of Human Evolution 63, 597–609.

Haile-Selassie, Y., 2010. Phylogeny of early *Australopithecus*: new fossil evidence from the Woranso-Mille (central Afar, Ethiopia). Philosophical Transactions of the Royal Society of London B 365, 3323–3331.

Haile-Selassie, Y., Latimer, B., Alene, M., Deino, A., Gibert, L., Melillo, S.M., Saylor, B., Scott, G.R., Lovejoy, C.O., 2010a. An early *Australopithecus afarensis* postcranium from Woranso-Mille, Ethiopia. Proceedings of the National Academy of Sciences 107, 12121–12126.

Haile-Selassie, Y., Saylor, B., Deino, A., Alene, M., Latimer, B.M., 2010b. New hominid fossils from Woranso-Mille (Central Afar, Ethiopia) and taxonomy of early *Australopithecus*. American Journal of Physical Anthropology 141, 406–417.

Hammond, A.S., Ward, C.V., 2013. *Australopithecus* and *kenyanthropus*. In: Begun, D.R. (Ed.), A Companion to Paleoanthropology. Wiley-Blackwell, West Sussex, pp. 434–456.

Harrison, T., 2011. Hominins from the Upper Laeotolil and Upper Ndolanya Beds, Laetoli. In: Harrison, T. (Ed.), Paleontology and Geology of Laetoli: Human Evolution in Context. Springer, New York, pp. 141–188.

Haüsler, M.F., Berger, L.R., 2001. Sts 441/465: a new fragmentary ilium of a small-bodied *Australopithecus africanus* from Sterkfontein, South Africa. Journal of Human Evolution 40, 411–418.

Heinrich, R.E., Rose, M.D., Leakey, R.E., Walker, A., 1993. Hominid radius from the middle Pliocene of Lake Turkana, Kenya. American Journal of Physical Anthropology 92, 139–148.

Herries, A.I.R., Pickering, R., Adams, Curnoe, D., Warr, G., Latham, A.G., Shaw, J., 2013. A multi-disciplinary perspective on the age of *Australopithecus* in Southern Africa. In: Reed, K.E., Fleagle, J.G., Leakey, R.E. (Eds.), The Paleobiology of Australopithecus. Springer, New York, pp. 21–40.

Holloway, R.L., 2008. The human brain evolving: a personal retrospective. Annual Review of Anthropology 37, 1–19.

Holloway, R.L., Yuan, M.S., 2004. Endocranial morphology of A.L. 444-2. In: Kimbel, W.H., Rak, Y., Johanson, D.C. (Eds.), The Skull of Australopithecus Afarensis. Oxford University Press, New York, pp. 123–135.

Howell, F.C., 1978. Hominidae. In: Maglio, V.J., Cooke, H.B.S. (Eds.), Evolution of African Mammals. Harvard University Press, Cambridge, pp. 154–248.

Johanson, D.C., White, T.D., Coppens, Y., 1978. A new species of the genus *Australopithecus* (Primates: Hominidae) from the Pliocene of Eastern Africa. Kirtlandia 28, 1–14.

Keyser, A.W., 2000. The Drimolen skull: the most complete australopithecine cranium and mandible to date. South African Journal of Science 96, 189–197.

Keyser, A.W., Menter, C.G., Moggi-Cecchi, J.M., Pickering, T.F., Berger, L.R., 2000. Drimolen: a new hominid-bearing site in Gauteng, South Africa. South African Journal of Science 96, 193–197.

Kibii, J.M., Churchill, S.E., Schmid, P., Carlson, K.J., Reed, N.D., de Ruiter, D.J., Berger, L.R., 2011. Partial pelvis of *Australopithecus sediba*. Science 333, 1407–1411.

Kimbel, W.H., 1988. Identification of the partial cranium of *Australopithecus afarensis* from Koobi Fora formation, Kenya. Journal of Human Evolution 17, 647–656.

Kimbel, W.H., Delezene, L.K., 2009. "Lucy" redux: a review of research on *Australopithecus afarensis*. Yearbook of Physical Anthropology 49, 2–48.

Kimbel, W.H., Johansen, D.C., Rak, Y., 1994. The first skull and other new discoveries of *Australopithecus afarensis* at Hadar, Ethiopia. Nature 368, 449–451.

Kimbel, W.H., Lockwood, C.A., Ward, C.V., Leakey, M.G., Rak, Y., Johanson, D.C., 2006. Was *Australopithecus anamensis* ancestral to A. afarensis? A case of anagenesis in the hominin fossil record. Journal of Human Evolution 51, 1–18.

Kimbel, W.H., Rak, Y., Johanson, D.C., 2004. The Skull of *Australopithecus afarensis*. Oxford University Press, New York.

Kimbel, W.H., White, T.D., Johanson, D.C., 1988. Variation, sexual dimorphism and the taxonomy of *Australopithecus*. In: Grine, F.E. (Ed.), Evolutionary History of the "Robust" Australopithecines. Aldine de Gruyter, New York, pp. 175–192.

Kivell, T.L., Kibii, J.M., Churchill, S.E., Schmid, P., Berger, L.R., 2011. *Australopithecus sediba* hand demonstrates mosaic evolution of locomotor and manipulative abilities. Science 333, 1411–1417.

Kullmer, O., Sandrock, O., Abel, R., Schrenk, F., Bromage, T.G., Juwayeyi, Y.M., 1999. The first *Paranthropus* form the Malawi Rift. Journal of Human Evolution 37, 121–127.

Lague, M.R., Jungers, W.L., 1996. Morphometric variation in Plio-Pleistocene hominid distal humeri. American Journal of Physical Anthropology 101, 401–427.

Larson, S.G., 2013. Shoulder morphology in early hominin evolution. In: Reed, K.E., Fleagle, J.G., Leakey, R.E. (Eds.), The Paleobiology of Australopithecus. Springer, New York, pp. 247–261.

Latimer, B., Ohman, J.C., Lovejoy, C.O., 1987. Talocrural joint in African hominoids: implications for *Australopithecus afarensis*. American Journal of Physical Anthropology 74, 155–175.

Leakey, L.S.B., 1959. A new fossil skull from Olduvai. Nature 184, 491–493.

Leakey, L.S.B., 1960. An alternative interpretation of the supposed giant deciduous hominid tooth from Olduvai: reply. Nature 185, 408.

Leakey, L.S.B., Leakey, M.D., 1964. Recent discoveries of fossil hominids in Tanganyika, at Olduvai and near Lake Natron. Nature 202, 5–7.

Leakey, M.D., 1981. Tracks and tools. Philosophical Transactions of the Royal Society of London 292, 95–102.

Leakey, M.D., 1987. Introduction. In: Leakey, M.D., Harris, J.M. (Eds.), Laetoli: A Pliocene Site in Northern Tanzania. Clarendon Press, Oxford, pp. 490–523.

Leakey, M.D., Hay, R.L., 1979. Pliocene footprints in the Laetoli beds at Laetoli, northern Tanzania. Nature 278, 317–323.

Leakey, M.G., Feibel, C.S., McDougall, I., Walker, A., 1995. New four-million-year-old hominid species from Kanapoi and Allia Bay, Kenya. Nature 376, 565–571.

Leakey, M.G., Feibel, C.S., McDougall, I., Ward, C.V., Walker, A., 1998. New specimens and confirmation of an early age for *Australopithecus anamensis*. Nature 393, 62–66.

Leakey, M.G., Spoor, F., Brown, F.H., Gathogo, P.N., Kiarie, C., Leakey, L.N., McDougall, I., 2001. New hominin genus from eastern Africa shows diverse middle Pliocene lineages. Nature 410, 433–440.

Leakey, R.E.F., Walker, A., 1988. New *Australopithecus boisei* specimens from east and west Lake Turkana, Kenya. American Journal of Physical Anthropology 76, 1–24.

Lockwood, C.A., Tobias, P.V., 2002. Morphology and affinities of new hominin cranial remains from Member 4 of the Sterkfontein Formation, Gauteng Province, South Africa. Journal of Human Evolution 42, 389–450.

Lockwood, C.A., Richmond, B.G., Jungers, W.L., Kimbel, W.H., 1988. Randomization procedures and sexual dimorphism in *Australopithecus afarensis*. Journal of Human Evolution 31, 537–548.

Lovejoy, C.O., 1981. The origin of man. Science 211, 341–350.

MacLatchy, L.M., DeSilva, J., Sanders, W.J., Wood, B., 2010. Hominini. In: Werdelin, L., Sanders, W.J. (Eds.), Cenozoic Mammals of Africa. University of California Press, Berkeley, pp. 471–540.

Manthi, F.K., Plavcan, J.M., Ward, C.V., 2012. New hominin fossils from Kanapoi, Kenya, and the mosaic nature of canine evolution in hominins. South African Journal of Science 108, 1–9.

McDougall, I., Brown, F., 2008. Geochronology of the pre-KBS Tuff sequence, Omo Group, Turkana Basin. Journal of the Geological Society 165, 549–562.

McHenry, H.M., 1991. Sexual dimorphism in *Australopithecus afarensis*. Journal of Human Evolution 20, 21–32.

McHenry, H.M., 1992. Body size and proportions in early hominids. American Journal of Physical Anthropology 87, 407–431.

McHenry, H.M., 1986. The first bipeds: a comparison of the *A. afarensis* and *A. africanus* postcranium and implications for the evolution of bipedalism. Journal of Human Evolution 15, 177–191.

Menter, C.G., Kuykendall, K.L., Keyser, A.W., Conroy, G.C., 1999. First record of hominid teeth from the Plio-Pleistocene site of Gondolin. South Africa. Journal of Human Evolution 37, 299–307.

van der Merwe, N., Masau, F., Bamford, M., 2008. Isotopic evidence for contrasting diets of early hominins *Homo habilis* and *Australopithecus boisei* of Tanzania. South African Journal of Science 104, 153–155.

Partridge, T.C., Granger, D.E., Caffee, M.W., Clarke, R.J., 2003. Lower Pliocene hominid remains from Sterkfontein. Science 300, 607–612.

Patterson, B., Howells, W.W., 1967. Hominid humeral fragment from early Pleistocene of northwestern Kenya. Science 156, 64–66.

Pickering, T.R., Clarke, R.J., Heaton, J.L., 2004. The context of Stw 573, an early hominid skull and skeleton from Sterkfontein Member 2: taphonomy and paleoenvironment. Journal of Human Evolution 46, 277–295.

Pickering, R., Kramers, J.D., Partridge, T., Kodolanyi, J., Pettke, T., 2010. U–Pb dating of calcite–aragonite layers in speleothems from hominin sites in South Africa by MC-ICP-MS. Quaternary Geology 5, 544–558.

Pickering, R., Dirks, P.H., Jinnah, Z., De Ruiter, D.J., Churchill, S.E., Herries, A.I., Woodhead, J.D., Hellstrom, J.C., Berger, L.R., 2011. *Australopithecus sediba* at 1.977 Ma and implications for the origins of the genus *Homo*. Science 333, 1421–1423.

Plavcan, J.M., 1997. Interpreting hominid behavior on the basis of sexual dimorphism. Journal of Human Evolution 32, 345–374.

Plavcan, J.M., Lockwood, C.A., Kimbel, W.H., Lague, M.R., Harmon, E.H., 2005. Sexual dimorphism in *Australopithecus afarensis* revisited: how strong is the case for a human-like pattern of dimorphism? Journal of Human Evolution 48, 313–320.

Prat, S., Brugal, J.-P., Roche, H., Texier, P.-J., 2003. Nouvelles decouvertes de dents d'hominidés dans le Membre Kaitio del a Formation de Nachukui (1.65-1.9 Ma) ouest du Lac Turkana (Kenya). Comptes Rendues Paleovol 2, 685–693.

Puech, P.-F., 1986. *Australopithecus afarensis* Garusi 1, diversité et spécialisation des premiers Hominidés d'après les caractères maxillo-dentaires. Comptes Rendues des l'Academie des Sciences, Paris, Série II 20 (303), 1819–1823.

de Ruiter, D.J., Churchill, S.E., Berger, L.R., 2013a. *Australopithecus sediba* from malapa, South Africa. In: Reed, K.E., Fleagle, J.G., Leakey, R.E. (Eds.), The Paleobiology of Australopithecus. Springer, New York, pp. 147–160.

de Ruiter, D.J., DeWitt, T.J., Carlson, K.B., Brophy, J.K., Schroeder, L., Ackerman, R.R., Churchill, S.E., Berger, L.R., 2013b. Mandibular remains support taxonomic validity of *Australopithecus sediba*. Science 340, 6129. http://dx.doi.org/10.1126/science.1232997.

Rak, Y., 1985. The Australopithecine Face. Academic Press, New York.

Rak, Y., Kimbel, W.H., 1991. On the squamosal suture of KNM-WT 17000. American Journal of Physical Anthropology 85, 1–6.

Rak, Y., Kimbel, W.H., 1993. Reply to Drs. Walker, Brown and Ward. American Journal of Physical Anthropology 90, 506–507.

Raichlen, D., Gordon, A.D., Harcourt-Smith, W.E.H., Foster, A.D., Haas Jr., W.R., 2010. Laetoli footprints preserve earliest direct evidence of human-like bipedal biomechanics. PLoS ONE 5, e9769.

Rayner, R.J., Moon, B.P., Masters, J.C., 1993. The Makapansgat australopithecine environment. Journal of Human Evolution 24, 219–231.

Reno, P.L., Meindl, R.S., McCollum, M.A., Lovejoy, C.O., 2003. Sexual dimorphism in *Australopithecus afarensis* was like that of modern humans. Proceedings of the National Academy of Sciences 100, 9404–9409.

Richmond, B.G., Hatala, K.G., 2013. Origin and evolution of human postcranial anatomy. In: Begun, D.R. (Ed.), A Companion to Paleoanthropology. Wiley-Blackwell, West Sussex, pp. 183–202.

Richmond, B.G., Jungers, W.L., 1995. Size variation and sexual dimorphism in *Australopithecus afarensis* and living hominoids. Journal of Human Evolution 29, 229–245.

Robinson, J.T., 1954. The genera and species of the Australopithecinae. American Journal of Physical Anthropology 12, 181–200.

Robinson, J.T., 1956. The dentition of the Australopithecinae. Transvaal Museum Memoirs 9, 1–179.

Robinson, J.T., 1968. The origin and adaptive radiation of the australopithecines. In: Kurth, G. (Ed.), Evolution und hominisation, second ed. Fischer, Stuttgart, pp. 150–175.

Robinson, J.T., 1972. Early Hominid Posture and Locomotion. University of Chicago Press, Chicago.

Ryan, A.S., Johanson, D.C., 1989. Anterior dental microwear in *Australopithecus afarensis*: comparisons with human and nonhuman primates. Journal of Human Evolution 18, 235–268.

Schmid, P., 2002. The gladysvale project. Evolutionary Anthropology 11, 45–48.

Schoeneman, P.T., 2013. Hominid brain evolution. In: Begun, D.R. (Ed.), A Companion to Paleoanthropology. Wiley-Blackwell, West Sussex, pp. 136–164.

Scott, R.S., Ungar, P.S., Bergstrom, T.S., Brown, C.A., Grine, F.E., Teaford, M.F., Walker, A., 2005. Dental microwear texture analysis reflects diets of living primates and fossil hominins. Nature 436, 693–695.

Sponheimer, M., Alemseged, Z., Cerling, T.E., Grine, F.E., Kimbel, W.H., Leakey, M.G., Lee-Thorp, J.A., Manthi, F.K., Reed, K.E., Wood, B.A., Wynn, J.G., 2013. Isotopic evidence of early hominin diets. Proceedings of the National Academy of Sciences 110, 10513–10518.

Sponheimer, M., Passey, B.H., de Ruiter, D.J., Guatelli-Steinberg, D., Cerling, T.E., Lee-Thorp, J.A., 2006. Isotopic evidence for dietary variability in the early hominin *Paranthropus robustus*. Science 314, 980–982.

Stern, J.T., 2000. Climbing to the top: a personal memoir of *Australopithecus afarensis*. Evolutionary Anthropology 9, 113–133.

Strait, D.S., Grine, F.E., 2004. Inferring hominoid and early hominid phylogeny using cranial characters: the role of fossil taxa. Journal of Human Evolution 47, 399–452.

Suwa, G., Wood, B.A., White, T.D., 1994. Further analysis of mandibular molar crown and cusp areas in Pliocene and early Pleistocene hominids. American Journal of Physical Anthropology 93, 407–426.

Suwa, G., Asfaw, B., Beyene, Y., White, T.D., Katoh, S., Nagaoka, S., Nakaya, H., Uzawa, K., Renne, P., Woldegabriel, G., 1997. The first skull of *Australopithecus boisei*. Nature 389, 489–492.

Suwa, G., Kono, R., Simpson, S.W., Asfaw, B., Lovejoy, C.O., White, T.D., 2009. Paleobiological implications of the *Ardipithecus ramidus* dentition. Science 326, 69–99.

Szalay, F.S., 1971. Biological level of organization of the Chesowanja robust australopithecine. Nature 234, 229–230.

Tague, R.G., Lovejoy, C.O., 1986. The obstetric pelvis of A.L. 288-1 (Lucy). Journal of Human Evolution 15, 237–255.

Teaford, M.F., Ungar, P.S., 2000. Diet and the evolution of the earliest human ancestors. Proceedings of the National Academy of Sciences 97, 13506–13511.

Tobias, P.V., 1967. Olduvai Gorge: The Cranium and Maxillary Dentition of Australopithecus (Zinjanthropus) Boisei, vol. 2. Cambridge University Press, Cambridge.

Toussaint, M., 2003. The third partial skeleton of a last Pliocene hominin (Stw 431) from Sterkfontein, South Africa. South African Journal of Science 99, 215–223.

Ungar, P.S., Grine, F.E., Teaford, M.F., 2008. Dental microwear and diet of the Plio-Pleistocene hominin *Paranthropus boisei*. PLoS ONE 3, e2044.

Ungar, P.S., Krueger, K.L., Blumenschine, R.J., et al., 2012. Dental microwear texture analysis of hominins recovered by the Olduvai Landscape Paleoanthropology Project, 1995–2007. Journal of Human Evolution 63, 429–437.

Walker, A., Brown, B., Ward, S.C., 1993. Squamosal suture of cranium KNM-WT 17000. American Journal of Physical Anthropology 90, 501–505.

Walker, A., Leakey, R.E., Harris, J.M., Brown, F., 1986. 2.5-Myr *Australopithecus boisei* from west of Lake Turkana, Kenya. Nature 322, 517–522.

Walker, J., Cliff, R.A., Latham, A.G., 2006. U-pb isotopic age of the Stw 573 hominid from Sterkfontein, South Africa. Science 314, 1592–1594.

Ward, C.V., 2002. Interpreting the posture and locomotion of *Australopithecus afarensis*: where do we stand? Yearbook of Physical Anthropology 45, 185–215.

Ward, C.V., 2013a. Early hominin posture and locomotion. In: Sponheimer, M., Reed, K.E., Ungar, P.S., Lee-Thorpe, J. (Eds.), Early Hominin Paleoecology. University of Chicago Press, Chicago, pp. 205–264.

Ward, C.V., 2013b. Postural and locomotor adaptations of *Australopithecus* species. In: Reed, K.E., Fleagle, J.G., Leakey, R.E. (Eds.), The Paleobiology of Australopithecus. Springer, New York, pp. 235–245.

Ward, C.V., Kimbel, W.H., Johanson, D.C., 2011. Complete fourth metatarsal and arches in the foot of *Australopithecus afarensis*. Science 331, 750–753.

Ward, C.V., Kimbel, W.H., Harmon, E.H., Johanson, D.C., 2012. New postcranial fossils of *Australopithecus afarensis* from Hadar, Ethiopia (1990–2007). Journal of Human Evolution 63, 1–51.

Ward, C.V., Leakey, M.G., Brown, B., Brown, F., Harris, J., Walker, A., 1999a. South Turkwel: a new Pliocene hominid site in Kenya. Journal of Human Evolution 36, 69–95.

Ward, C.V., Walker, A., Leakey, M.G., 1999b. The new hominid species *Australopithecus anamensis*. Evolutionary Anthropology 7, 197–205.

Ward, C.V., Plavcan, J.M., Manthi, F.K., 2010. Anterior dental evolution in the *Australopithecus anamensis-afarensis* lineage. Philosophical Transactions of the Royal Society of London 365, 3333–3344.

Ward, C.V., Plavcan, J.M., Manthi, F.K., 2013. New fossils of *Australopithecus anamensis* from Kanapoi, west Turkana, Kenya (2003–2008). Journal of Human Evolution 65, 501–524.

Ward, C.V., Walker, A., Leakey, M.G., 2001. Morphology of *Australopithecus anamensis* from Kanapoi and Allia Bay, Kenya. Journal of Human Evolution 41, 255–368.

White, T.D., 2002. Earliest hominids. In: Hartwig, W.C. (Ed.), The Primate Fossil Record. Cambridge University Press, Cambridge, pp. 407–417.

White, T.D., Falk, D., 1999. A quantitative and qualitative reanalysis of the endocast from the juvenile *Paranthropus* specimen L338y-6 from Omo, Ethiopia. American Journal of Physical Anthropology 110, 399–406.

White, T.D., Johanson, D.C., Kimbel, W.H., 1981. *Australopithecus africanus*: phyletic position reconsidered. South African Journal of Science 77, 445–470.

White, T.D., Suwa, G., Asfaw, B., 1994. *Australopithecus ramidis*, a new species of early hominid from Aramis, Ethiopia. Nature 371, 306–312.

White, T.D., Suwa, G., Hart, W.K., Walter, R.C., WoldeGabriel, G., 1993. New discoveries of *australopithecus* at Maka, Ethiopia. Nature 366, 261–265.

White, T.D., Suwa, G., Simpson, S., Asfaw, B., 2000. Jaws and teeth of *Australopithecus afarensis* from Maka, Middle Awash, Ethiopia. American Journal of Physical Anthropology 111, 45–68.

White, T.D., WoldeGabriel, G., Asfaw, B., Ambrose, S., Beyene, Y., Bernor, R.L., Boisserie, J.-R., Currie, B., Gilbert, H., Haile-Selassie, Y., Hart, W.K., Hlusko, L.J., Howell, F.C., Kono, R.T., Lehmann, T., Louchart, A., Lovejoy, C.O., Renne, P.R., Saegusa, H., Vrba, E.S., Wesselman, H., Suwa, G., 2006. Asai Issie, Aramis and the origin of *Australopithecus*. Nature 440, 883–889.

Wood, B., 1991. Koobi Fora Research Project. Hominid Cranial Remains, vol. 4. Clarendon Press, Oxford.

Wood, B., Constantino, P., 2007. *Paranthropus boisei*: fifty years of evidence and analysis. Yearbook of Physical Anthropology 50, 106–132.

Wood, B., Schroer, K., 2013. In: Begun, D.R. (Ed.), A Companion to Paleoanthropology. Wiley-Blackwell, West Sussex, pp. 457–478.

Wynn, J.G., Alemseged, Z., Bobe, R., Geraads, D., Reed, D., Roman, D.C., 2006. Geological and palaeontological context of a Pliocene juvenile hominin at Dikika, Ethiopia. Nature 443, 332–336.

Zipfel, B., DeSilva, J.M., Kidd, R.S., Carlson, K.J., Churchill, S.E., Berger, L.R., 2011. The foot and ankle of *Australopithecus sediba*. Science 333, 1417–1420.

Chapter 11

Early Pleistocene *Homo*

Scott W. Simpson
Department of Anatomy, Case Western Reserve University School of Medicine, Cleveland, OH, USA

SYNOPSIS

Today, it is easy to identify humans (*Homo sapiens*) vis-à-vis all other primates because the anatomical and behavioral gulf between us and our closest living relatives is great and obvious (e.g., the former's large brain size, technological sophistication, worldwide distribution, and language). In our practical experience, there is only one human species living today and this seems reasonable to us given that our behavioral and technological adaptations are so pervasive that there is not enough "ecological room" for another humanlike species and it is easy for us to consider that this uniqueness and suite of adaptations extended well into the past. However, this view is challenged as we learn more about the behavioral sophistication of other living apes, and as more ancient and anatomically primitive hominin fossils are recovered. While the appearance of modern humans (*H. sapiens*) is first documented in the fossil record at 195,000 years ago in southern Ethiopia (McDougall et al., 2008; Fleagle et al., 2008), the origins of the genus *Homo* extend back to 2.7–2.9 million years ago with a complex history that includes significant taxonomic diversity and marked behavioral and anatomical evolution. Here, the fossil evidence for Early Pleistocene (2.58–0.78 million years ago) *Homo* is reviewed.

DEFINING THE GENUS *HOMO*

Initially, the genus *Homo* was diagnosed based on the anatomy of modern humans (*Homo sapiens*), and prior to the recovery of fossil hominins, this definition was satisfactory (Stringer and Buck, 2014). The discovery of more historically ancient and anatomically primitive hominins required a concomitant reassessment of the genus's diagnosis, especially consideration of brain size and craniofacial shape. It was assumed, without much supporting data, that a brain volume of less than 750cc (Keith, 1948), or 800cc (Vallois, 1954), or 600cc (Montagu, 1961; Leakey et al., 1964) was insufficiently distinct from more primitive ancestors and could not support behaviors that are characteristically human in nature (e.g., speech, tool use/manufacture, and social complexity). As the fossil record has improved, brain size has proven to be an inviable generic diagnostic criterion in light of an increasingly rich fossil record in a lineage that exhibited such substantial evolution, parallelism, reversals, and variation in brain size (Spoor et al., 2015) (see chapter by Holloway in the present volume). For example, while the nomenclature and phyletic details of the Late Pleistocene hominin *Homo floresiensis* with a cranial capacity of 380–417cc (Brown et al., 2004) and a stature of about 1 m are unclear, its allocation to the genus *Homo* is widely accepted (e.g., Argue et al., 2006; Eckhardt et al., 2014).

A clade-based (phylogenetic) definition based on sharing a common ancestry from a single ancestor emphasizes the historical relationships of the phylogenetic unit relative to other such units (Wood, 2010). Taxonomic nomenclature needs to be consistent with the phyletic history of a group. For example, the phylogenetic proximity of separate lineages must be reflected in the taxonomic names. As noted above, the anatomically distinctive *H. floresiensis* is still assigned to the genus *Homo* because it shared a separate common ancestor with an independent phyletic history distinct from other Early Pleistocene hominins such as *Australopithecus africanus* and *A. (Paranthropus) boisei.* While this approach emphasizes its phylogenetic context, practically we must use phenotypic traits to make these assessments in the fossil record. Clearly, the earliest phylogenetic members of *Homo* did not exhibit the unique anatomical or behavioral traits of their descendants, yet those individuals were part of a discrete lineage distinct from other genetically independent and separately evolving units. Only where that lineage demonstrates clearly unique *Homo* anatomical characteristics can the fossil representatives be reliably assigned to our genus. Many of the criteria that have proven useful in this regard are associated with an unspecialized dentognathic system (smaller postcanine tooth crowns, enamel thickness thinner than that of "robust" australopiths, reduced prognathism, variable reduction in mandibular corpus robusticity, shorter mandibular corpora, and absence of *temporalis* muscle enlargement), and increasing brain size and its anatomical correlates (e.g., change in the relative position of the mandibular fossa and reduction of postorbital constriction). The dearth of taxonomically unambiguous postcranial elements in the fossil record precludes creating a reliable differential diagnosis for postcranial characters such as limb proportions, body mass and stature, and manual dexterity. While early *Homo* is generally thought to be the maker and user of stone tools and the author of the cut-marks on the c. 2.5 million-year-old (mya) animal remains (deHeinzelin et al., 1999; Semaw et al., 2003), this relationship needs to be more explicitly demonstrated as recent reports suggest that australopiths pursued stone tool-assisted carnivory by 3.3 mya (McPherron et al., 2010; Harmand et al., 2015).

HISTORY OF DISCOVERY OF EARLY *HOMO* FOSSILS

Eugène Dubois's discovery of fossil hominin remains in Java in 1891, in his search to discover evidence supporting Darwin's evolutionary proposals, represents perhaps one of the greatest exemplars of dedication to an idea and perseverance in the annals of human evolutionary studies (Shipman, 2001). Until that time, no truly ancient hominin fossils had been discovered and, in fact, hominin paleontology was not even a recognized field of research. The hominin fossils he discovered along the banks of the Trinil River in Java were found in association with extinct fauna suggesting an ancient age. Dubois named this species *Pithecanthropus erectus* ("ape-man that stands erect") based on the size and shape of the cranial vault (smaller endocranial volume than modern humans with a long and low cranial vault and prominent browridges) and the modern appearance of the recovered femora (see Ruff et al. (2015), who provide evidence that the iconic pathological Femur I—but not Femora II–IV—may be markedly younger than the calotte).

Subsequent research on fossils recovered from the Zhoukoudian caves near Beijing (then called Peking in the West), China in the early twentieth century (Black, 1926, 1928, 1929) recognized their similarity with those recovered in Java, and they too were proposed as human ancestors. Those fossils, now dated to between 0.4 and 0.78 mya (Shen et al., 2009), primarily craniodental remains, were assigned to the taxon *Sinanthropus pekinenensis* ("man from Peking, China") (Black, 1928, 1929). Reassessments of hominin taxonomy led to a reassignment of these Javan and Chinese samples first to *Pithecanthropus erectus* (Weidenreich, 1940) and then to *Homo erectus* (Mayr, 1950).

The recognition that the large-brained Piltdown fossil skull from England was fraudulent (Weiner et al., 1953; Weiner, 1955) removed the neurocentric model of *Homo* evolution from consideration, thus allowing the possibility of smaller-brained human ancestors. While primitive, small-brained, craniodentally specialized hominin fossils from Africa were known prior to the 1960s (e.g., *Australopithecus africanus* and *A. (Paranthropus) robustus*), a compelling fossil candidate of the earliest *Homo* had not been recognized at that time. The discovery of smaller-brained individuals lacking specialized masticatory anatomy in association with primitive stone tools in Early Pleistocene deposits at Olduvai Gorge in Tanzania by Louis and Mary Leakey beginning in the late 1950s (Tobias, 2009) provided evidence that two hominin species coexisted at Olduvai. These fossils were the earliest and most definitive evidence that *Homo* had a separate phylogenetic trajectory from the syntopic and synchronic *Australopithecus (P.) boisei*. These non-australopith specimens from Olduvai (OH 4, OH 6, OH 7, OH 8, OH 13) were assigned to a new species, *Homo habilis* (Leakey et al., 1964), and proposed to be uniquely ancestral to humans. The holotype for the species, OH 7 (Olduvai Hominid #7), was a juvenile that included fragmentary right and left parietal bones, a mandible, and a hand. This proposition was met with a variety of positive and negative responses. On the positive side, it had the right anatomy (craniodental material lacking obvious masticatory specializations such as a sagittal crest, robust mandibular corpus, tall mandibular ramus, postcanine megadontia, and anterior tooth size reduction) at the right time (earliest Pleistocene)[1] and in association with stone tools. However, its allocation to *Homo* was debated because the brain was considered to be too small (~650cc (Holloway et al., 2004)) (although endocranial volume for OH 7 has now been revised to 729–824cc (Spoor et al., 2015) for our genus and had yet to cross the "cerebral Rubicon" of brain size arbitrarily established at 700cc (see chapter by Holloway in this volume). One issue that remains concerning the early *Homo* fossils known from Olduvai Gorge is that the holotype of *H. habilis* (OH 7) is from a juvenile (Leakey et al., 1964; Tobias, 1991; Spoor et al., 2015)). Why this matters is that juvenile individuals do not necessarily exhibit the characteristic size and anatomical details expressed in adults and may be unreliable specific exemplars. The recovery of postcranial fossils, especially foot (OH 8) remains, reinforced the derived status of this hominin. However, the allocation of the OH 8 foot to *H. habilis* remains tentative, though numerous analyses have emphasized its full adaptation to bipedality (Susman, 2008; DeSilva et al., 2010).

Additional fossil collections from the Ethiopian–Kenyan Turkana Basin and South Africa continued to reinforce the presence of a gracile, smaller-brained hominin in the Early Pleistocene, although the magnitude of anatomical variation was increasing with additional finds, especially in dental size and proportions, and endocranial volume (especially fossils such as KNM-ER 1470, KNM-ER 1813, and the enigmatic KNM-ER 1805 (Prat, 2002)). The degree of anatomical variation in the growing sample of fossils was considered to exceed that of a single species and led to the naming of a new species, *Homo rudolfensis* (Alexeev, 1986; Wood, 1999) based largely on the KNM-ER 1470 cranium that emphasized its enlarged cranial capacity (742cc), reduced prognathism, and apparently enlarged postcanine tooth crowns. While there was little disagreement of which fossils were included in the "early" *Homo* group, the number of separate taxa and how they were distributed between the available taxa differed greatly (e.g., Antón et al., 2014; Spoor et al., 2015). This difficulty reflects the ongoing problem of classifying often-fragmentary fossils from phylogenetically closely related taxa into discrete and well-bounded groups.

Specimens from the karstic caves of South Africa have also been attributed to early *Homo*, including StW 53 from Sterkfontein Member 5 (Kuman and Clarke, 2000), Swartkrans Member 1 (Olson, 1978; Pickering et al., 2012), and Drimolen (Keyser et al., 2000; Moggi-Cecchi et al., 2010). These localities, dated between 1.4 and 2.0mya, also document the co-occurrence of *Homo* and *A. (P.) robustus.*

In summary, fossils attributed to *Homo* from the Early Pleistocene are known from a number of localities in Africa and Eurasia. These specimens document a series of species (*H. habilis, H. rudolfensis, H. erectus*) that have brains sizes larger than the ancestral australopiths and contemporary "robust" australopiths while also having a less specialized dentition and masticatory apparatus. Additional descriptions of the species are found below.

TAXONOMIC DIVERSITY IN THE GENUS *HOMO*

Here, the Early Pleistocene (2.58–0.78mya) diversity of the genus *Homo* is reviewed. There is a natural inclination to assign the fossil specimens to existing taxonomic bins, although this can be surprisingly difficult in practice due to the incomplete fossil record and marked anatomical variation. In addition, researchers can disagree on which traits should be emphasized for taxonomic assignment, and a single fossil specimen may have multiple taxonomic allocations (e.g., the KNM-ER 1802 mandible from Koobi Fora has been allocated to *H. habilis, H. rudolfensis,* and *H.* sp. *indet.*). Finally, it is unclear if the current number of recognized species is consistent with ancient biological diversity (Spoor et al., 2015). This is recognition of the complexity of the past, and the few—but increasing—data points to test these hypotheses.

While assignment of fossils to Linnaean units is standard in paleontology, other fossil organizing methods are possible, which incorporate time and space in addition to

1. In 1964, the absolute dating of the geological timetable was poorly established and the boundaries of the Pleistocene were not well dated. The application of radiometric dating techniques of volcanic materials led to the definition of the beginning of the Pleistocene epoch at 1.806 Ma (Lourens et al., 1996). More recently, the Pliocene–Pleistocene boundary was redefined as the Gauss-Matuyama paleomagnetic reversal that occurred at 2.588 Ma (Gibbard et al., 2010).

relying on morphology (Simpson, 1963; Howell, 1999; Asfaw et al., 2002; Antón et al., 2014). For example, Antón (Antón, 2012; Antón et al., 2014) organized the early (2.1–1.5 mya) *Homo* eastern African fossils into *H. erectus* and non-*H. erectus* groups. The latter group is subdivided into the following groups: "KNM-ER 1470" and "KNM-ER 1813" groups (named after the distinctive iconic crania), *H. habilis* (limited to the type specimen OH 7), and the enigmatic "unattributed non-*erectus Homo*." The morphology of OH 7 (larger neurocranium with smaller teeth) bears similarity to both the "1470" and "1813" groups, precluding a definitive assignment to either. The "1470" and "1813" groups are not synonymous with previously named *H. rudolfensis* and *H. habilis* species. However, species have been named—perhaps with some technical error—that use both KNM-ER 1470 and KNM-ER 1813 specimens as representatives—e.g., *H. rudolfensis* (Groves and Mazák, 1975) and *H. microcranous* (Ferguson, 1995), respectively. An advantage to this alternative approach is its flexibility and the recognition that biological processes operate within temporal and spatial contexts. That said, the species is used as a basic unit of organization below.

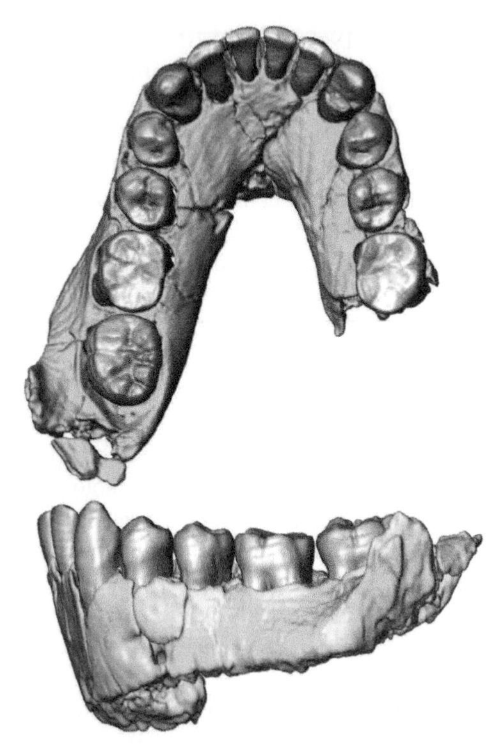

FIGURE 1 Computer-based reconstruction of the *Homo habilis* type specimen OH 7 including occlusal (top) and left lateral (bottom) views. (See color plate section). *Image used by permission of F. Spoor.*

Homo habilis

Non-australopith craniodental fossils were first recovered from Olduvai Gorge in Tanzania by Mary and Louis Leakey in the late 1950s and a new species, *H. habilis*, was named in 1964 (Leakey et al., 1964). In recognition of the larger brain and humanlike hand in the c. 1.84 mya type fossil OH 7 (Olduvai Hominid 7) and the presence of primitive stone tools, the team adopted the name "*habilis*" which means "able, handy, mentally skillful" for this species (Leakey et al., 1964). The *H. habilis* dentition lacks the large premolars and molars, and diminutive incisors and canines, of contemporaneous *Australopithecus* (Figure 1). The Olduvai *Homo* teeth are distinguishable from other *Homo* teeth by having slightly larger canines and incisors, and buccolingual narrowing of the postcanine teeth. The hand bones, while differing from modern humans in some anatomical details (e.g., robusticity and curvature of the proximal phalanges) are more similar to humans (presence of a volar pit on the terminal pollical phalanx, broad terminal phalanges) than to extant apes, suggesting a capacity for stone tool manufacture and use (Napier, 1962; Marzke and Shackley, 1987). An updated species diagnosis (Spoor et al., 2007) also identifies "reduced subnasal prognathism, thinner (than in *Australopithecus*) tooth enamel, lack of cranial keeling or bregmatic thickening, a mediolaterally narrower temporomandibular joint, and other cranial features" (MacLatchy et al., 2010). While the status of taxon *H. habilis* as either a valid taxon or one assignable to *Homo* was initially met with skepticism (Tobias, 2009), the recovery of additional material in the Turkana Basin of Kenya and Ethiopia, and reexamination of specimens from South Africa, recognized a reasonably coherent anatomically diagnosable species, thus affirming its distinctiveness from contemporary australopiths and the younger *H. erectus*. Part of the issue ("Is *H. habilis* a distinct species?") is a product of the increasing spatial distribution and morphological variation present in the fragmentary fossils. Through the 1990s, it was common to assign Early Pleistocene fossils not exhibiting characteristic *Australopithecus* masticatory or *H. erectus* cranial specializations to *H. habilis*, which had it becoming something of a "waste can" taxon. As more and better specimens, especially the 2.03–2.09-million-year-old KNM-ER 1470, KNM-ER 1813, and KNM-ER 1590 crania were discovered in the Turkana Basin in Kenya, the breadth of anatomical variation exceeded expectations for a single highly dimorphic species, providing evidence of greater taxonomic diversity. For example, the 1.86-million-year-old (Feibel et al., 2009) KNM-ER 1813 cranium is small brained (endocranial capacity of 513 cc (Holloway et al., 2004)), has smaller postcanine teeth with unreduced incisors, a slight post-toral sulcus, and a transversely rounded subnasal section (Figure 2). The 2.03–2.09-million-year-old (Joordens et al., 2013) cranium-labeled KNM-ER 1470 has a larger endocranial volume of 752 cc (Holloway et al., 2004), apparently larger postcanine teeth (inferred from the remaining alveoli and roots, since all of the crowns are missing), a transversely flat subnasal section,

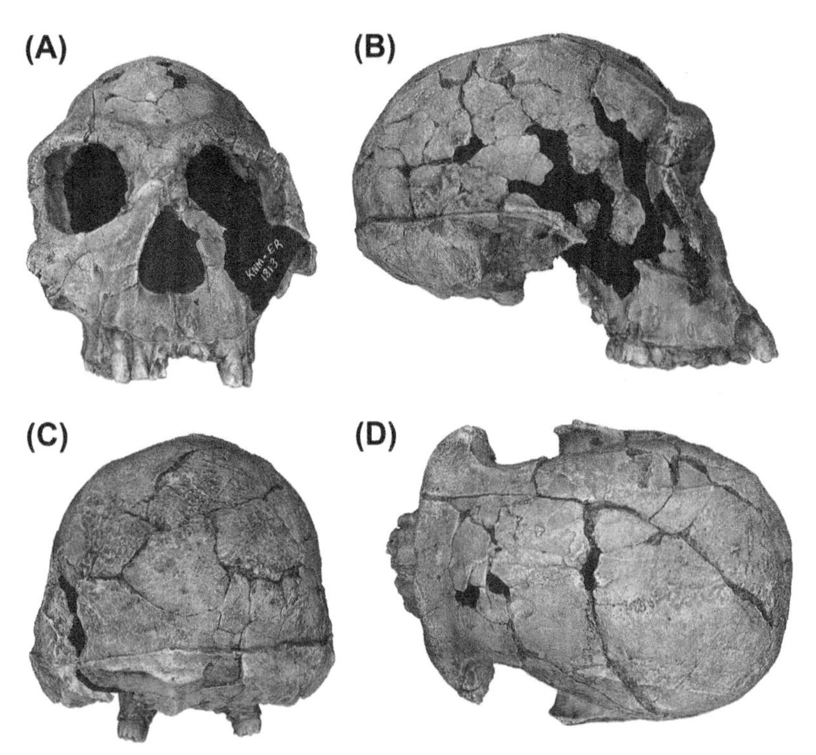

FIGURE 2 KNM-ER 1813 cranium (cast) attributed to *Homo habilis*, Turkana Basin, Kenya, 1.86 mya. (A) Facial view, (B) right lateral view, (C) posterior view, (D) superior view. (See color plate section).

and a taller midface. Some researchers identified this variation as compatible with a single species, perhaps as a consequence of sexual dimorphism within a single evolving lineage (Johanson and White, 1979; Tobias, 1991; Miller, 2000; Suwa et al., 2007). More recently, newly discovered fossils (Leakey et al., 2012) have been assigned to a more megadont, less prognathic, somewhat larger-brained species *H. rudolfensis*, a taxon originally named for KNM-ER 1470 (see discussion below), thus providing coherency for *H. rudolfensis* and distinctiveness from *H. habilis*.

Isolated postcranial bones are difficult to assign to taxon, and only those associated with diagnosable craniodental material are attributed to *H. habilis*. This limits the number of specimens to just a few (e.g., OH 7 (hand) (Leakey et al., 1964), OH 62 (partial skeleton) (Johanson et al., 1987), KNM-ER 3735 (partial skeleton) (Leakey and Walker, 1985)). The OH 7 hand includes shorter slightly curved—albeit robust—proximal phalanges, with phalangeal proportions that differ from those of earlier *Australopithecus*, which have longer, more-curved proximal phalanges. While the OH 8 foot and OH 35 tibia and fibula have been assigned to the same individual as OH 7 (Stern and Susman, 1982; Susman, 2008), the differences in developmental age and recovery location makes this association unlikely (DeSilva et al., 2010; Spoor et al., 2015). The OH 62 partial skeleton (Johanson et al., 1987) includes a palate, reasonably complete humerus, and partially complete ulna, radius, and femur. The palate is diagnostically not *Australopithecus* (e.g., broad palate, sharp nasal processes of the maxilla, the root of the zygoma is low), with teeth that are somewhat large and with complex premolar roots (Figure 3). The length of the humerus can be estimated with only a narrow range of possible values. The proximal femur is quite gracile and smaller in size than that of the 3.2-million-year-old A.L. 288-1 ("Lucy") *Australopithecus afarensis* individual who has a stature estimated to be about 105 cm. This comparison led to speculation that OH 62 had limb proportions and a height similar to those of Lucy. Unfortunately, the remaining portion of the OH 62 femur retains little anatomy that correlates with length. Thus, while the OH 62 femur may be more gracile than that of Lucy, its absolute length cannot be estimated reliably. Therefore, it is unclear whether it retained the short lower limbs of *Australopithecus* or the elongated lower limbs of later *Homo*.

A 2.5-million-year-old fossil partial skeleton (BOU-VP-12/1) was recovered from a similar stratigraphic level as that of the enigmatic *Australopithecus garhi* partial cranium (BOU-VP-12/130) from the Middle Awash project area in Ethiopia (Asfaw et al., 1999). The skeleton (BOU-VP-12/1) also has reasonably complete upper and lower limb bones that allow estimation of their length. This individual has a humerus that is about the same length as Lucy's, yet has a longer forearm (Asfaw et al., 1999). Most interesting, though, is that the femur is substantially longer than Lucy's and the BOU-VP-12/1 skeleton has a humero–femoral ratio similar to those of modern humans and *H. erectus*.

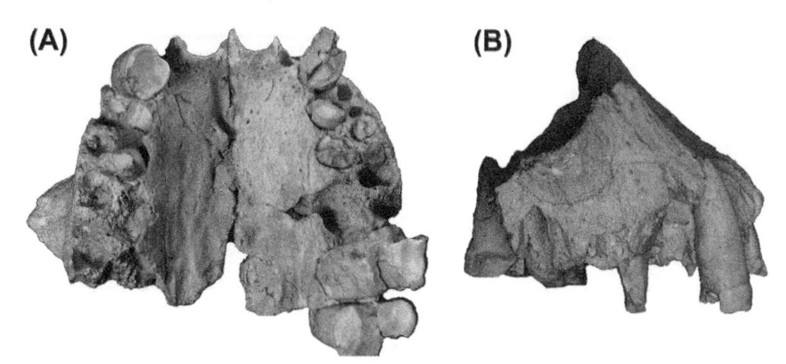

FIGURE 3 OH 62 maxilla (cast) attributed to *Homo habilis*, Olduvai Gorge, Tanzania, c. 1.8 mya. (A) Occlusal view, (B) right lateral view. (See color plate section).

Unfortunately, the partial skeleton does not retain any taxonomically diagnostic characters and cannot be assigned to a taxon. Additional fossils from this time period are necessary in order to resolve this issue.

Specimens attributed to *H. habilis* (or *H.* aff. *H. habilis*) have been recovered from deposits between 2.33 million years old (Hadar, Ethiopia (Kimbel et al., 1996, 1997)) and 1.65 million years old at Olduvai Gorge. More recently, Spoor et al. (2007) assigned a 1.44-million-year-old maxilla fragment (KNM-ER 42703), based on the fragment's dental size and proportion, to *H. habilis*, perhaps extending its temporal range. Following additional analyses, Spoor et al. (2015) have become more cautious about the taxonomic assignment of the KNM-ER 42703 specimen and no longer assign it to *H. habilis*. Thus, the OH 13 mandible, dated to 1.65 mya, is now widely regarded as the last evidence for the species.

Homo rudolfensis

The survey of the Early Pleistocene deposits around Lake Turkana (formerly known as Lake Rudolf) in Kenya, and to a lesser degree in Ethiopia, yielded fossils that were clearly not *Australopithecus*, but were dissimilar from the *H. habilis* specimens from Olduvai Gorge and other small-brained, small-toothed specimens from the Turkana Basin. Initially, these specimens were attributed to *H. habilis*, suggesting that species was highly variable in brain size, tooth crown size, and facial morphology. The key specimen is the significantly complete cranium (KNM-ER 1470) that has an endocranial volume estimated at 752 cc (Holloway et al., 2004), a less prognathic and transversely flat lower midface, and—although edentulous—evidence of large postcanine tooth crowns and complex roots. The recovery of additional larger-toothed—but not australopith—specimens from the Turkana Basin (e.g., KNM-ER 1802 (see below) and KNM-ER 1482) (Wood, 1992b) reinforced the observation that the KNM-ER 1470 specimen was distinctive enough from *H. habilis*, and suggested that two distinct phenotypes were present. This led to the naming of *H. rudolfensis* (Alexeev, 1986; Wood, 1991, 1992b, 1999; for additional history, see also Kennedy, 1999). Additional analyses of the craniodental materials further reinforced the distinctiveness of the two groups (Wood, 1985, 1991; Lieberman et al., 1988; Grine et al., 1996).

While clearly derived in cranial size and lacking evidence of masticatory muscle hypertrophy—i.e., sagittal and nuchal crests—these fossils were readily assignable to *Homo*. However, large postcanine tooth crowns and an orthognathic face are also found in *A. (P.) boisei* or *A. (P.) robustus*—species showing evidence of dietary specializations. Whether these similarities represent shared primitive retentions or adaptive convergences remains to be resolved. With the more recent discovery of additional craniodental remains, especially the 1.95–1.98-million-year-old KNM-ER 62000 face and 1.78–1.87-million-year-old KNM-ER 60000 mandible (Leakey et al., 2012), it is clear that the orthognathic face is not idiosyncratic to KNM-ER 1470 but is a reliable diagnostic character. Curiously, these new fossils suggest that some specimens previously attributed to *H. rudolfensis*, such as the 1.95–2.06-million-year-old (Joordens et al., 2013) KNM-ER 1802 mandible, may not be assignable to *H. rudolfensis*. While having the larger tooth crowns of this taxon, it appears that the KNM-ER1802 mandible belongs to a more prognathic individual (Leakey et al., 2012; Spoor et al., 2015), possibly *H. habilis* (see Antón et al., 2014).

Homo erectus

Homo erectus defies simple characterization due to its long duration (c. 1.8 mya to c. 125,000 years ago (kya)), its broad geographic distribution, and a fossil collection that has multiple crania, mandibles, and teeth but surprisingly few elements from other parts of the body. Part of this latter issue is a product of it being difficult to assign postcrania unambiguously to a taxon when multiple species are present. Fossils from Java, China, the Republic of Georgia, Turkey (Kappelman et al., 2008), Ethiopia, Eritrea (Abbate et al., 1998), Kenya, Tanzania, South Africa, Tunisia, and Morocco have been assigned to *H. erectus*, and some authors have included fossils from Spain, India, and Chad in this taxon too. These fossils span from about 1.4 to

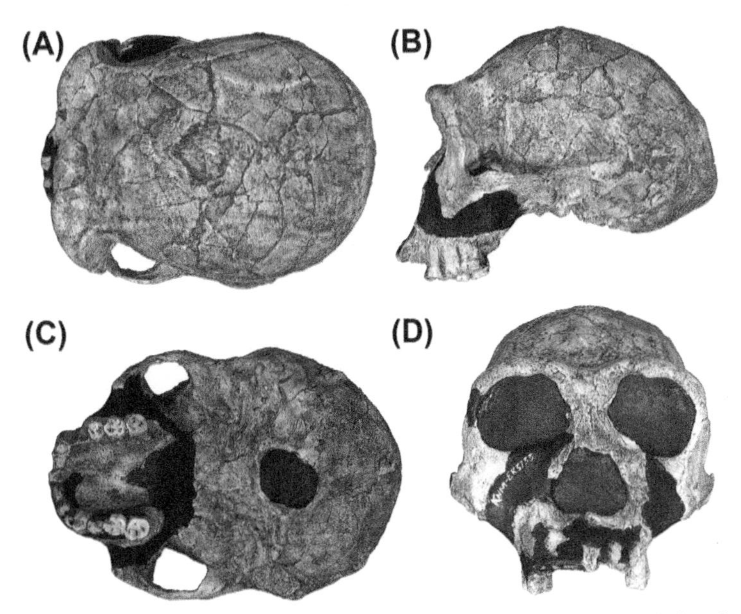

FIGURE 4 KNM-ER 3733 cranium (cast) attributed to *Homo erectus*, Turkana Basin, Kenya, c. 1.7 mya. (A) Superior view, (B) left lateral view, (C) basal view, (D) facial view. (See color plate section).

1.8 mya (Turkana Basin, Kenya (Lepre and Kent, 2010)); Sangiran Dome, Java (see Hyodo et al., 2011 for a younger chronology); Dmanisi, Republic of Georgia (Gabunia et al., 2000, 2002; Lordkipanidze et al., 2005, 2006, 2007, 2013) to as recently as 143 kya (range 546–143) in Java (Indriati et al., 2011). Among the earliest diagnostic *H. erectus* specimens from Africa is the cranium KNM-ER 3733, which is now dated to c. 1.7 mya (Lepre and Kent, 2010) (Figure 4). The youngest erectin from Africa is OH 12 from Olduvai Gorge (Antón, 2004) that is dated to about 0.78 mya (Tamrat et al., 1995). Recent redating of the Lantian, China specimen has pushed back the earliest appearance of *H. erectus* in China to c. 1.54–1.65 mya (Zhu et al., 2015). While the majority of fossils known for this species are neurocrania, the Dmanisi locality is producing postcrania (Lordkipanidze et al., 2007; Pontzer et al., 2010) in addition to well-preserved crania and skulls (Gabunia et al., 2000, 2002; Vekua et al., 2002; Lordkipanidze et al., 2005, 2006, 2013; Rightmire et al., 2006). With the preponderance of crania in the fossil record, analyses of *H. erectus* have focused on cranial morphology rather than the overall biology of the species (see Walker and Leakey, 1993; Antón, 2003, 2013; Antón et al., 2014). *H. erectus* was derived relative to earlier *Homo* with an absolutely and relatively larger brain (see chapter by Holloway) and larger body size. The Dmanisi sample expands dramatically the range of brain size of the species, with a male having an endocranial capacity of 546 ± 5 cc (Lordkipanidze et al., 2013). The diagnosis of *H. erectus* emphasize cranial morphology such as arching and projecting of the supraorbital tori, increased vault thickness especially at bregma, the formation of an occipital torus, a less prognathic face, and smaller postcanine teeth, especially M3 and premolars with less complex root systems.

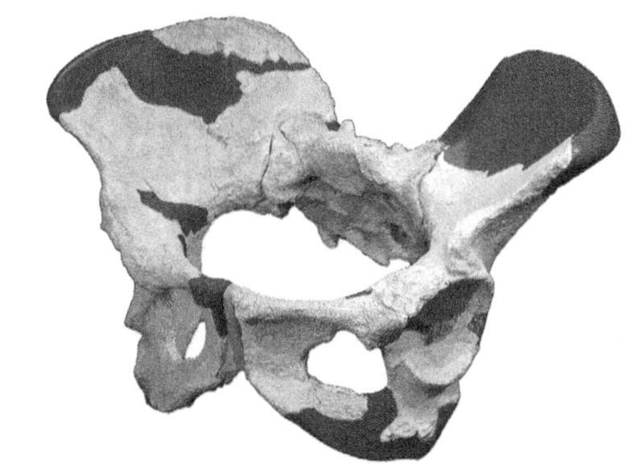

FIGURE 5 Reconstruction of the BSN49/P27 pelvis from Gona, Ethiopia (c. 0.9–1.4 mya) that is attributed to *Homo erectus*.

Our knowledge of the overall body plan of *H. erectus* increased substantially with the discovery of the significantly complete juvenile male skeleton (KNM-WT 15000) from Nariokotome, Kenya (Brown et al., 1985; Walker and Leakey, 1993) and the postcranial elements from Dmanisi (Lorkipanidze et al., 2007). The KNM-WT 15000 individual showed some distinctively *Homo* characteristics such as taller stature, modern *Homo sapiens*-like intra- and interlimb proportions, larger femoral head, and reduced toe length. *H. erectus* retained marked body size sexual dimorphism (Spoor et al., 2007; Lordkipanidze et al., 2013). The laterally flaring ilia and longer femoral necks are reminiscent of the *Australopithecus* condition, although the pelvis exhibits obstetric adaptations to birthing offspring with increasing fetal brain size (Simpson et al., 2008, 2014) (Figure 5).

Analyses of these fossils, which span a tremendous temporal and geographic range, have identified some regional anatomical themes that allow subspecific classifications of *H. erectus* (e.g., Kaifu et al., 2008). Howell (1999), using time, geographic location, and morphology, identified a series of paleodemes in Africa (Nariokotome, Olduvai/ LLK-II, Tighenif) and Eurasia (e.g., Dmanisi, Yunxian, Zhoukoudian, Sangiran (Glagahomba, Brangkal, Trinil)). This is a reasonable means of organizing disparate data to begin to understand the fossil samples as ancient biological populations. Despite this regionalization, unique anatomical features of crania suggest that gene flow existed, reinforcing the common specific allocation of these diverse fossils (Antón, 2003; Lordkipanidze et al., 2013; Rightmire, 1990, 2006: Asfaw et al., 2002; Baab, 2008; Rightmire, 2013). However, the biological history of the Chinese erectins may be more complex and exhibit greater diversity and regionalization than previously considered. The recent analysis of a series of teeth from the c. 412,000-year-old Longtan Cave, Hexian, China suggests that these teeth retain more primitive morphology (e.g., larger crowns, details of the occlusal surface and root form and robusticity) making them more similar to the >1.5-million-year-old teeth from Sangiran, Java than those from the c. 770,000-year-old (Shen et al., 2009) Zhoukoudian (China) Locality 1 (Xing et al., 2014).

The discovery of *H. erectus* fossils in eastern Asia provided reasonable evidence of a primitive human ancestor that was of a grade between apes and humans, hence Dubois's original generic nomen *Pithecanthropus*. When similar fossils were found elsewhere, especially in Africa, it was proposed that all specimens that had an enlarged brain, projecting supraorbital tori, a post-toral sulcus, and a long, low vault should be assigned to *H. erectus*. This included fossils from throughout Africa (South Africa, Tanzania, Kenya, Ethiopia, Sudan, Chad, Morocco), Europe (Greece, France, Italy, Spain), and Asia (Java, China, India, Turkey). Subsequent studies pared this list down, removing all fossils recovered from central and western Europe and fossils in Africa that are less than about 750 kya from *H. erectus*. In Africa—although the data are still quite meager—fossils younger than about 750 kya are now recognized as phylogenetically related, yet phenetically distinct by having larger brains and distinctive cranial architecture (Rightmire, 1990, 2013), and variously assigned to *H. rhodesiensis* or *H. heidelbergensis*. Therefore, while *H. erectus* in Asia has a broadly similar first appearance datum as those in Africa (at about 1.7 mya), the Asian lineage continued on Java until about 50 kya (Yokoyama et al., 2008) or 143 kya (range: 546–143 kya) (Indriati et al., 2011).

The Java deposits sample a time span of about 1.3 million years of *Homo erectus* evolution from the earliest samples recovered from Sangiran and Trinil through the Late Pleistocene fossils Sangiran and Trinil through the Late Pleistocene fossils from Sambungmacan and Ngandong. Numerous studies have identified the morphological continuity of the samples while recognizing that there appears to have been some anagenesis on the island populations (Kaifu et al., 2008; Santa Luca, 1980; Stringer, 1987; Antón et al., 2002, 2014; Baab, 2008; Baba et al., 2003; Zanolli, 2015). Comparison with the other major collection from Zhoukoudian, China that are dated to c. 600 kya suggests specific identity between Java and China that are distinguishable at the subspecific level (Santa Luca, 1980) (Figure 6).

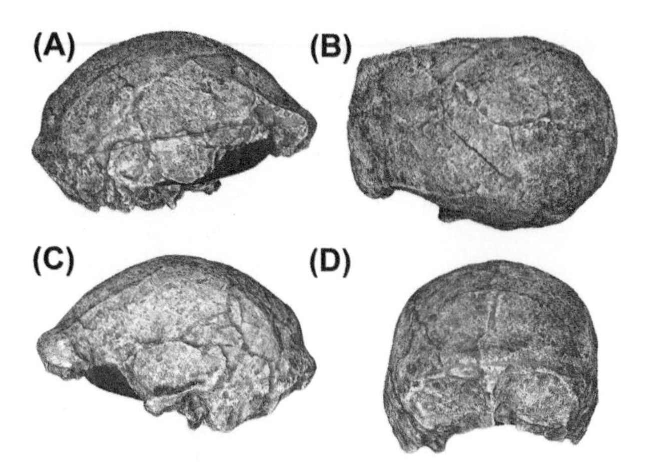

FIGURE 6 Ngandong XX craniums (cast) attributed to *Homo erectus*, Java, Indonesia. (A) Right lateral view, (B) superior view, (C) left lateral view, (D) posterior view. (See color plate section).

This contrasts with the African fossil record where *H. erectus* is replaced by a larger-brained hominin about 700 kya (Rightmire, 2009). So what is the phylogenetic relationship between the Asian and African specimens and were the samples part of a single biological species with an element of gene flow across the Old World?

Homo ergaster

This species was named by Groves and Mazák (1975) using the c. 1.5-million-year-old KNM-ER 992 mandible as its type. The origins of this naming stem from Groves and Mazák's recognition that species was distinguishable from *A. africanus* yet assignable to the genus *Homo*. As Antón (2003) noted, it was unfortunate that those authors did not compare the KNM-ER 992 mandible with Asian *H. erectus*, as this comparison would have shown marked anatomical similarity between the samples. There was little recognition of this taxon until Wood (1991) conducted his extensive analysis of the Koobi Fora craniodental materials where he, while recognizing the overall similarity in form between the African and Asian *H. erectus*, supported allocation of the Early Pleistocene African fossils to *H. ergaster*. The KNM-ER 992 mandible, while bearing some similarities to OH 13 that are attributed to *H. habilis*, is not significantly distinguishable from other African and Eurasian mandibles assigned to *H. erectus*; thus, the anatomical basis for defining a separate taxon requires additional evidence. The primary difference from Asian *H. erectus* is geographic location and age. While it is perhaps reasonable on biological grounds (e.g., lack of gene flow and genetic cohesion) to distinguish between the African and Javan fossils, it is difficult to distinguish them morphologically since they share many characteristics. This similarity is emphasized with the recovery of additional fossils from Dmanisi (Lordkipanidze et al., 2006, 2013; Rightmire et al., 2006), Turkey (Kappelman et al., 2008; Vialet et al., 2014), and Ethiopia (Asfaw et al., 2002)—the morphological distinctiveness of the African and Eurasian specimens at the level of species is increasingly difficult to demonstrate, as these fossils share characteristics in frontal and facial anatomy as well as reduced tooth dimensions. Rather than recognize the early African sample as a distinct species, it may be more useful to recognize regional patterns at the subspecific level, such as the identification of paleodemes (Howell, 1999; Asfaw et al., 2002), which are effectively equivalent to subspecies, with *H. ergaster* being consistent with the Nariokotome paleodeme—i.e., those specimens from Africa older than c. 1.4 mya (e.g., KNM-ER 3733, KNM-ER 3883, and KNM-WT 15000).

TAXONOMIC AND PHYLOGENETIC ISSUES: *HOMO HABILIS* AND *HOMO RUDOLFENSIS*

The recovery of additional hominin fossils from the Early Pleistocene has blurred rather than sharpened our understanding of the diversity of early *Homo*. The crucial period between 3 and 2 mya is poorly represented and the few fossils present (e.g., the LD 350-1 hemimandible (Figure 7) (Villmoare et al., 2015)), A.L. 666-1 maxilla (Kimbel et al., 1996, 1997), KNM-BC 1 Chemeron temporal (Sherwood et al., 2002), and 2.34-million-year-old KNM-WT 42718 isolated molar (Prat et al., 2005), while bearing *Homo* apomorphies, are imperfectly antecedent to later *Homo* such as *H. habilis* and *H. rudolfensis* to provide a clear demonstration of the phylogenetic and adaptive events occurring during this one-million-year span. The recovery of new fossils (KNM-ER 60000 and KNM-ER 62003 mandibles, KNM-ER 62000 maxilla (Leakey et al., 2012)) attributed to *H. rudolfensis* validates the presence of this contemporary of *H. habilis*, but creates additional taxonomic concerns for fossils previously attributed to *H. rudolfensis* on exclusionary rather than inclusive traits, such as the KNM-ER 1802 mandible (its teeth are too large for *H. habilis*, and it is too prognathic for *H. rudolfensis*), now create problems, since this mandible does not exhibit the characters expressed in the new mandible that emphasizes its very orthognathic face.

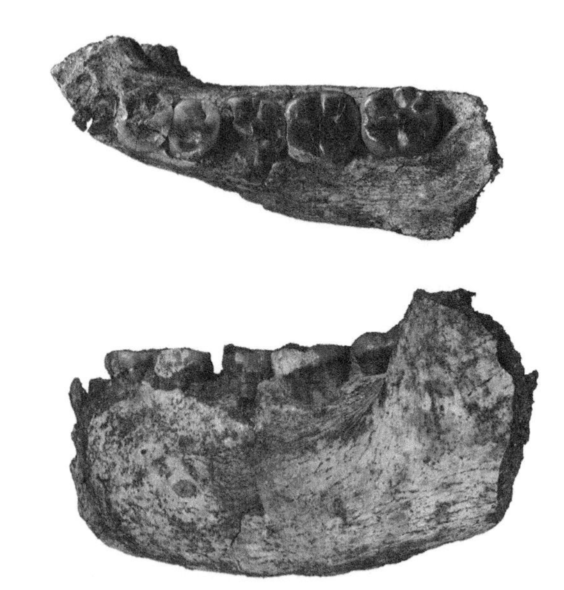

FIGURE 7 Occlusal and lateral views of the 2.8 mya early *Homo* from the Ledi-Geraru project area, Afar region, Ethiopia (Villmoare, et al., 2015). *Images provided by W.H. Kimbel.*

The naming of *H. habilis* (Leakey et al., 1964) appeared to fill a morphological and phylogenetic gap between *H. erectus* and *Australopithecus* by having a larger brain, less specialized dentition, a hand thought capable of making stone tools, and the right geologic age. However, the primitiveness (e.g., small brain, semicircular canal morphology, details of the hand anatomy, rate of ontogeny, and limb proportions) that distinguishes it from modern humans broadens the anatomical definition of the genus *Homo* to a degree that some consider untenably broad (Wood and Collard, 1999; Wood, 2009; Wood and Baker, 2011). See chapters by Ward and Hunt for further discussion.

Another concern with *H. habilis* is which fossils belong in the taxon (Antón et al., 2014; Spoor et al., 2015). The sample from Olduvai Gorge represents a reasonably coherent biological unit, although the species type specimen (the juvenile OH 7) has a larger endocranial capacity (729–824 cc (Spoor et al., 2015)) than other specimens from Olduvai Gorge (e.g., OH 24 (Leakey et al., 1971)) or specimens traditionally attributed to *H. habilis* such as KNM-ER 1813. When additional Early Pleistocene hominin fossils from the Turkana Basin and South Africa were discovered, often the initial taxonomic assignment of these fossils was to *H. habilis* based not on which characters the fossils exhibited, but on their geologic age (Early Pleistocene) and incompatibility with other taxa, notably *A. (P.) boisei* and *H. erectus*. Therefore, fossils were included in *H. habilis* based on exclusionary criteria rather than the sharing of common derived phena. A consequence of this is that the absence of a well-bounded diagnosis created the potential that multiple species could be included in a single taxon of *H. habilis*. For example, the range of variation known from the Olduvai Gorge sample could easily accommodate the small-brained, small-toothed, gracile KNM-ER 1813 cranium. However, the KNM-ER 1470 cranium with its larger brain, taller and flatter face, and probably larger teeth was most likely sampling a different biological unit. Confounding this, however, is the recognition that while the vault of the type specimen for *H. habilis*—the juvenile OH 7—bears some similarity to KNM-ER 1470, the dentognathic fossils do not. Thus, the lack of anatomical coherence, and poorly defined criteria used for inclusion into *H. habilis*, led to its reputation as a "waste can" taxon that masked the degree of taxonomic diversity and patterns of adaptive variation that existed in the Early Pleistocene. Careful analyses of the hominin remains resulted in the recognition that multiple morphotypes were discernable in specimens attributed to *H. habilis*, and a new species, *H. rudolfensis*, was recognized. Some researchers (Johanson et al., 1987; Suwa et al., 2007; Miller, 2000) considered that the existing data were insufficient to reject the null hypothesis of a single, albeit diverse, species, although additional finds (Leakey et al., 2012) have demonstrated that a distinct orthognathic species did exist that reflects the anatomical characters of KNM-ER 1470 (*H. rudolfensis*).

Another twist in the *H. habilis–H. rudolfensis* debate is that analysis of the 1.79–1.84-million-year-old well-preserved maxilla from Olduvai Gorge (OH 65) (Figure 8) (Blumenschine et al., 2003; Clarke, 2012) suggests that this specimen, which has morphological affinities to KNM-ER 1470 (which in turn has similarities in the size and shape of its parietals to OH 7 (the type specimen for *H. habilis*)), provides an anatomical link that allows these fossils to be accommodated into *H. habilis*. The consequence of this assignment is that the distinctiveness of *H. rudolfensis* is reduced and the taxon *H. rudolfensis* is thus unnecessary.

FIGURE 8 OH 65 maxilla attributed to *Homo habilis*, Olduvai Gorge, Tanzania, 1.79–184 mya. Occlusal view. (See color plate section). *Image provided by R. Blumenschine.*

More recently, Spoor et al. (2015) noted that the OH 65 maxilla, with its more modern parabolic arcade, would be incompatible with the reconstructed arcade for OH 7, suggesting that they belong in different taxa. In addition, the smaller-toothed and smaller-brained specimens that traditionally were assigned to *H. habilis* (e.g., OH 24, OH 62, KNM-ER 1813) would have to be reallocated to another taxon that Blumenschine et al. (2003) (Clarke, 2012) propose could be an as yet unnamed species of *Homo* or *Australopithecus*. The recovery of the platygnathic fossils attributed to *H. rudolfensis* from Koobi Fora weakens this contention (Leakey et al., 2012) by demonstrating that the face of *H. rudolfensis*, as first exemplified by KNM-ER 1470, is distinctive and dissimilar to other non-*rudolfensis* specimens.

While the recovery of these new *H. rudolfensis* fossils may resolve one issue, another is raised. A specimen from Koobi Fora, the c. 2.0-million-year-old KNM-ER 1802 mandible, has large postcanine tooth crowns, which results in a long postcanine tooth row, and a robust mandibular corpus. Similar anatomy is also seen in the 1.9–2.5-million-year-old UR-501 mandible from Malawi (Schrenk et al., 1993). Traditionally, these specimens had been assigned to *Homo* and grouped with KNM-ER 1470 (i.e., *H. rudolfensis*) (Wood, 1992a) rather than the more gracile specimens assigned to *H. habilis*. However, newly recovered dentognathic fossils (KNM-ER 60000 and KNM-ER 62003) (Leakey et al., 2012) that were assigned to *H. rudolfensis* are readily distinguishable from the KNM-ER 1802 mandible, suggesting that the latter specimen belongs to an as yet unnamed taxon, indicating a greater degree of taxic diversity than previously envisaged (Wood, 2012).

The c. 1.7-million-year-old Dmanisi locality in the Republic of Georgia has produced an impressive array of well-preserved fossils—especially crania—from a critical time period. While there was some initial uncertainty about the number of species present there due to the recovery of a large mandible (D2600) (Gabunia et al., 2002), further comparative analyses of the materials, as well as the recovery of the cranium (D4500) that articulates with the large mandible, now provide evidence of a coherent taxon of early *H. erectus* (Lordkipanidze et al., 2013; Rightmire et al., 2006). This larger sample broadens the range of anatomical variation of this deme as well as the diagnosis of *H. erectus*, with implications for the definition of other Early Pleistocene species *H. habilis* and *H. rudolfensis*. Lordkipanidze et al. (2013) have proposed that the range of anatomical variation in the Dmanisi sample and fossils now attributed to *H. habilis* and *H. rudolfensis* reflects intraspecies (or demic) levels of variation within a single evolving polymorphic lineage. Failure to reject this hypothesis presumes that all three species should then be consolidated into *H. erectus*, a position that has met with some skepticism (Spoor, 2013).

Homo habilis continues to "evolve" as additional analyses are conducted. The type specimen—OH 7—was subject to renewed analyses by Spoor et al. (2015), including restorations of the mandible and reconstruction of the parietal bones. Several very important conclusions were reached regarding the anatomy of the species, and a better understanding was reached of the patterns of variation in all of early *Homo*. The mandible appears to lack the parabolic dental arcade and reduced prognathism of later *Homo*, and its general conformation appears to be similar in some ways (e.g., more parallel sided dental rows) as observed in the much older *A. afarensis* (Spoor et al., 2015). The endocranial volume was revised upwards from the original estimates of 647–687 cc (Tobias, 1991; Holloway et al., 2004) to 729–824 cc (Spoor et al., 2015). The implication from this is that cranial capacity is highly variable within species and overlaps significantly between species such as *H. rudolfensis* and *H. erectus*, rendering it of reduced value as a reliable taxonomically valent character.

Finally, in their recent survey of early *Homo*, Antón et al. (2014) allocated these fossils into a series of groups as a means of organizing the fossils based on a combination of morphology and geological age. Major results of this analysis are that the amount of anatomical variation is great, that the named taxa do not adequately characterize the amount of variation, and that there may be greater taxonomic diversity than previously considered. For example, the three best-represented groups from eastern Africa are the "1470" group, the "1813" group, and *H. erectus*. The species *H. habilis* has a single fossil assigned to it—OH 7—since this damaged fossil is dissimilar from fossils in the other three groups. While this does not preclude that either the "1470" or "1813" groups may be subsumable into *H. habilis sensu stricto*, it is not clear which of these groupings will be preferred.

PHYLETIC ORIGINS AND THE DATING OF EARLY *HOMO*

The primary difficulty in recognizing the origins of the genus *Homo* is the lack of hominin fossils in the 2–3-million-year span. In addition, the earliest members of the *Homo* clade would be anatomically indistinguishable from their ancestors by the retention of a significant number of primitive traits and the absence of definitively derived characters. The earliest remains are from southern Ethiopia (Omo Valley Shungura Formation), the Afar region of Ethiopia (the Ledi-Geraru and Hadar project areas) and Kenya (Tulu Bor Formation, Koobi Fora). The recent announcement of a 2.8-million-year-old hemimandible (LD 350-1) from the Ledi-Geraru project area in the Afar region of Ethiopia (Villmoare et al., 2015; DiMaggio et al., 2015) becomes the oldest specimen attributed to the genus *Homo*. This specimen exhibits both derived characteristics (e.g., occlusal outline and accessory cusps on the molars, relative wear of the premolars, orientation of the mental foramen opening, and contours of the mandibular corpus) that distinguish it from *A. afarensis*, but retains some australopith characters such as the morphology of the anterior corpus and symphysis. At this time, there are no reports of associated stone tools or cut-marked/percussed faunal remains. The Omo remains are limited to a small sample of isolated teeth dated between 2.6 and 2.8 mya (Suwa, 1990; Suwa et al., 1996; Boisserie et al., 2008). These fossils' allocation to *Homo* sp. *indet.* is more a recognition that they do not exhibit the characteristic postcanine enlargement of the "robust" hominins and lack the distinctive morphology of the synchronic South African hominin *A. africanus* or the slightly older *A. afarensis* (Suwa et al., 1996), than that they exhibit *Homo* apomorphies. Another early fossil, the 2.42-million-year-old (Deino and Hill, 2002) temporal bone (KNM-BC one) recovered from the Chemeron Formation in Kenya (Hill et al., 1992), was in taxonomic limbo as it was allocated variously to early *Homo* or a robust *Australopithecus*. Its current assignment to *Homo* is now widely recognized due to the anatomical details and more medial position of the glenoid fossa relative to the temporal squama, suggesting both a smaller temporomandibular joint (which is broad in *Australopithecus*—the recovery of additional specimens from Dmanisi (Lordkipanidze et al., 2013), however, suggests that the position of the glenoid fossa may be variable in earliest *Homo*), and perhaps a slightly enlarged brain (Hill et al., 1992; Sherwood et al., 2002). A 2.3–2.4-million-year-old isolated first mandibular molar was recovered from West Turkana, Kenya that also has affinities with the genus *Homo* (Prat et al., 2005).

The oldest fossil assigned to a species in *Homo* (i.e., *Homo* aff. *H. habilis*) is a 2.33-million-year-old maxilla (A.L. 666-1) with dentition that was recovered in association with Oldowan stone tools from the lower Busidima Formation at Hadar in Ethiopia (Kimbel et al., 1996, 1997) (Figure 9). The contours of the maxilla are clearly not from an australopith, and while the premolar crowns are at the upper part of the *H. habilis* size range, the upper first and second molars are similar in size to other members

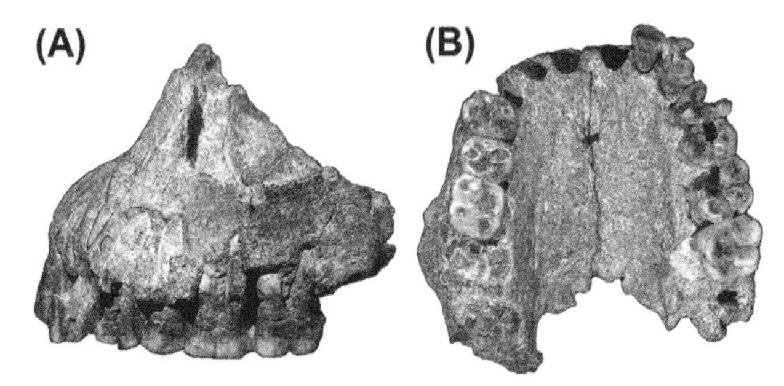

FIGURE 9 A.L. 666-1 maxilla attributed to *Homo* cf. *habilis*, Hadar, Ethiopia, 2.33 mya. (A) Left lateral view, (B) occlusal view. (See color plate section).

of the taxon (Kimbel et al., 1997). A mandible (UR 501) from the 2.5–2.3-million-year-old Chiwondo Beds in Malawi (Bromage et al., 1995; Schrenk et al., 1993, 2007) is assigned to another early *Homo* taxon, *H. rudolfensis,* based on its overall similarity to KNM-ER 1802, which was then assigned to that taxon. While these early specimens are allocated to different species, they both share the larger postcanine tooth crowns and more complex premolar roots than do the 2.5–2.4-million-year-old *Homo* specimens from Omo (Suwa et al., 1996).

The stem species from which *Homo* arose is still unclear, although *A. afarensis* is widely regarded as a possible ancestor (Johanson and White, 1979; Wood, 1992a) because of the morphology of the earliest nonaustralopith fossils, the generalized morphology of *A. afarensis,* its broad distribution in eastern Africa and perhaps Chad, and its temporal span. This possibility is supported by characteristics of the early (2.6–2.9 million years old) isolated teeth from Omo that (Suwa, 1990; Suwa et al., 1996) identified as exhibiting both *Homo* and early *Australopithecus* morphology. Furthermore, analyses of the recently announced Ledi-Geraru mandible (Villmoare et al., 2015) also document a derived dentition that retained elements of a more prognathic *A. afarensis* form. While *A. afarensis* may be "the right hominin at the right time" to be ancestral to *Homo*, other options need to be evaluated. The 2.5-million-year-old *A. garhi* (Asfaw et al., 1999) from the Hataye Formation on the Bouri peninsula in the Middle Awash (Ethiopia) study area has also been posited as an ancestor of later *Homo* due to the shape of its maxilla and relative dental proportions. While this is a possibility since it too has been posited to be in the "right place at the right time" (Asfaw et al., 1999, p. 634), the large size of the teeth, especially the canine, appear to preclude it as the unique ancestor of later *Homo*. Furthermore, the presence of more "*Homo*-like" hominins more than 2.5 million years old (e.g., the LD 350-1 mandible from Ledi-Geraru (Villmoare et al., 2015)) preclude this derived hominin as ancestral to *Homo*.

The recently discovered 1.95-million-year-old *A. sediba* from South Africa is also a candidate as an ancestor, although its geologic age (less than 2 million years) and craniodental and pedal morphology rule it out (Kimbel, 2013; Schroeder et al., 2014) as the unique ancestor of later *Homo*. Again, see chapter by Ward for complete discussion of all australopiths.

Kenyanthropus platyops is a poorly known species from the Turkana Basin in Kenya. This 3.5–3.2-million-year-old species is known from a small sample of isolated teeth and a distorted cranium (KNM-WT 4000). Recognizable and distinctive craniofacial anatomy exhibited by the cranium includes a lower midface that is more orthognathic and transversely flat (hence, its specific name *platyops* (= "flat face")), and tall zygomatic bodies. The reduced facial prognathism is thought to link it phyletically with the Pleistocene hominin *H. rudolfensis* that was until recently (Leakey et al., 2012) represented by the provocative but ambiguous collection of fossils exemplified by KNM-ER 1470 (Alexeev, 1986; Wood, 1991, 1992b). If *K. platyops* is uniquely ancestral to *H. rudolfensis* and unrelated to *H. habilis* (Schrenk et al., 2007; Leakey et al., 2001), this implies that the genus *Homo* is polyphyletic (Wood, 1991), with multiple lineages arising independently from australopith-grade ancestor(s), thus requiring a taxonomic revision of the Early Pleistocene gracile hominins.

An alternative (and much more ancient) proposed *Homo* ancestor is the c. 6-million-year-old hominin *Orrorin tugenensis* from the Baringo Basin in Kenya (see chapter by Hunt). While its dental remains are clearly primitive (large canines, small postcanine teeth with thinner enamel), the femora exhibit some derived anatomies, indicating it was bipedal and suggesting to the finders that it is more closely related with *Homo* than with either *Australopithecus* or *Paranthropus* (Pickford et al., 2002).

A natural inclination is to presume that all of the non-australopith species belong to the genus *Homo*—a lineage that had a singular origin about 2.8–3.0 mya—and that the phylogenetic diversification we now recognize arose subsequently from this ancestral population. Alternative possibilities exist, although these are inferred from the existing fossil record rather than being demonstrable at this time. For example, the amount of anatomical and taxonomic diversity we observe in the fossil record suggests that the degree of diversity arose quite early (>2.6 mya) (Spoor et al., 2015). If the genus *Homo* is monophyletic, then there is something distinctive about this lineage that leads to a rapid diversification in form. Alternatively, it could be that our estimates for the origins of a singular *Homo* lineage are in error, and that it became phylogenetically separate much earlier (>2.8–3.0 mya) than we now suppose. Finally, it could be that the genus *Homo* is polyphyletic and that taxa now assigned to *Homo* may have had separate origins from an australopith ancestor, requiring a reassessment of the history and nomenclature of the genus—changes that require new specimens and further analyses.

CONTEXT OF ORIGINS AND EXISTENCE

While some primates are known to eat meat (Butynski, 1982), especially chimpanzees (Boesch and Boesch, 1990) (see chapter by Watts), hominins do this on a regular basis, with the first evidence beginning about 2.58 mya (Semaw et al., 1997, 2003; deHeinzelin et al., 1999) or perhaps even earlier (McPherron et al., 2010). While it is perhaps intuitive that meat eating has a number of dietary and adaptive advantages, terrestrial carnivory requires a number of anatomical adaptations for the most part absent in primates (especially bipedal hominins), and exposed early hominins to a number of risks (e.g., predation and novel diseases) that can reduce survivorship. Similarly, while a large brain and a capacity for greater behavioral complexity seem to be obvious adaptations, these can create a demographic hurdle of a more protracted life history schedule, with consequences for survivorship and birth rates that only early *Homo* solved. In retrospect, keystone adaptations in the *Homo* lineage occurred serially over a span of time, indicating that these hominins were finding new ways to interact with the increasingly diverse habitats they encountered.

Based on analyses of archeological site location, there was a broadening of the habitats occupied by early *Homo*, and with the rise of *H. erectus* about 1.8 mya in Africa, these hominins increasingly occupied diverse habitats including more open and seasonal areas (Rogers et al., 1994). Even more remarkable is the rapid dispersal of these hominins to exploit the increasingly diverse ecological zones that they encountered in their rapid migration from Africa through easternmost Europe (Dmanisi (Gabunia et al., 2000; Lordkipanidze et al., 2013; Kappelman et al., 2008; Blain et al., 2014)) and ultimately into eastern Asia. Rapid dispersals such as this are evidence of rapid population growth and sufficient behavioral plasticity to adapt to novel habitats.

MUSCULOSKELETAL ADAPTATIONS IN EARLY *HOMO*

Body shape and limb proportions of Early Pleistocene *Homo* are still poorly known, although several specimens are quite informative, including the partial skeletons of *H. habilis*: OH 62 (Johanson et al., 1987), KNM-ER 3735 (Leakey and Walker, 1985); *H. erectus*: (KNM-WT 15000) (Brown et al., 1985; Walker and Leakey, 1993), 1.8-million-year-old postcranial elements from Dmanisi (Lordkipanidze et al., 2007), the 0.9–1.4-million-year-old Gona pelvis (BSN49/P27) (Simpson et al., 2008, 2014); or *Homo* sp. indet.: 1.9-million-year-old associated lower limb elements (KNM-ER 5881 (Ward et al., 2015), KNM-ER 3228 (Leakey, 1976)). There appears to have been a two-phase adaptation in the adoption of modern human limb proportions (Asfaw et al., 1999). Initially, there was a proportional elongation of the lower limb that is first exhibited in the 2.5-million-year-old taxonomically ambiguous BOU-VP-12/1 partial skeleton. This small-bodied individual clearly shows an elongation of the lower limb, attaining the humero–femoral ratio observed in modern humans with retention of a very long forearm. This appears to be the case in the 1.8-million-year-old *H. habilis* partial skeleton (OH 62) from Olduvai Gorge, Tanzania (Johanson et al., 1987). The second phase of relative antebrachial shortening appears by 1.53 mya with the appearance of the KNM-WT 15000 *H. erectus* skeleton.

While the elongation of the lower limb is undoubtedly related to the interplay between habitat preference and bipedal locomotion, the adaptive advantages of forearm shortening remain unclear. While the latter change appears to be near coincidental with the adoption of Acheulean stone tool manufacture, the functional and selective relationship is obscure.

The 0.9–1.4-million-year-old pelvis from the upper Busidima Formation at Gona, Ethiopia, provides information about body shape and obstetric capacity. While *H. erectus* has been widely considered essentially a modern human in shape and proportions below the neck, the broad bi-iliac breadth of the Gona pelvis indicates that elements of the laterally flaring ilia of *Australopithecus* were retained even as an increase in obstetric dimensions was occurring in *H. erectus*. These latter changes are a necessary coadaptation in a species where absolute brain size is increasing. Without these pelvic modifications, birthing of a large-brained offspring becomes quite challenging, compromising the health of both the mother and the baby (Trevethan, 1988; Rosenberg and Trevethan, 2002).

ASSOCIATION WITH STONE TOOLS

The earliest stone tools and evidence of stone tool use occur about 2.5–2.6 mya (Semaw et al., 1997, 2003; deHeinzelin et al., 1999) in Ethiopia, c. 2.34 mya at Lokalalei in Kenya (Tiercelin et al., 2010), about 1.78 mya in northern Africa (Sahnouni et al., 2013), and about 1.7-1.8 mya in Dmanisi (see chapter by Toth and Schick for a complete discussion of stone tools). These early tools of the Oldowan tradition are modified river cobbles flaked to produce a sharp edge (Semaw, 2000) where high-quality lithic sources are available, or using quartzite chunks when river cobbles are unavailable. Animal fossils have been recovered that have evidence of cutting and splintering due to the use of stone tools, providing the early and reliable evidence of larger-mammal carnivory by a hominin and indicating a fundamental shift in diet and behavior. These marks appear to be evidence of harvesting meat and viscera from animals and the breaking open of long bone shafts to access bone marrow. The recovery of even older cut-marked bones has been reported from the 3.4-million-year-old deposits at Hadar, Ethiopia—sediments from which *A. afarensis* is the only hominin yet known (McPherron et al., 2010). While visually compelling, other alternatives for their origins have been proposed (Domínguez-Rodrigo et al., 2012). More recently, stone tools dated to 3.3 mya from the Turkana Basin in Kenya have been reported (Harmand et al., 2015). If these observations are verified, then the origins of stone tool use and manufacture may no longer be considered a uniquely *Homo* adaptation.

This cobble tool tradition was the sole technological grade through the first half of the Early Pleistocene, although it is unclear whether evolution in Mode

I technology occurred during this time (Leakey, 1971; Kimura, 2002; Semaw et al., 2009; Stout et al., 2010). Beginning about 1.75 mya, the Oldowan toolkit was augmented by the development of the Acheulean tradition (Lepre et al., 2011; Beyene et al., 2013), and both continue into the Middle Pleistocene (Clark et al., 1994). The Acheulean period is characterized by the appearance of new tool types, including the development of bifacially flaked hand axes, cleavers, and trihedral picks (Lepre et al., 2011), with an increase in the frequency of manufacture from large flakes rather than cobbles (Beyene et al., 2013). The advantage of the Acheulean technology is that these larger-sized tools had more cutting edge than found in the Oldowan cobble tools. In addition, the adoption of distinctive tool types (e.g., hand axes, picks, and cleavers) suggests that these different forms reflect their use in different activities, and that there was a broadening or intensification of *H. erectus* extractive behaviors. Of additional interest is that hand axes continued to evolve in form over time, such that later tools were thinner, more symmetrical, and tended to be manufactured from large flakes rather than cobbles. This suggests that there was either a refinement in the shape to improve its function, or that the cultural norms for the idealized hand axe form were changing. In contrast, the trihedral picks did not change especially much over time, perhaps indicating that modifications to the tool did not enhance its function (Beyene et al., 2013). While each tool was capable of some degree of cutting and chopping, the details of their specific activities are difficult to discern.

The earliest appearance of the Acheulean stone tool tradition slightly antedates the appearance of *H. erectus* in Africa (c. 1.73 million years ago), raising the possibility that *H. habilis* manufactured the earliest Acheulean tools. The low resolution of the fossil record and the small temporal gap (30,000 years) between these two first-appearance data makes authorship of the Acheulean tools uncertain. Intuitively, it is most likely that the larger-brained *H. erectus* made these Acheulean tools and that their near-coincident appearance is linked, although this proposal requires the collection of additional data. It is important to note that the earliest *H. erectus* known from Eurasia apparently did not use Acheulean tools. Multiple explanations for this absence are possible, such as very small sample sizes from Dmanisi, a locality that may represent a carnivore-collected—and not archeologically associated—assemblage, and that Asian *H. erectus* "lost" this skill. Alternatively, it may be entertained that the origins of *H. erectus* and the Acheulean technique are unrelated.

FUTURE DIRECTIONS

The main issues in the study of the origins and evolution of the genus *Homo* are organizational (i.e., taxonomy and phylogeny) and paleobiological. The first refers to an accurate recounting of the diversity and phylogenetic relatedness of the different species, and the second relates to the behavioral, anatomical, and contextual issues of these biological species and their origins. Clearly, there was something different about early *Homo* in that they adopted behaviors that led to changing demographic parameters (Antón and Snodgrass, 2012) that resulted in expansion into new habitats within Africa and across Eurasia by 1.7 mya. They could manipulate the environment in ways that allowed the exploitation of new resources, using shaped stone tools to create a novel and unique niche for a primate. This is associated with an increase in brain size and decrease in tooth crown size. Limb proportions were also evolving, as early *Homo* increased the length of the lower limb and shortened the relative length of the forearm, indicating an anatomical dedication to terrestrial bipedality and the absence of any adaptations to arboreality. These adaptations had effects on other hominins on the landscape—the robust australopiths with their craniodental specializations (greatly enlarged postcanine tooth crowns and masticatory musculature, reduced anterior teeth, and a flatter taller face) and lack of significant brain size enlargement. Did competition between the two types of hominins result in resource competition and niche divergence, with the robust australopiths increasingly reliant on masticatory intensive herbivory while early *Homo* adopted a tool-facilitated omnivory? Isotopic analyses of hominin dental enamel provide good evidence that dietary preferences differed between the two lineages (Cerling et al., 2013), lending credence to models of niche separation. How did these novel behaviors result in changed demographic parameters that led to geographic expansion and a more protracted life history schedule?

The following questions still need to be addressed: when, where, and in what ecological context did *Homo* arise; are the Early Pleistocene species currently allocated to *Homo* part of a clade with a common origin or do they have multiple origins requiring a taxonomic reorganization to recognize these diverse origins; since stone tool use/manufacture and extractive carnivory arose about the same time as *Homo*—how are these events related; what were the body size dimensions and degree of sexual dimorphism (Plavcan, 2012); and how are the origins of Acheulean technology and *H. erectus* related? Did the niche competition between the more generalist *Homo* lead both to increased specialization of its masticatory apparatus and possibly to the demise of this more specialized robust australopith? During the Early Pleistocene, there appears to be greater diversity in Africa in the genus *Homo* than after about 1.5 mya—why? Also, what is the relationship between *H. erectus* and later hominins such as *H. heidelbergensis*, *H. rhodesiensis* (if a species distinct from *H. heidelbergensis*), and *H. antecessor* (see chapter by Ahern)?

Clearly, additional fossils are necessary to solve these and other issues about the phylogenetic and adaptive evolution of early *Homo*. Analyses of nuclear and mitochondrial DNA indicate that population diversity was marked in Eurasia by distinct Neanderthal and Denisovan groups that had some degree of interbreeding between themselves and with later-arriving modern humans (Reich et al., 2010; Prüfer et al., 2013). Are these genes the unique remnants of ancient *H. erectus*, providing us with evidence of interbreeding between the first inhabitants of Eurasia and the later-immigrating Neanderthal ancestors? Unfortunately, the age of the *H. erectus* fossils precludes recovery of DNA from their remains, so details of genetic relatedness and distinctiveness will always remain outside our grasp. Fortunately, some of these questions can be solved by the recovery and analyses of additional fossils and the application of new analytic techniques. The ultimate goal is to know as much about the history and biology of the earliest members of our genus as we do about any living species. Overall, continuing study of the evolution of early *Homo* has demonstrated more anatomical and adaptive diversity, and greater phylogenetic complexity, than previously considered. These should not be considered hindrances to future research; rather, they present interesting challenges as we unravel our earlier history.

REFERENCES

Abbate, E., Albianelli, A., Azzaroli, A., Benvenuti, M., Tesfamariam, B., Bruni, P., Cipriani, N., Clarke, R.J., Ficcarelli, G., Macchiarelli, R., Napoleone, G., Papini, M., Rook, L., Sagri, M., Tecle, T.M., Torre, D., Villa, I., 1998. A one-million-year-old *Homo erectus* cranium from the Danakil (Afar) Depression in Eritrea. Nature 393, 458–460.

Alexeev, V., 1986. The Origin of the Human Race. Progress Publishers, Moscow (Translation by H. Campbell).

Antón, S.C., 2003. Natural history of *Homo erectus*. Yearbook of Physical Anthropology 46, 126–170.

Antón, S.C., 2004. The face of Olduvai hominid 12. Journal of Human Evolution 46, 337–347.

Antón, S.C., 2013. *Homo erectus* and related taxa. In: Begun, D.R. (Ed.), A Companion to Paleoanthropology. Blackwell Publishing Ltd, Chichester, UK, pp. 497–516.

Antón, S.C., Márquez, S., Mowbray, K., 2002. Sambungmacan 3 and cranial variation in Asian *Homo erectus*. Journal of Human Evolution 43, 555–562.

Antón, S.C., 2012. Early Homo: Who, when, and where. Current Anthropology 53, s278–s298.

Antón, S.C., Potts, R., Aiello, L.C., 2014. Evolution of early *Homo*: an integrated biological perpective. Science 345, 1236828.

Antón, S.C., Snodgrass, J.J., 2012. Origins and evolution of genus *Homo*: new perspectives. Current Anthropology 53, s479–s496.

Argue, D., Donlon, D., Groves, C., Wright, R., 2006. *Homo floresiensis*: Microcephalic, pygmoid, *Australopithecus*, or *Homo*? Journal of Human Evolution 51, 360–374.

Asfaw, B., Gilbert, W.H., Beyene, Y., Hart, W.K., Renne, P.R., WoldeGabriel, G., Vrba, E.S., White, T.D., 2002. Remains of *Homo erectus* from Bouri, Middle Awash, Ethiopia. Nature 416, 317–320.

Asfaw, B., White, T., Lovejoy, O., Latimer, B., Simpson, S., Suwa, G., 1999. *Australopithecus garhi*: new species of early hominid from Ethiopia. Science 284, 629–635.

Baab, K., 2008. The taxonomic implications of cranial shape variation in *Homo erectus*. Journal of Human Evolution 54, 827–847.

Baba, H., Aziz, F., Kaifu, Y., Suwa, G., Kono, R.T., Jacob, T., 2003. *Homo erectus* calvarium from the Pleistocene of Java. Science 299, 1384–1388.

Beyene, Y., Katoh, S., WoldeGabriel, G., Hart, W.K., Uto, K., Sudo, M., Kondo, M., Hyodo, M., Renne, P.R., Suwa, G., Asfaw, B., 2013. The characteristics and chronology of the earliest Acheulean at Konso, Ethiopia. Proceedings of the National Academy of Science 110, 1584–1591.

Black, D., 1926. Tertiary man in Asia—the Chou Kou Tien discovery. Science 64, 586–587.

Black, D., 1928. Discovery of further hominid remains of Lower Quaternary age from the Chou Kou Tien deposit. Science 67, 135–136.

Black, D., 1929. *Sinanthropus pekinensis*: the recovery of further fossil remains of this early hominid from the Chou Kou Tien deposit. Science 69, 674–676.

Blain, H.-A., Agusti, J., Lorkipanidze, D., Rook, L., Delfino, M., 2014. Paleoclimatic and paleoenvironmental context of the Early Plesitocene hominins from Dmanisi (Georgia, Lesser Caucasus) inferred from the herpetofaunal assemblage. Quaternary Science Reviews 105, 136–150.

Blumenschine, R.J., Peters, C.R., Masao, F.T., Clarke, R.J., Deino, A.L., Hay, R.L., Swisher, C.C., Stanistreet, I.G., Ashley, G.M., McHenry, L.J., Sikes, N.E., van der Merwe, N.J., Tactikos, J.C., Cushing, A.E., Deocampo, D.M., Njau, J.K., Ebert, J.I., 2003. Late Pliocene *Homo* and hominid land use from western Olduvai Gorge, Tanzania. Science 299, 1217–1221.

Boesch, C., Boesch, H., 1990. Tool use and tool making in wild chimpanzees. Folia Primatologia 54, 86–99.

Boisserie, J.-R., Guy, F., Delagnes, A., Hlusko, L.J., Bibi, F., Beyene, Y., Guillemot, C., 2008. New palaeoanthropological research in the Plio-Pleistocene omo group, lower omo Valley, SNNPR (Southern Nations, Nationalities, and People regions), Ethiopia. Comptes Rendus Palevol 7, 429–439.

Bromage, T.G., Schrenk, F., Zonneveld, F.W., 1995. Paleoanthropology of the Malawi Rift: an early hominid mandible from the Chiwondo Beds, northern Malawi. Journal of Human Evolution 28, 71–108.

Brown, F., Harris, J., Leakey, R., Walker, A., 1985. Early *Homo erectus* skeleton from west Lake Turkana, Kenya. Nature 316, 788–792.

Brown, P., Sutikna, T., Morwood, M.J., Soejono, R.P., Jatmiko, W., Saptomo, E., Awe Due, R., 2004. A new small-bodied hominin from the Late Pleistocene of Flores, Indonesia. Nature 431, 1055–1061.

Butynski, T.M., 1982. Vertebrate predation by primates: a review of hunting patterns and prey. Journal of Human Evolution 11, 421–430.

Cerling, T.E., Manthi, F.K., Mbua, E.N., Leakey, L.N., Leakey, M.G., Leakey, R.E., Brown, F.H., Grine, F.E., Hart, J.A., Kaleme, P., Roche, H., Uno, K.T., Wood, B.A., 2013. Stable isotope-based diet reconstructions of Turkana Basin hominins. Proceedings of the National Academy of Sciences 110, 10501–10506.

Clark, J.D., de Heinzelin, J., Schick, K.D., Hart, W.K., White, T.D., WoldeGabriel, G., Walter, R.C., Suwa, G., Asfaw, B., Vrba, E., Haile-Selassie, Y., 1994. African *Homo erectus*: old radiometric ages and young Oldowan assemblages in the Middle Awash Valley, Ethiopia. Science 264, 1907–1910.

Clarke, R.J., 2012. A *Homo habilis* maxilla and other newly-discovered hominid fossils from Olduvai Gorge, Tanzania. Journal of Human Evolution 63, 418–428.

deHeinzelin, J., Clark, J.D., White, T., Hart, W., Renne, P., WoldeGabriel, G., Beyene, Y., Vrba, E., 1999. Environment and behavior of 2.5-million-year-old Bouri hominids. Science 284, 625–629.

Deino, A., Hill, A., 2002. ^{40}Ar/^{39}Ar geochronology and paleomagnetic stratigraphy of the Lukeino and lower Chemeron Formations at Tabarin and Kapcheberek, Tugen Hills, Kenya. Journal of Human Evolution 42, 117–140.

DeSilva, J.M., Zipfel, B., Van Arsdale, A.P., Tocheri, M.W., 2010. The Olduvai hominid 8 foot: Adult or subadult. Journal of Human Evolution 58, 419–423.

DiMaggio, E., Campisano, C.J., Rowan, J., Dupont-Nivet, G., Deino, A.L., Bibi, F., Lewis, M.E., Souron, A., Werdelin, L., Reed, K.E., Arrowsmith, J.R., 2015. Late Pliocene fossiliferous sedimentary record and the environmental context of early *Homo* from Afar, Ethiopia. Science. http://dx.doi.org/10.1126/science.aaa1415.

Domínguez-Rodrigo, M., Pickering, T.R., Bunn, H.T., 2012. Experimental study of cut marks made with rocks unmodified by human flaking and its bearing on claims of ~3.4-million-year-old butchery evidence from Dikika, Ethiopia. Journal of Human Evolution 39, 205–214.

Eckhardt, R.B., Henneberg, M., Weller, A.S., Hsü, K.J., 2014. Rare events in earth history include the LB1 human skeleton from Flores, Indonesia, as a developmental singularity, not a unique taxon. Proceedings of the National Academy of Sciences 111, 11961–11966.

Feibel, C.S., Lepre, C.J., Quinn, R.L., 2009. Stratigraphy, correlation, and age estimates for fossils from Area 123, Koobi Fora. Journal of Human Evolution 57, 112–122.

Ferguson, W.W., 1995. A new species of the genus *Homo* (Primates: Hominidae) from the Plio/Pleistocene of Koobi Fora, in Kenya. Primates 36, 69–89.

Fleagle, J.G., Assefa, Z., Brown, F.H., Shea, J.J., 2008. Paleoanthropology of the Kibish formation, southern Ethiopia: Introduction. Journal of Human Evolution 55, 360–365.

Gabunia, L., de Lumley, M.A., Vekua, A., Lordkipanidze, D., 2002. A new species of *Homo* represented by a fossil from the bottom part of the Pleistocene layer at Dmanisi, Georgia. Archaeol. Ethnol. Anthropol. Eurasia 4, 145.

Gabunia, L., Vekua, A., Lordkipanidze, D., Swisher 3rd, C.C., Ferring, R., Justus, A., Nioradze, M., Tvalchrelidze, M., Antón, S.C., Bosinski, G., Jöris, O., Lumley, M.A., Majsuradze, G., Mouskhelishvili, A., 2000. Earliest Pleistocene hominid cranial remains from Dmanisi, Republic of Georgia: taxonomy, geological setting, and age. Science 288, 1019–1025.

Gibbard, P.L., Head, M.J., Walker, M.J.C., Subcommission on Quaternary stratigraphy, 2010. Formal ratification of the Quaternary System/Period and the Pleistocene Series/Epoch with a base at 2.58 Ma. Journal of Quaternary Science 25, 96–102.

Grine, F.E., Jungers, W.L., Schultz, J., 1996. Phenetic affinities among early *Homo* crania from East and South Africa. Journal of Human Evolution 30, 189–225.

Groves, C.P., Mazák, V., 1975. An approach to the taxonomy of the Hominidae: gracile Villafranchian hominids of Africa. Časopis pro Mineralogii a Geologii 20, 225–247.

Harmand, S., Lewis, J.E., Feibel, C.S., Lepre, C.J., Prat, S., Lenoble, A., Boes, X., Quinn, R.L., Brenet, M., Arroyo, A., Taylor, N., Clément, S., Daver, G., Brugal, J.-P., Leakey, L., Kent, D.V., Mortlock, R.A., Wright, A.D., Roche, H., 2015. 3.3-million-year-old stone tools from Lomekwi 3, West Turkana, Kenya. Nature 521, 310–315.

Hill, A., Ward, S., Deino, A., Curtis, G., Drake, R., 1992. Earliest *Homo*. Nature 355, 719–722.

Holloway, R.L., Broadfield, D.C., Yuan, M.S., 2004. The Human Fossil Record, vol. 3. Wiley-Liss.

Howell, F. Clark, 1999. Paleo-demes, species clades, and extinctions in the Pleistocene hominin record. Journal of Anthropological Research 55, 191–243.

Hyodo, M., Masu'ura, S., Kamishima, Y., Kondo, M., Takeshita, Y., Kitaba, I., Danhara, T., Aziz, F., Kurniawan, I., Kumai, H., 2011. High-resolution record of the Matuyama-Brunhes transition contrsins the age of Javanese *Homo erectus* in the Sangiran dome, Indonesia. Proceedings of the National Academy of Science 108, 19563–19568. http://dx.doi.org/10.1073/pnas.1113106108.

Indriati, E., Swisher III, C.C., Lepre, C., Quinn, R.L., Suriyanto, R.A., Hascaryo, A.T., Grün, R., Feibel, C.S., Pobiner, B.L., Aubert, M., Lees, W., Antón, S.C., 2011. The age of the 20 meter Solo River terrace, Java, Indonesia and the survival of *Homo erectus* in Asia. PLoS One 6 (6), e21562. http://dx.doi.org/10.1371/journal.pone.0021562.

Johanson, D.C., Masao, F.T., Eck, G.G., White, T.D., Walter, R.C., Kimbel, W.H., Asfaw, B., Manega, P., Ndessokia, P., Suwa, G., 1987. New partial skeleton of *Homo habilis* from Olduvai Gorge, Tanzania. Nature 327, 205–209.

Johanson, D.C., White, T.D., 1979. A systematic assessment of early African hominids. Science 203, 321–330.

Joordens, J.C.A., Dupont-Nivet, G., Feibel, C.S., Spoor, F., Sier, M.J., van der Lubbe, J.H.J.L., Kellberg Nielsen, T., Knul, M.V., Davies, G.R., Vonhof, H.B., 2013. Improved age control on early *Homo* fossils from the upper Burgi Member at Koobi Fora, Kenya. Journal of Human Evolution 65, 731–745.

Kaifu, Y., Aziz, F., Indriati, E., Jacob, T., Kurniawan, I., Baba, H., 2008. Cranial morphology of Javanese *Homo erectus*: new evidence for continuous evolution, specialization, and terminal extinction. Journal of Human Evolution 55, 551–580.

Kappelman, J., Alçiçek, M.C., Kazanci, N., Schultz, M., Özkul, M., Şen, Ş., 2008. Brief Communication: first *Homo erectus* from Turkey and implications for migrations into temperate Eurasia. American Journal of Physical Anthropology 135, 110–116.

Keith, A., 1948. A New Theory of Human Evolution. Watts, London.

Kennedy, G.E., 1999. Is "*Homo rudolfensis*" a valid species? Journal of Human Evolution 36, 119–121.

Keyser, A., Menter, C.G., Moggi-Cecchi, J., Pickering, T.R., Berger, L.R., 2000. Drimolen: a new hominid-bearing site in Gauteng, South Africa. South African Journal of Science 96, 193–197.

Kimbel, W.H., 2013. Hesitation on hominin history. Nature 497, 573–574.

Kimbel, W.H., Johanson, D.C., Rak, Y., 1997. Systematic assessment of a maxilla of *Homo* from Hadar, Ethiopia. American Journal of Physical Anthropology 103, 235–262.

Kimbel, W.H., Walter, R.C., Johanson, D.C., Reed, K.E., Aronson, J.L., Assefa, Z., Marean, C.W., Eck, G.G., Bone, R., Hovers, E., Rak, Y., Vondra, C., Yemane, T., York, D., Chen, Y., Evensen, N.M., Smith, P.E., 1996. Late Pliocene *Homo* and Oldowan tools from the Hadar Formation (Kada Hadar Member), Ethiopia. Journal of Human Evolution 31, 549–561.

Kimura, Y., 2002. Examining time trends in the Oldowan technology at Beds I and II, Olduvai Gorge. Journal of Human Evolution 43, 291–321.

Kuman, K., Clarke, R.J., 2000. Stratigraphy, artefact industries and hominid associations for Sterkfontein, Member 5. Journal of Human Evolution 38, 827–847.

Leakey, L.S.B., Tobias, P.V., Napier, J.R., 1964. A new species of the genus *Homo* from Olduvai Gorge. Nature 202, 7–10.

Leakey, M.D., 1971. Olduvai Gorge, vol. 3. Cambridge University Press, Cambridge.

Leakey, M.D., Clarke, R.J., Leakey, L.S.B., 1971. New hominid skull from Bed 1, Olduvai Gorge, Tanzania. Nature 232, 308–312.

Leakey, M.G., Spoor, F., Brown, F.H., Gathogo, P.N., Kiarie, C., Leakey, L.N., McDougall, A., 2001. New hominin genus from eastern Africa shows diverse Middle Pliocene lineages. Nature 410, 433–440.

Leakey, M.G., Spoor, F., Dean, M.C., Feibel, G.S., Antón, S.C., Kiarie, C., Leakey, L.N., 2012. New fossils from Koobi Fora in northern Kenya confirm taxonomic diversity in early *Homo*. Nature 488, 201–204.

Leakey, R.E.F., 1976. New hominid fossils from the Koobi Fora formation in Northern Kenya. Nature 261, 574–576.

Leakey, R., Walker, A., 1985. Further hominids from the Plio-Pleistocene of Koobi Fora, Kenya. American Journal of Physical Anthropology 67, 135–163.

Lepre, C.J., Kent, D.V., 2010. New magnetostratigraphy for the Olduvai Subchron in the Koobi Fora Formation, northwest Kenya, with implications for early *Homo*. Earth and Planetary Science Letters 290, 362–374.

Lepre, C.J., Roche, H., Kent, D.V., Harmand, S., Quinn, R.L., Brugal, J.-P., Texier, P.-J., Lenoble, A., Feibel, C.S., 2011. An earlier origin for the Acheulian. Nature 477, 82–85.

Lieberman, D.E., Pilbeam, D.R., Wood, B.A., 1988. A probabilistic approach to the problem of sexual dimorphism in *Homo habilis*: a comparison of KNM-ER 1470 and KNM-ER 1813. Journal of Human Evolution 17, 503–511.

Lordkipanidze, D., Vekua, A., Ferring, R., Rightmire, G.P., Agusti, J., Kiladze, G., Mouskhelishvili, A., Nioradze, M., Ponce de León, M.S., Tappen, M., Zollikofer, C.P.E., 2005. Anthropology: the earliest toothless hominin skull. Nature 434, 717–718.

Lordkipanidze, D., Vekua, A., Ferring, R., Rightmire, G.P., Zollikofer, C.P., Ponce de León, M.S., Agusti, J., Kiladze, G., Mouskhelishvili, A., Nioradze, M., Tappen, M., 2006. A fourth hominin skull from Dmanisi, Georgia. Anatomical Record 288A, 1146–1157.

Lordkipanidze, D., Jashashvili, T., Vekua, A., Ponce de León, M.S., Zollikofer, C.P.E., Rightmire, G.P., Pontzer, H., Ferring, R., Oms, O., Tappen, M., Bukhsianidze, M., Agusti, J., Kahlke, R., Kiladze, G., Martinez-Navarro, B., Mouskhelishvili, A., Nioradze, M., Rook, L., 2007. Postcranial evidence from early *Homo* from Dmanisi, Georgia. Nature 449, 305–310.

Lordkipanidze, D., Ponce de León, M.S., Margvelashvili, A., Rak, Y., Rightmire, G.P., Vekua, A., Zollikofer, C.P.E., 2013. A complete skull from Dmanisi, Georgia, and the evolutionary biology of early *Homo*. Science 342, 326–331.

Lourens, L.J., Antonarakou, A., Hilgen, F.J., Van Hoof, A.A.M., Vergnaud-Grazzini, C., Zachariasse, W.J., 1996. Evaluation of the Plio-Pleistocene astronomical timescale. Paleoceanography 11391–11413.

MacLatchy, L.M., DeSilva, J., Sanders, W.J., Wood, B., 2010. Hominini. In: Werdelin, L., Sanders, W.J. (Eds.), Cenozoic Mammals of Africa. University of California Press, Berkeley, CA, USA, pp. 471–540.

Marzke, M.W., Shackley, M.S., 1987. Hominid hand use in the Pliocene and Pleistocene: evidence from experimental archaeology and comparative morphology. Journal of Human Evolution 15, 439–460.

Mayr, E., 1950. Taxonomic categories in fossil hominids. Cold Spring Harbor Symposium on Quantitative Biology 15, 109–118.

McDougall, I., Brown, F.H., Fleagle, J.G., 2008. Sapropels and the age of hominins Omo I and II, Kibish, Ethiopia. Journal of Human Evolution 55, 409–420.

McPherron, S.P., Alemseged, Z., Marean, C.W., Wynn, J.G., Reed, D., Geraads, D., Bobe, R., Béarat, H.A., 2010. Evidence for stone-tool-assisted consumption of animal tissues before 3.39 million years ago at Dikika, Ethiopia. Nature 466, 857–860.

Miller, J.M.A., 2000. Craniofacial variation in *Homo habilis*: an analysis of the evidence for multiple species. American Journal of Physical Anthropology 112, 103–128.

Moggi-Cecchi, J., Menter, C., Boccone, S., Keyser, A., 2010. Early hominin dental remains from the Plio-Pleistocene site of Drimolen, South Africa. Journal of Human Evolution 58, 374–405.

Montagu, A., 1961. The "Cerebral Rubicon": brain size and the achievement of hominid status. American Anthropologist 63, 377–378.

Napier, J., 1962. Fossil hand bones from Olduvai Gorge. Nature 196, 409–411.

Olson, T.R., 1978. Hominid phylogenetics and the existence of *Homo* in Member 1 of the Swartkrans Formation, South Africa. Journal of Human Evolution 7, 159–178.

Pickering, T.R., Heaton, J.L., Clarke, R.J., Sutton, M.B., Brain, C.K., Kuman, K., 2012. New hominid fossils from Member 1 of the Swartkrans formation, South Africa. Journal of Human Evolution 62, 618–628.

Pickford, M., Senut, B., Gommery, D., Treil, J., 2002. Bipedalism in *Orrorin tugenensis* revealed by its femora. Comptes Rendus Palevol 1, 1–13.

Plavcan, J.M., 2012. Body size variation, and sexual dimorphism in early *Homo*. Current Anthropology 53, S409–S423.

Pontzer, H., Rolian, C., Rightmire, G.P., Jashashvili, T., Ponce de León, M.S., Lordkipanidze, D., Zollikofer, C.P.E., 2010. Locomotor anatomy and biomechanics of the Dmanisi hominins. Journal of Human Evolution 58, 492–504.

Prat, S., 2002. Anatomical study of the skull of the Kenyan specimen KNM-ER 1805: a re-evaluation of its taxonomic allocation? Comptes Rendus Palevol 1, 27–33.

Prat, S., Brugal, J.-P., Tiercelin, J.-J., Barrat, J.-A., Bohn, M., Delagnes, A., Harmand, S., Kimeu, K., Kibunjia, M., Texier, P.-J., Roche, H., 2005. First occurrence of early Homo in the Nachukui formation (West Turkana, Kenya) at 2.3-2.4Myr. Journal of Human Evolution 49, 230–240.

Prüfer, K., et al., 2013. The complete genome sequence of a Neanderthal from the Altai mountains. Nature 505, 43–49.

Reich, D., Green, R.E., Kircher, M., Krause, J., Patterson, N., Durand, E.Y., Viola, B., Briggs, A.W., Stenzel, U., Johnson, P.L., Maricic, T., Good, J.M., Marques-Bonet, T., Alkan, C., Fu, Q., Mallick, S., Li, H., Meyer, M., Eichler, E.E., Stoneking, M., Richards, M., Talamo, S., Shunkov, M.V., Derevianko, A.P., Hublin, J.-J., Kelso, J., Slatkin, M., Pääbo, S., 2010. Genetic history of an archaic hominin group from Denisova Cave in Siberia. Nature 468, 1053–1060.

Rightmire, G.P., 1990. The Evolution of *Homo erectus*. Cambridge University Press, Cambridge.

Rightmire, G.P., 2009. Middle and later Pleistocene hominins in Africa and Southwest Asia. Proceedings of the National Academy of Sciences 106, 16046–16050.

Rightmire, G.P., 2013. *Homo erectus* and Middle Pleistocene hominins: brain size, skull form, and species recognition. Journal of Human Evolution 65, 223–252.

Rightmire, G.P., Lordkipanidze, D., Vekua, A., 2006. Anatomical descriptions, comparative studies and evolutionary significance of the hominin skulls from Dmanisi, Republic of Georgia. Journal of Human Evolution 50, 115–141.

Rogers, M.J., Harris, J.W.K., Feibel, C.S., 1994. Changing patterns of land use by Plio-Pliestocene hominids in the Lake Turkana basin. Journal of Human Evolution 27, 139–158.

Rosenberg, K., Trevethan, W., 2002. Birth, obstetrics and human evolution. British Journal of Obstetrics and Gynaecology 109, 1199–1206.

Ruff, C.B., Puymerail, L., Macchiarelli, R., Sipla, J., Ciochon, R.L., 2015. Structure and composition of the Trinil femora: functional and taxonomic implications. Journal of Human Evolution 80, 147–158.

Sahnouni, M., Rosell, J., van der Made, J., Vergès, J.M., Ollé, A., Kandi, N., Harichane, Z., Derradji, A., Medig, M., 2013. The first evidence of cut marks and usewear traces from the Plio-Pleistocene locality of El-Kherba (Ain Hanech), Algeria: implications for early hominin subsistence activities circa 1.8 Ma. Journal of Human Evolution 64, 137–150.

Santa Luca, A.P., 1980. The Ngandong Fossil Hominids: A Comparative Study of a Far Eastern Homo erectus Group, vol. 87. Yale University Press. 175 pp.

Schrenk, F., Bromage, T.G., Betzler, C., Ring, U., Juwayeyi, Y., 1993. Oldest *Homo* and Pliocene biogeography of the Malawi Rift. Nature 365, 833–836.

Schrenk, F., Kullmer, O., Bromage, T., 2007. The earliest putative *Homo* fossils. In: Henke, W., Tattersall, I. (Eds.), Handbook of Paleoanthropology. Springer-Verlag, pp. 1611–1631.

Schroeder, L., Roseman, C.C., Cheverud, J.M., Ackermann, R.R., 2014. Characterizing the evolutionary path(s) to early *Homo*. PLoS One 9, e114307.

Semaw, S., 2000. The world's oldest stone artefacts from Gona, Ethiopia: their implications for understanding stone technology and patterns of human evolution between 2.6-1.5 million years ago. Journal of Archaeological Science 27, 1197–1214.

Semaw, S., Renne, P., Harris, J.W.K., Feibel, C.S., Bernor, R.L., Fesseha, N., Mowbray, K., 1997. 2.5-million-year-old stone tools from Gona, Ethiopia. Nature 385, 333–336.

Semaw, S., Rogers, M.J., Quade, J., Renne, P.R., Butler, R.F., Dominguez-Rodrigo, M., Stout, D., Hart, W.S., Pickering, T., Simpson, S.W., 2003. 2.6-Million-year-old stone tools and associated bones from OGS-6 and OGS-7, Gona, Afar, Ethiopia. Journal of Human Evolution 45, 169–177.

Semaw, S., Rogers, M.J., Stout, D., 2009. The Oldowan-Acheulian transition: is there a 'Developed Oldowan: artifact tradition? In: Camps, M., Chauhan, P. (Eds.), Sourcebook of Paleolithic Transitions. Springer Science.

Shen, G., Gao, X., Gao, B., Granger, D.E., 2009. Age of Zhoukoudiean *Homo erectus* determined with $^{26}Al/^{10}Be$ burial dating. Nature 458, 198–200.

Sherwood, R.J., Ward, S.C., Hill, A., 2002. The taxonomic status of the Chemeron temporal (KNM-BC 1). Journal of Human Evolution 42, 153–184.

Shipman, P., 2001. The Man Who Found the Missing Link: Eugène Dubois and His Lifelong Quest to Prove Darwin Right. Simon & Schuster, N.Y.

Simpson, G.G., 1963. The meaning of taxonomic statements. In: Washburn, S.L. (Ed.), Classification and Human Evolution. Aldine, Chicago.

Simpson, S.W., Quade, J., Levin, N.E., Butler, R., Dupont-Nivet, G., Everett, M., Semaw, S., 2008. A female *Homo erectus* pelvis from Gona, Ethiopia. Science 322, 1089–1092.

Simpson, S.W., Quade, J., Levin, N.E., Semaw, S., 2014. The female *Homo* pelvis from Gona: response to Ruff (2010). Journal of Human Evolution 68, 32–35.

Spoor, F., 2013. Palaeoanthropology: small-brained and big-mouthed. Nature 502, 452–453.

Spoor, F., Gunz, P., Neubauer, S., Stelzer, S., Scott, N., Kwekason, A., Dean, M.C., 2015. Reconstructed *Homo habilis* type OH7 suggests deep-rooted species diversity in early *Homo*. Nature 519, 85–86.

Spoor, F., Leakey, M.G., Gathogo, P.N., Brown, F.H., Antón, S.C., McDougall, I., Kiarie, C., Manthi, F.K., Leakey, L.N., 2007. Implications of new early *Homo* from Ileret, east of Lake Turkana, Kenya. Nature 448, 688–691.

Stout, D., Semaw, S., Rogers, M.J., Cauche, D., 2010. Technological variation in the earliest Oldowan from Gona, Afar, Ethiopia. Journal of Human Evolution 58, 474–491.

Stringer, C.B., 1987. A numerical cladistics analysis for the genus *Homo*. Journal of Human Evolution 16, 135–146.

Stringer, C.B., Buck, L.T., 2014. Diagnosing *Homo sapiens* in the fossil record. Annals of Human Biology 41, 312–322.

Stern, J.T., Susman, R.L., 1982. Functional morphology of *Homo habilis*. Science 217, 931–934.

Susman, R.L., 2008. Brief communication: evidence bearing on the status of *Homo habilis* at Olduvai Gorge. American Journal of Physical Anthropology 137, 356–361.

Suwa, G., 1990. A Comparative Analysis of Hominid Dental Remains from the Shungura and Usno Formations, Omo Valley, Ethiopia (Ph.D. Dissertation). University of California, Berkeley.

Suwa, G., Asfaw, B., Haile-Selassie, Y., White, T., Katoh, S., WoldeGabriel, G., Hart, W.K., Nakaya, H., Beyene, Y., 2007. Early Pleistocene *Homo erectus* fossils from Konso, southern Ethiopia. Anthropological Science 115, 133–151.

Suwa, G., White, T.D., Howell, F.C., 1996. Mandibular postcanine dentition from the Shungura Formation Ethiopia: Crown morphology, taxonomic allocations, and Plio-Pleistocene hominid evolution. American Journal of Physical Anthropology 101, 247–282.

Tamrat, E., Thouveny, N., Taïeb, M., Opdyke, N.D., 1995. Revised magnetostratigraphy of the Plio-Pleistocene sedimentary sequence of the Olduvai Formation (Tanzania). Palaeogeography, Palaeoclimatology, Palaeoecology 114, 273–283.

Tiercelin, J.-J., Schuster, M., Roche, H., Brugal, J.-P., Thuo, P., Prat, S., Harmand, S., Davtian, G., Barrat, J.-A., Bohn, M., 2010. New considerations on the stratigraphy and environmental context of the oldest (2.34) Lokalalei archaeological site complex of the Nachukui Formation, West Turkana, northern Kenya Rift. Journal of African Earth Sciences 58, 157–184.

Tobias, P.V., 1991. The Skulls, Endocasts and Teeth of Homo habilis, Olduvai Gorge, vol. 4. Cambridge University Press, Cambridge.

Tobias, P.V., 2009. *Homo habilis*—a premature discovery: remembered by one of its founding fathers, 42 years later. In: Grine, F.E., et al. (Ed.), The First Humans: Origin and Early Evolution of the Genus *Homo*. Springer Science + Business Media B.V, pp. 7–15.

Trevethan, W.R., 1988. Fetal emergence patterns in evolutionary perspective. American Anthropologist 90, 674–681.

Vallois, H.V., 1954. La capacité crânniene chez les Primates supérieurs et le 'Rubicon cerebral'. Comptes rendus hebdomadaires des séances de l'Académie des sciences 238, 1349–1351.

Vekua, A., Lordkipanidze, D., Rightmire, G.P., Agusti, J., Ferring, R., Maisuradze, G., Mouskhelishvili, A., Nioradze, M., Ponce de León, M.S., Tappen, M., Tvalchrelidze, M., Zollikofer, C.P.E., 2002. A new skull of early *Homo* from Dmanisi, Georgia. Science 297, 85–89.

Vialet, A., Guipert, G., Alçiçek, M.C., de Lumley, M.A., 2014. La calotte crânienne de l'*Homo erectus* de Kocabaş (Bassin de Denizli, Turquie). L'anthropologie 118, 74–107.

Villmoare, B., Kimbel, W.H., Seyoum, C., Campisano, C.J., DiMaggio, E., Rowan, J., Braun, D.R., Arrowsmith, J.R., Reed, K.E., 2015. Early *Homo* from Ledi-geraru, Afar, Ethiopia. Science. http://dx.doi.org/10.1126/Science.aaa1343.

Walker, A., Leakey, R. (Eds.), 1993. The Nariokotome *Homo erectus* Skeleton. Harvard University Press, Cambridge, MA.

Ward, C.V., Feibel, C.S., Hammond, A.S., Leakey, L.N., Moffett, E.A., Plavcan, J.M., Skinner, M.M., Spoor, F., Leakey, M.G., 2015. Associated ilium and femur from Koobi Fora, Kenya, and postcranial diversity in early Homo. Journal of Human Evolution 81, 48–67. http://dx.doi.org/10.1016/j.jhevol.2015.01.005.

Weidenreich, F., 1940. Some problems in dealing with ancient man. American Anthropologist 42, 375–383.

Weiner, J.S., 1955. The Piltdown Forgery. Oxford University Press, Oxford.

Weiner, J.S., Oakley, K.P., le Gros Clark, W.E., 1953. The solution of the Piltdown hoax. Bulletin of the British Museum 2, 139–146.

Wood, B.A., 1985. Early *Homo* in Kenya, and its systematic relationships. In: Delson, E. (Ed.), Ancestors: The Hard Evidence. Alan R. Liss, New York, pp. 206–214.

Wood, B.A., 1991. Hominid Cranial Remains. Koobi Fora Research Project, vol. 4. Clarendon Press, Oxford.

Wood, B.A., 1992a. Origin and evolution of the genus *Homo*. Nature 355, 783–790.

Wood, B., 1992b. Early hominid species and speciation. Journal of Human Evolution 22, 351–365.

Wood, B.A., 1999. '*Homo rudolfensis*' Alexeev, 1986—fact or phantom? Journal of Human Evolution 36, 115–118.

Wood, B.A., 2009. Where does the genus *Homo* begin, and how would we know? In: Grine, F.E., et al. (Ed.), The First Humans: Origin and Early Evolution of the Genus *Homo*, 17 Vertebrate Paleobiology and Paleoanthropology. Springer Science+Business Media B.V.

Wood, B., 2010. Reconstructing human evolution: achievements, challenges, and opportunities. Proceedings of the National Academy of Science 107, 8902–8909.

Wood, B.A., 2012. Facing up to complexity. Nature 488, 162–163.

Wood, B.A., Baker, J., 2011. Evolution in the genus *Homo*. Annual Review of Ecology, Evolution, and Systematics 42, 47–69.

Wood, B.A., Collard, M., 1999. The human genus. Science 284, 65–71.

Xing, S., Martinón-Torres, M., Bermúdez de Castro, J.M., Zhang, Y., Fan, X., Zheng, L., Huang, W., Liu, W., 2014. Middle Pleistocene hominin teeth from Longtan Cave, Hexian, China. PLoS One e114265.

Yokoyama, Y., Falguères, C., Sémah, F., Jacab, T., Grün, R., 2008. Gamma-ray spectrometric dating of late *Homo erectus* skulls from Ngandong and Sambungmacan, Central Java, Indonesia. Journal of Human Evolution 55, 274–277.

Zanolli, C., 2015. Brief Communication: molar crown inner structural organization in Javanese *Homo erectus*. American Journal of Physical Anthropology 159, 148–157.

Zhu, Z.Y., Dennell, R., Huang, W.-W., Wu, Y., Rao, Z.-G., Qui, S.-F., Xie, J.-B., Liu, W., Fu, S.-Q., Han, J.-W., Zhou, H.Y., Ou Yang, T.-P., Li, H.M., 2015. New dating of the *Homo erectus* cranium from Lantian (Gongwangling), China. Journal of Human Evolution 78, 144–157. http://dx.doi.org/10.1016/j.jhevol.2014.10.001.

Chapter 12

Archaic *Homo*

James C.M. Ahern
Department of Anthropology, University of Wyoming, Laramie, WY, USA

SYNOPSIS

Archaic *Homo* refers to the human populations or species that were chronologically and anatomically intermediate between *Homo erectus* and modern *Homo sapiens*. Archaic *Homo* can be further subdivided into two chronological groupings, Early and Late Archaics. Anatomically, Archaic *Homo* exhibit larger braincases than *H. erectus* as well as greater body mass, among other features. Archaic *Homo* had controlled and regular use of fire and colonized temperate regions of Eurasia for the first time. In some regions, Archaic *Homo* populations appear to have directly contributed to the ancestry of modern humans, while the evidence for this in other regions is limited. The overall pattern of evolutionary relationships among Archaic *Homo*, as elucidated by the available fossil and paleogenomic evidence, indicates that not all archaics and modern humans were reproductively isolated from each other.

"Archaic *Homo*" refers to the populations or species of humans (genus *Homo*) that for the most part were chronologically and anatomically intermediate between Erectines (e.g., *Homo erectus*, *Homo ergaster*, and *Homo antecessor*; see the chapter by Simpson in this volume) and modern *Homo sapiens* (see the chapter by Holt in this volume). In 1856, the first Archaic *Homo* fossil to be recognized as such, the Feldhofer 1 skeleton, was discovered by workers in the Neander Valley ("Neander Thal") of western Germany. In the first half of the twentieth century, a scattering of additional archaic human fossils from outside Europe were found, such as Kabwe (1921) and Ngandong (1931–1933). However, the numerous discoveries of Neanderthal fossils in Europe have for better or worse shaped our understanding of what Archaic *Homo* were. It has only been in the last few decades that the fossil record has grown to provide a better perspective of non-Neanderthal populations of Archaic *Homo*.

Archaic *Homo*, as a biological and chronological category, has proven to be more of a heurism than an objectively distinct category. In most regions where Archaic *Homo* fossils have been found, the boundary between them and earlier Erectines is equivocal, with many fossils having been classified as *H. erectus* by some but Archaic *Homo* (e.g., archaic *Homo sapiens*, *Homo heidelbergeneis*, *Homo soloensis, Homo daliensis*, etc.) by others. The defining boundary on the other end is equally ambiguous, as illustrated by the continuing debate regarding a definition of "anatomically modern humans" that excludes all "archaics" while still including all Holocene humans (Stringer, 2012) (see the chapter by Holt). Nevertheless, the "archaic *Homo*" category continues to garner acceptance, if only as a heurism.

Chronologically, Archaic *Homo* existed from ~500 to 30 thousand years ago (kya). A simple chronological subdivision is Early Archaic *Homo* (~500–250 kya) and Late Archaic *Homo* (~250–30 kya in Eurasia, ~300–150 kya in Africa) (see Table 1). The term "Archaic *Homo*" is derived from the quasi-taxonomic term, "archaic *Homo* sapiens." This classification was commonly used in the 1960 and 1970s to distinguish robust, archaic-looking fossils with cranial capacities within the recent human range from both earlier *H. erectus* (with cranial capacities generally below the recent modern human range) and later " anatomically modern *Homo sapiens*" (which are generally more gracile) (Wolpoff, 1980). With the increasing impact of cladistics methodology and Gould and Eldredge's (1977) punctuated equilibrium theory of macroevolution, the "archaic *H. sapiens*" term found less relevance as "bushier" models of hominin evolution became more common (Cartmill and Smith, 2009). Thus, today, "Archaic *Homo*" tends to roughly mirror the earlier "archaic *H. sapiens*" in terms of the fossils it encompasses, while allowing for the possible recognition of multiple species.

The anatomical *definition* of Archaic *Homo* is largely one whereby traits fall within the range of modern living humans, but averages differ. For example (1) cranial capacities that fall within the living human range, although with population averages that tend to fall below those of the living human, (2) a braincase that is expanded relative to the cranial base in comparison with earlier hominins, (3) dental size that generally falls within the living human range, but with averages, especially of anterior dental size, that exceed the modern one, (4) thick cranial bone with well-developed cranial buttressing such as true supraorbital and nuchal/occipital tori, (5) prognathism that is exceptionally greater than that of most living humans, resulting in a general lack of chin development, and (6) postcranial robusticity that is on the high end of the living human range. Beyond these general characteristics, Archaic *Homo* anatomy shows regional variations that may reflect species-level taxonomic differences, or at the very least intraspecific variation that exceeds that seen among living humans. Body proportions and size vary, although archaic humans tend to have greater body mass than living humans. Archaic human stature tends to be similar to *H. erectus* but shorter than early modern humans. Behaviorally, archaic humans are usually associated with late Lower Paleolithic or more often Middle Paleolithic industries, with an occasional association with the early Upper Paleolithic. All archaic humans seem to have had controlled use of fire (Karkanas et al., 2007) and language (although perhaps not fully modern) (Dediu and Levinson, 2014) that were crucial to their colonization of temperate and cold climates.

GEOGRAPHICAL DISTRIBUTION

Archaic *Homo* fossils have been discovered on all Eastern Hemisphere continents, except Australia and Antarctica (see Figure 1). Europe has by far the best Archaic *Homo* record, especially of Late Archaic Neanderthals. Africa's record, although plagued by difficulties in accurately dating many fossils (Smith, 1992), has grown steadily to include numerous Early and Late Archaic fossils as well as the oldest reported anatomically modern humans (McDougall et al., 2008; White et al., 2003). Although the number of West Asian Archaic *Homo* individual sites is low, some of these sites (e.g., Shanidar, Iraq) have yielded numerous well-preserved skeletons of Late Archaics. Like West Asia, the East Asian Early Archaic *Homo* record is very limited, if present at all (see below). A scattering of late Middle Pleistocene and early Late Pleistocene fossils, mostly from China, represents Late Archaics in East Asia. Finally, the unclearly dated Javan fossils from Sambungmacan (Delson et al., 2001), Ngawi (Zeitoun and Widianto, 2001), and Ngandong (Weidenreich, 1951) represent Archaic *Homo* in Australasia.

CHRONOLOGICAL VARIATION

Some evolutionary trends appear to have been consistent among all Archaic *Homo* populations. Cranial

TABLE 1 Chronology and Classification of Pleistocene *Homo*, Including Archaic Humans. Lists of Fossils are Representative, Not Exhaustive. Dates are Approximate

Region		Erectines	Early Archaic *Homo*	Late Archaic *Homo*	Modern *H. sapiens*
Africa	*Dates*	1.8 mya to 500 kya	500–300 kya	300–150 kya	<200 kya
	Fossils	Koobi Fora, Olduvai, Daka, Olorgesailie, Swartkrans	Bodo, Kabwe, Ndutu	Eyasi, Florisbad, Ileret, Jebel Irhoud, Ngaloba	Aduma, Border Cave, Herto, Hofmeyr, Nazlet Khater, Omo
W. Asia	*Dates*	1.4 mya	Middle Pleistocene	120–45 kya	<110 kya
	Fossils	Ubeidiya	Zuttiyeh, Qesem	Amud, Dederiyeh, Kebara, Shanidar, Tabun	Qafzeh, Skhul
E. Asia	*Dates*	1.7[a] mya to 400 kya	400–300 kya	300–130 kya	<30 kya
	Fossils	Chenjiawo, Gongwangling, Nanjing, Yuanmou, Zhoukoudian locality 1	Hexian, Zhoukoudian H3[b]	Dali, Jinniushan, Maba	Tianyuan, Zhirendong, Zhoukoudian Upper Cave
Australasia	*Dates*	1.8 mya to early Middle Pleistocene	Late Middle Pleistocene[c]	Late Pleistocene[c]	<50 kya
	Fossils	Mojokerto, Sangiran	Nagwi, Ngandong?, Sambungmacan[c]	Liang-Bua[d], Ngandong?[c]	Keilor, Kow Swamp, Niah, Tabon, Willandra Lakes
Europe	*Dates*	1.8 mya to 500 kya	500–200 kya	200–30 kya	<40 kya
	Fossils	Ceprano, Dmanisi, Gran Dolina[e], Sima del Elefante	Arago, Boxgrove, Gran Dolina[d], Petralona, Sima de los Huesos, Swanscombe	Ehringsorf, Feldhofer, Guattari, Krapina, La Chapelle, La Ferrassie, Spy	Brassempouy, Kent's Cavern, La Quina-Aval, Mladeč, Peştera cu Oase

[a]*Zhu et al. (2008) report a paleomagnetic age estimate of 1.7 million years ago (mya) for the Yuanmou teeth, although this is significantly older than earlier age estimates for these fossils (Hyodo et al., 2002).*
[b]*Although these fossils are fairly late and show some more modern-like features than the earlier H. erectus fossils from China, they might be best classified as H. erectus.*
[c]*The Archaic* Homo *fossils from Java are very poorly dated. See the text for more discussion.*
[d]*The Liang-Bua remains from Flores Island, Indonesia are morphologically enigmatic and may represent a late-surviving, isolated population of H. erectus or perhaps a separate species of* Homo, *H. floresiensis (Brown et al., 2004). See the text.*
[e]*The Gran Dolina (Atapuerca, Spain) remains have been classified as* Homo antecessor *(Bermudez de Castro, 1997).*

capacity increases over time among all regional populations of Archaic *Homo* (Lee and Wolpoff, 2003), with Early Archaics falling within or just above the range of *H. erectus* and Late Archaics falling on the low end or completely overlapping with early modern *H. sapiens* (see Figure 2; also see the chapter by Holloway in the present volume). Other braincase dimensions reflect this trend. For example, while Early Archaics have upper braincases that are expanded compared with those of *H. erectus*, only among Late Archaics does upper braincase breadth generally exceed that of the lower braincase (see Figure 3). Additionally, while the most posterior point on Early Archaic braincases (*opisthocranion*) tends to coincide with the midpoint of the nuchal torus, the opisthocranion of many Late Archaics falls higher up on the occipital as it does in modern humans. Relatedly, the occipital plane increases in size over time among Archaic *Homo* relative to the nuchal plane.

In contrast to *H. erectus*, browridges among most Early Archaics are double-arched and dip inferiorly above the nasal region. Furthermore, browridge size reduces from Early to Late Archaics. Facial projection in front of the braincase reduces over time among Archaic *Homo*, with Early Archaics generally exhibiting total facial prognathism (Trinkaus, 1987) and Late Archaics exhibiting either overall reduced prognathism (Africa) or midfacial prognathism (see Figure 4). Dentally, the incisors increase in size slightly, while the canines decrease slightly.[1] Perhaps the most significant dental metric trend within Archaic *Homo* is a decrease in molar size. While Early Archaics have molars similar in size to late *H. erectus*, Late Archaics show a reduction in the direction of early modern humans (Figure 5) (Wolpoff, 1999).

1. However, the available dental data are mostly for European archaics. Thus, it is unclear if this trend characterizes all Archaic *Homo* in all regions.

FIGURE 1 Map of major Archaic *Homo* fossil sites. Circles represent Early Archaics (approximately 500–250kya), while diamonds represent Late Archaics (~250–30kya). Some of the fossils from these sites may better belong in other taxa, such as *Homo erectus* (e.g., Zhoukoudian H3, Yunxian, Hexian) or modern *Homo sapiens* (e.g., Laetoli (Ngaloba), Omo, Singa). The very recent (18kya) Liang-Bua fossils are anatomically very enigmatic, and may be best classified as their own species, *Homo floresiensis*.

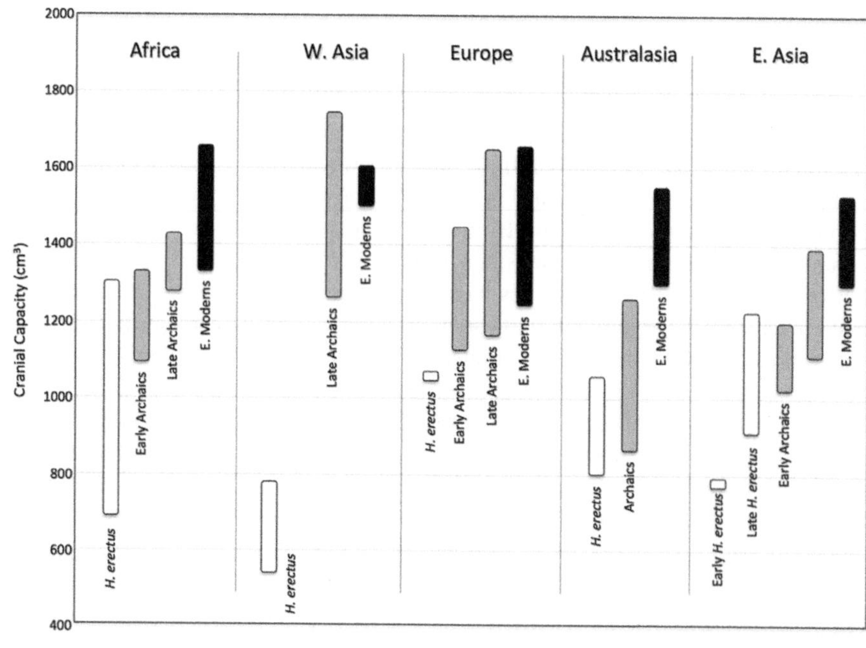

FIGURE 2 Cranial capacity ranges for Pleistocene *Homo* groups by region. There is an overall trend of increasing brain size during the Pleistocene in all regions. Specimens of Archaic *Homo* are characterized by possessing cranial capacities that usually fall within the range of living humans. Early Archaics generally exhibit cranial capacities that overlap with both *Homo erectus*, on the low end, and living humans, on the high end. Late Archaics, especially the Neanderthals of Europe and Western Asia, exhibit cranial capacities that fall in the upper half of living human variation, although are similar to the living humans of Western Eurasia, in this respect. Bars represent sample ranges. *Data from Holloway et al. (2004) and other sources. See the chapter by Holloway.*

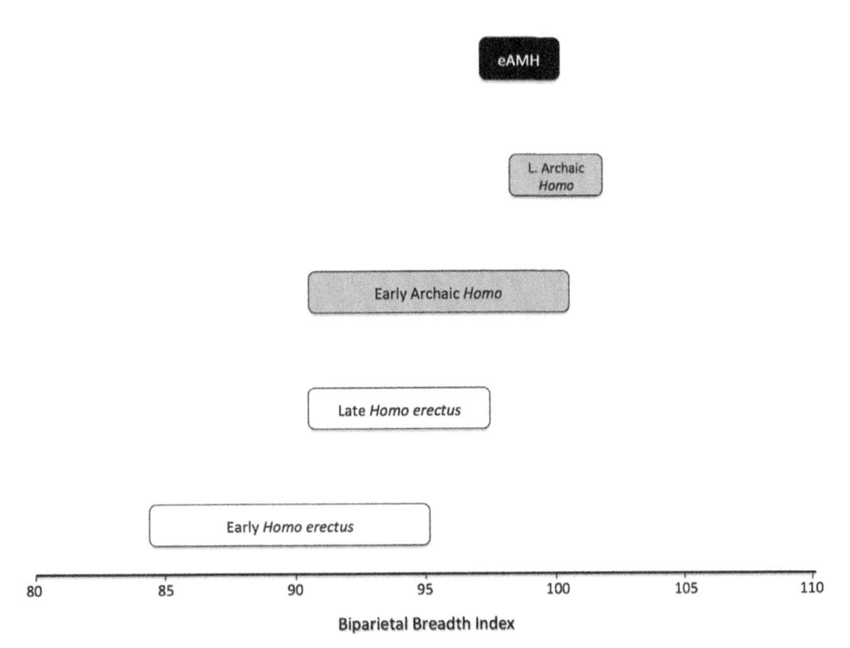

FIGURE 3 Ranges for relative upper braincase breadth, as measured by the biparietal index (100·(biparietal breadth/maximum cranial breadth)). "eAMH" stands for "early anatomically modern humans." The upper braincase expands over the course of *Homo* evolution. Bars represent sample ranges. *Data courtesy of F.H. Smith and M.H. Wolpoff.*

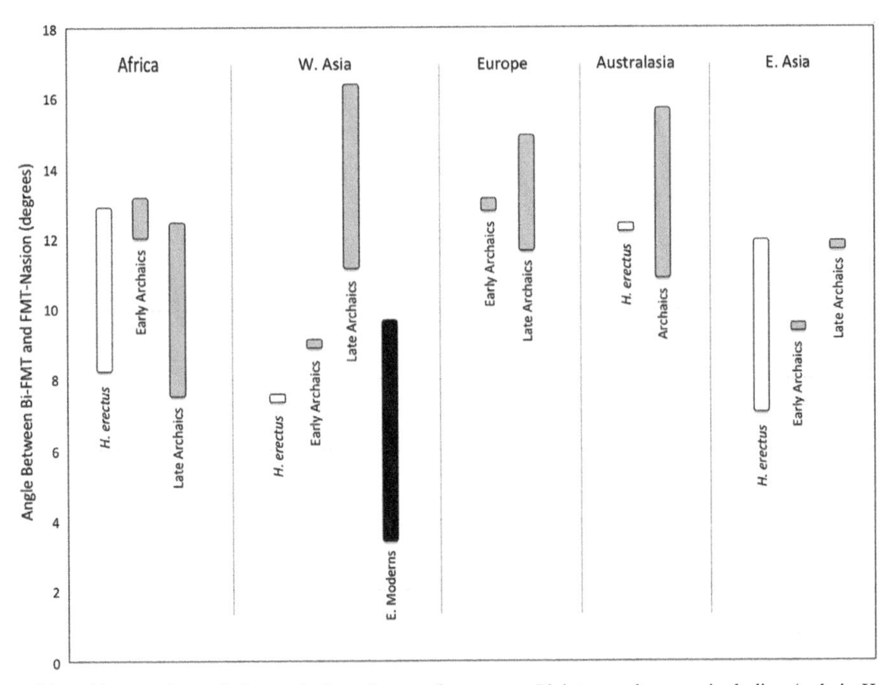

FIGURE 4 Projection of the mid-upper face relative to the lateral upper face among Pleistocene humans, including Archaic *Homo*, as measured by the angle between the bi-*frontomalare temporale* (bi-FMT) and FMT – *nasion* chords. Lower angle values correspond to less midfacial prognathism. Note than in all areas except Africa, Late Archaics show greater midfacial prognathism than Early Archaics. The African pattern of reduced midfacial prognathism potentially foreshadows the condition seen among modern humans. Bars represent sample ranges. *Data courtesy of M.H. Wolpoff.*

Postcranially, few temporal trends characterize all Archaic *Homo* populations. Archaic *Homo* had greater body mass on average than both *H. erectus* and early modern humans. Over time, Archaic *Homo* body mass appears to decrease (Froehle et al., 2013), although this trend is not statistically significant (see Figure 6). In terms of stature, archaic humans were similar to *H. erectus* but shorter than early modern humans (see Figure 7). Overall, Archaic *Homo* exhibited stockier bodies than *H. erectus* and early modern humans (Cartmill and Smith, 2009; Froehle et al., 2013).

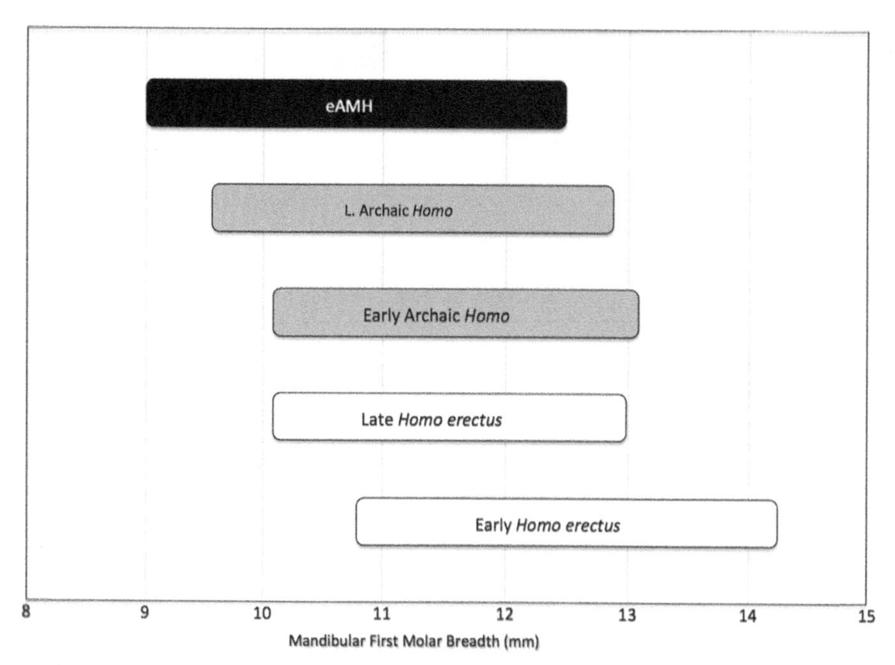

FIGURE 5 Sample ranges for first lower molar breadth (bucco-lingual) among Pleistocene *Homo*. "eAMH" = early anatomically modern humans. *Data courtesy of M.H. Wolpoff.*

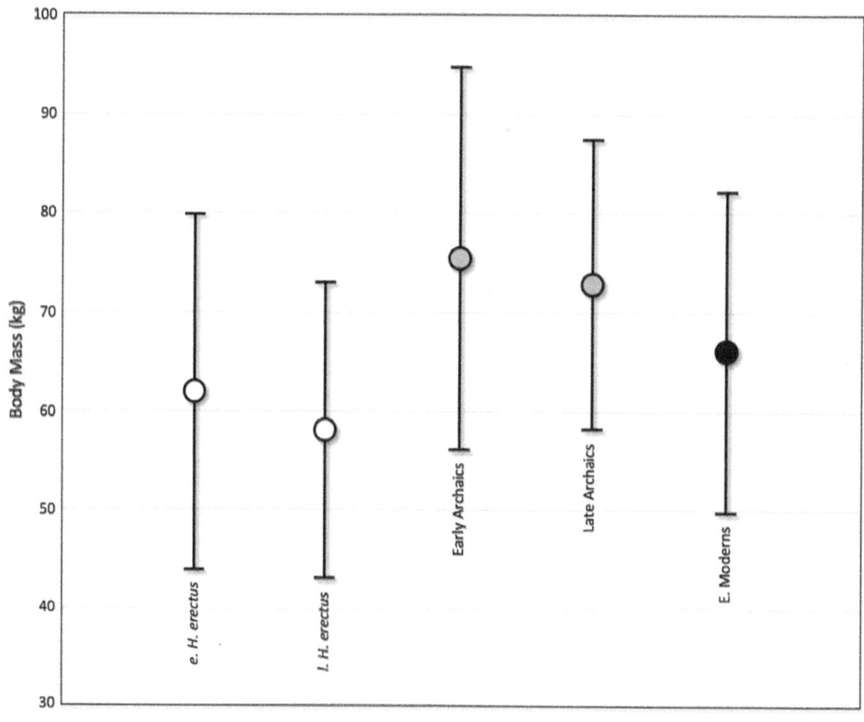

FIGURE 6 Pleistocene *Homo* estimated body mass in kilograms. The circles represent sample averages while the bars are ±2SD. Estimates are derived from postcranial measurements. *Data from Froehle et al. (2013).*

REGIONAL VARIATION

Our understanding of Archaic *Homo* and our interpretations of later human evolution in general have unfortunately been dominated by the extensive European fossil human record. This has largely been because paleoanthropology began in Europe, and Eurocentric perspectives have thus dominated explanations of human evolution. As paleoanthropology has become more cosmopolitan, the fossil records of other regions have grown. In particular, by the 1980s African and West Asian evidence became crucial to interpretations of later human evolution (cf., Bräuer, 1989; Stringer and Andrews, 1988). Furthermore,

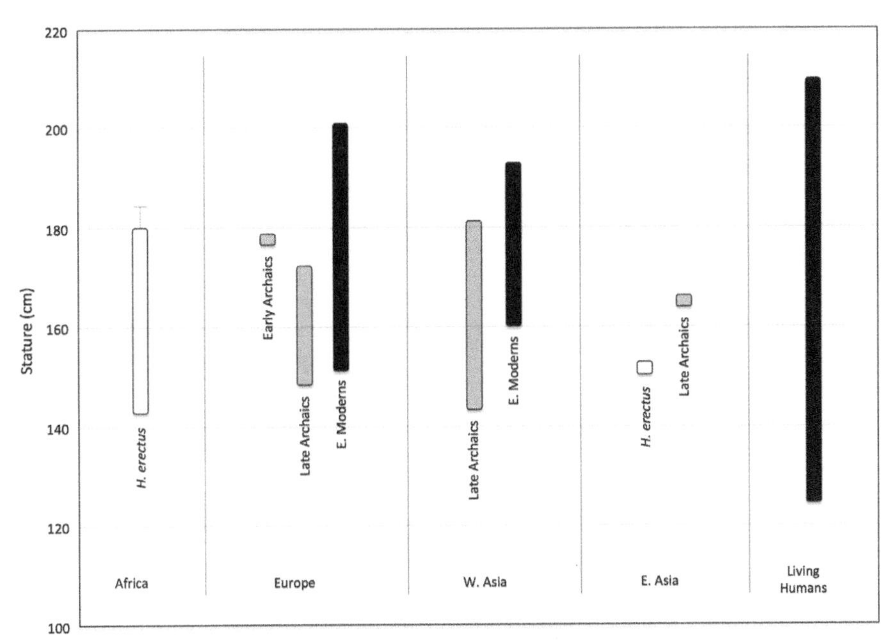

FIGURE 7 Sample ranges for Pleistocene *Homo* estimated stature in centimeters. *Data from Wolpoff (1999) and Feldesman et al. (1990).*

fossil records from East Asia and Australasia have also grown and have played a larger role in our interpretations (cf., Durband and Westaway, 2013; Rosenberg and Wu, 2013).

Europe

Hominins in the form of *H. erectus* or a similar species colonized far Eastern Europe by 1.8 million years ago (mya) (Gabunia et al., 2001) and Western Europe by 1.2 mya (Carbonell et al., 2008). By the middle of the Middle Pleistocene (Marine Isotope Stage [MIS] 12–13), Early Archaic *Homo* fossils appear. The largest sample of Early Archaic *Homo* fossils comes from Sima de los Huesos (Atapuerca, Spain), with over 5000 specimens representing at least 33 individuals (Arsuaga et al., 1997, 2014). Other important fossils come from England (Boxgrove and Swanscombe), France (Arago, Bau de l'Aubesier), Germany (Bilzingsleben, Mauer, Reilingen and Steinheim), Greece (Apidima and Petralona), and Hungary (Vértesszőllős). European Early Archaics are generally associated with Lower Paleolithic stone tool industries.

Anatomically, these European Early Archaics are characterized by a mixture of Erectine and Neanderthal (European Late Archaic) features (Arsuaga et al., 2014). Like other Early Archaic *Homo*, maximum cranial breadth remains low on the braincase (≈Erectines), but the upper portion of the braincase is expanded reflecting an increased cranial capacity. The 430 kya Sima de los Huesos fossils exhibit anterior vaults and faces that are more Neanderthal-like than the rest of the skull, which retains more Erectine-like anatomy (Arsuaga et al., 2014).

Late Archaic *Homo* in Europe is represented by numerous fossils of Neanderthals. Although the largest sample of Neanderthals from a single site comes from Central Europe (Krapina, Croatia), most well-preserved crania and skeletons derive from Western Europe (e.g., La Chapelle, La Ferrassie, Feldhofer, Spy). Anatomically, European Late Archaics are characterized by voluminous braincases that differ from modern humans by (1) being long and low, (2) being oval shaped when viewed from the rear, with maximum cranial breadth lying approximately halfway up the vault, (3) occipital bunning that involves flattening of the posterior parietals, a vertical occipital plane, and a horizontal nuchal plane, and (4) small mastoid processes that are often over-shadowed by large occipitomastoid crests (see Figure 8) (Cartmill and Smith, 2009).

European Late Archaics are generally associated with Middle Paleolithic assemblages (in particular, the Mousterian), although some late ones are associated with early Upper Paleolithic ones (Hublin, 2013; Ahern et al., 2013).

Recent work on the chronology of the Neanderthal–modern transition (Higham et al., 2014) indicates that Neanderthals persisted until around 39 kya and possibly even later (Higham et al., 2006), indicating that Neanderthals and early modern humans overlapped in Europe for 2600–5400 or more years. Anatomical (cf., Ahern et al., 2013) and paleogenomic (Green et al., 2010; see the chapter by Relethford in this volume) evidence indicates that these two groups interbred, although it is unclear if the interbreeding took place within or outside of Europe.

Africa

Erectines (e.g., Ergasters) appear in Africa after 2 mya and appear to have been widespread throughout the continent

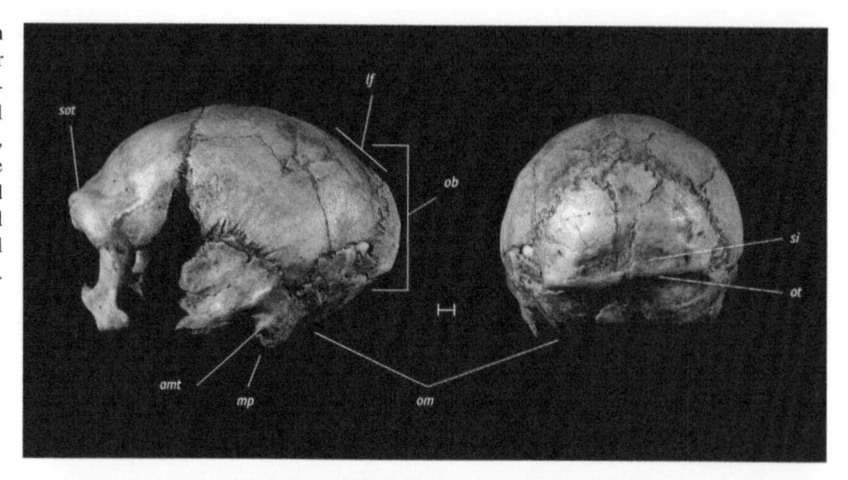

FIGURE 8 The La Quina V Neanderthal (European Late Archaic *Homo,* cast) in side (left) and rear (right) views. *amt:*anterior mastoid tubercle, *lf*: lambdoidal flattening, *mp*: mastoid process, *ob*: occipital bun, *om*: occipitomastoid crest, *ot*: occipital torus, *si*: suprainiac fossa, and *sot*: supraorbital torus. The anterior mastoid tubercle, a large occipitomastoid crest, an oblong suprainiac fossa, and an occipital bun with associated lambdoidal flattening are all Neanderthal characteristics. (See color plate section).

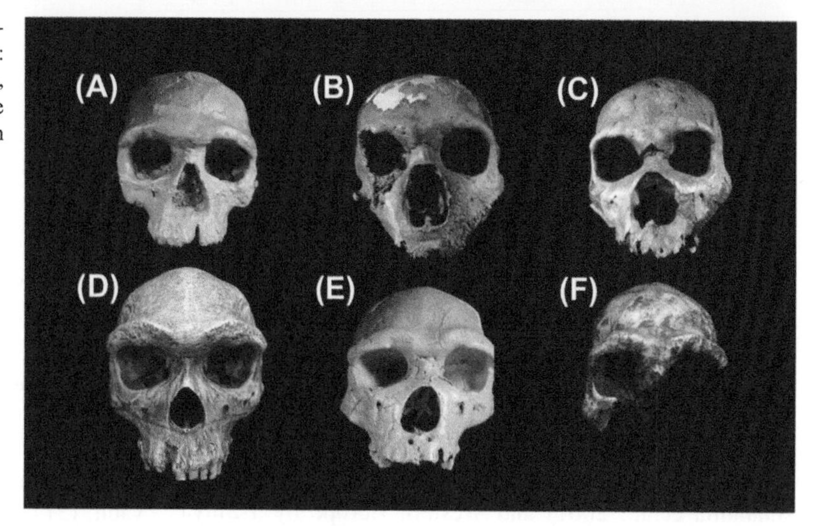

FIGURE 9 Anterior views of select Archaic *Homo* crania (all casts and approximately to scale). Late Archaics: (A) Jebel Irhoud (Morocco), (B) Guattari 1 (Italy), and (C) Shanidar 5 (Iraq). Early Archaics: (D) Kabwe (Zambia), (E) Arago 21 (France), and (F) Zuttiyeh (Israel). (See color plate section).

by 1.5–1.0 mya (Pearson, 2013). The few human fossils dating to 1.0–0.6 mya seem to indicate an evolution of Erectines into Early Archaic *Homo*. The Early Archaic fossil record is not extensive, although it includes some informative fossils such as those from Bodo (Ethiopia), Kabwe (Zambia, Figure 9), Ndutu (Tanzania), Saldanha (Namibia) and Salé (Morocco). The oldest of these, Bodo, dates to ~600 kya (Clark et al., 1994), indicating perhaps an earlier beginning of Archaic *Homo* in Africa than elsewhere. African Early Archaics are generally associated with the Early Stone Age.

Anatomically, the African Early Archaics exhibit a mixture of Erectine and African Late Archaic features (Cartmill and Smith, 2009; Pearson, 2013). Like Early Archaics from elsewhere, the African fossils have expanded upper braincases but with maximum breadth still down low near the base. Cranial buttressing (e.g., supraorbital and nuchal tori and cranial vault thickness) is extensive like in other Early Archaics. Interestingly, African Early Archaics exhibit midfacial prognathism that is not only greater than that of Erectines, but also greater than that of African Late Archaics (Figure 4). The anatomy of many postcranial fossils is modern-like, although some specimens show Erectine features (Pearson, 2013).

While African Early Archaic *Homo* are similarly robust as contemporary archaic populations elsewhere, African Late Archaic *Homo* appear decidedly more modern-like than contemporary Eurasian Late Archaics (Bräuer and Rimbach, 1990). The African Late Archaic fossil sample is not very extensive, with the most significant fossils coming from Eliye Springs (Kenya), Eyasi (Tanzania), Florisbad (South Africa), Ileret (Kenya), Jebel Irhoud (Morocco, Figure 9), and Ngaloba (Tanzania). Braincases in this group tend to be more globular and exhibit less robusticity than African Early Archaics, although the position of maximum cranial breadth ranges from low (e.g., Eliye Springs) to high on the parietals (e.g., Ngaloba). The facial skeleton also tends to be more modern-like than in contemporary Eurasian Archaics (Pearson, 2013). While some African Late Archaic crania exhibit true supraorbital tori (e.g., Jebel Irhoud 1), others (e.g., Florisbad) exhibit browridges that show a division into superciliary and lateral segments like most modern humans. Faces tend to be less prognathic than those of Eurasian Late Archaics and generally appear more modern-like. The mosaic of archaic and modern features among the African Late Archaic group has meant that some authors (Pearson, 2013; Smith, 1992) refer to them as "transitional." When archaeological context is known, African Late Archaics are associated with the Middle Stone Age.

The >160 kya fossils from Herto and Omo indicate that modern human anatomy appeared earliest in East Africa (Fleagle et al., 2003; White et al., 2003). A mosaic of archaic and modern features among some later fossils from other areas of Africa may indicate a demic diffusion of modern humans out of East Africa followed by admixture with resident Late Archaics in South and North Africa (Trinkaus, 2005; Smith et al., 2012).

West Asia

Few West Asian hominin fossils are known from before 200 kya. Given the presence of humans in far eastern Europe by 1.8 mya (Gabunia et al., 2001), early humans had likely colonized West Asia by then. However, the earliest fossil evidence for humans in the region comes from 'Ubeidiya (Israel) dated to 1.4 mya (Belmaker et al., 2002). Early Archaic *Homo* is represented by fragmentary fossils and isolated teeth from Qesem, Zuttiyeh (Figure 9), and Tabun Level E (all in Israel). The anterior craniofacial skeleton from Zuttiyeh comprises the best-preserved of the fossils. Its morphology is generally similar to that of other Early and Late Archaics from Africa and Europe, although it shares some affinities with early modern West Asians (Freidline et al., 2012). The teeth from Qesem are unusually small and show some anatomical similarities with West Asian early modern humans (Hershkovitz et al., 2011). West Asian Early Archaics are associated with the Lower Paleolithic.

In comparison with the Early Archaic *Homo* record, the West Asian fossil record of Late Archaic *Homo* is quite rich. Multiple partial skeletons are known from Amud (Israel), Dederiyeh (Syria), Kebara (Israel), Shanidar (Iraq) and Tabun (Israel). All of these remains are associated with the Middle Paleolithic. Additional and extensive human remains from the Middle Paleolithic sites of Qafzeh and Skhul either predate or overlap in time with West Asian Late Archaic *Homo*. However, the Qafzeh and Skhul fossils are generally viewed as those of early modern humans (Franciscus and Holliday, 2013). The West Asian Late Archaics are often classified as Neanderthals, although most if not all of the fossils lack at least some features common among European Neanderthals (Cartmill and Smith, 2009). This may indicate clinal variation among western Eurasian Late Archaics with far western populations being "hyper-Neandertal" and eastern populations being "hypo-Neandertal" (Voisin, 2006). Late Archaics persist in West Asia until at least 45 kya, with the oldest Upper Paleolithic-associated modern human remains dated to ~30 kya (Wolpoff, 1999).

South Asia

Archaic *Homo* is represented in South Asia by a single partial cranium from the Narmada Valley (Kennedy, 2000; Kennedy et al., 1991). Dating of this fossil is not clear, but it is likely from either the late Middle or early Late Pleistocene (MIS 5–7). Anatomically, braincase maximum breadth is low as in Erectines and many Early Archaics. Unlike Erectines, Narmada's occipital plane is larger than its nuchal plane, and the braincase is tall. Furthermore, its cranial capacity of 1260 cm^3 places it with other archaics. Associated with the Upper Paleolithic, modern humans appear by at least 38 kya in South Asia (Perera et al., 2011).

East Asia

Most of the East Asian hominin fossil record comes from China. Hominins appeared in East Asia by 1.7 mya. Fossils that could potentially be called Early Archaic *Homo* appear by ca. 200–600 kya as represented by those from Hexian, Yunxian, and Zhoukoudian H3 (Rosenberg and Wu, 2013). Although these fossils show some characteristics similar to other Archaic *Homo*, such as a broad upper braincase, they show their greatest similarities with East Asian Erectines such as those from Zhoukoudian locality 1 (Cartmill and Smith, 2009; Rosenberg and Wu, 2013).

The East Asian Late Archaic sample is more distinguishable from Erectines than is the small Early Archaic sample. Relatively well-preserved fossils from Dali, Jinniushan, Maba, and Nanjing all fall into this chronological grouping (~300–130 kya). The East Asian Late Archaic sample exhibits a great degree of anatomical variation. Some specimens (e.g., Dali) are similarly robust as Early Archaics from Africa and Europe. Others (e.g., Jinniushan) are very gracile with some aspects of anatomy being modern-like. Yet, still others (Nanjing and Salkhit) have Neanderthal-like features (Rosenberg and Wu, 2013). The broad, relatively flat faces seen in some of the Late Archaics, as well as other characteristics, may indicate evolutionary continuity between Late Archaics and modern humans in East Asia (Rosenberg and Wu, 2013). Anatomically modern humans appear by ~40 kya (Shang et al., 2007), although perhaps much earlier (Liu et al., 2010; Shen et al., 2002).

Australasia

Humans reached Australasia perhaps as early as 1.8 mya (Swisher et al., 1994). Up until the Late Pleistocene, humans appear to have been limited to the Asian side of Wallace's Line, although there is possible archaeological evidence of a much earlier colonization of Flores Island (Morwood et al., 1998). Most of the Archaic *Homo* fossil record comes from Java. Modern humans appear in Australasia by 30–50 kya.

The Australasian fossils that could possibly represent Archaic *Homo* are very poorly dated (Schwartz and Tattersall, 2003). Thus it is not possible to really distinguish between Early versus Late Archaic samples. Morphologically, some of the Archaic Javans (i.e., Sambungmacan) lie intermediate between earlier Erectines and the rest of the Archaic Javan sample (i.e., Ngandong and Ngawi). In general, the Javan Archaics exhibit the same fundamental characteristics as archaics from elsewhere, such as taller and broader braincases, larger cranial capacities, and greater cranial buttressing than Erectines. Differing from earlier Javan Erectines, the Javan Archaics share some features, such as straight supraorbital tori that are thickened laterally, flattened frontal bones, and prominent downward-pointing-at-the-midline nuchal tori (Wolpoff, 1999). While these features have been argued to be evolutionarily precedents to early modern Australasian anatomy (Thorne and Wolpoff, 1981), they may not actually be evidence of regional continuity (Durband and Westaway, 2013). Specifically, Durband and Westaway (2013) contend that cranial features seen in the Archaic Javan sample, such as supraorbital trigones, frontal/sagittal keeling, and a triangular-shaped projection at the midline of the nuchal tori are infrequent or perhaps absent among Pleistocene early modern Australians.

The remains of multiple individuals from Liang-Bua, Flores (Indonesia) have been assigned to their own species, *Homo floresiensis* (Brown et al., 2004). Remains attributed to this species are surprisingly recent (12–95 kya) (Morwood et al., 2005). Most of what is known of the anatomy of the Flores hominins is from a single partial skeleton, LB 1. LB 1 exhibits anatomy most similar to Erectines (Brown et al., 2004), although some details align it with even older hominins (Aiello, 2010). LB 1 stood only 1 m tall and had a cranial capacity of 380 cm^3, the smallest known for any hominin (Brown et al., 2004). Although the Flores remains are late in time, their primitive anatomy means that they are best not categorized as Archaic *Homo*.

GROWTH, DEVELOPMENT, AND ENERGETICS

Most of what is known about Archaic *Homo* growth, development, and energetics is from Neanderthals. This is due largely to the sheer size of the Neanderthal fossil sample, which includes many well-preserved individuals and many juveniles. Neanderthals and modern humans are different in some aspects of growth and development, although not appreciably more so than between the two species of chimpanzees, *Pan paniscus* and *Pan troglodytes* (Williams, 2013). Modern humans are characterized by a slowing down of facial growth resulting in small adult faces, compared with Archaic *Homo* (Williams, 2013).

The average estimated basal metabolic rate (BMR) differs significantly between archaic and modern humans ($t=2.437$, $p=0.02$). Early Archaics had higher BMR (male avg. = 1814 kcal/day, female avg. = 1349 kcal/day) than Late Archaics (male avg. = 1775 kcal/day, female avg. = 1430 kcal/day), and early modern humans had even lower energetic requirements (male avg. = 1652 kcal/day, female avg. = 1339 kcal/day) (Froehle et al., 2013).

LANGUAGE, BEHAVIOR, AND CULTURE

Generally, Early Archaic *Homo* fossils are associated with Lower Paleolithic ("Early Stone Age" in Africa) archaeological industries, while Late Archaics generally with Middle Paleolithic ("Middle Stone Age" in Africa) ones (see the chapter by Toth and Schick in this volume). Even though the earliest anatomically modern human fossils are also associated with the Middle Paleolithic, the complexity of the Upper Paleolithic is often used to contrast modern human behavior with that of archaics. The main technological difference between the Lower and Middle Paleolithic is in the number of distinctly different tool types, with the latter exhibiting more. The same applies to a comparison of the Middle and Upper Paleolithic, although the greater technological variation in the Upper Paleolithic is sometimes interpreted as stylistic and reflecting the development of ethnicities (cf., White et al., 1982). Stone tool manufacture also changes, especially from the Middle to Upper Paleolithic, with blade production being much more common in the latter period (see Toth and Schick).

Archaic humans had controlled and regular use of fire, which was essential to human occupation of temperate regions (Klein, 2009; Karkanas et al., 2007) and access to new food sources (Wrangham et al., 1999). They subsisted as foragers across a variety of environments and landscapes. There is clear evidence for cooperative hunting of large animals and dietary diversity not dissimilar to what is seen in the archaeological record of early modern humans (Stiner et al., 2009).

Symbolic thought is one of the key characteristics of living humans (Deacon, 1997). In the archaeological record, symbolic thought may be reflected in art and evidence of long-distance trade (Whallon, 1989). There is some evidence of symbolic behavior among Lower Paleolithic Early Archaic *Homo* (Bednarik, 1995, d'Errico and Nowell, 2000). For example, at Sima de los Huesos, all of the human remains may have accumulated because fellow humans were throwing bodies into the otherwise unused cave (Arsuaga et al., 1997). More evidence of symbolic behavior appears during the Middle Paleolithic of Eurasia and the Middle Stone Age of Africa. While some of this evidence has been linked to the earliest modern humans (Henshilwood et al., 2001), similar evidence is found in association with Neanderthals (d'Errico and Stringer, 2011; Zilhão et al., 2010). Although the general trend during human evolution of increasing evidence for symbolic behavior likely reflects some innate cognitive change, some of the differences among archaics and between archaics and modern humans may be explained better by differences in population density due to environmental variability (d'Errico and Stringer, 2011). As d'Errico and Stringer (2011) contend, evidence of complex symbolic behavior and adaptation to extreme environments fluoresces and then disappears at various times and among both archaic and early modern human populations.

Although archaeological evidence for symbolic thought implies symbolic communication or language, the limited symbolic evidence associated with Early Archaic *Homo* does not necessarily mean that these humans lacked language (but see the chapter by Leiberman). Fossil anatomy relevant to language production and processing might provide insights as to whether Archaic *Homo* had language. Much of the early anatomical work indicated that Neanderthals lacked the ability to speak the full range of modern human sounds (Lieberman, 1992). More recent evidence indicates that Archaic *Homo* had both modern-like speech (D'Anastasio et al., 2013; MartÌnez et al., 2008) and language-processing capabilities (Martínez et al., 2004, 2013). Paleogenomic evidence also supports Archaic *Homo* language capabilities (Ferretti, 2013).

EVOLUTIONARY RELATIONSHIPS AND TAXONOMY

Although some previous explanations for the evolution of the genus *Homo* interpreted all archaic humans as a phase in a unilineal process of human evolution (Wolpoff, 1980; Campbell, 1963), more recent interpretations have highlighted the distinctiveness and possibly unique evolutionary tendencies of at least some populations of archaic humans (Stringer et al., 1979; Tattersall, 1986, 1992). While the African fossil record (at least in East Africa) appears commensurate with a gradual evolution of Erectines into Archaic *Homo* and then into modern humans (by ~160 kya), the pattern of evolution during this time period in other regions is more ambiguous. The European (and possibly West Asian) Late Archaic *Homo* variant, the Neanderthals, seemingly evolved from earlier archaics in the region (Cartmill and Smith, 2009; Hublin, 1982; Arsuaga et al., 2014). However, the origins for these earlier European archaics are not as clear, with one possibility being a gradual evolution from the earliest Europeans (Erectines and/or Antecessors), and another possibility being a post-Erectine ancestor possibly from Africa that also gave rise to modern humans. Yet a third possibility is a combination of these two; i.e., there was an "out of Africa" event around 650 kya (Templeton, 2005) whereby Archaic *Homo* spread into Eurasia and intermixed with indigenous Erectines (Cartmill and Smith, 2009). The Australasian Archaics seem to be descended at least in part from Australasian Erectines. Less clear is the relationship between the Australasian Archaics and early modern humans. While some (Thorne and Wolpoff, 1981; Wolpoff et al., 1984) have argued for genetic continuity from the Javan Archaics to the early modern humans of Australia based on the persistence of regional anatomical features, others (Durband and Westaway, 2013; Habgood, 1989) do not find the evidence for continuity compelling. Similar disagreement can been seen in interpretations of the East Asian *Homo* fossil record, although a combination of continuity and extraregional gene flow seems to be most commensurate with the available evidence (Rosenberg and Wu, 2013).

In recent years, paleogenomic evidence has helped shed some light on the evolutionary relationships of archaic and modern humans. The overall pattern indicated by the ancient DNA (aDNA) evidence is that archaic and modern *Homo* admixed, perhaps throughout the settled Old World (Green et al., 2010; Reich et al., 2010, 2011). A hitherto unknown group of Archaic *Homo*, the Denisovans (and perhaps even another as yet unknown group (Callaway, 2013)), has been identified based solely on ancient DNA extracted from morphologically undiagnostic remains from Siberia (Reich et al., 2010; see chapters by Reletheford and Lao and Kayser in this volume). Interestingly, the mitochondrial DNA (mtDNA) from the Sima de los Huesos Early Archaics (Meyer et al., 2014) aligns them with Denisovans, although nuclear DNA, if successfully extracted, may demonstrate a more complex population history for European Early Archaics than the mtDNA evidence implies. Current evidence indicates that living humans carry both Neanderthal and Denisovan DNA, although the percentage of each varies by population (Green et al., 2010; Reich et al., 2011). Thus, the available paleogenomic evidence indicates complex population interactions inconsistent with a "many-species" model for Archaic *Homo* or a complete replacement scenario for the origin of modern humans (Hawks, 2013).

SUMMARY

Archaic *Homo* refers to the various fossil human populations that lie chronologically and anatomically intermediate between earlier Erectines and modern *H. sapiens*. The Archaic *Homo* group can be divided chronologically into Early Archaic *Homo* (~250–500 kya) and Late Archaic *Homo* (~30–250 kya). These chronological categories and even the term Archaic *Homo* itself should be regarded as heuristic.

While Early Archaics are generally associated with the Lower Paleolithic, Late Archaics are associated with the Middle Paleolithic. All archaics seemed to have had controlled use of fire, and this was important for the spread of hominins during the Middle and Late Pleistocene. Although there is some limited evidence for symbolic behavior among Early Archaics, clearer evidence of this is associated with Late Archaic *Homo*, especially in Africa and Europe. It is unclear whether Late Archaics had fully modern cognitive capabilities, given the fluorescence of complex behavior seen in the Upper Paleolithic associated with modern humans. However, environmental change leading to greater population densities for at least some early modern populations could explain increasing cultural complexity without evoking an innate cognitive change from the Late Archaic condition (d'Errico and Stringer, 2011).

In terms of temporal variation, archaic humans show an increase over time in terms of brainsize, stature, and body mass as well as dental changes. While Early Archaics from Africa and Western Eurasia are similar to each other, Late Archaics show greater regional differences. In Africa, Late Archaics appear much more similar to early modern humans than do Late Archaics of other regions. Although this is commensurate with an earliest appearance of modern humans in Africa, the persistence of regional Late Archaic anatomical and genomic characteristics into modern humans in Eurasia indicates a more complex evolutionary pattern than can be explained by a uniquely, recent African origin for modern humans.

REFERENCES

d'Anastasio, R., Wroe, S., Tuniz, C., Mancini, L., Cesana, D.T., Dreossi, D., Ravichandiran, M., Attard, M., Parr, W.C.H., Agur, A., Capasso, L., 2013. Micro-biomechanics of the kebara 2 hyoid and its implications for speech in Neanderthals. PLoS ONE 8, e82261.

Ahern, J.C.M., Janković, I., Voisin, J.L., Smith, F.H., 2013. Modern human origins in Central Europe. In: Smith, F.H., Ahern, J.C.M. (Eds.), The Origins of Modern Humans: Biology Reconsidered. Wiley-Blackwell, Hoboken, N.J.

Aiello, L.C., 2010. Five years of *Homo floresiensis*. Amercian Journal of Physical Anthropology 142, 167–179.

Arsuaga, J.L., Martínez, I., Arnold, L.J., Aranburu, A., Gracia-Téllez, A., Sharp, W.D., Quam, R.M., Falguères, C., Pantoja-Pérez, A., Bischoff, J., Poza-Rey, E., Parés, J.M., Carretero, J.M., Demuro, M., Lorenzo, C., Sala, N., Martinón-Torres, M., García, N., Alcázar De Velasco, A., Cuenca-Bescós, G., Gómez-Olivencia, A., Moreno, D., Pablos, A., Shen, C.-C., Rodríguez, L., Ortega, A.I., García, R., Bonmatí, A., Bermúdez De Castro, J.M., Carbonell, E., 2014. Neandertal roots: cranial and chronological evidence from Sima de los Huesos. Science 344, 1358–1363.

Arsuaga, J.L., Martínez, I., Gracia, A., Carretero, J.M., Lorenzo, C., García, N., Ortega, A.I., 1997. Sima de los Huesos (Sierra de Atapuerca, Spain). The site. Journal of Human Evolution 33, 109–127.

Bednarik, R.G., 1995. Concept-mediated marking in the lower Palaeolithic. Current Anthropology 36, 605–634.

Belmaker, M., Tchernov, E., Condemi, S., Bar-Yosef, O., 2002. New evidence for hominid presence in the lower Pleistocene of the Southern Levant. Journal of Human Evolution 43, 43–56.

Bermudez De Castro, J.M., 1997. A hominid from the lower Pleistocene of Atapuerca, Spain: possible ancestor to Neandertals and modern humans. Science 276, 1392–1395.

Bräuer, G., 1989. The evolution of modern humans: a comparison of the African and non-African evidence. In: Stringer, C.B., Mellars, P. (Eds.), The Human Revolution. Edinburgh University Press, Edinburgh.

Bräuer, G., Rimbach, K.W., 1990. Late archaic and modern *Homo* sapiens from Europe, Africa, and Southwest Asia: craniometric comparisons and phylogenetic implications. Journal of Human Evolution 19, 789–807.

Brown, P., Sutikna, T., Morwood, M., Soejono, R.P., Jatmiko, Saptomo, E.W., Due, R.A., 2004. A new small-bodied hominin from the Late Pleistocene of Flores, Indonesia. Nature 43, 1055–1061.

Callaway, E., 2013. Mystery humans spiced up ancients' sex lives. Nature 19. http://dx.doi.org/10.1038/nature.2013.14196.

Campbell, B., 1963. Quantitative taxonomy and human evolution. In: Washburn, S.L. (Ed.), Classification and Human Evolution. Aldine, Chicago.

Carbonell, E., Bermudez De Castro, J.M., Pares, J.M., Perez-Gonzalez, A., Cuenca-Bescos, G., Olle, A., Mosquera, M., Huguet, R., Van Der Made, J., Rosas, A., Sala, R., Vallverdu, J., Garcia, N., Granger, D.E., Martinon-Torres, M., Rodriguez, X.P., Stock, G.M., Verges, J.M., Allue, E., Burjachs, F., Caceres, I., Canals, A., Benito, A., Diez, C., Lozano, M., Mateos, A., Navazo, M., Rodriguez, J., Rosell, J., Arsuaga, J.L., 2008. The first hominin of Europe. Nature 452, 465–469.

Cartmill, M., Smith, F.H., 2009. The Human Lineage. Wiley-Blackwell, Hoboken, N.J.

Clark, J., De Heinzelin, J., Schick, K., Hart, W., White, T., Woldegabriel, G., Walter, R., Suwa, G., Asfaw, B., Vrba, E., Selassie, Y.H., 1994. African *Homo erectus*: old radiometric ages and young Oldowan assemblages in the Middle Awash valley, Ethiopia. Science 264, 1907–1910.

Deacon, T.W., 1997. The Symbolic Species: The Co-evolution of Language and the Brain. W.W. Norton and Co, New York.

Dediu, D., Levinson, S.C., 2014. The time frame of the emergence of modern language and its implications. In: Dor, D., Knight, C., Lewis, J. (Eds.), The Social Origins of Language. Oxford University Press, Oxford, pp. 184–195.

Delson, E., Harvati, K., Reddy, D., Marcus, L.F., Mowbray, K.M., Sawyer, G.J., Jacob, T., Márquez, S., 2001. The Sambungmacan 3 *Homo erectus* calvaria: a comparative morphometric and morphological analysis. Anatomical Record 262, 380–397.

Durband, A.C., Westaway, M.C., 2013. Perspectives on the origins of modern Australians. In: Smith, F.H., Ahern, J.C.M. (Eds.), The Origins of Modern Humans: Biology Reconsidered. Wiley-Blackwell, Hoboken, N.J.

d'Errico, F., Stringer, C.B., 2011. Evolution, revolution or saltation scenario for the emergence of modern cultures? Philosophical Transactions of the Royal Society B: Biological Sciences 366, 1060–1069.

d'Errico, F., Nowell, A., 2000. A new look at the Berekhat Ram figurine: implications for the origins of symbolism. Cambridge Archaeological Journal 10, 123–167.

Feldesman, M.R., Kleckner, J.G., Lundy, J.K., 1990. Femur/stature ratio and estimates of stature in Mid- and Late-Pleistocene fossil hominids. Amercian Journal of Physical Anthropology 83, 359–372.

Ferretti, F., 2013. Was there language outside Homo sapiens? A cognitive perspective. Journal of Anthropological Sciences = Rivista di antropologia: JASS/Istituto italiano di antropologia 91, 249–251.

Fleagle, J., Assefa, Z., Brown, F., Feibel, C., Mcdougall, I., Shea, J., 2003. The Omo I partial skeleton from the Kibish Formation. Amercian Journal of Physical Anthropology 36 (Suppl.), 95.

Franciscus, R.G., Holliday, T.W., 2013. Crossroads of the Old World: late hominin evolution in Western Asia. In: Smith, F.H., Ahern, J.C.M. (Eds.), The Origins of Modern Humans: Biology Reconsidered. Wiley-Blackwell, Hoboken, N.J.

Freidline, S.E., Gunz, P., Janković, I., Harvati, K., Hublin, J.J., 2012. A comprehensive morphometric analysis of the frontal and zygomatic bone of the Zuttiyeh fossil from Israel. Journal of Human Evolution 62, 225–241.

Froehle, A.W., Yokley, T.R., Churchill, S.E., 2013. Energetics and the origin of modern humans. In: Smith, F.H., Ahern, J.C.M. (Eds.), The Origins of Modern Humans: Biology Reconsidered. Wiley-Blackwell, Hoboken, N.J.

Gabunia, L., Anton, S., Lordkipanidze, D., Vekua, A., Justus, A., Iii, C.S., 2001. Dmanisi and dispersal. Evolutionary Anthropology 10, 158–170.

Gould, S.J., Eldredge, N., 1977. Punctuated equilibria: the tempo and mode of evolution reconsidered. Paleobiology 3, 115–151.

Green, R.E., Krause, J., Briggs, A.W., Maricic, T., Stenzel, U., Kircher, M., Patterson, N., Li, H., Zhai, W., Fritz, M.H.-Y., Hansen, N.F., Durand, E.Y., Malaspinas, A.-S., Jensen, J.D., Marques-Bonet, T., Alkan, C., Prufer, K., Meyer, M., Burbano, H.A., Good, J.M., Schultz, R., Aximu-Petri, A., Butthof, A., Hober, B., Hoffner, B., Siegemund, M., Weihmann, A., Nusbaum, C., Lander, E.S., Russ, C., Novod, N., Affourtit, J., Egholm, M., Verna, C., Rudan, P., Brajkovic, D., Kucan, Z., Gusic, I., Doronichev, V.B., Golovanova, L.V., Lalueza-Fox, C., De La Rasilla, M., Fortea, J., Rosas, A., Schmitz, R.W., Johnson, P.L.F., Eichler, E.E., Falush, D., Birney, E., Mullikin, J.C., Slatkin, M., Nielsen, R., Kelso, J., Lachmann, M., Reich, D., Paabo, S., 2010. A draft sequence of the Neandertal genome. Science 328, 710–722.

Habgood, P.J., 1989. An examination of regional features on Middle and early Late Pleistocene sub-Saharaan African hominids. South African Journal of Science 44, 17–22.

Hawks, J., 2013. The relevance of archaic genomes to modern human origins. In: Smith, F.H., Ahern, J.C.M. (Eds.), The Origins of Modern Humans: Biology Reconsidered. Wiley-Blackwell, Hoboken, N.J.

Henshilwood, C.S., D'errico, F., Marean, C.W., Milo, R.G., Yates, R., 2001. An early bone tool industry from the Middle Stone Age at Blombos Cave, South Africa: implications for the origins of modern human behaviour, symbolism and language. Journal of Human Evolution 41, 631–678.

Hershkovitz, I., Smith, P., Sarig, R., Quam, R., Rodríguez, L., García, R., Arsuaga, J.L., Barkai, R., Gopher, A., 2011. Middle Pleistocene dental remains from Qesem Cave (Israel). Amercian Journal of Physical Anthropology 144, 575–592.

Higham, T., Douka, K., Wood, R., Ramsey, C.B., Brock, F., Basell, L., Camps, M., Arrizabalaga, A., Baena, J., Barroso-Ruiz, C., Bergman, C., Boitard, C., Boscato, P., Caparros, M., Conard, N.J., Draily, C., Froment, A., Galvan, B., Gambassini, P., Garcia-Moreno, A., Grimaldi, S., Haesaerts, P., Holt, B., Iriarte-Chiapusso, M.-J., Jelinek, A., Jorda Pardo, J.F., Maillo-Fernandez, J.-M., Marom, A., Maroto, J., Menendez, M., Metz, L., Morin, E., Moroni, A., Negrino, F., Panagopoulou, E., Peresani, M., Pirson, S., De La Rasilla, M., Riel-Salvatore, J., Ronchitelli, A., Santamaria, D., Semal, P., Slimak, L., Soler, J., Soler, N., Villaluenga, A., Pinhasi, R., Jacobi, R., 2014. The timing and spatiotemporal patterning of Neanderthal disappearance. Nature 512, 306–309.

Higham, T., Ramsey, C.B., Karavanić, I., Smith, F.H., Trinkaus, E., 2006. Revised direct radiocarbon dating of the Vindija Upper Paleolithic Neandertals. Proceedings of the National Academy of Sciences USA. 103, 553–557.

Holloway, R.L., Broadfield, D.C., Yuan, M.S., 2004. The human fossil record. Brain Endocasts—the Paleoneurological Evidence, vol. 3, Wiley-Liss, Hoboken, N.J.

Hublin, J.-J., 1982. Les Anténéandertaliens: Presapiens ou Prénéandertaliens. Geobios, mémoire spécial 6, 345–357.

Hublin, J.-J., 2013. The makers of the upper paleolithic in western eurasia. In: Smith, F.H., Ahern, J.C.M. (Eds.), The Origins of Modern Humans: Biology Reconsidered. Wiley-Blackwell, Hoboken, N.J.

Hyodo, M., Nakaya, H., Urabe, A., Saegusa, H., Shunrong, X., Jiyun, Y., Xuepin, J., 2002. Paleomagnetic cates of hominid remains from Yuanmou, China, and other Asian sites. Journal of Human Evolution 43, 27–41.

Karkanas, P., Shahack-Gross, R., Ayalon, A., Bar-Matthews, M., Barkai, R., Frumkin, A., Gopher, A., Stiner, M.C., 2007. Evidence for habitual use of fire at the end of the Lower Paleolithic: site-formation processes at Qesem Cave, Israel. Journal of Human Evolution 53, 197.

Kennedy, K.A.R., 2000. Middle and late Pleistocene hominids of south Asia. In: Tobias, P.V., Raath, M., Moggi-Cecchi, J., Doyle, G. (Eds.), Humanity from African Naissance to Coming Millennia. Florence University Press, Firenze.

Kennedy, K.A.R., Chiment, J., Verma, K.K., Sonakia, A., 1991. Is the Narmada hominid an Indian *Homo erectus*? Amercian Journal of Physical Anthropology 86, 475–496.

Klein, R.G., 2009. The Human Career: Human Biological and Cultural Origins. The University of Chicago Press, Chicago.

Lee, S.-H., Wolpoff, M.H., 2003. The pattern of evolution in Pleistocene human brain size. Paleobiology 29, 186–196.

Lieberman, P., 1992. On Neanderthal speech and Neanderthal extinction. Current Anthropology 33, 409–410.

Liu, W., Jin, C.-Z., Zhang, Y.-Q., Cai, Y.-J., Xing, S., Wu, X.-J., Cheng, H., Edwards, R.L., Pan, W.-S., Qin, D.-G., An, Z.-S., Trinkaus, E., Wu, X.-Z., 2010. Human remains from Zhirendong, South China, and modern human emergence in East Asia. Proceedings of the National Academy of Sciences 107, 19201–19206.

Martìnez, I., Arsuaga, J.L., Quam, R., Carretero, J.M., Gracia, A., Rodrìguez, L., 2008. Human hyoid bones from the middle Pleistocene site of the Sima de los Huesos (Sierra de Atapuerca, Spain). Journal of Human Evolution 54, 118–124.

Martínez, I., Rosa, M., Arsuaga, J.-L., Jarabo, P., Quam, R., Lorenzo, C., Gracia, A., Carretero, J.-M., De Castro, J.-M.B., Carbonell, E., 2004. Auditory capacities in Middle Pleistocene humans from the Sierra de Atapuerca in Spain. Proceedings of the National Academy of Sciences USA 101, 9976–9981.

Martínez, I., Rosa, M., Quam, R., Jarabo, P., Lorenzo, C., Bonmatí, A., Gómez-Olivencia, A., Gracia, A., Arsuaga, J.L., 2013. Communicative capacities in Middle Pleistocene humans from the Sierra de Atapuerca in Spain. Quaternary International 295, 94–101.

Mcdougall, I., Brown, F.H., Fleagle, J.G., 2008. Sapropels and the age of hominins Omo I and II, Kibish, Ethiopia. Journal of Human Evolution 55, 409–420.

Meyer, M., Fu, Q., Aximu-Petri, A., Glocke, I., Nickel, B., Arsuaga, J.-L., Martinez, I., Gracia, A., De Castro, J.M.B., Carbonell, E., Paabo, S., 2014. A mitochondrial genome sequence of a hominin from Sima de los Huesos. Nature 505, 403–406.

Morwood, M.J., Brown, P., Jatmiko, Sutikna, T., Wahyu Saptomo, E., Westaway, K.E., Awe Due, R., Roberts, R.G., Maeda, T., Wasisto, S., Djubiantono, T., 2005. Further evidence for small-bodied hominins from the Late Pleistocene of Flores, Indonesia. Nature 437, 1012–1017.

Morwood, M.J., O'sullivan, P.B., Aziz, F., Raza, A., 1998. Fission-track ages of stone tools and fossils on the East Indonesian island of Flores. Nature 392, 173–176.

Pearson, O.M., 2013. Africa: the cradle of modern people. In: Smith, F.H., Ahern, J.C.M. (Eds.), The Origins of Modern Humans: Biology Reconsidered. Wiley-Blackwell, Hoboken, N.J.

Perera, N., Kourampas, N., Simpson, I.A., Deraniyagala, S.U., Bulbeck, D., Kamminga, J., Perera, J., Fuller, D.Q., Szabó, K., Oliveira, N.V., 2011. People of the ancient rainforest: Late Pleistocene foragers at the Batadomba-lena rockshelter, Sri Lanka. Journal of Human Evolution 61, 254–269.

Reich, D., Green, R.E., Kircher, M., Krause, J., Patterson, N., Durand, E.Y., Viola, B., Briggs, A.W., Stenzel, U., Johnson, P.L.F., Maricic, T., Good, J.M., Marques-Bonet, T., Alkan, C., Fu, Q., Mallick, S., Li, H., Meyer, M., Eichler, E.E., Stoneking, M., Richards, M., Talamo, S., Shunkov, M.V., Derevianko, A.P., Hublin, J.-J., Kelso, J., Slatkin, M., Pääbo, S., 2010. Genetic history of an archaic hominin group from Denisova Cave in Siberia. Nature 468, 1053–1060.

Reich, D., Patterson, N., Kircher, M., Delfin, F., Nandineni, M.R., Pugach, I., Ko, A.M., Ko, Y.C., Jinam, T.A., Phipps, M.E., Saitou, N., Wollstein, A., Kayser, M., Pääbo, S., Stoneking, M., 2011. Denisova admixture and the first modern human dispersals into Southeast Asia and Oceania. American Journal of Human Genetics 89, 516–528.

Rosenberg, K., Wu, X., 2013. A river runs through it: modern human origins in East Asia. In: Smith, F.H., Ahern, J.C.M. (Eds.), The Origins of Modern Humans: Biology Reconsidered. Wiley-Blackwell, Hoboken, N.J.

Schwartz, J., Tattersall, I., 2003. The Human Fossil Record. The Craniodental Morphology of the Genus Homo (Africa and Asia), vol. 2, Wiley Liss, New York.

Shang, H., Tong, H., Zhang, S., Chen, F., Trinkaus, E., 2007. An early modern human from Tianyuan Cave, Zhoukoudian, China. Proceedings of the National Academy of Sciences USA 104, 6573–6578.

Shen, G., Wang, W., Wang, Q., Zhao, J., Collerson, K., Zhou, C., Tobias, P.V., 2002. U-Series dating of Liujiang hominid site in Guangxi, Southern China. Journal of Human Evolution 43, 817–829.

Smith, F.H., 1992. Models and realitities in modern human origins: the African fossil evidence. Philosophical Transactions of the Royal Society of London 337, 243–250.

Smith, F.H., Hutchinson, V., Janković, I., 2012. Assimilation and modern human origins in the African peripheries. In: Reynolds, S.C., Gallagher, A. (Eds.), African Genesis: Perspectives on Hominin Evolution. Cambridge University Press, Cambridge.

Stiner, M.C., Barkai, R., Gopher, A., 2009. Cooperative hunting and meat sharing 400-200 kya at Qesem Cave, Israel. Proceedings of the National Academy of Sciences USA 106, 13207–13212.

Stringer, C.B., 2012. What makes a modern human. Nature 485, 33–35.

Stringer, C.B., Andrews, P., 1988. Genetic and fossil evidence for the origin of modern humans. Science 239, 1263–1268.

Stringer, C.B., Howell, F.C., Melentis, J.K., 1979. The significance of the fossil hominid skull from Petralona, Greece. Journal of Archaeological Science 6, 235–253.

Swisher, C.C., Curtis, G.H., Jacob, T., Getty, A.G., Suprijo, A., Widiasmoro, 1994. Age of the earliest known hominids in Java, Indonesia. Science 263, 1118–1121.

Tattersall, I., 1986. Species recognition in human paleontology. Journal of Human Evolution 15, 165–175.

Tattersall, I., 1992. Species concepts and species identification in human evolution. Journal of Human Evolution 22, 341–349.

Templeton, A.R., 2005. Haplotype Trees and modern human origins. Amercian Journal of Physical Anthropology 128, 33–59.

Thorne, A.G., Wolpoff, M.H., 1981. Regional continuity in Australasian Pleistocene hominid evolution. Amercian Journal of Physical Anthropology 55, 337–349.

Trinkaus, E., 1987. The Neandertal face: evolutionary and functional perspectives on a recent hominid face. Journal of Human Evolution 16, 429–443.

Trinkaus, E., 2005. Early modern humans. Annual Review of Anthropology 34, 207–230.

Voisin, J.-L., 2006. Speciation by distance and timing overlap: a new way of understanding Neanderthal evolution. In: Harvati, K., Harrison, T. (Eds.), Neanderthals Revisited: New Approaches and Perspectives. Springer, New York.

Weidenreich, F., 1951. Morphology of Solo man. Anthropological Papers of the American Museum of Natural History 43, 207–290.

Whallon, R., 1989. Elements of cultural change in the later Paleolithic. In: Mellars, P., Stringer, C.B. (Eds.), The Human Revolution. Edinburgh University Press, Edinburgh.

White, R., Arts, N., Bahn, P.G., Binford, L.R., Dewez, M., Dibble, H.L., Fish, P.R., Gamble, C., Meiklejohn, C., Ohel, M.Y., Pfeiffer, J., Straus, L.G., Weber, T., 1982. Rethinking the Middle/Upper Paleolithic transition (and comments and replies). Current Anthropology 23, 169–192.

White, T.D., Asfaw, B., Degusta, D., Gilbert, H., Richards, G.D., Suwas, G., Howell, F.C., 2003. Pleistocene *Homo sapiens* from middle Awash, Ethiopia. Nature 423, 742–747.

Williams, F.L., 2013. Neandertal craniofacial growth and development and its relevance for modern human origins. In: Smith, F.H., Ahern, J.C.M. (Eds.), The Origins of Modern Humans: Biology Reconsidered. Wiley-Blackwell, Hoboken, N.J.

Wolpoff, M., 1999. Paleoanthropology. McGraw-Hill, New York.

Wolpoff, M.H., 1980. Paleoanthropology. Knopf, New York.

Wolpoff, M.H., Wu, X., Thorne, A., 1984. Modern *Homo sapiens* origins: a general theory of hominid evolution involving the fossil evidence from East Asia. In: Smith, F.H., Spencer, F. (Eds.), The Origins of Modern Humans: A World Survey of the Fossil Evidence. Wiley-Liss, New York.

Wrangham, R.W., Jones, J.H., Laden, G., Pilbeam, D., Conklin, Äêbrittain, N., 1999. The raw and the stolen: cooking and the ecology of human origins. Current Anthropology 40, 567–594.

Zeitoun, V., Widianto, H., 2001. Phylogeny of the skull of Ngawi (Java, Indonesia). Journal of Human Evolution 40, A25.

Zhu, R.X., Potts, R., Pan, Y.X., Yao, H.T., Lü, L.Q., Zhao, X., Gao, X., Chen, L.W., Gao, F., Deng, C.L., 2008. Early evidence of the genus *Homo* in East Asia. Journal of Human Evolution 55, 1075–1085.

Zilhão, J., Angelucci, D.E., Badal-García, E., d'Errico, F., Daniel, F., Dayet, L., Douka, K., Higham, T.F.G., Martínez-Sanchez, M.J., Montes-Bernardez, R., Murcia-Mascarés, S., Pérez-Sirvent, C., Roldán-García, C., Vanhaeren, M., Villaverde, V., Wood, R., Zapata, J., 2010. Symbolic use of marine shells and mineral pigments by Iberian Neandertals. Proceedings of the National Academy of Sciences 107, 1023–1028.

Chapter 13

Anatomically Modern *Homo sapiens*

Brigitte M. Holt
Department of Anthropology, University of Massachusetts, Amherst, MA, USA

SYNOPSIS

Anatomically modern humans have relatively gracile skeletons and teeth, with crania exhibiting characteristic traits such as a vertical forehead, expanded parietal bones, a nonprojecting face, and the presence of a chin. Some of the distinctive characteristics of modern human behavior include more efficient hunting technology including prismatic blades and tools made of bone, ivory, and antler, broadening of diet, long-distance raw material procurement networks, use of symbolic elements such as ocher and personal decorations, and ceremonial burials. Genetic and fossil evidence shows that anatomically modern humans evolved in Africa around 200,000 years ago, and expanded out of Africa around 100,000 years ago. Analysis of fossil DNA shows that modern humans probably interbred with some local archaic populations, such as Neandertals.

HISTORY OF DISCOVERY

The first anatomically modern human skeleton was discovered in 1822–1823 at Goat's Hole Cave (Paviland) in England, by Rev. William Buckland. Buckland, who was also a professor of geology at Oxford, concluded that the Paviland skeleton was that of a Roman period woman, despite the presence of red ochre and mammoth ivory objects (Klein, 1999). In 1868, railway workers uncovered the bones of five to eight anatomically modern humans in the Cro-Magnon shelter in the southwest France village of Les Eyzies (Dordogne). French prehistorian Louis Lartet's subsequent recovery of associated stone tools and the bones of extinct fauna confirmed the antiquity of these remains. These finds represent the first well-accepted discovery of ancient anatomically modern humans, and the term Cro-Magnon is often treated as equivalent to "modern human." Numerous important discoveries of early anatomically modern humans made subsequently throughout the later part of the nineteenth and beginning of the twentieth century confirmed further the deep antiquity of humans.

EARLIEST FOSSIL EVIDENCE OF MODERN HUMANS

All fossils of modern humans older than 110,000 years ago (kya) are found in sub-Saharan Africa (Cartmill and Smith, 2009 and references therein; Table 1 and references therein). The earliest of these, Omo 1 and Herto 1 (Figure 1), both from Ethiopia, date to around 195 and 150 kya, respectively. Another handful of remains from South Africa (Klasies River Mouth), Sudan (Singa), and Tanzania (Mumba) range from about 150 to 100 kya. The earliest evidence of modern humans out of Africa comes from two sites in Israel, Skhul and Qafzeh. Dated between 90 and 110 kya, these two cave sites have yielded the largest early modern human sample known to date.

Modern humans appear to have made their way into Asia by around 63 kya (Cartmill and Smith, 2009). Preservation and depositional factors have long made accurate and precise dating of Asian fossils difficult. However, several recently excavated sites have yielded secure dates. Dates for the site of Tam Pa Ling in Laos range from around 46 kya to a maximum of 63 kya, followed by Ust'-Ishim in western Siberia, Tianyuandong in South China, and Fa Hein in Sri Lanka, at around 45, 40, and 36 kya, respectively. Other possible candidates for early modern humans in Asia and Australia are generally considered unreliable. A date of 100 kya for the site of Zhirendong in southern China appears to be secure, but the remains are too fragmentary for proper morphological evaluation. Between 65 and 45 kya, glacial conditions lowered sea levels by about 100 m, permitting modern humans to colonize the islands of the coast of Southeast Asia and Australia, an area known as Australasia. A modern human cranium recovered in 1958 on Borneo, the so-called "deep skull of Niah," is dated 35,200 kya. While dates for the earliest modern humans in Australia have long been controversial, several recent direct U-series and ESR dates for a skeleton from Lake Mungo (Willandra Lake) yielded an estimate of 62 kya. This date matches OSL dates taken from sediments associated with the burial.

Reconstructing the spread of modern humans into Europe remains one of the most hotly debated issues in paleoanthropology (Mellars, 2006; Benazzi et al., 2011; Higham et al., 2014). Reliably dated early modern human remains are scarce and often too fragmentary to be accurately diagnosed. In addition, the sample size of early modern humans in Europe has recently decreased as a result of improved radiocarbon dating techniques. This includes, ironically, the remains from Cro-Magnon that had stood as representative of early modern humans since their discovery in 1868. Previously believed to be around 40 kya old, Cro-Magnon has been redated to 27,680 kya. A handful of sites provide clear evidence of anatomically modern human remains in Europe by at least 35 kya. In Romania, a mandible and a partial cranium, Oase 1 and 2, respectively, from the site of Peştera cu Oase represent the oldest known modern human remains from Europe. In the Czech Republic, numerous skeletal elements were recovered from the cave complex at Mladeč, including three almost complete adult crania and a child's skull. Mladeč 1 and 2 are both dated directly at around 31 kya. Next, two crania from Romania, Muierii 1 and Cioclovina 1, date to between 30 and 29 kya.

These dates would suggest that modern humans reached the extreme southern portions of Asia at least 15,000 years before extending into western Eurasia. Such a late dispersal into Europe is difficult to reconcile with the overwhelming evidence of modern human behavior in Africa and the Near East by at least 100 kya (Henshilwood et al., 2004; Vanhaeren et al., 2006; Cartmill and Smith, 2009). Archeological evidence, in fact, suggests a much earlier arrival of modern humans into the eastern portion of Europe (Bar-Yosef, 2002; Hoffecker, 2009). Dates for the Aurignacian (see the chapter by Toth and Schick in this volume), an Upper Paleolithic industry generally considered a marker and proxy for the appearance of modern humans in Europe, are older than 35 kya (Hoffecker, 2009). Dates for industries falling into a category called Initial Upper Paleolithic (IUP), such as the South-Central European Bohunician and Bacho-Kirian, range between 46 and 43 kya. Technological similarities between these assemblages and other IUP industries from sites in the Near East suggest that these early Upper Paleolithic assemblages may document the first wave of modern human movement into Europe (Mellars, 2006; Hoffecker, 2009). A second wave may be represented

TABLE 1 Some Anatomically Modern Human Specimens[a]

Specimen	Location	Age	Dating Method	References
Omo Kibbish 1	East Africa	≤195,000, ≥150,000	40Ar/39Ar, paleoclimatic and stratigraphic correlation	McDougall et al. (2005)
Herto	East Africa	ca. 150,000–160,000	40Ar/39Ar, stratigraphic correlation	Clark et al. (2003)
Singa	East Africa	ca. 140,000–150,000	ESR, U-series	McDermott et al. (1996)
Mumba	East Africa	ca. 110,000–130,000	U-series	Bräuer and Mehlman (1988)
Klasies River, Mouth LBS	South Africa	ca. 100,000–120,000	OSL, U-series	Vogel (2001) and Feathers (2002)
Qafzeh	Southwest Asia	ca. 90,000–100,000	TL, ESR, U-series	Schwarcz et al. (1988) and McDermott et al. (1993)
Skhul	Southwest Asia	ca. 90,000–110,000	ESR, TL, U-series	Stringer et al. (1989), McDermott et al. (1993) and Mercier et al. (1993)
Aduma	East Africa	ca. 80,000–105,000	40Ar/39Ar, U-series	Haile-Selassie et al. (2004)
Bouri	East Africa	ca. 80,000–105,000	Stratigraphic correlation with Aduma	Haile-Selassie et al. (2004)
Klasies River, Mouth SAS	South Africa	ca. 65,000–90,000	OSL, U-series	Vogel (2001) and Feathers (2002)
Tam Pa Ling	Southeast Asia	63.600 ± 6000	U/Th	Demeter et al. (2012)
Lake Mungo (Willandra Lake Hominin 3)	Southeast Asia (Australia)	62,000 ± 2.6 61,000 ± 6	Direct ESR and U-series date, Indirect OSL date	Thorne et al. (1999)
Die Kelders 1	South Africa	ca. 60,000–70,000	OSL, ESR	Feathers and Bush (2000) and Schwarcz and Rink (2000)
Blombos	South Africa	ca. 65,000–70,000	Archeological correlation	Henshilwood et al. (2001)
Taramsa Hill	North Africa	ca. 50,000–80,000	OSL	Vermeersch et al. (1998)
Ust'-Ishim	Western Siberia	46,880–43,210	Direct AMS radiocarbon date	Fu et al. (2014)
Tianyuandong	Northeast Asia	42,000–39,000	Direct AMS radiocarbon date	Shang et al. (2007)
Nazlet Khater 1	North Africa	37,570 +350,–310	Direct AMS radiocarbon date	Vermeersch (2002)
Ksar 'Akil	Southwest Asia	37,000	Archeological correlation	Bergman and Stringer (1989)
Fa Hein	Southeast Asia	ca. 36,000		Kennedy et al. (1987)
Niah Cave	Southeast Asia	35,200 ± 2.6	U-series	Barker et al. (2007)
Peştera cu Oase 1	Eastern Europe	34,290, +970,–870	Direct AMS radiocarbon date	Trinkaus et al. (2003)
Mladeč 1-9	Central Europe	30,680 +380, –360 to 31,500 +420, –400	Direct AMS radiocarbon date	Wild et al. (2005)
Muierii 1	Eastern Europe	30,150 ± 800	Direct AMS radiocarbon date	Păunescu (2001)
Zhoukoudian Upper Cave	Northeast Asia	ca. 34,000–10,000	U-series, TL, AMS radiocarbon dates	Norton and Xing Gao (2008)
Cioclovina 1	Eastern Europe	29,000 ± 700	Direct AMS radiocarbon date	Păunescu (2001)
Cro-Magnon 1	Southwest Europe	27,680 ± 270	Direct AMS radiocarbon date	Henry-Gambier (2002)
Dolní Vestonicĕ 13–15	Central Europe	26,640 ± 110	Direct AMS radiocarbon date	Svoboda et al. (1996)

Continued

TABLE 1 Some Anatomically Modern Human Specimens[a]—cont'd

Specimen	Location	Age	Dating Method	References
Moh Khiew 1	Southeast Asia	25,800±600	AMS radiocarbon date	Matsumura and Pookajorn (2005)
Paviland 1	Northwest Europe	26,350±550, 25,840±280	Direct AMS radiocarbon date	Pettitt (2000)
Kostenki 4	Eastern Europe	21,020±180	Direct AMS radiocarbon date	Richards et al. (2001)

[a]*Specimens are listed from oldest to most recent.*

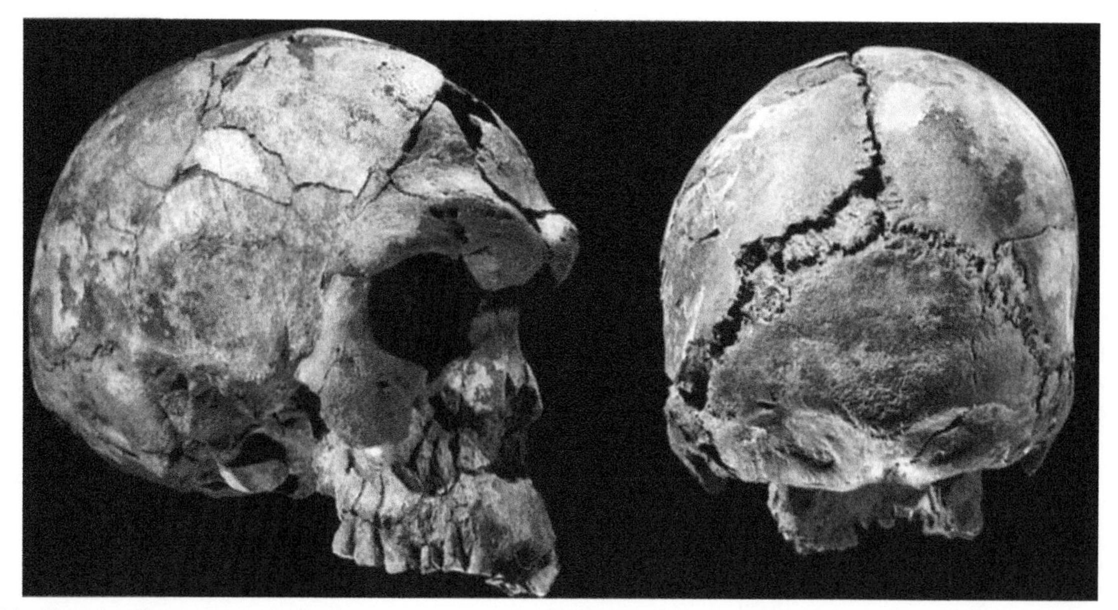

FIGURE 1 Herto, lateral and posterior views. This fossil cranium from Ethiopia dates to around 160kya and is one of the earliest anatomically modern human fossils known. Note the heavy brow ridges (an archaic trait) and the short and high cranium with vertical sides (anatomically modern traits). *From White et al. (2003); courtesy of Nature Publishing Group.*

by the Proto-Aurignacian, another early Upper Paleolithic industry found primarily in Southwest Europe (Italy, Bulgaria), and dating to at least 37 kya. The Proto-Aurignacian closely resembles the Ahmarian industry in the Near East (Hoffecker, 2009). At the Lebanese site of Ksar Akil (layer XVII), the Ahmarian is associated with a modern human skeleton dated at about 37 kya (Bar-Yosef, 2000). This supports a growing consensus that modern humans dispersed into Europe from the Near East at least by 50 kya, following the Mediterranean coastline (Bar-Yosef, 2002; van Andel et al., 2003; Mellars, 2006; Müller et al., 2011).

CLIMATIC CONDITIONS FOR THE SPREAD OF MODERN HUMANS

Deep-sea sediment and ice cores taken from Indian and Pacific Oceans, Antarctica, and Greenland provide detailed records of past global climates. Variation in the content of oxygen isotopes in these cores reflects the growth and decline of ice sheets. The isotope stages are referred to as Oxygen Isotope Stages or OIS. Integration of these climatic archives with various dating methods, such as ^{14}C and paleomagnetism, has produced a detailed account of climatic changes over the last 900,000 years. The appearance of modern humans in Africa and their expansion into Asia and Europe span OIS 6-3. OIS 6 (186–127 kya) corresponds to the penultimate glacial period (Cartmill and Smith, 2009). The subsequent long interglacial stage, OIS 5, comprises a series of warm and cold oscillations between 127 and 71 kya. A much shorter OIS 4 (71–57 kya) represents the beginning of the Last Glaciation and the first of two glacial maxima. OIS 3, a period central to the spread of modern humans, lasted from 57 to 24 kya. OIS 3 is characterized by an initial and brief warmer period, followed by a series of rapid and marked climatic oscillations.

ORIGINS OF MODERN HUMANS

The debate around modern human origins has long revolved around two extreme models: the Multiregional (MRM) and the Recent African Origin (RAO) models. The MRM proposes that, following expansion throughout the Old World around 2 mya, *Homo erectus* evolved into anatomically modern humans in multiple regions. Continuous gene flow between adjacent regions prevented extensive genetic divergence among African, European, and Asian populations. In contrast, the RAO argues that modern humans evolved in Africa after the initial spread of *H. erectus*. Equipped with better cognitive and technological abilities, as a new species, modern humans then expanded into Asia and Europe and replaced local archaic populations, such as Neandertals, around 100 kya. With new fossil, archeological and genetic data, and more precise dating techniques, these two models have been rejected by most as too simplistic, in favor of more nuanced and realistic intermediate models. The Assimilation model, for instance, accepts a single origin of modern humans in Africa, but argues that, rather than being completely replaced, some indigenous archaic populations interacted and interbred with modern humans (Cartmill and Smith, 2009). Mitochondrial and Y chromosome genetic data from present-day populations showing highest diversity in East and South African populations overwhelmingly support a single origin in Africa for modern humans, probably between 200 and 100 kya (Jobling et al., 2014). Analysis of complete mitochondrial DNA (mtDNA) sequences from six Neandertals reveals a substantial number of differences between Neandertal and modern humans. While the modern human sample varies by about 60 mutations, the Neandertal and modern human sample differs by about 200 mutations (Jobling et al., 2014). The ancient DNA data have also shown, however, good evidence that Neandertals contributed to the modern human gene pool (see chapters by Relethford, and Lao and Layser).

RELATIONSHIPS WITH ARCHAIC *HOMO*

Current evidence suggests that Neandertals and modern humans shared a common ancestor around 350 kya. The fossil record of separate Neandertal and modern human lineages emerge by at least 160 kya, and a number of African fossils dating to between 250 and 160 kya evince some aspects of modern human morphology, especially in the face. For instance, crania from the sites of Florisbad (South Africa), Jebel Irhoud (Morocco), and Ngaloba (Tanzania) have much lower nasal and upper facial height than Neandertals, and some exhibit traits typically associated with modern humans such as a canine fossa, reduced facial prognathism, and angled zygomatic bones. By 130 kya, geographic barriers produced by climate fluctuations appear to have isolated Neandertal populations in Europe from modern human populations to the south. Given that Neandertals were around until about 35 kya (see chapter by Ahern), they clearly must have come in contact with expanding modern human groups. The nature of that interaction has long been a source of heated debate. Recent genetic data show clear evidence of admixture between Neandertals and modern humans. Morphological data suggest, however, that the genetic admixture was relatively small.

TABLE 2 Some Cranial and Postcranial Features of Anatomically Modern Humans

Cranial
Large cranial capacity (recent human mean = $1349 \pm 77.5\ cm^3$
High, vertical forehead
Lateral parietal expansion, resulting in "house shape" posterior cranial outline
Orthognathic (nonprojecting) face (tucked below anterior edge of braincase)
Presence of canine fossa
Weak and arched supraorbital structures
Projecting mental eminence (chin)
Absence of retromolar space
Occipital lacking suprainiac fossa
Large and projecting mastoid processes
Postcranial
Relatively gracile long bones (including generally reduced muscular attachment scars and cortical thickness)
Presence of pilaster on posterior aspect of the femur

PALEOBIOLOGY OF ANATOMICALLY MODERN *HOMO SAPIENS*

Cranial Morphology

While most paleoanthropologists would agree that no concise definition encompasses the range of variability characterizing current and extant anatomically "modern" humans (AMH), the features listed in Table 2 are widely considered to distinguish "modern humans" from other hominins. Relative to archaic members of the genus *Homo*, such as *H. erectus* and Neandertals, AMH crania exhibit distinctively expanded parietal outlines and a steep vertical forehead, a short and high cranium, an orthognathic face, the presence of a chin, and general reduction of cranial thickness and superstructures such as brow ridges (Lahr, 1996; Lieberman, 1996; Cartmill and Smith, 2009; Figure 2). The etiology of modern human features is complex and evades unicausal explanations. As with other phenotypic traits, skeletal morphology results from the interaction of genetic and environmental influences. Comparative analyses of Neandertal and AMH ontogenetic patterns show that, while both groups follow essentially similar postnatal development trajectories, AMH exhibits reduced rates of cranial growth and shape change, differences

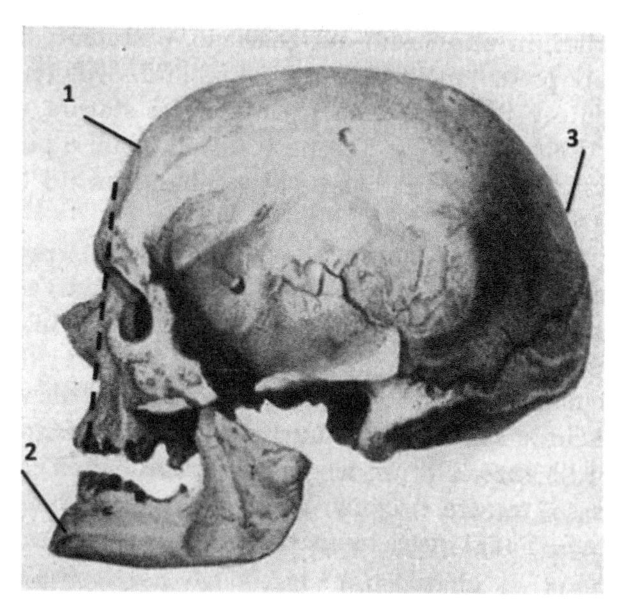

FIGURE 2 Cro-Magnon 1 (lateral view). This skull was discovered during the nineteenth century in the rockshelter of Cro-Magnon, Dordogne (France). Note the distinctive modern features: (1) steep and vertical forehead, (2) projecting mental eminence or chin, (3) rounded cranial contours, and face mostly positioned posterior to the anterior edge of the braincase (indicated by the dashed line). *Adapted from De Quatrefages and Hamy (1882).*

that may explain, for instance, why modern humans have a shorter face and cranial length than Neandertals (Ponce de Leon and Zollikofer, 2008). A recent alternative explanation invokes decreases in androgen testosterone in early anatomically modern humans as the cause of cranio-facial changes in robusticity and shape (Cieri et al., 2014). According to this model, higher population density evident in late Middle Stone Age and early Late Stone Age (LSA) would have required more social tolerance, maybe even resulting in possible selection of individuals with lower levels of testosterone and elevated levels of social behavior-mediating hormones, such as serotonin. Evidence from wolf and fox domestication, as well as comparative human data, highlights correlations between these hormonal changes and cranio-facial skeletal growth patterns. Domestication of foxes and wolves, for instance, results in shortened skulls and faces (Franciscus et al., 2013; Cieri et al., 2014). In humans, a low ratio of second to fourth digits length, a marker of elevated prenatal testosterone level, is associated with a thicker brow, broader faces, and more prominent chins.

Robusticity

Environmental influences such as activity levels, diet, and climate also played a crucial role in shaping skeletal morphology. Although the skeleton of modern humans is generally considered gracile relative to earlier hominins, this trend characterizes primarily more recent, Holocene populations (Trinkaus, 1997; Holt et al., 2000; Holt, 2003; Ruff, 2005; Shackelford, 2007). Femur robusticity values for Late Pleistocene modern humans, for instance, largely overlap those for Neandertals (Ruff et al., 1993). The mechanical loading patterns associated with rigorous lifestyles and reliance on largely unprocessed foods probably explains why Pleistocene modern humans have more robust skeletons than Holocene humans. There is a general reduction in cranio-dental size across Eurasian samples during the Upper Paleolithic, probably reflecting improving technology (Brace, 1962; Frayer, 1978, 1984; Kennedy et al., 1987; Hershkovitz et al., 1995). Relaxed selection for large anterior tooth size resulting from increased reliance on technology combined with lower levels of musculoskeletal stress to produce the gracile morphology typical of modern human crania. Continuing declines in levels of physical activity and increased reliance on technology during the Holocene resulted in further gracilization.

Although Upper Paleolithic populations are generally quite robust, some variability exists between early (EUP) and late (LUP) Upper Paleolithic populations. A number of indicators suggest, for instance, that EUP femora show more resistance to loads related to high mobility than LUP ones. EUP femora exhibit elevated anterior curvature, lower neck-shaft angle, and antero-posteriorly expanded midshaft cross-sections, all traits associated with high locomotor loads (Trinkaus, 1993; Holt, 2003). Recent urban populations, for instance, exhibit significantly lower anterior femur curvature than less industrialized groups, probably a reflection of sedentism and mechanization (Shackelford and Trinkaus, 2002). Asymmetry in Upper Paleolithic (UP) upper limb robusticity is quite high, even relative to recent marine foragers who otherwise have strong humeri, reflecting their extensive use of boats and kayaks over open waters (Churchill et al., 1996; Churchill, 2002; Schmitt et al., 2003). The high asymmetry measures in UP males, particularly in the EUP, probably reflect the repeated unimanual use of hunting weapons (Churchill et al., 2000; Schmitt et al., 2003). These changes in limb robusticity patterns characterize samples in many geographic areas, including Europe, Nile Valley, Mediterranean, and Asia, suggesting that the increase in population density, reduced mobility, and increased diet breadth that occurred worldwide as a result of Late Glacial Maximum-related declining temperatures and aridity affected human subsistence behavior in similar ways (Shackelford, 2007).

Climate has also played an important role. A strong correlation has been shown, for instance, between recent humans' body proportions and climate, with individuals in colder climates exhibiting short limbs and broad trunks while those living closer to the equator are characterized by long limbs and narrower trunks (Ruff, 1994). This may help explain why early modern humans living closer to the equator evolved a narrow trunk and pelvis. The combination of reduced mediolateral stresses resulting from a narrower pelvis and increased mechanical loads associated with long-distance mobility

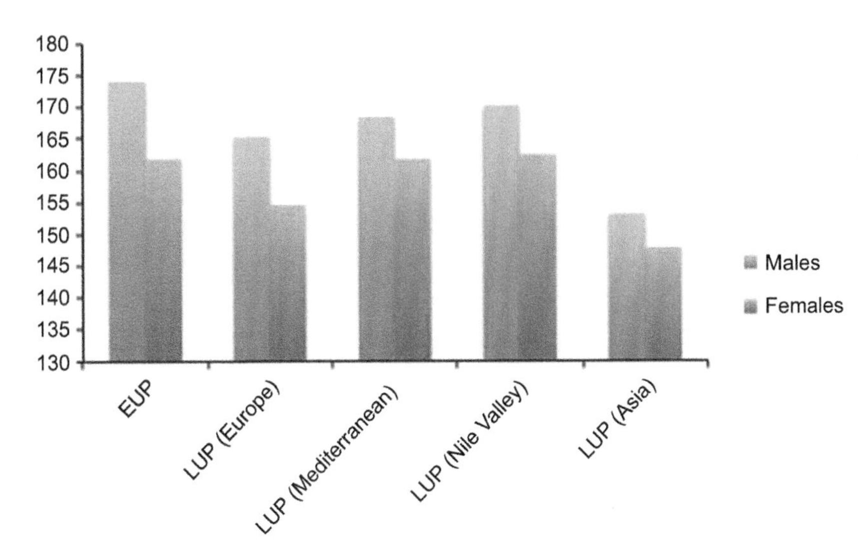

FIGURE 3 Graph of stature (in cm) for males and females from Early Upper Paleolithic group (EUP), Late Upper Paleolithic from Europe, Mediterranean, Nile Valley, and Southeast Asian groups (LUP). *Data for EUP and LUP Europeans are from Holt (2003), and data for LUP Mediterranean, Nile Valley, and Southeast Asia are from Shackelford (2007).*

contribute to the typical oval cross-sectional shape seen in early anatomically modern femora.

Stature

Phenotypic expression of stature depends on the complex interaction of genes and environmental factors. With a few exceptions such as the Near East Skhul/Qafzeh sample, the fragmentary state of the earliest AMH remains does not allow accurate stature reconstruction. Stature estimates for Skhul and Qafzeh show that early modern humans were much taller than Neandertals, averaging around 175 cm (). Another mostly complete female skeleton dated around 36 Holliday, 2000; Cartmill and Smith, 2009 kya from the site of Nazlet Khater (Egypt) measures around 1.61 cm (Crevecoeur, 2008). The larger and more complete UP European samples provide the best knowledge of stature in early anatomically modern humans. The EUP Gravettian people who inhabited glaciated Europe between 36 and 20 kya exhibit the tall stature and long limbs typical of many sub-Saharan populations, supporting an African or Near East origins for these populations (Holliday, 1997, 2000; Pearson, 2000). Stature decreases significantly after the Late Glacial Maximum. Comparisons of European data with samples from western Asia (Mediterranean), Africa (Nile Valley), and Southeast Asia show that Mediterranean and Nile valley LUP groups are taller than LUP Europeans, while Southeast Asians exhibit particularly low stature (Figure 3; Shackelford, 2007). Nutritional conditions may have played a role in the decrease in height between EUP and LUP (Formicola and Gianecchini, 1999). Isotope analysis suggests that the large herds populating glacial Europe afforded Gravettian groups a diet extremely rich in protein. The environmental degradation that ensued the Late Glacial Maximum forced populations to expand their diet to include low rank food sources such as aquatic resources. While there may have been a decline in availability of protein, neither the faunal evidence nor the biological stress markers and isotope data point to dramatic deterioration of life conditions and quality of diet.

Hand Morphology

Anatomically modern humans also exhibit some distinctive hand morphological features that suggest significant shifts in habitual manipulative behavior. Relative to Neandertals, modern humans exhibit a more concave articulation between the first and the second metacarpals and obliquely oriented facets between the second metacarpal and the capitate, resulting in a broader range of pronation and increased resistance to oblique loads (Niewohner, 2001). By contrast, Neandertal metacarpal morphology reflects resistance to powerful axial loads associated with a power grip. These adaptations to enhanced manipulation distinguish early modern humans such as Qafzeh and Skhul from Neandertals, and become more apparent in Late Upper Paleolithic humans and fully developed in recent human samples.

Life History

Relative to other primates, modern humans have prolonged periods of infant and childhood growth. These extended growth periods allow the time necessary for postnatal brain increase and for the acquisition of complex learning before reaching reproductive age. Dental development patterns provide important information about changes in growth trajectory in the human fossil record. Increasingly sophisticated methods based on X-ray synchrotron microtomography reveal exquisite details about tooth formation in children. Tooth development occurs through enamel accretion and the exact period of growth represented by each layer of enamel can be determined by counting the daily growth increments. Timing of crown formation time and age at death based

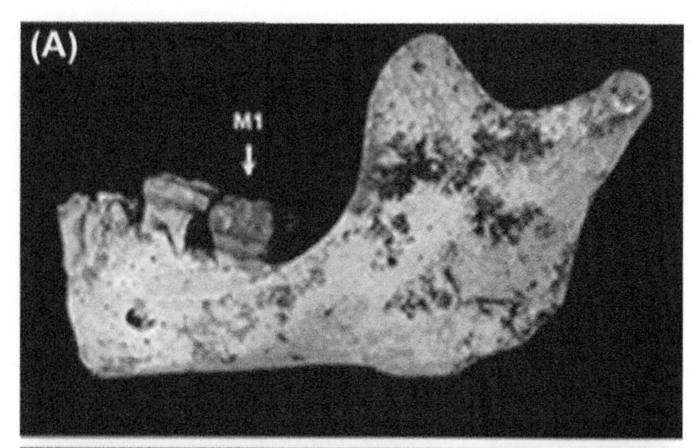

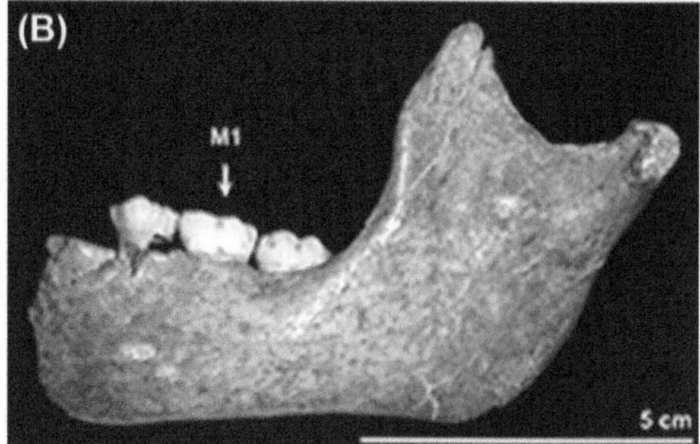

FIGURE 4 Comparison of degree of development in the anatomically modern human juvenile from Jebel Irhoud (A) and the Neandertal juvenile from Scladina (B) Both are 8 years old, but the Scladina child already has a second molar. (See color plate section). *From Smith et al. (2007); courtesy of the National Academy of Sciences (Figure 3)—Copyright (2013) National Academy of Sciences, USA.*

FIGURE 5 A rare example of Paleolithic dental enamel hypoplasia (Grotte des Enfants 6). (See color plate section). *Courtesy of the Musée d'Anthropologie Préhistorique de Monaco.*

on these incremental features can be used to estimate the speed of childhood maturation. The 160,000-year old modern human juvenile mandible from Jebel Irhoud exhibits a slow pattern of dental eruption and development, similar to recent populations, in contrast with the juvenile Neandertal from the site of Scladina (Belgium), dated between 80 and 127 kya, who matured faster (Figure 4; Smith et al., 2007). This extended developmental period implies that early modern human children enjoyed a prolonged childhood during which they acquired complex social knowledge.

Childbirth

Humans are unique in experiencing a difficult birthing process, owing to the necessary tradeoff between large neonatal brain size and the limits locomotion places on maternal pelvic dimensions. Efficient bipedal locomotion requires that the pelvic inlet be larger transversely than antero-posteriorly (front to back) (Rosenberg and Trevathan, 2001). The axis of the outlet, on the other hand, is oriented antero-posteriorly, probably an adaptation to the warm equatorial climate in which modern humans evolved and where a narrower body would have made heat dissipation easier (Franciscus, 2009). As a result of this twisted pelvic canal, the baby's head engages in the pelvis transversely, but about halfway through, it is forced to rotate 90° into an antero-posterior position by the presence of the ischial spines. The baby exits its mother's body facing backward, rotating one more time slightly sideways to pass the wide shoulders through the outlet. As a result of this complex "twisted" birthing process, humans rarely give birth unassisted, making obligate midwifery a cross-cultural norm (Rosenberg and Trevathan, 2001).

Health and Diseases

The overall good health of Paleolithic populations reflects the low pathogen load associated with a hunter and gatherer lifestyle (see chapters by Little, Morand, and Cook). Unlike Neolithic populations, late Pleistocene modern human populations rarely exhibit evidence of disease and malnutrition. Caries and periodontopathies are extremely infrequent, affecting a handful of individuals. Signs of infectious diseases in the postcranial skeleton are also extremely rare, generally limited to a few cases of periostitis, otitis, and bacterial infection (Trinkaus et al., 2005; Holt and Formicola, 2008; but see Cook). One notable exception to this generally healthy profile is found in one study of stress indicators in southwest France Upper Paleolithic samples pointing to an increase in the frequency of hypoplasias (Figure 5) and periostitis between EUP and LUP (Brennan, 1991; Holt and Formicola, 2008). Although the intensity of hypoplasias remains light in both periods, they may reflect nutritional stress and growth disruptions associated with decreased age at weaning. The severity of these stress indicators in the Upper Paleolithic, however, is much lower than that observed in more recent populations.

Some recurring pathologies involve minor cranial and limb fractures, many with extensive remodeling indicating that the individual survived the injury. While most of these injuries appear nonintentional, the rise in the number of wounds caused by projectiles during the LUP suggests increased conflict. Flint flakes, likely from javelin or arrows, were found embedded in the bones of two Late Epigravettian remains from Italy (Holt and Formicola, 2008). At Grotte des Enfants, the flint embedded in the 4th lumbar vertebra of Grotte des Enfants two appear to have caused the death of this child. At San Teodoro, an adult male survived the wound inflicted by a flake embedded in his hip bone. Besides these isolated instances, the large LUP sample from Jebel Sahaba in Sudan shows numerous signs of violence in the form of fractures and cut marks, providing more clear evidence of increased conflict during the latter part of the Upper Paleolithic (Anderson, 1968).

A surprisingly high number of cases of congenital diseases are found in Upper Paleolithic burials. The earliest evidence is presented by a congenital inner ear malformation affecting Mladeč 5 (Czech Republic), possibly resulting in hearing deficiency (Teschler-Nicola et al., 2006). At the site of Dolní Vestonicĕ (Moravia), one of the skeletons from a triple burial exhibits deformations in multiple limbs, possibly the result of a congenital disorder called chondrodysplasia calcificans punctate (Formicola et al., 2001). Despite these handicaps, upper and lower limb bones robusticity and articular changes attest to the active life of this individual into adulthood. In Russia, the deformed legs of a 10-year-old girl from the Sunghir double burial may be linked to the diabetic condition of the mother, thus providing indirect evidence of diabetes in the Gravettian (Formicola and Buzhilova, 2004). Finally, the dwarf from the LUP site of Romito (Italy) may represent the most intriguing evidence of inherited disorders in the Paleolithic. This young individual, buried with an adult female, is affected by an inherited form of acromesomelic dwarfism (Figure 6; Frayer et al., 1987; Frayer et al., 1988; Bachechi and Martini, 2002).

FIGURE 6 Double burial from Romito (Italy) when first discovered. The dwarf is on the right. *From Bachechi and Martini (2002); courtesy of F. Martini.*

Technology

The advent of modern human behavior is generally associated with the Upper Paleolithic in Europe and the LSA in Africa (see Chapters by Toth and Schick). Some of the distinctive characteristics of the UP/LSA include more efficient hunting technology such as long-distance projectiles, more diverse tool kits (including a preponderance of standardized prismatic blades and tools made of bone, ivory, and antler), broadening of diet, long-distance raw material procurement networks, ubiquitous use of symbolic elements such as ocher and personal decorations (ivory and shell beads and pendants), systematic use of portable and parietal art (carved figurines and painted or engraved cave and rock-shelter walls), and ceremonial burials. The elements of this "modernity behavioral package" form part of a broad and stable, but flexible, ecological niche (Marean et al., 2007; Shea, 2011).

Archeological evidence shows that most of these so-called hallmarks of modernity appear at least 30,000 years earlier in Africa than in Europe, and that their development may postdate the appearance of modern human anatomy by as much as 100,000 years. Advanced microlith technology appears earliest in South Africa, around 70 kya, in the Near East around 50 kya, and in Europe around 45 kya (Shea and Sisk, 2010; Brown et al., 2012). Developed for the production of more standardized tools that maximize cutting edge, prismatic blades (Figure 7) may reflect the intensification of resources and increased mobility in response to changing climate or higher population density. The earliest secure evidence of aerodynamic projectile points comes from early Upper Paleolithic contexts in the Near East and in European Upper Paleolithic assemblages around 45 kya (Shea and Sisk, 2010). African Middle Stone Age contexts contain numerous hypothetical stone projectile points, but their size and shape may be more compatible with their use as thrusting spears. The use of long-distance projectiles would have protected hunters from dangerous prey and help decrease the costs of pursuing potentially high return but high-risk food sources such as large terrestrial mammals (Shea, 2003). The earliest known microlithic technology used to create composite tool components as part of advanced projectile weapons comes from the site of Pinnacle Point in South Africa (Brown et al., 2012). There, heat-treated bladelets were used as early as 70 kya, although the tradition of heat treatment goes back to 162 kya, indicating the

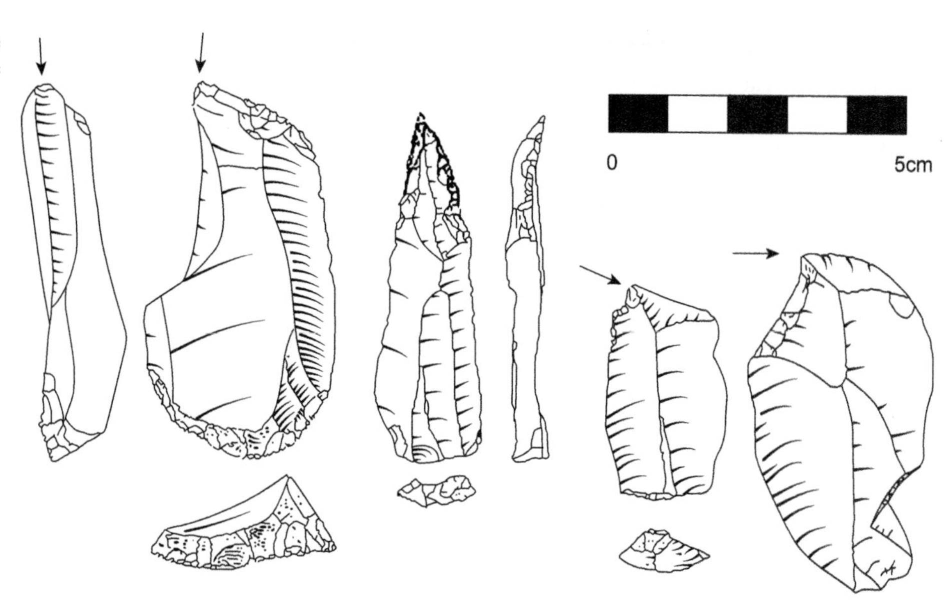

FIGURE 7 Examples of early blade tools from Initial Upper Paleolithic layers at Üçağızlı Cave (Turkey). *Adapted from Kuhn et al. (2009).*

early development in South Africa of modern human cognitive capacity to transmit cultural traditions.

Diet

While early modern humans regularly hunted large herbivores, faunal evidence from some LSA and UP sites suggests a shift to a broader dietary spectrum to include small game and fish (see the chapter by Little). Exploitation of low-ranked food items such as shellfish and fish and concomitant development of more specialized technology in the LSA may be tied to increased population pressure, necessitating intensification of resources and increased labor. The earliest dated evidence of marine food exploitation comes from Pinnacle Cave in South Africa. As early as 164kya, early modern human groups harvested shellfish on sandy beaches and rocky shores (Marean, 2010). Besides faunal and floral remains, carbon and nitrogen stable isotope ratios of human bone collagen provide important information about levels and source of dietary protein consumed over a number of years. Recent analyses of stable isotopes ratios of European Neandertal and modern human remains show that, while both groups incorporated large amounts of animal protein in their diet, modern humans exploited a larger spectrum of species. Isotope data from 14 EUP human remains ranging between 35 and 22kya, including the earliest known AMH in Europe (Oase 1), suggest that early modern humans obtained about 20–30% of their protein from marine resources such as fish or marine mammals (Richards and Trinkaus, 2009; Figure 8).

Mobility

The modern behavior package is generally associated with increased mobility involving long-distance exchange networks in lithics, raw materials, and marine shells. Upper Paleolithic raw material procurement distances often extended over several hundred kilometers (Bar Yosef, 2002). This increased mobility dovetailed with greater technological variability among UP/LSA lithic industries reflects a broader, more flexible and stable niche that probably included more reliable subsistence resources and enhanced social interaction. Mobility during the Upper Paleolithic period known as the Gravettian was particularly high. Studies of lithic procurement patterns show that groups often traveled distances in excess of 200km to obtain material necessary for tool production. Organization systems included complex social long-distance networks designed to cushion the impact of a highly seasonal environment and to ensure maintenance of social cohesion, mating networks, and exchange of information between small demes distributed across a vast territory (Gamble et al., 2004). The extensive use of portable art, such as beads and the so-called Venus figurines, probably functioned to extend social connections in absence of proximity. The striking resemblance in shape and size of these figurines and their geographic distribution across 2000km stress the maintenance of a common system of beliefs among distant populations, prompting some to refer to the EUP as an "open system."

Symbolic Behavior

All modern societies, regardless of cultural differences, use socially constructed symbols to mediate exchange of information. The meaning of these symbols is established by formal or simply agreed-on conventions independently of the physical characteristics of either sign or object (Deacon, 1997). The meaning is shared, learned, and transmitted across generations, establishing cultural continuity between and across generations and communities (Henshilwood

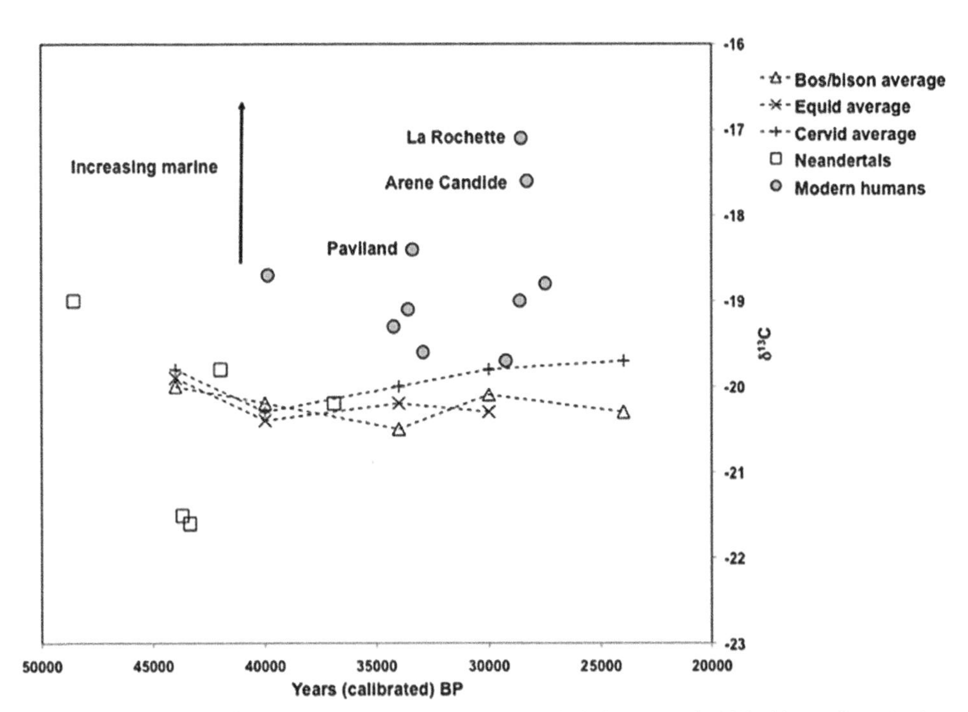

FIGURE 8 Carbon ($\delta_{13}C$) stable isotope values of early modern humans and Neanderthals compared with herbivores from Northern Europe from 50 to 20 kya, showing higher consumption of marine resources for modern humans. *From Richards and Trinkaus (2009); courtesy of the National Academy of Sciences (Figure 3).*

and Marean, 2003). This capacity for symbolically driven behavior cannot be decoupled from the evolution of anatomically modern humans. As such, the earliest occurrence of symbols lies at the core of research on the evolution of modernity.

The earliest securely dated evidence of symbolic behavior come from the Near East and Africa in the form of perforated marine shells (Figure 9) and engraved ochre fragments (Figure 10). At the Near East site of Skhul, several shell beads dated between 100 and 125 kya are associated with anatomically modern human burials (Vanhaeren et al., 2006), making them the oldest known symbolic objects. Engraved ochre and marine shell beads from Blombos Cave (South Africa) date to 74–84 kya (Henshilwood et al., 2001). Engraved ochre fragments are also found at Pinnacle Point Cave in layers dated at 164 kya (Henshilwood et al., 2002; Marean, 2010). Marine shell ornaments next appear systematically around 41 kya in East Africa, the Mediterranean basin, and central Europe, and thus appear to correspond with the geographic expansion of modern humans.

The appearance of self-ornamentation probably reflects the necessity to reinforce social ties. Portable objects such as shell beads are durable and plentiful and can be standardized to codify and transmit complex information across generations. Ornaments can function as information technology to express ethnicity and group identity in the context of increasingly complex social landscapes. While these ornaments appear early in the Middle Paleolithic and Middle Stone Age, their use becomes more systematic during the Gravettian in Europe. Between 28 and 20 kya, populations across 8000 km, from the Atlantic and Mediterranean shores to these Russian Plains, carved pendants in the shape of generally plump female so-called "Venus" figurines out of clay, stone, bone, antler, or ivory (Figure 11). A few of these figurines represent a woman continuing into a nonhuman being (Figure 12), a symbolic combination of elements that may reflect a complex system of beliefs in transformations and possible proto-shamanism practices (Mussi et al., 2000). The ubiquity and homogeneity of these figurines across such a large territory point to cultural homogeneity bridging populations across a vast depopulated territory.

Funerary Behavior

The earliest evidence of postmortem cultural modification by anatomically modern humans comes from the site of Herto (Ethiopia). Dated at around 160 kya, the three Herto crania show signs of defleshing cut marks probably made with sharp obsidian flakes (Clark et al., 2003). While these do not necessarily represent funerary treatment, ethnographic studies suggest that the cut marks could result from mortuary practices (White, 1992). Deliberate burials are known from numerous Middle Paleolithic and Middle Stone Age contexts. Many of these burials are associated with Neandertal remains, although none of these include ornaments. The earliest burials currently known date to around 110–90 kya, from the Near East sites of Skhul and Qafzeh where most of the anatomically modern human skeletons

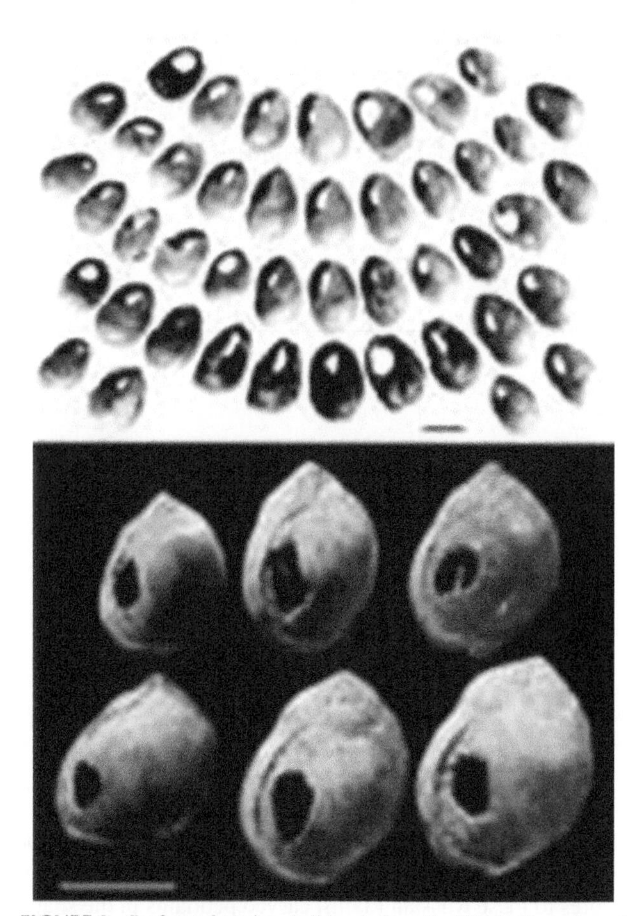

FIGURE 9 Perforated marine shell beads from the Middle Stone Age of Blombos Cave (South Africa). Scale bar: 5 mm. (See color plate section). *From Henshilwood et al. (2004); Courtesy of the American Association for the Advancement of Science.*

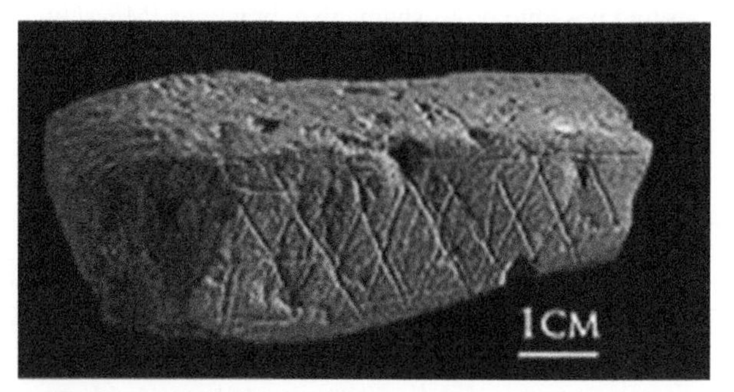

FIGURE 10 Engraved red ochre fragment from the Middle Stone Age of Blombos Cave (South Africa). (See color plate section).

were found in articulation, suggesting deliberate placement of the bodies. Skhul 5 and Qafzeh 11 were buried with a boar mandible and red deer antlers, respectively (Bar-Yosef and Vandermeersch, 1993; Shea, 2003). Shell beads have also been found at Skhul in association with skeletal remains (see Symbolic Behavior). At the 37,570-year-old site of Nazlet Khater (Egypt), a female skeleton was buried with a bifacial stone tool and covered with blocks.

FIGURE 11 Example of Upper Paleolithic (Gravettian) figurine from Willendorf (Austria). These so-called "Venus figures" are found throughout Europe between 28,000 and 20,000 years ago and often exhibit exaggerated sexual characteristics, such as enlarged buttocks, abdomen, and breasts. (See color plate section).

Although these objects probably form part of funerary rituals, the richest expression of such rituals comes much later, with the Upper Paleolithic burials from the Gravettian. Interestingly, the Aurignacian early phase of the Upper Paleolithic has yielded few human remains and none in formal burial contexts. Starting around 28 kya, burials become numerous and extraordinarily rich. These burials are generally single or double, even occasionally triple, with abundant red ochre sprinkled over the body. They invariably include rich and varied funerary objects carved from ivory, bone, shells, or stone. Among the most spectacular Gravettian burials, three are particularly noteworthy: the adolescent male (so-called "Principe") from Grimaldi (Italy) for its rich ornamentation, the double child burial from Sunghir (Russia) for its extraordinary ivory funerary objects, the triple burial from Dolní Vestonicě (Moravia) (Figure 13) comprising three young individuals, and the double burial from Grotta del Romito (Italy). In both Sunghir and Dolní Vestonicě, one individual exhibits congenital deformations. In the case of Romito, an adolescent affected by an

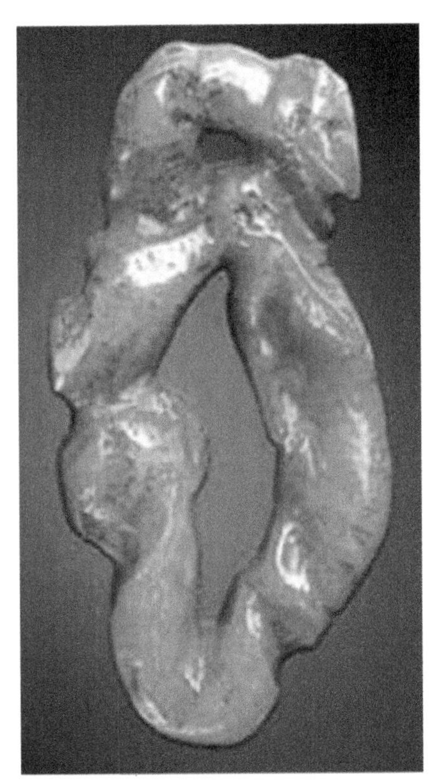

FIGURE 12 The "Doublet" figurine from the Upper Paleolithic (Gravettian) site of Balzi Rossi (Italy). Carved from steatite, this figure represents a woman (left) joined at the head, shoulder, and lower limb to a nonhuman being (right), possibly symbolizing a ritualistic transformation. (See color plate section). *Alexandre Louis Jullien Collection, copyright Pierre Bolduc.*

FIGURE 13 The triple burial from the Upper Paleolithic (Gravettian) site of Dolní Věstonice (Moravia). Two adolescent males flank a female exhibiting deformations from a congenital disorder. *From Formicola (2007); courtesy of the American Association of Physical Anthropologists.*

inherited form of acromesomelic dwarfism is buried with an adult female who holds him in one arm. The presence of these selected individuals in elaborate funerary contexts underscores the symbolic nature of these burials and raises the issue of social perception of diversity and the roles that their individuals played in Paleolithic society (Formicola, 2007; Holt and Formicola, 2008, and references therein).

Language

The ability to talk separates humans from animals (see the chapter by Lieberman). Numerous aspects of the human vocal tract and brain include adaptations for the production of speech. Uniquely specialized centers in the brain such as Broca's and Wernicke's areas play an important role in language comprehension, for instance. The human vocal tract has also been reshaped by natural selection for the production of complex sounds. The larynx, the complex structure of cartilage and muscles where sounds are produced, sits lower in the neck than in other animals, allowing air to pass through the mouth more easily. The combined action of lips, teeth, tongue, and vocal folds modify the air into myriad sounds such as vowels and consonants. There is little doubt that the capacity for speech evolved early in human evolution, probably well before the advent of anatomically modern humans. Because the anatomy of speech involves primarily soft tissues, such as cartilage and neural tissue, that do not fossilize, the fossil record affords few clues about the evolution of this hallmark of humanity. Evidence from remains of material culture has led some to argue that human speech was present as early as 164 kya. Transmission and sharing of symbolic meaning indicated by engraved ochre fragments from Blombos Cave and Pinnacle Cave in South Africa, for instance, may imply a capacity for syntactical language by 164 kya (Henshilwood, 2007). The recent discovery that Neandertals shared of a distinct human form of the FOXP2 gene responsible for speech production (Krause et al., 2007), however, suggests that adaptation to language evolved before Neandertals and AMH diverged.

REFERENCES

Anderson, J.E., 1968. Late Paleolithic skeletal remains from Nubia. The Prehistory of Nubia 2, 996–1040.

van Andel, T.H., Davies, W., Weniger, B., 2003. The human presence in Europe during the Last Glacial Period I: Migrations and the changing climate. In: van Andel, T.H., Davies, W. (Eds.), Neanderthals and Modern Humans in the European Landscape during the Last Glaciation. University of Cambridge Press, Cambridge, pp. 31–56. McDonald Institute for Archaeological Research.

Bachechi, L., Martini, F., 2002. Le sepulture. In: Martini, F. (Ed.), Grotta del Romito. Guide del Museo e Istituto Fiorentino di Preistoria, Firenze, pp. 29–36.

Barker, G., Barton, H., Bird, M., et al., 2007. The human revolution in lowland tropical Southeast Asia: the antiquity and behavior of anatomically modern humans at Niah Cave (Sarawak, Borneo). Journal of Human Evolution 52, 243–261.
Bar-Yosef, O., 2000. The Middle and early Upper Paleolithic in Southwest Asia and neighboring regions. In: Bar-Yosef, O., Pilbeam, D. (Eds.), The Geography of Neandertals and Modern Humans in Europe and the Greater Mediterranean. Peabody Museum, Harvard University, Cambridge, MA, pp. 107–1656.
Bar-Yosef, O., 2002. The Upper Paleolithic revolution. Annual Review of Anthropology 31, 363–393.
Bar-Yosef, O., Vandermeersch, B., 1993. Modern humans in the Levant. Scientific American 268, 94–100.
Benazzi, S., Douka, K., Fornai, C., et al., 2011. Early dispersal of modern humans in Europe and implications for Neanderthal behaviour. Nature 479, 525–528.
Bergman, C.A., Stringer, C.B., 1989. Fifty years after: Egbert, an early Upper Palaeolithic Juvenile from Ksar Akil, Lebanon. Paléorient 15, 99–111.
Brace, C.L., 1962. Cultural factors in the evolution of human dentition. In: Montagu, M.F.A. (Ed.), Culture and the Evolution of Man. Oxford University Press, New York, pp. 343–354.
Bräuer, G., Mehlman, M.J., 1988. Hominid molars from a Middle Stone Age level at the Mumba Rock Shelter, Tanzania. American Journal of Physical Anthropology 75, 69–76.
Brennan, M.U., 1991. Health and disease in the Middle and Upper Paleolithic of southwestern France. A bioarchaeological study. Ph.D. Dissertation, New York University.
Brown, K.S., Marean, C.W., Benjamin, J., et al., 2012. An early and enduring advanced technology originating 71,000 years ago in South Africa. Nature 491, 590–593.
Cartmill, M., Smith, F.H., 2009. The Human Lineage. Wiley-Blackwell, Hoboken.
Churchill, S.E., 2002. Of assegais and bayonets: reconstructing prehistoric spear use. Evolutionary Anthropology 11, 185–186.
Churchill, S.E., Weaver, A.H., Niewoehner, W.A., 1996. Late Pleistocene human technological and subsistence behavior: functional interpretations of upper limb morphology. Quaternaria Nova 6, 413–447.
Churchill, S.E., Formicola, V., Holliday, T.W., Holt, B., Schumann, B.A., 2000. The Upper Paleolithic population of Europe in an evolutionary perspective. In: Roebroeks, W., Mussi, M., Svoboda, J., Fennema, K. (Eds.), Hunters of the Golden Age: The Mid Upper Paleolithic of Eurasia (30,000–20,000 bp). Leiden University Press, Leiden, pp. 31–57.
Cieri, R.L., Churchill, S.E., Franciscus, R.G., Tan, J., Hare, B., 2014. Craniofacial feminization, social tolerance, and the origins of behavioral modernity. Current Anthropology 55, 419–443.
Clark, J.D., Beyenet, Y., WoldeGabriel, G., Hart, W.K., Renne, P.R., et al., 2003. Stratigraphic, chronological and behavioural contexts of Pleistocene *Homo sapiens* from Middle Awash, Ethiopia. Nature 423, 747–752.
Crevecoeur, I., 2008. Etude anthropologique du squelette du Paléolithique Supérieur de Nazlet Khater 2 (Egypte). Egyptian Prehistory Monographs 8. Leuven University Press.
Deacon, T., 1997. The Symbolic Species. Penguin, London.
Demeter, F., Shackelford, L.L., Bacon, A.-M., Duringer, P., Westaway, K., et al., 2012. Anatomically modern human in Southeast Asia (Laos) by 46 ka. Proceedings of the National Academy of Sciences of the United States of America 109, 14375–14380.
De Quatrefages, A., Hamy, T., 1882. Crania ethnica. Les crânes des races humaines, Paris.
Feathers, J.K., 2002. Luminescence dating in less than ideal conditions: case studies from Klasies River main site and Duinefontein, South Africa. Journal of Archaeological Science 29, 177–194.
Feathers, J.K., Bush, D.A., 2000. Luminescence dating of Middle Stone Age deposits at Die Kelders. Journal of Human Evolution 38, 91–119.
Formicola, V., 2007. From the Sunghir children to the Romito dwarf: aspects of the Upper Paleolithic funerary landscape. Current Anthropology 48, 446–453.
Formicola, V., Giannecchini, M., 1999. Evolutionary trends of stature in Upper Paleolithic and Mesolithic Europe. Journal of Human Evolution 36, 319–333.
Formicola, V., Buzhilova, A.P., 2004. Double child burial from Sunghir (Russia): pathology and inferences for Upper Paleolithic funerary practices. American Journal of Physical Anthropology 124, 189–198.
Formicola, V., Pontrandolfi, A., Svoboda, J., 2001. The Upper Paleolithic triple burial of Dolní Věstonice: pathology and funerary behavior. American Journal of Physical Anthropology 115, 372–379.
Franciscus, R.G., 2009. When did the modern human pattern of childbirth arise? New insights from an old Neandertal pelvis. Proceedings of the National Academy of Sciences of the United States of America 106, 9125–9126.
Franciscus, R.G., Maddux, S.D., Schmidt, K.W., 2013. Anatomically modern humans as a "self-domesticated" species: insights from ancestral wolves and descendant dogs. American Journal of Physical Anthropology 56 (Suppl.), 125.
Frayer, D.W., 1978. The evolution of the dentition in Upper Paleolithic and Mesolithic Europe. University of Kansas Publications in Anthropology. 10.
Frayer, D.W., 1984. Biological and cultural changes in the European Late Pleistocene and Early Holocene. In: Smith, F.H., Spencer, F. (Eds.), The origins of modern humans: a world survey of the fossil evidence. Alan R. Liss, Inc., New York, pp. 211–250.
Frayer, D.W., Horton, W.A., Macchiarelli, R., Mussi, M., 1987. Dwarfism in an adolescent from the Italian Late Upper Palaeolithic. Nature 330, 60–62.
Frayer, D.W., Macchiarelli, R., Mussi, M., 1988. A case of chondrodystrophic dwarfism in the Italian Late Upper Paleolithic. American Journal of Physical Anthropology 75, 549–565.
Fu, Q., Li, H., Moorjani, P., Jay, F., Slepchenko, S.M., et al., 2014. Genome sequence of a 45,000-year-old modern human from western Siberia. Nature 514, 445–449.
Gamble, C., Davies, W., Pettitt, P., Richards, M., 2004. Climate change and evolving human diversity in Europe during the last glacial. Philosophiscal Transactions of the Royal Society, London. B 359, 243–254.
Haile-Selassie, Y., Asfaw, B., White, T.D., 2004. Hominid cranial remains from Upper Pleistocene deposits at Aduma, Middle Awash, Ethiopia. American Journal of Physical Anthropology 123, 1–10.
Henry-Gambier, D., 2002. Les fossiles de Cro-Magnon (Les Eyzies-de-Tayac, Dordogne): nouvelle données sur leur position chronologique et leur attribution culturelle. Bulletin et Mémoire de la Société d'Anthropologie de Paris 14, 89–112.
Henshilwood, C.S., 2007. Fully symbolic sapiens behavior: innovation in the Middle Stone Age at Blombos Cave, South Africa. In: Mellars, P., Boyle, K., Bar-Yosef, O., Stringer, C. (Eds.), Rethinking the Human Revolution. McDonald Institute Monographs, Cambridge, pp. 123–132.

Henshilwood, C.S., d'Errico, F., Marean, C.W., Milo, R.G., Yates, R., 2001. An early bone tool industry from the Middle Stone Age, Blombos Cave, South Africa: implications for the origins of modern human behaviour, symbolism, and language. Journal of Human Evolution 41, 631–678.

Henshilwood, C.S., d'Errico, F., Yates, R., Jacobs, Z., Tribolo, C., et al., 2002. Emergence of modern human behaviour: Middle Stone Age engravings from South Africa. Science 295, 1278–1280.

Henshilwood, C.S., Marean, C.W., 2003. The origin of modern human behavior: critique of the models and their test implications. Current Anthropology 44, 627–651.

Henshilwood, C.S., d'Errico, F., Vanhaeren, M., van Niekerk, K., Jacobs, Z., 2004. Middle Stone Age shell beads from South Africa. Science 304, 404.

Hershkovitz, I.S.M., Frayer, D.W., Nadel, D., Wish-Baratz, S., Arensburg, B., 1995. Ohalo II H2: a 19,000-year-old skeleton from a water-logged site at the Sea at Galilee, Israel. American Journal of Physical Anthropology 96, 215–234.

Higham, T., Douka, K., Wood, R., et al., 2014. The timing and spatiotemporal patterning of Neanderthal disappearance. Nature 512, 306–309.

Hoffecker, J.F., 2009. The spread of modern humans in Europe. Proceedings of the National Academy of Sciences of the United States of America 10, 16040–16045.

Holliday, T., 2000. Evolution at the crossroad: modern human emergence in Western Asia. American Anthropologist 102, 54–68.

Holliday, T.W., 1997. Body proportions in Late Pleistocene Europe and modern human origins. Journal of Human Evolution 32, 423–447.

Holt, B., 2003. Mobility in Upper Paleolithic and Mesolithic Europe: evidence from the lower limb. American Journal of Physical Anthropology 122, 200–215.

Holt, B., Formicola, V., 2008. Hunters of the ice age: the biology of upper Paleolithic people. Yearbook of Physical Anthropology 51, 70–99.

Holt, B., Mussi, M., Churchill, S.E., Formicola, V., 2000. Biological and cultural trends in Upper Paleolithic Europe. Rivista Di Antropologia 78, 179–192.

Jobling, M., Hollox, E., Hurles, M., Kivisild, T., Tyler-Smith, C., 2014. Human Evolutionary Genetics, third ed. Garland Science, New York.

Kennedy, K.A.R., Roertgen, S.U., Chiment, J., Disotell, T., 1987. Upper Pleistocene fossil hominids from Sri Lanka. American Journal of Physical Anthropology 72, 441–461.

Klein, R.G., 1999. The Human Career. The University of Chicago Press, Chicago.

Krause, J., Lalueza-Fox, C., Orlando, L., Enard, W., Green, R.E., et al., 2007. The derived FOXP2 variant of modern humans was shared with Neandertals. Current Biology 17, 1908–1912.

Kuhn, S.L., Stiner, M.C., Gülec, E., Özer, I., Yılmaz, H., et al., 2009. The early Upper Paleolithic occupations at Üçağızlı Cave (Hatay, Turkey). Journal of Human Evolution 56, 87–113.

Lahr, M.M., 1996. The Evolution of Human Diversity: A Study of Cranial Variation. Cambridge University Press, Cambridge.

Lieberman, D.E., 1996. How and why recent humans grow thin skulls: experimental evidence for systemic cortical robusticity. American Journal of Physical Anthropology 101, 217–236.

Marean, C.W., 2010. Pinnacle Point Cave 13B (Western Cape Province, South Africa) in context: the Cape Floral kingdom, shellfish, and modern human origins. Journal of Human Evolution 59, 425–443.

Marean, C.W., Bar-Matthews, M., Bernatchez, J., Fisher, E., Goldberg, P., et al., 2007. Early human use of marine resources and pigment in South Africa during the Middle Pleistocene. Nature 449, 905–908.

Matsumura, H., Pookajorn, S., 2005. A morphometric analysis of the Late Pleistocene human skeleton from the Moh Kwiew Cave in Thailand. Homo 56, 93–118.

McDermott, F., Grün, R., Stringer, C., Hawkesworth, C.J., 1993. Mass-spectrometric U-series dates for Israeli Neanderthal/early modern hominid sites. Nature 363, 252–255.

McDermott, F., Stringer, C., Grün, R., Williams, C.T., Din, V.K., et al., 1996. New Late-Pleistocene uranium-thorium and ESR dates for the Singa hominid (Sudan). Journal of Human Evolution 31, 507–516.

McDougall, I., Brown, F.H., Fleagle, J.G., 2005. Stratigraphic placement and age of modern humans from Kibish, Ethiopia. Nature 433, 733–736.

Mellars, P., 2006. Why did modern human populations disperse from Africa ca. 60,000 years ago? A new model. Proceedings of the National Academy of Sciences of the United States of America 103, 9381–9386.

Mercier, N., Valladas, H., Bar-Yosef, O., Vandermeersch, B., Stringer, C., et al., 1993. Thermoluminescence date for the Mousterian burial site of Es-Skhul, Mt. Carmel. Journal of Archaeological Science 20, 169–174.

Müller, U.C., Pross, J., Tzedakis, P.C., Gamble, C., Kotthoff, U., et al., 2011. The role of climate in the spread of modern humans into Europe. Quaternary Science Reviews 30, 273–279.

Mussi, M., Cinq-Mars, J., Bolduc, P., 2000. Echoes from the mammoth steppes: the case of the Balzi Rossi. In: Roebroeks, W., Mussi, M., Svoboda, J., Fennema, K. (Eds.), Hunters of the Golden Age: The Mid Upper Palaeolithic of Eurasia (30,000–20,000 BP). Leiden University Press, Leiden, pp. 105–124.

Niewoehner, W.A., 2001. Behavioral inferences from the SkhulyQafzeh early modern human hand remains. Proceedings of the National Academy of Sciences of the United States of America 98, 2979–2984.

Norton, C.J., Gao, X., 2008. Zhoukoudian Upper Cave revisited. Current Anthropology 49, 732–745.

Păunescu, A., 2001. Paleoliticul Şi Mezoliticul Din Spaţiul Transilvan. Editura AGIR, Bucureşti.

Pearson, O.M., 2000. Activity, climate, and postcranial robusticity. Current Anthropology 41, 569–607.

Pettitt, P.B., 2000. The Paviland radiocarbon dating programme: reconstructing the chronology of faunal communities, carnivore activity and human occupation. In: Aldhouse-Green, S. (Ed.), Paviland Cave and the "Red Lady". West. Acad. Spec., Bristol, pp. 63–71.

Ponce de Leon, M.S., Zollikofer, C.P.E., 2008. Neanderthals and modern humans – chimps and bonobos: similarities and differences in development and evolution. In: Harvati, K., Harrison, T. (Eds.), Neanderthals Revisited: New Approaches and Perspectives. Springer, New York, pp. 71–88.

Richards, M.P., Trinkaus, E., 2009. Isotopic evidence for the diets of European Neanderthals and early modern humans. Proceedings of the National Academy of Sciences of the United States of America 106, 16034–16039.

Richards, M.P., Pettitt, P.B., Stiner, M.C., Trinkaus, E., 2001. Stable isotope evidence for increasing dietary breadth in the European mid-Upper Paleolithic. Proceedings of the National Academy of Sciences of the United States of America 98, 6528–6532.

Rosenberg, K.R., Trevathan, W., 2001. The evolution of human birth. Scientific American 285, 72–77.

Ruff, C.B., 1994. Morphological adaptation to climate in modern and fossil hominids. Yearbook of physical anthropology 37, 65–108.

Ruff, C.B., 2005. Mechanical determinants of bone form: insights from skeletal remains. Journal of Musculoskeletal and Neuronal Interactions 5, 202–212.

Ruff, C.B., Trinkaus, E., Walker, A., Larsen, C.S., 1993. Postcranial robusticity in *Homo*. I. Temporal trends and mechanical interpretation. American Journal of Physical Anthropology 91, 21–54.

Schmitt, D., Churchill, S.E., Hylander, W.L., 2003. Experimental evidence concerning spear use in Neandertals and early modern humans. Journal of Archaeological Science 30, 103–114.

Schwarcz, H.P., Rink, W.J., 2000. ESR dating of the Die Kelders 1 site, South Africa. Journal of Human Evolution 38, 121–128.

Schwarcz, H.P., Grün, R., Vandermeersch, B., Bar-Yosef, O., Valladas, H., et al., 1988. ESR dates for the hominid burial site of Qafzeh in Israel. Journal of Human Evolution 17, 733–737.

Shackelford, L.L., 2007. Regional variation in the postcranial robusticity of Late Upper Paleolithic humans. American Journal of Physical Anthropology 133, 655–668.

Shackelford, L.L., Trinkaus, E., 2002. Late Pleistocene human femoral diaphyseal curvature. American Journal of Physical Anthropology 118, 359–370.

Shang, H., Tong, H., Zhang, S., Chen, F., Trinkaus, E., 2007. An early modern human from Tianyuan Cave, Zhoukoudian, China. Proceedings of the National Academy of Sciences of the United States of America 104, 6573–6578.

Shea, J.J., 2003. Neandertals, competition, and the origin of modern human behavior in the Levant. Evolutionary Anthropology 12, 173–187.

Shea, J.J., 2011. *Homo sapiens* is as *Homo sapiens* was. Current Anthropology 52, 1–35.

Shea, J.J., Sisk, M.L., 2010. Complex projectile technology and *Homo sapiens* dispersal into Western Eurasia. Paleoanthropology 2010, 100–122.

Smith, T.M., Toussaint, M., Reid, D.J., Olejniczak, A.J., Hublin, J.J., 2007. Rapid dental development in a Middle Paleolithic Belgian Neanderthal. Proceedings of the National Academy of Sciences of the United States of America 104, 20220–20225.

Stringer, C.B., Grün, R., Schwarcz, H.P., Goldberg, P., 1989. ESR dates for the hominid burial site of Es-Skhul in Israel. Nature 338, 756–758.

Svoboda, J., Ložek, V., Vlček, E. (Eds.), 1996. Hunters between East and West: The Paleolithic of Moravia. Plenum Press, New York.

Teschler-Nicola, M., Czerny, C., Oliva, M., Schamall, D., Schultz, M., 2006. Pathological alterations and traumas in the human skeletal remains from Mladeč. In: Teschler-Nicola, M. (Ed.), The Mladeč Caves and Their Remains. Springer Verlag, Wien, pp. 473–487.

Thorne, A., Grün, R., Mortimer, G., Spooner, N.A., Simpson, J.J., et al., 1999. Australia's oldest human remains: age of the Lake Mungo 3 skeleton. Journal of Human Evolution 36, 591–612.

Trinkaus, E., 1993. Femoral neck-shaft angles of the Qafzeh-Skhul early modern humans, and activity levels among immature Near Eastern Middle Paleolithic hominids. Journal of Human Evolution 25, 393–416.

Trinkaus, E., 1997. Appendicular robusticity and the paleobiology of modern human emergence. Proceedings of the National Academy of Sciences of the United States of America 94, 13367–13373.

Trinkaus, E., Milota, S., Rodrigo, R., Mircea, G., Moldovan, O., 2003. Early modern human cranial remains from the Peştera cu Oase, Romania. Journal of Human Evolution 45, 245–253.

Trinkaus, E., Hillson, S.W., Franciscus, R.G., Holliday, T.W., 2005. Skeletal and dental paleopathology. In: Trinkaus, E., Svoboda, J. (Eds.), Early Modern Human Evolution in Central Europe: The People of Dolní Věstonice and Pavlov. University Press, Oxford, pp. 419–458.

Vanhaeren, M., d'Errico, F., Stringer, C., James, S.L., Todd, J.A., et al., 2006. Middle Paleolithic shell beads in Israel and Algeria. Science 312, 1785–1788.

Vermeersch, P.M., 2002. Two Upper Palaeolithic burials at Nazlet Khater. In: Vermeersch, P.M. (Ed.), Palaeolithic Quarrying Sites in Upper and Middle Egypt. Leuven University Press, Leuven, pp. 273–282.

Vermeersch, P.M., Paulissen, E., Stokes, S., Charlier, C., van Peer, P., et al., 1998. A Middle Palaeolithic burial of a modern human at Taramsa Hill, Egypt. Antiquity 72, 475–484.

Vogel, J.C., 2001. Radiometric dates for the Middle Stone Age in South Africa. In: Tobias, P.V., Raath, M.A., Moggi-Cecchi, J., Doyle, G.A. (Eds.), Humanity from African Naissance to Coming Millennia. Firenze University Press/Witwatersrand University Press, Florence/Johannesburg, pp. 261–268.

White, T.D., 1992. Prehistoric Cannibalism at Mancos 5MTUMR-2346. Princeton University Press, Princeton.

White, T.D., Asfaw, B., DeGusta, D., Gilbert, H., Richards, G.D., et al., 2003. Pleistocene *Homo sapiens* from Middle Awash, Ethiopia. Nature 423, 742–747.

Wild, E.M., Teschler-Nicola, M., Kutschera, W., Steier, P., Trinkaus, E., et al., 2005. First direct dating of Early Upper Paleolithic human remains from Mladeč. Nature 435, 332–335.

Chapter 14

Evolution of Tool Use

Nicholas Toth, Kathy Schick
The Stone Age Institute, Gosport, IN, USA; Indiana University, Bloomington, IN, USA

SYNOPSIS

The profound reliance of the human species on tools for its survival and adaptation is unique in the animal world. Prehistoric evidence for tool use as an adaptive strategy in human evolution extends back at least 3.3 million years, when stone tools began to be found at prehistoric sites in Africa in regions containing fossils of early bipedal ancestors. Archaeological research documents a long period of dependence on stone tools through a succession of ancestral species and a gradually accelerating pace of technological change, with emergence of large complex societies and rapid technological proliferation only within the past 10,000 years.

Humans and many of our extinct ancestral species have been coevolving with tools and technology for millions of years. In the kingdom Animalia, our species (anatomically modern *Homo sapiens*) is the consummate toolmaker and tool user; our very existence is dependent on our use of tools. Humans are unique in the history of life on Earth in their ability to adapt to a large and diverse range of environments and shape their world through the use of tools and technological systems. From its origins millions of years ago, the human lineage has greatly expanded its range, increased its population size, and increasingly affected global environmental and climatic patterns at an alarming and perilous pace. Understanding how we arrived at the modern human condition is the task not only of historians but of palaeoanthropologists that discover, study, and interpret the prehistoric fossil, archaeological, and palaeoenvironmental evidence.

The coevolution of technology and human biology has been an important focus of palaeoanthropology. This process has been dubbed the "biocultural feedback system" by noted anthropologist Sherwood Washburn (1960), "gene–culture coevolution" by sociobiologists Alan Lumsden and E.O Wilson (1983), and "techno-organic evolution" by the authors of this chapter (Schick and Toth, 1993).

Over a century-and-a-half palaeoanthropological investigations have taken place since Charles Darwin published his *Origin of Species by Means of Natural Selection* in 1859. Thousands of hominin fossils representing hundreds of individuals and dozens of taxa are now documented since the last common ancestor of chimpanzee and humans, perhaps 8 million years ago (mya); the archaeological record for tools and technology is documented for the past 3.3 million years. For overviews of the human evolutionary record, see the preceding seven chapters by Sept, Hunt, Ward, Simpson, Ahern, and Holt in the current volume. Also see, for example, Boyd and Silk (2012), Broadfield et al. (2010), Cartmill and Smith (2009), Delson et al. (2000), Foley and Lewin (2004), Henke and Tattersall (2007), Johanson and Edgar (2006), Klein (2009), Stringer and Andrews (2005), Roberts (2011), Scarre (2013), Schick and Toth (1993), Tattersall (2008), and Zimmer (2005).

EVOLVING HOMININ FORMS AND TECHNOLOGIES: A REVIEW

The backdrop for the evolution of technology includes a large number of evolving hominin species that predate the emergence of tools at 3.3 mya, and then continue to evolve and diversify over the past 3.3 million years of technological evolution (Table 1). (Further descriptions of technologies noted in this section will be provided in later sections of this chapter.) Between 7 and 4 mya, a number of hominin forms have been recovered from deposits in Africa that researchers consider to be closely related to our evolving lineage. These involve a number of species (*Sahelanthropus tchadensis*, *Orrorin tugenensis*, *Ardipithecus kadabba*, *Ardipithecus ramidis*) that appear to be apelike in many ways, but that show also some traits indicating evolutionary characteristics that may be ancestral to later hominins (see chapter by Hunt on Earliest Hominins). These are sometimes called the "pre-australopithecines," as they show possible relationships to the australopithecines that emerge approximately 4 mya.

The australopithecines, the various fossil species placed in the genus *Australopithecus*, have a span of almost 3 million years, and can be grouped by time into earlier and later forms (see chapter by Ward). The earlier australopithecines emerge in the fossil record a little more than 4 mya and continue evolving and diversifying until at least 2.5 mya. They are first evident in East Africa and then also in South Africa. Taxa include *Australopithecus anamensis*, *Australopithecus afarensis*, *Australopithecus garhi*, and *Australopithecus africanus*. Another genus, *Kenyanthropus*, dating to 3.5 mya, has also been proposed by some researchers (Leakey et al., 2001). The later australopithecines, sometimes put into the genus *Paranthropus*, forms with robust jaws and large cheek teeth, coexisted with evolving forms of *Homo* in both East and South Africa until their extinction between 1.5 and 1.0 mya. Our ancestral lineage almost certainly stems from the earlier australopithecines, with *A. afarensis*, a small-brained hominin with a well-developed adaptation to bipedal walking (Johanson, 2004; Johanson and Edgar, 2006; Tattersall, 2008), widely considered to be ancestral to later hominins. Very late in this span of early australopithecines, *A. garhi* (Asfaw et al., 1999), a still small-brained but nonrobust form, is found in deposits of similar age to early stone tools at Gona, Ethiopia (Semaw et al., 1997; Semaw, 2006), and in the same region, and so can be considered a candidate for the earliest toolmaker. There are also fossils in the Afar region of Ethiopia attributed to early *Homo*, including a mandible dating to 2.8 mya in the Ledi-Geraru region (Villmoare et al., 2015) and a fossil maxilla dated to 2.3 mya in the Hadar region (Kimbel et al., 1996). We currently do not have information regarding the morphology of the brain case or the cranial capacity of these fossils, but similar fossils would also be potential candidates for the toolmakers at Gona.

The robust australopithecine, *A. (P.) aethiopicus* (Kimbel et al., 1988), is also roughly contemporary with early stone tools, but would appear to be a less likely candidate for the maker of the first stone tools, as its massive chewing apparatus would indicate a biological adaptation with a strong emphasis on masticating hard,

TABLE 1 Hominin Forms in the Last 7 Million Years and Contemporary or Associated Technologies

Hominin Group	Species	Approximate Age	Associated/Contemporary Technological Tradition	Country/Region
Pre-Australopithecines	*Sahelonthropus tchadensis*	7 mya	None known	Chad (Central Africa)
	Orrorin tugenensis	6 mya	None known	Kenya (East Africa)
	Ardipithecus kadabba	5.6 mya	None known	Ethiopia (East Africa)
	Ardipithecus ramidus	4.4 mya	None known	Ethiopia (East Africa)
Earlier Australopithecines	*Australopithecus anamensis*	4.2–3.9 mya	None known	Kenya and Ethiopia (East Africa)
	Australopithecus afarensis	3.9–2.9 mya	(Earliest stone tools, Lomekwi, Kenya)	Ethiopia and Tanzania (East Africa)
	Australopithecus africanus	3–2 mya	None known	South Africa
	Australopithecus garhi	2.6 mya	(Early Oldowan)	Ethiopia (East Africa)
	Australopithecus aethiopicus	2.5 mya	(Early Oldowan)	Ethiopia and Kenya (East Africa)
Later Australopithecines	*Australopithecus boisei*	2.3–1.2 mya	(Oldowan/Acheulean)	Ethiopia, Kenya, Tanzania (East Africa)
	Australopithecus robustus	2–1.2 mya	(Oldowan/Acheulean)	South Africa
Early *Homo*	*Homo* sp.	2.3–2 mya	Early Oldowan	East Africa
	Homo rudolfensis	1.9 mya	Oldowan	Kenya (East Africa)
	Homo habilis	2.0–1.4 mya	Oldowan	East Africa, South Africa
	Homo ergaster/erectus	1.8 mya–200 kya	Oldowan/Acheulean	Africa, Eurasia
Later *Homo*	*Homo heidelbergensis*	600–250 kya	Later Acheulean/non-Acheulean	Africa, Eurasia
	Homo neandertalensis	300–30 kya	Middle Palaeolithic	Eurasia
	Homo helmei	260 kya	Middle Stone Age	South Africa
	Early *Homo sapiens*	160–50 kya	Middle Palaeolithic/Middle Stone Age	Africa, Eurasia
	Later *Homo sapiens*	50 kya to present	Upper Palaeolithic/Later Stone Age, Mesolithic, Neolithic, etc.	Africa, Eurasia, Australia, New Guinea, Americas

presumably lower energy foodstuffs. The later robust australopithecines in East Africa (*A. (P.) boisei*) and South Africa (*A. (P.) robustus*) carry on this biological adaptation until their ultimate extinction between 1.5 and 1 mya (Grine, 1988).

Starting about 2.8 mya, fossil remains have been found that have been placed in our genus, *Homo*, largely based on dental characteristics (see chapter by Simpson). The earliest fossils do not include evidence of cranial volume, so at present it is not known when the dramatic brain expansion evidence in the *Homo* lineage actually began. Starting approximately 2 mya, forms of *Homo* showing brain expansion appear in the fossil record in Africa (*Homo rudolfensis* and *Homo habilis*) and are presumed to be toolmakers responsible for the many Oldowan sites that have been found (Schick and Toth, 2006; Toth and Schick, 2010). By approximately 1.8 mya, a larger bodied and larger brained form appears in Africa, *Homo erectus* (also called *Homo ergaster*) and at the same time an evident expansion out of Africa is evidenced by fossils at the site of Dmanisi in the Republic of Georgia (Lordkipanidze et al., 2013). The presence of a large range of variation among the five hominin crania found at the Dmanisi site, in a single deposit and contemporary with one another, would indicate that hominins in this phase of human evolution constituted a quite variable species, and thus splitting Homo fossil samples into separate species based on morphological characteristics may not always be justifiable. *Homo erectus* is associated with both Oldowan and Acheulean sites in Africa and much of Eurasia, although the Acheulean is largely absent from eastern Europe through central and eastern Asia.

As multiple hominin species are contemporary with early Oldowan as well as Acheulean industries, both in East Africa and South Africa, the question naturally arises as to which species was or, perhaps, were responsible for early stone toolmaking. It cannot be ruled out, of course, that robust australopithecines made stone tools, at least on occasion, as they were present in the same regions during the time that early stone tool sites were produced. But, in addition to the noted emphasis among the robust australopithecines on a biological adaptation involving heavy mastication, stone tool sites continue on in time after their extinction and become even more numerous and widespread, as do evolving species of *Homo*. Thus, it is certain that the adaptation of *Homo* had incorporated a technological adaptation into its evolutionary strategy, and it would appear likely that *Homo* species were largely responsible for early stone tool industries.

In the past million years, *Homo* has undergone fairly rapid and dramatic evolutionary changes along with an accelerating pace of technological change. Between 1 mya and a few hundred thousand years ago, *H. erectus* evolved into a larger brained hominin with robust features of the cranium and face (see chapter by Simpson). In Africa, these larger brained forms included *Homo heidelbergensis*, *Homo helmei* (these two taxa were sometimes previously assigned to "archaic *Homo sapiens*"), and early anatomically modern humans (see chapter by Ahern). In Europe, these forms included *H. heidelbergensis* and *Homo neandertalensis* (Neanderthals). In East Asia, these forms are sometimes assigned to *H. heidelbergensis* and some more recent archaic form of *Homo* (species often not named). This later evolution of the genus *Homo* witnessed the technological transition from the later Acheulean to the Middle Palaeolithic (Middle Stone Age in sub-Saharan Africa) and then the Upper Palaeolithic (Later Stone Age in sub-Saharan Africa). (Note that the traditional British spelling of "Palaeolithic" is used throughout this chapter.)

By at least 160,000 years ago, fossils have been found in parts of Africa and the Near East that are considered to be early anatomically modern *H. sapiens* (AMHS), or early *H. sapiens* (see chapter by Holt). Early *H. sapiens* were generally associated with Middle Palaeolithic and Middle Stone Age technologies. Starting between 50,000 and 40,000 years ago, *H. sapiens* became much more widespread in the Old World, spreading through much of Eurasia (and coexisting with Neanderthals for some time in some regions), into the Pacific to Australia and ultimately to the Americas, and by this time are associated with Upper Palaeolithic or Later Stone Age technologies during the latter part of the Pleistocene. The emergence of food production, metalworking, complex societies, and escalating technological change are recent phenomena in our evolution (Scarre, 2013), all produced by modern *H. sapiens* since the end of the last Ice Age 10,000 to 12,000 years ago.

TOOLS IN THE ANIMAL WORLD

Although the definition of a "tool" may vary, in behavioral studies it usually means that an organism takes an extrasomatic object (not part of one's biology) and employs it toward some end (Schick and Toth, 1993; Shumaker et al., 2011). By this definition, a bird using a twig to probe for insects is an example of tool use, whereas a spider spinning a web (using its own biological repertoire) is not. An "artifact," in archaeological terminology, is an object that has been modified by a human or proto-human, whether it has been used or not. Waste flakes from stone tool manufacture, for example, are artifacts.

A range of animals are known to use tools in the wild. Interestingly, the wide range of animal species with tool-using behavior does not appear to correlate well with evolutionary history or neurological complexity, ranging from invertebrates such as digger wasps and octopus to fish, amphibians, reptiles, birds, nonprimate mammals, and primates (Shumaker et al., 2011; Sanz et al., 2013).

In nonprimate species it is likely that these tool-using behaviors are "instinctive" rather than learned, are common to the species, and have foundations in the genetics of the species. In primates, however, many instances of tool-related behaviors have been documented in a number of species that show good evidence of having been learned rather that innate and to be shared through learning within the primate group. Such acquired tool-using and toolmaking behavior has been documented among some monkeys such as capuchins (Shumaker et al., 2011), but is observable more typically among apes, especially chimpanzees.

CHIMPANZEE TOOL USE AS A MODEL FOR EARLY HOMININS

Chimpanzees have the richest tool repertoire of any nonhuman animal, and these tool behaviors are learned, often from infants watching their mothers and then going through an apprenticeship of trial and error until the tool behavior is mastered (McGrew, 1992, 2004). This includes termite and ant fishing with carefully shaped twigs, nut cracking with wood or stone hammers and anvils, and making sponges out of chewed leaves to collect water or for cleaning.

Wild chimpanzees in West and East Africa have been documented to have over three dozen cultural traits, many involving tool use (Whiten et al., 1999). Interestingly, these cultural traits demonstrate significant geographical clustering, so that chimpanzee groups situated closer to one another share more cultural traits, with the number of shared traits decreasing significantly among groups at greater distances from one another (Toth and Schick, 2009b; Whiten et al., 2009). This pattern of geographical clustering of shared cultural traits among chimpanzee groups may serve as a model for early hominin groups, such that geographic patterning in technological characteristics might be expected among groups in closer proximity to one another.

As our closest living relatives, the chimpanzees, show a diversity of tool-using and even sometimes toolmaking behaviors in the wild, it might be considered very likely that early hominin ancestors also utilized a variety of other materials before the prehistoric inception of stone tool manufacture and use. It is quite possible, or even likely in view of tool behaviors among chimpanzees, that before the advent of stone tools our ancestors also probed and manipulated materials in their environment with a variety of tools in materials other than stone, particularly in order to procure food or water. Unfortunately, materials such as wood, leaves, or bark that might readily be used or fashioned as tools are readily destroyed by natural forces, and so tools in materials other than stone would tend to be perishable and not likely preserved in the ancient prehistoric record. Thus, the beginning of a verified, documented record of tool use in the human lineage begins with the invention of flaked and battered stone tools.

In an experimental setting, modern bonobos ("pygmy chimpanzees") have learned to make and use flaked stone tools (Figure 1) and produce the rudiments of an Oldowan-like technology consisting of battered hammerstones, cobble cores, and flakes and fragments (Schick et al., 1999; Toth et al., 2006). These artifacts, however, do not show the level of skill and dexterity seen in the earliest archaeological sites (discussed below).

THE EARLIEST KNOWN STONE TOOLS

The earliest known stone tools are found from sites in the Great Rift Valley of East Africa. Although there are claims of cut marks from stone tools on 3.4 million-year-old animal bones from the site of Dikika in Ethiopia (McPherron et al., 2010), there are no known stone artifacts from this site, and it is possible that these marks were, in fact, tooth marks created by crocodiles (Njau, 2012), whose teeth are numerous at this locality.

FIGURE 1 Kanzi, a bonobo or "pygmy chimpanzee" (*Pan paniscus*), has learned to flake stone, producing Oldowan-like artifacts including battered hammerstones, simple cobble cores, and sharp-edged flakes and fragments.

At present, the earliest probable claim for stone artifacts is at Lomekwi 3, West Turkana, Kenya (Harmand et al., 2015a,b; Balter, 2015). They are found in a geological horizon believed to be 3.3 million years old based on paleomagnetic studies. The artifacts, simple cores, flakes, and fragments, are made on large lava cobbles, with 20 artifacts reported in situ and another 130 pieces from the surface. The next earliest documented and well-dated artifacts consist of stone tools from sites dated to between 2.6 and 2.5 mya found at the locality of Gona in Ethiopia in the Afar Rift (Semaw, 2006; Semaw et al., 1997; Toth et al., 2006).

These sites mark the beginnings of what is called the Palaeolithic, or "Old Stone Age," the period of human cultural development in which stone tools constituted the preeminent components of the known tool repertoire. However, whether these very early archaeological occurrences represent the beginnings of a continuous stone tool tradition only sporadically sampled in the prehistoric record, or whether these represent early, intermittent experiments in stone knapping without the establishment of a long-term tradition (i.e., "flashes in the pan"), is not known at this time.

THE OLDOWAN (EARLIER LOWER PALAEOLITHIC/EARLY STONE AGE)

The earliest stone tools are categorized as the "Oldowan," a tool industry named after the famous site of Olduvai Gorge in Tanzania (Hovers and Braun, 2009; Isaac, 1990;

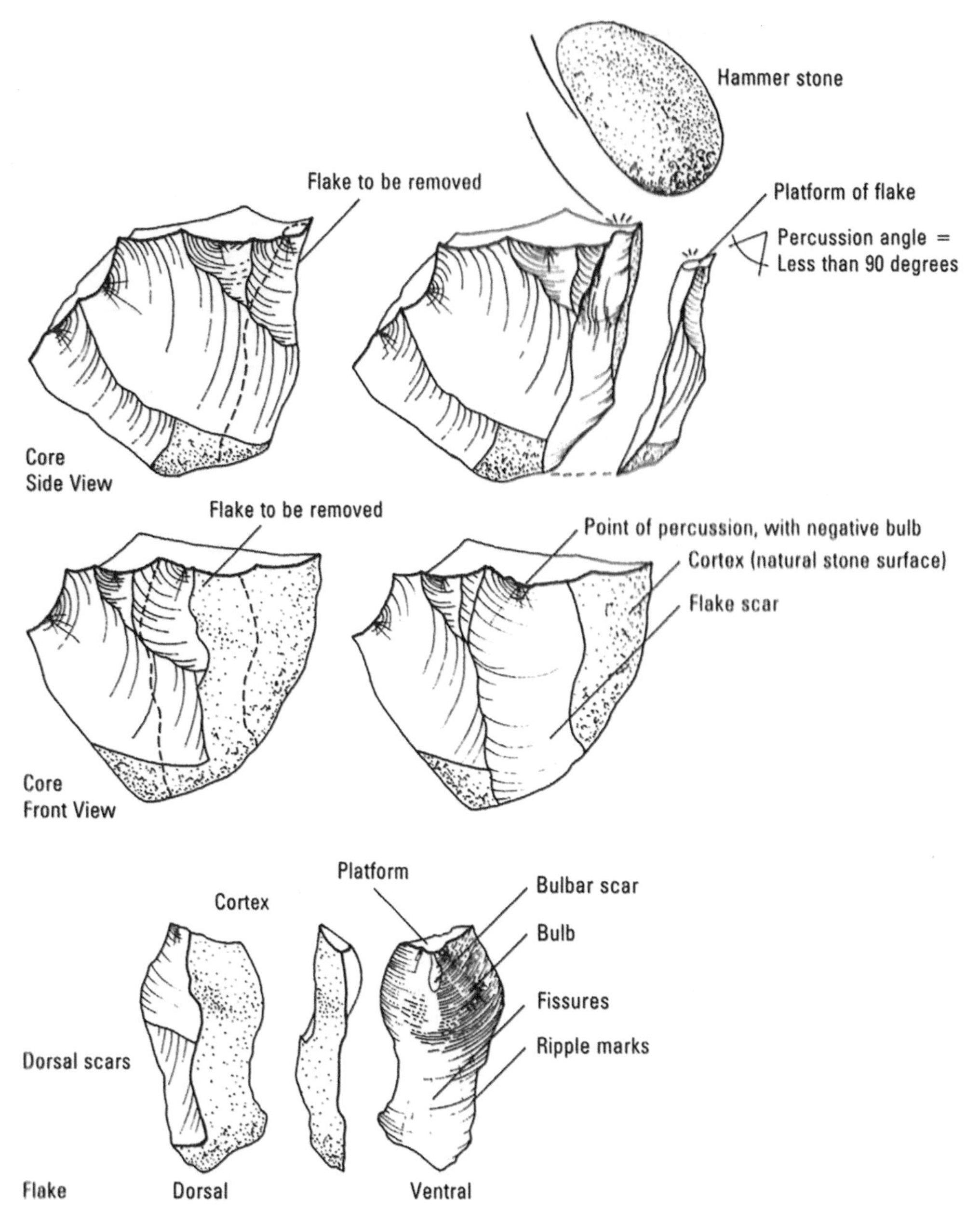

FIGURE 2 Controlled stone fracture (conchoidal fracture), showing a battered hammerstone, a core, and a flake.

Klein, 2009; Leakey, 1936, 1971; Schick and Toth, 1993, 2006, 2009; Toth, 1985; Toth and Schick, 2006, 2009a, 2009b, 2013). Starting approximately 2.3 mya, Oldowan sites become more numerous and widespread on the African continent, with the first evidence in Eurasia by c. 1.85 mya. Besides the earlier sites noted above (Lomekwi 3 and Gona), other major Oldowan sites include Hadar, Ethiopia (Kimbel et al., 1996), the Omo Valley, Ethiopia (Howell et al., 1987), West Turkana, Nachukui Formation, Kenya (Roche et al., 1999), Kanjera, Kenya (Plummer et al., 1999), East Turkana (Koobi Fora), Kenya (Isaac, 1997), Olduvai Gorge, Tanzania (Leakey, 1971) (all of the above in the East African Rift), Sterkfontein (Kuman, 1994) and Swartkrans (Clark, 1991) in South Africa, Ain Hanech and El-Kherba in Algeria (Sahnouni, 2006), and Dmanisi in the Republic of Georgia (Ferring et al., 2011).

Oldowan tools consist primarily of cobbles and chunks of stone that have been deliberately modified by percussive flaking, that is, hitting one rock forcefully against another and striking off stone flakes. Such artifacts typically exhibit evidence of deliberate, patterned conchoidal fracture induced through repeated sharp, percussive blows of one rock against another (Figure 2).

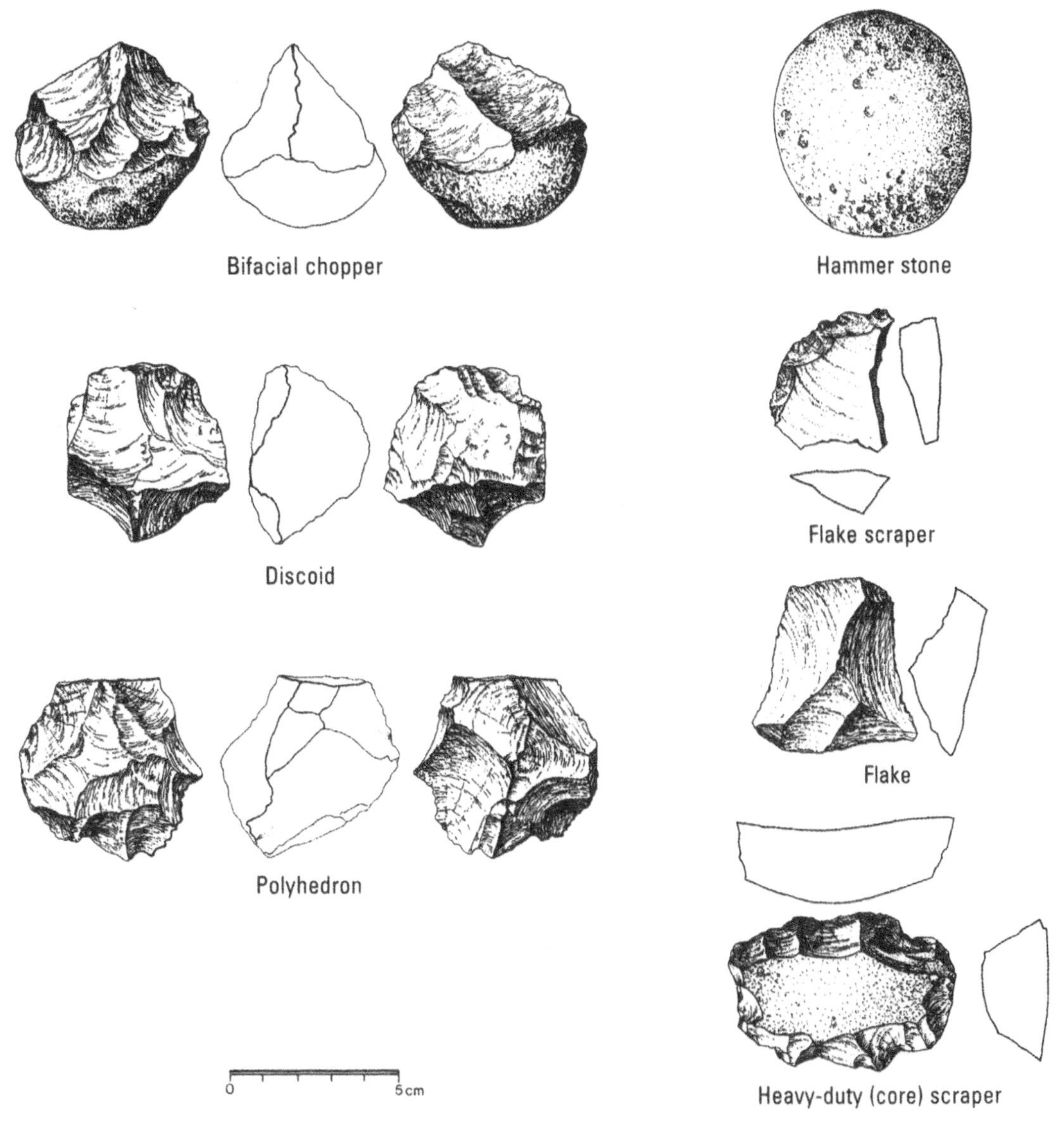

FIGURE 3 A range of Oldowan artifact forms, including a battered hammerstone, cores (chopper, discoid, polyhedron, heavy-duty scraper), flake, and retouched flake scraper.

Common elements of Oldowan stone tool assemblages include cores (pieces of stone that have been fractured through forceful blows with another stone), flakes (the often very sharp pieces of stone struck from these cores), and sometimes battered hammerstones that had been used to flake the cores (Figure 3).

Archaeologists have over the years developed a number of classification schemes to characterize the components of early stone technologies (Schick and Toth, 2006), usually employing terms that describe the shapes of the cores as well as possible functional qualities they might have had. Thus, such terms such as "chopper," "discoid," "polyhedron," and "core scraper" are often employed to describe cores in early stone tool assemblages. Experiments have demonstrated that all of the major tool categories can be produced as a by-product of removing stone flakes from cores, with the resultant shape influenced largely by a combination of the initial shape of the stone being flaked and how extensively the core was flaked (Toth, 1985). Thus, it is uncertain that early stone tool makers were deliberately producing cores with desired, targeted shapes, as they may well have been largely focused on producing quantities of flakes from stone cores. The Oldowan is also characterized by having retouched flakes (flakes chipped on their edge to resharpen or shape them), such as "light duty scrapers" and "awls." These retouched forms become more common as the Oldowan progresses.

Although for many years, an emphasis had been placed by archaeologists on the stone cores as the primarily tools within early stone assemblages, experimental archaeological studies have demonstrated that the flakes struck off from such cores also have superb utility as cutting tools (Toth, 1985). Functional studies of stone tool replicas have demonstrated how the different types of early stone artifacts may have been used, for example, the usefulness of stone "choppers" to cut branches into digging sticks, probes, or spears. Such experiments have also highlighted how a simple stone flake, produced by striking a stone hammer against another stone, can produce a keenly sharp cutting edge. Such stone flakes would effectively have provided early hominins with stone knives that could be used to procure meat resources, either from scavenging carcasses on the landscape and/or from (likely smaller) animals they might have been able to hunt and kill. As early hominins were not biologically well adapted to the acquisition of meat resources, lacking the sharp teeth and claws of carnivores, such toolmaking behavior would have opened up a new and vital niche for them.

It has been argued in the "expensive tissue hypothesis" that a dietary shift toward including more high-quality diet components such as meat resources may have been a critical factor in allowing for the evolutionary expansion of the human brain (Aiello and Wheeler, 1995). According to this model, a higher quality diet would have allowed an evolutionary shift in the body's energy budget, reducing energy demands for food processing in the gut and allowing a shift toward maintaining the very high energy demands of expanding brain tissue (see chapters by Little and Blumler, and Wiley in the current volume). Thus, stone tools could have been a critical factor in initially supporting the higher energy demands of brain expansion in our lineage. Interestingly, significant and rapid brain expansion is not observed in the human lineage until after the appearance of stone tools in the prehistoric record (Holloway et al., 2004; Toth and Schick, 2010). Improved adaptation through intelligent, inventive tool use may also have selected for enlarged brains, and the use of these tools to procure animal resources may have set the stage and allowed the rapid brain expansion that ensued. Of course, a circular feedback mechanism likely ensued, with the brain expansion also supporting enhanced tool-related behaviors and tool evolution, which would have been enhanced by further brain expansion, and so on.

It is open for speculation how stone tools were initially invented—that is, how early hominins discovered that hitting one rock against another would produce controlled fracture, and, moreover, that the resulting products could be employed for various tasks that could be useful in their lives and adaptation. This is not an intuitively obvious operation, and so it is very likely that it was discovered in the process of other tool-using tasks that involved the use of stones. One possible scenario is that stones accidentally fractured in the process of using stone hammers and anvils to break open hard objects, such as nuts or animal bones. The limb bones of dead animals can seasonally provide quantities of nutritious marrow, which is readily obtainable by animals such as hyenas biologically equipped with jaws and teeth that can induce fracture. A hominin using stone to venture into such a dietary niche could have accidentally discovered the basic principle of conchoidal fracture in the course of breaking open bones, and then went on to explore the properties of the artifacts so produced, for instance, in using stone flakes to cut remaining meat tissue or limbs from a carcass.

It is apparent, however, from the prehistoric record, that stone tools became incorporated within hominin behavior patterns starting approximately 3.3 mya. It is also clear that hominins began very often to transport stone around the landscape, carrying stone away from their original sources within stream beds and taking them to other locations where they were flaked and/or used, the locales that appear as "sites" in the archaeological record (Leakey, 1971; Schick and Toth, 1993, 2006). Sometimes significant concentrations of stone artifacts were produced on the landscape by such behaviors (Leakey, 1971; Schick, 1987), and some of these sites also contain numbers of animal bones and bone fragments.

In some instances, the formation of these large accumulations of stone artifacts ultimately involved transporting hundreds or even thousands of pounds of stone to a particular site, presumably through the efforts of a hominin group, and likely in some instances through repeated site visits (Schick, 1987). Flaked stone was also sometimes removed from such sites, presumably as a hominin group moved on from that location to another and carried potential tools with them for future use. Hominins did not take a random selection of stones locally available, as there is evidence of selectivity and preference in selection of stones better suited for flaking from within the local available sources of stone (Toth, 1985; Semaw et al., 2009).

Thus, starting at least 3.3 mya, the evolving human behavioral repertoire had solidly incorporated several new and significant activities revolving around stone tools: locating sources of stone on the landscape, selecting stone that would be better suited for toolmaking, transporting of stone to and from certain favorite locales, flaking stone, and using stone tools for various tasks. This marks a significant behavioral shift in our lineage, not only in the manufacture and use of stone tools, but also in the transport of large amounts of stone resources around the landscape and the evident decision-making and planning involved in such activities. These are behavior patterns unique to our lineage, as they have not been observed in any other species.

After its beginning at least 3.3 mya, Oldowan technology continued to develop over time, with some new tool categories added to the stone industries, including more tools made by modifying flakes (Leakey, 1971). Change was very slow, however, for several hundred thousands of years of Oldowan technology, without marked shifts in technological procedures. Stone tool sites became more numerous and widespread in Africa, particularly starting approximately 2 mya, with many Oldowan sites discovered first in East Africa and then in South Africa and North Africa. By at least 1.8 mya, hominins had spread out of Africa to western Asia, taking their Oldowan toolmaking tradition with them as evidenced by the site of Dmanisi in the Republic of Georgia (Lordkipanidze et al., 2013).

THE ACHEULEAN AND CONTEMPORANEOUS INDUSTRIES (LATER LOWER PALAEOLITHIC)

Starting nearly 1.8 mya, hominins developed new stone tools, called Acheulean technology, that mark a significant departure from Oldowan stone toolmaking (Klein, 2009, 2013; Lepre et al., 2011). The Acheulean tool industry is named after a site in France, St. Acheul, which is much younger than the first Acheulean sites in Africa but is where Acheulean tools had been discovered and recognized in the nineteenth century. These new technologies centered on large flaked tools shaped into forms such as handaxes, cleavers, and picks. These tools are usually rather elongate and often somewhat flat, and very often flaked on both sides (bifacially flaked). Very early Acheulean sites often contained rather thick, trihedral pick-shaped tools (Figure 4).

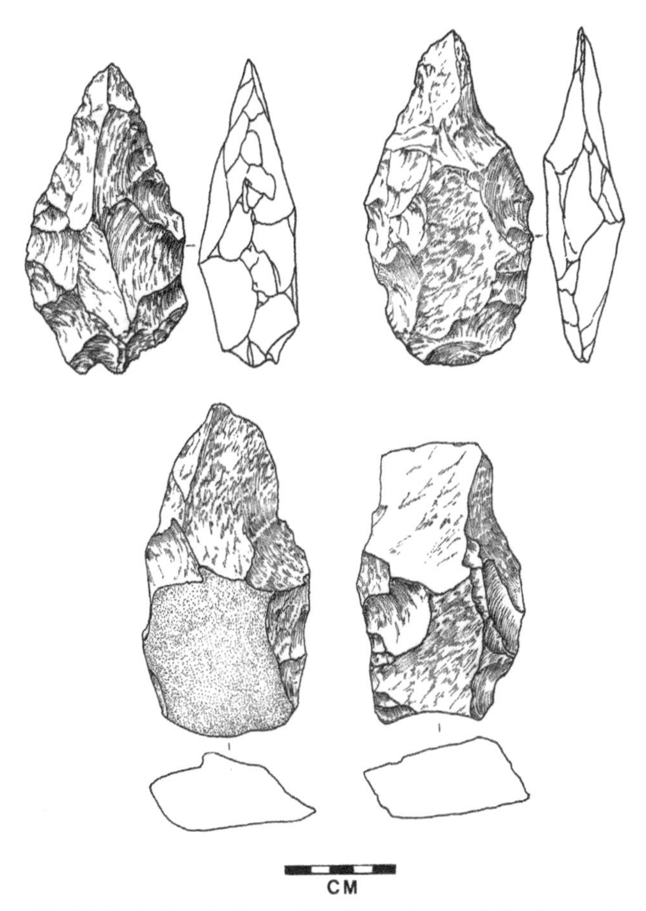

FIGURE 4 Early Acheulean artifact forms: three crude handaxes and a flake cleaver (bottom right).

The manufacture of Acheulean tools requires the use of larger pieces of stone than is characteristic of most Oldowan tool industries. In the early Acheulean, tools such as handaxes and cleavers were often made on large flakes that had been struck from very large boulder cores. Such tools can also be fashioned from large, relatively flat cobbles that have been flaked around much of their entire perimeter. The production of such large tools requires acquisition of large cores, which likely reflects ranging patterns that extended greater distances than those needed to acquire cobbles for the manufacture of Oldowan tools, which were generally available in streams more local to the sites. Large boulder cores are more readily available in more upstream, higher energy areas bordering a depositional basin, closer to the bedrock outcrops that serve as the ultimate source of the rock (Isaac, 1984). Thus, the development of Acheulean technology is evidence for hominins using more upstream parts of the environment, for which we have little direct evidence prior to this period.

The function of Acheulean tools has been a subject of great interest, speculation, and debate among archaeologists and others (Isaac, 1984; Schick and Toth, 1993). Some have argued that they represent cores for flake production rather than tool forms in themselves, while other arguments suggest they may represent multipurpose tools. Experiments in manufacturing Acheulean tools through replicating the technological procedures evidently followed render very unlikely the argument that they are simply cores. The production of these tool forms requires an elaborate series of steps and procedures that are far beyond those needed simply to produce useful flakes and, moreover, tend to minimize the size and number of useful flakes in the latter stages of production when final shaping is being undertaken. Based on experimental replication studies, the manufacture of Acheulean handaxe forms requires more than four times the number of cognitive decisions and procedures as those required by Oldowan toolmaking (Toth and Schick, 2009a). Acheulean tools such as handaxes, cleavers, and picks appear to represent deliberately shaped tools forms, purposefully and sometimes elaborately shaped into their forms for the functions they were to provide.

Experimental studies have revealed that handaxes and cleavers are remarkably useful and efficient tools for heavy-duty butchery of large animals, that is, in disarticulating limbs and other bones from animal carcasses (Schick and Toth, 1993; Toth and Schick, 2009a). Acheulean handaxes tend to show an emphasis by the tool makers in producing a sharp cutting edge toward the tip end of the tool. Moreover, such tools do not tend to show heavy damage on their sharper edges, indicating that they have been used on softer, more yielding materials rather than on harder materials that would have damaged their edges in use. As they tend to be sharpened along their edges, especially toward the tip end, handaxes provide a lengthy cutting edge along much of the tool's length, and meanwhile provide a good handle for gripping on the bottom end of the tool that also aids in applying pressure to the cutting operation. Moreover, when bifacially flaked, they essentially have a serrated edge, providing several cutting edges in tandem to allow for longer-lasting sharpness in cutting operations. Experiments have demonstrated that a handaxe not only has a longer cutting edge than a simple flake provides, but one that is useful for much longer (Toth and Schick, 2009a).

Acheulean technology represents an extremely long-lived tool tradition, lasting from almost 1.8 mya until approximately 250,000 years ago. As such, it is the longest tool tradition known, as subsequent tool industries became progressively shorter in duration and quicker to change over time. Interestingly, during the more than 1.5 million years of its duration, it also survived considerable expansion in geographic distribution of our lineage. The Acheulean tradition has been found from South Africa to the British Isles, from western Europe to the Indian subcontinent. The Acheulean did not, however, seem to spread for the most part throughout all of Eurasia, as it is largely absent from eastern Europe to central and far eastern Asia (Schick and Toth, 1993; Schick, 1994).

The manufacture of Acheulean tools such as handaxes and cleavers becomes more sophisticated over time, so that at sites later in the Acheulean period, by one million or half a mya, for example, Acheulean handaxes and cleavers can be very refined (Figure 5). Many such later Acheulean bifaces show remarkable symmetry, both in planform and in profile, and a great deal of finesse in their manufacture. The refinement in the methods of production and in the final product of many later Acheulean bifaces would appear to exceed considerations of function and may represent a developing sense of aesthetics in toolmakers (Schick and Toth, 1993) many hundreds of thousands of years prior to the evolution of *H. sapiens*.

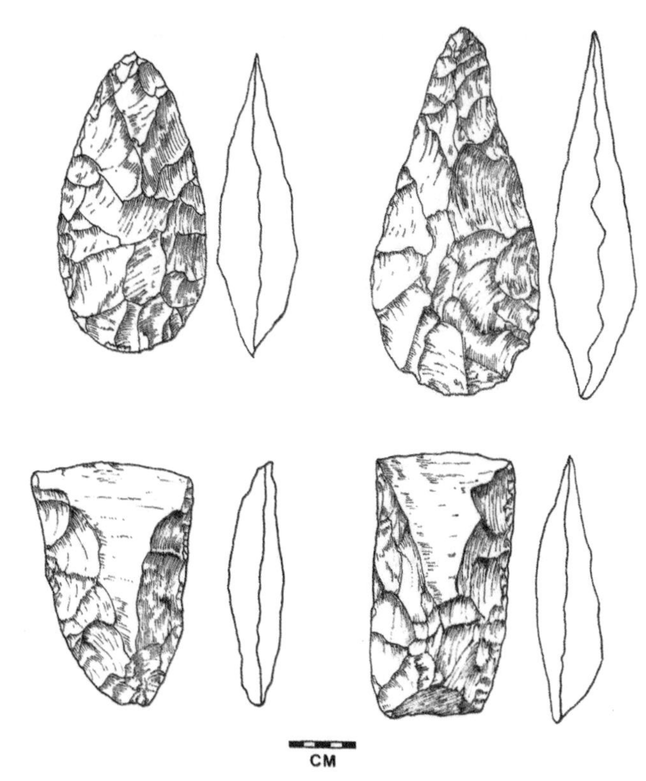

FIGURE 5 Later Acheulean artifact forms: an ovate handaxe (upper left), a lanceolate handaxe (upper right), and two cleavers made on large flakes (bottom).

THE MIDDLE PALAEOLITHIC/MIDDLE STONE AGE

Beginning around 300,000 years ago in some parts of Africa, and lasting in some parts of the Old World until approximately 30 thousand years ago (kya), this technological stage is usually called the Middle Palaeolithic in Europe, the Near East, and North Africa, and the Middle Stone Age in sub-Saharan Africa (Klein, 2009, 2013; Mellars, 1996). Hominins contemporaneous with this stage of technology include *H. neandertalensis* (Europe and the Near East), *H. helmei* and early *H. sapiens* (Africa), and archaic forms in East Asia (sometimes referred to as "Asian *heidelbergensis*" or "Asian archaic *H. sapiens*").

Technologies from this period are characterized by an abundance of flake tools (side scrapers, denticulates, backed knives, points) (Figure 6) that were made on flake blanks sometimes struck from special types of cores, notably radially flaked disc cores and Levallois cores. The Levallois method was used to carefully shape the morphology of the core in order to strike off one or more special flakes, such as large ovate flakes, convergent points, or occasionally blades (flakes that are at least twice as long as they are wide) (Figure 7). These "prepared core" technologies suggest higher cognitive skills and predetermination relative to earlier technologies.

Microscopic use-wear analysis on Middle Palaeolithic flint tools indicates that they were used for a variety of tasks, including animal butchery, hide scraping, and woodworking (Beyries, 1988). The morphologies of some of the retouched points suggest that they could have been hafted onto shafts for thrusting or throwing

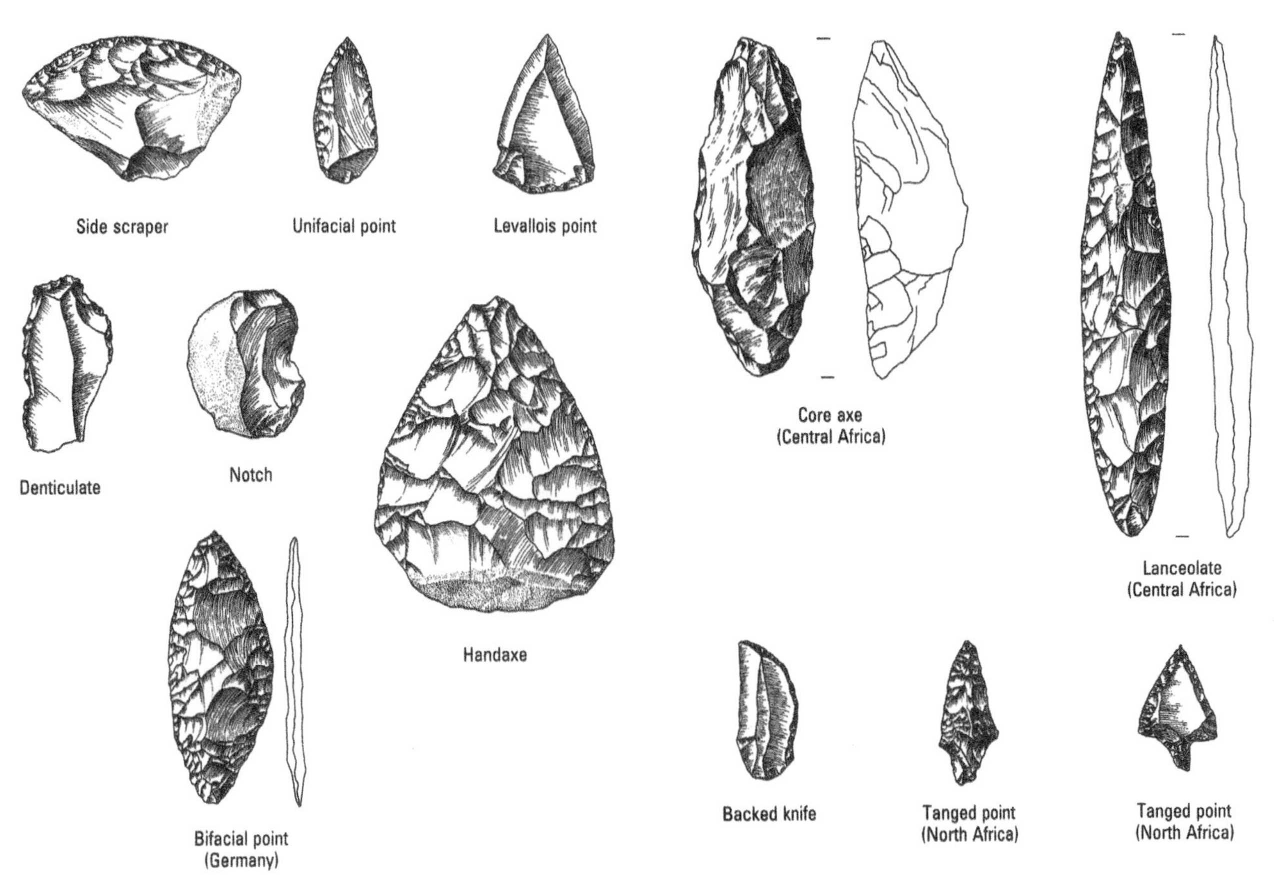

FIGURE 6 Middle Palaeolithic/Middle Stone Age artifact forms.

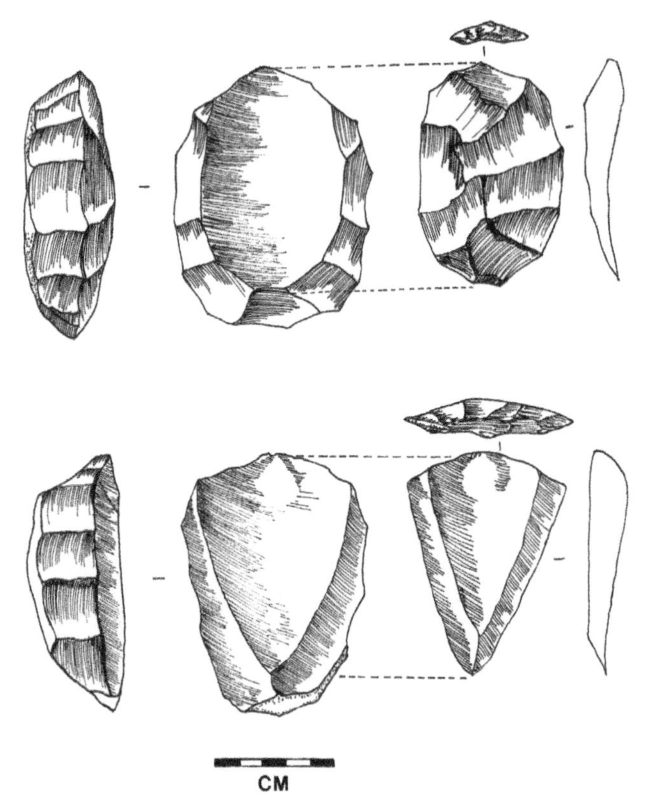

FIGURE 7 Examples of prepared cores. A Levallois "tortoise core" and flake (top), and a Levallois point core and Levallois point (bottom).

spears. In North Africa, tanged points and other tanged tools of the Aterian industry also suggest a hafting technology (Clark, 1982). Such technologies are indicative of "composite tools" that employ several elements together (e.g., stone point, wooden spear shaft, binding such as sinew or vegetable cordage, adhesive mastics such as gums, resin, or pitch).

Evidence of fire at habitation sites becomes much more common during this technological stage (Mellars, 1996; Roebroeks and Villa, 2011). Ethnographically, the major ways of hunter–gatherers producing fire are by creating high-temperature friction with the wooden drill/anvil technique, the wooden saw technique, or the thong technique (the latter usually creating friction with a supple strip of bamboo) (Spier, 1970; Hodges, 1995). Habitual manufacture and use of fire implies a sophisticated knowledge of the mechanical methods of production as well as the raw materials required (such as kindling of moss, dung, or some other material). The adaptive advantages of fire production (cooked food, warmth, predator avoidance) and its ultimate consequences are enormous, as the control of production and use of fire would have enabled human populations to spread to less hospitable geographic regions and climates, improve their ability to compete against predators on the landscape, and expand their access to food resources and their nutritional contents.

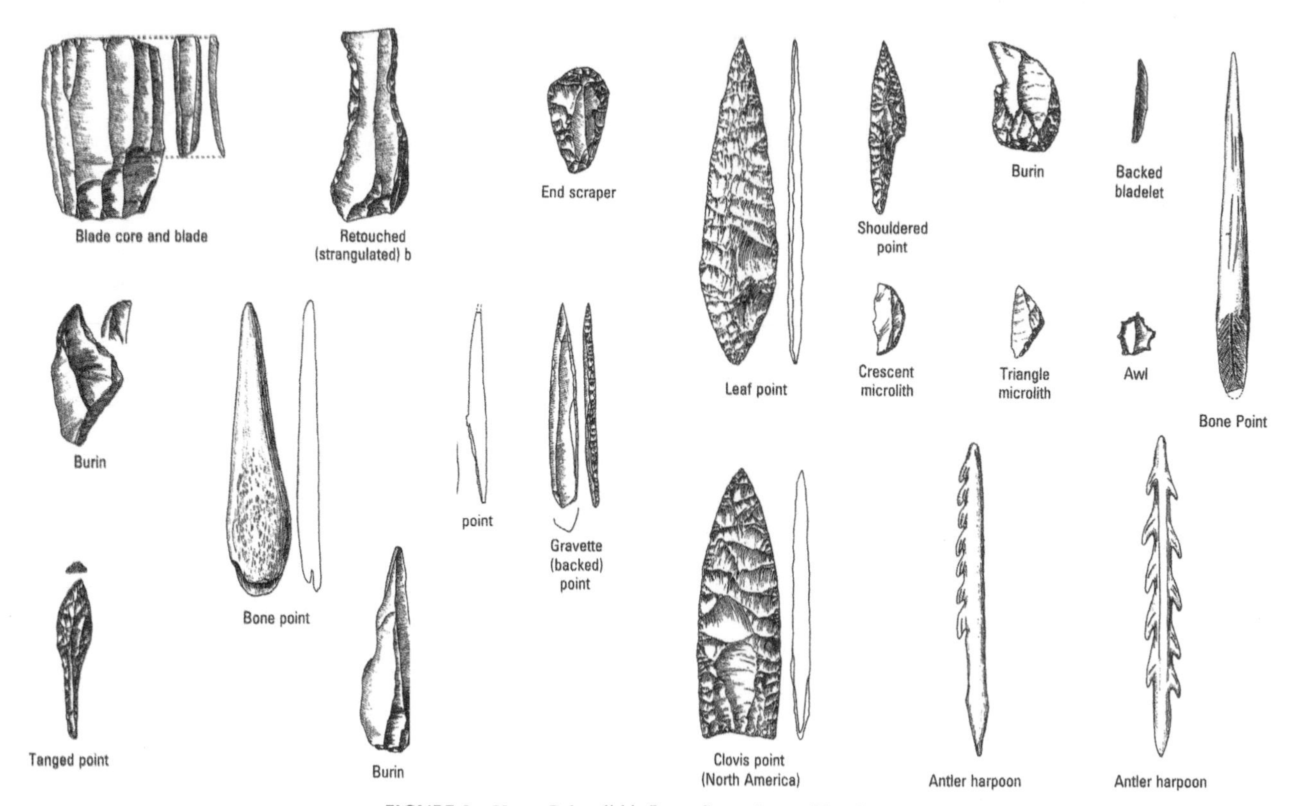

FIGURE 8 Upper Palaeolithic/Later Stone Age artifact forms.

It has been argued by some researchers (notably McBrearty and Brooks, 2000) that many of the behavioral traits that we consider to be characteristic of anatomically modern humans in the past 40,000 years have their roots in earlier periods, especially in the Middle Stone Age of Africa. They argue that traits such as blade technologies, hafting, bone tools, geometric crescents, abstract design, the use of ochre, and perforated beads can be found prior to 40,000 years ago. Although this is true, much of this evidence tends to be sporadic, with little long-term patterning or widespread distribution.

THE UPPER PALAEOLITHIC/LATE STONE AGE

Beginning around 50 kya, some anatomically modern hunter–gatherer groups began to have technologies that emphasized standardized blade technologies using a soft hammer or a bone or antler punch to produce long, thin blades (Klein, 2009; Pettitt, 2013). This technological stage is called the Upper Palaeolithic in Eurasia and North Africa and the Later Stone Age in sub-Saharan Africa. These blade technologies were a very efficient way of maximizing cutting edge relative to the mass of the stone. These blades could be used not only as cutting tools unmodified, but they also served as blanks for a wide range of other retouched tool forms, including end scrapers, burins (engraving tools), unifacial and bifacial spear points, backed knives, and pointed awls (Figure 8). In some of these industries flint was heated to make it easier to flake, and pressure flaking (pressing small flakes off with a fabricator of bone or antler) helped to shape delicate stone projectile points. Over time, there is a tendency to produce smaller geometric stone tools ("microliths") that served as elements of composite tools for points and knives. There is much more standardization of stone tools in this technological stage, as if the rules of production were becoming more codified in the cultures of these foragers.

Starting in this final period of the Palaeolithic, there is a record of accelerating use of materials other than stone for various purposes, such as bone, antler, and ivory for a variety of tools such as points, harpoons, needles, and spear throwers (Klein, 2009; Pettitt, 2013). In addition, Upper Palaeolithic peoples used fibers for mats and textiles (Adovasio et al., 1996), wood and animal bones for structures, shells and animal teeth for ornaments, and a variety of materials (ivory, limestone, calcite, clay, slate, ochers, and other pigments) for figurines and various types of artwork (White, 2003; Cook, 2013). Representative art in painting, engraving, and sculpture often

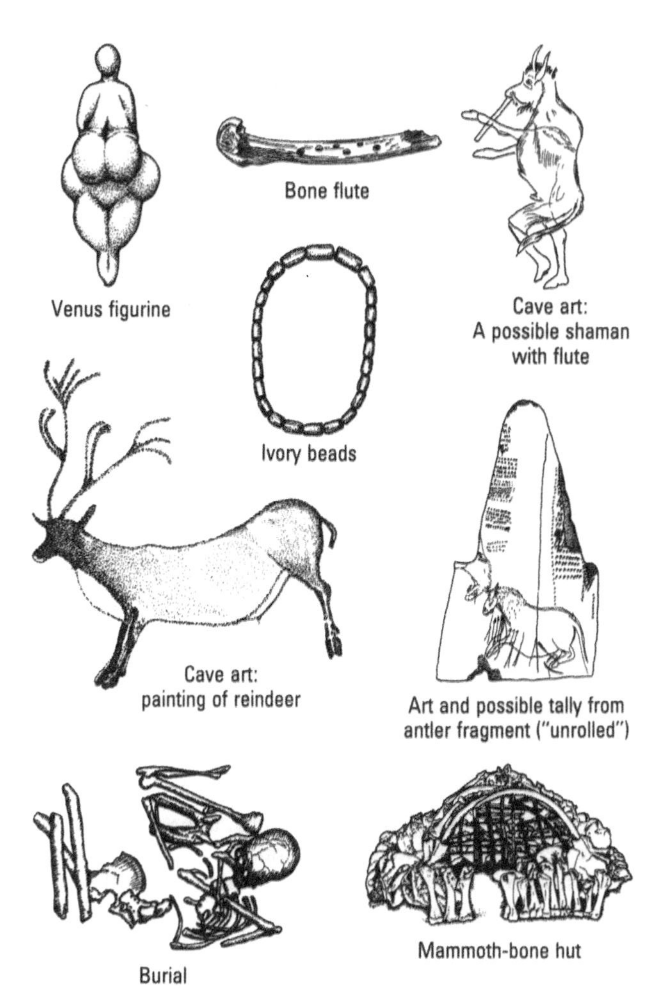

FIGURE 9 Upper Palaeolithic material culture and behaviors likely reflecting symbolism.

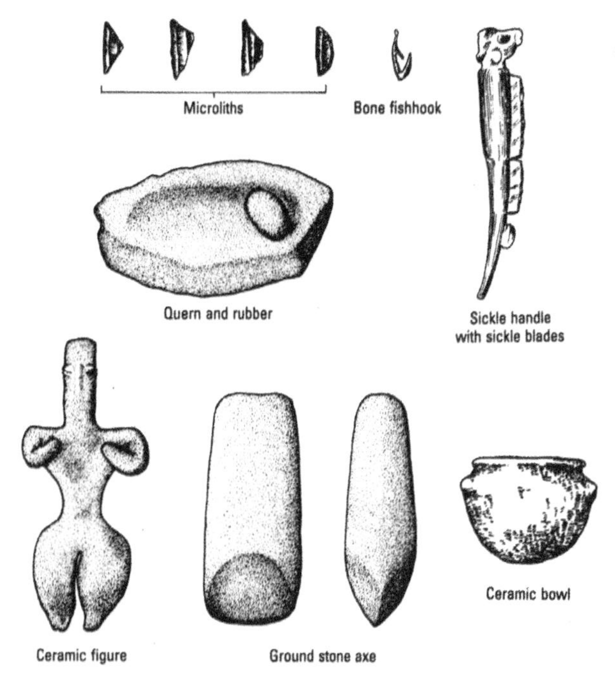

FIGURE 10 Neolithic artifact forms.

portrayed the animals of the artists' world. The famous caves of Chauvet and Lascaux in France and Altamira in Spain are classic examples of Upper Palaeolithic creativity (Clottes and Arnaud, 2003; White, 2003; Cook, 2013) It would appear that Upper Palaeolithic hominins were expressing themselves symbolically through material culture (as well as presumably in other forms of culture such as language, music, dance, and folklore) to a much greater degree than earlier hominins (Figure 9).

Architectural features, such as huts or tents, were much more common during this period, and hearths are often ringed with stones (Klein, 2009; Pettit, 2013). Long-distance trade of raw materials (e.g., flint, seashells, and amber) became much more common in this period (Mellars, 1990, 1996). It was during this time that modern humans spread to new continents: to Australia (crossing the Pacific waters from southeast Asia by boat) and to the Americas (spreading from East Asia over the land bridge that existed in the area of the Bering Straits during low sea levels or by following the southern coast by boat).

LATER DEVELOPMENTS

At the end of the Pleistocene and the retreat of the ice sheets, about 12,000 years ago, all of the world's human populations were foragers. In the Old World, the postglacial Holocene epoch witnessed many blade and bladelet technologies that tended to produce smaller stone tools (geometric microliths) that were used as composite tools for such implements as arrows and knives (Burdukiewicz, 2005). This technological stage is usually called the Mesolithic.

As the Holocene proceeded, hunter–gatherer groups slowly began to independently domesticate plants and animals in several areas of the world, including the Near East, Northern China, Southern China, sub-Saharan Africa, Mesoamerica, Peru, and New Guinea (see chapter by Little). These early farming communities in the Old World are called Neolithic. Societies tended to be more sedentary, with larger populations and more substantial architectural structures (including megaliths) and villages. New technological elements included ground stone axes, pottery, and grinding stones (mortars and pestles) (Figure 10). Long-distance trade of raw materials and manufactured goods (pottery, stone axes, etc.) is characteristic of this stage. As the Neolithic proceeded, there is increasing evidence of social stratification and more complex economic and political organization.

Not surprisingly, most of these areas of earliest food production and those to which early food production spread also saw the rise of even more complex societies ("states" or

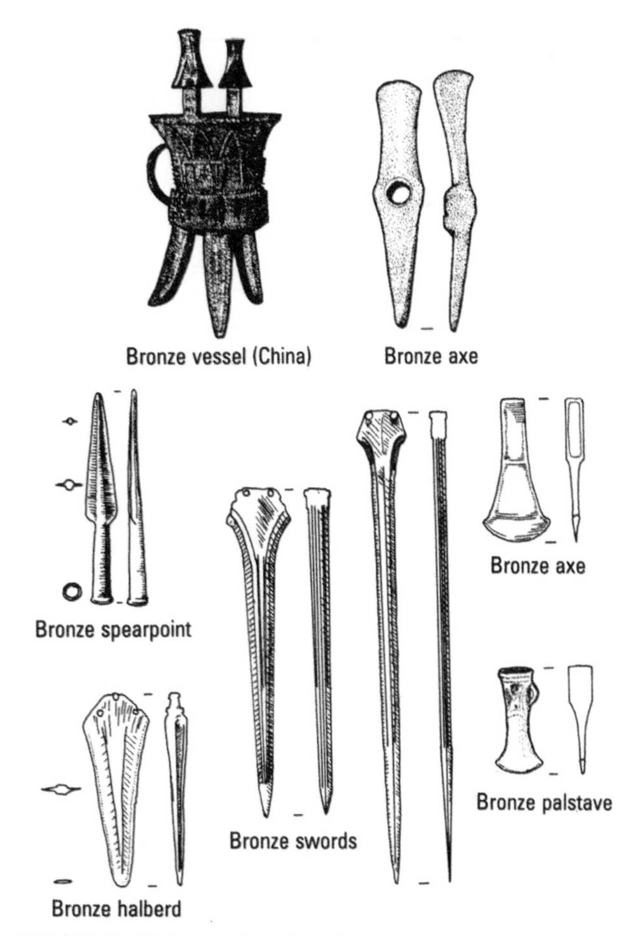

FIGURE 11 Early examples of metallurgy (Bronze Age).

FIGURE 12 Some material manifestations of complex societies: writing, monumental architecture, and currency.

"civilizations") often characterized by kingships, a standing army, and a priest class. In the Old World, these complex societies are associated with metallurgy (Tylecote, 2011), first using copper, then bronze, and then iron, wheeled vehicles, and horse riding (Scarre, 2013). In the Americas, these complex societies were beginning to experiment with metallurgy (Figure 11), but still had a strong reliance on stone tools at the time of European contact. Other traits of these societies often included writing, standardized systems of weights and measures, monumental architecture (such as palaces, temples, and tombs), and large-scale engineering such as organized systems of roads and water irrigation (Scarre, 2013; Figure 12).

Later civilizations led to the industrial revolution, including the use of fossil fuels to generate energy to power machines, as well as generate heat and light, and mass production of items with interchangeable, standardized components leading to the world we live in today. The challenges facing modern societies—overpopulation, overexploitation of resources, pollution, rise in carbon dioxide levels and world temperatures, etc.—are the end product of a long line of technological and adaptive patterns that can be traced back in the archaeological record 2.5 million years.

REFERENCES

Adovasio, J., Soffer, O., Klema, B., 1996. Upper Palaeolithic fibre technology: interlaced woven finds from Pavlov I, Czech Republic, c. 26,000 years ago. Antiquity 70, 526–534.

Aiello, L.C., Wheeler, P., 1995. The expensive tissue hypothesis. Current Anthropology 36 (3), 199–221.

Asfaw, B., White, T., Lovejoy, O., Latimer, B., Simpson, S., Suwa, G., 1999. *Australopithecus garhi*: a new species of early hominid from Ethiopia. Science 284, 629–635.

Balter, M., April 14, 2015. World's oldest stone tools discovered in Kenya. Sciencemag.org. http://dx.doi.org/10.1126/science.aab2487

Beyries, S., 1988. Functional variability of lithic sets in the Middle Paleolithic. In: Dibble, H., Montet-White, A. (Eds.), Upper Pleistocene Prehistory of Western Eurasia. University of Pennsylvania, Philadelphia, pp. 213–223.

Boyd, R., Silk, J., 2012. How Humans Evolved, sixth ed. Norton, New York.

Broadfield, D., Yuan, M., Schick, K., Toth, N., 2010. The Human Brain Evolving: Paleoneurological Studies in Honor of Ralph L. Holloway. The Stone Age Institute Press, Gosport, IN.

Burdukiewicz, J.M., 2005. Microlithic technology in the Stone Age. Journal of the Israel Prehistoric Society 35, 337–351.

Cartmill, M., Smith, F., 2009. The Human Lineage. Wiley-Blackwell, Hoboken.

Clark, J.D., 1982. The cultures of the Middle Palaeolithic/Middle Stone Age. In: Clark, J.D. (Ed.), The Cambridge History of Africa. From Earliest Times to c. 500 BC. Cambridge University Press, Cambridge, pp. 248–341.

Clark, J.D., 1991. Stone artifact assemblages from Swartkrans, Transvaal, South Africa. In: Clark, J.D. (Ed.), Cultural Beginnings: Approaches to Understanding Early Hominid Lifeways in the African Savanna. Dr. Rudolf Halbelt GMBH, Bonn, pp. 137–158.

Clottes, J., Arnaud, M., 2003. Chauvet Cave: The Art of Earliest Times. University of Utah Press, Salt Lake City.

Cook, J., 2013. Ice Age Art: The Arrival of the Modern Mind. The British Museum Press, London.

Darwin, C., 1859. The Origin of Species by Means of Natural Selection. John Murray, London.

Delson, E., Tattersall, I., Van Couvering, J.A., Brooks, A.S., 2000. Encyclopedia of Human Evolution and Prehistory, second ed. Garland Publishing, Inc, New York.

Ferring, R., Oms, O., Agusti, J., Berna, F., Nioradze, M., Shelis, T., Tappen, M., Vekua, A., Dhvania, D., Lordkipanidze, D., 2011. Earliest human occupations at Dmanisi (Georgian Caucasus) dated to 1.85–1.78 Ma. Proceedings of the National Academy of Sciences 108, 10432–10436.

Foley, R.A., Lewin, R., 2004. Principles of Human Evolution. Wiley-Blackwell, London.

Grine, F.E. (Ed.), 1988. Evolutionary History of the "Robust" Australopithecine. Aldine de Gruyter, New York.

Harmand, S., Lewis, J.E., Feibel, C.S., Lepre, C.J., Prat, S., Lenoble, A., Boes, X., Quinn, R.L., Brenet, M., Arroyo, A., Taylor, N., Clement, S., Daver, G., Brugal, J.-P., Leakey, L., Kent, D.V., Mortlock, R.A., Wright, J.D., Roche, H., 2015a. Early Tools from West Turkana, Kenya. Paleoanthropology Society Annual Meeting Abstract, San Francisco.

Harmand, S., Lewis, J.E., Feibel, C.S., Lepre, C.J., Prat, S., Lenoble, A., Boes, X., Quinn, R.L., Brenet, M., Arroyo, A., Taylor, N., Clement, S., Daver, G., Brugal, J.-P., Leakey, L., Mortlock, R.A., Wright, J.D., Lokorodi, S., Kirwa, C., Kent, D.V., Roche, H. 2015b. 3.3-million-year-old stone tools from Lomekwi 3, West Turkana, Kenya. Nature 521, 310–321.

Henke, W., Tattersall, I., 2007. Handbook of Paleoanthropology, vols. 1–3. Springer-Verlag, Berlin.

Hodges, H., 1995. Artifacts: An Introduction to Early Materials and Technology. Bristol Classical Press, Bristol.

Holloway, R.L., Broadfield, D.C., Yuan, M.S., 2004. The Human Fossil Record, Volume Three: Brain Endocasts: The Paleoneurological Evidence. Wiley-Liss, Hoboken.

Hovers, E., Braun, D., 2009. Interdisciplinary Approaches to the Oldowan. Springer, New York.

Howell, F.C., Haesaerts, P., de Heinzelin, J., 1987. Depositional environments, archaeological occurrences and hominids from Members E and F of the Shungura Formation (Omo basin, Ethiopia). Journal of Human Evolution 16, 665–700.

Isaac, G., 1984. The earliest archaeological traces. In: Clark, J.D. (Ed.), Cambridge History of Africa, vol. 1. Cambridge University Press, Cambridge, pp. 157–247.

Isaac, G., 1990. The Archaeology of Human Origins. Cambridge University Press, Cambridge.

Johanson, D., 2004. Lucy, thirty years later: an expanded view of *Australopithecus afarensis*. Journal of Anthropological Research 60, 465–486.

Johanson, D., Edgar, B., 2006. From Lucy to Language. Simon & Schuster, New York.

Kimbel, W.H., White, T.D., Johanson, D.C., 1988. Implications of KNM-WT 17000 for the evolution of "Robust *Australopithecus*.". In: Grine, F.E. (Ed.), Evolutionary History of the "Robust" Australopithecine. Aldine de Gruyter, New York, pp. 259–268.

Kimbel, W.H., Walter, R.C., Johanson, D.C., Reed, K.E., Aronson, J.L., Assefa, Z., Marean, C.W., Eck, G.C., Bobe, R., Hovers, E., Rak, Y., Vondra, C., Yemane, T., York, D., Chen, Y., Evensen, N.M., Smith, P.E., 1996. Late Pliocene *Homo* and Oldowan tools from the Hadar Formation (Kada Hadar Member), Ethiopia. Journal of Human Evolution 31, 549–561.

Klein, R.G., 2009. The Human Career: Human Biological and Cultural Origins, third ed. University of Chicago Press, Chicago.

Klein, R., 2013. Hominin dispersals in the Old World. In: Scarre, C. (Ed.), The Human Past: World Prehistory & the Development of Human Societies. Thames & Hudson, London, pp. 84–123.

Koobi Fora Research Project. In: Isaac, G. (Ed.), 1997. Plio-Pleistocene Archaeology, vol. 5. Clarendon Press, Oxford.

Kuman, K., 1994. The archaeology of Sterkfontein—past and present. Journal of Human Evolution 27, 471–495.

Leakey, L.S.B., 1936. Stone Age Africa. Oxford University Press, London.

Leakey, M.D., 1971. Olduvai Gorge. Excavations in Beds I and II, 1960–1963, vol. 3. Cambridge University Press, Cambridge.

Leakey, M.G., Spoor, F., Brown, F., Gathogo, P.N., Kiarie, C., Leakey, L.N., 2001. New hominin genus from eastern Africa shows diverse middle Pliocene lineages. Nature 410, 433–440.

Lepre, C.J., Roche, H., Kent, D.V., Harmand, S., Quinn, R.L., Brugal, J.P., Texier, P.J., Lenoble, A., Feibel, C.S., 2011. An earlier origin of the Acheulian. Nature 477, 82–85.

Lordkipanidze, D., Ponce de Leon, M.S., Margvelashvilli, A., Rak, Y., Rightmire, G.P., Vekua, A., Zollikofer, C.P.E., 2013. A complete skull from Dmanisi, Georgia, and the evolutionary biology of early *Homo*. Science 342, 326–331.

Lumsden, C.J., Wilson, E.O., 1983. Promethian Fire: Reflections into the Origins of Mind. Harvard University Press, Cambridge, Mass.

McBrearty, S., Brooks, A.S., 2000. The revolution that wasn't: a new interpretation of the origin of modern human behavior. Journal of Human Evolution 39, 453–563.

McGrew, W.C., 1992. Chimpanzee Material Culture: Implications for Human Evolution. Cambridge University Press, Cambridge.

McGrew, W.C., 2004. The Cultured Chimpanzee: Reflections on Cultural Primatology. Cambridge University Press, New York.

McPherron, S.P., Alemseged, Z., Marean, C.W., Wynn, J.G., Reed, D., Geraads, D., Bobe, R., Bearat, H.A., 2010. Evidence for stone tool-assisted consumption of animal tissues before 3.39 million year ago. Nature 446, 857–860.

Mellars, P., 1990. The Emergence of Modern Humans: An Archaeological Perspective. Cornell University Press, Ithaca, New York.

Mellars, P., 1996. The Neanderthal Legacy. Princeton University Press, Princeton.

Njau, J.K., 2012. Reading Pliocene bones. Science 336, 46–47.

Pettitt, P., 2013. The rise of modern humans. In: Scarre, C. (Ed.), The Human Past: World Prehistory & the Development of Human Societies. Thames & Hudson, London, pp. 124–173.

Plummer, T., Bishop, L.C., Ditchfield, P., Hicks, J., 1999. Research on late Pliocene Oldowan sites at Kanjera South, Kenya. Journal of Human Evolution 36, 151–170.

Roberts, A., 2011. Evolution: The Human Story. DK Publishing, New York.

Roche, H., Delagnes, A., Brugal, J., Feibel, C., Kibunjia, M., Mourre, V., Texier, P.-J., 1999. Early hominid stone tool production and technological skill 2.34 Myr ago in West Turkana, Kenya. Nature 399, 57–60.

Roebroeks, W., Villa, P., 2011. On the earliest evidence for habitual use of fire in Europe. Proceedings of the National Academy of Sciences 108, 5209–5214.

Sanz, C.M., Call, J., Boesch, C., 2013. Tool Use in Animals: Cognition and Ecology. Cambridge University Press, Cambridge.

Sahnouni, M., 2006. The North African Early Stone Age and the sites at Ain Hanech, Algeria. In: Toth, N., Schick, K. (Eds.), The Oldowan: Case Studies into the Earliest Stone Age. Stone Age Institute Press, Gosport, IN, pp. 77–111.

Scarre, C. (Ed.), 2013. The Human Past: World Prehistory & the Development of Human Societies. Thames and Hudson, London.

Schick, K., 1987. Modeling the formation of Early Stone Age artifact concentrations. Journal of Human Evolution 16, 789–807.

Schick, K., 1994. The Movius line reconsidered: perspectives on the earlier Paleolithic of Eastern Asia. In: Corruccini, R., Ciochon, R.L., Howell, F.C. (Eds.), Integrative Paths to the Past: Paleoanthropological Advances in Honor of F. Clark Howell. Prentice-Hall, Englewood Cliffs, New Jersey, pp. 569–596.

Schick, K., Toth, N., 1993. Making Silent Stones Speak: Human Evolution and the Dawn of Technology. Simon & Schuster, New York.

Schick, K., Toth, N., 2006. An overview of the Oldowan industrial complex: the sites and the nature of their evidence. In: Toth, N., Schick, K. (Eds.), The Oldowan: Case Studies into the Earliest Stone Age. Stone Age Institute Press, Gosport, IN, pp. 3–42.

Schick, K., Toth, N., 2009. The Cutting Edge: New Approaches to the Archaeology of Human Origins. Stone Age Institute Press, Gosport, IN.

Schick, K., Toth, N., Garufi, G., Savage-Rumbaugh, E.S., Rumbaugh, D., Sevcik, R., 1999. Continuing investigations into the stone tool-making and tool-using capabilities of a bonobo (*Pan paniscus*). Journal of Archaeological Science 26, 821–832.

Semaw, S., 2006. The oldest stone artifacts from Gona (2.6-2.5 Ma), Afar, Ethiopia: implications for understanding the earliest stages of stone knapping. In: Toth, N., Schick, K. (Eds.), The Oldowan: Case Studies into the Earliest Stone Age. The Stone Age Institute Press, Gosport, IN, pp. 43–76.

Semaw, S., Renne, P., Harris, J.W.K., Feibel, C.S., Bernor, R.L., Fesseha, N., Mowbray, K., 1997. 2.5 million-year-old stone tools from Gona, Ethiopia. Nature 385, 333–336.

Semaw, S., Rogers, M., Stout, D., 2009. Insights into late Pliocene lithic assemblage variability: the East Gona and Ounda Gona South Oldowan archaeology. In: Schick, K., Toth, N. (Eds.), The Cutting Edge: New Approaches to the Archaeology of Human Origins. Stone Age Institute Press, Gosport, IN, pp. 211–246.

Shumaker, R.W., Walkup, K.R., Beck, B.B., 2011. Animal Tool Behavior. John Hopkins University Press, Baltimore.

Spier, R., 1970. From the Hand of Man: Primitive and Preindustrial Technologies. Houghton Mifflin, New York.

Stringer, C., Andrews, P., 2005. The Complete World of Human Evolution. Thames and Hudson, London.

Tattersall, I., 2008. The Fossil Trail: How We Know What We Think We Know about Human Evolution. Oxford University Press, Oxford.

Toth, N., 1985. The Oldowan reassessed: a close look at early stone artifacts. Journal of Archaeological Science 12, 101–120.

Toth, N., Schick, K., 2006. The Oldowan: Case Studies into the Earliest Stone Age. Stone Age Institute Press, Gosport, IN.

Toth, N., Schick, K., 2009a. The importance of actualistic studies in Early Stone Age research. In: Schick, K., Toth, N. (Eds.), The Cutting Edge: New Approaches to the Archaeology of Human Origins. Stone Age Institute Press, Gosport, IN, pp. 267–344.

Toth, N., Schick, K., 2009b. The Oldowan: the tool making of early hominins and chimpanzees compared. Annual Review of Anthropology 38, 289–305.

Toth, N., Schick, K., 2010. Hominin brain reorganization, technological change, and cognitive complexity. In: Broadfield, D., Yuan, M., Schick, K., Toth, N. (Eds.), The Human Brain Evolving: Paleoneurological Studies in Honor of Ralph L. Holloway. Stone Age Institute Press, Gosport, IN, pp. 293–312.

Toth, N., Schick, K., 2013. African origins. In: Scarre, C. (Ed.), The Human Past: World Prehistory & the Development of Human Societies. Thames and Hudson, London, pp. 44–83.

Toth, N., Schick, K., Semaw, S., 2006. A comparative study of the stone tool-making skills of *Pan*, *Australopithecus*, and *Homo sapiens*. In: Toth, N., Schick, K. (Eds.), The Oldowan: Case Studies into the Earliest Stone Age. The Stone Age Institute Press, Gosport, IN, pp. 155–222.

Tylecote, R., 2011. A History of Metallurgy. Maney Publishing, Leeds.

Villmoare, V., Kimbel, W.H., Seyoum, C., Campisano, C.J., Dimaggio, E.N., Rowan, J., Braun, D.R., Ramon Arrowsmith, J., Reed, K.E., 2015. Early *Homo* at 2.8 Ma from Ledi-Geraru, afar, Ethiopia. Science 347, 1352–1355.

Washburn, S.L., 1960. Tools and human evolution. Scientific American 203 (9), 63–75.

White, R., 2003. Prehistoric Art: The Symbolic Journey of Humankind. Abrams, New York.

Whiten, A., Goodall, J., McGrew, W.C., Nishida, T., Reynolds, V., Sugiyama, Y., Tutin, C.E.G., Wrangham, R., Boesch, C., 1999. Culture in chimpanzees. Nature 399, 682–685.

Whiten, A., Schick, K., Toth, N., 2009. The evolution and cultural transmission of percussive technology; integrating evidence from palaeoanthropology and primatology. Journal of Human Evolution 57, 420–435.

Zimmer, C., 2005. Smithsonian Intimate Guide to Human Origins. Madison Press Books, Toronto.

Part IV

Genetics and Biology

Chapter 15

Contemporary Human Genetic Variation

John H. Relethford
Department of Anthropology, State University of New York College at Oneonta, Oneonta, NY, USA

SYNOPSIS

Many patterns of contemporary human genetic variation reflect the history of our species. Neutral genetic variation (not affected by natural selection) shows our African origin, later dispersion throughout the world, and some interbreeding with other human groups, such as the Neandertals. Some variation also reflects natural selection to different environments, disease, and diet, and much of these changes have occurred since the origin and spread of agriculture.

INTRODUCTION

Our species has spread out over almost the entire planet, a consequence of our cultural and biological adaptability. Globally, we see biological variation in our species, both genetic and phenotypic, and one of the goals of human biology and biological anthropology is to describe and analyze this variation. In pre-Darwinian times, the emphasis was on the enumeration and classification of human races, an approach that stressed variation between groups in a typological framework attempting to reduce human variation to a set number of discrete races. The fact that there have been centuries of debate over the number of human races (which often ranged from three to several dozen) is a reflection that the underlying model of discrete variation expected at a level of subspecies is not the most suitable means of analyzing human variation. Instead, genetic and phenotypic data for human populations show that although there are definite geographic patterns in human genetic variation, they do not conform well to the race model (e.g., Graves, 2001; Brace, 2005).

Following Darwin's insights and the subsequent growth of modern evolutionary thought, the study of human genetic variation eventually shifted to more analysis of local populations and an emphasis on explanation rather than on description. Explanations of evolutionary change and variation look at the mechanisms that alter allele frequencies over time: mutation, natural selection, genetic drift, and gene flow. Mutation is a random change in the underlying genetic code. Natural selection leads to changes in allele frequency based on differential survival and reproduction of genotypes. Genetic drift is the random fluctuation in allele frequencies because of sampling variation, which is greater in effect in smaller populations and leads to a loss in genetic diversity. Gene flow is the change in allele frequencies due to the movement of genes from one population to another, which acts to reduce genetic differences between populations.

As there have been literally thousands of studies of human variation for a wide range of traits and genetic markers, any brief review of contemporary human genetic variation must necessarily have a specific focus. The emphasis in this chapter is on several general findings of human genetic variation that are best interpreted in terms of our species' history of an African origin and subsequent dispersal. The focus here is initially on neutral genetic variation, where allele frequencies are subject to the joint action of mutation, drift, and gene flow for genes and DNA sequences that are not affected by natural selection. Neutral variation is a reflection of the overall history of human populations and this history is seen throughout our genome. On the other hand, natural selection affects specific genes or traits differently. Whereas the analysis of neutral variation is the study of the history of populations, the study of variation related to natural selection is the study of the history of specific traits or genes. The emphasis in this chapter is on patterns of neutral variation, with some secondary attention to the trait-specific actions of natural selection.

A BRIEF HISTORY OF HUMAN EVOLUTION

Contemporary genetic variation is understood best in light of our species' past. In order to set the stage for understanding present-day patterns of variation, we need to review briefly the fossil record for *hominin* evolution (hominin is the term for humans and human ancestors back to the time of a divergence from the African ape line). Fossil and genetic evidence suggests that the African ape and hominin lines diverged roughly 6–7 million years ago or so in Africa (Steiper and Young, 2006; Fabre et al., 2009). The earliest hominins, including species in the genera *Ardipithecus* and *Australopithecus*, lived in Africa and were bipeds with small brains and large faces (see chapters by Hunt and Ward in this volume). By 2 million years ago, the species *Homo erectus* emerged in Africa, with an increase in brain size, a reduction in the size of the face and teeth, and reliance on a stone tool technology (see the chapter by Simpson). By around 1.8 million years ago, *H. erectus* dispersed out of Africa into Southeast Asia and the eastern edge of Europe. By around 800,000 years ago or so, brain size increased more, and such specimens of *Homo* are often placed in the taxon *Homo heidelbergensis*, although others refer them to examples of early *Homo sapiens*. Fossils known collectively as Neandertals are found in Europe and the Middle East, with some features appearing between 100,000 and 200,000 years ago in Europe and "classic" Neandertals dating back to about 70,000 years ago (Harvati-Papatheodorou, 2013). Debate continues over whether they should be referred to as a subspecies of *H. sapiens* or a separate species (*Homo neanderthalensis*). Based on overall morphology, both specimens of *H. heidelbergensis* and the Neandertals are sometimes referred to collectively as "archaic" humans to contrast with anatomically modern humans (see chapter by Ahern).

There has been growing consensus that the morphology characteristic of anatomically modern humans appears first in Africa with a subsequent dispersion into Eurasia (see the chapter by Holt). Evidence of an earlier appearance of modern humans in Africa includes remains from Omo in Ethiopia at 195,000 years ago (McDougall et al., 2005), Herto in Ethiopia at 160,000 years ago (White et al., 2003), and the Klasies River Mouth in South Africa at 130,000 years ago (Grün et al., 1990). Some early expansion out of Africa is seen at the Skhul and Qafzeh sites in Israel dating back 92,000 years ago, followed by later expansions into Eurasia by about 60,000 years ago. Modern humans arrive in Australia sometime between 45,000 and 60,000 years ago and into Europe between 30,000 and 40,000 years ago. Modern humans later disperse into the Americas, most likely starting between 15,000 and 20,000 years ago.

The above review is necessarily brief, and does not include the fine details or nuances, particularly regarding the accumulating evidence of archaic admixture. Nonetheless, it is a useful starting point for addressing the general question regarding genetic variation in contemporary human populations. That is, what are the neutral genetic consequences of an initial appearance of modern humans in Africa, a subsequent dispersion of some populations into Eurasia and beyond, and the strong likelihood of admixture with preexisting archaic populations in Eurasia, such as the Neandertals (see the chapter by Lao and Kayser)? These questions help frame our interpretation of the patterns of neutral genetic variation that are outlined below.

OUR SPECIES HAS A RELATIVELY LOW LEVEL OF GENETIC VARIATION

There are several key features of neutral genetic variation in our species that can be explained in terms of our evolutionary history of an African origin and subsequent dispersion with admixture. One such feature is the relatively low level of genetic variation in our species. Mitochondrial DNA (mtDNA) diversity tends to be higher in apes than in humans, and autosomal DNA variation in apes is similar to, or larger than, that found in humans. Some of this variation points to likely different demographic histories (Fischer et al., 2006; Hawks, 2009). The reduced genetic variation in the human species is often attributed to one or more bottlenecks in our population history (where a reduction in population size would increase genetic drift and lead to a reduction in genetic diversity). One possibility favored by available evidence is a bottleneck accompanying a dispersion out of Africa (Garrigan and Hammer, 2006), where founder effect would lead to increased genetic drift and a reduction in diversity. Such a reduction in diversity could persist throughout later human evolution. Although the human species later increased dramatically in size, particularly after the origins of agriculture, our lower relative level of genetic diversity might still reflect a time when our numbers were small.

THERE IS HIGHER GENETIC VARIATION IN SUB-SAHARAN AFRICAN POPULATIONS THAN IN OTHER GEOGRAPHIC REGIONS

When genetic diversity is measured (either in terms of allele/haplotype frequencies or in terms of nucleotide diversity), variation in DNA markers shows a consistent pattern where genetic variation is higher in sub-Saharan Africa than elsewhere in the world. This pattern is seen in mitochondrial and autosomal DNA markers (e.g., Tishkoff et al., 2009; Xing et al., 2009), as well as phenotypic traits such as cranial measures and skin color (Relethford, 2008). Higher genetic (and phenotypic) variation in sub-Saharan Africa is typically considered a function of population history and the initial origin of modern humans in Africa. Here, the long period of existence of modern human populations in Africa prior to dispersal meant a long time for the accumulation of mutations in those populations, and hence higher levels of genetic diversity. Further, the occurrence of a bottleneck in population size accompanying the dispersion out of Africa would further mean increased genetic drift due to founder effect, which would reduce genetic variation in non-African populations. Founder effect is also consistent with the fact that DNA diversity outside of Africa is often a subset of the diversity found within Africa (e.g., Tishkoff and Gonder, 2007). Here, African populations accumulated genetic diversity over time, and as small founding populations left Africa, genetic drift caused the loss of some alleles, which led to non-African diversity being a subset of African diversity. As shown below, the hypothesis of founder effect and its influence on diversity are supported further by the geographic pattern of African and non-African genetic diversity.

GENETIC DIVERSITY DECLINES WITH DISTANCE OUT OF AFRICA

Genetic diversity outside of sub-Saharan Africa is structured geographically. Statistical analysis has shown that the farther a population is from East Africa in terms of likely routes of human dispersal, the lower its level of genetic diversity. This pattern has been seen for microsatellite DNA markers (short-repeated sequences of DNA) (Prugnolle et al., 2005; Ramachandran et al., 2005), SNPs (single-nucleotide polymorphisms) (Li et al., 2008), and craniofacial measures (Manica et al., 2007; von Cramon-Taubadel and Lycett, 2008; Betti et al., 2009). This stepwise decline in diversity out of East Africa is generally explained by the serial founder model (Ramachandran et al., 2005). Here, a small population initially disperses a short distance out of Africa. Because of founder effect, genetic diversity in the founders decreases. Over time, the population grows and then a small subset splits off again, causing a further reduction in genetic diversity due to drift. As shown by Ramachandran et al. (2005), a repeated series of such founder effects accompanying the spread of a species over space will give rise to the observed linear decline in genetic diversity out of Eastern Africa. The initial model considered these new populations as isolates, but a modified (and more realistic) serial founder model developed by DeGiorgio et al. (2009) shows that adding gene flow between neighboring populations also fits the observed diversity decline.

THERE ARE LOW LEVELS OF GENETIC DIFFERENTIATION BETWEEN GEOGRAPHIC REGIONS

Thus far, we have examined genetic variation within the species and genetic variation within different regional populations. Another focus of studies of contemporary genetic variation is the level of genetic variation between groups, indicating how much group allele frequencies diverge from one another. We are often interested in how much variation exists within groups and how much variation exists between groups. As a simple example, consider a gene with two alleles, *A* and *B*, in two different populations. Imagine that in both population 1 and population 2, the frequency of the *A* allele is 50% and the frequency of the *B* allele is 50%. There is variation within both populations, because there are two different alleles. However, there is no variation between the two populations because they have the same allele frequencies. Here, all of the variation is within groups, and there is no variation between groups. For another example, imagine allele frequencies of 100% *A* and 0% *B* in population 1, and 0% *A* and 100% *B* in population 2. In this example, there is no variation within either population because there is only one allele (*A* or *B*) in each population. However, there is variation between the groups because the allele frequencies are different in each population. In the real world, neither extreme would typically be found within a species for most genes, but instead we would see examples where there is variation both within and between groups, such as if we had allele frequencies of 80% *A* and 20% *B* in population 1, and 72% *A* and 28% *B* in population 2.

The level of genetic differentiation between populations is often quantified using a measure known as F_{ST}, which is the level of genetic variation between groups relative to the total amount of genetic variation present in the populations being studied. A number of studies have estimated F_{ST} among major geographic regions, such as sub-Saharan Africa, Europe, East Asia, and Australasia, among others. Studies of red blood cell genetic markers and autosomal DNA markers (microsatellite DNA) have typically found values of roughly 10% (Madrigal and Barbujani, 2007). Similar or slightly higher values of F_{ST} have been observed in studies of craniometric variation (Relethford, 2002). These estimates mean that roughly 10% of the variation found in our species is distributed between geographic regions while 90% is found within regions. This means that for most genes and other genetic markers (and many phenotypic traits), there is a great deal of overlap across the world, and there are few traits that are found exclusively in one region versus another (exceptions would include the high frequency of shovel-shaped incisors in East Asians and Native Americans, and the high frequency of the Duffy negative allele of the Duffy blood group in many African populations). Human genetic differentiation is thus less than what many would consider typical of species that are easily subdivided into subspecies or races (although this interpretation relies on models that may not be realistic of human demographic history—see Long and Kittles (2003) for a detailed critique).

On an evolutionary level, levels of F_{ST} tell us something about the impact of the three evolutionary forces that affect neutral genetic variation (mutation, drift, gene flow) and the amount of time since the initial divergence. In terms of our African origin and subsequent dispersal, the levels of F_{ST} observed in the human species could reflect insufficient time since dispersal into Eurasia and beyond to reach an equilibrium value of F_{ST} (because F_{ST} increases over time to asymptotically approach an equilibrium between genetic drift, which increases group differences, and gene flow, which reduces group differences). A low level of F_{ST} could also reflect increased gene flow, as humans have always migrated, both locally between groups as well as dispersing over long distances. Finally, the fact that the human species has increased over time, particularly since the development of agriculture, means that levels of F_{ST} would ultimately decline because F_{ST} is inversely proportional to population size. It is quite likely that all of these factors are important and further research is needed to understand the components of our species' demographic history and how they have affected genetic differences between human populations. Further, models that are more complex are needed to understand the interplay of drift and migration on patterns of between-group variation (Long and Kittles, 2003).

GENETIC DISTANCE IS CORRELATED WITH GEOGRAPHY

When we examine patterns of global genetic variation, the most noticeable observation is that our species' variation is geographically structured. We tend to resemble our neighbors more than we do populations farther away. Further, there is a definite pattern in this correlation with geography. Measures of genetic distance tend to increase over space, such that pairs of populations that are located farther apart are genetically more different. On a global level, many studies have shown a strong correlation between genetic distance (estimated in different ways, but always a measure of genetic differences between populations) and geographic distance. This correlation has been seen for red blood cell/white blood cell markers, DNA markers, and craniometric traits (e.g., Imaizumi et al., 1973; Cavalli-Sforza et al., 1994; Eller, 1999; Relethford, 2004a, b; Manica et al., 2005; Ramachandran et al., 2005). It is important to note that most of these global studies rely on a measure of geographic distance that considers likely travel routes, rather than simply measuring the straight-line distance between populations. For example, when measuring the geographic distance between indigenous populations in Africa and South America, we would not take the straight-line distance across the Atlantic Ocean because there is no archaeological evidence to support this route. Instead, we would consider a travel route out of Africa through Asia, across to the Americas at or near the Bering Straits, and then down into the North,

Central, and finally South America. In this way, geography reflects the history of population expansion in the human species.

What causes the correlation between genetics and geography? Numerous studies in anthropological genetics have also documented the correlation on a local and regional level (e.g., Jorde, 1980). In such cases, the observed correlation between genetics and geography reflects gene flow between neighboring populations under the classic isolation by distance model. The underlying principle of isolation by distance is that the level of gene flow between populations is constrained by the geographic distance between them, such that the farther two populations are apart from each other, the less gene flow will be shared and the higher the genetic difference between them. What is less clear is whether local gene flow can be used to explain the genetics–geography correlation over large continental distances, or whether there is some other factor. Ramachandran et al. (2005) suggested that the genetics–geography correlation could be explained by the serial founder model. As new populations split off to form new founding groups, these groups would necessarily be genetically related and geographically proximate. Their simulations show how a correlation between genetics and geography could occur under the assumptions of a serial founder model. However, as noted above, it may be more realistic to model human demographic history in terms of a combination of founder events, gene flow between local populations, and long-range dispersals (Relethford, 2004b). That populations tend to be similar genetically to their neighbors might be due to the facts that neighboring populations are ancestors or descendants, and that they are also those groups with which they are most likely to share mates. Further research is needed to help partition these effects in order to understand better the correlation between genetics and geography. In any event, it is likely that the dispersal of modern humans out of Africa, and the subsequent founder events, is at least a partial explanation of the genetics–geography correlation. As is the case with patterns of within-group variation, between-group variation is in part a reflection of our species' recent evolutionary history.

THE HUMAN GENOME SHOWS EVIDENCE OF ADMIXTURE FROM NEANDERTALS ... AND OTHERS

The debate over modern human origins was particularly intense from the late 1980s through the 1990s. Although there have been a variety of different models used to explain the origin of anatomically modern humans, much of the debate was polarized between two groups. One group saw a single origin for *Homo sapiens* in Africa, followed by dispersal throughout the rest of the planet (e.g., Stringer and Andrews, 1988). The other group argued for a multiregional explanation where there was no single point of origin in time and space for modern humans, but instead saw the transition to modernity as the coalescence of evolutionary changes across several continents connected by gene flow (e.g., Wolpoff et al., 1994). Among those who argued for an African origin, some proposed a complete replacement model, where other human groups such as the Neandertals became extinct without contributing anything to the genetic ancestry of living humans, and those who argued for some level of admixture (e.g., Smith et al., 1989).

Fossil evidence supports an earlier origin of modern humans in African than elsewhere, and genetic evidence has been consistent with this view as well (Relethford, 2008). There has been growing consensus that certain genetic and anatomical changes associated with modernity took place first in Africa. The debate continues over what happened next, as modern humans expanded out of Africa and moved into parts of the world where earlier more "archaic" human populations still lived. In Europe, for example, modern humans are found dating to a time when Neandertals still lived. As Neandertal populations are no longer with us, and have not been for 28,000 years (Smith et al., 1999), the question is whether there was any interbreeding between modern human and Neandertal populations before the latter disappeared as a group. In simpler terms, the question boils down to whether any (or all) modern humans have any Neandertal ancestry.

As questions of prehistoric admixture are difficult at best to address with fossil data, it is understandable that there was considerable excitement surrounding the extraction of a small sequence of mitochondrial DNA from a Neandertal fossil in 1997 (Krings et al., 1997). Since then, a total of 20 partial mtDNA sequences and seven complete mitochondrial genomes have been recovered from 27 Neandertals at 12 archaeological sites (Lalueza-Fox and Gilbert, 2011). Collectively, the mitochondrial DNA data suggested that Neandertals were a divergent group (and perhaps a different species), and there was no evidence of interbreeding with modern humans. However, the problem remained that because mtDNA is inherited intact without recombination it functions as a single locus, and therefore may not be representative of the history of the Neandertal lineage. Perhaps Neandertals interbred with early modern humans, but distinctive Neandertal mtDNA was lost because of genetic drift.

Further insight was gained with the completion of a draft sequence of the nuclear DNA genome of Neandertals, which found genetic variants shared by Neandertals and living Eurasians, but not with living Africans (Green et al., 2010). This pattern is best explained by interbreeding between Neandertals and early modern humans in the Middle East, after modern humans left Africa, but before they dispersed throughout Europe and Asia. Based on genetic and fossil evidence, this interbreeding likely took place between 50,000 and 100,000 years ago. The net result of this interbreeding is that Neandertals contributed between 1 and 4% of the ancestry of living Eurasians (but not Africans). Although the Neandertals as a group died out, some of their genes survived in living humans. These data argue against the idea of a complete replacement and instead support limited admixture with Neandertals.

The Neandertals were a regional population of earlier humans. Even though they interbred with modern humans to some extent, we still need information about the possible genetic contributions of other archaic human populations. Ancient DNA analysis has provided another piece of the puzzle with the sequencing of nuclear DNA from a finger bone found at Denisova Cave in Siberia. This population, known as the Denisovans, is genetically similar in some ways to the Neandertals, suggesting a common origin. Genomic comparisons reveal that 3–6% of the ancestry of living Melanesians comes from the Denisovans (Reich et al., 2010; Meyer et al., 2012). Denisovan DNA has also been detected in Australian aborigines and some populations in Southeast Asia, but not in other geographic regions. In particular, there is no evidence of Denisovan ancestry in mainland Southeast Asians, suggesting that living humans in that region may have moved there after the initial interbreeding with Melanesians (Reich et al., 2011). It seems likely that the Denisovans once occupied a large part of eastern Asia, but are now extinct, leaving only some ancestry in some living populations (see the chapter by Lao and Kayser).

These recent discoveries of ancient DNA reveal that the evolutionary relationship between archaic and modern humans is more complex than was once thought. As modern humans dispersed out of Africa, there appears to have been different instances of interbreeding with different archaic populations. Some living humans, such as those throughout much of Europe and Asia, retain a trace of Neandertal ancestry, whereas living Melanesians and Australian aborigines show evidence of both Neandertal and Denisovan ancestry. Further research will be needed to determine the existence of other ancestral lines. It seems likely that living human populations share a predominately African ancestry, with additional ancestry from one or more archaic groups at different times and places. It is now clear that although the dispersal of modern humans was perhaps the most important event in shaping neutral genetic variation in living humans, it was not the only one, and the earlier model of a *complete* replacement out of Africa is incorrect.

NATURAL SELECTION HAS FURTHER AFFECTED THE PATTERN OF CONTEMPORARY GENETIC VARIATION

An African origin and dispersal, combined in different places with admixture from archaic human populations, help explain much of our species' global variation for neutral genes. The view from neutral traits provides us the background for the expected patterns of variation in the absence of natural selection. For specific genes and traits, there are noticeable deviations from neutrality that provide us with a different pattern of human genetic variation. One example is human skin color, a trait whose distribution correlates strongly with latitude and the intensity of ultraviolet radiation (Jablonksi and Chaplin, 2000). Skin color is darkest for populations at or near the equator because darker skin protects against the harmful effects of excessive ultraviolet radiation, including sunburn, skin cancer, and folate deficiency, as well as enhancing the protective nature of the epidermis and resistance to infection (Elias et al., 2010) (see chapter by Elias and Williams). Lighter skin color in populations farther from the equator is typically considered an adaptive response to lowered levels of vitamin D production resulting from lower levels of ultraviolet radiation. Because of natural selection, human skin color shows a wide range of variation among continental populations that is atypical when compared with neutral traits (Relethford, 2002).

Other phenotypic traits show smaller deviations from the expected pattern from neutral traits. For example, global variation in craniometric size and shape corresponds to geographic patterns seen in neutral DNA markers (Relethford, 2004b; Roseman, 2004). However, several populations are outliers when comparing metric distances with genetic and geographic distances, such as the Buriat of Siberia and the Greenland Inuit (Roseman, 2004; Relethford, 2010). These populations live in the coldest climates, and their larger cranial size is consistent with Bergmann's rule predicting larger size of mammals in cold climates because of reduced surface area relative to volume.

A number of human genetic traits have distributions that have been shaped by natural selection to infectious disease. Notable examples include the evolution of different alleles for the hemoglobin protein and the Duffy blood group because of genetic adaptation to malaria (Mielke et al., 2011). Even the ABO blood group, whose global distribution can be described as neutral in some ways, shows evidence of potential selection relating to the incidence of infectious disease. What makes many of the links between infectious disease and different genes interesting in the context of human evolution is that many infectious diseases have likely been epidemic only with the increase in human population size that accompanied the transition to agriculture starting 12,000 years ago. If so, this means that much of the selection-related patterns of genetic variation that we see in human populations are of recent origin in evolutionary time and are related to cultural changes.

A classic example of the interrelationship of cultural and biological change is Livingstone's (1958) study of the hemoglobin *S* allele in western Africa. Although this allele is harmful in homozygotes (*SS*), individuals with the heterozygous genotype *AS* are resistant to the harmful effects of falciparum malaria. Livingstone shows that the transition to agriculture thousands of years ago in western Africa caused ecological changes that resulted in an increase in the mosquito population and the malarial parasite that it spreads. Mutation to the *S* allele provided genetic adaptation to malaria, resulting in its increase over time. Although the evolution of the *S* allele is now seen as more complex, likely involving independent origins in different parts of Africa and Asia (Mielke et al., 2011), the basic principle of biocultural evolution remains a good example of recent human evolution.

Another example of biocultural evolution is the evolution of lactase persistence alleles that have provided continued production of the lactase enzyme throughout life, thus allowing humans to digest milk sugar (lactose) into adulthood, rather than the normal mammalian pattern of lactase restriction after weaning (see chapters by Little and Wiley). Different lactase persistence mutations have appeared in recent human evolution, and have then been selected for in populations with a history of dairy farming. These changes have been tracked to the last 9000 years or less, coinciding with evidence for the domestication of cattle (Tishkoff et al., 2007). Recent molecular analyses have suggested that there are likely many other recent genetic changes in our species due to the increased number of mutations due to the ever-increasing population size of the human species combined with new adaptive possibilities due to rapid culture change (Hawks et al., 2007).

SUMMARY

The patterns of genetic variation that we see in the human species reflect our evolutionary past and present, and can be understood by considering the action of evolutionary forces (mutation, selection, drift, and gene flow) in terms of our species' origin, our geographic dispersals, and our biocultural adaptation to changing environmental conditions. The ever-increasing wealth of DNA markers provides us with ample evidence to characterize the pattern of neutral variation (not affected by natural selection). Here, a number of observations all support the view that much of neutral variation in modern humans reflects our recent African origin as a species 200,000 years ago with later dispersion out of Africa. However, we now see that our African origin and dispersion were not a complete replacement, and some of our current ancestry comes from interbreeding with other human populations (perhaps different species) that were already living outside of Africa, such as the Neandertals and Denisovans. Furthermore, some of our genome has also been affected by natural selection, and in many cases, selection has been very recent, since the time of the agricultural revolution. The recent advances in molecular genetics have provided data suggesting recent and continuing evolution of humans, including many relating to changes in human diet and disease.

REFERENCES

Betti, L., Balloux, F., Amos, W., Hanihara, T., Manica, A., 2009. Distance from Africa, not climate, explains within-population phenotypic diversity in humans. Proceedings of the Royal Society B 276, 809–814.

Brace, C.L., 2005. "Race" Is a Four Letter Word. Oxford University Press, New York.

Cavalli-Sforza, L.L., Menozzi, P., Piazza, A., 1994. The History and Geography of Human Genes. Princeton University Press, Princeton.

DeGiorgio, M., Jakobsson, M., Rosenberg, N.A., 2009. Explaining worldwide patterns of human genetic variation using a coalescent-based serial founder model of migration outward from Africa. Proceedings of the National Academy of Science 106, 16057–16062.

Elias, P.M., Menon, G., Wetzel, B.K., Williams, J.W., 2010. Barrier requirements as the evolutionary "driver" of epidermal pigmentation in humans. American Journal of Human Biology 22, 526–537.

Eller, E., 1999. Population substructure and isolation by distance in three continental regions. American Journal of Physical Anthropology 108, 147–159.

Fabre, P.-H., Rodrigues, A., Douzery, E.J.P., 2009. Patterns of macroevolution among Primates inferred from a supermatrix of mitochondrial and nuclear DNA. Molecular Phylogenetics and Evolution 53, 808–825.

Fischer, A., Pollack, J., Thalmann, O., Nickel, B., Pääbo, S., 2006. Demographic history and genetic differentiation in apes. Current Biology 16, 1133–1138.

Garrigan, D., Hammer, M.F., 2006. Reconstructing human origins in the genomic era. Nature Reviews Genetics 7, 669–680.

Graves Jr., J.L., 2001. The Emperor's New Clothes: Biological Theories of Race at the Millennium. Rutgers University Press, New Brunswick, NJ.

Green, R.E., Krause, J., Briggs, A.W., Maricic, T., Stenzel, U., Kircher, M., Patterson, N., Li, H., Zhai, W., Fritz, M.H.-Y., Hansen, N.F., Durand, E.Y., Malaspinas, A.-S., Jensen, J.D., Marques-Bonet, T., Alkan, C., Prüfer, K., Meyer, M., Burbano, H.A., Good, J.M., Schultz, R., Aximu-Petri, A., Butthof, A., Höber, B., Höffner, B., Siegemund, M., Weihmann, A., Nusbaum, C., Lander, E.S., Russ, C., Novod, N., Affourtit, J., Egholm, M., Verna, C., Rudan, P., Brajkovic, D., Kucan, Z., Gušic, I., Doronichev, V.B., Golovanova, L.V., Lalueza-Fox, C., de la Rasilla, M., Fortea, J., Rosas, A., Schmitz, R.W., Johnson, P.L.F., Eichler, E.E., Falush, D., Birney, E., Mullikin, J.C., Slatkin, M., Nielsen, R., Kelso, J., Lachmann, M., Reich, D., Pääbo, S., 2010. A draft sequence of the Neandertal genome. Science 328, 710–722.

Grün, R., Shackleton, N.J., Deacon, H.J., 1990. Electron-spin-resonance dating of tooth enamel from Klasies River Mouth. Current Anthropology 31, 427–432.

Harvati-Papatheodorou, K., 2013. Neanderthals. In: Begun, D.R. (Ed.), A Companion to Paleoanthropology. Wiley-Blackwell, Chichester, UK, pp. 538–556.

Hawks, J., 2009. Update to Eller et al.'s "Local extinction and recolonization, species effective size, and modern human origins" (2004). Human Biology 81, 825–828.

Hawks, J., Wang, E.T., Cochran, G.M., Harpending, H.C., Moyzis, R.K., 2007. Recent acceleration of human adaptive evolution. Proceedings of the National Academy of Sciences 104, 20753–20758.

Imaizumi, Y., Morton, N.E., Lalouel, J.M., 1973. Kinship and race. In: Morton, N.E. (Ed.), Genetic Structure of Populations. University Press of Hawaii, Honolulu, pp. 228–233.

Jablonksi, N.G., Chaplin, G., 2000. The evolution of human skin coloration. Journal of Human Evolution 39, 57–106.

Jorde, L.B., 1980. The genetic structure of subdivided human populations: a review. In: Mielke, J.H., Crawford, M.H. (Eds.), Current

Developments in Anthropological Genetics. Theory and Methods, vol. 1. Plenum Press, New York, pp. 135–208.
Krings, M., Stone, A., Schmitz, R.W., Krainitzki, H., Stoneking, M., Pääbo, S., 1997. Neandertal DNA sequences and the origin of modern humans. Cell 90, 19–30.
Lalueza-Fox, C., Gilbert, M.T.P., 2011. Paleogenomics of archaic humans. Current Biology 21, R1002–R1009. http://dx.doi.org/10.1016/j.cub.2011.11.021.
Li, J.Z., Absher, D.M., Tang, H., Southwick, A.M., Casto, A.M., Ramachandran, S., Cann, H.M., Barsh, G.S., Feldman, M., Cavalli-Sforza, L.L., Myers, R.M., 2008. Worldwide human relationships inferred from genome-wide patterns of variation. Science 319, 1100–1104.
Livingstone, F.B., 1958. Anthropological implications of sickle cell gene distribution in West Africa. American Anthropologist 60, 533–562.
Long, J.C., Kittles, R.A., 2003. Human genetic diversity and the nonexistence of human races. Human Biology 75, 449–471.
Madrigal, L., Barbujani, G., 2007. Partitioning of genetic variation in human populations and the concept of race. In: Crawford, M.H. (Ed.), Anthropological Genetics: Theory, Methods and Applications. Cambridge University Press, Cambridge, pp. 19–37.
Manica, A., Amos, W., Balloux, F., Hanihara, T., 2007. The effect of ancient population bottlenecks on human phenotypic variation. Nature 448, 346–349.
Manica, A., Prugnolle, F., Balloux, F., 2005. Geography is a better determinant of human genetic differentiation than ethnicity. Human Genetics 118, 366–371.
McDougall, I., Brown, F.H., Fleagle, J.G., 2005. Stratigraphic placement and age of modern humans from Kibish, Ethiopia. Nature 433, 733–736.
Meyer, M., Kircher, M., Gansauge, M.-T., Li, H., Racimo, F., Mallick, S., Schraiber, J.G., Jay, F., Prüfer, K., de Filippo, C., Sudmant, P.H., Alkan, C., Fu, Q., Do, R., Rohland, N., Tandon, A., Siebauer, M., Green, R.E., Bryc, K., Briggs, A.W., Stenzel, U., Dabney, J., Shendure, J., Kitzman, J., Hammer, M.F., Shunkov, M.V., Derevianko, A.P., Patterson, N., Andrés, A.M., Eichler, E.E., Slatkin, M., Reich, D., Kelso, J., Pääbo, S., 2012. A high-coverage genome sequence from an archaic Denisovan individual. Science 338, 222–226.
Mielke, J.H., Konigsberg, L.W., Relethford, J.H., 2011. Human Biological Variation, second ed. Oxford University Press, New York.
Prugnolle, F., Manica, A., Balloux, F., 2005. Geography predicts neutral genetic diversity of human populations. Current Biology 15, R159–R160. http://dx.doi.org/10.1016/j.cub.2005.02.038.
Ramachandran, S., Deshpande, O., Roseman, C.C., Rosenberg, N.A., Feldman, M.W., Cavalli-Sforza, L.L., 2005. Support for the relationship of genetic and geographic distance in human populations for a serial founder effect originating in Africa. Proceedings of the National Academy of Sciences 102, 15942–15947.
Reich, D., Green, R.E., Kircher, M., Krause, J., Patterson, N., Durand, E.Y., Viola, B., Briggs, A.W., Stenzel, U., Johnson, P.L.F., Maricic, T., Good, J.M., Marques-Bonet, T., Alkan, C., Fu, Q., Mallick, S., Li, H., Meyer, M., Eichler, E.E., Stoneking, M., Richards, M., Talamo, S., Shunkov, M.V., Derevianko, A.P., Hublin, J.-J., Kelso, J., Slatkin, M., Pääbo, S., 2010. Genetic history of an archaic hominin group from Denisova Cave in Siberia. Nature 468, 1053–1060.
Reich, D., Patterson, N., Kircher, M., Delfin, F., Madhusudan, R.N., Pugach, I., Ko, A.M., Ko, Y., Jinam, T.A., Phipps, M.E., Saitou, N., Wollstein, A., Kayser, M., Pääbo, S., Stoneking, M., 2011. Denisova admixture and the first modern human dispersals into Southeast Asia and Oceania. American Journal of Human Genetics 89, 1–13.
Relethford, J.H., 2002. Apportionment of global human genetic diversity based on craniometrics and skin color. American Journal of Physical Anthropology 118, 393–398.
Relethford, J.H., 2004a. Boas and beyond: migration and craniometric variation. American Journal of Human Biology 16, 379–386.
Relethford, J.H., 2004b. Global patterns of isolation by distance based on genetic and morphological data. Human Biology 76, 499–513.
Relethford, J.H., 2008. Genetic evidence and the modern human origins debate. Heredity 100, 555–563.
Relethford, J.H., 2010. Population-specific deviations of global human craniometric variation from a neutral model. American Journal of Physical Anthropology 142, 105–111.
Roseman, C.C., 2004. Detecting interregionally diversifying natural selection on modern human cranial form by using matched molecular and morphometric data. Proceedings of the National Academy of Sciences 101, 12824–12829.
Smith, F.H., Falsetti, A.B., Donnelly, S.M., 1989. Modern human origins. Yearbook of Physical Anthropology 32, 35–68.
Smith, F.H., Trinkaus, E., Pettitt, P.B., Karavanić, I., Paunović, M., 1999. Direct radiocarbon dates for Vindija G_1 and Velika Pećina Late Pleistocene hominid remains. Proceedings of the National Academy of Sciences 96, 12281–12286.
Steiper, M.E., Young, N.M., 2006. Primate molecular divergence dates. Molecular Phylogenetics and Evolution 41, 384–394.
Stringer, C.B., Andrews, P., 1988. Genetic and fossil evidence for the origin of modern humans. Science 239, 1263–1268.
Tishkoff, S.A., Gonder, M.K., 2007. Human origins within and out of Africa. In: Crawford, M.H. (Ed.), Anthropological Genetics: Theory, Methods and Applications. Cambridge University Press, Cambridge, pp. 337–379.
Tishkoff, S.A., Reed, F.A., Friedlaender, F.R., Ehret, C., Ranciaro, A., Froment, A., Hirbo, J.B., Awomoyi, A.A., Bodo, J.-M., Doumbo, O., Ibrahim, M., Juma, A.T., Kotze, M.J., Lema, G., Moore, J.H., Mortensen, H., Nyambo, T.B., Omar, S.A., Powell, K., Pretorious, G.S., Smith, M.W., Thera, M.A., Wambebe, C., Weber, J.L., Williams, S.M., 2009. The genetic structure and history of Africans and African Americans. Science 324, 1035–1044.
Tishkoff, S.A., Reed, F.A., Ranciaro, A., Voight, B.F., Babbitt, C.C., Silverman, J.S., Powell, K., Mortensen, H.M., Hirbo, J.B., Osman, M., Ibrahim, M., Omar, S.A., Lema, G., Nyambo, T.B., Ghori, J., Bumpstead, S., Pritchard, J.K., Wray, G.A., Deloukas, P., 2007. Convergent adaptation of human lactase persistence in Africa and Europe. Nature Genetics 39, 31–40.
von Cramon-Taubadel, N., Lycett, S.J., 2008. Human cranial variation fits iterative founder effect model with African origins. American Journal of Physical Anthropology 136, 108–113.
White, T.D., Asfaw, B., DeGusta, D., Gilbert, H., Richards, G.D., Suwa, G., Howell, F.C., 2003. Pleistocene *Homo sapiens* from middle Awash, Ethiopia. Nature 423, 742–747.
Wolpoff, M.H., Thorne, A.G., Smith, F.H., Frayer, D.W., Pope, G., 1994. Multiregional evolution: a world-wide source for modern human populations. In: Nitecki, M.H., Nitecki, D.V. (Eds.), Origins of Anatomically Modern Humans. Plenum Press, New York, pp. 175–200.
Xing, J., Watkins, W.S., Witherspoon, D.J., Zhang, Y., Guthery, S.L., Thara, R., Mowry, B.J., Bulayeva, K., Weiss, R.B., Jorde, L.B., 2009. Fine-scaled human genetic structure revealed by SNP microarrays. Genome Research 19, 815–825.

Chapter 16

Human Population Movements: A Genetic Perspective

Oscar Lao and Manfred Kayser
Department of Forensic Molecular Biology, Erasmus MC University Medical Center Rotterdam, Rotterdam, The Netherlands

SYNOPSIS

Movements of people from one place to another, also referred to as human migrations, played a fundamental role in the history of archaic and modern humans. Here, current evidence and hypotheses on two major prehistoric human migrations are described: the initial colonization of Eurasia by archaic and modern humans, and the Neolithic expansion of modern humans into Europe, and how they shaped human genetic diversity. Furthermore, current genetic evidence on admixture between modern and archaic humans is discussed, and showed based on two populations well-known for their particular migration history it is exemplified how genetic diversity can be used to trace population origins and migration history.

INTRODUCTION

Mythological and supernatural explanations aside, the natural origin and evolution of *Homo sapiens* is one of the main questions of humankind. The answer to this enigma is nourished by the direct and indirect traces that our ancestors left in the past. As the nature and tempo of evolution is extremely diverse, making sense of these messages from the past is usually complex and only affordable by studying them through numerous scientific disciplines. In some cases, the information provided by analyzing a particular characteristic using one scientific field can be contrasted to and corroborated by other scientific disciplines, for either the same or different items. In other cases, this information is irreplaceable and helps to fill the gap of otherwise unresolved aspects of human evolution. There is a large body of evidence of past human migrations since the first emergence of the genus *Homo;* some authors even consider members of *Homo* as migrants by nature (Crawford and Campbell, 2012). The causes underlying the movement of human individuals from one place to another, and consequent settlement histories, are not unique and include social and cultural features, technological improvements, and ecological factors such as climatic changes; some authors even propose that particular genetic changes and evolutionary pressures would enhance the capacity for migration (Crawford and Campbell, 2012). This chapter will briefly summarize current advances in the knowledge of human evolution and population history, focusing on the spread of archaic and modern humans out of the African continent and particularly more recent migratory waves, with a special emphasis on human genetic evidence.

OUT OF AFRICA MIGRATION OF ARCHAIC HUMANS

Anatomically modern humans are the only living species of the genus *Homo*. However, the fossil record contains substantial evidence of the past existence of other *Homo* species distributed in Africa and Eurasia (Cela-Conde and Ayala, 2007) (see the chapter by Simpson in this volume). The oldest fossil records of *Homo* are found in the African continent, pinpointing Africa as the birthplace of the genus (Fleagle et al., 2010). The earliest evidence of migration of hominins out of Africa is that of *Homo erectus* at ~1.8–1.7 million years ago (mya) (Anton, 2003); this evidence comes from paleontology. It has not been possible thus far to study the genetic diversity of *H. erectus* and useful material for study may be considered unlikely to ever be retrievable given the old age of the remains and current knowledge about DNA preservation over time (Allentoft et al., 2012). Many aspects of this initial Out of Africa diaspora (also known as the "Out of Africa I" hypothesis (Fleagle et al., 2010)) remain unknown. For example, the factors that caused *H. erectus* to expand outside the African continent, and which migratory routes the species followed out of Africa and across Eurasia, are only hypothesized (Fleagle et al., 2010). Moreover, there is no general agreement about the number of *Homo* species that existed during the Pleistocene, ranging from one to at least eight depending on the investigator (Stringer, 2002). It has even been proposed that morphological diversity in the fossil *Homo* record around 1.8 mya probably represents variation between demes of *H. erectus* (described in Asia), so that the previously described species *H. georgicus* (present in Caucasus) and *H. ergaster* (described in Africa) should be considered the chronospecies of a single *H. erectus* species (i.e., *H. e. georgicus* and *H. e. ergaster* respectively) (Lordkipanidze et al., 2013).

Independent of the taxonomic relationships among species or chronospecies, it has been suggested that climate changes including ice ages could have fueled the morphological diversification of the genus *Homo* (regardless of whether different species status is justified) by forcing population bottlenecks and massive migrations toward glacial and other refugia (Stewart and Stringer, 2012). For example, in Europe *H. heidelbergensis* expanded across Eurasia around 600,000 years ago and persisted there through entire ice age cycles, retreating to glacial refugia and eventually leading to a distinct human species known as *Homo neanderthalensis*. Neanderthal fossils can be found widespread all over Europe, and are also found in the Middle East and Western Asia (Cela-Conde and Ayala, 2007) until about 30,000 years ago (kya) (Stewart and Stringer, 2012) (see chapter by Ahern). A genetic characterization of Neanderthals is provided below.

OUT OF AFRICA MIGRATION OF ANATOMICALLY MODERN HUMANS

Whether *Homo sapiens* was the result from the evolution of the whole *Homo erectus* species or came about from a single subpopulation, either *H. erectus* or another descendant species, is still a matter of controversy among paleontologists. Several hypotheses explain the origin of anatomically modern humans (Stringer, 2002) (see Figure 1). Nevertheless, two of them—the Recent African Origin (RAO) model and the Multiregional model—can be considered the extremes of the currently considered plausible explanations of the development of humankind.

The multiregional or "Regional Continuity" hypothesis assumes a continuous genetic flow among the *Homo erectus* subpopulations in different parts of the world after the first diaspora out of the African continent (see Figure 1(D)), which prevented the mechanisms of speciation that could be expected under isolation and allowed the putative evolutionary changes that occurred in one subpopulation of *H. erectus* to spread to other subpopulations (Wolpoff et al., 2000). In contrast, the RAO model—also called the Out of

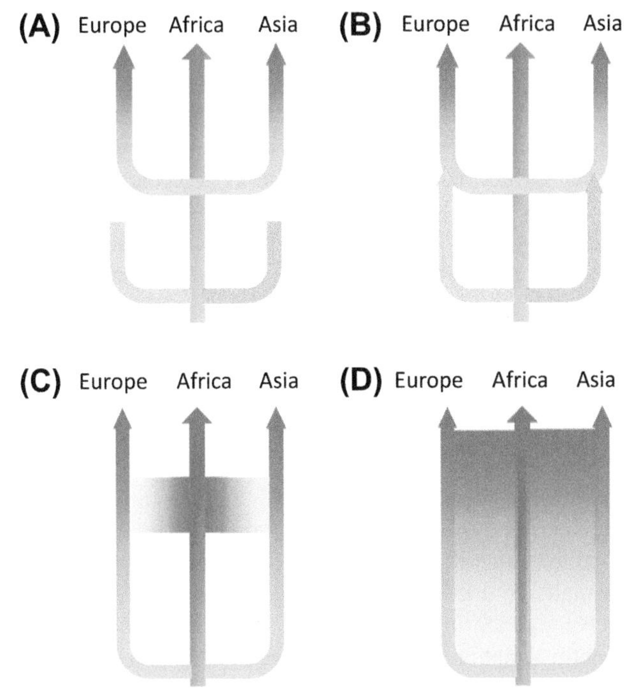

FIGURE 1 Models of evolution of anatomically modern humans. *(Adapted from Stringer, 2002.)* (A) Pure Recent African Origin (RAO) model: modern humans appear in Africa and spread worldwide, replacing archaic forms in Europe and Asia. (B) RAO model with gene flow between modern and archaic humans. (C) Assimilation model: modern human morphological characteristics appear in Africa and are spread to the archaic forms present outside of the African continent. (D) Multiregional hypothesis: after the first Out of Africa event, gene flow between populations in different continents allows evolutionary innovations that take place in one region to spread to other regions.

Africa model or "African Replacement" hypothesis (Excoffier, 2002)—considers *Homo sapiens* to have evolved from a unique African population descendant from *H. erectus*, which ~200 kya produced a second African diaspora replacing archaic *Homo* species (such as *Homo neanderthalensis*) (Stringer, 2002). Other models combining aspects from both the multiregional and the RAO hypotheses have been also proposed. These include an RAO with population substructure within Africa before the African diaspora started (Harding and McVean, 2004), recursive migrations out of the African continent before the RAO (Templeton, 2002), and the spread of the African modern human anatomy by way of diffusion rather than migration (Assimilation hypothesis (Smith, 1992); Figure 1(C)).

From a genetic point of view, it can be expected that each hypothesis leads to a distinctive signature in the amount and spatial distribution of the genetic variation of contemporary modern humans around the world (see the chapter by Relethford). Under a pure multiregional hypothesis, high levels of genetic differentiation are expected between human populations around the world (Barbujani and Colonna, 2010). In contrast, under the pure RAO hypothesis, the expected genetic differentiation of human populations around the world would be small. Further, under the RAO model it is expected that (1) the roots of the genetic diversity of humankind should be in Africa, (2) the genetic diversity of human populations within the African continent should be higher than that outside the African continent, and (3) genetic diversity of human populations should linearly decay with increased geographic distance from Africa due to the expected multiple bottlenecks that human populations likely have experienced during the migration and settlement history of the non-African world (Barbujani and Colonna, 2010).

Indeed, early analyses of the genetic variation present at the mitochondrial DNA (mtDNA), the maternally inherited part of the human genome (Pakendorf and Stoneking, 2005), using worldwide individuals showed that sub-Saharan African individuals tended to be placed at the oldest branches of the phylogenetic tree. This indicates that the origin of the current maternally inherited genetic diversity of modern humans points to Africa (Cann et al., 1987). Moreover, the time of the most recent common ancestor (TMRCA) of all humans was estimated with mtDNA diversity to be ~200 kya. Taking these observations together, the authors of this study concluded that the genetic variation present in mtDNA of modern humans supports the RAO hypothesis from the maternal genetic side (Cann et al., 1987). Subsequent studies have systematically shown that the origin of the genetic diversity of modern human mtDNA is recent and can be placed at the African continent (see van Oven and Kayser, 2009 for a comprehensive description of the current mtDNA phylogenetic tree). Further, analysis of mtDNA haplogroup frequencies suggests an expansion from East Africa toward the rest of the world (Oppenheimer, 2012). Similarly, analyses of the nonrecombining part of the Y chromosome (NRY), the paternally inherited part of the human genome, pinpoint a recent origin for modern humans in Africa and a spread out of Africa supporting the RAO hypothesis from the paternal genetic side (van Oven et al., 2013; Shi et al., 2010), although TMRCA was estimated as being substantially younger than in the case of mtDNA (i.e., ~60–140 kya (Wei et al., 2013; Hammer, 1995; Thomson et al., 2000; Cruciani et al., 2011)). However, a recent study analyzing sequence data estimated TMRCA in the Y chromosome to be older than it was by mtDNA (338 kya (Mendez et al., 2013)). Such discrepancies between TMRCA estimated by NRY or mtDNA have been explained by differences in the calibration of the mutation rates of Y and mtDNA chromosomes (Poznik et al., 2013). After controlling for these rates, the Y chromosome TMRCA has been estimated as 138 kya (120–156 kya) and mtDNA TMRCA as 124 kya (99–148 kya), thus agreeing in timing and suggesting a younger origin of humankind than those of previous studies.

Early analyses of alloenzymes (Lewontin, 1972) and further analyses using autosomal DNA markers (i.e., Romualdi et al., 2002) inherited from both parents equally showed that the extent of genetic differentiation among human populations around the world was relatively small compared with the genetic differentiation of individuals within populations, which is compatible with the hypothesis of a recent origin for modern humans. Further studies using different types of autosomal DNA markers showed that sub-Saharan African populations have the most genetic variation compared with non-African populations, and the decay in the genetic diversity was shown to be proportional to the geographic distance from Africa (i.e., (Ramachandran et al., 2005), Figure 2(A)). Moreover, it was demonstrated that non-African populations shared a subset of the autosomal haplotypic variation observed in African populations (Jakobsson et al., 2008), and the amount of linkage disequilibrium of a population correlates with its geographic distance from Africa (Jakobsson et al., 2008). Furthermore, population trees based on genetic distances showed that sub-Saharan African populations at the root (Li et al., 2008; Xing et al., 2010b; Pickrell and Pritchard, 2012) (Figure 2(B)), and modeling the spread of humans using autosomal short tandem repeat polymorphisms (STRs), assuming a single origin, fit well with the data (Liu et al., 2006). All these autosomal genetic findings are compatible with the RAO hypothesis. Therefore, all of the genetic evidence provided by DNA markers, following the three ways of inheritance, supports the RAO hypothesis.

It has been suggested that migration(s) out of Africa was accompanied by constant climate fluctuations (Petit et al., 1999). These climatic changes dramatically shaped the spatial distribution of animals and plants and provided or erased land pathways depending on the time period. On average, the climate in the past was cooler and drier, with a lower sea level, than nowadays (Jobling et al., 2004). Many details of the Out of Africa diaspora of modern humans, such as the number of migratory waves and the routes taken across Eurasia to Australia and the Americas, are still disputed among scholars from different disciplines. Genetic mtDNA and Y chromosome evidence suggest a common shared origin of populations outside of Africa, and has been interpreted as incompatible with the existence of two or more independent migratory events (Mellars, 2006). Similarly, a study using autosomal single-nucleotide polymorphisms (SNPs) in 73 Asian populations concluded that the data were consistent with a single wave of settlement of Asia that moved from south to north (Abdulla et al., 2009). However, another study of genome-wide SNP data of Southeast Asian and Oceanian populations argued in favor of two waves of settlement (Wollstein et al., 2010), as did an analysis of the genetic diversity in the human commensal bacterium *Helicobacter pylori* (Moodley et al., 2009). Also under debate is whether anatomically modern humans followed an early inland route or a later coastal route across South Asia (the so-called "coastal express" hypothesis) (Oppenheimer, 2012). The geographic distribution of mtDNA variation in modern humans has been interpreted as supporting the coastal express hypothesis (Macaulay et al., 2005). However, based on autosomal DNA sequence data, a third explanation, the "delayed expansion" hypothesis, has been proposed. According to this hypothesis, the modern human population(s) that migrated out of Africa remained in the Middle East for thousands of years before suddenly expanding across Eurasia (Xing et al., 2010a). A recent study analyzing autosomal SNPs and cranial features suggests multiple dispersals in which Australians, Papuans, and Melanesians remained relatively isolated after an early dispersal from Africa via a southern route (Reyes-Centeno et al., 2014).

GENETIC ADMIXTURE BETWEEN ARCHAIC AND MODERN HUMANS

The RAO hypothesis does not deny the possibility of inbreeding between anatomically modern humans and archaic *Homo* species living during the same time, although it assumes that if it happened, such admixture did not substantially contribute to the gene pool of modern humans as seen today (Stringer, 2002). A study considering current genetic variation of autosomal chromosomes in different populations, and modeling putative admixture between anatomically modern humans and archaic forms of *Homo*, concluded that the most likely demographic model was the one considering complete replacement (Fagundes et al., 2007). However, some authors argued that the size of the analyzed genomic fragments was not large enough to recover the fingerprint of putative archaic inbreeding (Garrigan and Hammer, 2008). Indeed, another study did detect signatures of introgression of archaic genetic variation at the human leukocyte antigen class I genes (Abi-Rached et al., 2011).

It has been suggested that in order to properly address this and related questions, genetic information from ancient DNA is essential (Pickrell and Reich, 2014). The outcomes of the first study of the genetic variation of the mtDNA of a Neanderthal individual were interpreted as evidence of replacement without admixture, given that the observed Neanderthal mtDNA variation did not overlap with that of modern humans from around the world (Krings et al., 1997). However, it was suggested that reaching such a conclusion was not reasonable given the small sample size of one Neanderthal individual used at that time (Nordborg, 1998). Notwithstanding that concern, subsequent studies of Neanderthal mtDNA considering more individuals, and eventually entire mtDNA genomes, did corroborate the strong genetic divergence between anatomically modern humans and Neanderthals (Fu et al., 2013), and further concluded the absence of maternal genetic admixture between both species (Lalueza-Fox and Gilbert, 2011).

New light on this question was shed by analyzing a large portion of the autosomal genome of Neanderthals and comparing its parts with the genomes of modern humans from different parts of the world, which became possible due to advances in DNA sequencing technologies (so-called next

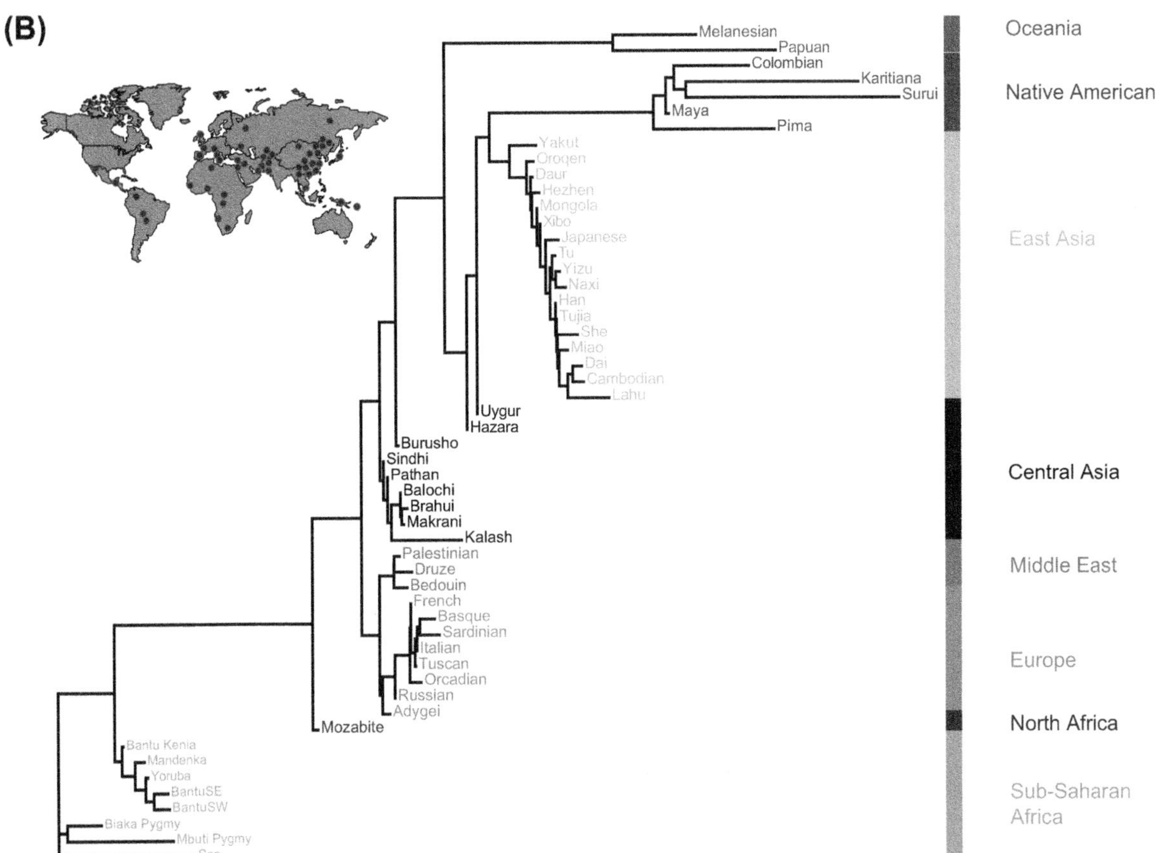

FIGURE 2 Genetic diversity in worldwide human populations. (A) Density map showing the genetic differentiation (estimated as Euclidean distance) within individuals from each population using 658 autosomal STRs in 5879 individuals from 250 populations (Pemberton et al., 2013). Sub-Saharan African chromosomes show the greatest genetic differentiation within individuals, and this level decays with the geographic distance from the African continent, with the lowest levels occurring in native South American populations. (B) Phylogenetic neighbor-joining tree (Saitou and Nei, 1987) showing the genetic affinities of 53 populations from the HGDP-CEPH panel (Cann et al., 2002) estimated by means of Fst distance (Weir and Cockerham, 1984) on ~650,000 autosomal SNPs in ~1000 individuals from ~50 worldwide populations (Li et al., 2008). Sub-Saharan African populations are separated from the other worldwide populations. Outside the African continent, populations are clustered by sampling continent. (See color plate section).

generation sequencing or NGS). The first draft of the Neanderthal genome showed that Neanderthals and anatomically modern humans split 270–400 kya. This study further suggested that there had been gene flow from Neanderthals into the ancestors of non-African modern humans but not African ones, so that 1%–4% of the current genomes of Eurasian and Oceanian people are derived from Neanderthals (Green et al., 2010). This finding was remarkable in that it not only presented genetic evidence for admixture but also showed admixture in human individuals from areas such as China and Papua New Guinea, where Neanderthal fossils are missing from the records and may not be expected to be found in the future given what is known about Neanderthals. This pattern would suggest that the admixture between Neanderthals and modern humans likely occurred after modern humans left Africa but before they spread around the world, pinpointing the Middle East as the most likely region of admixture, estimated at about 47–65 kya (Sankararaman et al., 2012). Analyses of Neanderthal introgression in the current human genome confirmed that the autosomal genome of East Asian and European populations carries an average of 1.38% and 1.15% Neanderthal ancestry, respectively; however, some genomic regions show inferred Neanderthal ancestry as high as 62% in East Asian populations and 64% in European populations, which could be mediated by selective pressures (Sankararaman et al., 2014). Interestingly, the estimated fraction of Neanderthal ancestry in the X chromosome is significantly smaller than in autosomal chromosomes, which could suggest male hybrid sterility (Sankararaman et al., 2014).

Additional evidence of genetic admixture between archaic and modern humans was found when analyzing the genetic variation present in the ancient genomic DNA from a hominin found in the Denisova cave in Siberia (Russia) dated ~30–50 kya (Krause et al., 2010). Mitochondrial DNA analysis showed that Denisovans diverged from the common lineage leading to modern humans and Neanderthals about one million years ago (Krause et al., 2010). Nevertheless, comparison using autosomal DNA showed that Denisovans are a sister group of the Neanderthals (Reich et al., 2010). Further comparisons with the genome of contemporaneous modern humans showed no traces of archaic Denisovan genetic variation in Eurasian individuals, but concluded that ~4.8% of the genome of current Melanesians from New Guinea shows signals of introgression of Denisovan genetic variation (see Figure 3). A subsequent

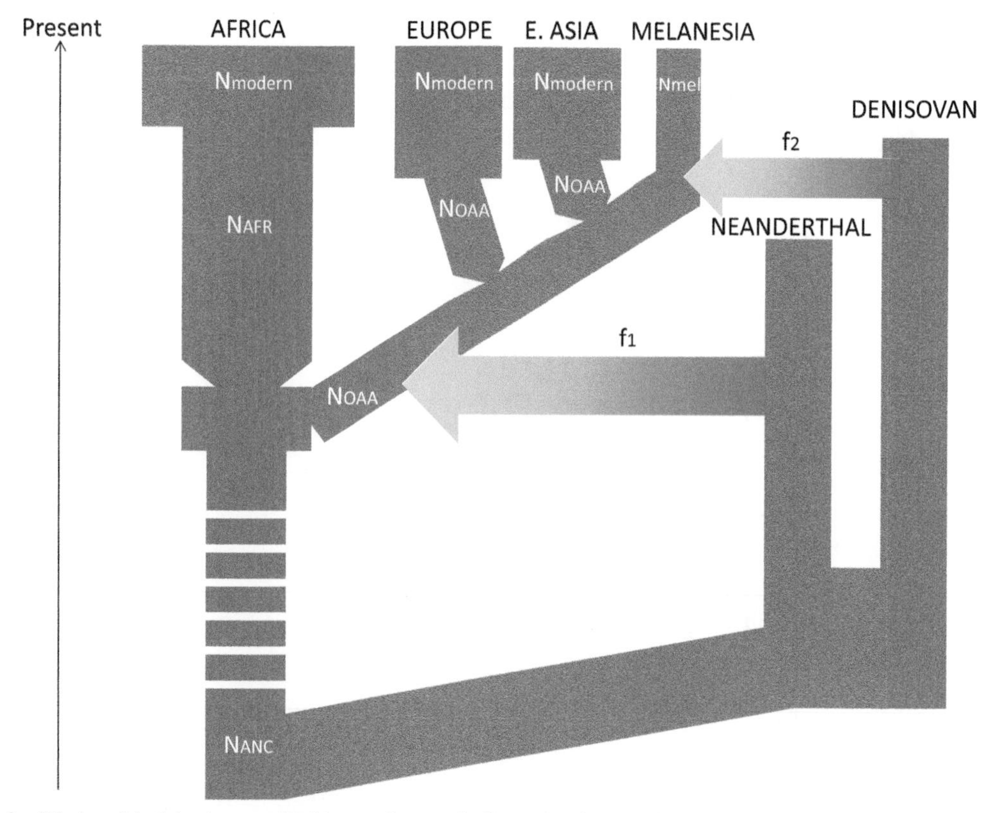

FIGURE 3 A simplified model of the demographic history of anatomically modern humans following Schaffner et al. (2005), and proposed relationships with archaic humans (Krause et al., 2010). Starting from an ancestral population with an effective population size N_{anc}, there was a split ~1 mya of the lineage that produced anatomically modern humans, Neanderthals, and Denisovans. During the Out of Africa event there was a bottleneck in human populations. Outside the African continent, Neanderthal inbreeding with modern humans occurred at the f1 rate. During human dispersal, Denisovans inbred with Melanesian populations at the f2 rate. Because of technological advances (see below), the human population increased exponentially (here modeled as a sudden population increase).

study (Reich et al., 2011) suggested that the most likely region of admixture between Denisovans and modern humans was Southeast Asia, where fossil evidence for Denisovans is notably absent thus far (as it currently is everywhere except the particular cave in Siberia). Recently, further evidence of Denisovan introgression into modern humans was detected at the *EPAS1* gene in Tibetans (Huerta-Sanchez et al., 2014). This gene has been associated with adaptation to high altitude (Yi et al., 2010) that in agreement with the observations in the Neanderthal case suggests that admixture with other hominin species could have helped modern humans to adapt to new environments.

TECHNOLOGICAL ADVANCES AND MIGRATION: THE CASE OF THE EUROPEAN FARMERS

For a long period after the worldwide dispersion of anatomically modern humans, humans were nomadic hunter-gatherers migrating under the availability of seasonal resources. Climatic changes, such as the last glacial maximum during 23–14 kya, produced population contractions/bottlenecks, and migrations of temperate-adapted animals and plants escaping from the advance of the ice sheets into glacial refugia that later expanded when the ice receded. However, from 12 kya to present, temperature has been unusually stable (Jobling et al., 2004).

The rise of sedentary communities, ceramics, and domesticated foods, known as the Neolithic Revolution, took place independently in different places in the world such as the Middle East, Central China, West Africa, the New Guinea Highlands, Mesoamerica, the Central Andes, and the Eastern Woodlands of the United States—and at different times (Jobling et al., 2004) (see chapters by Blumler and Little). This technological revolution had a dramatic impact not only on the social, ideological, and conceptual lifestyles of humans, but also on their demography (Bocquet-Appel, 2008). A question of interest is to what extent the Neolithic spread was indigenous and animated by imitation of information, language, and products (i.e., cultural diffusion) or was driven by an influx of dispersing human populations (i.e., demic diffusion). Because of the independent origin of the Neolithic in different parts of the world, it can be expected that the answer may depend on the geographic area analyzed (Diamond and Bellwood, 2003). Moreover, given the complexity of the Neolithic phenomenon, mixtures of the two hypotheses could be more likely than either of them alone (Diamond and Bellwood, 2003).

In the case of the European continent, the Neolithic way of life began and evolved in the Fertile Crescent (Middle East; see Figure 4(A)) for several millennia without having a notable influence on neighboring regions (Ozdogan, 2008). The wave-of-advance hypothesis (Ammerman and Cavalli-Sforza, 1971) and related ones (i.e., the time-delayed theory, Isern et al., 2012) propose that Neolithic culture and farmers moved from the Fertile Crescent in front of the advance toward Europe. Analysis of the timing of the first Neolithic settlements suggests an agropastoralism swept over Europe at a regular speed of 0.6–1.3 km/year, which has been interpreted as evidence for the demic diffusion model (Pinhasi et al., 2005) (see Figure 4(A)).

From a genetic point of view, earlier work on classical polymorphisms (e.g., blood groups) identified a gradient in the first dimension of a principal component analysis (PCA) (Cavalli-Sforza et al., 1994; Menozzi et al., 1978) and clinal patterns in the frequencies of different alleles (Sokal et al., 1991), which were interpreted as evidence of demic diffusion in Europe. However, some authors have claimed that PCA can produce statistical artifacts (Sokal et al., 1999; Novembre and Stephens, 2008), and the interpretation of PCA outputs is not straightforward (Arenas et al., 2013). When considering recent genome-wide SNP data, similar patterns in the PCAs are observed, with genetic data resembling sampling locations (Lao et al., 2008; Novembre et al., 2008) (see Figure 4(D)). Furthermore, Southern European populations show greater genetic diversity (see Figure 4(C)) and lower levels of linkage disequilibrium compared with those of Northern Europe (Lao et al., 2008), in line with the demic diffusion model. However, this finding could also suggest larger effective population sizes in southern populations due to more favorable climate conditions or a higher admixture rate with North African populations (Botigue et al., 2013). The estimated genetic contribution of Neolithic farmers to the patrilineal and matrilineal genetic pools of current Europeans depends on the genetic system analyzed and statistical method applied (Pinhasi et al., 2012). Studies of ancient mtDNA from skeletal remains of early European farmers and late European hunter-gatherers have revealed that current Europeans are more similar to Neolithic farmers than to Paleolithic hunter-gatherers in Europe (Fu et al., 2012). The latest analyses of genomic ancient DNA suggest a complex scenario where farming practices were brought to Northern Europe by a group of people genetically related to Southern Europe and distinct from resident hunter-gatherers (Skoglund et al., 2012), and that current Northern European populations would result from the admixture of people with these different genetic ancestries. Recent analyses of ancient DNA also suggest that the genomic landscape of current Europeans resulted from (at least) three components: Western European hunter-gatherers, ancient northern Eurasians related to Upper Paleolithic Siberians, and early European farmers (Lazaridis et al., 2014).

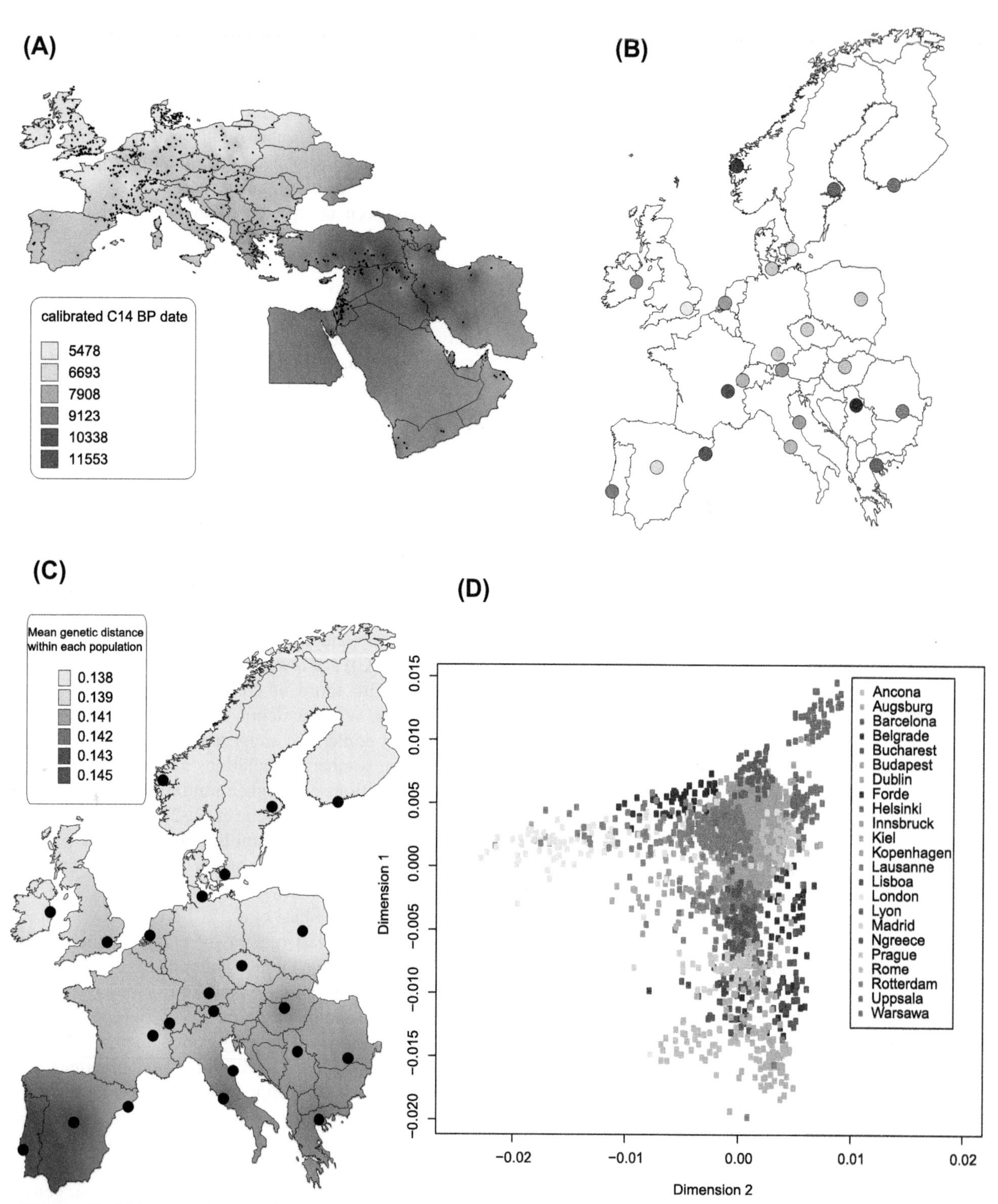

FIGURE 4 Neolithic sites and autosomal genetic variation in Europe. (A) Estimated C14 dates of Neolithic sites in Europe. *(Data taken from Pinhasi et al., 2005.)* The oldest Neolithic sites are found in the Fertile Crescent, while younger Neolithic sites are found as one moves away from the Middle East. (B) Geographic location of 23 European sampling sites/subpopulations containing 2457 individuals analyzed by Lao et al. (2008) with ~133,000 autosomal SNPs after linkage disequilibrium pruning. Each dot represents a sampling site or subpopulation. (C) Density map of the mean genetic diversity among individuals from the same European sampling site/subpopulation; within-population genetic differentiation decreases as latitude increases. (D) Plot of the first two dimensions computed by means of a classical multidimensional scaling analysis using a genetic distance (identical by state) matrix between pairs of European individuals; each dot is one individual. The relative position of each individual in the two dimensions resembles the sampling site location. (See color plate section).

HISTORIC MIGRATIONS INTO A PREVIOUSLY OCCUPIED TERRITORY: THE CASE OF THE ROMANI

Modern history is full of recorded human migrations, the details of which usually remain largely unknown—this is where the study of genetic variation in current populations can provide hints about the origin and details of historic migratory events. A case exemplifying the use of human population genetics to explore a migration in historic times is the history of the European Romani.

The Romani (also sometimes called Roma) people constitute a culturally and linguistically heterogeneous transnational group mainly living in different parts of Europe (~11 million people, Hancock, 2002) and also present in other regions of the world. Despite their cultural and geographic diversity, the Romani people consider themselves a unique group with a shared common origin (Hancock, 2002). First written documents suggest that the Romani people first arrived in Europe during the Middle Ages through the Balkans (see Figure 5(A)).

The relationships between the Romani newcomers and local non-Romani European populations were not always friendly, and the Romani people (traditionally represented by stereotypes such as nomadism, marginality, cultural difference, or illicit behavior) have been ignored, prosecuted, decimated, or enslaved depending on the time period and European region/country (Hancock, 2002). However, their origin remained unresolved for centuries, and fueled several legends including an Egyptian origin (the name "gypsy" refers to such a hypothetical origin). During the eighteenth century, the Indian origin hypothesis was proposed based on linguistic similarities of the Romani language with Sanskrit (Hancock, 2002), and was supported by subsequently accumulated linguistic evidence. However, the exact geographic origin of the Romani ancestors within India and the timing of when the Romani diaspora began remained disputed. From a genetic point of view, it has been shown by means of autosomal classical markers (Kalaydjieva et al., 2001), mtDNA, and Y chromosome DNA, that the Romani people are largely distinctive from non-Romani European populations and carry paternally (Rai et al., 2012) and maternally (Mendizabal et al., 2011) inherited haplogroups that are typical of South Asians such as those in Indian populations. Moreover, it was shown that European Romani carry a relatively high frequency of particular recessive Mendelian diseases that are rare or absent in other populations (Morar et al., 2004).

Analyses of genome-wide SNP data in Romani samples from across Europe and their comparison with worldwide populations including those from India and Europe in two independent studies (Moorjani et al., 2013; Mendizabal et al., 2012) have shown that European Romani carry a large proportion of non-Romani European ancestry, suggesting a more permissive admixture scenario than previously thought based on recorded relationships between the Romani European and their non-Romani European neighbors (Figure 5(D)). Furthermore, both studies concluded that the most likely geographic origin of the Romani is in Northwestern India (Figure 5(B)), a region that had been previously suggested based on linguistic evidence (Hancock, 2002). The Romani diaspora was estimated by Mendizabal et al. (2012) at 1.5 kya, with a strong reduction of the Proto-Romani effective population size and a demographic bottleneck at 0.95 kya, and it was revealed that the European dispersion of Romani populations took place from the Balkans and comprised additional population reduction and bottlenecks.

MOST RECENT MIGRATIONS INTO A NEWFOUND LAND: THE CASE OF THE POLYNESIANS

Described below is the most recent human migration into previously unoccupied territory, namely the settlement of the Pacific Islands. The archaeological record of Remote Oceania (comprising the islands south and east of the main Solomon Islands: Vanuatu, New Caledonia, Fiji, Micronesia, and Polynesia) provides evidence for a relatively recent human history in this part of the world. While Near Oceania (i.e., New Guinea and surrounding islands as well as the main Solomon Islands) was occupied by modern humans 40 kya or earlier, Remote Oceania was first settled by humans within the last 3 kya (Kirch, 2000). Archaeological and linguistic evidence and the topographic situation of Oceania suggest that the human migrations into Oceania were twofold: an early migration into Near Oceania and Australia about 50 kya, well before the two regions became separated by rising sea levels around 8 kya, likely a result of the first out of Africa route of migration of modern humans, and a much later migration wave into Remote Oceania. Currently, archaeologists and linguists widely agree that the later migration wave into Remote Oceania is associated with the expansion of Austronesian-speaking seafarers starting about 5.5 kya in Taiwan, arriving around 3.5 kya in the Bismarck Archipelago northeast of the New Guinea mainland, where the Lapita cultural complex and Proto-Oceanic language developed, and finally arriving in the most western parts of Remote Oceania around 3 kya, and across Polynesia 1.8–0.8 kya (Bellwood and Dizon, 2005; Blust, 1999; Gray et al., 2009; Spriggs, 2003). Because material remains associated with the Lapita cultural complex in Near Oceania were mostly found on the Bismarck Archipelago but appear to be largely absent from mainland New Guinea (Kirch, 2000), it was assumed for a while that the Austronesian expansion bypassed mainland New Guinea on its way toward Remote Oceania. This view was supported by mtDNA data showing that Polynesian maternal ancestry is mostly (>95%) from Asia,

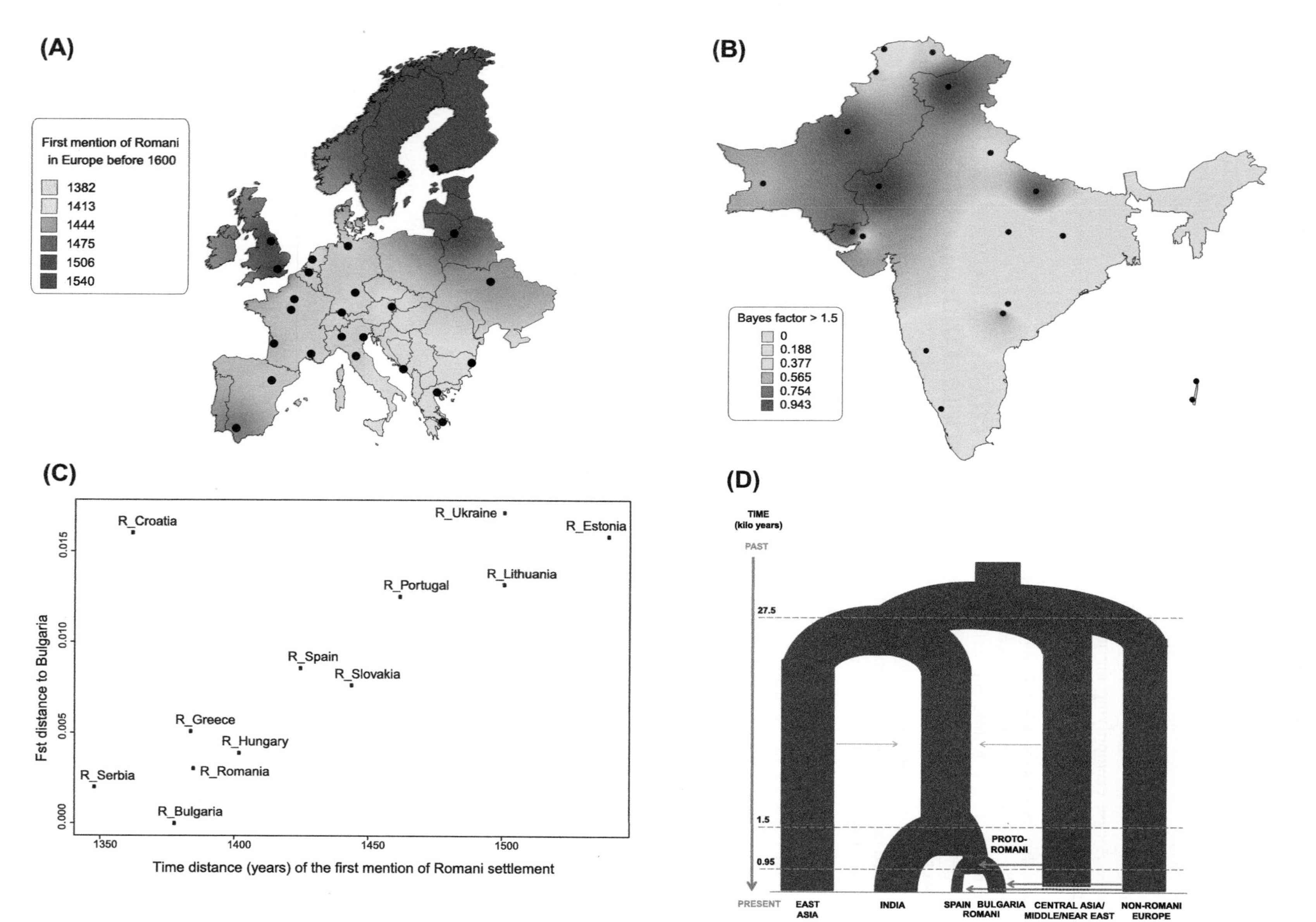

FIGURE 5 Analysis of the genetic variation of Romani populations in Europe using SNP data (Mendizabal et al., 2012). (A) Density map based on the first written mention of Romani in Europe before 1600 (Tcherenkov and Laederich, 2004). (B) Density map of Pakistan and India showing the number of times each population (a dot) is supported by statistical modeling as the homeland of the European Romani, compared with the other populations. The Northwestern India region (including Meghawal and Kashmiri Pandit populations) has the highest probability of representing the homeland of the European Romani. (C) Plot between the oldest historical records of the Romani settlements in each European country and the genetic distances (estimated by means of Fst Weir and Cockerham, 1984) between each Romani population and the Balkans (i.e., Bulgaria). (D) Inferred demographic history of the European Romani assuming the Out of Africa model and a generation time of 25 years. Branch width is proportional to the estimated effective population size; the presence of bottlenecks in the European Romani branch is indicated by red lines. Arrow width indicates migration rates, measured as the number of migrant chromosomes per generation from the donor population. (See color plate section).

while only a minor fraction of Polynesian mtDNA is of New Guinean origin (Melton et al., 1995; Sykes et al., 1995). This led Jared Diamond to propose the "Express Train" hypothesis of Polynesian origin (Diamond, 1988). However, earlier genetic work on autosomal globin genes, which are thought to be under selective constraint due to malaria resistance in Near Oceania, showed that the respective DNA variants are found in Polynesia despite the absence of malaria there; hence the Polynesian frequency of these variants must be explained solely by migration in the absence of selective pressure (Trent et al., 1988). Although this genetic evidence was ignored in the Express Train hypothesis, later Y chromosome DNA work provided overwhelming genetic evidence that the express train to Polynesia likely did not happen, because over 60% of Polynesian Y chromosomes were found to be of New Guinean origin, while ca. 40% were of Asian origin (Kayser et al., 2006, 2000). Y chromosome genetic data first led to the "Slow Boat" from Asia hypothesis of Polynesian origin (Kayser et al., 2000), under which Polynesian ancestors originated from East Asia, perhaps Taiwan, and migrated via Southeast Asia to New Guinea, where they mixed intensively with local New Guineans before they occupied the island world of Remote Oceania as a mixed group (see Figure 6). This genetic admixture likely was sex biased, with more Asian women and New Guinean men participating, which would be supported by the matrilineal culture and matrilocal residence pattern of the Austronesian arrivers and the polygyny of the New Guinea locals (Kayser et al., 2006, 2000; Hage and Marck, 2003). This hypothesis received further support from different genome-wide datasets (Kayser et al., 2008b; Wollstein et al., 2010), which revealed that about 85% of overall biparental genomic ancestry of Polynesians is from Asia while 15% is from New Guinea. Simulation studies using genome-wide data also revealed that this admixture between Asian and New Guinean ancestors of Polynesians happened about 3 kya, before the occupation of Polynesia about 1.8–0.8 kya (Wollstein et al., 2010). It is assumed that due to their boat technology (i.e., dugout canoes that did not allow for long sea crossings), people in Near Oceania before the Austronesian arrivers did not cross the seawater separating Near and Remote Oceania. However, the Austronesian arrivers, with their outrigger canoes and sophisticated navigation skills, managed not only to migrate all the way from Asia to Near Oceania, but also for the first time in the human history to cross the water border and enter Remote Oceania after mixing with local New Guineans, as genetic evidence suggests. Notably, no human genetic evidence so far has proposed the following Thor Heyerdahl's hypothesis (Heyerdahl, 1950) that Polynesia was occupied from South America.

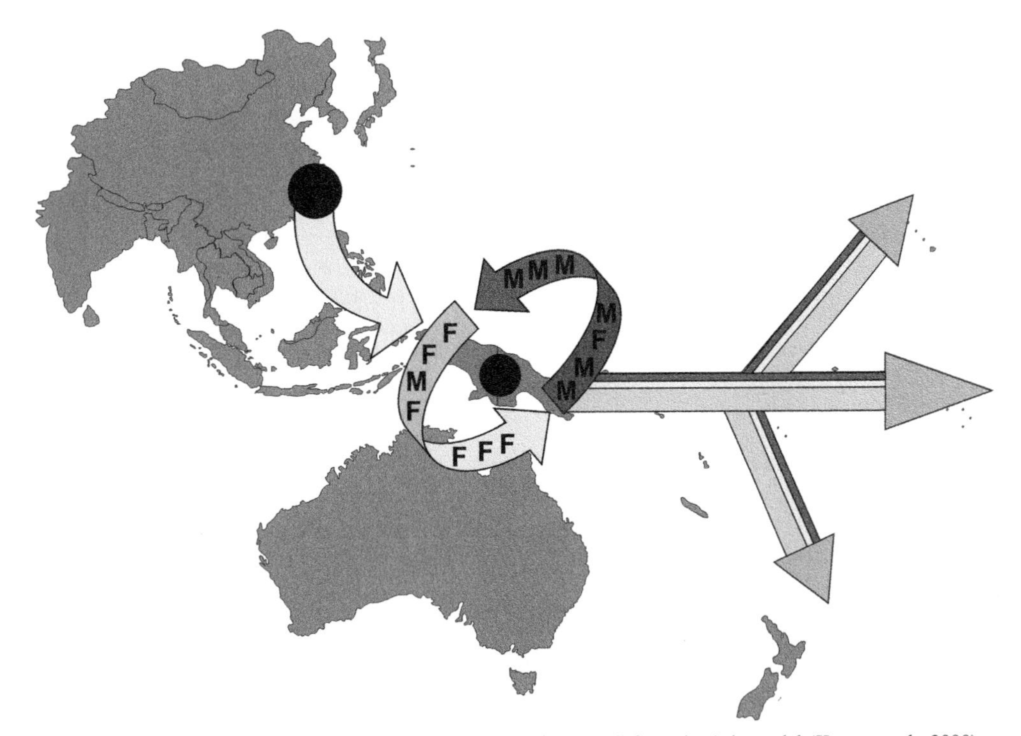

FIGURE 6 Migration history and origins of Polynesians proposed by the "Slow Boat" from the Asia model (Kayser et al., 2000) according to which Polynesian ancestors originated in East Asia, perhaps Taiwan, and migrated via Southeast Asia toward New Guinea (Kayser et al., 2006, Kayser et al., 2000), likely the Bismarck Archipelago (Kayser et al., 2008a), where they mixed in a sex-biased way that included more Asian women and New Guinean men (Kayser et al., 2006, Kayser et al., 2000, Kayser et al., 2008a, Kayser et al., 2008b; Wollstein et al., 2010), before the admixed population migrated further eastwards to Polynesia. (See color plate section).

CONCLUSIONS AND FUTURE PROSPECTS

Archaeological, fossil, and genetic records suggest that modern and archaic humans, such as Neanderthals and most likely the whole *Homo* genus, have been nomadic since the origin of the species and genus. Understanding the demographic processes that led to the current global distribution of humankind around the world is complex. However, new archaeological and fossil discoveries, as well as new statistical and technological developments in the field of human population genetics, constantly allow us to improve our understanding about the past of our species, including the migration waves that distributed modern humans around the world and how those migrators interacted with archaic humans present at the time. In many instances, human genetic evidence has strengthened hypotheses about the origin and migration history of our species as well as of certain regional populations, while in other cases human genetic evidence has allowed completely new hypotheses to be developed. Given the fast technological progress in human genetics both experimentally and analytically, currently ranging from single DNA polymorphisms, usually on the Y chromosome or the mtDNA, to hundreds of thousands of genome-wide SNPs genotyped in a single microarray experiment, in the future including whole-exome and even whole-genome data, and used in combination with the information from ancient DNA data, we are looking ahead to exciting new findings that we expect to be revealed in the area of human migration history research related to genetic evidence.

REFERENCES

Abdulla, M.A., Ahmed, I., Assawamakin, A., Bhak, J., Brahmachari, S.K., Calacal, G.C., Chaurasia, A., Chen, C.H., Chen, J., Chen, Y.T., Chu, J., Cutiongco-De La Paz, E.M., De Ungria, M.C., Delfin, F.C., Edo, J., Fuchareon, S., Ghang, H., Gojobori, T., Han, J., Ho, S.F., Hoh, B.P., Huang, W., Inoko, H., Jha, P., Jinam, T.A., Jin, L., Jung, J., Kangwanpong, D., Kampuansai, J., Kennedy, G.C., Khurana, P., Kim, H.L., Kim, K., Kim, S., Kim, W.Y., Kimm, K., Kimura, R., Koike, T., Kulawonganunchai, S., Kumar, V., Lai, P.S., Lee, J.Y., Lee, S., Liu, E.T., Majumder, P.P., Mandapati, K.K., Marzuki, S., Mitchell, W., Mukerji, M., Naritomi, K., Ngamphiw, C., Niikawa, N., Nishida, N., Oh, B., Oh, S., Ohashi, J., Oka, A., Ong, R., Padilla, C.D., Palittapongarnpim, P., Perdigon, H.B., Phipps, M.E., Png, E., Sakaki, Y., Salvador, J.M., Sandraling, Y., Scaria, V., Seielstad, M., Sidek, M.R., Sinha, A., Srikummool, M., Sudoyo, H., Sugano, S., Suryadi, H., Suzuki, Y., Tabbada, K.A., Tan, A., Tokunaga, K., Tongsima, S., Villamor, L.P., Wang, E., Wang, Y., Wang, H., Wu, J.Y., Xiao, H., Xu, S., Yang, J.O., Shugart, Y.Y., Yoo, H.S., Yuan, W., Zhao, G., Zilfalil, B.A., 2009. Mapping human genetic diversity in Asia. Science 326, 1541–1545.

Abi-Rached, L., Jobin, M.J., Kulkarni, S., Mcwhinnie, A., Dalva, K., Gragert, L., Babrzadeh, F., Gharizadeh, B., Luo, M., Plummer, F.A., Kimani, J., Carrington, M., Middleton, D., Rajalingam, R., Beksac, M., Marsh, S.G., Maiers, M., Guethlein, L.A., Tavoularis, S., Little, A.M., Green, R.E., Norman, P.J., Parham, P., 2011. The shaping of modern human immune systems by multiregional admixture with archaic humans. Science 334, 89–94.

Allentoft, M.E., Collins, M., Harker, D., Haile, J., Oskam, C.L., Hale, M.L., Campos, P.F., Samaniego, J.A., Gilbert, M.T., Willerslev, E., Zhang, G., Scofield, R.P., Holdaway, R.N., Bunce, M., 2012. The half-life of DNA in bone: measuring decay kinetics in 158 dated fossils. Proceedings of the Royal Society B: Biological Sciences 279, 4724–4733.

Ammerman, A.J., Cavalli-Sforza, L.L., 1971. Measuring the rate of spread of early farming in Europe. Man New Series 6, 674–688.

Anton, S.C., 2003. Natural history of Homo erectus (Suppl. 37) American Journal of Physical Anthropology 126–170.

Arenas, M., Francois, O., Currat, M., Ray, N., Excoffier, L., 2013. Influence of admixture and paleolithic range contractions on current European diversity gradients. Molecular Biology and Evolution 30, 57–61.

Barbujani, G., Colonna, V., 2010. Human genome diversity: frequently asked questions. Trends in Genetics 26, 285–295.

Bellwood, P., Dizon, E., 2005. The Batanes archaeological project and the Out of Taiwan hypothesis for Austronesian dispersal. Journal of Austronesian Studies 1, 1–32.

Blust, R., 1999. Subgrouping, circularity and extinction: some issues in Austronesian comparative linguistics. In: Zeitoun, E., Jen-Kuei Li, P. (Eds.), Selected Papers from the Eighth International Conference on Austronesian Linguistics. Academia Sinica, Taipei, pp. 31–94.

Bocquet-Appel, J., 2008. Explaining the neolithic demographic transition. In: Bocquet-Appel, J., Bar-Yosef, O. (Eds.), The Neolithic Demographic Transitions and Its Consequences. Springer.

Botigue, L.R., Henn, B.M., Gravel, S., Maples, B.K., Gignoux, C.R., Corona, E., Atzmon, G., Burns, E., Ostrer, H., Flores, C., Bertranpetit, J., Comas, D., Bustamante, C.D., 2013. Gene flow from North Africa contributes to differential human genetic diversity in southern Europe. Proceedings of the National Academy of Sciences United States of America 110, 11791–11796.

Cann, H.M., De Toma, C., Cazes, L., Legrand, M.F., Morel, V., Piouffre, L., Bodmer, J., Bodmer, W.F., Bonne-Tamir, B., Cambon-Thomsen, A., Chen, Z., Chu, J., Carcassi, C., Contu, L., Du, R., Excoffier, L., Ferrara, G.B., Friedlaender, J.S., Groot, H., Gurwitz, D., Jenkins, T., Herrera, R.J., Huang, X., Kidd, J., Kidd, K.K., Langaney, A., Lin, A.A., Mehdi, S.Q., Parham, P., Piazza, A., Pistillo, M.P., Qian, Y., Shu, Q., Xu, J., Zhu, S., Weber, J.L., Greely, H.T., Feldman, M.W., Thomas, G., Dausset, J., Cavalli-Sforza, L.L., 2002. A human genome diversity cell line panel. Science 296, 261–262.

Cann, R.L., Stoneking, M., Wilson, A.C., 1987. Mitochondrial DNA and human evolution. Nature 325, 31–36.

Cavalli-Sforza, L.L., Menozzi, P., Piazza, A., 1994. The History and Geography of Human Genes. Princeton University Press, Princeton (NJ).

Cela-Conde, C.J., Ayala, F.J., 2007. Human Evolution. Trails from the Past. Oxford University Press, Oxford.

Crawford, M.H., Campbell, C., 2012. Causes and Consequences of Human Migration. An Evolutionary Perspective. Cambridge University Press, Cambridge.

Cruciani, F., Trombetta, B., Massaia, A., Destro-Bisol, G., Sellitto, D., Scozzari, R., 2011. A revised root for the human Y chromosomal phylogenetic tree: the origin of patrilineal diversity in Africa. American Journal of Human Genetics 88, 814–818.

Diamond, J., 1988. Express train to Polynesia. Nature 336, 307–308.

Diamond, J., Bellwood, P., 2003. Farmers and their languages: the first expansions. Science 300, 597–603.

Excoffier, L., 2002. Human demographic history: refining the recent African origin model. Current Opinion in Genetical Development 12, 675–682.

Fagundes, N.J., Ray, N., Beaumont, M., Neuenschwander, S., Salzano, F.M., Bonatto, S.L., Excoffier, L., 2007. Statistical evaluation of alternative models of human evolution. Proceedings of the National Academy of Sciences United States of America 104, 17614–17619.

Fleagle, J.G., Shea, J.J., Grine, F.E., Baden, A.L., Leakey, R.E., 2010. Out of Africa I. The First Hominin Colonization of Eurasia. Springer.

Fu, Q., Mittnik, A., Johnson, P.L., Bos, K., Lari, M., Bollongino, R., Sun, C., Giemsch, L., Schmitz, R., Burger, J., Ronchitelli, A.M., Martini, F., Cremonesi, R.G., Svoboda, J., Bauer, P., Caramelli, D., Castellano, S., Reich, D., Paabo, S., Krause, J., 2013. A revised timescale for human evolution based on ancient mitochondrial genomes. Current Biology 23, 553–559.

Fu, Q., Rudan, P., Paabo, S., Krause, J., 2012. Complete mitochondrial genomes reveal neolithic expansion into Europe. PLoS One 7, e32473.

Garrigan, D., Hammer, M.F., 2008. Ancient lineages in the genome: a response to Fagundes et al. Proceedings of the National Academy of Sciences United States of America 105, E3 author reply E4.

Gray, R.D., Drummond, A.J., Greenhill, S.J., 2009. Language phylogenies reveal expansion pulses and pauses in Pacific settlement. Science 323, 479–483.

Green, R.E., Krause, J., Briggs, A.W., Maricic, T., Stenzel, U., Kircher, M., Patterson, N., Li, H., Zhai, W., Fritz, M.H., Hansen, N.F., Durand, E.Y., Malaspinas, A.S., Jensen, J.D., Marques-Bonet, T., Alkan, C., Prufer, K., Meyer, M., Burbano, H.A., Good, J.M., Schultz, R., Aximu-Petri, A., Butthof, A., Hober, B., Hoffner, B., Siegemund, M., Weihmann, A., Nusbaum, C., Lander, E.S., Russ, C., Novod, N., Affourtit, J., Egholm, M., Verna, C., Rudan, P., Brajkovic, D., Kucan, Z., Gusic, I., Doronichev, V.B., Golovanova, L.V., Lalueza-Fox, C., De La Rasilla, M., Fortea, J., Rosas, A., Schmitz, R.W., Johnson, P.L., Eichler, E.E., Falush, D., Birney, E., Mullikin, J.C., Slatkin, M., Nielsen, R., Kelso, J., Lachmann, M., Reich, D., Paabo, S., 2010. A draft sequence of the Neandertal genome. Science 328, 710–722.

Hage, P., Marck, J., 2003. Matrilineality and the melanesian origin of polynesian Y chromosomes. Current Anthropology 44, S121–S127.

Hammer, M.F., 1995. A recent common ancestry for human Y chromosomes. Nature 378, 376–378.

Hancock, I., 2002. We are the Romani People. University of Hertforshire Press, Great Britain.

Harding, R.M., McVean, G., 2004. A structured ancestral population for the evolution of modern humans. Current Opinion in Genetics and Development 14, 667–674.

Heyerdahl, T., 1950. The Kon-tiki Expedition: across the South Seas by Raft. Pocket Books, New York.

Huerta-Sanchez, E., Jin, X., Asan, Bianba, Z., Peter, B.M., Vinckenbosch, N., Liang, Y., Yi, X., He, M., Somel, M., Ni, P., Wang, B., Ou, X., Huasang, Luosang, J., Cuo, Z.X., Li, K., Gao, G., Yin, Y., Wang, W., Zhang, X., Xu, X., Yang, H., Li, Y., Wang, J., Wang, J., Nielsen, R., 2014. Altitude adaptation in Tibetans caused by introgression of Denisovan-like DNA. Nature 512, 194–197.

Isern, N., Fort, J., Vander Linden, M., 2012. Space competition and time delays in human range expansions. Application to the neolithic transition. PLoS One 7, e51106.

Jakobsson, M., Scholz, S.W., Scheet, P., Gibbs, J.R., Vanliere, J.M., Fung, H.C., Szpiech, Z.A., Degnan, J.H., Wang, K., Guerreiro, R., Bras, J.M., Schymick, J.C., Hernandez, D.G., Traynor, B.J., Simon-Sanchez, J., Matarin, M., Britton, A., Van De Leemput, J., Rafferty, I., Bucan, M., Cann, H.M., Hardy, J.A., Rosenberg, N.A., Singleton, A.B., 2008. Genotype, haplotype and copy-number variation in worldwide human populations. Nature 451, 998–1003.

Jobling, M.A., Hurles, M.E., Tyler-Smith, C., 2004. Human Evolutionary Genetics. Origins, Peoples & Disease. Garland Science, New York.

Kalaydjieva, L., Gresham, D., Calafell, F., 2001. Genetic studies of the Roma (Gypsies): a review. BMC Medical Genetics 2, 5.

Kayser, M., Brauer, S., Cordaux, R., Casto, A., Lao, O., Zhivotovsky, L.A., Moyse-Faurie, C., Rutledge, R.B., Schiefenhoevel, W., Gil, D., Lin, A.A., Underhill, P.A., Oefner, P.J., Trent, R.J., Stoneking, M., 2006. Melanesian and Asian origins of Polynesians: mtDNA and Y chromosome gradients across the Pacific. Molecular Biology and Evolution 23, 2234–2244.

Kayser, M., Brauer, S., Weiss, G., Underhill, P.A., Roewer, L., Schiefenhovel, W., Stoneking, M., 2000. Melanesian origin of Polynesian Y chromosomes. Current Biology 10, 1237–1246.

Kayser, M., Choi, Y., Van Oven, M., Mona, S., Brauer, S., Trent, R.J., Suarkia, D., Schiefenhovel, W., Stoneking, M., 2008a. The impact of the Austronesian expansion: evidence from mtDNA and Y chromosome diversity in the Admiralty Islands of Melanesia. Molecular Biology and Evolution 25, 1362–1374.

Kayser, M., Lao, O., Saar, K., Brauer, S., Wang, X., Nurnberg, P., Trent, R.J., Stoneking, M., 2008b. Genome-wide analysis indicates more Asian than Melanesian ancestry of Polynesians. American Journal of Human Genetics 82, 194–198.

Kirch, P.V., 2000. On the Road of the Winds: An Archaeological History of the Pacific Islands before European Contact. University of California press.

Krause, J., Fu, Q., Good, J.M., Viola, B., Shunkov, M.V., Derevianko, A.P., Paabo, S., 2010. The complete mitochondrial DNA genome of an unknown hominin from southern Siberia. Nature 464, 894–897.

Krings, M., Stone, A., Schmitz, R.W., Krainitzki, H., Stoneking, M., Paabo, S., 1997. Neandertal DNA sequences and the origin of modern humans. Cell 90, 19–30.

Lalueza-Fox, C., Gilbert, M.T., 2011. Paleogenomics of archaic hominins. Current Biology 21, R1002–R1009.

Lao, O., Lu, T.T., Nothnagel, M., Junge, O., Freitag-Wolf, S., Caliebe, A., Balascakova, M., Bertranpetit, J., Bindoff, L.A., Comas, D., Holmlund, G., Kouvatsi, A., Macek, M., Mollet, I., Parson, W., Palo, J., Ploski, R., Sajantila, A., Tagliabracci, A., Gether, U., Werge, T., Rivadeneira, F., Hofman, A., Uitterlinden, A.G., Gieger, C., Wichmann, H.E., Ruther, A., Schreiber, S., Becker, C., Nurnberg, P., Nelson, M.R., Krawczak, M., Kayser, M., 2008. Correlation between genetic and geographic structure in Europe. Current Biology 18, 1241–1248.

Lazaridis, I., Patterson, N., Mittnik, A., Renaud, G., Mallick, S., Kirsanow, K., Sudmant, P.H., Schraiber, J.G., Castellano, S., Lipson, M., Berger, B., Economou, C., Bollongino, R., Fu, Q., Bos, K.I., Nordenfelt, S., Li, H., De Filippo, C., Prufer, K., Sawyer, S., Posth, C., Haak, W., Hallgren, F., Fornander, E., Rohland, N., Delsate, D., Francken, M., Guinet, J.M., Wahl, J., Ayodo, G., Babiker, H.A., Bailliet, G., Balanovska, E., Balanovsky, O., Barrantes, R., Bedoya, G., Ben-Ami, H., Bene, J., Berrada, F., Bravi, C.M., Brisighelli, F., Busby, G.B., Cali, F., Churnosov, M., Cole, D.E., Corach, D., Damba, L., Van Driem, G., Dryomov, S., Dugoujon, J.M., Fedorova, S.A., Gallego Romero, I., Gubina, M., Hammer, M., Henn, B.M., Hervig, T., Hodoglugil, U., Jha, A.R., Karachanak-Yankova, S., Khusainova, R., Khusnutdinova, E., Kittles, R., Kivisild, T., Klitz, W., Kucinskas, V., Kushniarevich, A., Laredj, L., Litvinov, S., Loukidis, T., Mahley, R.W., Melegh, B., Metspalu, E., Molina, J., Mountain, J., Nakkalajarvi, K., Nesheva, D., Nyambo, T., Osipova, L., Parik, J., Platonov, F., Posukh, O., Romano, V., Rothhammer, F., Rudan, I., Ruizbakiev, R., Sahakyan, H., Sajantila, A., Salas, A., Starikovskaya, E.B., Tarekegn, A., Toncheva, D., Turdikulova, S., Uktveryte, I., Utevska, O., Vasquez, R., Villena, M., Voevoda, M., Winkler, C.A., Yepiskoposyan, L., Zalloua, P., et al., 2014. Ancient human genomes suggest three ancestral populations for present-day Europeans. Nature 513, 409–413.

Lewontin, R., 1972. The apportionment of human diversity. Evolutionary Biology 6, 8.

Li, J.Z., Absher, D.M., Tang, H., Southwick, A.M., Casto, A.M., Ramachandran, S., Cann, H.M., Barsh, G.S., Feldman, M., Cavalli-Sforza, L.L., Myers, R.M., 2008. Worldwide human relationships inferred from genome-wide patterns of variation. Science 319, 1100–1104.

Liu, H., Prugnolle, F., Manica, A., Balloux, F., 2006. A geographically explicit genetic model of worldwide human-settlement history. American Journal of Human Genetics 79, 230–237.

Lordkipanidze, D., Ponce De Leon, M.S., Margvelashvili, A., Rak, Y., Rightmire, G.P., Vekua, A., Zollikofer, C.P., 2013. A complete skull from Dmanisi, Georgia, and the evolutionary biology of early Homo. Science 342, 326–331.

Macaulay, V., Hill, C., Achilli, A., Rengo, C., Clarke, D., Meehan, W., Blackburn, J., Semino, O., Scozzari, R., Cruciani, F., Taha, A., Shaari, N.K., Raja, J.M., Ismail, P., Zainuddin, Z., Goodwin, W., Bulbeck, D., Bandelt, H.J., Oppenheimer, S., Torroni, A., Richards, M., 2005. Single, rapid coastal settlement of Asia revealed by analysis of complete mitochondrial genomes. Science 308, 1034–1036.

Mellars, P., 2006. Going east: new genetic and archaeological perspectives on the modern human colonization of Eurasia. Science 313, 796–800.

Melton, T., Peterson, R., Redd, A.J., Saha, N., Sofro, A.S., Martinson, J., Stoneking, M., 1995. Polynesian genetic affinities with Southeast Asian populations as identified by mtDNA analysis. American Journal of Human Genetics 57, 403–414.

Mendez, F.L., Krahn, T., Schrack, B., Krahn, A.M., Veeramah, K.R., Woerner, A.E., Fomine, F.L., Bradman, N., Thomas, M.G., Karafet, T.M., Hammer, M.F., 2013. An African American paternal lineage adds an extremely ancient root to the human Y chromosome phylogenetic tree. American Journal of Human Genetics 92, 454–459.

Mendizabal, I., Lao, O., Marigorta, U.M., Wollstein, A., Gusmao, L., Ferak, V., Ioana, M., Jordanova, A., Kaneva, R., Kouvatsi, A., Kucinskas, V., Makukh, H., Metspalu, A., Netea, M.G., De Pablo, R., Pamjav, H., Radojkovic, D., Rolleston, S.J., Sertic, J., Macek Jr., M., Comas, D., Kayser, M., 2012. Reconstructing the population history of European Romani from genome-wide data. Current Biology 22, 2342–2349.

Mendizabal, I., Valente, C., Gusmao, A., Alves, C., Gomes, V., Goios, A., Parson, W., Calafell, F., Alvarez, L., Amorim, A., Gusmao, L., Comas, D., Prata, M.J., 2011. Reconstructing the Indian origin and dispersal of the European Roma: a maternal genetic perspective. PLoS One 6, e15988.

Menozzi, P., Piazza, A., Cavalli-Sforza, L., 1978. Synthetic maps of human gene frequencies in Europeans. Science 201, 786–792.

Moodley, Y., Linz, B., Yamaoka, Y., Windsor, H.M., Breurec, S., Wu, J.Y., Maady, A., Bernhoft, S., Thiberge, J.M., Phuanukoonnon, S., Jobb, G., Siba, P., Graham, D.Y., Marshall, B.J., Achtman, M., 2009. The peopling of the Pacific from a bacterial perspective. Science 323, 527–530.

Moorjani, P., Patterson, N., Loh, P.R., Lipson, M., Kisfali, P., Melegh, B.I., Bonin, M., Kadasi, L., Riess, O., Berger, B., Reich, D., Melegh, B., 2013. Reconstructing Roma history from genome-wide data. PLoS One 8, e58633.

Morar, B., Gresham, D., Angelicheva, D., Tournev, I., Gooding, R., Guergueltcheva, V., Schmidt, C., Abicht, A., Lochmuller, H., Tordai, A., Kalmar, L., Nagy, M., Karcagi, V., Jeanpierre, M., Herczegfalvi, A., Beeson, D., Venkataraman, V., Warwick Carter, K., Reeve, J., De Pablo, R., Kucinskas, V., Kalaydjieva, L., 2004. Mutation history of the roma/gypsies. American Journal of Human Genetics 75, 596–609.

Nordborg, M., 1998. On the probability of Neanderthal ancestry. American Journal of Human Genetics 63, 1237–1240.

Novembre, J., Johnson, T., Bryc, K., Kutalik, Z., Boyko, A.R., Auton, A., Indap, A., King, K.S., Bergmann, S., Nelson, M.R., Stephens, M., Bustamante, C.D., 2008. Genes mirror geography within Europe. Nature 456, 98–101.

Novembre, J., Stephens, M., 2008. Interpreting principal component analyses of spatial population genetic variation. Nature Genetics 40, 646–649.

Oppenheimer, S., 2012. Out-of-Africa, the peopling of continents and islands: tracing uniparental gene trees across the map. Philosophical Transactions of the Royal Society of London B: Biological Sciences 367, 770–784.

Ozdogan, M., 2008. An alternative approach in tracing changes in demographic composition. In: Bocquet-Appel, J., Bar-Yosef, O. (Eds.), The Neolithic Demographic Transitions and its Consequences. Springer.

Pakendorf, B., Stoneking, M., 2005. Mitochondrial DNA and human evolution. Annual Review of Genomics and Human Genetics 6, 165–183.

Pemberton, T.J., Degiorgio, M., Rosenberg, N.A., 2013. Population structure in a comprehensive genomic data set on human microsatellite variation. G3 (Bethesda) 3, 891–907.

Petit, J.R., Jouzel, J., Raynaud, D., Barkov, N.I., Barnola, J.M., Basile, I., Bender, M., Chappellaz, J., Davis, M., Delaygue, G., Delmotte, M., Kotlyakov, V.M., Legrand, M., Lipenkov, V.Y., Lorius, C., Pépin, L., Ritz, C., Saltzman, E., Stievenard, M., 1999. Climate and atmospheric history of the past 420,000 years from the Vostok ice core, Antarctica. Nature 399, 429–436.

Pickrell, J.K., Pritchard, J.K., 2012. Inference of population splits and mixtures from genome-wide allele frequency data. PLoS Genetics 8, e1002967.

Pickrell, J.K., Reich, D., 2014. Toward a new history and geography of human genes informed by ancient DNA. Trends in Genetics 30, 377–389.

Pinhasi, R., Fort, J., Ammerman, A.J., 2005. Tracing the origin and spread of agriculture in Europe. PLoS Biology 3, e410.

Pinhasi, R., Thomas, M.G., Hofreiter, M., Currat, M., Burger, J., 2012. The genetic history of Europeans. Trends in Genetics 28, 496–505.

Poznik, G.D., Henn, B.M., Yee, M.C., Sliwerska, E., Euskirchen, G.M., Lin, A.A., Snyder, M., Quintana-Murci, L., Kidd, J.M., Underhill, P.A., Bustamante, C.D., 2013. Sequencing Y chromosomes resolves discrepancy in time to common ancestor of males versus females. Science 341, 562–565.

Rai, N., Chaubey, G., Tamang, R., Pathak, A.K., Singh, V.K., Karmin, M., Singh, M., Rani, D.S., Anugula, S., Yadav, B.K., Singh, A., Srinivasagan, R., Yadav, A., Kashyap, M., Narvariya, S., Reddy, A.G., Van Driem, G., Underhill, P.A., Villems, R., Kivisild, T., Singh, L., Thangaraj, K., 2012. The phylogeography of Y-chromosome haplogroup h1a1a-m82 reveals the likely Indian origin of the European Romani populations. PLoS One 7, e48477.

Ramachandran, S., Deshpande, O., Roseman, C.C., Rosenberg, N.A., Feldman, M.W., Cavalli-Sforza, L.L., 2005. Support from the relationship of genetic and geographic distance in human populations for a serial founder effect originating in Africa. Proceedings of the National Academy of Sciences United States of America 102, 15942–15947.

Reich, D., Green, R.E., Kircher, M., Krause, J., Patterson, N., Durand, E.Y., Viola, B., Briggs, A.W., Stenzel, U., Johnson, P.L., Maricic, T., Good, J.M., Marques-Bonet, T., Alkan, C., Fu, Q., Mallick, S., Li, H., Meyer, M., Eichler, E.E., Stoneking, M., Richards, M., Talamo, S., Shunkov, M.V., Derevianko, A.P., Hublin, J.J., Kelso, J., Slatkin, M., Paabo, S., 2010. Genetic history of an archaic hominin group from Denisova Cave in Siberia. Nature 468, 1053–1060.

Reich, D., Patterson, N., Kircher, M., Delfin, F., Nandineni, M.R., Pugach, I., Ko, A.M., Ko, Y.C., Jinam, T.A., Phipps, M.E., Saitou, N., Wollstein, A., Kayser, M., Paabo, S., Stoneking, M., 2011. Denisova admixture and the first modern human dispersals into Southeast Asia and Oceania. American Journal of Human Genetics 89, 516–528.

Reyes-Centeno, H., Ghirotto, S., Detroit, F., Grimaud-Herve, D., Barbujani, G., Harvati, K., 2014. Genomic and cranial phenotype data support multiple modern human dispersals from Africa and a southern route into Asia. Proceedings of the National Academy of Sciences United States of America 111, 7248–7253.

Romualdi, C., Balding, D., Nasidze, I.S., Risch, G., Robichaux, M., Sherry, S.T., Stoneking, M., Batzer, M.A., Barbujani, G.V., 2002. Patterns of human diversity, within and among continents, inferred from biallelic DNA polymorphisms. Genome Research 12, 602–612.

Saitou, N., Nei, M., 1987. The neighbor-joining method: a new method for reconstructing phylogenetic trees. Molecular Biology and Evolution 4, 406–425.

Sankararaman, S., Mallick, S., Dannemann, M., Prufer, K., Kelso, J., Paabo, S., Patterson, N., Reich, D., 2014. The genomic landscape of Neanderthal ancestry in present-day humans. Nature 507, 354–357.

Sankararaman, S., Patterson, N., Li, H., Paabo, S., Reich, D., 2012. The date of interbreeding between Neandertals and modern humans. PLoS Genetics 8, e1002947.

Schaffner, S.F., Foo, C., Gabriel, S., Reich, D., Daly, M.J., Altshuler, D., 2005. Calibrating a coalescent simulation of human genome sequence variation. Genome Research 15, 1576–1583.

Shi, W., Ayub, Q., Vermeulen, M., Shao, R.G., Zuniga, S., Van Der Gaag, K., De Knijff, P., Kayser, M., Xue, Y., Tyler-Smith, C., 2010. A worldwide survey of human male demographic history based on Y-SNP and Y-STR data from the HGDP-CEPH populations. Molecular Biology and Evolution 27, 385–393.

Skoglund, P., Malmstrom, H., Raghavan, M., Stora, J., Hall, P., Willerslev, E., Gilbert, M.T., Gotherstrom, A., Jakobsson, M., 2012. Origins and genetic legacy of Neolithic farmers and hunter-gatherers in Europe. Science 336, 466–469.

Smith, F., 1992. The role of continuity in modern human origins. In: Bräuer, G., Smith, F. (Eds.), Continuity or Replacement: Controversies in Homo sapiens Evolution. Balkema, Rotterdam, The Netherlands.

Sokal, R.R., Oden, N.L., Thomson, B.A.V., 1999. A problem with synthetic maps. Human Biology 71, 1–13 discussion 15–25.

Sokal, R.R., Oden, N.L., Wilson, C., 1991. Genetic evidence for the spread of agriculture in Europe by demic diffusion. Nature 351, 143–145.

Spriggs, M., 2003. Chronology of the neolithic transition in island southeast asia and the western Pacific: a view from 2003. Review of Archaeology 24, 57–80.

Stewart, J.R., Stringer, C.B., 2012. Human evolution out of Africa: the role of refugia and climate change. Science 335, 1317–1321.

Stringer, C., 2002. Modern human origins: progress and prospects. Philosophical Transactions of the Royal Society of London B: Biological Sciences 357, 563–579.

Sykes, B., Leiboff, A., Low-Beer, J., Tetzner, S., Richards, M., 1995. The origins of the Polynesians: an interpretation from mitochondrial lineage analysis. American Journal of Human Genetics 57, 1463–1475.

Tcherenkov, L., Laederich, S., 2004. The Rroma. History, Language, and Groups. Schwabe.

Templeton, A., 2002. Out of Africa again and again. Nature 416, 45–51.

Thomson, R., Pritchard, J.K., Shen, P., Oefner, P.J., Feldman, M.W., 2000. Recent common ancestry of human Y chromosomes: evidence from DNA sequence data. Proceedings of the National Academy of Sciences United States of America 97, 7360–7365.

Trent, R.J., Buchanan, J.G., Webb, A., Goundar, R.P., Seruvatu, L.M., Mickleson, K.N., 1988. Globin genes are useful markers to identify genetic similarities between Fijians and Pacific Islanders from Polynesia and Melanesia. American Journal of Human Genetics 42, 601–607.

van Oven, M., Kayser, M., 2009. Updated comprehensive phylogenetic tree of global human mitochondrial DNA variation. Human Mutation 30, E386–E394.

van Oven, M., Van Geystelen, A., Kayser, M., Decorte, R., Larmuseau, M.H., 2014. Seeing the Wood for the trees: a minimal reference phylogeny for the human Y chromosome. Human Mutation 35, 187–191.

Wei, W., Ayub, Q., Chen, Y., Mccarthy, S., Hou, Y., Carbone, I., Xue, Y., Tyler-Smith, C., 2013. A calibrated human Y-chromosomal phylogeny based on resequencing. Genome Research 23, 388–395.

Weir, B.S., Cockerham, C.C., 1984. Estimating F-statistics for the analysis of population structure. Evolution 38, 1358–1370.

Wollstein, A., Lao, O., Becker, C., Brauer, S., Trent, R.J., Nurnberg, P., Stoneking, M., Kayser, M., 2010. Demographic history of Oceania inferred from genome-wide data. Current Biology 20, 1983–1992.

Wolpoff, M.H., Hawks, J., Caspari, R., 2000. Multiregional, not multiple origins. American Journal of Physical Anthropology 112, 129–136.

Xing, J., Watkins, W.S., Hu, Y., Huff, C.D., Sabo, A., Muzny, D.M., Bamshad, M.J., Gibbs, R.A., Jorde, L.B., Yu, F., 2010a. Genetic diversity in India and the inference of Eurasian population expansion. Genome Biology 11, R113.

Xing, J., Watkins, W.S., Shlien, A., Walker, E., Huff, C.D., Witherspoon, D.J., Zhang, Y., Simonson, T.S., Weiss, R.B., Schiffman, J.D., Malkin, D., Woodward, S.R., Jorde, L.B., 2010b. Toward a more uniform sampling of human genetic diversity: a survey of worldwide populations by high-density genotyping. Genomics 96, 199–210.

Yi, X., Liang, Y., Huerta-Sanchez, E., Jin, X., Cuo, Z.X., Pool, J.E., Xu, X., Jiang, H., Vinckenbosch, N., Korneliussen, T.S., Zheng, H., Liu, T., He, W., Li, K., Luo, R., Nie, X., Wu, H., Zhao, M., Cao, H., Zou, J., Shan, Y., Li, S., Yang, Q., Asan, Ni, P., Tian, G., Xu, J., Liu, X., Jiang, T., Wu, R., Zhou, G., Tang, M., Qin, J., Wang, T., Feng, S., Li, G., Huasang, Luosang, J., Wang, W., Chen, F., Wang, Y., Zheng, X., Li, Z., Bianba, Z., Yang, G., Wang, X., Tang, S., Gao, G., Chen, Y., Luo, Z., Gusang, L., Cao, Z., Zhang, Q., Ouyang, W., Ren, X., Liang, H., Zheng, H., Huang, Y., Li, J., Bolund, L., Kristiansen, K., Li, Y., Zhang, Y., Zhang, X., Li, R., Li, S., Yang, H., Nielsen, R., Wang, J., Wang, J., 2010. Sequencing of 50 human exomes reveals adaptation to high altitude. Science 329, 75–78.

Chapter 17

Brain Evolution

Ralph L. Holloway
Department of Anthropology, Columbia University, New York, NY, USA

SYNOPSIS

The evolution of the human brain has been a combination of reorganization of brain components, and increases of brain size through both hyperplasia and hypertrophy during development, underlain by neurogenomic changes. Paleoneurology based on endocast studies is the direct evidence demonstrating volume changes through time, and if present, some convolutional details of the underlying cerebral cortex. Reorganizational changes include a reduction of primary visual cortex and relative enlargement of posterior association cortex and expanded Broca's regions, as well as cerebral asymmetries. The size of the hominid brain increased from about 450 ml at 3.5 million years ago to our current average volume of 1350 ml. These changes through time were sometimes gradual but not always.

INTRODUCTION

The evolution of the human brain has largely been a matter of integrating both the increases in the size of the brain and the organization of the brain through the past 3–4 million years ago (mya). Three lines of evidence are used by paleoneurologists to ascertain how these events might have occurred: (1) direct evidence from the brain endocasts of fossil hominids; (2) indirect evidence from comparative neuroscience, where variations in brain structures can be related to variations in behavior. The latter evidence is "indirect" because extant living animals are not ancestral to humans. Indeed, the last common ancestor for apes and the hominid line existed some 5–7 mya. (3) Newer neurogenomics evidence also promises to provide important clues to how and when certain aspects of brain changes occurred during human evolution.

The human brain evolved from an early hominid 3–4 mya, *Australopithecus afarensis*, having a size of roughly 400 ml to our present average of 1330 ml. These brain size increases, at different taxonomic levels, were mostly allometric, that is, related to body size, but not always. Integrated with these changes in brain size was reorganization of the cerebral cortex, as well as subcortical structures such as the hippocampus, amygdala, etc., to note a few important structures relating to aspects of social behavior that cannot be seen on endocasts.

At least three areas of the cerebral cortex were affected at different times: (1) a relative reduction of primary visual striate cortex (V1, PVC) and an attending relative increase in posterior parietal association cortex; (2) a change in Broca's region, resulting in a more human-like pattern; and (3) increasing degrees of cortical asymmetry, as well as increases in overall brain size, and number of neurons.

When did the human brain evolve, and how did it happen? Obviously, to answer this question will require a time machine, and thousands of generations of observations to ascertain both the variability and the direction of selection pressures in the past. We can, however, flesh out an initial understanding of how we got to be the animal *par excellence* that utilizes its brain for intelligent rationalizations based largely on the use of arbitrary symbol systems. The evidence consists of two components: (1) the "direct" evidence from the fossil record; and (2) the "indirect" evidence of the comparative neuroscientific record of extant living animals, particularly those most closely related to us, such as the chimpanzee and bonobo. There is a third possibility: since the human genome project has sequenced almost all of the genetic code, the future study of evolutionary neurogenomics might provide more data on the actual genetic history of our genus through time, as well as that of the great apes noted above. While the genetic code has been sequenced, the actual processes of how those sequences, involving metabolomics, transcriptomes, poteomes, and epigenetics interact is still unresolved. As this latter possibility is simply a speck in our eye at present, this article must concentrate on the evidence provided by the first two components.

LINES OF EVIDENCE

Direct Evidence

The term *paleoneurology* is used to describe evidence appraising the size and morphology of the casts made from the inside of actual fossil cranial remains. Occasionally, the casts are "natural," that is, where fine sediments have filled the inside of the cranial cavity, and became compacted through time. These casts often retain whatever morphological details were imprinted on the internal table of bone of the cranium. The famous australopithecine Taung child's skull, described by Dart (1925), is one of the best-known examples. Curiously, these natural endocasts are only found in the South African australopithecines, several of which exist, and date from about 3.0 mya to about 1.5 mya (see Figure 1). Most often, the paleoneurologist makes a cast of the inside of the fossil skull using rubber latex, or silicone rubber, and extracts this from the cranium. The partial cast is reconstructed by adding plasticine (modeling clay) to the missing regions. The whole is then measured by immersion into water, and the amount of water displaced is regarded as the volume of the once-living brain. Other measurements and observations are made on the original cast. More recently, "virtual" endocasts have been made from computed tomography (CT) scans of intact or partial crania, which have the advantage of being noninvasive, and as it is computer driven, there are different algorithms for measuring the size of the endocast and other metrics (Weber et al., 2012; see also Zollikofer and Ponce De Leon, 2013).

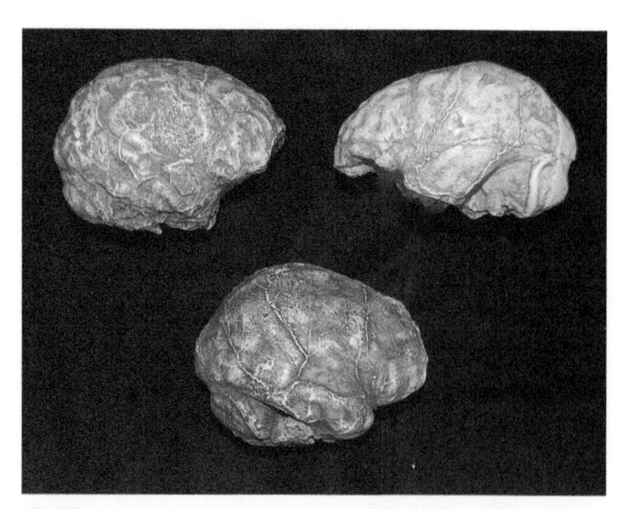

FIGURE 1 Casts of the Taung (left), STS60 (right), and SK1585 (bottom) "natural" endocasts of Australopithecines. (See color plate section).

During life the brain is surrounded by three dural tissues (dura mater, arachnoid tissue and its cerebrospinal fluid, and pia mater) that interface between the actual brain tissue (cerebral cortex, mostly) and the internal table of bone of the skull. The gyri and sulci (convolutions) of the once-pulsating cerebral cortex are rarely imprinted on the interior of the skull, and the degree of replication often varies in different regions, for example, sometimes the frontal lobe imprints more details than the parietal lobe, and also by age. The degree of replication also varies in different animals. Two extremely important considerations emerge from this: (1) the resulting imprints are never complete and thus "data poor," and never include subcortical structures; and (2) controversial interpretations of what the underlying brain once looked like are guaranteed (see Figures 2 and 3). Nevertheless, these endocranial casts do provide extremely important information regarding (1) size, (2) shape, (3) rough estimates of the lobar dimensions of the brain, and (4) cortical asymmetries that have relationships to hemispheric specializations, including handedness. In addition, (5) if the imprints of the underlying gyri and sulci are available, these can provide important information regarding the organization of the cerebral cortex, and whether the patterns of these are the same or different as in known extant primate brains. The infamous "lunate sulcus" is a good example, as it is a demarcation boundary between purely sensory PVC and multimodal association cortex. When the lunate sulcus appears in an anterior position, it is most similar to the condition known in modern apes. When it is found in a posterior position, it is more in a human-like condition. Ascertaining its correct position is essential in deciding whether or not such a fossil hominid had a brain organized along human or ape lines. In modern humans, the "lunate" is only partially homologous with that found in apes, and is usually fragmented (Allen et al., 2006). Figure 4(A) shows a comparison between a chimpanzee brain with lunate sulcus and that of the Taung child, *Australopithecus africanus*. Finally (6), meningeal arteries and veins that nourished the dura mater also imprint on the internal table of bone and these sometimes show patterns that are useful for deciding taxonomic issues; these have no known relationship to behavioral functions of the brain. (See also Grimaud-Hervé in Holloway et al., 2004, for further discussion and illustrations.) Figure 4(B) shows that a more recent *A. africanus* specimen from Sterkfontein, South Africa, Stw505, shows a clear lunate sulcus in a relatively posterior position compared to chimpanzee brains.

Indirect Evidence

This line of evidence is "data rich," providing comparative neurological information on living species, such as brain size (both absolute and relative, i.e., related to body size), the actual makeup of the brain from the gross to microscopic levels, including neural nuclei, fiber systems and interconnections, and distribution of neurotransmitters and neuroreceptors. Additionally, the brain can be studied ontogenetically, and neuroscientists can actually study the relationships between how the brain varies neurologically and how these variations relate to the behavioral variation. Figure 5 shows some different reorganization possibilities. Neurogenomic information will also add considerable details as to how living brains vary and operate, both within and between different species. This richness is simply lost to the paleoneurologist as it is not available as direct evidence. However, it is necessary to realize that the extant living species, for example, bonobo, chimpanzee, and macaque, are end points of their own evolutionary lines of development and are not our ancestors, however closely related (see the chapter by Martin in this volume). It is the blending and complementation of these two approaches that provide the best set of evidence for when and how our brains evolved. Another aspect of the comparative evidence is the question of how well we can explain species-specific behavior on the basis of what we know from comparative neurology. Consider the behavioral differences between chimpanzees and bonobos, gorillas and orangutans and try to explain these in terms of neuroanatomical details!

CHARACTERISTICS OF THE HUMAN BRAIN

Brain Size, Absolute and Relative

The human animal is obviously obsessed with size, and those who study the brain comparatively, perhaps more so. With an average brain weight of 1330 g and a body weight of 65,000 g (Tobias, 1971), the human species has the largest absolute brain size within the primate order, but is actually dwarfed by elephants and some of the whales, where brain weight can exceed 7500 g. Of course, the body weights are very much higher in elephants and whales. But even for its body weight, *Homo sapiens* does not have the largest relative brain weight (about 2% of body weight), being outdone by several monkeys, some rodents, and even some fish. Normal modern human brain size varies between roughly 900 and 2000 g, although some very small number of exceptions does occur, with sizes ranging from about 750 to 900 and 2000 to 2200 g. There exists human population variation as well as differences between the sexes. In general, Arctic peoples have larger brains than those living in the tropics and the smallest brains appear to be found among Ituri forest pygmies, also displaying small stature. Males in all populations for which good autopsy data have been gathered show brain sizes on the average of 100–150 g greater than females, an amount roughly the same as the range of modern human racial variation. It should be pointed out that these differences, and their possible relationship to cognitive skills, are highly controversial and

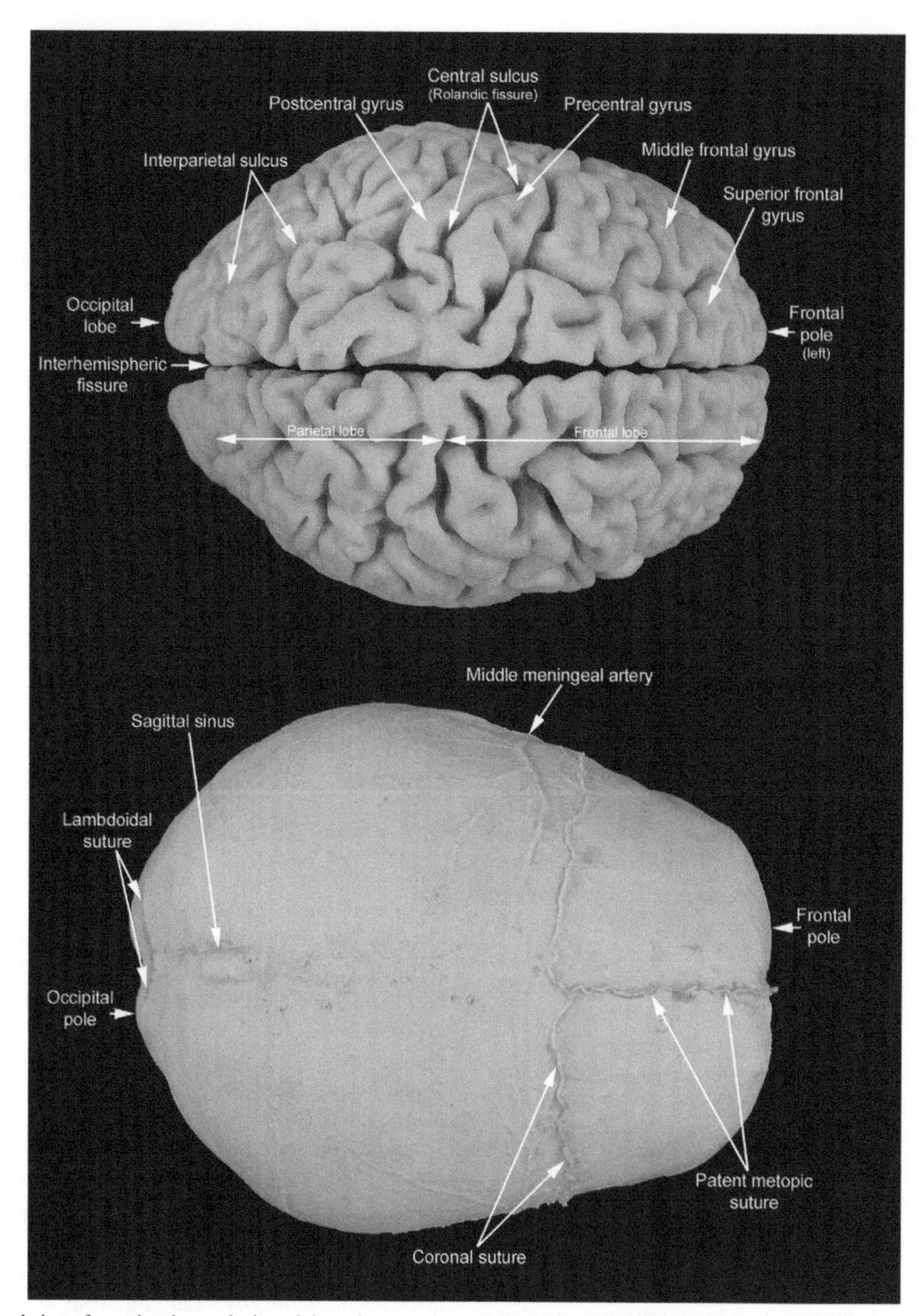

FIGURE 2 Dorsal view of a modern human brain and the endocast to demonstrate the loss of detail on the endocast surface.

simple correlations are deceptive (Holloway, 1996, 2008; Nyborg, 2003). Table 1 provides a listing of the major fossil hominid taxa and their respective brain sizes. Note that the range of values from the earliest australopithecine to modern *Homo* is roughly 1000 ml (grams and milliliters are used interchangeably, given that the specific gravity of neural tissue is approximately one) or about the same amount as the normal range of variation within our species.

Encephalization

Nevertheless, the human animal does come out on top of the evolutionary heap when its absolute brain and body weights are considered together. When the log (base 10) of brain weight is plotted against the log10 of body weight for a group of relevant taxa, the result is a linear line, where (log10) brain weight = a + b (log10) body weight. For a large array of primate data (e.g., Stephan et al., 1981), the slope

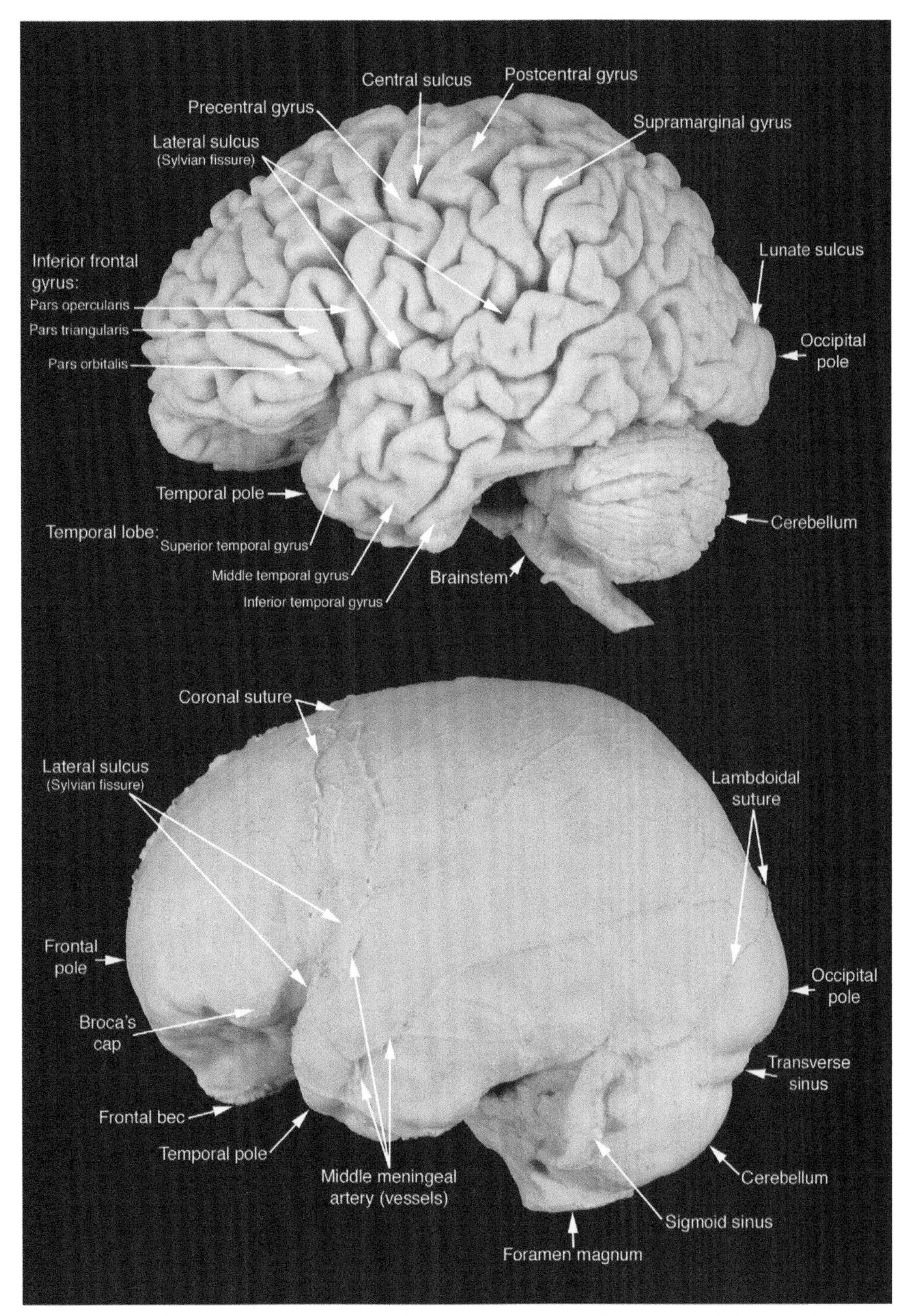

FIGURE 3 Lateral view of a modern human brain and the endocast to demonstrate the loss of detail on the endocast surface.

of the line (b in the equation above) is about 0.76, and the correlation coefficient is 0.98, indicating that the relationship is almost perfect. This relationship will naturally vary depending on the databases and the transformations used. This is known as an allometric equation, and these are used frequently in biology to assess the underlying relationships between the size of parts of the body and the whole. (See Figure 6, a graph of log brain weight against log body weight for most of the primates.) The slope sometimes has an interpretation suggesting functional relationships between the brain and other variables. For example, in the above example, the slope is 0.76, extremely close to 0.75 or 3/4, which often describes a metabolic relationship (Martin, 1983). The slope of 0.666, or 2/3, has been championed by some (e.g., Jerison, 1973) as indicating an important geometric relationship between volume and surface area. It is

(A)

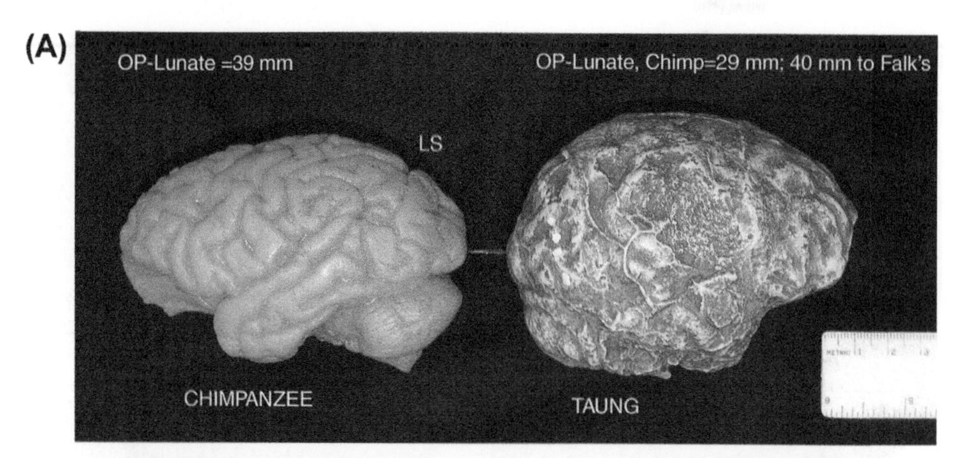

(B)

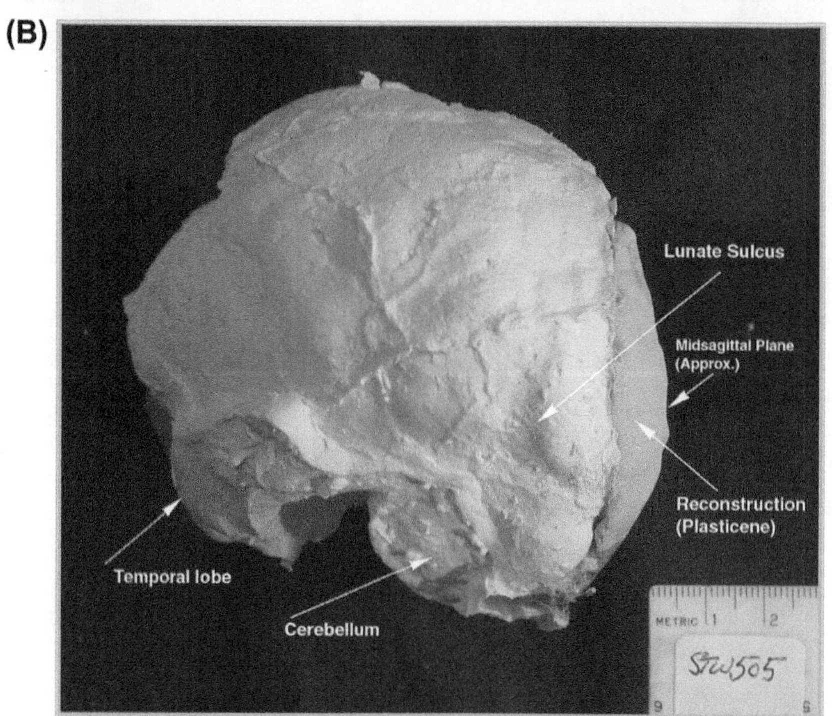

FIGURE 4 (A) Lateral view of a chimpanzee brain and the Taung *A. africanus* endocast. The lunate sulcus separates PVC from association cortex, and is in an anterior position in apes. The white dots on the Taung endocast show where the average chimpanzee lunate sulcus would fall, and that location violates the sulcal morphology on Taung. Placing it more anteriorly would be a monkey-like configuration. The Taung lunate sulcus would most probably be posterior, in a human-like position, which is near the lambdoid suture Holloway (1984). (B) Oblique view of the Stw 505 *A. africanus* specimen, showing a prominent lunate sulcus in a posterior position. This specimen makes it clear that at least some australopithecines had reduced primary visual cortices and expanded posterior parietal lobes, evidence showing that reorganization probably preceded brain size increases. (See color plate section).

important to realize that these slopes vary depending on the taxa examined. In general, as the taxa become more similar, the slope decreases. Species within a genus generally have a slope around 0.3; within a species the slope is smaller yet, about 0.2, and the correlation coefficient is also reduced (see also Martin and Isler, 2010).

Just as the human animal is curious, it is also vainglorious, always trying to find a measure that places it at the top. Thus we can fabricate a device, the "encephalization coefficient," or E.Q., which shows that relative to any database, the human animal is the most encephalized animal living. The point for *H. sapiens* shows a clear positive residual above the expected regression line, and in fact the human value is about three times that expected for a primate with its body weight. Table 2 provides a number of different equations based on differing databases, which give *H. sapiens* the highest value. (Actually, young immature dolphins will provide a higher number, but when compared to an immature human, the value is higher in the latter.) Two additional points should be made: (1) E.Q.s are relative to the databases used, and thus there is an inherent "relativity" to relative brain sizes; and (2) E.Q.s do not evolve,

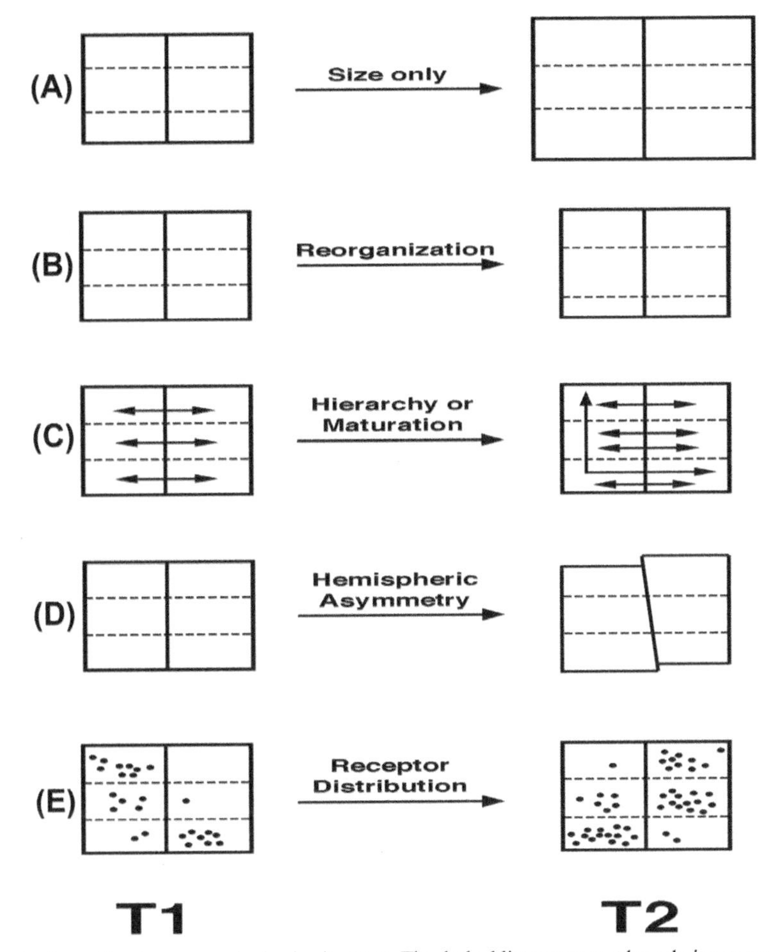

FIGURE 5 Types of reorganization without necessary brain size increase. The dashed lines represent boundaries among frontal, parietal, and occipital lobes, which if changed in relative positions from T1 to T2 would suggest reorganization.

only brain weight/body weight relationships do, and E.Q.s are simply a heuristic device enabling comparisons among taxa; they have no reality outside of the database chosen, or species within a taxa, and are not designed to discuss within-species variation. For example, female humans are "more" encephalized than males, given their smaller body sizes, have more body fat that is not innervated, and smaller brains, but the relationship might be simply a statistical artifact with no known gross behavioral manifestation given their nearly equal overall intelligence. It is more likely that small differences in neural reorganization might be related to behavioral differences such as language ability or math and spatiovisual manipulation rather than brain size or EQ.

I discuss below how the processes of hypertrophy and hyperplasia have been positively selected for in the course of the last 2–3 mya of hominid evolution. (Hypertrophy refers to increases in size of the neural components, e.g., neurons, dendritic branching, nuclei, and fiber tracts; hyperplasia refers to increased production of cells through mitotic division.) It is most probably the case that these processes are controlled by regulatory genes, and one of the major differences between ourselves and our closest nonhuman primate relative, the chimpanzee (brain size = ca. 385 g), relates to the schedules by which hyperplasia and hypertrophy are turned on and off during ontogenetic development (Holloway, 1979, 1995; Miller et al., 2012).

Brain Organization and Reorganization

It is well known that the brains of most animals are extremely similar to each other in terms of their overall organization, i.e., their neural nuclei and fiber systems. The human animal does not appear to show any different structures when compared to Old World monkeys, such as the macaque, or the Great Apes, including bonobo, chimpanzee, gorilla, and orangutan. Even the neural fiber tracts that are involved in human language appear in these primates (Deacon, 1997). One might ask, then, given the obvious species-specific repertoires that exist in all animals, how can these behaviors differ without differences in the underlying nervous systems? This is one of the major challenges of studying brain evolution, and in particular understanding what neural organizations account for the specificity of, for example, human behavior, the ability to use language composed of

TABLE 1 Some Fossil Hominid Brain Volumes

Group	Number	Location	Average Brain Volume	Range	Dating (mya)
A. afarensis	3	E. Africa	435	385–500+	3–4
A. africanus	8	S. Africa	440	420–500+	2–3
A. aethiopicus	1	E. Africa	410	Na	2.5
A. garhi	1	E. Africa	Ca. 450	Na	2.5
A. sediba	1	S. Africa	Ca. 420	Na	2–3
A. robustus	6	E. & S. Africa	512	500–530	1.6–2.0
H. rudolfensis	2	E. Africa	775	752–800	1.8
H. habilis	6	E. Africa	612	510–687	1.7–2.0
H. georgicus	3	Georgia, Europe	677	600–775	1.7
H. ergaster	2	E. Africa	826	804–848	1.6
H. erectus	2	E. Africa	980	900–1067	1.0–1.6
H. erectus	8	Indonesia	925	780–1059	1.0
H. erectus	8	China	1029	850–1225	0.6
Archaic *H. sapiens*	6	Indonesia (Solo)	1148	1013–1250	0.13
Archaic *H. sapiens*	6	Africa	1190	880–1367	0.125
Archaic *H. sapiens*	7	Europe	1315	1200–1450	0.5–0.25
H. sapiens (Neand.)	25	Europe, M. East	1415	1125–1740	0.09–0.03
H. sapiens sapiens	11	World	1506	1250–1600	0.025–0.01

Holloway, 1997.

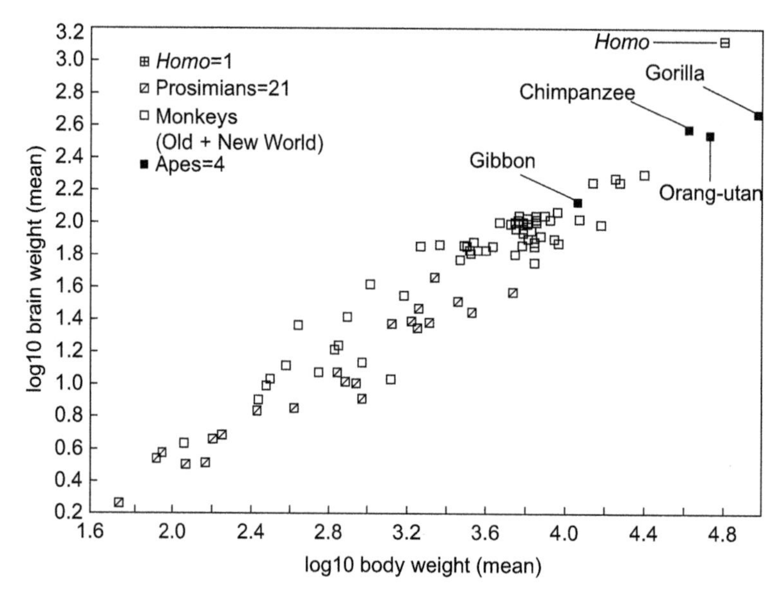

FIGURE 6 Graph showing human deviation from a plot of log brain weight against log body weight for primates, with *Homo* at the extreme upper right position.

TABLE 2 Some Examples of Encephalization Quotients

Species	Brain wt (g)	Body wt (g)	EQ Homo[a]	EQ Jerison[b]	EQ Primates[c]	EQ Stephan[d]
Lemur	23.3	1400	21	1.56 (22.6)	0.94 (32.7)	5.66 (19.6)
Baboon	201	25,000	28	1.97 (28.5)	0.90 (31.3)	7.94 (27.5)
Gorilla	465	165,000	23	1.56 (22.5)	0.61 (21.2)	6.67 (23.2)
Orang	370	55,000	31	2.15 (31.1)	0.91 (31.7)	8.90 (30.9)
Chimp	420	46,000	39	2.63 (28.1)	1.81 (41.1)	11.3 (39.3)
Human	1,330	65,000	100	6.91 (100)	2.87 (100)	28.8 (100)

Note: Each formula is based on a different set of data. The EQ Homo equation simply uses the average brain and body weight for *Homo sapiens*, and assumes an intercept where both brain and body weights are zero. The value of whichever animal is calculated is then given as a direct percentage of modern *Homo sapiens*. EQ Jerison is based on data for almost 200 mammals, while the EQ Primates is based on Martin's (1983) data set for primates only. The EQ Stephan equation is based on insectivores only. The numbers in the parentheses are the percentage of the *Homo sapiens* value.
[a]*Formulae: EQ Homo = Brain wt/1.0 Body* $wt^{0.64,906}$.
[b]*EQ Jerison = Brain wt/0.12 Body* $wt^{0.66}$.
[c]*EQ Primates = Brain wt/0.0991 Body* $wt^{0.76,237}$.
[d]*EQ Stephan = Brain wt/0.0429 Body* $wt^{0.63}$.
Holloway 1997.

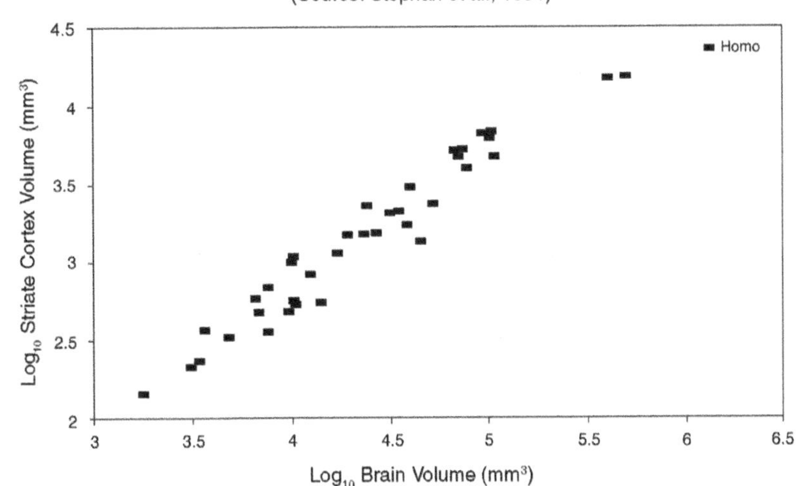

FIGURE 7 Graph showing log striate cortex (area 17) versus log brain volume, where the value for *Homo sapiens* (upper right) is 121% less than expected from the log–log regression. (See also Table 3, which shows other departures between actual and predicted values for different brain structures.)

arbitrary symbols. In other words, all mammals have a cerebral cortex, thalamus, cerebellum, hypothalamus, etc., and basically these structures possess almost identical divisions of nuclei and do the same neural tasks. Clearly, brain size alone will never explain species-specific behavior, and the relationships between neural nuclei and fiber tracts will only go so far in explaining behavioral differences.

Allometric equations showing the relationship between one bodily component and the whole are instructive here. If we were to plot the logs (base 10) of PVC against brain volume, we would find that the human PVC is 121% less than predicted, and similarly, the lateral geniculate nucleus of the thalamus is about 144% less than expected for a primate of our brain size (see Figure 7 showing striate cortex deviation for the human). In contrast, if one were to plot the amount of cerebral cortex against brain weight the result is a straight line, and the human point lies almost exactly on the line. In short, the human cerebral cortex is as large as would be expected for a primate of its brain size. But do portions of the cerebral cortex vary in size between different primates? In humans, the residuals noted above suggest that compared to chimpanzees, the amount of PVC is significantly smaller in humans, or alternatively put, the posterior association cortex of the parietal and temporal lobes is relatively larger in humans. Since there are no essential differences between chimpanzees and humans in their visual abilities and competencies, these differences reflect selection for expanded functioning of the association cortex in humans. This is precisely what is meant by "reorganization" (see Table 3 residuals).

TABLE 3 Residuals in percentage (Last Column) of Various Neural Structures for *Homo sapiens,* Based on Log–Log Allometric Equations for the Stephan et al. (1981) Primate Brain Structure Data Set. The Percentages Are the Deviations of Actual Volumes from Volumes Predicted on the Basis of Allometry

Human Brain Structre Residuals[a]							
Dependent variable	**Independent Variable**	**Number Species**	**Correl. Coeff. (*R*)**	**Actual Value (*A*)**	**Expected Value (*E*)**	**(*A*)/(*E*) Ratio**	**%Diff. (*A*)/(*E*) Homo**
Striate cortex	Brain weight (C)	37 19	0.971 0.977	22,866	50,598 38,097	0.45 0.60	−121.30 −66.60
Lateral geniculate	Brain weight (C)	37 19	0.978 0.982	416	1026 857	0.41 0.49	−146.60 −106.00
Cerebellum	Brain weight (C)	44 26	0.990 0.994	137,421	128,932 150,535	1.07 0.91	6.20 −9.50
Diencheph.	Brain weight (C)	44 26	0.995 0.998	33,319	51,512 47,899	0.65 0.70	−54.60 −43.70
Septum	Brain weight (C)	44 26	0.983 0.991	2610	2085 2201	1.25 1.19	20.10 15.70
Amygdala	Brain weight (C)	16 7	0.990 0.985	3015	4633 3753	0.65 0.80	−53.70 −24.50
Lateral geniculate	Thalamus	21 10	0.979 0.988	416 416	731 636	0.57 0.65	−75.72 −52.88

[a]*Note: Based on Stephan et al. (1981) data.*

When used in a comparative or evolutionary context, reorganization means changes in the sizes and proportions thereof of neural nuclei and their fiber tracts (see Figure 5). Given that chimpanzees and hominids last had a common ancestor some 5–7 mya, and that chimpanzees appear to have large PVC cortices, we infer that one aspect of human brain evolution has been some reorganization of the cerebral cortex, namely an increase in posterior association cortex (or, equally, a reduction in PVC) involved in polymodal cognitive tasks, where visual, auditory, and motor information are brought together in a synthetic whole. The trick, of course, is to demonstrate objectively when, where, and why these changes took place (Holloway, 1984). This example of PVC has been purposefully chosen because one of the sulcal landmarks of the cortex that defines the anterior border of PVC is the lunate sulcus, named for its crescentic shape, and there is some hope of identifying its position on some of the early hominid brain endocasts. In this regard, endocasts are most often frustratingly mute on other convolutional details.

Neuroanatomists have been trying for many decades to demonstrate the major differences between us and other primates, and aside from gross brain size, very little else has been shown as most of the differences can be explained as allometric scaling. The frontal lobe, and particularly its prefrontal portion, has been a favorite target and, indeed, Brodmann (1909) claimed that it was proportionally larger in humans, a view most recently championed by Deacon (1997). Unfortunately, other works have shown that the human brain has just as much frontal lobe as would be expected for a primate of its brain weight (Semendeferi et al., 1997; Uylings and van Eden, 1990), although the picture regarding the prefrontal cortex has yet to be determined objectively using cytoarchitectonic criteria, which is how prefrontal cortex is differentiated from the pure motor cortex behind it (Schenker et al., 2010; Sherwood et al., 2003; Rilling et al., 2008). Hominid brain endocasts do not, alas, provide any sulcal landmarks with enough reliability to determine the boundaries of prefrontal cortex, which is so important to impulse control, and higher cognitive functions such as planning and abstraction. Thus these regions cannot be accurately measured in a phylogenetic sequence. The Neandertals (see the chapter by Ahern) have frequently been described as having smaller frontal lobes; this is not based on objective measurements, but rather a perception that the large brow ridges on these humans were constraining frontal lobe development. Studying the Neandertal brain endocasts and comparing them to modern humans, I have failed to see any significant difference between these two groups, and (Bookstein et al., 1999) demonstrated that the prefrontal curvature was the same between these groups.

Similarly, regions such as "Broca's and Wernicke's areas," anterior and posterior association cortical regions involved in motor (Broca's) and receptive (Wernicke's) aspects of speech, are determinable on most fossil endocasts, and we can determine, for example, that Broca's region is more human-like on the brain cast of early *Homo*, some 1.8 mya. This is the famous KNM-ER 1470 endocast of *Homo rudolfensis* from Kenya, which had a brain volume of 752 ml, but which may not be a direct ancestor to our own line of *Homo*, but does show cerebral asymmetries similar to those found in modern *Homo* (see Figure 8). We know that Broca's regions in modern *Homo* are asymmetrical both in overall size and cytoarchitectonic divisions among areas 44, 45, and 47 of Brodmann (Amunts et al., 2010; Schenker et al., 2010). Interestingly, Neandertal endocasts show similar asymmetry in Broca's regions to modern humans (see Figure 9).

While the concept of reorganization has a heuristic value in directing our attention to changing quantitative relationships between different neural nuclei and fiber tracts, we cannot yet describe behavioral differences among closely related animals such as chimpanzee, gorilla, and orangutans, or different species of the genus *Macaca*, or indeed different breeds of dogs with their different temperaments, aptitudes, and sociality. We simply do not know what magic level of neural description is necessary to describe species-specific behavior. More recent research on prairie and mountain voles suggests that the difference in the females' ability to retrieve pups back to the nest depends on the distribution and number of neuroreceptors for the hormone oxytocin found in several nuclei of the brain, particularly

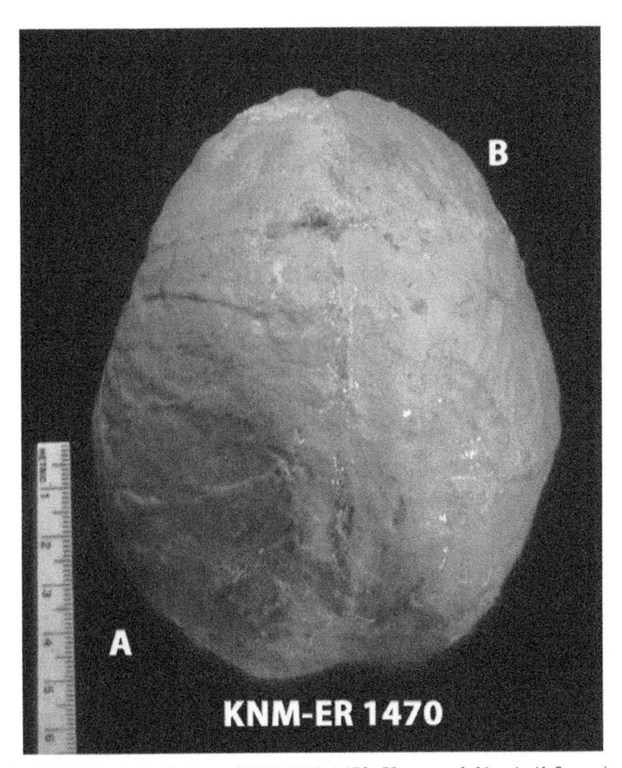

FIGURE 8 Dorsal view of KNM-ER 1470, *Homo rudolfensis* (1.8 mya), showing a typical *Homo* pattern of petalias, the left occipital projecting more posterior and being wider than the right side (A) and the right frontal being wider than the left (B). (See color plate section).

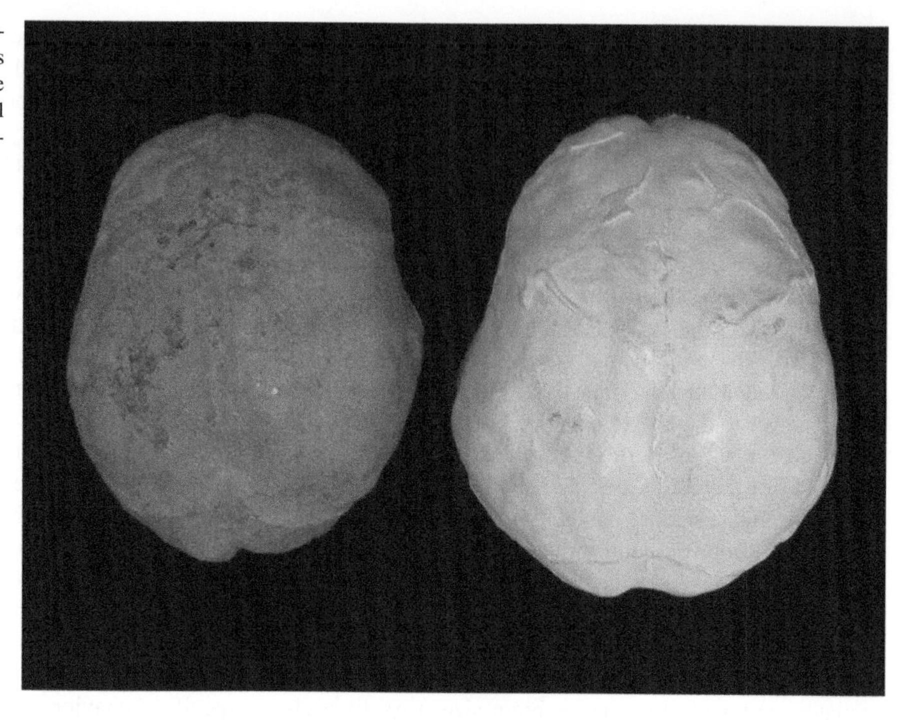

FIGURE 9 Neandertal cerebral asymmetries. Left is Monte Circeo and right is LaFerrassie. Both show a larger width of the right frontal lobe and a larger left occipital region (as in Figure 8). (See color plate section).

the thalamus. Otherwise, their brains appear identical (Insel and Shapiro, 1992). In addition, it is necessary to remember that the brain possesses aspects of plasticity that we did not appreciate except within the past decade, and that as the brain's organization unfolds ontogenetically, interactions with environmental stimuli are always occurring, and the brain builds its organization partly through its plasticity. It is difficult enough to study and understand such patterns in laboratory animals, let alone in our fossil ancestors! While the above suggests a somewhat pessimistic tone, we should remember that advances in noninvasive technology such as magnetic resonance imaging (MRI), functional magnetic resonance imaging, positron emission tomography, and tensor diffusion scanning have enormously increased our understanding of how the brain works, and how neural systems integrate and dissect data from the environment, always providing us with newer paradigms for further exploration about our brains and behavior, and in time, those of our closest relatives, the apes, in particular the bonobo and chimpanzee (see in particular Semendeferi et al., 2010).

Human Brain Asymmetry

The cerebral cortices of the human brain are usually asymmetrical, and tend to grow in a torque manner, reflecting minor differences in maturation rates. The hemispheres are seldom, if ever, equipotential in terms of functioning. Our left hemisphere is often characterized as "analytic" and involved with language tasks, while our right hemisphere appears to be most competent in visuospatial integration, and is often thought of as the "intuitive" or "gestalt" hemisphere. These characterizations, while crude, hold up fairly accurately for right-handers and many ambidextrals. From radiographic studies, it was possible for LeMay (1976) to ascertain different petalia patterns for right- and left-handed humans with a high degree of precision. These petalias are small extensions of cerebral cortex that extend farther in one part of a hemisphere than on the other side. For example, we speak of a left-occipital right-frontal torque pattern of petalias as occurring with high frequency in right-handed individuals. This means that the left occipital lobe bulges somewhat more posteriorly on the left hemisphere while the right hemisphere is somewhat broader in width in the frontal lobe. In true left-handers, who make up about 8–10% of human populations, the pattern is reversed, meaning that they exhibit a right-occipital left-frontal pattern. Petalia patterns for a large collection of apes indicated that while chimpanzees, gorillas, and orangutans sometimes demonstrated asymmetries, they did not show the particular torque pattern described above as frequently. The gorilla, incidentally, was the most asymmetrical of the apes (Holloway and de LaCoste-Lareymondie, 1982). On the other hand, brain asymmetries, particularly in the *planum temporale* (temporal cortex) of the chimpanzee, show a strong left-hemispheric size difference compared to the right (Gannon et al., 1998). This is simply puzzling as we do not have any evidence that chimpanzees use this structure in communication as do humans, and the fact that we share this difference with chimpanzees suggests that brain organizational features relating to complex cognitive functioning have been around for at least 5–7 mya. As our noninvasive scanning techniques become more sophisticated, we can expect to learn how these asymmetries function in animals other than ourselves. In fact, asymmetries appear in many animals and are hardly unique to primates. It is probably the degree of asymmetry that is important in attempting to distinguish humans from other primates (Balzeau and Gilissen, 2010).

Homin brain endocasts, when complete for both sides (unfortunately, this is very rare), allow the paleoneurologist to assess the cerebral asymmetries, and indeed, even australopithecines appear to show beginnings of the right-handed torque pattern found in humans, and, as one progresses through time, the petalia patterns become more accentuated in the modern human direction. If we add to these observations those of Toth's (1985) studies on the early stone tools of about 2 mya, which strongly suggest right-handedness, this underlines the fact that our early ancestors' brains, despite their small size (sometimes within the extant apes' range), were reorganized, and that they probably had some modes of cognition very similar to our own.

SYNTHESIS: PUTTING TOGETHER SIZE, ORGANIZATION, AND ASYMMETRY DURING HUMAN EVOLUTION

As noted earlier, human brain evolution has clearly been a process of integrating neurogenomic processes that led to increased size of the brain (hyperplasia and hypertrophy), and these neurogenomic changes also played roles in the reorganization (quantitative shifts) of neural nuclei, fiber tracts, and cortical cytoarchitectonics. In addition, it is probable that other changes occurred at the neurochemical level, involving neurotransmitters and receptor sites, but these are not well known from the comparative record, let alone the fossil one. This integration was sometimes gradual, sometimes "punctuated," at least based on the fossil hominid record currently available. The only reliable evidence from paleoneurology suggests that Brodmann's area 17 (PVC) was reduced early in hominid evolution, signs of the reduction being clear in *A. afarensis* some 3–3.5 mya. While this would have meant a relative increase in posterior parietal cortex (area 39) and peri- and parastriate cortex (areas 18 and 19, respectively), the vagaries of sulcal impressions do not allow for a totally unambiguous definition of these areas. Similarly, it is not possible at this time to measure and delineate remaining areas of the temporal cortex and superior parietal lobule unambiguously. What is suggested, however, is that visuospatial abilities were most probably cognitively enhanced early in hominid evolution. It is not until we come to *H. rudolfensis* ca. 1.8 mya that a case can be made for some frontal lobe reorganization in the third inferior frontal convolution, Broca's area. Thus it would appear that there was a gradient of cerebral reorganizational changes starting posteriorly, and progressing anteriorly. Table 4 outlines these changes.

Table 5 outlines the major size changes in the human brain during its evolutionary odyssey. Paleoneurological data simply are not detailed enough to integrate the two tables of size and reorganizational changes into one holistic sequence of events. Basically, the paleontological record supports an early reorganizational change resulting in an increase in posterior cortex associated with visuospatial processing, perhaps accompanied by a relatively small allometric increase in brain size from *A. afarensis* to *A. africanus*. This would correlate well with geological and paleontological evidence that shows that early hominids were expanding their ecological niches and becoming more diverse in their subsistence patterns in mixed habitats. We know this based on the fact that stone tool types are becoming standardized in form, tool inventories grow larger, and right-handedness is highly probable. With the evolution of *Homo*, we find strong evidence for a major increase in brain size, both allometric (related to body size) and nonallometric, and a reorganized frontal lobe, broader, and showing a more modern human-like Broca's area. This suggests that indeed there had been some strong and dramatic selection pressures for a somewhat different style of sociality, one most probably based on a primitive proto-language that had some arbitrary symboling elements as the standardization of stone tools (e.g., Acheulean hand axes) increases, suggesting social cohesion and control mediated through symbolically based

TABLE 4 Summary of Reorganizational Changes in the Evolution of the Human Brain

Brain Changes, Reorganizational	Taxon
1. Reduction of primary visual striate cortex, area 17, and a relative increase in posterior parietal and temporal cortex, Brodmann areas 37, 39, 40, as well as 5 and 7	*Australopithecus afarensis, Australopithecus africanus*
2. Reorganization of frontal lobe (3rd inferior frontal convolution, Broca's areas 44, 45, 47)	*Homo rudolfensis; early Homo*
3. Cerebral asymmetries, left-occipital right frontal petalias	Australopithecines and early *Homo*
4. Refinements in cortical organization to a modern *Homo sapiens* pattern	*Homo erectus* to present

Note: (4) is inferred, as brain endocasts cannot provide that level of detail necessary to demonstrate the refinements in cortical organization from surface features alone. Areas 18, 19 are peri- and parastriate cortex just anterior to area 17, and are included in posterior association cortex here.
Holloway 1997.

TABLE 5 Brain Size Changes in Hominid Evolution

Brain changes	Taxon	Time (mya)	Evidence
1. Small increase, allometric[a]	*A. afarensis* to *A. africanus*	3.5–2.5	Brain endocast increase from ca 400–450+ ml
2. Major increase, rapid, both allometric and nonallometric	*A. africanus* to *H. habilis, rudolfensis*	2.5–1.8	KNM–1470, 752 ml (300 ml increase)
3. Modest allometric increase in brain size to 800–1000 ml	*H. habilis* to *H. erectus*	1.8–0.5	*H. erectus* brain endocasts and postcranial bones
4. Gradual and modest size increase to archaic nonallometric FOXP2	*H. erectus* to *H. sapiens neanderthalensis*	0.5–0.075	Archaic *Homo H. sapiens*, Neandertal endocasts 1200–1700+ ml
5. Small reduction in brain size among modern allometric	*H. sapiens*	0.015–present	Modern endocranial volumes

[a]*Related to increase in body size only.*
Holloway 1997 and more recent endocast data Holloway et al., 2004.

communication (Holloway, 1981; Stout et al. 2006). Needless to say, this is but one speculative account of the evidence. But from about 1.8 to roughly 0.5 mya, there were minor allometric brain size increases to the earliest *H. erectus* hominids of Indonesia and China, where brain sizes ranged from 750 to 1250 ml in volume. We have very little evidence for body sizes, but we believe, on the basis of the KNM-WT 15,000 Nariokotome youth from Kenya at ca. 1.6 mya, that these did not differ significantly from our own.

This is also a time during which cerebral asymmetries are becoming more strongly pronounced. With the advent of Archaic *H. sapiens*, about 0.15–0.2 mya (see the chapter by Ahern), we find brain sizes well within modern human values, and no evidence for further allometric increases, except possibly for the Neanderthal humans in which it can be argued that their larger brain and body sizes (lean body mass: bone and muscle) were adaptations to colder conditions (see chapter by Leonard). If further changes took place in cerebral and/or subcortical organization, they are simply not apparent from a paleoneurological perspective. Yet, the Upper Paleolithic is the time when cave art makes its appearance, and one cannot help but wonder whether the explicit use of art involving symbolization might not also have been the time for the emergence of language. In fact, however, there is nothing in the direct fossil evidence, and in particular paleoneurology, which provides any evidence for such views.

Claims for a single mutation are extremely speculative, and while some genes have been identified, such as the FOXP2 (also in Neandertals), these also involve more general aspects of cognition. It is more likely that stone tool making and its underlying cognitive elements are very similar to language, if not partially homologous (Holloway, 1969, 1981, 2012). Finally, it would appear that there has actually been a small reduction in brain size, probably allometric in nature, from about 0.015 MY to the present (Henneberg, 1988; Hawks, 2012).

The totality of evidence shows that the brain has always been evolving during our evolutionary journey, with myriad changes taking place at different tempos during different times. As suggested (Holloway, 1997, p. 200):

In sum, the major underlying selectional pressures for the evolution of the human brain were mostly social. It was an extraordinary evolutionary 'decision' to go with an animal that would take longer to mature, reach sexual maturity later, and be dependent for its food and safety upon its caretakers (parents?) for a longer period of time. The benefits for the animal were many, including a longer learning period, a more advanced, larger, and longer-growing brain, and an increasing dependence on social cohesion and tool making and tool using to cope with the environments that they encountered. Needless to say, language abilities using arbitrary symbol systems were an important ingredient in this evolution.

The fossil record shows us that there was a feedback between the complexity of stone tools (which must be seen as a part of social behavior) and increasing brain size and the expansion of ecological niches. The 'initial kick,' however, the process that got the ball rolling, was a neuroendocrinological change affecting regulatory genes and target tissue–hormonal interactions that caused delayed maturation of the brain and a longer growing period, during which learning became one of our most important adaptations.

These ideas have been described elsewhere (Holloway, 1967, 1969, 1979, 1996, 2010), where more details may be found.

Finally, Figure 10 provides the often-seen relationship between time and endocranial volume, and as should be apparent, there is considerable overlap between fossil groups, and considerable variation within each taxon (e.g., *H. erectus*). Needless to say, such depictions cannot reveal

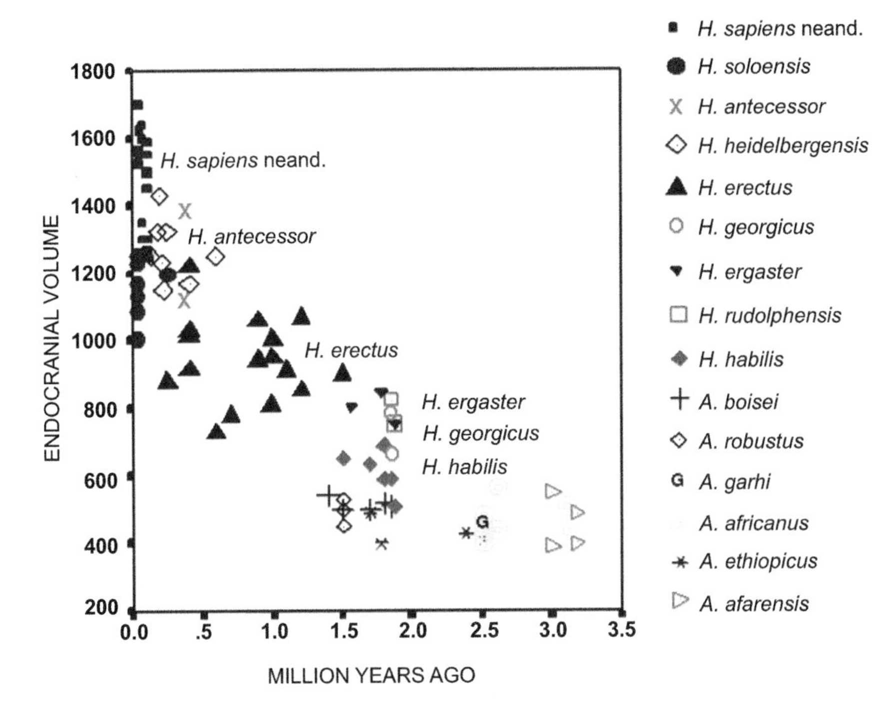

FIGURE 10 Endocranial volume plotted against time, showing an accelerated change in volume from *Homo erectus* on to anatomically modern *Homo* in the late Pleistocene. This figure cannot include times of reorganization events, changes in neurogenomic elements, or any of the finer-grained differences in morphology of the endocasts. It is important to observe overlap of endocranial volumes, as well as their variation within taxa.

the complex interactions between phases of reorganization, size increases through hypertrophy and hyperplasia, asymmetries in between left and right sides, different distributions of neuroreceptors and neurotransmitters, and the intricate interactions among natural selection, environmental challenges, mutation, drift, sensorimotor adaptations (think of the challenges of becoming fully bipedal), social behavior, communication skills, emotions, etc., all of which were operating during the whole of hominid brain evolution. I hope the point is obvious, that while we have learned much over the last century from fossil, comparative, and neurogenomic evidence, we remain almost totally ignorant of how it really happened.

AND TO THE FUTURE?

There appears to be two common presumptions about our future brain evolution. One is that our biological evolution has stopped. The second is that our brains will continue to grow in size, with bulging frontal lobes (sort of a cross between E.T. and X-Files ...), to handle our growing dependence on technology. What we have witnessed from the past fossil record is that our brains and bodies work largely in allometric fashion, and given the high metabolic cost of operating bigger brains (about 20–25% of our metabolic resources go to supporting our brains, which constitute only 2% of our total body weight), the second scenario seems highly unlikely. The first scenario is simply untrue, but it would require vast amounts of information from each generation of many living populations, something feasible perhaps, but not currently being collected. Furthermore, it is quite controversial whether brain size has any close relationship to intelligence, however intelligence is actually defined and measured. Recent research based on MRI determinations of brain volume and selected batteries of cognitive tests has shown correlations between test scores and brain volume ranging from 0.4 to 0.6 (Andreasen et al., 1993; Anderson, 2003; Davies et al., 2011). But if protein resources were to nosedive throughout the world for a significant period of time, selection would probably favor smaller body sizes in our species, and that could result in smaller brains, given an allometric relationship of roughly 0.3 between stature and brain size, at least in males (Holloway, 1980). While genetic engineering may well provide some respite between the ever-increasing mass of humanity, ecological and nutritive degradation, this too is likely to be nothing more than short-term fending off of the unstoppable future. These degradations are part and parcel of the human brain's capacity to ignore warnings that should curtail greed and stupidity. The paleontological record for most mammals suggests that at the taxonomic level of the genus (such as *Pan, Homo, Canis, Notocherus*, etc.), one finds a recognizable record of that genus spanning approximately 5–10 million years. Our genus has thus far been around for about two million years. We, as a genus, despite our largish highly encephalized brains, have another three million years to go if we wish to be as successful in the paleontological longevity game.

REFERENCES

Allen, J.S., et al., 2006. Looking for the Lunate Sulcus: A Magnetic Resonance Imaging Study in Modern Humans. Anat. Record Part A:288A:867–876.

Amunts, K., et al., September 2010. Broca's region: novel organizational principles and multiple receptor mapping. PLoS Biology 8 (9), e1000489.

Anderson, B., 2003. Brain imaging and g. In: Nyborg, H. (Ed.), The Scientific Study of General Intelligence: Tribute to Arthur R. Jensen. Pergamon Press, NY, pp. 29–40.

Andreasen, N.C., Flaum, M., Swayze, H.V., O'Leary, D.S., Alliger, R., Cohen, G., Ehrhardt, N., Yuh, W.T.C., 1993. Intelligence and brain structure in normal individuals. American Journal of Psychiatry 150, 130–134.

Balzeau, A., Gilissen, E., 2010. Endocranial shape asymmetries in *Pan paniscus, Pan troglodytes* and *Gorilla gorilla* assessed via skull based landmark analysis. Journal of Human Evolution 59, 54–69.

Bookstein, F., et al., 1999. Comparing frontal cranial profiles in archaic and modern *Homo* by morphometric analysis. Anatomical Record 6, 217–224.

Brodmann, K., 1909. Vergleichende Lokalizationzlehre der Grosshirnrinde. J.A. Barth, Leipzig, Germany.

Dart, R., 1925. *Australopithecus africanus*: the man-ape of South Africa. Nature 115, 195–199.

Davies, G., et al., 2011. Genome-wide association studies establish that human intelligence is highly heritable and polygenic. Molecular Psychiatry 16, 996–1005.

Deacon, T., 1997. The Symbolic Species: The Co-evolution of Language and the Brain. Norton, New York.

Gannon, P.J., Holloway, R.L., Broadfield, D.C., Braun, A.R., 1998. Asymmetry of chimpanzee planum temporale: humanlike pattern of Wernicke's brain language area homolog. Science 279, 220–222.

Hawks, John, 2012. Selection for Smaller Brains in Holocene Human Evolution. arXiv:1102.5604.

Henneberg, M., 1988. Decrease of human skull size in the Holocene. Human Biology 60, 395–405.

Holloway, R.L., 1967. The evolution of the human brain: some notes toward a synthesis between neural structure and the evolution of complex behavior. General Systems, XII, 13–19.

Holloway, R.L., 1969. Culture: a human domain. Current Anthropology 10, 395–412.

Holloway, R.L., 1979. Brain size, allometry, and reorganization: toward a synthesis. In Hahn, M.E., Jensen, C., Dudek, B.C. (Eds.), Development and Evolution of Brain Size: Behavioral Implications. New York: Academic Press, pp. 59–88.

Holloway, R.L., 1980. Within-species brain-body weight variability: a reexamination of the Danish data and other primate species. American Journal of Physical Anthropology 53, 109–121.

Holloway, R.L., 1981. Culture, symbols, and human brain evolution. Dialectical Anthropology 5, 287–303.

Holloway, R.L., 1984. The Taung endocast and the lunate sulcus: a rejection of the hypothesis of its anterior position. American Journal of Physical Anthropology 64, 285–287.

Holloway, R.L., 1995. Toward a synthetic theory of human brain evolution. In: Changeux, J.P., Chavaillon, J. (Eds.), Origins of the Human Brain. Clarendon Press, Oxford, UK, pp. 42–54.

Holloway, R.L., 1996. Evolution of the human brain. Chap. 4. In: Lock, A., Peters, C. (Eds.), Handbook of Human Symbolic Evolution. Oxford University Press, New York, pp. 74–116.

Holloway, R.L., 1997. Brain evolution. In: Dulbecco, R. (Ed.), Encyclopedia of Human Biology, vol. 2. Academic Press, New York, pp. 189–200.

Holloway, R.L., 2008. The human brain evolving: a personal retrospective. Annual Review of Anthropology 37, 1–19.

Holloway, R.L., 2012. Language and tool making are similar cognitive processes. Behavioral and Brain Sciences 35 (04), 326.

Holloway, R.L., Broadfield, D.C., Yuan, M.S., 2004. The human fossil record. In: Schwartz, J.H., Tattersall, I. (Eds.), Brain Endocasts: The Paleoneurological Evidence, vol. 3. Wiley-Liss, New York.

Holloway, R.L., 2010. Human brain endocasts, Taung, and the LB1 Hobbit brain. Chapt. 4. In: Broadfield, D., et al. (Ed.), The Human Brain Evolving: Paleoneurological Studies in Honor of Ralph L. Holloway. Stone Age Institute Press, Gosport, Indiana.

Holloway, R.L., de LaCoste-Lareymondie, M.C., 1982. Brain endocast asymmetry in pongids and hominids: some preliminary findings on the paleontology of cerebral dominance. American Journal of Physical Anthropology 58, 101–110.

Insel, T., Shapiro, L.E., 1992. Oxytocin receptors and maternal behavior. Annals of the New York Academy of Sciences 652, 448–451.

Jerison, H.J., 1973. Evolution of Brain and Intelligence. Academic Press, New York.

LeMay, M., 1976. Morphological cerebral asymmetries of modern man, fossil man, and nonhuman primates. Annals of the New York Academy of Sciences 280, 349–366.

Martin, R.D., Isler, K., 2010. The maternal energy hypothesis of brain evolution: an update. In: Broadfield, et al. (Ed.), The Human Brain Evolving: Paleoneurological Studies in Honor of Ralph L. Holloway. Stone Age Institute Press, Gosport, Indiana.

Martin, R.D., 1983. Human Evolution in an Ecological Context. American Museum of Natural History, New York. James Arthur lecture (1982).

Miller, D.J., et al., 2012. Prolonged myelination in human neocortical evolution. Proceedings National Academy of Sciences 109 (41), 16480–16485.

Nyborg, H. (Ed.), 2003. The Scientific Study of General Intelligence: Tribute to Arthur R. Jensen.

Rilling, J.K., et al., 2008. The evolution of the arcuate fasciculus revealed with comparative DTI. Nature Neuroscience 11, 382–384.

Schenker, N.M., et al., 2010. Broca's area homologue in chimpanzees (*Pan troglodytes*): probabilistic mapping, asymmetry, and comparison to humans. Cerebral Cortex 20, 730–742.

Semendeferi, K., Damasio, H., Frank, R., Van Hoesen, G.W., 1997. The evolution of the frontal lobes: a volumetric analysis based on three-dimensional reconstructions of magnetic resonance scans of human and ape brains. Journal of Human Evolution 32, 375–388.

Semendeferi, K., Barger, N., Schenker, N., 2010. Brain reorganization in humans and apes. In: Broadfield, D.C., et al. (Ed.), The Human Brain Evolving: Paleoneurological Studies in Honor of Ralph L. Holloway. Stone Age Institute Press, Gosport, Indiana.

Sherwood, C.C., et al., 2003. Variability in Broca's area homologue in African great apes: implications for language evolution. Anatomical Record 271A, 276–285.

Stephan, H., Frahm, H., Baron, G., 1981. New and revised data on volumes of brain structures in insectivores and primates. Folia Primatologia 35, 1–29.

Stout, D., 2006. Oldowan tool making and hominin brain evolution: theory and research using positron emission tomography (PET). In: Toth, N., Schick, K. (Eds.), The Oldowan: Case Studies into the Earliest Stone Age. Stone Age Institute Press, Gosport, Indiana.

Tobias, P.V., 1971. The Brain in Hominid Evolution. Columbia University Press, New York.

Toth, N., 1985. Archaeological evidence for preferential right-handedness in lower and middle Pleistocene, and its behavioral implications. Journal of Human Evolution 14, 607–614.

Uylings, H.B.M., van Eden, C.G., 1990. Qualitative and quantitative comparison of the prefrontal cortex in rats and primates, including humans. Progress in Brain Research 85, 31–62.

Weber, G.W., et al., 2012. Digital South African fossils: morphological studies using reference-based reconstruction and electronic preparation. In: Reynolds, S.C., Gallagher, A. (Eds.), African Genesis: Perspectives on Hominin Evolution. Cambridge, New York, pp. 298–316.

Zollikofer, C.P.E., Ponce De Leon, M.S., 2013. Pandora's growing box: inferring the evolution and development of hominin brains from endocasts. Evolutionary Anthropology 22, 20–23.

Chapter 18

Physiological Adaptations to Environmental Stressors

William R. Leonard
Department of Anthropology, Northwestern University, IL, USA

SYNOPSIS

Human biologists have long recognized the important role that environmental stressors play in shaping variation in human physiology and morphology. Among human populations around the world, body size and proportions are strongly influenced by differences in environmental temperature. Human populations living in cold, arctic climates are relatively heavy with large trunks and shortened limb lengths. Conversely, tropical populations are lighter and typically have more linear physiques. These patterns of variation are consistent with the classic "ecological rules" of Bergmann and Allen. Physiological adaptations to temperature stress involve strategies that regulate heat loss and production. Acclimation to cold stress involves increasing the efficiency of metabolic heat production and of heat delivery to peripheral tissues through earlier vasodilation. In contrast, acclimation to heat stress involves enhancing the capacity to dissipate heat production through greater vasodilation of peripheral blood vessels and increased sweat rates. Adaptations to high altitude hypoxia involve increasing the uptake and delivery of oxygen to peripheral tissues. Sea-level visitors to altitude respond by increasing the rate and depth of breathing and markedly increasing the oxygen carrying capacity of the blood. In contrast, native high altitude populations show a different set of responses including enlarged lung volumes and a modest increase in oxygen carrying capacity of the blood. The adaptations acquired during growth and development at altitude are critical for promoting complete functional adaptation to hypoxic stress.

INTRODUCTION

Over the last two million years, the human lineage (hominins) has expanded from Africa to colonize every major ecosystem on the planet, adapting to a wide range of environmental stressors (Antón et al., 2002). Adaptation to these diverse climates has selected for considerable variation in human physiology and morphology. For more than a century, anthropologists and physiologists have systematically studied the influence of adaptation to environmental stressors on human biological variation. Long-term studies have been carried out among indigenous human populations living in a range of extreme environments, including the arctic (Milan, 1980), both dry and humid tropical climes (Adolph, 1947; Hanna and Brown, 1983), and high altitude regions (Baker and Little, 1976). This research has documented the important role that environmental factors play in shaping human variation in (1) growth and development, (2) body size and composition, and (3) energy expenditure and metabolic function.

In the field of human biology, scholars have historically viewed adaptation (see the chapter by Futuyma in this volume) or adaptability as occurring along the three alternative, interacting pathways shown in Figure 1: (1) genetic, (2) functional/physiological, and (3) cultural/behavioral (cf. Frisancho, 1993; Lasker, 1969; Mazess, 1975). Genetic or Darwinian adaptations are the product of natural selection and are biologically heritable. In contrast, functional or physiological adaptations are those acquired during one's lifetime and are not genetically heritable. Physiological adaptations are mobilized to maintain homeostasis, allowing for normal biological function (e.g., maintenance of body temperature, oxygenation of the blood and tissues) in the face of environmental stressors. The process of responding to a single stressor, often in an experimental, laboratory context, is known as "acclimation." In contrast, "acclimatization" refers to longer term responses to environmental stressors in the natural world. Some physiological adaptations can only be developed if an individual is exposed to stress during growth; this process is known as developmental acclimatization.

Cultural/behavioral or technological adaptations are nonbiological strategies for adapting to environmental stressors. Although not biologically based, cultural adaptations are similar to genetic adaptations in that they can be transmitted across generations by teaching and learning (see the chapter by Walker). In addition, culture and technology not only serve as a buffer against environmental stressors, but also create new stressors to which human populations must respond (e.g., environmental pollution, social inequality in material and nutritional resources).

This chapter provides an overview of human physiological adaptations to major environmental stressors. First, the basic principles of human thermoregulation are discussed. Next, evidence for the influence of climate (environmental temperature) on worldwide human variation in body

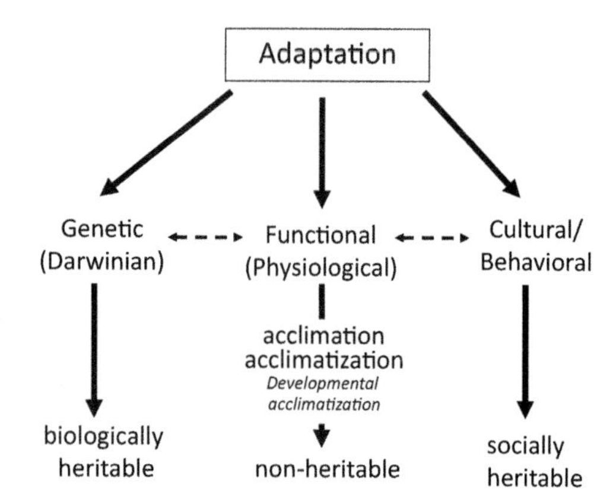

FIGURE 1 Schematic depiction of different types of adaptation to environmental stressors. Genetic (Darwinian) adaptations are the product of natural selection and are biologically heritable. Functional (physiological) adaptations are acquired during one's lifetime and are not genetically heritable. Physiological adaptations maintain homeostasis in the face of environmental stressors. Acclimation refers to responses to a single stressor, often in an experimental/laboratory setting. Acclimatization is the process of adaptation to multiple stressors in the natural world. Developmental acclimatization refers to responses that can only be acquired with exposure to the stressors during the growth period. Finally, cultural/behavioral adaptations are nonbiological responses to environmental stressors. These responses are "socially heritable," as they can be transmitted across generations by teaching and learning. Within human populations, these three adaptive domains interact to shape distinctive patterns of human variation.

morphology (body weight and proportions) is examined. From there, human physiological responses to the stresses of cold, heat, and high altitude hypoxia are reviewed. For each condition, the short-term responses of acclimation observed across all populations, and the distinctive suite of responses that are associated with lifelong acclimatization to these climatic extremes are examined.

PRINCIPLES OF THERMOREGULATION

Like other mammals, humans are homeotherms, who maintain relatively constant internal body (core) temperatures regardless of environmental temperatures. The maintenance of internal body temperature reflects a dynamic process of regulating both heat production and heat loss. Rates of heat production and energy use by the body are determined by a number of different biological (e.g., body size, composition) and environmental/lifestyle (e.g., temperature, diet, activity) factors. Basal metabolic rate (BMR) represents the minimum amount of energy that is required to keep a person alive, maintaining normal body temperature. As depicted in Figure 2, basal metabolism is measured under controlled temperatures (thermoneutral conditions), while a subject is lying in a relaxed and fasted state (at least 10 h after a meal) (see Schofield, 1985; McLean and Tobin, 1987). Humans (unclothed) have a thermoneutral range of 25–27 °C (77–81 °F). Within this temperature zone, the minimum amount of metabolic energy is spent to maintain body temperature (Erickson and Krog, 1956). At

FIGURE 2 Measurement of basal metabolic rate (BMR; kcal/day) in a Yakut subject from northeastern Siberia. BMR is measured under controlled, thermoneutral conditions, with subjects lying down, in a rested and fasted state. Basal (minimal) energy costs are determined by assessing breathing rates, and the concentrations of oxygen (O_2) and carbon dioxide (CO_2) in expired breath samples.

TABLE 1 World Health Organization's (2004) Predictive Equations for Estimating Basal Metabolic Rate (BMR; kcal/day) in Adults

Reference	Sex	Age (yr)	Equations[a]
FAO/WHO/ UNU (2004)	M	18–29	15.1(Wt)+692
		30–59	11.5(Wt)+873
		60 and older	11.7(Wt)+588
	F	18–29	14.8(Wt)+487
		30–59	8.1(Wt)+846
		60 and older	9.1(Wt)+659

[a]*Wt=weight (kg).*

temperatures outside of this range, the body utilizes additional energy to heat or cool itself.

In all mammals, the strongest determinant of metabolic heat production is body mass (see Kleiber, 1975; FAO/WHO/UNU, 2004). Table 1 shows the evidence for this, presenting the World Health Organizations predictive equations for estimating BMR based on age, sex, and body weight (FAO/WHO/UNU, 2004). Among adults, weight alone accounts for roughly 65–70% of the variation in BMR, with lean tissue (fat-free mass [FFM] = body weight – fat mass) being a much stronger predictor than fat mass (Johntone et al., 2005). In contrast, rates of heat loss from the body are strongly shaped by surface area (SA). The overall SA of the human body is determined by a number of factors, including weight, limb lengths, and trunk size. Consequently, the simple ratio of body SA to body weight provides a useful measure of heat loss capacity relative to heat production. The subsequent section on "Climate and Human Body Morphology" explores how variations in temperatures shape variations in SA to weight among human populations around the world.

TABLE 2 Mechanisms of Heat Loss from the Human Body (Unclothed) at Different Environmental Temperatures

Room Temperature	Radiation (%)	Convection (%)	Evaporation (%)
Comfortable (25 °C=77 °F)	67	10	23
Warm (30 °C=86 °F)	41	33	26
Hot (35 °C=95 °F)	4	6	90

Adapted from Frisancho (1993).

To maintain an equilibrium between heat production and heat loss in response to changing environmental temperatures, the body relies on alternative avenues of heat exchange: radiation, conduction, convection, and evaporation (Frisancho, 1993). Radiation refers to heat transfer from one object to another via electromagnetic waves (e.g., the loss of heat from the skin to the external air). Convection is the transfer of heat through the movement of air molecules, whereas conduction reflects heat transfer through direct physical contact. Evaporation is the loss of heat due to conversion of water into vapor. It is distinct from the three other modes in being an "active" form of heat transfer that always results in heat loss from the body.

Table 2 shows the relative contributions of each of the mechanisms of heat loss to the human body at different environmental temperatures. At a comfortable temperature (~25 °C), most of the metabolic heat is lost through radiation (67%), with evaporation (23%) and convection (10%) contributing much smaller proportions. As ambient temperatures rise, sweat rates and evaporative cooling increase dramatically to maintain thermal homeostasis. Thus, under very warm conditions (35 °C; 95 °F), fully 90% of heat dissipation is due to evaporation, while radiation (4%) and convection (6%) account for only small fractions (Frisancho, 1993).

CLIMATE AND BODY MORPHOLOGY

Humans are similar to other mammalian species in conforming to the classic ecological "rules" of Bergmann (1847) and Allen (1877) documenting the relationships between body morphology and climate. "Bergmann's rule" addresses the relationship between body weight (mass) and environmental temperature, noting that within a widely distributed species, body mass increases with decreasing average temperature (Bergmann, 1847). In contrast, "Allen's rule" considers the relationship between body proportionality and temperature (Allen, 1877).

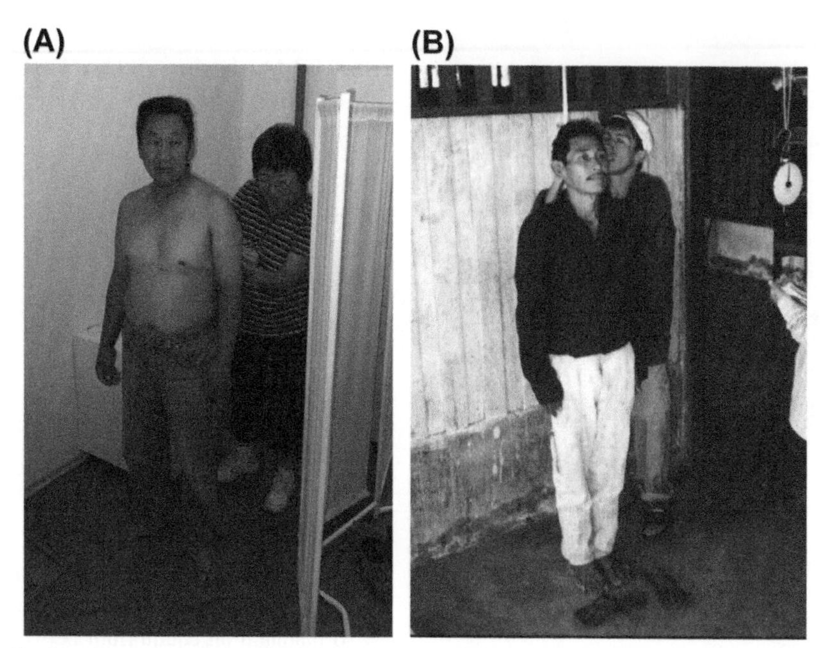

FIGURE 3 Comparison of body weight and morphology in adult men from populations of two different climatic extremes: (A) the Yakut of northeastern Siberia (arctic) and (B) the Tsimane' of lowland Bolivia (tropics). Body weights are significantly heavier in the Yakut (71 vs 62 kg; $P < 0.01$). The Yakut also have larger trunk and chest dimensions than the Tsimane, as reflected in their greater relative sitting height (RSH; 53 vs 49; $P < 0.01$). Difference in body morphology between the Yakut and the Tsimane' are consistent with Bergmann's and Allen's rules.

It finds that individuals of a species that are living in warmer climes have relatively longer limbs, whereas those residing in colder environments have relatively shorter extremities.

The physical basis of both of these ecological rules stems from the differences in the relationship between SA (cm^2) and volume (cm^3 proportional to mass [kg]) for organisms of different size (Schmidt-Nielson, 1984). Because volumetric measurements increase as the cube of linear dimensions, whereas SA increases as the square, the ratio of SA to volume (or mass) decreases as organisms increase in overall size. Since metabolic heat production is most strongly related to body mass, larger organisms are better suited to colder environments because they produce more heat and have relatively less SA through which to lose that heat. Conversely, small body size is better under warmer conditions, because these organisms will both produce less heat and have relatively greater SA for dissipating that heat. These morphological and physiological differences between organisms of different size are at the heart of the relationship described by Bergmann's rule.

Allen's rule, on the other hand, considers how changes in shape can alter SA to mass ratios. For organisms of the same size, the more elongated or linear the shape, the greater the SA to mass ratio. Thus, for organisms residing in tropical environs, a linear body plan—with less mass in the trunk and greater mass in the extremities—is better for facilitating heat dissipation. In contrast, for arctic-adapted organisms, a body build characterized by larger trunk size and shorter limbs will reduce metabolic heat loss by minimizing SA to mass.

Since the 1950s, numerous studies have demonstrated that contemporary human populations generally conform to the expectations of Bergmann's and Allen's rules, such that populations residing in colder climes are heavier and have shorter relative limb lengths, resulting in a decreased ratio of SA to body mass (see Schreider, 1950, 1957, 1964, 1975; Newman, 1953; Roberts, 1953, 1973, 1978; Barnicot, 1959; Baker, 1966; Walter, 1971; Stinson, 1990; Ruff, 1994; Katzmarzyk and Leonard, 1998). The most widely cited research on this topic is the classic work of Roberts (1953, 1978). Roberts (1953) paper on "Body Weight, Race and Climate" showed a significant negative correlation between body mass and mean annual temperature, indicating that humans appear to conform to Bergmann's rule. In subsequent work, Roberts (1973, 1978) showed that humans also conform to Allen's rule, such that populations living in colder regions have relatively shorter legs and larger relative sitting heights (RSH) (RSH = 100 × [sitting height]/[stature]) than those groups inhabiting warmer regions.

Figure 3 shows the differences in body mass and proportions between adult men from arctic and tropical populations, the Yakut (of northeastern Siberia) and Tsimane' (of lowland Bolivia), respectively. Yakut men have average body weights of 71 kg (160 lbs) and a mean RSH of 53. In contrast, Tsimane' men are significantly lighter (average weight = 62 kg (137 lbs)) relative longer leg length, as reflected in their RSH of 49.

Recently, Katzmarzyk and Leonard (1998) and Leonard and Katzmarzyk (2010) reexamined the influence of temperature on body mass and proportions in a worldwide sample of human populations studied after Roberts' initial work in 1953. These analyses confirmed many of Roberts' original

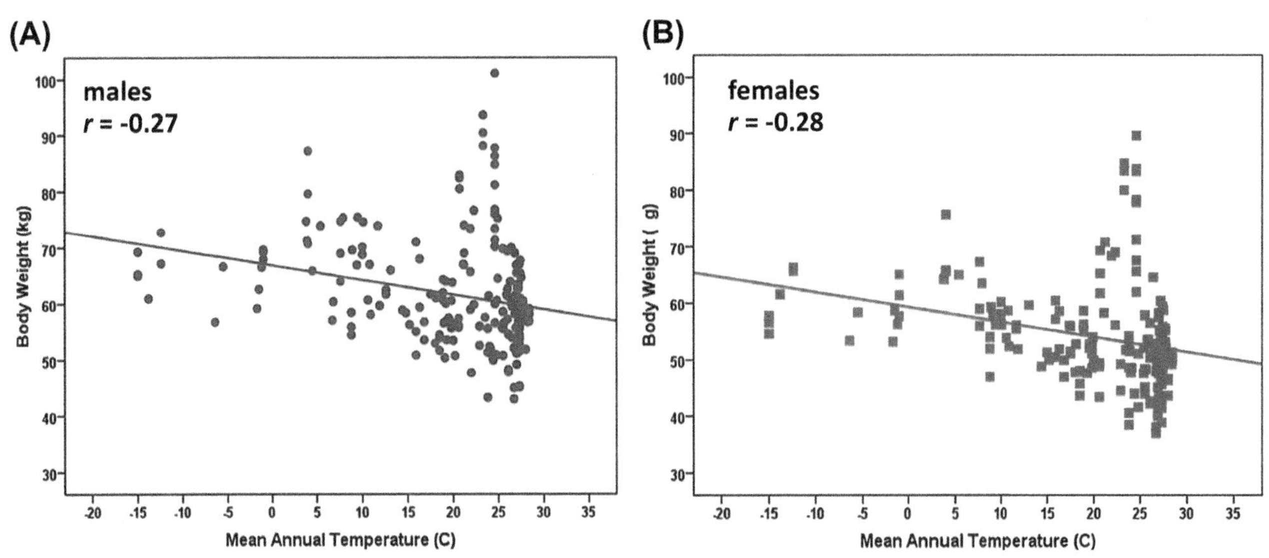

FIGURE 4 Relationship between body weight (kg) and mean annual temperature (°C) in a sample of 223 men and 198 women. In both sexes, weight is negatively associated with temperature ($r = -0.27$, males; $r = -0.28$ females; $P < 0.001$), consistent with the expectations of Bergmann's rule. *Adapted from Leonard and Katzmarzyk (2010).*

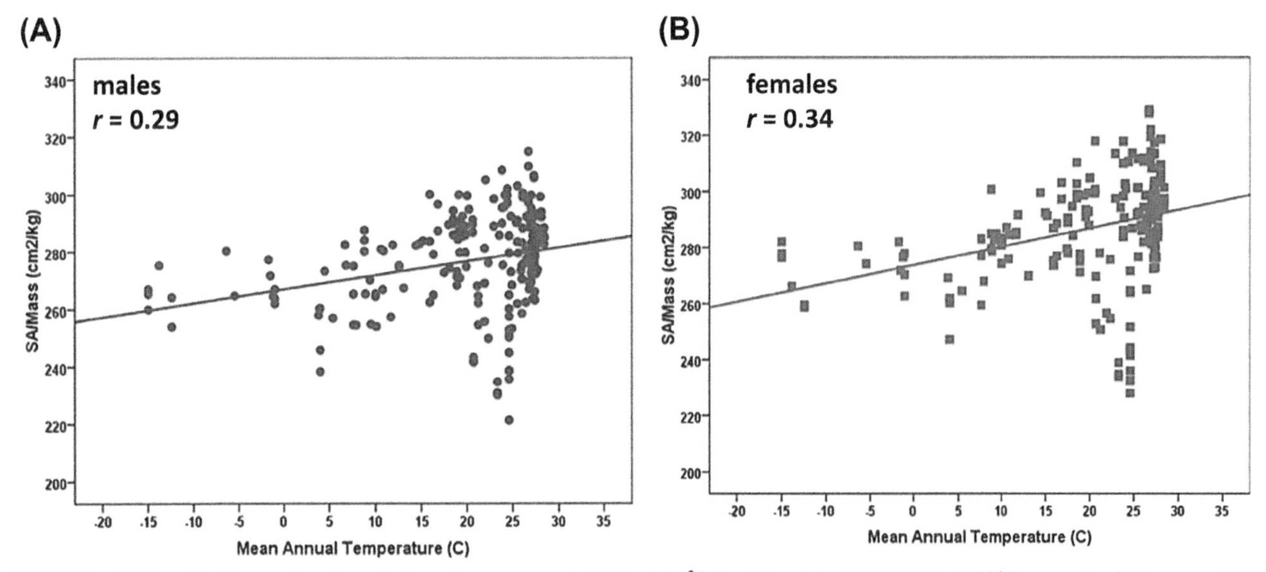

FIGURE 5 Relationship between the ratio of body surface area:mass (SA/Mass; cm^2/kg) and mean annual temperature (°C) in a sample of 222 men and 197 women. In both sexes, SA/Mass is positively related to temperature ($r = 0.29$ male; $r = 0.34$, females; $P < 0.001$). *Adapted from Leonard and Katzmarzyk (2010).*

findings. Specifically, we found that the inverse relationship between mass and temperature continues to persist for both men and women; however, the slope of the regression was significantly shallower than that reported by Roberts. Likewise, the relationship that we found between RSH and temperature was more modest than that of Roberts' sample. These differences partly reflect secular changes in growth and body size since 1950, and the development of improved technology that moderates extreme temperature exposure during development. These findings underscore the importance of both nutritional and temperature stresses in shaping human variation in body size and shape.

Figure 4 shows the relationship between body weight (kg) and mean annual temperature in the Leonard and Katzmarzyk (2010) sample of men and women. In both sexes, weight is significantly negatively correlated with mean annual temperature, as predicted by Bergmann's rule. The strength of association between body mass and temperature is comparable between the male and the female samples ($r = -0.27$ [males]; $r = -0.28$ [females]).

As shown in Figure 5, SA/mass (cm^2/kg) ratios are positively associated with mean annual temperature. This relationship indicates that populations of colder climes have body plans that minimize SA to mass to

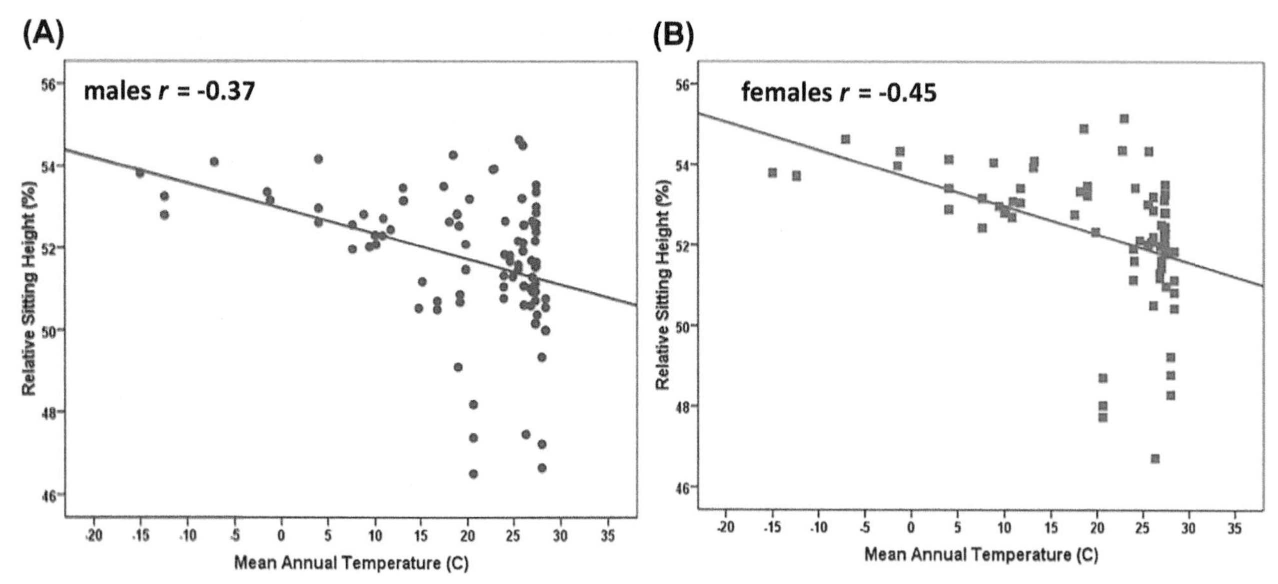

FIGURE 6 Relationship between the relative sitting height (RSH; %) and the mean annual temperature (°C) in a sample of 94 males and 74 females. In both sexes, RSH is inversely related to temperature ($r = -0.37$ males; $r = -0.45$, females; $P < 0.001$), consistent with the expectations of Allen's rule. *Adapted from Leonard and Katzmarzyk (2010).*

reduce metabolic heat loss, whereas those inhabiting hotter regions have physiques that maximize SA/mass to promote heat dissipation. The strength of the association is slightly stronger in females than in males (r = 0.34 vs 0.29).

Figure 6 shows the relationship between RSH and temperature in the Leonard and Katzmarzyk sample. Consistent with the expectations of Allen's rule, RSH is negatively correlated with temperature in both sexes ($r = -0.37$ in males; $r = -0.45$ in females), indicating that tropically adapted populations have a more linear body build, with relatively longer limbs and shorter trunks. In contrast, populations of cold, high latitude environments are characterized by a more stout body plan with shorter extremities and a relatively larger trunk.

PHYSIOLOGICAL ADAPTATIONS TO COLD STRESS

Adaptation to cold stress involves an integrated suite of hormonal, neurological, and vascular responses that both increase metabolic heat production and reduce heat loss to the environment. Our understanding of human responses to cold is based partly on controlled laboratory experiments and partly on field studies of populations residing in arctic and subarctic climates. This work has provided insights into the similarities and differences in short-term responses to acute cold exposure and the longer term (i.e., lifetime and transgenerational) responses to chronic cold stress observed among indigenous high latitude populations.

Acclimation to Cold Stress

At temperatures below the thermoneutral zone, the human body utilizes mechanisms to reduce metabolic heat loss and to increase heat production. On initial exposure to even mild cold stress, narrowing of the peripheral blood vessels to the skin helps to conserve heat and maintain internal (core) body temperature. This process of vasoconstriction is controlled by the sympathetic nervous system and results in a drop in skin temperature and a sharp reduction in the rate of heat loss to the environment. If the cold stress persists for sufficiently long, the continued drop in skin temperatures increases the risks of tissue damage (frostbite). In these circumstances, the body must ultimately respond by opening the peripheral blood vessels (vasodilation) to increase skin temperatures.

The alternation of vasoconstriction with vasodilation is a characteristic response to cold stress known as the "Lewis Hunting Phenomenon," and reflects the tradeoffs between minimizing heat loss and maintaining sufficient skin temperature to prevent frost bite (Lewis, 1930). Figure 7 shows the process of alternating vasodilation and vasoconstriction in response to placing one's hand in an ice bath. The key parameters that are measured to evaluate variation in cold responses in this protocol are: (a) onset time, the time from initial immersion to the onset of vasodilation; (b) T_{min}, the minimum skin temperature reached before initiation of vasodilation; (c) T_{max}, the maximum skin temperature reached following cold-induced vasodilation; and (d) T_{mean}, the average skin temperature during the test, excluding the onset time (Daanen, 2003). Note that skin temperature initially declines from roughly 32 to 3 °C (T_{min}) during the first 9–10 min of

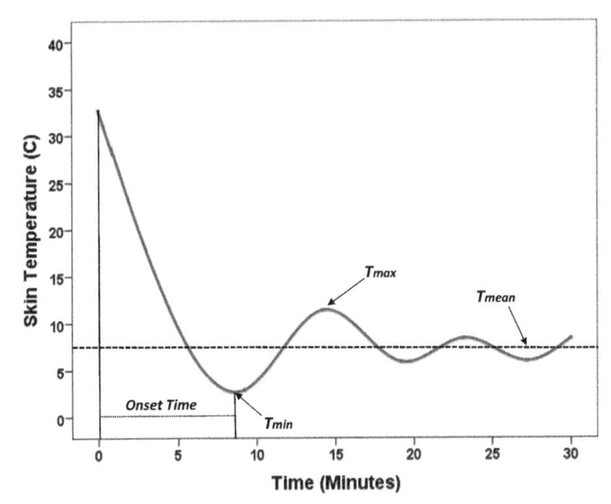

FIGURE 7 Changes in skin temperature (°C) in response to finger immersion in an ice bath. The fluctuations in skin temperature reflect the alternation of vasodilation and vasoconstriction ("Lewis Hunting Phenomenon") that is typical in response to this cold challenge. The onset time is the time from initial immersion to the start of vasodilation (9–10 min). T_{min} is the minimum skin temperature reached before initiation of vasodilation (3 °C); T_{max} is the maximum skin temperature reached following cold-induced vasodilation (11 °C), and T_{mean} is the average skin temperature during the test, excluding the onset time (7.2 °C). *Adapted from Daanen (2003).*

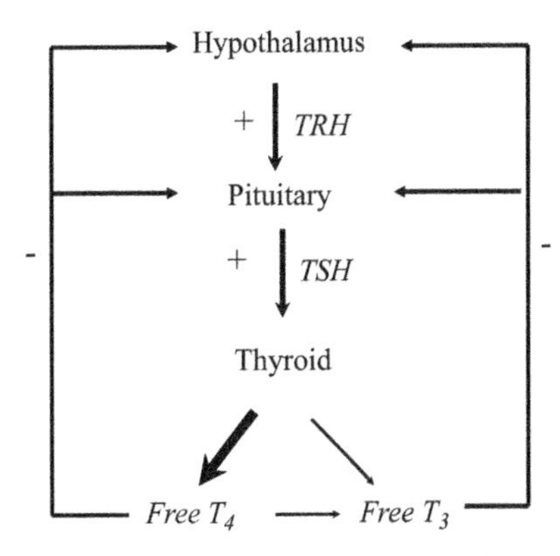

FIGURE 8 Schematic depiction of the regulation of thyroid metabolism. The thyroid gland secretes two types of thyroid hormone: tri-iodo thyronine (T_3) and thyroxine (T_4). Secretion of T_3 and T_4 is regulated by thyroid-stimulating hormone (TSH; or thyrotropin) from the pituitary. In turn, TSH is controlled by tryrotropin-releasing hormone (TRH), which is produced by the hypothalamus. Most of the thyroid hormone secreted by the thyroid gland are in the form of thyroxine. Thyroxine can be converted to T_3 in the peripheral tissues. Low levels of T_3 and T_4 in the bloodstream signal the hypothalamus and pituitary to stimulate hormone production in the thyroid gland. Conversely, adequate circulating levels of T_3 and T_4 signal the brain to shut down hormone production ("negative feedback").

the cold challenge. Vasodilation then results in an 8 °C rise in skin temperature over the next 5 min, to ~11 °C (T_{max}). Skin temperatures then fluctuate with smaller amplitudes (between 5.5 and 8.5 °C) for the remainder of the test, reflecting the changes in blood flow. The average skin temperature after the initial onset of vasodilation is 7.2 °C (T_{mean}).

When these initial vascular mechanisms of heat conservation are not sufficient to counteract heat loss, the body must respond by increasing heat production through shivering. Declines in skin temperature result in the stimulation of movement of the skeletal muscles to increase metabolic rates. While shivering is an effective way of increasing short-term heat production, it is metabolically inefficient due its high energetic costs (two to three times basal metabolic rates; Glickman et al., 1967) and rapid rates of heat loss. Consequently, as individuals acclimate to cold conditions, the time to initiation of shivering is delayed, and alternative avenues of "nonshivering thermogenesis" (NST) are enhanced.

NST refers to cold-induced increases in metabolic heat production that are not due to increased skeletal muscle activity. That is, rather that increasing metabolism through movement or activity, NST enhances heat production through neuroendocrine pathways that stimulate energy turnover at the cellular level. The hormones that are most critical for regulating NST are norepinephrine (NE), released from the sympathetic nervous system, and thyroid hormones, released from the thyroid gland.

Thyroid hormones play a critical role in regulating energy expenditure because they promote oxidative metabolism in most cells (Hadley, 1996). As shown in Figure 8, the thyroid gland secretes two types of thyroid hormone: tri-iodo thyronine (T_3) and thyroxine (T_4). Secretion of these hormones is regulated by thyroid-stimulating hormone (TSH; or thyrotropin) from the anterior pituitary. In turn, TSH is controlled by tryrotropin-releasing hormone (TRH) produced by the hypothalamus. Low levels of T_3 and T_4 in the bloodstream signal the hypothalamus and pituitary to stimulate hormone production in the thyroid gland. Conversely, adequate circulating levels of T_3 and T_4 signal the brain to shut down hormone production ("negative feedback").

The vast majority of the thyroid hormone secreted by the thyroid gland are in the form of thyroxine. Thyroxine can be converted to T_3 in the peripheral tissues. In the blood, most T_3 and T_4 are bound to proteins; however, it is only the free (or unbound) hormones that are biologically active (Hadley, 1996; Hardy, 1981).

Thyroid hormones increase energy expenditure, in part, by reducing metabolic efficiency at the cellular level. Although the specific molecular mechanisms are not fully resolved, current evidence indicates that thyroid hormones elevate cellular metabolic rates by uncoupling electron transport from the production of adenosine triphosphate (ATP) within the mitochondria. At the level of the mitochondrion, this "uncoupling" of oxidative phosphorylation results in reduced production of ATP (chemical energy to be used for cellular work) and increased heat production. Recent work suggests that thyroid hormones act partly by influencing transcription of genes coding for proteins that regulate the uncoupling process (e.g., "uncoupling proteins"; de Lange et al., 2001; Baccari et al., 2004).

With repeated exposure to cold, tissue uptake of free thyroid hormones dramatically increases, thus enhancing the capacity for metabolic heat production. Both laboratory and field studies have shown that prolonged exposure to cold stress results in declines in free T_3 and T_4 levels (Reed et al., 1990a,b; Pääkkönen and Leppäluoto, 2002). Under extreme arctic and Antarctic conditions, these changes are particularly dramatic, producing a distinctive set of physiological changes known as the "polar T3 syndrome" (see Reed et al., 1986, 1990a,b; Palinkas and Suedfeld, 2008). This phenomenon is discussed in more detail in the following section.

A primary site for NST is a special type of fat known as brown adipose tissue (BAT). Brown fat is distinct from normal (white) fat in having the ability to increase metabolic heat production at the cellular level. Under conditions of cold exposure, signals to increase metabolic rate are sent through the sympathetic nervous system to the BAT. The sympathetic nerves release NE that promotes increased heat production through enhanced fatty acid breakdown and greater uncoupling within the BAT (Cannon and Nedergaard, 2004, 2012).

In humans, BAT plays a critical role in regulating metabolic responses among infants. Because of their small size and high surface area to weight ratios, human infants are particularly vulnerable to cold stress. Consequently, BAT levels are highest early in life and help to protect against hypothermia and cold injury. Classic research done during the 1960s and 1970s showed that under conditions of even mild cold stress (15 °C), stimulation of BAT in infants produced a doubling of metabolic rates (Adamsons et al., 1965; Ahern and Hull, 1966; Brooke et al., 1973).

Until recently, it was generally thought that BAT was lost by adulthood and thus was only important in regulating responses to cold stress in infants. However, research by Nedergaard and colleagues (2007) documented the presence of significant amounts of BAT in adults using positron emission tomography (PET), a medical imaging technique that measures rates of blood flow and glucose update. Figure 9 shows a PET scan documenting the presence of metabolically active BAT in the neck, above the collar bone (the supraclavicular region).

Since 2007 there has been considerable research performed to investigate (a) variation in amounts of BAT found in adult humans (Cypress et al., 2009) and (b) the extent to which BAT can influence metabolic rates in adults (e.g., Muzik et al., 2013; van der Lans et al., 2013). It appears that BAT levels vary considerably in adults, ranging from none or trace amounts to maximum levels of approximately 170 g (Cypress et al., 2009). To date, research has shown clearly that BAT can be stimulated by cold to increase metabolic rates; however, results vary on how much extra energy this is. Among individuals with significant amounts of BAT, modest cold exposure results in a 15–20% (250–300 kcal/day) increase in metabolic rates (Muzik et al., 2013; van der Lans et al., 2013); however, individuals with little or no BAT show no increase in energy expenditure with cold exposure. These results suggest that stimulation of nonshivering thermogenesis through regular exposure to mild cold has the potential to increase daily energy budgets by significant amount (Lichtenbelt et al., 2014).

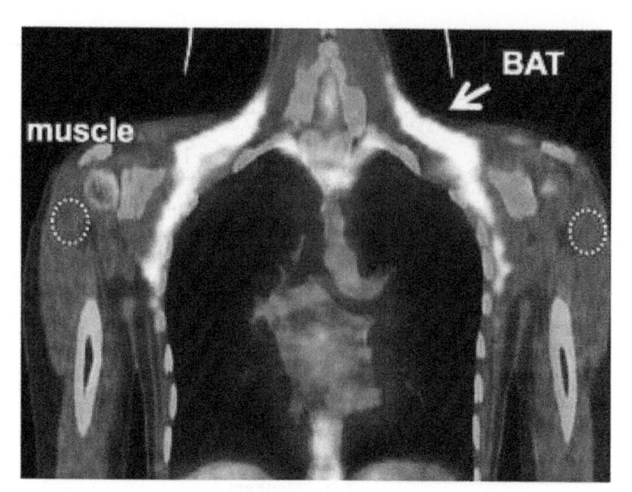

FIGURE 9 Positron emission tomography (PET) scan showing evidence of active brown adipose tissue (BAT) in the neck (supraclavicular) region of an adult female subject. The white/yellow colors in the neck indicate high levels of blood flow and heat production in the BAT in response to a mild, whole body cold challenge (15.5 °C). *Adapted from Muzik et al. (2013).*

In sum, the process of acclimation to cold stress involves the transition from relatively inefficient mechanical strategies for producing heat (shivering thermogenesis) to increased reliance on enhanced biochemical pathways for increasing metabolism (NST). The changes in response patterns are outlined in the flow diagrams in Figure 10. In the initial stages of cold acclimation (Figure 10(A)), the body favors vasoconstriction and delays vasodilation, attempting to limit heat loss. This response results in sharp declines in skin temperature that ultimately stimulate the shivering reflex as a mechanism for generating metabolic heat.

With greater acclimation, vasodilation occurs earlier, allowing for maintenance of higher peripheral temperatures and the dampening of the shivering responses (Figure 10(B)). This increased heat loss necessitates greater reliance on biochemical pathways for enhancing heat production. This is achieved through the increased production and uptake of thyroid hormones and the stimulation of BAT by NE.

Responses among Indigenous Arctic Populations

Indigenous populations of arctic (circumpolar) regions show enhanced physiological responses to cold stress. As noted above, cold-adapted populations such as the Inuit

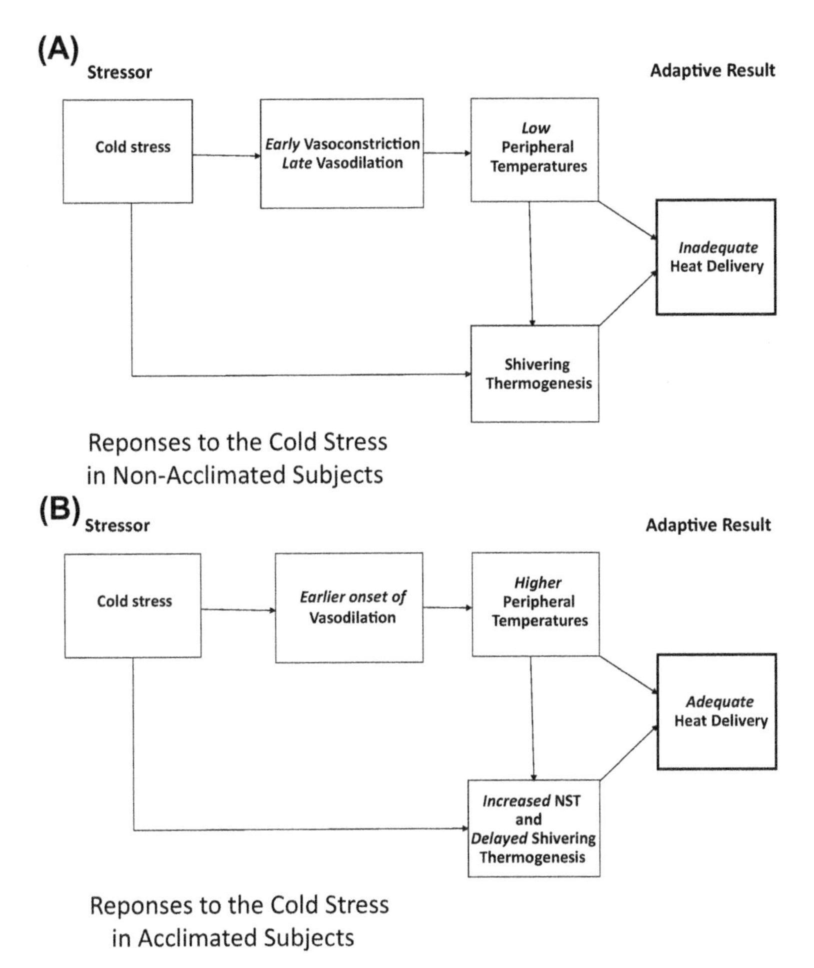

FIGURE 10 Schematic depiction of physiological responses to cold stress in (A) nonacclimated and (B) fully acclimated subjects. In nonacclimated subjects, delayed vasodilation results in a limited delivery of heat to the periphery, low skin temperature, and early onset of shivering as the primary method of heat production. Overall, these responses result in inadequate heat delivery and low tolerance to cold stress. With full acclimation, vasodilation occurs earlier, promoting greater heat delivery to the periphery and maintenance of higher skin temperatures. Higher skin temperatures are associated with delayed shivering and increased reliance on nonshivering thermogenesis (NST). This suite of responses promotes greater heat production and delivery, and better cold tolerance.

and Yakut have stocky body builds (heavy with large trunks and short arm and leg lengths) with a low surface area-to-weight ratios that reduce heat loss to the environment. These populations also show enhanced capacity for metabolic heat production in response to their cold and marginal climes. Research among North American Inuit populations has shown that when exposed to whole body cooling, these indigenous subjects display greater increases in metabolic rate and maintain higher core temperatures than their nonindigenous counterparts (Adams and Covino, 1958). Similarly, Inuit subjects also maintain higher skin temperatures and greater peripheral blood flow in response to localized cold challenges to the hands or feet (Adams and Covino, 1958; Brown and Page, 1953). The enhanced vascular responses appear to reflect developmental acclimatization, since Inuit children do not show the same pattern of increased peripheral skin temperature and blood flow that is observed in adults (Miller and Irving, 1962; Frisancho, 1993).

Indigenous circumpolar groups also have elevated BMR compared to populations from temperate and tropical climates. Early studies found that BMRs of Inuit populations of Alaska and Canada were 25–35% higher than reference values (Heinbecker, 1928, 1931; Crile and Quiring, 1939; Adams and Covino, 1958). However, these initial findings were questioned because of relatively small samples and lack of control for the potentially confounding effects of anxiety, diet, and body composition (see Rodahl, 1952). More recent research, with controlled measurement conditions, has largely confirmed the early studies in documenting elevated BMRs among indigenous arctic groups in both North America and Asia (Galloway et al., 2000; Leonard et al., 2002, 2005, 2014; Rode and Shephard, 1995; Snodgrass et al., 2005).

Figure 11 shows the relationship between BMR and FFM in a sample of 133 men and 179 women from indigenous populations of North America (Inuit) and Siberia (Evenki, Buryat, and Yakut). The BMRs of the arctic groups are substantially

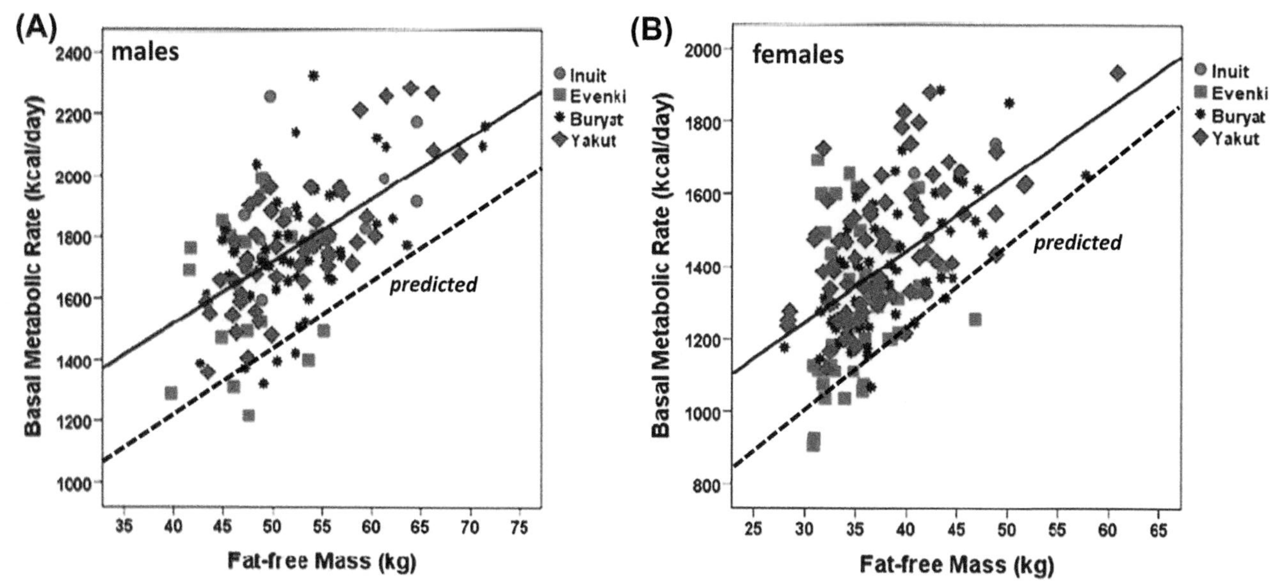

FIGURE 11 Basal metabolic rate (kcal/day) vs. fat-free mass (FFM, kg) among (A) men and (B) women from indigenous arctic populations (North American, Inuit; Siberian, Evenki, Buryat, Yakut). The solid lines denote the best fit regression lines for BMR versus FFM, while the dotted lines denote the predicted BMR values based on the equation of Cunningham (1991). Measured BMR values are significantly greater than those predicted based on FFM in both men and women (+18.2 and +18.3%, respectively; $P < 0.001$), indicating greater rates of metabolic heat production in these cold-adapted populations. (See color plate section).

elevated above those predicted from FFM based on the reference equation of Cunningham (1991). Among the men of the circumpolar groups, BMRs are, on average 1763 kcal/day, 18% higher than the mean predicted value of 1494 kcal/day ($P < 0.001$). Circumpolar women show similar metabolic increases (~18%), with average BMRs of 1396 kcal/day, compared to the predicted value of 1183 kcal/day ($P < 0.001$).

Recent work has also shown that populations living in extreme circumpolar environments show significant seasonal changes in BMR. Yakut men and women of northeastern Siberia experience a 6% increase in BMR during the winter, when temperatures regularly reach −50 to −60 °C (Leonard et al., 2014). This level of increase is similar to or greater than those reported by Kashiwazaki (1990) for populations of northern Japan (7%) and Plasqui et al. (2003) among a sample of Dutch students (4%).

Both the elevations and the seasonal changes in BMR observed in arctic populations appear to be attributable, in part, to alterations in thyroid function. The severe cold and short day lengths of arctic winters produce a distinctive set of physiological changes known as the polar T3 syndrome (see Reed et al., 1986, 1990a,b; Palinkas and Suedfeld, 2008). As shown in Figure 12, this suite of responses is characterized by increased rates of uptake of thyroid hormones exceeding production during the winter, leading to reduced circulating levels of free T_3 (Reed et al., 1990a,b; Harford et al., 1993; Levy et al., 2013). Recent work suggests that the wintertime declines in thyroid hormones are greater in indigenous, long-term arctic residents compared to short-term sojourners (Levy et al., 2013), thus providing a mechanism for the significant increases in wintertime BMRs.

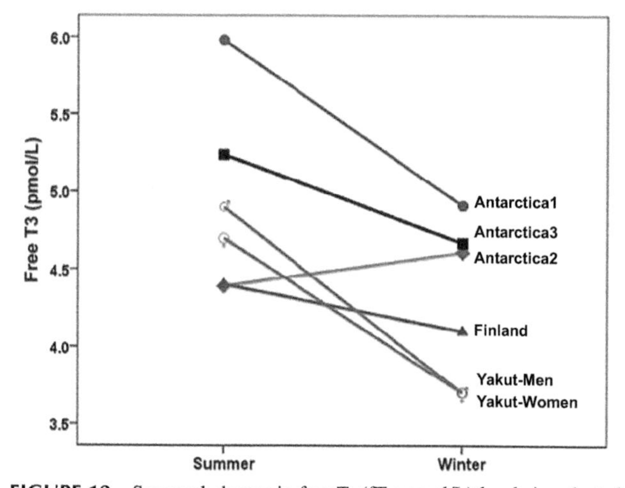

FIGURE 12 Seasonal change in free T_s (fT_3; pmol/L) levels in selected circumpolar groups. These groups are characterized by a pattern known as "Polar T_3 Syndrome"—marked wintertime declines in fT_3 levels associated with tissue update of thyroid hormones exceeding production. The seasonal changes are most pronounced in the indigenous, Yakut population, who show declines of 21 and 24% in men and women, respectively, as compared to the average declines of 8% in the three studies of Antarctic sojourners (all men) and declines of 7% in men of northern Finland. (See color plate section). *Adapted from Levy et al. (2013).*

In addition to alterations in thyroid function, preliminary evidence suggests that indigenous arctic populations may have elevated levels of BAT to enhance metabolic responses to cold. Levy et al. (2014) used thermal imaging to identify the presence of active BAT and measure the metabolic responses to cold stress in a sample of Yakut adults. They found that higher temperatures in the supraclavicular region were associated with increased metabolic rates.

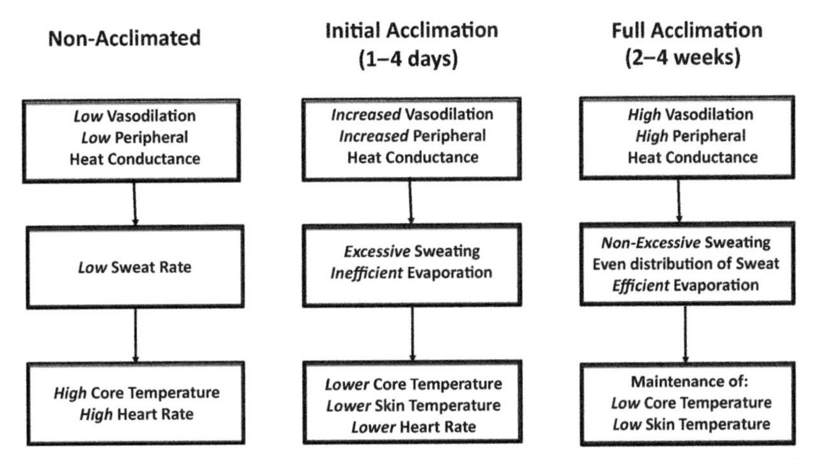

FIGURE 13 Schematic depiction of physiological responses to exercise during heat stress in nonacclimated, partially acclimated, and fully acclimated subjects. Nonacclimated subjects show low vasodilation, limited peripheral heat conductance, and low sweat rates. Together, these responses result in high core temperatures, elevated heart rates, and low tolerance to heat stress. During the initial stages of acclimation there is increased vasodilation, allowing for greater peripheral heat conductance. Sweat rates increase to excessive levels, resulting in inefficient evaporative cooling. Overall, core temperature, skin temperature, and heart rate are all reduced from the nonacclimated levels, producing greater heat tolerance. With full acclimation, maximal vasodilation is observed, producing high peripheral heat conductance. Sweat rates stabilize, with sweat being more evenly distributed over the body to enhance evaporation. These responses produce lower skin and core temperatures and result in maximal heat tolerance during exercise. *Adapted from Frisancho (1993).*

Recent genetic analyses have identified additional potential pathways for increasing metabolic heat production among indigenous circumpolar groups. Mishmar et al. (2003) and Ruiz-Pesini et al. (2004) examined global variation in mitochondrial DNA (mtDNA) genes associated with uncoupling of oxidative phosphorylation (e.g., ATP6). These studies found that the mtDNA lineages common to central Asia and Siberia were associated with greater uncoupling and increased metabolic heat production. They hypothesize that natural selection for mutations in mtDNA genes that increased uncoupling of oxidative phosphorylation was critical to the initial expansion of the human population into northern climes of Siberia and other parts of northern Eurasia. These findings imply that the metabolic influence of thyroid hormones is *greater* among indigenous Siberians who have genotypes associated with greater uncoupling in the mitochondria. Further research is needed to explore the influence of mtDNA haplotypes on variation in thyroid hormone levels and BMR.

PHYSIOLOGICAL ADAPTATIONS TO HEAT STRESS

Adaptation to heat stress predominantly involves physiological strategies to facilitate heat dissipation to the environment. Much of what we know about acclimation to heat stress is based on research in exercise and work physiology that has explored how heat tolerance shapes physical performance and health. Additionally, classic work in human biology and environmental physiology has documented a distinctive pattern of acclimatization among indigenous populations of the tropics.

Acclimation to Heat Stress

The classic studies of acclimation to heat stress have typically had subjects exercise in hot–dry or hot–humid conditions at standardized workloads for 1–2h per session for 10–24 days (e.g., Lind and Bass, 1963; Mitchell et al., 1976). During the exercise tests, a variety of physiological parameters are measured, including heart rates, sweat rates, skin temperature, core temperature, and rates of energy expenditure. Nonacclimated subjects show low sweat rates and limited ability to dissipate metabolic heat production through peripheral vasodilation. Consequently, they show rapid increases in core temperature and heart rate in response to exercise.

Figure 13 summarizes the physiological changes that occur at different stages of acclimation to heat stress. During the first 4 days of training, subjects show increased vasodilation and enhanced conductance of heat from the core to the periphery. At the same time sweat production increases dramatically; however, these excessive rates of sweating are relatively ineffective for cooling the body since only a small proportion of the sweat (roughly 10%) is evaporated (Frisancho, 1993). These high sweat rates also lead to excessive loss of water and electrolytes (McArdle et al., 2001). This partial level of acclimation is associated with lower skin and core temperatures and a reduction in heart rate.

Full acclimation to heat stress generally requires 2–4 weeks of training, and is associated with more efficient sweating and higher levels of vasodilation and peripheral heat conductance. Over the course of the training period, sweating capacity increases to almost twice baseline levels. Sweating is initiated at lower skin temperatures and is more evenly distributed over the body, thus facilitating higher rates of evaporation. Additionally, the sweat is more dilute in electrolytes, thus reducing sodium loss from the kidneys (McArdle et al., 2001). Together, the higher rates of peripheral heat conductance and more efficient sweating contribute to the maintenance of lower skin and rectal temperatures.

A number of factors contribute to variation in tolerance to heat stress; these include age, body size/shape, body fatness, and physical fitness levels. Both children (under the age of ~11 years) and older adults (greater than 46 years) have reduced heat tolerance compared to adolescents and younger adults (Lind et al., 1970; Wagner et al., 1972). Reduced heat tolerance in children is likely the result of incomplete development of the cardiorespiratory system. Similarly, increased sensitivity to heat stress in the older adults appears to be the consequence of declines in overall cardiovascular health (Frisancho, 1993).

Heavier individuals also have a more difficult time responding to heat stress, a pattern that reflects the differential changes in heat production and heat loss with increasing body mass. As discussed previously, metabolic heat production increases as a function of body weight (see Table 1), whereas the ability to dissipate heat—reflected in SA/weight ratios—*declines* with greater body mass. Consequently, heavier subjects are less able to handle increases in metabolic heat production, particularly under conditions of moderate to intense exertion.

Higher levels of body fatness also impair heat tolerance. Subcutaneous body fat provides greater insulation, and thus reduces the body's ability to dissipate heat from the core to the periphery. Reduced heat tolerance of overweight and obese subjects also reflects the fact that their SA/weight ratios are lower than their leaner counterparts.

Finally, greater physical fitness and cardiovascular health are associated with both improved heat tolerance and more rapid acclimation to heat stress. Aerobic fitness is commonly measured as "maximal oxygen consumption" (VO_{2max}). Acclimated subjects with higher VO_{2max} levels maintain lower core temperatures and have greater stamina while exercising under hot conditions (Wyndam et al., 1970). Higher aerobic capacity is associated with greater ability to expand blood volume to working muscles (Senay et al., 1976). This results in greater heat tolerance by enhancing the flow of warm blood from the core to the periphery during intense exercise.

Responses among Indigenous Populations of the Tropics

During the middle part of the twentieth century, several studies were carried out among indigenous populations of Africa (e.g., San, Bantu; Wyndham et al., 1964a,b) and Australia (Wyndham et al., 1964c) that documented their distinctive responses to heat stress. This research showed consistent differences between the indigenous and the acclimated nonindigenous subjects in tolerance to heat stress under exercise conditions. Figure 14 compares the physiological responses (core temperature, heart rate, and sweat rate) of indigenous Australians to those of acclimated, nonindigenous "control" subjects in responses to 4h of moderate intensity exercise under hot conditions (33 °C) (from Wyndham et al., 1964c). Overall, the indigenous subjects showed exercise capacity that was equal to or greater than that of their nonindigenous counterparts, despite having significantly lower sweat rates and maintaining higher core

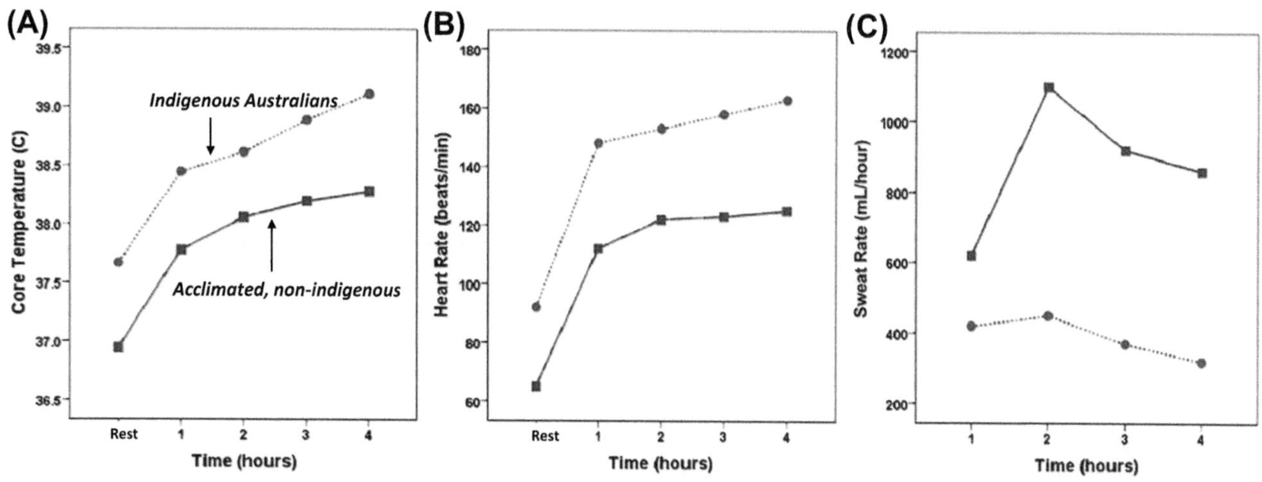

FIGURE 14 Physiological responses—(A) core temperature (C), (B) heart rate (beats/min), and (C) sweat rate (mL/h)—to 4h of moderate exercise under hot (33 °C) conditions in indigenous Australians and acclimated, nonindigenous subjects. The indigenous subjects showed exercise performance that was comparable to or better than that of the nonindigenous subjects despite maintaining higher core temperatures, higher heart rates, and much lower sweat rates. Over the duration of the test, sweat rates of the indigenous subjects averaged less than half that of their nonindigenous peers (406 vs. 890 mL/h; $P < 0.01$). The lower body weights (56.6 vs. 70.2 kg) and more linear body builds of the indigenous subjects allowed them to more effectively dissipate heat through passive, nonevaporative mechanisms. *Adapted from Wyndham et al. (1964c).*

temperatures. The concentrations of sodium and other minerals were also much lower in the indigenous subjects.

Frisancho (1993) has argued that this distinctive response pattern of indigenous tropical populations is the product of adaptation to both heat stress and limited nutritional resources during growth and development. These joint stressors contribute to the low adult body mass and linear body proportions that are typical of tropical populations throughout the world. Additionally, lifelong constraints on food and water availability appear to limit sweating capacity in these groups. Consequently, indigenous tropical populations rely much more heavily on passive, nonevaporative mechanisms of heat loss than their nonindigenous peers. Higher SA:weight ratios facilitate greater heat loss through radiation and convection, thus allowing for thermal homeostasis to be maintained with much lower sweat production.

Populations of the tropics also have reduced BMRs compared to temperate and circumpolar groups (Roberts, 1952, 1978; Henry and Rees, 1991; Soares et al., 1993; Froehle, 2008). Early work by Roberts (1952) provided systematic evidence of low and very low BMRs in tropical populations of India, Java, Australia, and South America. However, this research assessed variation in metabolic rates using simple ratios of BMR/weight or BMR/SA, approaches that have since been shown to produce spurious results (Poehlman and Toth, 1995; Weinsier et al., 1992).

More recent analyses have used standard regression approaches (BMR versus weight) to evaluate the evidence for greater metabolic economy among indigenous tropical populations (Henry and Rees, 1991; Soares et al., 1993; Froehle, 2008). Henry and Rees (1991) compiled 2822 BMR measurements from tropical populations around the world to develop a new set of age- and sex-specific predictive equations for these groups (see Table 3). These equations give BMR estimates that are approximately 8–11% below those derived from the standard WHO equations presented in Table 1.

Yet, even as there is a growing consensus that metabolic rates are reduced among indigenous populations from tropical climates, most researchers now believe that this pattern is largely the result of conditions of poverty and malnutrition, rather than adaptation to heat stress. Chronic undernutrition contributes to depressed metabolic rates through reductions in thyroid function and lean body mass (Soares and Shetty, 1991; Soares et al., 1992; Shetty, 1993; Ferro-Luzzi et al., 1997). Conditions of negative energy balance produce declines in T_3 production and uptake, which, in turn, result in marked reductions in cellular heat production (Keys et al., 1950; Shetty, 1996). Long-term nutritional stress also results in low BMI and reduced FFM. Thus, the low BMRs of tropical populations also reflect the fact that these groups are "undermuscled" for their weight.

TABLE 3 Henry and Rees (1991) Predictive Equations for Estimating Basal Metabolic Rate (BMR; kcal/day) in Adults of Tropical Populations

Reference	Sex	Age (year)	Equations[a]
Henry and Rees (1991)	M	18–29	13.4(Wt)+669
		30–60	11.0(Wt)+755
	F	18–29	11.5(Wt)+612
		30–60	11.5(Wt)+585

[a]*Wt=weight (kg).*

In sum, the distinctive physiological responses among indigenous tropical populations both enhance heat loss and reduce metabolic heat production. Long-term acclimatization to tropical climates clearly reflects adjustments to the joint effects of both heat stress and chronic undernutrition. During growth and development, these dual stressors result in the development of adult physiques (low body mass, linear body build) that facilitate passive avenues of heat loss, allowing for the maintenance of thermal homeostasis with relatively low rates of sweat production. Indigenous tropical populations also appear to have lower rates of metabolic heat production than populations of other climatic zones; however, these low BMRs seem to be primarily the consequence of nutritional rather than heat stress.

PHYSIOLOGICAL ADAPTATIONS TO HIGH ALTITUDE HYPOXIA

High altitude environments expose human populations to numerous environmental constraints, including cold, high solar radiation, and limited nutritional resources. However, the most pervasive and distinctive stressor at altitude is "hypobaric hypoxia," the reduction in oxygen availability due to lower atmospheric pressure at higher elevations. Unlike other ecological stressors, hypoxic stress cannot be ameliorated by cultural or behavioral responses alone. Consequently, human populations display a diverse range of physiological, developmental, and genetic adaptations to high altitude hypoxia.

Table 4 highlights the nature of hypoxic stress in high altitude environments. While the percentage of oxygen in the air is the same at all elevations (~20.9%), the pressure exerted on the air is lower at higher altitudes, making a given volume of air (e.g., 1 L) less dense. As a result, the body takes in fewer oxygen molecules with each breath in high altitude environments. The key measure of this reduced oxygen availability at altitude is the partial pressure of oxygen (PO_2), the portion of air pressure attributable to the O_2 component. PO_2 is calculated simply as: (20.9%)×(barometric pressure).

At sea level, atmospheric pressure is about 760 mm Hg, resulting in a PO_2 of 159 mm Hg. In moving to an elevation of 2500 m (8250 ft), the widely used "threshold" for

TABLE 4 Effects of Altitude on Barometric Pressure (mm Hg), Partial Pressure of Oxygen (PO_2; mm Hg), and Arterial Oxygen Saturation (SO_2; %)

Altitude (m)	Feet	Bar Pressure (mm Hg)	PO_2 (mm Hg)	SO_2 (%)
Sea level	0	760	159	98
1000	3280	674	141	
2500	8250	560	117	
3000	9840	526	110	90
4000	13,120	462	97	
6000	19,690	354	74	70
9000	29,530	230	48	20

defining high attitude environments, pressure drops to about 73% of sea-level values (117mm Hg). At 6000m (19,690ft), oxygen availability is less than half that of sea-level values (PO_2=74mm Hg).

Declining oxygen pressure at greater elevations results in a fewer of the body's hemoglobin molecules being bound with oxygen. Arterial oxygen saturation (SO_2) is the percentage of hemoglobin that is bound with oxygen. As shown in Table 4, at sea level ~98% of arterial hemoglobin is saturated with oxygen. When a sea-level resident ascends to altitude, SO_2 values decline, resulting in a reduced oxygen carrying capacity of blood. At an elevation of 3000m, SO_2 levels are about 90%, whereas elevations of 5500–6000m are associated with ~30% reductions in SO_2.

Given that oxygen is critical for cellular metabolism, hypoxic stress affects all systems of the body. Physiological responses to combat hypoxic stress and enhance oxygen uptake are typically divided into three major categories: (1) ventilatory, (2) diffusive, and (3) hematological. Ventilatory responses are those associated with changes in the rate and depth of breathing. Diffusive responses, in contrast, represent alterations in the transfer of oxygen from the lungs to the blood stream (pulmonary diffusion) or from the blood into the peripheral tissues (peripheral diffusion). Finally, hematological responses are changes in composition and chemistry of blood (especially the red blood cells) to alter its oxygen carrying capacity. Each of these three domains works in concert with one another to alleviate hypoxic stress; however, as outlined below, short-term and long-term residents to high altitude rely on a different "mix" of responses. Moreover, even among indigenous high altitude populations (e.g., Andean and Himalayan natives), there appears to be striking differences in physiological adaptations.

Acclimation to Hypoxic Stress

On the arrival to high altitude environments, nonacclimated, sea-level residents have a relatively limited range of physiological adaptations to hypoxic stress. With the initial exposure to hypoxia, chemoreceptors in the carotid artery (carotid body) detect changes in PO_2 in the bloodstream and relay that information to neurons in the brain. In turn, the brain signals elements of the sympathetic nervous system to stimulate pulmonary and vascular responses to increase oxygen update and delivery (Prabhakar, 2000).

Stimulation of the respiratory system involves increasing both the rate and the depth of breathing (ventilation). During initial exposure to elevations of 4300m, resting ventilation (liters/minutes) has been shown to increase by over 50% (from 6.8 to 10.5L/min; Moore et al., 1987). Similar proportional changes have been observed in resting heart rates, increasing from 70 to 105 beats/min (Frisancho, 1993).

These early responses to hypoxic stress involve making the body "work harder" at rest. Consequently, we find that BMR is increased by 17–27% during the initial stages of acclimation (Butterfield et al., 1992).

In addition to being energetically expensive, these initial responses are only partially effective at alleviating hypoxic stress. Thus, the first several days at altitude are typically associated with sluggishness, greatly reduced stamina, difficulty sleeping, and reduced mental acuity (Frisancho, 1993). Headaches are also common as a result of both hypoxia and hydration due to increased water loss at altitude.

Within 1–3 weeks of exposure to altitude, the secondary set of physiological responses begins to be fully mobilized, resulting in declines in resting ventilation and heart rate, although they remain higher than sea-level values. The secondary responses to hypoxia include the stimulation of increased red blood cell production (erythropoiesis) and the opening of capillaries in the peripheral tissues. Under hypoxic conditions the kidneys secrete the hormone erythropoietin, which promotes the production of red blood cells in the marrow of the long bones. During the first 2 weeks of exposure, erythropoietic activity is about three times that of sea-level values. These enhanced rates of red blood cell production result in marked increases in hemoglobin (Hb) levels, producing a condition known as polycythemia. After approximately 6 months at altitude, Hb levels stabilize at 18–20g/dL, substantially higher than the normal sea-level values (~12–16g/dL) (Frisancho, 1993). These higher Hb levels greatly increase the oxygen carrying capacity of the blood; however, they also make the blood thicker and more viscous, placing greater strain on the heart.

The opening of peripheral capillaries also occurs over the course of acclimation to enhance the diffusion of oxygen from the blood into the mitochondria of cells.

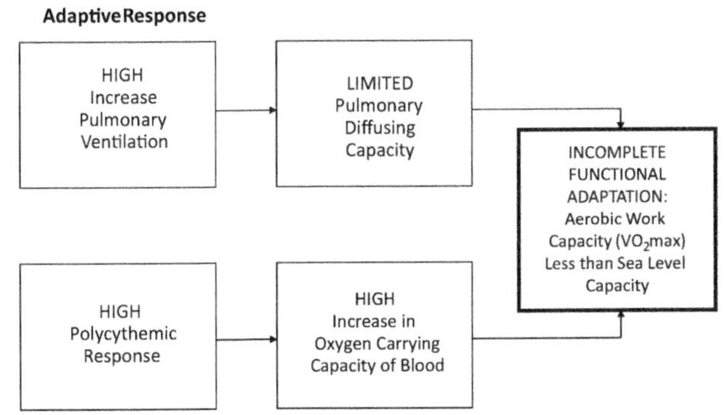

FIGURE 15 Schematic depiction of physiological responses to high altitude hypoxia in adult, sea-level sojourners to high altitude. These subjects rely primarily on increased pulmonary ventilation and a marked increase in red blood cell production to adjust to low oxygen availability. They have limited pulmonary diffusing capacity compared to high altitude natives, due to their smaller lung volumes. Overall, this response pattern results in partial functional adaptation to hypoxic stress, with subjects having lower aerobic capacities than at sea level. *Modified from Frisancho (1993).*

The rate of diffusion is related to PO_2 levels, which declines as oxygen passes through successive layers of tissues. Thus, increased capillarization reduces the distance that oxygen must travel, resulting in higher PO_2 levels and greater diffusion capacity (Frisancho, 1975). Adult sea-level sojourners to high altitude do not show significant increases in lung volume during the course of acclimation. As a consequence, pulmonary diffusing capacity remains relatively low, particularly in comparison with high altitude natives (see below).

Overall, acclimation to high altitude hypoxia among adult sea-level residents results in only partial aerobic adaptation. As shown in Figure 15, these individuals rely primarily on increased pulmonary ventilation and oxygen carrying capacity of the blood (polycythemia) and have limited pulmonary diffusing capacity relative to high altitude natives. This suite of responses results in reduced aerobic working capacity (VO_{2max}) compared to their original sea-level values. This is one of the primary differences in physical capacity between adult-acclimated subjects and those who are born and raised at altitude—those native to high altitude environments achieve adult VO_{2max} levels comparable to sea-level norms, while adult-acclimated sojourners do not (Frisancho, 1975).

Responses among Indigenous High Altitude Populations

Indigenous populations from high altitude regions around the world show a different set of adaptive responses to hypoxic stress than those observed among adult-acclimated sea-level sojourners. Pioneering research done among Quechua populations of highland Peru during the 1960s and 1970s documented the effects of hypoxic stress on growth and physiological function (Baker and Little, 1976). This work clearly demonstrated the important role of developmental acclimatization in shaping adaptive responses to high altitude hypoxia among native populations (Frisancho, 1993, 2013; Frisancho and Baker, 1970). Since the mid-1970s, extensive comparative research among other high altitude groups (e.g., Tibetans, Ethiopians) has shown remarkable diversity in adaptive patterns across populations (Beall, 2007, 2013; Moore et al., 1998). Hence, there is now substantial evidence to indicate that distinctive genetic adaptations to hypoxic stress have been selected for in different high altitude populations (see Beall, 2007; Bigham and Lee, 2014). It now appears that adaptations to hypoxic stress among indigenous high altitude populations reflect a mix of developmental and genetic responses.

All populations that grow up and develop under high altitude conditions show enlarged lung volumes in adulthood. This is evident in Figure 16, which shows variation in lung volume among adult men of Aymara, Quechua, and Tibetan populations living at similar altitudes (3600–3800m; data from Greksa et al., 1994; Frisancho et al., 1997; Droma et al., 1991). For each population, the measured lung volume is presented along with the predicted value from sea-level reference equations based on height and age (Johns Hopkins Pulmonary Function Laboratory: http://www.hopkinsmedicine.org/pftlab/predeqns.html; Goldman and Becklake, 1959). Total lung volume is divided into two component parts: (1) vital capacity, the volume of air that can be expelled from the lungs on forced exhalation after maximal inspiration, and (2) residual volume, the amount of air remaining in the lungs after forced exhalation. Note that total lung volumes in all high altitude groups are systematically greater than the predicted values based on sea-level norms. The Andean populations have lung volumes that are more than 30% greater than the predicted values (+33% for the Quechua and +46% for the Aymara), whereas the increases in the Tibetans are more modest at +12%. Additionally, relative to sea-level values, the proportional increases in residual volume are generally greater than those in vital capacity.

These expanded lung volumes are achieved through accelerated growth of lungs and chest dimensions during childhood and adolescence. This distinctive pattern of development is evident in Figure 17, which shows growth in (A) stature (height) and (B) chest circumference among Quechua boys between 6 and 19 years, compared to United States reference values (from Frisancho and Baker, 1970; Frisancho, 2008). Note that while the Quechua boys grow very slowly in height relative to their United States peers, their chest circumferences are larger than their United State counterparts across all ages. Indeed, physical development of indigenous high altitude populations is characterized by increased rates of growth of the organs associated with oxygen transport (e.g., heart, lungs), and declines in growth of the musculoskeletal

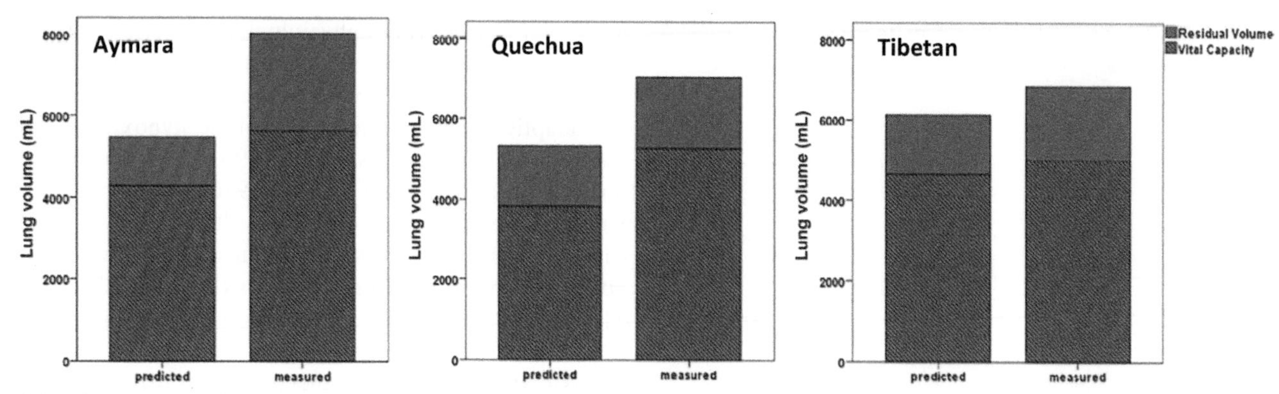

FIGURE 16 Variation in total lung volume (mL) among adult men of three high altitude populations: Aymara, Quechua, and Tibetan. For each population, the measured lung volume is presented along with the predicted value from sea-level reference equations (Johns Hopkins Pulmonary Function Laboratory: http://www.hopkinsmedicine.org/pftlab/predeqns.html). Total lung volume is composed of vital capacity (mL) and residual volume (mL). Men of all three high altitude groups have lung volumes that are larger than predicted based on sea-level norms. The increases relative to sea-level values are larger in the Andean populations (+46% in the Aymara; +33% in the Quechua) than in the Tibetans (+21%). *Data from Greksa et al. (1994), Frisancho et al. (1997), and Droma et al. (1991).*

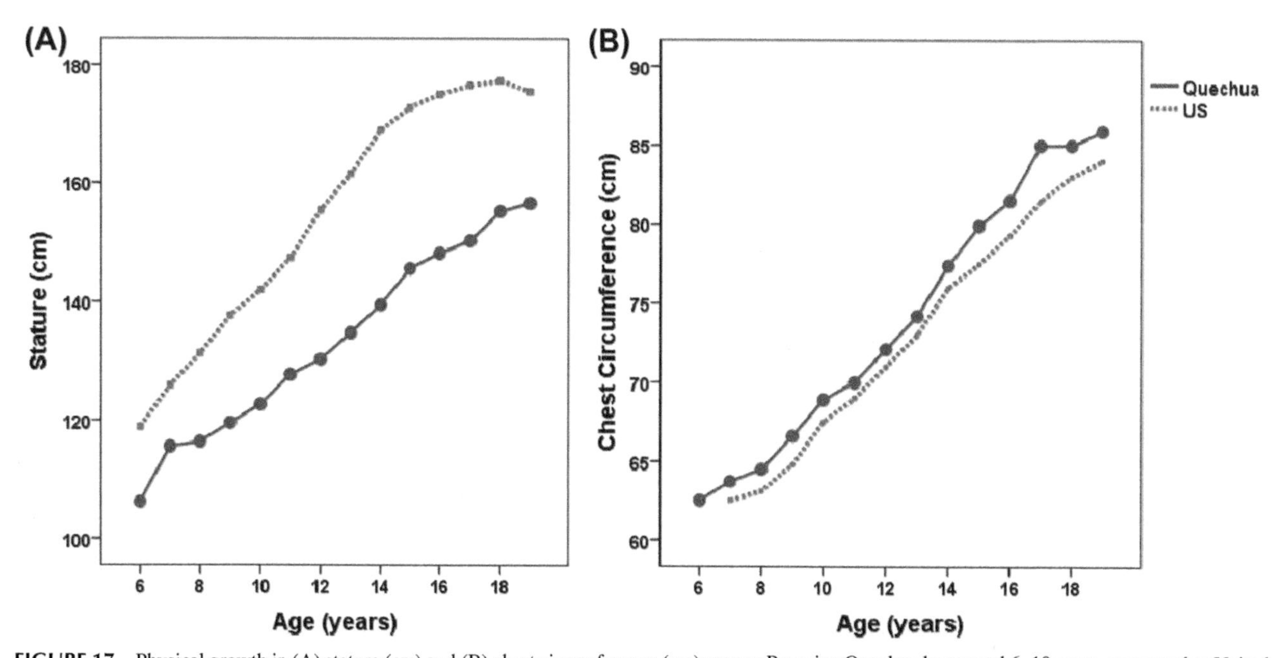

FIGURE 17 Physical growth in (A) stature (cm) and (B) chest circumference (cm) among Peruvian Quechua boys aged 6–19 years, compared to United States reference values. The Quechua boys have very slow rates of statural growth, falling well below United States norms. In contrast, growth in chest dimensions is accelerated in the Quechua boys. Despite their diminutive size, the chest dimensions of the Quechua boys are absolutely greater than those of their United States peers at all ages. The rapid rates of growth in chest dimensions and lung size in the Quechua reflect adaptation to hypoxic stress. The slow rates of growth in height are the result of both hypoxia and nutritional stress. *Data from Frisancho and Baker (1970) and Frisancho (2008).*

system (Frisancho, 1993). While the more rapid growth of lung volume and chest dimensions appears to be primarily attributable to hypoxic stress, the slow rates of growth in stature and body mass are the product of nutritional and hypoxic stress (Leonard, 1989; Leonard et al., 1990).

The development of large lung volumes in native high altitude populations results in greater surface area for diffusion of oxygen from the lungs into the blood. Within the lungs, oxygen exchange occurs in the alveoli, small balloon-like sacs at the ends of the bronchial tubes. Populations exposed to high altitude during growth appear to have more alveoli (with more surface area) and greater capillary density in the alveoli, both factors contributing to enhanced pulmonary diffusing capacity (Frisancho, 2013).

In addition, native high altitude populations do not show the exaggerated polycythemic response observed in adult-acclimated sojourners to altitude. Andean populations show moderate increases in Hb levels, while Tibetan populations have average Hb levels that are similar to sea-level values (Beall et al., 1998). These hematological responses

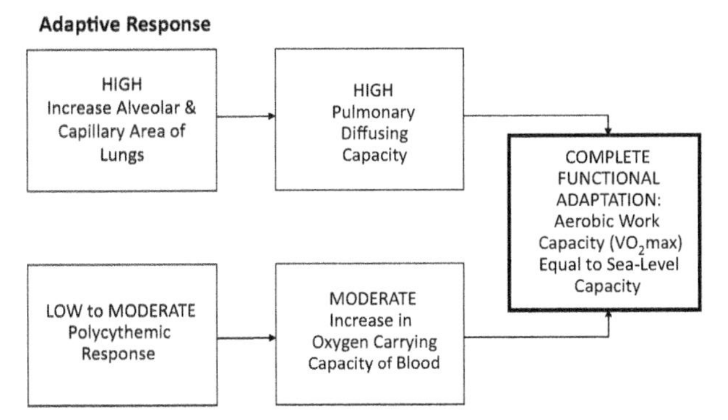

FIGURE 18 Schematic depiction of physiological responses to high altitude hypoxia in native high altitude populations. These subjects develop larger lung volumes during growth and development, and thus have high pulmonary diffusing capacity, facilitating the transfer of oxygen from the lungs into the bloodstream. Additionally, they show low-to-moderate increases in red blood cell production, resulting in a moderate rise in oxygen carrying capacity of the blood. Together, these pathways result in full functional adaptation to hypoxic stress, with subjects having aerobic capacities comparable to sea-level values. *Modified from Frisancho (1993).*

appear to produce modest increases in the oxygen carrying capacity of the blood, and thereby limit the stress on the heart and cardiovascular system due to increased viscosity of the blood.

Overall, the greater pulmonary diffusing capacity and the low-to-moderate hematological responses of highland natives allow them to achieve complete functional adaptation to hypoxic stress. These adaptive pathways are outlined in Figure 18. The larger lung volumes and enhanced pulmonary diffusing capacities permit populations raised at high altitude to have adult VO_{2max} levels comparable to sea-level values. Moore et al. (1998) found that Andean and Himalayan natives had aerobic capacities that were 12–16% higher than those of adult-acclimated sojourners.

Although all indigenous and lifelong high altitude residents are able to achieve sea-level aerobic capacities in adulthood, the specific constellation of adaptive responses differs across populations. To date, the differences between Andean (Quechua, Aymara) and Himalayan (Tibetan) populations have been most widely studied and are best understood (see Beall, 2007, 2013; Moore et al., 1998). These distinct adaptive patterns are summarized in Figure 19, which shows differences in (A) resting ventilation (L/min), (B) hypoxic ventilatory response (HVR, $\Delta L/min/\Delta SaO_2$), (C) Hb levels (g/dL), and (D) SaO_2 (%) in Aymara and Tibetan men and women living at similar elevations (3800–4065 m) (Beall et al., 1997a,b, 1998, 1999).

The most striking difference between the Andean and the Tibetan subjects is in their ventilatory responses. As shown in Figure 19(A), Tibetan men and women have significantly higher resting ventilation rates (L/min) than their Aymara counterparts. The rapid breathing rates observed in the Tibetans are higher than sea-level values, and are more similar to those observed among sea-level sojourners to altitude. In contrast, resting ventilation in the Aymara is higher than sea-level values, but substantially lower than those of the Tibetans and sea-level sojourners.

Aymara and Tibetan subjects also differ in their responses to an experimental hypoxic stress, a measure known as the HVR. Sea-level subjects display an exaggerated HVR when initially exposed to hypoxic stress, which decreases over time. Figure 19(B) shows that Tibetans have a sustained, moderate HVR, whereas the Aymara have a low or "blunted" HVR. Thus, Tibetans adapt to hypoxic stress with substantially higher breathing rates than their Aymara counterparts (Beall, 2007).

As shown in Figure 19(C), Hb levels are significantly higher in the Andean natives. Among men, Hb levels in the Aymara are 3.5 g/dL higher than in the Tibetans. The differences are similar in women, with the Aymara levels being 3.6 g/dL greater (Beall et al., 1998). Consequently, it appears that the Andean subjects have an increased erythropoietic response to hypoxic stress relative to the Tibetans (Beall et al., 1998).

The Andean subjects also show significantly higher oxygen saturation levels in the blood stream. As shown in Figure 19(D), Aymara men have SO_2 levels that are ~3% higher than their Tibetan counterparts, whereas Aymara women have levels that are ~2% greater. These results show that despite having higher rates of ventilation, the Tibetans are less able to transfer the oxygen from the lungs into the blood stream (Beall, 2007).

The differences in SaO_2 levels may be related to differences in pulmonary diffusing capacity between the Aymara and the Tibetans. While both groups show increased lung volumes over sea-level subjects, the degree of increase appears to be greater in the Aymara than in the Tibetans (see Figure 16). Additionally, Tibetans do not typically show the same "barrel-shaped" chest morphology that characterizes Quechua and Aymara highlanders (Beall, 1981). Further work is needed to more systematically compare variation in lung volume among Andean and Himalayan populations (Moore et al., 1998).

In sum, different native high altitude populations achieve complete functional adaptation to hypoxic stress through distinctive physiological pathways. Quechua and Aymara populations of the Andes have large lung volumes and expanded chest dimensions to greatly increase pulmonary diffusing capacity, and moderate-to-high Hb levels to increase the oxygen carrying capacity of the blood. In contrast, Tibetans have smaller lung volumes and lower Hb levels than Andean natives. Their adaptive strategy emphasizes higher ventilatory rates, both at rest and in response to a hypoxic challenge.

At present, the origin and nature of the differences between Andean and Tibetan populations remain unclear.

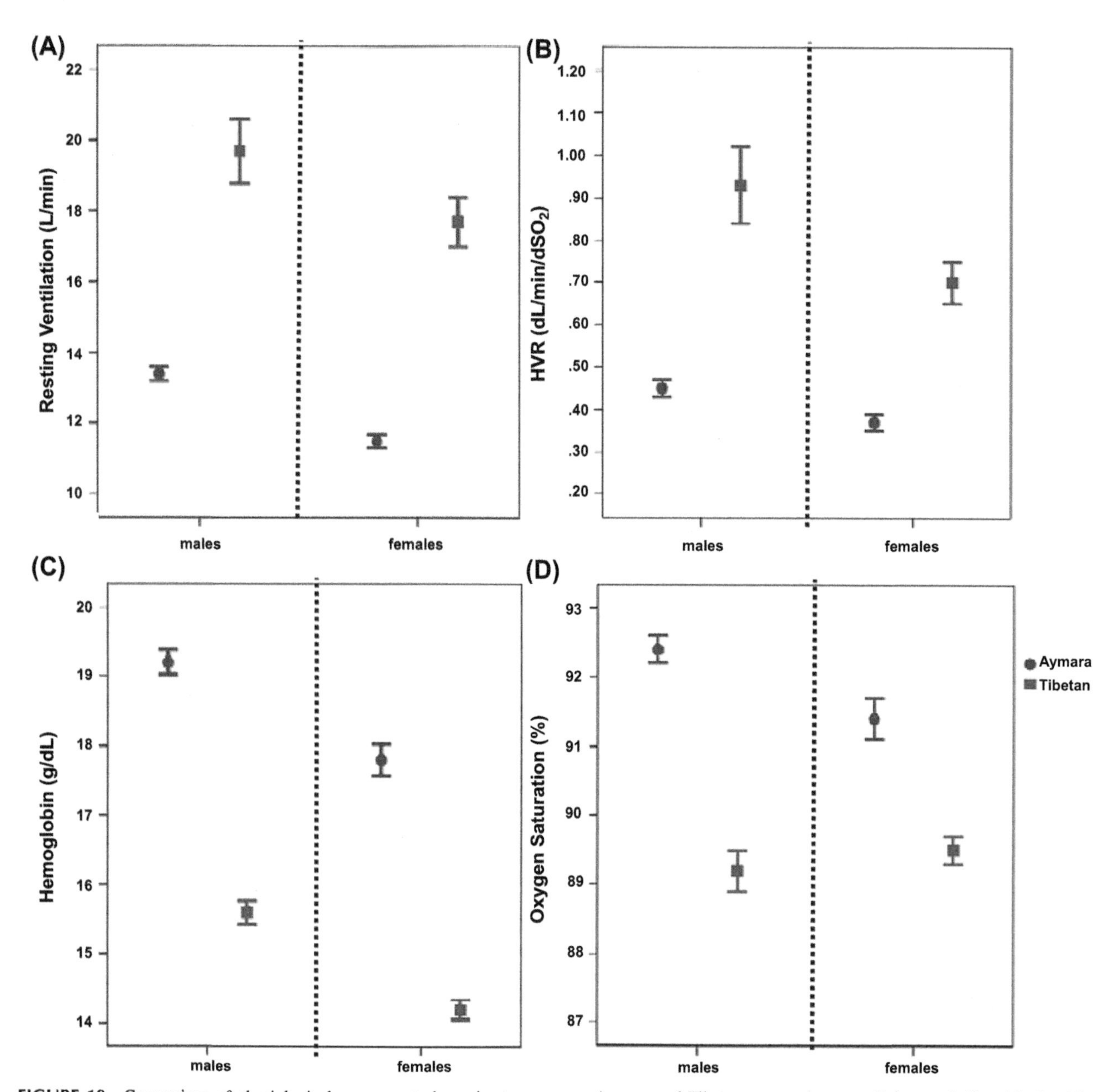

FIGURE 19 Comparison of physiological responses to hypoxic stress among Aymara and Tibetan men and women living at similar altitudes. The measures include (A) resting ventilation (L/min), (B) hypoxic ventilatory response (HVR; $\Delta L/min/\Delta SaO_2$), (C) hemoglobin (Hb) levels (g/dL), and (D) percentage oxygen saturation (SaO_2; %). Tibetan men and women have significantly higher resting ventilatory rates and HVRs than their Aymara counterparts ($P < 0.05$ for all sex-specific comparisons). In contrast, the Aymara have significantly higher Hb and SaO_2 levels ($P < 0.05$ for all sex-specific comparisons). Consequently, the Tibetan adaptive pattern emphasizes greater ventilatory responses and more moderate increases in pulmonary diffusion and red blood cell production. The Aymara, in turn, rely more heavily on enhanced pulmonary diffusion and show greater increases in red cell production. (See color plate section). *Data from Beall et al. (1997a,b, 1998, 1999).*

Some have suggested that the longer residence of Tibetan populations at altitude, ~25,000 versus 11,000 years for Andean populations, may have allowed for more time and opportunity for natural selection to operate (Moore et al., 1992). However, this explanation is problematic, given that we have abundant evidence of genetic adaptations in human populations arising in evolutionary time spans of under 10,000 years (see Lachance and Tishkoff, 2013). Thus, a more likely explanation is that in these genetically distinct high altitude populations, natural selection favored a different set of mutations, producing unique genetic adaptations in the two regions.

Ongoing research in statistical and molecular genetics is attempting to better understand the nature of the different responses observed in Andean and Tibetan populations. Quantitative genetic analyses have identified key traits in Tibetan populations (e.g., resting ventilation, SO_2 levels) whose variation is highly heritable, suggesting the opportunity for natural selection (Beall, 2007). Alternative approaches include the search for genetic loci (candidate genes) associated with adaptation to hypoxia. Particularly promising is research on transcription factors, proteins that bind to specific DNA sites and regulate the expression of different genes. The recently discovered hypoxia inducible factors (HIFs) are transcription factors that have been shown to regulate hundreds of genes involved with the maintenance of oxygen homeostasis in all metabolic processes (Semenza et al., 1991; Semenza, 2007). Thus, there is considerable interest in studying whether genes coding for transcription factors in the HIF pathway underlie the adaptive responses of both Andean and Tibetan populations (Bigham et al., 2009; Bighman and Lee, 2014).

SUMMARY

Human biologists have long recognized the important role that climatic and environmental stressors play in shaping variation in human physiology and morphology. Across the globe, human body size and proportions are strongly influenced by differences in environmental temperatures. Human populations conform to the classic "ecological rules" of Bergmann and Allen, as both body weight and RSH are inversely correlated with mean annual temperature.

Physiological adaptations to temperature stress involve strategies that influence patterns of heat loss and production. The process of acclimation to cold stress involves increasing efficiency of metabolic heat production and more effective delivery of warm blood to peripheral tissues through earlier vasodilation. Indigenous arctic populations show enhanced metabolic and vascular responses to cold. These populations display elevated BMRs and have the ability to maintain higher peripheral temperatures in response to acute cold challenges.

In contrast, acclimation to heat stress involves enhancing the capacity to dissipate metabolic heat production through greater vasodilation of peripheral blood vessels and increased sweat rates to promote evaporative cooling. Indigenous tropical populations maintain thermal homeostasis while exercising under hot conditions with much lower sweat rates than their nonindigenous counterparts. Their lower body weights and more linear physiques facilitate greater heat loss through passive, nonevaporative mechanisms (radiation and convection). Tropical populations also show reduced levels of metabolic heat productions; however, the reduced BMRs appear to be more strongly associated with nutritional stress than adaptation to heat stress.

Adaptations to high altitude hypoxia involve increasing the uptake and delivery of oxygen to peripheral tissues. Responses among sea-level visitors to altitude include increasing the rate and depth of breathing to bring more air into the lungs, and dramatically increasing Hb levels to raise the oxygen carrying capacity of the blood. These strategies are only partially successful, resulting in lower aerobic work capacities than at sea level.

Native high altitude populations show a different set of responses to hypoxic stress than those observed among adult-acclimated sea-level sojourners. Populations growing up under high altitude conditions have enlarged lung volumes and greater ability to diffuse oxygen from the lungs into the bloodstream. Additionally, they show low-to-moderate increases in Hb levels, rather than the exaggerated response observed among sea-level visitors. Overall, indigenous high altitude populations show complete functional adaptation to hypoxic stress, as evidenced by maximal working capacities that are comparable with sea-level norms.

REFERENCES

Adams, T., Covino, B.G., 1958. Racial variations to a standardized cold stress. Journal of Applied Physiology 12, 9–12.

Adamsons, K., Gandy, G.M., James, L.S., 1965. The influence of thermal factors upon oxygen consumption of the newborn human infant. Journal of Pediatrics 66, 495–507.

Adolph, E.F., 1947. Physiology of Man in the Desert. Interscience, New York.

Aherne, W., Hull, D., 1966. Brown adipose tissue and heat production in the newborn infant. Journal of Pathology and Bacteriology 91, 223–234.

Allen, J.A., 1877. The influence of physical conditions on the genesis of species. Radical Review 1, 108–140.

Antón, S.C., Leonard, W.R., Robertson, M.L., 2002. An ecomorphological model of the initial hominid dispersal from Africa. Journal of Human Evolution 43, 773–785.

Baccari, G.C., Monteforte, R., de Lange, P., Raucci, F., Farina, P., Lanni, A., 2004. Thyroid hormone affects secretory activity and uncoupling protein-3 expression in rat harderian gland. Endocrinology 145, 3338–3345.

Baker, P.T., 1966. Human biological variation as an adaptive response to the environment. Eugenics Quarterly 13, 81–91.

Baker, P.T., Little, M.A. (Eds.), 1976. Man in the Andes. Dowden, Hutchinson & Ross, Stroudsburg, PA.

Barnicot, N.A., 1959. Climatic factors in the evolution of human populations. Cold Spring Harbor Symposia on Quantitative Biology 24, 115–129.

Beall, C.M., 1981. Growth in a population of Tibetan origin at high altitude. Annals of Human Biology 8, 31–38.

Beall, C.M., 2007. Two routes to functional adaptation: Tibetan and Andean high-altitude native. Proceedings of the National Academy of Science 104, 8655–8660.

Beall, C.M., 2013. Human adaptability studies at high altitude: research designs and major concepts during fifty years of discovery. American Journal of Human Biology 25, 141–147.

Beall, C.M., Strohl, K.P., Blangero, J., Williams-Blangero, S., Decker, M.J., Brittenham, G.M., Goldstein, M.C., 1997a. Quantitative genetic analysis of arterial oxygen saturation in Tibetan highlanders. Human Biology 69, 597–604.

Beall, C.M., Strohl, K.P., Blangero, J., Williams-Blangero, S., Almasy, L.A., Decker, M.J., Worthman, C.M., Goldstein, M.C., Vargas, E., Villena, M., Soria, R., Alarcon, A.M., Gonzales, C., 1997b. Ventilation and hypoxic ventilatory response of Tibetan and Aymara high altitude natives. American Journal of Physical Anthropology 104, 427–447.

Beall, C.M., Brittenham, G.M., Strohl, K.P., Blangero, J., Williams-Blangero, S., Goldstein, M.C., Decker, M.J., Vargas, E., Villena, M., Soria, R., Alarcon, A.M., Gonzales, C., 1998. Hemoglobin concentration of high-altitude Tibetans and Bolivian Aymara. American Journal of Physical Anthropology 106, 385–400.

Beall, C.M., Almasy, L.A., Blangero, J., Williams-Blangero, S., Brittenham, G.M., Strohl, K.P., Decker, M.J., Vargas, E., Villena, M., Soria, R., Alarcon, A.M., Gonzales, C., 1999. Percent of oxygen saturation of arterial hemoglobin among Bolivian Aymara at 3900–4000 m. American Journal of Physical Anthropology 108, 41–51.

Bergmann, C., 1847. Uber die verhaltniesse der warmeokononomie der thiere zu ihrer grosse. Gottingen Studien 1, 595–708.

Bigham, A.W., Lee, F.S., 2014. Human high-altitude adaptation: forward genetics meets the HIF pathway. Genetics and Development 28, 2189–2204.

Bigham, A.W., Mao, X., Mei, R., Brutsaert, T., Wilson, M.J., Julian, C.G., Parra, E.J., Akey, J.M., Moore, L.G., Shriver, M.D., 2009. Identifying positive selection candidate loci for high-altitude adaptation in Andean populations. Human Genomics 4, 79–90.

Brooke, O.G., Harris, M., Salvosa, C.B., 1973. The response of malnourished babies to cold. Journal of Physiology 233, 75–91.

Brown, G.M., Page, J., 1953. The effect of chronic exposure to cold on temperature and blood flow of the hand. Journal of Applied Physiology 5, 221–227.

Butterfield, G.E., Gates, J., Fleming, S., Brooks, G.A., Sutton, J.R., Reeves, J.T., 1992. Increased energy intake minimizes weight loss in men at high altitude. Journal of Applied Physiology 72, 1741–1748.

Cannon, B., Nedergaard, J., 2004. Brown adipose tissue: function and physiological significance. Physiological Reviews 84, 277–359.

Cannon, B., Nedergaard, J., 2012. Yes, even human brown fat is on fire. Journal of Clinical Investigation 122, 487–489.

Crile, G.W., Quiring, D.P., 1939. Indian and Eskimo metabolism. Journal of Nutrition 18, 361–368.

Cunningham, J.J., 1991. Body composition as a determinant of energy expenditure: a synthetic review and a proposed general prediction equation. American Journal of Clinical Nutrition 54, 963–969.

Cypress, A.M., Lehman, S., Williams, G., Tal, I., Rodman, D., Goldfine, A.B., Kuo, F.C., Palmer, E.L., Tseng, Y.-H., Doria, A., Kolodny, G.M., Kahn, C.R., 2009. Identification and importance of brown adipose tissue in adult humans. New England Journal of Medicine 360, 1509–1517.

Daanen, H.A.M., 2003. Finger cold-induced vasodilation: a review. European Journal of Applied Physiology 89, 411–426.

Droma, T., McCullough, R.G., McCullough, R.E., Zhuang, J., Cymerman, A., Sun, S., Sutton, J.R., Moore, L.G., 1991. Increased vital and total lung capacities in Tibetan compared to Han residents of Lhasa (3658 m). American Journal of Physical Anthropology 86, 341–351.

Erickson, H., Krog, J., 1956. Critical temperature for naked man. Acta Physiologica Scandinavica 37, 35–39.

Ferro-Luzzi, A., Petracchi, C., Kuriyan, R., Kurpad, A.V., 1997. Basal metabolism of weight-stable chronically undernourished men and women: lack of metabolic adaptation and ethnic differences. American Journal of Clinical Nutrition 66, 1086–1093.

Food and Agriculture Organization, World Health Organization, and United Nations University (FAO/WHO/UNU), 2004. Human Energy Requirements. Report of Joint FAO/WHO/UNU Expert Consultation. World Health Organization, Geneva.

Frisancho, A.R., 1975. Functional adaptation to high altitude hypoxia. Science 187, 313–319.

Frisancho, A.R., 1993. Human Adaptation and Accommodation. University of Michigan Press, Ann Arbor.

Frisancho, A.R., 2008. Anthropometric Standards: An Interactive Nutritional Reference of Body Size and Body Composition for Children and Adults. University of Michigan Press, Ann Arbor.

Frisancho, A.R., 2013. Developmental functional adaptation to high altitude: review. American Journal of Human Biology 25, 151–168.

Frisancho, A.R., Baker, P.T., 1970. Altitude and growth: a study of the patterns of physical growth of a high altitude Peruvian Quechua population. American Journal of Physical Anthropology 32, 279–292.

Frisancho, A.R., Frisancho, H.G., Albalak, R., Villain, M., Vargas, E., Soria, R., 1997. Developmental, genetic, and environmental components of lung volumes at high altitude. American Journal of Human Biology 9, 191–203.

Froehle, A.W., 2008. Climate variables as predictors of basal metabolic rate: new equations. American Journal of Human Biology 20, 510–529.

Galloway, V.A., Leonard, W.R., Ivakine, E., 2000. Basal metabolic adaptation of the Evenki reindeer herders of Central Siberia. American Journal of Human Biology 12, 75–87.

Glickman, N., Mitchell, H.H., Keeton, R.W., Lambert, E.H., 1967. Shivering and heat production in men exposed to intense cold. Journal of Applied Physiology 22, 1–8.

Goldman, H.I., Becklake, M.R., 1959. Respiratory function tests; normal values at median altitudes and the prediction of normal results. American Review of Tuberculosis and Pulmonary Diseases 79, 457–467.

Greksa, L.P., Spielvogel, H., Caceres, E., 1994. Total lung capacity in young highlanders of Aymara ancestry. American Journal of Physical Anthropology 94, 477–486.

Hadley, M.E., 1996. Endocrinology, fourth ed. Prentice Hall, Upper Saddle River, NJ.

Hanna, J.M., Brown, D.E., 1983. Human heat tolerance: an anthropological perspective. Annual Review of Anthropology 12, 259–283.

Hardy, R.N., 1981. Endocrine Physiology. University Park Press.

Harford, R.R., Reed, H.L., Morris, M.T., Sapien, I., Warden, R., D'Alesandro, M.M., 1993. Relationship between changes in serum thyrotropin and total and lipoprotein cholesterol with prolonged Antarctic residence. Metabolism 42, 1159–1163.

Heinbecker, P., 1928. Studies on the metabolism of Eskimos. Journal of Biological Chemistry 80, 461–475.

Heinbecker, P., 1931. Further studies on the metabolism of Eskimos. Journal of Biological Chemistry 93, 327–336.

Henry, C.J.K., Rees, D.G., 1991. New predictive equations for the estimation of basal metabolic rate in tropical peoples. European Journal of Clinical Nutrition 45, 177–185.

Johnstone, A.M., Murison, S.D., Duncan, J.S., Rance, K.A., Speakman, J.R., 2005. Factors influencing variation in basal metabolic rate include fat-free mass, fat mass, age, and circulating thyroxine but not sex, circulating leptin, or triiodothyronine. American Journal of Clinical Nutrition 82, 941–948.

Kashiwazaki, H., 1990. Seasonal fluctuations of BMR in populations not exposed to limitation in food availability. Reality or illusion? European Journal of Clinical Nutrition 44 (Suppl. 1), 85–93.

Katzmarzyk, P.T., Leonard, W.R., 1998. Climatic influences on human body size and proportions: ecological adaptations and secular trends. American Journal of Physical Anthropology 106, 483–503.

Keys, A., Brožek, J., Henschel, A., Mickelsen, O., Taylor, H.L., 1950. The Biology of Human Starvation, 2 vols.University of Minnesota Press, Minneapolis.

Kleiber, M., 1975. The Fire of Life: An Introduction to Animal Energetics, second ed. Krieger, Huntington, N.Y.

Lachance, J., Tishkoff, S.A., 2013. Population genomics of human adaptation. Annual Review of Ecology, Evolution and Systematics 44, 123–143.

de Lange, P., Lanni, A., Beneduce, L., Moreno, M., Lombardi, A., et al., 2001. Uncoupling protein-3 is a molecular determinant of the regulation of resting metabolic rate by thyroid hormone. Endocrinology 142, 3414–3420.

Lasker, G.W., 1969. Human biological adaptability: the ecological approach in physical anthropology. Science 166, 1480–1486.

Leonard, W.R., 1989. Nutritional determinants of high-altitude growth in Nuñoa, Peru. American Journal of Physical Anthropology 80, 341–352.

Leonard, W.R., Katzmarzyk, P.T., 2010. Body size and shape: climatic and nutritional influences on human body morphology. In: Muehlenbein, M.P. (Ed.), Human Evolutionary Biology. Cambridge University Press, Cambridge, pp. 157–169.

Leonard, W.R., Leatherman, T.L., Carey, J.W., Thomas, R.B., 1990. Contributions of nutrition versus hypoxia to growth in rural Andean populations. American Journal of Human Biology 2, 613–626.

Leonard, W.R., Sorensen, M.V., Galloway, V.A., Spencer, G.J., Mosher, M.J., Osipova, L., Spitsyn, V.A., 2002. Climatic influence on basal metabolic rates among circumpolar populations. American Journal of Human Biology 14, 609–620.

Leonard, W.R., Snodgrass, J.J., Sorensen, M.V., 2005. Metabolic adaptation in indigenous Siberian populations. Annual Review of Anthropology 34, 451–471.

Leonard, W.R., Levy, S.B., Tarskaia, L.A., Klimova, T.M., Fedorova, V.I., Baltakhinova, M.E., Krivoshapkin, V.G., Snodgrass, J.J., 2014. Seasonal variation in basal metabolic rates among the Yakut (Sakha) of northeastern Siberia. American Journal of Human Biology 26, 437–445.

Levy, S.B., Leonard, W.R., Tarskaia, L.A., Klimova, T.M., Fedorova, V.I., Baltakhinova, M.E., Krivoshapkin, V.G., Snodgrass, J.J., 2013. Seasonal and socioeconomic influences on thyroid function among the Yakut (Sakha) of Eastern Siberia. American Journal of Human Biology 25, 814–820.

Levy, S.B., Leonard, W.R., Tarskaia, L.A., Klimova, T.M., Fedorova, V.I., Baltakhinova, M.E., Krivoshapkin, V.G., Snodgrass, J.J., 2014. An assessment of infrared thermal imaging as an indirect method for quantifying variation in brown adipose tissue using data from the Indigenous Siberian Health and Adaptation Project. American Journal of Human Biology 26, 270 (abstract).

Lewis, T., 1930. Observations upon the reactions of the vessels of the human skin to cold. Heart 15, 177–208.

Lichtenbelt, W.v., Kingma, B., van der Lans, A., Schellen, L., 2014. Cold exposure–an approach to increasing energy expenditure in humans. Trends in Endocrinology and Metabolism 25, 165–167.

Lind, A.R., Bass, D.E., 1963. Optimal exposure time for development of acclimatization to heat. Federation Proceedings 22, 704.

Lind, A.R., Humphreys, P.W., Collins, K.J., Foster, K., Sweetland, K.F., 1970. Influence of age and daily duration of exposure on responses of men to work in the heat. Journal of Applied Physiology 28, 50–56.

Mazess, R.B., 1975. Biological adaptation: aptitudes and acclimatization. In: Watts, E.S., Johnston, F.E., Lasker, G.W. (Eds.), Biosocial Interrelations in Population Adaptation. Mouton: the Hague, pp. 9–18.

McArdle, W.D., Katch, F.I., Katch, V.L., 2001. Exercise Physiology: Energy, Nutrition, and Human Performance, fifth ed. Lea & Febiger, Philadelphia.

McLean, J.A., Tobin, G., 1987. Animal and Human Calorimetry. Cambridge University Press, New York.

Milan, F.A. (Ed.), 1980. The Biology of Circumpolar Populations. Cambridge University Press, Cambridge.

Miller, L., Irving, L., 1962. Local reactions to air cooling in an Eskimo population. Journal of Applied Physiology 17, 449–455.

Mishmar, D., Ruiz-Pesini, E., Golik, P., Macaulay, V., Clarke, A.G., et al., 2003. Natural se lection shaped regional mtDNA variation in humans. Proceedings of the National Academy of Science 100, 171–176.

Mitchell, D.L., Senay, C., Wyndham, C.H., Van Rensberg, A.J., Rogers, G.G., Strydon, N.B., 1976. Acclimatization in a hot, humid environment: energy exchange, body temperature, and sweating. Journal of Applied Physiology 40, 768–778.

Moore, L.G., Cymerman, A., Huang, S.Y., McCullough, R.E., McCullough, R.G., Rock, P.B., Young, A., Young, P., Weil, J.V., Reeves, J.T., 1987. Propranolol blocks metabolic rate increase but not ventilatory acclimatization to 4300 m. Respiratory Physiology 70, 195–204.

Moore, L.G., Curran-Everett, L., Droma, T.S., Groves, B.M., McCullough, R.E., McCullough, R.G., Sun, S., Sutton, J.R., Zamudio, S., Zhuang, J.G., 1992. Are Tibetans better adapted? International Journal of Sports Medicine 13, S86–S88.

Moore, L.G., Neirmeyer, S., Zamudio, S., 1998. Human adaption to high altitude: regional and life-cycle perspectives. Yearbook of Physical Anthropology 41, 25–64.

Muzik, O., Mangner, T.J., Leonard, W.R., Kumar, A., Janisse, J., Granneman, J.G., 2013. ^{15}O PET measurement of blood flow and oxygen consumption in cold-activated human brown fat. Journal of Nuclear Medicine 54, 523–531.

Newman, M.T., 1953. The application of ecological rules to the racial anthropology of the aboriginal new world. American Anthropologist 55, 311–327.

Nedergaard, J., Bengtsson, T., Cannon, B., 2007. Unexpected evidence of brown adipose tissue in adult humans. American Journal of Physiology 293, E444–E452.

Pääkkönen, T., Leppäluoto, J., 2002. Cold exposure and hormonal secretion: a review. International Journal of Circumpolar Health 61, 265–276.

Palinkas, L.A., Suedfeld, P., 2008. Psychological effects of polar expeditions. Lancet 371, 153–163.

Plasqui, G., Kester, A.D.M., Westerterp, K.M., 2003. Seasonal variation in sleeping metabolic rate, thyroid activity, and leptin. Amercian Journal of Physiology 285, E338–E343.

Poehlman, E.T., Toth, M.J., 1995. Mathematical ratios lead to spurious conclusions regarding age- and sex-related differences in resting metabolic rate. American Journal of Clinical Nutrition 61, 482–485.

Prabhakar, N.R., 2000. Oxygen sensing by carotid body chemoreceptors. Journal of Applied Physiology 88, 2287–2295.

Reed, H.L., Burman, K.D., Shakir, K.M.M., O'Brian, J.T., 1986. Alterations in the hypothalamic-pituitary-thyroid axis after prolonged residence in Antarctica. Clinical Endocrinology 25, 55–65.

Reed, H.L., Brice, D., Shakir, K.M.M., Burman, K.D., D'Alesandro, M.M., O'Brian, J.T., 1990a. Decreased free fraction of thyroid hormones after prolonged Antarctic residence. Journal of Applied Physiology 69, 1467–1472.

Reed, H.L., Silverman, E.D., Shakir, K.M.M., Dons, R., Burman, K.D., O'Brian, J.T., 1990b. Changes in serum triiodothyronine (T3) kinetics after prolonged Antarctic residence: the polar T3 syndrome. Journal of Clinical Endocrinology and Metabolism 70, 965–974.

Roberts, D.F., 1952. Basal metabolism, race and climate. Journal of the Royal Anthropological Institute 82, 169–183.

Roberts, D.F., 1953. Body weight, race and climate. American Journal of Physical Anthropology 11, 533–558.
Roberts, D.F., 1973. Climate and Human Variability. An Addison-Wesley Module in Anthropology, No. 34.
Roberts, D.F., 1978. Climate and Human Variability, second ed. Cummings, Menlo Park, CA.
Rodahl, L.K., 1952. Basal metabolism of the Eskimos. Federation Proceedings 2, 130–137.
Rode, A., Shephard, R.J., 1995. Basal metabolic rate of Inuit. American Journal of Human Biology 7, 723–729.
Ruff, C.B., 1994. Morphological adaptation to climate in modern and fossil hominids. Yearbook of Physical Anthropology 37, 65–107.
Ruiz-Pesini, E., Mishmar, D., Brandon, M., Procaccio, V., Wallace, D.C., 2004. Effects of purifying and adaptive selection on regional variation in human mtDNA. Science 303, 223–226.
Schmidt-Nielson, K., 1984. Scaling: Why Animal Size is so Important. Cambridge University Press, Cambridge.
Schofield, C., 1985. An annotated bibliography of source material for basal metabolic rate data. Human Nutrition (Clinical Nutrition) 39C (Suppl. 1), 42–91.
Schreider, E., 1950. Geographical distribution of the body-weight/body-surface ratio. Nature 165, 286.
Schreider, E., 1957. Ecological rules and body-heat regulation in man. Nature 179, 915–916.
Schreider, E., 1964. Ecological rules, body-heat regulation, and human evolution. Evolution 18, 1–9.
Schreider, E., 1975. Morphological variations and climatic differences. Journal of Human Evolution 4, 529–539.
Semenza, G.L., 2007. Life with oxygen. Science 318, 62.
Semenza, G.L., Neifelt, M.K., Chi, S.M., Antonarakis, S.E., 1991. Hypoxia-inducible nuclear factors bind to an enhance element located on the 3' to the human erythropoietin gene. Proceedings of the National Academy of Science 88, 5680–5684.
Senay, L.C., Mitchell, D., Wyndham, C.H., 1976. Acclimatization in a hot, humid environment: body fluid adjustments. Journal of Applied Physiology 40, 786–796.
Shetty, P.S., 1993. Chronic undernutrition and metabolic adaptation. Proceedings of the Nutrition Society 52, 267–284.
Shetty, P.S., 1996. Metabolic adaptation in humans: Does it occur? In: Rosettta, L., Mascie-Taylor, C.G.N. (Eds.), Variability in Human Fertility. Cambridge University Press, Cambridge, pp. 125–147.
Snodgrass, J.J., Leonard, W.R., Tarskaia, L.A., Alekseev, V.P., Krivoshapkin, V.G., 2005. Basal metabolic rate in the Yakut (Sakha) of Siberia. American Journal of Human Biology 17, 155–172.
Soares, M.J., Kulkarni, R.N., Piers, L.S., Vaz, M., Shetty, P.S., 1992. Energy supplementation reverses changes in the basal metabolic rates of chronically undernourished individuals. British Journal of Nutrition 68, 593–602.
Soares, M.J., Francis, D.G., Shetty, P.S., 1993. Predictive equations for basal metabolic rates of Indian males. European Journal of Clinical Nutrition 47, 389–394.
Soares, M.S., Shetty, P.S., 1991. Basal metabolic rates and metabolic economy in chronicundernutrition. European Journal of Clinical Nutrition 45, 363–373.
Stinson, S., 1990. Variation in body size and shape among South American Indians. American Journal of Human Biology 2, 37–51.
van der Lans, A.A., Hoeks, J., Brans, B., Vijgen, G.H., Visser, M.G., Vosselman, M.J., Hansen, J., Jörgensen, J.A., Wu, J., Mottaghy, F.M., Schrauwen, P., Lichtenbelt, W.D.v, 2013. Cold acclimation recruits human brown fat and increases nonshivering thermogenesis. Journal of Clinical Investigation 123, 3395–3403.
Wagner, J.A., Robinson, S., Tzankoff, S.P., Marino, R.P., 1972. Heat tolerance and acclimatization to work in the heat in relation to age. Journal of Applied Physiology 33, 616–622.
Walter, H., 1971. Remarks on the environmental adaptation of man. Humangenetik 13, 85–97.
Weinsier, R.L., Shutz, Y., Bracco, D., 1992. Reexamination of the relationship of resting metabolic rate to fat-free mass and to the metabolically active components of fat-free mass in humans. American Journal of Clinical Nutrition 55, 790–794.
Wyndham, C.H., Morrison, J.F., Williams, C.G., Bredell, G.A.G., Von Raliden, M.J.E., Holdsworth, L.D., Van Graan, C.H., Van Rensburg, A.J., Munro, A., 1964a. Heat reactions of Caucasians and Bantus in South Africa. Journal of Applied Physiology 19, 598–606.
Wyndham, C.H., Strydom, N.B., Williams, C.G., Morrison, J.F., Bredell, G.A.G., Peter, J., Van Graan, C.H., Holdsworth, L.D., Van Rensburg, A.J., Munro, A., 1964b. Heat reactions of some Bantu tribesmen in southern Africa. Journal of Applied Physiology 19, 881–884.
Wyndham, C.H., McPherson, R.K., Munro, A., 1964c. Reactions to heat of aborigines and Caucasians. Journal of Applied Physiology 19, 1055–1058.
Wyndham, C.H., Strydom, N.B., Van Rensburg, A.J., Benade, A.J.S., Heyns, A.J., 1970. Relation between VO_2max and body temperature in hot, humid air conditions. Journal of Applied Physiology 29, 45–50.

Chapter 19

Evolution of Skin Color

Peter M. Elias[1] and Mary L. Williams[2]
[1]*Dermatology Service, Department of Veterans Affairs Medical Center and Department of Dermatology, University of California, San Francisco, CA, USA;* [2]*Departments of Dermatology and Pediatrics, University of California, San Francisco, CA, USA*

SYNOPSIS

Current theories for the development of epidermal pigmentation in hominins are problematic. Skin cancers occur beyond peak reproductive years, and minimal UV-B penetrates to deeper skin layers where folic acid circulates and eccrine glands reside. Generating an optimal permeability barrier was challenging for hominins who employed sweat to thermoregulate as they inhabited African savannas. Epidermal pigmentation provided hominins with a superior barrier. Latitude-dependent pigment dilution to facilitate cutaneous vitamin D3 (VD3) synthesis is also problematic, because: UV-B-irradiated skin, whether pigmented or nonpigmented, generates comparable VD3; extracutaneous storage of VD3 supplies year-round VD3 requirements; adequate dietary sources existed in the Upper Paleolithic; VD3 deficiency is rarely seen in pre-industrial age fossils; and evidence is lacking for mutations facilitating VD3 production/action. Pigment dilution instead largely served in metabolic conservation—the imperative to redirect scarce protein production toward more urgent requirements. Europeans of the far north instead developed loss-of-function filaggrin mutations leading to reduced *trans*-urocanic acid, the major UV-B absorber in stratum corneum, and to a corresponding increase in circulating VD3 levels.

WHEN AND WHY EPIDERMAL PIGMENTATION EVOLVED

Population genetic techniques show that the gene encoding the melanocortin 1 receptor (MC1R) stabilized in hominins around 1.2 million years ago (Harding et al., 2000; Rogers et al., 2004), providing an approximate date for the population of interfollicular epidermis by melanocytes. That epidermal pigmentation conferred significant survival benefits is shown by the strong conservation of not only *MC1R* (Harding et al., 2000), but also several other genes involved in pigment production (Graf et al., 2005; Lao et al., 2007; Parra, 2007), indicating strong evolutionary pressure to retain interfollicular pigmentation among humans residing in sub-Saharan Africa today. But melanin is a significant heat absorber (Hill, 1992; Blum, 1961), which would have posed substantial difficulties in thermoregulation. A most eloquent example of the role of epidermal pigmentation as an insulator is polar bear skin. Though hair follicles lack melanocytes and melanin, the epidermis underlying the white coat is widely populated with melanin (Figure 1). Therefore, in the heat of Africa, the benefits of pigmentation must have been quite robust to offset the formidable disadvantage of overheating due to the insulating effects of epidermal pigmentation.

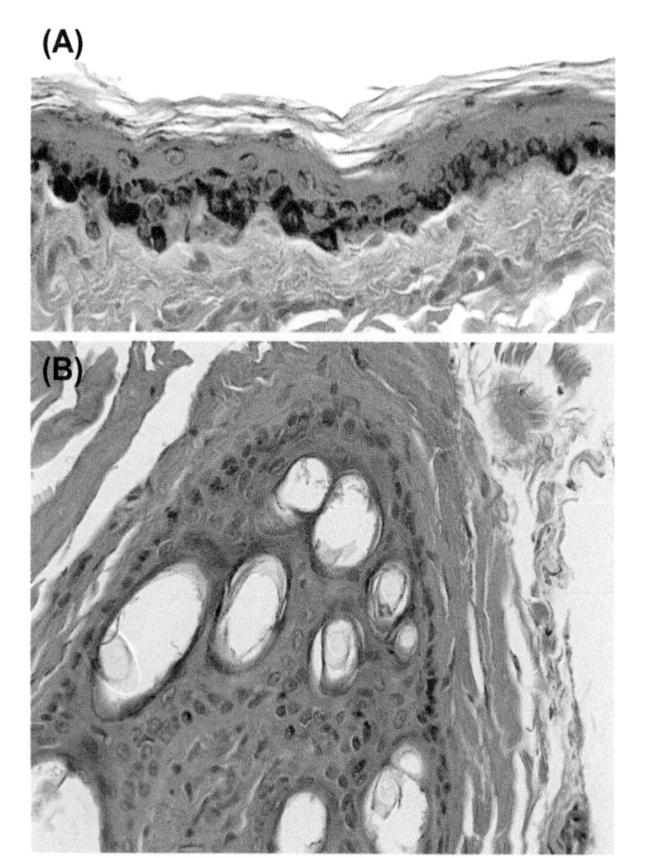

FIGURE 1 Polar bear skin: note dense carpet of melanocytes containing abundant melanin in interfollicular epidermis (A, arrows). (B) In contrast, the follicular epithelium lacks evidence of melanocytes, corresponding to the white fur of this species. (See color plate section). *Illustration courtesy of Elizabeth Maudlin, DVM, University of Pennsylvania.*

Protection Against Skin Cancer and Eccrine Gland Destruction

Among competing hypotheses put forward to explain the development of epidermal pigmentation, the "genotoxic hypothesis," though widely espoused, is untenable because the vast majority of skin cancers occur well past peak reproductive age. Though epidermal pigmentation protects against UV-B-induced sunburn, it likely did not evolve to protect against UV-B-induced destruction of nascent eccrine sweat glands, as proposed by Jablonski and Chaplin (2000) and Chaplin (2004), because these glands reside deep in the dermis where they are fully protected from UV-B exposure (Parrish et al., 1982). Though the heat of equatorial Africa can obstruct eccrine ducts, leading to heat intolerance and difficulties with thermoregulation, darkly pigmented skin remains just as prone to sweat gland dysfunction as lightly pigmented skin.

Protection Against Folate Degradation and Vitamin D Toxicity

In 1967, Loomis provided a map that showed latitude-dependent differences in skin coloration, and suggested that epidermal pigmentation evolved in ancestral hominins in order to protect against vitamin D intoxication, as first suggested by Murray (1934) (Figure 2). Then, he proposed oppositely that as humans moved northward out of equatorial Africa into regions with less exposure to ultraviolet light, pigmentation faded in order to augment cutaneous production of vitamin D (see also Chaplin and Jablonski, 2009; Yuen and Jablonski, 2010; Branda and Eaton, 1978; Jablonski, 2010; Neer, 1975). To generate vitamin D, a distal precursor of cholesterol, 7-dehydrocholesterol (7DHC), is first synthesized in the epidermis (Holick et al., 1980), then photoconverted to pre-vitamin D3 (pre-VD3), and finally thermally converted to vitamin D3 (VD3) (Holick et al., 1980). Yet even in light-skinned subjects, intense sun exposure never results in vitamin D intoxication (Holick et al., 1981), because excess pre-VD3 is shunted reversibly toward two biologically inactive metabolites, tachysterol and lumisterol. (Please see Part III, Hominins, Chapters 11 and 13 for further information).

Vitamin D is also often paired with folic acid (vitamin B9) in an antipode of "drivers" of skin coloration (Jablonski and Chaplin, 2010). According to this formulation, epidermal pigmentation evolved to protect folic acid and its metabolites tetrahydrofuran and *S*-methyl tetrahydrofolate from being destroyed by UV irradiation. Folic acid protects against the development of congenital neural anomalies, such as failure of spinal fusion (*spina bifida*) (Rayburn et al., 1996; Wilson et al., 2003, 2007). Hence, epidermal pigmentation could confer a considerable evolutionary advantage should it protect against photoinduced folic acid deficiency (Jablonski and Chaplin, 2010). Yet most neural tube defects are too mild to interfere with reproductive success (Jablonski and Chaplin, 2010), and the low overall prevalence of congenital anomalies (~1/2000 pregnancies) is likely too low to have "driven" natural selection (Rasmussen et al., 1998; Rayburn et al., 1996; Northrup and Volcik, 2000). While highly susceptible to photodegradation by UV-B and to a lesser extent by UV-A in vitro (Moan et al., 2012), another concern with the folate hypothesis is whether folic acid and its metabolites are vulnerable to UV-B-induced photodegradation in vivo. The blood vessels that transport folate and its active metabolites lie well beneath the epidermis, where likely too little UV-B penetrates to impact circulating folate levels (Anderson and Parrish, 1981). Though UV-A can penetrate to such depths, it is unlikely to degrade folate in vivo because intense doses of UV-A (and UV-B), when administered repeatedly in the treatment of patients with inflammatory skin diseases, do not provoke folic acid

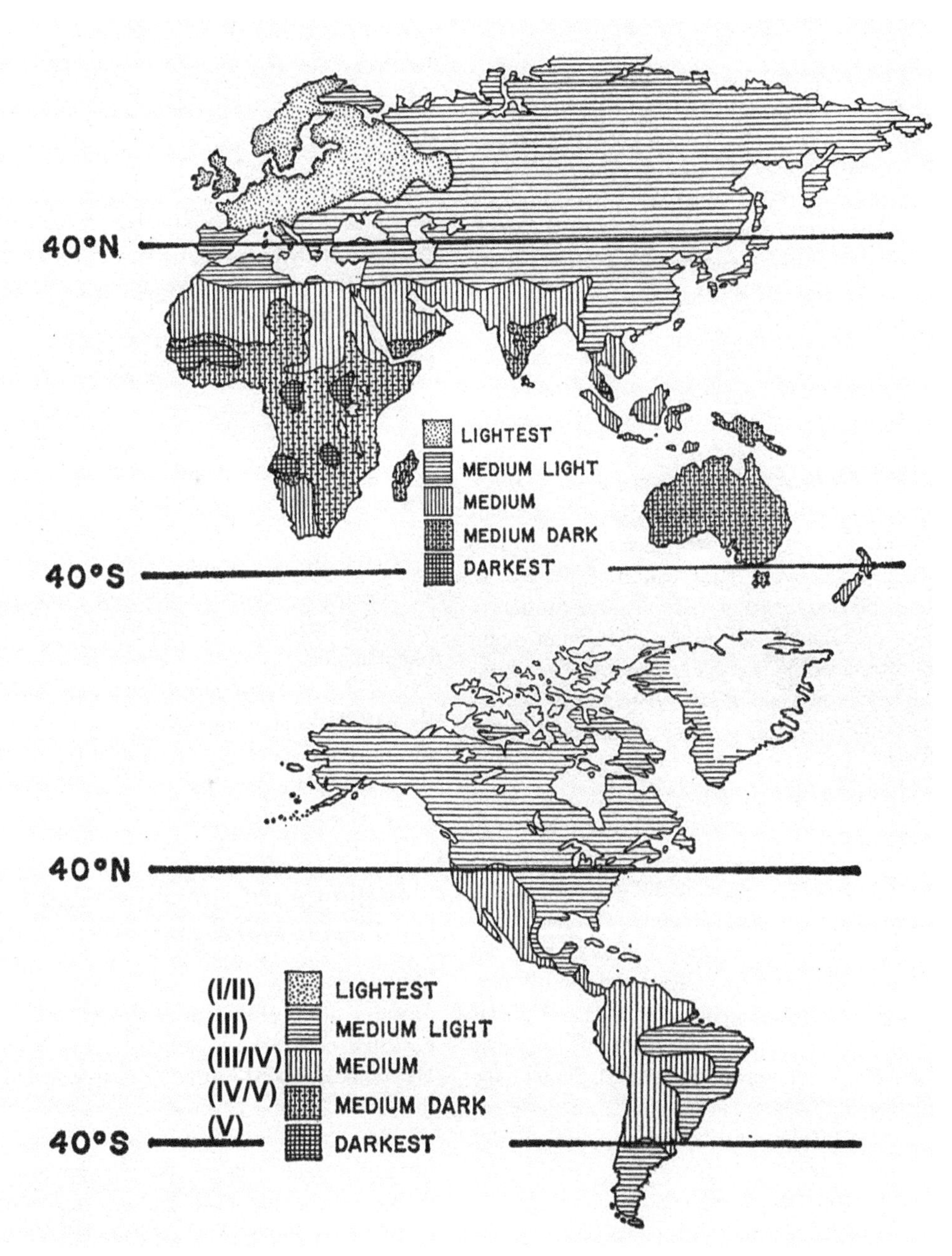

FIGURE 2 Distribution of human skin color before 1492. *(Modified from Loomis, 1967)*. Fitzpatrick-scale pigmentation from I (lightest) to VI (darkest) is shown.

deficiency (Juzeniene et al., 2010; Cicarma et al., 2010). Nor do folic acid levels decline following repeated sun exposure (Juzeniene et al., 2010; Cicarma et al., 2010). Therefore, it is highly unlikely that epidermal pigmentation developed to protect against UV-B-induced folic acid deficiency.

Barrier Requirements Likely Stimulated the Development of Epidermal Pigmentation

If pigmentation developed neither to prevent skin cancer nor to protect against eccrine gland destruction, vitamin D intoxication, or folic acid deficiency, then why did hominins become darkly pigmented? The answer almost certainly relates to the most critical function of the skin—the provision of a competent permeability barrier, which is a requirement for life in a desiccating terrestrial environment (previously proposed in (Elias et al., 2009, 2010)). Darkly pigmented human skin possesses a more competent skin barrier than does more lightly pigmented skin, a difference that correlates solely with pigment type rather than ethnicity, and is independent of latitude of residence (Reed et al., 1995; Gunathilake et al., 2009). Moreover, patients with vitiligo, whose pigment loss occurs in localized patches due to an absence of melanocytes, display reduced barrier function in depigmented regions (Liu et al., 2010). Finally, the skin of pigmented hairless mice (Skh2) displays a superior permeability barrier in comparison with the nonpigmented skin of albino (Skh1) hairless mice (Man et al., 2014). Mechanistic studies show that the reduced pH of darkly pigmented skin largely but not completely accounts for these differences, because exogenous acidification of the stratum corneum "resets" barrier function to levels seen in darkly pigmented subjects (Gunathilake et al., 2009; Man et al., 2014). Indeed, a reduced pH is highly beneficial for multiple epidermal functions, including permeability barrier homeostasis and antimicrobial defense (Fluhr and Elias, 2002; Elias, 2007).

Dark skin is also more resistant to all types of infections (Mackintosh, 2001). It is well known that darkly pigmented modern humans experience fewer skin infections than their cohabitating light-skinned neighbors, particularly when living in or visiting tropical latitudes (Wassermann, 1965; Mackintosh, 2001). Why is dark skin more resistant to infections? First, the more competent permeability barrier of normal pigmented skin generates a drier skin surface that is inimical to colonization by pathogenic microbes that prefer a moister environment (Elias, 2007). Second, the more-acidic surface pH of pigmented skin is hostile to the growth of bacterial pathogens (Korting et al., 1990, 1987). Cutaneous antimicrobial defense is pH-dependent (Elias, 2007) by several additional mechanisms including (1) increased cohesion of adjoining corneocytes (Gunathilake et al., 2009), which inhibits the penetration of pathogens; and likely; (2) increased production of antimicrobial lipids

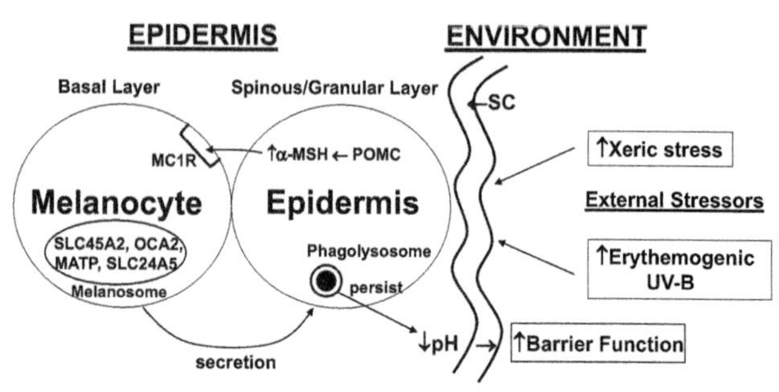

FIGURE 3 Epidermis–Melanocyte Cross Talk: Highly conserved genes in darkly pigmented skin acidify the stratum corneum, thereby improving barrier function in the face of extreme xeric stress and high amounts of erythemogenic UV-B irradiation. Whether environmental stress upregulates mechanisms that increase pigmentation (e.g., pro-opiomelanocortin), leading to increased α-melanocyte-stimulating hormone (α-MSH) is not known. Likewise, the impact of environmental stress on melanosome acidification and pigment production in melanocytes is unknown. SC = stratum corneum. *Modified from Figure 4, Elias et al. (2009).*

(i.e., acidic free fatty acids) that inhibit the growth of gram-positive bacteria and yeasts (Drake et al., 2008; Miller et al., 1988). Furthermore, pigmented epidermis produces increased quantities of non-melanin-derived antimicrobial peptides (Mackintosh, 2001; Man et al., 2014) (Figure 3), a highly conserved class of molecules that are found in epithelial barriers throughout the plant and animal kingdoms and are inimical to the growth of many disease-causing pathogens (Schroder and Harder, 2006; Nakatsuji and Gallo, 2012). Finally, melanin and its metabolites display potent antimicrobial activities (Montefiori and Zhou, 1991).

Large melanin granules are distributed evenly throughout the keratinocyte cytoplasm in darkly pigmented epidermis, and these more robust granules persist high into the outer nucleated layers and even into the stratum corneum, where some are discharged (with their load of protons) into the extracellular spaces (Man et al., 2014). Thus, the evolution of darkly pigmented skin also equipped hominins to withstand pathogens common in the tropics (Mackintosh, 2001; Wassermann, 1965).

Sweating was Critical for Heat Dissipation, but Placed Additional Stress on the Barrier

The climate that dominated sub-Saharan Africa at the time of pigment development in *H. erectus* was not only UV-B enriched, but also extremely arid (Blome et al., 2012; DeMenocal, 2004)—conditions that would have placed further stress on the permeability barrier. Because the vapor pressure at the skin surface is the primary determinant of transepidermal water loss, low environmental humidity steepens the gradient of water loss across the skin, inevitably imposing additional demands for a highly competent skin barrier. While the evolutionary development of eccrine sweating permitted hominins to hunt more actively on the savanna, the combination of sweating to dissipate heat and an inefficient (leaky) skin barrier would have quickly threatened these hunter-gatherers with dehydration. Hence, the subsequent development of a highly competent permeability barrier, assisted in part by the generation of interfollicular pigmentation, would have allowed movement by hominins over longer distances, even during midday hours.

While Erythemogenic UV-B is Toxic, Pigmentation Shifts the Dose–Response Curve toward the Beneficial Effects of Suberythemogenic UV-B

To further address the plausibility of this hypothesis, we must first examine the impact of UV-B irradiation on epidermal structure and function. Erythemogenic doses of UV-B damage DNA, induce keratinocyte cell death (apoptosis), and provoke inflammation (Young et al., 1998; Uchida et al., 2003; Parrish et al., 1982; Honigsmann, 2002; Anderson and Parrish, 1981). As an acute sunburn recedes, epidermal hyperproliferation propels layers of functionally incompetent keratinocytes through the outer epidermis, where they transiently compromise the permeability barrier (Haratake et al., 1997a, b; Holleran et al., 1997). Paradoxically, however, lower ("suberythemogenic") doses of UV-B instead benefit the permeability barrier function while enhancing cutaneous antimicrobial peptide production in parallel (Hong et al., 2008). Since even low doses of UV-B quickly become toxic in lightly pigmented humans, the endowment of hominin epidermis with dark pigmentation shifted the UV-B dose–response curve from a toxic range toward one that was beneficial.

BASIS FOR PIGMENT DILUTION IN MODERN HUMANS

An obvious feature of the northward dispersal of humans is a quasi-geographic reduction in pigmentation (Loomis, 1967; Murray, 1934; Chaplin and Jablonski, 2009) (Figure 2). But actual coloration varies greatly among Northerners. Using the Fitzpatrick scale of type I–VI pigmentation, native Inuits display medium-to-dark (type III/IV) rather than light pigmentation, as do both northern- and central-dwelling Asians, who display medium (type III) pigmentation. The retention of pigment presumably provided insulation (Hill, 1992), as did their clothing, and most Northerners obtained abundant VD3 from dietary sources (see also below). Recent population genetic data show that the reduction in skin pigmentation occurred sporadically and incompletely in European and Asian populations (Sturm, 2009; Olalde et al., 2014). Moreover, while modern humans reached Central Europe ~40 thousand years ago (kya), they reached Northern Europe only after the last ice sheets receded (<11 kya), and it is only these humans and their ancestors who display light pigmentation (type I/II) (Figure 2). Accordingly, recent molecular genetic studies suggest that the very light pigmentation of Northern Europeans did not develop until 5–6 kya (Norton et al., 2007; Norton and Hammer, 2008). Lighter pigmentation resulted from the accumulation of genetic polymorphisms in the melanocortin 1 receptor (Rana et al., 1999)] and in other genes that regulate either melanin synthesis or the acidification of melanosomal contents (Graf et al., 2005; Goding, 2007; Lao et al., 2007; Takeda et al., 2007; Soejima and Koda, 2007; Marconi et al., 2003; Lamason et al., 2005; Cook et al., 2009; McEvoy et al., 2006) (Figure 3). Because the polymorphisms associated with reduced pigment in Asian and European populations differ (Anno et al., 2008; Sturm, 2009), lighter pigmentation emerged independently in these populations (Alonso et al., 2008; Izagirre et al., 2006; Norton and Hammer, 2008; Norton et al., 2007; Sturm, 2009), suggesting that pigment dilution resulted from positive selection for lighter pigmentation (Sturm, 2009). Though these parallel genetic events suggest that pigment dilution occurred for a reason, what could that reason have been?

Cultural Theories for Pigment Dilution

While the requirement for vitamin D currently prevails as the theory for progressive skin lightening (Jablonski and Chaplin, 2013), other theories also have been forwarded. A now thoroughly discredited hypothesis posited that light skin could better tolerate cold temperatures (Post et al., 1975). This theory ignores the fact that relatively dark pigmentation persists in Inuits and Eurasians living at Arctic latitudes (Post et al., 1975), where the presence of pigmentation provides non-insignificant advantages by absorbing and retaining radiant heat (Hill, 1992). As noted above, this benefit is most vividly demonstrated in polar bear skin, which has deep epidermal pigmentation underneath pale light-reflecting fur (Figure 1), whereas the epidermis under the light stripes of zebras living in a hot environment lacks epidermal pigmentation (G.K. Menon, personal communication).

An even less credible theory to account for pigment dilution is that female hominins preferred reproduction with males who possess paler skin ("sexual selection" hypothesis) (Frost, 1988; van den Berghe and Frost, 1986; Aoki, 2002; Frost, 2007; Diamond, 2004). In Europe, pigmentation was considered a relatively unimportant cutaneous feature until the advent of slavery in the fifteenth century, followed by the emergence of theories of racial superiority late in the eighteenth century (Connor, 2004; Jablonski, 2012). There is no evidence that the Egyptians, Greeks, or Romans displayed pigment preferences (Jablonski, 2012), and it certainly is not known whether skin pigmentation influenced sexual selection in the Upper Paleolithic era, when lighter pigmentation evolved. While lighter skin pigmentation may be valued by males and/or females in some contemporary cultures, this theory presupposes that such a preference is itself hard-wired into our genomes and that such preferences would have been universal, not just evident in sporadic cultural samples. But most importantly, *dark skin could not have evolved in the first place if individuals innately preferred lighter-skinned mates*! Therefore, the "sexual selection theory" necessarily shifts the issue of skin pigmentation from evolutionary issues to a discussion instead of universal ideals of beauty (Etcoff, 1999). While "beauty" is a marker of health, and hence could serve as a measure of reproductive fitness, it is not immediately apparent how lighter skin connotes better health.

Most Pigment Dilution Did Not Occur from a Greater Need for Vitamin D

Though based for the most part on correlative evidence and a lack of more viable alternative hypotheses, evolutionary biologists have repeatedly proposed that lighter pigmentation evolved from an increased need for cutaneous vitamin D production (Jablonski and Chaplin, 2000; Jablonski, 2010; Hochberg and Templeton, 2010). Yet, Robins (2009) provided many arguments against this hypothesis, which we will buttress below with several additional points.

In the presence of sufficient UV-B exposure, most vitamin D is synthesized in the epidermis (Bikle, 2010). At latitudes that correspond to central Europe and subarctic North America, UV-B exposure in the late spring through early autumn months can generate sufficient vitamin D to prevent deficiency year-round, regardless of dietary intake (Holick et al., 1981; Chen et al., 2007). Indeed, exposure of only limited parts of the body such as the face and hands, a few times each week during the summer can generate substantial stores of vitamin D even in darkly pigmented populations living at temperate latitudes (Goding, 2007; Gilchrest, 2008; Brazerol et al., 1988; Marks et al., 1995; Rockell et al., 2008). The pale skin of populations in northern France, Germany, and the southern Baltic states may have evolved in part to facilitate VD3 production, because sufficient UV-B penetrates in summer months. But Europeans of the far north faced a more difficult challenge.

Vitamin D requirements can also be met by eating a diet enriched in oily fish and animals that eat fish (Bikle, 2010; Chen et al., 2007), as occurs in Arctic dwellers (Sharma et al., 2011). Although modern diets of highly refined grain-derived foods contain little vitamin D (Yuen and Jablonski, 2010; Jew et al., 2009), current dietary practices do not resemble those of the Upper Paleolithic, when few grains and cereal products were available. Indeed, a fish-enriched diet evolved in Central Europe over 20 kya, and the species of fish that inhabited inland waterways doubtlessly included some marine species (such as migrating Atlantic salmon) and oily freshwater fish, such as eels and catfish, that contain substantial vitamin D. Furthermore, Paleolithic humans likely wasted little in the food chain—the vitamin D–storing tissues such as fat, liver, kidney, bone marrow, and even skin and brain (Bikle, 2010) of wild game, likely would have been consumed.

Despite the many alluring examples of cave paintings from the Dordogne region of France, hunter-gatherers of this period were likely intimidated by these caves, which often were still inhabited by dangerous species of cave bears, lions, and hyenas. Hence, they chose to live an outdoor life that would have resulted in frequent exposure of their uncovered arms, legs, faces, and hands to UV-B radiation (Robins, 2009). With the addition of fish to a diet already enriched in fatty tissues, and with an outdoor life associated with continuous exposure of portions of their body surface, it is unlikely that hunter-gatherers of the Upper Paleolithic, even if still fully pigmented upon arrival in Central Europe, would have encountered difficulty meeting their vitamin D requirements.

There are still other serious problems with the vitamin D hypothesis. It is true that a severe deficiency in vitamin D can produce rickets in growing children and can narrow the pelvis, thus obstructing childbirth (Chaplin and Jablonski, 2009). But such severe cases of rickets are rare, and milder or later onset vitamin D deficiency likely would not have exerted deleterious effects on reproduction. Moreover, evidence for rickets in the fossil record and in disinterred bones from Medieval cemeteries is sparse and became prevalent only after the industrial revolution darkened the skies over Europe (Robins, 2009). Finally, the vitamin D hypothesis fails to explain why pigment was lost in sites that remain unexposed to light. If vitamin D requirements "drove" pigment dilution, pigmentation should have been lost preferentially on sun-exposed surfaces, such as the face and extremities, particularly since sufficient vitamin D can be generated with only limited exposure of these areas (Goding, 2007; Gilchrest, 2008; Holick, 1995). Moreover, the vitamin D hypothesis also fails to explain why hairs simultaneously became lightly pigmented, though hairs are not involved in vitamin D synthesis. Finally, population genetic studies have found few polymorphisms in genes that encode the VD3 synthetic pathway or VD3 receptor (Ahn et al., 2010). Hence, if sufficient VD3 is generated in all skin pigment types at temperate latitudes, pigment dilution necessarily served purposes unrelated to VD3 production.

Most importantly, the same doses of UV-B irradiation of either darkly pigmented or lightly pigmented skin produce comparable elevations of circulating 25-OH-vitamin D3 levels (Bogh et al., 2010). In fact, it seems naïve to assume that VD3 bioavailability would be regulated solely by skin pigmentation, because VD3 bioavailability can be influenced by multiple mechanisms including rates of 7DHC production and VD3-binding protein levels,

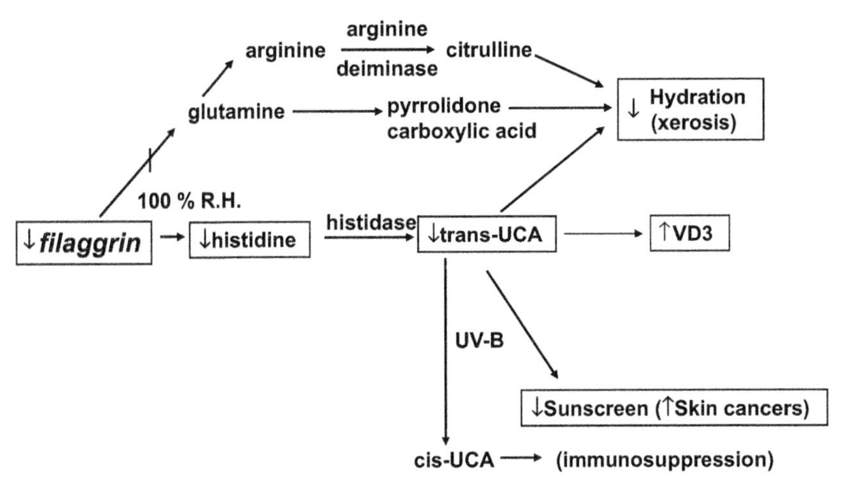

FIGURE 4 Downstream consequences of loss-of-function mutations in filaggrin.

as well as ligand affinity for the VD3 transporter; rates of transport from the blood into the liver and kidneys; differences in expression of the 1α- and 25-α hydroxylation enzymes; and VD3 receptor expression. It is likely that at least one of these mechanisms is upregulated in darkly pigmented skin to provide sufficient VD3 even with reduced UV-B penetration into the skin. In fact, the lower incidence of osteoporosis in darkly pigmented humans serves as eloquent evidence for increased bioavailability of VD3 independent of UV-B availability (Aloia, 2008; Vivanco-Munoz et al., 2012) (please also see Section 152 on Bone Disorders).

Non-pigment-based Mechanisms That Likely Enhance Vitamin D Bioavailability

While pigment dilution likely did not evolve to enhance VD3 generation in most Northerners, other mechanisms could have evolved to enhance its production. Not all incident UV-B is blocked by melanin—a substantial proportion (>35%) is absorbed by proteins and protein metabolites in the stratum corneum, even in darkly pigmented humans (Thomson, 1955). One protein has just emerged as a candidate to explain enhanced VD3 production—the stratum corneum structural protein filaggrin (FLG) (Thyssen et al., 2014). The link between FLG and UV-B bioavailability can be explained by the proteolytic processing of FLG into one of its constituent amino acids, histidine, followed by the deimination of histidine by the enzyme histidine ammonia-lyase (histidase), into the carboxylic acid *trans*-urocanic acid (t-UCA) (Scott, 1981; Brown and McLean, 2012) (Figure 4). The resulting t-UCA is a key endogenous sunscreen of the stratum corneum (Kripke, 1984) with an action spectrum in the UV-B spectrum (Brookman et al., 2002; Haralampus-Grynaviski et al., 2002; McLoone, Simics, Barton et al., 2005). Moreover, FLG knockdown results in subnormal levels of t-UCA and an increased susceptibility to UV-B-induced apoptosis (Mildner et al., 2010). Conversely, overexpression of caspase 14, which accelerates FLG breakdown, is photoprotective. Population genetics show that a substantial number (~10%) of normal Scottish, Irish, Dutch, and Scandinavians exhibit loss-of-function mutations in FLG (Irvine et al., 2011), with much lower prevalence in Central Europeans (c. 5%), Asians (<5%), Southern Europeans (<1%), and Africans (<1%) (Thyssen et al., 2014). Furthermore, while FLG mutations provoke defects in cutaneous barrier function that predispose to atopic dermatitis (Gruber et al., 2011; Scharschmidt et al., 2009; Fallon et al., 2009), Northern Europeans with FLG mutations, with or without AD, exhibit higher-than-normal circulating VD3 levels (Thyssen et al., 2012). Hence, a deficiency of t-UCA due to FLG deficiency would inevitably result in substantially more UV-B transmittance, likely leading to increased intracutaneous VD3 generation in extreme northern latitudes (Thyssen et al., 2014) (Figure 4).

CONSERVATION OF METABOLIC ENERGY

Then what could have been the benefit of reduced pigmentation in widely separated European and Asian populations? The most likely explanation is the ever-present imperative to conserve energy; that is, metabolic conservation (Elias and Williams, 2013). Briefly, when there is no biological advantage to expending metabolic energy in support of no-longer-critical functions (Gabay and Kushner, 1999), mutations that weed out energy-consuming processes become beneficial and favored by natural selection. Thus, a declining need to heavily pigment the epidermis favored the retention of mutations in genes that reduced pigment synthesis, thereby diverting energy toward the production of more urgently needed proteins (Elias and Williams, 2013). An eloquent demonstration of the metabolic cost of pigmentation occurs in children with protein malnutrition (i.e., kwashiorkor), who manifest marked lightening of the skin and hair (Latham, 1991). Analogously, adults on marginal diets stop menstruating, lactating, or making sperm, and children often stop growing in order to conserve limited caloric resources, not due to increased vitamin D requirements (Yuen and Jablonski, 2010) but rather to divert protein and energy supplies to other more critical purposes (e.g., growth and breast feeding). Finally, as noted above, pigment dilution occurred across the entire skin surface, not just in sun-exposed sites, reflecting broader priorities than cutaneous vitamin D production.

SUMMARY

Current theories to explain the basis for both gain of epidermal pigmentation and its subsequent loss are flawed. Pigmentation likely evolved to support permeability barrier function and antimicrobial defense, and became diluted as modern humans left Africa and diverted energy from pigment production toward other more critical purposes (e.g., metabolic conservation). Finally, FLG mutations in Northern Europeans may have emerged recently to favor cutaneous VD3 production.

REFERENCES

Ahn, J., Yu, K., Stolzenberg-Solomon, R., Simon, K.C., Mccullough, M.L., Gallicchio, L., Jacobs, E.J., Ascherio, A., Helzlsouer, K., Jacobs, K.B., Li, Q., Weinstein, S.J., Purdue, M., Virtamo, J., Horst, R., Wheeler, W., Chanock, S., Hunter, D.J., Hayes, R.B., Kraft, P., Albanes, D., 2010. Genome-wide association study of circulating vitamin D levels. Human Molecular Genetics 19, 2739–2745.

Aloia, J.F., 2008. African Americans, 25-hydroxyvitamin D, and osteoporosis: a paradox. American Journal of Clinical Nutrition 88, 545S–550S.

Alonso, S., Izagirre, N., Smith-Zubiaga, I., Gardeazabal, J., Diaz-Ramon, J.L., Diaz-Perez, J.L., Zelenika, D., Boyano, M.D., Smit, N., De La Rua, C., 2008. Complex signatures of selection for the melanogenic loci TYR, TYRP1 and DCT in humans. BMC Evolutionary Biology 8, 74.

Anderson, R.R., Parrish, J.A., 1981. The optics of human skin. Journal of Investigative Dermatology 77, 13–19.

Anno, S., Abe, T., Yamamoto, T., 2008. Interactions between SNP alleles at multiple loci contribute to skin color differences between caucasoid and mongoloid subjects. International Journal of Biological Sciences 4, 81–86.

Aoki, K., 2002. Sexual selection as a cause of human skin colour variation: Darwin's hypothesis revisited. Annals of Human Biology 29, 589–608.

Bikle, D.D., 2010. Vitamin D and the skin. Journal of Bone and Mineral Metabolism 28, 117–130.

Blome, M.W., Cohen, A.S., Tryon, C.A., Brooks, A.S., Russell, J., 2012. The environmental context for the origins of modern human diversity: a synthesis of regional variability in African climate 150,000-30,000 years ago. Journal of Human Evolution 62, 563–592.

Blum, H.F., 1961. Does the melanin pigment of human skin have adaptive value? an essay in human skin have adaptive value? An essay in human ecology and the evolution of race. Quarterly Review of Biology 36, 50–63.

Bogh, M.K., Schmedes, A.V., Philipsen, P.A., Thieden, E., Wulf, H.C., 2010. Vitamin D production after UVB exposure depends on baseline vitamin D and total cholesterol but not on skin pigmentation. Journal of Investigative Dermatology 130, 546–553.

Branda, R.F., Eaton, J.W., 1978. Skin color and nutrient photolysis: an evolutionary hypothesis. Science 201, 625–626.

Brazerol, W.F., Mcphee, A.J., Mimouni, F., Specker, B.L., Tsang, R.C., 1988. Serial ultraviolet B exposure and serum 25 hydroxyvitamin D response in young adult American blacks and whites: no racial differences. Journal of the American College of Nutrition 7, 111–118.

Brookman, J., Chacon, J.N., Sinclair, R.S., 2002. Some photophysical studies of cis- and trans-urocanic acid. Photochemical & Photobiological Sciences 1, 327–332.

Brown, S.J., Mclean, W.H., 2012. One remarkable molecule: filaggrin. Journal of Investigative Dermatology 132, 751–762.

Chaplin, G., 2004. Geographic distribution of environmental factors influencing human skin coloration. American Journal of Physical Anthropology 125, 292–302.

Chaplin, G., Jablonski, N.G., 2009. Vitamin D and the evolution of human depigmentation. American Journal of Physical Anthropology 139, 451–461.

Chen, T.C., Chimeh, F., Lu, Z., Mathieu, J., Person, K.S., Zhang, A., Kohn, N., Martinello, S., Berkowitz, R., Holick, M.F., 2007. Factors that influence the cutaneous synthesis and dietary sources of vitamin D. Archives of Biochemistry and Biophysics 460, 213–217.

Cicarma, E., Mork, C., Porojnicu, A.C., Juzeniene, A., Tam, T.T., Dahlback, A., Moan, J., 2010. Influence of narrowband UVB phototherapy on vitamin D and folate status. Experimental Dermatology 19, e67–72.

Connor, S., 2004. The Book of Skin. Cornell University Press, Ithaca, NY.

Cook, A.L., Chen, W., Thurber, A.E., Smit, D.J., Smith, A.G., Bladen, T.G., Brown, D.L., Duffy, D.L., Pastorino, L., Bianchi-Scarra, G., Leonard, J.H., Stow, J.L., Sturm, R.A., 2009. Analysis of cultured human melanocytes based on polymorphisms within the SLC45A2/MATP, SLC24A5/NCKX5, and OCA2/P loci. Journal of Investigative Dermatology 129, 392–405.

Demenocal, P.B., 2004. African climate change and faunal evolution during the Pliocene-Pleistocene. Earth and Planetary Science Letters 220, 3–24.

Diamond, J., 2004. Anthropology. The astonishing micropygmies. Science 306, 2047–2048.

Drake, D.R., Brogden, K.A., Dawson, D.V., Wertz, P.W., 2008. Thematic review series: skin lipids. Antimicrobial lipids at the skin surface. Journal of Lipid Research 49, 4–11.

Elias, P.M., 2007. The skin barrier as an innate immune element. Seminars in Immunopathology 29, 3–14.

Elias, P.M., Menon, G., Wetzel, B.K., Williams, J.J., 2009. Evidence that stress to the epidermal barrier influenced the development of pigmentation in humans. Pigment Cell & Melanoma Research 22, 420–434.

Elias, P.M., Menon, G., Wetzel, B.K., Williams, J.J., 2010. Barrier requirements as the evolutionary "driver" of epidermal pigmentation in humans. American Journal of Human Biology 22, 526–537.

Elias, P.M., Williams, M.L., 2013. Re-appraisal of current theories for the development and loss of epidermal pigmentation in hominins and modern humans. Journal of Human Evolution 64, 687–692.

Etcoff, N., 1999. Survival of the Prettiest: The Science of Beauty. Doubleday.

Fallon, P.G., Sasaki, T., Sandilands, A., Campbell, L.E., Saunders, S.P., Mangan, N.E., Callanan, J.J., Kawasaki, H., Shiohama, A., Kubo, A., Sundberg, J.P., Presland, R.B., Fleckman, P., Shimizu, N., Kudoh, J., Irvine, A.D., Amagai, M., Mclean, W.H., 2009. A homozygous frameshift mutation in the mouse Flg gene facilitates enhanced percutaneous allergen priming. Nature Genetics 41, 602–608.

Fluhr, J.W., Elias, P.M., 2002. Stratum corneum pH: Formation and function of the 'acid mantle'. Exogenous Dermatology 1, 163–175.

Frost, P., 1988. Human skin color: a possible relationship between its sexual dimorphism and its social perception. Perspectives in Biology and Medicine 32, 38–58.

Frost, P., 2007. Human skin-color sexual dimorphism: a test of the sexual selection hypothesis. American Journal of Physical Anthropology 133, 779–780 author reply 780–1.

Gabay, C., Kushner, I., 1999. Acute-phase proteins and other systemic responses to inflammation. The New England Journal of Medicine 340, 448–454.

Gilchrest, B.A., 2008. Sun exposure and vitamin D sufficiency. American Journal of Clinical Nutrition 88, 570S–577S.

Goding, C.R., 2007. Melanocytes: the new Black. International Journal of Biochemistry & Cell Biology 39, 275–279.

Graf, J., Hodgson, R., Van Daal, A., 2005. Single nucleotide polymorphisms in the MATP gene are associated with normal human pigmentation variation. Human Mutation 25, 278–284.

Gruber, R., Elias, P.M., Crumrine, D., Lin, T.K., Brandner, J.M., Hachem, J.P., Presland, R.B., Fleckman, P., Janecke, A.R., Sandilands, A., Mclean, W.H., Fritsch, P.O., Mildner, M., Tschachler, E., Schmuth, M., 2011. Filaggrin genotype in ichthyosis vulgaris predicts abnormalities in epidermal structure and function. American Journal of Pathology 178, 2252–2263.

Gunathilake, R., Schurer, N.Y., Shoo, B.A., Celli, A., Hachem, J.P., Crumrine, D., Sirimanna, G., Feingold, K.R., Mauro, T.M., Elias, P.M., 2009. pH-regulated mechanisms account for pigment-type differences in epidermal barrier function. Journal of Investigative Dermatology 129, 1719–1729.

Haralampus-Grynaviski, N., Ransom, C., Ye, T., Rozanowska, M., Wrona, M., Sarna, T., Simon, J.D., 2002. Photogeneration and quenching of reactive oxygen species by urocanic acid. Journal of the American Chemical Society 124, 3461–3468.

Haratake, A., Uchida, Y., Mimura, K., Elias, P.M., Holleran, W.M., 1997a. Intrinsically aged epidermis displays diminished UVB-induced alterations in barrier function associated with decreased proliferation. Journal of Investigative Dermatology 108, 319–323.

Haratake, A., Uchida, Y., Schmuth, M., Tanno, O., Yasuda, R., Epstein, J.H., Elias, P.M., Holleran, W.M., 1997b. UVB-induced alterations in permeability barrier function: roles for epidermal hyperproliferation and thymocyte-mediated response. Journal of Investigative Dermatology 108, 769–775.

Harding, R.M., Healy, E., Ray, A.J., Ellis, N.S., Flanagan, N., Todd, C., Dixon, C., Sajantila, A., Jackson, I.J., Birch-Machin, M.A., Rees, J.L., 2000. Evidence for variable selective pressures at MC1R. American Journal of Human Genetics 66, 1351–1361.

Hill, H.Z., 1992. The function of melanin or six blind people examine an elephant. Bioessays 14, 49–56.

Hochberg, Z., Templeton, A.R., 2010. Evolutionary perspective in skin color, vitamin D and its receptor. Hormones (Athens) 9, 307–311.

Holick, M.F., 1995. Environmental factors that influence the cutaneous production of vitamin D. American Journal of Clinical Nutrition 61, 638S–645S.

Holick, M.F., Maclaughlin, J.A., Clark, M.B., Holick, S.A., Potts Jr., J.T., Anderson, R.R., Blank, I.H., Parrish, J.A., Elias, P., 1980. Photosynthesis of previtamin D3 in human skin and the physiologic consequences. Science 210, 203–205.

Holick, M.F., Maclaughlin, J.A., Doppelt, S.H., 1981. Regulation of cutaneous previtamin D3 photosynthesis in man: skin pigment is not an essential regulator. Science 211, 590–593.

Holleran, W.M., Uchida, Y., Halkier-Sorensen, L., Haratake, A., Hara, M., Epstein, J.H., Elias, P.M., 1997. Structural and biochemical basis for the UVB-induced alterations in epidermal barrier function. Photodermatology, Photoimmunology & Photomedicine 13, 117–128.

Hong, S.P., Kim, M.J., Jung, M.Y., Jeon, H., Goo, J., Ahn, S.K., Lee, S.H., Elias, P.M., Choi, E.H., 2008. Biopositive effects of low-dose UVB on epidermis: coordinate upregulation of antimicrobial peptides and permeability barrier reinforcement. Journal of Investigative Dermatology 128, 2880–2887.

Honigsmann, H., 2002. Erythema and pigmentation. Photodermatology, Photoimmunology & Photomedicine 18, 75–81.

Irvine, A.D., Mclean, W.H., Leung, D.Y., 2011. Filaggrin mutations associated with skin and allergic diseases. The New England Journal of Medicine 365, 1315–1327.

Izagirre, N., Garcia, I., Junquera, C., De La Rua, C., Alonso, S., 2006. A scan for signatures of positive selection in candidate loci for skin pigmentation in humans. Molecular Biology and Evolution 23, 1697–1706.

Jablonski, N.G., 2010. The naked truth. Scientific American 302, 42–49.

Jablonski, N.G., 2012. Living Color: The Biological and Social Meaning of Skin Color. University of California Press.

Jablonski, N.G., Chaplin, G., 2000. The evolution of human skin coloration. Journal of Human Evolution 39, 57–106.

Jablonski, N.G., Chaplin, G., 2010. Colloquium paper: human skin pigmentation as an adaptation to UV radiation. Proceedings of the National Academy of Sciences of the United States of America 107 (Suppl. 2), 8962–8968.

Jablonski, N.G., Chaplin, G., 2013. Epidermal pigmentation in the human lineage is an adaptation to ultraviolet radiation. Journal of Human Evolution 65, 671–675.

Jew, S., Abumweis, S.S., Jones, P.J., 2009. Evolution of the human diet: linking our ancestral diet to modern functional foods as a means of chronic disease prevention. Journal of Medicinal Food 12, 925–934.

Juzeniene, A., Stokke, K.T., Thune, P., Moan, J., 2010. Pilot study of folate status in healthy volunteers and in patients with psoriasis before and after UV exposure. Journal of Photochemistry and Photobiology B 101, 111–116.

Korting, H.C., Hubner, K., Greiner, K., Hamm, G., Braun-Falco, O., 1990. Differences in the skin surface pH and bacterial microflora due to the long-term application of synthetic detergent preparations of pH 5.5 and pH 7.0. Results of a crossover trial in healthy volunteers. Acta Dermato-Venereologica 70, 429–431.

Korting, H.C., Kober, M., Mueller, M., Braun-Falco, O., 1987. Influence of repeated washings with soap and synthetic detergents on pH and resident flora of the skin of forehead and forearm. Results of a cross-over trial in health probationers. Acta Dermato-Venereologica 67, 41–47.

Kripke, M.L., 1984. Skin cancer, photoimmunology, and urocanic acid. Photodermatology 1, 161–163.

Lamason, R.L., Mohideen, M.A., Mest, J.R., Wong, A.C., Norton, H.L., Aros, M.C., Jurynec, M.J., Mao, X., Humphreville, V.R., Humbert, J.E., Sinha, S., Moore, J.L., Jagadeeswaran, P., Zhao, W., Ning, G., Makalowska, I., Mckeigue, P.M., O'donnell, D., Kittles, R., Parra, E.J., Mangini, N.J., Grunwald, D.J., Shriver, M.D., Canfield, V.A., Cheng, K.C., 2005. SLC24A5, a putative cation exchanger, affects pigmentation in zebrafish and humans. Science 310, 1782–1786.

Lao, O., De Gruijter, J.M., Van Duijn, K., Navarro, A., Kayser, M., 2007. Signatures of positive selection in genes associated with human skin pigmentation as revealed from analyses of single nucleotide polymorphisms. Annals of Human Genetics 71, 354–369.

Latham, M.C., 1991. The dermatosis of kwashiorkor in young children. Seminars in Dermatology 10, 270–272.

Liu, J., Man, W.Y., Lv, C.Z., Song, S.P., Shi, Y.J., Elias, P.M., Man, M.Q., 2010. Epidermal permeability barrier recovery is delayed in vitiligo-involved sites. Skin Pharmacology and Physiology 23, 193–200.

Loomis, W.F., 1967. Skin-pigment regulation of vitamin-D biosynthesis in man. Science 157, 501–506.

Mackintosh, J.A., 2001. The antimicrobial properties of melanocytes, melanosomes and melanin and the evolution of black skin. Journal of Theoretical Biology 211, 101–113.

Man, M.Q., Lin, T.K., Santiago, J.L., Celli, A., Zhong, L., Huang, Z.M., Roelandt, T., Hupe, M., Sundberg, J.P., Silva, K.A., Crumrine, D., Martin-Ezquerra, G., Trullas, C., Sun, R., Wakefield, J.S., Wei, M.L., Feingold, K.R., Mauro, T.M., Elias, P.M., 2014. Basis for enhanced barrier function of pigmented skin. Journal of Investigative Dermatology 134, 2399–2407.

Marconi, A., Terracina, M., Fila, C., Franchi, J., Bonte, F., Romagnoli, G., Maurelli, R., Failla, C.M., Dumas, M., Pincelli, C., 2003. Expression and function of neurotrophins and their receptors in cultured human keratinocytes. Journal of Investigative Dermatology 121, 1515–1521.

Marks, R., Foley, P.A., Jolley, D., Knight, K.R., Harrison, J., Thompson, S.C., 1995. The effect of regular sunscreen use on vitamin D levels in an Australian population. Results of a randomized controlled trial. Archives of Dermatology 131, 415–421.

Mcevoy, B., Beleza, S., Shriver, M.D., 2006. The genetic architecture of normal variation in human pigmentation: an evolutionary perspective and model. Human Molecular Genetics 15 (2), R176–R181.

Mcloone, P., Simics, E., Barton, A., Norval, M., Gibbs, N.K., 2005. An action spectrum for the production of cis-urocanic acid in human skin in vivo. Journal of Investigative Dermatology 124, 1071–1074.

Mildner, M., Jin, J., Eckhart, L., Kezic, S., Gruber, F., Barresi, C., Stremnitzer, C., Buchberger, M., Mlitz, V., Ballaun, C., Sterniczky, B., Fodinger, D., Tschachler, E., 2010. Knockdown of filaggrin impairs diffusion barrier function and increases UV sensitivity in a human skin model. Journal of Investigative Dermatology 130, 2286–2294.

Miller, S.J., Aly, R., Shinefeld, H.R., Elias, P.M., 1988. In vitro and in vivo antistaphylococcal activity of human stratum corneum lipids. Archives of Dermatology 124, 209–215.

Moan, J., Nielsen, K.P., Juzeniene, A., 2012. Immediate pigment darkening: its evolutionary roles may include protection against folate photosensitization. FASEB Journal 26, 971–975.

Montefiori, D.C., Zhou, J.Y., 1991. Selective antiviral activity of synthetic soluble L-tyrosine and L-dopa melanins against human immunodeficiency virus in vitro. Antiviral Research 15, 11–25.

Murray, F., 1934. Pigmentation, sunlight, and nutritional disease. American Anthropologist 36, 438–445.

Nakatsuji, T., Gallo, R.L., 2012. Antimicrobial peptides: old molecules with new ideas. Journal of Investigative Dermatology 132, 887–895.

Neer, R.M., 1975. The evolutionary significance of vitamin D, skin pigment, and ultraviolet light. American Journal of Physical Anthropology 43, 409–416.

Northrup, H., Volcik, K.A., 2000. Spina bifida and other neural tube defects. Current Problems in Pediatrics 30, 313–332.

Norton, H.L., Hammer, M., 2008. Sequence variation in the pigmentation candidate gene SLC24A5 and evidence for independent evolution of light skin in European and East Asian populations. In: 77th Annal Mtg of the Amer Assn of Phys Anthropologists, pp. 9–12 Columbus, OH.

Norton, H.L., Kittles, R.A., Parra, E., Mckeigue, P., Mao, X., Cheng, K., Canfield, V.A., Bradley, D.G., Mcevoy, B., Shriver, M.D., 2007. Genetic evidence for the convergent evolution of light skin in Europeans and East Asians. Molecular Biology and Evolution 24, 710–722.

Olalde, I., Allentoft, M.E., Sanchez-Quinto, F., Santpere, G., Chiang, C.W., Degiorgio, M., Prado-Martinez, J., Rodriguez, J.A., Rasmussen, S., Quilez, J., Ramirez, O., Marigorta, U.M., Fernandez-Callejo, M., Prada, M.E., Encinas, J.M., Nielsen, R., Netea, M.G., Novembre, J., Sturm, R.A., Sabeti, P., Marques-Bonet, T., Navarro, A., Willerslev, E., Lalueza-Fox, C., 2014. Derived immune and ancestral pigmentation alleles in a 7,000-year-old Mesolithic European. Nature 507, 225–228.

Parra, E.J., 2007. Human pigmentation variation: evolution, genetic basis, and implications for public health. American Journal of Physical Anthropology 45, 85–105.

Parrish, J.A., Jaenicke, K.F., Anderson, R.R., 1982. Erythema and melanogenesis action spectra of normal human skin. Photochemistry and Photobiology 36, 187–191.

Post, P.W., Daniels Jr., F., Binford Jr., R.T., 1975. Cold injury and the evolution of "white" skin. Human Biology 47, 65–80.

Rana, B.K., Hewett-Emmett, D., Jin, L., Chang, B.H., Sambuughin, N., Lin, M., Watkins, S., Bamshad, M., Jorde, L.B., Ramsay, M., Jenkins, T., Li, W.H., 1999. High polymorphism at the human melanocortin 1 receptor locus. Genetics 151, 1547–1557.

Rasmussen, L.B., Andersen, N.L., Andersson, G., Lange, A.P., Rasmussen, K., Skak-Iversen, L., Skovby, F., Ovesen, L., 1998. Folate and neural tube defects. Recommendations from a Danish working group. Danish Medical Bulletin 45, 213–217.

Rayburn, W.F., Stanley, J.R., Garrett, M.E., 1996. Periconceptional folate intake and neural tube defects. Journal of the American College of Nutrition 15, 121–125.

Reed, J.T., Ghadially, R., Elias, P.M., 1995. Skin type, but neither race nor gender, influence epidermal permeability barrier function. Archives of Dermatology 131, 1134–1138.

Robins, A., 2009. The evolution of light skin color: role of vitamin D disputed. American Journal of Physical Anthropology 139, 447–450.

Rockell, J.E., Skeaff, C.M., Williams, S.M., Green, T.J., 2008. Association between quantitative measures of skin color and plasma 25-hydroxyvitamin D. Osteoporosis International 19, 1639–1642.

Rogers, A., Iltis, D., Wooding, S., 2004. Genetic variation at the MC1R locus and the time since loss of human body hair. Current Anthropology 45, 105–124.

Scharschmidt, T.C., Man, M.Q., Hatano, Y., Crumrine, D., Gunathilake, R., Sundberg, J.P., Silva, K.A., Mauro, T.M., Hupe, M., Cho, S., Wu, Y., Celli, A., Schmuth, M., Feingold, K.R., Elias, P.M., 2009. Filaggrin deficiency confers a paracellular barrier abnormality that reduces inflammatory thresholds to irritants and haptens. Journal of Allergy and Clinical Immunology 124, 496–506 506 e1–6.

Schroder, J.M., Harder, J., 2006. Antimicrobial skin peptides and proteins. Cellular and Molecular Life Sciences 63, 469–486.

Scott, I.R., 1981. Factors controlling the expressed activity of histidine ammonia-lyase in the epidermis and the resulting accumulation of urocanic acid. Biochemical Journal 194, 829–838.

Sharma, S., Barr, A.B., Macdonald, H.M., Sheehy, T., Novotny, R., Corriveau, A., 2011. Vitamin D deficiency and disease risk among aboriginal Arctic populations. Nutrition Reviews 69, 468–478.

Soejima, M., Koda, Y., 2007. Population differences of two coding SNPs in pigmentation-related genes SLC24A5 and SLC45A2. International Journal of Legal Medicine 121, 36–39.

Sturm, R.A., 2009. Molecular genetics of human pigmentation diversity. Human Molecular Genetics 18, R9–R17.

Takeda, K., Takahashi, N.H., Shibahara, S., 2007. Neuroendocrine functions of melanocytes: beyond the skin-deep melanin maker. Tohoku Journal of Experimental Medicine 211, 201–221.

Thomson, M.L., 1955. Relative efficiency of pigment and horny layer thickness in protecting the skin of Europeans and Africans against solar ultraviolet radiation. Journal of Physiology 127, 236–246.

Thyssen, J.P., Bikle, D.D., Elias, P.M., 2014. Evidence that loss-of-function filaggrin gene mutations evolved in northern Europeans to favor intracutaneous vitamin D3 production. Evolutionary Biology 41, 388–396.

Thyssen, J.P., Thuesen, B., Huth, C., Standl, M., Carson, C.G., Heinrich, J., Kramer, U., Kratzsch, J., Berg, N.D., Menne, T., Johansen, J.D., Carlsen, B.C., Schwab, S., Thorand, B., Munk, M., Wallaschofski, H., Heickendorff, L., Meldgaard, M., Szecsi, P.B., Stender, S., Bonnelykke, K., Weidinger, S., Bisgaard, H., Linneberg, A., 2012. Skin barrier abnormality caused by filaggrin (FLG) mutations is associated with increased serum 25-hydroxyvitamin D concentrations. Journal of Allergy and Clinical Immunology 130, 1204–1207 e2.

Uchida, Y., Nardo, A.D., Collins, V., Elias, P.M., Holleran, W.M., 2003. De novo ceramide synthesis participates in the ultraviolet B irradiation-induced apoptosis in undifferentiated cultured human keratinocytes. Journal of Investigative Dermatology 120, 662–669.

Van Den Berghe, P., Frost, P., 1986. Skin color preference, sexual dimorphism and sexual selection: a case of gene culture co-evolution? Ethnic and Racial Studies.

Vivanco-Munoz, N., Jo, T., Gerardo, H.B., Juan, T., Clark, P., 2012. Physical activity and dark skin tone: protective factors against low bone mass in Mexican men. Journal of Clinical Densitometry 15, 374–379.

Wassermann, H.P., 1965. Human pigmentation and environmental adaptation. Archives of Environmental Health 11, 691–694.

Wilson, R.D., Davies, G., Desilets, V., Reid, G.J., Summers, A., Wyatt, P., Young, D., 2003. The use of folic acid for the prevention of neural tube defects and other congenital anomalies. Journal of Obstetrics and Gynaecology Canada 25, 959–973.

Wilson, R.D., Johnson, J.A., Wyatt, P., Allen, V., Gagnon, A., Langlois, S., Blight, C., Audibert, F., Desilets, V., Brock, J.A., Koren, G., Goh, Y.I., Nguyen, P., Kapur, B., 2007. Pre-conceptional vitamin/folic acid supplementation 2007: the use of folic acid in combination with a multivitamin supplement for the prevention of neural tube defects and other congenital anomalies. Journal of Obstetrics and Gynaecology Canada 29, 1003–1026.

Young, A.R., Chadwick, C.A., Harrison, G.I., Nikaido, O., Ramsden, J., Potten, C.S., 1998. The similarity of action spectra for thymine dimers in human epidermis and erythema suggests that DNA is the chromophore for erythema. Journal of Investigative Dermatology 111, 982–988.

Yuen, A.W., Jablonski, N.G., 2010. Vitamin D: in the evolution of human skin colour. Medical Hypotheses 74, 39–44.

Chapter 20

Human Growth and Development

Barry Bogin
Centre for Global Health & Human Development, School of Sport, Exercise & Health Sciences, Loughborough University, Loughborough, UK

SYNOPSIS

A biocultural approach is taken to the study of the evolution of human growth and development. The biocultural perspective of human development focuses on the constant interaction taking place during all phases of human development, between both genes and hormones within the body, and with the sociocultural environment that surrounds the body. The pattern of human postnatal growth and development—the stages of infancy, childhood, juvenility, and adolescence—is reviewed. Several hypotheses are discussed concerning how the new life stages of the human life cycle represent feeding and reproductive specializations, secondarily allowing for the human style of learning and cultural behavior.

Growth may be defined as a quantitative increase in size or mass. Measurements of height in centimeters or weight in kilograms indicate how much growth has taken place in a child. Development is defined as a progression of changes, either quantitative or qualitative, that lead from an undifferentiated or immature state to a highly organized, specialized, and mature state. Physical maturation is measured by functional capacity; for example, the maturation of bipedal walking results from changes with age in the skeletal, muscular, and motor skills of the infant and child.

GROWTH AND EVOLUTION

Human growth, development, and maturation have evolved, sometimes as discrete processes but more often as an integrated series of biological events. Biological anthropologists and human biologists have long been interested in how human growth, development, senescence, and aging differ from the corresponding processes in other apes, our closest phylogenic relatives, other nonhuman primates, and mammals. It is easy to document these differences, such as altricial offspring, slow and prolonged growth including childhood and adolescence stages, a late start to reproduction, menopause (see the chapter by Sievert in this volume), survival into the eighth and ninth decades, and a maximum life span of over 122 years (Crews and Bogin, 2010). Determining the evolutionary forces that produced these and other aspects of life history has not been as easy.

The evolution of the human pattern of growth may be understood by the study of growth and development in fossil species and by comparison of growth patterns in living species, especially the Primates. It is now clear that no living species of nonhuman primate has all the characteristics of human growth, in particular the human childhood and adolescent stages of life (these are defined and discussed below). This strongly suggests that the human pattern could only have evolved within the taxonomic group of the hominins, which includes the human species and extinct members of the genera *Homo, Australopithecus*, and perhaps older hominins that evolved by six to seven million years ago in Africa (see chapters by Hunt, Ward, and Simpson).

HUMAN VERSUS CHIMPANZEE GROWTH

Stages in the human life cycle are outlined in Table 1. In this entry, the focus is on human postnatal growth and development prior to adulthood, which may be divided into the stages of infancy, childhood, juvenility puberty, and adolescence (a more detailed explanation for the entire life cycle is found in Bogin, 2010; Bogin and Smith, 2012). Each stage may be defined by characteristics of dentition, changes related to methods of feeding, physical and mental competencies, and maturation of the reproductive system and sexual behavior. The most visually direct characteristic of each stage is its rate of growth. Shown in Figure 1 are rates and amounts of growth in height that take place between birth and adulthood for normal boys and girls. Growth in weight follows very similar curves. In Figure 1, the distance curve of growth—that is, the amount of growth achieved from year to year—is labeled on the right *y*-axis. The velocity curve, which represents the rate of growth during any one year, is labeled on the left *y*-axis. Below the velocity curve are symbols for each postnatal stage.

Presented in Figure 2 are the distance and velocity curves of body length growth for the chimpanzee. The figure is based on the longitudinal study of captive chimpanzee growth conducted by Hamada and Udono (2002). Postnatal growth of the chimpanzee has only two stages, infancy and juvenility prior to adulthood. The chimpanzees of this study were raised at two research institutes in Japan. As infants, five were nursed by their mothers and seven were bottle fed by human caregivers. After the nursing or bottle-feeding period, all were transferred to social groups with age peers, except for one infant chimp who remained with his mother. During their juvenile growth stage some of the chimpanzees were housed in social groups and some in individual cages. All chimpanzees were measured serially from infancy until adulthood. The measurements and routine medical examinations were performed at three-month or six-month intervals, with the chimpanzees anesthetized. These chimpanzees were given good care, but their captivity and treatment may have influenced their physical growth.

Infancy for both human beings and chimpanzees is characterized by the most rapid velocity of body growth of any of the postnatal stages, but also by a steep decline in velocity, or deceleration. Infancy of humans, chimpanzees, and other mammalian species is comparable in many other respects, such as feeding by maternal lactation and the appearance of deciduous teeth. However, in most mammals, including chimpanzees, infancy and lactation end with eruption of the first permanent molars. This occurs between the age of 48 and 60 months in chimpanzees, as indicated by the "W" (weaning, defined as cessation of nursing) in Figure 2. Note the change in chimpanzee growth velocity at the time of weaning. At this point the chimpanzee enters the juvenile growth stage. Juvenile mammals are largely responsible for their own care and feeding, but are still sexually immature.

In humans, by contrast, there is an interval of about three years between weaning, which takes place at a median age of 30–36 months in preindustrial societies, and eruption of the first permanent molars at about six years of age. This interval is the stage of life described here as childhood. The biological constraints of childhood, which include an immature dentition, a small digestive system, and a calorie-demanding brain that is both relatively large and growing rapidly, necessitate care and feeding that older individuals must provide.

TABLE 1 Stages in the Human Life Cycle and Life History

Stage	Growth Events/Duration (Approximate or Average)
Prenatal Development	
Fertilization	
First trimester	Fertilization to 12th week: embryogenesis
Second trimester	Fourth through sixth lunar month: rapid growth in length
Third trimester	Seventh lunar month to birth: rapid growth in weight and organ maturation
Birth	
Postnatal Development	
Neonatal period	Birth to 28 days: extrauterine adaptation, most rapid of postnatal growth and maturation
Infancy	Second month to end of lactation, usually by 36 months: rapid growth velocity, but with steep deceleration in growth rate, feeding by lactation to six months of age and then lactation with gradual introduction of complementary foods, deciduous tooth eruption, many developmental milestones in physiology, behavior, and cognition
Childhood	Years 3.0–6.9: moderate growth rate, dependency on older people for care and feeding, midgrowth spurt, eruption of first permanent molar and incisor, virtual completion of brain growth by end of stage
Juvenility	Years 7–10 for girls, 7–12 for boys: slower growth rate, capable of self-feeding, cognitive transition leading to learning of economic and social skills
Puberty	An event of short duration (days or a few weeks) at end of juvenility stage: reactivation in the hypothalamus of the GnRH pulse generator, dramatic increase in secretion of sex hormones from the ovaries/testes
Adolescence	The stage of development that lasts for 5–10 years after the onset of puberty: growth spurt in height and weight; permanent tooth eruption almost complete; development of secondary sexual characteristics; sociosexual maturation; intensification of interest in and practice of adult social, economic, and sexual activities
Adulthood	
Prime and transition	From 18–20 years of age for women to 45 years (end of childbearing) and from 21 to 25 years of age for men to about 55 years of age: commences with completion of skeletal growth, homeostasis in physiology, behavior, and cognition; loss of fecundity and menopause for women by age 50; fecundity for men may decline with age, but does not drop to zero at any age.
Old age and senescence	From end of childbearing years to death: decline in the function and repair ability of many body tissues or systems
Death	

The rate of body growth during childhood proceeds at a steady 5–6 cm per year. These growth rates are typical for healthy, well-nourished children. Indeed, the pattern of growth velocity from birth to adulthood is highly similar in all human populations, but growth rates and the total amount of growth will vary in relation to health and nutritional status. Many children experience a transient and small "spurt" in growth rate as they transition into the juvenile period. In many traditional human societies (such as hunter-gatherers, horticulturalists, and pastoralists), juveniles perform important work including food production and the care of children (i.e., "babysitting"). Juvenile growth rates decline until puberty. Here we define puberty as an event that takes place within the central nervous system resulting in a change in the regulation of hormone production and secretion, and initiating sexual maturation and the adolescent life stage (Bogin, 2010, 2011). The hormones responsible for sexual maturation also cause the adolescent growth spurt in stature and other skeletal dimensions. Growth of the skeleton ends at about 18–19 years for girls and 20–22 years for boys, and with this the adulthood or reproductive stage of life history begins.

Significant changes in motor control, cognitive function, and emotions are associated with infant, child, juvenile, and adolescent development—and are related to brain growth and maturation. Infancy and childhood are the times of most rapid postnatal brain growth in human beings. At birth, human brain size is about three times that of the chimpanzee—384 g versus 128 g. At adulthood, the difference increases to about 3.3 times—1352 versus grams 410 (Bogin

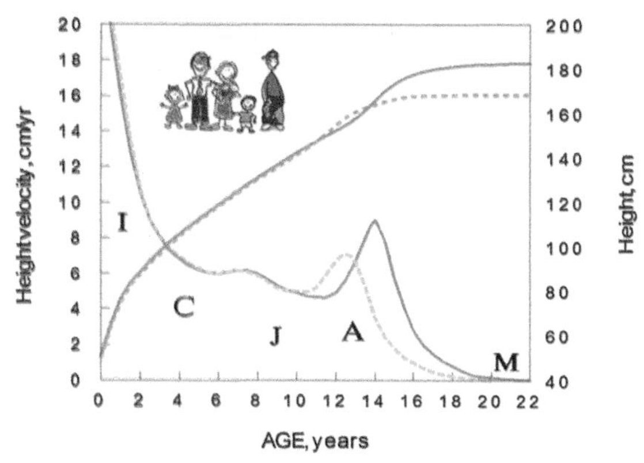

FIGURE 1 Distance and velocity curves of growth for healthy, well-nourished human beings. Boys, solid line; girls, dashed line. These are modal curves based on height data for Western European and North American populations. The stages of postnatal growth are abbreviated as follows: I, infancy; C, childhood; J, juvenility A, adolescence; M, mature adult. Weaning takes place at a mean age of 30–36 months—at the transition from infancy to childhood. *(Modified from Bogin, 1999.)* The distance curve (right y-axis) indicates the amount of height achieved at a given age. The velocity curve (left y-axis) indicates the rate of growth at a given age. Growth velocity during infancy is rapid with a steep deceleration. Childhood growth is relatively constant at about 6 cm per year. Growth rate slows during the juvenility stage and then accelerates during the first phase of adolescence—the adolescent growth spurt. Growth rates decline during the second phase until all growth in height stops at the onset of the adult stage. The image of the "family" is not meant to promote any particular type of family as desirable or normal; rather, the cartoon figures illustrate the stages of human life history between birth and adulthood from left to right—juvenile, adult male, adult female with infant, child, and adolescent. (See color plate section).

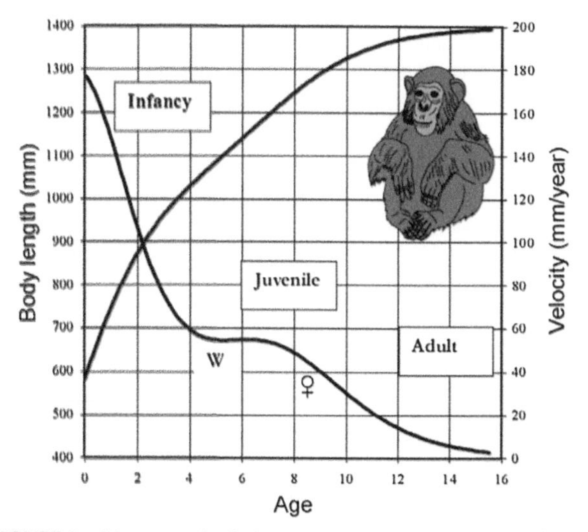

FIGURE 2 Distance and velocity curves for chimpanzee growth in body length (Hamada and Udono, 2002). In the wild, weaning (W) usually takes place between 48 and 60 months of age (Pusey, 1983).

and Smith, 2012). Most of the adult difference is achieved by the end of the human childhood growth stage, which has led to hypotheses that the evolution of childhood was to provide additional time for increased embodied capital, meaning larger brain size, greater learning, higher quality bodies, more complex social networks, and generally, human cultural behavior (Kaplan et al., 2000). An alternative hypothesis is that the primary selective pressure for a childhood stage was the reproductive advantages that accrued to the mother.

Chimpanzee and human mothers have very different reproductive patterns. Chimp mothers must feed and care for their infants for almost five years. No other chimpanzee helps, and the death of the mother usually results in the death of the infant. At five years of age, the young chimpanzee becomes a juvenile and must forage for food on its own and protect itself from dangers. Chimpanzee females must space their births at five-year intervals, as they cannot care for two infants simultaneously. This places constraints on chimpanzee reproductive success. In the wild, chimpanzee females are just able to produce and raise two offspring to adulthood. As a consequence, chimpanzee populations are stable, with equal numbers of births and deaths.

Human women in premodern and modern societies are almost never the sole caretakers of their infants. Other family members, including siblings, parents, husbands, and other children of the mother provide significant amounts of care to mothers and their infants (Hrdy, 2009). The major exceptions are the WEIRD societies—Western, Educated, Industrialized, Rich, and Democratic (Henrich et al., 2010). The social segregation, and often isolation, of mothers, children, adolescents, and the elderly from each other and working adults that is commonplace in North America, much of Western Europe, Japan, Australia, and other WEIRD nations is of very recent origin in human history. The typical human social condition for infant care for 99% of human evolutionary history (at least the past 125,000 years and maybe much longer) was cooperative breeding that eventually evolved into biocultural reproduction.

Cooperative breeding is found in some species of birds and mammals, such as wolves and hyenas, and it works to increase net reproductive output. Only two nonhuman primate groups are cooperative breeders, the marmosets and tamarins of South America. In those species, and in some but not all human groups, the cooperative breeders are close genetic relatives of the mother. By assisting the mother to care for her offspring, rather than selfishly caring only for themselves and their own offspring, the helpers increase their own inclusive fitness, meaning that they help to ensure that their genetic kin survive to reproductive age. Biologists define "fitness" of a species by the number of offspring produced that reach their own reproductive maturity.

Unlike the other cooperative breeding species, human societies define behavioral roles related to infant and child care on the basis of marriage and kinship categories. These often have greater social than genetic foundations. Hill et al. (2011) surveyed 32 present-day foraging societies, including the !Kung (Ju/'hoansi), Hadza, and Ache, and reported that human hunter-gatherer societies have a social structure that is unique among all primates. Hill and colleagues found that most individuals in residential groups were genetically unrelated, or at least not genetically related by descent from common parents or grandparents. This is due to the practice by both men and women of either dispersing from or remaining within their natal group for marriage. Migration to new groups dilutes genetic relationships and requires social kinship designations to structure relationships among people.

Human reproductive cooperation does more than increase biological fitness—it also enhances the technological, social, economic, and political "fitness" of the group. For this reason, the phrase biocultural reproduction seems to best describe the human practice (Bogin et al., 2014). The result of the human type of biocultural reproduction is that a human woman can successfully produce two or more infants in the time it takes a chimpanzee to produce one. It also means that 60%–97% of human infants survive to adulthood. Even in the traditional societies of hunters and gatherers, 50%–60% of live-born infants survived to adulthood. Chimpanzees, in contrast, successfully raise only 36% of their offspring to adulthood. In Darwinian evolutionary terms, these are significant biological advantages and explain, in part, why chimpanzee populations number in the thousands while humans number more than seven billion.

ADOLESCENCE

Puberty resets the central nervous system to resume positive feedback within the hypothalamic–pituitary–gonadal (HPG) axis. During fetal development and until about two to three years after birth, production of gonadotropin-releasing hormone by cells in the hypothalamus promotes the release of luteinizing hormone and follicle-stimulating hormone from cells in the anterior lobe of the pituitary gland (Bogin, 2011; Savage et al., 2011). These hormones stimulate the growth and development of the testes or ovaries, the release of androgens and estrogens, and sexual maturation of the fetus and infant. This process is paused in middle to late infancy when the HPG axis switches to negative feedback, by which circulating androgens and estrogens, as well as pituitary hormones, inhibit the hypothalamus (Ellison et al., 2012).

The change in regulation of the HPG axis from positive to negative feedback by the onset of the childhood growth stage, and the return to positive feedback at puberty, is common to many species of African monkeys, all apes, and human beings (Plant, 2008). Unique to humans is the shift from decelerating to accelerating skeletal growth that follows puberty (Bogin, 2010, 2011). Monkeys and apes have localized accelerations in growth, for example in the jaws, to accommodate projecting canine teeth in some species. Many nonhuman primate species have accelerations in body mass at puberty, especially those with male muscle development. Examples are gorillas and orangutans, as in these species adult males come to weigh more than twice as much as adult females. Nonhuman primates, however, do not have the human body-wide acceleration in the growth, development, and maturation of all long bones of the skeleton (such as the tibia, femur, and humerus), as well as the vertebrae, jaws, and other small bones. This systemic acceleration of skeletal biology following puberty is the human adolescent growth spurt. The spurt is clearly depicted in Figure 1. The spurt is absent in the chimpanzee (Figure 2) and other nonhuman primates (Bogin, 2010; Bogin and Smith, 2012).

There is more to human adolescence than just the skeletal growth spurt. In most species of primates, puberty is followed within a few months or a year by first reproduction. In humans, the delay between puberty and first reproduction is greater, usually on the order of 5–10 years. This interval is the human adolescent growth stage. In humans, the gonadal hormones responsible for the growth spurt also promote sexual maturation. In both sexes there is a sudden increase in the density of pubic hair and often other body hair. In boys, there may be an increased density and darkening of facial hair. The deepening of the voice (voice "cracking") is another sign of male puberty. In girls, a visible sign is the development of the breast bud, the first stage of breast development, which often precedes the appearance of dense pubic hair. The pubescent boy or girl, his or her parents, and relatives, friends, and sometimes everyone else in the social group can observe one or more of these signs of early adolescence.

Another notable feature of adolescence is the completion of permanent tooth eruption; the second permanent molar erupts at about 12 years of age, and the third molar at about 18–21 years. In addition, the years of the adolescence stage include the development of secondary sexual characteristics such as fat patterning and muscularity typical of each sex. At all ages from birth onward, girls on average have more body fat than boys. Just before the onset of puberty, at about nine years of age, the differences between the sexes are relatively small. Girls in the United States average about 30% body fat and 65% soft lean tissue (e.g., muscle and connective tissue) for total weight and boys average 25% body fat and 70% soft lean tissue. In both the girls and the boys the remaining 5% of body weight is made up of hard lean tissue such as bone. By 20 years of age, the differences become greater. Young women average 38% body fat and 60% soft lean tissue, and young men average 22% body fat and 71% soft lean tissue (Borrud et al., 2010). During adolescence, women tend to add body fat to breasts, buttocks, and thighs, while men tend to add muscle over the entire body. Another feature of adolescent development, common to both girls and boys, is the intensification of interest in and practice of adult social, economic, and sexual activities leading to sociosexual maturation in early adulthood.

EVOLUTION OF HUMAN ADOLESCENCE

Some theorists hypothesize that the adolescent stage of human growth evolved to provide the time to learn and practice the complex economic, social, and sexual skills required for effective food production, reproduction, and parenting. In this perspective, adolescence is a time for apprenticeship, working and learning alongside older and more experienced members of the social group. The benefits of skills acquired during adolescence are lower mortality of both first-time mothers and their offspring.

However, apprenticeship cannot be the primary cause for the evolution of adolescence. Learning for child care is an example. In most species of social mammals, the juveniles are often segregated from adults and infants. The ethnographic literature, however, documents that in human societies juvenile girls often are expected to provide significant amounts of child care for their younger siblings. Human girls enter adolescence with considerable knowledge of the needs of young children. Learning about child care, then, is not the reason why human girls experience adolescence. A similar case may be made for boys' learning of most skills needed for successful adulthood. More to the point, the additional 5–10 years of lack of fecundability associated with adolescence could not evolve by natural selection, since those individuals who "cheated" by terminating growth and reproductively maturing at an earlier age would begin reproducing sooner and would be at a reproductive advantage. All other primates do, in fact, begin reproducing at earlier ages than humans, and none of the nonhuman primates has a humanlike adolescent growth spurt or many of the other biological and behavioral features of human adolescence. Clearly, a juvenile primate does not need to pass through a lengthy period of adolescence, with apprenticeship-type learning, just to be reproductively successful (Bogin and Smith, 2012). As was the case for the evolution of human childhood, the evolution of human adolescence may be due in large part to the human style of biocultural reproduction. A multilevel model of selection for mating and parenting has been proposed for the evolution of human adolescence (Bogin, 2009). Multilevel models in evolutionary biology include selection at the level of the individual and at the level of the social group. Such models allow for time lags between the stage of life when selection takes place and the accrual of reproductive benefits later in life. The complex pattern of human individual growth and development, combined with equally complex human social and cultural behavior, seems to be better explained by multilevel evolutionary models than by simpler models, for example those focusing only on fertility or mortality of the adolescent.

Human mating and parenting are of course related, but they are not identical. Charles Darwin identified two types of biological selection—natural selection and sexual selection (see Futuyma)—and both are likely to be involved in the evolution of human adolescence. Sexual selection is all about opportunities for mating, while natural selection is in part about parenting (Darwin, 1871; see the chapter by Flinn, and chapter by Gangestad and Grebe). Darwin wrote of the many structures and instincts developed through sexual selection, including biological weaponry for offense and defense, often used to drive away rivals for mating opportunities; and ornaments, vocalizations, and glands to emit odors and attract mates. Some human examples are the waist-to-hip ratio and childlike voice pitch of women that may be alluring to men, and verbal skills for gossip and attack that may be used to drive away mating rivals (Locke and Bogin, 2006).

GIRLS AND BOYS: SEPARATE PATHS THROUGH ADOLESCENCE

The multilevel nature of the evolution of human adolescence may be seen by considering the trade-offs related to biocultural reproduction and the different sequence of biological and behavioral events experienced by adolescent girls and boys. The differences allow each sex to improve opportunities for mating and parenting. Mating will eventually lead to the birth of offspring, but producing offspring is only a small part of reproductive fitness. Rearing the young to their own reproductive maturity is a surer indicator of success. The developmental paths of girls and boys during adolescence may be key in helping each sex to both produce and rear its own young successfully.

The path that girls take gives them the outward appearance of sexual maturity many years before they are, in fact, fecund. In contrast, the path that boys take gives them an outward appearance of immaturity for several years after they are, in fact, fecund.

The order in which adolescent events occur in girls and boys may be expressed in terms of time before and after peak height velocity (PHV); that is, the maximum velocity during the adolescent growth spurt. The velocity curves for girls and boys depicted in Figure 1 are aligned on chronological age, and PHV takes place about two years earlier in girls than in boys. If the curves for girls and boys are aligned at PHV, the timing of events before and after PHV becomes easier to appreciate. The following analysis will use the Tanner Maturation Staging System for the development of secondary sexual characteristics (Tanner, 1962). This system is based on five stages. Prepubertal maturation is denoted as B1and PH1 for girls—the absence of breast development and pubic hair—or G1 and PH1 for boys—the absence of gonadal (testes or penis) enlargement and absence of pubic hair. The adult appearance is stage B5, PH5, or G5.

In both girls and boys, puberty begins with changes in the activity of the central nervous system, the HPG axis as described above. These events begin at the same relative age in both girls and boys—that is, three years before PHV. This is also the time when growth rates change from decelerating to accelerating. In girls, the order of events is:

1. The first outward sign of puberty is the development of the breast bud (B2) and wisps of pubic hair (PH2) between −2 and −1 years PHV;
2. A rise in serum levels of estradiol that leads to the laying down of fat on the hip buttocks, and thighs, at −1.5 to −1.0 years PHV;
3. Increased velocity of the adolescent growth spurt, noticeable between −2 and −1 years PHV;
4. Further growth of the breast and body hair (B3 and PH3) at about +3 months PHV. At this stage the adolescent girl has the outward appearance of fecundity.
5. Menarche (first menstruation) at about +1 to +1.3 years PHV;
6. Completion of breast and body hair development (B5 and PH5) between +2 and +3 years PHV;
7. Attainment of adult levels of ovulation frequency at about +4.5 years PHV. At this stage the adolescent girl is fecund;
8. About +6 years PHV, growth in height ends and adulthood begins.

The path of adolescent development in boys starts with:

1. A rise in serum levels of luteinizing hormone, and the enlargement of the testes and then penis (G2) at about −3 years PHV;
2. A rise in serum testosterone levels (T) that is closely followed by the appearance of pubic hair (PH2) at about −2 years PHV;
3. About a year later (−1 year PHV) motile spermatozoa may be detected in urine, stages G3 to G4 and PH3 to PH4. At this stage, the adolescent boy is fecund;
4. About +0.5 years PHV there is a deepening of the voice, and stages G5 and PH5 are achieved;
5. At about +2.5 years PHV boys undergo a spurt in muscular development and strength. At this stage the adolescent boy has the appearance of fecundity;
6. Between +4 and +9 years PHV, growth in height ends and adulthood begins.

The sex-specific order of adolescent events tends not to vary between early and late maturers. The normal age ranges for puberty are 8–13 years in girls and 9–14 years in boys (Tanner, 1962). The sequence of adolescent events is also essentially identical between well-nourished girls and boys and those who suffered from severe malnutrition in early life, between rural and urban dwellers, and between major geographic and ethnic groups (e.g., European, Asian, and African, Bogin and Smith, 2012).

The different paths of sexual development may allow girls and boys to best learn the physical, social, and emotional skills to be successful at mating (sexual selection) and parenting (natural selection). In human societies, adolescent girls gain knowledge of sexuality and reproduction because they look mature sexually and are treated as such several years before they actually become fecund. The dramatic changes of adolescence stimulate both girls and the adults around them to participate in adult social, sexual, and economic behavior. For the postmenarcheal adolescent girl, this participation may be "low risk" in terms of pregnancy. Even so, some may become pregnant, and there are other social and psychological risks to adolescent sexual behavior. Teenage mothers and their infants are at risk because of the reproductive and emotional immaturity of the mother (Cunnington, 2001). This often leads to a low-birth-weight infant, premature birth, and high blood pressure in the mother. The likelihood of these risks declines and the chance of successful pregnancy and birth increases markedly after 15 years of age, and reaches its nadir after 18 years of age (Bogin and Smith, 2012). Due to these biological and social risks, most human societies carefully regulate, according to age and sex, the onset and type of sexual behavior that is permitted by adolescent girls.

The adolescent development of boys is quite different from that of girls. Boys become fecund well before they assume the size and physical characteristics of men. Analysis of urine samples from boys 11–16 years old show that they begin producing sperm at a median age of 13.4 years. Yet, cross-cultural evidence indicates that few boys successfully father children until they are into their third decade of life. The explanation for the lag between sperm production and fatherhood is not likely to be a simple one of sperm performance, such as not having the endurance to swim to an egg cell in the woman's fallopian tubes. More likely is the fact that the average boy of 13.4 years is only beginning his adolescent growth spurt (Figure 1). Growth researchers have documented that in terms of physical appearance, physiological status, psychosocial development, and economic productivity, a 13-year-old boy is still more a juvenile than an adult.

The delay between sperm production and reproductive maturity is not wasted time in either a biological or a social sense. The obvious and the subtle psychophysiological effects of testosterone and other androgen hormones that are released after gonadal maturation may "prime" boys to be receptive to their future roles as men. Early in adolescence, sociosexual feelings including guilt, anxiety, pleasure, and pride intensify. At the same time, adolescent boys become more interested in adult activities, adjust their attitudes toward parental figures, and think and act more independently (Sisk and Zehr, 2005). In short, they begin to behave like men. Because their adolescent growth spurt occurs late in sexual development, young males can practice behaving like adults before they are actually the size of an adult and perceived as mature by other adults. The sociosexual antics of young adolescent boys are often considered to be more humorous than serious. Yet, they provide the experience to fine-tune their sexual and social roles before their lives or those of their offspring depend on them.

ADOLESCENT CONTRIBUTIONS TO THE REPRODUCTIVE SUCCESS OF ADULTS

Another aspect of the multilevel nature of human adolescence may be contributions that adolescents make to the reproductive success of older social group members. Adolescence delays the onset of reproduction for the adolescent, and allows the adolescent to channel food and work toward others. Human juveniles may hunt, gather, or produce some of their own food intake, but overall they require provisioning to achieve energy balance. In contrast, human adolescents are capable of producing sufficient quantities of food to exceed their own energy requirements. Some of the food that adolescents produce may be used to fuel their own growth and development, creating larger, stronger, and healthier bodies. Another portion of their production is shared with other members of the social group, including younger siblings, parents, and other immediate family members (defining families in the broad anthropological sense, Bogin, 2010; Bogin and Smith, 2012). Adolescent contributions enhance the fertility of adults and the survival of infants, children, and juveniles. The biological trade-off is the delay of years between puberty and first birth for adolescents. For their valuable services in food production, the adolescents receive care and protection to safeguard their health and survival. This is important because adolescents are immature in terms of sociocultural knowledge and experience.

RISKS OF CHILDHOOD AND ADOLESCENCE

The evolution of any new structure, function, or stage of development may bring about many biosocial benefits; however, it also incurs risks. Human childhood and adolescence come with their own set of specific risks. Children are dependent on older people for feeding, care, and protection. Neglect and abuse of children are common today and have been throughout recorded history. Predation on children by large carnivores and birds of prey was a real threat in the human past and still is today in some parts of the world. Childhood immaturity of the immune system means that viral and bacterial predators may also pose risks (Bogin and Smith, 2012). Immaturity and inexperience also mean that children are prone to accidents and other physical trauma.

Among the most common and serious threats to adolescents are psychiatric and behavioral disorders. The onset of such problems tends to peak during adolescence. Most mammalian species terminate all brain growth well before sexual maturation, but human adolescents show enlargement and pruning of some brain regions leading to structural changes in the cerebral cortex well into their third decade. Some scholars hypothesize that the increase in brain-related disorders may derive from these cortical changes, which affect the adolescent brain's sensitivity to reward. The reward system of the brain may lead adolescents toward risk-taking behavior. Whether risk taking is inherently biological or shaped by social stimuli is debated. What is clear is that adolescence is a time of life with a higher level of risk for certain diseases of the mind and body, and greater mortality, than was the case for the juvenility stage. Adolescent mortality often is associated with ritualized violence, such as serving as combatants in warfare, or being exposed to inherently dangerous but socially normative behaviors such as automobile driving, alcohol consumption, cigarette smoking, and sex, without appropriate instruction and regulation by society.

CONCLUSIONS

In this article, a biocultural approach is taken to the study of the evolution of human growth and development. The patterns of human postnatal growth and development—the stages of infancy, childhood, juvenility and adolescence—were reviewed. Several hypotheses were discussed concerning how the new life stages of the human life cycle represent feeding and reproductive specializations, secondarily allowing for the human style of learning and cultural behavior.

The biocultural perspective of human development focuses on the constant interaction taking place during all phases of human development, between both genes and hormones within the body, and with the sociocultural environment that surrounds the body. Research from anthropology, developmental psychology, endocrinology, primate behavior, and human biology shows how the biocultural perspective enhances our understanding of human development.

REFERENCES

Bogin, B., 1999. Patterns of Human Growth, second ed. Cambridge University Press, Cambridge.

Bogin, B., 2009. Childhood, adolescence, and longevity: a multilevel model of the evolution of reserve capacity in human life history. American Journal of Human Biology 21, 567–577.

Bogin, B., 2010. Evolution of human growth. In: Muehlenbein, M. (Ed.), Human Evolutionary Biology. Cambridge University Press, Cambridge, pp. 379–404.

Bogin, B., 2011. Puberty and adolescence: an evolutionary perspective. In: Brown, B.B., Prinstein, M.J. (Eds.), Encyclopedia of Adolescence, vol. 1. Academic Press, San Diego, pp. 275–286.

Bogin, B., Bragg, J., Kuzawa, C., 2014. Humans are not cooperative breeders but practice biocultural reproduction. Annals of Human Biology 41, 368–380.

Bogin, B., Smith, B.H., 2012. Evolution of the human life cycle. In: Stinson, S., Bogin, B., O'Rourke, D. (Eds.), Human Biology: An Evolutionary and Biocultural Perspective, second ed. Wiley-Blackwell, New York, pp. 515–586.

Borrud, L.G., Flegal, K.M., Looker, A.C., Everhart, J.E., et al., 2010. Body composition data for individuals 8 years of age and older: U.S. population, 1999–2004. Vital Health Statistics 11 (250). National Center for Health Statistics, Hyattsville, Maryland.

Crews, D.E., Bogin, B., 2010. Growth, development, senescence, and aging: a life history perspective. In: Larsen, C. (Ed.), A Companion to Biological Anthropology. Wiley-Blackwell, New York, pp. 124–152.

Cunnington, A.J., 2001. What's so bad about teenage pregnancy? Journal of Family Planning and Reproductive Health Care 27 (1), 36–41.

Darwin, C., 1871. The Descent of Man and Selection in Relation to Sex. John Murray, London.

Ellison, P.T., Reiches, M.W., Shattuck-Faegre, H., Breakey, A., Konecna, M., Urlacher, S., Wobber, V., 2012. Puberty as a life history transition. Annals of Human Biology 39 (5), 352–360.

Hamada, Y., Udono, T., 2002. Longitudinal analysis of length growth in the chimpanzee (*Pan troglodytes*). American Journal of Physical Anthropology 118, 268–284.

Henrich, J., Heine, S.J., Norenzayan, A., 2010. The weirdest people in the world? Behavioral and Brain Sciences 33, 61–83.

Hill, K.R., Walker, R.S., Bozicević, M., Eder, J., Headland, T., Hewlett, B., Hurtado, A.M., Marlowe, F., Wiessner, P., Wood, B., 2011. Co-residence patterns in Hunter-Gatherer societies show unique human social structure. Science 331, 1286–1289.

Hrdy, S., 2009. Mothers and Others: The Evolutionary Origins of Mutual Understanding. The Belknap Press of Harvard University Press, Cambridge.

Kaplan, H., Hill, K., Lancaster, J.B., Hurtado, A.M., 2000. A theory of human life history evolution: diet, intelligence, and longevity. Evolutionary Anthropology 9, 156–185.

Locke, J.L., Bogin, B., 2006. Language and life history: a new perspective on the development and evolution of human language. Behavioral and Brain Sciences 29, 259–325.

Plant, T.M., 2008. Hypothalamic control of the pituitary–gonadal axis in higher, primates: key advances over the last two decades. Journal of Neuroendocrinolgy 20, 719–726.

Pusey, A. 1983. Mother–offspring relationships in chimpanzees after weaning. Animal Behavior 31, 363–377.

Savage, M.O., Hwa, V., David, A., Rosenfeld, R.G., Metherell, L.A., December 12, 2011. Genetic defects in the growth hormone-IGF-I axis causing growth hormone insensitivity and impaired linear growth. Frontiers of Endocrinology (Lausanne) 2 (95). http://dx.doi.org/10.3389/fendo.2011.00095.

Sisk, C.L., Zehr, J.L., 2005. Pubertal hormones organize the adolescent brain and behavior. Frontiers of Neuroendocrinology 26 (3–4), 163–174.

Tanner, J.M., 1962. Growth at Adolescence, second ed. Blackwell, Oxford.

Chapter 21

Human Reproductive Ecology

Claudia R. Valeggia[1] **and Alejandra Núñez-de la Mora**[2]

[1]Department of Anthropology, Yale University, New Haven, CT, USA; [2]Instituto de Investigaciones Psicológicas, Universidad Veracruzana, Xalapa, Veracruz, Mexico

SYNOPSIS

Human reproductive ecology (HRE) studies human reproduction in the context of local ecological variables. The perspective of HRE is evolutionary, and its central goal is to provide models for explaining variation in reproductive patterns as adaptations to the environment. Reproductive function shows significant variation between and within individuals and populations. Human reproductive function has been shown to vary in response to ecological variables such as nutrition, physical activity, acute immunological challenges, psychological stress, and social context. HRE allows us to explore what the consequences of that variation are in terms of health, fertility, and ultimately, reproductive fitness.

INTRODUCTION

Human reproductive ecology (HRE) is a subfield of evolutionary biology that studies human reproduction in the context of local ecological variables. Historically, HRE started to take shape in the early 1990s as the result of a synergistic dialog between reproductive biologists and population scientists (Ellison, 1990; Campbell and Wood, 1994; Ellison, 1994a). As such, HRE is intrinsically interdisciplinary, drawing from biological, social, and environmental disciplines.

A central goal of HRE is to provide models for explaining variation in reproductive patterns. This focus on variability both separates HRE from other disciplines that also deal with human reproduction and defines its scope, breadth, and methods. Biomedical approaches to reproduction underscore the normal versus pathological dichotomy, work with mostly homogenous industrialized populations, and are based on research in clinical settings. Demographic approaches, on the other hand, emphasize universal patterns and attempt to measure and explain levels of fertility at a population level using large-scale retrospective surveys. In their quest to understand variation, human reproductive ecologists usually base their studies upon relatively small samples in a variety of ecological settings, and often incorporate prospective designs in which the individual becomes the unit of analysis. In addition, HRE models assume that variation in reproductive patterns is adaptive; that is, it is considered as an evolutionary response to the formative environment (see Figure 1).

During the last two decades, rapidly improving laboratory techniques have facilitated the collection of biological samples in a variety of settings and at a population scale (Ellison, 1988; O'Connor et al., 2003; McDade et al., 2007; Valeggia, 2007). Three of the most frequently applied techniques use saliva, urine, and dried blood spots for monitoring ovarian and testicular function in the field and also for associating reproductive outcomes with nutritional and immunological status.

These field-friendly techniques have allowed HRE researchers to show dramatic variation in reproductive hormone levels both within and between populations (Ellison and Lager, 1986; Ellison et al., 1993a, 2002; Bentley et al., 2000; Vitzthum et al., 2000, 2002; Nuñez de la Mora et al., 2007). The interpretation of these findings incorporates a strong evolutionary component and often uses life history theory as a rich framework within which reproductive ecology models can be developed and tested.

HUMAN LIFE HISTORY

A species' life history can be defined as a description of the timing of development, reproductive events, and mortality during the life course (Hill and Hurtado, 1996). For example, human life history is characterized by delayed maturation, relatively short interbirth intervals, and a long postreproductive life span, with females typically birthing comparatively large, altricial singleton offspring (Mace, 2000; Worthman and Kuzara, 2005). Following the same theoretical paradigm on which HRE rests, life history theory assumes that this schedule of development and reproduction represents an adaptive strategy, that is, species adjust their life history traits to local conditions so as to optimize reproductive success. The study of life history traits is ultimately the study of the trade-offs involved in the allocation of limited resources such as time and energy to biological functions such as growth, storage, maintenance, and reproduction (Stearns, 1992) (see Figure 2).

One key element of life histories is that certain traits appear to be linked, in the sense that they limit or constrain one another. For instance, current reproduction "competes" with current growth and body maintenance. Other typical examples of these trade-offs studied by human reproductive ecologists include current and future reproduction as well as quality versus quantity of offspring. The allocation of energy to different life history categories necessarily involves physiological trade-offs. For example, there is evidence of competition for nutrients between pregnant

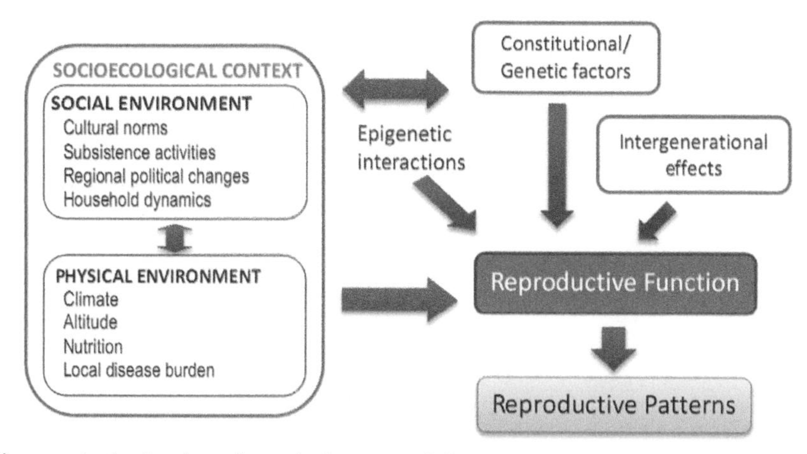

FIGURE 1 Factors affecting reproductive function and reproductive patterns in humans.

adolescent mothers—who are still growing—and their fetuses. Growing adolescents tend to give birth to smaller infants (Scholl et al., 1994, 1995; Wallace et al., 2004).

HRE evaluates reproductive patterns using evolutionary and life history models. Thus, it examines the interaction between socioecological variables and the possible physiological mechanisms underlying the trade-offs highlighted by life history theory. Interbirth intervals, for example, are life history traits that define the tempo of reproduction of a species. In natural fertility populations, the duration of interbirth intervals is variable and tends to reflect maternal nutritional status, which in turn is determined in great part by the ecology in which the mother lives (Ellison et al., 1993b; Ellison, 2001). In highly unpredictable environments, forgoing reproduction may imply increasing the chances of successful reproduction in the future (Wasser and Barash, 1983). This flexibility of response of female fecundity is interpreted to be an evolved trait that increases lifetime reproductive success.

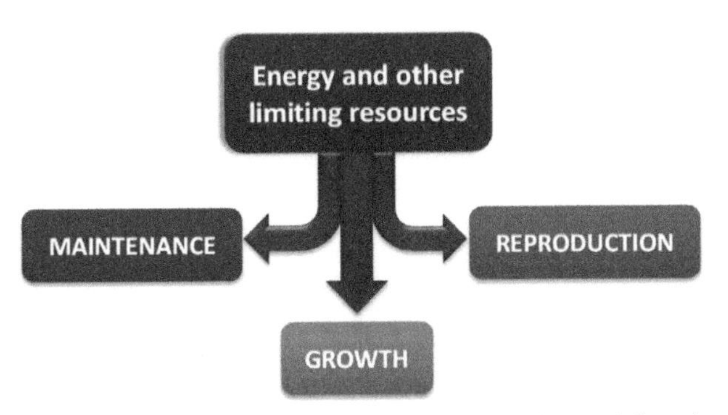

FIGURE 2 Life history theory studies the trade-offs that result from the differential allocation of energy and other limiting resources to different vital functions across an organism's life span.

HUMAN REPRODUCTIVE PHYSIOLOGY: THE BASICS

Like most mammals, the reproductive physiology of human females is organized in cycles that, in the absence of conception and subsequent pregnancy, recur regularly at roughly monthly intervals. On average, the ovarian cycle length of women is 28–29 days long. However, both within and between women, variation in cycle length is so substantial that this average should be used just for heuristic purposes.

Ovarian cycles are governed by the functioning of the major reproductive hormone axis—the hypothalamus–pituitary–ovarian (HPO) axis (see Figure 3(A)). Male reproductive endocrinology is regulated mainly by the hypothalamus–pituitary–testicular (HPT) axis (Figure 3(B)). A brief description of the HPO and HPT axes and their endocrine products as they relate to some of the proximate determinants of fertility is given below (Wood, 1994).

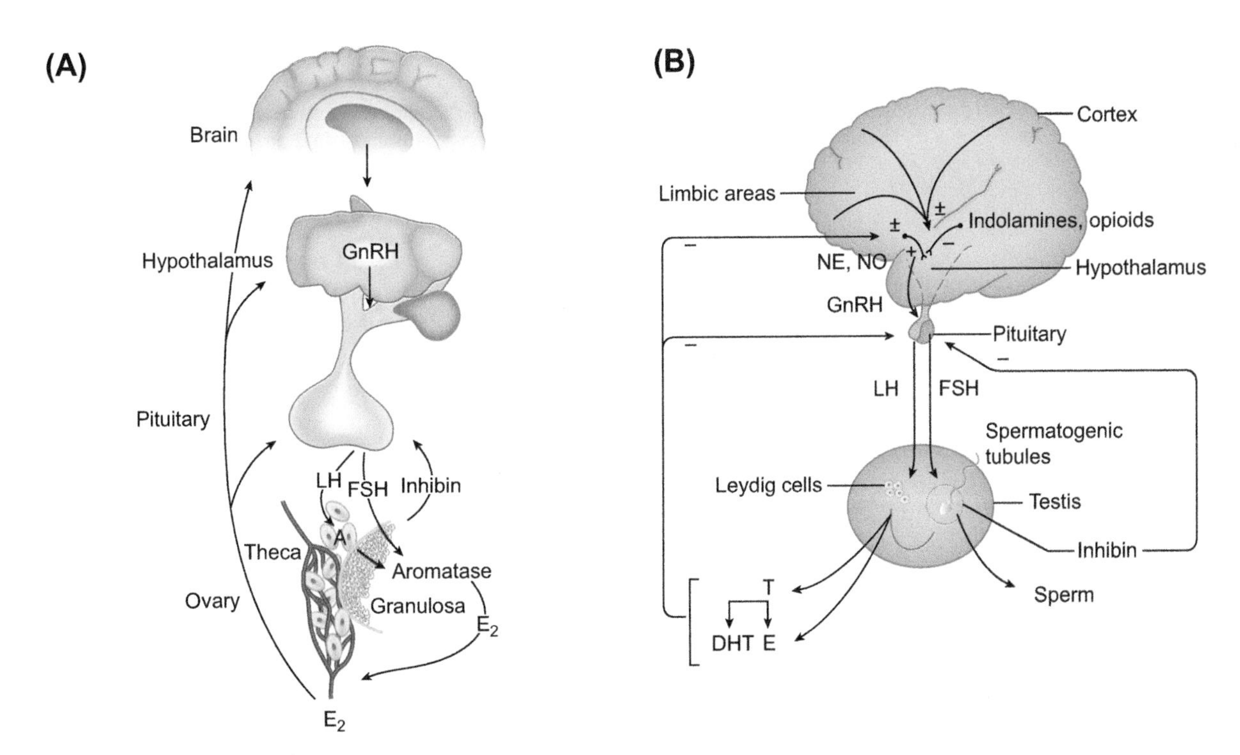

FIGURE 3 (A) The hypothalamic–pituitary–ovarian axis. GnRH=gonadotropin-releasing hormone; LH=luteinizing hormone; FSH=follicle-stimulating hormone; E_2=estradiol; (B) The hypothalamic–pituitary–testicular axis. DHT=dihydrotestosterone; E=estrogen; FSH=follicle-stimulating hormone; GnRH=gonadotropin-releasing hormone; LH=luteinizing hormone; NE=norepinephrine; NO=nitric oxide; T=testosterone. (See color plate section).

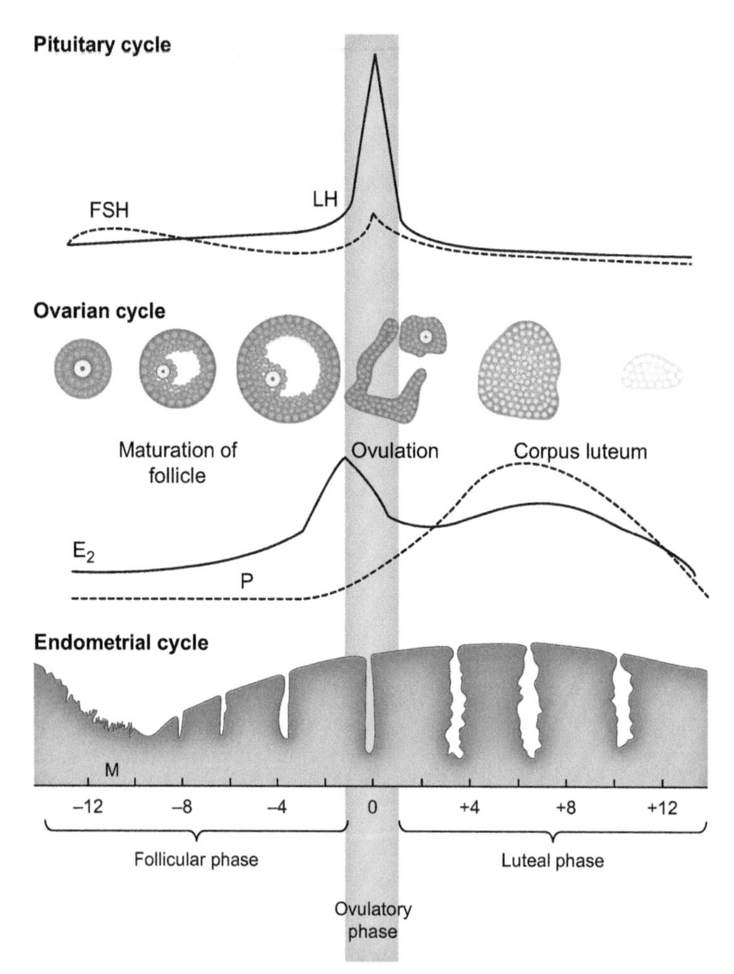

FIGURE 4 The idealized cyclic changes observed in gonadotropins, estradiol (E_2), progesterone (P), and uterine endometrium during the normal menstrual cycle. The data are centered on the day of the luteinizing hormone (LH) surge (day 0). Days of menstrual bleeding are indicated by M. FSH = follicle-stimulating hormone. (See color plate section). *Reprinted with permission from Endocrine and Metabolism Continuing Education Quality Control Program, 1982. Copyright American Association for Clinical Chemistry, Inc.*

The ovarian cycle has been described as biphasic: a follicular phase during which follicles grow in the ovary is followed by a luteal phase during which the reproductive system prepares for conception and implantation (see Figure 4). Ovulation (the release of a single mature egg from the follicle) is the event that marks the transition from one phase to the other, and it occurs at about the midportion of the cycle. Just before the beginning of each cycle, the pituitary gland, stimulated by GnRH, starts releasing follicle-stimulating hormone (FSH) and luteinizing hormone (LH). These hormones promote the growth of ovarian follicles. As follicles grow they begin secreting estradiol, which feeds back to the pituitary and the hypothalamus to inhibit the release of additional FSH and LH. The concentration of estradiol increases almost exponentially during the follicular phase. Conversely, the levels of FSH and LH remain low throughout most of the follicular phase. However, approximately one day before ovulation, there is a sharp surge of LH. This peak in LH concentration is the trigger for ovulation. LH and FSH levels drop back to baseline at the time of ovulation. After ovulation, the follicle cells remaining in the ovary develop into a specialized gland called corpus luteum, which secretes progesterone in large quantities and estradiol to a lesser extent. The postovulatory increase in progesterone is what defines the luteal phase. If there is no conception, the corpus luteum disintegrates and progesterone and estradiol levels drop, signaling the beginning of a new cycle. If conception occurs, the corpus luteum remains active and progesterone and estradiol levels increase dramatically as does the concentration of human chorionic gonadotropin, (hCG) secreted by the conceptus.

Male reproductive physiology is not organized in recognizable cycles. The hypothalamus releases pulses of GnRH that signal the pituitary to release LH and FSH. Increasing levels of LH stimulate the production and release of testosterone in the testes, while FSH (aided by testosterone) promotes the production of sperm. As is the case in females, feedback control mechanisms play an important role in this system. Testosterone inhibits the release of pituitary and hypothalamic hormones.

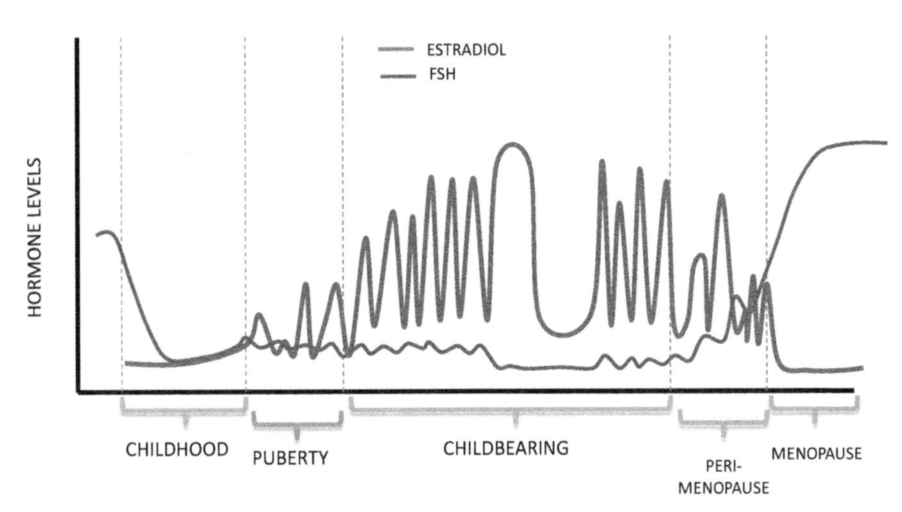

FIGURE 5 Schematic representation of the variation in estradiol and FSH (follicle-stimulating hormone) across the female life span. (See color plate section).

In women, each reproductive phase or status has its own characteristic hormone profile that can be monitored using the minimally invasive techniques mentioned above. Ovarian cycles are characterized by regular, measurable changes in ovarian hormones, but prepubertal and postmenopausal levels of ovarian hormones are noncyclic and very low (see Figure 5). Therefore, the beginning and the end of a woman's reproductive life can be established very precisely at a population level by estimating estradiol and progesterone levels in blood, urine, or saliva.

FEMALE REPRODUCTIVE ECOLOGY

Unlike their male counterparts, females' reproductive life spans are delimited on either side by biological events that mark the beginning and the end of reproductive potential: menarche and menopause, respectively. Within this period, females' lives are further punctuated by individual reproductive events, each typically comprising a pregnancy followed by a variable period of lactation. Despite the varied physiological requirements of each of these events, all are within the control of the HPO axis and its interactions with other vital functions.

Below is a brief description of the main physiological and developmental characteristics of each event that defines the female reproductive life span.

Menarche

Menarche, the first occurrence of menstruation, is a relatively late indicator of pubertal development that marks the beginning of menstrual cycling and is considered the start of a woman's reproductive span. In most girls, however, regular and/or hormonally robust ovulatory cycles are not well established until after a variable period (ranging from 1 to 5 years) of subfertility (Apter, 1980; Vihko and Apter, 1984). Recent estimates of the median age of menarche vary between 13 and 16 years (Adams Hillard, 2008). The first menstrual period typically occurs approximately two years after the onset of postnatal breast development (thelarche) and following the peak in pubertal growth spurt.

Although the precise details and mechanisms involved are not fully understood, the onset of menarche is associated with a gradual increase in estrogen levels derived from the hormone cascade triggered at puberty in the hypothalamus and affecting the rest of the HPO axis. Such increase in estrogenic activity elicits a series of physical changes that prepare the body for potential reproduction. Among these are the widening of the pelvis and hips and an increase in body fat deposition particularly in hips and tights. Linear growth after menarche is limited.

Almost all regions of the world have seen a secular trend toward earlier age at menarche that coincides with improvements in general living standards, particularly in nutrition, sanitation, and health care provision (Parent et al., 2003). Apart from this historical change, age at menarche varies substantially between and within populations as a function of genetic, energetic, social, and environmental factors and their interactions (Parent et al., 2003).

Twin and family studies show the strongest genetic association between mothers and daughters and heritability estimates ranging from 0.57 to 0.82 (Anderson et al., 2007; Morris et al., 2011; Dvornyk and Waqar-ul-Haq, 2012). Despite having a substantial genetic component, age at menarche remains a fairly plastic trait that responds to energetic and psychosocial conditions during early life. For instance, girls with larger body size (BMI, weight, and stature) (Cooper et al., 1996); faster growth rates during childhood (dos Santos Silva et al., 2002); higher childhood socioeconomic position (Wronka and Pawlinska-Chmara, 2005); family conflict and parental divorce (Wierson et al., 1993); presence of a stepfather (Ellis and Garber, 2000); and exposure to stressors such as war shortly before menarche (Prebeg and Bralic, 2000) are at higher risk of experiencing early menarche. In contrast, high level of physical activity and low level of nutritional intake, whether for athletic, body image purposes or daily subsistence, have a delaying effect (Moisan et al., 1991; Ellison, 1994a; Gluckman and Hanson, 2006).

From a life history perspective, variation in age at menarche in response to conditions experienced during childhood may help maximize reproductive fitness by adjusting an individual's reproductive schedule and the degree of investment to the available energetic and social resources. Furthermore, the trade-offs associated with each of the resulting strategies are likely to underlie the epidemiological association between age at menarche and adult health and reproductive outcomes (Eaton et al., 1994; Kuh and Hardy, 2002).

Variation in Adult Female Reproductive Function

Reproductive function in healthy human females, measured as average profiles of adult reproductive steroids, shows significant variation between and within individuals and populations (Ellison, 1993; Ellison, 1994b; Vitzthum et al., 2002; Jasienska and Jasienski, 2008) (see Figure 6).

Differences in baseline hormone levels are associated with genotype (Jasienska et al., 2006a), developmental conditions during fetal and childhood growth (Jasienska et al., 2006b,c; Nuñez de la Mora et al., 2007; Nuñez-De La Mora et al., 2008), age (Ellison, 1990; Lipson and Ellison, 1992) and nonenergetic variables such as qualitative dietary intake (Bagga et al., 1995; Schliep et al., 2013) and psychosocial stress (Edozien, 2006).

In addition, hormone levels vary along such baseline in response to acute energetic conditions. In situations of heightened energy expenditure and/or negative energy balance (Ellison, 2001), such as those related to seasonal food shortages (Bentley et al., 1998) and/or workloads in subsistence populations (Panter-Brick, 1993; Jasienska and Ellison, 1998, 2004), or to voluntary exercise (Ellison and Lager, 1986; Brooks et al., 1990; Morris et al., 1999; Pirke et al., 1999) and dieting in westernized societies (Schweiger et al., 1987), ovarian function is suppressed and the chances of conception are reduced (Bailey et al., 1992). This *normal*, temporary, and reversible modulation of ovarian function by ecological conditions is adaptive as it helps to optimize reproductive effort in an iteroparous organism over a lifetime of reproductive events (Ellison, 2003).

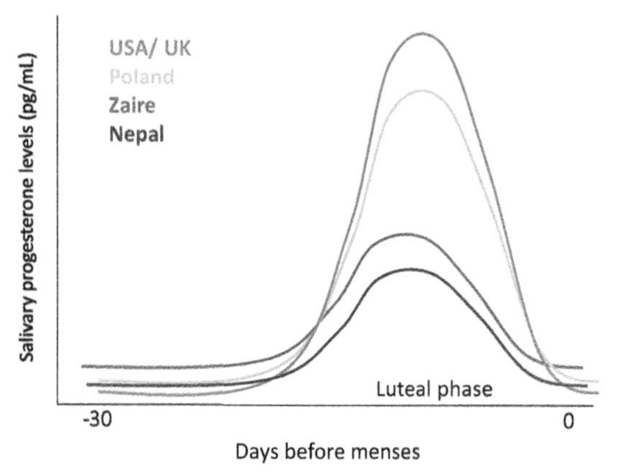

FIGURE 6 Interpopulation variation in ovarian function. (See color plate section). *Adapted from Ellison et al. (1993a).*

Pregnancy

Female reproduction is partitioned into gestation, parturition, lactation, and the resumption of menstrual cycling, all of which entail both time and energy costs. Compared with our closest primate relatives, human babies are born fatter, larger, and heavier relative to maternal size, which makes human babies more costly to produce (Kuzawa, 1998; Mace, 2000). Compared with other mammals of similar size, human and nonhuman primates have longer periods of gestation and lactation and a slower postnatal growth relative to that prenatally (Dufour and Sauther, 2002).

The energetic costs of pregnancy derive from fetal tissue growth, the growth and maintenance of supporting maternal tissues, the accumulation of maternal fat and the ever-increasing basal metabolic rate (BMR) of the growing fetus. By the sixth month of gestation, for instance, maternal daily energy expenditures may reach twice the prepregnancy BMR (Forsum and Löf, 2007). A range of physiological and behavioral coping strategies have evolved in human females that reduce and/or spare energy by increasing energy intake, decreasing physical activity and/or increasing metabolic efficiency through a reduction of BMR (Prentice and Goldberg, 2000; Dufour and Sauther, 2002). Although these strategies, which vary among individuals and populations, enable women to support reproduction on a limited energy supply, they represent a significant trade-off against long-term maternal health and survival (Lummaa, 2010).

In addition to direct energetic costs, gestation can also be evaluated in reproductive fitness terms. The short- and long-term trade-offs associated with reproductive investment in pregnancy often result in a conflict of interest between mother and offspring, who only share half their genes; what may benefit the mother may not necessarily be advantageous for the offspring and vice versa (Haig, 2008). Such parent–offspring genetic conflicts have been implied not only in pregnancy-related morbidity, but also in the evolution of human life history itself (Haig, 2010).

Given these high overall costs of gestation, human females have evolved a series of mechanisms to modulate investment according to the potential reproductive value of each event in any given circumstance (Wasser and Barash, 1983; Peacock, 1991). Along with the maternal age-related early spontaneous abortion of poor quality or abnormal fetuses (Forbes, 1997; Baird, 2009), investment in gestation varies in response to ecological conditions. For instance, in environments where maternal condition is compromised, the probability of conception is low (Bailey et al., 1992; Vitzthum et al., 2004), the risk of early pregnancy loss is high (Nepomnaschy et al., 2006; Vitzthum et al., 2009) as is the risk of preterm delivery (Pike, 2005).

The outcome of established pregnancies is equally affected by ecological and social variables (Pike, 2001); dietary intake, physical activity, disease burden and psychosocial stress impact fetal growth, development, and survival directly as well as through their negative effects on maternal condition and gestational weight gain (Prentice et al., 1987; Bonzini et al., 2007; Menezes et al., 2009; Schantz-Dunn and Nour, 2009; Loto and Awowole, 2012). Other aspects of the physical environment such as hypoxia at high altitudes (Moore et al., 2011) and heat stress (Bronson, 1995) have also been shown to affect pregnancy outcome.

In comparative terms, human gestational length and fetal growth are primarily constrained by maternal metabolism (Dunsworth et al., 2012). Within populations, the duration of pregnancy is affected by several factors including parity, sociodemographic characteristics, previous preterm delivery, cigarette smoking, maternal age, maternal energetic status and ethnicity (Shiono et al., 1986; Peacock et al., 1995; Roshni et al., 2004; Rayco-Solon et al., 2005). The range of the normal variation in the duration of term pregnancies at a population level has been poorly documented.

Lactation

In humans, the energetic demands associated with lactation are on a daily basis higher than for pregnancy, and are a consequence of milk synthesis and the maintenance of metabolically active mammary glands (Prentice and Prentice, 1988). Human females tend to increase energy intake to meet, in part, the cost of lactation, but in ecologies where that is not an option, women rely on similar strategies as those used during gestation to meet the remainder of the cost (Dufour and Sauther, 2002). In cases of extreme energetic constraints however, maternal depletion can occur with implications for both maternal and offspring health and survival (Tracer, 1991; Winkvist et al., 1992; Prentice and Goldberg, 2000).

The duration of human lactation depends on the socioecology of the population and thus it is highly variable across cultures, ranging from nil to more than five years (Sellen, 2001). Weaning from the breast is considered a process rather than an event and its beginning and duration are also extremely variable, occurring over a wide range of infant sizes (Sellen, 2007). Early and flexible weaning, another unique human life history trait, may have evolved to allow females to increase their reproductive success. By shifting from a costly direct and exclusive form of feeding (lactation) to a shared, less demanding but effective way of provisioning (safe and nutritionally adequate supplementary foods), women could recruit help from others in the group and direct their freed energy and time to a new pregnancy without risking the survival of their current offspring (Sellen, 2007). A consequence of this shift in maternal investment strategy is the characteristic human reproductive pattern of relatively short interbirth intervals and increased birth rates (Hill and Kaplan, 1999; Mace, 2000).

At a metabolic and physiological level, the variation in interbirth intervals depends on the duration of lactational amenorrhea, a period of suppressed ovarian function and subfecundity. There is strong evidence showing that the return to postpartum fecundity is largely determined by the dynamics of maternal energy availability, which are tightly linked to the ecological context in which lactation occurs (Ellison and Valeggia, 2003; Valeggia and Ellison, 2009).

Menopause

Menopause, the final cessation of menstruation, is the most prominent retrospective marker of reproductive aging and is declared after 12 consecutive months of amenorrhea (Utian, 1999) (see the chapter by Sievert in the present volume). It may be natural or initiated by surgical removal of the uterus and/or ovaries. Perimenopause, the transition from regular fertile cycles to infertility, usually occurs over a period of several years. It is characterized by subfecundity, increasingly irregular menstrual cycles, and dramatic variations in FSH, estradiol, and progesterone related to the dwindling pool of ovarian follicles (te Velde et al., 1998).

Age at natural menopause varies significantly within and between populations, with medians ranging from 49 to 52 years in industrialized ecologies and a few years younger in traditional societies (Sowers and La Pietra, 1995; Morabia and Costanza, 1998; Gold et al., 2001). Family studies indicate a strong relationship between mothers' and daughters' menopausal ages (Torgerson et al., 1997), and a number of genetic variants have been discovered in association with age at natural menopause, but menopause as a complex genetic trait remains to be comprehensively understood (de Bruin et al., 2001; Kok et al., 2005).

It has been postulated that in utero and perinatal conditions that impact growth could influence age at menopause by affecting the size of the initial ovarian reserve and the rate of oocyte depletion in postnatal life (Cresswell et al., 1997; te Velde et al., 1998). In line with these suggestions, not having been breastfed (Hardy and Kuh, 2002; Mishra et al., 2007), short stature at birth and poor early growth (Cresswell et al., 1997; Hardy and Kuh, 2002), adverse socioeconomic conditions in childhood (Hardy and Kuh, 2002; Mishra et al., 2007), and parental divorce (Hardy and Kuh, 2005) have all been associated with earlier menopause.

Many environmental and adult lifestyle factors including socioeconomic position, psychosocial stress, and reproductive health have been claimed to affect menopausal age, but inconsistencies exist for the majority of factors (see Kok et al., 2005; Mishra et al., 2009 for review) except cigarette smoking (Parente et al., 2008) and nulliparity (Cramer et al., 1995; Harlow and Signorello, 2000), which have consistently been shown to be related to early menopause.

The timing of menopause and the associated dramatic endocrinological, metabolic, and sociocultural changes have been shown to affect several aspects of women's health and wellbeing. For example, early age at menopause is associated with an increased risk of mortality, osteoporosis, cardiovascular disease, and mental health disorders, whereas later menopause increases the risk of reproductive cancers (see Kuh and Hardy, 2002). In addition, the decline in estrogen secretion in the perimenopausal period is associated to a varying degree with various somatic and psychological discomforts such as hot flashes, night sweats, and disturbed cognitive function. The experience and perception of these symptoms however, vary significantly among women across cultures (Melby et al., 2005; Sievert, 2006).

The fact that women may live several decades beyond the cessation of their child-bearing potential is one of human life history's most exceptional traits. From an evolutionary perspective, it has been suggested that the origin of menopause is related to the increased reproductive fitness benefits of stopping reproduction early and investing in existing offspring (Peccei, 1995, 2001) and grand-offspring (Hill, 1993; Hawkes et al., 1998; Lahdenpera et al., 2004; Shanley et al., 2007) over a long postreproductive period, compared with the benefits of continued reproduction, particularly as the costs associated with reproduction at advanced maternal age are high. Recently, a third hypothesis suggested that menopause and the minimal reproductive overlap between generations of women might be the outcomes of reproductive competition between generations (Cant and Johnstone, 2008).

Male Reproductive Ecology

The reproductive ecology of males has historically received less attention than that of females, in part because men's reproductive physiology was thought to be less responsive to environmental demands. However, there is mounting evidence that men's reproductive function does respond to environmental demands in ways that reflect both short- and long-term adaptive strategies (Bribiescas, 2001a,b; Bribiescas 2006). From a life history perspective, male reproductive ecology is best conceptualized by understanding energy-allocation processes that reflect competing needs for surviving, growing, and reproducing. Survival implicates expensive maintenance functions such as immunocompetence and metabolic regulation. Given that sperm production is relatively inexpensive, reproductive effort in males involves devoting energy mostly to competing for and maintaining mates, which necessitates increased musculoskeletal performance. This trade-off is physiologically reflected in the tight interaction among metabolic, immunological, and HPT axis functions, which men growing and living in different environments negotiate in various ways (Bribiescas, 2001a,b).

TESTICULAR FUNCTION DURING EARLY DEVELOPMENT

In human males, as in all placental mammals and marsupials, the presence of the SRY gene on the Y chromosome stimulates the development of testicular tissue in the first few weeks of embryonic life (Wallis et al., 2008). The developing testes start producing testosterone and other potent androgens that shape internal and external genitalia to produce a male phenotype. Testosterone production is intense during prenatal life and the first 6 months postpartum, but declines sharply during infancy and childhood (Rey et al., 2009). As has been postulated for females, the interaction among genetic, epigenetic, and environmental factors during these early developmental stages may program baseline testosterone levels and sperm production in later adulthood (Bribiescas, 2001a,b; Muehlenbein and Bribiescas, 2005; Magid, 2011).

Puberty

The exact physiological mechanisms leading to the activation of the HPT axis and establishment of mature testicular function remain unclear. The increased production of androgens by the adrenal gland during adrenarche would trigger changes at hypothalamic centers that eventually lead to the pulsatile release of GnRH and the establishment of the full HPT axis negative feedback loop (Campbell, 2006, 2011). During puberty, testosterone levels increase dramatically again, leading to the establishment of mature reproductive function (spermatogenesis) and the development of secondary sexual characteristics.

There is variation in the timing of pubertal onset in boys, but on average it occurs 1–2 years later than in girls—that is, around 12–13 years of age (Bogin, 1999). In part because of the lack of a visible marker of puberty such as menarche, studies of the population variation in age at puberty of boys are much more limited than those for girls (Zemel et al., 1993; Panter-Brick et al., 1996; Campbell et al., 2004). There is evidence that nutritional status and other environmental and lifestyle factors may be involved. Energy deficiency has been associated with later ages at puberty in Kenyan and Zambian boys, as measured by pubertal hormone changes (Campbell et al., 2005). Among British boys, faster early postnatal growth and weight gain were correlated with earlier sexual maturation trajectories (Ong et al., 2012).

Variation in Adult Male Reproductive Function

Within the context of HRE, male reproductive function is generally assessed by looking at circulating testosterone

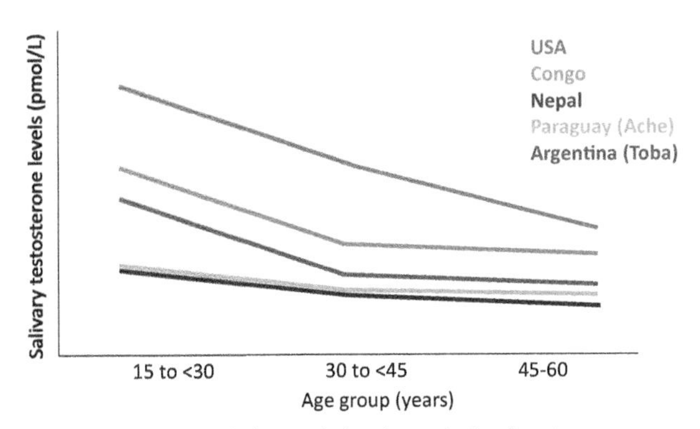

FIGURE 7 Interpopulation variation in testicular function across age groups. (See color plate section). *Adapted from Ellison et al. (2002) and Valeggia et al. (2009).*

levels. There is substantial individual and population variation in testosterone production and the sources of that variation, particularly the environmental ones, have received considerable attention (see Figure 7). In addition to genetic heterogeneity in hormone regulation (Xita and Tsatsoulis, 2010; Coviello et al., 2012), male reproductive function has been shown to vary in response to ecological variables such as nutrition, physical activity, acute immunological challenges, psychological stress, and social context (Bribiescas, 2001a,b; Muehlenbein and Bribiescas, 2010).

In general, minor-to-moderate changes in energy balance in an individual have little effect on adult testosterone levels or spermatogenesis (Bentley et al., 1993; Ellison and Panter-Brick, 1996; MacDonald et al., 2010; Muehlenbein and Bribiescas, 2010). It is only under extreme fasting or long-term, high-impact physical activity that male reproductive function is affected (Opstad, 1992; Gomez-Merino et al., 2003; Goto et al., 2013), and the effects seem to be easily reversible. However, at a population level, well-fed, sedentary men in industrialized societies tend to have higher circulating testosterone levels than non-Western men living in more calorie-restricted environments (Bribiescas, 2010).

Stressors other than energetic challenges have been shown to affect men's reproductive hormonal milieu. Mild immunological stressors, such as a common cold, were associated with significant reductions in testosterone levels (Muehlenbein and Bribiescas, 2005; Muehlenbein et al., 2010). At the same time, prolonged or chronic psychological stress has also been implicated in the regulation of the HPT axis (Chichinadze and Chichinadze, 2008); this response is likely to be mediated by glucocorticoids, which may suppress testicular function by affecting GnRH and LH production (Hardy et al., 2005; Hu et al., 2008).

Interestingly, testosterone levels may be increased at the initial states of acute stress (Chichinadze and Chichinadze, 2008). This relates to the dynamic two-way relationship between testosterone release and behavior. Human males show an increase in testosterone levels in the context of competition with other males (Booth et al., 1989; Bernhardt et al., 1998; Mazur and Booth, 1998)—which has been interpreted as a preparatory or dominance-seeking function—that leads to better physical and cognitive performance during the prospective match. In this context, testosterone has also been associated with aggressive behavior (Dabbs et al., 1987; Banks and Dabbs, 1996). However, there is no direct relationship between the level of circulating testosterone and the individual degree of aggressiveness (Pope et al., 2000). Instead, in humans testosterone has been postulated to have a "permissive effect"; i.e., testosterone allows the expression of aggression but not necessarily in a clear dose-dependent fashion (Sapolsky, 1998).

Testosterone levels often vary with mating and parenting status; pair-bonded relationships and fatherhood have been associated with suppressed testosterone production. For instance, several studies show that compared with single men matched for age, married men and fathers of young children have lower levels of circulating testosterone (Burnham et al., 2003; Gray, 2003; Gray et al., 2004, 2006, 2007; Gettler et al., 2011). Within the framework of life history theory, testosterone suppression would be the hormonal marker of the trade-off between mating and parenting efforts in men, or between current and future reproduction (Gray, 2010).

Male Reproductive Senescence

Unlike menopause, there is no abrupt cessation of reproductive function in men. However, there are distinct changes that characterize male reproductive aging (Bribiescas, 2006). Semen volume, sperm motility, and sperm morphology (but not sperm concentration) decline with age. This translates to lower fertility in older men, with pregnancy rates decreasing between 23% and 38% compared with those related to younger men (Kidd et al., 2001).

In most industrialized populations, men show a modest decline in testosterone levels with age. However, in non-Western populations, age has little to no effect on testosterone levels (Bribiescas, 2010). Interestingly, most testosterone variation is seen at younger ages, and populations tend to converge after the age of 50 (Bribiescas, 2006). From a human ecology perspective, these cross-population differences can be explained by different energy allocation strategies that are adapted to different ecological demands: men growing up in energetically demanding environments (low food availability, high pathogen load) would divert energy away from costly muscle building and toward immunological competence. Men living in more favorable ecologies would be able to afford higher testosterone levels without compromising other functions such as growth or immune function (Muehlenbein and Bribiescas, 2010).

CHALLENGES AND FUTURE DIRECTIONS OF THE FIELD OF HUMAN REPRODUCTIVE ECOLOGY

The main goal of HRE has been to understand how human reproduction responds to different environments and the consequences of variation in response in terms of health, reproductive success, and ultimately, reproductive fitness—in other words, to establish whether the flexibility and responsiveness of human reproductive function are adaptive.

As a collective, the findings of the past two decades provide a very persuasive argument in favor of the notion that much observed phenotypic variation in human reproductive function can be conceptualized as the product of natural selection rather than of developmental constraints or pathology (Ellison and Jasienska, 2007). One of the main challenges for future research is to demonstrate an actual relative advantage of such variation in terms of reproductive fitness using real data from contemporary populations. Some existing long-term epidemiological and clinical studies that include biomarker measurements provide an extraordinary opportunity for this (Stearns et al., 2010b).

Another aspect of reproductive variation that is instrumental to understanding the underlying evolutionary processes is its genetic basis. However, reproductive ecologists have so far done comparatively little work in this respect (Jasienska et al., 2006a; Iversen et al., 2012); much still needs to be learned about genetic variation, genotype–phenotype, and gene–environment interactions in relation to variation in reproductive function (Jasienska, 2008), a fact that identifies this area as ripe for future research. In this context, it is becoming increasingly clear that early developmental ecologies, likely acting through epigenetic processes, may program the reproductive trajectories followed later in life (Nuñez-de la Mora et al., 2007; Nuñez-De La Mora et al., 2008).

The trade-off concept, that individual differential allocation of finite energy among different life functions underlies variation in life history strategies, is central to the theoretical framework guiding much of HRE research. At a population level, trade-offs between reproductive effort and some other life history traits have been well described (Lummaa, 2010). At an individual level, however, the demonstration of trade-offs between physiological vital functions has been less straightforward (Clancy et al., 2013). The advent of new reliable methods for the simultaneous quantification of biomarkers associated with different vital functions (O'Connor et al., 2003; McDade et al., 2007; Valeggia, 2007; Salvante et al., 2012) is likely to remove some existing methodological constraints as well as boost research on trade-offs and on the dynamics of interactions between life functions. The development of suitable statistical techniques to manage the resulting data will need to follow suit (Stearns et al., 2010a).

Finally, perhaps one of the greatest challenges for HRE and many other disciplines relates to the biological and social consequences of the unprecedented change that most populations in the contemporary world are experiencing. The rapid economic, ecological, and cultural changes are generating new parameters and selective pressures, and as a result different trade-offs that can already be quantified as shifts in some human life history variables (Kirk et al., 2001; Byars et al., 2010; Courtiol et al., 2013). Understanding the variations in physiology, behavior, and life history patterns as humans experience these new biocultural environments is likely to engage human reproductive ecologists for decades to come.

REFERENCES

Adams Hillard, P.J., 2008. Menstruation in adolescents: what's normal, what's not. Annals of the New York Academy of Sciences 1135, 29–35.

Anderson, C.A., Duffy, D.L., et al., 2007. Estimation of variance components for age at menarche in twin families. Behavioural Genetics 37, 668–677.

Apter, D., 1980. Serum steroids and pituitary hormones in female puberty: a partly longitudinal study. Clinical Endocrinology 12 (2), 107–120.

Bagga, D., Ashley, J.M., et al., 1995. Effects of a very low fat, high fiber diet on serum hormones and menstrual function. Implications for breast cancer prevention. Cancer 76 (12), 2491–2496.

Bailey, R.C., Jenike, M.R., et al., 1992. The ecology of birth seasonality among agriculturalists in Central Africa. Journal of Biosocial Sciences 24, 393–412.

Baird, D.D., 2009. The gestational timing of pregnancy loss: adaptive strategy? American Journal of Human Biology 21 (6), 725–727.

Banks, T., Dabbs Jr., J.M., 1996. Salivary testosterone and cortisol in a delinquent and violent urban subculture. Journal of Social Psychology 136 (1), 49–56.

Bentley, G.R., Harrigan, A.M., et al., 1993. Seasonal effects on salivary testosterone levels among lese males of the Ituri Forest, Zaire. American Journal of Human Biology 5, 711–717.

Bentley, G.R., Harrigan, A.M., et al., 1998. Dietary composition and ovarian function among Lese horticulturalist women of the Ituri Forest, Democratic Republic of Congo. European Journal of Clinical Nutrition 52 (4), 261–270.

Bentley, G.R., Vitzthum, V.J., et al., 2000. Salivary Estradiol Levels from Conception and Nonconception Cycles in Rural Bolivian Women. Human Biology Association Abstracts (HBA Meetings). p. 279.

Bernhardt, P.C., Dabbs Jr., J.M., et al., 1998. Testosterone changes during vicarious experiences of winning and losing among fans at sporting events. Physiology and Behavior 65 (1), 59–62.

Bogin, B., 1999. Patterns of Human Growth. Cambridge University Press, Cambridge.

Bonzini, M., Coggon, D., et al., 2007. Risk of prematurity, low birthweight and pre-eclampsia in relation to working hours and physical activities: a systematic review. Occupatoinal and Environmental Medicine 64 (4), 228–243.

Booth, A., Shelley, G., et al., 1989. Testosterone, and winning and losing in human competition. Hormones and Behavior 23 (4), 556–571.

Bribiescas, R.G., 2001a. Reproductive ecology and life history of the human male. American Journal of Physical Anthropology 33, 148–176.

Bribiescas, R.G., 2001b. Reproductive physiology of the human male. In: Ellison, P. (Ed.), Reproductive Ecology and Human Evolution. Aldine de Gruyter, New York, pp. 106–133.
Bribiescas, R.G., 2006a. Men: Evolutionary and Life History. Harvard University Press, Cambridge, MA.
Bribiescas, R.G., 2006b. On the evolution, life history, and proximate mechanisms of human male reproductive senescence. Evolutionary Anthropology 15.
Bribiescas, R.G., 2010. An evolutionary and life history perspective on human male reproductive senescence. Annals of the New York Academy of Sciences 1204, 54–64.
Bronson, F.H., 1995. Seasonal variation in human reproduction: environmental factors. The Quarterly Review of Biology 70 (2), 141–164.
Brooks, S., Nevill, M.E., et al., 1990. The hormonal responses to repetitive brief maximal exercise in humans. European Journal of Applied Physiology and Occupational Physiolocy 60 (2), 144–148.
Burnham, T.C., Chapman, J.F., et al., 2003. Men in committed, romantic relationships have lower testosterone. Hormones and Behavior 44 (2), 119–122.
Byars, S.G., Ewbank, D., et al., 2010. Colloquium papers: natural selection in a contemporary human population. Proceedings of the National Academy of Sciences of the United States of America 107 (Suppl. 1), 1787–1792.
de Bruin, J.P., Bovenhuis, H., et al., 2001. The role of genetic factors in age at natural menopause. Human Reproduction 16 (9), 2014–2018.
Campbell, K.L., Wood, J.W., 1994. Introduction: what is human reproductive ecology, and why should we care about studying it? Annals of the New York Academy of Sciences 709, 1–8.
Campbell, B.C., Gillett-Netting, R., et al., 2004. Timing of reproductive maturation in rural versus urban Tonga boys, Zambia. Annals of Human Biology 31 (2), 213–227.
Campbell, B.C., Leslie, P., et al., 2005. Pubertal timing, hormones, and body composition among adolescent Turkana males. American Journal of Physical Anthropology 128, 896–905.
Campbell, B.C., 2006. Adrenarche and the evolution of human life history. American Journal of Human Biology 18 (5), 569–589.
Campbell, B.C., 2011. Adrenarche and middle childhood. Human Nature 22 (3), 327–349.
Cant, M.A., Johnstone, R.A., 2008. Reproductive conflict and the separation of reproductive generations in humans. Proceedings of the National Academy of Sciences of the United States of America 105 (14), 5332–5336.
Chichinadze, K., Chichinadze, N., 2008. Stress-induced increase of testosterone: contributions of social status and sympathetic reactivity. Physiology and Behavior 94 (4), 595–603.
Clancy, K.B., Klein, L.D., et al., 2013. Relationships between biomarkers of inflammation, ovarian steroids, and age at menarche in a rural polish sample. American Journal of Human Biology 25 (3), 389–398.
Cooper, C., Kuh, D., et al., 1996. Childhood growth and age at menarche. British Journal of Obstetrics and Gynaecology 103 (8), 814–817.
Courtiol, A., Rickard, I.J., et al., 2013. The demographic transition influences variance in fitness and selection on height and BMI in rural Gambia. Current Biology 23 (10), 884–889.
Coviello, A.D., Haring, R., et al., 2012. A genome-wide association meta-analysis of circulating sex hormone-binding globulin reveals multiple Loci implicated in sex steroid hormone regulation. PLoS Genetics 8 (7), e1002805.
Cramer, D.W., Xu, H., et al., 1995. Does "incessant" ovulation increase risk for early menopause? American Journal of Obstetrics and Gynecology 172 (2 Pt 1), 568–573.
Cresswell, J.L., Egger, P., et al., 1997. Is the age of menopause determined in-utero? Early Human Development 49 (2), 143–148.
Dabbs Jr., J.M., Frady, R.L., et al., 1987. Saliva testosterone and criminal violence in young adult prison inmates. Psychosomatic Medicine 49 (2), 174–182.
Dufour, D.L., Sauther, M.L., 2002. Comparative and evolutionary dimensions of the energetics of human pregnancy and lactation. American Journal of Human Biology 14 (5), 584–602.
Dunsworth, H.M., Warrener, A.G., et al., 2012. Metabolic hypothesis for human altriciality. Proceedings of the National Academy of Sciences of the United States of America 109 (38), 15212–15216.
Dvornyk, V., Waqar-ul-Haq, 2012. Genetics of age at menarche: a systematic review. Human Reproduction Update 18 (2), 198–210.
Eaton, S.B., Pike, M.C., et al., 1994. Women's reproductive cancers in evolutionary context. The Quarterly Review of Biology 69 (3), 353–367.
Edozien, L.C., 2006. Mind over matter: psychological factors and the menstrual cycle. Current Opinion in Obstetrics and Gynecology 18 (4), 452–456.
Ellis, B.J., Garber, J., 2000. Psychosocial antecedents of variation in girls' pubertal timing: maternal depression, stepfather presence, and marital and family stress. Child Development 71 (2), 485–501.
Ellison, P.T., Jasienska, G., 2007. Constraint, pathology, and adaptation: how can we tell them apart? American Journal of Human Biology 19, 622–630.
Ellison, P.T., Lager, C., 1986. Moderate recreational running is associated with lowered salivary progesterone profiles in women. American Journal of Obstetrics and Gynecology 154 (5), 1000–1003.
Ellison, P.T., Panter-Brick, C., 1996. Salivary testosterone levels among Tamang and Kami males of central Nepal. Human Biology 68, 955–965.
Ellison, P.T., Valeggia, C.R., 2003. C-peptide levels and the duration of lactational amenorrhea. Fertility and Sterility 80 (5), 1279–1280.
Ellison, P.T., Lipson, S.F., O'Rourke, M.T., Bentley, G.R., Harrigan, A.M., Panter-Brick, C., Vitzthum, V.J., 1993a. Population variation in ovarian function. Lancet 342 (8868), 433–434.
Ellison, P.T., Panter-Brick, C., et al., 1993b. The ecological context of human ovarian function. Human Reproduction 8 (12), 2248–2258.
Ellison, P.T., Bribiescas, R.G., et al., 2002. Population variation in age-related decline in male salivary testosterone. Human Reprodruction 17 (12), 3251–3253.
Ellison, P.T., 1988. Human salivary steroids: methodological considerations and applications in physical anthropology. Yearbook of Physical Anthropology 31, 115–142.
Ellison, P.T., 1990. Human ovarian function and reproductive ecology: new hypotheses. American Anthropologist 92, 933–952.
Ellison, P.T., 1993. Measurements of salivary progesterone. Annals of the New York Academy of Sciences 694, 161–176.
Ellison, P.T., 1994a. Advances in human reproductive ecology. Annual Review of Anthropology 23, 225–275.
Ellison, P.T., 1994b. Salivary steroids and natural variation in human ovarian function. Annals of the New York Academy of Sciences 709, 287–298.
Ellison, P.T., 2001. On Fertile Ground: A Natural History of Human Reproduction. Harvard University Press, Cambridge.
Ellison, P.T., 2003. Energetics and reproductive effort. American Journal of Human Biology 15 (3), 342–351.
Forbes, L.S., 1997. The evolutionary biology of spontaneous abortion in humans. Trends in Ecology and Evolution 12 (11), 446–450.
Forsum, E., Löf, M., 2007. Energy metabolism during human pregnancy. Annual Review of Nutrition 27, 277–292.

Gettler, L.T., McDade, T.W., et al., 2011. Longitudinal evidence that fatherhood decreases testosterone in human males. Proceedings of the National Academy of Sciences of the United States of America 108 (39), 16194–16199.

Gluckman, P.D., Hanson, M.A., 2006. Evolution, development and timing of puberty. Trends in Endocrinology and Metabolism 17 (1), 7–12.

Gold, E.B., Bromberger, J., et al., 2001. Factors associated with age at natural menopause in a multiethnic sample of midlife women. American Journal of Epidemiology 153 (9), 865–874.

Gomez-Merino, D., Chennaoui, M., et al., 2003. Immune and hormonal changes following intense military training. Military Medicine 168 (12), 1034–1038.

Goto, K., Shioda, K., et al., 2013. Effect of 2 days of intensive resistance training on appetite-related hormone and anabolic hormone responses. Clinical Physiological and Functional Imaging 33 (2), 131–136.

Gray, P.B., Campbell, B.C., et al., 2004. Social variables predict between-subject but not day-to-day variation in the testosterone of US men. Psychoneuroendocrinology 29 (9), 1153–1162.

Gray, P.B., Yang, C.F., et al., 2006. Fathers have lower salivary testosterone levels than unmarried men and married non-fathers in Beijing, China. Proceedings. Biological Sciences 273 (1584), 333–339.

Gray, P.B., Parkin, J.C., et al., 2007. Hormonal correlates of human paternal interactions: a hospital-based investigation in urban Jamaica. Hormones and Behavior 52 (4), 499–507.

Gray, P.B., 2003. Marriage, parenting, and testosterone variation among Kenyan Swahili men. American Journal of Physical Anthropology 122 (3), 279–286.

Gray, P.B., 2010. The evolution and endocrinology of human behavior: a focus on sex differences and reproduction. In: Muehlenbein, M.G. (Ed.), Human Evolutionary Biology. Cambridge University Press, Cambridge, UK, pp. 277–292.

Haig, D., 2008. Intimate relations: evolutionary conflicts of pregnancy and childhood. In: Stearns, S.C., Koella, J.C. (Eds.), Evolution in Health and Disease. Oxford University Press, New York, pp. 65–76.

Haig, D., 2010. Colloquium papers: Transfers and transitions: parent-offspring conflict, genomic imprinting, and the evolution of human life history. Proceedings of the National Academy of Sciences of the United States of America 107 (Suppl. 1), 1731–1735.

Hardy, R., Kuh, D., 2002. Does early growth influence timing of the menopause? Evidence from a British birth cohort. Human Reproduction 17 (9), 2474–2479.

Hardy, R., Kuh, D., 2005. Social and environmental conditions across the life course and age at menopause in a British birth cohort study. British Journal of Obstetrics and Gynaecology 112 (3), 346–354.

Hardy, M.P., Gao, H.B., et al., 2005. Stress hormone and male reproductive function. Cell and Tissue Research 322 (1), 147–153.

Harlow, B.L., Signorello, L.B., 2000. Factors associated with early menopause. Maturitas 35 (1), 3–9.

Hawkes, K., O'Connell, J.F., et al., 1998. Grandmothering, menopause, and the evolution of human life histories. Proceedings of the National Academy of Sciences of the United States of America 95 (3), 1336–1339.

Hill, K.R., Hurtado, A.M., 1996. Ache Life History: The Ecology and Demography of a Foraging People. Aldine de Gruyter, New York.

Hill, K., Kaplan, H., 1999. Life history traits in humans: theory and empirical studies. Annual Review of Anthropology 28, 397–430.

Hill, K.R., 1993. Life history and evolutionary anthropology. Evolutionary Anthropology 2 (3), 78–88.

Hu, G.X., Lian, Q.Q., et al., 2008. Rapid mechanisms of glucocorticoid signaling in the Leydig cell. Steroids 73 (9–10), 1018–1024.

Iversen, A., Thune, I., et al., 2012. Genetic polymorphism CYP17 rs2486758 and metabolic risk factors predict daily salivary 17beta-estradiol concentration in healthy premenopausal Norwegian women. The EBBA-I study. Journal of Clinical Endocrinology and Metabolism 97 (5), E852–E857.

Jasienska, G., Ellison, P.T., 1998. Physical work causes suppression of ovarian function in women. Proceedings. Biological Sciences 265 (1408), 1847–1851.

Jasienska, G., Ellison, P.T., 2004. Energetic factors and seasonal changes in ovarian function in women from rural Poland. American Journal of Human Biology 16, 563–580.

Jasienska, G., Jasienski, M., 2008. Interpopulation, interindividual, intercycle, and intracycle natural variation in progesterone levels: a quantitative Assessment and implications for population studies. American Journal of Human Biology 20, 35–42.

Jasienska, G., Kapiszewska, M., et al., 2006a. CYP17 genotypes differ in salivary 17-beta estradiol levels: a study based on hormonal profiles from entire menstrual cycles. Cancer Epidemiology, Biomarkers & Prevention 15 (11), 2131–2135.

Jasienska, G., Thune, I., et al., 2006b. Fatness at birth predicts adult susceptibility to ovarian suppression: an empirical test of the predictive adaptive response hypothesis. Proceedings of the National Academy of Sciences of the United States of America 103 (34), 12759–12762.

Jasienska, G., Ziomkiewicz, A., et al., 2006c. High ponderal index at birth predicts high estradiol levels in adult women. American Journal of Human Biology 18 (1), 133–140.

Jasienska, G., 2008. Determinants of variation in human female ovarian function: genetic polymorphism, developmental conditions and adult lifestyle. American Journal of Physical Anthropology 137 (Suppl. 47), 124.

Kidd, S.A., Eskenazi, B., et al., 2001. Effects of male age on semen quality and fertility: a review of the literature. Fertility and Sterility 75 (2), 237–248.

Kirk, K.M., Blomberg, S.P., et al., 2001. Natural selection and quantitative genetics of life-history traits in Western women: a twin study. Evolution 55 (2), 423–435.

Kok, H.S., Onland-Moret, N.C., et al., 2005. No association of estrogen receptor alpha and cytochrome P450c17alpha polymorphisms with age at menopause in a Dutch cohort. Human Reproduction 20 (2), 536–542.

Kuh, D., Hardy, R. (Eds.), 2002. A Life Course Approach to Women's Health. Oxford University Press, Oxford, UK.

Kuzawa, C.W., 1998. Adipose tissue in human infancy and childhood: an evolutionary perspective. American Journal of Physical Anthropology S27, 177–209.

Lahdenpera, M., Lummaa, V., et al., 2004. Menopause: why does fertility end before life? Climacteric 7 (4), 327–331 discussion 331–322.

Lipson, S.F., Ellison, P.T., 1992. Normative study of age variation in salivary progesterone profiles. Journal of Biosocial Sciences 24, 233–244.

Loto, O.M., Awowole, I., 2012. Tuberculosis in pregnancy: a review. Journal of Pregnancy 2012, 379271.

Lummaa, V., 2010. In: Frey, U.J., Störmer, C., Willführ, K.P. (Eds.), The Costs of Reproduction. Homo Novus—a Human without Illusions. Springer.

MacDonald, A.A., Herbison, G.P., et al., 2010. The impact of body mass index on semen parameters and reproductive hormones in human males: a systematic review with meta-analysis. Human Reproduction Update 16 (3), 293–311.

Mace, R., 2000. Evolutionary ecology of human life history. Animal Behaviour 59 (1), 1–10.

Magid, K.S., 2011. Reproductive Ecology and Life History of Human Males: A Migrant Study of Bangladeshi Men (Doctoral dissertation). Anthropology. London, University College London.

Mazur, A., Booth, A., 1998. Testosterone and dominance in men. Behavioral Brain Sciences 21 (3), 353–363 discussion 363–397.

McDade, T.W., Williams, S., et al., 2007. What a drop can do: dried blood spots as a minimally invasive method for integrating biomarkers into population-based research. Demography 44 (4), 899–925.

Melby, M.K., Lock, M., et al., 2005. Culture and symptom reporting at menopause. Human Reproduction Update 11 (5), 495–512.

Menezes, E.V., Yakoob, M.Y., et al., 2009. Reducing stillbirths: prevention and management of medical disorders and infections during pregnancy. BMC Pregnancy Childbirth 9 (Suppl. 1), S4.

Mishra, G., Hardy, R., et al., 2007. Are the effects of risk factors for timing of menopause modified by age? Results from a British birth cohort study. Menopause 14 (4), 717–724.

Mishra, G.D., Cooper, R., et al., 2009. Early life circumstances and their impact on menarche and menopause. Womens Health (London, England) 5 (2), 175–190.

Moisan, J., Meyer, F., et al., 1991. Leisure physical activity and age at menarche. Medicine and Science in Sports and Exercise 23 (10), 1170–1175.

Moore, L.G., Charles, S.M., et al., 2011. Humans at high altitude: hypoxia and fetal growth. Respiratory Physiology and Neurobiology 178 (1), 181–190.

Morabia, A., Costanza, M.C., 1998. International variability in ages at menarche, first livebirth, and menopause. World Health Organization Collaborative Study of Neoplasia and Steroid Contraceptives. American Journal of Epidemiology 148 (12), 1195–1205.

Morris, F.L., Payne, W.R., et al., 1999. Prospective decrease in progesterone concentrations in female lightweight rowers during the competition season compared with the off season: a controlled study examining weight loss and intensive exercise. British Journal of Sports Medicine 33 (6), 417–422.

Morris, D.H., Jones, M.E., et al., 2011. Familial concordance for age at menarche: analyses from the Breakthrough Generations Study. Paediatric and Perinatal Epidemiology 25, 306–311.

Muehlenbein, M.P., Bribiescas, R.G., 2005. Testosterone-mediated immune functions and male life histories. American Journal of Human Biology 17, 527–558.

Muehlenbein, M.P., Bribiescas, R.G., 2010. Male reproduction: physiology, behavior, and ecology. In: Muehlenbein, M.P. (Ed.), Human Evolutionary Biology. Cambridge University Press, Cambridge, UK, pp. 351–375.

Muehlenbein, M.P., Hirschtick, J.L., et al., 2010. Toward quantifying the usage costs of human immunity: Altered metabolic rates and hormone levels during acute immune activation in men. American Journal of Human Biology 22 (4), 546–556.

Nepomnaschy, P.A., Welch, K.B., et al., 2006. Cortisol levels and very early pregnancy loss in humans. Proceedings of the National Academy of Sciences of the United States of America 103 (10), 3938–3942.

Nuñez-de la Mora, A., Chatterton, R.T., et al., 2007. Childhood conditions influence adult progesterone levels. PLoS Medicine 4 (5), e167–e190.

Nuñez-De La Mora, A., Bentley, G.R., et al., 2008. The impact of developmental conditions on adult salivary estradiol levels: why this differs from progesterone? American Journal of Human Biology 20 (1), 2–14.

O'Connor, K.A., Brindle, E., et al., 2003. Urinary estrone conjugate and pregnanediol 3-glucuronide enzyme immunoassays for population research. Clinical Chemistry 49 (7), 1139–1148.

Ong, K.K., Bann, D., et al., 2012. Timing of voice breaking in males associated with growth and weight gain across the life course. Journal of Clinical Endocrinology and Metabolism 97 (8), 2844–2852.

Opstad, P.K., 1992. Androgenic hormones during prolonged physical stress, sleep, and energy deficiency. Journal of Clinical Endocrinology and Metabolism 74 (5), 1176–1183.

Panter-Brick, C., Todd, A., et al., 1996. Growth status of homeless Nepali boys: do they differ from rural and urban controls? Social Science and Medicine 43 (4), 441–451.

Panter-Brick, C., 1993. Seasonality of energy expenditure during pregnancy and lactation for rural Nepali women. American Journal of Clinical Nutrition 57 (5), 620–628.

Parent, A.S., Teilmann, G., et al., 2003. The timing of normal puberty and the age limits of sexual precocity: variations around the world, secular trends, and changes after migration. Endocrine Reviews 24, 668–693.

Parente, R.C., Faerstein, E., et al., 2008. The relationship between smoking and age at the menopause: a systematic review. Maturitas 61 (4), 287–298.

Peacock, J.L., Bland, J.M., et al., 1995. Preterm delivery: effects of socioeconomic factors, psychological stress, smoking, alcohol, and caffeine. British Medical Jounal 311, 531–535.

Peacock, N.R., 1991. In: Peacock, N.R. (Ed.), Rethinking the Sexual Division of Labor: Reproduction and Women's Work Among the Efe. Gender at the Crossroads of Knowledge: Feminist Anthropology in the Postmodern Era. University of California press, Berkeley, pp. 339–360.

Peccei, J.S., 1995. A hypothesis for the origin and evolution of menopause. Maturitas 21 (2), 83–89.

Peccei, J.S., 2001. A critique of the grandmother hypotheses: old and new. American Journal of Human Biology 13 (4), 434–452.

Pike, I.L., 2001. The evolutionary and ecological context of human pregnancy. In: Ellison, P.T. (Ed.), Reproductive Ecology and Human Evolution. Aldine Transactions, New Brunswick, pp. 39–58.

Pike, I.L., 2005. Maternal stress and fetal responses: evolutionary perspectives on preterm delivery. American Journal of Human Biology 17 (1), 55–65.

Pirke, K.M., Wurmser, H., et al., 1999. Early pubertal development and overweight in girls. Annals of the New York Academy of Sciences 892, 327–329.

Pope Jr., H.G., Kouri, E.M., et al., 2000. Effects of supraphysiologic doses of testosterone on mood and aggression in normal men: a randomized controlled trial. Archives of General Psychiatry 57 (2), 133–140 discussion 155–136.

Prebeg, Z., Bralic, I., 2000. Changes in menarcheal age in girls exposed to war conditions. American Journal of Human Biology 12 (4), 503–508.

Prentice, A.M., Goldberg, G.R., 2000. Energy adaptations in human pregnancy: limits and long-term consequences. American Journal of Clinical Nutrition 71 (Supl), S1226–S1232.

Prentice, A.M., Prentice, A., 1988. Energy costs of lactation. Annual Review of Nutrition 8, 63–79.

Prentice, A.M., Cole, T.J., et al., 1987. Increased birthweight after prenatal dietary supplementationof rural African women. American Journal of Clinical Nutrition 46, 912–925.

Rayco-Solon, P., Fulford, A.J., et al., 2005. Maternal preconceptional weight and gestational length. American Journal of Obstetrics and Gynecology 192 (4), 1133–1136.

Rey, R.A., Musse, M., et al., 2009. Ontogeny of the androgen receptor expression in the fetal and postnatal testis: its relevance on Sertoli cell maturation and the onset of adult spermatogenesis. Microscopy Research Techniques 72 (11), 787–795.

Roshni, R.P., Steer, P., et al., 2004. Does gestation vary by ethnic group? A London-based study of over 122 000 pregnancies with spontaneous onset of labour. International Journal of Epidemiology 33 (1), 107–113.

dos Santos Silva, I., De Stavola, B.L., et al., 2002. Prenatal factors, childhood growth trajectories and age at menarche. International Journal of Epidemiology 31 (2), 405–412.

Salvante, K.G., Brindle, E., et al., 2012. Validation of a new multiplex assay against individual immunoassays for the quantification of reproductive, stress, and energetic metabolism biomarkers in urine specimens. American Journal of Human Biology 24 (1), 81–86.

Sapolsky, R.M., 1998. The Trouble with Testosterone and Other Essays on the Human Predicament. Touchstone Books, New York.

Schantz-Dunn, J., Nour, N.M., 2009. Malaria and pregnancy: a global health perspective. Reviews in Obstetrics and Gynecology 2 (3), 186–192.

Schliep, K.C., Schisterman, E.F., et al., 2013. Energy-containing beverages: reproductive hormones and ovarian function in the BioCycle Study. American Journal of Clinical Nutrition 97 (3), 621–630.

Scholl, T., Hediger, M., et al., 1994. Maternal growth during pregnancy and the competition for nutrients. American Journal of Clinical Nutrition 60, 183–188.

Scholl, T., Hediger, M., et al., 1995. Maternal growth during adolescent pregnancy. Journal of the American Medical Association 274 (1), 26–27.

Schweiger, U., Laessle, R., et al., 1987. Diet-induced menstrual irregularities: effects of age and weight loss. Fertility and Sterility 48 (5), 746–751.

Sellen, D.W., 2001. Comparison of infant feeding patterns reported for nonindustrial populations with current recommendations. Journal of Nutrition 131, 2707–2715.

Sellen, D.W., 2007. Evolution of infant and young child feeding: implications for contemporary public health. Annual Review of Nutrition 27, 123–148.

Shanley, D.P., Sear, R., et al., 2007. Testing evolutionary theories of menopause. Proceedings. Biological Sciences 274 (1628), 2943–2949.

Shiono, P.H., Klebanoff, M.A., et al., 1986. Smoking and drinking during pregnancy. Their effects on preterm birth. Journal of the American Medical Association 255 (1), 82–84.

Sievert, L.L., 2006. Menopause: A Biocultural Perspective. Rutgers University Press, New Brunswick.

Sowers, M.R., La Pietra, M.T., 1995. Menopause: Its Epidemiology and Potential Association with Chronic Diseases. Epidemiological Reviews 17 (2), 287–302.

Stearns, S.C., Byars, S.G., et al., 2010a. Measuring selection in contemporary human populations. Nature Reviews Genetics 11 (9), 611–622.

Stearns, S.C., Nesse, R.M., et al., 2010b. Evolution in health and medicine Sackler colloquium: evolutionary perspectives on health and medicine. Proceedings of the National Academy of Sciences of the United States of America 107 (Suppl. 1), 1691–1695.

Stearns, S.C., 1992. The Evolution of Life Histories. Oxford University Press, Oxford.

Torgerson, D.J., Thomas, R.E., et al., 1997. Alcohol consumption and age of maternal menopause are associated with menopause onset. Maturitas 26 (1), 21–25.

Tracer, D.P., 1991. Fertility-related changes in maternal body composition among the Au of Papua New Guinea. American Journal of Physical Anthropology 85, 393–405.

Utian, W.H., 1999. The International Menopause Society menopause-related terminology definitions. Climacteric 2 (4), 284–286.

te Velde, E.R., Dorland, M., et al., 1998. Age at menopause as a marker of reproductive ageing. Maturitas 30 (2), 119–125.

Valeggia, C., Ellison, P.T., 2009. Interactions between metabolic and reproductive functions in the resumption of postpartum fecundity. American Journal of Human Biology 21 (4), 559–566.

Valeggia, C., Elwarch, C.L., Ellison, P.T., 2009. Testosterone, aging, and seasonality among Toba men of northern Argentina. American Journal of Physical Anthropology (Suppl. 48), 259.

Valeggia, C., 2007. Taking the lab to the field: minimally invasive techniques for monitoring reproductive hormones in population-scale research. Population Development Review 33 (3), 525–542.

Vihko, R., Apter, D., 1984. Endocrine characteristics of adolescent menstrual cycles: impact of early menarche. Journal of Steroid Biochemistry 20 (1), 231–236.

Vitzthum, V.J., Spielvogel, H., et al., 2000. Menstrual patterns and fecundity among non-lactating and lactating cycling women in rural highland Bolivia: implications for contraceptive choice. Contraception 62 (4), 181–187.

Vitzthum, V.J., Bentley, G.R., et al., 2002. Salivary progesterone levels and rate of ovulation are significantly lower in poorer than in better-off urban-dwelling Bolivian women. Human Reproduction 17 (7), 1906–1913.

Vitzthum, V.J., Spielvogel, H., et al., 2004. Interpopulational differences in progesterone levels during conception and implantation in humans. Proceedings of the National Academy of Sciences of the United States of America 101 (6), 1443–1448.

Vitzthum, V.J., Thornburg, J., et al., 2009. Seasonal modulation of reproductive effort during early pregnancy in humans. American Journal of Human Biology 21 (4), 548–558.

Wallace, J.M., Aitken, R.P., et al., 2004. Nutritionally mediated placental growth restriction in the growing adolescent: consequences for the fetus. Biology of Reproduction 71 (4), 1055–1062.

Wallis, M.C., Waters, P.D., et al., 2008. Sex determination in mammals–before and after the evolution of SRY. Cellular and Molecular Life Sciences 65 (20), 3182–3195.

Wasser, S.K., Barash, D.P., 1983. Reproductive suppression among female mammals: implications for biomedicine and sexual selection theory. Quarterly Review of Biology 58 (4), 513–538.

Wierson, M., Long, P.J., et al., 1993. Toward a new understanding of early menarche: the role of environmental stress in pubertal timing. Adolescence 28 (112), 913–924.

Winkvist, A., Rasmussen, K.M., et al., 1992. A new definition of maternal depletion syndrome. American Journal of Public Health 82 (5), 691–694.

Wood, J.W., 1994. Dynamics of Human Reproduction: Biology, Biometry, Demography. Aldine de Gruyter, New York.

Worthman, C.M., Kuzara, J., 2005. Life history and the early origins of health differentials. American Journal of Human Biology 17 (1), 95–112.

Wronka, I., Pawlinska-Chmara, R., 2005. Menarcheal age and socio-economic factors in Poland. Annals of Human Biology 32 (5), 630–638.

Xita, N., Tsatsoulis, A., 2010. Genetic variants of sex hormone-binding globulin and their biological consequences. Molecular and Cellular Endocrinology 316 (1), 60–65.

Zemel, B., Worthman, C.M., et al., 1993. Differences in endocrine status associated with urban-rural patterns of growth and maturation in Bundi (Gende-speaking) adolescents of Papua New Guinea. In: Schell, L.M., Smith, M.T., Bilsborough, A. (Eds.), Urban Ecology and Health in the Third World. Cambridge University Press, Cambridge, pp. 39–60.

Chapter 22

Human Senescence

Lynnette L. Sievert
Department of Anthropology, Machmer Hall, UMass Amherst, Amherst, MA, USA

SYNOPSIS

Senescence can be defined as age-related, deleterious changes that result in the reduced probability of reproduction and increased likelihood of death. The evolution of human senescence can be explained by reference to the brief life expectancy of our ancestors. The deleterious alleles that currently cause degeneration and disease during our long years of postreproductive life were infrequently expressed in our evolutionary and historical past. Because those alleles were rarely expressed, they were not exposed to the forces of natural selection. Three theories of senescence are described, along with theories that address the evolution of female postreproductive life.

INTRODUCTION

Worldwide, there were almost 525 million people aged 65 or older in 2010. That number is expected to rise to over 700 million people by 2020 and over 975 million people by 2030 (United Nations Population Division, 2011). Table 1 shows the number of people aged 65 or more in select countries, along with population projections to 2020 and 2030. The increase in the number of people aged 65 or older is not limited to wealthy nations. As shown in Figure 1, by 2030, 30% of Japan's population may be aged 65 or older, up from 23% in 2010, while 12% of Mexico's population may be aged 65 or older, up from 6% in 2010.

The increase in the percentage of people aged 65 or older would be unremarkable were it not for the burden of deleterious physiological and cognitive changes that often accompanies chronological aging. Worldwide, as more people survive to later ages, there is an increase in the prevalence of chronic degenerative diseases. These diseases are by turns inconvenient, uncomfortable, costly, and tragic. Why is the latter half of the human life span characterized by somatic and reproductive senescence? Why, as Shaw observed, is youth wasted on the young?

It was not always this way. Life expectancy was probably 15–20 years for our hominin ancestors, Australopithecus (see the chapter by Ward in the present volume) and early *Homo* (see the chapter by Simpson). Fossil remains from Paleolithic Neanderthals (see the chapter by Ahern) suggest that nearly half of the population died prior to the age of 20, and only 10% survived beyond the age of 40 years (Trinkaus, 1995). Within our own *Homo sapiens* (see the chapter by Holt), prehistoric foragers had a mortality pattern similar to Neanderthals, with a survivorship of just 17% beyond the age of 40 (Kennedy, 2003). Figure 2 compares life expectancy at birth across a number of hominin, historical, and contemporary populations.

As shown in Figure 2, the estimated life expectancy at birth in Western Europe during the nineteenth century ranged from 29 to 42 years. Infant and child mortality is one of the causes of low life expectancy; however, even if one considers life expectancy at age 10, life expectancy in Western Europe during the nineteenth century was still less than 50 years. Low

TABLE 1 The Estimated Number of People Aged 65 or Older and the Total Population (Medium Variant) in Four Countries and the World

		2010	2020	2030
Bangladesh	Ages 65+	6,819,000	8,509,000	13,747,000
	Total	148,692,000	167,256,000	181,863,000
	Percent 65+	**4.6%**	**5.1%**	**7.6%**
Japan				
	Ages 65+	28,707,000	35,432,000	36,404,000
	Total	126,536,000	124,804,000	120,218,000
	Percent 65+	**22.7%**	**28.4%**	**30.3%**
Mexico				
	Ages 65+	7,201,000	10,463,000	15,826,000
	Total	113,423,000	125,928,000	135,398,000
	Percent 65+	**6.3%**	**8.3%**	**11.7%**
United States				
	Ages 65+	40,534,000	54,655,000	72,000,000
	Total	301,384,000	337,102,000	361,680,000
	Percent 65+	**13.4%**	**16.2%**	**19.9%**
World				
	Ages 65+	524,364,000	717,797,000	976,111,000
	Total	6,895,889,000	7,656,528,000	8,321,380,000
	Percent 65+	**7.6%**	**9.4%**	**11.7%**

United Nations, Population Division, 2011.

life expectancy characterized our species until the turn of the past century. Over the last 100 years, life expectancy has increased dramatically to a current life expectancy at birth of 78.7 years in the United States (Murphy et al., 2013).

When life expectancies are low, age-related chronic degenerative diseases are few; however, since the time of the Neanderthals, there were always some individuals at risk for osteoarthritis and other age-related disorders. Menopause was most likely experienced by a small number of female Neanderthals as well (Trinkaus, 1995). Why, in the course of evolving a uniquely human pattern of life history characterized by long periods of childhood, adolescent growth, and postreproductive life, was there no selection against senescent change?

The answer lies in the brief life expectancy of our ancestors. The deleterious alleles that currently cause degeneration and disease during our long years of postreproductive life were infrequently expressed in our evolutionary and historical past. Because those alleles were rarely expressed, they were not exposed to the forces of natural selection. They were not selected against; therefore, the deleterious alleles that cause the senescent changes associated with aging are part of our gene pool. The increase in life expectancy that came about due to cultural and environmental adaptations uncovered those deleterious alleles. The alleles are not new, but they are now expressed in human populations with alarming frequency. As shown in Figure 3, the prevalence of cancer increases most dramatically after the ages at which most of our hominin ancestors died.

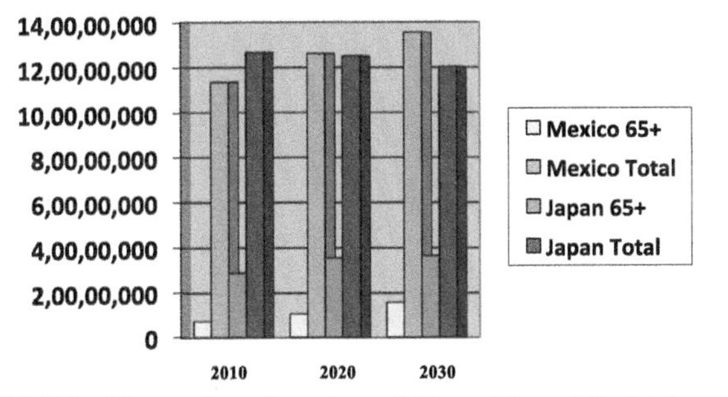

FIGURE 1 The number of people aged 65 or older and the total population (medium variant) in 2010, 2020, and 2030 in Mexico and Japan (United Nations, Population Division, 2011). (See color plate section).

Definition of Senescence

Terms like "aging" and "senescence" are familiar but imprecise. There is no agreed-upon definition for either term, and some people use the terms interchangeably. Various criteria have been suggested to define the age-related changes of senescence, including cumulative, deleterious, progressive (gradual), irreversible, intrinsic (within the individual), multifactorial (involving environmental and genetic factors), and universal to the species. The criterion of universality is a point of particular disagreement among researchers. Almost all researchers agree, however, that age-related changes, both structural and functional, decrease the ability of an individual to withstand stress, reduce the probability of reproduction, and increase the likelihood that an individual will die.

From an evolutionary perspective, one of the most important considerations is the timing of age-related changes. Two points of view place the beginning of senescence at the end of development, just after puberty. According to Arking (2006), senescence begins just after the minimum point in age-specific mortality. In the United States, death rates are lowest among children aged five to nine (see Figure 4), and therefore senescence begins around

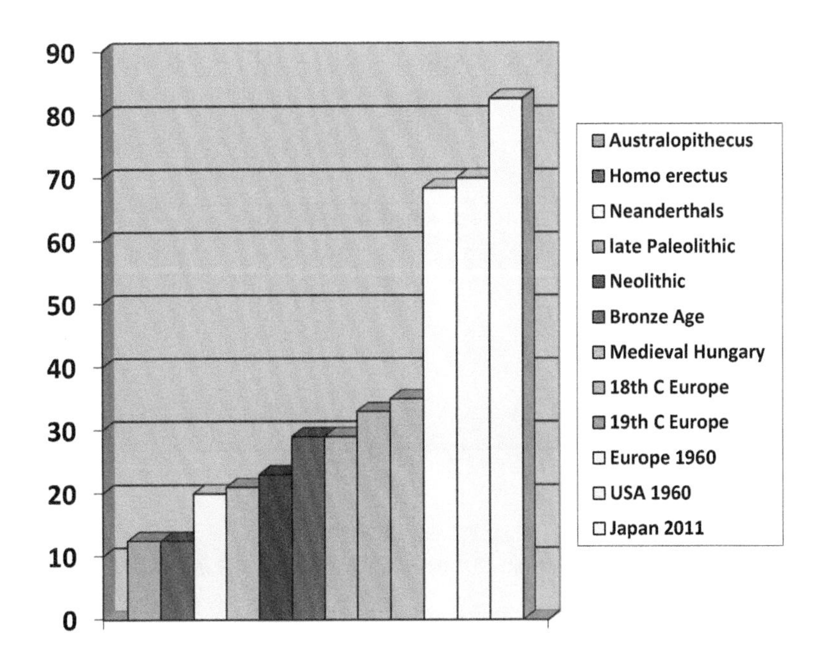

FIGURE 2 Life expectancy at birth, in years, across a number of hominin, historical, and contemporary populations. (See color plate section). *Data from Kennedy (2003) and World Bank.*

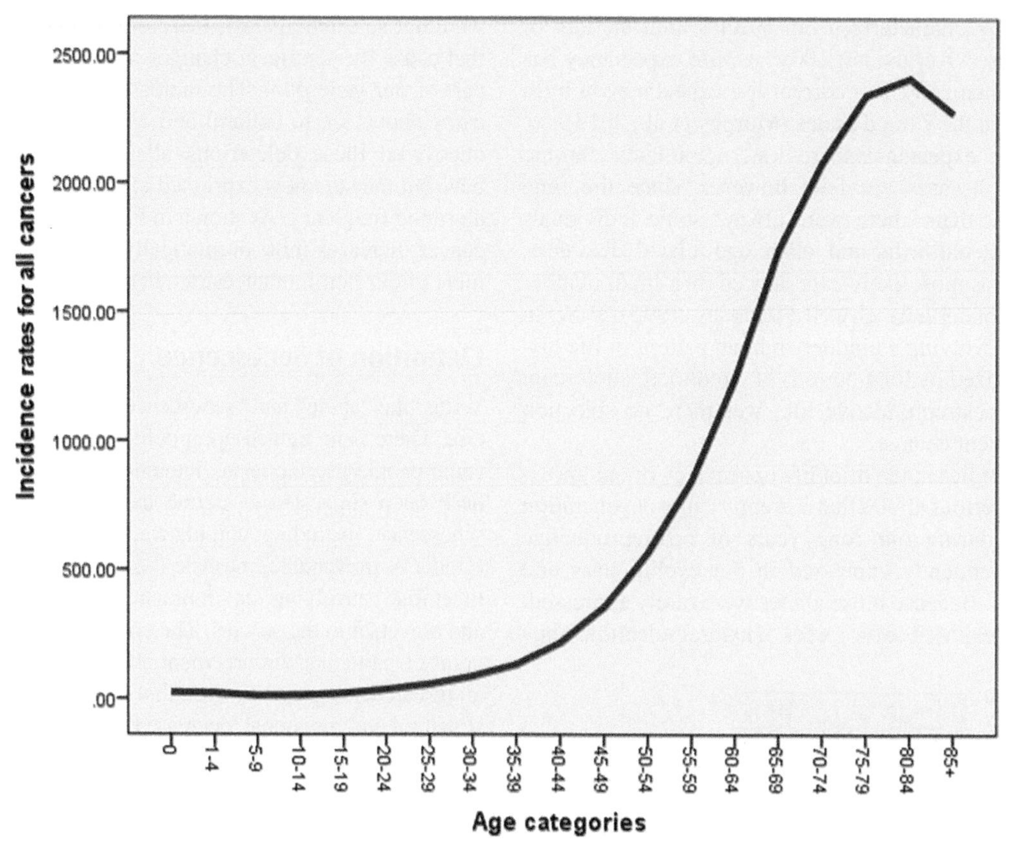

FIGURE 3 Age-specific SEER Incidence Rates, 2006–2010, for all cancers by age at diagnosis. Rates are per 100,000 and age-adjusted to the 2000 US standard population (SEER Cancer Statistics Review, 1975–2010).

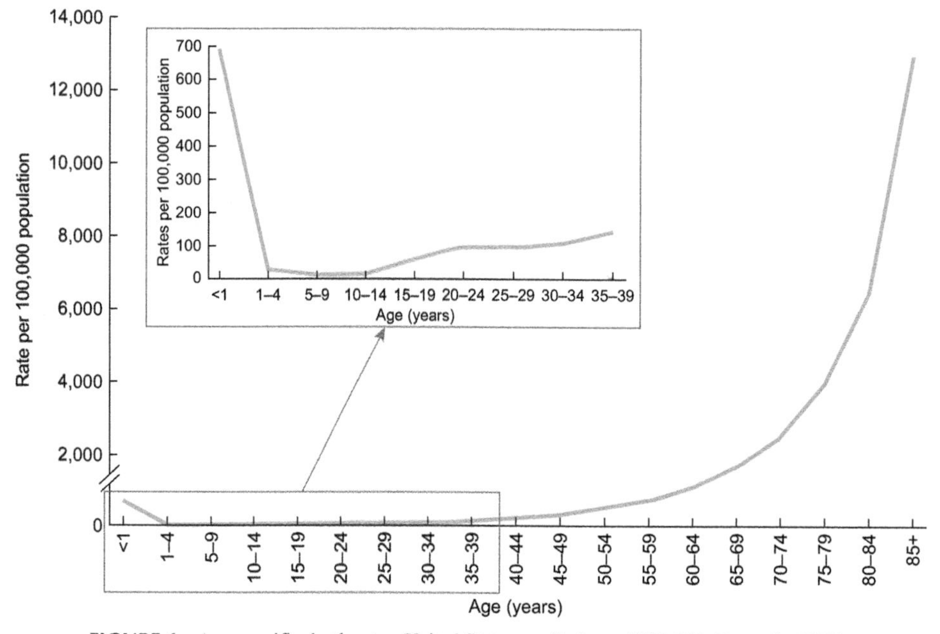

FIGURE 4 Age-specific death rates: United States, preliminary 2007 (Miniño et al., 2009).

the age of 10. In contrast, Crews (2003) focuses not on mortality but on fertility, and argues that senescence begins just after the attainment of maximum reproductive potential—the point at which organisms are able to bear and rear offspring with maximum efficiency. In humans, maximum reproductive potential is around the age of 19. This age may seem early to us now, but is rather late in the context of the life expectancy of our hominin ancestors. Other researchers narrow the definition of senescence to mean only the changes that occur during the postreproductive period.

The causes of senescence are multifactorial—that is, both environmental and genetic. Senescence is also polygenic, meaning that multiple genetic factors are involved. There is no single known biological mechanism responsible for age-related changes across the many levels of senescence, from modifications in DNA to alterations in cells and organ systems. Age-related changes that appear to be senescent include reduced repair of damaged DNA, accumulation of lipofuscin in cells, decreased permeability of cell membranes, declining hormone production, lowered immune function, loss of respiratory and circulatory reserve capacity, progressive dementia, atrophy of muscle, and loss of bone (Arking, 2006; Crews, 2003). All somatic systems are susceptible to disease and breakdown at about the same age. This is in part because senescent changes accumulate over time before causing problems, and also because senescence cannot be effectively selected against after the end of the reproductive period.

Loss of reproductive function in women is the result of senescent changes, but menopause is different from the somatic changes listed above. For one thing, menopause is universal among all female humans who live to the age of 60. The loss of ovarian follicles, which results in menopause, appears to be programmed. And, in humans, the timeline for reproductive senescence is notably disconnected from the timeline for somatic senescence. For these reasons, the evolution of senescence in general and menopause in particular will be addressed separately.

THE EVOLUTION OF SENESCENCE

Biomedical researchers, geneticists, and clinicians work to understand and treat the proximate mechanisms underlying senescent changes. Evolutionary biologists try to understand the ultimate questions about why humans and other living things are vulnerable to senescent changes. Much progress has been made toward understanding the mechanisms underlying age-related changes; however, theories about the evolution of senescence have remained relatively stable. It appears that forces of evolution shape a similar pattern of life history, including senescence, for all sexually reproducing species. Why is senescence part of our life history?

The evolution of senescence has been of interest to biologists since the nineteenth century. Most reviews start with the work of August Weismann (1891) and Peter Medawar (1946, 1952), then move to newer understandings based on the work of George Williams (1957) and William Hamilton (1966). More recently, theories of aging and senescence have been advanced by Thomas Kirkwood (1977, 1987), Michael Rose (1991), Caleb Finch (1990), Stephen Stearns (1992), Steven Austad (1997), and others. Currently, three major theories explain the evolution of senescence: (1) the disposable soma theory, (2) the theory of mutation accumulation and late-acting genes, and (3) antagonistic pleiotropy. These theories have different histories and proponents, but are not mutually exclusive.

Disposable Soma Theory

This theory has been attributed to the German evolutionary biologist August Weismann (1891), who suggested that sexually reproducing organisms invest in reproduction at the expense of maintaining the soma (body). Thomas Kirkwood (1977) proposed an idea similar to the disposable soma theory, making an analogy with products that are manufactured to last for only as long as needed. According to this theory, available resources (e.g., energy) are limited, and these resources must be allocated within an organism for growth, maintenance, and/or reproduction. Once growth and sexual maturation are achieved, natural selection favors investment in reproductive success. However, increasing age is accompanied by the loss of reproductive potential.

Natural selection works by acting on existing variation within a population. Some individuals have more offspring than others, so their alleles are passed along more successfully compared with the alleles of other individuals, generation after generation. Natural selection acts through differential fertility; therefore, as individuals age and reproductive potential declines, natural selection loses its strength. For example, natural selection selects for increasing bone density in young women so they have the mineral reserve necessary to meet the requirements of fetuses during pregnancy and the needs of babies during lactation. Over many generations, women with greater bone density at the ages of 20–30 gave birth to healthier children who went on to reproduce healthier children, carrying forward the alleles for good bone density at the ages of 20–30. In contrast, women with greater bone density at 50 did not have a reproductive advantage. There was no pressure in our evolutionary past to maintain high levels of bone mineral density in older women, and the result is that at menopause bone density falls quickly. In other words, extra bone density was disposable after the age of 50. Our ancestors died before the loss of bone mineral density could result in osteoporosis and fractures; however, we now reap the legacy of the decrease in the force of natural selection with age (see Figure 5). According to the CDC, there were 281,000 hospital admissions for hip fractures among people aged 65 and older in the United States in 2007.

Maintenance of the soma is needed to ensure a sound body, but only until the age when most individuals have died from accidental causes. Therefore, once reproductive maturity has been reached, natural selection is more strongly focused on reproduction than somatic maintenance and repair. This leads to degeneration in somatic tissues that appears to result from "wear and tear." In this theory of senescence, natural selection sacrifices continued vigor of the body, exchanging it for increased reproduction (Rose, 1991) by investing as little as possible in somatic maintenance.

There have been a number of critiques of the disposable soma theory (Blagosklonny, 2010). For example, a number of animal models have demonstrated that caloric restriction results in increased longevity. This seems to be at odds with what would be predicted by the disposable soma theory, if aging results from an evolved balance between resources devoted to maintenance and those devoted to reproduction. One might expect limited resources to compromise somatic maintenance. On the other hand, adaptive plasticity may adjust the optimal balance between resources devoted to maintenance or reproduction in response to changing environments (Shanley and Kirkwood, 2000).

Mutation Accumulation and Late-Acting Genes

This second theory of senescence is generally credited to Peter Medawar (1952). New mutations caused by UV light, mistakes in cell replication, and environmental toxins appear in all populations. Most mutations are harmful and selected against—e.g., individuals with deleterious alleles die before having a chance to reproduce, or have few surviving offspring. However, Medawar realized that deleterious mutations that act late in life would not be selected against, because most of our ancestors who carried those mutations would die at an earlier age from disease, predation, or other environmental hazards. If individuals died from external causes prior to the expression of a deleterious allele, natural selection could not act to decrease the frequency of the allele within the population; therefore, deleterious alleles accumulated in individuals within the population. The accumulation of deleterious alleles is not important until the environment changes and life spans lengthen. When this happens, the accumulated late-acting deleterious mutations are expressed.

The expected life span in many human populations has increased dramatically over the last 100 years because of behavioral and cultural change, such as improved sanitation. Levels of infant and child mortality have been reduced, mortality due to infectious diseases has fallen, and a larger number of individuals now live to older ages. According to this theory of senescence, alleles with late-life deleterious effects have been accumulating, mostly undetected,

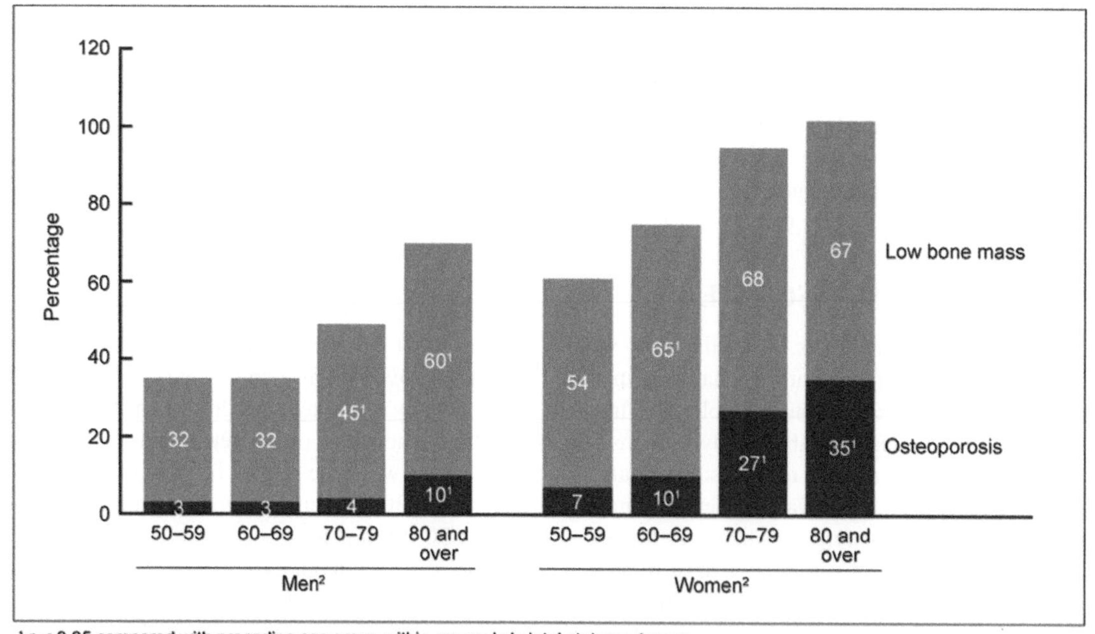

[1] $p < 0.05$ compared with preceding age group within sex and skeletal status category.
[2] $p < 0.05$ for trend by age group within sex for both osteoporosis and low bone mass.

FIGURE 5 Osteoporosis or low bone mass at the femur neck or lumbar spine, by age in adults 50 years and older (Looker et al., 2012).

generation after generation, and have now been "uncovered" by longer life spans. The visible effect of late-acting deleterious mutations is senescence.

Alleles that predispose an individual to prereproductive mortality, infertility, and low relative fitness are selected against without regard to their late-life benefits (see Table 2). Accumulated deleterious alleles are expressed after maturation, and after the opportunity to mate, give birth, and rear multiple offspring. If it looks as if aging happens all at once—for example, the similar ages of onset for cardiovascular diseases, diabetes, and cancers in the United States—this is because multiple alleles express age-specific effects at about the same time. Deleterious alleles are more likely to show up at later ages because they were not expressed, or were not deleterious, earlier in life. The increasing frequency of chronic degenerative diseases across populations is due to the short life span characteristic of our evolutionary history, the accumulation of late-acting deleterious alleles, and our present, novel environment that allows for a longer life.

Antagonistic Pleiotropy

George C. Williams (1957) first developed the idea that pleiotropy might be an important determinant of senescence. Pleiotropy simply means that single genes have multiple effects. With regard to aging, Williams proposed that the same genes could have beneficial effects early in life but detrimental effects late in life. Because the effects are counteracting, the theory has been called "antagonistic pleiotropy." This theory is not completely separate from the previous theory of mutation accumulation and late-acting genes. That theory addresses the timing of senescence, while antagonistic pleiotropy explains how the same allele can confer both early positive and later negative effects. As with the mutation accumulation hypothesis, alleles with antagonistic pleiotropy produce dysfunction when the life span extends beyond maturation and reproduction.

One example of antagonistic pleiotropy is testosterone activity in men. Testosterone contributes to secondary sex characteristics, the regulation of puberty, and reproductive success early in life. These are positive effects. Unfortunately, the same testosterone activity may result in compromised immune function, prostate cancer, and an increased risk of cardiovascular disease later in life (Muehlenbein and Bribiescas, 2005; Finch and Rose, 1995). Some researchers see menopause as an example of antagonistic pleiotropy. Multiple ovarian follicles develop to produce the estrogen needed during the reproductive period. Unfortunately, this requirement for periodic elevations in the level of estrogen results in depletion of follicles from the ovary and eventually menopause. First this, then unfortunately that, is the sentence structure that describes antagonistic pleiotropy.

In Summary

The force of natural selection is diminished with age. This, according to Rose (1991), is the essential evolutionary theory of aging and senescence. Mutations that produce deleterious effects at later ages may remain in the population gene pool unaffected by natural selection if they are not expressed, or if they are associated with positive or neutral effects early in life. Positive effects early in the life span can be due to an allele's own effects or the effects of other genes with which it is linked. Across time, environmental changes brought about a lengthened life span in *Homo sapiens*. Late-acting deleterious genes are now expressed in what looks like synchronized age-related changes. When the same genes have beneficial effects early but deleterious effects later in the life span, this is known as antagonistic pleiotropy.

TABLE 2 Alleles Associated with or Linked to Alleles Associated with Early-Life Benefits Are Selected Regardless of Late-Life Effects. Alleles Not Associated with or Linked to Alleles Associated with Early-Life Benefits Can Be Maintained in a Population by Random Genetic Drift

Allele Associated with Prereproductive Survival	Allele Associated with Fertility	Allele Associated with High Relative Fitness	Allele Associated with Late-Life Survival	Consequences
Beneficial	Beneficial	Beneficial	Beneficial	Natural selection for allele(s)
Beneficial	Beneficial	Beneficial	**Deleterious**	Natural selection for allele(s) = **senescence**
Deleterious	Beneficial	Beneficial	Beneficial	Natural selection against allele(s)
Beneficial	**Deleterious**	Neutral	Beneficial	Natural selection against allele(s)
Neutral	Neutral	Neutral	**Deleterious**	Maintained in the population by genetic drift = **senescence**

REPRODUCTIVE SENESCENCE

Male Reproductive Senescence

There is a decline in male fertility with age that generally begins between the ages of 30 and 35. Men do not, however, experience a universal and age-related cessation of reproductive function. Unlike women who run out of viable eggs, sperm is found in the ejaculate of almost half of men aged 80–90. Luciano Pavarotti was 67 at the birth of his last child, Pablo Picasso was 68, Jacques Cousteau was 72, Rupert Murdoch was 72, Tony Randall was 78, and Saul Bellow was 84. Internet sources may be unreliable; however, each of these men was clearly past the age at which a woman's menstruation ceases.

While women experience a universal decline in estrogen levels due to the loss of ovarian follicles (described below), the decline in male testosterone is population specific. Several large studies of men in the United States found that testosterone levels decline by about 1% per year after the age of 40; however, studies of testosterone in non-Western populations, including urban Japan, suggest that the declines are not universal, and that stable levels of testosterone can be maintained well past the age of 80 (Bribiescas, 2006; Ellison et al., 2002).

There appears to be agreement among researchers that, with age, there is a loss of circadian rhythm in testosterone and a blunting of the morning peak. With increasing age, levels of sex-hormone binding globulin (SHBG) increase. Some of the age-related decline in bioavailable testosterone is due to the increase in SHBG (Feldman et al., 2002). There are also age-related structural changes in the testis, accessory sexual glands, and penis (Bribiescas, 2006; Sievert, 2001). The ability to produce and maintain an erection can be affected by age, hypertension, diabetes, and smoking. The National Health and Social Life Survey found that, compared with men aged 18–29, men aged 50–59 were three times as likely to experience erection problems and low sexual desire (Laumann et al., 1999). Still, some males continue to produce sperm to very old ages, and the National Survey of Sexual Health and Behavior found that 19% of men aged 80 and older reported engaging in intercourse a few times per month (Schick et al., 2010). In contrast, only 4% of women aged 80 years and older reported the same frequency of sexual activity, and not one would have been able to ovulate a viable egg.

Female Reproductive Senescence

Menopause is the cessation of menses due to the loss of ovarian follicular activity. It is a human female universal, experienced by all women who live beyond 60 years of age. Menopause generally occurs around the age of 50, even though humans have a maximum potential life span of 122 years. Other female animals also experience postreproductive life, most famously killer whales and short-finned pilot whales. Among primates, a study of Japanese macaques found that every female monkey ceased reproduction by the age of 25, and only 3% lived to the age of 26 (Pavelka and Fedigan, 1999). Among chimpanzees in the Mahale mountains of Tanzania, six females were observed to be postreproductive for 5 years or longer (Nishida el al. 2003).

As a general rule, female mammals stop forming new eggs around the time of birth, and the number of primordial follicles is set for the rest of the female's life. There is disagreement about whether some mammals can form new eggs after birth. If the ability is present, it is expressed for a relatively short amount of time during the postnatal period.

Across the life span, the female ovary contains follicles in a number of stages of development. These follicles surround the developing egg and produce hormones. In mammals, undeveloped eggs and follicles in all stages of development are lost through the process of atresia (degeneration) that continues across the life span. Figure 6 compares the number of eggs formed in human, monkey, cow, and rat ovaries. The same pattern is seen across mammalian species—a peak in egg production followed by a steady decline in numbers until none remain. Only a very few eggs will ovulate (about 400 in humans). At least 34 species of mammals experience the complete loss of follicles prior to death (Cohen, 2004)—this loss of follicles can result in menopause and the start of postreproductive life.

In contrast to the mammalian pattern of egg production and loss, fish, amphibians, and possibly all reptiles have stem cells that can produce new eggs throughout the female's life. Eggs are produced during each breeding season through mitosis, and almost all eggs are ovulated. Menopause is a mammalian characteristic because of the pattern of making eggs all at once during fetal and neonatal development (depending on the species), coupled with the loss of eggs through the process of atresia across the life span until follicle reserves are depleted. In mammals, the timing of menopause is determined by (1) the number of eggs formed in the female ovary during fetal development or soon after birth; (2) the rate of egg loss across the life span through ovulation and atresia; and (3) the threshold number of ovarian follicles needed to maintain reproductive function through the production of hormones.

Why, in human females, does menopause occur at a relatively early age in the life span? Why are we capable of such a long period of postreproductive life? A number of adaptationist scenarios have been suggested. For example, menopause prevents the use of old and potentially damaged eggs, prevents the loss of offspring due to mortality among older mothers, allows postmenopausal grandmothers to invest in their extended family, conserves energy, and reduces the risk of reproductive cancers.

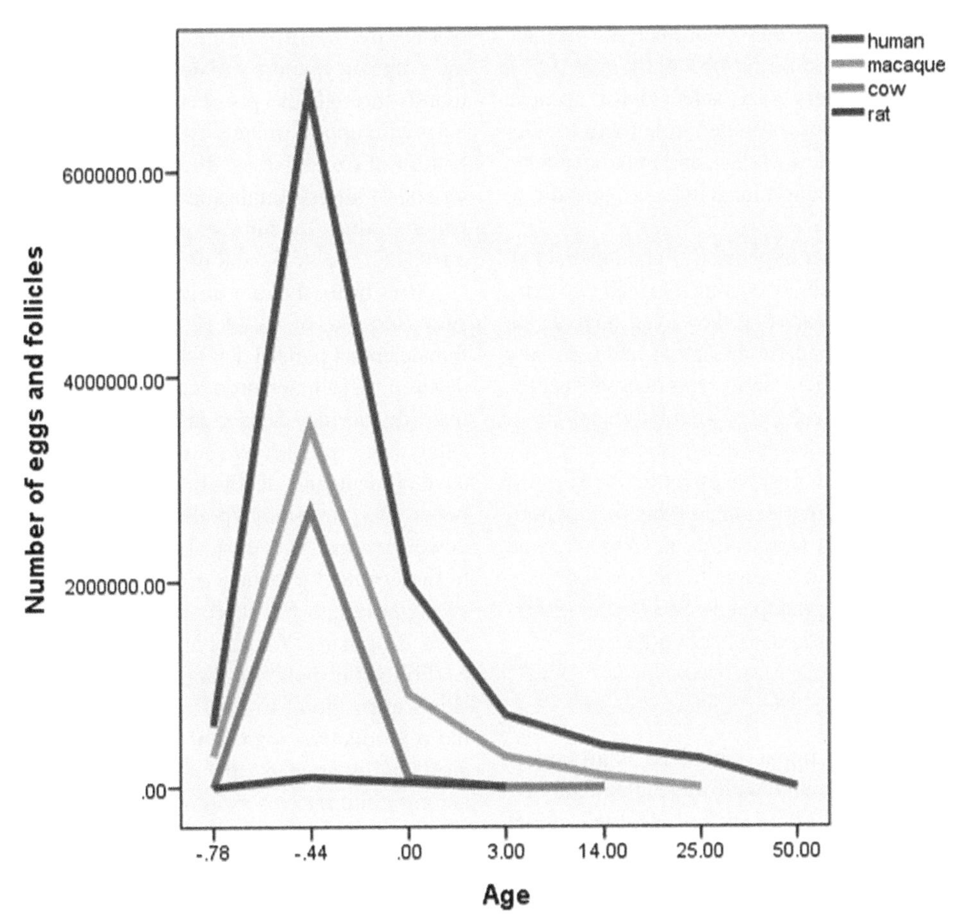

FIGURE 6 The rise and decline in the number of eggs and ovarian follicles in human, rhesus macaque, cow, and rat ovaries during the prenatal period, at birth, and at ages 3, 14, 25, and 50 (Baker, 1986; Erickson, 1966a,b; Hansen et al., 2008; Ryan, 1981). (See color plate section).

Aging Eggs

Menopause may have been selected for to ensure that old eggs are not fertilized. Because eggs develop when the female is still a fetus, those eggs can wait 15–50 years in the ovary before they are ovulated and fertilized. This wait time is related to an increase in the risk of chromosomal abnormalities (Pellestor et al., 2003). As women age, the number of eggs in the ovary declines, so there are fewer primordial follicles to recruit for development. As a result, there is an increased risk of chromosomal abnormalities in the eggs that are selected to develop and ovulate. Menopause functions to prevent the ovulation and fertilization of abnormal oocytes.

Aging Mothers

Menopause ensures that mothers are young enough to survive pregnancy, the difficulties of giving birth, and the infancy and childhood of their offspring. The demands of motherhood are heightened by the extreme helplessness of human infants (Williams, 1957). In our ancestral past, it is likely that offspring survivorship was low when mothers died in the first five years of the child's life (Hill and Hurtado, 1991). Because of menopause, women cease childbearing early enough to allow the last child to remain dependent until it can survive on its own. This scenario implies that maternal death has a greater impact than paternal death on the survival of the youngest offspring.

From a different perspective, older women are at greater risk for pregnancy complications compared with younger women. Untreated, these complications result in death. Menopause protects older women from the demands of pregnancy and childbirth.

The Grandmother Hypothesis

This adaptationist scenario has been advanced most notably by the work of Kristen Hawkes (2003), although many authors now argue that menopause and postreproductive aging were selected for by the evolutionary benefits gained through grandmothering (Voland et al., 2005). During human evolution, females increased their reproductive success, measured in terms of inclusive fitness, by investing in their grandchildren. It is argued that the reproductive success of our postmenopausal ancestors was greater when they invested in their extended family than when, after a certain point in the life span, they continued to produce children of their own (Hamilton, 1966; Williams, 1957). Postmenopausal grandmothers were selected for because they produced surplus calories, provided infant and toddler care, served as reservoirs of knowledge, and may have been the first birth attendants when hominins moved from solo to assisted births.

In a test of the grandmother hypothesis, Lahdenperä et al. (2004) showed that postreproductive women gained two extra grandchildren for every 10 years that they lived beyond the age of 50. Other studies in Japan, rural Gambia, and Germany have also demonstrated an increase in reproductive success among offspring with the presence of grandmothers. However, some of these same studies have also shown that only maternal grandmothers have a positive effect on the survival of grandchildren, whereas paternal grandmothers and grandfathers have a negative effect (Jamison et al., 2002; Voland and Beise, 2002). It is difficult to argue for the evolutionary selection of just one type of grandparent. See Peccei (2001) for other concerns about the grandmother hypothesis.

Energy Conservation

Hall (2004) suggests that menopause evolved as an opportunity to save energy associated with the high metabolic costs of reproduction. Menopause may have evolved so that energy spent in menstrual cycling could instead be used to support offspring survival.

Prevention of Reproductive Cancers

Lifelong exposure to periodic elevation in the levels of estrogen associated with monthly menstrual cycles increases the risk of breast and endometrial cancers. Without menopause, women would continue to be exposed to high, cyclic levels of estrogen into the seventh or eighth decade of life. Menopause is therefore adaptive within the context of contemporary industrialized societies, because it lowers the risk of breast cancer and other estrogen-dependent cancers that can directly reduce reproductive success.

Menopause as a By-Product of Other Adaptations

In addition to the above adaptive scenarios, a "by-product scenario" has been advanced to argue that menopause was not selected for directly, but is simply the by-product of a conserved mammalian pattern of egg production and follicular atresia coupled with a lengthened life span (Sievert, 2011). In this scenario, menopause and postreproductive life are neutral by-products of natural selection for human longevity.

As Figure 6 illustrates, millions of eggs are produced by mitosis at the beginning of human female life. This production continues until seven million eggs are present in the ovaries by the fifth month in utero. For the remainder of the pregnancy, approximately one million eggs are lost per month through the process of atresia until the neonate is born with approximately two million ovarian follicles. This pattern of excessive production and loss of eggs is characteristic of all mammals and may have been selected for as a way to reduce the number of eggs that have developmental errors and chromosomal mutations.

After birth, the rate of loss of ovarian follicles begins to slow to a rate of about 12,000 per month, and the human female enters puberty with about 400,000 follicles. Waves of developing follicles produce hormones during prereproductive life and may help to initiate regular cycles. In terms of antagonistic pleiotropy (described above), waves of follicle development and degeneration may be necessary prior to puberty to prime the hypothalamic–pituitary–ovarian axis—however, there is a cost. This loss of follicles contributes to the eventual exhaustion of ovarian reserves, resulting in menopause and postreproductive life in long-lived species such as humans (Wood et al., 2001).

Following puberty, the rate of loss of ovarian follicles slows even more, to approximately 700 per month across the reproductive span. The rate of loss varies across species, but it appears that the rate of follicular depletion among chimpanzees is almost identical to that of humans (Jones et al., 2007). As follicles mature, they produce hormones such as estrogen. These hormones help to maintain regular cycles during the early reproductive years—another example of antagonistic pleiotropy (Wood et al., 2001). The development of many ovarian follicles maintains regular menstrual cycles. This is the benefit. However, the loss of nonovulatory follicles can be viewed as a deleterious trait at later ages when older women run out of the ovarian follicles needed for fertility.

The by-product scenario illustrates that if a female produced only the eggs necessary for ovulation, she may not be able to produce as many offspring because (1) some of the eggs would be defective, (2) there would be no extra follicles to produce hormones during the prepubertal period when developing follicles may be necessary to prime the hypothalamic–pituitary–ovarian axis, and (3) there would be no accessory follicles available to produce the hormones necessary for fertile ovulatory cycles.

It appears that the system evolved to provide viable eggs until at least 40 or 45 years of age. This age far exceeds the life span of most of our hominin ancestors (Kennedy, 2003). When our life span extended beyond 50 years of age, menopause became a common event. Menopause is not "new," because the capacity for menopause has always existed for any individual female who lived beyond her egg supply. The pattern of egg production and loss is highly conserved. It is the evolution of human longevity that explains the appearance of menopause and a long reproductive life.

FUTURE DIRECTIONS

Biomarkers of aging provide one way to measure the accumulation of damage to the soma across time (Butler et al., 2004; Levine, 2013). Biomarkers include measures of metabolic function (glycated hemoglobin and total cholesterol), cardiac function (systolic blood pressure), lung function (forced expiratory volume), kidney function (serum creatinine and serum urea nitrogen), liver function (serum albumin and serum alkaline phosphatase), and immune function (C-reactive protein, cytomegalovirus optical density, and interleukin-6), as well as cell blood count, telomere length, and measures of oxidative damage. Eventually, genomics, proteomics, and metabolomics will identify new biomarkers of somatic and reproductive aging that can be measured in early adult life, perhaps before the age of 30, to predict the survival and reproductive potential of each individual (Arking, 2006). That said, knowing more about the mechanisms of aging may not fundamentally change our understanding of why we senesce. Molecular insights, however, may narrow the range or increase the depth of evolutionary explanations. More information will most certainly change the way we understand and treat the currently unavoidable and unfortunate changes associated with senescence (Rose, 2005).

REFERENCES

Arking, R., 2006. The Biology of Aging: Observations and Principles, third ed. Oxford University Press, New York.

Austad, S.N., 1997. Why We Age: What Science Is Discovering about the Body's Journey through Life. J. Wiley & Sons, New York.

Baker, T.G., 1986. Gametogenesis. In: Dukelow, W.R., Erwin, J. (Eds.), Comparative Primate Biology, Vol. 3: Reproduction and Development. Alan R. Liss, New York, pp. 195–213.

Blagosklonny, M.V., 2010. Why the disposable soma theory cannot explain why women live longer and why we age. Aging 2, 884–887.

Bribiescas, R.G., 2006. Men: Evolutionary and Life History. Harvard University Press, Cambridge, MA.

Butler, R.N., Sprott, R., Warner, H., Bland, J., Feuers, R., Forster, M., Fillit, H., Harman, S.M., Hewitt, M., Hyman, M., Johnson, K., Kligman, E., McClearn, G., Nelson, J., Richardson, A., Sonntag, W., Weindruch, R., Wolf, N., 2004. Biomarkers of aging: from primitive organisms to humans. Journals of Gerontology. Series A, Biological Sciences and Medical Sciences 59, 560–567.

Cohen, A.A., 2004. Female post-reproductive lifespan: a general mammalian trait. Biological Reviews 79, 733–750.

Crews, D.E., 2003. Human Senescence: Evolutionary and Biocultural Perspectives. Cambridge University Press, Cambridge, UK.

Ellison, P.T.R., Bribiescas, R.G., Bentley, G.R., Campbell, B.C., Lipson, S.F., Panter-Brick, C., Hill, K.R., 2002. Population variation in age-related decline in male salivary testosterone. Human Reproduction 17, 3251–3253.

Erickson, B.H., 1966a. Development and radio-response of the prenatal bovine ovary. Journal of Reproduction and Fertility 11, 97–105.

Erickson, B.H., 1966b. Development and senescence of the postnatal bovine ovary. Journal of Animal Science 25, 800–805.

Feldman, H.A., Longcope, C., Derby, C.A., Johannes, C.B., Araujo, A.B., Coviello, A.D., Bremner, W.J., McKinlay, J.B., 2002. Age trends in the level of serum testosterone and other hormones in middle-aged men: longitudinal results from the Massachusetts male aging study. Journal of Clinical Endocrinology and Metabolism 87, 589–598.

Finch, C.E., 1990. Longevity, Senescence and the Genome. University of Chicago Press, Chicago, IL.

Finch, C.E., Rose, M.R., 1995. Hormones and the physiological architecture of life history evolution. The Quarterly Review of Biology 70, 1–52.

Hall, R., 2004. An energetics-based approach to understanding the menstrual cycle and menopause. Human Nature 15, 83–99.

Hamilton, W.D., 1966. The moulding of senescence by natural selection. Journal of Theoretical Biology 12, 12–45.

Hansen, K.R., Knowlton, N.S., Thyer, A.C., Charleston, J.S., Soules, M.R., Klein, N.A., 2008. A new model of reproductive aging: the decline in ovarian non-growing follicle number from birth to menopause. Human Reproduction 23, 699–708.

Hawkes, K., 2003. Grandmothers and the evolution of human longevity. American Journal of Human Biology 15, 380–400.

Hill, K., Hurtado, A.M., 1991. The evolution of premature reproductive senescence and menopause in human females: an evaluation of the "grandmother hypothesis". Human Nature 2, 313–350.

Jamison, C.S., Cornell, L.L., Jamison, P.L., Nakazato, H., 2002. Are all grandmother's equal? A review and a preliminary test of the "grandmother hypothesis" in Tokugawa Japan. American Journal of Physical Anthropology 119, 67–76.

Jones, K.P., Walker, L.C., Anderson, D., Lacreuse, A., Robson, S.L., Hawkes, K., 2007. Depletion of ovarian follicles with age in chimpanzees: similarities to humans. Biology of Reproduction 77, 247–251.

Kennedy, G.E., 2003. Palaeolithic grandmothers? Life history theory and early Homo. Journal of the Royal Anthropological Institute 9, 549–572.

Kirkwood, T.B.L., 1977. Evolution of ageing. Nature 270, 301–304.

Kirkwood, T.B.L., 1987. Immortality of the germ-line versus disposability of the soma. In: Woodhead, A.D., Thompson, K.H. (Eds.), Evolution of Longevity in Animals: A Comparative Approach. Plenum Press, New York, pp. 209–218.

Lahdenperä, M., Lummaa, V., Helle, S., Tremblay, M., Russell, A.F., 2004. Fitness benefits of prolonged post-reproductive lifespan in women. Nature 428, 178–181.

Levin, M.E., 2013. Modeling the rate of senescence: can estimated biological age predict mortality more accurately than chronological age? Journal of Gerontology A: Biological Sciences and Medical Sciences 28, 667–674.

Looker, A.C., Borrud, L.G., Dawson-Hughes, B., Shepherd, J.A., Wright, N.C., 2012. Osteoporosis or Low Bone Mass at the Femur Neck or Lumbar Spine in Older Adults: United States, 2005–2008. NCHS Data Brief, No. 93. National Center for Health Statistics, Hyattsville, MD. http://www.cdc.gov/nchs/data/databriefs/db93.htm.

Laumann, E.O., Paik, A., Rosen, R.C., 1999. Sexual dysfunction in the United States: prevalence and predictors. Journal of the American Medical Association 281, 537–544.

Medawar, P.B., 1946. Old age and natural death. Modern Quarterly 1, 30–56.

Medawar, P.B., 1952. An Unsolved Problem of Biology. H. K. Lewis, London, UK.

Miniño, A.M., Xu, J., Kochanek, K.D., Tejada-Vera, B., 2009. Death in the United States, 2007. NCHS Data Brief, No 26. National Center for Health Statistics, Hyattsville, MD. http://www.cdc.gov/nchs/data/databriefs/db26.htm.

Muehlenbein, M.P., Bribiescas, R.G., 2005. Testosterone-mediated immune functions and male life histories. American Journal of Human Biology 17, 527–558.

Murphy, S.L., Xu, J., Kochanek, K.D., 2013. Deaths: final data for 2010. National Vital Statistics Reports 61 (4).

Nishida, T., Corp, N., Hamai, M., Hasegawa, T., Hiraiwa-Hasegawa, M., Hosaka, K., Hunt, K.D., Itoh, N., Kawanaka, K., Matsumoto-Oda, A., Mitani, J.C., Nakamura, M., Norikoshi, K., Sakamaki, T., Turner, L., Uehara, S., Zamma, K., 2003. Demography, female life history, and reproductive profiles among the chimpanzees of Mahale. American Journal of Primatology 59, 99–121.

Pavelka, M.S.M., Fedigan, L.M., 1999. Reproductive termination in female Japanese monkeys: a comparative life history perspective. American Journal of Physical Anthropology 109, 455–464.

Peccei, J.S., 2001. A critique of the grandmother hypothesis: old and new. American Journal of Human Biology 13, 434–452.

Pellestor, F., Andreo, B., Arnal, F., Humeau, C., Demaille, J., 2003. Maternal aging and chromosomal abnormalities: new data drawn from in vitro unfertilized human oocytes. Human Genetics 112, 195–203.

Rose, M.R., 1991. Evolutionary Biology of Aging. Oxford University Press, New York, NY.

Rose, M.R., 2005. The Long Tomorrow: How Advances in Evolutionary Biology Can Help Us Postpone Aging. Oxford University Press, New York.

Ryan, R.J., 1981. Follicular atresia: some speculations of biochemical markers and mechanisms. In: Schwartz, N.,B., Hunzicker-Dunn, M. (Eds.), Dynamics of Ovarian Function. Raven Press, New York, pp. 1–11.

Schick, V., Herbenick, D., Reece, M., Sanders, S.,A., Dodge, B., Middlestadt, S.E., Fortenberry, J.D., 2010. Sexual behaviors, condom use, and sexual health of Americans over 50: implications for sexual health promotion for older adults. Journal of Sexual Medicine 7 (Suppl. 5), 315–329.

SEER Cancer Statistics Review, 1975–2010. National Cancer Institute, U.S. National Institutes of Health. US Department of Health and Human Services. http://seer.cancer.gov/csr/1975_2010/browse_csr.php?section=2&page=sect_02_table.07.html.

Shanley, D.P., Kirkwood, T.B.L., 2000. Calorie restriction and aging: a life history analysis. Evolution 54, 740–750.

Sievert, L.L., 2001. Aging and reproductive senescence. In: Ellison, P. (Ed.), Reproductive Ecology and Human Evolution. Aldine de Gruyter, Hawthorne, New York, pp. 267–292.

Sievert, L.L., 2011. The evolution of post-reproductive life: adaptationist scenarios. In: Mascie-Taylor, C.G.N., Rosetta, L. (Eds.), Reproduction and Adaptation. Cambridge University Press, Cambridge, UK, pp. 149–170.

Stearns, S.C., 1992. The Evolution of Life Histories. Oxford University Press, Oxford, UK.

Trinkaus, E., 1995. Neanderthal mortality patterns. Journal of Archaeological Science 22, 121–142.

United Nations, Department of Economic and Social Affairs, Population Division, 2011. World Population Prospects: The 2010 Revision. United Nations, New York.

Voland, E., Beise, J., 2002. Opposite effects of maternal and paternal grandmothers on infant survival in historical Krummhörn. Behavioral Ecology and Sociobiology 52, 435–443.

Voland, E., Chasiotis, A., Schiefenhovel, W., 2005. Grandmotherhood: The Evolutionary Significance of the Second Half of Female Life. Rutgers University Press, New Brunswick, New Jersey.

Weissmann, A., 1891. Essays on Heredity. Clarendon Press, Oxford, UK.

Williams, G.C., 1957. Pleiotropy, natural selection, and the evolution of senescence. Evolution 11, 398–411.

Wood, J.W., Holman, D.J., O'Connor, K.A., 2001. Did menopause evolve by antagonistic pleiotropy? In: Homo unsere Herkunft und Zukunft. Proceedings 4.Kongress der Gesellschaft für Anthropologie (GfA). Cuvillier Verlag, Göttingen, pp. 483–490.

Part V

Lifeways

Chapter 23

Hunter-Gatherers

Michael A. Little[1] and Mark A. Blumler[2]

[1]*Department of Anthropology, Binghamton University, State University of New York, Binghamton, NY, USA;* [2]*Department of Geography, State University of New York, Binghamton, NY, USA*

SYNOPSIS

Human hunter-gatherer (H-G) populations date back to earliest Paleolithic times more than 200,000 years ago, although nonhuman primate ancestral hunting probably dates back to earlier times. H-G populations/societies constituted all human social groups until the Neolithic Revolution (10,000–12,000 years ago), when agriculture and husbandry developed and then became the dominant pattern of subsistence. H-G numbers declined after the Neolithic and then further during the age of exploration and colonization in the sixteenth through nineteenth centuries, when indigenous groups became marginalized and the H-G population showed further declines. Living H-G populations make up only a tiny fraction of the seven billion people on the planet. Existing hunter-gatherers are largely acculturated and remain in enclaves in tropical Africa, South America, and Southeast Asia. These remaining populations show considerable variation in subsistence, diet and nutrition, demographic characteristics, disease patterns, reproduction, child growth, and overall health. Because of the small-scale population units that characterize most H-G societies, they have been essentially free from most epidemic diseases, but have been subject to heavy parasite loads and infections. Death rates are high and life expectancies at birth are low, and natural selection has probably operated in a major way via differential mortality. Other evolutionary forces may have operated through genetic drift because of the small bands in which hunter-gatherers lived. The use of contemporary or historical H-G populations as models of prehistoric hunter-gatherers is still a debated issue.

EARLY HUNTING AND GATHERING SUBSISTENCE

All terrestrial mammals procure food through the processes of hunting (predation), gathering and foraging of plants, fishing, scavenging, and other means of acquiring animal or plant nutrients. Our ancestral primates were almost exclusively herbivores or insectivores, and most nonhuman primates today collect or catch the same foods as their ancestors. There are some exceptions, notably the common chimpanzee (*Pan troglodytes*), which has been observed on many occasions to participate in organized hunts of monkeys and small mammals in arboreal or forest environments (Goodall, 1968, 1986; Teleki, 1973; see the chapter by Watts). Sanford et al. (1994) reported in detail on the Tanzanian Gombe National Park chimpanzee predation on red colobus monkeys (*Colobus badius*), where adolescent and adult males would work together in an organized band fashion to trap, capture, kill, and eat the monkeys. This is a clear example of division of labor by sex in a nonhuman primate species and reflects quite closely the pattern of human hunters being almost exclusively males. Sanford et al. (1994) did report that slightly more than 10% of the kills were made by adult female chimpanzees. Between 1982 and 1991 in Gombe, chimpanzees were able to kill 429 largely immature mammals, including monkeys, bushbucks, and bushpigs. Of these kills, about 80% were red colobus monkeys. The meat acquired from these predatory kills amounted to a relatively small amount of the largely herbivorous/frugivorous diet at Gombe, but it did supplement the protein intakes of the male hunter-chimps. There also are indications in Uganda that chimpanzee predation of the red colobus is not sustainable and is leading to a decline in some colobus populations (Teelen, 2008; Watts and Amsler, 2013). In addition to Gombe, common chimpanzee predation has been observed in the Tai National Park, Ivory Coast and the Mahale Mountain National Park, Tanzania (Boesch and Boesch, 1989; Uehara, 1997). Of the other apes, bonobo chimpanzees (*Pan paniscus*), our closest primate relatives, have been observed hunting occasionally and eating prey, principally solitary and terrestrial ungulates (Hohmann and Fruth, 2008; Surbeck et al., 2009). These observations suggest strongly that our early hominin ancestors already had a prehominin tradition of hunting for small mammals, and our ape precursors were adding meat to their diets five million years ago (ya) or earlier.

During the Lower Paleolithic (roughly 2.5 million to 300,000 ya), Middle Paleolithic (300,000 to 40,000 ya), and Upper Paleolithic (40,000 to 10,000–12,000 ya), H-G technology was characterized by use of flaked stone tools (see chapter by Toth and Schick). These periods included species of hominins such as *Homo (Australopithecus) habilis*, *Homo erectus*, and *Homo heidelbergensis* during the Lower Paleolithic, archaic and anatomically modern *Homo sapiens* during the Middle Paleolithic, and with the gradual disappearance of archaic humans such as the Neanderthals, the increasingly wide distribution of cognitively modern *H. sapiens* during the Upper Paleolithic. Tool manufacture and hunting techniques became successively more sophisticated over these stages in human prehistory, and there is abundant evidence that Paleolithic hominins hunted for both large game and smaller animals, and also scavenged (Smith, 1999).

Throughout the evolution of the hominin lineage, a complex of behaviors evolved that incorporated meat into a largely vegetarian primate diet. These behaviors also selected for anatomical, physiological, and neurological transformations that facilitated the incorporation of foods into what became an omnivorous diet. It is believed by many that meat provided the concentrated energy to fuel the evolution of an expanded brain (Aiello and Wheeler, 1995; Leonard and Robertson, 1997). This allowed humans, in an evolutionary framework, to develop complex cultural behaviors that increased the success of food procurement. An omnivorous diet also required evolutionary dental changes and digestive modifications in the intestines associated with such dietary alterations. The increased metabolic needs of an expanded brain during hominin evolution from *Homo (Au.) habilis* to *H. erectus* to *H. sapiens* are thought to have been compensated for by a dramatically reduced intestinal system (Aiello and Wheeler, 1995). Energetic needs of the expanded brain were provided by a high-quality, energy-dense, and easy-to-digest food, mainly provided by hunted mammals. An evolutionary scenario might be envisioned as early ape hunting (as in chimpanzees and bonobos), which selected for reduced gut size because of the energy dense meat and in turn provided the energy needed for brain expansion (a form of relaxed selection allowing for brain expansion in early hominins).

Although there is no full agreement on how this occurred in our Paleolithic ancestors, the transition from a largely plant diet to an omnivorous diet (including a significant proportion of meat) probably took place at the time of the branching of hominoids and hominins. Hominin hunting and gathering became more successful with the invention of flaked stone tools, as noted, somewhere around two and a half million ya (see chapter by Toth and Schick). By the time of the second major hominin movement out of Africa about 100,000 or more ya, tool manufacture and use allowed humans considerable success in hunting large and small mammals, and this success also contributed to population increase and further expansion. Increasing sophistication in food procurement took place throughout the Paleolithic Age with the development of elaborate and well-designed stone, bone, and wooden tools. At the same time, advanced social and cultural behaviors contributed to increasing success in hunting, fishing, and gathering of food. By about 12,000 to 15,000 ya, human efforts to procure food still largely depended on hunting and gathering and nomadic movement in search of food. The Mesolithic and later Neolithic Revolution of plant and animal domestication introduced new ways of food acquisition that allowed for greater stability in the food supply and also contributed to the establishment of a more sedentary or settled way of life (see chapters by Little and by Blumler). These new patterns of food production through domestication and selection of favorable qualities of plants and animals began to be practiced in many parts of the Old and New Worlds. However, many societies continued to employ the traditional hunting and gathering practices as primary or singular forms of subsistence, and these traditional practices persisted in many parts of the world up until quite recently.

HUNTER-GATHERERS DURING THE HISTORICAL ERA

Lee and Daly (1999a, p. 3) defined foraging, or the hunter-gatherer (H-G) society's means of food procurement, as "subsistence based on hunting of wild animals, gathering of wild plant foods, and fishing, with no domestication of plants, and no domesticated animals except the dog." Another definition of hunting and gathering societies by Panter-Brick et al. (2001, p. 3) is more concise, where these societies have "subsistence activities entailing negligible control over the gene pool of food resources." These statements are intended to define hunter-gatherers from both Paleolithic times and from the period when agriculture and pastoralism first arose (the Neolithic) up to the present—that is, when hunter-gatherers were likely to be in contact or trading with cultivators. Despite the generality of the definitions identifying hunting-gathering as a "type" of subsistence, it is acknowledged that H-G populations display a remarkable degree of variation in general subsistence patterns, and that this variation certainly existed in the distant past as well as in historical times (Lee and Daly, 1999b; Lee and DeVore, 1968; Kelly, 2013). Such variation in subsistence and other attributes of culture have been represented broadly through time in both evolutionary context and geographic space.

We have evidence for variation in prehistoric Paleolithic H-G populations from archaeology; evidence for historical H-G populations from travel accounts, particularly during the Age of Exploration, and abundant evidence for twentieth and twenty-first century H-G populations from ethnographic writings (Lee and DeVore, 1976; Lee and Daly, 1999b). These are quite different classes of populations superimposed on basic evolutionary and biogeographic variability. Because of traditions and limited technology, Paleolithic populations were absolutely dependent on H-G subsistence. Historical H-G populations were subject to acculturation from neighboring populations practicing cultivation, in which exchanges of foods and technology were likely to have occurred (see Lee and Daly, 1999b). Modern and recent H-G populations are subject to influences from neighbors practicing cultivation and animal husbandry, as well as Western and now global technologies, foods, and values, not the least of which is tourism. A good example is the Pygmy populations from west and central Africa. They are dependent to varying degrees on trade with their Bantu cultivator neighbors and have continued these relationships for hundreds, perhaps thousands, of years (Turnbull, 1965; Cavalli-Sforza, 1986b). They are nevertheless primarily hunter-gatherers, but in a reciprocal economic relationship with Bantu cultivators.

There has been considerable debate among anthropologists on whether historical or contemporary H-G populations can serve as analogous models of Paleolithic populations. This use of analogy (Cummings, 2013), where contemporary peoples are substituted for Paleolithic populations in either theoretical evolutionary frameworks or archaeological reconstructions, was criticized very early on by Franz Boas (1896). Over the twentieth century, ethnographic analogy has often been criticized (Freeman, 1968) but has been used commonly, and some attempts have been made to conduct experiments to study simulated H-G settlement remains as they might have been deposited at archaeological sites (Binford, 1980). To some extent, the debate is an unproductive one, since the cultural diversity of at least Upper Paleolithic H-G populations was matched or exceeded by historical H-G diversity, so there is no archetype culture of hunter-gatherers.

The historical period can be thought of as the period following 1500 CE, when H-G populations were still widely distributed throughout the world and major influences from Western European exploration and colonization began. It is estimated that before 1500 CE, "hunter-gatherers still occupied almost one-third of the world's landmass, including all of Australia, the northwestern half of North America, and the southern part of South America, as well as smaller parts of sub-Saharan Africa and south, southeast, and northeast Asia" (Smith, 1999, p. 389). In Australia, an estimated half-million people procured food exclusively by H-G means, and early populations had been doing so for as many as 60,000 years (Peterson, 1999). In North America, northwest Pacific coast populations reached high levels of art, architecture, material culture, and stratified social classes, and built elaborate cultures (Kwakiutl, Tsimshian, Bella Coola, Nootka, Haida) on the basis of rich marine (whaling) and freshwater (salmon) food resources (Boas, 1921; Feit, 1999). Other societies in North America, such as the California coast Chumash and the Great Basin Paiute, had simpler but still quite sophisticated technology, and they had very limited resources (Feit, 1999). South American H-G societies included the plains hunters of Argentina, the tropical forest dwellers in Amazonia, and the coastal seafood-gathering peoples of the south Chilean archipelago (Rival, 1999). Asia was represented by cultures from

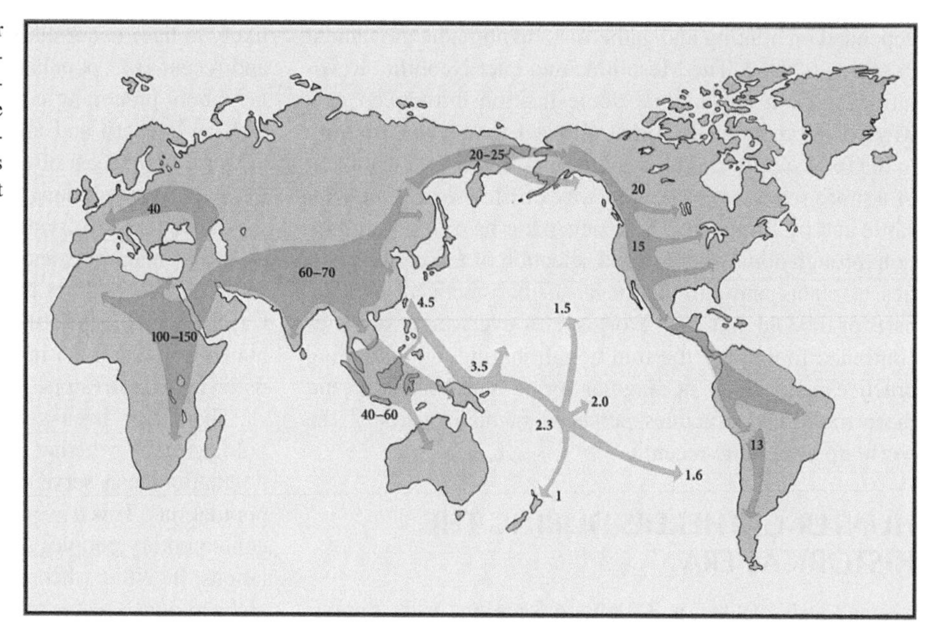

FIGURE 1 Approximations of some major migratory routes of hunter-gatherers following the Out of Africa exodus of anatomically modern *Homo sapiens* in the Middle Paleolithic. Values are in thousands of years ago. Arctic migrations in Eurasia and North America are not shown. (See color plate section).

the Siberian Evenki in the north to the many cultures in south and southeast Asia (Schnirelman, 1999; Bird-David, 1999; Endicott, 1999). Finally, Africa, the birthplace of anatomically modern *H. sapiens*, had a number of H-G populations, the best known of which were the numerous Pygmy groups in central Africa and the San (Bushmen) of South Africa, Namibia, and Botswana (Hitchcock, 1999). In a recent genetic study, Henn et al. (2011) found extremely high levels of genetic diversity in ‡Khomani Bushmen of South Africa, suggesting that the origin of modern humans is southern rather than eastern Africa. Further, they noted that African H-G populations tend to have high genetic diversity supporting the contention that these populations are descendants of the stem populations leading to modern humans[1] (see the chapter by Hawks).

MIGRATION, BIOGEOGRAPHY, AND CONTEMPORARY POPULATIONS

A map showing some migratory routes of human H-G populations out of Africa and later into other regions of the Old World and New World is shown in Figure 1 (also see Figure 2 in Morand chapter). Some of the most recent migrations of seafaring Pacific populations were cultivators who also subsisted on marine fish and shellfish. After the glacial maximum around 18,000 ya, as glaciers retreated, Paleolithic H-G populations had moved into ice-free areas in the Eurasian north, and some populations were beginning to cross either the Bering Land Bridge or the southern coastal route into North America (Schurr, 2004). Following the Pleistocene glacial retreats around 10,000–12,000 ya, global climate warming, and the subsequent rise of agriculture, Holocene hunter-gatherers were still widely distributed throughout Africa, Asia, Europe, Australia, and the newly colonized parts of North and South America (Bellwood, 2013). The earliest site in South America along the Chilean coast at Monte Verde dates to about 14,700 ya. This date and other sources of evidence suggest that: (1) a Bering migration occurred sometime after 20,000 ya, probably around 16,000 ya; (2) this was a single wave migration; and (3) this was from a central Asian population, perhaps west of Lake Baikal (Zegura et al., 2009).

As plant and animal domestication diffused from the Near East and Asia to other parts of the world during the millennia that followed the glacial retreats, there were still many areas occupied by hunter-gatherers that were marginal to agriculture and husbandry. With the continued expansion and migration of agriculturalists and diffusion of agricultural subsistence, H-G populations were forced into areas that were not amenable to cultivation. Lee and DeVore (1968) estimated that by 1500 CE, there were still 3.5 million hunter-gatherers in the world, and they constituted about one percent of the world's population. Nearly 500 years later during the latter half of the twentieth century, hunter-gatherers had declined to about 0.001% of the world's population. This resulted from an absolute increase in world population numbers and geographic expansion coupled with a dramatic decline in H-G populations. At the time of the "Man the Hunter" conference in 1965, Murdock (1968) identified 27 major H-G societies. He excluded three subsistence categories (that others might have included in a more comprehensive framework): *mounted hunters* (e.g.,

1. The fact that "stem" or "founding" populations are identified as those with the greatest genetic diversity is based on the following principle. As populations migrate beyond their home areas, only segments (samples) of the population with limited diversity emigrate, while the majority remains. This pattern of sequential migrations (serial founder effect) or "demic" expansion leads to successive reductions in genetic diversity, so that the most distant populations from a purported origin are likely to have the least genetic diversity (Henn et al., 2012).

FIGURE 2 Map showing the location of major hunting-gathering groups discussed here. (See color plate section).

US Plains Indians) who employ domestic horses or camels in pursuit of their livelihood; *sedentary fisherman* who live in settled villages and exploit fish and/or shellfish; and *incipient tillers* who are mixed farmers, hunters, and gatherers, many of whom get the bulk of their food from H-G activities. Six of the 27 H-G populations were from Africa, five from Asia and Southeast Asia, one from Oceania (Australians), nine from North America, and seven from South America. Murdock's choices were largely determined by those societies that had been studied by anthropologists.

The most recent assessment of hunter-gatherers was compiled by Lee and Daly (1999b) more than 30 years after the "Man the Hunter" conference. The contributors to *The Cambridge Encyclopedia of Hunters and Gatherers* (Lee and Daly, 1999b) provided brief ethnographies of 53 known historical and contemporary H-G societies by geographic area (North America, South America, North Eurasia, Africa, South Asia, Southeast Asia, and Australia). The majority of these societies no longer practice an H-G way of life, especially those societies that have survived into the twenty-first century. A number of societies, however, still retain a semblance of their H-G past and their subsistence continues with the practice of hunting, gathering, and fishing as primary or secondary means of food procurement. Some of these H-G populations are shown in Figure 2 and are listed in the glossary.

HUNTER-GATHERERS AND EVOLUTION

Evolutionary processes in H-G populations can be linked to geography, climate, resource availability, technology, social practices, and any number of environmental, cultural, and stochastic factors. These processes are highly variable and somewhat unpredictable. However, many attributes of hunter-gatherers that have evolutionary implications are held in common. First, since largely nomadic H-G populations are usually found in bands with an average size of 50 or fewer individuals (Forde, 1934; Steward, 1936), these bands are subject to microevolution through considerable genetic drift because of their small population sizes. Diets tend to be diverse with a balance of animal and plant foods, limited fat and salt intakes, and substantial fiber intake. Yet seasonal fluctuations in resources and food availability are common, and periods of decreased dietary intake may occur. Both a small-band social and subsistence community and a nomadic hunting pattern will influence disease, first by limiting the spread of epidemic diseases that are more common in nucleated agricultural societies, and second by limiting disease transmission from human waste, since H-G bands abandon and move their campsites frequently. Fertility and mortality rates of hunter-gatherers must have been closely balanced throughout Paleolithic times, since the estimated rates of population increase are extremely low. Pennington (2001) summarized the demographic literature and estimated very slow population growth rates (less than 0.004% per year) during the Paleolithic, with a population expansion during the modern hominin movement out of Africa, and another expansion at the beginning of the Neolithic that occurred with the rise of agriculture.

The sections that follow treat essential attributes of human population biology with evolutionary perspective for hunter-gatherers and include demography, diet and nutrition, child growth and adult size, and disease and morbidity.

Demographic Characteristics

Population structure (numbers, density, sex, and age), fertility, mortality, and migration are fundamental demographic variables. They are also the fundamental variables of microevolution at the population level. For example, differential fertility and differential mortality are the principal ways that selection operates; genetic drift occurs via small population sampling variation (see the chapter by Futuyma); and kinship practices and physical migration are fundamental mechanisms for gene flow. Hence, knowledge of H-G demography is essential to gaining insights into their evolutionary processes.

One of the earliest demographic studies of contemporary hunter-gatherers was conducted by Joseph Birdsell (1953) with Australian Aborigines. Based on data that had been collected as a part of the Harvard-Adelaide Universities Anthropological Expedition in 1938–1939 (Tindale, 1940), Birdsell (1953) found a strong curvilinear relationship between tribal territorial area and rainfall. About 65% of the variance in tribal area (and population density) was explained by the amount of rainfall—a large tribal area (low population density) was associated with low rainfall. With a relatively uniform tribal size (about 500) for the estimated precontact Australian Aborigines, Birdsell suggested that this demonstrated an equilibrium density system and that the Australian H-G populations existed at the approximate carrying capacity of the environment (Birdsell, 1968). Scientists today are cautious about the concept of carrying capacity (or an environment of maximum sustainability), "equilibrium systems," and what has been identified as "environmental determinism." Nevertheless, Birdsell's research was a pioneering demographic effort that cast light on conditions that probably existed for Paleolithic populations. The logic is that among hunter-gatherers (assuming uniform and limited material culture and technology), population density is based on available food resources.

Considerable demographic research has been conducted on H-G populations since Birdsell's early studies of Australian Aborigines. In an early review of demographic research in anthropology (Baker and Sanders, 1972), it was noted that only 4 of the 35 papers in the *Man the Hunter* book (Lee and DeVore, 1968) employed demographic data, and largely on the population size and density in Australian Aboriginal populations and the health and mortality of several other H-G populations. Perhaps the earliest comprehensive work on H-G demography was conducted by Nancy Howell (1976, 1979) who studied Dobe !Kung (Zhũ/twasi) Bushmen on the Botswana side of the northern Kalahari Desert between 1967 and 1969. During that period, and with earlier data from the Kalahari Research Group, she was able to accumulate a substantial sample of 840 individuals, from which she gathered data on fertility, mortality, and migration. The !Kung had a moderate completed fertility of between four and five births, which is relatively low for a noncontraceptive population (Campbell and Wood, 1988) and an infant mortality (from birth to 1 year of age) of about 20%, also moderate. Howell (1979) applied Crow's (1958) *Index of Opportunity for Selection* to the !Kung fertility and overall mortality data and found that selection in this population was relatively low because of the limited variability in fertility and moderate mortality (selection intensity tends to be highest in human populations with high variation in fertility as well as high mortality).

In Pennington's (2001) comparative review of H-G demography, she surveyed 25 populations from Africa, Asia, Australasia, North America, and South America which were studied from the 1950s through the 1980s. Fertility (total fertility rates or completed fertility) ranged from 2.6 in Efe Pygmies from the Ituri Forest in the Congo to 8.0 in the Ache in eastern Paraguay. The low fertility rate in Efe Pygmies is generally attributed to sexually transmitted diseases and may be atypical for H-G populations. Excluding the Efe fertility, the mean total fertility for the remaining populations equates to 5.4 live births/woman, which places the !Kung fertility only slightly below the mean for all H-G populations surveyed by Pennington. Among those H-G populations with good mortality data, life expectancy at birth (e_0) ranged between 24 and 37 years (Pennington, 2001). With moderate fertility and relatively low life expectancies at birth, selection opportunities for hunter-gatherers are likely to have been most effective through differential mortality. However, this pattern might be an oversimplification based on the degree of variation in both fertility and age-related mortality among H-G populations. For example, there are indications of much higher young and adult mortality rates among Venezuelan Hiwi hunter-gatherers (Hill et al., 2007), conditions that may more closely represent Paleolithic times.

Diet and Nutrition

In general, both historical and prehistoric H-G populations have had diets that are relatively high in protein, fiber, and micronutrients (vitamins and minerals), and relatively low in fat and salt (Eaton et al., 1996; Eaton and Eaton, 1999; Jenike, 2001; see the chapter by Wiley). However, within this nutritional generalization, what really characterizes H-G diet and nutrition is the remarkable diversity of food intake—a function of the diversity and resource density in any given environment. It is also a function of the remarkable omnivorous capacity of humans. For Arctic and subarctic populations such as the Inuit, Inupiat, and Inland Cree, sea mammals, caribou and other land mammals, fish, and wild fowl are primary food sources contributing to high intakes of protein and fat. Carbohydrates are limited to seasonal berries and other plants. Some northern European subarctic populations (Saami, Evenki) hunted and then domesticated reindeer, while others engaged in hunting and fishing (Shnirelman, 1999). In the cool rainforests of the northwest Pacific coast of North America, H-G populations were privileged with abundant sea mammal and salmon food resources, and achieved high levels of ethnic diversity and cultural complexity in art, architecture, rituals, and social stratification (Feit, 1999). Semiarid lands such as the Mojave Desert of the southwestern United States, the Kalahari Desert of southern Africa, and the dry savanna of Tanzania are represented by Shoshone, Ju/'hoansi (!Kung Bushmen), and Hadza hunter-gatherers. Despite the impoverished nature of the food resources, numerous species

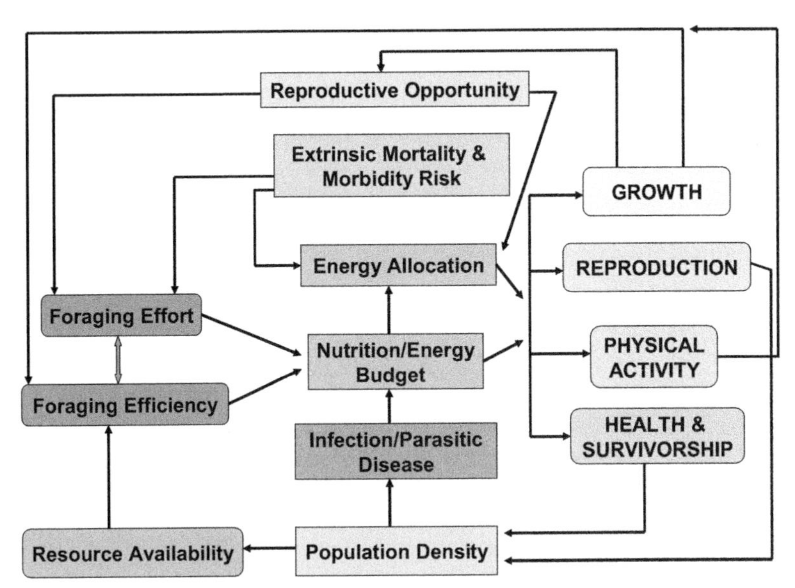

FIGURE 3 A diagram representing relationships among the environment, nutrition, energy intake and expenditure, and other variables associated with foraging in hunter-gatherers. (See color plate section). *Redrawn and modified from Jenike (2001).*

of edible plants in these areas provided staple foods that were supplemented with hunted animals. The wet tropics of the Congo Basin and Southeast Asia are the habitations of Twa, Aka, Bambuti and other Pygmy groups, the Agta and Aeta (Philippine Negritos), and Semang (Malayan Negritos). Pygmies are often divided into bow and net hunters, but are also characterized as maintaining symbiotic relationships with Bantu farmers, whereby hunted and gathered forest foods are exchanged for cultivated foods. Agta hunt, fish, gather plants, and also occasionally cultivate fields of manioc and other crops, while Semang trade with neighboring farmers. It is generally the case that tropical forest hunters have access to some cultivated foods to supplement their H-G subsistence. Pagezy (1988) identified some of the major food items consumed by the somewhat acculturated Ntomba Twa Pygmies in the Lake Tumba region of the Congo: (1) lake, river, and swamp fish, (2) various insect larvae, (3) termites, (4) small game, (5) shrimp, (6) meat from animal husbandry, (7) various wild plants, (8) wild mushrooms, fruits, and tubers, and (9) honey. Game foods are either exchanged for cultivated crops with the associated Oto Bantu people or retained in cases where crops are now cultivated directly by the Twa.

There is marked seasonality in food supply in Arctic, temperate, and both wet and dry tropical areas. Seasonal movements of land and sea mammals, seasonal availability of game and fish (and their nutritional content), seasonal production of plants (fruits and seeds), and seasonal reproduction of insects (e.g., termites) are common in all environments. This has affected H-G food procurement and has clearly produced seasonal hunger and selective evolutionary pressures in areas of reproduction, child growth, immune capacity, and infectious and nutritional disease, as well as in the capacity to perform work associated with subsistence (Harrison, 1988; Stini, 1988). Jenike (2001) compiled seasonal energy intake from four H-G studies (Ache, Efe, Hadza, and Hiwi) and found that lean-season energy intake (in kcal) ranged from 49% to 74% of the intake during the high season. Weight losses associated with seasonal hunger have been widespread and documented for a number of groups, including the Kalahari San (Wilmsen, 1978) and African nomadic pastoralists (Little, 1989; Little et al., 1999).

Marshall Sahlins (1968) presented the provocative idea that hunter-gatherers or foragers were "the original affluent society" based on the limited time devoted to immediate hunting, gathering, and fishing each day. The idea has since been debunked with accumulated data on physical activities as well as data acquired about the work required to maintain an H-G economy (Kelly, 2013). Kelly (2013) found that individuals in some H-G populations forage for more than 8 h per day, whereas in other H-G societies foraging is done only every other day. Jenike (2001) compiled physical activity levels (PAL[2]) for several H-G groups and compared them with other subsistence groups (horticulturalist and agriculturalist). Most H-G men had moderate activity levels (PAL = 1.7–2.1), while those of women were lower (PAL = 1.5–1.9). Figure 3 shows a model that illustrates some highly interactive relationships among the environment, nutrition, energy intake and expenditure, and other variables for forager groups.

Child Growth, Body Size, and Life History

Body sizes of hunter-gatherers show considerable variation geographically, although within that variation the shortest-statured populations in the world are tropical forest hunter-gatherers and horticulturalists. Walker et al. (2006) compiled

2. Physical activity level (PAL) is the total (24 h) daily energy expenditure (TDEE) divided by the basal energy expenditure (BMR). A TDEE/BMR = 2.0 indicates that the daily energy expenditure (and food energy requirement) is twice that of the basal requirement of 1.

growth rates and adult sizes for 22 "small-scale" societies, most of which were made up of hunter-gatherers. For the H-G groups, adult height ranged from short-statured African Pygmies of 144 and 136cm (males and females), to relatively tall Australian Aborigines from Arnhem Land of 171 and 156cm, and from Argentine Toba of 172 and 157cm. The tallest among the 22 societies were the Turkana (175 and 166cm), who are pastoralists and as tall as average North Americans (Little et al., 1983). As noted, the tropical forest H-G and horticulturalist populations tend to include the shortest people of all the societies around the world. Table 1 is a list of heights and weights of a number of largely H-G populations. The shortest were Mbuti Pygmies, with men's average stature at 144cm (4′ 9″) and women's at 136cm (4′ 3″). Also relatively short, but much taller than the Mbuti or Lese, are the Twa at 160cm (5′ 3″) for men and 150cm (4′ 11″) for women. However, the Twa, who are from the Lake Tumba region in the Congo, live in the same villages as the Oto Bantu, have a long history of intermarriage, and represent an admixed "Pygmoid" population (Pagezy, 1988). Other H-G populations are also short-statured and range from the shortest Southeast Asian males (150+cm, or 5′ 0″) to the slightly taller Bushmen, Hadza, and Aché. Body mass indices for these H-G populations tend to be low (lean) and range between 18 and 24.

Hypotheses on why Pygmies and other tropical forest hunter-gatherers ("Pygmoid" and "Negrito" populations) are so short compared with worldwide average statures have been varied (Bailey, 1991a; Becker et al., 2011). One hypothesis bears on temperature regulation, in that short stature tends to maximize surface area per weight relationships and should give Pygmies an advantage in hot-humid tropical forests. Another suggests that small body size is advantageous in reducing individual and population food energy requirements. A third has even suggested that dense jungle can be more easily navigated by small-sized individuals. Finally, there may be a connection to island populations, wherein humans and other mammals are small because of limited territory; in the case of tropical forest dwellers, this may relate to the concept of dwelling in enclaves. These are difficult hypotheses to test, and the real answer may be in some combination of the causes. Nevertheless, the pattern of Pygmy growth has been established, and growth in childhood is not remarkably different from that in other populations. First, birth weights are slightly lower than for the neighboring Bantu (Bailey, 1991b), but by three years of age, Aka Pygmy children are close to the middle of a worldwide sample of like-aged children (van de Koppel and Hewlett, 1986). However, Pygmies begin to show faltering growth during puberty and adolescence, with growth slowing down beginning at about eight years of age and becoming flat by 12.5 years in girls and by 14.5 years in boys. Accordingly, Pygmies show minimal or no adolescent growth spurt (van den Koppel and Hewlett, 1986).

Studies of Pygmy growth and development have uncovered endocrine and genetic causes for their short stature. Merimee and Remoin (1986) gathered data on insulin-induced human growth hormone (hGH) and arginine-induced hGH, as well as associated hormones from 22 Pygmies in the

TABLE 1 Heights and Weights of Short-Statured Hunter-Gatherer and Horticulturalist Populations

	Males		Females		
Populations	Ht (cm)	Wt (kg)	Ht (cm)	Wt (kg)	References
Aché	158	60	149	54	Hill and Hurtado (1996)
Aeta	150	40	140	38	Migliano (2005)
Agta	153	45	144	38	Headland (1989)
Batak	153	47	143	41	Eder (1987)
Hadza	161	54	150	48	Hiernaux and Hartono (1980)
Hiwi	154	56	145	48	Hurtado and Hill (1987)
Ju'/hoansi San	159	55	149	50	Tanaka (1980)
!Kung	160	49	150	41	Lee (1979)
Onge	148	–	138	–	Stock and Migliano (2009)
Pygmy, Efe	143	43	136	38	Bailey and Peacock (1988)
Pygmy, Mbuti	144	42	136	37	Cavalli-Sforza (1986a)
Pygmy, Twa	160	46	150	42	Pagezy (1982, 1984, 1988)
Yanomamo	152	52	142	45	Walker et al. (2006)

Central African Republic. For both tests they found no differences in hGH levels between the Pygmies and a European control group; however, the Pygmies' plasma glucose responses to insulin and plasma insulin responses to arginine were similar to those of hGH-deficient dwarfs. Further studies by Merimee and Remoin (1986) demonstrated that Pygmies had peripheral tissue insensitivity to hGH due to a deficiency of somatomedin or insulin-like growth factor I (IGF-I), a hormone which affects skeletal growth. Recent work by Bozzola et al. (2009) has supported these early studies, having found a marked reduction in hGH receptor gene expression, and that Pygmies lack both an adolescent growth spurt and an associated pubescent serum IGF-I surge. Finally, Becker et al. (2011) provided convincing evidence that short Pygmy stature is under genetic control through admixture studies of more than 1100 Pygmy and Bantu adults in central Africa.

With the demonstration that Pygmy hunter-gatherers' short stature is controlled by a genetic insensitivity to hGH and other hormonal pathways, the question remains: Why are Pygmies short in stature? The most recent explanation is associated with a life history trade-off between accelerated growth, early sexual maturation, and early reproduction as compensation for high mortality at relatively young and older ages (Migliano et al., 2007). This was hypothesized by Migliano (2005) and Walker et al. (2006), and tested by Migliano et al. (2007). Both Walker's and Migliano's research employed comparative data from Asian, African, and South American H-G populations. Walker et al. (2006, p. 295) suggested it is "selective pressure for accelerated development in the face of higher mortality" that has led to the short stature of many H-G peoples. Migliano et al. (2007, p. 20216) feel "that human pygmy populations and adaptations evolved independently as the result of a life history tradeoff between the fertility benefits of larger body size against the costs of late growth cessation, under circumstances of significant young and adult mortality." Moreover, Migliano et al. (2007, p. 20216) emphasize that "human pygmies do not appear to have evolved through positive selection for small stature—this was a by-product of selection for early onset of reproduction." There has been some criticism of this work (see Becker et al., 2010; and a response by Migliano et al., 2010), but fundamentally the relationship of high mortality (short life expectancy), early maturation, and resultant short stature in some H-G societies remains a working hypothesis in the fields of life history and evolutionary theory.

Disease and Morbidity

The relatively high mortality at young and middle ages in many H-G populations, coupled with the consequent low life expectancy, implies considerable illness or morbidity, accidents, homicide, warfare, famine, or other causes of death. Rates of microbial and parasitic infection appear to be high, especially among tropical forest hunter-gatherers. Tropical forests, which are characterized as having high biodiversity, have high parasite and infectious microorganism biodiversity as well. Cavalli-Sforza's (1986b) project on Pygmies of central Africa found very high levels of intestinal, blood, and skin parasites associated with spleen and liver enlargement, a high prevalence of nonvenereal yaws, and common occurrences of respiratory and gastrointestinal infections. He noted, however, that "the superficial impression one received is that Pygmies are exceptionally healthy and fit" (Cavalli-Sforza, 1986b, p. 421). This impression is inconsistent with a combined infant, child, and adolescent mortality that is greater than 50%. Froment (2001, p. 252), however, supports these observations when he states: "Many foraging peoples … look very healthy but they experience very high childhood mortality." This suggests, then, very high selection intensity based on differential early mortality, rather than differential fertility.

In a review of the health and evolution of hunter-gatherers, Froment (2001) noted that a very low population density (<1 person/km^2), which characterizes most H-G populations, is below the threshold for the transmission of many infectious diseases. A low population density limits the transmission of vector-borne parasites such as malaria and other parasites (*host dilution*). That is, parasite transmission is limited in hunter-gatherers when there are no neighboring populations with higher population densities. On the negative side, close contact with game animals for food increases the risk to hunter-gatherers of zoonotic disease transmission to a greater extent than in farmers (see chapter by Morand). Surveys of hunter-gatherers (Australia, Ju/'hoansi/Bushmen, Hadza, Pygmies) by Froment (2001) showed high levels of parasites, including giardia, hookworm amoebas, and ascaris. Yaws prevalence was low in savanna Hadza, but high in many Cameroon tropical forest Pygmies (55–85%) versus Bantu farmers (37%).

Detailed data on mortality and morbidity were gathered on the Casiguran Agta by Headland (1989), who spent many years studying these "Negrito" hunter-gatherers of eastern Luzon in the Philippines. Census data indicated that the population was in decline and had dropped 40%, from about 1000 people in 1936 to 609 people in 1984. Among the problems that the people suffered were high infant mortality (342/1000 births) and a 49% incidence of infant, child, and adolescent deaths before 15 years of age. Life expectancy at birth was 20.7 years for males and 22.0 years for females, but the total fertility rate was quite high at 6.3 live births per woman. Maternal deaths in childbirth were extraordinarily high at 14% of all births. The Casiguran Agta are probably a good example of acculturative population decline resulting from loss of food resources and territory, foreign intrusion, increased infectious disease, and malnutrition. Another case of disruptive acculturation is in the Philippine Aeta from western Luzon. This formerly H-G population also suffered from high parasite loads. Bernstein and Dominy (2013) analyzed Aeta women's breast milk on the grounds that maternal exposures to disease can influence breast milk immune constituents. They found that Aeta had higher levels of transforming growth factor-β_2 (TGF-β_2, an antiinflammatory cytokine), compared with levels in Ilocano rice farmers, suggesting that the Aeta had an elevated exposure to pathogens.

Hill and Hurtado's (1996, pp. 171–177) comprehensive treatment of Aché hunter-gatherers included cause-of-death data for nearly 850 Aché born since 1890. Their data were partitioned into the precontact or forest period (1890–1971), the contact period (1971–1977), and the reservation period (1978–1993). Before 1971, the major causes of death were violence, illness and disease, and accidents. Remarkably, within-group violence was the major cause of death among all age groups including infants and young children. When the authors compared illness, accidents, and violence in early !Kung (before 1973) and early Yanomamo (1970–1974), *illness* was the major cause of death for both groups: 52–88% mortality for the !Kung and 74% mortality for the Yanomamo. These death statistics can be compared with the precontact Aché, where illness constituted only about 24% of deaths while *violence* constituted about 55% of deaths. Much of the violence in Aché during the precontact period resulted from warfare with neighboring Guarani and Paraguayan peasants. Yet, many violent deaths in infants and children were within-Aché homicides. Of the Aché infectious disease mortality, gastrointestinal rather than acute respiratory diseases affected many children, and staphylococcal skin infections were common at all ages. Mortality in the Aché presents a picture that may not be atypical for precontact H-G populations in the past, but we cannot be certain. Questions continue to remain on whether any contemporary H-G population can be used as a model for prehistoric peoples.

HUNTER-GATHERERS IN EVOLUTIONARY PERSPECTIVE: SUMMARY

The evolution of the human lineage was almost certainly driven by environmental changes, resulting in changes in posture, locomotion and manual dexterity, expansion of the brain, and concomitant changes in behavior. A major behavior transition was linked to patterns of food acquisition. From ape origins and plant gathering to the rise of hominins, hunting for meat became an important way to procure food, particularly the energy-dense high-protein food that could fuel the needs of a large brain and allow for even further expansion of this complex organ. As social organization and technology became more and more sophisticated, so did hunting techniques and successes. Hunting in some populations became the primary means of food procurement, while in others, hunting was subordinate to gathering plant foods in its contribution to the diet. Dating back to Paleolithic times, human societies in the Old World, and later in the New World, showed remarkable variation in hunting methods and procedures, and equally remarkable variation in the animals hunted and fished and the plants that were gathered. These patterns of food procurement persisted through time and enabled a slow expansion of human populations up to Neolithic times, when cultivation of plants and animal husbandry arose. This in turn led to a dramatic geographic and population expansion. Despite this expansion into more successful means of food procurement through plant and animal domestication, subsistence by way of hunting, fishing, and gathering persisted in many parts of the globe.

Contemporary H-G societies are in decline, and most are linked now to the greater agricultural societies that dominate the planet. A few hunter-gatherers who persist in marginal areas of tropical savannas and forests, and Arctic populations in North America, Europe, and Asia, still hunt and trap land animals and sea mammals, but these ethnic minorities are also involved in trade relations, and many are subsidized by governments. The 53 H-G societies in Lee and Daly's (1999b) *Cambridge Encyclopedia* have almost certainly been reduced in the years ensuing its publication. Individuals from many societies around the world still engage in hunting and fishing for sport and food—nevertheless, the primary H-G subsistence societies of cognitively modern humans that have existed for more than 100,000 years will not survive into the next century.

REFERENCES

Aiello, L.C., Wheeler, P., 1995. The expensive tissue hypothesis: the brain and the digestive system in human and primate evolution. Current Anthropology 36, 199–221.

Arcand, B., 1999. The Cuiva. In: Lee, R.B., Daly, R. (Eds.), The Cambridge Encyclopedia of Hunters and Gatherers. Cambridge University Press, Cambridge, pp. 97–100.

Bahuchet, S., 1988. Food supply uncertainty among the Aka Pygmies (Lobaye, Central African Republic). In: de Garine, I., Harrison, G.A. (Eds.), Coping with Uncertainty in Food Supply. Clarendon Press, Oxford, pp. 118–149.

Bailey, R.C., 1991a. The Behavioral Ecology of Efe Pygmy Men in the Ituri Forest, Zaire. Anthropological Papers, Museum of Anthropology. University of Michigan, Ann Arbor. No. 86.

Bailey, R.C., 1991b. The comparative growth of Efe Pygmies and African farmers from birth to age 5 years. Annals of Human Biology 18, 113–120.

Bailey, R.C., Peacock, N.R., 1988. Efe Pygmies of northeast Zaïre: subsistence strategies in the Ituri forest. In: de Garine, I., Harrison, G.A. (Eds.), Coping with Uncertainty in Food Supply. Clarendon Press, Oxford, pp. 88–117.

Baker, P.T., Sanders, W.S., 1972. Demographic studies in anthropology. Annual Review of Anthropology 1, 151–178.

Becker, N.S.A., Verdu, P., Hewlett, B., Pavard, S., 2010. Can life history trade-offs explain the evolution of short stature in human Pygmies? A response to Migliano et al. (2007). Human Biology 82, 17–27.

Becker, N.S.A., Verdu, P., Froment, A., Le Bomin, S., Pagezy, H., Bahuchet, S., Heyer, E., 2011. Indirect evidence for the genetic determination of short stature in African Pygmies. American Journal of Physical Anthropology 145, 390–401.

Bellwood, P., 2013. First Migrants: Ancient Migration in Global Perspective. Wiley Blackwell, Oxford.

Bernstein, R.M., Dominy, N.J., 2013. Mount Pinatubo, inflammatory cytokines, and the immunological ecology of Aeta hunter-gatherers. Human Biology 85, 231–250.

Binford, L., 1980. Willow smoke and dogs' tails: hunter-gatherer settlement systems and archaeological site formation. American Antiquity 45, 4–20.

Bird-David, N., 1999. Introduction: southeast Asia. In: Lee, R.B., Daly, R. (Eds.), The Cambridge Encyclopedia of Hunters and Gatherers. Cambridge University Press, Cambridge, pp. 231–237.

Birdsell, J.B., 1953. Some environmental and cultural factors influencing the structuring of Australian aboriginal populations. American Naturalist 87, 171–207.

Birdsell, J.B., 1968. Some predictions for the Pleistocene based on equilibrium systems among recent hunter-gatherers. In: Lee, R.B., DeVore, I. (Eds.), Man the Hunter. Aldine, Chicago, pp. 229–240.

Boas, F., 1896. The limitations of the comparative method of anthropology. Science 4 (103), 901–908.

Boas, F., 1921. Ethnology of the Kwakiutl. Annual Report of the Bureau of American Ethnology, vol. 35 Smithsonian Institution, Washington, DC.

Boesch, C., Boesch, H., 1989. Hunting behavior of wild chimpanzees in the Tai National Park. American Journal of Physical Anthropology 78, 547–573.

Bozzola, M., Travaglino, P., Marziliano, N., Meazza, C., Pagani, S., Grasso, M., Tauber, M., Diegoli, M., Pilotto, A., Disabella, E., Tarantino, P., Brega, A., Arbustini, E., 2009. The shortness of Pygmies is associated with severe under-expression of the growth hormone receptor. Molecular Genetics and Metabolism 98, 310–313.

Burch, E.S., Csonka, Y., 1999. The caribou inuit. In: Lee, R.B., Daly, R. (Eds.), The Cambridge Encyclopedia of Hunters and Gatherers. Cambridge University Press, Cambridge, pp. 56–60.

Campbell, K.L., Wood, J.W., 1988. Fertility in traditional societies. In: Diggory, P., Potts, M., Teper, S. (Eds.), Natural Human Fertility: Social and Biological Determinants. Macmillan Press, Houndmills, Basingstoke, Hampshire, England, pp. 39–69.

Cavalli-Sforza, L.L., 1986a. Anthropometric data. In: Cavalli- Sforza, L.L. (Ed.), African Pygmies. Academic Press, Orlando, FL, pp. 81–93.

Cavalli-Sforza, L.L., 1986b. African Pygmies: an evaluation of the state of research. In: Cavalli-Sforza, L.L. (Ed.), African Pygmies. Academic Press, Orlando, FL, pp. 361–426.

Chagnon, N., 1968. Yanomamö: The Fierce People. Holt, Rinehart & Winston, New York.

Crow, J.F., 1958. Some possibilities for measuring selection intensities in man. Human Biology 30, 1–13.

Cummings, V., 2013. The Anthropology of Hunter-Gatherers: Key Themes for Archaeologists. Bloomsbury, London.

Eaton, S.B., Eaton III, S.B., 1999. Hunter-gatherers and human health. In: Lee, R.B., Daly, R. (Eds.), The Cambridge Encyclopedia of Hunters and Gatherers. Cambridge University Press, Cambridge, pp. 449–456.

Eaton, S.B., Eaton, S.B., Konner, M.J., Shostak, M., 1996. An evolutionary perspective enhances understanding of human nutritional requirements. Journal of Nutrition 126, 1732–1740.

Eder, J.F., 1987. On the Road to Tribal Extinction. University of California Press, Berkeley.

Eder, J.F., 1999. The Batak of peninsular Malaysia. In: Lee, R.B., Daly, R. (Eds.), The Cambridge Encyclopedia of Hunters and Gatherers. Cambridge University Press, Cambridge, pp. 294–297.

Endicott, K., 1999. Introduction: southeast Asia. In: Lee, R.B., Daly, R. (Eds.), The Cambridge Encyclopedia of Hunters and Gatherers. Cambridge University Press, Cambridge, pp. 275–283.

Feit, H., 1999. Introduction: North America. In: Lee, R.B., Daly, R. (Eds.), The Cambridge Encyclopedia of Hunters and Gatherers. Cambridge University Press, Cambridge, pp. 23–30.

Forde, C.D., 1934. Habitat, Economy and Society: A Geographical Introduction to Ethnology. Methuen, London.

Fowler, C.S., 1999. The Timbisha Shoshone of Death Valley. In: Lee, R.B., Daly, R. (Eds.), The Cambridge Encyclopedia of Hunters and Gatherers. Cambridge University Press, Cambridge, pp. 66–70.

Freeman Jr., L.G., 1968. A theoretical framework for interpreting archaeological materials. In: Lee, R.B., DeVore, I. (Eds.), Man the Hunter. Aldine, Chicago, pp. 262–267.

Froment, A., 2001. Evolutionary biology and health of hunter-gatherer populations. In: Panter-Brick, C., Layton, R.H., Rowley-Conwy, P. (Eds.), Hunter-Gatherers: An Interdisciplinary Perspective. Cambridge University Press, Cambridge, pp. 239–266.

Goodall, J., 1968. Behavior of free-living chimpanzees of the Gombe Stream area. Animal Behavior Monographs 1, 163–311.

Goodall, J., 1986. The Chimpanzees of Gombe: Patterns of Behavior. Harvard University Press, Cambridge, MA.

Griffin, P.B., Griffin, M.B., 1999. The Agta of eastern Luzon, Philippines. In: Lee, R.B., Daly, R. (Eds.), The Cambridge Encyclopedia of Hunters and Gatherers. Cambridge University Press, Cambridge, pp. 289–293.

Harrison, G.A., 1988. Seasonality and human population biology. In: de Garine, I., Harrison, G.A. (Eds.), Coping with Uncertainty in Food Supply. Clarendon Press, Oxford, pp. 26–31.

Headland, T.N., 1989. Population decline in a Philippine Negrito hunter-gatherer society. American Journal of Human Biology 1, 59–72.

Henn, B.M., Gignoux, C.R., Jobin, M., Granka, J.M., Macpherson, J.M., Kidd, J.M., Rodriguez- Botigué, L., Ramachandran, S., Hon, L., Brisbin, A., Lin, A.A., Underhill, P.A., Comas, D., Kidd, K.K., Norman, P.J., Parham, P., Bustamante, C.D., Mountain, J.L., Feldman, M.W., 2011. Hunter-gatherer genomic diversity suggests a southern African origin for modern humans. Proceedings of the National Academy of Sciences 108 (13), 5154–5162.

Henn, B.M., Cavalli-Sforza, L.L., Feldman, M.W., 2012. The great human expansion. Proceedings of the National Academy of Sciences 109 (44), 17758–17764.

Hiernaux, J., Hartono, D.B., 1980. Physical measurements of the adult Hadza of Tanzania. Annals of Human Biology 7, 339–346.

Hill, K., Hurtado, A.M., 1996. Ache Life History: The Ecology and Demography of a Foraging People. Aldine de Gruyter, New York.

Hill, K., Hurtado, A.M., Walker, R.S., 2007. High adult mortality among Hiwi hunter-gatherers: implications for human evolution. Journal of Human Evolution 52, 443–454.

Hitchcock, R.K., 1999. Introduction: Africa. In: Lee, R.B., Daly, R. (Eds.), The Cambridge Encyclopedia of Hunters and Gatherers. Cambridge University Press, Cambridge, pp. 175–184.

Hohmann, G., Furth, B., 2008. New records on prey capture and meat eating by bonobos at Lui Kotale, Salonga National Park, Democratic Republic of Congo. Folia Primatologica 79, 103–110.

Howell, N., 1976. The population of the Dobe area !Kung. In: Lee, R.B., DeVore, I. (Eds.), Kalahari Hunter-Gatherers: Studies of the !Kung San and Their Neighbors. Harvard University Press, Cambridge, pp. 137–151.

Howell, N., 1979. Demography of the Dobe !Kung. Academic Press, New York.
Hurtado, A.M., Hill, K., 1987. Early dry season subsistence ecology of the Cuiva foragers of Venezuela. Human Ecology 15, 163–187.
Jenike, M.R., 2001. Nutritional ecology: diet, physical activity and body size. In: Panter-Brick, C., Layton, R.H., Rowley-Conwy, P. (Eds.), Hunter-Gatherers: An Interdisciplinary Perspective. Cambridge University Press, Cambridge, pp. 205–238.
Kaare, B., Woodburn, J., 1999. The Hadza of Tanzania. In: Lee, R.B., Daly, R. (Eds.), The Cambridge Encyclopedia of Hunters and Gatherers. Cambridge University Press, Cambridge, pp. 200–204.
Kelly, R.L., 2013. The Lifeways of Hunter-Gatherers: The Foraging Spectrum. Cambridge University Press, Cambridge.
Lee, R.B., 1979. The !Kung San. Cambridge University Press, Cambridge.
Lee, R.B., Daly, R., 1999a. Foragers and others. In: Lee, R.B., Daly, R. (Eds.), The Cambridge Encyclopedia of Hunters and Gatherers. Cambridge University Press, Cambridge, pp. 1–19.
Lee, R.B., Daly, R. (Eds.), 1999b. The Cambridge Encyclopedia of Hunters and Gatherers. Cambridge University Press, Cambridge.
Lee, R.B., DeVore, I. (Eds.), 1968. Man the Hunter. Aldine, Chicago.
Lee, R.B., DeVore, I. (Eds.), 1976. Kalahari Hunter-Gatherers: Studies of the !Kung San and Their Neighbors. Harvard University Press, Cambridge.
Leonard, W.R., Robertson, M.L., 1997. Comparative primate energetics and hominid evolution. American Journal of Physical Anthropology 102, 265–281.
Little, M.A., 1989. Human biology of African pastoralists. Yearbook of Physical Anthropology 32, 215–247.
Little, M.A., Galvin, K., Mugambi, M., 1983. Cross-sectional growth of nomadic Turkana pastoralists. Human Biology 55, 811–830.
Little, M.A., Dyson-Hudson, R., Dyson-Hudson, N., Winterbauer, N.L., 1999. Environmental variations in the South Turkana ecosystem. In: Little, M.A., Leslie, P.W. (Eds.), Turkana Herders of the Dry Savanna: Ecology and Biobehavioral Response of Nomads to an Uncertain Environment. Oxford University Press, Oxford, pp. 316–330.
Merimee, T.J., Remoin, D.L., 1986. Growth hormone and insulin-like growth factors in the Western Pygmy. In: Cavalli-Sforza, L.L. (Ed.), African Pygmies. Academic Press, Orlando, FL, pp. 167–177.
Migliano, A.B., 2005. Why Are Pygmies Small? Ontogenetic Implications of Life History Evolution (Ph.D. dissertation). University of Cambridge, Cambridge.
Migliano, A.B., Vinicius, L., Lahr, M.M., 2007. Life history trade offs explain the evolution of human pygmies. Proceedings of the National Academy of Sciences 104, 20216–20219.
Migliano, A.B., 2010. Why are Pygmies so short? A defense of Migliano's hypothesis. Human Biology 82, 109–113.
Migliano, A.B., Romero, I.G., Metspalu, M., Leavesley, M., Pagani, L., Antao, T., Huang, D.-W., Sherman, B.T., Siddle, K., Scholes, C., Hudjashov, G., Kaitokai, E., Babalu, A., Belatti, M., Cagan, A., Hopkinshaw, B., Shaw, C., Nelis, M., Metspalu, E., Mägi, R., Lempicki, R.A., Villems, R., Lahr, M.M., Kivisild, T., 2013. Evolution of the Pygmy phenotype: evidence of positive selection from genome-wide scans in African, Asian, and Melanesian Pygmies. Human Biology 85, 251–384.
Murdock, G.P., 1968. The current status of the world's hunting and gathering peoples. In: Lee, R.B., DeVore, I. (Eds.), Man the Hunter. Aldine, Chicago, pp. 13–20.
Pagezy, H., 1982. Seasonal hunger as experienced by the Oto and Twa of a Ntomba Village in the equatorial forest (Lake Tumba, Zaïre). Ecology of Food and Nutrition 12, 139–153.
Pagezy, H., 1984. Seasonal hunger as experienced by the Oto and Twa women of a Ntomba village in the equatorial forest (Lake Tumba, Zaïre). Ecology of Food and Nutrition 15, 13–27.
Pagezy, H., 1988. Coping with uncertainty in food supply among the Oto and Twa living in the equatorial flooded forest near Lake Tumba, Zaïre. In: de Garine, I., Harrison, G.A. (Eds.), Coping with Uncertainty in Food Supply. Clarendon Press, Oxford, pp. 175–209.
Pandya, V., 1999. The Andaman islanders of the Bay of Bengal. In: Lee, R.B., Daly, R. (Eds.), The Cambridge Encyclopedia of Hunters and Gatherers. Cambridge University Press, Cambridge, pp. 243–247.
Panter-Brick, C., Layton, R.H., Rowley-Conwy, P., 2001. Lines of enquiry. In: Panter-Brick, C., Layton, R.H., Rowley-Conwy, P. (Eds.), Hunters-Gatherers: An Interdisciplinary Perspective. Cambridge University Press, Cambridge, pp. 1–11.
Pennington, R., 2001. Hunter-gatherer demography. In: Panter-Brick, C., Layton, R.H., Rowley-Conwy, P. (Eds.), Hunters-Gatherers: An Interdisciplinary Perspective. Cambridge University Press, Cambridge, pp. 170–204.
Peterson, N., 1999. Introduction: Australia. In: Lee, R.B., Daly, R. (Eds.), The Cambridge Encyclopedia of Hunters and Gatherers. Cambridge University Press, Cambridge, pp. 317–323.
Rival, L.M., 1999. Introduction: South America. In: Lee, R.B., Daly, R. (Eds.), The Cambridge Encyclopedia of Hunters and Gatherers. Cambridge University Press, Cambridge, pp. 77–85.
Sahlins, M.D., 1968. Notes on the original affluent society. In: Lee, R.B., DeVore, I. (Eds.), Man the Hunter. Aldine, Chicago, pp. 85–89.
Sanford, C.B., Wallis, J., Matama, H., Goodall, J., 1994. Patterns of predation by chimpanzees on red colobus monkeys in Gombe National Park. American Journal of Physical Anthropology 94, 213–228.
Schurr, T.G., 2004. The peopling of the new world: perspectives from molecular anthropology. Annual Review of Anthropology 33, 551–583.
Shnirelman, V.A., 1999. Introduction: North Eurasia. In: Lee, R.B., Daly, R. (Eds.), The Cambridge Encyclopedia of Hunters and Gatherers. Cambridge University Press, Cambridge, pp. 119–125.
Smith, A.B., 1999. Archaeology and evolution of hunters and gatherers. In: Lee, R.B., Daly, R. (Eds.), The Cambridge Encyclopedia of Hunters and Gatherers. Cambridge University Press, Cambridge, pp. 384–390.
Steegmann Jr., A.T. (Ed.), 1983. Boreal Forest Adaptations: The Northern Algonkians. Plenum, New York.
Steward, J.H., 1936. The economic and social basis of primitive bands. In: Kroeber, A.L. (Ed.), Essays in Honor of. University of California Press, Berkeley, pp. 331–350.
Stini, W.A., 1988. Food, seasonality, and human evolution. In: de Garine, I., Harrison, G.A. (Eds.), Coping with Uncertainty in Food Supply. Clarendon Press, Oxford, pp. 32–51.
Stock, J.T., Migliano, A.B., 2009. Stature, mortality, and life history among indigenous populations of the Andaman Islands 1871–1986. Current Anthropology 50, 713–725.
Surbeck, M., Fowler, A., Deimel, C., Hohmann, G., 2009. Evidence for the consumption of arboreal, diurnal primates by bonobos (*Pan paniscus*). American Journal of Primatology 71, 171–174.
Tanaka, J., 1980. The San, Hunter-Gatherers of the Kalahari. University of Tokyo Press, Tokyo.
Teelen, S., 2008. Influence of chimpanzee predation on the red colobus population at Ngogo, Kibale National Park, Uganda. Primates 49, 41–49.
Teleki, G., 1973. The Predatory Behavior of Wild Chimpanzees. Bucknell University Press, Lewisburg, PA.

Tindale, N.B., 1940. Results of the Harvard-Adelaide Universities anthropological expedition, 1938–1939: distribution of Australian aboriginal tribes: a field survey. Transactions of the Royal Society of South Australia 64, 140–231.

Turnbull, C.M., 1965. The Mbuti Pygmies of the Congo. In: Gibbs Jr., J.L. (Ed.), Peoples of Africa. Holt, Rinehart and Winston, New York, pp. 281–317.

Uehara, S., 1997. Predation on mammals by the chimpanzee (*Pan troglodytes*): an ecological review. Primates 38, 193–214.

van de Koppel, J.M.H., Hewlett, B.S., 1986. Growth of Aka Pygmies and Bagandus of the Central African Republic. In: Cavalli-Sforza, L.L. (Ed.), African Pygmies. Academic Press, Orlando, FL, pp. 95–102.

Walker, R., Gurven, M., Hill, K., Migliano, A., Chagnon, N., De Souza, R., Djurovic, G., Hames, R., Hurtado, A.M., Kaplan, H., Kramer, K., Oliver, W.J., Valeggia, C., Yamauchi, T., 2006. Growth rates and life histories in twenty-two small-scale societies. American Journal of Human Biology 18, 295–311.

Watts, D.P., Amsler, S.J., 2013. Chimpanzee-red colobus encounter rates show a red colobus population decline associated with predation by chimpanzees at Ngogo. American Journal of Primatology 75, 927–937.

Wilmsen, E.N., 1978. Seasonal effects of dietary intake on Kalahari San. Federation Proceedings 37, 65–71.

Worl, R., 1999. Inupiat Arctic whalers. In: Lee, R.B., Daly, R. (Eds.), The Cambridge Encyclopedia of Hunters and Gatherers. Cambridge University Press, Cambridge, pp. 61–65.

Zegura, S.L., Karafet, T.M., Hammer, M.F., 2009. The peopling of the Americas as viewed from the Y chromosome. In: Peregrine, P.N., Peiros, I., Feldman, M. (Eds.), Ancient Human Migrations: A Multidisciplinary Approach. University of Utah Press, Salt Lake City, pp. 127–148.

Chapter 24

Pastoralism

Michael A. Little
Department of Anthropology, Binghamton University, State University of New York, Binghamton, NY, USA

SYNOPSIS

Pastoralism (keeping domestic herbivores) is a fundamental subsistence pattern that dates back over 10,000 years to the global warming that ended the Pleistocene Epoch. Excluding dogs, the earliest domesticated animals were goats, sheep, and cattle. Livestock provide meat, milk, and other food products that are dietary staples for many populations. Livestock are also disease reservoirs of both ancient and emerging diseases that affect humans. Humans have adapted genetically to milk sugar (lactose) and to some of the zoonotic diseases transmitted by livestock. This demonstrates that livestock and humans have coevolved because of their intimate association over the past 100 centuries.

PASTORALISM AS SUBSISTENCE

Pastoralism refers to the practice of keeping livestock as a means of either primary or secondary subsistence. Pastoralist societies usually keep animals that require pasture, but not necessarily animals that can be kept in permanent or semipermanent surroundings (e.g., swine, fowl, rabbits, and guinea pigs). Ordinarily, pastoral societies manage ruminant animals that have natural herding tendencies and that inhabited grasslands, either tropical savanna or temperate prairie lands, in their wild ancestral forms. These environments are often marginal to agriculture but appropriate for providing forage for both wild animals and livestock. In a mixed grassland and forest environment, pastoralists will often burn the scrubland and woodland growth to maintain a grassland environment for their herds. Subsistence practices range in a continuum from a mix of plant cultivation and herding (agropastoralism) to an almost exclusive practice of herding animals. Movement patterns range from sedentary use of rich pasturelands (little migration) to transhumance (seasonal migration) to seminomadic (permanent homestead) to nomadic pastoralism (frequent homestead movement).

Nomadic pastoralism is often practiced in semiarid ecosystems, where green pastures are either patchy and widely dispersed or dramatically limited by seasonal drought. In transhumant, seminomadic, and nomadic pastoralists, "spatial mobility is regularly employed as a survival strategy" (Dyson-Hudson and Dyson-Hudson, 1980, p. 15). Rather than bringing feed to the livestock (with limited grazing), as is a modern Western practice, traditional non-Western subsistence pastoralists move the animals to where green forage can be found. As Dyson-Hudson and Dyson-Hudson (1980, p. 51) noted livestock mobility "does not demand large capital investment, does not entail huge inputs of fuel energy, and does not require a heavy dependence on foods suitable for human consumption [such as grains] to support livestock during periods when plant growth is limited." It is quite likely that transhumant, seminomadic, and nomadic pastoralism linked to livestock mobility date back to the earliest herding peoples in the Neolithic Era.

Traditional or native pastoralists, who have been practicing this form of subsistence for hundreds or thousands of years, can be found on nearly all continents. For example, horses are kept in Central Asia; sheep and goats are commonly herded in the Mediterranean and Near East; cattle are kept in Europe, the Near East, South Asia, and Africa; water buffalo are a domestic stock in Southeast Asia, China, and India; dromedary camels and donkeys are herded in Africa and the Near East; reindeer are kept in northern Europe and Siberia; yaks are herded in the Himalayas; and camelids (llamas and alpacas) are kept in the Andes of South America. The primary products from these domesticated animals are meat, milk, hides, wool, and dung for fuel and fertilizer. Other important, but secondary, products include bone, horn, sinew, and hooves. Many of these domestic animals are also used for transport of goods and people and as draft animals.

PREHISTORY OF PASTORALISM

The earliest domestication of any animal was the dog, with dates from the late Paleolithic and Mesolithic eras as much as 15,000 years ago (Clutton-Brock, 1995). DNA evidence suggests that the presence of genetic variation in living dog breeds indicates a close affinity with Middle Eastern gray wolves (vonHoldt et al., 2010). Dogs are thought to have been used primarily to support hunting activities, but were likely to have been eaten as well. It is probable that dogs and some livestock species were domesticated independently on several occasions in the past, since a sole diffusion model of domestication for individual species is likely to be too simple a pattern.

How any single wild animal species was originally domesticated is highly speculative. However, based on a variety of information sources, the domestication process involved capturing animals who lived in close proximity to humans, managing young animals, taming them in captivity, selectively breeding them for desired traits such as docility and size, occasionally adding wild animals to the breeding stock, and continuing to selectively breed the animals for size, body composition, morphology, and manageable behavior. It is likely that tractable animal behavior was the most desirable characteristic of many of the early wild forms that were domesticated. Recent experiments with the domestication of the silver fox demonstrated how a wild, aggressive animal can be transformed into a behaviorally gentle pet through selective breeding over many generations (Trut, 1999). The process of domestication is an evolving relationship between animals and people where ultimately a mutualistic or symbiotic association develops; that is, both the domestic species and its human host benefit from the association (Zeder et al., 2006). The animals benefit by being cared for, fed, watered, and protected from predators; the people benefit by having a more secure and dependable food supply.

Melinda Zeder (2012) developed a sophisticated model for three alternate pathways to early domestication in the distant past. The first is the *commensal pathway*, where animals are drawn to human settlements or camps by attraction to food refuse, and become accustomed to humans through close association. Dogs, cats, and guinea pigs may have begun the domestication process by this pathway. The second is the *prey pathway*, where animals that were actively hunted may have been initially managed (game management) according to hunted species that had behaviors amenable to such management. Cattle, goats, sheep, camelids, reindeer, and other social herding animals (livestock) are likely candidates for this pathway. The third is the *directed pathway*, where animals may have already begun the domestication process via one of the other pathways, or where a more deliberate effort has been employed to capture, domesticate, and use these animals. Several examples given by Zeder (2012) are the domestication of horses, donkeys, and camels. Recent patterns of animal domestication all fall within this third pattern of directed domestication, whereas for the other two pathways domestication was likely not a planned, stepwise, or directed process that involved human agency alone. A detailed treatment of these domestication pathways and the species taxa that were likely to have been domesticated in each pathway is given by Larson and Fuller (2014).

Each of these pathways effectively led to conditions such that by the early-to-mid-nineteenth century when Darwin (1859) first used the model of "artificial selection" (selective animal and plant breeding) to conceptualize "natural selection," there were literally hundreds, and perhaps thousands, of domestic breeds of livestock that had been created by humans around the globe, by the very process of selection that Darwin had described.

Prehistoric domestic species can be identified largely by changes in the morphology, skeleton, and teeth of animal remains found in association with humans from archaeological sites. Also, as determined from bone assemblages, domestic herd demography is always different from wild herd demography in that domestic herds are likely to have no or few postreproductive females and a predominance of young female animals (Zeder, 2008). As noted above, more recently, ancient DNA analysis of animal bones and teeth of both domestic and ancestral wild species can be compared to determine the degree of genetic difference, approximate dates of domestication, and patterns of migration (Pereira et al., 2009; Warmuth et al., 2012). These analytical procedures are being used more and more not only to identify domestication events and wild ancestral species, but also to trace domestic lineages.

The domestication of livestock species began between 10,000 and 12,000 years ago in the Near East. The two earliest domesticated species were goats (*Capra hircus*) and sheep (*Ovis aries*) around 11,000 years ago, while the earliest evidence for cattle (*Bos taurus*) is 10,000 years ago (Zeder, 2008). The wild ancestors of goats and sheep were *Capra aegagrus* and *Ovis orientalis*, respectively. Two species of domestic cattle were derived from the wild form called the aurochs or *Bos primigenius*, and these were Eurasian cattle (*B. taurus*) and zebu cattle (*Bos indicus*). This wild aurochs ancestor of domestic cattle (see Figure 1) became extinct in 1627 with the death of the last female in Poland. Both animal domestication and plant cultivation began in the Near East at about the same time following the climatic warming trend and glacial retreats at the end of the Pleistocene Epoch. The end of the Pleistocene and beginning of the Holocene Epoch, about 10,000–12,000 years ago, ushered in a warming climatic trend and the retreat of the last major continental glacier (Würm/Wisconsin). It was then that human populations were able to expand into lands freed by the glacial retreat, and the warming trend allowed humans to experiment with plant cultivation and ultimately domestication. This period is known in prehistory as the Neolithic Revolution (named also for a new form of stone tool manufacture).

FIGURE 1 Painting of an Aurochs (*Bos primigenius*) by Gilles Tosello. (See color plate section). *With permission, ©Gilles Tosello.*

A major new source of food was acquired when animals began to be exploited for milk. The late Andrew Sherratt (1981) identified milk, wool, and other products from herds as constituting the "secondary products revolution." He believed that this occurred several thousand years after the Neolithic Revolution. More recent evidence from the Near East and southeastern Europe has identified milk residues in pottery that push dates for milk use (and milking animals) back to between 8000 and 9000 years ago (Evershed et al., 2008). Even more recently, Dunne et al. (2012) have found residues of alkanoic acids of milk fat in Saharan Africa, indicating the presence of dairying about 7000 years ago on that continent. Another food extraction practice, not well documented in prehistory, is bleeding live animals. Today this is employed only in East Africa, where many pastoral peoples bleed animals to provide food to supplement milk and meat when those foods are in scarce supply (Little, 1989).

BIOGEOGRAPHY OF PASTORALISM

There have been many independent livestock domestication events, mostly in the Near East and Asia. Based on mtDNA analysis, Loftus et al. (1994) identified two major evolutionary clades of cattle, one lineage constituting all

European and African breeds (*Bos taurus*), and the other including all Indian breeds (*Bos indicus*). These wild lineages may have separated as much as 1.7–2.0 million years ago (Hiendleder et al., 2008). These data were interpreted as representing two independent domestication events from two subspecies of the ancestral aurochs. This conclusion was supported by McTavish et al. (2013) from DNA single-nucleotide polymorphisms (SNPs) of New World cattle breeds, some of which were hybridized from representatives of the two domestic species. Cattle derived from these two domestic lineages are now found throughout the world, including Asia, Europe, Africa, Australia, North America, and South America. Europeans brought cattle to the New World in 1493 and to Australia in the late 1700s.

The prehistoric domestication of cattle and other livestock species is given in Table 1 (also see Figure 3 in the chapter by Morand in this volume). As noted, goats, sheep, and taurine cattle were all domesticated in the Near East between 10,000 and 11,000 years ago (Zeder, 2008). These livestock diffused outward to many of the marginal coastal areas of the Mediterranean basin (both European and African areas) between 7000 and 9000 years ago, and then later to other parts of Europe. Zebu cattle were domesticated independently, probably in the Indus Valley, about 9000 years ago. Horses (*Equus caballus*) were domesticated from the wild species (*Equus ferus*) several thousand years later in the middle regions of the Western Eurasian steppes (Warmuth et al., 2012). The domestic water buffalo (*Bubalus bubalis*) was derived from the wild buffalo (*Bubalus amee*) between 5000 and 6000 years ago in India (river buffalo) and later in China (swamp buffalo) (Kumar et al., 2007). The camelids, llama (*Lama glama*) and alpaca (*Lama pacos*), were domesticated in the Andes from the wild guanaco (*Lama guanicoe*) and vicuña (*Vicugna vicugna*) (Mengoni Goñalons and Yacobaccio, 2006), and are the only large New World domesticated mammals. Donkeys (*Equus asinus*) are the only large mammal domesticated in Africa, and were derived from the African wild ass (*Equus africanus*). Old-world camels are of two species—*Camelus bactrianus* (two-humped) and *Camelus dromedarius* (one-humped)—and were almost certainly products of independent domestication events. More recently domesticated species, at about 2000 years ago or less, include the yak (*Bos grunniens*) and the reindeer (*Rangifer tarandus*). Yaks are found throughout the vast area of the Tibetan Plateau, while domestic reindeer are herded across the northern circumpolar regions from northwestern Europe by the Saami and to northeastern Siberia by the Samoyeds.

FOOD, DIET, AND CUISINE

In the subsistence transition from hunting and gathering to pastoralism, the food supply became more regular and stable. This enabled human populations to increase in number along with the expansions of their animal herds. An increase in population was also associated with the rise of agriculture and associated agropastoralism. With early pastoralists, meat was the first important food provided by animals kept near human settlements. The success of goat and sheep pastoralism in the Near East was the result of having animals of relatively small body size to care for. The slaughter of a single animal (goat or lamb) could provide food for a small group of people or a family without waste or spoilage and without significantly

TABLE 1 Domestication of Principal Livestock Species

Common Name	Species	Date, Years Ago	Place	References
Goats	*Capra* sp.	11,000	Near East (SW Asia)	Zeder (2008)
Sheep	*Ovis* sp.	11,000	Near East	Zeder (2008)
Cattle	*Bos taurus* *Bos indicus*	10,000 9000	Near East India	Zeder (2008)
Horses	*Equus caballus*	6000	Eurasian steppes (Ukraine, Kazakhstan)	Clutton-Brock (1992) and Olsen (2006)
Water buffalo	*Bubalus bubalis*	5000–6000 4000	India (river buffalo) China (swamp buffalo)	Olsen (1993)
Camelids	*Llama* sp.	6000	Central Andes	Novoa and Wheeler (1984) and Mengoni Goñalons and Yacobaccio (2006)
Donkeys	*Equus asinus*	5000	NE Africa	Clutton-Brock (1992)
Camels	*Camelus* sp.	5000	Arabia (dromedary) central Asia (bactrian)	Clutton-Brock (1999)
Yak	*Bos grunniens*	>2000	Tibetan plateau	Olsen (1990)
Reindeer	*Rangifer tarandus*	<2000	Northern Eurasia	Aikio (1989)

depleting herd resources. Today for subsistence pastoralists who keep several species of domestic stock, small animals such as sheep and goats are commonly slaughtered for meat, while large animals such as cattle or camels are kept for their milk and only slaughtered for special ritual occasions and feasts (Galvin and Little, 1999). Meat could be preserved by drying or salting, and provided by culling unproductive (older or sickly) animals from the herd. This enabled early pastoral populations to procure food throughout the year and during difficult times such as droughts. Other food provided by the herds included blood from freshly slaughtered animals and fat from the animal carcass and bone marrow. One other unusual form of food production from livestock is still employed by pastoral populations in parts of East Africa. This is the practice of bleeding live animals, generally female animals that are not pregnant or producing milk or male animals that are mature and healthy (Little, 1989; Galvin and Little, 1999). It is carried out for large animals, such as cattle, camels, and donkeys, by carefully shooting a blunt arrow into the jugular vein. For small animals such as sheep and goats, a facial vein is opened with a sharp knife. The amounts of blood taken are small, but in the aggregate may help a family to survive during periods of hunger. Blood taken from live animals in this fashion is usually considered a supplemental or emergency food.

When milk offtake from animals, or *dairying*, was discovered some 8000–9000 years ago, it ushered in a new and revolutionary form of food production with a variety of new foods and preservation techniques. Keeping livestock for meat and other products involves a simpler pattern of management than that required for maintaining milking herds. Careful attention must be given to maintain a balance between the amount of milk taken for human consumption and the amount left for the nursing of young animals; hence, milk production must be balanced with the needs for herd reproduction. Keeping herds of milch (milking) animals requires special attention to watering and grazing, division of human labor by specific tasks, and maintaining the number of human laborers to carry out these tasks. Since many subsistence pastoral units are composed of families, expanding herds must be matched by growing families who provide the basic labor for herd management. Thus, herd reproduction and population expansion are closely linked to the maximization of human reproduction and fertility (Leslie et al., 1999).

Although fresh milk is a common form of dietary consumption by pastoral peoples, there is a remarkable variety of food products that can be derived from raw milk. The fat fraction of milk will rise to the top and can be skimmed off and churned into butter, or heated to remove the protein to produce ghee (a pure form of milk fat or oil that can be preserved without spoiling for several weeks or more). The energy value of ghee is high at 9 kcal/g. When there is a surplus, Turkana pastoral nomads preserve whole milk by pouring a thin layer on hides that are placed in the sun to dry. When the milk is dried, it is scraped off and stored in a container for later use (Little et al., 2001). The simplest form of food storage is human consumption of excess or surplus food. When livestock are highly productive in seasons or years of abundant vegetation, and milk is available in great quantities, people will consume large amounts of milk and store the excess calories as body fat. This stored energy can then be drawn on during later periods of limited food availability (Little et al., 1999a).

A variety of other products make up the cuisine of some pastoral groups. Yogurt, cheeses, buttermilk, and other fermented products allow longer storage times and change the nutritional content of the original fresh milk. Some kinds of processing remove or reduce the content of lactose (milk carbohydrate or sugar) from milk, thereby concentrating the protein and fat or removing the fat from the original product. Few pastoral populations feed entirely on animal food products, since many groups procure other food items through trade or hunting and gathering (Casimir, 1991; Little and Leslie, 1999). Despite this dietary diversity, there is always the likelihood that some pastoral populations have in the past had diets so specialized that there may have been nutritional deficiencies. However, as noted, such deficiencies are less common in pastoralists who supplement their diets with traded, cultivated, or gathered foods.

Among contemporary pastoralists from Africa, there is substantial variation in dietary intakes. This is a function of both climate and the subsistence traditions of pastoralism and agropastoralism. People in arid and semiarid climatic zones are unable to cultivate crops as a reliable source of food and hence depend more heavily on livestock products, trade, and some hunting and gathering. Figure 2 illustrates dietary intakes for 11 herding populations ranked according to milk consumption. Most of these populations exploit the semiarid savannas of East Africa and the Sahel farther west. The values given in Figure 2 are annual percentage values for various food products consumed by different ethnic pastoralists, and do not show either the substantial seasonal variation in food intakes experienced by the different groups or the variation in intakes by age and sex. Nor does the figure indicate the number of food calories consumed by each population. All of these factors produce even more variation in dietary intake and also reflect variations in subsistence practices. Thus, there is no archetypical diet of even African pastoralists. To use the Turkana as an example, milk constitutes about 62% of the annual caloric intake of the people and is clearly their most important staple food. However, during the season in which milk is being produced in large quantities by many of the livestock after giving birth to young, milk may constitute up to 90% of the diet (Galvin, 1985). This high milk production of livestock follows the rainy season vegetation flush in particularly good years of production. Dietary

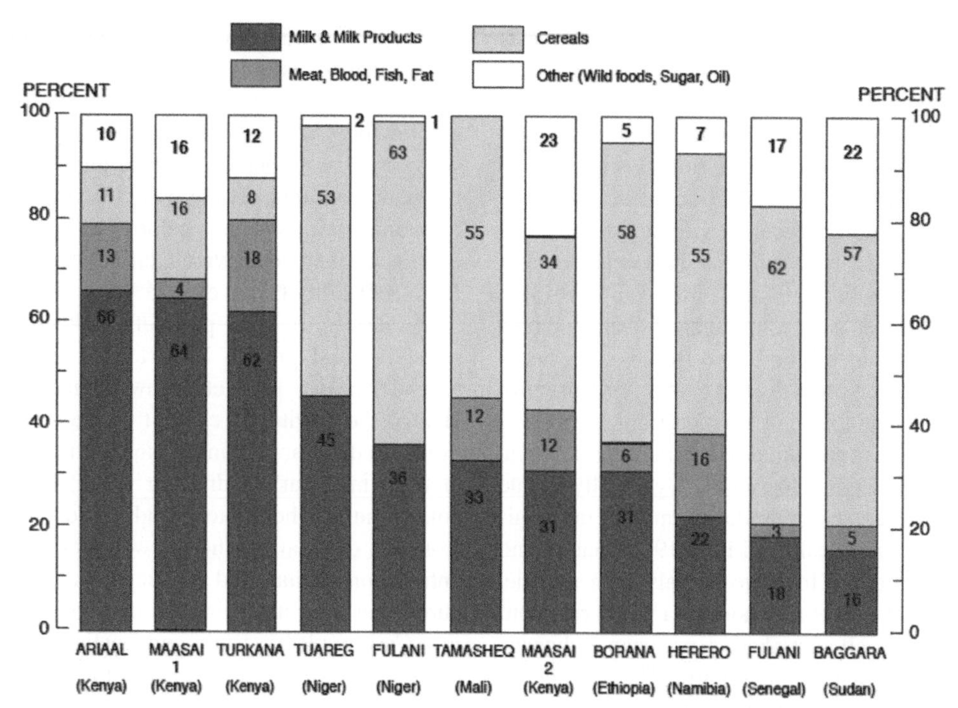

FIGURE 2 (African populations). Intakes of milk and other food categories in 11 pastoral populations from Africa. (See color plate section). *Modified from Galvin (1992). Data are drawn from Ariaal (Fratkin, 1991), Baggara (Holter, 1988), Borana (Galvin et al., 1994), Herero (O'Keefe et al., 1988), Maasai 1 and 2 (Nestel, 1985, 1986), Senegal Fulani (Bénéfice et al., 1984), Tamasheq (Wagenaar-Brouwer, 1985), Tuareg and Niger Fulani (Bernus, 1988), and Turkana (Galvin, 1985).*

intake of milk is also higher in children and adolescents than in adults and higher in adult women than in adult men. Children are given priority over adults in the distribution of milk, and women have a higher milk intake than men because they have greater access to animals since they do most of the milking. Despite the fact that there is no typical pastoral dietary pattern, milk is a staple food and is the preferred food of most African pastoralists (Galvin, 1992). At an even broader level, Casimir (1991, pp. 227–234) reported on 57 pastoral populations from Africa and Asia and found a remarkable variety of diets. Milk consumption was highest in East Africa and Central Asia.

Pastoralists' dependence on milk as a staple food is a remarkable example of how food production, subsistence, and culture can drive evolutionary processes through *natural selection* and genetic change. The ability of adult humans to digest milk easily (that is, milk sugar or lactose) is a recent evolutionary event that almost certainly dates back to the earliest dairying practices in prehistory. Durham (1991, p. 282) characterized it, in the true Darwinian sense, as a "coevolutionary relationship" between dairying as a cultural tradition and adult lactose absorption as a biological process.

MILK AND THE EVOLUTIONARY BASIS FOR LACTOSE TOLERANCE

Lactose is the principal carbohydrate (sugar) in milk. It is an indigestible disaccharide that requires the enzyme lactase (more correctly, lactase-phlorizin hydrolase or LPH) to break it down into its digestible monosaccharides, glucose and galactose. The enzyme lactase is produced in the lining of the small intestine. Nursing infants have the ability to digest lactose in mother's milk, but in a majority of populations around the world, infants lose this ability as they move past weaning into early childhood (see the chapter by Wiley). This is true not only for most human populations, but also for all other mammalian species; hence, it is the mammalian norm. Humans become intolerant of milk sugar or suffer from what is called *lactose intolerance* or *hypolactasia*; in other words, they are no longer able to digest milk sugar because intestinal lactase production ceases. This condition, which is under genetic control, persists into adulthood, such that individuals who have hypolactasia and who consume milk suffer from acute intestinal distress, diarrhea, and poor absorption of food. There is strong evidence that, as with other mammals, this is the normal condition for humans; or rather, adult humans in our evolutionary past prior to the Neolithic Revolution, when our subsistence was principally hunting and gathering. Today, this condition of adult hypolactasia or lactose intolerance has continued in humans for whom milk is not a part of the daily diet. For others, notably Northern and Central European and some African populations, there is an adult genetic ability to produce the lactase enzyme and to digest lactose: this is known as *lactase persistence*.

Simoons (1970) and McCracken (1971) first identified this genetically-based variation in the ability of adults to digest milk and proposed a hypothesis to explain the basis for the variation. The hypothesis, which has been well supported

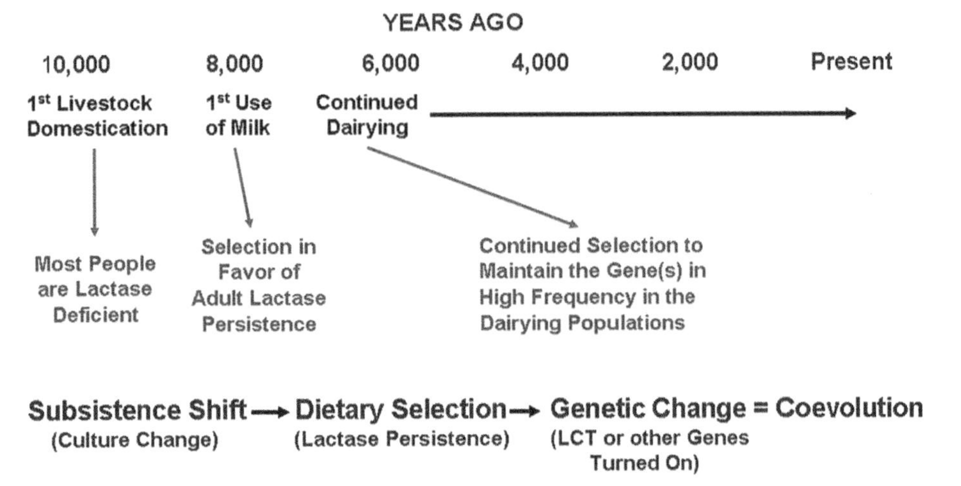

FIGURE 3 A diagram illustrating the proposed patterns of cultural and biological change associated with adult lactase persistence.

over the past four decades, is that the earliest dairy milk production during the Neolithic Era served to select for a rare gene that enabled adults to produce the enzyme lactase (lactase persistence) to digest lactose (Page and Bayless, 1981; Scrimshaw and Murray, 1988; Durham, 1991; Wiley, 2008). This selective process probably was based on the value of animal milk as a highly nutritious food that contributed to the dietary needs of people who were harvesting that product from their livestock. The frequencies of lactase persistence are highest in Northern and Central European populations and in sub-Saharan African pastoralist populations (Little, 1989; Durham, 1991, pp. 234–235). West African agriculturalist Bantu and Bantu-derived populations (including African-Americans in the United States) have low frequencies of lactase persistence (Little, 1989; Scrimshaw and Murray, 1988). Other adult populations, including Aboriginal Australians, hunter-gatherers from around the world, Asians, Pacific Islanders, and Native Americans, lack the ability to digest lactose and are intolerant of milk in varying degrees.

The genetics of lactase persistence is complex. Lactase production is under the control of the dominant lactase gene (LCT) that is "downregulated" (turned off) sometime during childhood. This prevents the synthesis of lactase and leads to lactase nonpersistence (the physiological state) and lactose intolerance (the digestive state). Current evidence suggests that lactase persistence in adults results from an "enhancer effect" of other genes on the same chromosome that facilitates lactase production in the intestine (Järvelä, 2005; Ingram et al., 2009). Tishkoff et al. (2007) found other genes—SNPs—that were correlated with lactase persistence in East African adults from Tanzania, Kenya, and the Sudan. They suggested that an independent selection and convergent evolution occurred among the North European, Central European, and East African pastoral populations in which lactase persistence is most prevalent.

The predominant hypothesis for the establishment and maintenance of adult lactase persistence is that it is a genetic adaptation to drinking milk from domestic animals, and that milk contributes essentially to food resources. There are several other hypotheses that suggest different selective pressures contributing to the worldwide genetic variation seen in human populations. Holden and Mace (2002) tested three hypotheses, including the predominant one. The second hypothesis they tested is based on the uptake of calcium being promoted by lactose, which would compensate for lowered vitamin D production (also important in calcium uptake) in northern temperate climates (with limited solar radiation to stimulate vitamin D synthesis), particularly Europe (Flatz and Rotthauwe, 1973). The third hypothesis centers on pastoralism in semiarid environments and proposes that milk drinking provides both fluid (hydration) and electrolytes in dry environments where much herding occurs. Holden and Mace (2002) used a variety of tests (correlation, phylogeny relationships, and estimates of gene flow) to determine the validity of the three hypotheses, and found that the "food resource" hypothesis was supported while the other two were not. There are good reasons for this, in that milk is a staple food in some living East African pastoralists, and some members of these populations would simply not survive were they unable to drink milk (Little et al., 2001). The "hydration" hypothesis might even be rejected on practical grounds, since watering livestock is a time-consuming and labor-intensive activity and it is more efficient to just drink water than to process it through other animals. In a later study by Gerbault et al. (2009), there was some support for the "lactose promotion of calcium uptake" hypothesis, because they found a southwest-to-northeast Europe gradient in lactase persistence. An illustration of the relationships for the generally accepted lactase persistence hypothesis is shown in Figure 3.

HEALTH, DISEASE, AND PASTORALISM

Some of the health benefits of pastoralism include a stable and accessible food supply, a food supply that is high in protein and has a broad spectrum of other nutrients, a life that involves considerable physical activity, and with nomadic and transhumant subsistence, a pattern of household movement where human and other wastes are dispersed over a wide area. This last practice reduces pollution and the spread of infectious disease (Western and Dunne, 1979). With small dispersed populations where livestock are owned and managed at the household level, epidemic diseases are less likely to become established than they are in larger-scale nucleated agricultural societies. A serious problem faced by early domestic livestock during the Neolithic that persists to the present is livestock disease. A well-supported assumption is that wild ruminants have adapted to animal diseases and have some degree of resistance to animal diseases found in grasslands. With domestication, there is often a reduction in resistance to animal diseases that has occurred in part from selective breeding for traits other than disease resistance. Livestock diseases are always a threat to pastoralists' survival.

An example can be given for contemporary East African nomadic Turkana pastoralists who keep camels, cattle, sheep, goats, and donkeys. They have limited access to veterinary services because of their nomadic movement and relative isolation from population centers. Also, animals often experience high mortality when various diseases strike domestic herds. Three diseases that have high mortality (when untreated) are viral ovine rinderpest (66% mortality), caprine pleuropneumonia (62% mortality), and mange (28%–32% mortality) (Bett et al., 2009). In 1979–1981 there was a disastrous drought in the Turkana District, Kenya, followed by particularly heavy chilling monsoon rains that led to high livestock losses due to starvation, poor nutritional health, and infection (both caprine and bovine pleuropneumonia) resulting from poor health (Little et al., 1999b). Livestock mortality from this one period of drought and rains was extraordinary, with some Turkana losing 90% of their small stock and 70% of their cattle and camels (McCabe, 2004, p. 217).

One of the most devastating cattle diseases is rinderpest virus (RPV). Evans-Pritchard (1940, pp. 68–69), in his classic ethnography on the Sudanese Nuer, noted the disastrous effects of RPV on these pastoralists and their cattle when it was first introduced during the late 1890s by Arab raiders. The effects of RPV and bovine pleuropneumonia on the Nuer were felt in human population loss from poor nutrition and disease, increased armed conflict with neighboring Dinka, declines in bride wealth, and changes in kinship patterns. Although it is not valid to use contemporary populations as accurate models of prehistoric populations, it is quite clear that livestock disease can wreak havoc on the structure, subsistence, and very survival of pastoral societies, and that this has certainly happened in the past as well as in the present.

Among pastoralists, there are also the problems of "zoonotic diseases" (those diseases transmitted to humans from other vertebrates—wild and domestic animals) (see the chapter by Morand and Sattenspiel) and parasites, many of which are also zoonotic. Rinderpest is a zoonotic livestock virus that probably underwent a mutation to jump the species barrier to humans sometime in the past. For example, measles or rubella virus (MeV) is a member of the *Morbillivirus* genus and is closely related to RPV. Furuse et al. (2010) have provided phylogenetic evidence that MeV diverged from RPV in about the eleventh or twelfth century CE. As Tomley and Shirley (2009) observed, zoonotic livestock disease has a double-barreled effect by acting on the livestock as a food supply and by transmission of the disease to humans. The World Health Organization (WHO) reported that "at least 61% of all human pathogens are zoonotic, and have represented 75% of all emerging pathogens during the past decade" (WHO, 2007). Many of these zoonotic pathogens are from vertebrates other than pastoral livestock—dogs, cats, pigs, fowl, wild birds, bats, and other wildlife. Among cattle there are a number of zoonotic diseases that have been transmitted to humans. Some are quite common, such as: cryptosporidiosis (protozoal parasite), *Escherichia coli* (bacterium), leptospirosis (bacterium), giardiasis (protozoal), ringworm (fungal), and salmonellosis (bacterium). Other rarer but very dangerous zoonotic diseases are anthrax (bacterium), brucellosis (bacterium), Q fever (bacterium), and tuberculosis (bacterium) (Pelzer and Currin, 2009). Several of these diseases are included by WHO in their list of "Seven Neglected Endemic Zoonoses," a majority of which are found in the Third World (Maudlin et al., 2009).

There is no evidence available for the effects of zoonotic diseases on prehistoric pastoral populations and limited evidence for even contemporary traditional pastoralists. It is generally agreed, however, that farmwork in Western societies is dangerous not only because of accidents, but also due to the risk of exposure to zoonotic diseases from farm animals (LeJeune and Kersting, 2010). This risk may be extrapolated back to the earliest pastoralists who lived in very close association with their animals, who had little knowledge of animal and human diseases, and who lived under less than optimal sanitary conditions. There are numerous examples of zoonotic diseases infecting children at petting farms, fairs, and zoos (Valkenburgh and Heuvelink, 2006), and veterinary students working with calves (Preiser et al., 2003). Zoonotic tuberculosis, usually transmitted from cows to humans by consumption of raw milk, was prevalent in Europe before pasteurization. It is again on the rise because of farm-family consumption of fresh milk and sales of unpasteurized "organic" milk (de la Rua-Domenech, 2006). Most African and Asian subsistence pastoralists consume raw milk that is neither boiled nor pasteurized. A study of urban dairy farming households in Nairobi, Kenya, found that women were particularly at high risk of contracting Cryptosporidiosis because they milked the animals, drank raw milk, and had more daily contact with animal dung than men or children (Kimani et al., 2012). Among the Kenyan and Tanzanian

Maasai herders, women milk cows and construct huts with cow dung plaster that they spread with their hands!

In the modern world, livestock numbers are increasing at about the same rate as human population numbers. At present, there are roughly 1.4 billion cattle (with nearly 50% in Asia and Africa), two billion small ruminants (1.1 billion sheep and 0.9 billion goats), and 14 million domestic camels (with dromedaries about 90%) (Tomley and Shirley, 2009). Hence, the major species of livestock populations constitute a number equal to more than 50% of the human population number, and when pigs are added, the number is well over 60%. Household livestock holdings by African and Asian pastoralists have little disease control and are likely to be foci for newly emerging zoonotic diseases. Also, because of the proximity of wildlife to domestic stock, the likelihood of zoonotic transfer from wildlife to livestock to humans is likely to be increased as pressure on land resources intensifies.

Another source of zoonotic disease is caused by prions that lead to a number of related transmissible spongiform encephalopathies (TSEs) (Belay, 1999). There are a variety of forms of these nervous degenerative diseases that are always lethal and appear to be most common in domestic or near-domestic species. The earliest recorded disease is *scrapie* in sheep and goats, which has been known for centuries. It is believed that *bovine spongiform encephalopathy* (BSE, or mad cow disease), which reached epidemic proportions in the United Kingdom in the late 1980s, was transmitted to cattle through their feed, which contained ground carcasses and bone from infected sheep. BSE was then transmitted from cows directly to humans and crossed the so-called species barrier. It was estimated that more than a million cattle were infected and millions of human residents were exposed, hence setting up natural conditions for prions to switch hosts from cattle to humans. *Creutzfeldt–Jakob* disease (CJD) is the human counterpart of the animal encephalopathies, and the form transmitted from cattle to humans is a variant form of CJD. Another prion encephalopathy, *chronic wasting disease*, was discovered in a captive mule deer in Colorado in 1960 (Belay and Schonberger, 2005). It is now present in wild cervids (elk, deer, and moose) in several US states, and although it has not yet crossed the cervid–human species barrier, it has the potential to do so (Garruto et al., 2008). There are many frightening aspects of these TSEs, including a slow incubation period (often years), lack of immune response, lack of a cure, and uniform fatality. As zoonotic diseases, they are among the most dangerous in their effects on both livestock and their human managers.

COEVOLUTION OF LIVESTOCK AND THEIR HUMAN HOSTS

Gerbault et al. (2011) identified the practice of livestock keeping and milk production as an example of "human niche construction." This concept of "human niche construction" might be broadened to include all of pastoralism, agropastoralism, and in fact any organized framework of human subsistence (or culture). The pastoral niche, established during the Neolithic Era, was a new and ingenious way to harness a reliable food supply, but it involved risks and hard work in dealing with wild animals at its initial stages of organization. It also led to unique selective pressures on and evolutionary changes in both the livestock and their human hosts. This coevolutionary process involved the artificial selection of livestock traits in behavior and biology, and the Darwinian selection of human traits that allowed, for example, the digestion of lactose. The coevolutionary process also contributed to the human biobehavioral evolution of tolerance to a variety of animal foods and the development of cuisine, and the fine-tuning of the human immune system to zoonotic diseases. This coevolutionary process of reciprocal species' (human and livestock) selection and adaptation is ongoing. With a current global human population of about seven billion and a livestock population equal to half this human number, these ongoing coevolutionary processes will continue into the future.

REFERENCES

Aikio, P., 1989. The changing role of the reindeer in the life of the Sámi. In: Clutton-Brock, J. (Ed.), The Walking Larder: Patterns of Domestication, Pastoralism, and Predation. Unwin Hyman, London, pp. 169–184.

Belay, E.D., 1999. Transmissible spongiform encephalopathies in humans. Annual Review of Microbiology 53, 283–314.

Belay, E.D., Schonberger, L.B., 2005. The public health impact of prion diseases. Annual Review of Public Health 26, 191–212.

Bénéfice, E., Chevassus-Agnes, S., Barral, H., 1984. Nutritional situation and seasonal variations for pastoralist populations of the Sahel (Senegalese Ferlo). Ecology of Food and Nutrition 14, 229–247.

Bernus, E., 1988. Seasonality, climatic fluctuations, and food supplies. In: de Garine, I., Ainsworth Harrison, G. (Eds.), Coping with Uncertainty in Food Supply. Oxford University Press, Oxford, pp. 318–336.

Bett, B., Jost, C., Allport, R., et al., 2009. Using participatory epidemiological techniques to estimate the relative incidence and impact on livelihoods of livestock diseases amongst nomadic pastoralists in Turkana South District, Kenya. Preventative Veterinary Medicine 90, 194–203.

Casimir, M.J., 1991. Flocks and Food: A Biocultural Approach to the Study of Pastoral Foodways. Böhlau Verlag, Cologne.

Clutton-Brock, J., 1992. Horse Power: A History of the Horse and the Donkey in Human Societies. Harvard University Press, Cambridge.

Clutton-Brock, J., 1995. Origins of the dog: domestication and early history. In: Serpell, J. (Ed.), The Domestic Dog: Its Evolution, Behavior and Interactions with People. Cambridge University Press, Cambridge, pp. 7–20.

Clutton-Brock, J., 1999. A Natural History of Domesticated Mammals, second ed. Cambridge University Press, Cambridge.

Darwin, C.R., 1859. On the Origin of Species by Means of Natural Selection or the Preservation of Favoured Races in the Struggle for Life. John Murray, London.

Dunne, J., Evershed, R.P., Salque, M., et al., 2012. First dairying in green Saharan Africa in the fifth millennium BC. Nature 486, 390–394.

Durham, W.H., 1991. Coevolution: Genes, Culture and Human Diversity. Stanford University Press, Stanford, California.

Dyson-Hudson, R., Dyson-Hudson, N., 1980. Nomadic pastoralism. Annual Review of Anthropology 9, 15–61.

Evans-Pritchard, E.E., 1940. The Nuer: A Description of the Modes of Livelihood and Political Institutions of a Nilotic People. Oxford University Press, Oxford.

Evershed, R.P., Payne, S., Sherratt, A.G., et al., 2008. Earliest date for milk use in the Near East and southeastern Europe linked to cattle herding. Nature 455, 528–531.

Flatz, G., Rotthauwe, H.W., 1973. Lactose nutrition and natural selection. Lancet II, 76–77.

Fratkin, E., 1991. Surviving Drought and Development: Ariaal Pastoralists of Northern Kenya. Westview Press, Boulder.

Furuse, Y., Suzuki, A., Oshitani, H., 2010. Origin of measles virus: divergence from rinderpest virus between the 11th and 12th centuries. Virology Journal 7, 52–55.

Galvin, K., 1985. Food Procurement, Diet, Activities and Nutrition of Ngisonyoka, Turkana Pastoralists in an Ecological and Social Context. Ph.D. dissertation in anthropology. State University of New York, Binghamton.

Galvin, K.A., 1992. Nutritional ecology of pastoralists in dry tropical Africa. American Journal of Human Biology 4, 209–221.

Galvin, K.A., Coppock, D.L., Leslie, P.W., 1994. Diet, nutrition, and the pastoral strategy. In: Fratkin, E., Galvin, K.A., Roth, E.A. (Eds.), African Pastoralist Systems: An Integrated Approach. Rienner, Boulder, pp. 113–131.

Galvin, K.A., Little, M.A., 1999. Dietary intake and nutritional status. In: Little, M.A., Leslie, P.W. (Eds.), Turkana Herders of the Dry Savanna: Ecology and Biobehavioral Response of Nomads to an Uncertain Environment. Oxford University Press, Oxford, pp. 124–145.

Garruto, R.M., Reiber, C., Alfonso, M.P., et al., 2008. Risk behaviors in a rural community with a known point-source exposure to chronic wasting disease. Environmental Health 7, 31–38.

Gerbault, P., Liebert, A., Itan, et al., 2011. Evolution of lactase persistence: an example of human niche construction. Philosophical Transactions of the Royal Society B: Biological Sciences (London) 366, 863–877.

Gerbault, P., Moret, C., Currat, M., Sanchez-Mazas, A., 2009. Impact of selection and demography on the diffusion of lactase persistence. PLoS One 4 (7), e6369.

Hiendleder, S., Lewalski, H., Janke, A., 2008. Complete mitochondrial genomes of *Bos taurus* and *Bos indicus* provide new insights into intra-species variation, taxonomy and domestication. Cytogenetic and Genome Research 120, 150–156.

Holden, C., Mace, R., 2002. Pastoralism and the evolution of lactase persistence. In: Leonard, W.R., Crawford, M.H. (Eds.), Human Biology of Pastoral Populations. Cambridge University Press, Cambridge, pp. 280–307.

Holter, U., 1988. Food consumption of camel nomads in the North West Sudan. Ecology of Food and Nutrition 21, 95–115.

vonHoldt, B.M., Pollinger, J.P., Lohmueller, K.E., et al., 2010. Genome-wide SNP and haplotype analyses reveal a rich history underlying dog domestication. Nature 464, 898–902.

Ingram, C.J., Mulcare, C.A., Itan, Y., et al., 2009. Lactose digestion and the evolutionary genetics of lactase persistence. Human Genetics 124, 579–591.

Järvelä, I.E., 2005. Molecular genetics of adult-type hypolactasia. Annals of Medicine 37, 179–185.

Kimani, V.N., Mitoko, G., McDermott, B., et al., 2012. Social and gender determinants of risk of cryptosporidiosis, an emerging zoonosis, in Dagoretti, Nairobi, Kenya. Tropical Animal Health and Production 44 (Suppl. 1), S17–S23.

Kumar, S., Nagarajan, M., Sandhu, J.S., et al., 2007. Phylogeography and domestication of Indian River buffalo. BMC Evolutionary Biology 7, 186.

Larson, G., Fuller, D.Q., 2014. The evolution of animal domestication. Annual Review of Ecology, Evolution, and Systematics 45, 115–136.

LeJeune, J., Kersting, A., 2010. Zoonoses: an occupational hazard for livestock workers and a public health concern for rural communities. Journal of Agricultural Safety and Health 16, 161–179.

Leslie, P.W., Little, M.A., Dyson-Hudson, R., et al., 1999. Synthesis and lessons. In: Little, M.A., Leslie, P.W. (Eds.), Turkana Herders of the Dry Savanna: Ecology and Biobehavioral Response of Nomads to an Uncertain Environment. Oxford University Press, Oxford, pp. 355–373.

Little, M.A., 1989. Human biology of African pastoralists. Yearbook of Physical Anthropology 32, 215–247.

Little, M.A., Dyson-Hudson, R., Dyson-Hudson, N., et al., 1999a. Environmental variations in the South Turkana ecosystem. In: Little, M.A., Leslie, P.W. (Eds.), Turkana Herders of the Dry Savanna: Ecology and Biobehavioral Response of Nomads to an Uncertain Environment. Oxford University Press, Oxford, pp. 318–330.

Little, M.A., Dyson-Hudson, R., McCabe, J.T., 1999b. Ecology of South Turkana. In: Little, M.A., Leslie, P.W. (Eds.), Turkana Herders of the Dry Savanna: Ecology and Biobehavioral Response of Nomads to an Uncertain Environment. Oxford University Press, Oxford, pp. 45–65.

Little, M.A., Gray, S.J., Campbell, B.C., 2001. Milk consumption in African pastoral peoples. In: de Garine, I., de Garine, V. (Eds.), Drinking: Anthropological Approaches. Berghahn Book, New York, pp. 66–86.

Little, M.A., Leslie, P.W. (Eds.), 1999. Turkana Herders of the Dry Savanna: Ecology and Biobehavioral Response of Nomads to an Uncertain Environment. Oxford University Press, Oxford.

Loftus, R.T., MacHugh, D.E., Bradley, D.G., et al., 1994. Evidence for two independent domestications of cattle. Proceedings of the National Academy of Sciences of the United States of America 91, 2757–2761.

Maudlin, I., Eisler, M.C., Welburn, S.C., 2009. Neglected and endemic zoonoses. Philosophical Transactions of the Royal Society B: Biological Sciences (London) 364, 2777–2787.

McCabe, J.T., 2004. Cattle Bring Us to Our Enemies: Turkana Ecology, Politics, and Raiding in a Disequilibrium System. University of Michigan Press, Ann Arbor, MI.

McCracken, R.D., 1971. Lactase deficiency: an example of dietary evolution. Current Anthropology 12, 479.

McTavish, E.J., Decker, J.E., Schnabel, R.D., et al., 2013. New World cattle show ancestry from multiple independent domestication events. Proceedings of the National Academy of Sciences of the United States of America 110, E1398–E1406.

Mengoni Goñalons, G.L., Yacobaccio, H.D., 2006. The domestication of South American camelids: a view from the south-central Andes. In: Zeder, M.A., Bradley, D.G., Emshwiller, E., Smith, B.D. (Eds.), Documenting Domestication: New Genetic and Archaeological Paradigms. University of California Press, Berkeley, pp. 228–244.

Nestel, P.S., 1985. Nutrition of Maasai Women and Children in Relation to Subsistence Food Production. Ph.D. dissertation in nutrition. University of London, London.

Nestel, P.S., 1986. A society in transition: developmental and seasonal influences on the nutrition of Maasai women and children. Food and Nutrition Bulletin 8, 2–18.

Novoa, C., Wheeler, J.C., 1984. Llama and alpaca. In: Mason, I. (Ed.), Evolution of Domesticated Animals. Longman, London, pp. 116–128.

O'Keefe, S.J.D., Rund, J.E., Marot, N.R., et al., 1988. Nutritional status, dietary intake and disease patterns in rural Hereros, Kavangos and Bushmen in South West Africa/Namibia. South African Medical Journal 73, 643–648.

Olsen, S.J., 1990. Fossil ancestry of the yak. Its cultural significance and domestication in Tibet. Proceedings of the Philadelphia Academy of Natural Sciences 142, 73–100.

Olsen, S.J., 1993. Evidence of early domestication of the water buffalo in China. In: Clason, A., Payne, S., Uerpmann, H.-P. (Eds.), Skeletons in Her Cupboard: Festschrift for Juliet Clutton-Brock. Oxbow Monograph 34, Oxford, pp. 151–156.

Olsen, S.L., 2006. Early horse domestication on the Eurasian steppe. In: Zeder, M.A., Bradley, D.G., Emshwiller, E., Smith, B.D. (Eds.), Documenting Domestication: New Genetic and Archaeological Paradigms. University of California Press, Berkeley, pp. 245–269.

Page, D.M., Bayless, T.M. (Eds.), 1981. Lactose Digestion: Clinical and Nutritional Implications. Johns Hopkins University Press, Baltimore.

Pelzer, K.D., Currin, N., 2009. Zoonotic Diseases of Cattle. Publication 400–460, Virginia Cooperative Extension. Virginia Polytechnic Institute, Blacksburg, VA.

Pereira, F., Queirós, S., Gusmão, L., et al., 2009. Tracing the history of goat pastoralism: new clues from mitochondrial and Y chromosome DNA in North Africa. Molecular Biology and Evolution 26, 2765–2773.

Preiser, G., Preiser, L., Madeo, L., 2003. An outbreak of cryptosporidiosis among veterinary science students who work with calves. Journal of American College Health 51, 213–215.

de la Rua-Domenech, R., 2006. Human *Mycobacterium bovis* infection in the United Kingdom: incidence, risks, control measures and review of the zoonotic aspects of bovine tuberculosis. Tuberculosis (Edinburgh) 86, 77–109.

Scrimshaw, N.S., Murray, E.B., 1988. The acceptability of milk and milk products in populations with a high prevalence of lactose intolerance. American Journal of Clinical Nutrition 48 (Suppl. 4), 1079–1159.

Sherratt, A.G., 1981. Plough and pastoralism: aspects of the secondary products revolution. In: Hodder, I., Isaac, G., Hammond, N. (Eds.), Pattern of the Past: Studies in Honor of David Clark. Cambridge University Press, Cambridge, pp. 261–305.

Simoons, F.J., 1970. Primary adult lactose intolerance and the milk-drinking habit. A problem in biological and cultural interrelations. II. A culture historical hypothesis. American Journal of Digestive Diseases 15, 695–710.

Tishkoff, S.A., Reed, F.A., Ranciaro, A., et al., 2007. Convergent adaptation of human lactase persistence in Africa and Europe. Nature Genetics 39, 31–40.

Tomley, F.M., Shirley, M.W., 2009. Livestock infectious diseases and zoonoses. Philosophical Transactions of the Royal Society B Biological Sciences (London) 364, 2637–2642.

Trut, L.N., 1999. Early canid domestication: the farm-fox experiment. American Scientist 87, 160–169.

Valkenburgh, S.M., Heuvelink, A.E., 2006. Children's farms in the Netherlands: hygiene and zoonotic pathogens (in Dutch). Tijdschrift voor Diergeneeskd 131, 224–227.

Wagenaar-Brouwer, M., 1985. Preliminary findings on the diet and nutritional status of some Tamasheq and Fulani sgroups in the Niger Delta of Central Mali. In: Hill, A.G. (Ed.), Population, Health and Nutrition in the Sahel: Issues in the Welfare of Selected West African Communities. Kegan Paul International, London, pp. 226–252.

Warmuth, V., Eriksson, A., Bower, M.A., et al., 2012. Reconstructing the origin and spread of horse domestication in the Eurasian steppe. Proceedings of the National Academy of Sciences of the United States of America 109, 8202–8206.

Western, D., Dunne, T., 1979. Environmental aspects of settlement site decisions among pastoral Maasai. Human Ecology 7, 75–98.

WHO, 2007. Zoonoses and Veterinary Public Health: The Control of Neglected Zoonotic Diseases. WHO http://www.who.int/zoonoses/control_neglected_zoonoses/en/index.html.

Wiley, A.S., 2008. Cow's milk consumption and health: an evolutionary Perspective. In: Trevathan, W.R., Smith, E.O., McKenna, J.J. (Eds.), Evolutionary Medicine and Health: New Perspectives. Oxford University Press, New York, pp. 116–133.

Zeder, M.A., 2008. Domestication and early agriculture in the Mediterranean Basin: origins, diffusion, and impact. Proceedings of the National Academy of Sciences of the United States of America 105, 11597–11604.

Zeder, M.A., 2012. The domestication of animals. Journal of Anthropological Research 68, 161–190.

Zeder, M.A., Bradley, D.G., Emshwiller, E., et al., 2006. Documenting domestication; bringing together plants, animals, archaeology, and genetics. In: Zeder, M.A., Bradley, D.G., Emshwiller, E., Smith, B.D. (Eds.), Documenting Domestication: New Genetic and Archaeological Paradigms. University of California Press, Berkeley, pp. 1–12.

Chapter 25

Agriculturalism

Mark A. Blumler
Department of Geography, State University of New York, Binghamton, NY, USA

SYNOPSIS

Farming began in several regions of the globe at approximately the same time, suggesting an environmental rather than a sociocultural cause. The spread of agriculture was an exercise in niche construction, essentially transferring (modified) whole ecosystems from the source regions elsewhere. Crop domestication entailed significant morphological, but relatively little genetic, change, while humans took over the competitive, protective, and dispersal functions that wild plants must handle themselves. Dietary changes due to agriculture presumably have been great, and these changes produced an evolutionary mismatch, to which we have responded genetically, microbially, and/or culturally.

INTRODUCTION

In the original, Hebrew version of the Biblical story of Eden, it is not an apple that causes Adam and Eve's downfall, but an unspecified fruit. According to the Talmudic commentaries, that fruit was wheat, an interpretation echoed in the Koran and in the Black Book of the Yezidis. Archaeobotanical investigations suggest that emmer wheat was the plant around which early Near Eastern agriculture centered (Blumler, 1998b; Nesbitt, 2002), and when one considers that Cain and Abel were respectively a herder and a farmer, the Eden story becomes a parable about the transition from hunter-gathering to agro-pastoralism. This view of agriculture as a fall from hunter-gatherer grace has many adherents today, notably among environmentalists who worry that we are overpopulating the planet, and also among those who promote the "paleo diet" to which they argue we are adapted (Cordain, 2001). Advocates of the paleo diet argue that we suffer from evolutionary mismatch that we are poorly adapted to our current diet and lifestyle because our genes still reflect adaptation as hunter-gatherers (see the chapters by Wiley and Low et al. in this volume). In contrast, Zuk (2013) argues that reconstructions of preagricultural diets and lifestyles are "paleofantasies" given the fragmentary nature of the fossil evidence and the diversity of modern hunter-gatherer diets, and because we are not genetically identical to our ancestors: we have evolved, in some respects perhaps dramatically, in response to agriculture and the changes it wrought. This chapter covers the origin and spread of agriculture, beliefs about how it came into being, the evolution of crop plants, coevolutionary interactions, and the impacts on humans, including the degree to which we may be mismatched today.

Agriculture is the cultivation of domesticated plants for food. Cultivation can vary from a huge agribusiness field to a pot on a patio. Today, farming is practiced almost everywhere that it can be, although it is disappearing from lands that are urbanizing. At the same time, the definition of "arable" is continuing to expand, as humans find ways to farm lands formerly thought unsuitable. For example, the cerrado of Mato Grasso, Brazil, was largely restricted to cattle ranching until recently because of its infertile soils; now much of it is in soy after massive fertilization. Although farming was always primarily about food, some crops from very early times were put to other uses such as clothes/textiles (flax, cotton) and building construction (grain straw for roofs, and mixed with clay for adobe). Today, crops such as corn and wheat have many industrial uses, besides serving as primary staples (Head et al., 2012).

Without agriculture, we never would have seen so much that followed: cities, civilization, the Industrial Revolution, and perhaps a generalized increase in social complexity and hierarchy (though some hunter-gatherer societies with abundant resources did achieve considerable complexity, including, as in the case of Pacific Northwest groups, widespread slavery). Much of the paleo diet discourse presumes that all of this is unnatural. But we are not the only farmers: several species of ants, termites, and beetles cultivate fungi (Farrell et al., 2001; Schultz and Brady, 2008). The fungi are not known to occur outside cultivation, and thus, qualify as domesticated species (Diamond, 1998).

If fully domesticated, a species will be so altered from its wild form that it no longer can survive without human assistance (Blumler and Byrne, 1991). A classic example is maize: after maturity, its seeds remain attached to the cob, which remains attached to the stalk. Unless harvested and sown, the seeds cannot reach the ground. If the stalk falls over some seeds may contact the soil, but even if they germinate they will do so in a mass, competing with each other. In contrast, the wild ancestor, teosinte, produces seeds that detach individually from the plant when ripe. Our agricultural plants and animals vary in their degree of domestication, and not all are completely helpless without our assistance. Pigs, for instance, are notorious for their ability to survive in the wild. Some herbs, such as spearmint, are little altered from the wild condition, and probably could thrive without humans. Even cabbage frequently establishes along coastal cliffs, reflecting its derivation from a wild Mediterranean sea-cliff species. Several feral "cabbage" populations are scattered along the Pacific Coast of California. In an unpublished investigation, Berkeley geneticist Herbert Baker (personal communication, 1994) cultivated seeds from these plants and found that one population was feral broccoli, another was kale, still another green cabbage, and so on (despite their striking morphological differences, all are the same species, *Brassica oleracea*). In establishing in the wild, they had converged on a common morphology, little distinct from the wild phenotype in its native habitat. This example also illustrates that crop evolution, while giving rise to striking morphological changes, may do so through alterations in relatively few genes (cf. Paterson, 2002).

From an evolutionary perspective, crop domestication is fascinating (Hancock, 2005; Purugganan and Fuller, 2011). Darwin (1859, 1883) drew on what was then known about crop variation to support his theory of evolution. More recently, crop geneticists have played a major role in the study of introgressive hybridization, and the recognition of its importance in plant evolution (Hancock, 2005). Increasingly, domestication is being studied to shed light on fundamental theoretical questions about evolution (Hancock, 2005; Purugganan and Fuller, 2009). Crop evolution has given us examples of recent speciation, *contra* the beliefs of Creationists: for instance, bread wheat speciated after agriculture began (Dvorak et al., 1998). If one considers the major evolutionary trends of the past 10,000 years (Table 1), agriculture figures prominently. In addition to fostering the evolution of domesticates and "pests," it is largely responsible for massive

TABLE 1 Major Evolutionary Trends of the Past 10,000 years
Extinction of megafauna (Martin and Wright, 1967)
Extinctions and contractions in range due to human impacts
Evolution of domesticates, including ornamentals
Evolution of weeds and other "pests"
Invasions, many if not most of them due to the spread of agriculture, leading to biotic homogenization

invasions of alien species, which in turn are a major cause of extinction of native species (Blumler, 2011).

AGRICULTURAL ORIGINS

In present evidence, humans began farming in several regions of the globe at approximately the same time (Table 2 and Figure 1). The consensus origin date for each region is provided to the extent that it can be concluded that there is a consensus. The timing is established best for the Fertile Crescent, albeit with some variance in opinion (Nesbitt, 2002; Ozdogan, 2002; Zeder, 2011; Zeder et al., 2006; Riehl et al., 2013). For Mexico, Smith's (1997) report of squash (*Cucurbita pepo*) domestication is supported by evidence for maize domestication within a millennium or so thereafter (Matsuoka et al., 2004; Piperno et al., 2009). Similarly, for China, genetic and archaeological studies support early domestication of rice (MacNeish and Libby, 1995; Huang et al., 2012), and foxtail and broomcorn millets (Crawford, 2009; Lu et al., 2009; Yang et al., 2012). Many archaeologists regard North (millet) and Central (rice) China as separate centers of agricultural origin, but given their proximity, they are combined here. The consensus date for New Guinea (Denham and Haberle, 2008) rests on admittedly limited archaeological evidence. Finally, the Andean region is clearly ancient, but the date given here reflects the interpretation of the authors of a single study (Dillehay et al., 2007), who found domesticated plants in western Peru that must have originated earlier on the east side of the Andes. As Blumler (1992b) points out, the wild progenitors of many Andean crops are located on the Bolivian side of the range, while the archaeology concentrated on the western (Peruvian) side. Consequently, the picture is somewhat confused (see also Piperno, 2011). The comparative timing of origins matters in that it bears on highly contentious debates over the causes of agricultural origins, and diffusion versus independent invention (Blumler, 1992a, 1996, 2002), as discussed below.

The wild progenitors of the earliest crops, for the most part, were large-seeded annuals, or plants with starchy tubers (Blumler, 1992b). They were mesophytes: fast-growing plants adapted to fertile soil conditions (Blumler, 1994). Subsequent domestications, often of edible weeds growing in the agricultural fields, also tended to be of mesophytic species. For instance, many of our vegetables are from the mustard and spinach families. Many members of these families have lost the ability to form associations with mycorrhizal fungi (Wang and Qiu, 2006). In contrast, most plants characteristically have a coevolutionary relationship with mycorrhizae, in which the plant sends sugar to the fungus, and the fungus decomposes organic matter and passes mineral nutrients, particularly phosphorus, to the plant. This is beneficial especially to plants on poor soil. The absence of mycorrhizal associates for many mustards and spinaches suggests that they "expect" to grow in soils with excellent nutrient availability (Trappe, 1987). They presumably can grow faster than other species under such conditions, since no sugar must be diverted to a fungal partner.

COMPETING HYPOTHESES

Rare events, such as the initiation of farming, can be modeled as having a Poisson distribution. As such, the likelihood of several independent origins occurring synchronously in widely separated regions is highly unlikely if the primary cause is endogenous, that is, the result of sociocultural developments. Synchronous origins would be more plausibly triggered by some global environmental change. Similarly, synchronicity of widely separated regions tends to rule out diffusion from one to the other. Given how ancient the Mexican center is, it is unlikely that its origin was influenced by Old World farmers (but see Erickson et al., 2005). Two regions, the Mississippi Valley and the Sahel, may have independently undertaken farming, but given their proximity to much earlier agricultural centers (Mexico and the Fertile Crescent, respectively), they also may have adopted agriculture as a consequence of diffusion.

The literature on the transition from hunter-gathering to agriculture is necessarily speculative. Despite remarkable advances both in the cytogenetic study of crops and in the collection and interpretation of archaeobotanical information, the empirical evidence remains incomplete, and legitimately subject to varying interpretations. The best evidence comes from the Fertile Crescent (Zohary et al., 2012), though even there it is not nearly as clear as one would wish (e.g., Fuller et al., 2012; Abbo et al., 2013). From Mexico and China there is some good, though limited, evidence. The evidence from the other centers is very equivocal.

Some old ideas, such as that farming was invented as part of the march of "progress," are no longer accepted. The publication of *Man the Hunter* (Lee and DeVore, 1968), which demonstrated that the hunter-gathering !KungSan work few hours in comparison to members of modern societies, let alone traditional farmers, was highly influential. Consequently, Malthusian, population pressure hypotheses (e.g., Cohen, 1977)

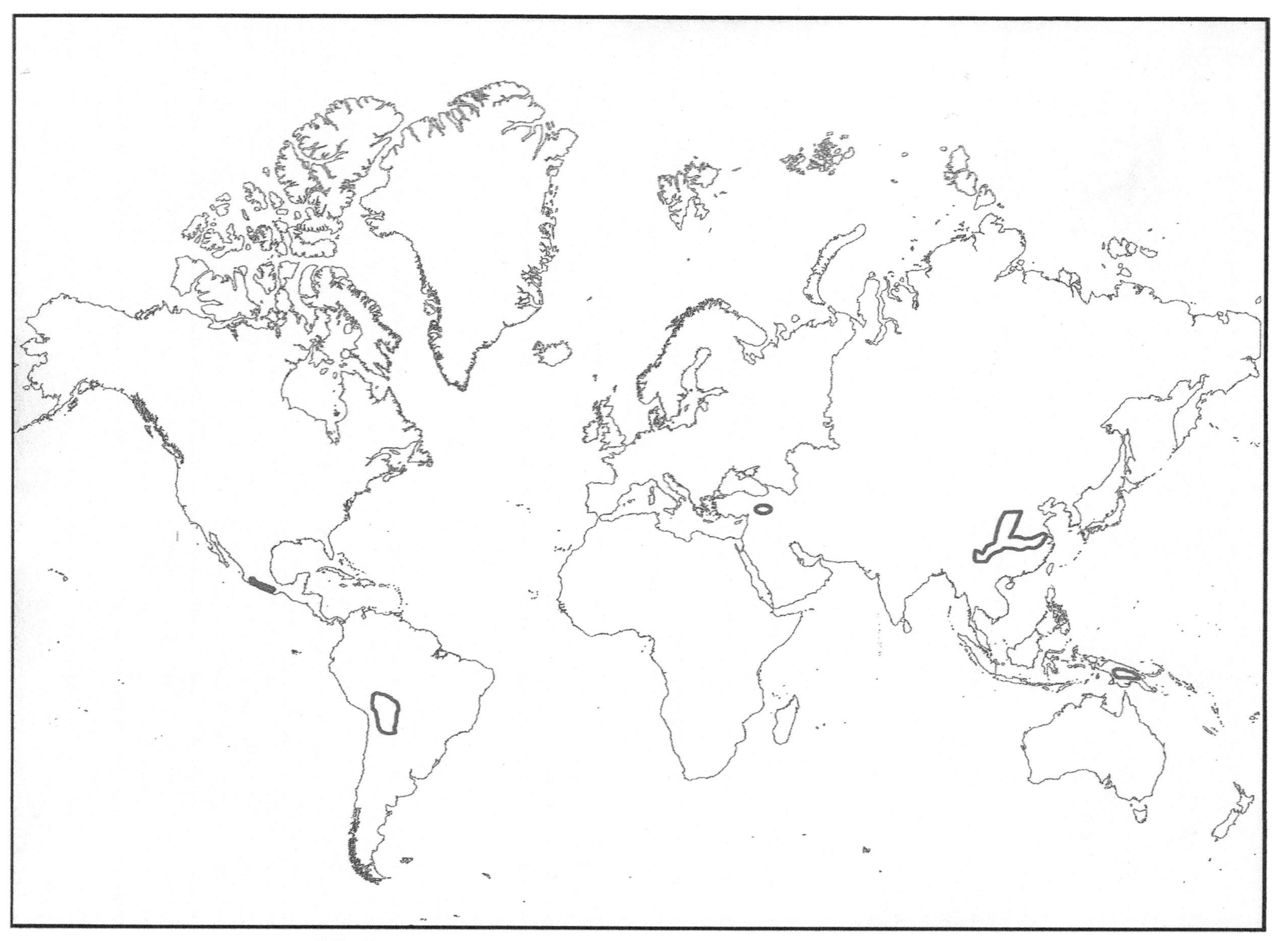

FIGURE 1 Centers of agricultural origin, based on current evidence. (See color plate section).

TABLE 2 Early Agriculture Centers, Consensus Date of Origin, and Their Major Early Crops

Fertile Crescent	11,000 BP	Caprines, emmer, einkorn, lentils, peas, chickpeas, flax
China	10,500 BP	Rice, millets
Mexico	10,000 BP	Squash, maize
New Guinea	10,000 BP	Taro, sugar, bananas
Central Andes	10,000 BP	Squash, quinoa
Sahel	7000 BP	Sorghum
Mississippi Valley	5000 BP	Sunflower, squash

became popular, and remain so today (Cohen, 2009; Lambert, 2009). Cohen and Armelagos (1984) gathered evidence that Neolithic farmers in the Near East suffered from malnutrition and that Neolithic women frequently developed joint problems as a result of the physical act of grinding grain for long periods of time. In contrast, Gage (2005) showed that there is little evidence this was other than a temporary effect related to the initial adoption of farming.

It is difficult to reconcile population pressure hypotheses with the evidence for very early origin in Mexico, which had only recently been peopled, and where population density clearly was very low. Kent Flannery, who dominated agricultural origin theorizing for a long time, and who was inclined toward scenarios involving human manipulation of environment with population pressure, was aware of this problem. He suggested that although Mexico was not densely populated when farming began, it had reached the point where people had "settled in" (Flannery, 1986, p. 11). Alternatively, perhaps the elimination of the native megafauna through overhunting (Martin and Wright, 1967; Flannery, 2001), more or less coincident with the Younger Dryas climatic episode (see below), created a depletion of food resources that necessitated experimentation with cultivation.

Numerous scholars have advocated some version of a gradual intensification or "continuum" hypothesis, in which proto-agricultural practices such as replanting of tubers grade into horticulture and eventually, full-fledged farming (e.g., Harris, 1977; Denham and Haberle, 2008; Zeder and Smith, 2009). These ideas derive in part from so-called "evolutionary" models in anthropology, which are not really in accordance with evolutionary theory, and are more properly termed developmental (Orlove, 1980; Blumler, 1996). Since they propose a gradual, incremental growth of preagricultural practices, they also do not align well with the current consensus that New World origin is almost as early as in the Fertile Crescent and China. Moreover, the early cereal domesticates in the Fertile Crescent and China seem poorly suited to a horticultural system (in contrast, the New Guinean crops, and squashes in Mexico and Peru/Bolivia, do seem appropriate). In contrast to the gradualist paradigm, Ozdogan (2002, p. 156) concluded: "The rate of change in the Neolithic of the Near East can only be compared with that of the industrial revolution..."

Another set of hypotheses involves environmental change, and has become more popular as the concern over global warming has increased appreciation of the magnitude of climate change in the past. Several scholars have argued that farming was not feasible during the glacial periods of the Pleistocene, when it was not only colder, but drier, and far more variable as well (Sherratt, 1997; Richerson et al., 2001; Bettinger et al., 2009). Sage (1995) pointed out that the low CO_2 levels in the atmosphere during the Pleistocene would have posed a problem for plants, which obtain their carbon from the air. Byrne (1987) argued that climates became more seasonal during the Pleistocene–Holocene transition, and that farming began in regions with seasonal drought. The first domesticates, annual plants and those with starchy tubers, are well-adapted to seasonal drought. Blumler (1992b, 1998a) and cf. Diamond (1997) showed that large-seeded annual grasses are strongly associated with regions of seasonal drought, and that such plants contributed disproportionately to early agriculture. The Fertile Crescent has the most extreme seasonality of rainfall in the world, both Mexico and the central Andes have pronounced seasonal drought, and China has a monsoon climate; but New Guinea does not fit the pattern very well. It does not have a seasonal drought, and none of its early domesticates were annual seed crops.

The timing of agricultural origins, which has been pushed back repeatedly so that now it follows only shortly after the Younger Dryas, has given rise to the suggestion that farming was a response to severe, natural, environmental disruption (Moore and Hillman, 1992; Moore et al., 2000; Bar-Yosef, 2002; Harris, 2003; Wells, 2010). The Younger Dryas was a dramatic, probably global, reversal back to glacial-like conditions that lasted about 1500 years, and ended with extremely rapid warming—in Greenland, temperature may have increased 7 °C in fewer than 10 years (Alley et al., 1993)! It also appears to have been drier, and one study has suggested that the Dead Sea dried up completely (Yechieli et al., 1993). Since the Dead Sea is 400 km deep, and receives only about 50 mm of rainfall per year, the aridity necessary to eliminate it is difficult to conceive. In a sense, then, Gordon Childe's (1951) oasis hypothesis, long fallen out of favor, is now back in play.

Finally, there are theories specific to the Fertile Crescent, emphasizing trade networks (Runnels and van Andel, 1988), or religion (Ozdogan, 2002). The obsidian trade extended over hundreds of kilometers, including to Greek islands; the major deposits are located in southeast Turkey (Wright and Gordus, 1969), close to where agriculture apparently began (Lev-Yadun et al., 2000; Zeder et al., 2006). Religious temples existed in southeast Turkey before cultivation is attested there, and Ozdogan (2002) reported that the society was already highly stratified. Thus, the old, Marxist view that organized religion was an outgrowth of agriculture is also called into question (Blumler, 1993b).

While each scholar has her/his predilection, it should be emphasized that much depends on the origin dates, and that these are subject to change. For instance, some scholars had suggested that agriculture came to the Mississippi Valley from Mexico, given the early radiocarbon dates initially reported from the famous site of Tehuacan (Byers, 1967). But then the more accurate accelerator mass spectrometry radiocarbon dating technique was applied, and the site turned out to be several thousand years younger than formerly thought (Long et al., 1989; Kaplan and Lynch, 1999). Fritz (1994) noted, correctly, that this strengthened the case for independent origin in the Mississippi Valley. Subsequently, however, Mexican origin dates were pushed back again, and are now even earlier than before. Even the sunflower, long believed to be a North American domesticate, has its earliest date from a site in Mexico (Pope et al., 2001; Lentz et al., 2008); this suggests either that it was domesticated in Mexico, or more likely, that there were maritime trade relations between Mexico and the Mississippi Valley. Smith (1992, 1996) and others have argued that Mexican crops came to the Mississippi late, via the American Southwest. But a maritime route would shorten the necessary time for dispersal. Given that the sunflower find is located near the Gulf of Mexico, and since there are other indications of interactions between the two regions, a maritime connection seems possible (Blumler, 1998a). The sunflower find is dated long after Mexican farming began, and about the time that the earliest agriculture is attested in the Mississippi Valley (cf. Smith, 2006).

Currently, all five (six, if one separates north and central China) early centers cluster chronologically. The coincidence in timing suggests some global impact on humans, most likely the Younger Dryas. It is difficult to reconcile other theories with this synchronicity. But since further revision of the dates is not only possible but likely, it would be premature to rule out other explanations.

Once it appeared, agriculture was self-reinforcing because it enabled population growth and craft specialization (technological advances) (Diamond, 1997). Farming increased the food supply by removing vegetation that humans could not digest, and replacing it with plants that humans could. The adoption and diffusion of farming raised human carrying capacity dramatically. One estimate is that a global hunter-gatherer population of one to five million increased to several hundred million as farming spread (Zuk, 2013).

THE SPREAD OF AGRICULTURE

Agriculture spread (diffused) from the origin centers, with the diffusion being most dramatic in the case of the Fertile Crescent (Figure 2) (Diamond, 1997; cf. Bellwood, 2009). As agriculture spread, farmers moved into the territory already occupied by hunter-gatherers. The latter would have had three choices: (1) fight; (2) retreat; or (3) become farmers themselves. While they might have won some battles initially, in the long run they would have lost to the superior numbers of the farmers. On the other hand, if they retreated they would have eventually run out of land. Thus, in the long run their only option, other than extinction, would be to become farmers. (I am presuming here that the territory in question is arable.) Diamond (1997) argued that hunter-gatherers would have mostly rejected agriculture, and while at first that would have been so, ultimately they would have folded into the system or been wiped out. Diamond and others also have related the spread of language groups such as Indo-European to the invention of agriculture, but this is contested by linguists (Mallory, 1989; Haarmann, 1999; Anthony, 2007). There is no reason to think that the original farmers were the ones who also spread the technology; more likely it was a group or groups on the edge of a given agricultural origin center who became mobile. Certainly some of the examples that Diamond (2002) offers, such as the Bantu, the Polynesians, and the Korean migration into Japan, seem to fit the latter scenario. In any case, a similar process is playing out today, which we call globalization: traditional societies, such as hunter-gatherers, slash and burn farmers, and nomadic pastoralists, are being forced to join the world system whether they wish to or not.

As farming spread from the origin centers, it was increasingly likely to encounter environments to which the crops were not well-adapted, creating an evolutionary mismatch. In some cases, this mismatch could not be overcome: Near Eastern cereals could not penetrate Africa south of Egypt and the Ethiopian highlands, because they require cool weather during the growing season (Figure 2). Near Eastern crops were also delayed initially in the spread north from the Mediterranean region into Europe, perhaps due to a need to adjust phenologically to day-length differences, or to winter cold, although eventually they succeeded (Colledge et al., 2005). But in northwest Europe, which has a cool, rainy climate, there continued to be a mismatch for Near Eastern plants during ripening, when they "expected" hot, dry conditions. Wheat and barley have upright spikes that are subject to fungal attack if rain falls on them; rain can also cause sprouting within the ear. These are not problems in the very dry early summer of the Fertile Crescent. In contrast, oat species produce hanging spikelets that shed rain water. Thus, they are better adapted to cool, wet conditions postanthesis. Although wild oat is a dominant species across the Mediterranean, the domesticates (e.g., common oat and sand oat) are not Near Eastern in origin (Vavilov, 1926). The Near Eastern grains can grow in northwest Europe, and over time presumably evolved to be better adapted, yet were beset with problems such as ergot, especially during cooler, wetter epochs such as the Little Ice Age (1350–1850). Matossian (1989)

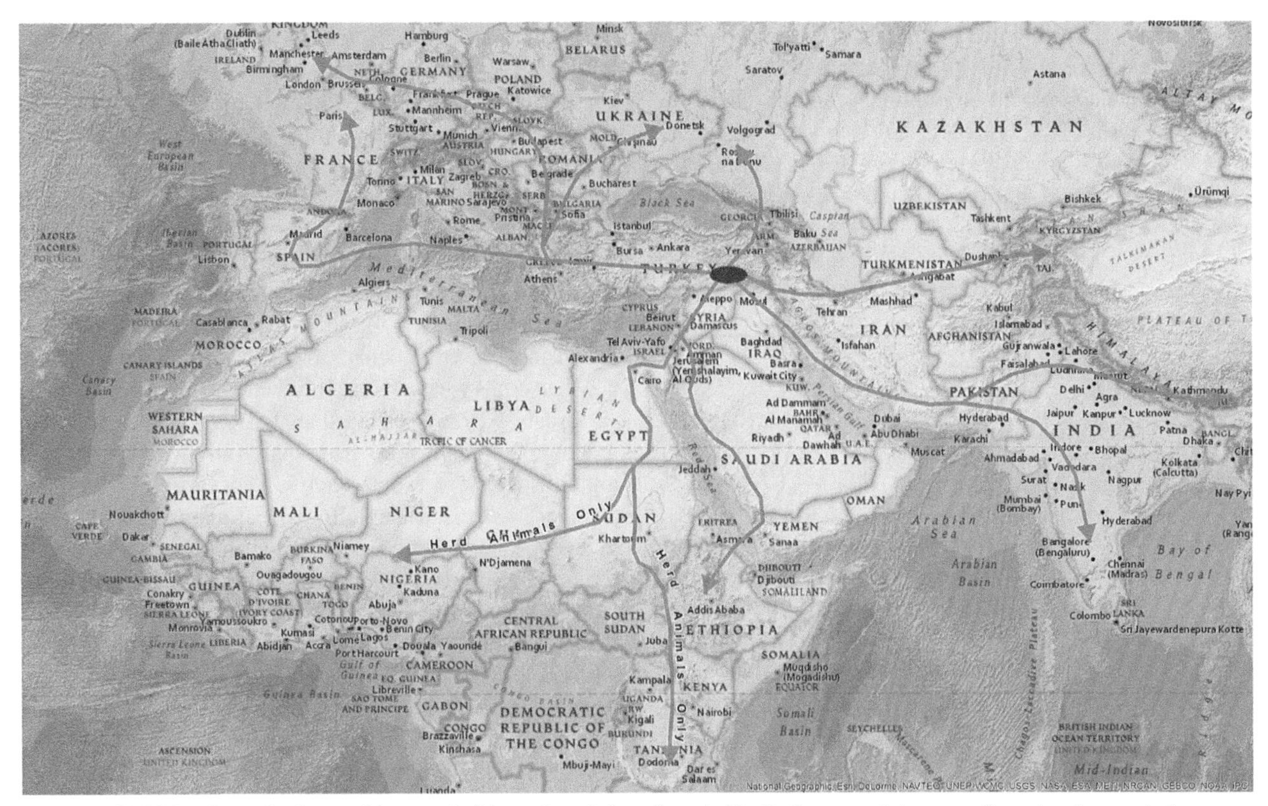

FIGURE 2 Generalized map of the spread of domesticated plants from the Fertile Crescent origin center. (See color plate section).

argued that many of the accusations of witchcraft during the Middle Ages were directed at victims of ergot poisoning, or their caregivers, and reflected a lack of understanding of the disease. Her thesis, if correct, would illustrate that our domesticated plants have had complex influences on our societies. Much has been made of the characteristically low cereal yields in northwest Europe during the Middle Ages; they probably should not be taken as representative of yields in the Near Eastern Neolithic, however.

One way to address evolutionary mismatch is through "niche construction" (Odling-Smee et al., 2003). The beaver is the classic example, which literally constructs a habitat favorable for itself. Today, dense wild cereal stands are characteristic of annual grasslands on hard limestone and basalt in the Fertile Crescent (Zohary, 1969; Blumler, 1993c, 1999). Near Eastern farmers attempted to create the equivalent of such stands in other habitats, sowing their grains where the wild plants normally would not thrive, and nurturing them at the expense of the natural dominants of the vegetation. Sherratt (1980) pointed out that cultivation took place initially on hydromorphic soil. Such soil is not part of the natural habitat for wild Near Eastern cereals and legumes (Zohary, 1969; Blumler, 1999, 2002). In all likelihood, perennials would be more competitive on such soils than on the soils that the wild cereals naturally dominate (cf. Blumler, 1993c). Blumler and Waines (2009) showed that evolutionary changes in phenology and growth habit (upright, prostrate, etc.) would have been necessary to adapt cereals to Near Eastern hydromorphic soils.

Initially, farmers recreated the annual grassland/wild cereal plant community in other environments within the Fertile Crescent; subsequently, as agriculture spread out of the region, into places with climates more favorable to perennials, plowing served to remove these competitors, thus in effect enabling the Fertile Crescent annual grassland to spread. Many of the natural associates of wild cereals came along as weeds, and some of them such as wild oat, black mustard, and wild carrot, were later domesticated. Thus, ultimately an entire plant community was able to expand its range, albeit with significant deletions and additions of species.

A striking illustration is the "rabi" system in India (Figure 2). Supposedly from the Arabic word for spring, reflecting the season at which harvest occurs, rabi is also an Indo-European word for mustard (as in broccoli rabe, rape seed oil, etc.), one of the most important of the rabi crops, as well as a major rabi weed. Lowland India is tropical, but it cools off slightly in the fall especially in the north; the monsoon rains come in summer. But the Near Eastern crops, so important in Indian agriculture, are adapted to winter rain and summer drought. As they moved east, farmers could find close climatic analogs as far as the Indus, but not beyond. Instead, they learned to dry fallow in summer, enabling the soil to accumulate monsoon moisture, and then planted the Near Eastern grains and legumes in fall as the temperature cooled slightly. The plants grow on the stored moisture, and normally will produce a good crop so long as there is a small amount of additional rainfall during the winter and spring (e.g., Howard and Howard, 1909). The better-known system characteristic of Egypt's Nile Valley in ancient times—indeed, up until the construction of the Aswan Dam—is a similar example of ecological niche construction to favor the cereals (and unintentionally, their associated weeds), and consequently, farmers.

Some Consequences of Agriculture

Perhaps the most significant consequence of agriculture, and the population growth that followed, was the emergence of serious epidemic diseases (McNeill, 1976; Diamond, 1997). "Crowd" diseases such as measles require a minimum human population size to remain established. Cholera spreads through water polluted with infected feces, which only becomes an issue when people return to the same water source repeatedly because they are sedentary. (Sedentism is generally thought to be an outgrowth of agriculture, except in the Fertile Crescent where permanent villages such as Jericho may have preceded farming.)

Many of the major epidemic diseases are zoonotic (see chapters by Little, Sattenspiel, and Morand), spreading from domesticated animals to humans: examples include smallpox, tuberculosis, diphtheria, and influenza (Crosby, 1973; McNeill, 1976; Diamond, 1997). Farmers lived in such close proximity to domesticated animals, sometimes sleeping in the same building with them, that it is not surprising that some diseases were able to adapt to humans. Still other diseases, such as malaria, were present before agriculture, but became more prevalent during periods of forest clearance or expansion of swamps due to accelerated soil erosion off hill slopes (Sallares and Gomzi, 2001).

As Crosby (1973; see also Diamond, 1997) pointed out, animal domestication, occurring as it did primarily in the Old World, by giving rise to a host of epidemic diseases ultimately was a major reason that Europeans were able to conquer much of the world after 1492; it also explains why they were not able to maintain control of Africa and Asia, but did retain control over the Americas and Australia. In the former, the inhabitants were just as disease resistant as Europeans, if not more so, while in the latter, isolated from events in the Old World for many millennia, the native populations suffered enormous mortality from the diseases that Europeans brought with them (Crosby, 1973).

Finally, vegetarianism was not an option for hunter-gatherers because their primary protein sources, dairy and legumes, are postagricultural, on the whole (see discussion regarding legumes). In meat-poor regions such as South India, vegetarianism became a widely adopted practice.

THE COLUMBIAN EXCHANGE

Prior to 1492, people mostly grew crops derived from the agricultural origin center closest to them, typically with the addition of a small number of locally derived species. After 1492, there was a massive exchange of organisms—pathogens, crops, domestic animals, weeds, and "varmints" such as rats—between continents, which Crosby (1973) famously dubbed the Columbian Exchange. The European colonial powers consciously introduced crops all over the globe, with the result that many species found favorable habitat overseas. The classic example was the introduction of the potato from Peru to Ireland, where it was much better adapted than the Near Eastern grains. One net result was a global increase in agricultural productivity, because farmers now had crops that were matched better to their environments.

Crosby (1986) proposed that European crops, weeds, grazing animals, and pathogens all worked together mutualistically to replace native ecosystems in the "neo-Europes"—overseas regions with climates similar to that of Europe. His thesis is overstated, but not entirely without merit (Blumler, 1993a); it is more applicable, however, to the "neo-Fertile Crescents" than to neo-Europes. Construction of the Fertile Crescent wild cereal/annual grassland ecological niche overseas after 1492 caused rapid and massive impacts on places with winter-rain, summer-dry climates such as California. There, while a significant portion of the native vegetation was lost to agriculture, even the remaining wildlands suffered as both the crops (for example, radish (Panetsos and Baker, 1967)) and the associated weeds were able to become invasive and dominate over huge areas (Blumler, 1995, 2011).

Ecological imperialism in a different sense is illustrated by the mint and parsley families. Both are diverse in the Mediterranean and adjacent Near East, and many members of both became culinary herbs (Table 3). Both families are diverse in other parts of the world too, yet few of these are readily available in stores. For instance, Mexico and California have hundreds of species of sage (*Salvia*), but the commonly used species is native to the Mediterranean region, where the genus is less diverse. In effect, these seasoning herbs have come along for the ride, hitchhiking on the spectacular diffusion of the Near Eastern agropastoralist system.

Another significant outcome of the Columbian Exchange was the diffusion of the tropical plantation system and slavery, though not its origin: sugar plantations with slaves were widespread across the Mediterranean already before 1492 (Lewis, 1990). The colonial powers very quickly discovered that the key to plantation success was to grow the crop overseas from the region where it had originated. By moving the crop to another continent or island, they enabled it to escape its native insect pests and pathogens. Most plantation crops, such as sugar, cotton, indigo, and tobacco, were not nutritious; plantation sugar, in particular, may mark the inception of the modern problem of overproduction of cheap "junk food" that we tend to overeat, to the detriment of our health (see chapters by Wiley, Low et al., and Dressler).

TABLE 3 Members of the Mint and Parsley Families Native to the Mediterranean/Near East That Are Used as Food or Seasoning

Mints	Parsleys
Spearmint	Carrot
Peppermint	Parsley
Lemon mint, etc.	Celery
Sage	Cilantro (coriander)
Rosemary	Anise
Thyme	Dill
Oregano	Fennel
Marjoram	Caraway
Summer savory	Cumin
Winter savory	Asafoetida
Lavender	Chervil
Lemon balm	Lovage
Catnip	Za'atar

"SCIENTIFIC BREEDING," AND THE INDUSTRIAL REVOLUTION

The rise of scientific breeding in the nineteenth century further extended the Columbian Exchange as agronomists searched the globe for crop varieties that would grow in their home environments, and hybridized foreign germplasm with existing lines to produce new varieties. Subsequently, yields skyrocketed as fossil fuels were applied to agriculture first through machinery such as tractors and then through other inputs such as artificial fertilizers, herbicides, pesticides, and pumped irrigation water (Blumler, 2008, 2009). Smil (2006) argued cogently that the Haber-Bosch process (using methane to manufacture artificial fertilizer from nitrogen gas) was the most important invention of the twentieth century because of its dramatic effect on agricultural yields. These practices constituted a "Green Revolution," though the term was only coined when the practices were extended to developing countries after World War II. Until then, agriculture had relied entirely on renewable resources; since fossil fuels are not renewable, it is uncertain whether current yields are sustainable in the long run. Much hope rests with further plant breeding, but so far the much debated genetically modified organisms feature little besides the insertion of herbicide resistance genes. In short, our crop plants remain surprisingly similar to their wild progenitors physiologically, though not morphologically (Blumler, 1994, 2008; cf. Bamakhramah et al., 1984; Amir and Sinclair, 1994).

CROP EVOLUTION

There are several excellent, illuminating treatments of aspects of crop domestication (e.g., Harlan, 1975; Harlan et al., 1973; Hancock, 1992, 2005; Doebley et al., 2006; Purugganan and Fuller, 2009). These are incomplete, however, and since archaeologists typically do not know a great deal about botanical physiology and genetics, there has been little critical feedback within the agriculture origins literature (Blumler, 1991b; Blumler and Byrne, 1991). Crop genetics and evolution are complicated, and the agricultural origins literature sometimes oversimplifies in the interest of searching for generalities. For instance, there has been much discussion of the rate at which domesticated traits such as indehiscence (retention of the seed on the stalk) would have evolved, with some arguing for rapid (Zohary, 1969; Iltis, 1983; Hillman and Davies, 1990; Paterson, 2002) and others for protracted evolution (Harlan, 1975; Purugganan and Fuller, 2011). The reality is that either is possible, depending on circumstances, species, and trait (Blumler and Byrne, 1991). On the other hand, the agronomic literature related to crop breeding is enormous, with an emphasis on the pragmatic; therefore, informed historical perspective tends to be scattered and buried within the vast literature. Below are some generalities that emphasize aspects of crop evolution that relate to human diet and the evolutionary mismatch question.

Crop Choice

Diamond (1997) argued that agriculture began wherever there were domesticable wild plants and/or animals. This is clearly untenable: for instance, the wild progenitors of many early African domesticates grow both in the Sahel and south of the equator, but domestication only occurred in the Sahel. Wild onions grow all over the Northern Hemisphere, but only species located in or near centers of origin were domesticated. Nonetheless, it is probably true that only certain species were domesticable initially, while others could be added to the system as it developed. Zohary and Spiegel-Roy (1975) noted that in the Near East the earliest domesticated woody species—the vine, the olive, the date, and the fig—have in common that they can be propagated vegetatively, without employing complicated techniques such as grafting. Woody species are outcrossers, and as such, cannot be expected to breed true. Vegetative propagation allowed humans to maintain the variants that they wanted. Tree crops such as almonds that were grown from seed were domesticated later (Zohary and Spiegel-Roy, 1975).

Phylogenetic constraint clearly played a role. Many of our domesticates came from certain characteristic families: grasses for cereals; legumes; mints and parsleys for seasoning (and mints for medicinal teas); mustard and spinach families for vegetables; the poisonous nightshade family for vegetables, including staples such as potatoes and eggplant;

the squash family for cucumber, melons, and gourds as well as the squashes themselves; and the rose family for most of our temperate fruits (apple, pear, peach, apricot, plum, cherry, quince, strawberry, raspberry, blackberry, including more than one species in several cases, plus additional species from east Asia (loquat, etc.)). The aforementioned almond also is a rose family member.

Vavilov (1917) suggested that some crops originated as weeds: as a crop was taken into a new, and not so favorable, environment, the weed might be better adapted to that environment and overwhelm the crop. Then the farmers would begin to eat the weed. Vavilov pointed out that rye is cold tolerant and frequently a weed in wheat fields, and suggested that it became a crop as wheat moved north. He made a similar argument regarding oats (Vavilov, 1926). Vavilov called this "secondary domestication." It is likely that many crops arose in this manner.

Crop Mimics

Some secondary domesticates began as crop mimics (Barrett, 1983; Hancock, 2005). These are weeds that hide from the humans attempting to eliminate them by resembling the crop. Perhaps the best known example is false flax, *Camelina sativa*, a member of the mustard family that has two distinct weed forms, mimicking the oilseed and the fiber varieties, respectively, of domesticated flax. The oilseed form of false flax is also a minor cultivated crop. The wild progenitor of the crop has the potential to be a particularly good mimic, and may pick up useful genes from the crop through hybridization. In the case of mimics unrelated to the crop, seeds often are larger than those of the wild ancestor, enabling the weed to mimic the large seeds of the domesticate with which it grows (for example, darnel (*Lolium temulentum*), corn cockle (*Agrostemma githago*), and common wild oat (*Avena fatua*)). Since traditional seed cleaning entailed running the harvest through sieves of different sized meshes, the mimic would stay with the crop while the smaller-seeded wild form would be eliminated (Hillman, 1984). Also, some mimics have become indehiscent, and are harvested unintentionally with the crop, and then sown again with it, also unintentionally. In this case, the mimic is entirely dependent on humans for its survival, though not yet used as food. The ultimate stage is when people begin intentionally planting the mimic, usually under conditions that it tolerates better than the crop. False flax is an example; another is sand oat (*A. strigosa*). Until recently, sand oat was cultivated on sandy soils on which other grains yielded poorly. It existed not only as a crop but also as a crop mimic in common oat (*A. sativa*), and as a wild plant. The crop and crop mimic are indistinguishable. The crop and crop mimic are becoming rare today, close to if not entirely extinct except in seed banks, as agricultural systems continue to change. The spread of mechanical seed cleaning technology is eliminating many seed mimics, so much so that some, such as corn cockle, now are classified as endangered species (Cockroft, 2014).

Unconscious (Automatic) versus Conscious Evolution

Some traits would have developed as a result of repeated cultivation, harvest, and replanting, whether those traits were selected intentionally or not (Table 4). Other characteristic changes were probably consciously selected by farmers (Blumler and Byrne, 1991; Blumler, 1994). The trends listed in Table 4 represent generalizations only; there are several exceptions to many of them. The literature on this topic tends to overemphasize a few tendencies, for instance, it has long been asserted that under domestication perennials evolve the annual habit. While this is certainly the case for a few species, such as cotton, most if not all of the cereal progenitors and most of the legumes were annuals prior to cultivation and domestication. In part, this reflects the fact that annual-dominated vegetation is much easier to manipulate with basic tools than that which is vegetated with perennials, and so cultivation is more likely to arise there than where there is a dense, perennial sod (Blumler, 1992b, 1998a). Similarly, Harlan et al. (1973) and Purugganan and Fuller (2009) asserted that cultivation produced selection for increased seedling vigor and therefore, larger seeds. This is unlikely; if anything, wild progenitor seedlings appear to be more vigorous than their domesticated descendants (Chapin et al., 1989; Blumler, 1992b, 1994, 1998a). Seed size did increase under cultivation in many species, but the initial increase was probably phenotypic (not genetic), reflecting the comparatively good growing conditions in the cultivated field. Subsequently, farmers may have consciously selected for larger seeds, as is generally true of the plant part harvested for food.

Much has been made of the breeding of improved varieties as a part of the Green Revolution, but in fact, the changes have been minor. They have primarily involved adjusting the crop to the exigencies of mechanical harvest and abundant supplies of nutrients and water. The former require uniform height and maturity, while the latter can cause traditional varieties to lodge, not because they grow poorly, but because they grow too well (Blumler, 2008).

Loss of the Competitive, Protective, and Dispersal Functions

Wild plants must compete with their neighbors, protect their seeds from predators, and disperse their seeds to suitable germination sites. For domesticated crops, these functions are largely accomplished by the farmer. Wheat is an example. Figure 3 shows a diaspore (seed dispersal unit) of wild emmer wheat. Typically, it tightly encloses (and protects) two grains, which are elongated to fit into the enclosing

TABLE 4 Evolutionary Trends under Domestication

Automatic (unconscious)
Indehiscence
Uniform maturity
Loss of competitive ability (upright growth)
Increased self-fertilization
Probably resulting at least in part from conscious farmer selection
Uniform germination (loss of dormancy)
Increased size of the desired plant part
Increased harvest index (proportion of the crop biomass harvested for food)
Loss of antiherbivore defense (increased palatability)
Hypothesized changes that are without empirical support
Increased seedling vigor
Increased adaptation to fertile soils
Consistent physiological changes

Modified from Blumler (1994).

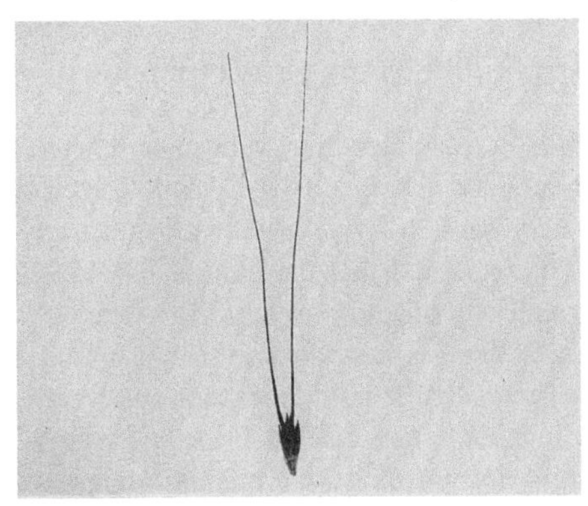

FIGURE 3 A diaspore of wild emmer wheat (× 1/2).

glumes. Zohary (1965) described the diaspore as arrow shaped. Recurved barbs along the awns protect it from birds and also ensure that it can only move in the direction the arrow is pointing. The heavy "arrowhead" ensures that it tends to fall near the mother plant, and then creeps over the ground, thanks in part to the hygroscopic awns which spread apart and draw close again with changes in humidity. If it encounters a crack in the soil or adjacent to a rock, it burrows in and plants itself (see Blumler, 1991b, 1999).

Under cultivation, this structure was no longer needed. Selection favored mutations that caused the diaspore to open up, so that the seed could become more plump, and even "naked" (not enclosed within hulls) (Blumler, 1998b). This made for easier processing of the harvested grain, and increased harvest index since photosynthate (products of photosynthesis) that had gone to creating the diaspore could now be diverted for seed filling. But it was dependent on humans protecting the ripening grain from birds and other granivores. It is the enormous amount of effort required to protect cultivated plants, from competition (by plowing and weeding) and predation, and to disperse (plant) the seeds, that causes farming to be hard work relative to hunting and gathering.

Speciation

Most crop evolution has occurred at the intraspecific level; for example, emmer and durum wheat though morphologically very different from wild emmer are still fully interfertile with it and each other (Zohary, 1969; Harlan, 1975). But there have been a few cases of speciation under domestication. The most important example is bread wheat, which is an allopolyploid derivative of emmer and a closely related goatgrass, *Aegilops squarrosa*, that apparently arose more than once in cultivated fields (Dvorak et al., 1998; Giles and Brown, 2006; Gu et al., 2006). A more recent example is triticale, another allopolyploid, in this case of bread wheat and rye, intentionally created by crop geneticists. Genetic interactions among polyploid crops and their wild or weedy relatives can be very complex, and allow for rapid evolution (Zohary, 1965; Harlan, 1975; Paterson, 2002; Hancock, 2005).

Coevolution

Crop domestication is a coevolutionary process. In evolutionary biology, coevolution occurs when a genetic change in species A leads to a genetic change in species B, which then causes further genetic change in species A. Domesticated plants and animals clearly underwent significant genetic changes as a result of their adoption into a farming system, but it is less certain that humans have always responded with genetic changes. As Zuk (2013) pointed out, our coevolutionary response can be (1) genetic; (2) microbial; and/or (3) cultural. Perhaps the best studied example is the digestion of milk, which in most mammals ceases after early childhood. Consequently, the enzyme lactase, which digests lactose, is no longer active in most adult mammals. Some pastoralists have evolved lactose tolerance, with lactase remaining active throughout their adult lives (see chapters by Little and Wiley). But as Zuk pointed out, there are also groups of people who appear to have intestinal flora that digest lactose. This would be an example of a coevolutionary response that is microbial. Finally, in many societies, dairy consumption is primarily as cheese or yogurt, which contain less lactose. This would reflect a cultural response since it depends on human behavior (i.e., humans intentionally create the conditions that enable the appropriate microbes to interact with the milk and break down the lactose).

Palatability

Since agriculture began, humans have consciously selected for increased yield and for palatability. Palatability tends to mean more sugar, more calories, and fewer toxins (which often have a bitter taste), but also less fiber, and fewer antioxidants (Table 5). There are also many exceptions to these tendencies. For instance, anthocyanin, an antioxidant, decreased in some crops such as several cereals, but increased in others such as purple cabbage. Bitterness was selected against in most plants that defended themselves with cyanogenic glycosides, such as almond. But bitter almond is still used as a flavoring, and many seasoning herbs such as oregano and hops remain bitter.

Increasing caloric content and yield often entailed increasing starch in the plant part harvested and eaten. Harlan et al. (1973) and others have stated that the protein content of grains decreased under cultivation/domestication, but it appears that this is only true on a percentage basis. In tetraploid wheat the total amount of protein in the domesticated grain is often about the same as, or even larger than, its wild ancestor, but the seed size has increased due to the opening up of the investing structures, as described above. The extra volume in the grain floret typically fills with starch, after the protein has first been transported into the developing grain. While protein content of cereals is less on a percentage basis than in their wild ancestors, today they are typically supplemented with legumes, which are high in protein. Legumes were much less available to Paleolithic peoples, because in the wild they are usually toxic (except sometimes when still green). Because of their association with nitrogen-fixing bacteria, legumes usually have more nitrogen than they need. The excess is used typically to produce alkaloids and other poisons such as prussic acid. Although legumes became important in all agricultural origin centers eventually, the timing of domestication was highly variable. In the Fertile Crescent, domesticated legumes were about as early as cereals. In China, it appears that no legume was domesticated for many millennia after the appearance of cereal agriculture (Lee et al., 2011). This is somewhat paradoxical, in that the Near East also had dairy as a source of protein, while China did not.

TABLE 5 Palatability-Related Changes in Domesticated Plants

Protein: decreases
Starch: increases
Sugar: increases
Fiber: decreases
Antioxidants: decreases
Bitterness: decreases

Palatability for humans reflects dietary preferences or cravings that go back to hunter-gatherer times or before, and which presumably would have been adaptive then, as evolutionary psychologists have pointed out (Tooby and Cosmides, 1990). This is also the basis of the paleo diet perspective (Eaton and Konner, 1985; Lindeberg, 2010). We crave salt, fat, and sugar. Unless along a seacoast, it was difficult to obtain enough salt, while fat was generally the limiting item in the hunter-gatherer diet. Most game animals are lean, and there are relatively few sources of edible oilseeds in nature. Sugar craving encouraged humans to eat wild fruits. Occasionally a hunter-gatherer band might encounter a beehive, in which case they might be able to feast on honey for only a few days. With agriculture, we gained the ability to satiate ourselves, and then some! Sugar, fat, and salt are all now cheap and abundant, with effects on our health that are not good. In contrast, hunter-gatherers would have consumed more than sufficient fiber and antioxidants, so it is less likely that they would have developed cravings for these substances. If anything, one might expect selection to have favored mild avoidance of high fiber and high antioxidant foods, which is consonant with the general reduction in these substances in the human diet since agriculture.

EVALUATING THE PALEO DIET HYPOTHESIS

Proponents of the paleo diet differ somewhat; here the perspective of Lindeberg (2010) is emphasized. His book featured a comprehensive review of the nutrition literature. In it, he demonstrated that the consensus beliefs of nutritionists frequently have no or very little empirical support. For instance, there is no evidence that insoluble fiber, the kind present in cereals with the exception of oats, is of any benefit. Lindeberg stressed that from an evolutionary perspective it is not normally in a plant's interest to allow its seeds to be eaten, and so they tend to defend them with toxic compounds. In the case of cereals, these compounds are concentrated in the bran, which raises the question (though Lindeberg did not discuss it) of whether whole grains are perhaps not really as nutritious as refined cereals. In this regard, Jasny's (1940) classic study of the reasons people choose one grain over another made the then entirely reasonable assumption that all humans would prefer white flour when they could get it. Entirely reasonable because at that time only monks given to asceticism ate whole grains in preference to refined breads, white rice, etc.

Lindeberg (2010) stated categorically that grains and beans were not eaten in Paleolithic times. This appears extreme: there is some evidence for consumption of grains then (Henry et al., 2010), and both grains and bean family members were or are eaten by modern hunter-gatherers. Wild grass seeds were harvested in significant quantities by some Australian Aborigines (Head et al., 2012), and also by native peoples in the Great Basin and California. California Indians also feasted on several species of native clover during the spring (clovers are legumes). The native clovers are less well-defended than introduced, Old World clovers, but even so, at times individuals would eat so much that they suffered severely from bloat (Chesnut, 1897). California Indians also ate native vetches (broad bean, pea, and lentil are domesticated vetches of Near Eastern origin) but they ate them when the pods were still immature (much as Americans and east Asians tend to eat peas). In the eastern United States, one of the favored wild foods was the groundnut (*Apios americana*), another member of the bean family. British settlers learned about this resource from the native peoples, and often depended on it during the first year of settlement out on the frontier, until their crops began to produce. Of course, these examples are from the ethnographic present, and Paleolithic peoples probably did not have the same diet; but what they did eat remains somewhat speculative.

Lindeberg also emphasized that Paleolithic peoples ate a diverse diet, which would have served to mitigate the impacts of plant poisons. Given that the toxic compounds tend to differ from species to species, a mixed feeding strategy can ensure that only small quantities of any one toxin are ingested. This is an entirely reasonable argument; however, the literature on diet breadth is conflicting. Some believe that diet simplified under agriculture, with its emphasis on a few cereal grains or tubers; but others argue that hunter-gatherers made relatively little use of seasonings, spices, and the like. Certainly, recent times have seen a global narrowing of the food base, with a few staples such as maize, rice, and wheat becoming more and more predominant (Harris, 2009). This has especially affected the urban poor. In contrast, the rich and even the middle class today typically have much greater choice of vegetables and fruits (though not necessarily of grains and tubers) than ever before. And they have access to fresh vegetables and fruits year-round, which was not the case even 50 years ago. Harris (2009) found that hunter-gatherers are reported to consume from 50 to over 100 plant species. But the Japanese have a folk saying that one should eat at least 25 different foods each day. Thus, it is by no means clear that diet breadth has narrowed for everyone.

Because of the supposed lack of grains in the Paleolithic diet, Lindeberg (2010) argued that starch is a food to which we are poorly adapted. On the other hand, humans have evolved increased copies of the amylase (starch digestion) gene in recent times (Perry et al., 2007). Lindeberg asserted that the amount of evolution that has occurred was insufficient to enable us to deal with the very large increase in the amount of starch we consume, but this rests on his belief that we did not consume grains at all prior to the Neolithic. In addition, in east and southeast Asia people have selected mutant forms of several cereals that are high in amylopectin; the best known example is "sticky rice," also called "sweet rice" (Sakamoto, 1996; Hunt et al., 2012). While so far the nutritional research on the sticky cereals has been limited, it is worth noting that traditionally they are fed to invalids. In short, agriculture was associated with a shift to grains or tubers as staples, and along with the increase in proportion of starch in the domesticated grains that ensures that we are eating more starch today—but how much more is unclear.

In any case, refined cereals are useful in another circumstance. Early childhood mortality is the major source of mortality in many countries, and a major cause of this mortality (directly or indirectly) is often dysentery. While dysentery probably has been with us for a long while (see the chapter by Morand), it is almost certain that its prevalence increased with the increase in population density associated with farming. The traditional food of choice for individuals suffering from dysentery is starch, the purer (i.e., the more highly refined) the better.

Antinutritional compounds are present in many other domesticates, in addition to the seed crops, as Lindeberg (2010) pointed out. For instance, while the levels of cyanogenic glycosides are generally reduced sufficiently that we can ingest domesticated crops without being fatally poisoned, he reviewed evidence that if ingested in small quantities, the cyanogenic glycosides are goitrogenic. This is likely to become particularly problematic when a crop is consumed as a staple. For instance, cabbage is goitrogenic, and the traditional method of boiling it for a long period of time before consumption may have been desirable when it was eaten in large quantities—as it often was.

While a mixed Paleolithic diet would have reduced the deleterious effects of plant toxins, humans also developed processing techniques that reduced the levels of these compounds. These techniques ranged from cooking to fermentation to the ingestion of clays that adsorb toxins (Johns, 1990; Johns and Kubo, 1988). Fermentation is a particularly widespread and interesting case. Yogurt and cheese have already been noted. An example involving legumes is soy, which in its native east Asia is seldom cooked and consumed in the whole, ripe state. Instead, for the most part it is consumed only unripe (edamame, bean sprouts), or after fermentation or similar processing (tofu, soy sauce, miso, black bean sauce, tempeh, natto).

CONCLUSIONS

To return to the evolutionary mismatch question, Lloyd et al. (n.d.) point out that, in a sense, evolutionary mismatch is always with us, because our environment is always changing if only because other species are evolving. Moreover, the diet of Paleolithic peoples was not constant, but also must have changed over time. Zuk rightfully claims that paleo-diet enthusiasts are greatly overstated and that their viewpoint likely exaggerates the persistence of hunter-gatherer adaptations. On the other hand, Lindeberg (2010), Lloyd et al. (n.d.), and others are making a genuine contribution to nutrition research by attempting to put it on an evolutionary footing. Moreover, it is unquestionably true that today we have an abundance of cheap, readily available sugar, salt, fat, and other high calorie foods, and that we tend to overeat them, because of cravings that were adaptive in ancient times, when such foods were not available in quantity. Agriculture certainly started the process of making such foods readily available to us, and it did so because we selected for palatability, that is, according to our cravings. But for most of the 10,000 years of agriculture, there were countervailing forces that encouraged us to continue to eat bitter herbs, for instance, and that militated against drastic overconsumption of calories and sugars. In addition, we evolved, as, for instance, the amylase gene copy data suggest (Perry et al., 2007). There is certainly a serious evolutionary mismatch today, but much of this derives from the Industrial Revolution, which began only a short time ago. This is an evolutionary mismatch of our own creation.

REFERENCES

Abbo, S., Lev-Yadun, S., Heun, M., Gopher, A., 2013. On the 'lost' crops of the neolithic Near East. Journal of Experimental Botany 64, 815–822.

Alley, R.B., Meese, D.A., Shuman, C.A., Gow, A.J., Taylor, K.C., Grootes, P.M., White, J.W.C., Ram, M., Waddington, E.D., Mayewski, P.A., Zielinski, G.A., 1993. Abrupt increase in Greenland snow accumulation at the end of the Younger Dryas event. Nature 362, 527–529.

Amir, J., Sinclair, T.R., 1994. Cereal grain yield: biblical aspirations and modern experience in the Middle East. Agronomy Journal 86, 362–364.

Anthony, D.W., 2007. The Horse, the Wheel, and Language: How Bronze-Age Riders from the Eurasian Steppes Shaped the Modern World. Princeton University Press, Princeton, NJ.

Bamakhramah, H.S., Halloran, G.M., Wilson, J.H., 1984. Components of yield in diploid, tetraploid and hexaploid wheats (*Triticum* spp.). Annals of Botany 54, 51–60.

Bar-Yosef, O., 2002. The role of the Younger Dryas in the origin of agriculture in West Asia. In: Yasuda, Y. (Ed.), The Origins of Pottery and Agriculture. Roli/Lustre, New Delhi, pp. 39–54.

Barrett, S.C.H., 1983. Crop mimicry in weeds. Economic Botany 37, 255–282.

Bellwood, P., 2009. The dispersals of established food-producing populations. Current Anthropology 50, 621–626.

Bettinger, R., Richerson, P., Boyd, R., 2009. Constraints on the development of agriculture. Current Anthropology 50, 627–631.

Blumler, M.A., 1991b. Modelling the origins of legume domestication and cultivation. Economic Botany 45, 243–250.

Blumler, M.A., 1992a. Independent inventionism and recent genetic evidence on plant domestication. Economic Botany 46, 98–111.

Blumler, M.A., 1992b. Seed Weight and Environment in Mediterranean-type Grasslands in California and Israel (Ph. D. dissertation). University of California, Berkeley.

Blumler, M.A., 1993a. 'EcologicalImperialism': a botanical perspective. In: Proceedings of the 27th International Geographical Congress, Washington, DC, August, 1992, pp. 141–142.

Blumler, M.A., 1993b. On the tension between cultural geography and anthropology: commentary on Christine Rodrigue's 'Early animal domestication.'. The Professional Geographer 45, 359–363.

Blumler, M.A., 1993c. Successional pattern and landscape sensitivity in the Mediterranean and Near East. In: Thomas, D.S.G., Allison, R.J. (Eds.), Landscape Sensitivity. John Wiley & Sons, Chichester, pp. 287–305.

Blumler, M.A., 1994. Evolutionary trends in the wheat group in relation to environment, Quaternary climate change and human impacts. In: Millington, A.C., Pye, K. (Eds.), Environmental Change in Drylands. John Wiley & Sons, Chichester, pp. 253–269.

Blumler, M.A., 1995. Invasion and transformation of California's valley grassland, a Mediterranean analogue ecosystem. In: Butlin, R., Roberts, N. (Eds.), Human Impact and Adaptation: Ecological Relations in Historical Times. Blackwell, Oxford, pp. 308–332.

Blumler, M.A., 1996. Ecology, evolutionary theory, and agricultural origins. In: Harris, D.R. (Ed.), The Origins and Spread of Agriculture and Pastoralism in Eurasia. UCL Press, London, pp. 25–50.

Blumler, M.A., 1998a. Evolution of caryopsis gigantism and the origins of agriculture. Research in Contemporary and Applied Geography: A Discussion Series 22 (1–2), 1–46.

Blumler, M.A., 1998b. Introgression of durum into wild emmer and the agricultural origin question. In: Damania, A.B., Valkoun, J., Willcox, G., Qualset, C.O. (Eds.), The Origins of Agriculture and Crop Domestication. ICARDA, Aleppo, pp. 252–268.

Blumler, M.A., 1999. Edaphic ecology of the wild cereals. Research in Contemporary and Applied Geography: A Discussion Series 23 (3–4), 1–58.

Blumler, M.A., 2002. Changing paradigms, wild cereal ecology, and agricultural origins. In: Cappers, R.T.J., Bottema, S. (Eds.), The dawn of farming in the Near East, studies in early Near Eastern production, subsistence and environment, 6. Ex Oriente, Berlin, pp. 95–111.

Blumler, M.A., 2008. The fossil fuel revolution: a great, and largely unrecognized, experiment. The Paennsylvania Geographer 46 (2), 3–21.

Blumler, M.A., 2009. The frontier is dead, long live the frontier: responses to the official "closing" of the American settlement frontier. The Paennsylvania Geographer 47 (2), 3–19.

Blumler, M.A., 2011. Invasive species, in geographical perspective. In: Millington, A.C., Blumler, M.A., Schickoff, U. (Eds.), Handbook of Biogeography. Sage Publications, London, pp. 510–527.

Blumler, M.A., Byrne, R., 1991. The ecological genetics of domestication and the origins of agriculture. Current Anthropology 32, 23–54.

Blumler, M.A., Waines, J.G., 2009. On the potential for spring sowing in the ancient Near East. In: Fairbairn, A.S., Weiss, E. (Eds.), From Foragers to Farmers: Papers in Honour of Gordon C. Hillman. Oxbow Monographs, Oxford, pp. 19–26.

Byers, D. (Ed.), 1967. Environment and Subsistence, Volume 1: The Prehistory of the Tehuacan Valley. University of Texas Press, Austin.

Byrne, R., 1987. Climatic change and the origins of agriculture. In: Manzanilla, L. (Ed.), Studies in the Neolithic and Urban Revolutions: The V. Gordon Childe Colloquium, pp. 21–34 British Archaeological Reports, International Series 349.

Chapin, F.S., Groves, R.H., Evans, L.T., 1989. Physiological determinants of growth rate in response to phosphate supply in wild and cultivated *Hordeum* species. Oecologia 79, 96–105.

Chesnut, V.K., 1897. Plants Used by the Indians of Mendocino County, California. US Department of Agriculture, Washington DC.

Childe, V.G., 1951. Man Makes Himself. Watts, London.

Cockroft, S., 2014. The Flower That Can Kill: Deadly British Plant Thought to Be Extinct Is Discovered by a Lighthouse. Daily Mail 7/15/14. http://www.dailymail.co.uk/news/article-2693783/Deadly-British-plant-thought-extinct-discovered-lighthouse-Public-warned-not-touch-corn-cockle-kill.html (last accessed 23.12.14.).

Cohen, M.N., 1977. The Food Crisis in Prehistory: Overpopulation and the Origins of Agriculture. Yale University Press, New Haven CT.

Cohen, M.N., 2009. Introduction: rethinking the origins of agriculture. Current Anthropology 50, 591–595.

Cohen, M.N., Armelagos, G.J., 1984. Paleopathology at the Origins of Agriculture. Academic Press, New York.

Colledge, S., Connelly, J., Shennan, S., 2005. The evolution of Neolithic farming from SW Asian origins to NW European limits. European Journal of Archaeology 8, 137–156.

Cordain, L., 2001. The Paleo Diet: Lose Weight and Get Healthy by Eating the Foods You Were Designed to Eat. Wiley, New York.

Crawford, G.W., 2009. Agricultural origins in North China pushed back to the Pleistocene-Holocene boundary. Proceedings of the Natlional Academy of Sciences of the United States of America 106, 7271–7272.

Crosby, A.W., 1973. The Columbian Exchange: Biological and Cultural Consequences of 1492. Greenwood Publishing Co, Westport, CT.

Crosby, A.W., 1986. Ecological Imperialism: The Biological Expansion of Europe, 900-1900. Cambridge University Press, Cambridge.

Darwin, C., 1859. On the Origin of Species by Means of Natural Selection, or the Preservation of Favoured Races in the Struggle for Life. John Murray, London.

Darwin, C., 1883. The Variation of Animals and Plants under Domestication. D. Appleton and Co, New York.

Denham, T.P., Haberle, S.G., 2008. Agricultural emergence and transformation in the Upper Wahgi Valley during the Holocene: theory, method and practice. Holocene 18, 499–514.

Diamond, J., 1997. Guns, Germs, and Steel. Norton, New York.

Diamond, J., 1998. Ants, crops, and history. Science 281, 1974–1975.

Dillehay, T.D., Rossen, J., Andres, T.C., Williams, D.E., 2007. Preceramic adoption of peanut, squash, and cotton in northern Peru. Science 316, 1890–1893.

Doebley, J.F., Gaut, B.S., Smith, B.D., 2006. The molecular genetics of crop domestication. Cell 127, 1309–1329.

Dvorak, J., Luo, M.-C., Yang, Z.-L., 1998. Genetic evidence on the origin of *Triticum aestivum* L. In: Damania, A.B., Valkoun, J., Willcox, G., Qualset, C.O. (Eds.), The Origins of Agriculture and Crop Domestication. ICARDA, Aleppo, pp. 235–251.

Eaton, S., Konner, M., 1985. Paleolithic nutrition.A consideration of its nature and current implications. New England Journal of Medicine 312, 283–289.

Erickson, D., Smith, B.D., Clarke, A.C., Sandweiss, D.H., Tuross, N., 2005. An Asian origin for a 10,000 year-old domesticated plant in the Americas. Proceedings of the National Academy of Sciences of the United States of America 102, 18315–18320.

Farrell, B.D., Sequeira, A.S., O'Meara, B.C., Normark, B.B., Chung, J.H., Jordal, B.H., 2001. The evolution of agriculture in beetles (Curculionidae: Scolytinae and Platypodinae). Evolution 55, 2001–2007.

Flannery, K.V., 1986. The research problem. In: Flannery, K.V. (Ed.), Guila Naquitz: Archaic Foraging and Early Agriculture in Oaxaca, Mexico. Academic Press, Orlando, FL, pp. 3–18.

Flannery, T., 2001. The Eternal Frontier. Atlantic Monthly Press, New York.

Fritz, G.J., 1994. Are the first farmers getting younger? Current Anthropology 35, 305–309.

Fuller, D.Q., Willcox, G., Allaby, R.G., 2012. Early agriculture pathways: moving outside the 'core area' hypothesis in Southwest Asia. Journal of Experimental Botany 63, 617–633.

Gage, T.B., 2005. Are modern environments really bad for us?: revisiting the demographic and epidemiological transitions. Yearbook of Physical Anthropology 48, 96–117.

Giles, R., Brown, T., 2006. *GluDy* allele variation in *Aegilops tauschii* and *Triticum aestivum*: implications for the origins of hexaploid wheats. Theoretical and Applied Genetics 112, 1563–1572.

Gu, Y.Q., Salse, J., Coleman-Derr, D., Dupin, A., Crossman, C., Lazo, G.R., Huo, N., Belcram, H., Ravel, C., Charmet, G., Charles, M., Anderson, O.D., Chalhoub, B., 2006. Types and rates of sequence evolution at the high-molecular-weight glutenin locus in hexaploid wheat and its ancestral genomes. Genetics 174, 1493–1504.

Haarmann, H., 1999. On the problem of primary and secondary diffusion of Indo-Europeans and their languages. Journal of Indo-European Studies 26, 3–4.

Hancock, J.F., 1992. Plant Evolution and the Origin of Crop Species. Prentice Hall, Englewood Cliffs NJ.

Hancock, J.F., 2005. Contributions of domesticated plant studies to our understanding of plant evolution. Annals of Botany 96, 953–963.

Harlan, J.R., 1975. Crops & Man. American Society of Agronomy, Madison, WI.

Harlan, J.R., de Wet, J.M.J., Price, E.G., 1973. Comparative evolution of cereals. Evolution 27, 311–325.

Harris, D.R., 1977. Alternative pathways toward agriculture. In: Reed, C.A. (Ed.), Origins of Agriculture. Mouton, The Hague, pp. 179–243.

Harris, D.R., 2003. Climate change and the beginnings of agriculture: the case of the Younger Dryas. In: Rothschild, L., Lister, A. (Eds.), Evolution on Planet Earth: The Impact of the Physical Environment. Elsevier, London, pp. 379–394.

Harris, D.R., 2009. Agriculture, cultivation, and domestication: exploring the conceptual framework of early food production. In: Denham, T.P., Iriarte, J., Vrydaghs, L. (Eds.), Rethinking Agriculture: Archaeological and Ethnoarchaeological Perspectives. Left Coast Press, Walnut Creek, CA, pp. 16–35.

Head, L., Atchison, J., Gates, A., 2012. Ingrained: A Human Bio-geography of Wheat. Ashgate, Burlington, VT.

Henry, A.G., Brooks, A.S., Piperno, D.R., 2010. Microfossils in calculus demonstrate consumption of plants and cooked food in Neanderthal diets (Shanidar III, Iraq; Spy I and II, Belgium). Proceedings of the National Academy of Sciences of the United States of America 108, 486–491.

Hillman, G.C., 1984. Traditional husbandry and processing of archaic cereals in modern times: Part I, the glume-wheats. Bulletin on Sumerian Agriculture 1, 114–152.

Hillman, G.C., Davies, M.S., 1990. Domestication rates in wild wheats and barley under primitive cultivation. Biological Journal of the Linnean Society 39, 39–78.

Howard, A., Howard, G.L.C., 1909. Wheat in India. Thacker, Spink, and Co, Calcutta.

Huang, X., Kurata, N., Wei, X., Wang, Z.X., Wang, A., Zhao, Q., Zhao, Y., Liu, K., Lu, H., Li, W., Guo, Y., Lu, Y., Zhou, C., Fan, D., Weng, Q., Zhu, C., Huang, T., Zhang, L., Wang, Y., Feng, L., Furuumi, H., Kubo, T., Miyabayashi, T., Yuan, X., Xu, Q., Dong, G., Zhan, Q., Li, C., Fujiyama, A., Toyoda, A., Lu, T., Feng, Q., Qian, Q., Li, J., Han, B., 2012. A map of rice genome variation reveals the origin of cultivated rice. Nature 490, 497–503.

Hunt, H.V., Moots, H.M., Graybosch, R.A., Jones, H., Parker, M., Romanova, O., Jones, M.K., Howe, C.J., Trafford, K., 2012. Waxy phenotype evolution in the allotetraploid cereal broomcorn millet: mutations at the *GBSSI* locus in their functional and phylogenetic context. Molecular Biology and Evolution 30, 109–122.

Iltis, H.H., 1983. From teosinte to maize: the catastrophic sexual transmutation. Science 222, 886–894.

Jasny, N., 1940. Competition Among Grains. Food Research Institute, Stanford University, Palo Alto, CA.
Johns, T., 1990. With Bitter Herbs They Shall Eat It: Chemical Ecology and the Origins of Human Diet and Medicine. University of Arizona Press, Tucson.
Johns, T., Kubo, I., 1988. A survey of traditional methods employed for the detoxification of plant foods. Journal of Ethnobiology 8, 81–129.
Kaplan, L., Lynch, T.F., 1999. *Phaseolus* (Fabaceae) in archaeology: AMS radiocarbon dates and their significance for pre-Columbian archaeology. Economic Botany 53, 261–272.
Lambert, P.M., 2009. Health versus fitness: competing themes in the origins and spread of agriculture? Current Anthropology 50, 603–608.
Lee, G.A., Crawford, G.W., Liu, L., Sasaki, Y., Chen, X.X., 2011. Archaeological soybean (*Glycine max*) in East Asia: does size matter? PLoS One 6: e26720.
Lee, R.B., DeVore, I., 1968. Man the Hunter. Aldine Press, Chicago.
Lentz, D.L., Pohl, M.D., Alvarado, J.L., Tarighat, S., Bye, R., 2008. Sunflower (*Helianthus annuus* L.) as a pre-Columbian domesticate in Mexico. Proceedings of the National Academy of Sciences 105, 6232–6237.
Lev-Yadun, S., Gopher, A., Abbo, S., 2000. The cradle of agriculture. Science 288, 1602–1603.
Lewis, B., 1990. Race and Slavery in the Middle East. Oxford University Press, New York.
Lindeberg, S., 2010. Food and Western Disease. Wiley-Blackwell, Chichester.
Lloyd, E.A., Wilson, D.S., Sober, E., n.d. Evolutionary Mismatch and What to Do About It: A Basic Tutorial. http://evolutioninstitute.org/sites/default/files/articles/Mismatch%20updated%20March%2026%202014.pdf (last accessed 23.12.14.).
Long, A., Benz, B.F., Donahue, D.J., Jull, A.J.T., Toolin, L.J., 1989. First direct AMS dates on early maize from Tehuacan, Mexico. Radiocarbon 31, 1035–1040.
Lu, H., Zhang, J., Liu, K.B., Wu, N., Li, Y., Zhou, K., Ye, M., Zhang, T., Zhang, H., Yang, X., Shen, L., Xu, D., Li, Q., 2009. Earliest domestication of common millet (*Panicum miliaceum*) in East Asia extended to 10,000 years ago. Proceedings of the National Academy of Sciences of the United States of America 106, 7367–7372.
MacNeish, R.S., Libby, J.G., 1995. Origins of Rice Agriculture: The Preliminary Report of the Sino-american Jiangxi Project.
Mallory, J.P., 1989. In Search of the Indo-europeans: Language, Archeology and Myth. Thames & Hudson, London.
Martin, P.S., Wright Jr., H.E., 1967. Pleistocene Extinctions: The Search for a Cause. Yale University Press, New Haven, CT.
Matossian, M.K., 1989. Poisons of the Past. Yale University Press, New Haven, CT.
Matsuoka, Y., Vigouroux, Y., Goodman, M.M., Sanchez, G.,J., Buckler, E., Doebley, J., 2004. A single domestication for maize shown by multilocus microsatellite genotyping. Proceedings of the National Academy of Sciences of the United States of America 99, 6080–6084.
McNeill, W.H., 1976. Plagues and Peoples. Anchor, Garden City, NJ.
Moore, A.M.T., Hillman, G.C., 1992. The Pleistocene to Holocene transition and human economy in Southwest Asia: the impact of the Younger Dryas. American Antiquity 57, 482–494.
Moore, A.M.T., Hillman, G.C., Legge, A.J., 2000. Village on the Euphrates: From Foraging to Farming at Abu Hureyra. Oxford University Press, New York.
Nesbitt, M., 2002. When and where did domesticated cereals first occur in Southwest Asia? In: Cappers, R.T.J., Bottema, S. (Eds.), The Dawn of Farming in the Near East, Studies in Early Near Eastern Production, Subsistence and Environment, 6. Ex Oriente, Berlin, pp. 113–132.
Odling-Smee, F.J., Laland, K.N., Feldman, W., 2003. Niche Construction. Princeton University Press, Princeton, NJ.
Orlove, B.S., 1980. Ecological anthropology. Annual Review of Anthropology 9, 235–273.
Ozdogan, M., 2002. Redefining the Neolithic of Anatolia: a critical review. In: Cappers, R.T.J., Bottema, S. (Eds.), The Dawn of Farming in the Near East: Studies in Early Near Eastern Production, Subsistence and Environment, 6. Ex Oriente, Berlin, pp. 151–158.
Panetsos, C.A., Baker, H.G., 1967. The origin of variation in "wild" *Raphanus sativus* (Cruciferae) in California. Genetica 38, 243–274.
Paterson, A.H., 2002. What has QTL mapping taught us about plant domestication? New Phytologist 154, 591–608.
Perry, G.H., Dominy, N.J., Claw, K.G., Lee, A.S., Fiegler, H., Redon, R., Werner, J., Villanea, F.A., Mountain, J.L., Misra, R., Carter, N.P., Lee, C., Stone, A.C., 2007. Diet and the evolution of human amylase gene copy number variation. Nature Genetics 39, 1256–1260.
Piperno, D.R., 2011. The origins of plant cultivation and domestication in the New World Tropics. Current Anthropology 52, S453–S470.
Piperno, D.R., Ranere, A.J., Holst, I., Iriarte, J., Dickau, R., 2009. Starch grain and phytolith evidence for early ninth millennium B.P. maize from the Central Balsas River Valley, Mexico. Proceedings of the National Academy of Sciences of the United States of America 106, 5019–5024.
Pope, K.O., Pohl, M.E.D., Jones, J.G., Lentz, D.L., von Nagy, C., Vega, F.J., Quitmyer, I.R., 2001. Origin and environmental setting of ancient agriculture in the lowlands of Mesoamerica. Science 292, 1370–1373.
Purugganan, M.D., Fuller, D.O., 2009. The nature of selection during plant domestication. Nature 457, 843–848.
Purugganan, M.D., Fuller, D.O., 2011. Archaeological data reveal slow rates of evolution during plant domestication. Evolution 65, 171–183.
Richerson, P.J., Boyd, R., Bettinger, R.L., 2001. Was agriculture impossible during the Pleistocene but mandatory during the Holocene? A climatic change hypothesis. American Antiquity 66, 387–412.
Riehl, S., Zeidl, M., Conard, N.J., 2013. Emergence of agriculture in the foothills of the Zagros Mountains of Iran. Science 341, 65–67.
Runnels, C., van Andel, T.H., 1988. Trade and the origins of agriculture in the eastern Mediterranean. Journal of Mediterranean Archaeology 1, 83–109.
Sage, R.F., 1995. Was low atmospheric CO_2 during the Pleistocene a limiting factor for the origin of agriculture? Global Change Biology 1, 93–106.
Sakamoto, S., 1996. Glutinous-endosperm starch food culture specific to Eastern and Southeastern Asia. In: Ellen, R., Kikuchi, K. (Eds.), Redefining Nature. Ecology, Culture and Domestication. Oxford University Press, Oxford, pp. 215–231.
Sallares, R., Gomzi, S., 2001. Biomolecular archaeology of malaria. Ancient Biomolecules 3, 195–213.
Schultz, T.R., Brady, S.G., 2008. Major evolutionary transitions in ant agriculture. Proceedings of the National Academy of Sciences of the United States of America 105, 5435–5440.
Sherratt, A., 1980. Water, soil and seasonality in early cereal cultivation. World Archaeology 11, 313–330.
Sherratt, A., 1997. Climatic cycles and behavioural revolutions: the emergence of modern humans and the beginning of farming. Antiquity 71, 271–287.
Smil, V., 2006. Transforming the Twentieth Century: Technological Innovations and Their Consequences. Oxford University Press, Oxford.
Smith, B.D., 1992. Rivers of Change: Essays on Early Agriculture in Eastern North America. Smithsonian Institution Press, Washington, DC.
Smith, B.D., 1996. The Emergence of Agriculture. Scientific American Library, New York.

Smith, B.D., 1997. The initial domestication of *Cucurbita pepo*in the Americas 10,000 years ago. Science 276, 932–934.

Smith, B.D., 2006. Eastern North America as an independent center of plant domestication. Proceedings of the National Academy of Sciences of the United States of America 103, 12223–12228.

Tooby, J., Cosmides, L., 1990. The past explains the present: emotional adaptations and the structure of ancestral environments. Ethology and Sociobiology 11, 375–424.

Trappe, J.M., 1987. Phylogenetic and ecologic effects of mycotrophy in the angiosperms from an evolutionary standpoint. In: Safir, G.R. (Ed.), Ecophysiology of VA Mycorrhizal Plants. CRC, Boca Raton FL.

Vavilov, N.I., 1917. On the origin of cultivated rye. Bulletin of Applied Botany 10, 561–590.

Vavilov, N.I., 1926. Studies on the origin of cultivated plants. Bulletin of Applied Botany 16, 1–245.

Wang, B., Qiu, Y.L., 2006. Phylogenetic distribution and evolution of mycorrhizas in land plants. Mycorrhiza 16, 299–363.

Wells, S., 2010. Pandora's Seed: The Unforeseen Cost of Civilization. Random House, New York.

Wright, G., Gordus, A., 1969. Distribution and utilization of obsidian from Lake Van sources between 7500 and 3500 B.C. American Journal of Archaeology 73, 75–77.

Yang, X., Wan, Z., Perry, L., Lu, H., Wang, Q., Zhao, C., Li, J., Xi, E., Yu, J., Cui, T., Wang, T., Li, M., Ge, Q., 2012. Early millet use in northern China. Proceedings of the National Academy of Sciences of the United States of America 109, 3726–3730.

Yechieli, Y., Margaritz, M., Levy, Y., Weber, U., Kafri, U., Woelfli, W., Bonani, B., 1993. Late Quaternary geological history of the Dead Sea area, Israel. Quaternary Research 39, 59–67.

Zeder, M.A., 2011. The origins of agriculture in the Near East. Current Anthropology 52, S221–S235.

Zeder, M.A., Emshwiller, E., Bradley, D., Smith, B.D., 2006. Documenting Domestication: New Genetic and Archaeological Paradigms. University of California Press, Berkeley.

Zeder, M.A., Smith, B.D., 2009. A conversation on agricultural origins: talking past each other in a crowded room. Current Anthropology 50, 681–691.

Zohary, D., 1965. Colonizer species in the wheat group. In: Baker, H.G., Stebbins, G.L. (Eds.), The Genetics of Colonizing Species. Academic Press, New York, pp. 403–419.

Zohary, D., 1969. The progenitors of wheat and barley in relation to domestication and agricultural dispersal in the old world. In: Ucko, P.J., Dimbleby, G.W. (Eds.), The Domestication and Exploitation of Plants and Animals. Duckworth, London, pp. 47–66.

Zohary, D., Hopf, M., Weiss, E., 2012. Domestication of Plants in the Old World, fourth ed. Oxford University Press, Oxford.

Zohary, D., Spiegel-Roy, P., 1975. Beginning of fruit growing in the old world. Science 187, 319–327.

Zuk, M., 2013. Paleofantasy: What Evolution Really Tells Us About Sex, Diet, and How We Live. W. W. Norton, New York.

Part VI

Health

Chapter 26

Evolutionary and Developmental Origins of Chronic Disease

Felicia M. Low[1], Peter D. Gluckman[1] and Mark A. Hanson[2]

[1]Liggins Institute, The University of Auckland, Auckland, New Zealand; [2]Institute of Developmental Sciences, University of Southampton, Southampton General Hospital, Southampton, UK

SYNOPSIS

Multiple developmental pathways affect the development of obesity and associated morbidities. Early-life cues may evoke processes of developmental plasticity that while adaptive (in the evolutionary sense) in origin may have longer-term adverse consequences when a mismatch between the early- and later-life environments arises, resulting in an increased susceptibility to disease. The inability of our evolved metabolic physiology to cope with the evolutionary novelty of the modern obesogenic environment is a further factor. Using obesity and related chronic diseases as exemplars, we examine how evolutionary and developmental histories influence predisposition to disease, and review the evidence for the role of epigenetic mechanisms.

EVOLUTIONARY PERSPECTIVE ON HUMAN DISEASE

In evaluating the etiology of human disease, medical science can offer illuminating insights into the genetic, molecular, and physiological underpinnings of how pathology has arisen—that is, the "proximate" explanations for disease. However, such explanations do not provide any context for or understanding of why disease has arisen in the first place, or why susceptibility to disease may vary among individuals or populations. Incorporating an evolutionary perspective to consider "ultimate" explanations can offer much more meaningful insights into human health and disease, and indeed can lay a path toward further basic and applied research (Gluckman et al., 2011b; Stearns, 2012).

The multidisciplinary and expanding field of evolutionary medicine draws together concepts from evolutionary biology, biomedicine, genetics, epidemiology, and public health to provide a heuristic approach for understanding and improving human health, particularly from the ultimate perspective. One of its fundamental principles is that selection acts to maintain or promote Darwinian fitness (Nesse and Williams, 1995; Gluckman et al., 2009); that is, once the peak reproductive period has passed, there is little selective pressure for energetic resources to be devoted to maintaining health and longevity. This becomes a liability in long-lived species such as humans, in whom longevity has increased markedly over the past few centuries, since little negative selection pressure has been exerted against diseases that manifest later in life. Modeling studies based on hunter-gatherer and contemporary human populations have shown that the largest single contributor to human fitness is survival to puberty (Jones, 2009). On the other hand, a significant determinant of health in middle age, and also longevity, is the integrity of cellular and DNA repair processes; their decline contributes to morbidity and mortality, as encapsulated by the concept of antagonistic pleiotropy.

Another important consideration is that contemporary human populations now tend to live in highly novel—in evolutionary terms—nutritional, technological, and social environments. Throughout most of our evolutionary past, social groups were small and consumed very different diets, but it was under these conditions that selective processes operated to shape much of our biology. Yet temporal constraints on selection and other evolutionary processes limit the ability of our adaptive capacity to match the rapid rate of environmental change, thus impairing responses to challenges such as a rich diet and increasing vulnerability to disease. Finally, it is important to correct a common misconception that our evolutionary or developmental histories are themselves generally *causative* of disease. Rather, they serve to modulate our susceptibility to disease in specific environments, and whether adverse health consequences occur is dependent on an individual's later-life environment. Note that in this chapter, adaptation is used in its evolutionary sense, namely a change in phenotype that enhances Darwinian fitness.

We have previously identified several pathways by which evolutionary processes can influence disease risk (Gluckman et al., 2009). In this chapter we shall draw on two of these non-mutually exclusive pathways—evolutionarily mismatched or novel environments, and life-history-associated factors linking to developmental mismatch—in explaining how variations in disease risk can arise. The case of obesity and its associated chronic diseases is used as an illustrative example.

DEVELOPMENTAL ORIGINS OF HEALTH AND DISEASE

In recent decades, a large body of experimental data in animals and epidemiological and clinical studies has amassed, showing that a wide array of early-life influences can modulate the risk of noncommunicable disease in later life. This phenomenon had, in fact, been noted in the literature as early as the 1930s following the advent of cohort analysis approaches, which led to an important report highlighting the potential role of childhood factors in accounting for adult mortality rates (Kermack et al., 1934). However, the notion that adult disease risk has a developmental component has only relatively recently become formalized as "Developmental Origins of Health and Disease" (DOHaD; Gluckman and Hanson, 2006b); this followed thorough and systematic epidemiological investigations by several investigators in the 1980s, including David Barker and colleagues, which showed a negative correlation between birth weight—an approximate indicator of fetal growth and intrauterine nutrition—and the risk of adult cardiovascular disease–related mortality (Barker et al., 1989; Osmond et al., 1993). Developmental influences have subsequently been implicated in other pathological conditions including diabetes, osteoporosis, breast cancer, mental disorders, poor respiratory health, and atopy (reviewed in Godfrey, 2006).

Barker, together with Hales, developed an initial theoretical model known as the "thrifty phenotype hypothesis." They proposed that nutritional deprivation prompts the fetus to develop insulin resistance that restricts growth and favors survival to birth, but that this would lead to a predisposition to diabetes later (Hales and Barker, 1992). However, this precise hypothesis has not been confirmed experimentally, as growth-restricted neonates do not become insulin resistant until several years after birth (Mericq et al., 2005), and indeed may first go through a period of increased insulin sensitivity. Furthermore, other assumptions implicit in the model, including that poor fetal growth reflected in low birth weight is the causal factor, and that undernutrition is required to be severe for phenotypic effects to be observed, have met with opposing empirical evidence requiring the development of modified explanatory frameworks. For example, variation in the balance of maternal nutrition has effects on childhood carotid intima–media thickness, a risk marker of atherosclerosis, in a birth weight–independent manner (Gale et al., 2006), while the association between birth weight and the incidence of type 2 diabetes or impaired glucose tolerance in adulthood shows gradation over the normal range of birth weights (Godfrey, 1998). These observations indicate the operation of early-life effects over a normative range of developmental exposures.

However, the thrifty phenotype hypothesis became the basis from which others developed conceptual models that addressed these limitations and that incorporated emerging understandings in evolutionary and developmental biology. It is now clear that multiple developmental pathways affect later disease risk. Broadly, these can be defined as those having an evolved origin because of their potential adaptive advantage as seen in the predictive adaptive response (PAR) paradigm, and those that arise because of changes in the environment of the fetus due to probable evolutionary novelty and for which an adaptive explanation is inappropriate, as in the situations of maternal obesity, gestational diabetes, and infant overfeeding (see later). While all these relationships are encapsulated in the DOHaD paradigm, mechanistically they must be considered separately.

The DOHaD paradigm was slow to be accepted: it was unclear what its mechanistic explanation was in either theoretical or biological terms, and how important it was in explaining either individual or population-based risk. The early studies had inappropriately focused on the extremes of birth weight, thus narrowing their perceived relevance. More recently, however, each of these issues has been addressed and will be expanded here.

DEVELOPMENTAL PLASTICITY AND PREDICTIVE ADAPTIVE RESPONSES

While severe insults such as toxicants may lead to disruption of the developmental program, resulting in teratogenesis and necessitating phenotypic accommodation, organisms are well equipped to cope with environmental challenges of lesser severity operating over a range of timescales. On a short timescale, homeostasis serves as a rapid and reversible mechanism to cope with immediate challenges. At the other end of the spectrum, selection acts at the population level over multiple generations, whereby responses to environmental changes become reflected in genomic changes; for example, the increase in the amylase gene copy number in populations with a history of starch consumption (Perry et al., 2007). On an intermediate timescale, the processes of "developmental plasticity"—responses of a potentially adaptive nature that enable the organism to adjust its phenotypic development to match its environment—operate over a life course, although there can be transgenerational echoes. They can be evoked by cues such as altered maternal nutrition, maternal stress, and predator exposure. Importantly, these are evolved adaptive responses to cues operating within the normal range of ecological conditions.

Such potentially adaptive plasticity may lead to a range of phenotypes manifest at maturity. These may be discontinuous phenotypes (morphs) reflecting polyphenism (Mayr, 1963) or a continuous range in phenotype as determined by the extent to which specific traits are evident, known as reaction norms (Schlichting and Pigliucci, 1998). An example of the former is the development of female honey bee larvae into highly fecund queens or sterile worker morphs depending on whether the larval diet consisted exclusively of royal jelly (Foret et al., 2012). Reaction norms can be manifest in many aspects such as physiological set points (e.g., body composition and hypothalamic–pituitary–adrenal axis responses), life history (e.g., timing of puberty), and behavior (e.g., parenting and exercise propensity). Thus, a single genotype has the potential to give rise to multiple phenotypes, depending on the developmental environment.

Adaptive responses may result in effects that occur soon after exposure to the inducing cue, in which case they are broadly classified as immediate adaptive responses. An example is growth restriction by the fetus to counteract a deprived nutritional environment, whereby the adaptation increases chances of survival to birth, albeit incurring a trade-off of higher risk of morbidity and mortality in infancy. However, many plastic responses prompt effects that only manifest later in the life course for anticipated fitness benefit—the aforementioned "predictive adaptive responses" or PARs (Gluckman et al., 2005; Bateson et al., 2014). Such developmental anticipation is ubiquitous across taxa. For example, it can be seen in the development of an antipredator protective helmet in *Daphnia* offspring whose mothers were exposed to remains of other killed *Daphnia* (Agrawal et al., 1999), and in rats that had received low levels of neonatal maternal care that displayed enhanced learning and memory in response to high-stress situations (Champagne et al., 2008). The distinction between immediate responses and PARs is not absolute, and they may occur concurrently, as in the case of the growth-restricted fetus demonstrating stunting in childhood in anticipation of postnatal nutritional scarcity.

Adopting an anticipatory strategy carries the inherent possibility that predictions may turn out to be wrong—for example, if the cue is not an accurate reflection of the environment, or if the environment changes substantially after anticipatory responses have been made and the developmental trajectory has been established, creating a situation of "mismatch" in later life. This scenario is widely reported in societies where nutrition or other aspects of lifestyle change between generations, for example through socioeconomic improvement. Despite the potential for PARs to become maladaptive should such discordance arise, they have not been eliminated by negative selection, because inaccurate predictions do not incur a major fitness cost as long as reproductive success is largely unaffected. Thus, if the consequence of inaccurate prediction is disease in middle life, there will be little negative selection. Furthermore, in species such as *Homo sapiens* that have low offspring numbers and for which environmental change has generally taken place over a timescale of generations, modeling studies suggest that fitness advantage may still be obtained even if predictions are not entirely accurate (Jablonka et al., 1995). It is also likely that there has been an evolved bias in favor of anticipation of poor conditions due to the differential fitness costs of making inaccurate bidirectional predictions. To use food availability to illustrate: predicting a plentiful nutritional environment but encountering sparse nutritional conditions will leave the individual phenotypically ill-equipped to cope with the unforeseen deficit—empirical evidence shows that this incurs a greater cost to fitness than predicting a poor nutritional environment but experiencing abundant nutrition. This asymmetry suggests that humans have evolved with PARs biased toward the latter. The retention of PARs through evolution also suggests that humans have been exposed to varied environmental conditions, and hence diets, in different locales over their evolutionary history.

Empirical Support for the PAR Hypothesis

One way to test the validity of the PAR hypothesis is by determining whether the primed phenotype can be reversed by applying a manipulation (e.g., hormonal or nutritional) during the early-life window of development to alter responses. One animal study providing experimental support for this approach is the maternal undernutrition rat model, wherein offspring of dams exposed to a hypocaloric diet during pregnancy (30% of ad libitum of a standard diet) grow up to display a range of characteristics consistent with an energy-conserving integrated phenotype, including hyperphagia, reduced activity, and a dietary preference for fat. In addition, they tend to be obese, hypertensive, and insulin- and leptin-resistant—features that become exacerbated with consumption of a postnatal high-fat diet (Vickers et al., 2000, 2003). This adult phenotype suggests that the pups have responded to the early cues by predicting a postnatal low-nutrition environment and modifying their developmental trajectory accordingly. At least in rodents, the key energy-regulating pathways continue to develop after birth, presenting an opportunity for their adjustment by pre- and postnatal interventions. In support of this, injection of the satiety adipokine leptin during the neonatal period can restore the adult phenotype to that of a pup adequately nourished in utero (Vickers et al., 2005). Similarly, leptin administration in intrauterine growth-restricted piglets can reverse several aspects of the induced phenotype, such as poor postnatal growth rate and decreased lean mass (Attig et al., 2008). The precise molecular mechanisms by which these interventions achieve phenotypic reversal remain unclear, although it is known that leptin regulates neuronal activity in the arcuate nucleus to suppress food intake (Vickers and Sloboda, 2012), and that it has neurotrophic actions on the infant rat (Bouret et al., 2004).

We have provided the first direct evidence that responses made in utero have fitness-promoting effects later in life in humans (Forrester et al., 2012). We conducted a clinical study on Jamaican survivors of marasmus or kwashiorkor, two syndromes of severe childhood malnutrition that can occur in a population undergoing nutritional stress. Marasmus arises from insufficient total caloric intake, in contrast to kwashiorkor, which is characterized by insufficient protein consumption but adequate caloric intake. The latter syndrome presents with a higher mortality rate, as well as patterns of protein and lipid metabolism that appear to be less efficient at exploiting existing tissue stores (Jahoor et al., 2008). In this cohort of survivors, those children developing marasmus were found to have had a significantly lower birth weight (333 g difference) than those developing kwashiorkor. This suggests that the more constrained fetal environment experienced by marasmics, as manifest in lower birth weight, had prompted them to predict a low-nutrition postnatal environment and hence develop the appropriate physiological phenotype to cope with it. The presence of a survival advantage among marasmics provides a clear demonstration of the adaptive significance of PAR.

EPIGENETICS AS AN UNDERPINNING MECHANISM

It is now becoming increasingly apparent that epigenetic processes, which are the mitotically stable molecular mechanisms that modulate gene expression without altering DNA sequence, are involved in effecting developmental plasticity. Epigenetic mechanisms are responsible for some of the most fundamental cellular processes such as cell differentiation, transposon silencing, and gene dosage control, including parental imprinting and X chromosome inactivation (Allis et al., 2007). These are achieved via multiple pathways, including the attachment of a methyl group to specific cytosine residues, known as DNA methylation; covalent modifications such as acetylation and methylation at specific amino acid residues of histone proteins to modulate chromatin compaction or decondensation; and the activity of regulatory noncoding RNA strands in mediating chromatin remodeling as well as transcriptional and posttranscriptional activities (Jones, 2012; Bannister and Kouzarides, 2011; Morris and Mattick, 2014). Additionally, enzymatic alteration of RNA sequence through RNA editing, which is particularly prevalent in the brain, has been proposed to underlie a form of neural plasticity (Mattick, 2010). The reversibility of the chemical modifications involved, and their receptivity to external cues, allows epigenetic mechanisms to function as a nexus between environmental exposures and transcriptional regulation.

Nonhuman Animal Evidence

Several animal models demonstrating early-life determinants of metabolic disease risk have helped to shed light on the underlying mechanistic basis (Seki et al., 2012). In the maternal undernutrition rat model, epigenetic analysis of hepatic tissue in fetally undernourished offspring revealed hypermethylation at the promoter encoding the lipid metabolism regulator PPARα, which was also associated with decreased expression of its gene (Gluckman et al., 2007b). Upregulation of hepatic 11β-HSD2, a glucocorticoid inactivator, was also observed. Importantly, such effects were completely abolished in leptin-treated pups, thus lending support to the epigenetic basis of the primed phenotype.

In another model, where rat dams are fed a pregnancy diet in which protein content is reduced by ≥50%, offspring become hypertensive and demonstrate vascular endothelial dysfunction and altered lipid metabolism in adulthood (Langley and Jackson, 1994; Erhuma et al., 2007). Curiously, in this model hypomethylation at the promoters of the *Ppara* and *GR* genes in hepatocytes, and corresponding increases in gene expression, were seen in offspring (Lillycrop et al., 2005). This may reflect the very specific epigenetic responses to different intrauterine nutritional perturbations. The aberrant epigenetic regulation at *GR* appears to be modulated by decreased expression of Dnmt1, the DNA methyltransferase responsible for maintaining extant patterns of DNA methylation (Lillycrop et al., 2007). Other epigenetic analyses have suggested a role for the renin–angiotensin system, showing differential methylation at the rat adrenal gland angiotensin II receptor, type 1b (*Agtr1b*) promoter (Bogdarina et al., 2007), and at the mouse brain angiotensin-converting enzyme-1 (*Ace-1*) gene promoter (Goyal et al., 2010).

Notably, even a moderate reduction in food intake (to 70% of ad libitum) in baboons (*Papio* sp.) that had only minor effects on overall growth and weight was sufficient to elicit hypomethylation at the promoter of *PCK1*, which encodes the rate-limiting enzyme in gluconeogenesis (Nijland et al., 2010). This underscores how even environmental challenges that are not severe can induce alterations in fetal development.

Human Evidence

While animal experimental studies have provided a wealth of information on associations between early-life events and epigenetic changes, there is a relative paucity of human data, owing to obvious ethical constraints. Much of the current human evidence has been derived from retrospective studies based on natural experiments involving adverse circumstances such as famine. A major contributor to this body of literature is a cohort of individuals whose mothers were exposed to the Dutch Hunger Winter famine of 1944–45 during pregnancy. A heightened risk of developing coronary heart disease, obesity, and a poorer lipid profile in adulthood, compared with unexposed controls, has been observed among those affected during early pregnancy (Roseboom et al., 2006). Interestingly, impaired glucose tolerance was seen irrespective of the time of exposure in gestation, while microalbuminuria and obstructive airways disease were linked to exposure in mid-gestation, suggesting that there are certain ontogenetic windows in fetal development during which traits relating to risks of specific pathologies are established, and beyond which trajectories become canalized. Analysis of genomic DNA from these individuals nearly 60 years post exposure uncovered differences in methylation levels at gene loci implicated in growth and metabolic and cardiovascular disease (Heijmans et al., 2008; Tobi et al., 2009), underscoring the enduring molecular effects of transient environmental exposure.

Although studies of famine do give insights, they represent extreme scenarios that are unlikely to be experienced by most of the population, and the small size of the effects observed has been thought to limit interpretation. However, we have demonstrated that even relatively unremarkable variations in maternal nutrition in normal pregnancies are linked to altered offspring epigenetic states at birth. In a cohort of children whose umbilical cord tissues were analyzed at birth, lower maternal carbohydrate intake in early pregnancy was strongly associated with higher levels of methylation at one CpG of *RXRA*, encoding the retinoid X receptor alpha protein involved in fat metabolism and insulin sensitivity (Godfrey et al., 2011). This is in accordance with prior research implicating such a dietary pattern in neonatal adiposity (Godfrey et al., 1997). Of particular interest, *RXRA* methylation levels were positively correlated with adiposity in children at six or nine years of age in two independent cohorts, demonstrating for the first time an association between the epigenetic profile at birth and clinically relevant variation in later phenotype. The methylation–childhood adiposity relationship was sustained throughout the normal range of maternal nutritional states and birth weights. Thus, in contrast with observations from famine-based studies, these results provide compelling evidence that even subtle variations in the prenatal environment in uncomplicated pregnancies can modulate an offspring's later risk of obesity and predisposition to metabolic disease through epigenetic mechanisms. These results strongly suggest that developmental plasticity is a significant determinant of interindividual variations in disease risk. Furthermore, famine-induced methylation differences, while statistically robust, were small (absolute within-pair difference ≤2.4%; (Tobi et al., 2009)); however, in the cohorts of normal pregnancies, *RXRA* methylation could, remarkably, predict 25% of variance in childhood adiposity. Links between epigenetic marks in cord blood and child body composition have also been reported (Relton et al., 2012; Perkins et al., 2012).

Subsequent studies, albeit of smaller scale, have investigated the effects of prenatal nutrition on neonatal or later-life epigenetic marks with possible physiological relevance. For example, a pilot study of Scottish mothers specifically advised to adopt a low-carbohydrate/high-meat diet during pregnancy uncovered differential methylation at the promoter regions of *HSD2* and *GR*, and the *H19* imprinting control region—genes involved in blood pressure, glucocorticoid action, and fetal growth, respectively—in offspring at adulthood (Drake et al., 2012). This aligns with separate analyses of offspring adult phenotype showing a link between a similarly unbalanced maternal diet and risk of hypertension, hypercortisolemia, and potentiated hypothalamic–pituitary–adrenal axis stress responses.

While ongoing technological developments such as whole-genome scanning have boosted the field of human developmental epigenetics, a number of challenges remain. To gain better appreciation of the functional significance of changes in epigenetic marks, epigenetic epidemiology studies need to be cognizant of the importance of validation through within- and between-cohort replication, such as that reported by Godfrey et al. (2011). Other studies have reported small effect sizes that are of uncertain translational value. Further, many technical issues such as tissue heterogeneity or nonnormal distribution of epigenetic marks may complicate data interpretation (Niculescu, 2012). As new human epigenetic data emerge, careful methodological scrutiny will be essential in order to avoid errors through inappropriate extrapolation.

EVOLUTIONARY AND DEVELOPMENTAL MISMATCH: THE CASE OF OBESITY AND RELATED CHRONIC DISEASES

Over recent decades, the global rates of overweight and obesity, as well as related chronic conditions including type 2 diabetes and cardiovascular disease, have increased rapidly (Ng et al., 2014; International Diabetes Federation, 2013), and this is forecast to impose heavy burdens on public health, the economy, and society even in economically prosperous countries (Wang et al., 2011). This predicament is likely to worsen in the light of the decreasing age of disease onset, and indeed several alarming statistics have been published describing the growing prevalence of diabetes, hypertension, and cardiometabolic risk factors in children and adolescents (Adair et al., 2014; Xi et al., 2014; May et al., 2012).

The relentless increase in obesity rates suggests that public health initiatives formulated to combat these epidemics by targeting adult lifestyle choices to change the energy-intake:energy-expenditure ratio are inadequate, and that a reassessment of the current paradigm for combating the trend is needed. We have argued for the importance of incorporating both evolutionary and developmental perspectives into the conceptual framework for understanding the basis of obesity and its associated diseases, and developing more holistic prevention approaches (Gluckman and Hanson, 2008; Gluckman et al., 2011a).

Evolutionary Mismatch

The ready availability of the modern-day high-fat, high-calorie, and high-glycemic index diet presents a vastly different nutritional environment from what prevailed throughout most of our evolutionary history, during which humans and our hominin ancestors survived as hunter-foragers, with selective processes operating to shape our physiology and metabolism according to the dietary and physical requirements of such a lifestyle. Much of the nutritional change has occurred relatively recently in evolutionary terms—over the past several thousand to several hundred years, with particularly rapid change in recent decades. The speed at which our genotype has evolved has not kept pace with such change, consequently leading to a metabolic mismatch between our physiology and our contemporary environment (see Figure 1) (Gluckman and Hanson, 2006a). While there is still some debate on whether decreased energy expenditure today is also a contributing factor (Pontzer et al., 2012), the implications of this mismatch have become a well-established public health issue in many high income countries, and are increasingly affecting low-middle income countries.

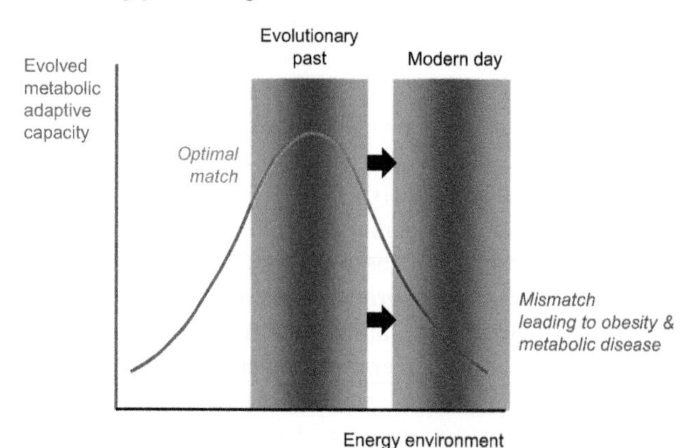

FIGURE 1 Energetic mismatch is the basis of metabolic disease: The typical energy environment of most of our evolutionary past (left gradient block) corresponds to the optimal level to which our physiology and metabolism have evolved to respond (green (light gray in print versions) line). Our physiology is less well adapted to the modern-day nutritional transition to a higher energy environment (right gradient block), and the resultant mismatch manifests as an increased risk of obesity and associated metabolic disease (red (gray in print versions) line).

It is important to note that even though humans as a species are susceptible to evolutionary mismatch, there is considerable variation in an individual's or population group's vulnerability to disease. For example, modeling work has shown that among individuals with a normal BMI of 24, those of South Asian descent have a greater than five-fold incidence rate of diabetes compared with white subjects (Chiu et al., 2011). Early hopes that such variation had a genetic explanation had to be tempered by findings from genome-wide association studies showing relatively modest effect sizes for known variants (McCarthy, 2010). There has been a growing consensus that holding a narrow gene-centric view of disease determinants is no longer tenable, particularly in light of the recent milestones of the Encyclopedia of DNA Elements (ENCODE) project in assigning potential regulatory roles to much of the human genome, suggesting that clues to the genetic basis of disease may lie within noncoding loci (The ENCODE Project Consortium, 2012). It is becoming evident that much of the phenotypic variance in disease risk has a developmental component with an underlying epigenetic basis.

Developmental Mismatch

In maximizing adaptation of the organism to the nutritional environment, selection may act both directly on traits regulating metabolism and indirectly through the processes of developmental plasticity, to enable phenotypic adjustments to best match the current or predicted environment. Thus, metabolic mismatch could also arise if developmental cues are misinterpreted—for example, because of placental inadequacy or dysfunction that limits nutrient transfer and erroneously signals a low-nutrition environment, or encountering an environment in later life that differs from that experienced in early life. The latter is well demonstrated by studies in the maternal undernutrition rat model: offspring primed by maternal undernutrition to anticipate a low-nutrition environment become obese when food supply is freely available, but crucially, metabolic dysfunction is amplified when the degree of mismatch is increased via exposure to a hypercaloric diet (Vickers et al., 2000).

There are several circumstances in which developmental mismatch has potential broad impact on human health. The first relates to the rapid nutritional transition occurring in many low-middle income countries as a result of economic advancements bringing about improved standards of living, often over the timeframe of a single generation. Thus, those whose early-life experiences were obtained prior to the transition are unlikely to have developed the appropriate phenotype for coping with an abundant nutrient environment and will therefore be at elevated risk of obesity and metabolic disease. Epidemiological evidence also points to rural-to-urban migrants, or individuals emigrating to more prosperous countries, as being at similar risk (Ebrahim et al., 2010; Shan et al., 2011). Short stature, or stunting, has an incidence rate of about 30% among children under five in certain low-middle income regions of the world. There is a strong association between stunting and visceral adiposity in children living in countries undergoing the nutritional transition (Popkin et al., 1996), particularly manifest in girls at the time of puberty, and concerns over the potential economic and social consequences of this are rising (Hoffman and Klein, 2012).

A change of environment need not be a prerequisite for mismatch to occur. Studies have found that few women planning a pregnancy adhere to appropriate nutritional and lifestyle guidelines (Inskip et al., 2009); this has been observed even among women undergoing in vitro fertilization treatment—a group that would generally be characterized as motivated to ensure the best outcome for their conceptus (Redward et al., 2012). More widely, in the general population many women consume unbalanced diets even in affluent societies, and change their diets and lifestyle little even when pregnant. Given that minor dietary variations during pregnancy could affect offspring disease risk, and that many pregnancies are unplanned, there may be utility in implementing DOHaD-based nutritional education programs for adolescents as a preventative strategy (Bay et al., 2012), although such programs may be more effective if conducted outside the school classroom (Grace et al., 2012).

Maternal Constraint

A possibly important contributor to developmental mismatch is the processes of maternal constraint. This concept refers to restrictions on nutrient transfer imposed by the mother during gestation, via placental–fetal endocrine interactions, to prevent fetal size at birth from exceeding her pelvic dimensions (Gluckman and Hanson, 2004). Such a limitation is necessary due to the evolutionary compromise needed to balance a large brain and bipedal posture with minimizing the risk of dystocia (Hanson and Godfrey, 2008). In some low-middle income countries where medical care is not rapidly accessible, the risk of damaging obstructed labor, sometimes leading to obstetric fistula and fetal death, is a major concern. As predicted by the concept of maternal constraint, such problems are more commonly associated with primiparous than subsequent pregnancies, shorter women, and those carrying a male fetus (Muleta et al., 2010). However, because maternal constraint operates in all pregnancies, even in high income countries, all fetuses to a degree predict a lower postnatal nutritional plane than is in fact the case, producing a potential mismatch against the backdrop of the modern-day high-nutrient environment.

Another consequence of this phenomenon is that first-borns are subject to a greater degree of maternal constraint compared with their siblings. From a proximate viewpoint, this greater constraint may be a result of lower vasodilative capacity of uterine blood vessels in the first pregnancy. However, an ultimate perspective would suggest that for much of our evolutionary history, firstborns were probably conceived fewer than four years postmenarche, before the age at which the woman's pelvis reaches its maximal diameter. Firstborns are about 100–150g lighter at birth (Ward, 1993), and in support of the predictive and mismatch models, have been found to have increased fat mass through childhood, adolescence, and young adulthood in comparison with subsequent siblings (Celi et al., 2003; Reynolds et al., 2010). Unsurprisingly, the risk of developing metabolic syndrome is elevated in these individuals (Gaillard et al., 2014). The wider impact of such adverse health consequences becomes apparent in view of the declining fertility rate in many European and Asian countries. It is also pertinent to note that the overwhelming majority of children born in China over the past few decades and to the present time are firstborn.

Excessive Nutrition and Gestational Diabetes Mellitus

Much of the early work in the field of nutrition-related DOHaD, inspired by Barker and colleagues' research, focused on nutrient deprivation as a developmental cue. However, maternal factors such as high maternal prepregnancy body mass index, excessive gestational weight gain (GWG), and gestational diabetes mellitus (GDM) are rapidly becoming pervasive. There is now a burgeoning body of research investigating how excessive nutrition in early life may impact later-life health. For example, mice dams exposed to a high-fat diet from weaning through to pregnancy and lactation have offspring predisposed toward weight gain, hypertension, fatty liver, and disrupted lipid and glucose metabolism in adulthood, even when raised on a control postnatal diet (Elahi et al., 2009). Even restricting high-fat maternal feeding to the pregnancy or lactation periods alone led to elevated blood pressure, obesity, and hyperinsulinemia in control diet–fed offspring (Khan et al., 2005). Mild overnutrition before weaning, in rats induced through litter reduction, was sufficient to lead to greater adiposity and to interfere with glucose homeostasis and gene expression of the hypothalamic leptin receptor (Rajia et al., 2010). In humans, preconceptional overweight or obesity and excessive GWG are risk factors for being born large for gestational age (Black et al., 2012), and are predictors of neonatal adiposity and overweight in early adulthood. Similarly, infant formula feeding appears to be associated with greater adiposity (Stettler et al., 2005); formula-fed infants tend to be poorer at self-regulating energy intake than breastfed infants (Li et al., 2010). There is also considerable evidence that GDM is linked to heightened risk of obesity in late childhood (Crume et al., 2011)—this is presumably induced by fetal hyperinsulinemia leading to greater adipogenesis.

The full mechanistic basis for how the apparently opposite developmental influences of maternal under- and overnutrition can lead to a similar adult phenotype in the offspring, characterized by obesity and metabolic dysfunction, remains to be elucidated. However, while the former is likely to reflect the evolved mechanisms of developmental plasticity, the latter is likely to represent evolutionary novelty. Both may involve epigenetic mechanisms that have been well described as a response to a wide variety of developmental exposures, including those that are toxic (Knopik et al., 2012). An increasing number of studies are identifying physiologically relevant genes at which epigenetic changes occur in response to early-life overnutrition. Adipose tissue in offspring of mice fed a high-fat diet during pregnancy showed methylation changes or histone modifications at genes encoding the adipocyte-derived hormones adiponectin and leptin (Masuyama and Hiramatsu, 2012). Employing an obesity-resistant rat strain to exclude the effects of gestational adiposity or diabetes, Strakovsky et al. (2011) found that a gestational high-fat diet altered multiple histone marks at hepatic *Pck1*, leading to increased gene expression and suggesting an impact on gluconeogenic function in offspring. Interestingly, there is also evidence that a maternal high-fat diet may bias offspring dietary preference toward sugar and fat via epigenetic modulation within the mesocorticolimbic reward system (Vucetic et al., 2010).

Because early-life overnutrition as experienced by humans in the modern-day environment is essentially an evolutionary novelty, it is likely that the PAR model does not apply to the physiological responses to maternal obesity or GDM, as PARs have not evolved to operate under nutritionally surfeit conditions. Indeed, we have proposed that GDM represents an exaggerated form of an otherwise normal physiological response to pregnancy (Ma et al., 2013). Maternal insulin resistance occurs to a degree in all pregnancies to promote nutrient transfer to the fetus. However, there is an absence of a mechanism to limit placental glucose transfer, even if maternal hyperglycemia reaches extreme levels. This suggests that GDM was not a frequent negative selector in our evolutionary past, and that the evolution of the placenta favored fetal buffering against under- rather than overnutrition. Furthermore, mild fetal hyperglycemia stimulates insulin production to promote somatic growth and adiposity, and hence fitness—the greater fatness of human neonates relative to other mammals is posited to protect the energy-intensive human brain in case of nutritional scarcity (Kuzawa, 2010).

Maternal obesity and GDM create particular challenges because of the potential for intergenerational transmission of the metabolic phenotype to offspring (Dabelea and Crume, 2011). Some populations may be at particular risk—rates of GDM are high among Canadian First Nations people, and a modeling approach has shown that GDM may explain up to 30% of cases of type 2 diabetes within this population (Osgood et al., 2011).

TRANSGENERATIONAL INHERITANCE

The multiple ways by which determinants of disease risk that arise from early-life events can become transgenerationally transmitted have been well canvassed (Gluckman et al., 2007a; Danchin et al., 2011). One of these is the possibility of direct *trans*-meiotic passage of developmentally induced epigenetic marks. A growing number of animal studies are seeking experimental evidence for this by focusing on male-line transmission or by investigating effects beyond the F_2 generation to eliminate confounding contributions of maternal physiological or behavioral effects as well as grand maternal effects. Because the only paternal contribution to offspring in experimental settings is sperm, experimentally induced phenotypic effects may be attributed to epigenetic reprogramming in the paternal germline. Such modifications need to survive erasure during spermatogenesis in order to be passed on to the next generation, and molecular plausibility has been provided by observations that specific sperm histones at developmental loci are preserved during human spermiogenesis (Hammoud et al., 2009). In addition, although the amount of cytoplasm in sperm is small, it nonetheless contains some noncoding micro-RNAs that may be another mode of intergenerational transmission of the epigenetic mark and may affect the early stages of embryonic development (Krawetz et al., 2011).

Both male and female progeny of male mice raised on a protein-restricted diet showed changes in a raft of hepatic genes known to be involved in lipid and cholesterol metabolism (Carone et al., 2010). Among the many but small differences in hepatic DNA methylation levels detected was hypermethylation at an intergenic CpG island upstream of *Ppara*. Male rats fed a high-fat diet and presenting with obesity, glucose intolerance, and insulin resistance sired female offspring that were glucose intolerant due to abnormal insulin secretion and disrupted β-cell function (Ng et al., 2010). Another study examining F_2 offspring of mice that had experienced GDM found hypermethylation at certain differentially methylated regions of the imprinted genes *Igf2* and *H19* in pancreatic islets, along with decreased expression of these genes (Ding et al., 2012). Of note, *Igf2* and *H19* gene expression was found to be downregulated in sperm of F_1 males, supporting the premise that germ cell epigenetic modifications play a role in transgenerational transmission.

There is currently very limited human evidence for the multigenerational effect of early-life environmental influences. Hints of this have come from epidemiological analysis of a historical Swedish cohort, showing that the paternal grandfather's prepubertal food supply was linked to his grandson's mortality risk ratio; effects were only seen with paternal grandparents, suggesting operation through the male line (Pembrey et al., 2006). The potential for heritable, developmentally induced epigenetic changes to become genomically fixed and impact evolutionary processes has been flagged as an important research avenue (Bateson and Gluckman, 2011).

CONCLUDING REMARKS

By emphasizing a more contextual interpretation of health and disease, evolutionary principles serve as important complements to traditional medical explanations in understanding how variations in disease risk between individuals or population groups can arise. The overwhelming evidence that early-life factors are important modulators of disease risk in later life argues for the inclusion of development as a third component of the gene–environment interaction model. However, the impact of the DOHaD paradigm clearly extends far beyond conceptual arguments, as a deeper understanding of the developmental origins of obesity and its associated diseases may pave the way for appropriate clinical interventions to be applied. Multiple animal studies have demonstrated that a shift toward a normal phenotype and epigenotype may be achieved by intervening early in the life course.

While there are caveats in interpreting human developmental epigenetic data, these cannot negate the considerable therapeutic potential for epigenetic biology in tackling the epidemics of obesity and metabolic disease, for example through early prognostic identification of individuals most at risk. Whatever interventional strategies are devised, it is unequivocal that consideration of developmental factors needs to be given greater weight.

REFERENCES

Adair, L.S., Gordon-Larsen, P., Du, S.F., Zhang, B., Popkin, B.M., 2014. The emergence of cardiometabolic disease risk in Chinese children and adults: consequences of changes in diet, physical activity and obesity. Obesity Reviews 15, 49–59.

Agrawal, A.A., Laforsch, C., Tollrian, R., 1999. Transgenerational induction of defences in animals and plants. Nature 401, 60–63.

Allis, C.D., Jenuwein, T., Reinberg, D., 2007. Epigenetics. CSHL Press, New York.

Attig, L., Djiane, J., Gertler, A., Rampin, O., Larcher, T., Boukthir, S., Anton, P.M., Madec, J.Y., Gourdou, I., Abdennebi-Najar, L., 2008. Study of hypothalamic leptin receptor expression in low-birth-weight piglets and effects of leptin supplementation on neonatal growth and development. American Journal of Physiology–Endocrinology and Metabolism 295, E1117–E1125.

Bannister, A.J., Kouzarides, T., 2011. Regulation of chromatin by histone modifications. Cell Research 21, 381–395.

Barker, D.J.P., Winter, P.D., Osmond, C., Margetts, B., Simmonds, S.J., 1989. Weight in infancy and death from ischaemic heart disease. Lancet 2, 577–580.

Bateson, P., Gluckman, P., 2011. Plasticity, Robustness, Development and Evolution. Cambridge University Press, Cambridge.

Bateson, P., Gluckman, P.D., Hanson, M.A., 2014. The biology of developmental plasticity and the predictive adaptive response hypothesis. Journal of Physiology 592, 2357–2368.

Bay, J.L., Mora, H.A., Sloboda, D.M., Morton, S.M., Vickers, M.H., Gluckman, P.D., 2012. Adolescent understanding of DOHaD concepts: a school-based intervention to support knowledge translation and behaviour change. Journal of Developmental Origins of Health and Disease FirstView, 1–14.

Black, M.H., Sacks, D.A., Xiang, A.H., Lawrence, J.M., 2012. The relative contribution of prepregnancy overweight and obesity, gestational weight gain, and IADPSG-defined gestational diabetes mellitus to fetal overgrowth. Diabetes Care 36, 56–62.

Bogdarina, I., Welham, S., King, P.J., Burns, S.P., Clark, A.J.L., 2007. Epigenetic modification of the renin-angiotensin system in the fetal programming of hypertension. Circulation Research 100, 520–526.

Bouret, S.G., Draper, S.J., Simerly, R.B., 2004. Trophic action of leptin on hypothalamic neurons that regulate feeding. Science 304, 108–110.

Carone, B.R., Fauquier, L., Habib, N., Shea, J.M., Hart, C.E., Li, R., Bock, C., Li, C., Gu, H., Zamore, P.D., Meissner, A., Weng, Z., Hofmann, H.A., Friedman, N., Rando, O.J., 2010. Paternally induced transgenerational environmental reprogramming of metabolic gene expression in mammals. Cell 143, 1084–1096.

Celi, F., Bini, V., Giorgi, G.D., Molinari, D., Faraoni, F., Stefano, G.D., Bacosi, M.L., Berioli, M.G., Contessa, G., Falorni, A., 2003. Epidemiology of overweight and obesity among school children and adolescents in three provinces of central Italy, 1993–2001: study of potential influencing variables. European Journal of Clinical Nutrition 57, 1045–1051.

Champagne, D.L., Bagot, R.C., van Hasselt, F., Ramakers, G., Meaney, M.J., de Kloet, E.R., Joëls, M., Krugers, H., 2008. Maternal care and hippocampal plasticity: evidence for experience-dependent structural plasticity, altered synaptic functioning, and differential responsiveness to glucocorticoids and stress. The Journal of Neuroscience 28, 6037–6045.

Chiu, M., Austin, P.C., Manuel, D.G., Shah, B.R., Tu, J.V., 2011. Deriving ethnic-specific BMI cutoff points for assessing diabetes risk. Diabetes Care 34, 1741–1748.

Crume, T.L., Ogden, L., Daniels, S., Hamman, R.F., Norris, J.M., Dabelea, D., 2011. The impact of in utero exposure to diabetes on childhood body mass index growth trajectories: the EPOCH study. The Journal of Pediatrics 158, 941–946.

Dabelea, D., Crume, T., 2011. Maternal environment and the transgenerational cycle of obesity and diabetes. Diabetes 60, 1849–1855.

Danchin, É., Charmantier, A., Champagne, F.A., Mesoudi, A., Pujol, B., Blanchet, S., 2011. Beyond DNA: integrating inclusive inheritance into an extended theory of evolution. Nature Reviews Genetics 12, 475–486.

Ding, G.-L., Wang, F.-F., Shu, J., Tian, S., Jiang, Y., Zhang, D., Wang, N., Luo, Q., Zhang, Y., Jin, F., Leung, P.C.K., Sheng, J.-Z., Huang, H.-F., 2012. Transgenerational glucose intolerance with *Igf2/H19* epigenetic alterations in mouse islet induced by intrauterine hyperglycemia. Diabetes 61, 1133–1142.

Drake, A.J., McPherson, R.C., Godfrey, K.M., Cooper, C., Lillycrop, K.A., Hanson, M.A., Meehan, R.R., Seckl, J.R., Reynolds, R.M., 2012. An unbalanced maternal diet in pregnancy associates with offspring epigenetic changes in genes controlling glucocorticoid action and fetal growth. Clinical Endocrinology 77, 808–815.

Ebrahim, S., Kinra, S., Bowen, L., Andersen, E., Ben-Shlomo, Y., Lyngdoh, T., Ramakrishnan, L., Ahuja, R.C., Joshi, P., Das, S.M., Mohan, M., Smith, G.D., Prabhakaran, D., Reddy, K.S., 2010. The effect of rural-to-urban migration on obesity and diabetes in India: a cross-sectional study. PLoS Medicine 7, 12.

Elahi, M.M., Cagampang, F.R., Mukhter, D., Anthony, F.W., Ohri, S.K., Hanson, M.A., 2009. Long-term maternal high-fat feeding from weaning through pregnancy and lactation predisposes offspring to hypertension, raised plasma lipids and fatty liver in mice. British Journal of Nutrition 102, 514–519.

Erhuma, A., Salter, A.M., Sculley, D.V., Langley-Evans, S.C., Bennett, A.J., 2007. Prenatal exposure to a low-protein diet programs disordered regulation of lipid metabolism in the aging rat. American Journal of Physiology–Endocrinology and Metabolism 292, E1702–E1714.

Foret, S., Kucharski, R., Pellegrini, M., Feng, S., Jacobsen, S.E., Robinson, G.E., Maleszka, R., 2012. DNA methylation dynamics, metabolic fluxes, gene splicing, and alternative phenotypes in honey bees. Proceedings of the National Academy of Sciences 109, 4968–4973.

Forrester, T.E., Badaloo, A.V., Boyne, M.S., Osmond, C., Thompson, D., Green, C., Taylor-Bryan, C., Barnett, A., Soares-Wynter, S., Hanson, M.A., Beedle, A.S., Gluckman, P.D., 2012. Prenatal factors contribute to emergence of kwashiorkor or marasmus in response to severe undernutrition: evidence for the predictive adaptation model. PLoS One 7, e35907.

Gaillard, R., Rurangirwa, A.A., Williams, M.A., Hofman, A., MacKenbach, J.P., Franco, O.H., Steegers, E.A.P., Jaddoe, V.W.V., 2014. Maternal parity, fetal and childhood growth, and cardiometabolic risk factors. Hypertension 64, 266–274.

Gale, C.R., Jiang, B., Robinson, S.M., Godfrey, K.M., Law, C.M., Martyn, C.N., 2006. Maternal diet during pregnancy and carotid intima-media thickness in children. Arteriosclerosis, Thrombosis, and Vascular Biology 26, 1877–1882.

Gluckman, P., Hanson, M., 2006a. Mismatch: Why Our World No Longer Fits Our Bodies. Oxford University Press, Oxford.

Gluckman, P.D., Beedle, A.S., Hanson, M.A., 2009. Principles of Evolutionary Medicine. Oxford University Press, Oxford.

Gluckman, P.D., Hanson, M., Zimmet, P., Forrester, T., 2011a. Losing the war against obesity: the need for a developmental perspective. Science Translational Medicine 3, 93cm19.

Gluckman, P.D., Hanson, M.A., 2004. Maternal constraint of fetal growth and its consequences. Seminars in Fetal and Neonatal Medicine 9, 419–425.

Gluckman, P.D., Hanson, M.A., 2006b. Developmental Origins of Health and Disease. Cambridge University Press, Cambridge.

Gluckman, P.D., Hanson, M.A., 2008. Developmental and epigenetic pathways to obesity: an evolutionary-developmental perspective. International Journal of Obesity 32, S62–S71.

Gluckman, P.D., Hanson, M.A., Beedle, A.S., 2007a. Non-genomic transgenerational inheritance of disease risk. Bioessays 29, 145–154.

Gluckman, P.D., Hanson, M.A., Spencer, H.G., 2005. Predictive adaptive responses and human evolution. Trends in Ecology and Evolution 20, 527–533.

Gluckman, P.D., Lillycrop, K.A., Vickers, M.H., Pleasants, A.B., Phillips, E.S., Beedle, A.S., Burdge, G.C., Hanson, M.A., 2007b. Metabolic plasticity during mammalian development is directionally dependent on early nutritional status. Proceedings of the National Academy of Sciences of the United States of America 104, 12796–12800.

Gluckman, P.D., Low, F.M., Buklijas, T., Hanson, M.A., Beedle, A.S., 2011b. How evolutionary principles improve the understanding of human health and disease. Evolutionary Applications 4, 249–263.

Godfrey, K., 2006. The 'developmental origins' hypothesis: epidemiology. In: Gluckman, P.D., Hanson, M.A. (Eds.), Developmental Origins of Health and Disease. Cambridge University Press, Cambridge, pp. 6–32.

Godfrey, K.M., 1998. Maternal regulation of fetal development and health in adult life. European Journal of Obstetrics and Gynecology and Reproductive Biology 78, 141–150.

Godfrey, K.M., Barker, D.J., Robinson, S., Osmond, C., 1997. Maternal birthweight and diet in pregnancy in relation to the infant's thinness at birth. British Journal of Obstetrics and Gynaecology 104, 663–667.

Godfrey, K.M., Sheppard, A., Gluckman, P.D., Lillycrop, K.A., Burdge, G.C., McLean, C., Rodford, J., Slater-Jefferies, J.L., Garratt, E., Crozier, S.R., Emerald, B.S., Gale, C.R., Inskip, H.M., Cooper, C., Hanson, M.A., 2011. Epigenetic gene promoter methylation at birth is associated with child's later adiposity. Diabetes 60, 1528–1534.

Goyal, R., Goyal, D., Leitzke, A., Gheorghe, C.P., Longo, L.D., 2010. Brain renin-angiotensin system: fetal epigenetic programming by maternal protein restriction during pregnancy. Reproductive Sciences 17, 227–238.

Grace, M., Woods-Townsend, K., Griffiths, J., Godfrey, K., Hanson, M., Galloway, I., Azaola, M.C., Harman, K., Byrne, J., Inskip, H., 2012. Developing teenagers' views on their health and the health of their future children. Health Education 112, 543–559.

Hales, C.N., Barker, D.J., 1992. Type 2 (non-insulin-dependent) diabetes mellitus: the thrifty phenotype hypothesis. Diabetologia 35, 595–601.

Hammoud, S.S., Nix, D.A., Zhang, H., Purwar, J., Carrell, D.T., Cairns, B.R., 2009. Distinctive chromatin in human sperm packages genes for embryo development. Nature 460, 473–478.

Hanson, M.A., Godfrey, K.M., 2008. Commentary: maternal constraint is a pre-eminent regulator of fetal growth. International Journal of Epidemiology 37, 252–254.

Heijmans, B.T., Tobi, E.W., Stein, A.D., Putter, H., Blauw, G.J., Susser, E.S., Slagboom, P.E., Lumey, L.H., 2008. Persistent epigenetic differences associated with prenatal exposure to famine in humans. Proceedings of the National Academy of Sciences of the United States of America 105, 17046–17049.

Hoffman, D.J., Klein, D.J., 2012. Growth in transitional countries: the long-term impact of under-nutrition on health. Annals of Human Biology 39, 395–401.

Inskip, H.M., Crozier, S.R., Godfrey, K.M., Borland, S.E., Cooper, C., Robinson, S.M., Southampton Women's Survey Study Group, 2009. Women's compliance with nutrition and lifestyle recommendations before pregnancy: general population cohort study. BMJ 338, b481.

International Diabetes Federation, 2013. IDF Diabetes Atlas, Sixth ed. International Diabetes Federation, Brussels.

Jablonka, E., Oborny, B., Molnar, I., Kisdi, E., Hofbauer, J., Czaran, T., 1995. The adaptive advantage of phenotypic memory in changing environments. Philosophical Transactions of the Royal Society of London–Series B: Biological Sciences 350, 133–141.

Jahoor, F., Badaloo, A., Reid, M., Forrester, T., 2008. Protein metabolism in severe childhood malnutrition. Annals of Tropical Paediatrics 28, 87–101.

Jones, J.H., 2009. The force of selection on the human life cycle. Evolution and Human Behavior 30, 305–314.

Jones, P.A., 2012. Functions of DNA methylation: islands, start sites, gene bodies and beyond. Nature Reviews Genetics 13, 484–492.

Kermack, W., McKendrick, A., McKinlay, P., 1934. Death rates in Great Britain and Sweden: some general regularities and their significance. Lancet 223, 698–703.

Khan, I.Y., Dekou, V., Douglas, G., Jensen, R., Hanson, M.A., Poston, L., Taylor, P.D., 2005. A high-fat diet during rat pregnancy or suckling induces cardiovascular dysfunction in adult offspring. American Journal of Physiology 288, R127–R133.

Knopik, V.S., Maccani, M.A., Francazio, S., McGeary, J.E., 2012. The epigenetics of maternal cigarette smoking during pregnancy and effects on child development. Development and Psychopathology 24, 1377–1390.

Krawetz, S.A., Kruger, A., Lalancette, C., Tagett, R., Anton, E., Draghici, S., Diamond, M.P., 2011. A survey of small RNAs in human sperm. Human Reproduction 26, 3401–3412.

Kuzawa, C.W., 2010. Beyond feast–famine: brain evolution, human life history, and the metabolic syndrome. In: Muehlenbein, M.P. (Ed.), Human Evolutionary Biology. Cambridge University Press, Cambridge, pp. 518–527.

Langley, S.C., Jackson, A.A., 1994. Increased systolic blood pressure in adult rats induced by fetal exposure to maternal low protein diets. Clinical Science 86, 217–222.

Li, R., Fein, S.B., Grummer-Strawn, L.M., 2010. Do infants fed from bottles lack self-regulation of milk intake compared with directly breastfed infants? Pediatrics 125, e1386–e1393.

Lillycrop, K.A., Phillips, E.S., Jackson, A.A., Hanson, M.A., Burdge, G.C., 2005. Dietary protein restriction of pregnant rats induces and folic acid supplementation prevents epigenetic modification of hepatic gene expression in the offspring. Journal of Nutrition 135, 1382–1386.

Lillycrop, K.A., Slater-Jefferies, J.L., Hanson, M.A., Godfrey, K.M., Jackson, A.A., Burdge, G.C., 2007. Induction of altered epigenetic regulation of the hepatic glucocorticoid receptor in the offspring of rats fed a protein-restricted diet during pregnancy suggests that reduced DNAmethyltransferase-1 expression is involved in impaired DNA methylation and changes in histone modifications. British Journal of Nutrition 97, 1064–1073.

Masuyama, H., Hiramatsu, Y., 2012. Effects of a high-fat diet exposure in utero on the metabolic syndrome-like phenomenon in mouse offspring through epigenetic changes in adipocytokine gene expression. Endocrinology 153, 2823–2830.

Ma, R.C.W., Chan, J.C.N., Tam, W.H., Hanson, M.A., Gluckman, P.D., 2013. Gestational diabetes, maternal obesity and the NCD burden. Clinical Obstetrics and Gynecology 56, 633–641.

Mattick, J.S., 2010. RNA as the substrate for epigenome-environment interactions. Bioessays 32, 548–552.

May, A.L., Kuklina, E.V., Yoon, P.W., 2012. Prevalence of cardiovascular disease risk factors among US adolescents, 1999–2008. Pediatrics 129, 1035–1041.

Mayr, E., 1963. Animal Species and Evolution. Harvard University Press, Cambridge, Mass.

McCarthy, M.I., 2010. Genomics, type 2 diabetes, and obesity. New England Journal of Medicine 363, 2339–2350.

Mericq, V., Ong, K.K., Bazaes, R.A., Pena, V., Avila, A., Salazar, T., Soto, N., Iniguez, G., Dunger, D.B., 2005. Longitudinal changes in insulin sensitivity and secretion from birth to age three years in small- and appropriate-for-gestational-age children. Diabetologia 48, 2609–2614.

Morris, K.V., Mattick, J.S., 2014. The rise of regulatory RNA. Nature Reviews Genetics 15, 423–437.

Muleta, M., Rasmussen, S., Kiserud, T., 2010. Obstetric fistula in 14,928 Ethiopian women. Acta Obstetricia et Gynecologica Scandinavica 89, 945–951.

Ng, M., Fleming, T., Robinson, M., Thomson, B., Graetz, N., Margono, C., Mullany, E.C., Biryukov, S., Abbafati, C., Abera, S.F., Abraham,

J.P., Abu-Rmeileh, N.M.E., Achoki, T., AlBuhairan, F.S., Alemu, Z.A., Alfonso, R., Ali, M.K., Ali, R., Guzman, N.A., Ammar, W., Anwari, P., Banerjee, A., Barquera, S., Basu, S., Bennett, D.A., Bhutta, Z., Blore, J., Cabral, N., Nonato, I.C., Chang, J.-C., Chowdhury, R., Courville, K.J., Criqui, M.H., Cundiff, D.K., Dabhadkar, K.C., Dandona, L., Davis, A., Dayama, A., Dharmaratne, S.D., Ding, E.L., Durrani, A.M., Esteghamati, A., Farzadfar, F., Fay, D.F.J., Feigin, V.L., Flaxman, A., Forouzanfar, M.H., Goto, A., Green, M.A., Gupta, R., Hafezi-Nejad, N., Hankey, G.J., Harewood, H.C., Havmoeller, R., Hay, S., Hernandez, L., Husseini, A., Idrisov, B.T., Ikeda, N., Islami, F., Jahangir, E., Jassal, S.K., Jee, S.H., Jeffreys, M., Jonas, J.B., Kabagambe, E.K., Khalifa, S.E.A.H., Kengne, A.P., Khader, Y.S., Khang, Y.-H., Kim, D., Kimokoti, R.W., Kinge, J.M., Kokubo, Y., Kosen, S., Kwan, G., Lai, T., Leinsalu, M., Li, Y., Liang, X., Liu, S., Logroscino, G., Lotufo, P.A., Lu, Y., Ma, J., Mainoo, N.K., Mensah, G.A., Merriman, T.R., Mokdad, A.H., Moschandreas, J., Naghavi, M., Naheed, A., Nand, D., Narayan, K.M.V., Nelson, E.L., Neuhouser, M.L., Nisar, M.I., Ohkubo, T., Oti, S.O., Pedroza, A., Prabhakaran, D., Roy, N., Sampson, U., Seo, H., Sepanlou, S.G., Shibuya, K., Shiri, R., Shiue, I., Singh, G.M., Singh, J.A., Skirbekk, V., Stapelberg, N.J.C., Sturua, L., Sykes, B.L., Tobias, M., Tran, B.X., Trasande, L., Toyoshima, H., van de Vijver, S., Vasankari, T.J., Veerman, J.L., Velasquez-Melendez, G., Vlassov, V.V., Vollset, S.E., Vos, T., Wang, C., Wang, X., Weiderpass, E., Werdecker, A., Wright, J.L., Yang, Y.C., Yatsuya, H., Yoon, J., Yoon, S.-J., Zhao, Y., Zhou, M., Zhu, S., Lopez, A.D., Murray, C.J.L., Gakidou, E., 2014. Global, regional, and national prevalence of overweight and obesity in children and adults during 1980–2013: a systematic analysis for the global burden of disease study 2013. Lancet 384, 766–781.

Ng, S.-F., Lin, R.C.Y., Laybutt, D.R., Barres, R., Owens, J.A., Morris, M.J., 2010. Chronic high-fat diet in fathers programs β-cell dysfunction in female rat offspring. Nature 467, 963–966.

Nesse, R.M., Williams, G.C., 1995. Why We Get Sick: The New Science of Darwinian Medicine. Times Books, New York.

Niculescu, M.D., 2012. Challenges in nutrition-related DNA methylation studies. Biomolecular Concepts 3, 151–160.

Nijland, M.J., Mitsuya, K., Li, C., Ford, S., McDonald, T.J., Nathanielsz, P.W., Cox, L.A., 2010. Epigenetic modification of fetal baboon hepatic phosphoenolpyruvate carboxykinase following exposure to moderately reduced nutrient availability. The Journal of Physiology 588, 1349–1359.

Osgood, N.D., Dyck, R.F., Grassmann, W.K., 2011. The inter- and intragenerational impact of gestational diabetes on the epidemic of type 2 diabetes. American Journal of Public Health 101, 173–179.

Osmond, C., Barker, D.J.P., Winter, P.D., Fall, C.H.D., Simmonds, S.J., 1993. Early growth and death from cardiovascular disease in women. British Medical Journal 307, 1519–1524.

Pembrey, M.E., Bygren, L.O., Kaati, G., Edvinsson, S., Northstone, K., Sjöström, M., Golding, J., the Alspac study team, 2006. Sex-specific, male-line transgenerational responses in humans. European Journal of Human Genetics 14, 159–166.

Perkins, E., Murphy, S.K., Murtha, A.P., Schildkraut, J., Jirtle, R.L., Demark-Wahnefried, W., Forman, M.R., Kurtzberg, J., Overcash, F., Huang, Z., Hoyo, C., 2012. Insulin-like growth factor 2/H19 methylation at birth and risk of overweight and obesity in children. The Journal of Pediatrics 161, 31–39.

Perry, G.H., Dominy, N.J., Claw, K.G., Lee, A.S., Fiegler, H., Redon, R., Werner, J., Villanea, F.A., Mountain, J.L., Misra, R., Carter, N.P., Lee, C., Stone, A.C., 2007. Diet and the evolution of human amylase gene copy number variation. Nature Genetics 39, 1256–1260.

Pontzer, H., Raichlen, D.A., Wood, B.M., Mabulla, A.Z.P., Racette, S.B., Marlowe, F.W., 2012. Hunter-gatherer energetics and human obesity. PLoS One 7, e40503.

Popkin, B.M., Richards, M.K., Montiero, C.A., 1996. Stunting is associated with overweight in children of four nations that are undergoing the nutrition transition. The Journal of Nutrition 126, 3009–3016.

Rajia, S., Chen, H., Morris, M.J., 2010. Maternal overnutrition impacts pffspring adiposity and brain appetite markers—modulation by postweaning diet. Journal of Neuroendocrinology 22, 905–914.

Redward, A., Cutfield, W.S., Peek, J., Young, N., 2012. The lifestyle habits and dietary intake of women undergoing in vitro fertilisation (IVF) treatment. In: Annual Scientific Meeting of the Fertility Society of Australia. Auckland.

Relton, C.L., Groom, A., St Pourcain, B., Sayers, A.E., Swan, D.C., Embleton, N.D., Pearce, M.S., Ring, S.M., Northstone, K., Tobias, J.H., Trakalo, J., Ness, A.R., Shaheen, S.O., Davey Smith, G., 2012. DNA methylation patterns in cord blood DNA and body size in childhood. PLoS One 7, e31821.

Reynolds, R.M., Osmond, C., Phillips, D.I.W., Godfrey, K.M., 2010. Maternal BMI, parity, and pregnancy weight gain: influences on offspring adiposity in young adulthood. Journal of Clinical Endocrinology and Metabolism 95, 5365–5369.

Roseboom, T., de Rooij, S., Painter, R., 2006. The Dutch famine and its long-term consequences for adult health. Early Human Development 82, 485–491.

Schlichting, C.D., Pigliucci, M., 1998. Phenotypic Evolution: A Reaction Norm Perspective. Sinauer Associates, Sunderland.

Seki, Y., Williams, L., Vuguin, P.M., Charron, M.J., 2012. Epigenetic programming of diabetes and obesity: animal models. Endocrinology 153, 1031–1038.

Shan, G., Wei, D., Wang, C., Zhang, J., Wang, B., Ma, M., Pan, L., Yu, T., Xue, F., Wu, Z., 2011. Trends of overweight and obesity in Yi people between 1996 and 2007: an Yi migrant study. Biomedical and Environmental Sciences 24, 467–474.

Stearns, S.C., 2012. Evolutionary medicine: its scope, interest and potential. Proceedings of the Royal Society B: Biological Sciences 279, 4305–4321.

Stettler, N., Stallings, V.A., Troxel, A.B., Zhao, J., Schinnar, R., Nelson, S.E., Ziegler, E.E., Strom, B.L., 2005. Weight gain in the first week of life and overweight in adulthood: a cohort study of European American subjects fed infant formula. Circulation 111, 1897–1903.

Strakovsky, R.S., Zhang, X., Zhou, D., Pan, Y.-X., 2011. Gestational high fat diet programs hepatic phosphoenolpyruvate carboxykinase gene expression and histone modification in neonatal offspring rats. The Journal of Physiology 589, 2707–2717.

The ENCODE Project Consortium, 2012. An integrated encyclopedia of DNA elements in the human genome. Nature 489, 57–74.

Tobi, E.W., Lumey, L.H., Talens, R.P., Kremer, D., Putter, H., Stein, A.D., Slagboom, P.E., Heijmans, B.T., 2009. DNA methylation differences after exposure to prenatal famine are common and timing- and sex-specific. Human Molecular Genetics 18, 4046–4053.

Vickers, M.H., Breier, B.H., Cutfield, W.S., Hofman, P.L., Gluckman, P.D., 2000. Fetal origins of hyperphagia, obesity, and hypertension and postnatal amplification by hypercaloric nutrition. American Journal of Physiology 279, E83–E87.

Vickers, M.H., Breier, B.H., McCarthy, D., Gluckman, P.D., 2003. Sedentary behavior during postnatal life is determined by the prenatal environment and exacerbated by postnatal hypercaloric nutrition. American Journal of Physiology 285, R271–R273.

Vickers, M.H., Gluckman, P.D., Coveny, A.H., Hofman, P.L., Cutfield, W.S., Gertler, A., Breier, B.H., Harris, M., 2005. Neonatal leptin treatment reverses developmental programming. Endocrinology 146, 4211–4216.

Vickers, M.H., Sloboda, D.M., 2012. Strategies for reversing the effects of metabolic disorders induced as a consequence of developmental programming. Frontiers in Physiology 3, 242.

Vucetic, Z., Kimmel, J., Totoki, K., Hollenbeck, E., Reyes, T.M., 2010. Maternal high-fat diet alters methylation and gene expression of dopamine and opioid-related genes. Endocrinology 151, 4756–4764.

Wang, Y.C., McPherson, K., Marsh, T., Gortmaker, S.L., Brown, M., 2011. Health and economic burden of the projected obesity trends in the USA and the UK. The Lancet 378, 815–825.

Ward, W.P., 1993. Birth Weight and Economic Growth: Women's Living Standards in the Industrializing West. University of Chicago Press, Chicago.

Xi, B., Li, H., Li, S., Mi, J., 2014. Recent prevalence of hypertension among Chinese children and adolescents based on 2010 China national blood pressure references. International Journal of Cardiology 174, 870–871.

Chapter 27

Modernization and Disease

William W. Dressler
Department of Anthropology, University of Alabama, Tuscaloosa, AL, USA

SYNOPSIS

Modernization describes a process in which societies relying on traditional patterns of subsistence, with long-established and usually kin-based social structures, are transformed by interaction with market-oriented industrial societies. Modernization is broadly associated with an increase in the risk of chronic diseases, especially cardiovascular disease. Anthropological research over the past 40 years has focused on unraveling the factors associated with that risk. This has been a fruitful area of research that has contributed substantially to the development and refinement of a biocultural perspective.

INTRODUCTION

The study of change is an integral part of human biology, behavior, and culture. While the evolution of the human species has long been a major focus, change occurring in a shorter time frame has also been an emphasis in various investigations. The study of change in the modern world intensified after the Second World War. While there has hardly ever been a time in the history of human beings when groups existed in pristine isolation from one another, the increased impact of capitalist industrial societies on less technologically sophisticated traditional societies struck the imagination of anthropologists and other social scientists in the late 1940s and the early 1950s. There appeared to be a process at work in which societies (nations) followed a linear trajectory of economic change toward industrialization, which in turn provided opportunities for wage labor (vs. household-based subsistence) requiring increased formal education, followed by fundamental changes in social interaction and beliefs. This process was referred to as "modernization." Migration and urbanization tended to accompany modernization. Also, since the transition to industrialized economies took place first in Western Europe and North America, the term "Westernization" was sometimes used as a synonym for modernization.

The question was then posed: are there biological implications of this social and cultural transformation? In the 1950s, a number of biological anthropologists began to examine modernization within the emerging human adaptability paradigm; that is, how did humans as individuals and groups respond biologically to rapidly changing social and physical environments?

The study of modernization and disease proved to be a fruitful pathway in the further development of this biocultural perspective. The aim of this chapter is to trace this path. First, classic research on modernization and disease will be presented. Second, problems with the entire concept of modernization will be discussed along with alternate ways of conceptualizing processes of change. Third, biocultural theories developed in the context of the study of modernization and disease will be examined. And fourth, prospects for the future will be discussed.

MODERNIZATION AND DISEASE: BASIC FINDINGS

The concept of modernization tends to be used in two senses in biocultural discussions. The first sense is descriptive. Global historical trends, whether measured over decades or centuries, have involved an elaboration of technologies enabling greater energy capture from the environment along with increasing information storage and transfer (Wolf, 1982). This technological change has had variable effects on social organization and belief systems. Modernization as a term is a convenient shorthand gloss to allude to these processes. Even those who are critical of theories of modernization tend to slip into this usage.

The second sense of the term modernization is more precise theoretically and is not unlike older anthropological notions of unilineal progressive evolution. In this sense, modernization is a process in which societies pass from largely agricultural-based economies to those with a capitalist industrial base. These technological changes are in turn accompanied by changes in social structure and ideology. These include a shift from traditional foundations of social structure defined in terms of kinship to individualistic, voluntaristic forms of social interaction, as well as an increased emphasis on a scientific rationalism as opposed to an ideology rooted in tradition. The underlying sense of this specific modernization theory is progress from the shackles of tradition to the freedom of modernity, along with a near inevitability of the process (Rostow, 1960).

Anthropologists have used the term in both of these senses. They have also expanded the scope of the concept by examining how modernization influences specific patterns of social interaction and behavior that in turn have direct impacts on health. These include changes in work patterns, patterns of energy expenditure, diet, and psychosocial stress.

The basic theory of modernization has been criticized extensively, but for now it is useful to examine some of the results from research inspired by these ideas. Ingrid Waldron et al. (1982) provided an interesting synopsis of results of cross-cultural studies of blood pressure in relation to subsistence economy that conveniently summarizes findings in global terms. Blood pressure is a biological outcome that has been studied extensively in this literature. There are three good reasons for this. First, blood pressure is a variable influenced by a wide variety of factors ranging from body mass to nutrient intake to psychosocial stress; therefore, as an outcome it is unusually suited to examining the mix of causal factors thought to influence human biology which are also implicated in the process of modernization. Second, while its measurement is not without technical difficulties, it can be measured reliably and noninvasively in the field. Third, the invariance of what is being measured across cultures can be assumed.

There have been a considerable number of studies of blood pressure in a wide variety of communities, ranging from foragers to members of urban industrial society. Waldron and associates collated community studies of blood pressure from 82 different reports and integrated these data with sociocultural data from the Human Relations Area Files. Community average blood pressures (adjusted for age) were then examined in relation to subsistence economy, which is often used as an indicator of sociocultural complexity. The results of this analysis are summarized in Figure 1. Clearly, as societies increase in sociocultural complexity, community average blood pressures increase. Interestingly, the rise

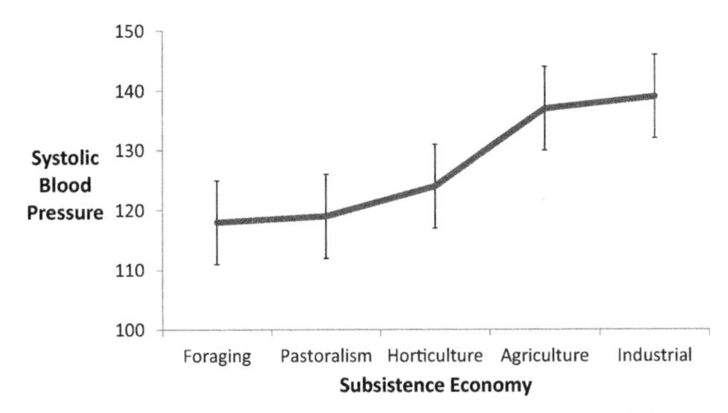

FIGURE 1 Community average blood pressure in relation to subsistence economy. *Adapted from Waldron et al. (1982).*

in blood pressure is not constant across types of subsistence economy; rather, a sharp increase separates foragers, pastoralists, and horticulturalists from agriculturalists and members of industrial societies (see chapters by Blumler, Little, and Little and Blumler in this volume).

From the late 1950s through the early 1980s, there were a number of focused studies of modernization and disease. Many of these used blood pressure and related indicators of cardiovascular disease risk (e.g., serum lipids) as outcome variables (see Dressler, 1999 for a review). Two basic research designs have been employed in modernization research. In the first, two or more communities that could be arrayed along a continuum of modernization were selected, and random samples of individuals within these communities were chosen and studied. Cross-community comparisons provided estimates of the effects of modernization. In the second type of research design, migrants from a less modern community were followed as they moved to a more modern—usually urban—community. They were then compared with long-term residents of the urban area, as well as to members of their home communities who remained.

Perhaps the most complete, sophisticated, and well known of the first type of study is the Penn State Samoa Project led by Paul Baker (Baker et al., 1986) that was conducted primarily in the 1970s. These investigators selected three types of communities as foci of study. The first was located in the Independent State of Samoa (previously known as and often still referred to as "Western Samoa"), an independent nation. The Western Samoa community represented the most traditional. The community was dependent in part for subsistence on a traditional mix of horticulture and fishing. What development that occurred took place in the agricultural sector by the growing of coconuts and copra for export. Social organization emphasized large extended families that also served as local residence groups. A *matai* or chief headed the extended family, representing it in interaction with other communities. The *matai* also served a distributive/redistributive function, allocating productive resources (e.g., land, tools, and implements) to family members, and then redistributing products to the family. While Samoa had long been visited by Christian missionaries, Western religions were adapted to the traditional Samoan way of life (Bindon, 2004).

The most modernized of the communities was an urban neighborhood in Pago Pago, the capital of American Samoa. American Samoa became a protectorate of the United States in the early twentieth century. American Samoa underwent extensive development as a result of a US naval base that was established there. This led to population growth along with opportunities for wage labor on the naval base, in the growing governmental bureaucracy, and with nascent industries such as fish canneries. This development required more extensive formal education, and life in the urban area changed the role of the *matai* in family affairs.

Finally, a community that was considered intermediate in its level of modernization, between these two extremes, was also chosen in American Samoa. It was distant from the urban area, but still offered increased opportunities for formal education and wage labor. A variety of parameters were compared across these communities. Figure 2 presents data on blood pressure.

Clearly, as community modernization increases, average sample blood pressure increases as well. Controlling for age and skinfold measures (indicators of fat mass) made no difference in the results (McGarvey and Baker, 1979).

The Tokelau Island Migrant Study is a notable example from migration studies (Joseph et al., 1983). The Tokelau Islands make up an archipelago that lies to the north of Samoa. The traditional way of life on Tokelau was much the same as for Western Samoa: horticulture, fishing, large extended families, and a chiefdom form of political organization. In 1966 a major hurricane struck Tokelau, and over half the population migrated to New Zealand in its aftermath. There they began work largely as unskilled factory laborers. Follow-up with these migrants was made in the late 1970s, and their average blood pressures were compared with their nonmigrant counterparts. After adjustment for age and body mass index, migrant men had blood pressures 7–10 mm Hg higher than those of nonmigrants. For women, migrants had blood pressures 3–4 mm Hg higher

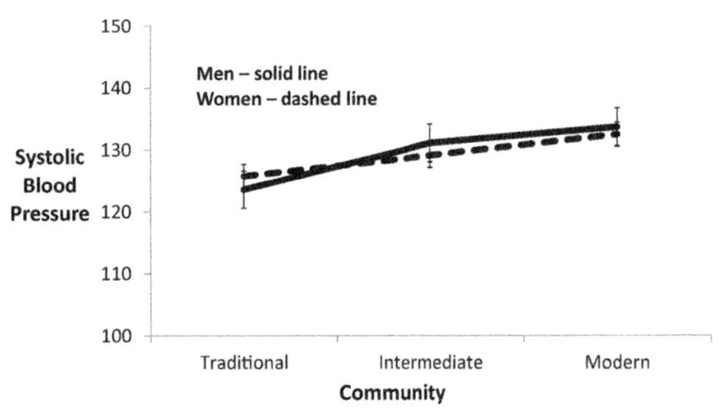

FIGURE 2 Average systolic blood pressure by community modernization. *Adapted from McGarvey and Baker (1979).*

than those of nonmigrants. For women, these are detectable and statistically reliable differences, although the biological implications are probably small; for men, however, blood pressure differences of that magnitude more than double the risk of a myocardial infarction or heart attack.

These studies can be taken as representative of biological research on modernization and migration (see Dressler (1999) for a more extensive review; see also p. 554: Physiological Adaptations to Environmental Stressors). A wide variety of health outcomes have been studied including cardiovascular disease, diabetes, obesity, and psychiatric disorders, across a wide variety of populations. Results are generally consistent. Increasing modernization and migration to more modernized communities are associated with changes in human biology, especially in terms of increasing risk of disease.

CRITIQUES OF STUDIES OF MODERNIZATION

Research on modernization and disease is solid but opaque; that is, there is little insight into the processes involved. Researchers expended considerable energy examining a variety of factors to account for the association. For example, modernization is associated with both increasing obesity and a shift from locally grown to processed foods in the diet (Bindon, 1982; Bindon and Baker, 1985; also see the chapter by Wiley); however, controlling for body mass, diet, and many other factors only explained a portion of the differences between communities. This led many to speculate that modernization was a socially and psychologically stressful experience, and that this stress process accounted for the impact of modernization. While this is a very important hypothesis, it awaited elaboration both theoretically and empirically (Dressler, 1999).

Furthermore, the entire concept of modernization, both in its descriptive form and in its stronger usage in a theory of modernization, has been extensively criticized. Theories of modernization imply that development occurring in a society takes place in some kind of pristine vacuum, divorced from other centers of power and influence. This is clearly not the case, and alternative theories to describe processes of global change have been offered. World systems theory emphasizes the fact that societies do not move through stages on their own, but rather are part of a global economic and information system (Goodman and Leatherman, 1998; Wallerstein, 1974). A key feature of world systems theory is the distinction drawn between the core and the periphery in the world system. The core, residing in urban centers in Europe and North America, is defined in terms of its control over capital and information. The periphery is made up of those developing societies that are the focus of studies of modernization and disease. The periphery is dependent on the core for capital, material, and information. A concrete example of this would be the introduction of outboard motors for use on fishing canoes in Samoa. The outboard motor certainly increases the productivity of an individual canoe, given the ease with which the crew can move from one productive fishing area to another. Yet the motors, the parts for them, the gas to run them, and the knowledge to repair them are not parts of the local economic system. It is in this sense that Pelto and Pelto (1975) described modernization as a process of "de-localization"—local social and economic systems become dependent on distant centers of economic and political power. Furthermore, world systems theorists argue that this dependency is an enforced and continuing part of this process.

A related critique of modernization theory, which is also a critique of world systems theory, is that it fails to account for local culture and local history. Modernization theory posited a particular trajectory for societies and individuals within those societies, the ultimate outcome being a Western-style development. Wolf (1982) argued that developmental and modernization processes are not inevitable, but rather are transformed at the local level as modernizing forces come into contact with specific societies having particular institutional arrangements and individual social actors.

There were two researchers in particular, however, coming out of the tradition of modernization research, who offered more complex, nuanced, and above all socioculturally informed foundations for new approaches in anthropology, specifically exploring how the process of modernization generated specific kinds of stresses that in turn led to disease. The anthropologist Norman Scotch (1963) carried out a study of Zulu migrants to urban areas in South Africa, comparing them with persons who remained in rural areas. Urban migrants had higher blood pressures compared with rural residents; but more importantly, Scotch found differing social patterns associated with blood pressure within each community. For example, residents who had converted to Christianity had higher blood pressures in the rural community but lower blood pressures in the urban community. Scotch interpreted these differing patterns in terms of the "fit" between the behavior pattern or characteristic and the overall culture of the community. Converts to Christianity living in the urban area fit into the dominant pattern of church affiliation, while in the rural area they stood out by their rejection of traditional belief systems. He argued that the stresses associated with these kinds of incongruities led to the risk of high blood pressure.

The social epidemiologist John Cassel (1976) offered a related novel hypothesis to account for the health effects of migration. Cassel argued that migrants to a novel cultural setting arrive, in essence, with a set of cultural models (although he did not use that terminology) in which they have been socialized and that guide their behaviors and understanding in mundane social interaction. These cultural models are likely incongruent with the prevailing cultural models guiding social behavior in the new setting. Where the cultural models of migrants and the receiving community are inconsistent, there is likely to be confusion, uncertainty, and perhaps conflict in social interaction. This can be stressful in and of itself. Furthermore, the process of learning new norms and values takes a toll on migrants. These stresses can combine to increase the risk of disease.

The critique of modernization theory, and the novel hypotheses offered by Scotch and Cassel, provided the impetus for new approaches to the study of modernization and disease.

NEW APPROACHES TO THE STUDY OF MODERNIZATION AND DISEASE

There are three general models that build on conventional studies of modernization and disease, insights derived from the critique of modernization theory, and the novel hypotheses offered to account for the relationship of modernization and disease. These are (1) modernization and psychological/physiological stress; (2) cultural consonance; and (3) the political economy of health. All three of these models use the term modernization in a broad descriptive sense, but propose hypotheses of greater specificity within the context of this general process.

Modernization and Stress

Theories of stress and homeostasis have evolved considerably since Walter Cannon (1932). Hans Selye (1956) expanded on and popularized these ideas (see the chapter by Worthman). In contemporary thinking, understanding the stress process requires consideration of both the factors that challenge or place adaptive demands on individuals—or "stressors"—and the resources that individuals have to respond to those demands—or "resistance resources"(Dressler, 2011; Ice and James, 2012). Anthropologists adapted this general model to fit the realities of communities in the process of modernization. This required attention to local ethnography both to identify stressors generated in the modernization process and to identify resistance resources available in sociocultural systems.

Dressler (1982) studied modernization and blood pressure on the West Indian island of St. Lucia. One of the "small islands" in the eastern Caribbean, St. Lucia experienced slow but sustained economic growth after the Second World War as a result of the development of the banana industry. While the majority of the profits from this growth went to large landowners, it created more work for agricultural laborers, and through marketing cooperatives small-scale growers were able to participate to a limited degree in this international economy. A predictable corollary of this kind of development is a growing availability of consumer goods characteristic of the North American middle class (e.g., appliances, radios, stereos, manufactured furniture). Furthermore, higher social status tends to be accorded those persons observed to accumulate these consumer goods. This occurred in St. Lucia, and set up a situation in which individuals aspired to a higher material lifestyle or greater accumulation of these goods. The problem in a slowly developing economy, however, is that aspirations to this lifestyle can outpace the ability to easily achieve it, resulting in a situation in which scarce economic resources are diverted to material lifestyles, putting economic strain on the household as a whole.

At the same time, a material lifestyle was not the only marker of status in the community. Moving beyond unskilled labor to a more prestigious occupation, such as being a shop clerk or even a teacher, was also an indicator of higher social status. But in this slowly growing economy, few of these higher-status occupational positions are available. A nearly inevitable result of the modernization process is one in which some individuals aspire to higher-status lifestyle, but in the context of meager economic resources and lower-status occupations. This inconsistency between lifestyle and occupation—or "lifestyle incongruity"—was hypothesized to be a chronic stressor characteristic of modernization.

At the same time, traditional social support systems that had developed in St. Lucia during the long days of slavery and its aftermath remained intact. A developmental cycle of domestic groups is found throughout the West Indies, in which it is common for persons to have children with two or more partners early in life and then to live with and ultimately marry one of those partners, with the household comprising their children in common and the woman's children from other relationships. The male partner in this situation is then likely to have offspring resident in other households (called "outside children" in St. Lucia). This leads to complex linkages among households. On top of these social network links, there is a strong moral obligation to, at some level, support one's children. The end result is a system of redistribution in which resources, including food and money, flow among households as men support all their children and women seek support for their children. Coupled with other links among households formed by the close attachment of adult siblings, this leads to households having varying access to resources.

Figure 3 shows lifestyle incongruity in relation to systolic blood pressure, broken down by social resources, in St. Lucia. Higher lifestyle incongruity is associated with higher blood pressure if an individual has access to few other social resources; if an individual has more social resources, however, the effect of lifestyle incongruity on blood pressure is moderated. This is a classic "buffering effect" of social support on a stressor (Cassel, 1976).

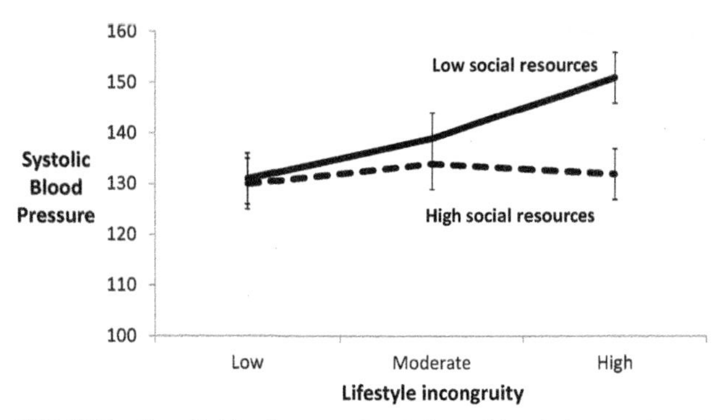

FIGURE 3 Systolic blood pressure in relation to lifestyle incongruity and social resources in St. Lucia (Dressler, 1982).

Janes (1990) found a very similar pattern in his study of Samoan migrants to northern California. Rather than looking at incongruities in lifestyle, however, he examined the inconsistency between an individual's occupational status and his or her educational level, reasoning that in adapting to American society, these were more important parameters. He also looked at social supports, especially the degree to which individuals maintained a close relationship with adult siblings and other family members. He found that for men, status inconsistency was associated with higher blood pressure, while more social resources were associated with lower blood pressure. For women, a slightly different pattern emerged. Stressors occurring within the family were the most important correlates of blood pressure, and these were moderated by greater access to social resources, indicating that the migration experience can affect men and women differently.

McDade (2002) examined lifestyle incongruity and immune function among Samoan adolescents. As a measure of immune function, he assessed antibodies to the Epstein–Barr virus (EBV). EBV is nearly ubiquitous in humans and is kept in check by an adequately functioning immune system; however, the experience of psychosocial stress can suppress the functioning of the immune system, leading to a proliferation of EBV and hence the immune system's production of antibodies. McDade reasoned that lifestyle incongruity at the household level could influence adolescent stress levels and hence immune function. Furthermore, he conceptualized lifestyle incongruity, in a novel way, as the discrepancy between the accumulation of a modern lifestyle and achievement of a traditional status, that of *matai*. As noted earlier, the *matai* is the leader of an extended family group and represents the family in various ways. McDade examined the health effects of having achieved a modern material lifestyle, but not having the traditional status position of *matai*. He found that any inconsistency between lifestyle and *matai* status at the household level was associated with stimulation of immune function in adolescents, and that the association was strongest in the most modernized communities. Furthermore, this effect was exacerbated among adolescents who reported higher levels of social interaction with others (McDade, 2001; see also Sorensen et al., 2009).

Cultural Consonance

Cultural consonance refers to the degree to which individuals approximate, in their own beliefs and behaviors, the prototypes for belief and behavior that are encoded in shared cultural models (Dressler, 2005, 2012). While the theory of cultural consonance has multiple origins in anthropology, it is certainly consistent with John Cassel's theories of migration. As noted above, Cassel argued that migrants are confronted with the task of learning new cultural models in a novel social setting in order to interpret others' behaviors and guide their own behaviors. This suggests that new migrants would have low cultural consonance in relation to dominant cultural models.

For example, Copeland (2011) studied cultural models for the management of HIV/AIDS among female migrants to Nairobi, Kenya from the countryside. These mostly impoverished women are considered poor candidates for antiretroviral treatment, but have developed a cultural model for the management of HIV outside of the biomedical establishment. This cultural model represents a schema or blueprint for how to manage HIV-positive status in lieu of access to biomedical treatment. The model combines information from the health education literature supplied by local nongovernmental organizations with some traditional understanding of illness treatment as well as strategies that women living in this circumstance have developed and in turn pass on to others. Copeland (2011) found this to be a shared cultural model; furthermore, a woman's understanding of the cultural model increases with her time of residence in Nairobi. Finally, the degree to which a woman can successfully adhere to the model—her cultural consonance in the model—is associated with higher immunocompetence and fewer opportunistic infections (Copeland, 2013).

The theory of cultural consonance extends to any social context. In any community, there will be more or less widely shared cultural models regarding salient domains within that community. There will be variation in how well individuals understand that cultural domain, but more importantly there will be variation in the degree to which individuals can put their understanding into practice, or cultural consonance. The individual with low cultural consonance in a domain with high cultural consensus understands what the culturally valued behaviors are, but is unable to implement them in his or her own life. This is likely to be a chronically stressful experience associated with poor health outcomes.

Dressler and associates have examined these processes in a variety of settings, especially urban Brazil. Using a two-stage method, these investigators first examine the degree of sharing, or cultural consensus, regarding the importance of beliefs and behaviors in a particular cultural domain. For example, in urban Brazil they selected the domains of lifestyle (material goods and leisure time activities that constitute having a good life); social support (who one can turn to for help in times of felt need); family life (characteristics of a good family); and national identity (what makes Brazilians unique). In each of these domains, there was a shared understanding of the importance of elements of the domain (e.g., what are the most important lifestyle items necessary for having a good life, or the most important characteristics of a good Brazilian family?). They then developed individual-level measures of cultural consonance, assessing the degree to which individuals actually engaged in those behaviors.

They found that low cultural consonance was associated with higher blood pressure (Dressler and Bindon, 2000; Dressler et al., 1998, 2005), higher body mass index (Dressler et al., 2008, 2012), and more symptoms of depression (Dressler et al., 2007a,b). Finally, they found that cultural consonance interacts with a single-nucleotide polymorphism in a receptor for serotonin (a neurotransmitter) in the prediction of depressive symptoms (Dressler et al., 2009). For persons with a specific variant of the gene for the serotonin receptor, the effect of cultural consonance is amplified. That is, if they have low cultural consonance, their level of depressive symptoms is considerably higher than for persons with other gene variants; if they have high cultural consonance, their level of depressive symptoms is considerably lower than for persons with other gene variants. The polymorphism appears to function as an amplifier of culturally meaningful experience.

Political Economy and Health

A political economy of health grew out of researchers taking the critique of conventional modernization theory seriously and attempting to incorporate into their research insights drawn from that critique. Goodman and Leatherman (1998) outline several features of a biological anthropology informed by political economy. These features include (1) an emphasis on the social relations of production by which individuals and groups gain access to scarce resources; (2) tracing explicitly the links between local and global processes; (3) how local history and historical contingency affect biological outcomes; (4) how groups and individuals are active agents in shaping their environments; and (5) how the control of knowledge helps to influence the process (ibid. 19–20). This work is very clear in its attempt to incorporate the critique of modernization, and especially to trace how differences in power impact biological outcomes.

The most completely developed of these research programs is probably Leatherman's (1996) work in the Peruvian Andes. The Andean project was a part of the human adaptability paradigm, with a focus principally on adaptation to high altitude. As the work progressed, it became apparent that altitude and hypoxia had less to do with biological adaptation than the oppressive and exploitive conditions under which highland peasants labored (Thomas, 1998). In order to examine these political economic factors in health more closely, Leatherman compared three highland communities. The area had undergone substantial change in the 1960s as a result of land reform. While the traditional *haciendas* (large farms controlled by a single owner) were broken up, peasant farmers did not necessarily enjoy greater access to land. Farming cooperatives were created, one of which was studied by Leatherman. A second cooperative emphasizing alpaca herding was also included. The third community was the largest town in the district, which grew in population as some peasants were displaced from the land.

A major focus of research was child growth and development. Using height-for-age as a measure of growth, Leatherman (1996) found that the highest rate of stunting (very low height-for-age) was found among the members of the farming cooperative, despite the fact that they lived at the lowest altitude. Members of the herding cooperative had the lowest rate of stunting, despite living at the highest altitude. Leatherman points out that the conventional adaptability paradigm would predict higher rates of stunting at higher altitudes due to increased hypoxia, while the data show the opposite. He argues that this can be accounted for by the changing nature of the social relations of production in the communities. Land reform was accompanied by a "rational" allocation of land, ending the informal arrangements that had traditionally been in place. This meant reduced areas for planting for many households, and along with an increased capitalization of the local agricultural market meant that they needed more ready cash for their production, which was often unavailable in the farming cooperative. The herding cooperative, on the other hand, was able to adjust to the greater need for capital, resulting in greater dietary diversity and less evidence of stunting among children.

This same pattern was evident in work days lost to illness among adults; the herding cooperative reported fewest work days lost, followed by the town and the farming community. Leatherman argues that this increased level of illness then feeds back on household productivity and interacts with the loss of traditional work patterns. Households with sicker members are forced to hire temporary labor to maintain their production levels, which in turn saps their limited resources, which diminishes dietary quality, which in turn leads to more illness. In other words, there is a kind of vicious circle in which changing national and international policies impact the local level and households' abilities to adjust.

The ultimate point of this work is again that communities and households living in places as remote as the Peruvian highlands must be understood against a backdrop of political economic change occurring in distant centers of power, both in the nation and internationally, and that this political economic environment can be a more potent influence on health and well-being than the physical environment to which they must adapt.

Godoy and associates (Godoy et al., 2009) have carried out a comprehensive political economic investigation of the Tsimané of lowland Bolivia. The Tsimané are a large group who traditionally practiced horticulture supplemented by hunting and gathering. Changing global markets increased the interest of developers in exploiting the forest and riverine resources of the Tsimané territory, leading to the construction of roads and increased interaction with developers. One effect of road construction has been to increase the access of some Tsimané communities to local market towns, while leaving others quite isolated. This has meant that Tsimané have increased their participation in the market as both producers and consumers, and have increased access to wage labor as a function of development.

Godoy et al. (2009) found mixed effects of increasing market involvement on the Tsimané. Significant improvement over time in diet, body mass index, and psychological distress was observed. At the same time, the overall health status, as assessed by self-reported recent sicknesses, declined (Godoy et al., 2009). Two things distinguish this study. First, five annual rounds of data collection were carried out, making this a true panel study (a longitudinal study in which the same participants are followed over time). Second, very sophisticated time-series econometric models were employed in the data analysis.

Similar findings emerge from a recent study among the Shuar, a group practicing horticulture in the Ecuadorian Amazon (Liebert et al., 2013). Shuar have a history of resistance to colonizers, both pre-Columbian and Spanish. As such they maintained their way of life apart from outside influences nearly into the twentieth century. In the past decades, due to the unique geography of the area they inhabit, segments of Shuar have been drawn into market participation, while other segments have not. Liebert et al. (2013) compared these groups with respect to cardiovascular disease risk factors.

These investigators also found mixed effects of market integration on human biology. Overall, greater market integration did not appear to have any large effects on cardiovascular risk factors, nor were the effects all in the same direction. For example, Shuar in the more market-integrated areas tended to have somewhat higher serum lipid values, while those in the less integrated areas had higher diastolic blood pressure. While these effects are detectable, the impact on health is unclear given the small magnitude of the differences (Liebert et al., 2013).

DISCUSSION

Studies of modernization and disease have a history rich in results, controversy, and theoretical development. Early studies, such as the Penn State Samoa Project (Baker et al., 1986) and the Tokelau Island Migration Study (Joseph et al., 1983), provided findings that were both powerful and provocative. The contrasts between populations that were following traditional subsistence patterns with kin-based social structures relatively intact, and populations engaged in wage labor and social structures that were much transformed, were dramatic. From these studies, modernization appeared to have substantial effects on human biology and the risk of disease. The problem was that the construct and measurement of modernization proved to be a rather blunt instrument. So many changes in economic activities, diet, physical activity, social interaction, and beliefs were entailed in these contrasts that the mediators of change could not be identified.

The relative insensitivity of modernization as a concept was further revealed by the theoretical critique of the research (Goodman and Leatherman, 1998). Fundamentally, this theoretical critique emphasized that groups arrayed along a continuum of modernization could not be regarded as existing in a political economic vacuum; rather, they were subject to forces far beyond their control. Yet at the same time, each modernizing group brought with it its own history and trajectory of sociocultural development, which brought to bear on the modernization process varying influences, all of which were meaningful with respect to adaptation.

This theoretical critique and the insights provided by early thinkers led to the development of new models for investigating societies in transition. The nascent consensus regarding models of psychosocial stress in industrial societies led to their adaptation for the study of modernizing societies, with some considerable success. In the studies selected here as exemplars of this approach (Dressler, 1982; Janes, 1990; McDade, 2002), it is noteworthy that the factors influencing disease risk involved an interaction between the change in society brought about by the impact of capitalist industrial societies at the local level, especially in terms of changing lifestyles and the status accorded them, and the nature of social status and social resources that were specific to those societies. So, in the case of St. Lucia (Dressler, 1982) and Samoan migrants to California (Janes, 1990), the stresses of status incongruity were buffered by the strength of traditional social resources. In McDade's (2002) study of immunocompetence in Samoa, the intersection of status as defined by a capitalist consumer mentality and status as defined traditionally proved to be a key mediator of biological response. These studies showed that a unilinear model of modernization assumed to be associated with a uniform pattern of stress was inadequate, and that explanation depended on the intersection of global and local factors.

The concept of cultural consonance was derived in part from studies of modernization and especially suppositions regarding migration (Cassel, 1976). These ideas brought a renewed emphasis on culture as meaning to this area. While cultural consonance as a concept can be applied in any setting regardless of the change or population movements occurring, the idea that modernization creates a situation in which the cultural models that people use to interpret the world and guide their behavior can lose their relevance, and that new models must be learned in the process of adaptation, provides a clear example of the importance of the process. What is also important, however, is how people can put their understanding into practice, which ultimately influences their biological response (Copeland, 2013; Dressler et al., 2005).

Finally, political economic models emerged both from a critique of modernization (Goodman and Leatherman, 1998) and from the critique of the entire concept of adaptation (Singer, 1986; Thomas, 1998). Basically, these models focus much more attention on the imbalances of power and systems of inequality brought about by change, focusing carefully on who benefits and who does not in the process, with disease as the outcome.

Studies of modernization and disease thus laid the foundation for the development of these contemporary biocultural models in anthropology. This intellectual trajectory demonstrates the importance of a cumulative scientific approach in human biology and the value of anthropology's historic emphasis on a holistic perspective on the human species.

REFERENCES

Baker, P.T., Hanna, J.M., Baker, T.S., 1986. The Changing Samoans: Behavior and Health in Transition. Oxford University Press, New York.

Bindon, J., 1982. Breadfruit, banana, beef, and beer—modernization of the Samoan diet. Ecology of Food and Nutrition 12, 49–60.

Bindon, J., Baker, P., 1985. Modernization, migration and obesity among Samoan adults. Annals of Human Biology 12, 67–76.

Bindon, J.R., 2004. Samoa. In: Ember, C.R., Ember, M. (Eds.), Encyclopedia of Medical Anthropology, vol. 2. Kluwer Academic/Plenum Publishers, New York, pp. 929–936.

Cannon, W.B., 1932. The Wisdom of the Body. W.W. Norton, New York.

Cassel, J.C., 1976. The contribution of the social environment to host resistance. American Journal of Epidemiology 104, 107–123.

Copeland, T.J., 2011. Poverty, nutrition, and a cultural model of managing HIV/AIDS among women in Nairobi, Kenya. Annals of Anthropological Practice 35, 81–97.

Copeland, T.J., February 20–23, 2013. To keep this disease from killing you: cultural competence, consonance, and health among HIV-positive women in Nairobi, Kenya. In: Paper Presented at the Annual Meeting of the Society for Anthropological Sciences, Mobile, AL.

Dressler, W., 1999. Modernization, stress, and blood pressure: new directions in research. Human Biology 71, 583–605.

Dressler, W.W., 1982. Hypertension and Culture Change: Acculturation and Disease in the West Indies. Redgrave Publishing, South Salem, NY.

Dressler, W.W., 2005. What's cultural about biocultural research? ETHOS 33 (1), 20–45.

Dressler, W.W., 2011. Culture and the stress process. In: Singer, M., Erickson, P. (Eds.), A Companion to Medical Anthropology. Wiley-Blackwell, New York, pp. 119–134.

Dressler, W.W., 2012. Cultural consonance: Linking culture, the individual and, health. Preventive Medicine 55, 390–393.

Dressler, W., Bindon, J., 2000. The health consequences of cultural consonance: cultural dimensions of lifestyle, social support, and arterial blood pressure in an African American community. American Anthropologist 102, 244–260.

Dressler, W., Balieiro, M., Dos Santos, J., 1998. Culture, socioeconomic status, and physical and mental health in Brazil. Medical Anthropology Quarterly 12, 424–446.

Dressler, W., Balieiro, M., Ribeiro, R., Dos Santos, J., 2005. Cultural consonance and arterial blood pressure in urban Brazil. Social Science and Medicine 61, 527–540.

Dressler, W., Balieiro, M., Ribeiro, R., Dos Santos, J., 2007a. Cultural consonance and psychological distress: examining the associations in multiple cultural domains. Culture, Medicine and Psychiatry 31, 195–224.

Dressler, W., Balleiro, M., Ribeiro, R., Dos Santos, J., 2007b. A prospective study of cultural consonance and depressive symptoms in urban Brazil. Social Science and Medicine 65, 2058–2069.

Dressler, W., Balieiro, M., Ribeiro, R., Dos Santos, J., 2009. Cultural consonance, a 5HT2A receptor polymorphism, and depressive symptoms: a longitudinal study of gene x culture interaction in urban Brazil. American Journal of Human Biology 21, 91–97.

Dressler, W.W., Oths, K.S., Ribeiro, R.P., Balieiro, M.C., Dos Santos, J.E., 2008. Cultural consonance and adult body composition in urban Brazil. American Journal of Human Biology 20, 15–22.

Dressler, W.W., Oths, K.S., Balieiro, M.C., Ribeiro, R.P., Dos Santos, J.E., 2012. How culture shapes the body: cultural consonance and body mass in urban brazil. American Journal of Human Biology 24, 325–331.

Godoy, R., Reyes-García, V., Gravlee, C.C., et al., 2009. Moving beyond a snapshot to understand changes in the well-being of native Amazonians: panel evidence (2002–2006) from Bolivia. Current Anthropology 50, 563–573.

Goodman, A.H., Leatherman, T.L., 1998. Building a New Biocultural Synthesis: Political-economic Perspectives on Human Biology. University of Michigan Press, Ann Arbor.

Ice, G.H., James, G.D., 2012. Measuring Stress in Humans. Cambridge University Press, Cambridge.

Janes, C., 1990. Migration, Social Change, and Health: A Samoan Community in Urban California. Stanford University Press, Stanford, CA.

Joseph, J.G., Prior, I.A.M., Salmond, C.E., Stanley, D., 1983. Elevation of systolic and diastolic blood pressure associated with migration: the Tokelau Island migrant study. Journal of Chronic Disease 67, 507–516.

Leatherman, T.L., 1996. A biocultural perspective on health and household economy in southern Peru. Medical Anthropology Quarterly 10, 476–495.

Liebert, M.A., Snodgrass, J.J., Madimenos, F.C., Cepon, T.J., et al., 2013. Implications of market integration for cardiovascular and metabolic health among an indigenous Amazonian Ecuadorian population. Annals of Human Biology 40, 228–242.

McDade, T., 2001. Lifestyle incongruity, social integration, and immune function in Samoan adolescents. Social Science and Medicine 53, 1351–1362.

McDade, T., 2002. Status incongruity in Samoan youth: a biocultural analysis of culture change, stress, and immune function. Medical Anthropology Quarterly 16, 123–150.

McGarvey, S., Baker, P., 1979. Effects of modernization and migration on Samoan blood pressures. Human Biology 51, 461–479.

Pelto, P.J., Pelto, G.H., 1975. Intra-cultural diversity: some theoretical issues. American Ethnologist 2, 1–18.

Rostow, W.W., 1960. The Stages of Economic Growth: A Non-communist Manifesto. Cambridge University Press, Cambridge.

Selye, H., 1956. The Stress of Life. McGraw-Hill, New York.

Scotch, N., 1963. Sociocultural factors in epidemiology of Zulu hypertension. American Journal of Public Health Nations Health 53, 1205–1213.

Singer, M., 1986. Developing a critical perspective in medical anthropology. Medical Anthropology Quarterly 17, 128–137.

Sorensen, M.V., Snodgrass, J.J., Leonard, W.R., McDade, T.W., et al., 2009. Lifestyle incongruity, stress and immune function in indigenous Siberians: the health impacts of rapid social and economic change. American Journal of Physical Anthropology 138, 62–69.

Thomas, B., 1998. The evolution of human adaptability paradigms: toward a biology of poverty. In: Goodman, A., Leatherman, T. (Eds.), Building a New Biocultural Synthesis: Political-Economic Perspectives on Human Biology. University of Michigan Press, Ann Arbor, MI, pp. 43–74.

Waldron, I., Nowotarski, M., Freimer, M., et al., 1982. Cross-cultural variation in blood pressure: a quantitative analysis of the relationships of blood pressure to cultural characteristics, salt consumption and body weight. Social Science and Medicine 16, 419–430.

Wallerstein, I., 1974. The Modern World-System, Vol. I: Capitalist Agriculture and the Origins of the European World-economy in the Sixteenth Century. Academic Press, New York.

Wolf, E.R., 1982. Europe and the People without History. University of California Press, Los Angeles and Berkeley.

Chapter 28

Modern Human Diet

Andrea S. Wiley
Department of Anthropology, Indiana University, Bloomington, IN, USA

SYNOPSIS

Humans have myriad adaptations to diet. This chapter considers these adaptations in relation to contemporary human diets and biology, highlighting the ways in which these diets have changed over our evolutionary and recent history and the ways in which those changes have had consequences for biological function. Specific topics include diet and the evolution of bipedalism and the large brain, genetic adaptations to agricultural diets, the globalization and industrialization of diet, and the consequences for human biology. Paleolithic-inspired diets are also reviewed, as well as vegetarian diets and caloric restriction (CR) as dietary means to reduce the risk of chronic diseases.

INTRODUCTION

Accessing sufficient food is a fundamental problem that all organisms must solve. For animals, food must be acquired to provide energy for the physiological activities needed to support life. These include the life history processes of growth and reproduction that are essential to an organism's fitness. It is no surprise then that organisms' biologies are shaped by diet, with adaptations for accessing food, extracting nutrients, and detoxifying potentially harmful food components. Evidence for these adaptations can be found throughout the body. The sensory organs are attuned to finding and evaluating food sources, and contribute to preferences for foods rich in valuable nutrients; the musculoskeletal system facilitates physical access; the dentition allows for the initial breakdown of foods in preparation for digestion; the digestive system allows for efficient extraction of nutrients; and the circulatory system delivers nutrients to cells, which in turn have metabolic capabilities to convert food energy into fuel for the organism's activities. Liver enzymes convert toxins to less harmful products, and the immune system mounts a response to potentially life-threatening pathogens. The respiratory system plays a role in providing oxygen, which is necessary for energy production and removing carbon dioxide, a by-product of metabolism. Thus to understand many aspects of contemporary human biology, it is essential to know something about diets over the course of human evolution and more recent history, and the ways these have impacted both our biological structures and their functioning.

This chapter outlines the major trends in human dietary evolution, briefly reviewing the major hypotheses for how diet has shaped some unique species-wide human characteristics. I next consider the major changes human diets have sus tained over the past 10,000 years, and describe how agricultural diets produced lasting effects on human biological variation, particularly in milk and possibly also starch digestion, as well as biological consequences of the more recent global exchange of foods, industrialization of food production and processing, and the nutrition transition. Diets have changed markedly since our early evolution as a species, and the relevance of preagricultural Paleolithic diets to contemporary human health conditions are highlighted, with a focus on current dietary trends such as "Paleo" and vegetarian diets, as well as an evaluation of CR on chronic disease and aging.

THE ROLE OF DIET IN HOMININ EVOLUTION

Changes in food acquisition and diet composition have been proposed as major shapers of two of the characteristics that make *Homo sapiens* unique as a species: bipedalism and a brain that is three times the size of that of our closest living relatives, the African apes. The transition from quadrupedalism and suspensory arboreal locomotion to walking on two legs occurred at least two million years prior to the enlargement of the brain, and so the role of diet is likely to be different in the evolution of these two traits.

Bipedalism

The earliest members of the lineage that likely led to *Homo* lived in Africa beginning more than four million years ago (mya) (see the chapter by Hunt in this volume on Early Hominins). These were small bipedal apes, who descended from a broad primate ancestry that included heavy reliance on fruits as food (Milton, 1999). With life in a more terrestrial context that included open grassland, forest, and riverine environments, the australopithecines faced new dietary challenges that are reflected in changes in the dentition. In particular, their molar teeth were flat, broad, and thickly enameled, well suited for chewing hard, brittle foods such as fruits and nuts, or soft mashable foods, but not meat, tough fruits, or leaves (Teaford and Ungar, 2000). Stable isotope analysis suggests that they consumed grasses and sedges, as well as their roots and underground storage organs, as part of an overall drying trend during this period (Klein, 2013).

While these data indicate a shift away from fruit eating to a more diverse set of foods found in a terrestrial context, the ways in which this dietary shift is related to the evolution of bipedalism remain opaque. One advantage to bipedalism as a form of terrestrial locomotion is that it is more energetically efficient than moving around on all four limbs at walking speeds (for a complete discussion on bipedalism, see the chapter by Hunt in this volume). This allows for tremendous caloric savings when moving across larger spaces, which is typical of contemporary human foragers, who range much more widely than apes (Leonard and Robertson, 1992; Leonard et al., 2007; although see Halsey and White, 2012). Bipedalism would also allow for more efficient upright feeding (Hunt, 1994) and carrying of prized food resources to safe places for consumption (Hewes, 1964). Experimental evidence with chimpanzees and bonobos suggests that these are the two situations that promote bipedal locomotion among apes who are more habitually quadrupedal when on the ground (Videan and McGrew, 2002). Bipedalism may have facilitated acquisition of more diverse terrestrial foods, but it is important to note that there were many species of early African hominids whose phylogenetic relationship to each other and to our genus *Homo* is uncertain, and their diets appear to have been quite diverse (Sponheimer et al., 2013). That said, these early hominids were not likely subsisting on meat or utilizing tools beyond the types of crude implements (termite wands, stones for cracking nuts) used by contemporary chimpanzees.

Brain and Body Size

Homo sapiens has a brain size that is about three times that of our closet living relatives among the apes, and even after adjusting for our larger body size, our brain-to-body size ratio is about 2.5 times that of the apes (Leonard and Robertson, 1994). Evidence of brain case enlargement is used to assign fossils to the genus *Homo*, and first appears around 2 mya in Africa. One significant feature of large brains is that they are metabolically costly. About 20% of our basal metabolic rate (BMR; the energy we need just to stay alive) goes to supporting our large brain; this is in contrast to other primates, whose brains use about 8–10% of their BMR (Leonard and Robertson, 1994). Despite having a large and costly brain, human metabolic needs per kilogram of body mass are not substantially different from that of mammals in general, suggesting that energy needs have been redistributed among the organs that require energy. In other words, other organs must have reduced their energy needs in order to compensate for the greater needs of the large brain. At the same time, a larger body size does require more energy in absolute terms.

The fact that our caloric needs per unit of body mass did not increase suggests that there were trade-offs among those organs that require large amounts of energy such as the brain, heart, kidney, liver, digestive tract, and skeletal muscle, with the first five accounting for the majority of the BMR. The "expensive tissue hypothesis" (Aiello and Wheeler, 1995) proposes that while humans deviate from the typical primate by having a much larger brain, they have a much smaller gut than predicted for their body size. Thus the larger brain was made possible by a trade-off with a smaller gut. The latter, in turn, would be possible only if *Homo* was consuming a nutrient-dense diet, since this smaller gut has a reduced capacity to ferment fibrous plant foods. High-fiber foods are relatively low in energy, and require prolonged fermentation by bacteria in either the foregut (i.e., the stomach) or hindgut (the cecum and/or colon). In terms of proportions of the gastrointestinal system, humans have an unusually long small intestine, but a small colon and none of the specialized fermentation chambers listed above (Milton, 1999). Thus acquisition of a high-quality diet, the large brain, and smaller digestive tract would have likely coevolved together.

Some researchers have contested Aiello and Wheeler's hypothesis, noting that humans may not deviate in overall gut volume compared to other primates' fat-free body mass rather than total body mass (Navarrete et al., 2011). Indeed, relative to fat-free mass, the human basal metabolic rate is actually substantially higher than other primates, providing further support to a need for an energy-rich diet in the course of human evolution. Leonard et al. (2003) have suggested that the trade-off for the large brain took the form of loss of muscle, as humans have less muscle mass and more fat than other primates. Fat has low metabolic costs and serves as an energy store, both of which would have been useful to a large-brained species. Others have suggested that energetic savings from bipedal locomotion or a slower life history may have compensated for the brain's high energy needs (Navarrete et al., 2011).

Acquisition of more high-quality food could have been made possible with elaborated stone tool use and inclusion of meat in the diet. While the earliest stone tools would have been of little use in hunting, as they were hand tools rather than projectiles, they could have been employed in butchering or preparing hard plant foods (e.g., cracking nuts, extracting roots and tubers, chopping), making extraction of calories and nutrients more efficient.

While there is broad agreement that the evolution of *Homo* is characterized by greater omnivory, there is considerable debate as to the importance of meat in the diet, especially in relation to the evolution of the larger brain, complex behavior, and a longer, slower life history. Many researchers have proposed that it was the inclusion of meat that provided an additional nutrient-dense food to support a large brain and the other traits typical of *Homo* (Bunn, 2007; see selections in Stanford and Bunn, 2001). It is important to note that evidence for meat eating is overrepresented in the archaeological record, as animal bones are much more likely to be preserved than are plant foods, so it is difficult to ascertain its proportion in the diet. Studies of contemporary hunter gatherers suggest a range of meat in the diet, from ~30 to 80% of total calories (Kaplan et al., 2000). Others maintain that the addition of meat allowed plant nutrients (primarily glucose, which can only be acquired from plant foods as animals do not store glucose except in very small amounts) to be freed up for use by the brain, as glucose is its primary energy source (Milton, 2003). Another alternative is that protein and fat from animals were necessary to grow the large brain (Kuipers et al., 2010), or that only animals (especially fish and shellfish) can provide sufficient docosahexaenoic acid, an ω-3 fatty acid essential to the growth of the brain during development, and thus early *Homo* must have relied heavily on those foods (Broadhurst et al., 2002; although see Carlson and Kingston, 2007). Overall, the relative proportions of plant and animal foods in the diets remain unknown, and undoubtedly varied by location and season, especially at the higher latitudes, when *Homo* expanded its range into Europe and Asia, but omnivory and dietary flexibility became hallmarks of our evolution (Hladik and Pasquet, 2002).

At some point during the evolution of *Homo* our ancestors learned how to control and utilize fire. The earliest date for controlled use of fire is a matter of controversy, but current evidence suggests that it may be around 1 mya in South Africa (Berna et al., 2012), but much later (~400,000 ya) in Europe (Roebroeks and Villa, 2011). This opened up a new niche for them, as fire provided warmth, protection, light, and also the ability to cook foods. The significance of cooking should not be underestimated. It allowed for the detoxification of some plant secondary compounds and reduction in pathogen exposure, both of which expanded dietary diversity (Stahl et al.,

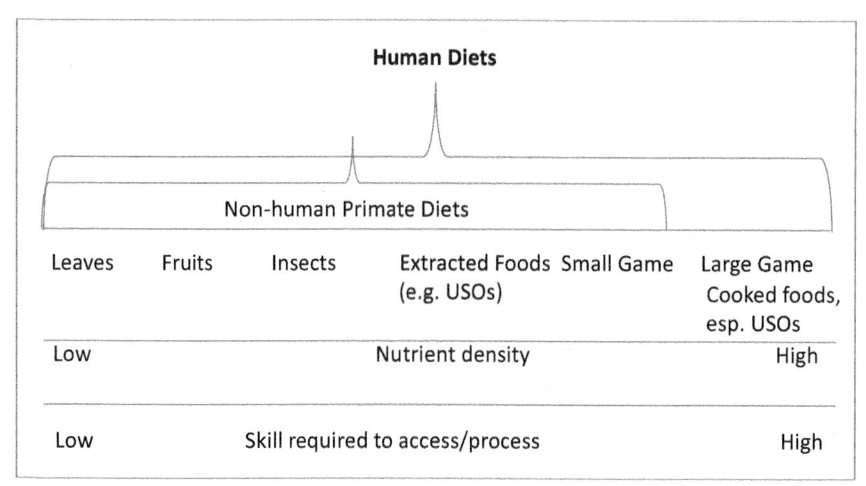

FIGURE 1 A comparison of the diversity of human and nonhuman primate diets. *Adapted from Kaplan, H., Hill, K., et al., 2000. A theory of human life history evolution: diet, intelligence, and longevity. Evolutionary Anthropology Issues News and Reviews 9 (4), 156–185.*

1984; Wrangham, 2009; Wrangham and Carmody, 2010). Cooking also enhances the digestibility of carbohydrates in tubers, which could have been a major advantage for the growing brain. Furthermore, it softened foods, reducing stresses on the dentition and jaw. This offered up new dietary possibilities for feeding the very young with soft mashed and safer foods, perhaps making it possible to wean earlier and involve others in feeding infants, thereby freeing the mother to have another offspring sooner (Bogin, 2006). Indeed food sharing and provisioning of the young are rare among primates, but is elaborated among all human societies (Hrdy, 2009). Thus cooking provides an alternative hypothesis to those emphasizing meat in the evolution of human intelligence, and is perhaps more compelling because it provided more of the very fuel that the brain required and is a behavior that is truly unique to our species. The two arguments are not mutually exclusive, and meat eating appears to predate the earliest documented uses of fire, as does the cranial expansion associated with early *Homo*. In sum, greater omnivory, foods higher in nutrient density, overall increases in foods, and technologies (including cooking) that increased efficiency in nutrient extraction and reduced chewing stresses, along with a larger body and brain size all typified the evolution of our genus (see Figure 1).

AGRICULTURAL TRANSITION: DIETARY AND EVOLUTIONARY CONSEQUENCES

The transition from foraging to food production occurred at several sites in the Old and New Worlds starting around 8000 years before the common era (see the chapter by Blumler). This transition is typified by domestication of plants and animals, a more limited array of staple crops, mostly seeds and tubers, and permanent settlements. This marked a radical change in human diets, with reduced dietary diversity and increased vulnerability to food catastrophes caused by weather or pests (Cohen, 1989). In addition, settlements facilitated the spread of infectious disease, and introduced zoonotic diseases from domesticated animals and greater temporal variability in physical activity and dietary patterns. While the causes of this transition likely differed by site and stemmed from multiple processes, the consequences for human diet and health were remarkable. Jared Diamond described agriculture as "the worst mistake in the history of the human race" (Diamond, 1987), and evidence suggests a decline in health among agricultural populations compared to hunter–gatherers (Cohen and Armelagos, 1984; Bocquet-Appel and Bar-Yosef, 2008; Mummert et al., 2011) (see chapters by Little and Blumler, Blumbler, and Dressler). Agriculture did allow for production of a surplus, the specialization of labor, and overall larger population size. As a consequence, diets for most contemporary humans stem from cultivation of a relatively narrow array of plant and animal species, and Solomon Katz has argued that food processing techniques became especially important in this context, as many appear to be adaptive insofar as they enhance nutrient availability or reduce toxicity (Katz, 1987). Increasing numbers of humans are no longer directly involved in food production and must rely on the producers or the market for access.

There is evidence that the changes in subsistence stemming from the agricultural transition have resulted in evolutionary changes in human populations engaging in specific agricultural practices. These have contributed to population differences in genes associated with food digestion. Two examples are described below; other possible genetic changes related to dietary modifications associated with agriculture are described elsewhere (Luca et al., 2010).

Starch Digestion

Grains and tubers are a major staple of many agricultural populations. These foods are rich sources of starch, which is made up of strands of glucose, some with elaborate branching structures. Only glucose molecules bound by alpha glycosidic bonds are digestible by humans (and other animals); those with glucose molecules connected by beta bonds are collectively referred to as fiber, and are resistant to digestive enzymes. Their energy can only be captured if they are fermented into short-chain fatty acids by gut bacteria, whose importance to human nutrition and a wide array of physiological functions is just beginning to be appreciated. Cooking does not change the digestibility of fiber, but as noted above, heat enhances the digestibility of starches, especially those in tubers, by increasing their solubility.

Digestion of starch begins in the mouth with the salivary enzyme amylase, which breaks up the polymers of glucose into two-glucose units called maltose. The amount of maltose depends on how long food stays in the mouth. After swallowing, stomach acids deactivate salivary amylase; in the small intestine amylase from the pancreas carried out the conversion of starch to maltose. In order to be absorbed and utilized for energy, maltose in turn must be converted into single glucose molecules by the enzyme maltase, which is produced in the cells lining the small intestine.

In humans, the gene for salivary (α-) amylase in on chromosome 1. Researchers have discovered that humans appear to have more copies of the gene for salivary amylase (AMY1) than our ape relatives (Perry et al., 2007), but also that some seed- and unripe fruit-eating Old World monkey species (e.g., baboons and geladas) appear to have higher copy numbers than humans (Mau et al., 2010). Across individuals AMY1 copy number correlates with amylase production and speed of oral digestion of starch into maltose (Perry et al., 2007; Mandel et al., 2010). Individual variation in copy number is high, but the high average copy number in our species relative to other apes suggests that recent evolutionary history of high-starch foods such as underground storage organs might have selected for greater amylase activity through gene duplication. In addition, it appears that human populations that currently have high-starch intake have higher average AMY1 copy number than those whose subsistence strategies do not include reliance on high-starch diets (Perry et al., 2007). Thus there may have been later selection for greater salivary amylase production among those populations consuming a high-starch diet from cultivated grains or tubers.

Whether this selection has occurred awaits further systematic investigation among more extant human populations. It is also the case that salivary amylase is only the first step in starch digestion; the rate-limiting step is actually intestinal maltase, which is essential for the final cleaving of maltose into readily absorbed glucose subunits. Interestingly, domesticated dogs show an increase in pancreatic amylase copy number (AMY2B) and mutations in the maltase–glucoamylase gene (MGAM) that appear to increase maltase activity (Axelsson et al., 2013). These observations suggest domestic canine adaptations to a starch-rich diet akin to that consumed by its human companions. Whether there are positive relationships between salivary amylase and maltase remains unknown.

Variation in Adult Milk Digestion

The best documented example is that of population variation in milk digestion, which is well described at the genetic, phenotypic, and behavioral levels. The production and consumption of milk and milk products date to the domestication of animals, around 8000 years before the common era (Burger et al., 2007; Craig et al., 2005) (see the chapter by Little), but the use of domesticated mammals for dairying was limited to nomadic pastoralist groups and the settled large-scale populations of Europe and South Asia. Prior to European colonization, there were virtually no domesticated mammals, and hence no dairying tradition, among Native American, Australian, or Oceanic populations. In East and Southeast Asia, the use of domesticated animals for dairying was never institutionalized, and in sub-Saharan Africa, while pastoralist groups have made extensive use of their animals for milk, most settled agricultural groups did not milk their mammalian domesticates (Simoons, 1981). This variation between populations in dairy production and consumption historically mapped onto biological differences in the ability to digest the milk sugar lactose throughout life, suggesting some coevolutionary interactions among culture, diet, and biology.

Human biological variation in relation to milk centers on the milk sugar lactose. Lactose is a disaccharide unique to mammalian milk, and while not all mammals produce milk with lactose in it, the majority of milk-producing domesticates do contain this sugar. It cannot be absorbed in the small intestine directly, but must be cleaved into its component sugars glucose and galactose, which can then be absorbed, enter the body's circulatory system, and used for energy. This process requires lactase, a specialized enzyme found along the cells that line the upper small intestine. In general, infant mammals produce lactase in order to digest the lactose they are ingesting while nursing. However, lactase production diminishes over time and eventually stops all together, usually around the time of weaning (see Figure 2). Mammals living in the wild never consume milk again after weaning; since the lactase appears to have no other function, it would be wasteful of scarce nutrients to continue to produce it. Those mammals whose milk does not contain lactose (e.g., the marine mammals) also do not produce lactase, suggesting that the enzyme was eliminated by natural selection because it was not needed.

Among humans, the gene for the lactase enzyme (LCT) is on chromosome 2, but a regulatory region upstream from the lactase gene itself is the locus of variation in adult lactase activity (Enattah et al., 2002; Ingram et al., 2007, 2009).

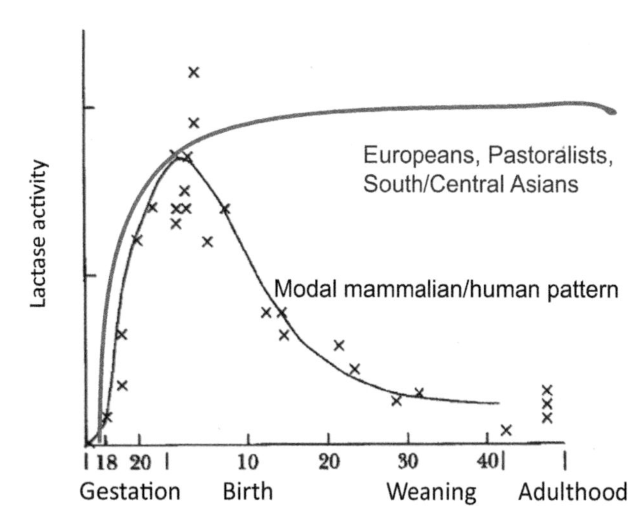

FIGURE 2 Life course variation in lactase activity among mammalian species and human populations.

Some sub-Saharan African, European, Middle Eastern, and Central and South Asian populations have shown relatively high frequencies of mutations in this region that act to keep lactase activity throughout life, while all other humans have the ancient mammalian DNA sequence, which results in lactase being turned off. Individuals who continue to produce lactase in adulthood are lactase persistent, while those whose lactase activity diminishes in childhood are lactase impersistent. The mutations for lactase persistence appear to be dominant to those that result in lactase impersistence, and thus only one copy of the allele is needed for phenotypic expression of lactase persistence. The decline of lactase production is independent of the presence of lactose in the small intestine (Sahi, 1994a,b); continued milk drinking does not cause lactase production to remain and cessation of milk drinking does not cause it to stop. Lactase impersistent individuals often experience the gastrointestinal symptoms of lactose intolerance when they consume milk or lactose-rich milk products.

All groups that have high frequencies of the mutations for lifelong lactase activity share a history of dairy animal domestication. It is easy to hypothesize that mutations that allowed people to digest milk throughout their lives would have been advantageous in dairying populations, as they would have been able to utilize a nutrient-rich food and thus been better nourished and presumably healthier, and ultimately had more offspring who also had the mutation (Ingram et al., 2009; Simoons, 1978, 2001; Ingram et al., 2007). Evidence suggests that the mutation for ongoing lactase production spread rapidly after domestication (Burger et al., 2007), and it was under very strong selection. At present there is no consensus on the exact nature of the selective force that enabled the spread of the lactase persistence trait—it could have been access to an energy- and protein-rich food, or a specific component such as calcium or lactose itself. Sugars are rare in pastoralist diets, and milk is the only way carbohydrates can be obtained from animal foods. Lactose may also substitute for vitamin D in enhancing calcium absorption; vitamin D synthesis is limited in high-latitude regions such as northern Europe where UV-B rays are insufficient for much of the year. Milk is also a source of water, important to pastoralist groups living in arid regions. Or, given that milk is produced to support the very rapid growth of infants, and that bovine infants grow more rapidly and to larger body sizes than humans, it is possible that humans who continued to consume milk grew more rapidly or reached sexual maturity earlier, thus enhancing fitness (Wiley, 2011a, 2012). In all likelihood different selective forces shaped the evolution of lactase persistence in populations living in different environments. But most humans maintain the ancestral mammalian pattern of lactase impersistence. Despite this, milk consumption is recommended in many contemporary dietary guidelines, and milk consumption is growing most rapidly in East Asia, where lactase impersistence is the norm (Wiley, 2011b).

INDUSTRIALIZATION OF THE DIET AND CONSEQUENCES FOR HUMAN BIOLOGY

Two historical trends that relied on agricultural production of a surplus further shaped contemporary human diets. One is the globalization of diet (see the chapter by Dressler), which began with European exploration and subsequent colonization of the New World and regions of Africa, Asia, and Oceania beginning in the fifteenth century. Combined with more efficient forms of transport, this allowed for commodities to circulate across oceans, transforming both production and consumption in the colonies and in Europe. Sugar is a classic example of an Old World food that was cultivated in New World colonies, and then exported back to Europe, becoming an intrinsic part of the diet (e.g., in tea, coffee, and sweets of various kind, cf. Mintz, 1985). The second related trend was the industrialization of diet, which began in the nineteenth century with technological innovations that created efficiencies in production, processing, preservation, and transport of food. Details of these historical processes are beyond the scope of this chapter (see: Crosby, 1986; Pollan, 2006; Mintz, 1985; Grew, 1999; Nutzenadel and Trentmann, 2008; Phillips, 2006), but they appear to have contributed to a surfeit of calories available in industrialized countries (Nestle and Nesheim, 2012), which, along with declines in infectious disease, contributed to increases in weight, height and life expectancy (Komlos, 1994; Steckel, 2009). Here we focus on the consequences for human diet and nutrition around the globe over the course of the twentieth century and now into the twenty-first century.

A major outcome of these trends includes the homogenization of diets across human populations, often referred to as the spread of the Western diet, reflecting the preponderance

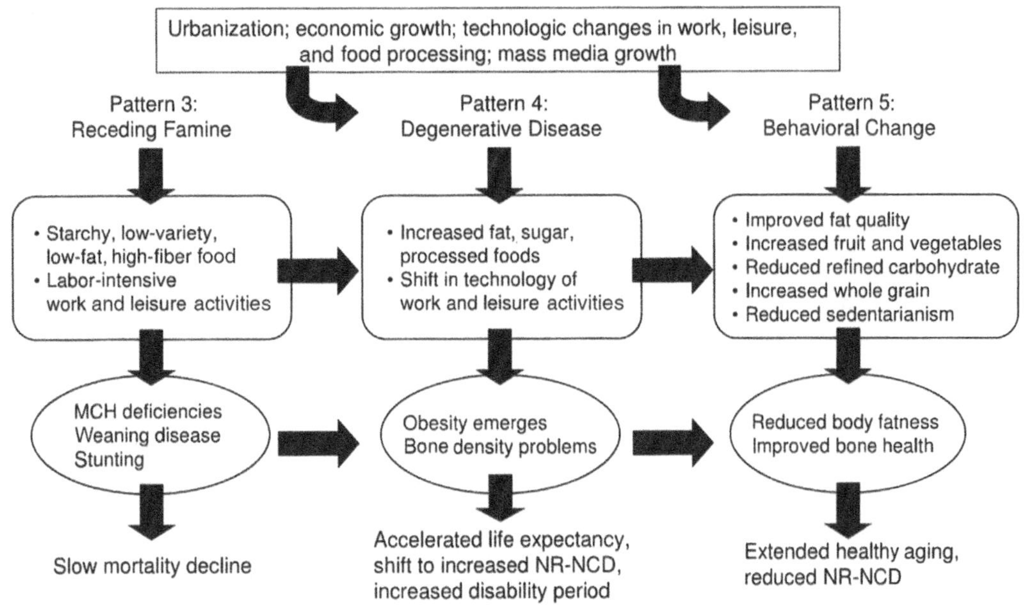

FIGURE 3 Patterns of the nutrition transition. *From Popkin, B.M., 2006. The American Journal of Clinical Nutrition 84, 289–298; NR-NCD, nutrition-related non-communicable disease; MCH, maternal and child health. ©2006 by American Society for Nutrition.*

of foodstuffs that were crafted in Western countries such as those in Europe and North America. These include highly processed foods that are high in salt, refined grains, sugar (or high-fructose corn syrup, commonly used in sweetened beverages), and fat as well as meat and dairy products, which are marketed by rapidly spreading multinational fast food industries. These foods are products of industrial agriculture, which generate large surpluses of grains (including corn, which can be turned into the sweetener corn syrup) and plant oils that find their way into a wide variety of processed foods. Collectively these changes in food production and consumption have been termed the Nutrition Transition (cf. Caballero, 2002), although this is just one in the series of nutritional transitions described by Barry Popkin (2004) (see Figure 3). Alterations in diet are coincident with other social changes: urbanization, reductions in occupations requiring high levels of energy expenditure, greater life expectancy, declining fertility, and increasing income disparities (Popkin, 2006).

These changes appear to have had a major impact on human biology, with rates of obesity, metabolic disorders, and other noncommunicable diseases swiftly rising in developing countries to match or exceed those of industrialized countries where they have been increasing since the mid-twentieth century. Humans now are taller and heavier than at any time in the evolutionary or historical past, and this may put them at risk for a variety of chronic diseases (metabolic disorders such as diabetes, cancer, and cardiovascular diseases). It should be noted that while there is an overall trend toward overnutrition, undernutrition, measured by stunted height and underweight, remains endemic to impoverished populations of South Asia and sub-Saharan Africa. Many developing countries are experiencing a "double burden" of obesity and undernutrition, with policies and public health infrastructures incapable of addressing these issues at the national level (Kennedy et al., 2006).

The causes of the dramatic rise in weight among human populations are not well established, although most scholars and policymakers consider it to be an increase in calories consumed coupled with a reduction in calories expended, leading to excess energy storage as fat. There are other hypotheses that do not invoke changes in diet and activity, including exposure to endocrine disruptors, positive assortative mating, and reductions in sleep (McAllister et al., 2009).

PALEOLITHIC PRESCRIPTIONS

One way in which biological anthropologists have come to understand this unprecedented growth in body height and weight and its attendant chronic diseases is by viewing our current diet and activity patterns as "mismatched" to those of our evolutionary past (see the chapter by Low et al.). That is, our biology is well adapted to the environment that prevailed during the course of our evolution as a species (the environment of evolutionary adaptedness; EEA), but our current dietary environment and many aspects of our lifestyle are quite different. The major transition between the EEA and the current diets is considered to be agriculture, and its introduction of grains, legumes, and dairy in the diet, although it should be emphasized that rates of chronic disease do not appear to rise until well into the twentieth century, suggesting that more recent dietary changes must be taken into account. Obesity, metabolic and cardiovascular diseases, and some forms of cancer could be considered consequences of this "mismatch." In 1988, S. Boyd Eaton, Melvin Konner, and Marjorie Shostak outlined this approach in an article entitled "Stone Agers in the Fast Lane" (Eaton et al., 1988a), and later expanded it in a popular book entitled, *The Paleolithic Prescription* (Eaton et al., 1988b). More recently it has been advocated by Loren Cordain (Cordain et al., 2005), who has popularized and commercialized the "Paleo Diet" (cf. www.thepaleodiet.com). This perspective has antecedents in James Neel's "thrifty gene" hypothesis, which proposed that our ancestors faced high levels of variability in access to food, thus making efficient absorption and metabolism adaptive. However, in the context of routine access to high calorie foods, these efficiencies could be deleterious, contributing to metabolic disorders such as Type II diabetes (Neel, 1962).

Drawing heavily on evidence from contemporary hunter–gatherer populations, especially the !Kung of South Africa who appear to have very low rates of chronic disease (note that this is not a consequence of their shorter life expectancy; the latter is most reduced by infant and child mortality), they outlined the ways in which diet and nutrition varied between them and contemporary Americans. This comparison was updated in 2010 (Konner and Eaton, 2010), and some of the differences in diet can be found in Table 1. Overall, subsistence based on hunting and gathering would mean consumption of wild plant and animal foods, focusing on fruits, nuts, tubers, and a variety of vertebrate and invertebrate animal products. Notably absent from that list are dairy products, grains, and legumes, which are routinely consumed in many contemporary diets, especially in industrialized societies, and hence most Paleolithic-inspired diets

TABLE 1 Widely Agreed-On Qualitative Differences between Average Ancestral (Hunter–Gatherer) Diets and Contemporary Western Diets

	Ancestral (Hunter–Gatherer)	Contemporary Western
Total energy intake	More	Less
Caloric density	Very low	High
Dietary bulk	More	Less
Total carbohydrate intake	Less	More
Added sugars/refined carbohydrates	Very little	Much more
Glycemic load	Relatively low	High
Fruits and vegetables	Twice as much	Half as much
Antioxidant capacity	Higher	Lower
Fiber	More	Less
Soluble:insoluble	Roughly 1:1	<1 insoluble
Protein intake	More	Less
Total fat intake	Roughly equal	
Serum cholesterol-raising fat	Less	More
Total polyunsaturated fat	More	Less
ω-6:ω-3	Roughly equal	Far more ω-6
Long-chain essential fatty acids	More	Less
Cholesterol intake	Equal or more	Equal or less
Micronutrient intake	More	Less
Sodium:potassium	<1	>1
Acid–base impact	Alkaline or acidic	Acidic
Milk products	Mother's milk only	High, lifelong
Cereal grains	Minimal	Substantial
Free water intake	More	Less

Konner, M, and Eaton, S.B., 2010. Nutrition in Clinical Practice 25, 594–602.

are low in carbohydrates. Wild plant foods are much higher in fiber and phytochemicals than cultivated plant foods (which have been domesticated both for increased productivity and palatability), and wild animals are leaner than domesticated, and often grain-fed, animals such as cows, pigs, and chickens. Also note in Table 1 that sodium, carbohydrates, and saturated fat would have been much lower in the EEA diet, while fiber, micronutrients, protein, and ω-3 fatty acids are likely to have been more common. On the whole, the diet would have contained more foods deemed to be "healthy" and fewer of those identified as risk factors for chronic disease (cf. American Heart Association Diet and Lifestyle Guidelines, www.heart.org). However, it is at odds with contemporary Dietary Guidelines for Americans (United States Department of Health and Human Services and United States Department of Agriculture, 2010) as well as dietary guidance from many countries that emphasize grains and low-fat milk and dairy products.

Principles of this evolutionary approach to diet and nutrition have been tested empirically, with evidence suggesting that adherence to a diet more akin to that of the EEA can has a positive impact on biomarkers associated with chronic disease risk (O'dea, 1984; Jönsson et al., 2009; Frassetto et al., 2009; Lindeberg et al., 2003). At the same time, there is debate about what the relative proportions of meat and plant foods should be. As noted earlier, this reflects uncertainty about dietary proportions from the archaeological record of preagricultural *H. sapiens*, and geographic and temporal variability. It is also the case that diet must be considered in light of physical activity; current reports suggest though, that total energy expenditure (controlling for body size, age, and sex) is not significantly greater among contemporary hunter–gatherers than Western adults or subsistence farmers (Pontzer et al., 2012), and most physical activity is characterized as light to moderate rather than vigorous (Gurven et al., 2013).

It is notable that the two common contemporary dietary components absent in Paleolithic diets—grains and milk—are both implicated in intolerance syndromes. Lactose intolerance from milk consumption has already been noted, but there has recently been a surge in diagnosis of individuals reporting gluten intolerance in Western countries. Gluten is the main protein in grains such as wheat, rye, and barley, and gluten intolerance has three different sources: allergy (characterized by antibodies to gluten), autoimmune (celiac disease, characterized by gut inflammation), and gluten sensitivity, which has none of the other diagnostic features, but is associated with symptoms that improve with a gluten-free diet (Sapone et al., 2012). These gluten intolerances appear to be increasing in frequency, for reasons that are not well understood, but they are probably related to the overall increase in allergies and autoimmune disorders in industrialized countries and not just better detection (Lohi et al., 2007). These syndromes are likely to increase in frequency with the nutrition transition, as populations previously subsisting on rice, corn, or other staples adopt diets with more wheat products (e.g., bread).

While some individuals avoid these postagriculture products in an attempt to enhance their health, there are other dietary patterns worth considering from an evolutionary perspective. One is the purposeful limitation of animal products in the diet, from lacto–ovo vegetarianism (including dairy and eggs) to veganism (eschewing any animal product including honey). As noted before, our evolutionary history has been characterized by omnivory, and thus there is no justification for vegetarianism as the "natural" diet of *H. sapiens*. At the same time, there is good evidence that individuals who consume little or no meat have lower body mass index and lower risk of various chronic diseases, cancers, and all-cause mortality (Fraser, 2009; Tantamango-Bartley et al., 2013; Orlich et al., 2013; Huang et al., 2012, provide a meta-analysis), although study results differ in part because what is defined as a vegetarian diet is quite variable. The mechanisms underlying this association are not known with any certainty: vegetarian diets provide more fiber, phytochemicals, and micronutrients and are associated with reduced inflammation, circulating cholesterol (only animal foods provide meaningful amounts of cholesterol in the diet) and platelet aggregation (Barnard et al., 2009; Boeing et al., 2012). Many phytochemicals and micronutrients have antioxidant properties. Despite the benefits associated with plant-based diets, global surveys indicate that only a few countries' citizens even come close to meeting the World Health Organization's recommendation of five servings of fruits and vegetables (~400 g) per day (Hall et al., 2009; Boeing et al., 2012).

While different diets may promote or harm human health in various ways, the only dietary approach that consistently seems to retard some diseases of aging or extend life expectancy in adulthood is dramatic CR imposed on nutrient sufficiency. It has been known for most of the twentieth century that laboratory animals on restricted calorie diets (up to 40% calorie reduction) have lives that are markedly longer than those fed a standard diet (Colman and Anderson, 2011). Currently, two long-term studies of two groups of rhesus monkeys given CR diets (30% reduction in calories) starting in adulthood have shown that while CR improves some biomarkers of health risk, only one CR group had a longer lifespan than the non-CR monkeys (Mattison et al., 2012; Colman et al., 2009). The researchers are trying to ascertain what contributed to the different results and what mechanisms might link CR to healthier old ages.

Among humans, the Okinawan population has had the longest life expectancy, with older individuals showing low prevalence of age-related disease and reportedly having a long-term modest CR diet (<1800 kcal) that provided adequate micronutrients (Willcox et al., 2007). Energy expenditure was relatively high from farming, and BMI was low. The diet was based on sweet potatoes and legumes, and rich in antioxidants (Willcox et al., 2007). Studies have not been able to point to specific dietary composition factors, and younger Okinawans appear to be consuming more calories. Overall it appears to be quite difficult for humans to adhere voluntarily to such a low calorie diet, which probably reflects adaptations to avoid starvation and undernutrition, both of which impair fitness.

CONCLUSION

Human evolution has been typified by changes in diet relative to our closest living relatives. These have been marked by increased omnivory and a broader array of more terrestrial foods with the commitment to bipedalism. Technological changes such as stone tools and controlled use of fire enhanced food acquisition and processing, opening up new dietary niches, and these are likely to be related to the evolution of the large brain of *H. sapiens*. In turn, high-quality (i.e. nutrient dense) food and the ability to transform it into that easily consumed by the very young as transitional foods, and when such foods could be provided by someone other than the mother, breastfeeding could have been shortened and fertility enhanced, setting the stage for demographic expansion of our species, which is unusual for its routine engagement in cooking and food sharing.

With the transition to food production, major changes occurred in the realm of diet, including reliance on more grains and tubers and less dietary variety. One of the best described examples of human population genetic variation stems from the domestication of animals and use of their milk. More recently, global exploration, trade, and colonialism ushered in a new era of global dietary homogenization, wherein many foods are now consumed well beyond their local origins. Industrial processing of foods and movement away from subsistence agriculture have added to the spread of the "Western" diet, and exploit human taste preferences for fat, sweet, and salt. At present the consequences for human biology have not been positive, and health worries have prompted attempts to conform to a diet more typical of our ancestors, with the constraint that almost all foods available to humans are highly domesticated.

REFERENCES

Aiello, L.C., Wheeler, P., 1995. The expensive-tissue hypothesis: the brain and the digestive system in human and primate evolution. Current Anthropology 36, 199–221.

Axelsson, E., Ratnakumar, A., Arendt, M.-L., Maqbool, K., Webster, M.T., Perloski, M., Liberg, O., Arnemo, J.M., Hedhammar, Å., Lindblad-Toh, K., 2013. The genomic signature of dog domestication reveals adaptation to a starch-rich diet. Nature 495, 360–364.

Barnard, N.D., Katcher, H.I., Jenkins, D.J.A., Cohen, J., Turner-Mcgrievy, G., 2009. Vegetarian and vegan diets in type 2 diabetes management. Nutrition Reviews 67, 255–263.

Berna, F., Goldberg, P., Horwitz, L.K., Brink, J., Holt, S., Bamford, M., Chazan, M., 2012. Microstratigraphic evidence of in situ fire in the Acheulean strata of Wonderwerk Cave, Northern Cape province, South Africa. Proceedings of the National Academy of Sciences 109, E1215–E1220.

Bocquet-Appel, J.-P., Bar-Yosef, O. (Eds.), 2008. The Neolithic Demographic Transition and its Consequences. Springer-Verlag, New York.

Boeing, H., Bechthold, A., Bub, A., Ellinger, S., Haller, D., Kroke, A., Leschik-Bonnet, E., Müller, M., Oberritter, H., Schulze, M., Stehle, P., Watzl, B., 2012. Critical review: vegetables and fruit in the prevention of chronic diseases. European Journal of Nutrition 51, 637–663.

Bogin, B., 2006. Modern human life history: the evolution of human childhood and fertility. In: Hawkes, K., Paine, R.R. (Eds.), The Evolution of Human Life History. Santa Fe: School of American Research Press.

Broadhurst, C.L., Wang, Y., Crawford, M.A., Cunnane, S.C., Parkington, J.E., Schmidt, W.F., 2002. Brain-specific lipids from marine, lacustrine, or terrestrial food resources: potential impact on early African Homo sapiens. Comparative Biochemistry and Physiology Part B: Biochemistry and Molecular Biology 131, 653–673.

Bunn, H.T., 2007. Meat made us human. In: Ungar, P.S. (Ed.), Evolution of the Human Diet: The Known, the Unknown, and the Unknowable. Oxford University Press, New York.

Burger, J., Kirchner, M., Bramanti, B., Haak, W., Thomas, M.G., 2007. Absence of the lactase-persistence-associated allele in early Neolithic Europeans. Proceedings of the National Academy of Sciences 104, 3736–3741.

Caballero, B., 2002. The Nutrition Transition: Diet and Disease in the Developing World. Academic Press, New York.

Carlson, B.A., Kingston, J.D., 2007. Docosahexaenoic acid, the aquatic diet, and hominin encephalization: difficulties in establishing evolutionary links. American Journal of Human Biology 19, 132–141.

Cohen, M.N., 1989. Health and the Rise of Civilization. Yale University Press, New Haven, CT.

Cohen, M.N., Armelagos, G.J. (Eds.), 1984. Paleopathology at the Origins of Agriculture. Academic Press, New York.

Colman, R.J., Anderson, R.M., 2011. Nonhuman primate calorie restriction. Antioxidants and Redox Signaling 14, 229–239.

Colman, R.J., Anderson, R.M., Johnson, S.C., Kastman, E.K., Kosmatka, K.J., Beasley, T.M., Allison, D.B., Cruzen, C., Simmons, H.A., Kemnitz, J.W., 2009. Caloric restriction delays disease onset and mortality in rhesus monkeys. Science 325, 201–204.

Cordain, L., Eaton, S.B., Sebastian, A., Mann, N., Lindeberg, S., Watkins, B.A., O'keefe, J.H., Brand-Miller, J., 2005. Origins and evolution of the Western diet: health implications for the 21st century. The American Journal of Clinical Nutrition 81, 341–354.

Craig, O.E., Chapman, J., Heron, C., Willis, L.H., Bartosiewicz, L., Taylor, G., Whittle, A., Collins, M., 2005. Did the first farmers of central and eastern Europe produce dairy foods? Antiquity 79, 882–894.

Crosby, A.W., 1986. Ecological Imperialism: The Biological Expansion of Europe 900-1900. Cambridge University Press, New York.

Diamond, J., 1987. The worst mistake in the history of the human race. Discover 8, 64–66.

Eaton, S.B., Konner, M., Shostak, M., 1988a. Stone agers in the fast lane: chronic degenerative diseases in evolutionary perspective. The American Journal of Medicine 84, 739–749.

Eaton, S.B., Shostak, M., Konner, M., 1988b. The Paleolithic Prescription: A Program of Diet & Exercise and a Design for Living. Harper & Row, New York.

Enattah, N.S., Sahi, T., Savilahti, E., Terwilliger, J.D., Peltonen, L., Varvela, I., 2002. Identification of a variant associated with adult-type hypolactasia. Nature Genetics 30, 233–237.

Fraser, G.E., 2009. Vegetarian diets: what do we know of their effects on common chronic diseases? The American Journal of Clinical Nutrition 89, 1607S–1612S.

Frassetto, L.A., Schloetter, M., Mietus-Synder, M., Morris, R., Sebastian, A., 2009. Metabolic and physiologic improvements from consuming a paleolithic, hunter–gatherer type diet. European Journal of Clinical Nutrition 63, 947–955.

Grew, R., 1999. Food in Global History. Westview Press, Boulder, CO.

Gurven, M., Jaeggi, A.V., Kaplan, H., Cummings, D., 2013. Physical activity and Modernization among Bolivian Amerindians. PloS One 8, e55679.

Hall, J.N., Moore, S., Harper, S.B., Lynch, J.W., 2009. Global variability in fruit and vegetable consumption. American Journal of Preventive Medicine 36, 402–409 e5.

Halsey, L.G., White, C.R., 2012. Comparative energetics of mammalian locomotion: humans are not different. Journal of Human Evolution 63, 718–722.

Hewes, G.W., 1964. Hominid bipedalism: independent evidence for the food-carrying theory. Science 146, 416–418.

Hladik, C.M., Pasquet, P., 2002. The human adaptations to meat eating: a reappraisal. Human Evolution 17, 199–206.

Hrdy, S.B., 2009. Mothers and Others. Harvard University Press, Cambridge, MA.

Huang, T., Yang, B., Zheng, J., Li, G., Wahlqvist, M.L., Li, D., 2012. Cardiovascular disease mortality and cancer incidence in vegetarians: a meta-analysis and systematic review. Annals of Nutrition and Metabolism 60, 233–240.

Hunt, K.D., 1994. The evolution of human bipedality: ecology and functional morphology. Journal of Human Evolution 26, 183–202.

Ingram, C.J., Elamin, M.F., Mulcare, C.A., Weale, M.E., Tarekegn, A., Raga, T.O., Bekele, E., Elamin, F.M., Thomas, M.G., Bradman, N., Swallow, D.M., Itan, Y., Tishkoff, S.A., Reed, F.A., Ranciaro, A., Voight, B.F., Babbitt, C.C., Silverman, J.S., Powell, K., Mortensen, H.M., Hirbo, J.B., Osman, M., Ibrahim, M., Omar, S.A., Lema, G., Nyambo, T.B., Ghori, J., Bumpstead, S., Pritchard, J.K., Wray, G.A., Deloukas, P., 2007. A novel polymorphism associated with lactose tolerance in Africa: multiple causes for lactase persistence? Human Genetics 120, 779–788.

Ingram, C.J., Mulcare, C.A., Itan, Y., Thomas, M.G., Swallow, D.M., 2009. Lactose digestion and the evolutionary genetics of lactase persistence. Human Genetics 124, 579–591.

Jönsson, T., Granfeldt, Y., Ahrén, B., Branell, U.-C., Pålsson, G., Hansson, A., Söderström, M., Lindeberg, S., 2009. Beneficial effects of a Paleolithic diet on cardiovascular risk factors in type 2 diabetes: a randomized cross-over pilot study. Cardiovascular Diabetology 8, 1–14.

Kaplan, H., Hill, K., Lancaster, J., Hurtado, A.M., 2000. A theory of human life history evolution: diet, intelligence, and longevity. Evolutionary Anthropology Issues News and Reviews 9, 156–185.

Katz, S.H., 1987. Food and biocultural evolution: a model for the investigation of contemporary nutrition problems. In: Johnston, F.E. (Ed.), Nutritional Anthropology. Allen Liss, New York.

Kennedy, G., Nantel, G., Shetty, P., 2006. Assessment of the double burden of malnutrition in six case study countries. FAO Food and Nutrition Paper 84, 1–20.

Klein, R.G., 2013. Stable carbon isotopes and human evolution. Proceedings of the National Academy of Sciences 110, 10470–10472.

Komlos, J. (Ed.), 1994. Stature, Living Standards, and Economic Development: Essays in Anthropometric History. University of Chicago Press, Chicago.

Konner, M., Eaton, S.B., 2010. Paleolithic nutrition twenty-five years later. Nutrition in Clinical Practice 25, 594–602.

Kuipers, R.S., Luxwolda, M.F., Janneke Dijck-Brouwer, D., Eaton, S.B., Crawford, M.A., Cordain, L., Muskiet, F.A., 2010. Estimated macronutrient and fatty acid intakes from an East African Paleolithic diet. British Journal of Nutrition 104, 1666–1687.

Leonard, W.R., Robertson, M.L., 1992. Nutritional requirements and human evolution: a bioenergetics model. American Journal of Human Biology 4, 179–195.

Leonard, W.R., Robertson, M.L., 1994. Evolutionary perspectives on human nutrition: the influence of brain and body size on diet and metabolism. American Journal of Human Biology 6, 77–88.

Leonard, W.R., Robertson, M.L., Snodgrass, J.J., 2007. Energetic models of human nutritional evolution. In: Ungar, P.S. (Ed.), Evolution of the Human Diet: The Known, the Unknown, and the Unknowable. Oxford University Press, New York.

Leonard, W.R., Robertson, M.L., Snodgrass, J.J., Kuzawa, C.W., 2003. Metabolic correlates of hominid brain evolution. Comparative Biochemistry and Physiology-Part A: Molecular & Integrative Physiology 136, 5–15.

Lindeberg, S., Cordain, L., Eaton, S.B., 2003. Biological and clinical potential of a Palaeolithic diet. Journal of Nutritional and Environmental Medicine 13, 149–160.

Lohi, S., Mustalahti, K., Kaukinen, K., Laurila, K., Collin, P., Rissanen, H., Lohi, O., Bravi, E., Gasparin, M., Reunanen, A., Mäki, M., 2007. Increasing prevalence of coeliac disease over time. Alimentary Pharmacology & Therapeutics 26, 1217–1225.

Luca, F., Perry, G.H., Di Rienzo, A., 2010. Evolutionary adaptations to dietary changes. Annual Review of Nutrition 30, 291–314.

Mandel, A.L., Des Gachons, C.P., Plank, K.L., Alarcon, S., Breslin, P.A., 2010. Individual differences in AMY1 gene copy number, salivary α-amylase levels, and the perception of oral starch. PloS One 5, e13352.

Mattison, J.A., Roth, G.S., Beasley, T.M., Tilmont, E.M., Handy, A.M., Herbert, R.L., Longo, D.L., Allison, D.B., Young, J.E., Bryant, M., Barnard, D., Ward, W.F., Qi, W., Ingram, D.K., De Cabo, R., 2012. Impact of caloric restriction on health and survival in rhesus monkeys from the NIA study. Nature 489, 318–321.

Mau, M., Südekum, K.H., Johann, A., Sliwa, A., Kaiser, T.M., 2010. Indication of higher salivary α-amylase expression in hamadryas baboons and geladas compared to chimpanzees and humans. Journal of Medical Primatology 39, 187–190.

Mcallister, E.J., Dhurandhar, N.V., Keith, S.W., Aronne, L.J., Barger, J., Baskin, M., Benca, R.M., Biggio, J., Boggiano, M.M., Eisenmann, J.C., 2009. Ten putative contributors to the obesity epidemic. Critical Reviews in Food Science and Nutrition 49, 868–913.

Milton, K., 1999. Nutritional characteristics of wild primate foods: do the diets of our closest living relatives have lessons for us? Nutrition 15, 488–498.

Milton, K., 2003. The critical role played by animal source foods in human (Homo) evolution. The Journal of Nutrition 133, 3886S–3892S.

Mintz, S., 1985. Sweetness and Power: The Place of Sugar in Modern History. Viking, New York.

Mummert, A., Esche, E., Robinson, J., Armelagos, G.J., 2011. Stature and robusticity during the agricultural transition: evidence from the bioarchaeological record. Economics & Human Biology 9, 284–301.

Navarrete, A., Van Schaik, C.P., Isler, K., 2011. Energetics and the evolution of human brain size. Nature 480, 91–93.

Neel, J.V., 1962. Diabetes mellitus: a "thrifty" genotype rendered detrimental by "progress"? American Journal of Human Genetics 14, 353–362.

Nestle, M., Nesheim, M., 2012. Why Calories Count: From Science to Politics. University of California Press, Berkeley.
Nutzenadel, A., Trentmann, F., 2008. Food and Globalization. Berg, New York.
O'dea, K., 1984. Marked improvement in carbohydrate and lipid metabolism in diabetic Australian Aborigines after temporary reversion to traditional lifestyle. Diabetes 33, 596–603.
Orlich, M., Singh, P., Sabaté, J., 2013. Vegetarian diets associated with lower mortality, including CVD-related death. Journal of the American Medical Association 173, 1230–1238.
Perry, G.H., Dominy, N.J., Claw, K.G., Lee, A.S., Fiegler, H., Redon, R., Werner, J., Villanea, F.A., Mountain, J.L., Misra, R., 2007. Diet and the evolution of human amylase gene copy number variation. Nature Genetics 39, 1256–1260.
Phillips, L., 2006. Food and globalization. Annual Review of Anthropology 35, 37–57.
Pollan, M., 2006. The Omnivore's Dilemma: A Natural History of Four Meals. Penguin, New York.
Pontzer, H., Raichlen, D.A., Wood, B.M., Mabulla, A.Z., Racette, S.B., Marlowe, F.W., 2012. Hunter–gatherer energetics and human obesity. Plos One 7, e40503.
Popkin, B.M., 2004. The nutrition transition: an overview of world patterns of change. Nutrition Reviews 62, S140–S143.
Popkin, B.M., 2006. Global nutrition dynamics: the world is shifting rapidly toward a diet linked with noncommunicable diseases. The American Journal of Clinical Nutrition 84, 289–298.
Roebroeks, W., Villa, P., 2011. On the earliest evidence for habitual use of fire in Europe. Proceedings of the National Academy of Sciences 108, 5209–5214.
Sahi, T., 1994a. Genetics and epidemiology of adult-type hypolactasia. Scandinavian Journal of Gastroenterology 29, 7–20.
Sahi, T., 1994b. Hypolactasia and lactase persistence: historical review and the terminology. Scandinavian Journal of Gastroenterology 29, 1–6.
Sapone, A., Bai, J.C., Ciacci, C., Dolinsek, J., Green, P.H., Hadjivassiliou, M., Kaukinen, K., Rostami, K., Sanders, D.S., Schumann, M., Ullrich, R., Villalta, D., Volta, U., Catassi, C., Fasano, A., 2012. Spectrum of gluten-related disorders: consensus on new nomenclature and classification. BMC Medicine 10, 13.
Simoons, F.J., 1978. The geographic hypothesis and lactose malabsorption. American Journal of Digestive Diseases 23, 963–980.
Simoons, F.J., 1981. Geographic patterns of primary adult lactose malabsorption: a further interpretation of evidence for the Old World. In: Paige, D.M., Bayless, T.M. (Eds.), Lactose Digestion: Clinical and Nutritional Implications. The Johns Hopkins University Press, Baltimore.
Simoons, F.J., 2001. Persistence of lactase activity among northern Europeans: a weighing of evidence for the calcium absorption hypothesis. Ecology of Food and Nutrition 40, 397–469.
Sponheimer, M., Alemseged, Z., Cerling, T.E., Grine, F.E., Kimbel, W.H., Leakey, M.G., Lee-Thorp, J.A., Manthi, F.K., Reed, K.E., Wood, B.A., Wynn, J.G., 2013. Isotopic evidence of early hominin diets. Proceedings of the National Academy of Sciences 110, 10513–10518.
Stahl, A.B., Dunbar, R., Homewood, K., Ikawa-Smith, F., Kortlandt, A., Mcgrew, W., Milton, K., Paterson, J., Poirier, F., Sugardjito, J., 1984. Hominid dietary selection before fire [and Comments and Reply]. Current Anthropology 25, 151–168.
Stanford, C.B., Bunn, H.T. (Eds.), 2001. Meat-eating and Human Evolution. Oxford University Press, New York.
Steckel, R.H., 2009. Heights and human welfare: recent developments and new directions. Explorations in Economic History 46, 1–23.
Tantamango-Bartley, Y., Jaceldo-Siegl, K., Fan, J., Fraser, G., 2013. Vegetarian diets and the incidence of cancer in a low-risk population. Cancer Epidemiology Biomarkers and Prevention 22, 286–294.
Teaford, M.F., Ungar, P.S., 2000. Diet and the evolution of the earliest human ancestors. Proceedings of the National Academy of Sciences 97, 13506–13511.
United States Department of Health and Human Services & United States Department of Agriculture, 2010. Dietary Guidelines for Americans. [Online]. USDHHS/USDA. Available http://health.gov/dietaryguidelines/dga2010/DietaryGuidelines2010.pdf (accessed 03.01.14).
Videan, E.N., Mcgrew, W.C., 2002. Bipedality in chimpanzee (Pan troglodytes) and bonobo (Pan paniscus): testing hypotheses on the evolution of bipedalism. American Journal of Physical Anthropology 118, 184–190.
Wiley, A.S., 2011a. Milk intake and total dairy consumption: associations with early menarche in NHANES 1999–2004. PLoS One 6, e14685.
Wiley, A.S., 2011b. Re-imagining Milk. Routledge, New York.
Wiley, A.S., 2012. Cow milk consumption, insulin-like growth factor-I, and human biology: a life history approach. American Journal of Human Biology 24, 130–138.
Willcox, B.J., Willcox, D.C., Todoriki, H., Fujiyoshi, A., Yano, K., He, Q., Curb, J.D., Suzuki, M., 2007. Caloric restriction, the traditional Okinawan diet, and healthy aging. Annals of the New York Academy of Sciences 1114, 434–455.
Wrangham, R., Carmody, R., 2010. Human adaptation to the control of fire. Evolutionary Anthropology: Issues, News, and Reviews 19, 187–199.
Wrangham, R.W., 2009. Catching Fire: How Cooking Made Us Human. Basic Books, New York.

Chapter 29

Diversity and Origins of Human Infectious Diseases

Serge Morand
CNRS ISEM–CIRAD AGIRs, Centre d'Infectiologie Christophe Mérieux du Laos, Vientiane, Lao PDR, France

SYNOPSIS

Human diseases are not distributed at random geographically or temporally. The present chapter reviews how pathogens (infectious agents that cause disease) of humans have changed over evolutionary and historical times. Modern humans have been parasitized by the infectious agents of our earlier hominin ancestors, or by acquiring them from wild or domesticated animal species according to several epidemiological transitions: (1) the "out of Africa" source where pathogens followed the dispersal and expansion of modern humans and earlier hominins from Africa, as humans acquired new parasites and pathogens during the geographic expansion process (but also immune genes to cope with these new parasites by interbreeding with archaic humans); (2) the "domestication" source where pathogens spread from domesticated animals and then dispersed more globally; and (3) the "globalization" source, with expansion of pathogens in relation to historical and recent trade routes. Ongoing environmental changes result in both the emergence of new infectious diseases and the loss in parasite diversity, with striking consequences that include the rise of autoimmune diseases as well as zoonotic pathogens.

THE DIVERSITY OF INFECTIOUS DISEASES IN SPACE

Humans are parasitized by a large number of pathogenic species, and we are likely the most infected species on earth. More than 1400 parasite species have been listed as pathogenic in humans (Cleaveland et al., 2001) and, among them, at least 60% are zoonotic in origin (Taylor et al., 2001). Understanding the ecological, historical, and biogeographical associations between humans and parasites has been the subject of numerous studies (May, 1958; Cockburn, 1963; McNeil, 1976; Diamond, 1997; Ashford and Crewe, 1998; Wolfe et al., 2007; Perrin et al., 2010), with some recent studies that further take into account the evolutionary history of animal domestication (Morand et al., 2014b; see chapters by Little and Sattenspiel in this volume).

Human pathogens are not randomly distributed geographically (Dunn et al., 2010), as the density of these pathogens increases with lower latitudes (Guernier et al., 2004; Dunn et al., 2010). Interestingly, this pattern follows a similar trend observed for general biodiversity with, for example, richness in bird and mammalian species being highest in tropical zones (Schipper et al., 2008). Hence, among countries, a positive relationship is observed between bird and mammal richness and human infectious diseases: A country with high bird and/or mammal biodiversity has a greater diversity of human pathogens (Dunn et al., 2010; Morand et al., 2014c) (Figure 1).

Cultural diversity also appears to mirror biological diversity (Maffi, 2005). For example, countries with high cultural diversity (e.g., higher number of spoken languages) have greater diversity of infectious diseases (Fincher and Thornhill, 2008). Biodiversity, cultural diversity, and the diversity of human pathogens are entangled in their actual geographical distributions due to evolutionary and ecological processes.

The aim of the present chapter is to review how the communities of human pathogens have changed over evolutionary and historical times. Reviews on this subject have emphasized how humans have gained their parasites either through descent (i.e., inherited from a common ancestor) or from either wild or domesticated animal species according to three major hypotheses (Wolfe et al., 2007; Morand, 2012): (1) the "out of Africa" source where parasites followed the dispersal and expansion of modern humans out of Africa, and where humans acquired new pathogens during the geographic expansion process; (2) the "domestication" source where parasites were transmitted from domesticated animals and then dispersed globally; and (3) the "globalization" source, reflecting the distribution of pathogens in relation to historical and more recent trade routes.

Ongoing global changes (e.g., climate change, biodiversity loss, land use change, biological invasion, etc.) are dramatically modifying the epidemiological environment (Daily and Ehrlich, 1996), with decreases in the burden of parasitic infection in the developed countries and increases of emerging infectious diseases in others (Keesing et al., 2010; Morand et al., 2014c). These environmentally driven epidemiological changes may even have consequences for the emergence of noncommunicable diseases, in particular autoimmune diseases (Muehlenbein, 2010; Parker et al., 2012).

ORIGINS OF INFECTIOUS DISEASES IN NONHUMAN PRIMATES

Early human populations were likely parasitized by pathogens from hominin and other primate ancestors, through descent or from habitat overlap. The study of cospeciation between primates and their parasites has evidenced the presence of many host-specific parasites in humans wordwide, including pinworms (Hugot, 1999), fungi (Hugot et al., 2003), lice (Reed et al., 2007), and some viruses (Switzer et al., 2005). Davies and Pedersen (2008) have shown that primate species that are closely related phylogenetically as well as those that live within the same geographic region likely share more pathogens than more distantly related primate hosts, or those separated geographically. Host shifts among closely related primates living in sympatry is even a likely explanation for the sharing of helminths, bacteria, and protozoa between human and nonhuman primates (Davies and Pedersen, 2008). It is likely that geographic overlap has been critical, perhaps even more important than phylogenetic relatedness, in facilitating host shifts of viruses (Davies and Pedersen, 2008). More recently, Gómez et al. (2013) have reinvestigated the patterns of primate host–pathogen sharing, identifying how a node (a host species) within a network is connected to other nodes through the sharing of pathogens. Both geographic distribution and population density of primate hosts are key factors in pathogen sharing.

THE FIRST EPIDEMIOLOGICAL TRANSITION: OUT OF AFRICA

Anatomically modern *Homo sapiens* likely left Africa beginning around 150,000 years ago (Callaway, 2011; Henn et al., 2012; see chapters by Holt, and Lao and Kayser). Humans dispersed out of Africa toward the Middle–East and then independently to Europe and Asia (Henn et al., 2012), and probably in two major waves to Asia (Rasmussen et al., 2011). Dispersals to the Americas occurred around 15,000 years ago by peoples of East Asian ancestry who crossed the Bering Straits in two major migrations (Schurr and Sherry, 2004). Finally, the islands of the Western Pacific were populated by people who originated in Taiwan around 5500 years before present (Gray et al., 2009) (Figure 2; also see Figure 1 in Little and Blumler).

Human microbes and parasites have been used as markers to retrace human dispersals (Wirth et al., 2005;

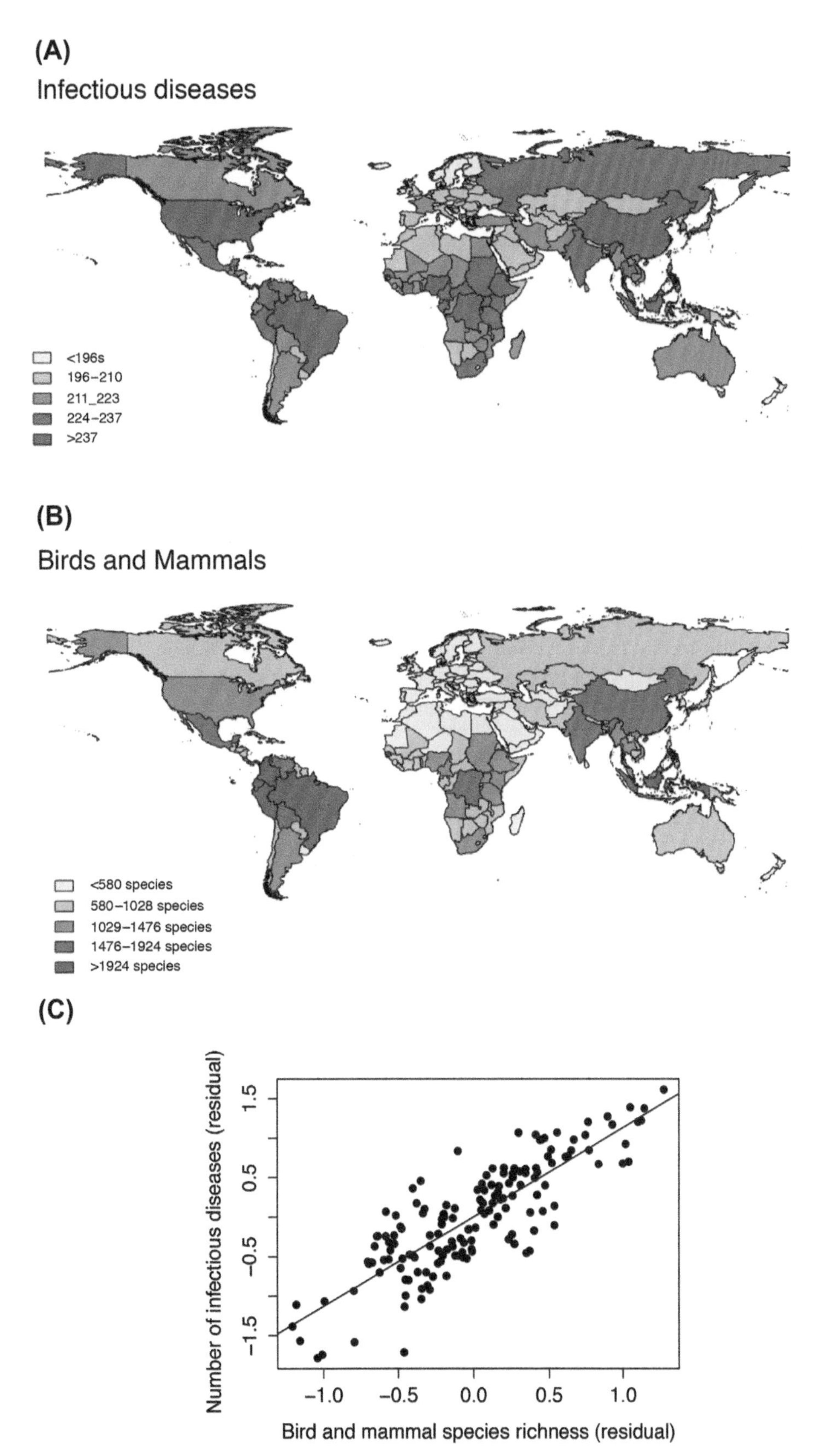

FIGURE 1 Richness maps of (A) human infectious diseases and (B) bird and mammal species with (C) positive relationship between both richness values correcting for potential confounding factors (such as nation size). *The data are from Bird Life International, UCN, and GIDEON database; redrawn from Morand et al. (2014c).*

Achtman, 2008; Dominguez-Bello and Blaser, 2011; Reinhard et al., 2013). These include human polyomavirus JCV (Zheng et al., 2003), human T cell lymphotropic virus I (Miura et al., 1994), hepatitis G virus (HGV or GBV-C) (Muerhoff et al., 2005), *Haemophilus influenzae* (Musser et al., 1990), *Mycobacterium tuberculosis* (Kremer et al., 1999), *Mycobacterium leprae* (Monot et al., 2005), *Histoplasma capsulatum* (Kasuga et al., 1999), *Streptococcus mutans* (Caufield et al., 2007), and *Helicobacter pylori* (Ghose et al., 2002; Falush et al., 2003; Wirth et al., 2004) among others. All are hypothesized to have accompanied humans during their ancient and recent dispersals, and they may help us to understand better the human evolutionary history of our species (Wirth et al., 2005).

For example, phylogenetic analyses of HGV or GBV-C, including sequences from chimpanzee isolates, suggest an ancient African origin of the virus in humans (Muerhoff et al., 2005), with this virus following the dispersion of humans out of Africa as well as from Asia to the Americas (Loureiro et al., 2002). Similarly, phylogeographic studies of the *M. tuberculosis* complex illustrate a cluster of six lineages, likely reflecting ancient human dispersals (Gagneux and Small, 2007; also see the chapter by Cook in this volume). The *M. tuberculosis* complex is supposed to be part of a highly diversified proto-species that has infected hominins since their origins and was subject to an extreme genetic bottleneck following the emergence of the modern humans and their dispersal in and out of Africa (Gutierrez et al., 2005; Achtman, 2008). Similarly, phylogeographic studies support an African origin of *M. leprae*, the agent of leprosy, followed by early dispersion of modern humans (Monot et al., 2009).

Phylogenetic studies of the parasite *Taenia* suggest that it infected humans before the domestication of cattle and swine by Neolithic farmers (Hoberg et al., 2001). Therefore, earlier human populations likely became infected from consuming raw meat from hunting or possibly scavenging. *Taenia* could have then accompanied early human dispersals out of Africa, and later swine and cattle could have acquired their *Taenia* infections during early domestication (Hoberg et al., 2001).

During their migrations to Eurasia, modern humans encountered other hominins such as Neanderthals and Denisovans. Genomic analyses show that interbreeding occurred between these different hominin species (Abi-rached et al., 2011; see chapters by Relethford, and Lao and Kayser). Moreover, it has been suggested that this interbreeding was beneficial to our species by introducing helpful immune genes, allowing us to colonize Eurasia and new epidemiological environments in which Neanderthals and Denisovans coevolved with their pathogens previously. Abi-rached et al. (2011) found that allele variants of a group of immune genes (the human leukocyte antigens, HLAs) in Neanderthals and Denisovans also occur in modern humans in Europe and Asia, and they are absent in current African populations. HLA genes are known to evolve rapidly, suggesting that their presence in European and Asian populations is the result of interbreeding between species after the

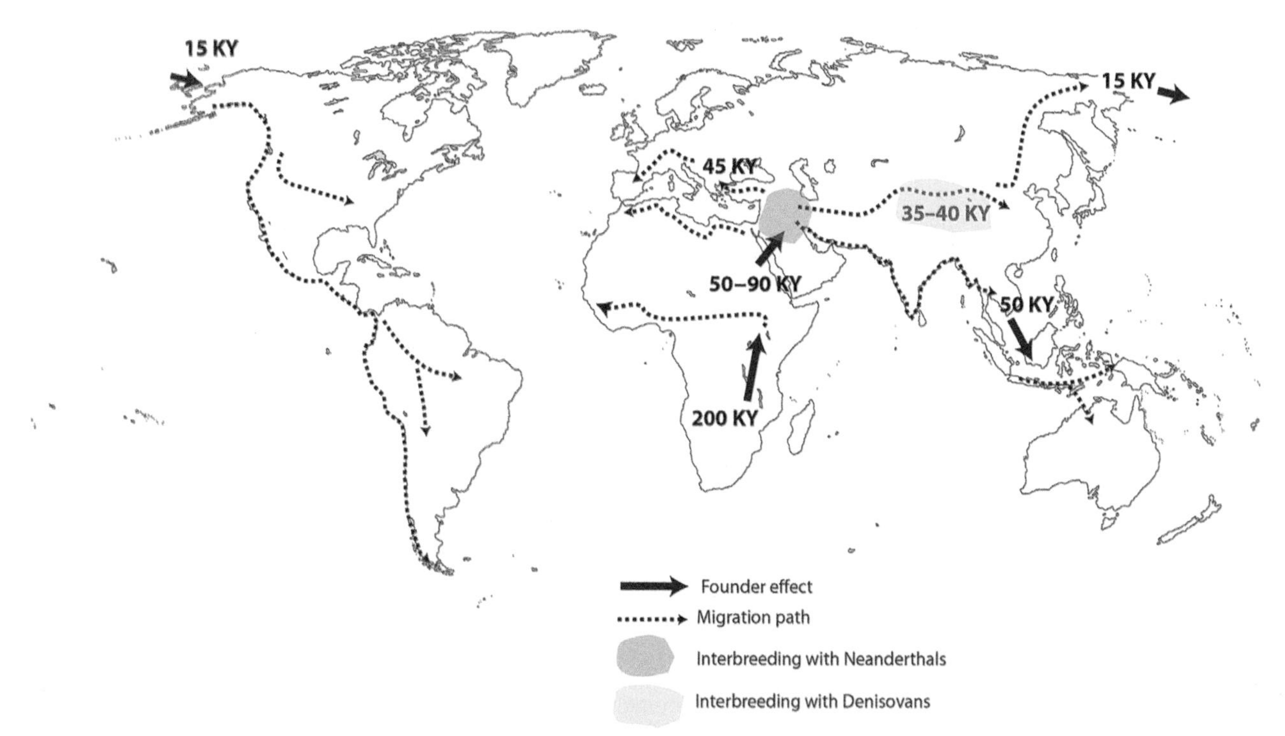

FIGURE 2 Migration pathways of modern human *Homo sapiens* with potential localizations of interbreeding with Neanderthals and Denisovans. *Redrawn after Henn et al. (2012); Abi-rached et al. (2011); Callaway (2011).*

migration of modern humans out of Africa. HLA variants are involved in innate immunity, and some of these ancient HLA genes are still common in modern Asia, providing immunity against Epstein Barr virus. Furthermore, high HLA gene polymorphism is observed in tropical regions with high virus diversity (Prugnolle et al., 2005), while neutral genetic diversity decreases farther away from Africa (Handley et al., 2007; Henn et al., 2012).

THE SECOND EPIDEMIOLOGICAL TRANSITION: ANIMAL DOMESTICATION

Archeological studies in tropical and subtropical areas of Asia, northern and central Africa, and Central America suggest rapid and large-scale domestication of animals beginning around 12,000 years ago when an intense humid phase and equable climates were detected in paleosurveys (Gupta, 2004; see the chapter by Little in this volume). Most animal domestication originates from the Middle East, Central, Southwest, Southern, and East Asia with fewer species domesticated in Africa, Western Europe, and the New World (Diamond, 1997; Gupta, 2004; Naderi et al., 2008; Driscoll et al., 2009; VonHoldt et al., 2010) (Figure 3; also see the table in the chapter by Little). Animal domestication has led to significant changes in human nutrition and health. For example, the health status of early Neolithic populations exhibited significant deterioration in comparison with the hunter–gatherer populations that preceded them in the Southern Levant Horwitz and Smith, 2000; see chapters by Little, Blumler, and Little and Blumler for more details of human lifestyle transitions.

McNeil (1976) first hypothesized that there was a positive relationship between the length of time since domestication and the number of parasite species shared by humans and those domesticates. Morand et al. (2014b) reinvestigated this hypothesis using more accurate information on the dates and origins of domestication (Driscoll et al., 2009) along with several independent datasets on the pathogens of humans and their domestic mammals. They found that the number of pathogens shared between domesticated animals and humans increases with time since domestication, even when controlling for the total number of pathogens hosted by domesticated animals as well as phylogenetic relationships among the domesticated animals (Morand et al., 2014b), confirming McNeil's (1976) hypothesis.

To visualize better the overall interactions between the pathogens of humans and domestic animals, Morand et al. (2014b) used a network-based approach (like those used widely in epidemiology and disease ecology to study transmission heterogeneity; Bansal et al., 2007) to determine which domestic mammals are the primary sources of human pathogens. This was based largely on the concept of centrality, for which the centrality of a host is a good estimate of its potential to be a source of pathogens to other species. A central host is the one with a high value of centrality, or is infected by many pathogens that also infect many other hosts in the network. Domesticated mammalian hosts that were associated with humans a long time ago are central hosts within these networks, indicating pathogen transmission from these domesticates to humans as well as other domesticated mammals (Figure 4).

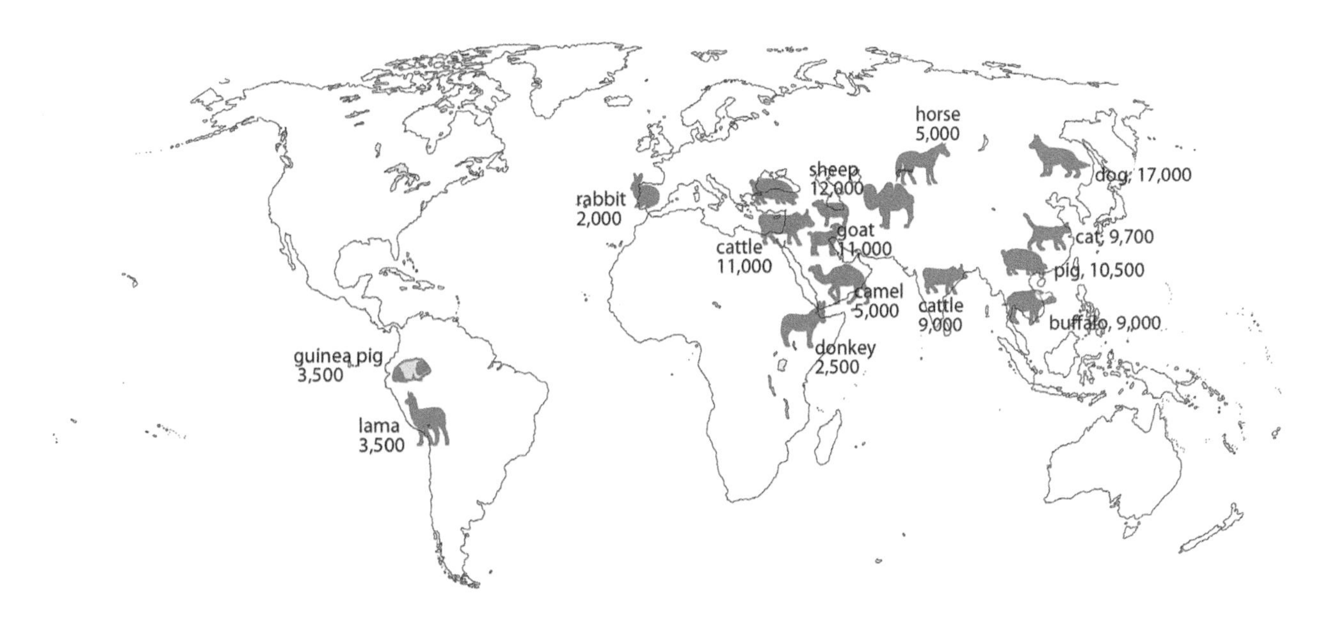

FIGURE 3 Map of localizations of the principal domesticated mammals with dates since their domestication (in thousand years before present). *Data from Driscoll et al. (2009).*

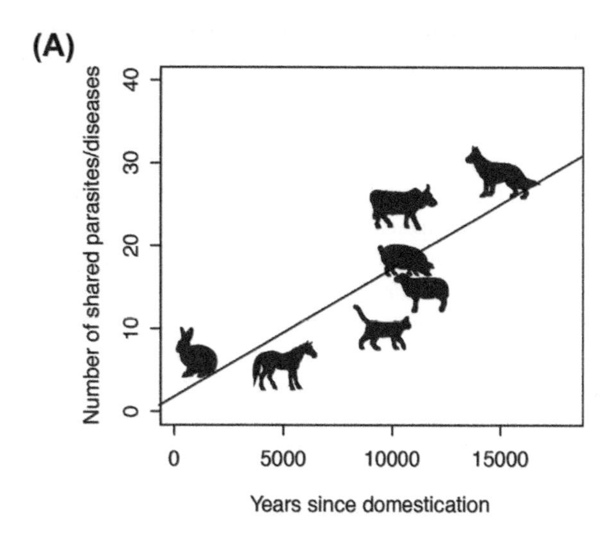

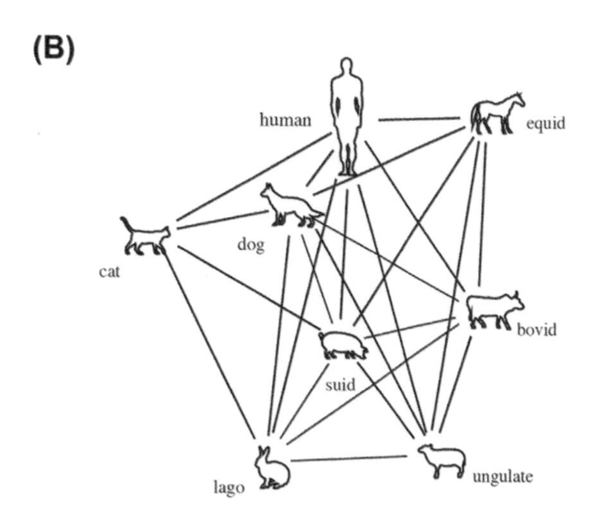

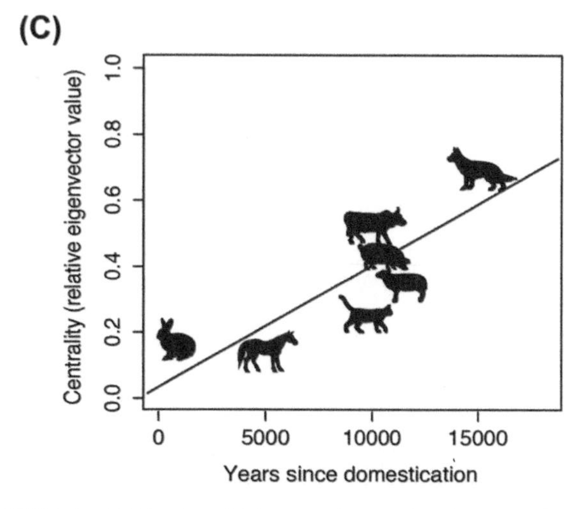

FIGURE 4 (A) Number of infectious diseases shared between domestic mammals and humans in relation to domestication time. (B) Network depicting the pattern of shared diseases by humans and domestic animal species. The links among host nodes depict shared diseases. (C) Relationship between centrality of domestic animals from network analysis based on shared diseases and domestication time. *Redrawn from Morand et al. (2014b); data from GIDEON database.*

This finding emphasizes that the time of geographic overlap is important for building these communities of shared pathogens, as also highlighted by Wolfe et al. (2007). However, this type of network analysis does not take into account the source of the donor and receiver of pathogens. Indeed, although the results of several phylogenetic studies show that cattle and swine were the sources of pathogens for humans (e.g., *Ascaris* from wild boar at the very start of their domestication; Criscione et al., 2007), they were also recipients of pathogens from humans. Phylogenetic studies suggest that *Mycobacterium bovis* that infects cattle actually originated from a human *M. tuberculosis* strain (Smith et al., 2009).

THE THIRD EPIDEMIOLOGICAL TRANSITION: FIRST GLOBALIZATIONS

Several studies have investigated the burden of infectious diseases in hunter–gatherer societies compared with pastoralist and sedentary societies, emphasizing an increase of the burden of infectious diseases due largely to the increase of population size from hunter–gatherer bands to sedentary societies (Araujo et al., 2003). Smallpox, measles, and mumps did not likely affect early hunter–gatherer groups, while sedentary agricultural societies with higher population density became more easily infected (Cockburn, 1963, 1971; Araujo et al., 2003). The opening of new landscapes by clearance of forests for agricultural purposes favored the spread of biting arthropods and thus the transmission of malaria, yellow fever, and dengue fever (Carter and Mendis, 2002). Moreover, intensive contacts with domestic animals and the colonization of human settlements by new commensal animals (e.g., rats) would have favored host shifts of several new pathogens to humans (Kosoy et al., 2015). Although the transition from hunting and gathering to agriculture in Europe had increased infection from zoonotic diseases, the same transition cannot be said for the Americas (Reinhard et al., 2013). Very few zoonotic diseases have emerged in historical times in the Americas with some notable exceptions like Chagas disease (which seems to be associated with the domestication of the guinea pig; Araujo et al., 2003), although new zoonotic diseases have recently emerged such as New World hantaviruses first observed in United States in the region of Four Corners in 1993 (where rodents act as reservoirs) and subsequently in other regions of the Americas causing hantavirus pulmonary syndrome (Jonsson et al., 2010).

What has really changed with the emergence of agrarian societies is the importance of trade routes. Human societies that are engaged in trading activities exchanged not only goods but also infectious diseases (McNeil, 1976; Diamond, 1997). Plague, syphilis, smallpox, and leprosy are strongly associated with trade as well as wars and the resulting human displacements. The most famous example is that of *Yersinia pestis*, the agent of two ancient pandemics: Justinian's plague (541–767 AD) and the Black Death

(1346–1800s). The last pandemic was related to shipping from Hong Kong, which carried infected rats and fleas to the entire globe in 1894 (Achtman, 2008). According to phylogeographic studies, Africa, Central Asia, and East Asia were all centers of origin for these pandemic plagues (Achtman, 2008).

Leprosy (*M. leprae*) was thought to have originated in Europe or the Middle East and then spread to China via the Silk Road and thereafter to the Far East, yet comparative genomic analyses now suggest an African origin with dispersal patterns among early humans via trade routes (Monot et al., 2009). Syphilis (*Trepanoma pallidum* subspecies *pallidum*) arose in Africa and Southeast Asia as nonveneral infections (*T. pallidum* subspecies *pertenue* I and II) that spread to the Middle East and Eastern Europe as endemic syphilis, and then later to the Americas as yaws (Harper et al., 2008). These results support the great Columbian exchange theory of the origin of syphilis, as an American *T. pallidum* strain was reintroduced to the Old World giving rise to the progenitor of modern syphilis-causing strains (Harper et al., 2008).

Dispersal of domestic animals from their domestication centers may have also favored the spread of zoonotic diseases. Rosenthal et al. (2008) investigated the diversification of the nematode *Trichinella spiralis* and found that European lineages of T. *spiralis* originated several thousand years ago, when pigs were first domesticated there. It is hypothesized that Europeans introduced *T. spiralis* to America via infected pigs and/or rats, and their hypothesis is supported by lower genetic diversity observed in European lineages of *T. spiralis* compared with Asian lineages (also in accordance with lower genetic diversity of European wild boar in comparison with Asian wild boar) (Rosenthal et al., 2008). Similarly, according to molecular clock analyses, anthrax (*Bacillus anthracis*) diversified more than 17,000 years ago and then spread globally with the dispersal of domesticated cattle (Van Ert et al., 2007).

THE FOURTH EPIDEMIOLOGICAL TRANSITION: RECENT EMERGENCES AND THE HOMOGENIZATION OF INFECTIOUS DISEASES

The number of emerging infectious diseases has dramatically increased during the last century (Wilcox and Gubler, 2005; Jones et al., 2008). Concurrently, the number of reported infectious disease outbreaks has also dramatically increased from 1954 to 2014 (Morand and Waret-Szkuta, 2012; Morand et al., 2014a). The explanatory factors reported for these trends are generally associated with ongoing global changes such as climate change, global trade, land use change, and biodiversity loss, which are still occurring on an unprecedented scale (Chivian and Berstein, 2004; Patz et al., 2004; Wilcox and Colwell, 2005).

Biodiversity loss also applies to the parasites themselves. More than half of all biodiversity are parasites Morand et al., 2015 yet the rate of parasite extinction in the wild is far from being estimated accurately (Dunn et al., 2010). In developed countries, a large decline of parasite burdens and the extinction of some human infectious diseases have been observed in the last century (Armstrong et al., 1999). Although the total number of human pathogens worldwide has not changed very much in the past several decades, the number of global outbreaks has (Morand et al., 2014a; see Figure 5). This homogenization of global parasite distribution began around 1960 (Smith et al., 2007), with a striking decrease of the modularity of the country–pathogen network (Poisot, Nunn and Morand, personal communication). There is also a loss in pathogen genetic diversity, probably linked with a loss in livestock genetic resources (Rosenthal, 2009). For example, broadly distributed parasites of livestock such as *Trichinella, Taenia*, and *Toxoplasma* show almost uniform population genetic diversity, reflecting both global trade and global circulation of parasites (Rosenthal, 2009).

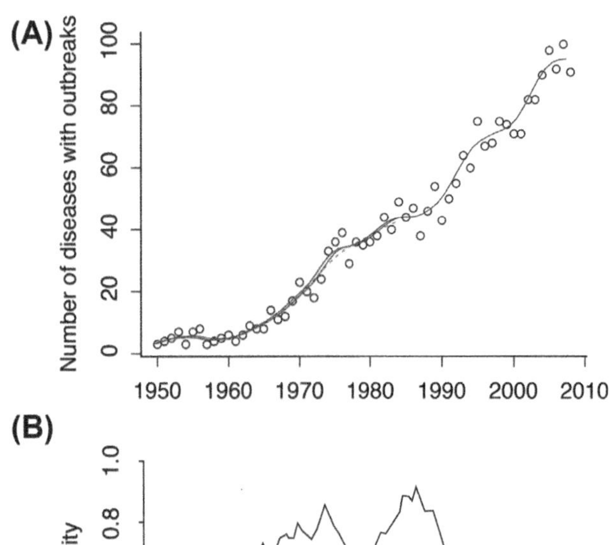

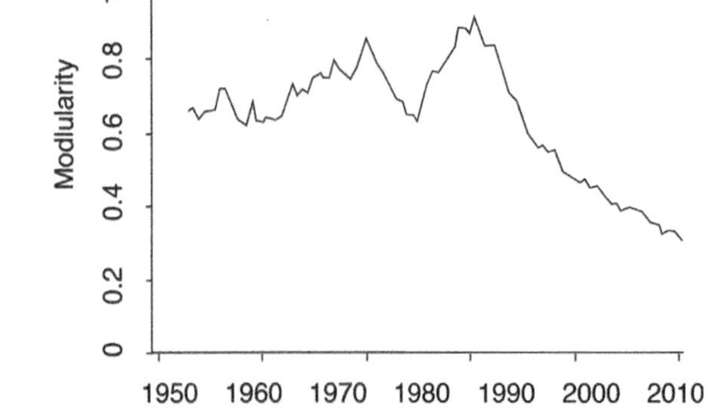

FIGURE 5 (A) Increase in the number of infectious diseases presenting outbreaks globally and from 1954 to 2014. *(Redrawn from Morand et al. (2014a); data from GIDEON database.)* (B) Change in values of modularity of country–infectious diseases over time. The decrease of modularity values since the beginning of the 1960s depicts a global homogenization of infectious diseases. *(Using the GIDEON database; Poisot, Nunn, and Morand, personal communication; data from GIDEON database.)*

CONCLUDING REMARKS

Human pathogen communities evolved as a result of phylogenetic inheritance, animal domestication, and globalization mechanisms, with significant changes in biogeography resulting from our interactions with other animals. Human evolution has produced population densities and distributions that offer exceptional conditions for pathogen coevolution, particularly through our interactions with nonhuman animal species. In response to this diversification of pathogens, humans have evolved elaborate immune responses (Muehlenbein, 2010). Topics, such as conflict between organisms and evolutionary mismatch between current conditions and those we originally evolved in, represent different "evolutionary explanations" for disease. These and more form the basis of the topic of evolutionary medicine, and readers are encouraged to review Nesse and Williams (1996); Ewald (1996), and Stearns and Koella (2008), among others. Understanding evolutionary and ecological principles allows us to formulate both explanations for disease and potential methods of prevention and treatment.

Interestingly, as parasite biodiversity declined in developed countries, new health problems emerged. Two last examples will illustrate this point. First, control and eradication of smallpox (which nobody obviously regrets) led to the abandonment of the smallpox vaccine. Smallpox vaccination provided protection against other related viruses, and the unexpected result of abandonment of the smallpox vaccination was to favor new infections with related viruses such as monkeypox and other orthopoxvirus viruses (Vorou et al., 2008). Second, the decline in parasite biodiversity has led to the emergence of some autoimmune diseases. The ulcers caused by *H. pylori* appear to be linked to the disappearance of parasitic nematodes and tapeworms in many developed countries, and resultant changes in inflammatory responses that now lead to ulceration (Hunter and McKay, 2004; Weinstock et al., 2004). Changes in hygiene have altered our microbiome in ways that change our likelihood of allergies and autoimmunity (Strachan, 1989; De Silva et al., 2003; Parker et al., 2012). This ironically takes place while we are simultaneously facing new pandemic threats, illustrating how we must remain cognizant about how alterations in our relationships with other species greatly affect both the human individual and our global population.

REFERENCES

Abi-rached, L., Jobin, M.J., Kulkarni, S., McWhinnie, A., Dalva, K., Gragert, L., Babrzadeh, F., Gharizadeh, B., Luo, M., Plummer, F.A., Kimani, J., Carrington, M., Middleton, D., Rajalingam, R., Beksac, M., Marsh, S.G.E., Maiers, M., Guethlein, L.A., Tavoularis, S., Little, A.-M., Green, R.E., Norman, P.J., Parham, P., 2011. The shaping of modern human immune systems by multiregional admixture with archaic humans. Science 334, 89–94.

Achtman, M., 2008. Evolution, population structure, and phylogeography of genetically monomorphic bacterial pathogens. Annual Review of Microbiology 62, 53–70.

Armstrong, G.L., Conn, L.A., Pinner, R.W., 1999. Trends in infectious disease mortality in the United States during the 20th century. Journal of the American Medical Association 281, 61–66.

Araujo, A., Jansen, A.M., Bouchet, F., Reinhard, K., Ferreira, L.F., 2003. Parasites, the diversity of life, and paleoparasitology. Memorias Instituto Oswaldo Cruz 98, S5–S11.

Ashford, R.W., Crewe, W., 1998. The Parasites of Homo sapiens: An Annotated Checklist of the Protozoa, Helminths and Arthropods for Which We Are Home. Liverpool School of Tropical Medicine, Liverpool.

Bansal, S., Grenfell, B.T., Meyers, L.A., 2007. When individual behaviour matters: homogeneous and network models in epidemiology. Journal of the Royal Society Interface 4, 879–891.

Callaway, E., 2011. Ancient DNA reveals secrets of human history. Nature 476, 136–137.

Carter, R., Mendis, K.N., 2002. Evolutionary and historical aspects of the burden of malaria. Clinical Microbiology Reviews 15, 564–594.

Caufield, P.W., Saxena, D., Fitch, D., Li, Y., 2007. Population structure of plasmid-containing strains of *Streptococcus mutans*, a member of the human indigenous biota. Journal of Bacteriology 189, 1238–1243.

Chivian, E., Bernstein, A.S., 2004. Embedded in nature: human health and biodiversity. Environmental Health Perspectives 112, 12–13.

Cleaveland, S., Laurenson, M.K., Taylor, L.H., 2001. Diseases of humans and their domestic mammals: pathogen characteristics, host range and the risk of emergence. Philosophical Transaction of the Royal Society of London B 356, 991–999.

Cockburn, T.A., 1963. The Evolution and Eradication of Infectious Diseases. Johns Hopkins Press, Baltimore.

Cockburn, T.A., 1971. Infectious disease in ancient populations. Current Anthropology 12, 45–62.

Criscione, C.D., Anderson, J.D., Sudimack, D., Peng, W., Jha, B., Williams-Blangero, S., Anderson, T.J.C., 2007. Disentangling hybridization and host colonization in parasitic roundworms of humans and pigs. Proceedings of the Royal Society of London B 274, 2669–2677.

Daily, G.C., Ehrlich, P.R., 1996. Global change and human susceptibility to disease. Annual Review of Energy and Environment Vol. 21, 125–144.

De Silva, N.R., Brooker, S., Hotez, P.J., Montresor, A., Engels, D., Savioli, L., 2003. Soil-transmitted helminth infections: updating the global picture. Trends in Parasitology 19, 547–551.

Davies, T.J., Pedersen, A.B., 2008. Phylogeny and geography predict pathogen community similarity in wild primates and humans. Proceedings of the Royal Society of London B 275, 1695–1701.

Diamond, J., 1997. Guns, Germs and Steel: The Fates of Human Societies. Norton, New York.

Dominguez-Bello, M.G., Blaser, M.J., 2011. The human microbiota as a marker for migrations of individuals and populations. Annual Reviews of Anthropology 40, 451–474.

Driscoll, C., Macdonald, D.W., O'Brien, S.J., 2009. From wild animals to domestic pets, an evolutionary view of domestication. Proceedings of the National Academy of Sciences of the USA 106, 9971–9978.

Dunn, R.R., Davies, T.J., Harris, N.C., Galvin, M.C., 2010. Global drivers of human pathogen richness and prevalence. Proceedings of the Royal Society of London B 277, 2587–2595.

Dunn, R.R., Harris, N.C., Colwell, R.K., Koh, L.P., Sodhi, N.S., 2009. The sixth mass coextinction: are most endangered species parasites and mutualists? Proceedings of the Royal Society of London B 276, 3037–3045.

Ewald, P.W., 1996. Evolution of Infectious Disease. Oxford University Press, New York.

Falush, D., Wirth, T., Linz, B., Pritchard, J.K., et al., 2003. Traces of human migrations in *Helicobacter pylori* populations. Science 299, 1582–1585.

Fincher, C., Thornhill, R., 2008. A parasite-driven wedge: infectious diseases may explain language and other biodiversity. Oikos 9, 1289–1297.

Gagneux, S., Small, P.M., 2007. Global phylogeography of *Mycobacterium tuberculosis* and implications for tuberculosis product development. Lancet Infectious Diseases 7, 328–337.

Ghose, C., Perez-Perez, G.I., Dominguez-Bello, M.G., Pride, D.T., Bravi, C.M., Blaser, M.J., 2002. East Asian genotypes of *Helicobacter pylori* strains in Amerindians provide evidence for its ancient human carriage. Proceedings of the National Academy of Sciences of the USA 99, 15107–15111.

Gómez, J.M., Nunn, C.L., Verdú, M., 2013. Centrality in primate-parasite networks reveals the potential for the transmission of emerging infectious diseases to humans. Proceedings of the National Academy of Sciences of the USA 110, 7738–7741.

Gray, R.D., Drummond, A.J., Greenhill, S.J., 2009. Language phylogenies reveal expansion pulses and pauses in Pacific settlements. Science 323, 479–483.

Guernier, V., Hochberg, M.E., Guégan, J.-F., 2004. Ecology drives the worldwide distribution of human diseases. PLoS Biology 2, 740–746.

Gupta, A.K., 2004. Origin of agriculture and domestication of plants and animals linked to early Holocene climate amelioration. Current Science 87, 54–59.

Gutierrez, M.C., Brisse, S., Brosch, R., Fabre, M., Omais, B., et al., 2005. Ancient origin and gene mosaicism of the progenitor of *Mycobacterium tuberculosis*. PLoS Pathogens 1, e5.

Handley, L.J.L., Manica, A., Goudet, J., Balloux, F., 2007. Going the distance: human population genetics in a clinal world. Trends in Genetics 23, 432–439.

Harper, K.N., Ocampo, P.S., Steiner, B.M., George, R.W., Silverman, M.S., et al., 2008. On the origin of the treponematoses: a phylogenetic approach. PLoS Neglected Tropical Diseases 2, e148.

Henn, B.M., Cavalli-Sforza, L.L., Feldman, M.W., 2012. The great human expansion. Proceedings of the National Academy of Sciences of the USA 109, 17758–17764.

Hoberg, E.P., Alkire, N.L., de Queiroz, A., Jones, A., 2001. Out of Africa: origins of the *Taenia* tapeworms in humans. Proceedings Royal Society of London B 268, 781–787.

Horwitz, L.K., Smith, P., 2000. The contribution of animal domestication to the spread of zoonoses: a case study from the Southern Levant. Anthropozoologica 31, 77–84.

Hugot, J.-P., 1999. Primates and their pinworm parasites: the Cameron hypothesis revisited. Systematic Biology 48, 523–546.

Hugot, J.-P., Demanche, C., Barriel, V., Dei-Cas, E., Guillot, J., 2003. Phylogenetic systematics and evolution of the Pneumocystis parasite on primates. Systematic Biology 52, 735–744.

Hunter, M.M., McKay, D.M., 2004. Helminths as therapeutic agents for inflammatory bowel disease. Aliment Pharmacology and Therapeutics 19, 167–177.

Jones, K.E., Patel, N.G., Levy, M.A., et al., 2008. Global trends in emerging infectious diseases. Nature 451, 990–994.

Jonsson, C.B., Figueiredo, L.T.M., Vapalahti, O., 2010. A global perspective on hantavirus ecology, epidemiology, and disease. Clinical Microbiology Reviews 23, 412–441.

Kasuga, T., Taylor, J.W., White, T.J., 1999. Phylogenetic relationships of varieties and geographical groups of the human pathogenic fungus *Histoplasma capsulatum* Darling. Journal of Clinical Microbiology 37, 653–663.

Keesing, F., Belden, L.K., Daszak, P., et al., 2010. Impacts of biodiversity on the emergence and transmission of infectious diseases. Nature 468, 647–652.

Kosoy, M., Khlyap, L., Cosson, J-F., Morand, S., 2015. Aboriginal and invasive rats of genus *Rattus* as hosts of infectious agents. Vector-Borne and Zoonotic Diseases, 15, 3–12.

Kremer, K., van Soolingen, D., Frothingham, R., Haas, W.H., Hermans, P.W.M., Martín, C., Palittapongarnpim, P., Plikaytis, B.B., Riley, L.W., Yakrus, M.A., Musser, J.M., van Embden, J.D.A., 1999. Comparison of methods based on different molecular epidemiological markers for typing of *Mycobacterium tuberculosis* complex strains: interlaboratory study of discriminatory power and reproducibility. Journal of Clinical Microbiology 37, 2607–2618.

Loureiro, C.L., Alonso, R., Pacheco, B.A., Uzcategui, M.G., Villegas, L., León, G., De Saéz, A., Liprandi, F., López, J.L., Pujol, F.H., 2002. High prevalence of GB virus C/hepatitis G virus genotype 3 among autochthonous Venezuelan populations. Journal of Medical Virology 68, 357–362.

Maffi, L., 2005. Linguistic, cultural and biological diversity? Annual Review Anthropology 29, 599–617.

May, J.M., 1958. The Ecology of Human Disease. M.D. Publications, New York.

McNeil, W.H., 1976. Plagues and People. Anchor Press, New York.

Miura, T., Fukunaga, T., Igarashi, T., Yamashita, M., Ido, E., Funahashi, S., Ishida, T., Washio, K., Ueda, S., Hashimoto, K., 1994. Phylogenetic subtypes of human T-lymphotropic virus type I and their relations to the anthropological background. Proceedings of the National Academy of Sciences of the USA 91, 1124–1127.

Monot, M., Honoré, N., Garnier, T., Araoz, R., et al., 2005. On the origin of leprosy. Science 308, 1040–1042.

Monot, M., Honoré, N., Garnier, T., Zidane, N., et al., 2009. Comparative genomic and phylogeographic analysis of Mycobacterium leprae. Nature Genetics 41, 1282–1289.

Morand, S., 2012. Phylogeography helps with investigating the building of human parasite communities. Parasitology 139, 1966–1974.

Morand, S., Jittapalapong, S., Supputamongkol, Y., Abdullah, M.T., Huan, T.B., 2014a. Infectious diseases and their outbreaks in Asia-Pacific: biodiversity and its regulation loss matter. PLoS One 9, e90032.

Morand, S., Krasnov, B.R., Littlewood, T. (Eds.), 2015. Parasite Diversity and Diversification: Evolutionary Ecology Meets Phylogenetics. Cambridge: Cambridge University Press.

Morand, S., McIntyre, K.M., Baylis, M., 2014b. Domesticated animals and human infectious diseases of zoonotic origins: domestication time matters. Infection Genetics Evolution 24, 76–87.

Morand, S., Owers, K., Bordes, F., 2014c. Biodiversity and emerging zoonoses. In: Akio Yamada, A., Kahn, L.H., Kaplan, B., Monath, T.P., Woodall, J., Conti, L. (Eds.), Confronting Emerging Zoonoses: The One Health Paradigm. Tokyo: Springer, pp. 27–41.

Morand, S., Waret-Szkuta, A., 2012. Determinants of human infectious diseases in Europe: biodiversity and climate variability influences. Bulletin Epidémiologique Hebdomadaire 12-13, 156–159.

Muehlenbein, M.M., 2010. Evolutionary medicine, immunity, and infectious disease. In: Muehlenbein, M.M. (Ed.), Human Evolutionary Biology. Cambridge University Press, Cambridge, pp. 459–490.

Muerhoff, A.S., Leary, T.P., Sathar, M.A., Dawson, G.J., Desai, S.M., 2005. African origin of GB virus C determined by phylogenetic analysis of a complete genotype 5 genome from South Africa. Journal of General Virology 86, 1729–1735.

Musser, J.M., Kroll, J.S., Granoff, D.M., et al., 1990. Global genetic structure and molecular epidemiology of encapsulated Haemophilus influenzae. Reviews of Infectious Diseases 12, 75–111.

Naderi, S., Rezaei, H.R., Pompanon, F., et al., 2008. The goat domestication process inferred from large-scale mitochondrial DNA analysis of wild and domestic individuals. Proceedings of the National Academy of Sciences of the USA 105, 17659–17664.

Nesse, R.M., Williams, G.C., 1996. Why We Get Sick: The New Science of Dalwinian Medicine. Vintage Books, New York.

Patz, J.A., Daszak, P., Tabor, G.M., et al., 2004. Unhealthy landscapes: policy recommendations on land use change and infectious disease emergence. Environmental Health Perspectives 112, 1092–1098.

Parker, W., Perkins, S.E., Harker, M., Muehlenbein, M.M., 2012. A prescription for clinical immunology: the pills are available and ready for testing. A review. Current Medical Research and Opinion 28, 1193–1202.

Perrin, P., Herbreteau, V., Hugot, J., Morand, S., 2010. Biogeography, humans and their parasites. In: Morand, S., Krasnov, B.R. (Eds.), The Biogeography of Host–Parasite Interactions. Oxford University Press, Oxford, pp. 41–57.

Prugnolle, F., Manica, A., Charpentier, M., Guégan, J.F., Guernier, V., Balloux, F., 2005. Pathogen-driven selection and worldwide HLA class I diversity. Current Biology 15, 1022–1027.

Rasmussen, M., Guo, X., Wang, Y., Lohmueller, K.E., et al., 2011. An Aboriginal Australian genome reveals separate human dispersals into Asia. Science 334, 94–98.

Reed, D.L., Light, J.E., Allen, J.M., Kirchman, J.J., 2007. Pair of lice lost or parasites regained: the evolutionary history of anthropoid primate lice. BMC Biology 5, 7.

Reinhard, K.J., Ferreira, L.F., Bouchet, F., Sianto, L., Dutra, J.M.F., Iniguez, A., Leles, D., Le Bailly, M., Fugassa, M., Pucu, E., Araújo, A., 2013. Food, parasites, and epidemiological transitions: a broad perspective. International Journal of Paleopathology 3, 150–157.

Rosenthal, B.M., 2009. How has agriculture influenced the geography and genetics of animal parasites? Trends in Parasitology 25, 67–70.

Rosenthal, B.M., LaRosa, G., Zarlenga, D., Dunams, D., et al., 2008. Human dispersal of Trichinella spiralis in domesticated pigs. Infection Genetics and Evolution 8, 799–805.

Schipper, J., Chanson, J.S., Chiozza, F., et al., 2008. The status of the world's land and marine mammals: diversity, threat, and knowledge. Science 322, 225–230.

Schurr, T., Sherry, S.T., 2004. Mitochondrial DNA and Y chromosome diversity and the peopling of the Americas: evolutionary and demographic evidence. American Journal of Human Biology 16, 420–439.

Smith, N.H., Hewinson, R.G., Kremer, K., Brosch, R., Gordon, S.V., 2009. Myths and misconceptions: the origin and evolution of *Mycobacterium tuberculosis*. Nature Review Microbiology 7, 537–544.

Smith, K.F., Sax, D.F., Gaines, S.D., Guernier, V., Guégan, J.-F., 2007. Globalization of human infectious disease. Ecology 88, 1903–1910.

Stearns, S.C., Koella, J.C., 2008. Evolution in Health and Disease, second ed. Oxford University Press, New York.

Strachan, D.P., 1989. Hay fever, hygiene, and household size. British Medical Journal 299, 1259–1260.

Switzer, W.M., Salemi, M., Shanmugam, V., Gao, F., Cong, M., Kuiken, C., Bhullar, V., Beer, B.E., Vallet, D., Gautier-Hion, A., Tooze, Z., Villinger, F., Holmes, E.C., Heneine, W., 2005. Ancient co-speciation of simian foamy viruses and primates. Nature 434, 376–380.

Taylor, L.H., Latham, S.M., Woolhouse, M.E.J., 2001. Risk factors for human disease emergences. Philosophical Transaction of the Royal Society of London B 356, 983–989.

Van Ert, M.N., Easterday, W.R., Huynh, L.Y., et al., 2007. Global genetic population structure of *Bacillus anthracis*. PLoS ONE 2, e461.

Von Holdt, B.M., Pollinger, J.P., Lohmueller, K.E., et al., 2010. Genome-wide SNP and haplotype analyses reveal a rich history underlying dog domestication. Nature 464, 898–902.

Vorou, R.M., Papavassiliou, V.G., Pierroutsakos, I.N., 2008. Cowpox virus infection: an emerging health threat. Current Opinion Infectious Diseases 21, 153–156.

Weinstock, J.V., Summers, R., Elliott, D.E., 2004. Helminths and harmony. Gut 53, 7–9.

Wilcox, B.A., Colwell, R.R., 2005. Emerging and reemerging infectious diseases: biocomplexity as an interdisciplinary paradigm. EcoHealth 2, 244–257.

Wilcox, B.A., Gubler, D.J., 2005. Disease ecology and the global emergence of zoonotic pathogens. Environmental Health Preventive Medicine 10, 263–272.

Wirth, T., Meyer, A., Achtman, M., 2005. Deciphering host migrations and origins by means of their microbes. Molecular Ecology 14, 3289–3306.

Wirth, T., Wang, X., Linz, B., Novick, R.P., Lum, J.K., Blaser, M., Morelli, G., Falush, D., Achtman, M., 2004. Distinguishing human ethnic groups by means of sequences from *Helicobacter pylori*: lessons from Ladakh. Proceedings of the National Academy of Sciences of the USA 101, 4746–4751.

Wolfe, N.D., Panosian Dunavan, C., Diamond, J., 2007. Origins of major human infectious diseases. Nature 447, 279–283.

Zheng, H.-Y., Sugimoto, C., Hasegawa, M., Kobayashi, N., Kanayama, A., Rodas, A., Mejia, S.M., Nakamichi, S.J., Guo, J., Kitamura, T., Yogo, Y., 2003. Phylogenetic relationships among JC virus strains in Japanese/Koreans and native Americans speaking Amerind or Na-Dene. Journal of Molecular Evolution 56, 18–27.

Chapter 30

Coevolution of Humans and Pathogens

Lisa Sattenspiel
Department of Anthropology, University of Missouri-Columbia, Columbia, MO, USA

SYNOPSIS

Coevolution can be defined as evolutionary changes that occur within two or more organisms as a response to interactions between them and the resulting mutual selective pressures that those interactions cause. This chapter discusses coevolutionary mechanisms between human hosts and their pathogens and gives four examples: the relationship between ABO blood groups and disease, genetic variants that have arisen in response to malaria, the evolution of new strains of the influenza A virus, and the evolution of the CCR5-Δ32 variant and associated resistance to HIV infection.

As is the case for all other organisms, human life goes hand in hand with the lives of many different kinds of pathogens. Some of these pathogens have only been associated with humans for a short time; others have infected humans for millennia. The focus of this chapter is on evolutionary relationships between humans and their pathogens. Although there has been great interest in the media in newly evolved and reemerging pathogens in recent years, this chapter will concentrate on pathogens that have been infecting humans for long periods. The chapter consists of two major parts: an overview of basic concepts and questions related to human–pathogen coevolution, and a discussion of several examples of human–pathogen interactions that provide clear evidence of evolution of host and/or pathogen in response to each other.

FIGURE 1 John Tenniel's illustration of Alice and the Red Queen racing as fast as they can only to end up right where they started, as depicted in Lewis Carroll's *Through the Looking Glass, and What Alice Found There*. *Source: http://commons.wikimedia.org/wiki/File:Alice_queen2.jpg.*

THE IMPORTANCE OF COEVOLUTION BETWEEN HOSTS AND PATHOGENS

Coevolution can be defined as evolutionary changes that occur within two or more organisms as a response to interactions between them, and the mutual selective pressures that those interactions cause (Van Blerkom, 2003). Haldane (1949) was one of the first scientists to emphasize the importance of infectious diseases as agents of selection within humans as well as other organisms. Although this work largely consisted of speculations derived from the work of others rather than discussions of his own experimental work, Haldane's ideas have inspired much thinking on disease and evolution (Lederberg, 1999).

Haldane (1949) recognized that disease can be either a selective advantage or a disadvantage to a host when competing with other species. For example, some wild African ungulates have evolved the ability to tolerate nonlethal trypanosome infections and thus outcompete introduced cattle that die from the infections. More often the situation is one of disadvantages, as in the risk of predation in a species that is highly susceptible to a pathogen and has a high prevalence of infected individuals. Such a species may not survive and reproduce as well when faced with a predator, such as a cat, that chases after its prey. Haldane further pointed out that individuals who possess rare phenotypes may be more likely to be resistant to pathogens than individuals with common phenotypes, and therefore frequency-dependent natural selection in favor of rare types in an environment of parasitism may foster the evolution of new alleles (increased levels of genetic variation) in both hosts and pathogens. This idea is at the core of a prominent hypothesis in studies of the coevolution of host and pathogens, the Red Queen hypothesis.

Van Valen (1973) used Alice's interactions with the Red Queen in Lewis Carroll's *Through the Looking Glass and What Alice Found There* to illustrate the nature of all kinds of species interactions (Figure 1). In that story, Alice moves backward when she wants to go forward, and if she wants to go quickly, she comes to an abrupt stop. She wants to talk to the Red Queen, but in order to do so, she must walk away from her. Finally, at one point she and the Red Queen start running somewhere (although Alice is not quite sure where), and at the end of a long hard run, Alice observes that after all that running, they are still right where they started.

Van Valen used this story as a metaphor for the impact of species interactions on rates of extinction over long spans of time, but others have used the same story to explain coevolutionary interactions between hosts and pathogens (see, for example, Bell, 1982; Lively and Dybdahl, 2000; Lively, 2010). This latter version of the Red Queen hypotheses is also used to illustrate how host–pathogen interactions may help to explain the maintenance of sexual reproduction. Sexual reproduction provides increased opportunities for new (and rare) offspring phenotypes that pathogens would not have evolved to infect, which could allow more opportunity for hosts to escape the negative effects of the parasitism. It is important to note that changes in the genetic structure of hosts, however, often stimulate the pathogen to evolve new strategies to infect the host (Muehlenbein, 2010).

The Red Queen hypothesis for host–pathogen coevolution and the maintenance of sexual reproduction is attractive and logically compelling, but empirical evidence to support it is rare, largely because it is difficult to track oscillations in genotype and phenotype frequencies as well as linkage disequilibrium effects over time (Salathé et al., 2008). In addition, it has been questioned on theoretical grounds, but it is still being studied and remains one of the primary hypotheses related to host–pathogen coevolution.

ENSURING PATHOGEN PERSISTENCE OVER TIME: MODES OF TRANSMISSION

In order for any species to survive and evolve, it is necessary for at least some of the genetic material possessed by individuals within a species be passed from one generation to the next. Human pathogens require that at least part, although not necessarily all, of their life cycle be spent in the human host, but since all living organisms have a finite life span, in order for the pathogen species to remain viable over time it must eventually find a way to move to another host of the same or different species. Pathogens use many different strategies, called modes of transmission, to facilitate this task. These modes of transmission ensure that a pathogen species can survive over time and influence the potential for evolution of not only the pathogen strains, but the human host as well.

Modes of transmission of human pathogens can be conveniently divided into direct modes of transmission where a pathogen spreads directly from one human host to another, and indirect modes of transmission that require an intermediate host or agent to facilitate transmission of the pathogen between human hosts. The primary direct modes for human diseases are respiratory, fecal–oral, sexual, vertical (or congenital), and direct physical contact; indirect modes include vehicle-borne (water-borne, foodborne, and soil-borne transmission as well as needle sharing and contact with fomites), vector-borne (which involve the action of a living vector such as a mosquito, tick, or flea), and complex patterns of transmission, which involve multiple host species and/or environmental stages. Table 1 gives examples of common human diseases spread by each of these modes of transmission. Some human pathogens are typically diseases of other animals that occasionally infect humans (often through vector-borne or complex modes of transmission). Such diseases are called zoonotic diseases; outbreaks of these diseases are called zoonoses.

Modes of transmission are the mechanisms that infectious organisms use to ensure an unbroken chain of infection from one generation to the next, and knowledge of these mechanisms is essential to fully understand how and why coevolution may have occurred between humans and one of their pathogens. As illustrated by the examples discussed below, the strategies used by pathogens to spread from one host to another have strongly affected how and to what degree humans and their pathogens have evolved in response to one another.

TABLE 1 Common Modes of Transmission of Human Pathogens

Mode of Transmission	Common Human Diseases That Spread by the Mode
Respiratory/Droplet	Influenza, strep throat, measles, mumps, smallpox
Fecal–oral	Cholera, rotavirus, giardia, typhoid
Sexual	Syphilis, gonorrhea, *chlamydia*, HIV
Vertical/Congenital	Rubella, HIV, syphilis, herpes
Direct physical contact	Yaws, impetigo, warts, athlete's foot
Water-borne	Cholera, typhoid, cryptosporidiosis, *shigella*
Foodborne	Salmonella, tapeworm, hepatitis B, *E. coli*, botulism
Soil-borne	Tetanus, hookworm, trichuriasis, ascaris
Needle sharing	HIV, hepatitis B, hepatitis C
Fomite	Influenza, methicillin-resistant *S. aureus* (MRSA)
Vector-Borne	
Mosquito	Malaria, yellow fever, dengue fever, West Nile
Flea	Plague, murine typhus, tularemia, tapeworm
Tick	Lyme disease, epidemic typhus, ehrlichiosis
Fly	Leishmaniasis, onchocerciasis, African trypanosomiasis
Lice	Epidemic typhus, trench fever, louse-borne relapsing fever
Other	Chagas disease, scrub typhus, rickettsialpox
Complex patterns	Schistosomiasis, dracunculiasis, echinococcosis

MECHANISMS OF HOST–PATHOGEN COEVOLUTION

The Red Queen hypothesis posits that interactions between pathogens and their hosts will generally be disadvantageous for common host types. In other words, due to increased mortality caused by pathogens common host types will not as readily pass their genes on to future generations as rare types will. As a result, pathogens will evolve strategies that allow infection of the rare types (since that ensures pathogen reproduction), and then the host will generate even newer rare types that resist pathogen infection. Ultimately in a situation such as this, the host and pathogen coevolve in such a way that the host develops resistance to the pathogen while the pathogen becomes less and less virulent. This is the traditional view of host–pathogen coevolution, and it has been observed in some situations. For example, such a situation occurred with myxomatosis and rabbits after the pathogen, which was native to northern hemisphere rabbits, was introduced into Australia in an attempt to help control a population explosion of rabbits that had been introduced into the continent from Europe (Fenner and Ratcliff, 1965).

More recent work has shown that host–pathogen coevolution does not always lead to the evolution of reduced virulence in the pathogen and/or greater resistance of the host. What seems to be important in determining whether these kinds of changes will occur is the mode of transmission of the pathogen.

One of the most important virulence-influencing aspects of the mechanism a pathogen uses to get from one host to another is whether that mechanism requires a mobile host (Ewald, 1994). Pathogens that can only spread to new hosts if the present host physically carries the pathogen to a location where the host can facilitate its spread (e.g., pathogens such as the common cold virus that spread through a close physical contact between ill and susceptible hosts) cannot afford to make the host so sick that movement becomes impossible. In other words, pathogen strains that make hosts too sick to move around will fail to find enough new hosts to keep their chain of transmission going and will be selected against in favor of milder strains. More virulent pathogen strains become more common if their mode of transmission allows transmission to occur even when hosts are immobile. For example, Ewald (1994) suggests that this may be the case for cholera, which causes profuse watery diarrhea and vomiting that can spread pathogen-laden fluids for some distance even when a host is immobilized by the disease.

Other situations that may facilitate transmission of virulent pathogens include whether the pathogen only occasionally infects a specific host (Ebert and Bull, 2008); whether the pathogen has recently evolved to infect a new host, as in the case of avian influenza of humans; the host's recovery rate or age and geographic distribution (Koella and Turner, 2008); and the amount of competition among different pathogen strains (Nowak and May, 1994). Galvani (2003) has pointed out that virulent pathogen strains are likely to evolve whenever the benefits they receive through the exploitation of the host exceed the costs they incur as a consequence of the damage they cause to the host.

EXAMPLES OF PATHOGENS THAT HAVE COEVOLVED WITH HUMANS

Many examples of evolutionary relationships between humans and their pathogens have been documented and studied, and help to elucidate the nature and the timing of important adaptations humans have made to their environments throughout the history of the species. The examples in this section illustrate the process of coevolution between humans and specific pathogens and how knowledge of that process has provided insights into the details of events that occurred during human evolution. Examples include the influence of infectious diseases on the distribution of ABO blood groups worldwide, how malaria has stimulated the evolution of hemoglobin variants, the evolution of influenza viruses, and the evolution of CCR5-Δ32, a recently evolved allele that protects humans from HIV infection.

ABO Blood Groups and Infectious Diseases

The ABO blood group system was the first biochemical genetic polymorphism detected in humans. Frequencies of the alleles in this genetic system vary markedly across space, and evolutionary explanations have been proposed to explain this variation since the system's discovery by Karl Landsteiner in the early 1900s. The ABO system was first recognized as a consequence of studies to determine why blood transfusions were sometimes successful and at other times unsuccessful. Landsteiner (1900) showed that human blood could be classified into different types on the basis of whether the mixing of two samples of blood led to clumping (agglutination). This agglutination occurred when donors and recipients had certain types of unmatched blood, and resulted from an antigen–antibody response that occurred when the recipient produced antibodies to fight off the foreign antigens present on a donor's blood cells. Recognition of the ABO blood types, and subsequent efforts to match them when giving transfusions, greatly improved the success rates of the procedures.

The molecular basis of the ABO blood system was first proposed by Yamamoto et al. (1990). The ABO gene codes for a glycosyltransferase (an enzyme) that facilitates the addition of a sugar to the end of a precursor antigen, the H antigen, on the surface of red blood cells. The system consists of three alleles, A, B, and O. Presence of the A allele causes the glycosyltransferase to attach the sugar *N*-acetyl D-galactosamine to the H antigen, converting it to the A antigen; the B allele's presence causes the glycosyltransferase to attach a different sugar, D-galactose, converting it to the B antigen. The O allele causes the glycosyltransferase gene to become inactivated; thus, individuals who possess only O alleles (i.e., have blood type O) do not alter the H antigen and so have neither A nor B antigens on the surface of their red blood cells (Anstee, 2010; Yamamoto et al., 1990). The A, B, and H antigens are not limited to just red blood cells, but are found in a variety of other cells in the body (Anstee, 2010).

As a result of the medical importance of the ABO blood group system, abundant data are available on the worldwide frequencies of the A, B, and O alleles. At the worldwide level, the O allele is most common and occurs about 63% of the time, the A allele occurs 21% of the time, and the B allele occurs about 16% of the time. However, there is extensive geographic variation. For example, the B allele is most common among individuals of Asian ancestry, but it is nearly absent among pure-blooded indigenous North and South Americans, even though many other genetic traits indicate that these populations originated in Asia (Brues, 1977; Cavalli-Sforza et al., 1994).

The high degree of genetic variability in the ABO system in all human populations outside of the New World is too high to be explained by recurrent mutation (see Futuyma in this volume). Thus, hypotheses about the roles of other evolutionary forces have been repeatedly proposed. For example, the ubiquitous presence of the O allele in most pure-blooded indigenous North and South Americans is often suggested to be a consequence of founder effect and subsequent genetic drift occurring in the undoubtedly small groups that migrated into the New World from Asia. It is important to note, however, that the evidence for multiple migrations from Asia to the New World reduces the chance that founder effect and genetic drift alone are responsible, since these processes would be just as likely to result in high frequencies of either the A or the B allele in different groups. Matson and Schrader (1933) observed that Blackfeet and Blood Indians in northwestern North America have a much higher frequency of the A allele than do other Amerindian groups, possibly because genetic drift in these groups resulted in an increase of the A allele relative to the O allele rather than extinction of the A and B alleles (Szathmary and Auger, 1983).

Natural selection has also often been proposed as an explanation for the distribution of ABO blood groups. Often the selection arguments have posited that interactions of human groups with infectious diseases may be responsible for the observed variation, although noninfectious diseases are also commonly implicated (Table 2).

TABLE 2 Known and Proposed Associations between ABO Blood Type and Human Disease

Disease/Pathogen Implicated	Selected References
I. Associations with Infectious Diseases	
Helicobactor pylori, peptic ulcer	Alkout et al. (2000) and Edgren et al. (2010)
Malaria	Cserti and Dzik (2007) and Zerihun et al. (2011)
Cholera	Glass et al. (1985) and Harris et al. (2005)
E. coli	Blackwell et al. (2002)
Norovirus[a]	Hutson et al. (2002) and Tan and Jiang (2005, 2010)
Smallpox[b]	Pettenkofer et al. (1962) and Chakravartti et al. (1966)
Plague[b]	Vogel et al. (1960) and Pettenkofer et al. (1962)
II. Associations with Other Diseases	
Arterial and venous thromboembolism	Anstee (2010)
Salivary and gastric cancer	Aird et al. (1953) and Edgren et al. (2010)
Exocrine pancreatic cancer[c]	Iodice et al. (2010) and Wang et al. (2012)

[a]Resistance to norovirus infection depends on human histo-blood group antigens, which are related to secretor and Lewis alleles in addition to ABO alleles.
[b]The associations of smallpox and plague with ABO blood groups are among the first proposed relationships between blood groups and disease, but the idea is now controversial (Anstee, 2010).
[c]Wang et al. (2012) suggested that interactions between hepatitis B virus and ABO blood type may play a role in the development of pancreatic cancer.

Many of the arguments related to infectious diseases hinge on the recognition that some kinds of infectious pathogens possess antigens on the surface of their cells that are similar in structure to the A or B antigens on human red blood cells. Persons possessing the A and/or B antigens on their red blood cells do not mount antibody responses against those cells (which otherwise would destroy their own blood) and so the argument is that they also cannot recognize pathogens that contain similar antigens. For example, the smallpox virus possesses an antigen similar to the A antigen on red blood cells. Thus, the immune system of individuals who have A or AB blood type may not be able to recognize the smallpox virus as well as individuals who have B or O blood type, and may have been selected against during smallpox epidemics. Although this is an attractive hypothesis, it remains controversial, largely because results of studies looking at whether there was an observed relationship in the postulated direction have been mixed (Mielke et al., 2011). A similar argument related to the red blood cell H antigen described above was proposed to explain a relatively low frequency of the O allele in parts of Europe and Asia, and additional relationships between infectious diseases and the frequency of the ABO alleles in different locations have been proposed (Mielke et al., 2011). The observed relationships between malaria, cholera, *Helicobactor pylori*, and norovirus mentioned in Table 2 do seem to be solid, but in general the associations have not been adequate to fully assess whether infectious diseases have played important roles in generating the observed distributions of the ABO alleles.

Genetic Adaptations to Malaria

According to the World Health Organization, about half the world's population (3.3 billion people) are at risk of malaria. In 2010, there were about 219 million cases of malaria worldwide and about 660,000 deaths, with about 90% of all deaths occurring in Africa (World Health Organization, 2013). These numbers are astounding in and of themselves, but they represent a 25% reduction globally and a 33% reduction in Africa since 2000. Malaria remains one of the most serious of all infectious diseases of humans. Almost certainly because of its long-standing severity, there is more evidence for the evolution of genetic adaptations that reduce the effects of malaria than for any other infectious disease.

The human genetic adaptations to malaria are of several types. The best-known adaptation is the evolution of hemoglobin S (Hb S). This variant of normal hemoglobin, which causes sickling of red blood cells, reaches its highest frequencies in parts of Africa which are especially at high risk for malaria (Figure 2). Individuals who are homozygous for Hb S develop sickled red blood cells that do not carry sufficient oxygen to the tissues of the body and cause a severe

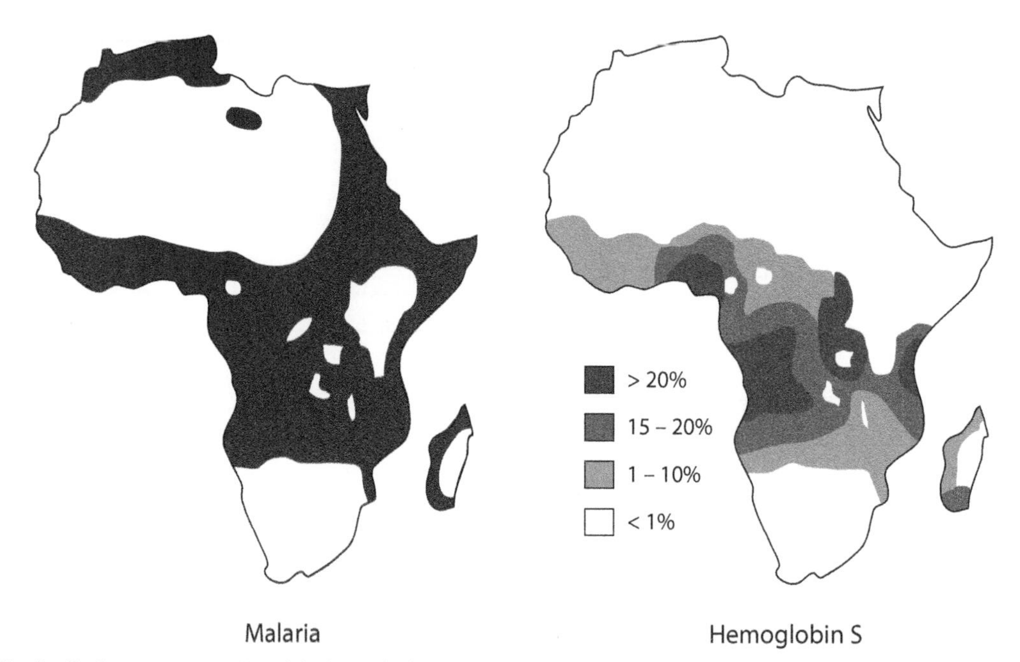

FIGURE 2 The distribution of malaria (left) and the hemoglobin S (Hb S) allele (right) in Africa. Note the absence of Hb S in regions that do not experience malaria. (See color plate section). *Adapted from the original figure of Anthony Allison, dedicated by him to the public domain and made available at http://commons.wikimedia.org/wiki/File:Malaria_versus_sickle-cell_trait_distributions.png.*

anemia; such individuals tend to die at a young age and have reduced fertility if they do survive to maturity. Individuals who are homozygous for the normal hemoglobin (Hb A) are at greatest risk of dying from malaria. Heterozygotes who have both types of hemoglobin tend to survive and reproduce the best; they have only a mild anemia, but the presence of sickled cells tends to protect them against malaria, because in such cells the survival and/or reproduction of the plasmodium that causes the disease is significantly reduced. This heterozygote advantage leads to balancing selection that ensures that both the Hb A and Hb S alleles are maintained in the population (see the chapter by Relethford).

Livingstone (1958) suggested that the spread of Hb S throughout Africa could be tied to the spread of rice-based and yam agriculture. He argued that virgin forest in the parts of Africa affected by malaria would not possess the proper conditions for reproduction of the mosquito vector, but when humans cleared the forest to make way for farmland, they reduced shade cover and increased the amount of standing water that mosquitoes would need to reproduce. He also correlated population movements associated with the spread of agriculture with the distribution of the Hb S allele and illustrated the key role that human behaviors played in the spread of and adaptation to malaria in Africa (see chapters by Blumler and Lao and Kayser).

Hemoglobin S is only one of several hemoglobin variants that appear to confer some resistance to malaria. Hemoglobin E (Hb E) is the second most common variant and is especially common in malarial regions of Southeast Asia. Another common variant in West Africa is hemoglobin C.

Besides hemoglobin variants, malaria has been implicated in the evolution of α- and β-thalassemia, the Duffy trait, and G6PD deficiency. The thalassemias result when either of the hemoglobin chains (the α-and β-chains) that make up normal adult human hemoglobin do not get synthesized by the body. They reach highest frequencies in the Mediterranean region, but different variants are known from other malarial regions of the world as well. The Duffy trait is one of the best genetic markers for African ancestry. Two common Duffy antigens, Fy^a and Fy^b, occur on the surface of red blood cells. These antigens have been shown to be receptors for one species of human malaria, *Plasmodium vivax* (Chaudhuri and Pogo, 1995; Pogo and Chaudhuri, 1997). A high proportion of Africans and those with African ancestry lack these antigens, which increases resistance to malaria by interfering with the introduction of the pathogen into red blood cells. G6PD (glucose-6-phosphate-dehydrogenase) is a red cell enzyme that performs a variety of functions in the body including protection against harmful by-products of cell metabolism (Mielke et al., 2011). It is an X-linked trait with the variant allele resulting in a lack of production of the enzyme. The variant is most often found in African, Middle Eastern, and South Asian people. G6PD-deficient individuals appear to be at an advantage in a malarial environment, although it is not completely clear whether the advantage holds only for males with the allele (who have only the variant allele) or for heterozygous females as well.

All the examples discussed here involve the development of human adaptations to malaria, but a coevolutionary framework implies that the pathogen evolves in response to the host as well. Is there evidence for such an effect with malaria? Mackinnon and Read (2004) addressed this very question. They analyzed data on *Plasmodium falciparum* infection in humans to see if virulence of the pathogen would stabilize when the cost to transmission caused by the death of hosts outweighed the benefits. Their analyses indicated that at least for young children, the hypothesized virulence trade-off was apparent. They further suggested that when assessing the response of populations to new medical interventions such as vaccines or drugs, careful attention should be paid on how the virulence of a pathogen might be affected.

Interacting Species and the Evolution of Influenza Viruses

One of the most compelling examples of coevolution between humans and their pathogens is that of influenza A viruses (IAVs), the most common viruses implicated in human influenza. Influenza B and C viruses also cause disease in humans, but the discussion in this chapter is limited to IAVs. Work on understanding the evolution of IAVs has been ongoing at least since the 1918–19 influenza pandemic, and much is now known about this process. Yet there are still significant questions about how and when new strains of human influenza virus evolve and so prediction remains inadequate, in large part because multiple genes are involved as well as multiple mechanisms that different virus strains have used to solve similar problems (Taubenberger and Kash, 2010). What is clear is that the primary natural hosts of IAVs are wild aquatic waterfowl and shorebirds (Webster et al., 1992), and that complex interactions between domesticated birds, swine, and humans are the most common sources for new human IAV strains (Taubenberger and Kash, 2010).

IAV strains are designated by the type of two different glycoproteins present on the surface of the virus, hemagglutinin (HA) and neuraminidase (NA). For example, the 1918–19 virus is designated as H1N1, meaning it is Type 1 hemagglutinin and Type 1 neuraminidase. Since 2005, the major avian influenza strain that has infected humans and circulated in Asia is designated as H5N1. Table 3 lists the IAV strains that have been observed to infect humans, only three of which appear to be capable of infecting humans on a regular basis—H1N1, H2N2, and H3N2. It is too soon to tell whether H5N1 or other influenza strains will join this select group.

The evolution of IAVs involves two primary kinds of evolutionary processes. Antigenic drift occurs when selectively advantageous mutations evolve and result in small changes in the HA and/or NA glycoproteins. These result in regularly occurring but relatively minor changes in the nature of the virus. Larger evolutionary changes of the virus occur when genetic reassortment occurs between different virus strains cocirculating in a host. Commonly this reassortment involves multiple species. For example, swine influenza usually evolves when strains derived from domesticated or wild birds reassort within swine. Many common strains presently circulating in swine are derived from the 1918 pandemic IAV (Taubenberger et al., 2001), but since 1998, strains derived as a consequence of "triple" reassortment of swine, human, and avian IAVs have been causing disease in the United States and globally (Olsen, 2002). It is a common view that antigenic shifts lead to worldwide epidemics or pandemics while antigenic drift is responsible for more modest seasonal outbreaks of flu, but this view is

TABLE 3 Influenza A Strains Observed in Humans

I. Adapted to Humans; Sustained Circulation with Regular Outbreaks

Strain	*Primary Host*
H1N1	Humans, pigs, birds
1918 pandemic strain (1918–57, 1977–2010)	
2009 pandemic strain (2009–present)	
H2N2	Humans, birds
1957 pandemic strain (1957–68)	
H3N2	Humans, pigs, birds
1968 pandemic strain (1968–present)	

II. Observed in Humans, but without Sustained Transmission

Strain	*Primary Host*	*Selected References*
H1N2	Pigs	Xu et al. (2002) and Choi et al. (2002)
H3N8	Horses, dogs	Dowdle (1999) and Morens and Taubenberger (2011)
H5N1	Poultry	Peiris et al. (2007) and Taubenberger and Kash (2010)
H7N2	Poultry	de Wit and Fouchier (2008)
H7N3	Poultry	Tweed et al. (2004)
H7N7	Poultry	Fouchier et al. (2004) and Taubenberger and Kash (2010)
H7N9	Poultry	Chen et al. (2013), Gao et al. (2013) and Morens et al. (2013)
H9N2	Poultry	Lin et al. (2000), Cheng et al. (2011) and Morens et al. (2013)
H10N7	Poultry	Arzey et al. (2012)

probably too simplistic and has recently been called into question (Morens et al., 2009).

The H1N1/2009 pandemic strain appears to have been derived primarily from a swine IAV. Research has shown that the human strain carries genes from several different swine viruses, and that initial transmission to humans predated recognition of the outbreak by several months. Furthermore, the reassortment of swine lineages that eventually gave rise to the new human strain may have occurred years before it emerged in humans (Smith et al., 2009).

Human–animal interactions have played a very important role in the evolution of new influenza strains (see the chapter by Morand). For example, some new strains have been traced to a cycle involving swine herding, duck domestication, and fish farming in Asia. According to Scholtissek and Naylor (1988), a common agricultural practice in Thailand was to keep domesticated chickens in cages above swine, which then ate the chicken feces. The swine were in pens directly above fish ponds, and defecated into them to feed the fish and/or fertilize the ponds. Humans could be exposed to viral strains through their interactions with any components of this complex agricultural system. Similar systems involving ducks, swine, and fish were also used in other places (Scholtissek and Naylor, 1988). Even today, humans are infected sporadically by various avian IAVs as a consequence of intensive and intimate contact with domesticated birds (Schoub, 2012). Since many of the farms where such contact occurs also raise swine and other domesticates, there is always potential for the development of new strains of the virus. Luckily, the transformation of a bird virus to a form that allows efficient human–human direct transmission appears to be a very complex process (Taubenberger and Morens, 2009). Thus, although bird flu cases occur on a regular basis, genetic changes that lead to the evolution of brand new and potentially deadly human strains occur only rarely.

Why Does the CCR5-Δ32 Allele Reach Such High Frequencies in European Populations?

One of the newest pathogen-related genetic variants identified in human populations is the CCR5-Δ32 allele. This is a variant allele for the gene that codes for the CCR5 receptor on immune system cells. The receptor itself, in combination with CD4, provides a "doorway" through which HIV-1 can enter immune system cells (Alkhatib et al., 1996; Choe et al., 1996; Deng et al., 1996; Doranz et al., 1996; Dragic et al., 1996; Stephens et al., 1998). In the CCR5-Δ32 variant, 32 base pairs have been deleted, which changes the structure of the receptor and prevents its expression on the cell surface (Galvani and Slatkin, 2003). Individuals who are homozygous for this variant are almost completely resistant to HIV-1 infection; heterozygous individuals show partial resistance and slower disease progression (Dean et al., 1996; Galvani and Slatkin 2003; Huang et al., 1996; Liu et al., 1996; Michael et al., 1997; Samson et al., 1996).

The frequency of the CCR5-Δ32 allele varies in a clinal fashion across Europe, with the highest frequencies (>15%) occurring in areas around the Baltic and White Seas in Northern Europe and in Central Russia near Novosibirsk. The frequency decreases from north to south throughout Europe, although there are pockets of high frequency in the northern coasts of France and parts of Russia (Balanovsky et al., 2005). The allele is virtually absent in African, Asian, Middle Eastern, and Native American populations (Galvani and Slatkin, 2003; Stephens et al., 1998) (see the chapter by Relethford).

It is clear that HIV-1 cannot be responsible for the marked geographic distribution of the CCR5-Δ32 allele, because the virus has not been present in human populations for long enough to generate this diversity. In addition, regions with high frequencies of the CCR5-Δ32 allele are not the world's major hotspots for HIV-1 prevalence. So what is responsible for the allele's distribution?

One of the major recent technological advances that can be used to address questions like this is the ability to rapidly determine the genetic code underlying the expression of different genes and to compare this information across different populations. Several molecular studies have been done on the CCR5-Δ32 allele, and all suggest that the variant has a single origin that occurred relatively recently; nonetheless, different methods give widely differing estimates of the time of origin of the allele. Using the coalescence theory on a sample of over 4000 individuals, Stephens et al. (1998) estimated that the deletion originated about 700 years ago, with an estimated range of 275–1875 years. Libert et al. (1998) used microsatellite analysis and estimated that the variant evolved a few thousand years ago. Sabeti et al. (2005) used single nucleotide polymorphism (SNP) data and estimated an age of 5075 years, while ancient DNA evidence suggests the allele appeared in Germany at least 2900 years ago (Hummel et al., 2005). A recent study of Scandinavian Mesolithic DNA has pushed the date back to about 7000 years ago (Lidén et al., 2006). These results clearly show that one must be cautious in drawing conclusions about where, when, and how the CCR5-Δ32 allele evolved, at least until there is some consensus on which methods are best used for the type of data and time period involved.

Given modern understanding of how the forces of evolution influence allele frequencies over time and the general agreement about the recent evolution of the CCR5-Δ32 allele, it is almost certain that exceptionally strong selection pressures were needed in order to drive the frequency of the CCR5-Δ32 allele to observed levels in the short time indicated by even the earliest of these dates. Since epidemics of infectious diseases, particularly new diseases, tend to kill substantial portions of populations, they are commonly suggested as possible agents of selection when trying to explain data such as that for CCR5-Δ32.

Numerous scenarios have been proposed in the literature. Perhaps the earliest hypothesis was put forth by Stephens et al. (1998), who suggested that bubonic plague may have been the responsible pathogen. Recall that Stephens, Reich, Goldstein et al. derived a date of origin of about 700 years ago, which places the evolution of CCR5-Δ32 at about AD 1300, right around the time of the Black Death in Europe. They further cited research indicating that the bacterium that causes plague, *Yersinia pestis*, causes diminished immune defenses, although it was not known at the time whether the mechanisms involved the CCR5 receptor.

The plague hypothesis is no longer accepted, however. Galvani and Slatkin (2003) showed that even if the CCR5-Δ32 allele was a fully dominant allele (which would mean that heterozygotes were indistinguishable from homozygotes for the allele), bubonic plague would not be able to generate sufficient selective pressure to account for the current allele frequencies in Europe. Mecsas et al. (2004) showed that there was no difference in either growth rate of the plague *bacillus* or the survival time when comparing normal and CCR5-deficient mice infected with plague. Elvin et al. (2004) also failed to observed a difference in survival between normal and CCR5-deficient mice with plague, although they were cautious about the significance of experiments using twentieth-century *Yersinia pestis* strains when assessing events that would have happened in the fourteenth century. They also cautioned that there may be differences in outcome when the *bacillus* is injected, as in the experimental studies, instead of naturally transmitted by fleas in the environment. Faure and Royer-Carenzi (2008) pointed out that the plague hypothesis does not fit historical evidence indicating that the epidemic spread from east to west (not from north to south, the direction of the cline in allele frequencies). Furthermore, the observed gradient of Black Death mortality sloped in the opposite direction from what would be expected if it were the selective force responsible for the increase in the frequency of CCR5-Δ32. In other words, if plague were the selective factor, then the descendants of survivors living in the regions with the highest mortality during the Black Death should have the highest frequency of the "protective" allele, but these descendants have some of the lowest frequencies of the allele in all of Europe.

One of the problems associated with determining the cause of the cline in allele frequencies is that, as described above, the timing estimates vary markedly from study to study. Different hypotheses have been proposed for each of these different estimates. For example, using the date of 700 years ago proposed by Stephens et al. (1998), Galvani and Slatkin (2003) proposed that smallpox may be the causal pathogen responsible for selective pressures on the CCR5-Δ32 allele. They used an age-structured population model to assess whether smallpox and plague could generate appropriate selection coefficients and found that smallpox, but not plague, had the possible capacity to do so. The essential differences between smallpox and plague are that (1) smallpox epidemics were a regular feature of life at the time and represented weaker but more constant selection than plague epidemics, and (2) smallpox tended to kill young individuals disproportionately, whereas plague killed individuals of all ages. Because high rates of death among young individuals influence both survivorship and future fertility, diseases like smallpox that possess such age-biased mortality toward the young tend to be more potent selective forces. In addition to these arguments, Galvani and Slatkin (2003) pointed out that poxviruses tend to use a similar mechanism to that of HIV-1 to enter immune system cells, and so a mutation such as that occurring with CCR5-Δ32 would almost certainly have resulted in increased resistance to smallpox infection.

Smallpox has also been used to explain the possible origin of CCR5-Δ32 at earlier postulated times. Holtby et al. (2012) suggested that the frequency of the allele increased in Northern Europe when agriculture first spread to the region around 7000 years ago. They argued that local residents possessing the CCR5-Δ32 allele would be more resistant to smallpox that was carried into the region by migrants coming from the farming regions where smallpox itself originated (see chapters by Lao and Kayser, and Blumler). Klitz et al. (2001) suggested that CCR5-Δ32 spread from Northern European populations to Ashkenazi Jews, and that carriers of the allele in both groups were favored during repeated smallpox epidemics beginning in the eighth century AD.

The specific diseases implicated at other possible times of origin of the CCR5-Δ32 allele are less certain. Faure and Royer-Carenzi (2008) suggested that the spread of pathogens during the Roman expansion may have led to the origin of the CCR5-Δ32 allele, and they propose that some as yet unidentified zoonotic disease may have been important at this time. Lucotte (2001) and Lucotte and Dieterlen (2003) suggested that the allele evolved in Scandinavia some time before the Vikings were a dominant power, and then spread throughout Europe and Russia as a consequence of their dispersal patterns. They did not implicate a particular pathogen in this process, however.

Clearly some kind of strong selective force must have been operating to increase the frequency of CCR5-Δ32 in European populations, and good arguments have been made for the potential role of pathogens in this process. Uncertainties remain, and it may be that other factors are at play, but the stories of CCR5-Δ32 described here illustrate nicely the possible mechanisms by which humans and their pathogens coevolve.

CONCLUSIONS

As the examples above illustrate, pathogens can be potent forces influencing genetic change in humans. Particularly when they first evolve to infect humans, pathogens may cause very high mortality among individuals of all ages that can lead to rapid evolutionary change, especially when the affected humans vary in their susceptibility or resistance to the pathogen. Yet as the models of Galvani and Slatkin (2003) suggest, pathogens such as smallpox, which have been circulating within human groups for some time, may be equally important or even more potent evolutionary forces than newly evolved pathogens, primarily because they provide constant levels of selection and tend to target very young individuals, thus affecting both survivorship and future fertility.

At the same time, genetic changes occur readily in pathogens that infect humans, especially given the flexibility of human immune responses. Many pathogens have generation times on the order of days or hours, and as a result they have abundant opportunity for mutations to evolve that may allow them to bypass existing host defenses. The development of antibiotic resistance is a clear example of evolutionary processes working on the pathogens to help them survive and reproduce in the face of human strategies designed to prevent such events.

Space limitations allow consideration, in this chapter, of only a selection of possible human–pathogen evolutionary interactions. They have been chosen because of their historical importance and because they illustrate the variety of ways that humans and pathogens interact within evolutionary scenarios. Other prominent examples include the question of whether human tuberculosis evolved from bovine tuberculosis as a consequence of animal domestication or if instead the evolutionary pathway took the opposite direction; how, when, and where venereal syphilis evolved; and how the human gut microbiota influences a variety of disease processes, including infection with pathogenic microorganisms and chronic conditions that may or may not involve pathogens.

By the middle of the twentieth century, in response to the impressive health advances associated with the development of sanitation systems, personal hygiene, antibiotic use, and many other improvements, the Western medical establishment was misled into thinking that humans had largely learned to control the survivorship and reproduction of pathogens to keep them from causing significant levels of illness and death. It is now abundantly clear that this optimistic view was premature—pathogens have always been and always will be a part of human life, and humans and their pathogens will always influence each other's evolution.

REFERENCES

Aird, I., Bentall, H.H., Roberts, J.A.F., 1953. A relationship between cancer of stomach and the ABO blood groups. British Medical Journal 1, 799–801.

Alkhatib, G., Combadiere, C., Broder, C.C., et al., 1996. CC CKR5: a RANTES, MIP-1α, MIP-1ß receptor as a fusion cofactor for macrophage-tropic HIV-1. Science 272, 1955–1958.

Alkout, A.M., Blackwell, C.C., Weir, D.M., 2000. Increased inflammatory responses of persons of blood group O to *Helicobacter pylori*. The Journal of Infectious Diseases 181, 1364–1369.

Anstee, D.J., 2010. The relationship between blood groups and disease. Blood 115, 4635–4643.

Arzey, G.G., Kirkland, P.D., Arzey, K.E., et al., 2012. Influenza virus A (H10N7) in chickens and poultry abattoir workers, Australia. Emerging Infectious Diseases 18 (5), 814–816.

Balanovsky, O., Pocheshkhova, E., Pshenichnov, A., et al., 2005. Is spatial distribution of the HIV-1-resistant CCR5-Δ32 allele formed by ecological factors? Journal of Physiological Anthropology and Applied Human Science 24, 375–382.

Bell, G., 1982. The Masterpiece of Nature: The Evolution and Genetics of Sexuality. University of California Press, Berkeley.

Blackwell, C.C., Dundas, S., James, V., et al., 2002. Blood group and susceptibility to disease caused by *Escherichia coli* O157. The Journal of Infectious Diseases 185, 393–396.

Brues, A.M., 1977. Peoples and Races. Macmillan, New York.

Cavalli-Sforza, L.L., Menozzi, P., Piazza, A., 1994. The History and Geography of Human Genes. Princeton University Press, Princeton.

Chakravartti, M.R., Verma, B.K., Hanurav, T.V., Vogel, F., 1966. Relation between smallpox and the ABO blood groups in a rural population of West Bengal. Humangenetik 2, 78–80.

Chaudhuri, A., Pogo, A.O., 1995. The Duffy blood group system and malaria. In: Cartron, J.-P., Rouger, R. (Eds.), Blood Cell Biochemistry. Molecular Basis of Human Blood Group Antigens, vol. 6. Plenum, New York, pp. 243–265.

Chen, Y., Liang, W., Yang, S., et al., 2013. Human infections with the emerging avian influenza A H7N9 virus from wet market poultry: clinical analysis and characterisation of viral genome. The Lancet. http://dx.doi.org/10.1016/S0140-6736(13)60903-4.

Cheng, V.C.C., Chan, J.F.W., Wen, X., et al., 2011. Infection of immunocompromised patients by avian H9N2 influenza A virus. Journal of Infection 62, 394–399.

Choe, H., Farzan, M., Sun, Y., et al., 1996. The ß-chemokine receptors CCR3 and CCR5 facilitate infection by primary HIV-1 isolates. Cell 85, 1135–1148.

Choi, Y.K., Goyal, S.M., Farnham, M.W., Joo, H.S., 2002. Phylogenetic analysis of H1N2 isolates of influenza A virus from pigs in the United States. Virus Research 87, 173–179.

Cserti, C.M., Dzik, W.H., 2007. The ABO blood group system and *Plasmodium falciparum* malaria. Blood 110, 2250–2258.

de Wit, E., Fouchier, R.A.M., 2008. Emerging influenza. Journal of Clinical Virology 41, 1–6.

Dean, M., Carrington, M., Winkler, C., et al., 1996. Genetic restriction of HIV-1 infection and progression to AIDS by a deletion allele of the CCR5 structural gene. Science 273, 1856–1862.

Deng, H., Liu, R., Ellmeier, W., et al., 1996. Identification of a major coreceptor for primary isolates of HIV-1. Nature 381, 661–666.

Doranz, B.J., Rucker, J., Yi, Y., et al., 1996. A dual-tropic primary HIV-1 isolate that uses fusin and the ß-chemokine receptors CKR-5, CKR-3, and CKR-2b as fusion cofactors. Cell 85, 1149–1158.

Dowdle, W.R., 1999. Influenza A virus recycling revisited. Bulletin of the World Health Organization 77, 820–828.

Dragic, T., Litwin, V., Allaway, G.P., et al., 1996. HIV-1 entry into $CD4^+$ cells is mediated by the chemokine receptor CC-CKR5. Nature 381, 667–673.

Ebert, D., Bull, J.J., 2008. The evolution and expression of virulence. In: Stearns, S.C., Koella, J.C. (Eds.), Evolution in Health and Disease, second ed. Oxford University Press, New York, pp. 153–168.

Edgren, G., Hjalgrim, H., Rostgaard, K., et al., 2010. Risk of gastric cancer and peptic ulcers in relation to ABO blood type: a cohort study. American Journal of Epidemiology 172, 1280–1285.

Elvin, S.J., Williamson, E.D., Scott, J.C., et al., 2004. Ambiguous role of CCR5 in *Y. pestis* infection. Nature 430, 417.

Ewald, P.W., 1994. Evolution of Infectious Disease. Oxford University Press, New York.

Faure, E., Royer-Carenzi, M., 2008. Is the European spatial distribution of the HIV-1-resistant CCR5-Δ32 allele formed by a breakdown of the pathocenosis due to the historical Roman expansion? Infection, Genetics, and Evolution 8, 864–874.

Fenner, F., Ratcliff, F.N., 1965. Myxomatosis. Cambridge University Press, Cambridge.

Fouchier, R.A.M., Schneeberger, P.M., Rozendaal, F.W., et al., 2004. Avian influenza A virus (A7N7) associated with human conjunctivitis and a fatal case of acute respiratory distress syndrome. Proceedings of the National Academy of Sciences, USA 101, 1356–1361.

Galvani, A.P., 2003. Epidemiology meets evolutionary ecology. Trends in Ecology and Evolution 18, 132–139.

Galvani, A.P., Slatkin, M., 2003. Evaluating plague and smallpox as historical selective pressures for the CCR5-Δ32 HIV-resistant allele. Proceedings of the National Academy of Sciences, USA 100, 15276–15279.

Gao, R., Cao, B., Hu, Y., et al., 2013. Human infection with a novel avian-origin influenza A (H7N9) virus. New England Journal of Medicine 368, 1888–1897.

Glass, R.I., Holmgren, J., Haley, C.E., et al., 1985. Predisposition for cholera of individuals with O blood group. Possible evolutionary significance. American Journal of Epidemiology 121, 791–796.

Haldane, J.B.S., 1949. Disease and evolution. La Ricerca Scientifica Supplemento 19, 1–11.

Harris, J.B., Khan, A.I., LaRocque, R.C., et al., 2005. Blood group, immunity, and risk of infection with *Vibrio cholerae* in an area of endemicity. Infection and Immunity 73, 7422–7427.

Holtby, I., Scarre, C., Bentley, R.A., Rowley-Conwy, P., 2012. Disease, CCR5-Δ32 and the European spread of agriculture? A hypothesis. Antiquity 86, 207–210.

Huang, Y., Paxton, W.A., Wolinsky, S.M., et al., 1996. The role of a mutant CCR5 allele in HIV-1 transmission and disease progression. Nature Medicine 2, 1240–1243.

Hummel, S., Schmidt, D., Kremeyer, B., Herrmann, B., Oppermann, M., 2005. Detection of the CCR5-Δ32 HIV resistance gene in Bronze Age skeletons. Genes and Immunity 6, 371–374.

Hutson, A.M., Atmar, R.L., Graham, D.Y., Estes, M.K., 2002. Norwalk virus infection and disease is associated with ABO histo-blood group type. The Journal of Infectious Diseases 185, 1335–1337.

Iodice, S., Maisonneuve, P., Botteri, E., Sandri, M.T., Lowenfels, A.B., 2010. ABO blood group and cancer. European Journal of Cancer 46, 3345–3350.

Klitz, W., Brautbar, C., Schito, A.M., Barcellos, L., Oksenberg, J.R., 2001. Evolution of the CCR5-Δ32 mutation based on haplotype variation in Jewish and Northern European population samples. Human Immunology 62, 530–538.

Koella, J.C., Turner, P., 2008. Evolution of parasites. In: Stearns, S.C., Koella, J.C. (Eds.), Evolution in Health and Disease, second ed. Oxford University Press, New York, pp. 229–237.

Landsteiner, K., 1900. Zur Kenntnis der antifermentativen, lytischen und agglutinierenden Wirkungen des Blutserums und der Lymphe. Zentralblatt für Bakteriologie 27, 357–362.

Lederberg, J., 1999. J. B. S. Haldane (1949) on infectious disease and evolution. Genetics 153, 1–3.

Libert, F., Cochaux, P., Beckman, G., et al., 1998. The Δccr5 mutation conferring protection against HIV-1 in Caucasian populations has a single and recent origin in Northeastern Europe. Human Molecular Genetics 7, 399–406.

Lidén, K., Linderholm, A., Götherström, A., 2006. Pushing it back. Dating the CCR5-Δ32 bp deletion to the Mesolithic in Sweden and its implications for the Meso/Neo transition. Documenta Praehistorica 33, 29–37.

Lin, Y.P., Shaw, M., Gregory, V., et al., 2000. Avian-to-human transmission of H9N2 subtype influenza A viruses: relationship between H9N2 and H5N1 human isolates. Proceedings of the National Academy of Sciences, USA 97, 9654–9658.

Liu, R., Paxton, W.A., Choe, S., et al., 1996. Homozygous defect in HIV-1 coreceptor accounts for resistance of some multiply-exposed individuals to HIV-1 infection. Cell 86, 367–377.

Lively, C.M., 2010. A review of Red Queen models for the persistence of obligate sexual reproduction. Journal of Heredity 101 (Supp. 1), S13–S20.

Lively, C.M., Dybdahl, M.F., 2000. Parasite adaptation to locally common host genotypes. Nature 405, 679–681.

Livingstone, F.B., 1958. Anthropological implications of sickle cell gene distribution in West Africa. American Anthropologist 60, 533–562.

Lucotte, G., 2001. Distribution of the CCR5 gene 32-basepair deletion in West Europe. A hypothesis about the possible dispersion of the mutation by the Vikings in historical times. Human Immunology 62, 933–936.

Lucotte, G., Dieterlen, F., 2003. More about the Viking hypothesis of origin of the Δ32 mutation in the CCR5 gene conferring resistance to HIV-1 infection. Infection, Genetics and Evolution 3, 293–295.

Mackinnon, M.J., Read, A.F., 2004. Virulence in malaria: an evolutionary viewpoint. Philosophical Transactions of the Royal Society of London, B. Biological Sciences 359, 965–986.

Matson, G.A., Schrader, H.G., 1933. Blood grouping among the "Blackfeet" and "Blood" tribes of American Indians. Journal of Immunology 25, 155–163.

Mecsas, J., Franklin, G., Kuziel, W.A., et al., 2004. CCR5 mutation and plague protection. Nature 427, 606.

Michael, N.L., Chang, G., Louie, L.G., et al., 1997. The role of viral phenotype and CCR-5 gene defects in HIV-1 transmission and disease progression. Nature Medicine 3, 338–340.

Mielke, J.H., Konigsberg, L.W., Relethford, J.H., 2011. Human Biological Variation, second ed. Oxford University Press, New York.

Morens, D.M., Taubenberger, J.K., 2011. Pandemic influenza: certain uncertainties. Reviews in Medical Virology 21, 262–284.

Morens, D.M., Taubenberger, J.K., Fauci, A.S., 2009. The persistent legacy of the 1918 influenza virus. New England Journal of Medicine 361, 225–229.

Morens, D.M., Taubenberger, J.K., Fauci, A.S., 2013. Pandemic influenza viruses—hoping for the road not taken. New England Journal of Medicine 368, 2345–2348.

Muehlenbein, M.P., 2010. Evolutionary medicine, immunity, and infectious disease. In: Muehlenbein, M.P. (Ed.), Human Evolutionary Biology. University Press, Cambridge, pp. 459–490.

Nowak, M.A., May, R.M., 1994. Superinfection and the evolution of parasite virulence. Proceedings of the Royal Society of London, Series B 255, 81–89.

Olsen, C.W., 2002. The emergence of novel swine influenza viruses in North America. Virus Research 85, 199–210.
Peiris, J.S.M., de Jong, M.D., Guan, Y., 2007. Avian influenza virus (H5N1): a threat to human health. Clinical Microbiology Reviews 20, 243–267.
Pettenkofer, H.J., Stöss, B., Helmbold, W., Vogel, F., 1962. Severe smallpox scars seen in A+AB. Nature 193, 445–446.
Pogo, A.O., Chaudhuri, A., 1997. The Duffy blood group system and its extension in nonhuman primates. In: Blancher, A., Klein, J., Socha, W.W. (Eds.), Molecular Biology and Evolution of Blood Group and MHC Antigens in Primates. Springer-Verlag, Berlin, pp. 219–235.
Sabeti, P.C., Walsh, E., Schaffner, S.F., et al., 2005. The case for selection at CCR5-Δ32. PLoS Biology 3, e378.
Salathé, M., Kouyos, R.D., Bonhoeffer, S., 2008. The state of affairs in the kingdom of the Red Queen. Trends in Ecology and Evolution 23, 439–445.
Samson, M., Libert, F., Doranz, B.J., et al., 1996. Resistance to HIV-1 infection in Caucasian individuals bearing mutant alleles of the CCR-5 chemokine receptor gene. Nature 382, 722–725.
Scholtissek, C., Naylor, E., 1988. Fish farming and influenza pandemics. Nature 331, 215.
Schoub, B.D., 2012. Zoonotic diseases and human health: the human influenza example. Onderstepoort Journal of Veterinary Research 79 (2). Art. #489, 4 pages http://dx.doi.org/10.4102/ojvr.v79i2.489.
Smith, G.J.D., Vijaykrishna, D., Bahl, J., et al., 2009. Origins and evolutionary genomics of the 2009 swine-origin H1N1 influenza A epidemic. Nature 459, 1122–1126.
Stephens, J.C., Reich, D.E., Goldstein, D.B., et al., 1998. Dating the origin of the CCR5-Δ32 AIDS-resistance allele by the coalescence of haplotypes. American Journal of Human Genetics 62, 1507–1515.
Szathmary, E.J.E., Auger, F., 1983. Biological distances and genetic relationships within Algonkians. In: Steegman Jr., A.T. (Ed.), Boreal Forest Adaptations. Plenum Press, New York, pp. 289–315.
Tan, M., Jiang, X., 2005. Norovirus and its histo-blood group antigen receptors: an answer to a historical puzzle. Trends in Microbiology 13, 285–293.
Tan, M., Jiang, X., 2010. Norovirus-host interaction: multi-selections by human histo-blood group antigens. Trends in Microbiology 19, 382–388.
Taubenberger, J.K., Kash, J.C., 2010. Influenza virus evolution, host adaptation, and pandemic formation. Cell Host and Microbe 7, 440–451.
Taubenberger, J.K., Morens, D.M., 2009. Pandemic influenza—including a risk assessment of H5N1. Revue Scientifique et Technique (International Office of Epizootics) 28, 187–202.
Taubenberger, J.K., Reid, A.H., Janczewski, T.A., Fanning, T.G., 2001. Integrating historical, clinical and molecular genetic data in order to explain the origin and virulence of the 1918 influenza virus polymerase genes. Philosophical Transactions of the Royal Society of London, B. Biological Sciences 356, 1829–1839.
Tweed, S.A., Skowronski, D.M., David, S.T., et al., 2004. Human illness from avian influenza H7N3, British Columbia. Emerging Infectious Diseases 10, 2196–2199.
Van Blerkom, L.M., 2003. Role of viruses in human evolution. Yearbook of Physical Anthropology 46, 14–46.
Van Valen, L., 1973. A new evolutionary law. Evolutionary Theory 1, 1–30.
Vogel, F., Pettenkofer, H.J., Helmbold, W., 1960. Über die Populationsgenetik der ABO-Blutgruppen. Acta Genetica 10, 267–294.
Wang, D., Chen, D., Ren, C., et al., 2012. ABO blood group, hepatitis B viral infection and risk of pancreatic cancer. International Journal of Cancer 131, 461–468.
Webster, R.G., Bean, W.J., Gorman, O.T., Chambers, T.M., Kawaoka, Y., 1992. Evolution and ecology of influenza A viruses. Microbiological Reviews 56, 152–179.
World Health Organization, 2013. Malaria Fact Sheet. http://www.who.int/mediacentre/factsheets/fs094/en/.
Xu, X., Smith, C.B., Mungall, B.A., et al., 2002. Intercontinental circulation of human influenza A(H1N2) during the 2001-2002 influenza season. The Journal of Infectious Diseases 186, 1490–1493.
Yamamoto, F., Clausen, C., White, T., Marken, J., Hakomori, S., 1990. Molecular genetic basis of the histo-blood group ABO system. Nature 345, 229–233.
Zerihun, T., Degarege, A., Erko, B., 2011. Association of ABO blood group and *Plasmodium falciparum* malaria in Dore Bafeno Area, Southern Ethiopia. Asian Pacific Journal of Tropical Biomedicine 2011, 289–294.

Chapter 31

Paleopathology

Della C. Cook
Department of Anthropology, Indiana University, Bloomington, IN, USA

SYNOPSIS

Paleopathology is the study of diseases in ancient organisms. The lives of hominid fossils, prehistoric people, and people from the recent past are better understood if evidence for health and disease derived from their remains is considered. Some diseases can be diagnosed from ancient skeletons and mummies. Others can be identified through ancient DNA techniques or through paleoparasitology. Bioarchaeology seeks to place evidence for health and disease in cultural, historical, ecological, and archaeological contexts. Osteobiography uses the techniques of paleopathology to describe the life experiences both of persons known from history and of the anonymous ancient dead. New technologies and new media promise to improve communication in this interdisciplinary field.

Paleopathology (palaeopathology, traditional British spelling) is the study of ancient health and disease. The field is interdisciplinary, stretching from art history to paleontology and from parasitology to genomics. Whereas most contributors to paleopathology have disciplinary homes in medical sciences or in anthropology, specialists in many areas offer new and intriguing perspectives on the ancient diseases.

How ancient is ancient? The questions that paleopathologists address range from the origins of disease in deep time to scientific evidence of diseases that may be invisible in early twentieth century medical records. As a discipline, paleopathology has a diverse history (Buikstra and Roberts, 2012). The earliest efforts addressed relatively recent humans: Medieval and early modern investigators of saints' remains sometimes looked for injuries and other individual features that could confirm the identity of a skeleton. For example, the skeleton identified as Saint Stephen in 415 AD was expected to show evidence of stoning (Bynum, 1995).

As paleontology emerged as a discipline, anatomists who described fossils found that they needed to distinguish the effects of diseases on bones from the normal features of dinosaurs and cave bears. Johann Friedrich Esper in the eighteenth century, and Georges Cuvier and Rudolf Virchow in the nineteenth century made important contributions to distinguishing normal differences among fossil species from disease processes in individual fossils. In 1872, Virchow argued that the first Neandertal fossil suffered from rickets and arthritis, and that these diseases accounted for the features that made his skeleton look more primitive than it otherwise would, in hindsight an incorrect inference. It is thus no accident that the word paleopathology was coined in the late nineteenth century almost simultaneously by two physicians working with very different material: American Robert Shufeldt investigating Pleistocene birds in 1892, and British Marc Armand Ruffer studying Egyptian mummies shortly thereafter (Buikstra and Roberts, 2012).

The earliest evidence for disease as the focus of paleopathology was an obsession of one of the field's founders, paleontologist and anatomist Roy Lee Moodie. He searched for evidence of disease in fossil marine invertebrates and even speculated about the antiquity of plant diseases, reserving for the paleopathology of humans a small part of his 1923 work (Moodie, 1923). More recent research on deep time issues focuses on the evolutionary origins of parasitism and the role of infectious disease in extinction. Epidemic disease as a force in the demographic collapse of simple societies under the onslaught of European colonialism interested Charles Darwin (Darwin, 1874), and it continues to play this role today in studies of historical demography (see chapters by Sattenspiel, Dressler, and Morand in this volume).

FOSSIL HUMANS AND DISEASE

After an eclipse since Moodie's era, paleopathology has recently reemerged as a lively topic in paleontology, including that of humans. Several studies of the ecology of fossil animals have employed paleopathology to ask how species interacted. Inferences about activities as diverse as avoiding predators, capturing prey, competing for mates, and sunning for warmth abound in recent literature about ancient animals. In hominid paleopathology there has been striking attention to predation. Early hominids were prey to crocodiles at Olduvai Gorge, Tanzania (Njao and Blumenschine, 2012), and the Taung child may have been eaten by an eagle (Berger, 2006). Damage consistent with various large cats has been described in australopithecines from Sterkfontein and Swartkrans in South Africa and the *Orrorin* fossils from Kenya, as well as in *Homo ergaster* fossils from Dmanisi in Georgia (Hart and Sussman, 2005). Distinguishing predation from other causes of injury can be difficult. A few injuries in early hominids are attributed to interpersonal violence, for example, the healed head wound in the Maba hominid from China (Wu et al., 2011) and rib injuries in one of the Shanidar Neandertals (Churchill et al., 2009). Fine cutmarks on the facial skeleton of the Bodo *Homo heidelbergensis* from Ethiopia point to cultural practices such as defleshing or mask making (White, 1986). More controversially, the excess of skull caps with respect to other bones has been used to argue for mortuary practices or cannibalism in the Zhoukoudian *Homo erectus* remains from China almost since their discovery, whereas the most recent reassessment points to predator damage consistent with the giant hyenas that also occupied the cave (Boaz et al., 2004).

Reports of chronic diseases of the skeleton in fossil humans range from developmental defects to arthritis and from tooth loss to tumors. The Qafzeh 12 Neandertal child from Israel has an enlarged skull and growth retardation suggestive of hydrocephalus (Tillier et al., 2001). Periosteal new bone cloaking a femur and humerus has been described as yaws in an adolescent *H. erectus* from Kenya (Rothschild et al., 1995), although the same lesions were earlier attributed to hypervitaminosis A (Walker et al., 1982). This interesting case merits new study with advanced imaging and careful differential diagnosis. Childhood anemia has been described in several fossils (Domínguez-Rodrigo et al., 2012; Garcia et al., 2009). Intervertebral disc lesions in an isolated australopithecine vertebra from Sterkfontein in South Africa have been attributed to brucellosis, a common infectious disease in herd animals (D'Anastasio et al., 2009). Scheuerman's disease, a curvature of the spine that begins in adolescence and reflects weight bearing, has been reported in the "Lucy" *Australopithecus* from Ethiopia (Cook et al., 1983; Figure 1), and DISH (diffuse idiopathic skeletal hyperostosis), a fusion of the spine common in older, obese

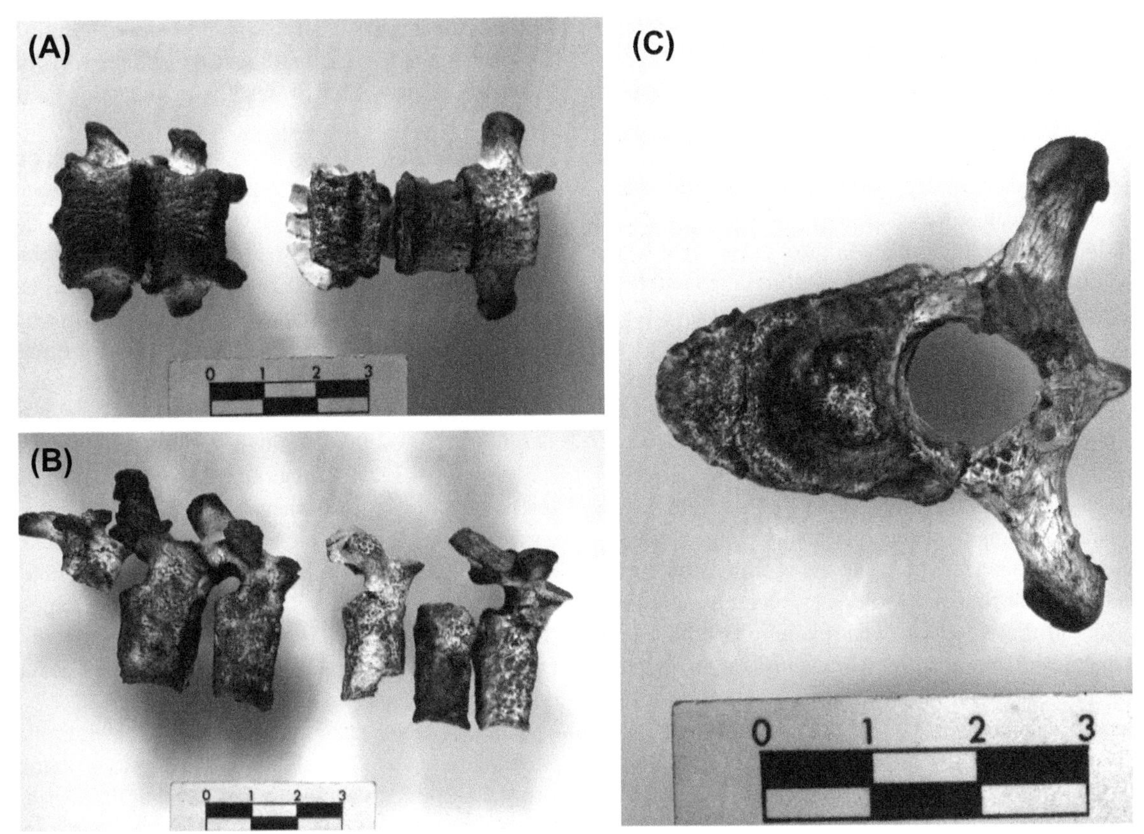

FIGURE 1 Vertebrae of the 3.5 million-year-old "Lucy" australopithecine showing Scheuermann's disease or adolescent kyphosis, associated with lifting heavy weights in modern people. (A) anterior, (B) right side, (C) upper surface of TV6. The light-colored anterior third of the vertebral body is pathological new bone. (See color plate section).

people, has been reported in the Shanidar 1 Neandertal from Iraq and in the Kiik-Koba Neandertal (Trinkaus et al., 2008). The Kanam mandible from Kenya has thick, irregular bone in the inner side of the mandible that has been diagnosed as osteosarcoma, Burkitt's lymphoma, and most recently as a healed fracture of the chin (Phelan et al., 2007). Arthritis in the Olduvai 8 hominid foot from Tanzania has been used to show that it was incorrectly associated with a juvenile *Homo habilis* and is perhaps the foot of *Paranthropus* instead (Weiss, 2012). In this long list of diseased fossils, several conditions—anemia, DISH, brucellosis, and hypervitaminosis A—have been used to infer meat eating.

The survival of ancient persons with disabilities, for example, the Neandertal child from Sima de los Huesos in Spain with premature suture closure and probable mental deficits (Gracia et al., 2009), the Shanidar 1 Neandertal from Iraq with an arm amputated above the elbow early in life, and the 1.7 million-year-old (mya) *H. ergaster* from Dmanisi, Georgia, who had lost all his teeth, has been seen as evidence for provisioning and social support for the disabled (Lordkipanidze et al., 2005). A lively debate about compassion as a fundamental aspect of human nature has surrounded these claims (Hublin, 2009).

Virchow's problem of distinguishing normal from pathological features in fossil humans has been revived in the controversy surrounding the subfossil Flores hominid from Indonesia: Are unusual features of this small-brained person evidence for the population characteristics of a small-brained, late-surviving extinct species, or did this person suffer from microcephaly (Argue et al., 2006; Villa et al., 2012), cretinism (Obendorf et al., 2008), Down syndrome (Henneberg et al., 2014), growth hormone insensitivity (Hershkovitz et al., 2007), or some other disabling condition? In the decade since *Homo floresiensis* was discovered, dozens of papers have been published on each side.

DISEASES IN THE PAST

Much as in medicine, many paleopathologists are somewhat specialized in a particular disease. Only a few infectious diseases leave clear signatures on the human skeleton, and most of these are chronic conditions consistent with long survival. A very large portion of the literature on paleopathology is for this reason devoted to just two groups of infectious diseases: the treponematoses and the mycobacterioses (Aufderheide and Rodrigues Martin, 1998; Ortner, 2011; Grauer, 2012).

The treponematoses are a group of very closely related bacterial diseases that include pinta, yaws, bejel, and syphilis. The first was a skin disease of the American tropics. Yaws and bejel were once widespread in the moist tropics and in temperate regions, respectively, while the syphilis is still a common, worldwide, sexually transmitted disease. All but pinta can cause skin and bone lesions if untreated, particularly affecting the forehead, nose, palate, and lower legs. Syphilis can cause devastating mental illness and heart disease, stillbirth, blindness, and deafness (Powell and Cook, 2005). Yaws can infect African apes, and there are several more distantly related treponematoses that are associated with periodontal disease in humans (Maixner et al., 2014) and with communicable diseases of rabbits, swine, and monkeys. When and how the treponematoses diverged from one another, whether the human treponematoses are really separate strains or species, or just one organism differently expressed under different living conditions, how their epidemiology relates to colonial contact between the Old World and the New, and what evolutionary pressures led to sexual and maternal/fetal transmission in syphilis are hotly contested questions, persisting even as useful genomic data are brought to bear on them. Among many recent discoveries are a few convincing cases of treponematosis from the Old World (Ortner, 2011), as well as evidence for social discrimination against people with severe lesions in the New (Smith et al., 2011).

The mycobacterioses include leprosy, or Hansen's disease, and tuberculosis. Both affect the respiratory system and are spread via close contact and shared housing. Each can progress to cause characteristic destructive lesions on the face and hands in leprosy and of the spine and joints in tuberculosis. Until quite recently, there was general agreement that leprosy and tuberculosis were Old World diseases prior to the voyages of Columbus, and that human tuberculosis—and perhaps leprosy—evolved from bovine tuberculosis quite recently as a consequence of the domestication of cattle. Leprosy, an organism confined to people, was once seen as a much older disease that was gradually replaced by more infectious tuberculosis. These scenarios have been demolished. A major turning point was the discovery that modern human TB strains as well as bovine tuberculosis evolved quite recently from human-infecting organisms that are now common only in Africa (Brosch et al., 2002), but comparable insights from paleopathology have been equally revolutionary.

Good evidence for tuberculosis in the Americas was discovered in the mid-twentieth century, but it was dismissed due to mimicry by other infections among advocates of the bovine model. In the latter half of the twentieth century paleopathologists looked systematically for evidence of the mycobacterioses in remains from ancient hospitals, monasteries, and leprosaria in the Old World, and from Precolumbian sites in the New World (Ortner, 2011). Autopsied cadaver skeletons from several institutions were combed for evidence, leading to the discovery of facial lesions of leprosy and rib lesions of tuberculosis that had gone undetected in clinical medicine. *Mycobacterium* resists treatment and persists in the body to break out again when the victim is immune compromised, in part because it has a dense outer coat of wax-like material that resists degradation. Fortuitously for paleopathologists, this waxy coat protects the bacterial DNA, and it differs in chemical composition among *Mycobacterium* species, providing an independent tool for species identification (Redman et al., 2009). Ancient DNA studies have confirmed that there is abundant Precolumbian evidence for New World tuberculosis (Figure 2)—at least some of it resembling a mycobacterial disease of seals (Bos et al., 2014)—and that evidence for bovine tuberculosis in ancient humans is surprisingly hard to find. Several ancient sites in Europe and Africa have shown long-term coexistence of leprosy and tuberculosis, and a few cases of coinfection have come to light (Donoghue et al., 2005). Thus, paleopathological ancient DNA studies have complicated evolutionary scenarios for this group of diseases.

Ancient DNA techniques have also revolutionized our understanding of several infectious diseases that are not easily diagnosed in ancient bones. Chagas disease, a vector-borne parasitic disease endemic to South America that can

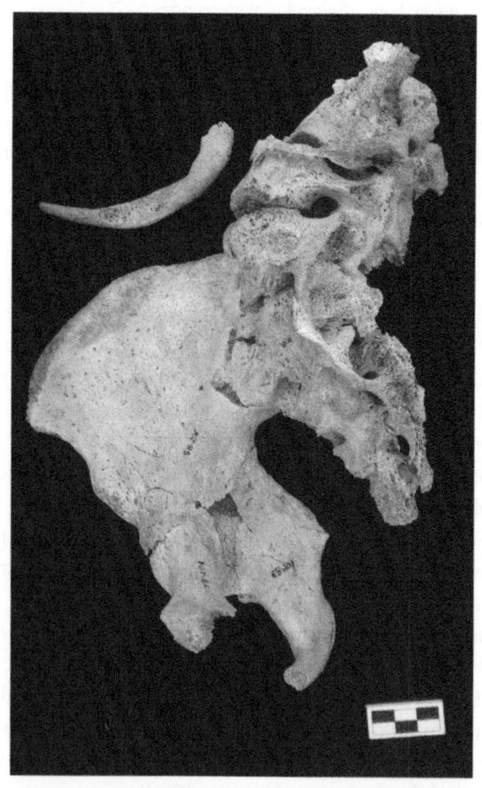

FIGURE 2 Tuberculosis in the spine and hip bone of a young Mississippian woman from Illinois, 1000 AD. (See color plate section).

cause heart disease and megacolon (fecal impaction resulting from failure of the nerves that control peristalsis), can now be studied in mummified and skeletonized remains (Aufderheide et al., 2004; Lima et al., 2008; Reinhard et al., 2003). Even though they are even less visible in bones and mummified tissue, leishmaniasis, plague, and malaria can be identified through ancient DNA techniques, permitting fascinating new discoveries about their evolution and epidemiology. The recent discovery that skeletons recovered Medieval plague pits harbor fragments of the DNA of plague bacteria and that the sixth to eighth century AD Plague of Justinian matches DNA evidence for an increase in diversity of *Yersinia pestis*—the organism that causes plague—demonstrates the growing relevance of paleopathology for disease history (Bos et al., 2012).

Tooth decay is a communicable disease because babies get their oral flora, including *Streptococcus mutans*, from their mothers shortly after birth. We no longer think of dental caries as a serious disease, but a carious tooth can progress to osteomyelitis, systemic infection, and death, and did so not infrequently before the mid-twentieth century (Figure 3). The sugar and starch content and stickiness and abrasiveness of foods influence caries formation, so cooking technologies complicate interpretation. Caries and tooth loss rates have been used as a proxy for carbohydrate content of the diet in ancient groups for whom other data are not available (Cucina et al., 2011), and sex differences have been shown to reflect both dietary differences and reproductive biology (Lukacs and Largaespada, 2006). These studies offer epidemiological insights that complement present-day research (Russell et al., 2013). In contrast, periodontal disease is relatively uncommon in the ancient world. A study from Medieval England (Dewitte and Bekvalac, 2011) reveals the same association with increased mortality that was discovered quite recently in modern people.

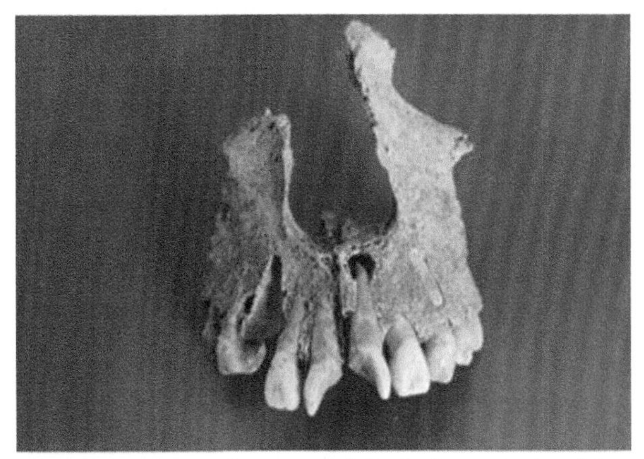

Fragmento de maxilar com padrão de desgaste dentário

FIGURE 3 Decorative notching of the central incisors that has resulted in an abscessed tooth in an enslaved young woman from Cemeterio Pretos Novos, Rio de Janeiro, Brazil. (See color plate section).

Paleoparasitology is a specialty within paleopathology as well as the passion of a small but active community of parasitologists (Faulkner and Reinhard, 2014). When environmental conditions permit, fecal matter is preserved in situ in mummies or separately in dry or cold localities such as caves. In both cases, coprolites yield abundant information about ancient parasites as well as ancient meals. More limited data can be obtained from latrine deposits, and some parasite life stages are extremely resistant to decomposition. Fleas and lice can be recovered from mummified tissue and from clothing (Fornaciari et al., 2009).

It is not surprising that ancient people suffered more parasite infestations than modern people do. However, parasite loads can be related to ancient water and waste disposal systems, to population density, and to various cultural practices. Several parasites, for example *Diphyllobothrium*, a fish tapeworm, give evidence of culinary habits. Others, for example, *Schistosoma*, a parasite whose life cycle involves snails and humans, reflect technologies such as irrigation. This parasite was discovered by Ruffer in kidney tissue from 20th Dynasty Egyptian mummies in 1910. Its eggs have been recovered from French latrine deposits dating to the fifteenth–sixteenth century, outside the present distribution, a finding attributed to the African slave trade (Bouchet et al., 2002). The discovery of a second French *Schistosoma* egg in the ninth century dental calculus (Charlier et al., 2013) could indicate that the ancient distribution of this devastating parasite may have been more extensive than it is today.

Icelandic monks suffered from hydatid cyst (Kristjánsdóttir and Collins, 2011), a zoonotic tapeworm they acquired from the sheep and dogs that accompanied them to Iceland, as did Alaskan Iniut, their dogs and the caribou they hunted (Ortner, 2011). Parasite diversity and distribution thus reflect the patterns of population movement in the spread of humans throughout the world, as well as recent historical contacts. The coevolution of *Taenia* tapeworms and their human and nonhuman hosts have generated interesting scenarios for the origins of omnivory in humans, with domestic cattle and pigs eventually replacing their wild kin in the life cycle of the parasite. The coevolution of head lice, body lice, and pubic lice with their ape and human hosts reflects both the loss of body hair in deep time in Africa and the invention of clothing more recently in several locations (Kittler et al., 2003). Even the lice found in ancient mummies tell a fascinating story. The question of so-called "diseases of civilization" is amenable to evaluation through paleopathology. Cancers of various kinds have been diagnosed in human remains from all time periods, although their relative rarity and age dependence makes for statistical difficulties in determining whether present-day rates are higher or lower than they were in the past. It is clear that claims that cancer did not exist in the past—because ancient humans did not reach sufficiently advanced ages, or because ancient environments were free of cancer-causing pollutants—are misinformed speculations (David and Zimmerman, 2010). Paleopathologists have also found ample evidence of the many degenerative diseases that affect bone, and there is even bone evidence for ancient diabetes, pseudogout, and gout. Medieval monks, mocked since Chaucer's time for their luxurious habits, have been a particularly fruitful subject for research of this kind, and several monastery cemeteries have revealed high rates of DISH (Rogers and Waldron, 2001) and diets surprisingly rich in meat or dairy (Gregoricka and Sheridan, 2013). Gallstones, bladder stones, dental caries, and other miseries are amply documented in many ancient remains.

Atherosclerosis—calcified lipid deposits in arteries—has been studied systematically through detailed radiographic survey of 52 Egyptian mummies in the Horus Study. The frequency of calcification suggests that Egyptian elites suffered from cardiovascular disease at much the same level as that of modern sedentary populations, but the initial focus on the elite status of these mummies was premature. When the researchers added Peruvian, Basketmaker, and Aleutian mummies to the study, all four groups showed similar patterns, even though the three new groups consisted of nonelite farmers and hunter–gatherers (Wann and Thomas, 2014; Thompson et al., 2013). Sample sizes for the last two groups are very small, and more work is needed before lifestyle and diet (rather than age and genetic predisposition) can be demonstrated as important factors in cardiovascular disease.

BIOARCHAEOLOGY

Bioarchaeology is the study of human remains with regard to their archaeological context, coined in this sense by anthropologist Jane Buikstra in 1976. Its focus is broader than paleopathology, as it includes various tools for assessing genetic relationships from the skeleton (Buikstra and Beck, 2006). Bioarchaeology emerged from the "New Archaeology," a movement that stressed ecological and systems approaches to understanding the past. While its roots are Americanist, there are similar efforts in the Old World to provide an anthropologically rich interpretation of human remains from archaeological sites. For paleopathology, important contextual aspects at the individual level include mortuary practices and grave offerings that may reflect social distinctions a person enjoyed in life. Important contextual aspects at the group level include trade, marriage exchange, population distribution, architecture, subsistence, and environmental factors that play important roles in community and individual health. Comparative studies evaluate the health costs and benefits of changes in technology and environment through time. This application of paleopathology focuses on relatively common and easily quantified conditions—survivorship, dental caries and tooth loss, stature, trauma, bone density, bone lesions that reflect anemia, and markers of disturbed bone and enamel formation—that are amenable to statistical evaluation. The landmark 1984 publication *Paleopathology at the Origins of Agriculture* has been recently reissued for a new generation of workers in the field (Cohen and Armelagos, 2013). It demonstrated that paleopathology was a useful tool for quantifying the health consequences of technological change.

More recently there have been two ambitious extensions of this approach. "Biohistory" is largely the province of economic historians, who have adopted stature—from contemporary records or from skeletal remains—as a proxy for health. A more nuanced and interdisciplinary effort is the *Backbone of History Project,* which attempted systematic analysis of health markers across time, technology, and observers. A summary volume for comprehensive, parallel studies of New World populations has appeared (Steckel and Rose, 2002). It is clearly limited by a lack of standardization of data between observers. A more ambitious European project has resulted in many methodological and substantive studies, but a summary volume is still in the planning stages at this time. These large-scale studies have stimulated interesting innovations in method. In the European case a renewed interest in the paleopathology of domestic animals can perhaps be credited to the larger focus on archaeological context.

Recent elaborations on this theme explore correlates of some of the signs and symptoms that are seen in bones. For example, cribra orbitalia and porotic hyperostosis, skeletal changes that expand the bone marrow space and are used by paleopathologists to evaluate anemia, are much better understood as a result of recent work. While iron deficiency anemia related to diet or parasitism remains the most popular explanation, specific nutritional deficiencies such as scurvy (Geber and Murphy, 2012), vitamin B12 and folate deficiency (Walker et al., 2009), associations with infectious disease such at tuberculosis (Blom et al., 2005), or systemic infection (Turner and Armelagos, 2012) have been proposed, and the link between the two indicators (Walker et al., 2009) and the rigor of scoring criteria have been challenged (Wapler et al., 2004).

Mortality related to pregnancy and childbirth has proved remarkably difficult to study in past populations, despite its importance in human evolution. While fetal bones can be recovered when remains are well preserved, burials of pregnant women with the fetus in situ are rarer than one would expect, and there are many cultural practices that may render deaths during pregnancy invisible (Lewis, 2007). Women dying of complications following childbirth are even more difficult to identify, and the published examples are from mummies (Aufderheide, 2003). Detailed study of sex ratios among young adults may provide a way out of this dilemma by focusing on mortality related to the first pregnancy (Pfeiffer et al., 2014).

It is perhaps inevitable that paleopathologists have become interested in trauma in a time of war. There are many fascinating studies of battlefield remains (Fiorato et al., 2000), complemented by contextualized studies of evidence for healed and unhealed injuries (Martin et al., 2012) that aim at teasing out structural state from interpersonal violence, and all of these from sports injuries (Judd, 2006). Some of this literature critiques facile generalizations about human propensity for violence in the popular books of Jared Diamond and Steven Pinker (Fry and Soderberg, 2013). Careful study of human remains yields a complex, nuanced view of social structure and violence, for example, in prehistoric California hunter–gatherers (Andrushko et al., 2010). Ancient civilizations once characterized as peaceful and egalitarian have produced human remains that bear witness to interpersonal violence, for example, at Harappa in the Indus Valley (Schug et al., 2012).

Structural violence is the subject of several recent studies of enslaved Africans and on the urban poor, both from excavation of cemeteries associated with asylums, poorhouses, prisons, and similar institutions and from cadaver collections made in the late nineteenth and early twentieth centuries that consist of the remains of the indigent (Figure 4). The health costs of nineteenth century lead mining are visible in people buried at the Colorado Mental Health Institute. High bone lead levels are consistent with mental retardation and neuropathy resulting in institutionalization (Bower et al., 2007). The bodies of inmates of Kilkenny Union Workhouse show evidence of scurvy and appalling infectious disease mortality during the Irish Potato Famine of 1845–1852 (Geber and Murphy, 2012). Cadaver collections assembled from the dissected bodies of the urban poor tell a surprising story of race differences in trauma and infectious disease experience in nineteenth to early twentieth century America (De La Cova, 2010, 2011). These cadaver collections were assembled for other purposes, primarily for the study of age and sex differences in the skeleton, and had not been previously studied for their historical context. A Portuguese cadaver collection of the same era with occupation data is providing a basis for developing new techniques in paleopathology and testing old ones (Redman et al., 2009; Villotte et al., 2010).

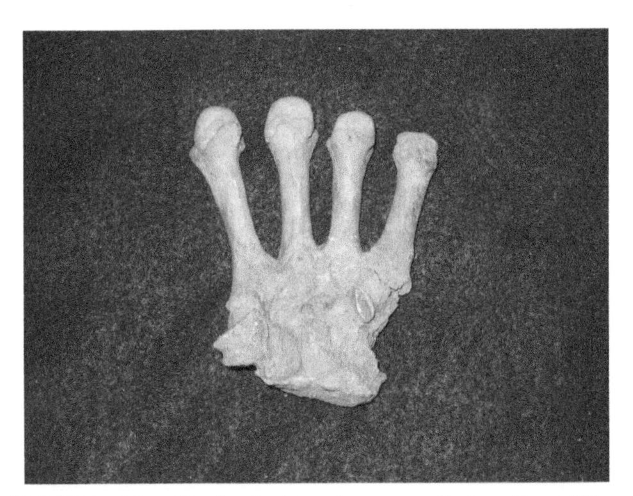

FIGURE 4 Man with a fused wrist, perhaps a complication of immobilization after an injury that would have posed a substantial disability, Milwaukee Poorhouse. (See color plate section). *Photo courtesy of Sean Dougherty.*

OSTEOBIOGRAPHY

Osteobiography is a genre of science writing in which findings on a skeleton or mummy serve as a means of exploring the life experience of an ancient person. The word was coined by physician and anthropologist Frank Saul in describing three Archaic burials from Sheep Rock Shelter in Pennsylvania (1968), and later in ancient Maya from Guatemala. It has come to mean a contextualized case study of ancient remains in which information about chronic disease, habitual activities, body modification, healed trauma, last meals, and circumstances of death is integrated with texts, burial furnishings, and other historical or archaeological information. The related term pathobiography describes a similar genre in the literature.

Anonymous prehistoric people, and more recent persons whose names are lost to history, as well as the ancient rich and famous, are potential subjects for osteobiography. Kennewick Man, a 9000-year-old hunter from Washington State, Ötzi the Ice Man from Chalcolithic Italy, and Juanita the sacrificed 15-year-old Llullaillaco Ice Maiden from the Peruvian Andes are examples of the anonymous group, although they have been named by their osteobiographers. Kennewick Man passed at least two decades of adult life with a projectile point fragment imbedded in his hip with little long-term consequence for mobility or activity (Owsley and Jantz, 2014). Ötzi suffered a lethal injury from a spear and a recently broken incisor, both reflecting interpersonal violence; his catalog of other oral health problems includes periodontal disease (Seiler et al., 2013; Maixner et al., 2014). Children sacrificed to the Inca mountain gods around 1400 AD were recruited from several regions and ate a special diet high in meat and maize in the year prior to their deaths; diet and coca consumption were placed precisely through analysis of hair strands from the scalp outward (Wilson et al., 2007).

The rich and famous are well represented by Tutankhamun and other Egyptian royalty, Richard III of England, and much of the Medici family. King Tut had a club foot and aseptic necrosis of his toes, and like many members of his family he suffered from malaria, but, he did not have Marfan's syndrome or evidence of several other syndromes that his unusual portraits have suggested in the past (Hawass et al., 2010). Even his widely discussed fracture may be excavation damage, and many of the more interesting diagnoses that have been proposed in the past century relate to artistic conventions and are contradicted by evidence from the mummy (Rühli and Ikram, 2014). Richard III, dead in battle in 1485, had a medically unimportant scoliosis that has been misrepresented to posterity as debilitating hunchback (Appleby et al., 2014). He had roundworms, and he drank a great deal of wine during his rule (Lamb et al., 2014). Famous subjects of lesser means include Baroque cellist Luigi Boccherini, who suffered from arthritis and bone deformities, perhaps as a consequence of his many hours of practice (Ciranni and Fornaciari, 2003), and Joseph Merrick, the nineteenth century "Elephant Man" who was afflicted with the medical rarity *Proteus* syndrome (Huntley et al., 2015), among many others.

Osteobiography has controversial relatives in other disciplines. Historians have debated the pitfalls of retrospective diagnosis, and there has been similar discussion of reductionism regarding pathobiography in the literature. Although it is clearly wrong to view a person as just the product of his illness, paleopathology does add to our understanding of the lives of the dead. A recent collection of illuminating case studies stresses illness as an aspect of osteobiography (Stodder and Palkovich, 2011), and concepts from the social science literature on disability have enriched the contextual interpretation of the lives of an adolescent with chondrodystrophic dwarfism from the Upper Paleolithic (Tilley, 2015) and a young man with possible cerebral palsy drowned aboard the Mary Rose in 1545 (Drew, 2015).

INTERDISCIPLINARITY AND DIFFICULTIES IN COMMUNICATION

The past of paleopathology has become a lively topic among its practitioners (Buikstra and Roberts, 2012), just as most historians of medicine have taken a radical turn away from looking at historical sources for evidence of modern diseases as they are understood in modern medicine. In archaeology there has been a similar turn away from what one can learn from old bones and mummies. This is particularly unfortunate, because paleopathologists, radiologists, geneticists, and others are developing new techniques that were unimaginable just a few years ago. In many countries, ethical objections of the long-term curation of human remains are resulting in repatriation and destruction of the collections on which research in paleopathology depends. On the brighter side, several databank projects have been initiated, for example, the longstanding Manchester Museum Mummy Project, that promise to conserve resources for the future.

Because paleopathology is interdisciplinary, finding all the interesting literature can be difficult. Much paleopathology has appeared incidentally in appendices of archaeological reports and in systematic publications on fossils. Articles that focus on paleopathology can be found occasionally in the journals of both these fields, as well as in general science journals. They are a bit more common in physical anthropology journals. A search for the keyword paleopathology in PubMed (accessed April 13, 2015) yielded 1902 articles in medical periodicals, but such searches miss many articles. Finding literature and colleagues has become easier through the efforts of three organizations: the Paleopathology Association, the Paleopathology Club of the International Academy of Pathology, and the International Council for Archaeozoology (ICAZ) Animal Palaeopathology Working Group. Two new journals promise better communication among specialists: *International Journal of Paleopathology* and *Yearbook of Mummy Studies.*

QUESTIONS FOR THE FUTURE

Will paleopathologists find new ways to characterize extinct ancient pathogens that are independent from modern diagnostic models?

Will a comprehensive paleoepidemiology emerge from the synthesis of ancient DNA studies and diagnosis of lesions in bone and mummified tissue?

Can we quantify trauma in a way that permits better testing of hypotheses about human propensity for violence?

Can public opinion be influenced to stop the wholesale destruction of skeletal collections and the failure to curate newly discovered ancient human remains?

REFERENCES

Appleby, J., Mitchell, P.D., Robinson, C., Brough, A., Rutty, G., Harris, R.A., Thompson, D., Morgan, B., 2014. The scoliosis of Richard III, last Plantagenet King of England: diagnosis and clinical significance. Lancet 383, 1944.

Andrushko, V.A., Schwitalla, A.W., Walker, P.L., 2010. Trophy-taking and dismemberment as warfare strategies in prehistoric central California. American Journal of Physical Anthropology 141, 83–96.

Argue, D., Donlon, D., Groves, C., Wright, R., 2006. *Homo floresiensis*: microcephalic, pygmoid, *Australopithecus*, or *Homo*? Journal of Human Evolution 51, 360–374.

Aufderheide, A.C., 2003. The Scientific Study of Mummies. Cambridge University Press, Cambridge.

Aufderheide, A.C., Rodriguez Martin, C., 1998. The Cambridge Encyclopedia of Human Paleopathology. Cambridge University Press, Cambridge.

Aufderheide, A.C., Salo, W., Madden, M., Streitz, J., Buikstra, J., Guhl, F., Arriaza, B., Renier, C., Wittmers Jr., L.E., Fornaciari, G., Allison, M., 2004. A 9,000-year record of Chagas' disease. Proceedings of the National Academy of Sciences of the United States of America 101, 2034–2039.

Berger, L.R., 2006. Brief communication: predatory bird damage to the Taung type-skull of *Australopithecus africanus* Dart 1925. American Journal of Physical Anthropology 131, 166–168.

Blom, D.E., Buikstra, J.E., Keng, L., Tomczak, P.D., Shoreman, E., Stevens-Tuttle, D., 2005. Anemia and childhood mortality: latitudinal patterning along the coast of pre-Columbian Peru. American Journal of Physical Anthropology 127, 152–169.

Boaz, N.T., Ciochon, R.L., Xu, Q., Liu, J., 2004. Mapping and taphonomic analysis of the *Homo erectus* loci at Locality 1 Zhoukoudian, China. Journal of Human Evolution 46, 519–549.

Bos, K.I., Harkins, K.M., Herbig, A., Coscolla, M., Weber, N., Comas, I., Forrest, S.A., Bryant, J.M., Harris, S.R. Schuenemann, V.J., Campbell, T.J., Majander, K., Wilbur, A.K., Guichon, R.A., Wolfe Steadman, D.L., Cook, D.C., Niemann, S., Behr, M.A., Zumarraga, M., Bastida, R., Huson, D., Nieselt, K., Young, D., Parkhill, J., Buikstra, J.E., Gagneux, S., Stone, A.C., Krause, J., 2014. Pre-Columbian mycobacterial genomes reveal seals as a source of New World human tuberculosis. Nature 514 (7523), 494–497.

Bos, K., Stevens, P., Nieselt, K., Poinar, H.N., DeWitte, S.N., Kraus, J., 2012. *Yersinia pestis*: new evidence for an old infection. PLoS ONE 7 (11), e49803.

Bouchet, F., Harter, S., Paicheler, J.C., Aráujo, A., Ferreira, L.F., 2002. First recovery of *Schistosoma mansoni* eggs from a latrine in Europe (15–16th centuries). Journal of Parasitology 88, 404–405.

Bower, N.W., McCants, S.A., Custodio, J.M., Ketterer, M.E., Getty, S.R., Hoffman, J.M., 2007. Human lead exposure in a late 19th century mental asylum population. Science of the Total Environment 372, 463–473.

Brosch, R., Gordon, S.V., Marmiesse, M., Brodin, P., Buchrieser, C., Eiglmeier, K., Garnier, T., Gutierrez, C., Hewinson, G., Kremer, K., Parsons, L.M., Pym, A.S., Samper, S., van Soolingen, D., Cole, S.T., 2002. A new evolutionary scenario for the *Mycobacterium tuberculosis* complex. Proceedings of the National Academy of Sciences of the United States of America 99 (6), 3684–3689.

Buikstra, J.E., Beck, L.A. (Eds.), 2006. Bioarchaeology: The Contextual Study of Human Remains. Elsevier, Amsterdam.

Buikstra, J.E., Roberts, C.A. (Eds.), 2012. Global History of Paleopathology: Pioneers and Prospects. Oxford University Press, New York.

Bynum, C.W., 1995. The Resurrection of the Body in Western Christianity, 200-1336. Columbia University Press, New York.

Charlier, P., Abadie, I., Cavard, S., Brun, L., 2013. Ancient calculus egg. British Dental Journal 215, 489–490.

Churchill, S.E., Franciscus, R.G., McKean-Peraza, H.A., Daniel, J.A., Warren, B.R., 2009. Shanidar 3 Neandertal rib puncture wound and palaeolithic weaponry. Journal of Human Evolution 57, 163–178.

Ciranni, R., Fornaciari, G., 2003. Luigi Boccherini and the Barocco cello: an 18th century striking case of occupational disease. International Journal of Osteoarchaeology 13, 294–302.

Cohen, M.N., Armelagos, G.J. (Eds.), 2013. Paleopathology at the Origins of Agriculture, second ed. University Press of Florida, Gainesville.

Cook, D.C., Buikstra, J.E., DeRousseau, C.J., Johanson, D.C., 1983. Vertebral pathology in the Afar australopithecines. American Journal of Physical Anthropology 60, 83–101.

Cucina, A., Cantillo, C.P., Sosa, T.S., Tiesler, V., 2011. Carious lesions and maize consumption among the Prehispanic Maya: an analysis of a coastal community in northern Yucatan. American Journal of Physical Anthropology 145, 560–567.

D'Anastasio, R., Zipfel, B., Moggi-Cecchi, J., Stanyon, R., Capasso, L., 2009. Possible brucellosis in an early hominin skeleton from Sterkfontein, South Africa. PLoS One 4 (7), e6439.

Darwin, C., 1874. The Descent of Man and Selection in Relation to Sex, second ed. A.L. Burt Company, New York.

David, A.R., Zimmerman, M.R., 2010. Cancer: an old disease, a new disease or something in between? Nature Reviews: Cancer 10, 728–733.

De La Cova, C., 2010. Cultural patterns of trauma among 19th-century-born males in cadaver collections. American Anthropologist 112, 589–606.

De La Cova, C., 2011. Race, health, and disease in 19th-century-born males. American Journal of Physical Anthropology 144, 526–537.

Dewitte, S.N., Bekvalac, J., 2011. The association between periodontal disease and periosteal lesions in the St. Mary Graces cemetery, London, England A.D. 1350–1538. American Journal of Physical Anthropology 146, 609–618.

Domínguez-Rodrigo, M., Pickering, T.R., Diez-Martín, F., Mabulla, A., Musiba, C., Trancho, G., Baquedano, E., Bunn, H.T., Barboni, D., Santonja, M., Uribelarrea, D., Ashley, G.M., Martínez-Ávila, M.S., Barba, R., Gidna, A., Yravedra, J., Arriaza, C., 2012. Earliest porotic hyperostosis on a 1.5-million-year-old hominin, Olduvai Gorge, Tanzania. PLoS One 7, e46414.

Donoghue, H.D., Marcsik, A., Matheson, C., Vernon, K., Nuorala, E., Molto, J.E., Greenblatt, C.L., Spigelman, M., 2005. Co-infection of *Mycobacterium tuberculosis* and *Mycobacterium leprae* in human archaeological samples: a possible explanation for the historical decline of leprosy. Proceedings Biological Sciences (The Royal Society) B 272 (1561), 389–394.

Drew, R., 2015. Lives of deprivation of lives of industry: possible cerebral palsy on the Mary Rose. The Medieval History Journal 18, 1–21.

Faulkner, C.T., Reinhard, K.J., 2014. A retrospective examination of paleoparasitology and its establishment in the Journal of Parasitology. Journal of Parasitology 100, 253–259.

Fiorato, V., Boylston, A., Knüsel, C., 2000. Blood Red Roses: The Archaeology of a Mass Grave from the Battle of Towton AD 1461. Oxbow Books, Oxford.

Fornaciari, G., Giuffra, V., Marinozzi, S., Picchi, M.S., Masetti, M., 2009. "Royal" pediculosis in renaissance Italy: lice in the mummy of the King of Naples Ferdinand II of Aragon (1467–1496). Memorias do Instituto Oswaldo Cruz 104 (4), 671–672.

Fry, D.P., Soderberg, P., 2013. Lethal aggression in mobile forager bands and implications for the origins of war. Science 341, 270–273.

Geber, J., Murphy, E., 2012. Scurvy in the great Irish famine: evidence of vitamin C deficiency from a mid-19th century skeletal population. American Journal of Physical Anthropology 148, 512–524.

Gracia, A., Arsuaga, J.L., Martínez, I., Lorenzo, C., Carretero, J.M., Bermúdez de Castro, J.M., Carbonell, E., 2009. Craniosynostosis in the Middle Pleistocene human Cranium 14 from the Sima de los Huesos, Atapuerca, Spain. Proceedings of the National Academy of Sciences of the United States of America 106, 6573–6578.

Grauer, A.L. (Ed.), 2012. A Companion to Paleopathology. Wiley-Blackwell, Chichester.

Gregoricka, L.A., Sheridan, S.G., 2013. Ascetic or affluent? Byzantine diet at the monastic community of St. Stephen's, Jerusalem from stable carbon and nitrogen isotopes. Journal of Anthropological Archaeology 32, 63–73.

Hart, D., Sussman, R.W., 2005. Man the Hunted: Primates, Predators, and Human Evolution. Westview Press, Cambridge, MA.

Hawass, Z., Gad, Y.Z., Ismail, S., Khairat, R., Fathalla, D., Hasan, N., Ahmed, A., Elleithy, H., Ball, M., Gaballah, F., Wasef, S., Fateen, M., Amer, H., Gostner, P., Selim, A., Zink, A., Pusch, C.M., 2010. Ancestry and pathology in King Tutankhamun's family. Journal of the American Medical Association 303, 638–647.

Henneberg, M., Eckhardt, R.B., Chavanaves, S., Hsü, K.J., 2014. Evolved developmental homeostasis disturbed in LB1 from Flores, Indonesia, denotes Down syndrome and not diagnostic traits of the invalid species Homo floresiensis. Proceedings of the National Academy of Sciences United States of America 111, 11967–11972.

Hershkovitz, I., Kornreich, L., Laron, Z., 2007. Comparative skeletal features between *Homo floresiensis* and patients with primary growth hormone insensitivity (Laron Syndrome). American Journal of Physical Anthropology 134, 198–208.

Hublin, J.J., 2009. The prehistory of compassion. Proceedings of the National Academy of Sciences United States of America 106, 6429–6430.

Huntley, C., Hodder, A., Ramachandran, M., 2015. Clinical and historical aspects of the Elephant Man: exploring the facts and the myths. Gene 555, 63–65.

Judd, M.A., 2006. Continuity of interpersonal violence between Nubian communities. American Journal of Physical Anthropology 131, 324–333.

Kittler, R., Kayser, M., Stoneking, M., 2003. Molecular evolution of *Pediculus humanus* and the origin of clothing. Current Biology 13, 1414–1417.

Kristjánsdóttir, S., Collins, C., 2011. Cases of hydatid disease in medieval Iceland. International Journal of Osteoarchaeology 21, 479–486.

Lamb, A.L., Evans, J.E., Buckley, R., Appleby, J., 2014. Multi-isotope analysis demonstrates significant lifestyle changes in King Richard III. Journal of Archaeological Science 50, 559–565.

Lewis, M.E., 2007. The Bioarchaeology of Children. Cambridge University Press, Cambridge.

Lima, V.S., Iniguez, A.M., Otsuki, K., Fernando Ferreira, L., Araújo, A., Vicente, A.C., Jansen, A.M., 2008. Chagas disease in ancient hunter-gatherer population, Brazil. Emerging Infectious Diseases 14, 1001–1002.

Lordkipanidze, D., Vekua, A., Ferring, R., Rightmire, G.P., Agusti, J., Kiladze, G., Mouskhelishvili, A., Nioradze, M., Ponce de Leon, M.S., Tappen, M., Zollikofer, C.P., 2005. The earliest toothless hominin skull. Nature 434 (7034), 717–718.

Lukacs, J.R., Largaespada, L.L., 2006. Explaining sex differences in dental caries prevalence: saliva, hormones, and "life-history" etiologies. American Journal of Human Biology 18, 540–555.

Maixner, F., Thomma, A., Cipollini, G., Widder, S., Rattei, T., Zink, A., 2014. Metagenomic analysis reveals presence of *Treponema denticola* in a tissue biopsy of the Iceman. PLoS One 18; 9 (6), e99994.

Martin, D.L., Harrod, R.P., Pérez, V.R. (Eds.), 2012. The Bioarchaeology of Violence. University Press of Florida, Gainesville.

Moodie, R.L., 1923. Paleopathology: An Introduction to the Study of Ancient Evidences of Disease. University of Illinois Press, Urbana.

Njau, J.K., Blumenschine, R.J., 2012. Crocodylian and mammalian carnivore feeding traces on hominin fossils from FLK 22 and FLK NN 3, late Pliocene, olduvai gorge, Tanzania. Journal of Human Evolution 63, 408–417.

Obendorf, P.J., Oxnard, C.E., Kefford, B.J., 2008. Are the small human-like fossils found on Flores human endemic cretins? Proceedings: Biological Sciences: the Royal Society 275 (1640), 1287–1296.

Ortner, D.J., 2011. What skeletons tell us. The story of human paleopathology. Virchows Archiv 459, 247–254.

Owsley, D., Jantz, R. (Eds.), 2014. Kennewick Man: The Scientific Investigation of an Ancient American Skeleton. Texas A&M University Press, College Station.

Pfeiffer, S., Doyle, L.E., Kurki, H.K., Harrington, L., Ginter, J.K., Merritt, C.E., 2014. Discernment of mortality risk associated with childbirth in archaeologically derived forager skeletons. International Journal of Paleopathology 7, 15–24.

Phelan, J., Weiner, M.J., Ricci, J.L., Plummer, T., Gauld, S., Potts, R., Bromage, T.G., 2007. Diagnosis of the pathology of the Kanam mandible. Oral Surgery, Oral Medicine, Oral Pathology, Oral Radiology, and Endodontology 103, e20.

Powell, M.L., Cook, D.C. (Eds.), 2005. The Myth of Syphilis: A Natural History of North American Treponematosis. University Press of Florida, Gainesville.

Redman, J.E., Shaw, M.J., Mallet, A.I., Santos, A.L., Roberts, C.A., Gernaey, A.M., Minnikin, D.E., 2009. Mycocerosic acid biomarkers for the diagnosis of tuberculosis in the Coimbra Skeletal Collection. Tuberculosis (Edinburgh) 89, 267–277.

Reinhard, K., Fink, T.M., Skiles, J., 2003. A case of megacolon in Rio Grande valley as a possible case of Chagas disease. Memorias do Instituto Oswaldo Cruz 98 (Suppl. 1), 165–172.

Rogers, J., Waldron, T., 2001. DISH and the monastic way of life. International Journal of Osteoarchaeology 11, 357–365.

Rothschild, B.M., Hershkovitz, I., Rothschild, C., 1995. Origin of yaws in the Pleistocene. Nature 378 (6555), 343–344.

Rühli, F.J., Ikram, S., February 2014. Purported medical diagnoses of Pharaoh Tutankhamun, c. 1325 BC-. Homo 65 (1), 51–63.

Russell, S.L., Gordon, S., Lukacs, J.R., Kaste, L.M., 2013. Sex/gender differences in tooth loss and edentulism: historical perspectives, biological factors, and sociologic reasons. Dental Clinics of North America 57, 317–337.

Saul, F.P., 1968. The human skeletal remains of the Sheep Rock Shelter: an osteobiographic analysis. Occasional Papers in Anthropology, Pennsylvania State University 5, 199–220.

Schug, G.R., Gray, K., Mushrif-Tripathy, V., Sankhyan, A.R., 2012. A peaceful realm? Trauma and social differentiation at Harappa. International Journal of Paleopathology 2, 136–147.

Seiler, R., Spielman, A., Zink, A., Rühli, F., 2013. Oral pathologies of the Neolithic Iceman, c.3,300 BC. European Journal of Oral Sciences 121 (3 Pt 1), 137–141.

Smith, M.O., Betsinger, T.K., Williams, L.L., 2011. Differential visibility of treponemal disease in pre-Columbian stratified societies: does rank matter? American Journal of Physical Anthropology 144, 185–195.

Stodder, A.L., Palkovich, A.M. (Eds.), 2011. The Bioarchaeology of Individuals. University Press of Florida, Gainesville.

Steckel, R.H., Rose, J.C. (Eds.), 2002. The Backbone of History: Health and Nutrition in the Western Hemisphere. Cambridge University Press, Cambridge.

Thompson, R.C., Allam, A.H., Lombardi, G.P., Wann, L.S., Sutherland, M.L., Sutherland, J.D., Soliman, M.A., Frohlich, B., Mininberg, D.T., Monge, J.M., Vallodolid, C.M., Cox, S.L., Abd el-Maksoud, G., Badr, I., Miyamoto, M.I., el-Halim Nur el-Din, A., Narula, J., Finch, C.E., Thomas, G.S., 2013. Atherosclerosis across 4000 years of human history: the Horus study of four ancient populations. Lancet 381, 1211–1222.

Tilley, L., 2015. Accommodating difference in the prehistoric past: revisiting the case of Romito 2 from a bioarchaeology of care perspective. International Journal of Paleopathology 03 (8), 64–74.

Tillier, A.M., Arensburg, B., Duday, H., Vandermeersch, B., 2001. Brief communication: an early case of hydrocephalus: the Middle Paleolithic Qafzeh 12 child (Israel). American Journal of Physical Anthropology 114, 166–170.

Trinkaus, E., Maley, B., Buzhilova, A.P., 2008. Brief communication: paleopathology of the Kiik-Koba 1 Neandertal. American Journal of Physical Anthropology 137, 106–112.

Turner, B.L., Armelagos, G.J., 2012. Diet, residential origin, and pathology at Machu Picchu, Peru. American Journal of Physical Anthropology 149, 71–83.

Villa, C., Persson, L., Alexandersen, V., Lynnerup, N., 2012. A small skull from Flores dated to the 20th century. HOMO—Journal of Comparative Human Biology 63, 12–20.

Villotte, S., Castex, D., Couallier, V., Dutour, O., Knüsel, C.J., Henry-Gambier, D., 2010. Enthesopathies as occupational stress markers: evidence from the upper limb. American Journal of Physical Anthropology 142, 224–234.

Walker, A., Zimmerman, M.R., Leakey, R.E., 1982. Possible case of hypervitaminosis A in *Homo erectus*. Nature 296 (5854), 248–250.

Walker, P.L., Bathurst, R.R., Richman, R., Gjerdrum, T., Andrushko, V.A., 2009. The causes of porotic hyperostosis and cribra orbitalia: a reappraisal of the iron-deficiency-anemia hypothesis. American Journal of Physical Anthropology 139, 109–125.

Wann, S., Thomas, G.S., 2014. What can ancient mummies teach us about atherosclerosis? Trends in Cardiovascular Medicine 24, 279–284.

Wapler, U., Crubézy, E., Schultz, M., 2004. Is cribra orbitalia synonymous with anemia? Analysis and interpretation of cranial pathology in Sudan. American Journal of Physical Anthropology 123, 333–339.

Weiss, E., 2012. Olduvai Hominin 8 foot pathology: a comparative study attempting a differential diagnosis. Homo 63, 1–11.

White, T.D., 1986. Cutmarks on the Bodo cranium: a case of prehistoric defleshing. American Journal of Physical Anthropology 69, 503–509.

Wilson, A.S., Taylor, T., Ceruti, M.C., Chavez, J.A., Reinhard, J., Grimes, V., Meier-Augenstein, W., Cartmell, L., Stern, B., Richards, M.P., Worobey, M., Barnes, I., Gilbert, M.T., 2007. Stable isotope and DNA evidence for ritual sequences in Inca child sacrifice. Proceedings of the National Academy of Sciences of the United States of America 104, 16456–16461.

Wu, X.J., Schepartz, L.A., Liu, W., Trinkaus, E., 2011. Antemortem trauma and survival in the late Middle Pleistocene human cranium from Maba, South China. Proceedings of the National Academy of Sciences United States of America 108 (49), 19558–19562.

Part VII

Behavior and Culture

Chapter 32

Evolutionary Biology of Human Stress

Carol M. Worthman
Department of Anthropology, Emory University, Atlanta, GA, USA

SYNOPSIS

"Stress" is a modern word for an ancient phenomenon, related to evolved adaptations that detect threats or damage to well-being and to mobilize responses to them. As such, it lies at the heart of adaptation at the phenotypic level, so that the study of stress has become a major venue for tracking not only the sources of human health disparities, but also adaptive processes and life history evolution. This is especially true for humans, whose evolution has been shaped by sociocultural specializations integrally related to cognition and behavior along with their corresponding biology. Therefore, stress responses are being intensely studied, and are increasingly well characterized. The multidisciplinary field of stress research has elucidated molecular, genetic, systemic, organismic, cognitive, behavioral, developmental, life course, and intergenerational processes involved in meeting the continuous daily challenges of living by detecting and responding to demands and threats. This exciting, rapidly moving field is revealing how cognition and behavior are modulated by neuroendocrine-driven adaptive mechanisms that integrate genetic, epigenetic, and experience-derived cues. Such insights have shed light on old questions or revised existing views of concepts like homeostasis and phenotypic adaptation, biological inheritance, development, and pathogenesis. Studies of stress will continue to shape our understanding of human adaptive design as well. They have enriched our knowledge of the human condition. Most profoundly, stress research has revealed the power of social factors, from personal to structural, in human functioning and well-being. This work, more than any other, demonstrates our common stake in issues of social structure, culture, and allocation of resources at the roots of human welfare and health disparities.

"Stress" is a modern word for an ancient phenomenon, as old perhaps as life itself. Throughout evolution, organisms have faced daily challenges to survival and welfare, and have evolved myriad ways adaptively to meet their needs under diverse and often uncertain conditions. Most basic of these needs are to secure resources and avoid harm: stress concerns threats or damage to the interests and well-being of an organism and its efforts to detect, avoid, or manage them. As such, it lies at the heart of adaptation at the phenotypic level, so that the study of stress has become a major venue for tracking not only the sources of human health disparities, but also adaptive processes and life history evolution. This is especially true for humans, whose evolution has been shaped by sociocultural specializations integrally related to cognition and behavior along with their corresponding biology. As you will see here, the phenomena are complex and their study yields novel insights into the human condition.

Stress adaptations aim to solve the fundamental problem of how to adjust internal conditions and action to meet external conditions and demands. The highly complex stress system operates both continually at basal conditions and mobilizes physiological and behavioral responses to challenge. Developmental sensitivity to stressors encountered from gestation onward provides another layer of adaptability to this highly responsive system.

EVOLVING CONCEPTS OF STRESS AND ADAPTATION

Stress springs from the relationship between an organism and its environment, particularly the fit of its capacities and needs with existing demands and resources. Definitions of stress have evolved as empirical research has traced its causes, pathways, and effects (reviewed in Goldstein, 2010). The concept arose as medical physiology began to regard bodily functions dynamically, in terms of negotiating the ongoing need to both keep the organism alive and accommodate external perturbations and demands. In the nineteenth century, Bernard's insight that maintenance of an internal environment (*milieu intérieur*) (Bernard, 1865) was a crucial feature in organismic design paved the way for Cannon's twentieth century formulation of homeostasis as the regulatory means the body uses to maintain internal conditions within tolerable ranges (Cannon, 1929). Cannon concentrated first on feedback mechanisms for temperature or glucose regulation, and then included responses to psychosocial challenges represented in "fight or flight" responses. This work identified a neuroendocrine axis comprising the sympathetic nervous system and the adrenal medulla as the bases for a rapid, generalized response to meet homeostatic challenges.

The word "stress" itself was popularized by Selye's mid-twentieth century definition as a *response to demand*, specifically a suite of nonspecific responses by the body to any present or perceived challenge (Selye, 1973). His model emphasized endocrine orchestration of the response as mediated by cortisol. Stress responses also are costly to the organism; therefore, the more and more frequent the demands to be met, the greater the cumulative burden of meeting them, and the greater the risk of impairment. A trade-off is established in which stress responses apparently are beneficial for well-being at low, intermittent levels of demand but potentially harmful at high, chronic levels. Yet intense chronic stressors may be those most threatening to well-being: stress responses enable the organism to survive harsh conditions at the expense of long-term survival. Chronic stress leads to cumulative risk as the costs from such trade-offs mount up. Incomplete understandings of stress have fostered ambiguous use of terminology, leading to the present scientific distinction among *stress* as a demand–response condition between organism and environment; *stressor* as the demand on the organism, whether perceived or actual; and *stress response* as the changes in the organism triggered by demand, in biological, cognitive, and behavioral terms. These terms fail to capture the dimensions of mismatch and burden, which are expressed by allostasis and allostatic load. *Allostasis* comprises the suite of dynamic adjustments required to maintain well-being under varying internal and external conditions. *Allostatic load* refers to cumulative wear and tear from allostasis, and refers to the costs of phenotypic adaptation. Thus, allostatic load forms a basis for linking stress to both adaptation and health (McEwen and Wingfield, 2003). Indeed, in both popular and scientific literature, "stress" commonly refers to burden now defined as allostatic load. In this chapter also, unless otherwise stated, stress will be used in the sense of allostatic load, but with a focus on psychosocial demands.

Life History

Conceptualizations of stress are grounded in attempts to understand fundamental adaptive strategies that organisms have evolved to meet threats to well-being, in terms of survival and reproduction. Indeed, physiologic and cognitive–behavioral adaptations to stress exhibit highly conserved elements. But specific needs or critical resources, and thus the potential adaptive challenges, vary among species. Life history theory aims to explain species-specific suites of characteristics such as the timing of puberty and reproduction or adult body size and longevity that make up its distinctive life course (Charnov, 1993; Stearns, 2000). These characteristics represent a series of evolved solutions to the basic challenges of growing up, surviving, and reproducing. Limited resources, energy, and time must be allocated among the key domains of growth, reproduction, and maintenance (the costs of staying alive), setting up trade-offs for optimizing the balance of costs and benefits for assigning such limited resources to one domain versus another under ambient selection pressures driving mortality and fecundity. For example, if mortality risk is low and early death unlikely, investment in slow maturation and later reproduction are favored when larger size and more experience improve reproductive outcomes.

Each species has a distinctive life history strategy that comprises a set of evolved mechanisms or algorithms for handling allocation trade-offs and is adapted to the environments it can expect to face in a lifetime (Brommer, 2000). Thus, the evolved designs for implementing a life history strategy set priorities on resource demands and allocate how they are met across the range of expectable operating conditions. Features of human life history include singleton births, slow development with intense early, prolonged dependency, late maturation and initiation of reproduction, large adult body size, and long life expectancy, along with a corresponding set of mechanisms for resource allocation that mediate trade-offs and the range of phenotypic plasticity (reaction norms) (Hill and Kaplan, 1999).

Such features establish the critical conditions and resources necessary for successfully enacting this life history, and set the range of tolerance or phenotypic plasticity under different environmental conditions. Indeed, the tempo and phenotypic outcomes of human life history exhibit considerable plasticity: rates of growth and timing of maturation vary markedly in relation to conditions of nutrition and health. A prominent example of such plasticity in the timing of maturation is reflected in the large reaction norm for population median age at menarche, documented from as low as 12 to as high as 18 (Eveleth and Tanner, 1990). In life history terms, the conditions of maternal health, juvenile morbidity rates, and nutrition determine productivity (net energy availability) and the cost of maintenance. These, in turn, set the energy available for growth or reproduction to drive growth rates and maturational timing. Ambient conditions also influence age at maturity indirectly by signaling reliability of resource availability and level of mortality risk. As we shall see, stress response systems appear to play a central role in linking risk/threat detection with graded responses that mediate life history trade-offs.

Endocrine Architecture of Life History

If a life history strategy comprises an adaptive set of evolved mechanisms for getting a life and getting through it with optimum relative reproductive success, then what are these mechanisms? A vast literature indicates that the primary proximate means for implementing life history are physiologic mechanisms operating via the neuroendocrine system (Finch and Rose, 1995). Neuroendocrine actions both determine net energy availability (productivity) by regulating uptake and maintenance costs and mediate resource partitioning among growth, reproduction, and maintenance (Worthman, 2003). They negotiate resource allocation by networks of axial cross talk, or the competing and combined actions of centralized (e.g., somatotrophic or hypothalamo–pituitary–gonadal (HPG) axes (also referred to as the hypothalamic–pituitary–gonadal axis)) and distributive (e.g., immune, insulin) regulatory systems. Moreover, they manage short- and long-term resource allocation; for example, energy use in aging is downregulated by the hypothalamo–pituitary–thyroid axis (Worthman, 1999). Likewise, neuroendocrine actions schedule key life history events by acting as pacemakers for growth, developmental transitions, reproductive effort, and aging; the onset of puberty is triggered by the HPG axis, for instance. Most pertinent to this chapter, neuroendocrine systems operate at the interface between individual and environment, organizing and implementing internal (physiologic, cognitive) and external (behavioral) responses to ambient demands such as psychosocial stress, thermal load, or workload. Note also that facultative adjustment of life history parameters, such as variation in age at menarche, is mediated by the pacemaking actions of neuroendocrine mechanisms.

Each of the foregoing points furthermore supports the view that the neuroendocrine architecture for carrying out the life history strategy also constitutes the architecture of resource allocation. Indeed, the elements of this architecture represent the evolved adaptive mechanisms that effect life history, implement trade-offs, and generate plasticity. This observation relates directly to the evolutionary biology of the stress process, but first a review of that process is in order.

HOW STRESS WORKS

As a response that arises when demands must be met at some cost to the organism, stress might be seen as inherent to being alive. Indeed, the systems that mediate stress are intimately involved in juggling everyday demands and have deep phylogenetic roots. Stress responses are crucial to adaptation but become problematic when the costs measurably impair the ability to survive or pursue life goals (growth, reproduction, integrity of other systems). Therefore, a high premium rests on the ability not only to detect a real or potential threat but also to gauge whether and how much to respond. The steps involved in the stress process are as follows: stressor → detection–assessment → response → feedback.

Stressors

Stressors can be virtually anything that presents a challenge to the organism, about which it must do something either to keep things as they are (stay alive) or to pursue a goal (grow up, reproduce). The availability and quality of vital resources are major potential stressors, as are challenges to physical integrity. Vital resources include food, water, and space. Among the classic physical stressors are workload, illness, malnutrition, and climate. Signals that these resources are threatened are potent stressors that trigger neural alarm systems. We focus here on psychosocial stressors, because access to vital resources in humans is mediated wholly through the social world they inhabit. Resources such as relationships and information are crucial for securing vital resources and avoiding or mitigating harm, and are particularly important among social species such as humans. Intriguingly, many of the same neural alarm regions that respond to threats to physical well-being also are triggered by threats to social well-being (Eisenberger and Cole, 2012). The adaptive niche that humans inhabit involves an unusually expansive set of vital needs that include meaning, purpose, and social capital (e.g., networks, reputation, respect). Humans engage in extensive sharing and reciprocity, habitually giving or taking benefit with expectation of return but without specifying type or time of repayment. Distinctive cognitive capacities—memory, language, simulation and projection—support such behaviors. The dramatic cortical expansion during human evolution was driven in part by the exponentially expanded computational load required to cultivate relationships, recall personal histories, and keep track of exchanges among all members of even a small social group. Moreover, humans possess interrelated, possibly unique capacities for mental simulation to remember the past, imagine the future, and project the thoughts and feeling of others (theory of mind, empathy). Such simulations occupy much of conscious experience and are central to social cognition, decision making, and production of meaning (Gilbert, 2006).

Hence, although humans are wholly dependent on sociality, the cognitive–behavioral demands of life in social groups provide myriad grounds for stress. The cognitive adaptations that support intense sociality in humans—language, memory, simulation—create new types of stressors, both present and imagined. A verbal twist, perceived slight, or even an unmet expectation can catalyze stress. Capacities for simulation open wide possibilities for stressors that are remote, even imaginary, arising from recall (e.g., recrimination, regret, rumination), projection or empathy (toward, e.g., people, creatures, inanimate objects), and even concepts (e.g., honor, environmentalism). These capacities can generate stressors such as shared loss or pain, envy or fear of envy, and anticipation (future contingencies, possibilities, consequences) (Sapolsky, 1998).

For all these reasons, social threat, along with unpredictability and lack of control, is the stressor that most reliably triggers an acute stress response (Dickerson and Kemeny, 2004). Moreover, people who experience social marginalization, loneliness or isolation, or poor relationships are more likely to display the clear hallmarks of chronic stress and suffer its toll on mental and physical health (Cole et al., 2010).

Detection–Assessment

Recognition is a key step in the stress process (McEwen and Gianaros, 2010). Stressors are defined by host conditions, whereby the stressor is sensed and recognized as a challenge requiring a response. Clearly, if there are no means to detect the stressor, there is no stimulus for response—and thus no stress—even if the threat exists (atomic radiation and humans, for example). Vigilance, or scanning frequency, plays a role in cue detection. Such vigilance can be neurocognitive, as in visual scanning rates and threshold for threat detection. It also can be physiological, as in threshold for inflammatory response or pain sensitivity. However, the challenges posed by stressors are graded (low to high) and come up against other priorities. Given the costs of stress responses and the finite time and resources available, filtering mechanisms for gauging cost/benefit and weighing response options are crucial adaptive capacities. Knowing what to ignore can be as important as being able to respond. Many of the stressors and the regulatory responses they trigger involve bottom-up biogenic challenges and lie well outside of awareness (Pecoraro et al., 2006), Regulation of core elements of well-being such as hydration, blood sugar, and oxygenation are detected, assessed, and managed by sensory, neural, endocrine, and chemical mechanisms throughout the body including the central nervous system. Only a minute fraction of this action enters consciousness, due to potent gating and attentional mechanisms that determine what does.

The phylogenetically ancient limbic system is a key player in assessment, including perception of challenge or threat (Catani, 2013). This set of subcortical brain structures (e.g., amygdala, hippocampus, anterior cingulate cortex) is the seat of emotion processing where incoming stimuli are both vetted for immediate action and screened for referral to higher cortical regions based on salience and valence (positive or negative, with strong sensitivity to negative) via both inbuilt unconscious pathways and preconscious memory-linked processing (Pessoa, 2008). The vast realm of unconscious activity stretches across a continuum from being inaccessible to being readily available to conscious awareness, or from unconscious and preconscious processing. Stress-related un- and preconscious processes include many other regions (thalamus, insula) involved in self-relevant sensory processing, feed forward to conscious experience, and influence the likelihood and nature of a response (Arnsten, 2009). Consequently, stress comprises unconscious, preconscious, and conscious elements depending on the level of response involved. Notably, conscious recognition and reaction to an external stressor commonly are preceded and shaded by both preconscious processing through emotion, self-referential, and memory systems and mobilization of autonomic response.

The human capacity for internal generation of stress is both powerful and possibly unique (Sapolsky, 1998). The aforementioned capacities for simulation involved in recall, empathy, scenario building, and future projection enable humans to experience anything from mild stress under terrible circumstances, to profound distress under apparently benign conditions. Perception and emotion regulation are decisive factors. The immediate reasons for stress responses and the differential impact of stressors on individuals are inherently variable and rarely lie solely in the present. Perception relies on meanings—significance, value, implications—to evaluate circumstances or events, which opens the door for influences from culture (models, values, norms), history (previous experiences), and private idiosyncrasy (Dressler, 2011).

Whether the stimuli are biogenic or psychogenic, conditions assessed as stressors that reliably provoke a stress response are those that are perceived as unpredictable, uncontrollable, or socially threatening (Dickerson and Kemeny, 2004). The impoverished graduate student who envisions a glowing postgraduate future or the similarly impoverished artist or social activist who regards the work as intrinsically rewarding and socially valuable does not express the same level of stress as those similarly impoverished but who feel marginalized, oppressed, or trapped. The sense of self, not unique to but elaborated in humans, also contributes to these processes. The concepts of eustress, or a positive response to a stressor, contrasted with toxic stress, or stressors that induce harm, aim to capture an important distinction in terms of optimal levels and types of stimulation or change as opposed to those that degrade performance or well-being (Aschbacher et al., 2013).

Response

We all know how stress feels: the mouth goes dry, eyes widen, skin goes clammy or sweaty, we breathe faster and harder, the heart thuds rapidly, muscles tense, blood pounds in our ears, and the stomach tightens and churns. The stress response is a graded sequence of events that may involve body, brain, and behavior (McEwen and Gianaros, 2010). We have seen that body and brain commonly act together in stressor recognition. The brain, in turn, orchestrates stress responses by the periphery. Concurrently, perceived stress may come to consciousness (or be generated there in the first place) to mobilize a cognitive cascade that shapes the intensity and duration of the response. Un-, pre-, and conscious cognition also can drive a behavioral response. These steps will be discussed in turn.

The Brain Talks to the Body

The brain's mobilization of peripheral stress responses commences in limbic and cortical circuits including the amygdala, hippocampus, and prefrontal cortex, and proceeds largely via two interacting effector systems running through the hypothalamus and brain stem, the sympathetic nervous system (SNS) and the hypothalamo–pituitary adrenal (HPA) axis (also commonly referred to as the hypothalamic–pituitary–adrenal axis) (Figure 1) (Chrousos, 2009). Stress mediators unleashed by these systems organize generalized effects at every functional level, from the cardiorespiratory, metabolic, gastrointestinal, and immune systems; the growth, thyroid, and reproductive axes; anger/fear and reward circuits; and cognitive–executive processing (Table 1). Together they represent a reprioritization of resources away from long-term processes of maintenance, growth, or

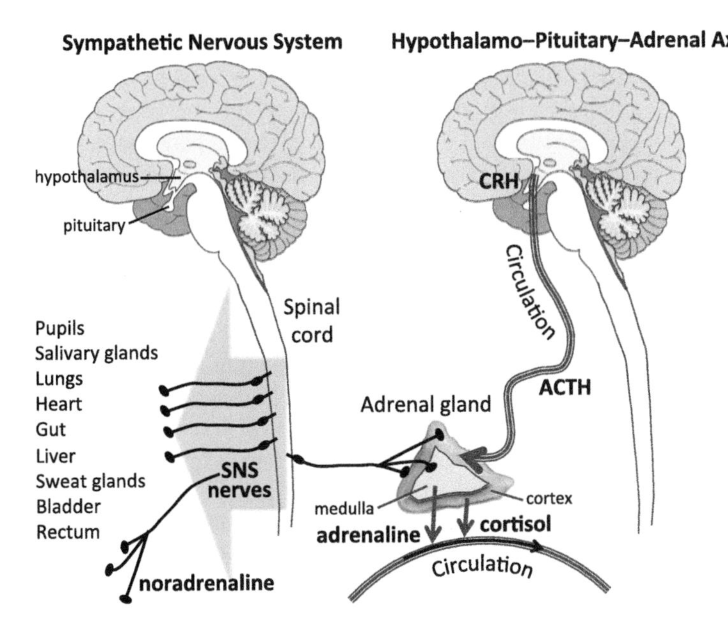

FIGURE 1 Schematic of two arms of the stress response system, the sympathetic nervous system and the hypothalamo–pituitary–adrenal axis.

TABLE 1 Actions of the Sympathetic Nervous System and Hypothalamo–Pituitary–Adrenal Axis during the Stress Response

System	Sympathetic Nervous System	Hypothalamo–Pituitary–Adrenal Axis
Sensory	Pupil dilation	
Systemic	(+) heart rate, respiration rate	(–) vascular tone
	(+) vasoconstriction, water retention	(–) vasoconstriction, water retention (contingent)
Metabolic	(–) salivation	(+) energy mobilization
	(–) appetite, gut activity	(+) appetite
	(+) liver glucose release	(–) glucose uptake, use
Immunologic	(+) immune activation	(–) proinflammatory cytokines and antiviral agents
Endocrine	(+) adrenal catecholamine release	(–) hypothalamo–pituitary–gonadal axis
Central	(+) cerebral blood flow, glucose use	(–) cerebral glucose transport, use
	(+) memory formation	(–) memory formation, neural integrity
Structural		(–) bone formation
		(–) growth and development

reproduction, toward acute physical and cognitive mobilization for action (Sapolsky, 1998). The two arms operate at different speeds and contribute to a temporally graded physiologic, cognitive, and behavioral response. In the early response phase, sympathetic activation proceeds through two routes, most immediately by its direct innervation of viscera to produce coordinated effects throughout the body, including pupil dilation, reduced salivation (dry mouth), increased heart rate, slowed digestion, and increased sweating. Almost as swiftly, sympathetic enervation to the adrenal medulla results in the release of adrenaline into circulation through the adrenal medulla. Together, these fast-acting catecholamines acutely increase peripheral readiness (increased heart and respiration rate, glucose availability) and cognitive arousal (attention, alertness, vigilance).

The more slowly responding HPA axis uses endocrine signaling along a pathway from release by the hypothalamus of cortisol releasing hormone, CRH, and arginine vasopressin, AVP, to stimulate release by the anterior pituitary of adrenocorticotropic hormone, which in turn triggers release by the adrenal cortex of cortisol that peaks in circulation 15–30 min after initiation of the response (Chrousos, 1992). Graded widespread actions of cortisol coordinate shifts in body and brain toward coping, recovery and, hence, adaptation to a stressor. As a steroid, cortisol passes unhindered through lipid membranes and thus reaches every cell of the body. Endocrine action relies on receptors, and cortisol works through a system of two nuclear receptors (glucocorticoid and mineralocorticoid) that operate synergistically to set up a signaling cascade adjustable from basal to stressful conditions, are widespread in tissues, and activate gene networks responsible for a coordinated adaptive stress response (Sapolsky, 1996). Indeed, cortisol might be considered less simply a stress hormone and more accurately a "traffic controller" that reorganizes the body's priorities for resource allocation (Worthman and Kuzara, 2005). This hormone directly influences the activity (transcription) of as much as one-fifth of expressed genes and thereby exerts widely pleiotropic effects across many systems central to immediate function and survival (respiration and blood supply, metabolism, immune activity), intermediate actions (energy storage, repair, immunity), and long-term projects (growth and maturation, reproduction) (Table 1) (de Kloet et al., 2005).

A potent inbuilt moderator of stress responsivity, the activity of the parasympathetic nervous system, generally counterbalances that of the sympathetic nervous system (Porges, 2007). Indeed, enhanced tonic parasympathetic tone has been associated with resilience to challenge, a muted response to stressors, and enhanced emotion regulation and attention regulation. Low parasympathetic tone potentiates the excitatory actions of the SNS, essentially increasing physiologic "vigilance" and stress responsivity, and acute parasympathetic withdrawal has the immediate effect of derepressing sympathetic action and its systemic effects. Thus, factors that influence vagal tone thereby affect stress responsiveness.

The immune system is both a target and a trigger for the stress response. Thus, it is integral to stress, although the stress–immune relationship is complicated by the need to minimize potential harm or costs unleashed by immune activation. Immune responses to stress hormones have deep evolutionary roots in neuroendocrine–immune connections that help coordinate activities across cells, organs, and systems, and reallocate resources to both meet current challenges and maintain optimal immunity (Adamo, 2012). However, the overall effects of neuroendocrine stress responses are finely modulated by the antagonistic effects of HPA and SNS on immune activation: cortisol suppresses inflammatory activity while noradrenaline and adrenaline released by SNS nerves fibers and adrenal medulla, respectively, excite inflammatory and dampen antiviral immune activity (Irwin and Cole, 2011). Suppression of "adaptive" cellular immunity during acute stress appears to be of questionable adaptive value, compromising resistance to infection just when the risks of damage would seem greatest. Yet it clearly represents a shunting of resources away from more remote infectious risk toward damage repair through concurrent activation of innate immunity via proinflammatory actions (Raison et al., 2006). Chronic stress, by contrast, is associated with increased inflammatory activity despite elevated cortisol levels as immune activity becomes desensitized to cortisol (Adamo, 2012). In this case, given the energy costs, nutrient demands, and risks of self-harm from inflammatory responses, loss of sensitivity to cortisol regulation by proinflammatory agents during chronic stress suggests a trade-off shift toward immediate survival at the expense of longevity (Demas et al., 2011).

And the Body Talks Back

Although the brain initiates the stress response, it also is a target of that response, at two levels, regulatory feedback and cognition. Feedback is crucial because cortisol action is relatively slow, but it is powerful and lingering, mediating both acute and late-recovery phases of the stress response. Once mounted, therefore, the response also needs to be terminated if the stimulus is withdrawn or modulated if it is not. The very power and reach of the stress response set up serious cost–benefit trade-offs that put a premium on ability to regulate the strength and duration of the response, and to mitigate or reverse its effects. The HPA is controlled by negative feedback loops at several levels of regulation, including the anterior pituitary, the hypothalamus, and several locations in the brain, prominently the hippocampus. Both the actions and the feedback regulation of cortisol operate through receptors, so factors that affect receptor number, specificity, or distribution exert powerful effects on HPA activity and efficacy (Cole et al., 2007). For instance, early trauma and chronic stress each have been associated with alterations in feedback sensitivity and resultant dysregulation of the HPA via epigenetic effects (Heim et al., 2002; Strüber et al., 2014).

Cross talk among nervous, endocrine, and immune systems not only links stress with immune activity but also mediates cognitive and behavioral effects of stress (Demas et al., 2011). Proinflammatory cytokines act directly on brain activity, such that stress-mediated immune changes result in feedback to cognitive and regulatory centers in the brain with consequent effects on behavior. Sickness behavior exemplifies these effects: the cognitive states and behaviors we associate with illness—lassitude, loss of appetite, social withdrawal, poor concentration, depression, irritability—are adaptive reactions induced by effects of inflammatory responses to infection on the central nervous system (Irwin and Cole, 2011). Similarly, changes in mood associated with stress are mediated in part by actions of proinflammatory cytokines in the brain (Raison et al., 2006).

In addition to the cytokine-mediated effects of stress responses, corticoid receptors are ubiquitous in neurons and glia, which makes the brain a target organ for cortisol (Sapolsky et al., 2000). Stress-induced peripheral neuroendocrine responses potently influence attention, learning, and memory. Acute elevations of CRH and AVP prompt increased attention, arousal, and vigilance. Basal levels of cortisol play a permissive role in the formation of long-term memory in the hippocampus, whereas high cortisol levels impair it. Nevertheless, arousing situations leave more durable memories. It appears that event- or situation-related cortisol facilitates memory, while elevated cortisol before memory formation interferes with retention. Stressor controllability inhibits stress-induced cortical activation (Sapolsky et al., 2000). Thus, modulated receptor activity mediated by intensity and controllability of the stressor permit subtle modulations of cortisol's effects in the brain. Such modulatory capacity enhances adaptability of the stress response (Wolf and Linden, 2012).

STRESS MODERATORS AND BUFFERS

Many factors have been found to moderate or buffer stress. This is unsurprising. Given their power and costliness, stress responses must be finely tuned to meet the challenge, neither more nor less than is required. For our social primate relatives, and particularly for humans who are so dependent on social conditions and relationships, the social world is rife with sources of stress. Yet that world also is a prime source for stress modulators or buffering. Thus, we may regard the systems involved in stress responses as subject to social regulation.

Increasingly detailed evidence documents the role of social factors in mitigating the impact of adversity and promoting well-being by activating safety/reward-related systems that counterbalance threat/harm circuits involved in stress (Eisenberger and Cole, 2012). The perception of social connection, presence of social ties, and experience of social belonging and support all have been found to diminish vulnerability to stressors. Being connected to others indexes social security and hence safety, and has been linked to activation of neural reward systems, inhibition of threat responsivity, and promotion of care for others (Norman et al., 2012). Two prime sources for buffering are brain regions that detect degree of safety from incoming stimuli and inhibit stress responses, and brain and endocrine pathways regulating caregiving behavior have been found to decrease threat responses and promote adaptive prosocial behavior during stress. Notably, neuroendocrine signatures of social bonding involve hormones (endogenous opioids and oxytocin) that reduce stress (Taylor et al., 2008). Thus, signals to "tend and befriend" can buffer or neutralize signals for "fight or flight" (Taylor et al., 2000).

Individual differences in vulnerability to stress also arise from genetic and epigenetic sources, indicating that such variation has adaptive bases (Wolf and Linden, 2012). Humans have been found to have unusually high genetic diversity at loci related to neuroendocrine responsiveness to stressors (reviewed in Worthman (2009)). The differences in biological sensitivity to context fostered by such genetic variation suggest niche partitioning, or occupation of different physical and social locations within the societal matrix to access a subset of available resources and reduce global intraspecific resource competition. Niche partitioning is a powerful adaptive force among humans, whose social-cultural matrix presents tremendous potential for niche diversity and construction observed even in small foraging bands and elaborated in large settled societies (Worthman and Brown, 2005). Furthermore, several genes linked to sensitivity and reactivity to stressors are known to yield different phenotypes, depending on rearing conditions. For example, children with an allele that alters neuronal reuptake of serotonin, a central neurotransmitter that influences mood and behavior, are more likely to have prosocial skills and low stress reactivity if they grow up in supportive conditions, but more likely to have behavioral problems and high stress reactivity if they grow up in disrupted, emotionally unsupportive conditions (Ellis and Bjorklund, 2012). Such findings demonstrate biological bases for context-dependent canalization of life history strategies via genetically informed, context-sensitive processes of psychobehavioral development. Underlying genetic polymorphisms in humans may be maintained by different strategies, one yielding high returns under good conditions, the other yielding less outstanding but more reliable returns even under poor conditions.

DEVELOPMENTAL AND INTERGENERATIONAL PROCESSES

Stress provides valuable information to the individual, by signaling a mismatch between an organism's current states and the demands it faces to sustain well-being, acquire resources, or pursue long-term goals. We have seen that responses to this signal adaptively reallocate resources to meet immediate challenges at the expense of competing demands or life history pursuits. Stress initially was cast in terms of such acute, generalized processes of stressor–assessment–response, and the adverse effects from chronic activation of the response. Now, stress is understood to comprise a graded range of adaptive responses from acute and momentary to long-term, life course adjustments (Hunter and McEwen, 2013). Table 2 traces the impact of exposures/experiences on stress response through development and aging, and by chronicity of exposure. Accumulating evidence suggests that such adjustments are life course organismic adaptations designed to predict, anticipate, and modulate responses to potential stressors. Neuroendocrine mechanisms drive facultative adjustments in life history based on cues about expectable environments the person will face in the future, acting through both developmental and epigenetic pathways. Thus, although the stress response comprises ancient and highly conserved elements (Denver, 2009), plasticity and sensitivity to experiences throughout the life course and across generations contribute to wide individual differences in vulnerability and response to stress.

Developmental Processes

A vast literature documents the influence of stress on cognitive, emotional, and social development (Shonkoff et al., 2009). Gestation and infancy are formative periods in the development of stress responsivity. Development of the fetal HPA is heavily informed by the placenta, the pipeline between mother and fetus that mediates organizational effects of maternal stress not only on postnatal stress responses but also on outcomes such as later development, metabolism, reproduction, and cardiovascular function (Braun et al., 2013). During the first years of life, the HPA axis reorganizes, settles into a diurnal activity pattern, and establishes regulation of stress reactivity and response (Adam et al., 2007). Poor early rearing conditions or disruptions of caregiver relationships have lasting effects on socioemotional and cognitive outcomes as well as mental and physical health. Harsh or uncertain early circumstances such as family dysfunction, harsh parenting, maltreatment, or neglect are associated with changes in operation and balance of stress systems (nervous, endocrine, and immune) (Danese and McEwen, 2012). One might expect that early adversity would exacerbate sensitivity and responses to stress, yet on the contrary, early life stress more commonly is associated with reduced cortisol reactivity (Gunnar et al., 2009). Dampened stress responses may be adaptive by reducing the biological costs of contending with what promises to be a stressful environment. Evidence suggests that such sensitivity sharpens attention and preparation for social challenges (caregiver or other relationship threats, social instability, emotional insecurity) in the world in which the child is growing up (Flinn et al., 2011).

TABLE 2 Graded Timelines of Stress Responses across the Life Course

Factors Influencing Stress Responses	Organizational Effects[a]				
Period of Life	**Mild Stress**	**Severe Stress**	**Social Buffering[b]**	**Poststress Resilience[c]**	**Sensitivity to Stress[e]**
Gestation	+	+++	+	——	+
Early infancy (0–12 months)	—	+++	+++	——	++
Infant, early childhood	—	++	+++	—	+
Middle-to-late childhood	—	+	+++	+	
Puberty	—	+	++	+++	+
Adulthood			+++	++	
Aging			+++	—	+
Stress responses / **Stress duration**	**HPA**	**SNS**	**Immune**	**Poststress resilience**	**Behavior[d]**
Acute	+	+	—+	+	+—Transient
Chronic	—	++	+	—	— Persistent

—Reduced/negative.
+Increased/positive.
Blank: neither notably reduced nor increased.
[a]*Long-term effects on stress responsivity (+increased, — reduced).*
[b]*Capacity to reduce stress responses.*
[c]*Prognosis for recovery later in life.*
[d]*Cognitive and behavioral effects.*
[e]*Vulnerability to stressors.*
Barrientos et al. (2012); Hunter and McEwen (2013).

Epigenetic mechanisms during development have been discovered to forge a direct link of the social environment with long-term effects on stress responsiveness and behavior that have consequences for later health (Szyf et al., 2008). The human infant is highly dependent on caregivers for survival; to a large degree, then, early stress neurobiology and maturation may be said to be socially regulated by proximal social ecology (caregiver qualities and relationships) (Gunnar and Donzella, 2002). Indeed, a suite of temperamental, maturational, and behavioral characteristics in mice has been found to vary in relation to specific qualities of early maternal care that cue epigenetic effects (or modifications of gene function) on stress systems that establish those characteristics in offspring (Meaney et al., 2007). Similar effects have been observed in early-maltreated adults (Szyf et al., 2008). Such shifts demonstrate adjustments in life history strategy induced by social experiences.

Although psychosocial stress in the first five years, during the period of most rapid brain development, exerts powerful effects on psychobehavioral outcomes, later stress exposures and social regulation of stress systems remain potent through childhood and adolescence (Adam et al., 2007). Thus, for example, Finnish adults who had been removed from their parents, evacuated, and returned after World War II experienced greatly elevated risk for stress-related disorders (depression, hypertension, diabetes) as adults (Alastalo et al., 2009). The impact on those who were evacuated aged 7–11 years was equivalent to those evacuated under age 5. Later stressors can compound the health consequences of exposure to early harsh conditions. For instance, heightened stress responsiveness of women exposed to early life stress is exacerbated by further trauma in adulthood and associated with depression (Heim et al., 2002).

Transgenerational Effects

Recognition of transgenerational effects of stress and discovery of their modes of inheritance have perturbed the established anti-Lamarkian view that acquired characteristics cannot be inherited. Transgenerational effects begin during gestation: maternal stress in pregnancy has been associated with altered stress responses in the offspring from early infancy (Thayer and Kuzawa, 2014) on into adulthood (Entringer et al., 2009). As noted above, disruptions in maternal–child relationship are linked to enduring alterations in stress responsivity and behavior of the child via a suite of epigenetic mechanisms (Zhang et al., 2013). However, those disruptions often have their source in the mother's own early experiences that have shaped her cognitive, emotional, and social capacities. Suggestively, persistent transgenerational effects have been observed in rodent studies finding that exposure to stressors alters offspring stress response generations later (Hunter and McEwen, 2013; Meaney et al., 2013). Transmission of early rearing environments via their effects on maternal caregiving has been shown to reliably reproduce maternal phenotype in rodents, including aspects of life history strategy such as maturation rate, temperament, and behavior (Hunter and McEwen, 2013). As noted above, such transmission relies on epigenetic mechanisms operating during development.

Gender Differences

An evolutionary perspective might predict sex differences in sources of and responses to stress, and indeed are observable across the lifespan (Davis and Pfaff, 2014). Regulation of HPA activity differs by sex already at birth. Thus, men are less prone than women to stress-related disorders (depression, autoimmunity) and more prone to metabolic and immune dysregulaion (Bourke et al., 2012). Such gender differences are due in part to interactions of gonadal steroid hormones with cortisol action in target tissues of brain and immune system (Muehlenbein and Bribiescas, 2005). For instance, these hormones alter HPA regulation during chronic stress, leading to greater sensitization to chronic stress as well as more powerful effects of early life stress on HPA regulation in females. Thus, gonadal hormone-mediated differences in regulation of HPA activity and cortisol action yield a range of sex differences in effects of stress on physical (immune, metabolic) and cognitive (mood, behavior regulation) function. These differences also relate to sex differences in health consequences of psychosocial stress (Davis and Pfaff, 2014). Such differences also trace back to gestational factors as well. Maternal psychosocial stress early in pregnancy is linked to increased physiological and behavioral stress reactivity in male offspring (Bale, 2011). The sexes also differ in mental health effects of early stressors: males show greater increases in risk for neuropsychiatric disorders (autism, schizophrenia, anxiety) than females, who experience more increased risk for affective problems (Davis and Pfaff, 2014).

STRESS AND HEALTH

Psychosocial stress inherently incurs health costs. The resources that might be allocated to other developmental, reproductive, productive, and maintenance demands must be diverted to meet the challenge triggering the stress response. Evaluative and feedback processes regulating the response are tightly honed to gauge the level of threat in view of available resources and act accordingly. Nevertheless, the physical, mental, and social burdens imposed by stress take a toll on mental and physical health, particularly when those burdens are severe, recurrent, unpredictable, and/or timed in a sensitive period (Strüber et al., 2014). When mounting costs to meet urgent needs erode everyday function, health impairment is inevitable. Refer again to Table 1 and note the wide range of functions—cardiovascular, metabolic, immune, neuroendocrine, cognitive, and even skeletal—that are affected in stress responses.

Effects of stress conduce to immune dysregulation and increased inflammatory activity; conversely, positive affect is associated with reduced inflammation (Steptoe et al., 2008). Inflammation, in turn, is implicated in all the aforementioned mental and physical health issues. Indeed, inflammation may be the single common cause for many chronic mental and physical diseases. We therefore may not be surprised at the pronounced effects of serious and/or chronic stressors on cardiovascular (blood pressure, heart disease, stroke), metabolic (obesity, diabetes), and mental (depression, anxiety, memory impairment) health, child growth, vulnerability to disease (cancer, infectious illness), and life expectancy (Sapolsky, 1998). Moreover, the developmental impact of pronounced stress—stunted growth, poor social functioning (aggression, withdrawal, hypervigilance)—is associated with impairments in school performance, resilience, social relations, and adult mental and physical health (Danese and McEwen, 2012).

STRESS AND LIFE HISTORY

Over 60 years of stress research has yielded extraordinary insights into the roots of human function and well-being, arguably because core adaptive processes are involved. These studies have elucidated molecular, genetic, systemic, organismic, cognitive, behavioral, developmental, life course, and intergenerational processes involved in meeting the continuous daily challenges of living by detecting and responding to demands and threats. Such insights have shed light on old questions or revised existing views of concepts like homeostasis and phenotypic adaptation, biological inheritance, development, and pathogenesis. Most profoundly, stress research has revealed the power of social factors, from personal to structural, in functioning and well-being. For humans, as intensely social beings, the social world structures demands and the resources to meet them. Accordingly, stress responses are acutely tuned to the ever-shifting conditions of that world, and geared to adjust perception and behavior as well as function across a graded response curve—momentary, cumulative, lifetime, and even transgenerational. Social subordination (Hackman et al., 2010), marginalization or isolation (Eisenberger and Cole, 2012), disparagement (O'Connor et al., 2009), and early poverty (Ziol-Guest et al., 2012), for instance, all induce marked responses along this gradient. The accumulating evidence suggests that these responses relate to life history strategy, reallocating resources and adjusting tradeoffs between immediate and long-term needs according to accumulating and predicted information about environmental demands and supports. Findings linking early or chronic stress to changes in perception, behavior, and biology at the cost of long-term health and survival also accord with life history theory.

Mounting discoveries about the elegant mechanisms that detect challenge and reallocate resources across a graded time course have convinced many scientists that evolved adaptive mechanisms are involved. Nevertheless, how good is our evidence that reallocation of mental, physical, and behavioral resources by stress responses promotes adaptation and fitness, in terms of improved survival and reproductive success? These powerful responses clearly meet perceived threats and needs, but at a cost. Do those demonstrable immediate and long-term costs to function and health represent adaptive designs that improve fitness, or are they merely unavoidable insults to the organism (Ellison and Jasienska, 2007)? Recall that stress serves as a signal to the organism about the quality of environments in which it must function. When they reliably predict future conditions, such signals can be used to guide development or adjust ongoing function at any age, as seen in Table 2. Stress research is revealing how cognition and behavior are modulated by neuroendocrine-driven adaptive mechanisms that integrate genetic, epigenetic, and experience-derived cues (Wolf and Linden, 2012). This tremendous integrative work doubtless will continue to shape our understanding of human adaptive design as well.

Although questions about the evolutionary biology of human stress remain, one profound lesson stands out in the evidence. Social factors are integral to human welfare, built into how we become human, experience our lives, and function in the world. Because social conditions are such a potent determinant of environmental quality and experience, we can predict that development of stress regulation and the thresholds for stress responses at any age will be sensitive to social cues. And indeed, evidence shows that to be the case (Flinn et al., 2011; Hunter and McEwen, 2013), so decisively so that biomedical and public health understandings of disease have changed (Hertzman and Boyce, 2010). Both now highlight social factors as sources of both illness and wellness, mental and physical. Consequently, our emerging understanding of biological design for sensitivity to context and the resultant effects on function and health within and across generations bear directly on policy issues and questions of social justice.

REFERENCES

Adam, E.K., Klimes-Dougan, B., Gunnar, M.R., 2007. Social regulation of the adrenocortical response to stress in infants, children, and adolescents: Implications for psychopathology and education. In: Coch, D., Dawson, G., Fischer, K.W. (Eds.), Human Behavior, Learning, and the Developing Brain: Atypical Development. Guilford Press, New York, NY US, pp. 264–304.

Adamo, S.A., 2012. The effects of the stress response on immune function in invertebrates: an evolutionary perspective on an ancient connection. Hormones and Behavior 62 (3), 324–330.

Alastalo, H., Raikkonen, K., Pesonen, A.K., Osmond, C., Barker, D.J., Kajantie, E., Eriksson, J.G., 2009. Cardiovascular health of Finnish war evacuees 60 years later. Annals of Medicine 41 (1), 66–72.

Arnsten, A.F., 2009. Stress signalling pathways that impair prefrontal cortex structure and function. Nature Reviews Neuroscience 10 (6), 410–422.

Aschbacher, K., O'Donovan, A., Wolkowitz, O.M., Dhabhar, F.S., Su, Y., Epel, E., 2013. Good stress, bad stress and oxidative stress: insights from anticipatory cortisol reactivity. Psychoneuroendocrinology 38 (9), 1698–1708.

Bale, T.L., 2011. Sex differences in prenatal epigenetic programming of stress pathways. Stress 14 (4), 348–356.

Barrientos, R.M., Frank, M.G., Watkins, L.R., Maier, S.F., 2012. Aging-related changes in neuroimmune-endocrine function: implications for hippocampal-dependent cognition. Hormones and Behavior 62 (3), 219–227.

Bernard, C., 1865. Introduction à l'étude de las médecine expérimentale. J.B. Bailière, Paris.

Bourke, C.H., Harrell, C.S., Neigh, G.N., 2012. Stress-induced sex differences: adaptations mediated by the glucocorticoid receptor. Hormones and Behavior 62 (3), 210–218.

Braun, T., Challis, J.R., Newnham, J.P., Sloboda, D.M., 2013. Early-life glucocorticoid exposure: the hypothalamic-pituitary-adrenal axis, placental function, and long-term disease risk. Endocrine Reviews 34 (6), 885–916.

Brommer, J.E., 2000. The evolution of fitness in life-history theory. Biological Reviews 75, 377–404.

Cannon, W.B., 1929. Organization for physiological homeostasis. Physiological Reviews 9, 399–431.

Catani, M., Dell'acqua, F., Thiebaut de Schotten, M., 2013. A revised limbic system model for memory, emotion and behaviour. Neuroscience and Biobehavioral Reviews 37 (8), 1724–1737.

Charnov, E.L., 1993. Life History Invariants: Some Explorations of Symmetry in Evolutionary Ecology. Oxford University Press, Oxford.

Chrousos, G.P., 1992. Regulation and dysregulation of the hypothalamic-pituitary-adrenal axis. The corticotropin-releasing hormone perspective. Endocrinology and Metabolism Clinics of North America 21 (4), 833–858.

Chrousos, G.P., 2009. Stress and disorders of the stress system. Nature Reviews Endocrinology 5 (7), 374–381.

Cole, S.W., Arevalo, J.M.G., Takahashi, R., Sloan, E.K., Lutgendorf, S.K., Sood, A.K., Seeman, T.E., 2010. Computational identification of gene-social environment interaction at the human IL6 locus. PNAS 107 (12), 5681–5686.

Cole, S.W., Hawkley, L.C., Arevalo, J.M., Sung, C.Y., Rose, R.M., Cacioppo, J.T., 2007. Social regulation of gene expression in human leukocytes. Genome Biology 8 (9), R189.

Danese, A., McEwen, B.S., 2012. Adverse childhood experiences, allostasis, allostatic load, and age-related disease. Physiology and Behavior 106 (1), 29–39.
Davis, E.P., Pfaff, D., 2014. Sexually dimorphic responses to early adversity: implications for affective problems and autism spectrum disorder. Psychoneuroendocrinology 49, 11–25.
de Kloet, E.R., Joels, M., Holsboer, F., 2005. Stress and the brain: from adaptation to disease. Nature Reviews Neuroscience 6 (6), 463–475.
Demas, G.E., French, S.S., Adamo, S.A., 2011. Neuroendocrine-immune crosstalk in vertebrates and invertebrates: implications for host defence (electronic resource). Functional ecology 25 (1), 29–39.
Denver, R.J., 2009. Structural and functional evolution of vertebrate neuroendocrine stress systems. Annals of the New York Academy of Sciences 1163, 1–16.
Dickerson, S.S., Kemeny, M.E., 2004. Acute stressors and cortisol responses: a theoretical integration and synthesis of laboratory research. Psychological Bulletin 130, 355–391.
Dressler, W.W., 2011. Culture and the stress process. In: Singer, M., Erickson, P.I. (Eds.), A Companion to Medical Anthropology. Blackwell, Chichester, UK, pp. 119–134.
Eisenberger, N.I., Cole, S.W., 2012. Social neuroscience and health: neurophysiological mechanisms linking social ties with physical health. Nature Neuroscience 15 (5), 669–674.
Ellis, B.J., Bjorklund, D.F., 2012. Beyond mental health: An evolutionary analysis of development under risky and supportive environmental conditions: an introduction to the special section. Developmental Psychology 48 (3), 591–597.
Ellison, P.T., Jasienska, G., 2007. Constraint, pathology, and adaptation: How can we tell them apart? American Journal of Human Biology 19 (5), 622–630.
Entringer, S., Kumsta, R., Hellhammer, D.H., Wadhwa, P.D., Wüst, S., 2009. Prenatal exposure to maternal psychosocial stress and HPA axis regulation in young adults. Hormones and Behavior 55 (2), 292–298.
Eveleth, P.B., Tanner, J.M., 1990. Worldwide variation in human growth, 2nd ed. edn. Cambridge University Press, New York.
Finch, C.E., Rose, M.R., 1995. Hormones and the physiological architecture of life history evolution. Quarterly Review of Biology 70 (1), 1–52.
Flinn, M.V., Nepomnaschy, P.A., Muehlenbein, M.P., Ponzi, D., 2011. Evolutionary functions of early social modulation of hypothalamic-pituitary-adrenal axis development in humans. Neuroscience and Biobehavioral Reviews 35 (7), 1611–1629.
Gilbert, D.T., 2006. Stumbling on Happiness. Knopf, New York.
Goldstein, D.S., 2010. Adrenal responses to stress. Cellular and Molecular Neurobiology 30 (8), 1433–1440.
Gunnar, M.R., Donzella, B., 2002. Social regulation of the cortisol levels in early human development. Psychoneuroendocrinology 27, 199–220.
Gunnar, M.R., Frenn, K., Wewerka, S.S., Van Ryzin, M.J., 2009. Moderate versus severe early life stress: associations with stress reactivity and regulation in 10-12-year-old children. Psychoneuroendocrinology 34 (1), 62–75.
Hackman, D.A., Farah, M.J., Meaney, M.J., 2010. Socioeconomic status and the brain: mechanistic insights from human and animal research. Nature Reviews Neuroscience 11 (9), 651–659.
Heim, C., Newport, D.J., Wagner, D., Wilcox, M.M., Miller, A.H., Nemeroff, C.B., 2002. The role of early adverse experience and adulthood stress in the prediction of neuroendocrine stress reactivity in women: a multiple regression analysis. Depression and Anxiety 15, 117–125.
Hertzman, C., Boyce, T., 2010. How experience gets under the skin to create gradients in developmental health. Annual Review of Public Health 31, 329–347.
Hill, K.R., Kaplan, H., 1999. Life history traits in humans: theory and empirical studies. Annual Review of Anthropology 28, 397–430.
Hunter, R.G., McEwen, B.S., 2013. Stress and anxiety across the lifespan: structural plasticity and epigenetic regulation. Epigenomics 5 (2), 177–194.
Irwin, M.R., Cole, S.W., 2011. Reciprocal regulation of the neural and innate immune systems. Nature Reviews Immunology 11 (9), 625–632.
McEwen, B.S., Gianaros, P.J., 2010. Central role of the brain in stress and adaptation: links to socioeconomic status, health, and disease. Annals of the New York Academy of Sciences 1186, 190–222.
McEwen, B.S., Wingfield, J.C., 2003. The concept of allostasis in biology and biomedicine. Hormones and Behavior 43, 2–15.
Meaney, M.J., Aitken, D.H., Bodnoff, S.R., Iny, L.J., Tatarewicz, J.E., Sapolsky, R.M., 2013. Early postnatal handling alters glucocorticoid receptor concentrations in selected brain regions. Behavioral Neuroscience 127 (5), 637–641.
Meaney, M.J., Szyf, M., Seckl, J.R., 2007. Epigenetic mechanisms of perinatal programming of hypothalamic-pituitary-adrenal function and health. Trends in Molecular Medicine 13 (7), 269–277.
Muehlenbein, M.P., Bribiescas, R.G., 2005. Testosterone-mediated immune functions and male life histories. American Journal of Human Biology 17 (5), 527–558.
Norman, G.J., Hawkley, L.C., Cole, S.W., Berntson, G.G., Cacioppo, J.T., 2012. Social neuroscience: the social brain, oxytocin, and health. Social Neuroscience 7 (1), 18–29.
O'Connor, M.F., Bower, J.E., Cho, H.J., Creswell, J.D., Dimitrov, S., Hamby, M.E., Irwin, M.R., 2009. To assess, to control, to exclude: effects of biobehavioral factors on circulating inflammatory markers. Brain, Behavior, and Immunity 23 (7), 887–897.
Pecoraro, N., Dallman, M.F., Warne, J.P., Ginsberg, A.B., Laugero, K.D., la Fleur, S.E., Akana, S.F., 2006. From Malthus to motive: how the HPA axis engineers the phenotype, yoking needs to wants. Progress in Neurobiology 79 (5–6), 247–340.
Pessoa, L., 2008. On the relationship between emotion and cognition. Nature Reviews Neuroscience 9 (2), 148–158.
Porges, S.W., 2007. The polyvagal perspective. Biological Psychology 74 (2), 116–143.
Raison, C.L., Capuron, L., Miller, A.H., 2006. Cytokines sing the blues: inflammation and the pathogenesis of depression. Trends in Immunology 27 (1), 24–31.
Sapolsky, R.M., 1996. Stress, glucocorticoids, and damage to the nervous system: the current state of confusion. Stress 1 (1), 1–19.
Sapolsky, R.M., 1998. Why Zebras Don't Get Ulcers: An Updated Guide to Stress, Stress-related Diseases, and Coping, second ed. W.F. Freeman, New York.
Sapolsky, R.M., Romero, L.M., Munck, A.U., 2000. How do glucocorticoids influence stress responses? Integrating permissive, suppressive, stimulatory, and preparative actions. Endocrine Reviews 21, 55–89.
Selye, H., 1973. The evolution of the stress concept. American Scientist 61 (6), 692–699.
Shonkoff, J.P., Boyce, W.T., McEwen, B.S., 2009. Neuroscience, molecular biology, and the childhood roots of health disparities: building a new framework for health promotion and disease prevention. JAMA 301 (21), 2252–2259.

Stearns, S.C., 2000. Life history evolution: successes, limitations, and prospects. Naturwissenschaften 87, 476–486.

Steptoe, A., O'Donnell, K., Badrick, E., Kumari, M., Marmot, M., 2008. Neuroendocrine and inflammatory factors associated with positive affect in healthy men and women–the Whitehall II study. American Journal of Epidemiology 167 (1), 96–102.

Strüber, N., Strüber, D., Roth, G., 2014. Impact of early adversity on glucocorticoid regulation and later mental disorders. Neuroscience and Biobehavioral Reviews 38, 17–37.

Szyf, M., McGowan, P., Meaney, M.J., 2008. The social environment and the epigenome. Environmental and Molecular Mutagenesis 49, 46–60.

Taylor, S.E., Burklund, L.J., Eisenberger, N.I., Lehman, B.J., Hilmert, C.J., Lieberman, M.D., 2008. Neural bases of moderation of cortisol stress responses by psychosocial resources. Journal of Personality and Social Psychology 95 (1), 197–211.

Taylor, S.E., Klein, L.C., Lewis, B.P., Gruenewald, T.L., Gurung, R.A., Updegraff, J.A., 2000. Biobehavioral responses to stress in females: tend-and-befriend, not fight-or-flight. Psychological Review 107 (3), 411–429.

Thayer, Z.M., Kuzawa, C.W., 2014. Early origins of health disparities: maternal deprivation predicts maternal evening cortisol in pregnancy and offspring cortisol reactivity in the first few weeks of life. American Journal of Human Biology 26 (6), 723–730.

Wolf, C., Linden, D.E.J., 2012. Biological pathways to adaptability–interactions between genome, epigenome, nervous system and environment for adaptive behavior. Genes, Brain, and Behavior 11 (1), 3–28.

Worthman, C.M., 1999. The epidemiology of human development. In: Panter-Brick, C., Worthman, C.M. (Eds.), Hormones, Health, and Behavior: A Socio-ecological and Lifespan Perspective. Cambridge University Press, Cambridge, pp. 47–104.

Worthman, C.M., 2003. Energetics, sociality, and human reproduction: life history theory in real life. In: Wachter, K.W., Bulatao, R.A. (Eds.), Offspring: Human Fertility Behavior in Biodemographic Perspective. National Academies Press, Washington, DC, pp. 289–321.

Worthman, C.M., 2009. Habits of the heart: life history and the developmental neuroendocrinology of emotion. American Journal of Human Biology 211, 772–781.

Worthman, C.M., Brown, R.A., 2005. A biocultural life history approach to the developmental psychobiology of male aggression. In: Stoff, D.M., Susman, E.J. (Eds.), Developmental Psychobiology of Aggression. Cambridge University Press, New York, pp. 187–222.

Worthman, C.M., Kuzara, J.L., 2005. Life history and the early origins of health differentials. American Journal of Human Biology 17, 95–112.

Zhang, T.Y., Labonte, B., Wen, X.L., Turecki, G., Meaney, M.J., 2013. Epigenetic mechanisms for the early environmental regulation of hippocampal glucocorticoid receptor gene expression in rodents and humans. Neuropsychopharmacology 38 (1), 111–123.

Ziol-Guest, K.M., Duncan, G.J., Kalil, A., Boyce, W.T., 2012. Early Childhood Poverty, Immune-Mediated Disease Processes, and Adult Productivity. Proceedings of the National Academy of Sciences of the United States of America 09 (Suppl. 2), 17289–17293.

Chapter 33

Aggression, Affiliation, and Parenting

Mark V. Flinn
Department of Anthropology, University of Missouri, Columbia, MO, USA

SYNOPSIS

This chapter examines how and why humans evolved the seemingly oppositional traits of lethal coalitional aggression and intensive, loving parental care. Coalitions are one of the most striking aspects of human behavior. All societies recognize alliances among communities, usually based in part on kinship. Intergroup competition is ubiquitous, often deadly and fueled by revenge, and critical to general human welfare. The ontogeny of coalitions is a key aspect of childhood; children in all cultures develop friendship cliques and engage in group-against-group team play. Parenting is also a fundamental component of human sociality. The extent and duration of offspring care is extraordinary—and unique in the huge informational transfer via language and culture. The emotional, cognitive, and behavioral components of human family relationships are biologically embedded in neurological and physiological mechanisms. Analysis of affiliative and stress hormonal axes can provide clues into the emotional and cognitive systems that underpin these core aspects of human sociality.

Humans are the species that raises children.

Melvin Konner (1991), p. 421

Why are humans alone at the pinnacle of the particular direction of rapid evolutionary change that led to the combination of traits such as a huge brain, complex intellect, upright posture, concealed ovulation, menopause, virtual hairlessness, a physically helpless but mentally precocial baby, and above all our tendency and ability to cooperate in social groups of millions?"

Richard D. Alexander, 1990b, p. 1

No hunt needs quite so much forethought or ability to communicate complex instructions as does a war, nor do such drastic demographic consequences hinge on the outcome.

William D. Hamilton, 1976, p. 165

The human family seems to follow a typical mammalian pattern: intense maternal care, including breastfeeding of an altricial (relatively undeveloped at birth) offspring, with some support from an assortment of other relatives—fathers, siblings, aunts, and the like. Beyond the shared mammal/primate commonality, however, humans exhibit a suite of highly unusual traits. We are the only species characterized by the combination of stable breeding bonds, extensive caregiving by males residing in a multimale group, lengthy childhood, extended bilateral kin recognition, grandparenting, and controlled exchange of mates among kin groups (Chapais, 2008, 2014; Flinn, 2011; Hawkes, 2003; Hrdy, 2005). These characteristics are important for the theoretical and pragmatic understanding of family relationships and the supporting cognitive, emotional, and physiological mechanisms; they also provide critical insight into the puzzle of human evolution.

BRAIN, CHILDHOOD, AND PARENTING

Information processing (intelligence) and social communication (language) are core human adaptations. By all measures, the human brain that enables these abilities is an astonishing organ (see chapter by Holloway in the present volume). Its cortex comprises about 30 billion neurons of 200 different types, each of which is interlinked by about a thousand synapses, resulting in more than 10^{15} connections working at rates of up to 10^{10} interactions per second (Edelman, 2006; Koch, 1999; Williams and Herrup, 1988). Quantifying the transduction of these biophysical actions into specific mental activities (i.e., thoughts and emotional feelings) is difficult, but it is likely that humans have more information-processing capacity than any other species (Roth and Dicke, 2005).

The human brain evolved at a rapid pace: Hominin cranial capacity tripled (from an average of about 450–1350 cc) in less than 2 million years (Lee and Wolpoff, 2003)—roughly 100,000 neurons and supportive cells per generation. Structural changes, such as increased convolutions, thickly myelinated cortical neurons, lateral asymmetrics, increased von Economo neurons, expansion of the neocortex, and enhanced integration of the cerebellum also were significant (Allman, 1999; Amodio and Frith, 2006; Rilling, 2013; Schoenemann, 2006; Sherwood et al., 2006; Spoctor et al., 2010; see chapter by Holloway). In comparison with most other parts of the human genome, selection on genes involved with brain development was especially intense (Gilbert et al., 2005).

The human brain has high metabolic costs: About 50% of a child's and 20% of an adult's, energetic resources are used to support neurological activity (Aiello and Wheeler, 1995; Elia, 1992; Holliday, 1986; Leonard et al., 2007). Thoughts are not free; the high levels of glucose and other energetic nutrients required to fuel human cognition involve significant trade-offs. Although the increase in energetic resources allocated to the brain was accompanied by a corresponding decrease in digestive tissue, this does not explain what the selective pressures for enhanced information processing were, nor why the resources were not reallocated to direct reproductive function (Aiello and Wheeler, 1995). The obstetric difficulties associated with birthing a large-headed infant generate additional problems (Rosenberg and Trevathan, 2002). The selective advantages of increased intelligence must have been high to overcome these costs.

The human brain, in short, is a big evolutionary puzzle. It is developmentally and metabolically expensive. It evolved rapidly and consistently. And it enables unusual human cognitive abilities, such as language, empathy, foresight, consciousness, mental time travel, creativity, and theory of mind (ToM) (see chapter by Vonk and Aradhye). The advantages of a larger brain may include enhanced information-processing capacities to contend with ecological pressures that involve sexually dimorphic activities, such as hunting and complex foraging (Kaplan and Robson, 2002). There is little evidence, however, of sufficient domain-specific enlargement of those parts of the brain associated with selective pressures from the physical environment, including subsistence activities (Adolphs, 2003; Geary and Huffman, 2002). Indeed, human cognition has little to distinguish itself in the way of specialized ecological talents. Our remarkable aptitudes for tool use and other technical behaviors depend primarily on more general aptitudes for social learning and fluid intelligence (Geary, 2005). A large brain may have been sexually selected because intelligence was an attractive trait for mate choice (Miller, 2000). However, there is little sexual dimorphism in encephalization quotient or intelligence psychometrics (Jensen, 1998), nor is there a clear reason why brains would have been a target for sexual selection driven by mate choice uniquely and consistently among hominins (and no other species...).

The human brain did not evolve as an isolated trait. Concomitant changes in other traits provide clues to what selective pressures were important during hominin evolution.

Changes in life history patterns accompanied the evident increases in information processing and communication during the Pleistocene (Dean et al., 2001). Gestation (pregnancy) was lengthened, but the resultant infant was even more altricial (Rosenberg, 2004). Human infants must be carried, fed, and protected for a long period in comparison with those of other primates (see chapter by Valeggia and Nunez de la Mora). And yet, humans have shorter interbirth intervals than other hominoids (Bogin et al., 2007). Human childhood and adolescence are also exceptionally lengthy (Bogin, 1999; Del Giudice, 2009; Leigh, 2004; Smith, 1994; see chapter by Bogin). This extension of the juvenile period appears costly in evolutionary terms. The delay of reproduction until at least 15 years of age involves prolonged exposure to extrinsic causes of mortality and longer generation intervals. Parental and other kin investment continues for an unusually long time, often well into adulthood and perhaps even after the death of the parents. Like the big brain, human life history is an evolutionary puzzle (Hill and Hurtado, 1995; Mace, 2000; Muehlenbein and Flinn, 2011; Quinlan et al, 2003).

Of course, the child must accumulate energetic resources necessary for physical somatic growth. Whether the lengthening of the human juvenile period was an unavoidable response to an increasing shortage of calories, however, is uncertain. Other hominoids (chimpanzees, gorillas, and orangutans) grow at similar overall rates, but mature earlier (Leigh, 2004). Increased body fat is associated with earlier puberty for girls, although psychological and genetic factors are also important (Walvoord, 2010), and the relation is not significant for boys (Lee et al., 2010). Moreover, low birth weight is associated with earlier puberty in some conditions (Karaolis-Danckert et al., 2009). The peculiarities of the human growth curve are also difficult to explain from a simple model of food scarcity—the general timing of growth spurts does not appear to be linked to a pattern of caloric surpluses. Hence, although it is clear that human female growth and reproductive maturation are sensitive to fat accumulation (Ellison, 2001; Sloboda et al., 2007), the lengthening of the juvenile period during human evolution seems likely to have involved more than simple energetic constraints on growth.

The life history stage of human childhood appears to be an adaptation, at least in part, for the function of enabling cognitive development, including complex social skills and emotional regulation (Alexander, 1987, 1990a; Bogin, 1999; Bjorklund and Pelligrini, 2002; Del Giudice et al., 2009; Flinn, 2004; Konner, 2010). The human child is an extraordinarily social creature, motivated by and highly sensitive to interpersonal relationships (Gopnik et al., 1999). Learning, practice, and experience are imperative for social success. The information-processing capacity used for human social interactions is considerable, and perhaps significantly greater than that involved with foraging, locomotion, toolmaking, and other subsistence skills (Rilling et al., 2002; Roth and Dicke, 2005; Schoenemann, 2006).

The child needs to master complex dynamic social tasks, such as developing appropriate cognitive and emotional responses during interactions with peers and adults in the local community (Bugental, 2000). The learning environments that facilitate and channel these aspects of human mental phenotypic plasticity appear to take on a special importance (Posner, 2005). Much of the data required for the social behavior necessary to be successful as a human cannot be "preprogrammed" into specific, detailed, fixed responses. Social cleverness in a fast-paced, cumulative cultural environment must contend with dynamic, constantly shifting strategies of friends and enemies, and hence needs information from experiential social learning and creative scenario building (Flinn, 1997; 2006a; Flinn and Alexander, 2007; del Giudice, 2009).

In brief, human childhood may be viewed as a life history stage necessary for acquiring the information and practice to build and refine the mental algorithms critical for negotiating the social relationships that are key to success in our species (Flinn et al., 2009; Geary and Flinn, 2001; Joffe, 1997; Muehlenbein and Flinn, 2011) Mastering the social environment presents special challenges for the human child. Social competence is difficult because the targets (other children and adults) are constantly changing and similarly equipped with ToM and other cognitive abilities. Selection for flexible cognitive problem solving would also enhance complementary development of more sophisticated ecological skills, such as hunting and complex extractive foraging (Kaplan et al., 2000).

Human social relationships are especially complex because they involve extensive coalitions. We are extraordinarily cooperative, most exceptionally and importantly in regard to competition with other groups (Alexander, 1979, 2006; Bowles, 2009). Humans are unique in being the only species that engages in group-against-group play (Alexander, 1990b), including team sports. This trait is cross-culturally universal, emerges early in child development, and often is the object of tremendous collective effort.

The family environment is a primary source and mediator of the ontogeny of information-processing abilities, including social competencies and group cooperation. Human biology has been profoundly affected by our evolutionary history as unusually social creatures, immersed in networks of family, kin, and dynamic intercommunity coalitions.

THE HUMAN FAMILY

All human societies recognize kinship as a key organizational principle (Brown, 1991). All languages have kinship terminologies and concomitant expectations of obligations and reciprocity (Fortes, 1969; Murdock, 1949). Human kinship systems appear unique in their universal recognition of both bilateral (maternal and paternal) and multigenerational structures, with a general trend for coresidence of male kin, but with a dozen or more major variants (Flinn, 1981; Flinn and Low, 1986; Murdock, 1949). These aspects of human kinship link families into broader cooperative systems and provide additional opportunities for alloparental care during the long social childhood. Three species-distinctive characteristics stand out as unusually important in this regard: (1) fathering—that is, extensive and specific investment by males; (2) grandparenting; and (3) networks of kinship that extend among communities and involve affinal (ties by marriage) and consanguineous (ties by blood) relationships.

Fathers

Mammals that live in groups with multiple males, such as chimpanzees (Pan troglodytes), usually have little or no paternal care, because the nonexclusivity of mating relationships obscures paternity (Alexander, 1974; Clutton-Brock, 1991). In contrast, it is common for human fathers to provide protection, information, food, and social status for their children (Gray and Anderson, 2010). Paternal care in humans appears to be facilitated by relatively stable pair-bonds (see chapter by Gangestad and Grebe), not only involves cooperation between mates that often endures over the life span, but also requires an unusual type of cooperation among coresiding males—that is, respect for each other's mating relationships.

The relatively exclusive mating relationships that are characteristic of most human societies (Flinn and Low, 1986) generate natural factions within the group. Mating relationships also can create alliances in human groups, linking two families or clans together (e.g., Chagnon, 1966; Macfarlan et al., 2014). By way of comparison, in chimpanzee communities, it is difficult for even the most dominant male to monopolize an estrous female; usually, most of the males in a community mate with most of the females (Goodall, 1986; Mitani et al., 2010) (see chapter by Watts). Chimpanzee males in effect "share" a common interest in the community's females and their offspring. Human groups, in contrast, are composed of family units, each with distinct reproductive interests. Human males do not typically share mating access to all the group's females; consequently, there are usually reliable cues identifying children are their genetic offspring, and are those of other males (for exceptions see Beckerman and Valentine, 2002; Walker et al., 2010). Because humans live in multimale groups, yet often maintain stable and exclusive mating relationships, the potential for fission along family lines is high. Still, human groups overcome this inherent conflict between family units to form large, stable coalitions.

This unusual tolerance among coresidential males and their families stands in contrast to the norm of polygamous mate competition in group-living nonhuman primates. Selection pressures favoring such tolerance are uncertain, but likely involve the importance of both male parental investment (Alexander, 1990b; Geary and Bjorklund, 2000) and male coalitions for intraspecific conflict (Alexander, 1990b; Bernhard et al., 2006; Flinn et al., 2005a,b; Gavrilets and Vose, 2006; Wrangham, 1999).

The advantages of intensive parenting, including paternal protection and other care, require a most unusual pattern of mating relationships: moderately exclusive pair-bonding in multimale groups. No other primate (or mammal) that lives in large, cooperative multi-reproductive-male groups has such extensive male parental care targeted at specific offspring. Competition for females in multimale groups usually results in low confidence of paternity (e.g., among bonobos and chimpanzees). Males forming exclusive "pair bonds" in multimale groups would provide cues of to nonparental males of their nonpaternity, and hence place the offspring of bond-forming males in greater danger of infanticide (Hrdy, 1999). Paternal care is most likely to be favored by natural selection in conditions in which males can identify their offspring with sufficient probability to offset the costs of investment, although reciprocity with mates is also likely to be involved (Geary and Flinn, 2001; Smuts and Smuts, 1993). Humans exhibit a unique "nested family" social structure, involving complex reciprocity among males and females to restrict direct competition for mates among group members (see chapter by Gangestad and Grebe).

It is difficult to imagine how this system could be maintained in the absence of another unusual human trait: concealed or "cryptic" ovulation (Alexander and Noonan, 1979). Human groups tend to be male philopatric (males tending to remain in their natal groups), resulting in extensive male kin alliances, useful for competing against other groups of male kin (LeBlanc, 2003; Wrangham and Peterson, 1996). Females also have complex alliances, but usually are not involved directly in the overt physical aggression characteristic of intergroup relations (Campbell, 2002; Geary and Flinn, 2002).

Relationships among human brothers and sisters are lifelong, even when residence is in different communities, in contrast with the absence of significant ties or apparent kin recognition after emigration in other hominoids. Parents, grandparents, and other kin may be especially important for the child's mental development of social and cultural maps because they can be relied upon as landmarks who provide relatively honest information. From this perspective, the evolutionary significance of the human family in regard to child development is viewed more as a nest from which social skills may be acquired than simply as an economic unit centered on the sexual division of labor (Flinn and Leone, 2006, 2009; Flinn and Ward, 2005).

In summary, the care-providing roles of fathers are unusually important in humans, particularly in regard to protection and social power, but are flexible components of the human family and linked with the roles of other relatives, including grandparents. In addition to the effects of direct parental care, paternity provides the basis for critical bilateral kinship links that extend across communities and generations. The neuroendocrine mechanisms that underpin human paternal and grandparental psychology are not well studied, but likely involve the common mammalian affiliative hormones oxytocin (OT) and arginine vasopressin (AVP), with additional influence from the hypothalamic–pituitary–gonadal and hypothalamic–pituitary–adrenal systems (Gray and Campbell, 2009).

Grandparents

Grandparents and grandoffspring share genes that are 25% identical by descent, a significant opportunity for kin selection. Few species, however, live in groups with multiple overlapping generations of kin. Fewer still have significant social relationships among individuals two or more generations apart. Humans appear rather exceptional in this regard. Grandparenting is cross-culturally ubiquitous and pervasive (Murdock, 1967; Sear et al., 2000; Voland et al., 2005). Our life histories allow for significant generational overlaps, including an apparent extended postreproductive stage facilitated by the unique human physiological adaptation of menopause (Alexander, 1974, 1987; Hawkes, 2003; see chapter by Sievert).

The significance of emotional bonding between grandparents and grandchildren is beyond doubt. The evolved functions are uncertain, but likely involve the exceptional importance of long-term, extensive, and intensive investment for the human child. The emotional and cognitive processes that guide grandrelationships must have evolved because they enhanced the survival and eventual reproductive success of grandchildren. Leaving children with grandparents and other alloparental care providers allows parents to pursue productive activities that would otherwise be risky or difficult when encumbered with childcare. In addition to the physical basics of food, protection, and hygienic care, psychological development of the human child is strongly influenced by the dynamics of the social environment (Dunn, 2004; Hetherington, 2003a,b; Konner, 1991, 2010). Grandparents may have knowledge and experience that are important and useful for helping grandchildren and other relatives survive (Sear et al., 2000; Sear and Mace, 2008) and succeed in social competition (Coe, 2003; Voland et al., 2005). Humans are unusual in the critical role of kin in alloparental care and group coalitions (Hrdy, 2009).

Extended Kinship and Control of Mating Relationships

> *The direct application of theory from evolutionary biology to human marriage behavior and mating strategies is... not possible until the theory is modified to take into consideration the interdependency of individuals... and how their interdependency—coalition alliances—structures human mating behavior.*
>
> Napolean Chagnon, 1979, p. 88

Human communities are composed of families embedded in complex kin networks. The importance of kinship in traditional societies is paramount; social power is primarily contingent upon support from relatives. Complex kinship alliances are arguably the most distinguishing social behavioral characteristic of humans in preindustrial cultures, and yet it is rarely discussed in evolutionary psychology or evolutionary economics. Reciprocity in all its various guises (for review see Alexander, 2006) is inextricably bound up with kinship in traditional societies, perhaps most importantly in regard to the control of mating within the institution of marriage. The vast majority of nonindustrial cultures in the Ethnographic Atlas (Murdock, 1967) have rules and preferences specifying what categories of relatives are appropriate for mating/marriage; these rules and preferences involve issues of resultant kin ties in addition to inbreeding avoidance (for ethnographic examples see Chagnon, 1966; Gough, 1959). It is worth emphasizing that humans are unique in the regulation of mating relationships by kin groups. The reason for controlling who mates with whom is that humans are unique in the great importance of kinship ties for alliances among groups (Fortes, 1969; Levi-Straus, 1949/1969). Mates in traditional societies are usually obtained via strategic negotiation between kin groups. No other species exhibits systematic preferences and prohibitions for mating relationships between specific types of cousins.

If human ancestors had intergroup relations similar to those of chimpanzees (see Mitani et al., 2010; Wrangham, 1999; Wrangham and Peterson, 1996), it would have been difficult to make even the first steps toward cooperative alliances among males (and females, including sisters) in different communities. An adult male attempting to establish a relationship with another group likely would be killed as he entered their range. Somehow, our ancestors overcame such obstacles to taking the first steps toward the core human adaptation of intercommunity alliances. It is possible that our ancestors did not have hostile intergroup relations; this seems unlikely, however, on both empirical (LeBlanc, 2003) and theoretical (Alexander, 1990b) grounds. Surely, the most potent factor driving the evolution of the psychological, social, and cultural mechanisms enabling the formation of increasingly large and complex coalitions was competition with other such coalitions (Alexander, 1990b; Flinn et al., 2005a,b; Gavrilets and Vose, 2006).

Recognition of kinship among individuals residing in different communities is key to intergroup cooperation. Humans are different from other hominoids in the coevolutionary development of (1) stable and moderately exclusive breeding bonds, (2) bilateral kin recognition and relationships, and (3) reciprocity and kin links among coresident families (Alexander, 1990b; Chapais, 2008, 2014; Flinn et al., 2007). In short, the family was a critical building block for the evolution of more complex communities, such as the band, with flexible residence choice with kin in multiple communities and apparent intentional cultivation of ties with relatives in multiple locations.

Hard evidence for the evolutionary trajectory of human family, kinship, and intergroup relations is scarce and indirect; neurobiology and physiology, however, provide some important clues.

NEUROLOGICAL AND PHYSIOLOGICAL MECHANISMS

Neuroendocrine systems may be viewed as complex sets of mechanisms designed by natural selection to communicate information among cells and tissues. Steroid and peptide hormones, associated neurotransmitters, and other chemical messengers guide the behaviors of mammals in many important ways (Ellison and Gray, 2009; Lee et al., 2009; Panksepp, 2009). For example, analysis of patterns of hormone levels in naturalistic contexts can provide important insights into the evolutionary functions of the neuroendocrine mechanisms that guide human behaviors. Here, the basic neuroendocrine mechanisms that underpin affiliative human family relationships and aggressive coalitional behavior are discussed.

Hormonal Basis for Attachment and Family Love

Some of the most precious human feelings are stimulated by close social relationships: a mother holding her newborn infant; brothers reunited after a long absence; lovers entangled in each other's arms. Natural selection has produced human neurobiological mechanisms, in concert with endocrine systems, that generate potent sensations during our interactions with these most evolutionarily significant individuals. We share with our primate relatives the same basic hormones and neurotransmitters that underlie these mental gifts. But our unique evolutionary history has modified us to respond to different circumstances and situations; we are rewarded and punished for somewhat different stimuli than were our phylogenetic cousins. Chimpanzees and humans share the delight—the sensational reward—when biting into a ripe, juicy mango. But the endocrine, neurological, and associated emotional responses of a human father to the birth of his child (e.g., Gray et al., 2002; Storey et al., 2000) are likely to be quite different from those of a chimpanzee male. Happiness for a human (Buss, 2000) has many unique designs, such as romantic love (Fisher et al., 2006), that involve modifications of the neurological receptors and processors of shared endogenous messengers from our phylogenetic heritage.

Attachments or bonding are central in the lives of social mammals. Basic to survival and reproduction, these interdependent relationships are the fabric of the social networks that permit individuals to maintain cooperative relationships over time. Although attachments can provide security and relief from stress, close relationships also exert pressures on individuals to which they continuously respond. It should not be surprising, therefore, that the neuroendocrine mechanisms underlying attachment and stress are intimately related to one another (see chapter by Worthman). And although more is known about the stress response systems than the affiliative systems, we are beginning to gain some important insights into the neuroendocrine mechanisms that enable human relationships (Lee et al., 2009; Panksepp, 2004).

The mother–offspring relationship is at the core of mammalian life, and it appears that some of the biochemistry at play in the regulation of this intimate bond was also selected to serve in primary mechanisms regulating bonds between mates, paternal care, the family group, and even larger social networks (Fisher et al., 2006; Hrdy, 1999; Wynne-Edwards, 2001). Although a number of hormones and neurotransmitters are involved in attachment and other components of relationships, the two peptide hormones OT and AVP appear to be primary (Carter, 2002; Curtis and Wang, 2003; Heinrichs and Domes, 2008; Heinrichs et al., 2009; Lee et al., 2009; Seltzer et al., 2010; Insel, 2000), with dopamine, cortisol, and other hormones and neurotransmitters having mediating effects.

The hypothalamus is the major brain site where OT and AVP (closely related chains of nine amino acids) are produced. From there, they are released into the central nervous system, as well as transported to the pituitary where they are stored until secreted into the bloodstream. OT and AVP act on a wide range of neurological systems, and their influence varies among mammalian species and by stage of development. The neurological effects of OT and AVP appear to be key mechanisms (e.g., Bartels and Zeki, 2004) involved in the evolution of human family behaviors. The effects of OT and AVP in humans are likely to be especially context dependent, because of the variable and complex nature of family relationships.

Along with OT and AVP, prolactin, estrogen, testosterone, and progesterone are involved in parental care among mammals (Insel, 2000). The roles of these hormones vary across species and between males and females. The effects of these hormones are influenced by experience and context. Among rats, for example, estrogen and progesterone appear to prime the brain during pregnancy for parental behavior. Estrogen has been found to activate the expression of genes that increase the receptor density for OT and prolactin, thus increasing their influence postnatally (Insel, 2000).

OT is most well known for its role in regulating birth and lactation, but along with AVP, it has also been found to play a central role in maternal care and attachment (Fleming et al., 1999). Just prior to birth, an increase in OT occurs that is seen as priming maternal care. An injection of OT in virgin rats has been found to induce maternal care, whereas an OT antagonist administered to pregnant rats interferes with the development of maternal care (Carter, 2002).

Among mammals, hormonal activation initially stimulates maternal behavior among new mothers. Once she has begun to care for her offspring, however, hormones are not required for maternal behavior to continue. Olfactory and somatosensory stimulation from interactions between offspring and mother are, however, usually required for parental care to continue (e.g., Fleming et al., 1999). The stimulation from suckling raises OT levels in rodents and breastfeeding women, which then results in not only milk letdown but also a decrease in limbic hypothalamic–pituitary (anterior)–adrenal (cortex) system (HPA) activity and a shift in the autonomic nervous system from a sympathetic tone to a parasympathetic tone. This results in a calmness seen as conducive to remaining in contact with the infant. It also results in a shift from externally directed energy toward the internal activities of nutrient storage and growth (Uvnas-Moberg, 1998).

Experience also affects the neuroendocrine systems involved in the expression of maternal care. The HPA system of offspring during development is influenced by variations in maternal care, which then influences offspring maternal behavior as adults. Such changes involve the production of, and receptor density for, stress hormones and OT (Champagne and Meaney, 2001; Fleming et al., 1999).

HPA-modulated hormones and maternal behavior are related in humans during the postpartum period (Fleming et al., 1999). During this time, cortisol appears to have an arousal effect, focusing attention on infant bonding. Mothers with higher cortisol levels were found to be more affectionate, more attracted to their infant's odor, and better at recognizing their infant's cry during the postpartum period.

Functional magnetic resonance imaging (fMRI) studies of brain activity involved in maternal attachment in humans indicate that the activated regions are part of the reward system and contain a high density of receptors for OT and AVP (Bartels and Zeki, 2004; Fisher et al., 2006). These studies also demonstrate that the neural regions involved in attachment activated in humans are similar to those activated in nonhuman animals. Among humans, however, neural regions associated with social judgment and assessment of the intentions and emotions of others exhibited some deactivation during attachment activities, suggesting possible links between psychological mechanisms for attachment and management of social relationships. Falling in love with a mate and affective bonds with offspring may involve temporary deactivation of psychological mechanisms for maintaining an individual's social "guard" in the complex reciprocity of human social networks. Dopamine levels are likely to be important for both types of relationship but may involve some distinct neural sites. It will be interesting to see what fMRI studies of attachment in human males indicate, because that is where the most substantial differences from other mammals would be expected. Similarly, fMRI studies of attachment to mothers, fathers, and alloparental care providers in human children may provide important insight into the other side of parent–offspring bonding (Swain, 2011).

Paternal care is uncommon among mammals (Clutton-Brock, 1991). For evolutionary reasons noted earlier, it is found among some rodent and primate species, including humans. The extent and types of paternal care vary among species. The hormonal influence in parental care among males appears to differ somewhat from that found among females. AVP appears to function as a male "booster" to OT (Insel, 2000). Along with prolactin and OT, AVP prepares the male to be receptive to, and care for, infants (Bales et al., 2004).

Paternal care is more common in monogamous than polygamous mammals and is often related to hormonal and behavioral stimuli from the female. In the monogamous California mouse, disruption of the pair-bond does not affect maternal care but does diminish paternal care (Gubernick, 1988). In some other species with biparental care, however, paternal care is not as dependent on the presence of the female (Insel, 2000). Experience also plays a role in influencing hormonal activation and paternal behavior. Among tamarins, experienced fathers have higher levels of prolactin than do first-time fathers (Ziegler and Snowdon, 1997).

Androgens, including testosterone, also appear to be involved in the regulation of paternal behavior. For example, human fathers tend to have lower testosterone levels when they are involved in childcare activities (Berg and Wynne-Edwards, 2001; Fleming et al., 2002; Gray and Campbell, 2009; Kuzawa et al., 2009), although the relation of these levels with the key paternal role of offspring protection is uncertain. Human males stand out as very different from our closest relatives, the chimpanzees, in the areas of paternal attachment and investment in offspring. Investigation of the neuroendocrine mechanisms that underpin male parental behavior may provide important insights into these critical evolutionary changes.

Like male parental care, bonding between mates is also uncommon among mammals but has been selected for when it has reproductive advantages for both parents (Clutton-Brock, 1991; Carter, 2002; Young et al., 2002). Monogamy is found across many mammalian taxa, but most of the current knowledge related to the neuroendocrine basis of this phenomenon has been obtained from the comparative study of two closely related rodent species. The prairie vole (*Microtus ochrogaster*) mating pair nests together and provides prolonged biparental care, whereas their close relative, the meadow vole (*Microtus pennsylvanicus*) does not exhibit these behaviors (Young et al., 2002).

As with other social behaviors in rodents, OT and AVP have been found to be central in the differences that these related species exhibit with respect to pair-bonding. Pair-bonding occurs for the prairie vole following mating. Vaginocervical stimulation results in a release of OT and the development of a partner preference for the female (Carter, 2002). For the male, it is an increase in AVP following mating, and not just OT, that results in partner preference. Injections of exogenous OT in the female and AVP in the male prairie vole result in mate preference even without mating. This does not occur with meadow voles (Young et al., 2002).

The receptor density for OT and AVP in specific brain regions might provide the basis for mechanisms underlying other social behaviors. Other neurotransmitters, hormones, and social cues also are likely to be involved, but slight changes in gene expression for receptor density, such as those found between the meadow and prairie voles in the ventral palladium (located near the nucleus accumbens, an important component of the brain's reward system), might demonstrate how such mechanisms could be modified by selection. The dopamine D2 receptors in the nucleus accumbens appear to link the affiliative OT and AVP pair-bonding mechanisms with positive, rewarding mental states (Aragona et al., 2006; Curtis and Wang, 2003). The combination results in the powerful addiction that parents have for their offspring.

Given the adaptive value of extensive biparental care and prolonged attachment found in the mating pair and larger family network, it is not surprising that similar neurohormonal mechanisms active in the maternal–offspring bond would also be selected to underlie these other attachments. Although some variation exists among species and between males and females, the same general neuroendocrine systems active in pair-bonding in other species are found in our own species (Lee et al., 2009; Panksepp, 2004; Wynne-Edwards, 2003). Androgen response to pair-bonding appears complex (e.g., van der Meij et al., 2008), but similar to parent–offspring attachment in that pair-bonded males tend to have lower testosterone levels in nonchallenging conditions (Alvergne et al., 2009; Gray and Campbell, 2009). Moreover, males actively involved in caretaking behavior appear to have temporarily diminished testosterone levels (Gray et al., 2007).

Hormonal mechanism for another key human adaptation—bonding among adult males to form coalitions (or a "band of brothers")—is less well studied. Social effects, such as victories against outsiders, produce elevations in testosterone, but defeating friends does not (Flinn et al., 2009, 2013; Fuxjager et al., 2009; Gleason et al., 2009; Wagner et al., 2002). Human males, moreover, may differentially respond to females contingent on whether the females are in a stable breeding bond with a close friend; males have lower testosterone after interacting with the wives of their relatives and friends (Flinn et al., 2012). Involvement of the affiliative neuropeptides (OT and AVP) in relationships among adult males is unknown (but see de Dreu, 2012).

The challenge before human evolutionary biologists and psychologists is to understand how these general neuroendocrine systems have been modified and linked with other special human cognitive systems (e.g., Allman 1999; Ponzi et al., 2014; Ulrich-Lai and Herman, 2010) to produce the unique suite of human family behaviors. Analysis of hormonal responses to social stimuli may provide important insights into the selective pressures that guided the evolution of these key aspects of the human mind.

CONCLUSIONS

Human childhood is a life history stage that appears necessary and useful for acquiring the information and practice to build and refine the mental algorithms critical for negotiating the social coalitions that are key to success in our species (del Giudice et al., 2009; cf. Bissonnette et al., 2015). Mastering the social environment presents special challenges for the human child. Social competence is difficult because the target is constantly changing and is similarly equipped with ToM and other cognitive abilities. Parenting is a primary source and mediator of the ontogeny of social competencies.

Social competence is developmentally expensive in time, instruction, and parental care. Costs are not equally justified for all expected adult environments. The human family may help children adjust development in response to environmental exigencies for appropriate trade-offs in life history strategies. An evolutionary developmental perspective of the family can be useful in these efforts to understand this critical aspect of a child's world by integrating knowledge of physiological causes with the logic of adaptive design by natural selection (Flinn, 2006b). Human biology has been profoundly affected by our evolutionary history as unusually social creatures, including, perhaps, a special reliance upon cooperative fathers, grandparents, and kin residing in other groups. Indeed, the mind of the human child may have design features that enable its development as a group project, guided by the multitudinous informational contributions of its ancestors and codescendants (Flinn and Coe, 2007).

The core human adaptations involving affiliation, emotional warmth, and love—stable breeding bonds, biparental care, altricial infancy, prolonged childhood, complex social intelligence, and extended kinship networks—seem difficult to reconcile with the pervasiveness of intergroup aggression. Humans have extraordinary potential for in-group cooperation, perhaps driven in part by an evolutionary history of out-group competition.

REFERENCES

Adolphs, R., 2003. Cognitive neuroscience of human social behavior. Nature Reviews, Neuroscience 4 (3), 165–178.

Aiello, L.C., Wheeler, P., 1995. The expensive-tissue hypothesis: the brain and the digestive system in human and primate evolution. Current Anthropology 36, 199–221.

Alexander, R.D., 1974. The evolution of social behavior. Annual Review of Ecology and Systematics 5, 325–383.

Alexander, R.D., 1979. Darwinism and Human Affairs. University of Washington Press, Seattle.

Alexander, R.D., 1987. The Biology of Moral Systems. Aldine Press, Hawthorne, NY.

Alexander, R.D., 1990a. Epigenetic rules and Darwinian algorithms: the adaptive study of learning and development. Ethology and Sociobiology 11, 1–63.

Alexander, R.D., 1990b. How Humans Evolved: Reflections on the Uniquely Unique Species. Museum of Zoology (Special Publication No. 1). The University of Michigan, Ann Arbor.

Alexander, R.D., 2006. The challenge of human social behavior. Evolutionary Psychology 4, 1–32.

Alexander, R.D., Noonan, K.M., 1979. Concealment of ovulation, parental care, and human social evolution. In: Chagnon, N.A., Irons, W. (Eds.), Evolutionary Biology and Human Social Behavior: An Anthropological Perspective. Duxbury Press, North Scituate, MA, pp. 436–453.

Allman, J., 1999. Evolving Brains. Scientific American Library, New York.

Alvergne, A., Faurie, C., Raymond, M., 2009. Variation in testosterone levels and male reproductive effort: insight from a polygynous human population. Hormones and Behavior 56 (5), 491–497.

Amodio, D.M., Frith, C.D., 2006. Meeting of minds: the medial frontal cortex and social cognition. Nature Reviews Neuroscience 7 (4), 268–277.

Aragona, B.J., Liu, Y., Yu, Y.J., Curtis, J.T., Detwiler, J.M., Insel, T.R., et al., 2006. Nucleus accumbens dopamine differentially mediates the formation and maintenance of monogamous pair bonds. Nature Neuroscience 9, 133–139.

Bales, K.L., Kim, A.J., Lewis-Reese, A.D., Carter, C.S., 2004. Both oxytocin and vasopressin may influence alloparental behavior in male prairie voles. Hormones and Behavior 45 (5), 354–361.

Bartels, A., Zeki, S., 2004. The neural correlates of maternal and romantic love. NeuroImage 21, 1155–1166.

Beckerman, S., Valentine, P. (Eds.), 2002. Cultures of Multiple Fathers: The Theory and Practice of Partible Paternity in South America. University of Florida Press, Gainesville.

Berg, S.J., Wynne-Edwards, K.E., 2001. Changes in testosterone, cortisol, and estradiol levels in men becoming fathers. Mayo Clinic Proceedings 76, 582–592.

Bernhard, H., Fischbacher, U., Fehr, E., 2006. Parochial altruism in humans. Nature 442 (7105), 912–915.

Bissonnette, A., Perry, S., Barrett, L., Mitani, J., Flinn, M.V., Gavrilets, S., De Waal, F.B., 2015. Coalitions in theory and reality: a review of pertinent variables and processes. Behaviour 152 (1), 1–56. http://dx.doi.org/10.1163/1568539x-00003241.

Bjorklund, D.F., Pellegrini, A.D., 2002. The Origins of Human Nature: Evolutionary Developmental Psychology. APA Press, Washington, DC.

Bogin, B., 1999. Patterns of Human Growth, second ed. Cambridge University Press, Cambridge.

Bogin, B., Silva, M.I.V., Rios, L., 2007. Life history trade-offs in human growth: Adaptation or pathology? American Journal of Human Biology 19 (5), 631–642.

Bowles, S., 2009. Did warfare among ancestral hunter-gatherers affect the evolution of human social behaviors? Science 324, 1293–1298.

Brown, D.E., 1991. Human Universals. Temple University Press, Philadelphia.

Buss, D.M., 2000. The evolution of happiness. American Psychology 55, 15–23.

Campbell, A., 2002. A Mind of Her Own: The Evolutionary Psychology of Women. Oxford University Press, London.

Carter, C.S., 2002. Neuroendocrine perspectives on social attachment and love. In: Caciooppo, J.T., Berntson, G.G., Adolphs, R., Carter, C.S., Davidson, R.J., McClintock, M.K., et al. (Eds.), Foundations in Social Neuroscience. MIT Press, Cambridge, MA, pp. 853–890.

Chagnon, N.A., 1966. Yanomamo. Holt, Rinehart & Winston, New York.

Chagnon, N.A., 1979. Mate competition, favoring close kin, and village fissioning among the Yanomamo Indians. In: Chagnon, N.A., Irons, W.G. (Eds.), Evolutionary Biology and Human Social Behavior. Duxbury Press, pp. 86–131. N. Scituate, MA.

Champagne, F.A., Meaney, M.J., 2001. Like mother, like daughter: evidence for non-genomic transmission of parental behavior and stress responsivity. Progress in Brain Research 133, 287–302.

Chapais, B., 2008. Primeval Kinship: How Pair-bonding Gave Birth to Human Society. Harvard University Press, Cambridge, MA.

Chapais, B., 2014. Complex kinship patterns as evolutionary constructions, and the origins of sociocultural universals. Current Anthropology 55 (6), 751–783.

Clutton-Brock, T.H., 1991. The Evolution of Parental Care. Princeton University Press, Princeton, NJ.

Coe, K., 2003. The Ancestress Hypothesis: Visual Art as Adaptation. Rutgers University Press, New Brunswick.

Curtis, T.J., Wang, Z., 2003. The neurochemistry of pair bonding. Current Directions in Psychological Science 12 (2), 49–53.

Dean, C., Leakey, M.G., Reid, D., Schrenk, F., Schwartz, G.T., Stringer, C., Walker, A., 2001. Growth processes in teeth distinguish modern humans from Homo erectus and earlier hominins. Nature 414, 628–631.

De Dreu, C.K.W., 2012. Oxytocin modulates cooperation within and competition between groups: an integrative review and research agenda. Hormones and Behavior 61 (3), 419–428.

Del Giudice, M., 2009. Sex, attachment, and the development of reproductive strategies. Behavior and Brain Science 32, 1–21.

Del Giudice, M., Angeleri, R., Manera, V., 2009. The juvenile transition: a developmental switch point in human life history. Developmental Reviews 29, 1–31.

Dunn, J., 2004. Understanding children's family worlds: family transitions and children's outcome. Merrill-Palmer Quarterly 50 (3), 224–235.

Edelman, G.M., 2006. Second Nature: Brain Science and Human Knowledge. Yale University Press, New Haven, CT.

Elia, M., 1992. Organ and tissue contribution to metabolic rate. In: Kinner, J.M., Tucker, H.N. (Eds.), Energy Metabolism: Tissue Determinants and Cellular Corollaries. Raven Press, New York, pp. 61–79.

Ellison, P.T., 2001. On Fertile Ground, a Natural History of Human Reproduction. Harvard University Press, Cambridge, MA.

Ellison, P.T., Gray, P.B. (Eds.), 2009. Endocrinology of Social Relationships. Harvard University Press, Cambridge, MA.

Fisher, H., Aron, A., Brown, L.L., 2006. Romantic love: a mammalian system for mate choice. Philosophical Transactions of the Royal Society B–Biological Sciences 361, 2173–2186.

Fleming, A.S., Corter, C., Stallings, J., Steiner, M., 2002. Testosterone and prolactin are associated with emotional responses to infant cries in new fathers. Hormones and Behavior 42, 399–413.

Fleming, A.S., O'Day, D.H., Kraemer, G.W., 1999. Neurobiology of mother-infant interactions: experience and central nervous system plasticity across development and generations. Neuroscience and Biobehavioral Reviews 23, 673–685.

Flinn, M.V., 1981. Uterine and agnatic kinship variability. In: Alexander, R.D., Tinkle, D.W. (Eds.), Natural Selection and Social Behavior: Recent Research and New Theory. Blackwell Press, New York, pp. 439–475.

Flinn, M.V., 1997. Culture and the evolution of social learning. Evolution and Human Behavior 18 (1), 23–67.

Flinn, M.V., 2004. Culture and developmental plasticity: evolution of the social brain. In: MacDonald, K., Burgess, R.L. (Eds.), Evolutionary Perspectives on Child Development. Sage, Thousand Oaks, CA, pp. 73–98.

Flinn, M.V., 2006a. Cross-cultural universals and variations: the evolutionary paradox of informational novelty. Psychological Inquiry 17, 118–123.

Flinn, M.V., 2006b. Evolution and ontogeny of stress response to social challenge in the human child. Developmental Review 26, 138–174.

Flinn, M.V., Alexander, R.D., 2007. Runaway social selection. In: Gangestad, S.W., Simpson, J.A. (Eds.), The Evolution of Mind. Guilford Press, New York, pp. 249–255.

Flinn, M.V., Coe, K.C., 2007. The linked red queens of human cognition, coalitions, and culture. In: Gangestad, S.W., Simpson, J.A. (Eds.), The Evolution of Mind. Guilford Press, New York, pp. 339–347.

Flinn, M.V., Geary, D.C., Ward, C.V., 2005a. Ecological dominance, social competition, and coalitionary arms races: why humans evolved extraordinary intelligence. Evolution and Human Behavior 26 (1), 10–46.

Flinn, M.V., Leone, D.V., 2006. Early trauma and the ontogeny of glucocorticoid stress response: grandmother as a secure base. Journal of Developmental Processes 1 (1), 31–68.

Flinn, M.V., Leone, D.V., 2009. Alloparental care and the ontogeny of glucocorticoid stress response among stepchildren. In: Bentley, G., Mace, R. (Eds.), Substitute Parents. Biosocial Society Symposium Series, Berghahn Books, Oxford, pp. 212–231.

Flinn, M.V., Low, B.S., 1986. Resource distribution, social competition, and mating patterns in human societies. In: Rubenstein, D., Wrangham, R. (Eds.), Ecological Aspects of Social Evolution. Princeton University Press, Princeton NJ, pp. 217–243.

Flinn, M.V., Muehlenbein, M.P., Ponzi, D., 2009. Evolution of neuroendocrine mechanisms linking attachment and life history: social endocrinology of the human child. Behavioral and Brain Sciences 32 (1), 27–28.

Flinn, M.V., Nepomnaschy, P.A., Muehlenbein, M.P., Ponzi, D., 2011. Evolutionary functions of early social modulation of hypothalamic-pituitary-adrenal axis development in humans. Neuroscience and Biobehavioral Reviews 35 (7), 1611–1629.

Flinn, M.V., Quinlan, R.J., Ward, C.V., Coe, M.K., 2007. Evolution of the human family: cooperative males, long social childhoods, smart mothers, and extended kin networks (Chapter 2). In: Salmon, C., Shackelford, T. (Eds.), Family Relationships. Oxford University Press, Oxford, pp. 16–38.

Flinn, M.V., Ward, C.V., 2005. Evolution of the social child (Chapter 2). In: Ellis, B., Bjorklund, D. (Eds.), Origins of the Social Mind: Evolutionary Psychology and Child Development. Guilford Press, London, pp. 19–44.

Flinn, M.V., Ward, C.V., Noone, R., 2005b. Hormones and the human family (Chapter 19). In: Buss, D. (Ed.), Handbook of Evolutionary Psychology. Wiley, New York, pp. 552–580.

Fortes, M., 1969. Kinship and the Social Order. Aldine, Chicago.

Fuxjager, M.J., Mast, G., Becker, E.A., Marler, C.A., 2009. The "home advantage" is necessary for a full winner effect and changes in post-encounter testosterone. Hormones and Behavior 56, 214–219.

Gavrilets, S., Vose, A., 2006. The dynamics of Machiavellian intelligence. Proceedings of the National Academy of Sciences 103 (45), 16823–16828.

Geary, D.C., 2005. The Origin of Mind: Evolution of Brain, Cognition, and General Intelligence. American Psychological Association, Washington, DC.

Geary, D.C., Bjorklund, D.F., 2000. Evolutionary developmental psychology. Child Development 71 (1), 57–65.

Geary, D.C., Flinn, M.V., 2001. Evolution of human parental behavior and the human family. Parenting: Science and Practice 1, 5–61.

Geary, D.C., Flinn, M.V., 2002. Sex differences in behavioral and hormonal response to social threat. Psychological Review 109 (4), 745–750.

Geary, D.C., Huffman, K.J., 2002. Brain and cognitive evolution: forms of modularity and functions of mind. Psychological Bulletin 128 (5), 667–698.

Gilbert, S.L., Dobyns, W.B., Lahn, B.T., 2005. Genetic links between brain development and brain evolution. Nature Reviews Genetics 6 (7), 581–590.

Gleason, E.D., Fuxjager, M.J., Oyegbile, T.O., Marler, C.A., 2009. Testosterone release and social context: when it occurs and why. Frontiers in Neuroendocrinology 30 (4), 460–469.

Goodall, J., 1986. The Chimpanzees of Gombe: Patterns of Behavior. Belknap Press of Harvard University Press, Cambridge, MA.

Goodson, J.L., Kabelik, D., 2009. Dynamic limbic networks and social diversity in vertebrates: from neural context to neuromodulatory patterning. Frontiers in Neuroendocrinology 30 (4), 429–441.

Gopnik, A., Meltzoff, A.N., Kuhl, P.K., 1999. The Scientist in the Crib: Minds, Brains, and How Children Learn. William Morrow & Co, New York.

Gough, E.K., 1959. The Nayars and the definition of marriage. Journal of the Royal Anthropological Institute 89, 23–34.

Gray, P.B., Anderson, K.G., 2010. Fatherhood: Evolution and Human Paternal Behavior. Harvard University Press, Cambridge, MA.

Gray, P.B., Campbell, B.C., 2009. Human male testosterone, pair bonding and fatherhood. In: Ellison, P.T., Gray, P.B. (Eds.), Endocrinology of Social Relationships. Harvard University Press, Cambridge.

Gray, P.B., Kahlenberg, S., Barrett, E., Lipson, S., Ellison, P.T., 2002. Marriage and fatherhood are associated with lower testosterone in males. Evolution and Human Behavior 23, 193–201.

Gray, P.B., Parkin, J.C., Samms-Vaughan, M.E., 2007. Hormonal correlates of human paternal interactions: a hospital-based investigation in urban Jamaica. Hormones and Behavior 52, 499–507.

Gray, P.B., Yang, C.J., Pope Jr., H.G., 2006. Fathers have lower salivary testosterone levels than unmarried men and married non-fathers in Beijing, China. Proceedings of the Royal Society of London B: Biological Sciences 273, 333–339.

Guatelli-Steinberg, D., Reid, D.J., Bishop, T.A., Larsen, C.S., 2005. Anterior tooth growth periods in Neanderthals were comparable to those of modern humans. Proceedings of the National Academy of Sciences 102 (40), 14197–14202.

Gubernick, D.J., 1988. Reproduction in the California mouse, *Peromyscus californicus*. Journal of the Mammal 69, 857–860.

Hamilton, W.D., 1976.

Hawkes, K., 2003. Grandmothers and the evolution of human longevity. American Journal of Human Biology 15, 380–400.

Heinrichs, M., Dawans, B.V., Domes, G., 2009. Oxytocin, vasopressin, and human social behavior. Frontiers in Neuroendocrinology 30 (4), 548–557.

Heinrichs, M., Domes, G., 2008. Neuropeptides and social behaviour: effects of oxytocin and vasopressin in humans. Progress in Brain Research 170, 337–350.

Hetherington, E.M., 2003a. Intimate pathways: changing patterns in close personal relationships across time. Family Relations: Interdisciplinary Journal of Applied Family Studies 52 (4), 318–331.

Hetherington, E.M., 2003b. Social support and the adjustment of children in divorced and remarried families. Childhood: A Global Journal of Child Research 10 (2), 217–236 (San Diego: Sage Publications, USA).

Hill, K., Hurtado, A., 1995. Ache Life History: The Ecology and Demography of a Foraging People. Aldine de Gruyter, Hawthorne, NY.

Holliday, M.A., 1986. Body composition and energy needs during growth. In: Falkner, F., Tanner, J.M. (Eds.), Human Growth: A Comprehensive Treatise. Plenum Press, New York.

Hrdy, S.B., 1999. Mother Nature: A History of Mothers, Infants, and Natural Selection. Pantheon, New York.

Hrdy, S.B., 2005. Evolutionary context of human development: the cooperative breeding model. Dahlem Workshop 92. In: Carter, C.S., Ahnert, L. (Eds.), Attachment and Bonding: A New Synthesis. MIT Press, Cambridge, MA.

Hrdy, S.B., 2009. Mothers and Others: The Evolutionary Origins of Mutual Understanding. Harvard University Press, Cambridge.

Insel, T.R., 2000. Toward a neurobiology of attachment. Review of General Psychology 4 (2), 176–185.

Jensen, A.R., 1998. The G Factor: The Science of Mental Ability. Praeger, New York.

Joëls, M., Baram, T.Z., 2009. The neuro-symphony of stress. Nature Reviews Neuroscience 10 (6), 459–466.

Joffe, T.H., 1997. Social pressures have selected for an extended juvenile period in primates. Journal of Human Evolution 32, 593–605.

Kaplan, H., Hill, K., Lancaster, J., Hurtado, A.M., 2000. A theory of human life history evolution: diet, intelligence and longevity. Evolutionary Anthropology 9, 156–183.

Kaplan, H.S., Robson, A.J., 2002. The emergence of humans: the coevolution of intelligence and longevity with intergenerational transfers. Proceedings of the National Academy of Sciences 99 (15), 10221–10226.

Karaolis-Danckert, N., Buyken, A.E., Sonntag, A., Kroke, A., 2009. Birth and early life influences on the timing of puberty onset: results from the DONALD (DOrtmund Nutritional and Anthropometric Longitudinally Designed) Study. American Journal of Clinical Nutrition 90 (6), 1559–1565.

Koch, C., 1999. Biophysics of Computation. Information Processing in Single Neurons. Oxford University Press, New York.

Konner, M., 1991. Childhood. Little, Brown, Boston.

Konner, M., 2010. The Evolution of Childhood: Relationships, Emotion, Mind. Harvard University Press, Cambridge, MA.

Kuzawa, C.W., Gettler, L.T., Muller, M.N., McDade, T.W., Feranil, A.B., 2009. Fatherhood, pairbonding and testosterone in the Philippines. Hormones and Behavior 56 (4), 429–435.

Leblanc, S.A., 2003. Constant Battles: The Myth of the Peaceful, Noble Savage. St. Martin's Press, New York.

Lee, H.-J., Macbeth, A.H., Pagani, J., Young 3rd, W.S., 2009. Oxytocin: the great facilitator of life. Progress in Neurobiology 88 (2), 127–151.

Lee, J.M., Kaciroti, N., Appugliese, D., Corwyn, R.F., Bradley, R.H., Lumeng, J.C., 2010. Body mass index and timing of pubertal initiation in boys. Archives of Pediatrics and Adolescent Medicine 164 (2), 116–123.

Lee, S.H., Wolpoff, M.H., 2003. The pattern of evolution in Pleistocene human brain size. Paleobiology 29, 186–196.

Leigh, S.R., 2004. Brain growth, cognition, and life history in primate and human evolution. American Journal of Primatology 62, 139–164.

Leonard, W.R., Snodgrass, J.J., Robertson, M.L., 2007. Effects of brain evolution on human nutrition and metabolism. Annual Review of Nutrition 27, 311–327.

Levi-Strauss, C., 1949/1969. Bell, J.H., von Sturmer, J.R., Rodney Needham (Trans.), 1969. In: Needham, R. (Ed.), Les Structures élémentaires de la parenté (The elementary structures of kinship). Beacon Press, Boston.

Mace, R., 2000. Evolutionary ecology of human life history. Animal Behaviour 59, 1–10.

Macfarlan, S.J., Walker, R.S., Flinn, M.V., Chagnon, N.A., 2014. Lethal coalitionary aggression and long-term alliances among Yanomamö men. Proceedings of the National Academy of Sciences 111 (47), 16662–16669. http://dx.doi.org/10.1073/pnas.1418639111.

Miller, G.E., 2000. The Mating Mind: How Sexual Choice Shaped the Evolution of Human Nature. Doubleday, New York.

Mitani, J.C., Watts, D.P., Amsler, S.J., 2010. Lethal intergroup aggression leads to territorial expansion in wild chimpanzees. Current Biology 20 (12), R507–R508.

Muehlenbein, M.P., Flinn, M.V., 2011. Patterns and processes of human life history evolution. In: Flatt, T., Heyland, A. (Eds.), Mechanisms of Life History Evolution (Chapter 12). Oxford University Press, Oxford.

Murdock, G.P., 1949. Social Structure. Macmillan, New York.

Murdock, G.P., 1967. Ethnographic Atlas. University of Pittsburgh Press, Pittsburgh.

Panksepp, J., 2004. Affective Neuroscience: The Foundations of Human and Animal Emotions. Oxford University Press, New York.

Panksepp, J., 2009. Carving "natural" emotions: "Kindly" from bottom-up but not top-down. Journal of Theoretical and Philosophical Psychology 28 (2), 395–422.

Ponzi, D., Muehlenbein, M.P., Sgoifo, A., Geary, D.C., Flinn, M.V., 2014. Day-to-day variation of salivary cortisol and dehydroepiandrosterone (DHEA) in children from a rural Dominican community. Adaptive Human Behavior and Physiology 1, 12–24. http://dx.doi.org/10.1007/s40750-014-0002-4.

Posner, M.I., 2005. Genes and experience shape brain networks of conscious control. Progress in Brain Research 150, 173–183.

Quinlan, R.J., Quinlan, M.B., Flinn, M.V., 2003. Parental investment and age at weaning in a Caribbean village. Evolution and Human Behavior 24 (1), 1–17.

Rilling, J.K., 2013. The neural and hormonal bases of human parental care. Neuropsychologia 51 (4), 731–747.

Rilling, J., Gutman, D., Zeh, T., Pagnoni, G., Berns, G., Kilts, C., 2002. A neural basis for social cooperation. Neuron 35 (2), 395–405.

Rosenberg, K., 2004. Living longer: Information revolution, population expansion, and modern human origins. Proceedings of the National Academy of Sciences 101 (30), 10847–10848.

Rosenberg, K., Trevathan, W., 2002. Birth, obstetrics and human evolution. BJOG: An International Journal of Obstetrics and Gynecology 109 (11), 1199–1206.

Roth, G., Dicke, U., 2005. Evolution of the brain and intelligence. Trends in Cognitive Sciences 9 (5), 250–257.

Schoenemann, P.T., 2006. Evolution of the size and functional areas of the human brain. Annual Review of Anthropology 35, 379–406.

Sear, R., Mace, R., McGregor, I.A., 2000. Maternal grandmothers improve the nutritional status and survival of children in rural Gambia. Proceedings of the Royal Society London B 267, 1641–1647.

Sear, R., Mace, R., 2008. Who keeps children alive? A review of the effects of kin on child survival. Evolution and Human Behavior 29 (1), 1–18.

Seltzer, L.J., Ziegler, T.E., Pollak, S.D., 2010. Social vocalizations can release oxytocin in humans. Proceedings of the Royal Society London B. 277 (1694), 2661–2666. http://dx.doi.org/10.1098/rspb.2010.0567.

Sherwood, C.C., Stimpson, C.D., Raghanti, M.A., Wildman, D.E., Uddin, M., Grossman, L.I., et al., 2006. Evolution of increased glia–neuron ratios in the human frontal cortex. Proceedings of the National Academy of Sciences 103 (37), 13606–13611.

Sloboda, D.M., Hart, R., Doherty, D.A., Pennell, C.E., Hickey, M., 2007. Age at menarche: Influences of prenatal and postnatal growth. Journal of Clinical Endocrinology and Metabolism 92, 46–50.

Smuts, B.B., Smuts, R.W., 1993. Male aggression and sexual coercion of females in nonhuman primates and other mammals: evidence and theoretical implications. Advances in the Study of Behavior 22, 1–63.

Spocter, M.A., Hopkins, W.D., Garrison, A.R., Bauernfeind, A.L., Stimpson, C.D., Hof, P.R., Sherwood, C.C., 2010. Wernicke's area homologue in chimpanzees (Pan troglodytes) and its relation to the appearance of modern human language. Proceedings of the Royal Society London B 277, 2165–2174. http://dx.doi.org/10.1098/rspb.2010.0011.

Storey, A.E., Walsh, C.J., Quinton, R.L., Wynne-Edwards, K.E., 2000. Hormonal correlates of paternal responsiveness in new and expectant fathers. Evolution and Human Behavior 21, 79–95.

Swain, J.E., 2011. The human parental brain: in vivo neuroimaging. Progress in Neuro-Psychopharmacology and Biological Psychiatry 35 (4), 1242–1254.

Ulrich-Lai, Y.M., Herman, J.P., 2010. Neural regulation of endocrine and autonomic stress responses. Epub ahead of print, May 13, 2009 Nature Reviews Neuroscience. http://dx.doi.org/10.1038/nrn2647.

Uvnas-Moberg, K., 1998. Oxytocin may mediate the benefits of positive social interaction and emotions. Psychoneuroendocrinology 23, 819–835.

van der Meij, L., Buunk, A.P., van de Sande, J.P., Salvador, A., 2008. The presence of a woman increases testosterone in aggressive dominant men. Hormones and Behavior 54, 640–644.

Voland, E., Chasiotis, A., Schiefenhövel, W., 2005. Grandmotherhood: The Evolutionary Significance of the Second Half of Female Life. Rutgers University Press, New Brunswick, NJ.

Wagner, J.D., Flinn, M.V., England, B.G., 2002. Hormonal response to competition among male coalitions. Evolution and Human Behavior 23 (6), 437–442.

Walker, R.S., Flinn, M.V., Hill, K., 2010. The evolutionary history of promiscuous mating and partible paternity in lowland South America. Proceedings of the National Academy of Sciences.

Walvoord, E.C., 2010. The timing of puberty: Is it changing? Does it matter? Journal of Adolescent Health online.

Williams, R.W., Herrup, K., 1988. The control of neuron number. Annual Review of Neuroscience 11, 423–453.

Wrangham, R.W., 1999. Evolution of coalitionary killing. Yearbook of Physical Anthropology 42, 1–30.

Wrangham, R.W., Peterson, D., 1996. Demonic Males. Houghton Mifflin Company, New York.

Wynne-Edwards, K.E., 2001. Hormonal changes in mammalian fathers. Hormones and Behavior 40, 139–145.

Wynne-Edwards, K.E., 2003. From dwarf hamster to daddy: the intersection of ecology, evolution, and physiology that produces paternal behavior. In: Slater, P.J.B., Rosenblatt, J.S., Snowden, C.T., Roper, T.J. (Eds.), Advances in the Study of Behavior, vol. 32. Academic Press, San Diego, pp. 207–261.

Young, L., Wang, Z., Insel, T.R., 2002. Neuroendocrine bases of monogamy. In: Cacioppo, J.T., Berntson, G.G., Adolphs, R., Carter, C.S., Davidson, R.J., McClintock, M.K., et al. (Eds.), Foundations in Social Neuroscience. MIT Press, Cambridge, MA, pp. 809–816.

Ziegler, T.E., Snowdon, C.T., 1997. Role of prolactin in paternal care in a monogamous New World primate, *Saguinus oedipus*. The integrative neurobiology of affiliation. Annals of the New York Academy of Sciences 807, 599–601.

Chapter 34

Human Mating Systems

Steven W. Gangestad, and Nicholas M. Grebe
Department of Psychology, University of New Mexico, Albuquerque, NM, USA

SYNOPSIS

A variety of different mating systems—distributions of sexual activity within populations—exist in the animal world. Much recent theory in evolutionary biology has clarified the circumstances giving rise to mating systems. An issue of long-standing interest concerns the mating systems of ancestral hominins, giving rise to human mating adaptations. Multiple lines of evidence suggest that humans evolved the capacity to pair-bond, with monogamous or, less commonly, polygynous mating arrangements prevailing and men investing in offspring. At the same time, human mating systems have likely been variable (as they now are, within foraging societies), contingent on ecological circumstances.

MATING SYSTEMS: BASIC CONCEPTS AND UNDERSTANDINGS

The concept of a *mating system* is a core one within reproductive biology. A mating system concerns the distribution of sexual activity, and ultimately reproduction, within a population, defining which males mate with which females and the conditions under which those matings occur. In brief, mating systems dictate who mates with whom, and when. Mating systems are evolved outcomes; they emerge from the costs and benefits of males and females affecting mate choice and competition for mating opportunities, which in turn select for male and female adaptations. The details of a mating system can thereby offer telltale signs of ancestral selection that forged reproductive behavior. But mating systems also affect subsequent evolved outcomes: They impose selective pressures on males and females, in that they affect the reproductive strategies and tactics most beneficial, under various circumstances, to male and female individuals.

A related but distinct concept is that of a *breeding system*. A breeding system specifies not only how sexual unions occur, but also who produces offspring and who invests energetic resources into their care. An example is a communal breeding system, within which adult individuals care for offspring who are not their own. Because a large amount of the total energetic effort that contributes to reproduction often goes toward parental care, knowledge of the breeding system in which a mating system is embedded is often required in order to derive implications of the mating system for selection on male and female phenotypes.

A BASIC CLASSIFICATION OF MATING SYSTEMS

A Traditional View

Within plants, the primary distinction between different kinds of mating systems concerns the extent and circumstances of outcrossing as opposed to selfing. Many plants either self-fertilize (the same individual donating both male and female gametes) or reproduce asexually in the absence of fertilization. Some animals too are asexual, others hermaphroditic. The vast majority of animals, however, reproduce through outcrossing, whereby two different individuals, one male and one female, provide the gametes that define the DNA of the developing offspring. In broad strokes, reproductive biologists traditionally classify animal mating systems involving outcrossing as one of three types.

Monogamy: males and females have exclusive sexual access to one another.

Polygamy: sexual exclusivity occurs, but is not monogamous. Three subtypes of polygamy can be discriminated: (1) *Polygyny*: individual males have exclusive sexual access to two or more females; (2) *Polyandry*: individual females have exclusive sexual access to two or more males; (3) *Polygynandry*: two or more males have exclusive sexual access to two or more females, and vice versa.

Promiscuity: males within a group mate with any female, and vice versa.

Mating structures are often seasonal such that, for instance, male and female pairs coreside for a breeding season but then form new pairs in the next season. In other cases, male–female pairings extend over multiple seasons (see Davies et al., 2012).

Variants with Stable and Unstable Unions

The three traditional categories do not exhaustively cover all possibilities. Monogamy and polygamy assume forms of sexual exclusivity. Promiscuity assumes no sexual exclusivity. In the late 1980s and early 1990s, avian biologists discovered that many birds with pair-bonding, such that male and female pairs share a nest and cooperate to raise a brood, nonetheless exhibit high rates of extra-pair paternity (EPP)—paternity by males other than social partners. Indeed, the mean EPP rate among passerine birds is about 10% (e.g., Griffith et al., 2002; Westneat and Stewart, 2003). An *extra-pair mating system* is one in which most sexual activity occurs within structured units (e.g., individual male–female units, individual male–multifemale units, and so on), but some mating—"extra-pair" mating—takes place outside of these units. In recognition of the social unit in which most mating occurs, systems with pair-bonding and extra-pair sex are said to reflect *social monogamy with EPC* (extra-pair copulation). In essence, this structure of mating is a form of polygynandry, though with the proviso that extra-pair mateships are typically much more fluid and fleeting than in-pair mateships.

Naturally, some systems entail very little EPC, whereas others may entail a great deal. Social monogamy with EPC hence lays on a continuum of degree of sexual exclusivity. The level of EPC is thought to generally be reflected in the EPP rate.

Mixed Systems

Some population exhibit mixed mating systems. Two kinds of mixtures exist. First, different forms of sexual exclusivity may simultaneously exist. Hence, in many avian species with polygyny, such that some males mate with two females, monogamous mating arrangements occur in the same population. Second, different forms of exclusivity may exist temporally or spatially within the same population or species. As noted below, both kinds of mixtures are observed in humans.

Sexual Selection

Sexual selection operates on phenotypes that affect mating success, due to differential access to quantity or quality of mates (see Davies et al., 2012). In turn, differential access to mates can result from (1) differential ability to compete with other individuals of the same sex to gain access to mates (intrasexual competition); (2) differential ability to appeal to the mate preferences of members of the other sex (intersexual choice); (3) differential ability to gain access to matings through coercive tactics (intersexual coercion; see Arnqvist and Rowe, 2005) (see Davies et al., 2012). Mating systems arise, in part, through sexual selection processes; for instance, whether males care for offspring in monogamous pair-bonds, or refrain from care and exclusively engage in competition for mates, depends on the relative success of each strategy (Kokko and Jennions, 2008). The intensity of sexual selection refers to the degree of differential reproductive success that is due to differences in mating success covarying with phenotypic features. For example, when a few members of a sex copulate with many individuals while many others do not copulate with anyone, and success covaries with phenotypic traits, sexual selection is relatively intense.

WHAT SELECTION PRESSURES GIVE RISE TO MATING SYSTEMS?

Traditionally, it was widely assumed by sexual selection theorists that of possible mating systems, those that involve some degree of polygyny—mating by males with multiple females—most readily emerge from selection. The reason is the "Darwin-Bateman effect:" the number of offspring increases as a function of the number of matings with different females by males, but there is no similar increase as a function of the number of matings with different males (in the production of a single offspring or brood) experienced by females (see Parker and Birkhead, 2013). Ultimately, it has been thought that the effect is due to anisogamy, differences in the size of the gametes and the initial investment in offspring by the sexes. In placental mammals, internal fertilization and gestation prevents production of new offspring after conception of another until birth (or some period of lactation) or fetal loss. Hence, according to this paradigm, whereas males can benefit from seeking multiple mates, females reap very little if any benefit. Selection hence may be expected under many conditions to favor male multiple mating, but only rarely to favor female multiple mating.

Polygamy can take on different forms defined by the male features that foster access to multiple females (Emlen and Oring, 1977). In resource-defense polygyny, males control resources valued by females because they fuel offspring production. In male-dominance polygyny, such as in leks, males are chosen by females for their ability to compete with other males, but they offer no material resources to females to be directed toward investment in offspring.

Monogamy, in this view, typically evolves because, due to ecological conditions, males cannot defend the resources that permit them to attract multiple mates. Hence, in this view, most birds (over 90%) are socially monogamous because resources are dispersed to such a degree that males cannot typically command resources for more than one female's offspring at a time (Emlen and Oring, 1977).

THE POLYANDRY "REVOLUTION" WITHIN BEHAVIORAL BIOLOGY

Over the past three decades, behavioral ecologists have increasingly recognized the benefits of a female mating with multiple males, so much so that Parker and Birkhead (2013) refer to the change as a "revolution." Both direct benefits (increasing female reproductive success) and indirect benefits (increasing genetic fitness of offspring) have been recognized. Direct benefits include increased rates of fertilization, increased attainment of resources offered by males, and reduction in male aggression toward offspring due to "paternity confusion" (multiple mating that does not permit males within a group to rule out their own paternity of an offspring). Indirect benefits include attaining a sire with DNA compatible with the mother's (where selection occurs "cryptically"—within the reproductive tract of the female), diversification of offspring (bet-hedging), and in the context of extra-pair mating, attaining a sire with greater genetic fitness ("intrinsic good genes") than an in-pair male. As a result, female multiple mating is not a rare anomaly; it is commonplace.

Extra-Pair Mating

The benefits with the greatest empirical support are direct ones (see discussion of chimpanzees below). The most controversial are "good genes" benefits attained through multiple mating. As noted above, many socially monogamous birds engage in EPC. What are its benefits?

One question is whether any female benefits drive the evolution of EPC in these species. Males, once again, can benefit from multiple mating. Female EPC, then, may be due to male assertion. In some instances, the costs of resisting copulation (e.g., physical harm) outweigh its benefits. EPC can commonly arise, as a result, exclusively due to male sexual interests and female tolerance. Arnqvist and Kirkpatrick (2005) present data consistent with this scenario of "sexual conflict" (see also Parker and Birkhead, 2013).

At the same time, in some avian species females clearly do solicit EPC (e.g., see Parker and Birkhead, 2013). A "good genes" hypothesis is one potential explanation, and Açkay and Roughgarden (2007) reviewed its empirical tests. As roughly half the studies found empirical support for a "good genes" account, they concluded that support is weak, a view echoed by other authors (Arnqvist and Kirkpatrick, 2005; Parker and Birkhead, 2013).

Others have argued that data are inconclusive but not necessarily negative (Eliassen and Kokko, 2008; Griffith, 2007; Whittingham and Dunn, 2010). Açkay and Roughgarden (2007) counted a study as supportive only if a statistically robust result was achieved. Even well-conducted studies can yield inconclusive data when true effects exist due to chance alone. And EPC can arise from multiple causes. When EPC arising from male assertion is combined, in the same review, with EPC arising from female choice for genetic benefits, mixed results are expected.

The jury on avian female EPC functioning to attain "good genes" in a sire, then, is still out. One outstanding question concerns the conditions under which adaptive female EPC evolves. Why in only some species do females appear to derive genetic benefits from EPC (e.g., see Jennions and Petrie, 2000)?

NEW MODELS OF SEXUAL SELECTION

In recent years, theorists have also questioned the traditional view's perspective on male interests. The traditional view is that, because of the Darwin-Bateman effect, males should possess promiscuous sexual motivations and, where ecological circumstances permit, compete for the resources or standing that permit multiple mating. Yet, the fact that males may reproductively benefit from attaining multiple mates need not imply that every male benefits from putting forth the effort to compete for multiple mates. Males can benefit from providing care for existing offspring as well. When the net benefits from providing care to offspring exceed the net benefits from competing for mates, males should be selected to care. These conditions are most likely to exist when (1) competing for mates is costly (e.g., males risk injury) relative to the costs of caring, (2) the net benefits to competing are relatively weak (a male is not competitive, or many males compete), and (3) the benefits to care are relatively great (males are able to discriminate their own offspring, and offspring benefit from complementary care from males and females) (Kokko and Jennions, 2008). This framework offers a different perspective on why pair-bonding with biparental care is so common in birds. The traditional view focuses on male inability to defend the resources supportive of multiple mates and their offspring (e.g., due to resource dispersion). More recent theory focuses instead on the substantial benefits to male care (such as the complementarity in aerial species, with one parent defending a nest of offspring that can't yet fly, while the other forages) and the relatively poor gain from competing (Kokko and Johnstone, 2002; other circumstances favoring the evolution of male care are discussed by Kokko and Jennions, 2008).

HUMAN MATING SYSTEMS

We now turn to mating systems in humans. The primary question is not what mating systems humans have created. Human groups, large and small, have created and can create a vast array of different mating systems. Rather, the primary question is, what mating systems recurrently existed in ancestral human groups, such that male and female adaptations for mating were shaped by selection imposed by those mating systems?

Mating in Hominoids and the Ancestral State

One issue is a phylogenetic one. Whatever mating systems evolved in humans, they evolved from a mating system in an ancestral species that the human lineage shares with related ape species. Humans share a common ancestor with chimpanzees and bonobos that existed more than five million years ago (see chapter by Hunt in this volume). These three species share a common ancestor with gorillas that existed several million years previously. What mating system did these ancestral species possess?

This issue has been vigorously debated recently. Each of our closest relatives—chimpanzees, bonobos, and gorillas—has distinct mating systems. If all were the same, the parsimonious inference would be that our common ancestor had that mating system. The fact that they differ leaves open the possibility that our common ancestor had any of the three. In general, however, scholars tend to favor either a chimpanzee or bonobo mating system as the basic ancestral model (e.g., Dixson, 2009; Chapais, 2008).

Chimpanzee Mating

Both chimpanzees and bonobos live in groups. In both species, the mating system is promiscuous (see chapter by Watts and Aronsen). Females are sexually active about 10 days per ~30-day cycle, during which they possess hindquarter sexual swellings (a feature likely not prominent in our own ancestry; e.g., Pagel and Mead, 2006). During their sexual phase, females will mate with all adult males in the group, typically multiple times. The favored hypothesis for the function of promiscuous mating is Hrdy's (1979): promiscuous mating confuses paternity. If no male can rule out his own paternity, each is less likely to aggress against the subsequent offspring. Recent studies suggest that female sexual interests change across the sexual period. Only a two- to three-day window within it is conceptive; on the remaining days, females are not fertile. During "extended sexuality" (the nonfertile phase), females are most promiscuous. During the fertile phase, in contrast, females tend to prefer particular males (in one study, younger males who were rising in status; Stumpf and Boesch, 2005). One explanation is that females confuse paternity during nonconceptive (extended) sexuality, but they bias sireship toward males who can offer genetic benefits or certain forms of direct benefits (e.g., physical protection) when fertile. And indeed, male reproductive success is very skewed, with one or two males typically siring many more offspring than the majority of other adult males, despite promiscuity (e.g., Stumpf and Boesch, 2005).

Bonobo Mating

The mating system of bonobos is similarly promiscuous. Female coalitions, however, are considerably more powerful in bonobo than chimpanzee society. Female–female sexual exchanges regulate female status hierarchies and coalitional activity (Hohmann and Fruth, 2000). No study has yet examined changes in female sexual activity across the cycle. Despite widespread promiscuity, however, male reproductive skew is strong, suggesting similar bias in mating with certain males at peak fertility (Gerloff et al., 1999). (An alternative view is that the skew is due to dominant males being able to hoard matings at peak fertility, a possibility that better fits chimpanzees than bonobos; Muller et al. 2009b; Wobber et al., 2013; Surbeck et al., 2012).

As Dixson (2009) notes, humans have diverged considerably from chimpanzees in a variety of respects. Testes size and ejaculate volume in chimpanzees is consistent with the high degree of sperm competition (multiple males' sperm residing in the female reproductive tract vying for conception) evident in their mating system. Human testes and sperm volume are consistent with lower levels of sperm competition (e.g., Gomendio et al., 1998). Nonetheless, human mating adaptations may possess vestiges of chimpanzee mating, as illustrated below (Gangestad and Garver-Apgar, 2013).

Human Marital Systems

Marriage is a near-universal institution in human societies. (Purportedly, the Na, an ethnic minority living in the Himalayan foothills in China, lack any such institution. Rather, brothers and sisters live together for life. Siblings help women care for offspring. Fathers do not (see Hua, 2001)). The Standard Cross-Cultural Sample (SCCS) is a collection of 186 modern and historical human societies selected by Murdock and White (1969) because they are, purportedly, weakly redundant representations of human culture, not closely deriving from common cultures or possessing similarities due to horizontal cultural diffusion. Within the SCCS, over 80% of societies permit polygyny. Fewer than 20% are completely monogamous, and 1% are characterized by a nonzero level of polyandry. As agriculture, herding, and other relatively recent means of production may alter mating arrangements, Marlowe (2003b) examined mating arrangements in the SCCS's 36 foraging groups (that attain <10% of their diet from cultivated foods or domesticated animals), and found that about 90% exhibit a nonzero level of polygyny.

Even in most societies that permit polygyny, however, monogamy is the norm. In about two-thirds of foraging societies, for instance, the percentage of polygynously married women is 12% or less (Marlowe, 2003b). In only 13% of these societies does the rate reach 50%. Most marital arrangements across human foraging societies, then, are monogamous unions, though many societies permit polygamy, and there is much cross-cultural variability in rates of polygyny.

Reproductive Skew

Mating systems give rise to the nature and level of reproductive skew. Monogamous mating systems generate variation in reproductive success nearly equal across the sexes. Polygynous ones generate greater male than female variation. Brown et al. (2009) examined variance in reproductive success across 18 traditional or preindustrial societies. On average, male variance exceeded female variance (median ratio $\approx$ 1.70). Yet, consistent with degree of polygyny varying widely across societies, so does this ratio; values range from <1 to nearly 5.

THE HUMAN CASE: DO MALES POSSESS ADAPTATIONS FOR CARE?

Marital arrangements, of course, need not be direct reflections of mating arrangements. And, indeed, a question intensely debated within evolutionary anthropology over the past few decades is whether human pair-bonding is central to the mating system. A fundamental issue in this regard concerns male care. Do males possess adaptations for providing care for offspring? Did ancestral conditions that favored male care, at least contingently, exist recurrently in human groups, leading selection to shape male features that promoted care? Pair-bonding need not imply biparental care, but the two are strongly associated in the animal world.

Hunting-As-Paternal-Effort Views

In most primate species, adult individuals of both sexes are largely responsible for their own subsistence, and mothers harvest the overwhelming majority of calories consumed by offspring during pregnancy and lactation. In contrast, in most human foragers, men generate more calories than they consume: 64% of the total calories produced in 95 foraging societies on which sufficient information is available (Marlowe, 2001). The primary activity through which men generate surplus calories in foraging societies is hunting (broadly defined to include any activity aimed to harvest animal meat, including fishing). Though women forage and extract roots (and, in a minority of societies, produce more calories than men), only rarely do they hunt a lot (for an exception, see Hart et al., 1987). Whereas chimpanzees obtain about 95% of their calories from collected foods requiring no extraction (e.g., fruits and leaves), only about 8% of calories consumed by modern hunter–gatherers are from foods requiring no extraction, with a large proportion (30–80%) derived from vertebrate meat (Kaplan et al., 2000).

Women reproductively benefit from the male-generated surplus. The level of male contribution to the diet varies considerably across foraging societies (~40%–90+%). Women's offspring production covaries positively with the male contribution to subsistence (Marlowe, 2001). Specifically, men's contribution to subsistence predicts shorter interbirth intervals (see Ellison, 2001).

A traditional anthropological view is that male surplus food production evolved as paternal care (e.g., Lancaster and Lancaster, 1983; Lovejoy, 1981; Westermarck, 1929). In this view, the nuclear family is a key economic unit in the evolution of human mating relations. For subsidies generated by male hunting to function as parental effort, nutrients that men generate must flow from them to mates (and then to offspring), or directly to offspring.

Hunting-As-Mating-Effort Views

The male-hunting-as-parental-effort theory, as critics claim, faces a fundamental difficulty: nuclear families are not, in fact, potent economic units in foraging societies (Hawkes, 1991, 2004; Hawkes et al., 1991, 2001). In the Hadza of Tanzania and the Ache of Paraguay, for instance, hunters have little control over the distribution of meat they generate. Instead, meat is shared widely across community members. According to Hawkes (2004), then, men's hunting functions as (that is, evolved for) mating effort to compete for access to mates through "showing off" (Hawkes et al., 1991) rather than as parental effort. Men garner prestige through successful hunting exploits, and prestige translates into mating opportunities (including with other men's wives) (see Kaplan and Hill, 1985; Marlowe, 2003a).

Of course, male hunting subsidizes the diets of women and their offspring. But these subsidies, in this view, are not generated directly by women's own mates or by children's own fathers. Rather, they are generated through the efforts of men in general to gain mates. In economists' terms, the surplus calories generated by male hunting that benefit women and offspring are "positive externalities" of men showing off—windfalls they enjoy, not benefits the men's efforts were designed to achieve.

Hawkes et al. (2001) do argue that the diets of women and their children are subsidized through the efforts of family members, but not primarily husbands. Rather, maternal kin—most importantly, mothers' mothers (i.e., children's grandmothers)—directly subsidize the diets of women of reproductive age and their offspring (Hawkes, 2004). More generally, Hrdy (2009) has argued that humans have evolved to be communal breeders, with many adults (often female kin) sharing the responsibility of caring for offspring. (On the negative effects of communal breeding on male investment in offspring, see Wu et al., 2013.)

A Blended View

A blended view is possible: historically, men may have benefited from hunting in currencies of enhanced viability of offspring *and* mating opportunities (e.g., Gurven and Hill, 2009). Patterns of Hadza foraging rates and activities support it (Marlowe, 2003a). Overall, married Hadza women produce as many calories as married Hadza men. Women with young children, however, do not, as their childcare interferes with effective foraging. In such circumstances, their husbands forage more; in couples with an infant less than one year of age, men produce almost 70% of the calories. Hadza men adjust their work efforts (and perhaps the prey items they target) in response to the direct food production of wives, as it varies with the presence or absence of young children, which the hunting-as-mating-effort view cannot explain (see also Marlowe, 1999.) Across the SCCS, pair-bond stability predicts older ages at weaning (Quinlan and Quinlan, 2008; see also Quinlan et al., 2003). Lactation interferes with women's ability to produce food, and male subsidy permits women to nurse longer.

Earlier, the great variability in degree of polygyny across human foraging societies was noted. As expected, if men contribute to household consumption, societies with high levels of monogamy are also the ones in which men produce relatively many calories (Marlowe, 2003b).

EXTRA-PAIR PATERNITY: SEXUAL OR SOCIAL MONOGAMY?

Again, marital systems pertain to social arrangements. They need not imply sexual exclusivity. Though most offspring may be produced by in-pair partners, some may be produced through extra-pair matings. A question of interest to evolutionary anthropologists is, how common is extra-pair paternity (EPP)?

The answer involves guesswork. Anderson (2006) reviewed studies estimating nonpaternity, largely in developed countries (see also Voracek et al., 2008). Nearly all samples give biased estimates, depending on how they were selected. High paternity confidence samples estimate an EPP rate averaging just 2%. Samples of men with suspicions estimate it to be, on average, about 30%. Once again, tremendous variation exists—in high paternity confidence samples, from <1% (in Switzerland and Germany; see also Wolf et al., 2012) to about 12% (in Monterrey, Mexico).

More recently, Scelza (2011) reported a level of 17% EPP (based on women's reports) in the Himba of Namibia. All occurred within arranged marriages. For some other traditional societies, such as the Tiwi (e.g., Hart et al., 1987) and Canela (Crocker, 1990), ethnographies report very frequent EPC. Within developed countries, EPP rates have declined in the past century; the advent of effective birth control may reduce incidence of conceptive EPC, suppressing the EPP rate (Voracek et al., 2008). In traditional settings, then, one guess is that EPP rates have on average been low (5% or less) but variable (in some societies >10%).

As discussed earlier, the benefits that give rise to female extra-pair mating are debated; some arguing that benefits lie exclusively with males who initiate them, with others open to the possibility that females too, in certain circumstances, benefit.

ADAPTATIONS FOR MATING

Thus far, patterns of human marriage and reproduction have been described. In general, the evidence is consistent with human pair-bonding and biparental investment, albeit with modest levels of polygyny, nonzero rates of EPC, and notable variability. We now turn to an important source of evidence of human mating systems: male and female adaptations for mating. Recurrent mating systems through recent human evolution favored certain male and female features over others. Those features, in some instances, serve as telltale signatures of the mating system that led them to evolve. We focus our discussion on several of them. (For others, see, e.g., Dixson, 2009.)

Male Adaptations for Pair-Bonding and Parenting

Both sexes have the capacity for romantic love, a capacity that can be found across cultures (e.g., Jankowiak and Fischer, 1992; see also Fisher et al., 2005). The precise function of romantic love is not clear. One view is that romantic love motivates the individual experiencing the emotion to protect the pair-bond (Fisher, 2004). Another view is that romantic love and the behavior it promotes act as signals to the loved target that the lover intends to be loyal and faithful (Frank, 1988; Gangestad and Thornhill, 2007). These views are not mutually exclusive; specific features of romantic love may have been shaped by each function.

Mutual Mate Choice

In species in which males and females pair-bond, "mutual mate choice" typically evolves: Members of both sexes are advantaged through preference for some mates over others (e.g., Kokko and Johnstone, 2002). In many instances, choice for mates that exhibit good parenting qualities should be preferred. Studies of mate preferences strongly point to mutual mate choice in modern human societies. In seeking a long-term mate, both men and women are equally "choosy" (e.g., Kenrick et al., 1990). And in Buss's (1989) classic study of mate preferences in 39 cultures, both men and women on average rated "kindness and understanding" as the top preference.

We can look to specific forms of mate preference for particularly compelling examples of adaptation for mutual choice. Specific major histocompatibility complex (MHC) genes code for cell-surface markers that function to "declare" that a cell is uninfected (when the MHC molecule presents only self-peptides) or infected (when the MHC molecule binds a non-self-peptide structure "visible" to the immune system). All else equal, it pays to mate with someone who possesses alleles different from one's own, as then only advantaged heterozygotic offspring are conceived (see Penn and Potts, 1999; Geise and Hedrick, 2003; Thursz et al., 1997; Wegner et al., 2003; though some have argued that intermediate levels of heterozygosity across MHC loci are advantaged in some species such as stickleback fish; Wegner et al., 2003). MHC appears to be detectable through signatures in scent, and possibly by visual stimuli. In a variety of species (see Penn and Potts, 1999), females prefer the scent of males who possess different MHC from their own. Studies on humans strongly suggest that we too are most sexually attracted to scents of others who possess nonshared MHC alleles. Consistent with *mutual* mate choice, preferences exist in *both* sexes (4 of 5 studies of normally ovulating women: Wedekind et al., 1995; Wedekind and Füri 1997; Santos et al., 2005; Tal 2009; cf. Thornhill et al., 2003; and 3 of 4 studies of men: Wedekind and Füri, 1997; Thornhill et al., 2003; Tal, 2009; cf. Santos et al., 2005).

Discriminative Care

Men should not invest in offspring unconditionally. As paternity is not 100% certain, men's parental efforts may be contingent on cues of paternity, such as self-resemblance. In a large Western sample, men assisted offspring they report likely to be their own genetic offspring more than those they suspect reflect EPP (Anderson et al., 2007). Ingenious behavioral studies have examined possible psychological underpinnings of discriminative parenting. In one design, a digital photograph is taken of a participant. The participant's own face, or alternatively that of another participant, is digitally combined with the face of a small child to create two composite images of child faces—one "self-resembling" and one not. Participants choose which of the two children they would be more likely to invest in (e.g., spend time with, or adopt). Men prefer the self-resembling one, an effect not due to conscious recognition of self-resemblance (Platek et al., 2002, 2003, 2005; DeBruine, 2004). Indeed, men respond to self-resembling child faces with more overall brain activation than women, despite women exhibiting stronger brain responses to presentations of children's faces in general (Platek et al., 2004).

Oxytocin

Oxytocin is a peptide hormone produced by the posterior pituitary gland that functions, among other things, as an important neurotransmitter that facilitates maternal focus on offspring well-being (e.g., Gordon et al., 2011). It is also associated with consortship formation in some nonhuman primates (e.g., Moscovice and Ziegler, 2012). Only recently have investigators examined levels of oxytocin in romantically involved couples. Schneiderman et al. (2012) found that both men and women in newly involved relationship couplings exhibited high serum levels of oxytocin, especially if they evidenced "interactive reciprocity" (e.g., through affectionate touch).

Male Testosterone and Care

Hormones may be thought of as messengers in distributed communication networks (e.g., Finch and Rose, 1995). Across vertebrate taxa, physiological actions of the androgen testosterone (T) have been shaped to facilitate male mating effort by channeling energetic resources to features particularly useful in male–male competition (e.g., muscles, sensitivity to dominance ranks and cues of social hierarchy; e.g., Mazur and Booth, 1998), and due to necessary trade-offs, away from other targets of allocation (e.g., repair and immune function; see Bribiescas, 2001; Ellison, 2003; Muehlenbein and Bribiescas, 2005). In species in which males exert parental effort, a modification may often have evolved: T may also modulate allocations of effort to mating versus parenting (in shorthand, competing vs caring). In some species in which males invest in offspring (e.g., marmosets and some birds), male T levels drop after the birth/hatching of the mate's offspring (e.g., Nunes et al., 2001; for a review, see Muller and Wrangham, 2001). This feature of T has independently evolved several times (e.g., in birds at least once, and in mammals several times).

In some birds, males who have already mated (and perhaps fathered offspring in a season) may benefit from seeking additional mates, particularly in pair-bonding bird species in which females engage in EPCs at higher rates. When females are relatively "faithful" to their partners, few additional mating possibilities are available, and male efforts to seek additional matings are on average less successful. Across bird species, male T levels covary positively with the total EPP rate (Garamszegi et al., 2005), but not with overall levels of polygyny. This pattern is consistent with T in male birds, having evolved in response to conditions that affect the relative value of exerting mating versus parental effort.

In pair-bonding birds and some mammals, then, reductions in male T following birth of their social mate's offspring may reflect adaptation for exerting parental effort. Mounting evidence suggests that men's T does too; on average, men's T levels drop when they become mated or have offspring (e.g., Booth and Dabbs, 1993; Gray et al., 2002, 2004, 2006; Mazur and Michalek, 1998; Burnham et al., 2003; Berg and Wynne-Edwards, 2001; Storey et al., 2000; Gettler et al., 2011). Some evidence indicates that the association depends partly on changes in T following changes in mating or paternal status (e.g., Mazur and Michalek, 1998; cf. Van Anders and Watson, 2006). Among fathers, those with lower levels of T and smaller testes evidence a pattern of brain activation when looking at photos of their own children indicative of greater paternal involvement; men with higher levels of T evidence lower paternal involvement (Mascaro et al., 2013); for additional evidence that T functions to modulate mating and parental effort, see also Fleming et al., 2002.

The effect of mating status on men's T is moderated by men's interest in pursuing extra-pair relationships with women other than their primary partners (McIntyre et al., 2006; Edelstein et al., 2011). Men with little interest in and history of extra-pair relationships reveal the typical drop in T when mated, as compared with being single. Men with interest in extra-pair relationships, in contrast, showed no difference: their T levels were just as high when in relationships as when single (McIntyre et al., 2006; Edelstein et al., 2011).

Muller et al. (2009a) examined differences in T levels as a function of paternity in two neighboring Tanzanian groups that differ in paternal involvement. Hadza forager men engage in substantial amounts of paternal care, whereas Dotoga pastoralist fathers rarely engage in direct paternal care but instead invest substantially in efforts to find new mates. As expected, Hadza men's T levels vary as a function of fatherhood; Datoga men's do not. Gray (2003) found that when polygynously mated, Kenyan Swahili men's T levels remain high, perhaps because maintaining multiple mates requires sustained mating effort.

Female Estrus and Extended Sexuality

In most vertebrate species, females are sexually receptive during their fertile reproductive phase, referred to as estrus. For decades, scholars had thought that women had evolved to lose estrus, replaced by continuous, unchanging sexual interests across their reproductive cycles. In recent years, however, research has demonstrated that women's sexual interests do change across the cycle (e.g., Gangestad and Thornhill, 2008; Thornhill and Gangestad, 2008; Gildersleeve and Haselton, in press). When fertile, women are attracted more to certain male features (in broad strokes, masculine features—e.g., more masculine faces and voices). And, should their primary partners lack those features, they report greater attraction to men other than their partners, but only when fertile (e.g., Gangestad et al., 2005; Haselton and Gangestad, 2006; Larson et al., 2012). One study found that women's sexual fantasies are more likely to include multiple men when fertile (Nummi and Pellikka, 2011). One possibility is that women's estrus was not fully lost, and was modified by selection to facilitate adaptive EPC when fertile. Nonetheless, another possibility cannot be ruled out: that women have retained, at least in vestigial form, estrous sexual interests of an ancestor shared with chimpanzees, not functionally modified in the context of pair-bonding (e.g., Gangestad and Garver-Apgar, 2013).

Women do possess, in addition to estrus, a long phase of extended sexuality. They are sexually receptive and proceptive throughout the cycle, most of which is nonconceptive (see Martin, 2007). Unlike chimpanzees, however, extended sexuality in women does not function to confuse paternity. The little work focused on identifying the distinctive features of extended sexuality performed to date suggests that, to the contrary, sexual interests during the nonconceptive phase function, at least in part, to foster intimacy with valued pair-bond partners (e.g., Grebe et al., 2013; Sheldon et al., 2006). Women's extended sexuality, then, possesses telltale signs of selection favoring pair-bonding.

CONCLUSIONS

What can we conclude about the nature of human mating systems, as they have shaped human adaptations for reproduction?

First, humans likely evolved systems in which pair-bonding has been important, and in which men often exert considerable effort to invest in offspring (e.g., Chapais, 2008; Dixson, 2009; Thornhill and Gangestad, 2008). Social monogamy is the most frequent mating arrangement, especially when male contribution to subsistence is great, but most human societies permit polygyny. Both men and women possess adaptations that serve as telltale signs of selection in favor of pair-bonding and paternal investment.

Second, social monogamy need not imply sexual monogamy. And indeed, nonzero though typically low EPP rates characterize most human groups. Female benefits to EPC ancestrally are not fully understood. They may or may not include genetic benefits.

Third, despite these general patterns, wide variations across human groups in degrees of monogamy, sex differences in reproductive skew, and EPP rates exist. Some variation derives from flexibility in who invests heavily in offspring: Reliance on paternal contributions fosters monogamy, and we suspect relatively low EPP rates; heavy reliance on maternal kin (arrangements of communal breeding) likely leads to increased EPP rates.

Finally, human mating systems likely have vestiges of the mating systems of our common ancestors with close relatives, including chimpanzees and bonobos. Evolutionary biologists have intensely debated the order in which we proceeded through specific primate-like mating systems, as well as the events and adaptations that eventually spurred a transition to distinctly human mating systems. Though no clear consensus exists, ongoing research can help illuminate the functional framework of a remarkably flexible mating system. While human mating systems vary, they do so in systematic, structured ways.

REFERENCES

Akcay, E., Roughgarden, J., 2007. Extra-pair paternity in birds: review of the genetic benefits. Evolutionary Ecology Research 9, 855–868.

Anderson, K.G., 2006. How well does paternity confidence match actual paternity? Evidence from worldwide nonpaternity rates. Current Anthropology 47, 513–520.

Anderson, K.G., Kaplan, H., Lancaster, J., 2007. Confidence of paternity, divorce, and investment in children by Albuquerque men. Evolution and Human Behavior 28, 1–10.

Arnqvist, G., Kirkpatrick, M., 2005. The evolution of infidelity in socially monogamous passerines: the strength of direct and indirect selection on extrapair copulation behavior in females. American Naturalist 165, S26–S37.

Arnqvist, G., Rowe, L., 2005. Sexual Conflict. Princeton University Press, Princeton, NJ.

Berg, S.J., Wynne-Edwards, K.E., 2001. Changes in testosterone, cortisol, and estradiol levels in men becoming fathers. Mayo Clinic Proceedings 76, 582–592.

Booth, A., Dabbs, J.M., 1993. Testosterone and mens' marriages. Social Forces 72, 463–477.

Bribiescas, R.G., 2001. Reproductive ecology and life history of the human male. Yearbook of Physical Anthropology 44, 148–176.

Brown, G.R., Laland, K.N., Borgerhoff Mulder, M., 2009. Bateman's principles and human sex roles. Trends in Ecology and Evolution 24, 297–304.

Burnham, J.C., Chapman, J.F., Gray, P.B., McIntyre, M.H., Lipson, S.F., Ellison, P.T., 2003. Men in committed, romantic relationships have lower testosterone. Hormones and Behavior 44, 119–122.

Buss, D.M., 1989. Sex differences in human mate preferences: evolutionary hypotheses tested in 37 cultures. Behavioral and Brain Sciences 12, 1–49.

Chapais, B., 2008. Primeval Kinship: How Pair-bonding Gave Birth to Human Society. Harvard University Press, Cambridge, MA.

Crocker, W.H., 1990. The Canela (Eastern Tibira): I. Ethnographic Introduction. Smithsonian Institution Press, Washington, DC.

DeBruine, L.M., 2004. Resemblance to self increases the appeal of child faces to both men and women. Evolution and Human Behavior 25, 142–154.

Dixson, A.F., 2009. Sexual Selection and the Origins of Human Mating Systems. Oxford University Press, New York.

Davies, N.B., Krebs, J.R., West, S.A., 2012. An Introduction to Behavioral Ecology, fourth ed. Wiley-Blackwell, Oxford, UK.

Edelstein, R.S., Chopik, W.A., Kean, E.L., 2011. Sociosexuality moderates the association between testosterone and mating status in men and women. Hormones and Behavior 60, 248–255.

Eliassen, S., Kokko, H., 2008. Current analyses do not resolve whether extra-pair paternity is male or female driven. Behavioral Ecology and Sociobiology 62, 1795–1804.

Ellison, P.T., 2001. On Fertile Ground: A Natural History of Reproduction. Harvard University Press, Cambridge, MA.

Ellison, P.T., 2003. Energetics and reproductive effort. American Journal of Human Biology 15, 342–351.

Emlen, S.T., Oring, L.W., 1977. Ecology, sexual selection, and the evolution of mating systems. Science 197, 215–223.

Finch, C.E., Rose, M.R., 1995. Hormones and the physiological architecture of life history evolution. Quarterly Review of Biology 70, 1–52.

Fisher, H., 2004. Why We Love: The Nature and Chemistry of Romantic Love. Henry Holt, New York.

Fisher, H., Aron, A., Brown, L.L., 2005. Romantic love: an fMRI study of a neural mechanism for mate choice. Journal of Comparative Neurology 493, 58–62.

Fleming, A.S., Corter, C., Stallings, J., Steiner, M., 2002. Testosterone and prolactin are associated with emotional responses to infant cries in new fathers. Hormones and Behavior 42, 399–413.

Frank, R.H., 1988. Passions within Reason: The Strategic Role of the Emotions. Norton, New York.

Gangestad, S.W., Garver-Apgar, C.E., 2013. The nature of female sexuality: insights into the dynamics of romantic relationships. In: Simpson, J.A., Campbell, L. (Eds.), The Oxford Handbook of Relationships. Oxford University Press, New York.

Gangestad, S.W., Thornhill, R., 2007. The evolution of social inference processes: the importance of signaling theory. In: Forgas, J.P., Haselton, M.G., von Hippel, W. (Eds.), Evolutionary Psychology and Social Cognition. Psychology Press, New York, pp. 33–48.

Gangestad, S.W., Thornhill, R., 2008. Human oestrus. Proceedings of the Royal Society of London B 275, 991–1000.

Gangestad, S.W., Thornhill, R., Garver-Apgar, C.E., 2005. Women's sexual interests across the ovulatory cycle depend on primary partner fluctuating asymmetry. Proceedings of the Royal Society of London B 272, 2023–2027.

Garemszegi, L.Z., Eens, M., Hurtrez-Boussès, S., Møller, A.P., 2005. Testosterone, testes size and mating success in birds: a comparative study. Hormones and Behavior 47, 389–409.

Geise, A.R., Hedrick, P.W., 2003. Genetic variation and resistance to a bacterial infection in endangered Gila topminnow. Animal Conservation 6, 369–377.

Gerloff, U., Hartung, B., Fruth, B., Hohmann, G., Tautz, D., 1999. Intercommunity relationships, dispersal pattern and paternity success in a wild living community of Bonobos (*Pan paniscus*) determined from DNA analysis of faecal samples. Proceedings of the Royal Society of London B 266, 1189–1195.

Gettler, L.T., McDade, T.W., Kuzawa, C.W., 2011. Cortisol and testosterone in Filipino young adult men: evidence for co-regulation of both hormones by fatherhood and relationship status. American Journal of Human Biology 23, 609–620.

Gildersleeve, K.A. Haselton, M.G. Do women's mate preferences change across the ovulatory cycle?: A meta-analytic review. Psychological Bulletin, in press.

Gomendio, M., Harcourt, A.H., Roldan, E.R.S., 1998. Sperm competition in mammals. In: Birkhead, T.R., Møller, A.P. (Eds.), Sperm Competition and Sexual Selection. Academic Press, New York, pp. 667–756.

Gordon, I., Martin, C., Feldman, R., Leckman, J.F., 2011. Oxytocin and social motivation. Developmental Cognitive Neuroscience 1, 471–493.

Gray, P.B., 2003. Marriage, parenting, and testosterone variation among Kenyan Swahili men. American Journal of Physical Anthropology 122, 279–286.

Gray, P.B., Chapman, J.F., Burnham, T.C., McIntyre, M.H., Lipson, S.F., Ellison, P.T., 2004. Human male pairbonding and testosterone. Human Nature 15, 119–131.

Gray, P.B., Kahlenberg, S.M., Barrett, E.S., Lipson, S.F., Ellison, P.T., 2002. Marriage and fatherhood are associated with lower testosterone in males. Evolution and Human Behavior 23, 193–201.

Gray, P.B., Yang, C.F.J., Pope, H.G., 2006. Fathers have lower salivary testosterone levels than unmarried men and married non-fathers in Beijing, China. Proceedings of the Royal Society of London B 273, 333–339.

Grebe, N.M., Gangestad, S.W., Garver-Apgar, C.E., Thornhill, R., 2013. Women's luteal phase proceptivity and the functions of extended sexuality. Psychological Science.

Griffith, S.C., Owens, I.P.F., Thuman, K.A., 2002. Extra-pair paternity in birds: a review of interspecific variation and adaptive function. Molecular Ecology 11, 2195–2212.

Griffith, S.C., 2007. The evolution of infidelity in socially monogamous passerines: neglected components of direct and indirect selection. American Naturalist 169, 274–281.

Gurven, M., Hill, K., 2009. Why do men hunt?: A reevaluation of "man the hunter" and the sexual division of labor. Current Anthropology 50, 51–74.

Hart, C.W.M., Pilling, A.R., Goodale, J.C., 1987. The Tiwi of North Australia, third ed. Holt, Rinehart, and Winston, New York.

Haselton, M.G., Gangestad, S.W., 2006. Conditional expression of women's desires and male mate retention efforts across the ovulatory cycle. Hormones and Behavior 49, 509–518.

Hawkes, K., 1991. Showing off: tests of an hypothesis about men's foraging goals. Ethology and Sociobiology 12, 29–54.

Hawkes, K., 2004. Mating, parenting, and the evolution of human pairbonds. In: Chapais, B., Berman, C.M. (Eds.), Kinship and Behavior in Primates. Oxford University Press, Oxford, U.K., pp. 443–473.

Hawkes, K., O'Connell, J.F., Blurton Jones, N.G., 1991. Hunting patterns among the Hadza: big game, common goals, foraging goals and the evolution of the human diet. Philosophical Transactions of the Royal Society of London B 334, 243–251.

Hawkes, K., O'Connell, J.F., Blurton Jones, N.G., 2001. Hunting and nuclear families – some lessons from the Hadza about men's work. Current Anthropology 42, 681–709.

Hohmann, G., Fruth, B., 2000. Use and function of genital contacts among female bonobos. Animal Behaviour 60, 107–120.

Hrdy, S.B., 1979. Infanticide among animals, a review, classification and examination of the implications for the reproductive strategies of females. Ethology and Sociobiology 1, 3–40.

Hrdy, S.B., 2009. Mothers and Others. Harvard University Press, Cambridge, MA.

Hua, C., 2001. A Society without Fathers or Husbands: The Na of China. MIT Press, Cambridge, MA.

Jankowiak, W.R., Fischer, E.F., 1992. A cross-cultural perspective on romantic love. Ethnology 31, 148–155.

Jennions, M.D., Petrie, M., 2000. Why do females mate multiply? A review of the genetic benefits. Biological Reviews 75, 21–64.

Kaplan, H., Hill, K., 1985. Hunting ability and reproductive success among male Ache foragers: preliminary tests. Current Anthropology 26, 131–133.

Kaplan, H., Hill, K., Lancaster, J., Hurtado, A.M., 2000. A theory of human life history evolution: diet, intelligence, and longevity. Evolutionary Anthropology 9, 156–185.

Kenrick, D.T., Sadalla, E.K., Groth, G., Trost, M.R., 1990. Evolution, traits, and the stages of human courtship: qualifying the parental investment model. Journal of Personality 58, 97–116.

Kokko, H., Jennions, M.D., 2008. Parental investment, sexual selection and sex ratios. Journal of Evolutionary Biology 21, 919–948.

Kokko, H., Johnstone, R.A., 2002. Why is mutual mate choice not the norm? Operational sex ratios, sex roles and the evolution of sexually dimorphic and monomorphic signalling. Philosophical Transactions of the Royal Society of London B 357, 319–330.

Lancaster, J.B., Lancaster, C.S., 1983. Parental investment: the hominid adaptation. In: Ortner, D. (Ed.), Parental Care in Mammals. Plenum, New York, pp. 347–387.

Larson, C.M., Pillsworth, E.G., Haselton, M.G., 2012. Ovulatory shifts in women's attractions to primary partners and other men: further evidence of the importance of primary partner sexual attractiveness. PLoS One 7, e44456.

Lovejoy, C.O., 1981. The origins of man. Science 211, 341–350.

Marlowe, F., 1999. Male care and mating effort among Hadza foragers. Behavioral Ecology and Sociobiology 46, 57–64.

Marlowe, F., 2001. Male contribution to diet and female reproductive success among foragers. Current Anthropology 42, 755–760.

Marlowe, F.W., 2003a. A critical period for provisioning by Hadza men: implications for pair bonding. Evolution and Human Behavior 24, 217–229.

Marlowe, F.W., 2003b. The mating system of foragers in the standard cross-cultural sample. Cross-Cultural Research 37, 282–306.

Martin, R.D., 2007. The evolution of human reproduction: a primatological perspective. American Journal of Physical Anthropology Supplement 45, 59–84.

Mascaro, J.S., Hackett, P.D., Rilling, J.K., 2013. Testicular volume is inversely correlated with nurturing-related brain activity in human fathers. Proceedings of the National Academy of Science USA.

Mazur, A., Booth, A., 1998. Testosterone and dominance in men. Behavioral and Brain Sciences 21, 353–397.

Mazur, A., Michalek, J., 1998. Marriage, divorce, and male testosterone. Social Forces 77, 315–330.

McIntyre, M.H., Gangestad, S.W., Gray, P.E., Chapman, J.F., Burnham, T.C., O'Rourke, M.T., Thornhill, R., 2006. Romantic involvement often reduces men's testosterone levels—but not always: the moderating effect of extra-pair sexual interest. Journal of Personality and Social Psychology 91, 642–651.

Moscovice, L.R., Ziegler, T.E., 2012. Peripheral oxytocin in female baboons relates to estrous state and maintenance of sexual consortships. Hormones and Behavior 62, 592–597.

Muehlenbein, M.P., Bribiescas, R.G., 2005. Testosterone-mediated immune functions and male life histories. American Journal of Human Biology 17, 527–558.

Muller, M.N., Marlowe, F.W., Bugumba, R., Ellison, P.T., 2009a. Testosterone and paternal care in East African foragers and pastoralists. Proceedings of the Royal Society B 276, 347–354.

Muller, M.N., Wrangham, R.W., 2001. The reproductive ecology of male hominoids. In: Ellison, P.T. (Ed.), Reproductive Ecology and Human Evolution. Aldine, New York, pp. 397–427.

Muller, M.N., Kahlenberg, S.M., Wrangham, R.W., 2009b. Male aggression against females and sexual coercion in chimpanzees. In: Muller, M.N., Wrangham, R.W. (Eds.), Sexual Coercion in Primates and Humans: An Evolutionary Perspective on Male Aggression Against Females. Harvard University Press, Cambridge, MA, pp. 184–217.

Murdock, G.P., White, D.R., 1969. Standard cross-cultural sample. Ethnology 9, 329–369.

Nummi, P., Pellikka, J., 2011. Do female sexual fantasies reflect adaptations for sperm competition? Annales Zoologici Fennici 49, 93–102.

Nunes, S., Fite, J.E., Patera, K.J., French, J.A., 2001. Interactions among paternal behavior, steroid hormones, and parental experience in male marmosets (*Callithrix kuhlii*). Hormones and Behavior 39, 70–82.

Pagel, M., Meade, A., 2006. Bayesian analysis of correlated evolution of discrete characters by reversible-jump Markov chain Monte Carlo. American Naturalist 167, 808–825.

Parker, G.A., Birkhead, T.R., 2013. Polyandry: history of a revolution. Philosophical Transactions of the Royal Society B 368.

Penn, D.J., Potts, W.K., 1999. The evolution of mating preferences and major histocompatibility complex genes. American Naturalist 153, 145–164.

Platek, S.M., Burch, R.L., Panyavin, I.S., Wasserman, B.H., Gallup Jr., G.G., 2002. Reaction to children's faces—resemblance affects males more than females. Evolution and Human Behavior 23, 159–166.

Platek, S.M., Critton, S.R., Burch, R.L., Frederick, D.A., Myers, T.E., Gallup Jr., G.G., 2003. How much paternal resemblance is enough? Sex differences in hypothetical investment decisions, but not in the detection of resemblance. Evolution and Human Behavior 24, 81–87.

Platek, S.M., Keenan, J.P., Mohamed, F.B., 2005. Sex differences in the neural correlates of child facial resemblance: an event-related fMRI study. Neuroimage 25, 1336–1344.

Platek, S.M., Raines, D.M., Gallup, G.G., Mohamed, F.B., Thomson, J.W., Myers, T.E., Panyayin, I.S., Levin, S.L., Davis, J.A., Fonteyn, L.C.M., Arigo, D.R., 2004. Reactions to children's faces: males are more affected by resemblance than females are, and so are their brains. Evolution and Human Behavior 25, 394–405.

Quinlan, R.J., Quinlan, M.B., 2008. Human lactation, pair-bonds and alloparents: a cross-cultural analysis. Human Nature 19, 87–102.

Quinlan, R.J., Quinlan, M.B., Flinn, M.V., 2003. Parental investment and age of weaning in a Caribbean village. Evolution and Human Behavior 24, 1–16.

Santos, P.S.C., Schinemann, J.A., Gabardo, J., Bicalho, M.D., 2005. New evidence that the MHC influences odor perception in humans: a study with 58 southern Brazilian students. Hormones and Behavior 47, 384–388.

Scelza, B.A., 2011. Female choice and extra-pair paternity in a traditional human population. Biology Letters 7, 889–891.

Schneiderman, I., Zagoora-Sharon, O., Leckman, J.F., Feldman, R., 2012. Oxytocin during the initial stages of romantic attachment: relation to couples' interactive reciprocity. Psychoneuroendocrinology 37, 1277–1285.

Sheldon, M.S., Cooper, M.L., Geary, D.C., Hoard, M., DeSoto, M.C., 2006. Fertility cycle patterns in motives for sexual behavior. Personality and Social Psychology Bulletin 32, 1659–1673.

Storey, A.E., Walsh, C.J., Quinton, R.L., Wynne-Edwards, K.E., 2000. Hormonal correlates of paternal responsiveness in new and expectant fathers. Evolution and Human Behavior 21, 79–95.

Stumpf, R.M., Boesch, C., 2005. Does promiscuous mating preclude female choice? Female sexual strategies in chimpanzees (*Pan troglodytes verus*) of the Taï National Park, Côte d'Ivoire. Behavioral Ecology and Sociobiology 57, 511–524.

Surbeck, M., Deschner, T., Schubert, G., Weltring, A., Hohmann, G., 2012. Mate competition, testosterone and intersexual relationships in bonobos, *Pan paniscus*. Animal Behaviour 83, 659–669.

Tal, I., 2009 (unpublished dissertation data). Department of Psychology, University of New Mexico.

Thornhill, R., Gangestad, S.W., 2008. The Evolutionary Biology of Human Female Sexuality. Oxford University Press, New York.

Thornhill, R., Gangestad, S.W., Miller, R., Scheyd, G., Knight, J., Franklin, M., 2003. MHC, symmetry, and body scent attractiveness in men and women. Behavioral Ecology 14, 668–678.

Thurz, M.R., Thomas, H.C., Greenwood, B.M., Hill, A.V.S., 1997. Heterozygote advantage for HLA class-II type in hepatitis B virus infection. Nature Genetics 17, 11–12.

Van Anders, S.M., Watson, N.V., 2006. Relationship status and testosterone in North American heterosexual and non-heterosexual men and women: cross-sectional and longitudinal data. Psychoendocrinology 31, 715–723.

Voracek, M., Haubner, T., Fisher, M.L., 2008. Recent decline in nonpaternity rates: a cross-temporal meta-analysis. Psychological Reports 103, 799–811.

Wedekind, C., Füri, S., 1997. Body odor preference in men and women: do they aim for specific MHC combinations or simply heterozygosity? Proceedings of the Royal Society of London B 264, 1471–1479.

Wedekind, C., Seebeck, T., Bettens, F., Paepke, A.J., 1995. MHC-dependent mate preferences in humans. Proceedings of the Royal Society of London B 260, 245–249.

Wegner, K.M., Reusch, T.B.H., Kalbe, M., 2003. Multiple parasites are driving major histocompatibility complex polymorphism in the wild. Journal of Evolutionary Biology 16, 224–232.

Westermarck, E., 1929. Marriage. Jonathan Cape and Harrison Smith, New York.

Westneat, D.F., Stewart, I.R.K., 2003. Extra-pair paternity in birds: causes, correlates, and conflict. Annual Review of Ecology, Evolution and Systematics 34, 365–396.

Whittingham, L.A., Dunn, P.O., 2010. Fitness benefits of polyandry for experienced females. Molecular Ecology 19, 2328–2335.

Wobber, V., Hare, B., Lipson, S., Wrangham, R., Ellison, P., 2013. Different ontogenetic patterns of testosterone production reflect divergent reproductive strategies in chimpanzees and bonobos. Physiology and Behavior 116, 44–53.

Wolf, M., Musch, J., Enczmann, J., Fischer, J., 2012. Estimating the prevalence of nonpaternity in Germany. Human Nature 23, 208–217.

Wu, J.J., He, Q.Q., Deng, L.L., Wang, S.C., Mace, R., Ji, T., Tao, Y., 2013. Communal breeding promotes a matrilineal social system where husband and wife live apart. Proceedings of the Royal Society B 280, 20130010.

Chapter 35

Evolution of Cognition

Jennifer Vonk and Chinmay Aradhye
Department of Psychology, Oakland University, Rochester, MI, USA

SYNOPSIS

Researchers have rigorously investigated the cognitive capacities of various nonhuman primate species in an attempt to elucidate the factors that define human uniqueness and that are responsible for the emergence of these traits. It is proposed that humans can be distinguished from other primates by their heightened capacity for abstraction, which permits reasoning about unobservable social and physical forces and prompts the capacity to seek causal explanations for events. It is suggested that these capacities arose as a result of the distinctly cooperative and communicative societies that humans occupy as well as the need to outcompete more physically formidable species.

EVOLUTION OF COGNITION

No human would pay any more attention to this order of mammals than to any other group of living creatures were it not for our unique interest in understanding our own place in biological nature.

Friderun Ankel-Simons (2007, p. xiii)

Researchers have adopted multiple approaches to studying the evolution of the human mind. One particularly productive approach has been to compare various aspects of human cognition with those of other extant species that have diverging evolutionary paths from a common ancestor and occupy different environmental niches (see also Heyes, 2012a). This comparison allows us to understand the contributions of environmental factors to the expression of particular cognitive traits. For instance, two popular hypotheses for the emergence of primate cognitive specializations have been the social intelligence hypothesis, which emphasizes the importance of group living and social complexity (Dunbar, 1998, 2009; Humphrey, 1976; Jolly, 1966), and the foraging/ecological hypothesis, which focuses on the challenges of finding food that may be difficult to procure or patchily distributed (Milton, 1981). Researchers have compared the presence or complexity of cognitive traits such as spatial memory, episodic memory, causal reasoning, and problem solving in species living in social groups with the presence or complexity of those traits in solitary species (Croney and Newberry, 2007; Dunbar, 2001), in species such as folivores, for whom food is readily available, and frugivores, who must forage over long distances for temporarily available resources (Milton, 1999). This research has extended beyond comparisons of different primate species to include, for example, caching and noncaching bird species (Shettleworth, 2009).

Although it is useful to compare ecological, structural, and behavioral characteristics across taxa in order to understand human evolution, a particularly important group for study includes members of the order Primates (see chapters by Ward, Simpson, Ahern, and Holt in the present volume). It is well known that the closest extant relatives of humans are chimpanzees (*Pan troglodytes*) and bonobos (*Pan paniscus*), with whom we shared a common ancestor about four to seven million years ago (Chatterjee et al., 2009; Dawkins, 2005; Steiper and Young, 2006; Wilson and Reeder, 2005). The estimated timeline for primate phylogenetic relatedness appears in Figure 1. Along with a closer phylogenetic relationship, primates share important common features such as vertical flat faces allowing stereoscopic vision and prehensile hands and feet. Some adaptations, such as gripping capacity and depth perception, were designed to assist with arboreal lifestyles (Relethford, 2010; Ridley, 2004). However, along with the evolutionary history of our structural capacities such as brain size and complexity (see Aiello and Wheeler, 1995; Chittka and Niven, 2009; Dunbar, 1998, 2001; Gazzaniga, 2009), many of our more unique and recent cognitive capacities can be traced back to the time that our ancestors left the trees and inhabited the forests. For instance, adaptations for bipedality (see the chapter by Hunt) are seen in the feet and the leg bones, back vertebrae, length of arms, position of the skull on the backbone, fully opposable thumbs that facilitate tool use, and teeth and jaws for acquiring food, chewing, etc. (for a detailed overview see Fleagle, 2013).

Particularly important for understanding comparative behavior is the architecture of the human brain (see the chapter by Holloway). Although the human brain has gone through significant increases in absolute size since splitting from the hominin (great ape) ancestry (Relethford, 2010), many similarities exist among nonhuman primate, hominid, and human brains (see Gazzaniga, 2009) that help us understand certain important aspects of the framework within which complex brains might have evolved (de Waal, 2001; Pusey, 2001). For instance, human and nonhuman primate (especially great ape) brains share the basic makeup of the brain, the location of the cortices, and the different lobes and their functions. Differences lie in the percentage of the brain area devoted to a particular lobe, and perhaps the strength of the neuronal networks (Gazzaniga, 2009; Semendeferi et al., 2002). Therefore, calculated comparisons between differing lifestyles of closely related species and humans, in addition to brain mechanisms and structures, provide remarkable insights into the evolution of our core cognitive faculties, such as the realization that complex human social and cultural behavior might have something to do with our denser frontal lobes (Dunbar, 2003).

For decades, scientists in various fields have tried to understand the unique place that humans (*Homo sapiens*) occupy in their natural environment by asking questions such as "What makes us special?" Most unequivocally, the candidate for uniqueness has been some variation of the concept of "human intelligence" (for discussions on technical and social intelligence see Byrne, 2002; for neuroanatomical uniqueness of humans see Gazzaniga, 2009; for discussion on cognitive skills see Penn and Povinelli, 2007a; Vonk and Povinelli, 2012; for ecological cognitive adaptations see Pinker, 2010; Sapolsky, 2011). The question of why human intelligence is so distinct from that of other species was the one that bifurcated Darwin's and Wallace's understanding of the limits of natural selection (Gross, 2010). Darwin claimed that the manner in which human intelligence was unique was by degree and not any specific type (Darwin, 1871), whereas Wallace famously proposed that human intelligence could not have been a product of sheer natural selection at all, by which he implied that a force other than natural selection was responsible for human intelligence (Wallace, 1864), an idea quite contradictory to Darwin's understanding of the history and development of life. "Darwin's bulldog" and an authority in

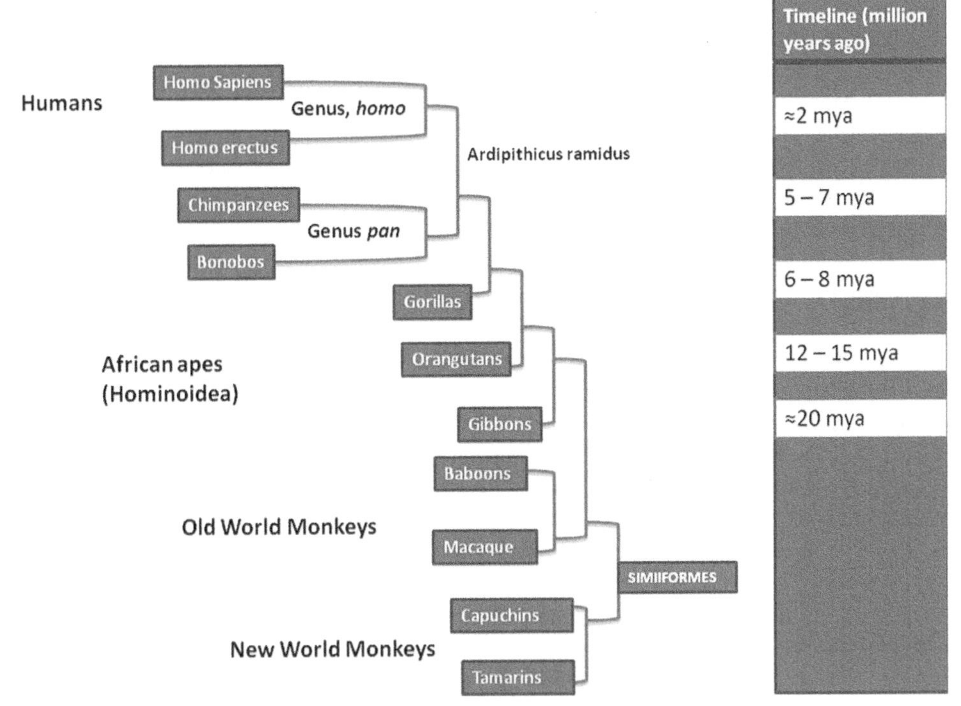

FIGURE 1 A timeline of phylogenetic distance within the primate order. *Updated and adapted from information provided in Wilson and Reeder (2005).*

neuroanatomy, Thomas Huxley, held the position that there was no structure or organization in the human brain that was absent from the nonhuman primate brain (Huxley, 1863), perhaps in an attempt to overcompensate for the unwillingness of his opposition to accept the phylogenetic relationship between humans and nonhuman primates. Today, each of these three claims has been falsified by new evidence; currently the contributing primary neuroanatomical predictors of intelligence are absolute brain size, neuronal connectivity, and encephalization quotient (EQ), for which humans are remarkably unique. There has been much debate about whether relative or absolute brain size is more critical in accounting for cognitive differences (Yopak et al., 2010; although see also Azevedo et al., 2009). Clearly size is not of sole importance—convolutions, EQ, and cortical complexity are also critical (Kudo and Dunbar, 2001), as is the ratio of forebrain to absolute brain size (Lefebvre et al., 1997). Barton's (2012) recent work suggests a coevolution of the neocortex and cerebellum, not just in primates but in mammals in general. Such an analysis suggests that the evolution of human cognition is part of a larger trend toward more powerful, embodied modes of thought. Several specialists have indicated that differences may be as much a matter of reorganization as one of any real quantitative differences in matter (Gazzaniga, 2009; Holloway, 1968; this volume; Preuss, 2001). How exactly neuroanatomical factors correspond to observable differences in intelligent behavior and/or potential is still unknown (see Gazzaniga, 2009, Schoenemann, 2006; Willemet, 2013). In addition, it has proven difficult to link environmental changes to periods of punctuated changes in brain size evolution, lending credence to the notion that the evolution of human cognition was influenced more strongly by social or cognitive factors (such as language) than by environmental ones (Shultz et al., 2012). Lefebvre et al. (1997) suggested that an examination of behavioral innovation in relation to relative forebrain size may be a useful direction for research, following an exploration in various bird species. There have been several attempts to identify specific systems, operations within those systems, and even basic structural brain differences that may explain special human abilities, but a definitive understanding of the core cognitive differences between humans and nonhuman primates has yet to yield a unanimously agreed-upon candidate (Gazzaniga, 2009; Wasserman and Zentall, 2006; Willemet, 2013).

For many years the small subset of psychologists studying primate cognition has been largely divided into two disparate camps—those who believe that advances in research techniques reveal evidence for a decreasing divide between nonhuman and human minds (de Waal and Ferrari, 2010; Tomasello et al., 2003a, b; Whiten and Erdal, 2012) and those who believe that existing evidence is consistent with a striking gap between human and nonhuman cognition (Penn et al., 2008; Penn and Povinelli, 2007a, b). Indeed, the growing neurological data support such a disparity (see Premack, 2007 for a review). Members of the latter group also tend to believe that the current tests are not diagnostic with regard to the questions being investigated, because of either the experimental tasks themselves (Penn et al., 2008; Penn and Povinelli, 2007a, b; Povinelli and Vonk, 2003, 2004), or failure to take individual subject history into account (Boesch, 2007). Many feel that the current corpus of data has been misinterpreted in favor of one view over the other (Heyes, 1998; Povinelli and Vonk, 2003, 2004). To date, there is little consensus over the extent to which human and other primate minds are similar or different. This disagreement extends to the study of other nonprimate animal minds, in which investigators have increasingly claimed evidence for continuity between nonprimates and humans with regard to capacities such as theory of mind (Kaminski et al., 2008), episodic memory (Clayton et al., 2001), prosocial sentiments (de Waal, 2009), and metacognition (Foote and Crystal, 2007).

A multitude of proposals have been made regarding the dimension(s) on which a possible disparity could be revealed. Most readers will be more than familiar with the types of cognitive capacities traditionally deemed to be unique to humans. These capacities include, among others, language, the ability to represent and reason about mental states (theory of mind) and relations, metacognition, self-awareness, causal reasoning, tool use (see the chapter by Toth and Schick) imitation, culture, abstraction, mental time travel, and aspects of morality such as other-regarding preferences and concepts of equity/fairness (see the chapter by Allchin). As noted above, however, researchers have been strongly motivated to find evidence against such proposals in order to champion the notion of cognitive similarities. Furthermore, there has been discussion of late as to whether human cognitive uniqueness is domain general or domain specific (Shettleworth, 2012a). A domain-general view supports the notion that a more general cognitive capacity such as superior working memory or greater representational ability (Penn et al., 2008) underlies differences in multiple domains.

CONSCIOUSNESS

Long following Descartes's famous claim that thinking on our existence provided proof of that very existence (Descartes, 2006), researchers have been driven to determine whether the ability to reflect on one's own existence might be shared with other species. Although no researcher has produced a valid methodology for assessing precisely this capacity, researchers have skirted around the edges—assessing the capacity for metacognitive awareness and episodic memory.

Metacognition entails the ability to reflect on one's own cognitions—memory, awareness, deficits, etc. Researchers have used a variety of techniques to study metacognition in nonhuman animals. Most commonly they have explored the extent to which animals selectively seek information when they had not been given access to needed information (Call and Carpenter, 2001; Hampton, 2009; Hampton et al., 2004; Terrace, 2005) and the extent to which they know when to opt out of making an uninformed response (Beran et al., 2009; Hampton, 2009; Smith et al., 1997). Monkeys, dolphins, and pigeons all appear to use the "uncertainty response" selectively, but there has been much debate about potential alternative explanations that rest on associations indicating memory strength or other functions (Beran et al., 2010; Kornell, 2009; Smith, 2009; Smith et al., 2003) rather than true metacognitive ability. For instance, animals may have learned that it is more advantageous to choose the uncertainty option under conditions that also predict poor memory strength, such as longer delays or a greater number of alternatives, without reflecting on the reason for this association. To their credit, researchers have ingeniously continued to advance new methodologies in this domain, continuing to support the idea that at least some components of metacognition are likely shared within the animal kingdom. Importantly, Smith et al. (2012b) pointed out that one need not expect the entire suite of metacognitive abilities that has evolved in humans to also be present in nonhumans before rejecting the possibility that some components of the system might exist. In fact, there have been many proposals for mosaic-like (Calvin, 1996; Dunbar and Barrett, 2007; Shettleworth, 2012a) or modular cognition (Leslie, 1987; Penn et al., 2008; Pinker, 1989, 2007)—both models might suppose aspects of a suite of abilities to be present or absent in different species. Frith (2012) has proposed that implicit metacognition might be shared broadly, whereas explicit metacognition may be unique to humans. Again, one needs to consider which capacities make sense in a given species. For what purpose would those capacities have evolved?

Episodic memory, as defined by Tulving (2002), refers to the ability to reflect on past, present, and future with "autonoetic" awareness. That is, episodic memory goes beyond the memory for details of events by also containing the subjective quality of having experienced the event. Others have more recently referred to this ability as "mental time travel" (Suddendorf et al., 2009), and in animals researchers have focused on "episodic-*like*" memory, with the acknowledgment that in the absence of language it is difficult to find clear evidence for true episodic memory. The most convincing evidence for episodic-like memory in nonhumans comes from Clayton and colleagues' work with scrub jays (Clayton et al., 2001). The initial experiment revealed that jays searched for different items at different time intervals following caching. After short intervals they selectively retrieved preferred food items, but at longer intervals, when the preferred items would have decayed, they reversed their preferences and retrieved less preferred but nondecayed items (Clayton and Dickinson, 1998). This finding indicated that the jays were aware of what they had hidden when and where. These results provoked use of the term "what-where-when" memory. Since then, Crystal and colleagues (Babb and Crystal, 2006; Foote and Crystal, 2007) have determined that rats may also make use of such a memory capacity. As Pattison et al. (2013) have pointed out that it makes sense to consider not just whether an animal would have a capacity for episodic memory but why it would be adaptive for that species to have this capacity and what it would mean for that species if that ability were lacking. Shettleworth (2012b) has also pointed out that it makes little sense for an animal to evolve such a capacity in order to reflect on its past history. After all, animals do not write memoirs; the best use of such a system is to allow an animal to apply lessons learned from past experiences toward predicting the future. Why then would humans have expanded on this capacity to reflect, in the thoughtful manner that they do, about what it means to exist? We suspect that this reflective tendency coevolved, along with the ability to reason about other minds and the ability to communicate those thoughts to others, with the goal of teaching others about our own experiences and building a collective culture. Taken together, these aspects are unique to human cultures.

LANGUAGE

Along with the ability to reflect on our own conscious existence, the ability to communicate abstract ideas regarding events both in the distant past and projected into the future was heralded early on as a hallmark of human uniqueness. As other purportedly unique human capacities such as tool use were demonstrated at increasing rates in the animal kingdom (Beck, 1980; Bentley-Condit and Smith, 2010; Roffman et al., 2012), researchers also sought to wrench language from its special position on the list of uniquely human capacities. Despite innovative and promising research programs from various ape language labs, including the work of Gardner & Gardner (1969); Hayes & Hayes (1952); Savage-Rumbaugh (1986); Patterson (1981); and Miles (1983), using multiple methods such as sign language and lexigram keyboards, the field of ape language studies was almost singlehandedly dismissed by the work of Herbert Terrace in the 1970s (Terrace et al., 1979). Terrace showed that many of the findings championed as evidence for humanlike language capacities could be explained by simpler associative mechanisms. Although Savage-Rumbaugh's impressive accomplishments with the bonobo Kanzi have been highly illuminating with regard to the thought processes of these close relatives, it is clear that human language incorporates generativity and flexibility unparalleled elsewhere in the animal kingdom. Researchers once focused on language as the dividing factor enabling the capacity for abstraction, theory of mind, and various other important capacities. However, the focus has shifted in more recent years, as language is no longer considered ubiquitous in explaining all other potential differences between humans and other closely related species. Indeed, language may be a fairly recent advance, and a commonly held position today is that the human ability to produce language has resulted from major selection pressures over our evolutionary past. These pressures may have been exerted in the forms of needing to deal with causal contingencies, exchanging information with members of the group and dealing with social challenges in general (see Pinker and Bloom, 1995).

The effect of selection pressures on language ability can be traced with a comparative approach. For instance, one obvious difference between the socialities of humans (*H. sapiens*) and nonhuman primates is that ancestral *human* societies were much more complex in terms of their number, associations, and networks (Dunbar, 2001). Perhaps due to simpler social structures, hierarchies, and social demands, monkeys and apes even today show a much smaller inventory of calls and concepts, and even those are far less flexible than human language allows (Hammerschmidt and Fischer, 2008). On the other hand, nonhuman primates still show remarkable contingency-based signaling (Cheney and Seyfarth, 1999) and are not totally devoid of semantics (Gouzoules et al., 1998; Zuberbühler et al., 1999). This general lack of "complexity" of language is on the whole thought to be the result of a weaker theory of mind possessed by nonhuman primates, which forces them to be unsure about the intentions of other group members (Cheney and Seyfarth, 2005) and less able to determine the need to communicate complex information.

Nonhuman primates seem to be more restricted to vocal production than comprehension. In terms of comprehension capacities, nonhuman primates seem to grasp a much wider scope of associations between external stimuli and sounds, even those produced by members of other species (Fichtel, 2004; Seyfarth and Cheney, 1990). This may be a result of the gradual development of communication abilities. However, in comparison with human infants, nonhuman primates seem to learn far fewer new calls and sounds to add to their inventory even after being fostered by another species (Owren et al., 1993), and as a result their general capacity to produce varying sounds for varying messages is limited. Significant discretion is shown, however, in particular calls that have been learned within a group. For example, it was found that vervet monkeys (*Cercopithecus aethiops*) choose to give a predator-detection-alarm call only in the presence of their kin members (Cheney and Seyfarth, 1992). Female diana monkeys (*Cercopithecus diana*) keep track of which predator call has been made before by a male member and sound a predator alarm only if the other predator threat is detected (Zuberbühler et al., 1999). These studies show that nonhuman primates may not be as endowed in their linguistic capacity as humans, but show significant complexity in signaling and communication.

In humans, the study of the emergence of language, the psychological basis for it, and the social structure for which it has been utilized has thrown much light on human social cognition and social dynamics (for detailed discussion, see Pinker, 2007; Pinker and Bloom, 1995) (see chapter by Leiberman). For example, most of the basic contents of human languages can be categorized under objects, substances, space, and time, and in addition their relationships and properties such as motion, agency, and causation (Pinker, 2008, 2010). Human survival issues therefore seem to be centered on manipulating objects, spaces, relationships, and properties, and communicating those to their counterparts, thus giving rise to a complex sociopolitical dimension (Pinker, 2010). Along similar lines, Seyfarth and Cheney (1999) regard it as an important commonality between the language development of both human and nonhuman primate infants that both learn to assign external stimuli to particular sounds, although the production of sounds themselves involves a very different developmental progression. This difference has led some thinkers to argue that language is not an instinct at all, but rather a learned skill that is an extension of the motor system operating a functional language system comprising different parts of the brain. For comparative theoretical views, see Lieberman (2002) and Pinker (2007).

Language has been postulated as a major developmental advance, drastically restructuring the nature of human thought itself (Piaget, 1974). There has been debate about whether language is necessary for abstraction (Premack, 1983) or has emerged as the result of abstraction and the need to communicate ever more complex ideas (Chomsky, 2006; Corballis, 2011). Without question, language and abstraction are intimately related. The use of language can highlight commonalities or differences between members of a category or items that share a non-perceptually obvious relationship (Kotovsky and Gentner, 1996). The ability to communicate at such a level has clearly allowed humans to relay ideas, beliefs, and knowledge to others and to create institutions through which all members can share common goals and ideologies—something that clearly cannot be expressed in other animal groups despite the shared existence of in-groups and out-groups.

ABSTRACTION

Although it is hardly contested that many other species are capable of representing complex concepts, including the honeybee's capacity to differentiate between "same" and "different" (Giurfa et al., 2001), a capacity it shares with various mammals and birds (Wasserman et al., 2001; Wright et al., 1983), the more revealing issue of what precisely supports such representations is more contentious. Researchers have examined whether animals form concepts and classify objects on the basis of abstract theoretical principles or strictly physical characteristics. For instance, do they create categories on the basis of directly perceivable characteristics such as shape, color, taste, and smell or on the basis of more abstract nonperceivable qualities such as "goodness" and "badness"? It has been argued that whereas many animals form concepts comprising a diversity of features and can thus be described as abstract, none have demonstrated the capacity for concepts that cannot be described using observable features (Vonk and Povinelli, 2012).

Researchers have also examined the process by which animals wield the concepts they represent in order to solve problems in the social and physical domains. Bringing these two related lines of inquiry together, they have explored the extent to which primates make causal inferences about unobservable forces as opposed to forming associations and making predictions about physically observable characteristics of the objects, individuals, and events they encounter (Povinelli, 2004; Povinelli and Vonk, 2003, 2004). At play are two distinctions between (1) *what kind of* concepts primates can reason about—unobservable theoretical entities versus concrete observable features, and (2) the *processes by which* primates might reason about such concepts—through flexible inferential processes or rigid behaviorist associations (Vonk and Povinelli, 2012).

In both the human and nonhuman literature, a longstanding debate has ensued between those who believe that learning primarily depends on abstract representations and those who believe that it depends on specific modality-dependent sensorimotor representations. Contrary to the popular viewpoint, humans themselves may rely on learning mechanisms that are dependent on perceptual characteristics rather than abstract representations. For instance, the perception of entropy (perceptual irregularity in this case) may underlie decisions about sameness and difference (Fagot et al., 2001). This general point has been made often (Andrews, 2007; Heyes, 2012b; Povinelli, 2004) and corroborated by many empirical findings including recent research into artificial grammar learning (Conway and Christiansen, 2006), physical causal reasoning (Silva et al., 2005), and concept formation (Ashby and Maddox, 2005). However, this point often seems to be lost in the deluge of research reports that champion the existence of the most abstract humanlike representations in animal minds. What has been lacking until recently (Barrett, 2012) is the understanding that an animal's cognitive capacities must be tightly tied to its perceptual and physiological adaptations. Just as it made little sense to study language by exploring vocal production in species that were physically incapable of producing humanlike sounds, it makes little sense to explore concepts by presenting visual discriminations to animals lacking visual acuity, color vision, or depth perception. However, without much regard to ecological or physiological differences, researchers have studied two main areas in which highly abstract concept representations might be expressed—reasoning about mental states (theory of mind) and reasoning about unobservable physical causes.

THEORY OF MIND

The term "theory of mind," coined by Premack and Woodruff (1978), implies a representational theory of what it means to have a mind. However, the term has been applied more colloquially to refer to any system that represents and reasons about underlying mental states such as beliefs, thoughts, and goals. Elsewhere, paradigms most commonly heralded as evidence for the existence of at least some components of the theory of mind system (i.e., an understanding of seeing) in chimpanzees have been critically evaluated (Heyes, 1998; Karin-Darcy and Povinelli, 2002; Penn and Povinelli, 2007b; Povinelli and Vonk, 2003, 2004). Alternative mechanisms—other than mental state attribution—by which the subjects of the relevant experiments may have operated have been proposed in detail previously, so we will not reiterate those arguments in full here. To summarize briefly, however, it has been argued that the ability to represent and reason about mental states evolved subsequently to the ability to reason about observable behavioral predictors of actions and events (Penn and Povinelli, 2007b). This new system was tightly woven alongside the more ancient system for reasoning about behaviors and allowed for an additional level of explanatory power. This has been labeled "the reinterpretation hypothesis." Notably, it is not more parsimonious to interpret an animal's behavior as involving the reading of mental states, because that understanding is necessarily predicated on an understanding of all the observable manifestations of the assumed underlying mental state (Povinelli and Vonk, 2004). That is, understanding mental states depends *first* upon being able to predict behaviors from behaviors!

In humans one has clear evidence of reasoning about mental states, because humans can use language to express such thoughts. In nonhumans and preverbal human infants, one must rely on an individual's behavior to make inferences about its thought processes. In the absence of language, one must create scenarios and stimuli in visual (sometimes auditory) form that represent varying contents of different minds with access to different information, and then interpret the resulting behavior as being indicative of reasoning about an individual's behavior or mental states. In order to provide the relevant information about underlying mental states in a nonverbal test, the researcher must necessarily provide observable cues as to that underlying mental state, thus making it impossible to disentangle those instances when the organism reasons about the observable behaviors alone from those that involve reasoning about the mental state that has presumably given rise to those behaviors, as both necessarily lead to identical outcomes (Povinelli and Vonk, 2003, 2004).

Members of Tomasello's research group have repeatedly argued that chimpanzees might differentially make use of the ability to reason about mental states in competitive versus cooperative paradigms (Brauer et al., 2006a, b; Hare, 2001; Hare et al., 2005; Hare and Tomasello, 2004; Hermann and Tomasello, 2006; Melis et al., 2006). Whereas this hypothesis is intuitively appealing, why should the capacity to reason about mental states exist only in these constrained situations? This constraint would be particularly puzzling because the adaptive value of such a response is clearly to allow one to reason flexibly and generalize across a wide variety of circumstances. In any case, if chimpanzees can exhibit these capabilities in only the sorts of situations that they encounter routinely in their everyday lives, it is certainly a more limited capability than the one that humans regularly draw upon. Such a scenario also causes one to question whether the apparent ability to reason about mental states instead reflects a highly canalized ability to respond efficiently that has been shaped by evolutionary pressures only in the kinds of settings that chimpanzees might routinely encounter in nature (see also Premack, 2007). Thus, the response is not representative of abstract inferential abilities but rather is sculpted by evolution and hardwired. Humans occupy a variety of ecological niches and exploit a vast and varied diet. In order to exploit animals as both prey and possible service animals, while also outcompeting more formidable predators, it would be adaptive for humans to evolve a system to flexibly predict animals' behavior in as yet unexperienced situations. Humans may be among the few species that occupy a niche between that of predator and prey. Perhaps it is for this reason that theory of mind evolved in a species that has continuously relied on intellect over force to survive in many difficult environments.

CAUSAL REASONING

Both mental states and causal forces in the physical domain represent certain classes of unobservable theoretical entities. As such, evidence for reasoning about either class of unobservables can be construed as evidence for causal reasoning. Although data from the social realm are currently unconvincing, modularity in cognition may be possible, such that an animal's abilities in the physical realm could differ from those in the social realm. Indeed, many researchers have proposed such a dichotomy in the cognition of humans with autism spectrum disorders (ASD) (Baron-Cohen, 1991; David et al., 2008; Leekam et al., 1997), and some have suggested parallels between the cognition of individuals with ASD and that of nonhumans in this regard (Gomez, 2009; Machado and Bachevalier, 2003; Winslow, 2005). Furthermore, some have argued that individuals who do not live in large social groups and thus do not require the ability to reason about the motives and goals of other individuals may yet evolve advanced abilities for reasoning about the behavior of prey or predators, or events that might predict resource availability (Vonk and Beran, 2012; Vonk et al., 2012).

Premack (2007) has distinguished between causal illusion and causal reasoning. He describes causal illusion as follows: "Any goal directed-act that is followed by a desired item will cause the illusion that the goal-directed act produced the desired item" (p. 13,862). When coupled with associative learning, this cognitive mechanism can mimic causal reasoning. However, as Premack points out, animals that know that a larger rock is more likely to break an object, compared with a smaller one, will not necessarily infer that the rock crushed an object if they come across the rock lying next to a broken object. Coming from Premack, this point is especially weighty given that it was Premack's early work with the famous chimpanzee Sarah that suggested that chimpanzees recognize agents of change in a causal fashion (Premack, 1976). However, Premack (2007) now interprets these data as being indicative of an understanding of physical action, which he distinguishes from true causal reasoning. This point epitomizes the argument that primates may lack the explanatory drive that characterizes much of human cognition (Andrews, 2005; Vonk, 2005). Few empirical papers have attempted to test this hypothesis. The one existing study (Povinelli and Dunphy-Lelii, 2001) suggests that chimpanzees, unlike three- to five-year-old human children, do not seek causal explanations when objects do not respond as expected.

Hauser and Spaulding (2006) presented wild rhesus macaques with an experiment fashioned after Premack's (1976) famous early experiments with Sarah. Using an expectancy violation looking time method, they reported that wild rhesus macaques were able to distinguish between possible and impossible physical transformations using objects that would be completely foreign to them. The monkeys were shown possible or impossible physical transformations (e.g., an apple apparently cut by a knife vs. a glass of water) and looked longer at impossible events. In agreement with Premack's later assessment (2007) of his own early research design, we suggest that the performance of these macaques, while impressive, indicates at best the ability to reason about physical action and not causal forces. Furthermore, although looking time/violation of expectation experiments are widely accepted in studies of both human infants and nonhumans, they are always somewhat speculative as to what underlying process the effects truly indicate.

An inability to reason about unobservable theoretical constructs such as mental states, and physical forces such as gravity, is likely connected to a failure to question events in an explanatory fashion (Povinelli and Dunphy-Lelii, 2001). Thus, animals that do not reason about unobservables do not reason about causal forces at all. We extend this argument to propose that other animals may additionally fail to reason about why events occur even when triggering events are entirely observable, such as when a cat is given ear mite treatment and subsequently walks through the house in front of other cats with ears laid back. Even though the other cats have witnessed their owner filling the cat's ears with drops, they will not subsequently discount the cat's laid-back ears as a result of the medication, and instead will still react as though the ear position signifies an imminent attack. The other cats will respond by hissing, and the medicated cat will appear surprised by their reaction. This is likely because cats automatically respond to the behavioral signs of impending aggression without reasoning about the underlying causes of the observable behavior. They do not distinguish between internal aggression and external causes such as medicine in the ears. Thus, nonhumans may fail to seek explanations regardless of whether precipitating causes are of an observable or unobservable nature (Andrews, 2005; Vonk, 2005). Indeed, Vonk and Subiaul (2009) found that even chimpanzees failed to reason in a causal manner about events for which the precipitating causes were clearly visible in a paradigm that asked them to predict which experimenters could offer them food. Thus the question of whether primates reason about unobservables is not the same as the question of whether they reason causally. We propose that nonhumans do neither, and that most purported "evidence" for causal reasoning in animals can be explained by their proclivity to make accurate predictions based on contingencies between observable events. That is, they can predict or make forward attributions but may not engage in backward reasoning. Thus, many of the same sorts of arguments applied above in the realm of reasoning about mental states will apply again in the physical domain.

AN ECONOMIC MIND

We have emphasized the importance of social reasoning and group living in the grand scheme of intellectual differences among taxa. An important outcome of social living is able to share resources with group members (Adolphs, 1999). Trade and commercial development have been the backbone of the increasingly larger human civilization, and humans seem to have complex economic and trade systems that require considerable cognitive effort to learn and understand. However, it is well known that today's international economic structure was once a small and simple barter-based system. Therefore, it is of interest to evaluate whether other primates that engage in resource sharing possess the ability to think in terms of basic economic gains and losses in order to optimize their sharing strategies accordingly. For instance, human players generally show a high sensitivity to fairness and reject anything that seems like an unfair offer in an ultimatum game (Camerer, 2003), although culture does seem to influence people's ideas of fairness (Oosterbeek et al., 2004). A study across 15 cultures found that subjects in remote hunter-gatherer cultures were willing to accept lower offers from the proposers and generally seemed to not care about fairness (Henrich et al., 2006). Jensen et al. (2007) found similar results in chimpanzees. These findings have been interpreted as the ability to maximize gains, which humans forego in the interest of fairness. Others have found that capuchin monkeys (Brosnan and De Waal, 2003) and chimpanzees (Brosnan et al., 2010) reject the offer of unequal reward for effort. However, nonhuman primates seem to be concerned only about offers that are unfair to themselves, not offers that are unfair to their partners. This result is in contrast to human behavior, in which humans are generally concerned with equity and the welfare of others, even when they could benefit from an unfair exchange (Brosnan, 2006). It is important to keep in mind, however, that a complete understanding of context effects on performance in ultimatum games and unequal exchanges is still under question (Camerer, 2003).

Most experiments on nonhuman primate neuroeconomics have been conducted with the rhesus macaque (*Macaca mulatta*), an Old World monkey that among primates is second only to humans in its widespread distribution around the world—indicating its flexible and resilient nature. Experiments studying the neurological activity in the brains of macaques while they play economic games have indicated that they may be influenced by recent past choice and recent past reward in the same way that human players are (Lau and Glimcher, 2005). This finding highlights the role of reinforcement learning in economic decision-making (for a detailed review, see Sanfey and Dorris, 2009). Brown capuchin monkeys (*Sapajus paella)* have also been studied in behavioral economic paradigms (Santos and Chen, 2009), supporting the host of findings that capuchins' decision-making shows remarkable resemblance to human economic decision-making. Results from capuchins are predictable by the tenets of behavioral economics, such as price theory and prospect theory, even without formal training or learned behavioral strategies (Chen et al., 2006). Brosnan et al. (2011) and Brosnan (2013) have also found convergence in the decision-making of capuchins, chimpanzees, and humans when task parameters were made as similar as possible across species. Comparative research continues to raise interesting questions about the role of innate origins of human economic thinking.

COOPERATION

Closely related to the topic of behavioral economics is the study of cooperation and prosocial behavior (see chapters by Flinn and Walker). It has been frequently argued that humans are alone in possessing concern for others' welfare (Silk et al., 2005). Although nonhumans frequently cooperate, it is usually assumed that cooperative acts confer advantages upon the actor and can be construed as basically selfish acts, as the same benefits may not be conferred when working alone. The same is likely true of many cooperative human endeavors. However, it appears that humans alone risk their lives to assist strangers, and engage in other highly prosocial acts (see Hollis and Nowbahari, 2013, who study rescue behavior in antlions, and Wilkinson's (1984) work on vampire bats). Although initial research efforts painted a picture of nonhuman primates as selfish (yet not spiteful) (Jensen et al., 2006; Silk et al., 2005; Vonk et al., 2008), some follow-up investigations called this conclusion into question (see especially Horner et al., 2011; Lakshminarayanan and Santos, 2008), and the reviews have outlined many factors that mediate findings of prosocial behavior (Cronin, 2012; Silk, 2009). Most of the studies have focused on our closest living relatives, although it is important for researchers to continue to probe the existence of prosocial sentiments in other more distantly related species—both those that share conditions common to most humans, such as social living, and those that do not.

CONCLUSIONS

In general, we are in need of much more extensive comparative studies (see Vonk and Shackelford, 2012) so that our explorations are not limited to species that are more human-like, more social, or of a higher profile. Only by broadly comparative studies can we truly elucidate evolutionary forces that may be related to clusters or constellations of traits. What we can conclude based on existing evidence is that humans have evolved sophisticated cognitive abilities, which when shared even within the entire animal kingdom are quantitatively more advanced, flexible, and reflective than those exhibited by even our closest living relatives. When examining the factors that have differed in our evolutionary history, it seems clear that humans developed mental resources to outcompete other species that were bigger, faster, stronger, and more specialized at exploiting particular resources. In contrast, humans have become generalists, finding ingenious ways to exploit numerous habitats, as well as other organisms, through a keen and flexible intellect (see also Heyes, 2012a). A need to understand competitors, as well as prey, likely led to an evolved system for reading other minds—leading to complex concepts that had to be communicated to conspecifics using an advanced language. This view is consistent with that of Heyes (2012a), who suggests that the evolution of technical skills (such as toolmaking) coevolved alongside the evolution of social skills (such as cooperation), perhaps due to a domain-general cognitive process that supported advancements in multiple domains. Thus, humans stand alone in terms of their advanced theory of mind, language, and abstraction capabilities—all of which are closely connected and thus confer a competitive advantage. Humans, like any other species, must be studied with a clear understanding of their relationships to other species and the environments in which they have evolved. Furthermore, specific cognitive capacities should not be studied in isolation without attention to relationships with other sometimes seemingly unrelated adaptations.

REFERENCES

Adolphs, R., 1999. Social cognition and the human brain. Trends in Cognitive Sciences 3, 469–479.

Aiello, L.C., Wheeler, P., 1995. The expensive-tissue hypothesis: the brain and the digestive system in human and primate evolution. Current Anthropology 36, 199–221.

Andrews, K., 2005. Chimpanzee theory of mind: looking in all the wrong places? Mind & Language 20, 521–536.

Andrews, K., 2007. Critter psychology: on the possibility of nonhuman animal folk psychology. In: Folk Psychology Re-assessed. Springer, Netherlands, pp. 191–209.

Ankel-Simons, F., 2007. Primate Anatomy: An Introduction. Academic Press, Burlington, MA.

Ashby, F.G., Maddox, W.T., 2005. Human category learning. Annual Review of Psychology 56, 149–178.

Azevedo, F.A.C., Carvalho, L.R.B., Grinberg, L.T., Farfel, J.M., Ferretti, R.E.L., Leite, R.E.P., Herculano-Houzel, S., 2009. Equal numbers of neuronal and nonneuronal cells make the human brain an isometrically scaled-up primate brain. The Journal of Comparative Neurology 513, 532–541. http://dx.doi.org/10.1002/cne.21974.

Babb, S.J., Crystal, J.D., 2006. Episodic-like memory in the rat. Current Biology 16, 1317–1321.

Baron-Cohen, S., 1991. The theory of mind deficit in autism: how specific is it? British Journal of Developmental Psychology 9, 301–314.

Barrett, L., 2012. Why behaviorism isn't Satanism. In: Vonk, J., Shackelford, T.K. (Eds.), Oxford Handbook of Comparative Evolutionary Psychology. Oxford University Press, New York, NY, pp. 17–38.

Barton, R.A., 2012. Embodied cognitive evolution and the cerebellum. Philosophical Transactions of the Royal Society B: Biological Sciences 367, 2097–2107.

Beck, B.B., 1980. Animal Tool Behavior: The Use and Manufacture of Tools by Animals. Garland STPM Pubishing.

Bentley-Condit, V., Smith, E.O., 2010. Animal tool use: current definitions and an updated comprehensive catalog. Behaviour 147, 185–221.

Beran, M.J., Smith, J.D., Coutinho, M.V., Couchman, J.J., Boomer, J., 2009. The psychological organization of "uncertainty" responses and "middle" responses: a dissociation in capuchin monkeys (*Cebus apella*). Journal of Experimental Psychology: Animal Behavior Processes 35, 371.

Beran, M.J., Couchman, J.J., Coutinho, M.V.C., Boomer, J., Smith, J.D., 2010. Metacognition in Nonhumans: Methodological and Theoretical Issues in Uncertainty Monitoring. Springer Science, New York, NY, pp. 21–35.

Boesch, C., 2007. What makes us human (*Homo sapiens*)? The challenge of cognitive cross-species comparison. Journal of Comparative Psychology 121, 227.

Brauer, J., Call, J., Tomasello, M., 2006a. Are apes really inequity averse? Proceedings of the Royal Society B: Biological Sciences 273, 3123–3128.

Brauer, J., Kaminski, J., Riedel, J., Call, J., Tomasello, M., 2006b. Making inferences about the location of hidden food. Social dog, causal ape. Journal of Comparative Psychology 106, 38–47.

Brosnan, S.F., 2006. Nonhuman species' reactions to inequity and their implications for fairness. Social Justice Research 19, 153–185. http://dx.doi.org/10.1007/s11211-006-0002-z.

Brosnan, S.F., 2013. Justice-and fairness-related behaviors in nonhuman primates. Proceedings of the National Academy of Sciences, 110, 10416–10423.

Brosnan, S.F., De Waal, F.B., 2003. Monkeys reject unequal pay. Nature 425, 297–299.

Brosnan, S.F., Parrish, A., Beran, M.J., Flemming, T., Heimbauer, L., Talbot, C.F., Wilsone, B.J., 2011. Responses to the assurance game in monkeys, apes, and humans using equivalent procedures. PNAS Proceedings of the National Academy of Sciences of the United States of America 108, 3442–3447. http://dx.doi.org/10.1073/pnas.1016269108.

Brosnan, S.F., Talbot, C., Ahlgren, M., Lambeth, S.P., Schapiro, S.J., 2010. Mechanisms underlying responses to inequitable outcomes in chimpanzees, *Pan troglodytes*. Animal Behaviour 79, 1229–1237. http://dx.doi.org/10.1016/j.anbehav.2010.02.019.

Byrne, R.W., 2002. Social and technical forms of primate intelligence. In: de Waal, F.B. (Ed.), Tree of Origin: What Primate Behavior Can Tell Us about Human Social Evolution. Harvard University Press, Cambridge, MA, pp. 145–172.

Call, J., Carpenter, M., 2001. Do apes and children know what they have seen? Animal Cognition 3, 207–220.

Calvin, W.H., 1996. The Cerebral Code: Thinking a Thought in the Mosaics of the Mind. The MIT Press, Cambridge, MA, US.

Camerer, C., 2003. Behavioral Game Theory: Experiments in Strategic Interaction. Princeton University Press, Princeton, NJ.

Chatterjee, H.J., Ho, S.W.H., Barnes, I., Groves, C., 2009. Estimating the phylogeny and divergence times of primates using a supermatrix approach. BMC Evolutionary Biology 9, 259. http://dx.doi.org/10.1186/1471-2148-9-259.

Chen, K., Lakshminarayanan, V., Santos, L., 2006. The evolution of our preferences: evidence from capuchin monkey trading behavior. Journal of Political Economy 114, 517–537.

Cheney, D.L., Seyfarth, R.M., 1992. How Monkeys See the World: Inside the Mind of Another Species. University of Chicago Press, Chicago, IL.

Cheney, D.L., Seyfarth, R.M., 1999. Recognition of other individuals' social relationships by female baboons. Animal Behaviour 58, 67–75.

Cheney, D.L., Seyfarth, R.M., 2005. Constraints and preadaptations in the earliest stages of language evolution. Linguistic Review 22, 135.

Chittka, L., Niven, J., 2009. Are bigger brains better? Current Biology 19, R995–R1008.

Chomsky, N., 2006. Language and Mind. Cambridge University Press, Cambridge, UK.

Clayton, N.S., Dickinson, A., 1998. Episodic-like memory during cache recovery by scrub jays. Nature 395, 272–274.

Clayton, N.S., Griffiths, D.P., Emery, N.J., Dickinson, A., 2001. Elements of episodic-like memory in animals. Philosophical Transactions of the Royal Society of London. Series B: Biological Sciences 356, 1483–1491.

Conway, C.M., Christiansen, M.H., 2006. Statistical learning within and between modalities pitting abstract against stimulus-specific representations. Psychological Science 17, 905–912.

Corballis, M.C., 2011. The Recursive Mind. Princeton University Press, Princeton, NJ.

Croney, C.C., Newberry, R.C., 2007. Group size and cognitive processes. Applied Animal Behaviour Science 103, 215–228.

Cronin, K.A., 2012. Prosocial behaviour in animals: the influence of social relationships, communication and rewards. Animal Behaviour 84, 1085–1093.

Darwin, C., 1871. The Descent of Man in Relation to Sex. Murray, London.

David, N., Gawronski, A., Santos, N.S., Huff, W., Lehnhardt, F., Newen, A., Vogeley, K., 2008. Dissociation between key processes of social cognition in autism: impaired mentalizing but intact sense of agency. Journal of Autism and Developmental Disorders 38, 593–605. http://dx.doi.org/10.1007/s10803-007-0425-x.

Dawkins, R., 2005. The Ancestor's Tale: A Pilgrimage to the Dawn of Evolution. Mariner Books.

Descartes, R., 2006. Discourse on Method (I. Maclean, Trans.). Oxford University Press, Oxford, UK. Original work published 1637.

de Waal, F.B., 2001. Apes from Venus: bonobos and human social evolution. In: In de Waal, F.B. (Ed.), Tree of Origin: What Primate Behavior Can Tell Us about Human Social Evolution. Harvard University Press, Cambridge, MA, pp. 39–68.

de Waal, F.B., 2009. Primates and Philosophers: How Morality Evolved: How Morality Evolved. No. 148. Princeton University Press.

de Waal, F.B., Ferrari, P.F., 2010. Towards a bottom-up perspective on animal and human cognition. Trends in Cognitive Sciences 14, 201–207.

Dunbar, R.I.M., 1998. The social brain hypothesis. Evolutionary Anthropology 6, 178–190.

Dunbar, R.I., 2001. Brains on two legs: group size and the evolution of intelligence. In: De Wall, F. (Ed.), Tree of Origin: What Primate Behavior Can Tell Us about Human Social Evolution. Harvard University Press, Cambridge, MA, pp. 173–192.

Dunbar, R.I., 2003. The social brain: mind, language, and society in evolutionary perspective. Annual Review of Anthropology 163–181.

Dunbar, R.I.M., 2009. The social brain hypothesis and its implications for social evolution. Annals of Human Biology 36, 562–572.

Dunbar, R., Barrett, L., 2007. Evolutionary psychology in the round. In: Dunbar, R., Barrett, L. (Eds.), Oxford Handbook of Evolutionary Psychology. Oxford University Press, New York, NY, pp. 3–9.

Fagot, J., Wasserman, E.A., Young, M.E., 2001. Discriminating the relation between relations: the role of entropy in abstract conceptualization by baboons (*Papio papio*) and humans (*Homo sapiens*). Journal of Experimental Psychology: Animal Behavior Processes 27, 316–328. http://dx.doi.org/10.1037/0097-7403.27.4.316.

Fichtel, C., 2004. Reciprocal recognition of sifaka (*Propithecus verreauxi verreauxi*) and redfronted lemur (*Eulemur fulvus rufus*) alarm calls. Animal Cognition 7, 45–52.

Fleagle, J.G., 2013. Primate Adaptation and Evolution, third ed. Academic Press.

Foote, A.L., Crystal, J.D., 2007. Metacognition in the rat. Current Biology 17, 551–555.

Frith, C.D., 2012. The role of metacognition in human social interactions. Philosophical Transactions of the Royal Society B: Biological Sciences 367, 2213–2223.

Gardner, R.A., Gardner, B.T., 1969. Teaching sign language to a chimpanzee. Science 165, 664–672.

Gazzaniga, M.S., 2009. Human. Harper Collins E-Books.

Giurfa, M., Zhang, S., Jenett, A., Menzel, R., Srinivasan, M.V., 2001. The concepts of 'sameness' and 'difference' in an insect. Nature 410, 930–933.

Gomez, J.C., 2009. Embodying meaning: insights from primates, autism, and brentano. Neural Networks 22, 190–196. http://dx.doi.org/10.1016/j.neunet.2009.01.010.

Gouzoules, H., Gouzoules, S., Tomaszycki, M., 1998. Agonistic screams and the classification of dominance relationships: are monkeys fuzzy logicians? Animal Behaviour 55, 51–60.

Gross, C., 2010. Alfred Russell Wallace and the evolution of the human mind. The Neuroscientist 16, 496–507.

Hammerschmidt, K., Fischer, J., 2008. Constraints in primate vocal production. In: The Evolution of Communicative Creativity: From Fixed Signals to Contextual Flexibility, pp. 93–119.

Hampton, R.R., 2009. Multiple demonstrations of metacognition in nonhumans: converging evidence or multiple mechanisms? Comparative Cognition & Behavior Reviews 4, 17–28. http://dx.doi.org/10.3819/ccbr.2009.40002.

Hampton, R.R., Zivin, A., Murray, E.A., 2004. Rhesus monkeys (*Macaca mulatta*) discriminate between knowing and not knowing and collect information as needed before acting. Animal Cognition 7, 239–246.

Hare, B., 2001. Can competitive paradigms increase the validity of experiments on primate social cognition? Animal Cognition 4, 269–280.

Hare, B., Tomasello, M., 2004. Chimpanzees are more skilful in competitive than in cooperative cognitive tasks. Animal Behaviour 68, 571–581.

Hare, B., Plyusnina, I., Ignacio, N., Schepina, O., Stepika, A., Wrangham, R., Trut, L., 2005. Social cognitive evolution in captive foxes is a correlated by-product of experimental domestication. Current Biology 15, 226–230.

Hauser, M., Spaulding, B., 2006. Wild rhesus monkeys generate causal inferences about possible and impossible physical transformations in the absence of experience. Proceedings of the National Academy of Sciences 103, 7181–7185.

Hayes, K.J., Hayes, C., 1952. Imitation in a home-raised chimpanzee. Journal of Comparative and Physiological Psychology 45, 450.

Heyes, C.M., 1998. Theory of mind in nonhuman primates. Behavioral and Brain Sciences 21, 101–114.

Heyes, C., 2012a. New thinking: the evolution of human cognition. Philosophical Transactions of the Royal Society B: Biological Sciences 367, 2091–2096.

Heyes, C., 2012b. Simple minds: a qualified defence of associative learning. Philosophical Transactions of the Royal Society B: Biological Sciences 367, 2695–2703.

Henrich, J., McElreath, R., Barr, A., Ensminger, J., Barrett, C., Bolyanatz, A., Ziker, J., 2006. Costly punishment across human societies. Science 312, 1767–1770.

Herrmann, E., Tomasello, M., 2006. Apes' and children's understanding of cooperative and competitive motives in a communicative situation. Developmental Science 9, 518–529.

Holloway, R.L., 1968. The evolution of the primate brain: some aspects of quantitative relations. Brain Research 7, 121–172.

Hollis, K.L., Nowbahari, E., 2013. A comparative analysis of precision rescue behaviour in sand- dwelling ants. Animal Behavior 85, 537–545.

Horner, V., Carter, J.D., Suchak, M., de Waal, F.B.M., 2011. Spontaneous prosocial choice by chimpanzees. PNAS Proceedings of the National Academy of Sciences of the United States of America 108, 13847–13851. http://dx.doi.org/10.1073/pnas.1111088108.

Humphrey, N., 1976. The social function of intellect. In: Bateson, P.P.G., Hinde, R.A. (Eds.), Growing Points in Ethology. Cambridge University Press, Cambridge, UK, pp. 303–317.

Huxley, T.H., 1863. Evidence as to Man's Place in Nature. Williams and Norgate, London, U.K.

Jensen, K., Call, J., Tomasello, M., 2007. Chimpanzees are rational maximizers in an ultimatum game. Science 318 (5847), 107–109.

Jensen, K., Hare, B., Call, J., Tomasello, M., 2006. What's in it for me? Self-regard precludes altruism and spite in chimpanzees. Proceedings of the Royal Society B: Biological Sciences 273, 1013–1102.

Jolly, A., 1966. Lemur social behavior and primate intelligence. Science 153, 501–506.

Kaminski, J., Call, J., Tomasello, M., 2008. Chimpanzees know what others know, but not what they believe. Cognition 109, 224–234.

Karin-D'Arcy, R.M., Povinelli, D.J., 2002. Do chimpanzees know what each other see? A closer look. International Journal of Comparative Psychology 15, 21–54.

Kornell, N., 2009. Metacognition in humans and animals. Current Directions in Psychological Science 18, 11–15.

Kotovsky, L., Gentner, D., 1996. Comparison and categorization in the development of relational similarity. Child Development 67, 2797–2822.

Kudo, H., Dunbar, R.I.M., 2001. Neocortex size and social network size in primates. Animal Behaviour 62, 711–722.

Lakshminarayanan, V.R., Santos, L.R., 2008. Capuchin monkeys are sensitive to others' welfare. Current Biology 18, R999–R1000.

Lau, B., Glimcher, P.W., 2005. Dynamic response-by-response models of matching behavior in rhesus monkeys. Journal of the Experimental Analysis of Behavior 84, 555.

Leekam, S., Baron-Cohen, S., Perrett, D., Milders, M., Brown, S., 1997. Eye-direction detection: a dissociation between geometric and joint attention skills in autism. British Journal of Developmental Psychology 15, 77–95.

Lefebvre, L., Whittle, P., Lascaris, E., Finkelstein, A., 1997. Feeding innovations and forebrain size in birds. Animal Behaviour 53, 549–560.

Leslie, A.M., 1987. Pretense and representation: the origins of "theory of mind". Psychological Review 94, 412–426.

Lieberman, P., 2002. Human Language and Our Reptilian Brain: The Subcortical Bases of Speech, Syntax, and Thought. Harvard University Press.

Machado, C.J., Bachevalier, J., 2003. Non-human primate models of childhood psychopathology: the promise and the limitations. Journal of Child Psychology and Psychiatry 44, 64–87. http://dx.doi.org/10.1111/1469-7610.00103.

Melis, A.P., Hare, B., Tomasello, M., 2006. Engineering cooperation in chimpanzees: tolerance constraints on cooperation. Animal Behaviour 72, 275–286.

Miles, H.L., 1983. Apes and language: the search for communicative competence. In: de Luse, J., Wilder, H.T. (Eds.), Language in Primates: Perspectives and Implications. Springer, New York, pp. 43–61.

Milton, K., 1981. Distribution patterns of tropical plant foods as an evolutionary stimulus to primate mental development. American Anthropologist 83, 534–548.

Milton, K., 1999. A hypothesis to explain the role of meat-eating in human evolution. Evolutionary Anthropology Issues News and Reviews 8, 11–21.

Oosterbeek, H., Sloof, R., Van De Kuilen, G., 2004. Cultural differences in ultimatum game experiments: evidence from a meta-analysis. Experimental Economics 7, 171–188.

Owren, M.J., Dieter, J.A., Seyfarth, R.M., Cheney, D.L., 1993. Vocalizations of rhesus (*Macaca mulatta*) and Japanese (*M. Fuscata*) macaques cross-fostered between species show evidence of only limited modification. Developmental Psychobiology 26, 389–406.

Patterson, F.G., 1981. Ape language. Science 211, 86–87.

Pattison, K.F., Laude, J.R., Zentall, T.R., 2013. The case of the magic bones: dogs' memory of the physical properties of objects. Learning and Motivation 44.

Penn, D.C., Holyoak, K.J., Povinelli, D.J., 2008. Darwin's mistake: explaining the discontinuity between human and nonhuman minds. Behavioral and Brain Sciences 31, 109–129.

Penn, D.C., Povinelli, D.J., 2007a. Causal cognition in human and nonhuman animals: a comparative, critical review. Annual Review of Psychology 58, 97–118.

Penn, D.C., Povinelli, D.J., 2007b. On the lack of evidence that non-human animals possess anything remotely resembling a 'theory of mind'. Philosophical Transactions of the Royal Society B: Biological Sciences 362, 731–744.

Piaget, J., 1974. Understanding Causality. Norton, New York.

Pinker, S., 1989. Language Acquisition. The MIT Press, Cambridge, MA, US.

Pinker, S., 2007. The Language Instinct: How the Mind Creates Language. Harper Perennial Modern Classics.

Pinker, S., 2008. The Stuff of Thought: Language as a Window into Human Nature. Penguin Group USA.

Pinker, S., 2010. The cognitive niche: coevolution of intelligence, sociality, and language. Proceedings of the National Academy of Sciences 107, 8993–8999.

Pinker, S., Bloom, P., 1995. Natural language and natural selection. In: Barkow, J.H., Cosmides, L., Tooby, J. (Eds.), The Adapted Mind: Evolutionary Psychology and the Generation of Culture. Oxford University Press, New York, pp. 451–494.

Povinelli, D.J., 2004, Winter. Behind the ape's appearance: escaping anthropocentrism in the study of other minds. Daedalus 133, 29–41.

Povinelli, D.J., Dunphy-Lelii, S., 2001. Do chimpanzees seek explanations? Preliminary comparative investigations. Canadian Journal of Experimental Psychology 55, 185–193.

Povinelli, D.J., Vonk, J., 2003. Chimpanzee minds: suspiciously human? Trends in Cognitive Sciences 7, 157–160.

Povinelli, D.J., Vonk, J., 2004. We don't need a microscope to explore the chimpanzee's mind. Mind & Language 19, 1–28.

Premack, D., 1976. Intelligence in Ape and Man. Lawrence Erlbaum Associates, Hillsdale, NJ.

Premack, D., 1983. The codes of man and beast. Behavioral and Brain Science 6, 125–137.

Premack, D., 2007. Human and animal cognition: continuity and discontinuity. Proceedings of the National Academy of Sciences 104, 13861–13867.

Premack, D., Woodruff, G., 1978. Does the chimpanzee have a theory of mind? Behavioral and Brain Sciences 1, 515–526.

Preuss, T.M., 2001. The discovery of cerebral diversity: an unwelcome scientific revolution. In: Falk, D., Gibson, K. (Eds.), Evolutionary Anatomy of the Primate Cerebral Cortex. Cambridge University Press, Cambridge, pp. 138–164.

Pusey, A.E., 2001. Of genes and apes: chimpanzee social organization and reproduction. In: In de Waal, F.B. (Ed.), Tree of Origin: What Primate Behavior Can Tell Us about Human Social Evolution. Harvard University Press, Cambridge, MA, pp. 39–68. 9–38.

Relethford, J., 2010. The Human Species: An Introduction to Biological Anthropology. McGraw-Hill Companies.

Ridley, M., 2004. Evolution. Blackwell Publishing, Malden, Massachusetts. isbn:978-1-4051-3459.

Roffman, I., Savage-Rumbaugh, S., Rubert-Pugh, E., Ronen, A., Nevo, E., 2012. Stone tool production and utilization by bonobo-chimpanzees (*Pan paniscus*). Proceedings of the National Academy of Sciences 109, 14500–14503.

Sanfey, A., Dorris, M., 2009. Games in humans and non-human primates: scanners to single units. Neuroeconomics: Decision Making and the Brain 63–80.

Santos, L.R., Chen, M.K., 2009. The evolution of rational and irrational economic behavior: evidence and insight from a non-human primate species. Neuroeconomics: Decision Making and the Brain 81–93.

Sapolsky, R., March 2, 2011. Are Humans Just Another Primate? (video file) Retrieved from http://www.youtube.com/watch?v=YWZAL64E0DI.

Savage-Rumbaugh, E.S., 1986. Ape Language: From Conditioned Response to Symbol. Columbia University Press.

Schoenemann, P.T., 2006. Evolution of the size and functional areas of the human brain. Annual Review of Anthropology 35, 379–406.

Shultz, S., Nelson, E., Dunbar, R.I.M., 2012. Hominin cognitive evolution: identifying patterns and processes in the fossil and archaeological record. Philosophical Transactions of the Royal Society B 367, 2130–2140. http://dx.doi.org/10.1098/rstb.2012.0115.

Semendeferi, K., Lu, A., Schenker, N., Damásio, H., 2002. Humans and great apes share a large frontal cortex. Nature Neuroscience 5, 272–276.

Seyfarth, R., Cheney, D., 1990. The assessment by vervet monkeys of their own and another species' alarm calls. Animal Behaviour 40, 754–764.

Seyfarth, R.M., Cheney, D.L., 1999. Production, usage, and response in nonhuman primate vocal development. In: The Design of Animal Communication, pp. 390–417.

Shettleworth, S.J., 2009. Cognition, Evolution, and Behavior. Oxford University Press, USA.

Shettleworth, S.J., 2012a. Modularity, comparative cognition and human uniqueness. Philosophical Transactions of the Royal Society B: Biological Sciences 367, 2794–2802.

Shettleworth, S.J., 2012b. Darwin, Tinbergen, and the evolution of comparative cognition. In: Vonk, J., Shackelford, T.K. (Eds.), Oxford Handbook of Comparative Evolutionary Psychology. Oxford University Press, New York, NY, pp. 529–546.

Silk, J.B., 2009. Nepotistic cooperation in non-human primate groups. Philosophical Transactions of the Royal Society B: Biological Sciences 364, 3243–3254.

Silk, J.B., Brosnan, S.F., Vonk, J., Henrich, J., Povinelli, D.J., Richardson, A.S., Lambeth, S.P., Mascaro, J., Schapiro, S.J., 2005. Chimpanzees are indifferent to the welfare of unrelated group members. Nature 437, 1357–1359.

Silva, F.J., Page, D.M., Silva, K.M., 2005. Methodological-conceptual problems in the study of chimpanzees' folk physics: how studies with adult humans can help. Animal Learning & Behavior 33, 47–58.

Smith, J.D., Shields, W.E., Schull, J., Washburn, D.A., 1997. The uncertain response in humans and animals. Cognition 62, 75–97.

Smith, J.D., Shields, W.E., Washburn, D.A., 2003. The comparative psychology of uncertainty monitoring and metacognition. Behavioral and Brain Sciences 26, 317–339.

Smith, J.D., 2009. The study of animal metacognition. Trends in Cognitive Sciences 13, 389–396.

Smith, J.D., Couchman, J.J., Beran, M.J., 2012a. The highs and lows of theoretical interpretation in animal-metacognition research. Philosophical Transactions of the Royal Society B: Biological Sciences 367, 1297–1309.

Smith, J.D., Coutinho, M.V.C., Boomer, J., Beran, M.J., 2012b. Metacognition across species. In: Vonk, J., Shackelford, T.K. (Eds.), Oxford Handbook of Comparative Evolutionary Psychology. Oxford University Press, New York, NY, US, pp. 271–294.

Steiper, M.E., Young, N.M., 2006. Primate molecular divergence dates. Molecular Phylogenetics and Evolution 41, 384–394.

Suddendorf, T., Addis, D.R., Corballis, M.C., 2009. Mental time travel and the shaping of the human mind. Philosophical Transactions of the Royal Society B: Biological Sciences 364, 1317–1324.

Terrace, H.S., 2005. The Missing Link in Cognition: Origins of Self-reflective Consciousness. Oxford University Press, New York, NY, US.

Terrace, H.S., Petitto, L.A., Sanders, R.J., Bever, T.G., 1979. Can an ape create a sentence? Science 206, 891–902.

Tomasello, M., Call, J., Hare, B., 2003a. Chimpanzees understand psychological states–the question is which ones and to what extent. Trends in Cognitive Sciences 7, 153–156.

Tomasello, M., Call, J., Hare, B., 2003b. Chimpanzees versus humans: it's not that simple. Trends in Cognitive Sciences 7, 239–240.

Tulving, E., 2002. Episodic memory: from mind to brain. Annual Review of Psychology 53, 1–25.

Vonk, J., 2005. Causality in non-humans: empirical questions. In: Invited Paper Published Online for Interdisciplines: Causality, May 23, 2005.

Vonk, J., Beran, M.J., 2012. Bears "count" too: quantity estimation and comparison in black bears (*Ursus americanus*). Animal Behaviour 84, 231–238.

Vonk, J., Brosnan, S.F., Silk, J.B., Henrich, J., Richardson, A.S., Lambeth, S.P., Povinelli, D.J., 2008. Chimpanzees do not take advantage of very low cost opportunities to deliver food to unrelated group members. Animal Behaviour 75, 1757–1770.

Vonk, J., Jett, S.E., Mosteller, K.W., 2012. Concept formation in American black bears (*Ursus americanus*). Animal Behaviour 84, 953–964.

Vonk, J., Povinelli, D.J., 2012. Similarity and difference in the conceptual systems of primates: the unobservability hypothesis. In: Wasserman, E., Zentall, T. (Eds.), Oxford Handbook of Comparative Cognition, second ed. Oxford University Press, New York, NY.

Vonk, J., Shackelford, T.K., 2012. The Oxford Handbook of Comparative Evolutionary Psychology. Oxford University Press, New York, NY.

Vonk, J., Subiaul, F., 2009. Do chimpanzees know what others can and cannot do? Reasoning about 'capability'. Animal Cognition 12, 267–286.

Wallace, A.R., 1864. The origin of human races and the antiquity of man deduced from the theory of "natural selection". Journal of the Anthropological Society of London 2, clviii–clxxxvii.

Wasserman, E.A., Fagot, J., Young, M.E., 2001. Same-different conceptualization by baboons (*Papio papio*): the role of entropy. Journal of Comparative Psychology 115, 42–52.

Wasserman, E.A., Zentall, T.R., 2006. Comparative Cognition: Experimental Explorations of Animal Intelligence. Oxford University Press, USA.

Whiten, A., Erdal, D., 2012. The human socio-cognitive niche and its evolutionary origins. Philosophical Transactions of the Royal Society B: Biological Sciences 367, 2119–2129.

Wilkinson, G.S., 1984. Reciprocal food sharing in the vampire bat. Nature 308, 181–184.

Willemet, R., 2013. Reconsidering the evolution of brain, cognition and behaviour in birds and mammals. Frontiers in Comparative Psychology, 4, 396.

Wilson, D.E., Reeder, D.M. (Eds.), 2005. Mammal Species of the World: A Taxonomic and Geographic Reference, vol. 2. John Hopkins University Press.

Winslow, J.T., 2005. Neuropeptides and non-human primate social deficits associated with pathogenic rearing experience. International Journal of Developmental Neuroscience 23, 245–251. http://dx.doi.org/10.1016/j.ijdevneu.2004.03.003.

Wright, A.A., Santiago, H.C., Urcuioli, P.J., Sands, S.F., 1983. Monkey and pigeon acquisition of same/different concept using pictorial stimuli. Quantitative Analyses of Behavior 4, 295–317.

Yopak, K.E., Lisney, T.J., Darlington, R.B., Collin, S.P., Montgomery, J.C., Finlay, B.L., 2010. A conserved pattern of brain scaling from sharks to primates. PNAS Proceedings of the National Academy of Sciences of the United States of America 107, 12946–12951. http://dx.doi.org/10.1073/pnas.1002195107.

Zuberbühler, K., Cheney, D.L., Seyfarth, R.M., 1999. Conceptual semantics in a nonhuman primate. Journal of Comparative Psychology 113, 33.

Chapter 36

Evolution of Language

Philip Lieberman
Department of Cognitive, Linguistic, and Psychological Sciences, Brown University, Providence, RI, USA

SYNOPSIS

Evidence from fossils as well as from the archaeological record and its biological bases allows us to rule out implausible scenarios and permit reasonable inferences concerning the evolution of language. Human language serves as a medium of both thought and communication. Specialized anatomy has evolved to enhance the robustness of human speech. Specialized neural capacities are implicated in speech production and confer the cognitive capacities that make language possible. It is unlikely that any single factor, such as social interaction, could have led to the evolution of human linguistic capabilities. Natural selection rules out the *universal grammar* posited by Noam Chomsky.

EVIDENCE FROM COMPARATIVE STUDIES

Present-day apes and humans share a common ancestor. Thus any aspect of language that apes can master most likely was present in early hominin languages. If exposed to human language early in life, chimpanzees can acquire active vocabularies of about 150 words and master simple syntax (Gardner and Gardner, 1969; Savage-Rumbaugh et al., 1985). It is thus improbable that any hominin species was limited to "protolanguage" lacking words. However, no nonhuman species can talk. Apes instead use sign language and other manual systems to signify words, lending plausibility to the idea of gestures playing a greater role in the early stages of the evolution of language (Hewes, 1973).

Human Speech

Apart from not having to direct one's visual attention to individuals who are communicating, and freeing one's hands, speech allows humans to transmit information at rates that exceed the fusion frequency of the auditory system—the rate at which sounds merge into a meaningless buzz. Alvin Liberman and his colleagues in 1967 showed that this process, which they termed "encoding," derived from the physiology of speech production and perceptual processes that are still being studied.

The lungs provide the source of energy for speech production, generally on the outward expiratory flow of air. As Charles Darwin pointed out in 1859, the lungs of most terrestrial species evolved from the swim bladders of fish. Swim bladders enabled fish to hover by storing air extracted from water in elastic swim bladders, enlarging their body size so as to displace water at a given depth. Human lungs retain this elastic property. During quiet inspiration the diaphragm, intercostal, and abdominal muscles expand the lungs; during expiration, the elastic recoil of the lungs expels air. The duration of inspiration and expiration is almost equal. The alveolar (lung) air pressure during expiration impinging on the vocal cords of the larynx starts at a high level and falls as elastic lung sacks deflate. The fundamental frequency of phonation (Fo)—perceived as the "pitch" of a speaker's voice—is determined by the magnitude of alveolar air pressure and the tension placed on the vocal cords of the larynx. Human vocal cords are complex structures whose evolution can be traced back to lungfish. A series of adaptations changed their role from the sealing of the lungs from the intrusion of water to phonation (Negus, 1949).

The pattern of muscular control during speech is quite clear. The diaphragm is immobilized, and the duration of expiration is keyed to the length of the sentence that the speaker intends to produce. Alveolar air pressure is maintained at an almost uniform level until the end of each expiration by means of complex muscular commands. A speaker must "program" a set of instructions to the intercostal and abdominal muscles so that they "hold back" against the force generated by the elastic recoil force of the lungs, which is high at the start of the expiration and gradually falls as the lung volume falls. The intercostal and abdominal muscles contain muscle "spindles" that monitor the force that they produce. The diaphragm contains few spindles, which may account for its being immobilized during speech and singing (Bouhuys, 1974). Speakers generally anticipate the length of the sentence that they intend to produce, taking in more air before the start of a long sentence during spontaneous speech (Lieberman and Lieberman, 1973). Whether and when hominins controlled alveolar air pressure and keyed the duration of an expiration to a sentence in like manner is unknown.

The fundamental frequency of phonation and amplitude of the speech signal (its intonation) cue the ends of sentences, usually remaining more or less level and abruptly falling at the end of a sentence. In many dialects of English and other languages, Fo remains level or rises for yes–no questions (Armstrong and Ward, 1926; Pike, 1945; Lieberman, 1967). Local modulation of Fo contours differentiate words in tone languages such as the Chinese languages (Tseng, 1981). Independent studies—for example, Cheyney and Seyforth (1990)—show that nonhuman primates can signal referential information by means of calls that have different Fo contours. This again points out the implausibility of any stage in early hominin communication that exclusively relied on manual gestures.

Mammalian infants and their caretakers, including humans, signal attention through Fo modulations (Fernald et al., 1989). Some of the neural circuits involved in controlling phonation can be traced back to therapsids, mammal-like reptiles who lived in the age of the dinosaurs. The anterior cingulate cortex (ACC) of the mammalian paleocortex forms part of a neural circuit involving the basal ganglia, subcortical structures deep within the brain (Alexander et al., 1986). The soft tissue of the therapsid brain has not survived, and the inference that therapsids had an ACC is based on their fossil remains, which have the three middle ear bones found in all present-day mammals. The initial function of the hinge bones of the reptilian jaw was to open the jaw wide. In the course of evolution, the final mammalian transition involved migration of the former jaw bones into the middle ear, where they serve as a mechanical amplifier that increases auditory acuity. Being better able to hear enhances the possibility of a mother keeping in contact with her suckling infants and the inference is that, as is the case for living mammals, the therapsid brain possessed an ACC.

Lesion studies show that female mice do not pay attention to their infants when neural circuits involving the ACC are disrupted (Maclean and Newman, 1988). General problems in maintaining attention occur when the ACC to basal neural circuits are degraded in humans. Patients become apathetic and do not attend to events in the flow of life when the ACC to basal ganglia circuit is degraded (Cummings, 1993). Maclean and Newman showed that the ACC has another role in mother–infant interaction—it controls the laryngeal "mammalian isolation cry," the sound that keeps parents awake for months. Basal ganglia lesions in adult humans, as well as Parkinson disease (PD), also disrupt basal ganglia circuits to the ACC, resulting in mutism (Cummings, 1993) and aberrant patterns of laryngeal control during speech (Lieberman et al., 1992; Pickett et al., 1998). Thus, it is almost a certainty that early hominins possessed language in which words and probably simple syntax were communicated vocally, perhaps using Fo variations.

The Supralaryngeal Vocal Tract and Speech Encoding

Figure 1 presents a sketch of a midsagittal view of an adult supralaryngeal airway. Speech encoding devolves from the physical constraints governing the changing of the shape of the airway above the larynx—the supralaryngeal vocal tract (SVT). The SVT determines sound quality in a manner analogous to a pipe organ. In a pipe organ, a source of acoustic energy with a wide frequency spectrum is filtered by pipes that allow energy to pass through them in narrow ranges of frequency, producing particular musical notes. The acoustic energy generated by the larynx is the "source" of energy for phonated vowels such as the vowels and initial consonants of the words *bit* and *map.* Acoustic energy occurs at the fundamental frequency of phonation and at its

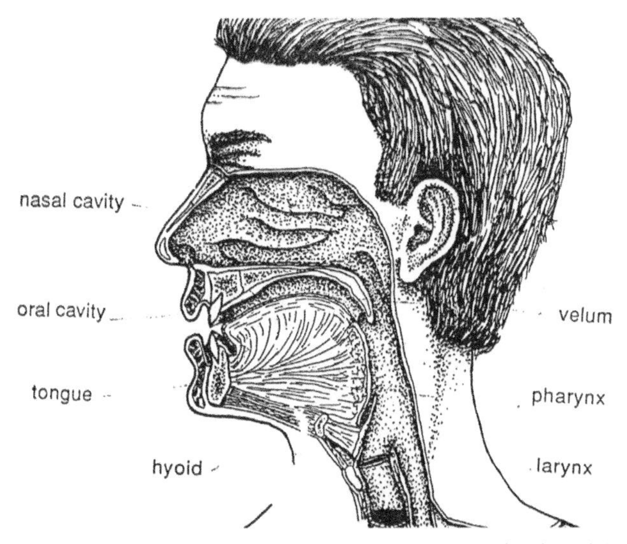

FIGURE 1 The supralaryngeal airway. The human tongue after six to eight years of age has a posterior circular contour when at rest. Approximately half of the tongue is positioned in the oral cavity, and half in the pharynx. When raised, the velum seals off the nasal cavity during the production of sounds other than "nasal" consonants such as [m] and [n], and nasalized vowels in languages such as Portuguese. The larynx is positioned low in the neck in contrast to its high position in infants and nonhuman primates.

harmonics—integral multiples of Fo. For any given shape of the SVT, maximum acoustic energy will pass through it at "formant frequencies." The vowel of the word "see"—in phonetic notation [i]—could have local energy maxima at 270, 2300, and 3000 Hz for a speaker whose overall SVT length is 17 cm. The same speaker's [a] vowel (the vowel of the word "ma") would be 730, 1100, and 2400 Hz. A speaker's consonants and vowels are distinguished by their formant frequency patterns, as well as different durations and the timing between tongue and lip movements and laryngeal phonation (Fant, 1960).

Movable type is often used as a metaphor to describe the speech signal; we hypothetically string together *phonemes*, sounds approximated by the letters of the alphabet, to form words. The phonemes [t], [a], and [b], for example, can be rearranged to form tab, bat, at, and ba. Attempts were made in the 1960s to build devices that would talk by isolating phonemes and then stringing them together. It was thought that it would be possible to isolate phonemes from tape recordings. When a person spoke the word *too* there should have been a segment of tape that contained the sound [t], before a segment of tape that contained the sound [u] (the vowel of the word *too*). However, when the segment of recording tape that was supposed to correspond to the "phoneme" [t] in the word *too* was isolated and linked to a vowel [i] segmented from the word *tea*, the result was incomprehensible. The reason for this failure became evident—the positions assumed by the lips, tongue, jaw, and larynx for the phoneme [t] are affected by those necessary to produce the vowel [u] in a different manner than for the vowel [i], yielding different overall "encoded" formant frequency patterns for the words *too* and *tea.*

Encoding is a general effect. For example, when producing a word such as bit, the positions of the tongue, jaw, lips, and larynx that determine the formant frequencies of [b] must move to the different positions necessary for [I] (the vowel of the word "bit") and then to [t]. As they move, albeit rapidly, there must be a transition between each SVT shape. The perceptual unit that listeners pay attention to is the encoded word, a consonant–vowel syllable being the minimal unit that they then perceptually decode, taking account of the constraints of speech production at some internal level (Liberman et al., 1967). In short, the formant frequency patterns of the hypothetical independent phonemes posited by linguists are always melded together into syllables and words. This seeming "deficiency" explains why speech is the medium for language.

Speech is special in that it allows humans to transmit information at a rate that exceeds that of any other acoustic signal, by means of a complex perceptual process. The minimal units transmitted are encoded words, at a rate below seven units per second. The perceived speech signal can then be perceptually decoded into sequences of phonemes. In a sense, phonemes are abstract speech, motor control, and perception instruction sets. Chinese orthography, which codes words, better approximates the speech signal than alphabetic systems do. Computerized speech-recognition systems use algorithms that match the incoming signal to probable word templates.

One of the problems encountered in speech-recognition systems is how to account for the effect of speakers' differing SVT lengths. Since the length of the SVT varies from person to person as well as during the years of childhood and adolescence for the same individual, the absolute values of the formant frequency pattern vary. For example, the formant frequencies of an [i] are 1.5 times higher for a child whose SVT length is 11.3 cm long than for an adult whose SVT is 17 cm long. This difference would appear to create a problem for listeners, but both of the two different formant frequency patterns would be perceived as examples of an [i] owing to a speech-specific process of perceptual normalization in which listeners internally estimate the length of a speaker's SVT (Nearey, 1978). Listeners can estimate SVT length after hearing a short stretch of speech or by "reverse engineering" a known phrase such as a person saying *hello*, but Nearey showed that the vowel [i] (of the word *see*) was an optimal signal for SVT normalization. The data of the Peterson and Barney (1952) study aimed at developing speech recognition for telephone dialing specified [i] as an optimal cue for SVT normalization. Two errors in 10,000 trials occurred for [i] in this study, where listeners had to identify monosyllabic words that differed with respect to the vowel and speaker. The words uttered by 10 different speakers were presented in quasi-random order to listeners in this experiment; the listeners had to immediately adjust for different speakers' voices. Hillenbrandt et al. (1995) reported similar results.

The Unique Human Tongue and Supralaryngeal Vocal Tract

Charles Darwin pointed out the unique difficulties that humans faced while eating:

> *The strange fact that every particle of food and drink which we swallow has to pass over the orifice of the trachea, with some risk of falling into the lungs*
>
> Darwin (1859, p. 191).

In the twentieth century Victor Negus's studies of comparative anatomy pointed out the species-specific nature of the human tongue that was the cause of this problem. Negus concluded that the adult human larynx (opening to the trachea) was carried down into the pharynx because it "is closely apposed to the tongue" (Negus, 1949, pp. 25–26). Choking on food, owing to the low human larynx, remains the fourth-leading cause of accidental death in the United States (http://www.nsc.org/library/report_injury_usa.htm). Negus thought that the position and shape of the human tongue in some manner facilitated speech communication. That supposition has been validated by computer-modeling studies that calculate the range of formant frequencies that can be produced by adult human and nonhuman SVTs (e.g., Lieberman et al., 1969, 1972; Carre et al., 1995; De Boer, 2010).

The initial 1969 study calculated the formant frequency patterns of the vowels that a rhesus macaque's SVT could produce. The monkey's tongue was positioned as far as possible to yield the SVT shape used by adult human speakers to yield the "point" vowels [i], [u], and [a]. These vowels delimit the range of vowels used in human languages (Greenberg, 1963). The computer-modeling technique showed that the monkey's vowel space did not include these point vowels. Newborn infants have SVTs similar to those of nonhuman primates (Negus, 1949; Crelin, 1969). Lieberman et al. (1972) used similar techniques to model the SVTs of chimpanzees and human newborn infants. Crelin concluded that the vocal tract of La Chapelle-aux-Saints Neanderthal was similar to that of a large human newborn, and it too was modeled (Lieberman and Crelin, 1971). Cineradiographic data of the newborn infant cry (Truby et al., 1965) guided the jaw, tongue, lip, and laryngeal maneuvers in these studies.

At birth in humans, most of the tongue is positioned in the mouth and its shape is flat, as is the case for other mammals. The proportion of the tongue in the oral "horizontal" (SVTh) part of the infant oral cavity relative to the part of the tongue in the "vertical" pharynx (SVTv)—SVTh/SVTv—is 1.5 when the larynx is positioned at its lowest point during the "forceful" cries pictured in Truby et al. (1965). The human tongue does not attain its adult 1:1 SVTh/SVTv proportions and almost circular posterior midsagittal shape until six to eight years of age. The developmental process by which the species-specific human vocal tract is formed is complex and sometimes is not complete until 10 years of age (Lieberman and McCarthy, 1999; Lieberman et al., 2001). The length of the oral cavity is first shortened in humans by developmental processes that move the hard palate back on the base of the skull (Lieberman, 2011). The shape and position of the tongue then gradually change from those of the newborn tongue, which is flat and positioned almost entirely in the oral cavity. The human tongue descends down into the pharynx and achieves its posterior rounded contour, carrying the larynx down with it. In contrast, the nonhuman primate tongue throughout life is long, rectangular, and positioned primarily in the oral cavity. In fetal development and shortly after birth, the chimpanzee larynx drops slightly owing to an increase in the distance between the larynx and hyoid (Nishimura, 2005), whereas the human growth pattern involves the descent and shaping of the tongue. Tongue shape and SVTh/SVTv proportions in nonhuman primates remain almost constant from birth onwards.

The human tongue's oral and pharyngeal proportions and shape explain why only adultlike human SVTs can produce the vowels [i], [u], and [a], and why these vowels contribute to the robustness of human vocal communication. Stevens (1972) showed that the species-specific human SVT can produce the 10-to-1 midpoint area function discontinuities needed to produce the vowels [i], [u], and [a], which Stevens termed "quantal." Stevens employed both computer modeling and physical models (wooden tubes that could be shifted to change the position of the 10:1 changes in SVT cross-sectional area). The quantal vowels are perceptually salient owing to convergence of two formant frequencies that yields spectral peaks. Their formant frequency patterns also do not shift when tongue position varies ~1 cm about the midpoint. Speakers can thus be sloppy and still produce the "same" vowel.

The Speech Capacities of Other Species

The limits on the passive vocabulary of chimpanzees have not been determined, but some dogs can learn at least the primary meanings of 200 words (Kaminski et al., 2004). The neural basis for vocal tract normalization also appears to derive from other species that estimate the size of conspecifics from their vocalizations. Fitch and Reby (2001) documented larynx lowering in deer that serves the purpose of increasing the length of the vocal tract, thereby producing lower formant frequencies. The lowered formant frequencies serve to signal to conspecifics that the animal is larger than it actually is. However, the animal's tongue remains anchored in its mouth while the vocal tract elongates. The larynx transiently descends by increasing the distance between the hyoid bone and larynx. This maneuver does not change the shape of the SVT—its cross-sectional area as a function of distance. The cineradiographs of other mammals vocalizing in Fitch (2000), contrary to his claims, show that though transient larynx lowering occurs, the animals cannot produce quantal vowels because their tongues are still positioned in their mouths. Thus the dynamic changes discussed in Fitch (2010) do not increase the phonetic range of their vocalizations.

It is imperative to point out that speech and language are possible without the ability to produce quantal vowels. Lieberman and Crelin (1971) concluded that Neanderthals (see the chapter by Ahern in this volume) lacked this capacity. However, the archaeological record demonstrated that Neanderthals must have possessed language to transmit their stoneworking technology. In light of the selective advantage of speech's information transfer rate, Neanderthals undoubtedly talked.

NEURAL CIRCUITS AND THE EVOLUTION OF HUMAN LANGUAGE

Studies comparing the vocal repertoires of nonhuman primates with SVT-modeling studies consistently show that these animals are unable to make full use of their phonetic potential. They could communicate to each other or mimic human speech, albeit with reduced intelligibility since their SVTs would enable them to produce only nonquantal vowels in addition to most consonants. However, they cannot talk. Current studies point to a link between speech motor control capabilities and the cognitive capacities that distinguish humans from other species, deriving from neural circuits linking subcortical and cortical structures of the human brain.

Broca's Area

The traditional answer to the question of why humans can talk is that Broca's area of the cortex instantiates the "faculty of language" often noted by Noam Chomsky and his colleagues. In 1861 Paul Broca published his study of a stroke victim whose speech was limited to a syllable that sounded like *tan* (Broca, 1861). Broca's postmortem observations were confined to the cortical surface of this patient's brain. Broca's patient's brain was preserved in alcohol, and a magnetic resonance imaging (MRI) of the preserved brain shows that it had massive damage to the basal ganglia and other subcortical structures, pathways connecting cortical and subcortical neural structures, and other cortical areas (Dronkers et al., 2007). However, in the decades that followed Broca's publication, "Broca's area" was taken to be the brain's speech and language "organ." A few neurologists demurred, pointing to postmortem examinations showing that language and speech are disrupted only when subcortical brain damage is present. Neuroimaging studies using CT scans and MRIs resolved the issue. Patients who had suffered strokes that damaged the cortex, sparing subcortical brain structures, recovered. Conversely, aphasia resulted when subcortical structures were damaged (e.g., Alexander et al., 1987). Stuss and Benson (1986) concluded that aphasia never occurs absent subcortical damage.

Circuits Linking Cortex and the Basal Ganglia

Invasive tracer studies of the brains of monkeys and other animals first mapped out a class of circuits that linked areas of the motor cortex with the subcortical basal ganglia. Other cortical–basal ganglia circuits connected areas of prefrontal cortex, through the basal ganglia and other subcortical structures, to temporal and parietal cortical regions of the brain (e.g., Alexander et al., 1986). Noninvasive diffusion tensor imaging confirmed similar cortical-to-basal-ganglia circuits in humans (Lehericy et al., 2004). Studies of neurodegenerative diseases identified some of the cognitive and linguistic operations performed by these neural circuits. In PD, depletion of the neurotransmitter dopamine degrades the basal ganglia (Jellinger, 1990). Patients have difficulty sequencing the submovements that carry out internally directed motor acts such as walking (e.g., Harrington and Halland (1991)). Speech motor control deteriorates (Lieberman et al., 1992).

Cognitive inflexibility and difficulties performing cognitive acts that require planning or selecting criteria occur (e.g., Lange et al., 1992). Patients so afflicted are unable to change the direction of a thought process or action (Flowers and Robertson, 1985). Focal brain damage limited to the basal ganglia results in similar speech production and cognitive deficits. Bilateral basal ganglia lesions in the subject studied by Pickett et al. (1998), for example, resulted in severe deficits in sequencing the laryngeal, lingual, and lung motor activities necessary to produce articulate speech. The subject also had extreme difficulties comprehending distinctions in meaning conveyed by syntax, and was unable to change the criteria by which she had to sort cards on the "odd man out" test that Flowers and Robertson (1985) devised to test PD patients' cognitive flexibility.

Comparative studies of the architecture of the frontal regions of the brains of monkeys and humans have been conducted over the course of more than a century. One explanation for why we, but no other primates, can talk and command complex syntax might rest on structural cortical differences. However, this does not seem to be the case—Petrides (2005) concludes that the basic architectonic organization (the distribution of neurons in the frontal layers of the cortex) is the same in both humans and monkeys. Nor do humans possess unique cortical-to-laryngeal neural circuits that enable them to talk (Lieberman, 2012).

Neuroimaging

Studies using functional magnetic resonance imaging (fMRI), which tracks the level of neural activity in a particular region of the brain by monitoring oxygen depletion, can infer whether a particular neural structure is active during cognitive tasks. Some of the local operations performed by the different components of neural circuits in humans are becoming apparent (e.g., Duncan and Owen, 2000; Petrides, 2005). Dorsal posterior motor cortex areas control fine motor control. Ventrolateral prefrontal cortex is active during tasks that involve actively selecting and retrieving information stored in other regions of the brain according to specific criteria. Dorsolateral prefrontal cortex is active while monitoring motor or cognitive events during a task that takes into account earlier events in working memory. Studies using fMRI can reveal the roles played by different basal ganglia structures and cortical areas in subjects performing linguistic and cognitive tasks. The Wisconsin card sorting task (WCST) is an instrument that measures cognitive flexibility—a person's ability to form cognitive criteria and shift from one criterion to another. The usual form of the WCST entails subjects sorting cards, each with an image that differs with respect to shape, color, and number. Monchi et al. (2001) reported the involvement of a cortical–striatal loop involving the ventrolateral prefrontal cortex, the caudate nucleus, and the thalamus during WCST criterion sorting shifts. Another cortical–striatal loop including the posterior prefrontal cortex and the putamen was observed during the execution of a sorting criterion set-shift. Dorsolateral prefrontal cortex was involved whenever subjects made any decision as they performed card sorts, to apparently monitor whether their responses were consistent with the chosen criterion. Another fMRI study showed that the caudate nucleus uses this information when a novel action needs to be planned. Similar activation patterns were apparent when subjects were sorting words instead of images and had to match words on the basis of semantic similarity, phonetic similarity of the start of the syllable, or rhyme (Simard et al., 2011). The neural circuits involved are thus not domain specific and operate solely on visual criteria. Studies ranging from electrophysiologic recordings of neuronal activity in the basal ganglia of mice and other animals as they learn tasks (Graybiel, 1995; Mirenowicz and Schultz, 1996; Jin and Costa, 2010) to studies of PD patients (Lang et al., 1992; Monchi et al., 2006) and birds (Brainard and Doupe, 2000) also show that the basal ganglia play a critical role in associative learning and in planning and executing motor acts including speech, and in songbirds, their songs.

THE IMPLAUSIBILITY OF UNIVERSAL GRAMMAR

Noam Chomsky's basic premise is that all human brains contain genetically transmitted "universal grammar" (UG) that specifies the details of syntax of all human languages. Chomsky stated that

> *language is as much an organ of the body as the eye or heart or the liver. It's strictly characteristic of the species, has a highly intricate structure, developed more or less independently of experience in very specific ways, and so on.*
>
> Chomsky (1976, p. 57).

In his subsequent publications, the specific operations that the hypothetical language organ carries out have changed, but UG remains an innate species-specific entity devoted to language. The syntactic rules, speech sounds, and syllable structures of the particular language that a child hears are acquired because the "parameters and principles" that activate them are innate and hence genetically transmitted. It is as though all human brains have identical preloaded software that automatically selects the appropriate instruction set for whatever language a child encounters in the first years of life. The biological implausibility of the Chomskian enterprise rests on Charles Darwin's central insight concerning evolution—the interplay among variation, natural selection, and the ecosystem.

Chomsky must postulate identical UG in every dead, living, or yet-to-be-born human because UG is the causal agent for any human acquiring any language. This entails that every person who ever was or will be born has identical UG owing to the knowledge that any normal child can acquire any human language. If variation existed in UG, some children would be unable to acquire particular languages because critical features were missing from their UG as a result of that variation. A child raised in a German-speaking setting, for example, might not have the parameters necessary to activate German syntax, but instead might be able to activate Chinese. We would find thousands of children who could not acquire their native language. Such a result, however, is not supported by empirical evidence.

Moreover, natural selection does not exist in Chomskian linguistics. Chomsky notes:

> *It is perfectly safe to attribute this development [of innate language structures] to 'natural selection,' so long as we realize that there is no substance to this assertion, that it amounts to no more than a belief that there is some naturalistic explanation for these phenomena*
>
> Chomsky (1972, p. 97).

Similar statements dismissing the role of natural selection in evolution are made throughout Chomsky (2012). However, it is apparent that natural selection continues to act on humans. As Darwin pointed out, natural selection acts on any aspect of biology that contributes to the survival of individuals and their children. An example is lactase tolerance in adult populations that rely heavily on mammalian milk for nutrition (Tishkoff et al., 2007) (see chapters by Little and Wiley in this volume). Overall, language is much more critical to human survival than adult lactase tolerance. Virtually all aspects of human life are results of cultural aggregation and are culturally transmitted. And the medium of transmission is language. Thus if your ancestors had been using any particular language for an extended period, natural selection would have acted to optimize UG. According to Chomsky, his hypothetical UG is *necessary* to "acquire" any language. The United States is an optimal "experiment-in-nature," since the ancestors of most Americans did not speak English before they arrived here. If UG existed, your UG would have been optimized for your ancestral language, rendering it deficient for acquiring English. The Chinese language, for example, has a long attested time depth, and we would expect Americans with Chinese-speaking ancestors to be unable to speak English fluently. Since that is not the case, we must conclude that Chomsky's hypothetical universal grammar does not exist.

FULLY HUMAN LINGUISTIC AND COGNITIVE CAPABILITY

Brain Size

The first anatomical study of an ape, Tyson's 1699 dissection of an orangutan, revealed many anatomical affinities between great apes and humans, but the ape brain was much smaller. Much attention has since been focused on humans having big brains, because brains require lots of biological support. Current assessments show that we have a scaled-up primate brain that has about three times as many neurons, which are the basic computing elements of all brains, as there are in a chimpanzee brain (Herculano-Houzel, 2009) (see the chapter by Holloway). The size of most parts of the mammalian brain scale up in proportion to overall brain size (Finlay and Darlington, 1995), but the human brain differs in that the posterior temporal cortex is disproportionately larger than would be expected (Semendeferi et al., 1997, 2002). The temporal cortex is part of our long-term information storage system, which may explain its larger size. Working memory, which keeps information in short-term memory during cognitive processing, also appears to involve accessing information through neural circuits linking the prefrontal cortex to the temporal cortex and other structures (Badre and Wagner, 2006; Postle, 2006).

The human prefrontal cortex has long been associated with "higher" cognition (see the chapter by Vonk and Aradhye) and when linked with the basal ganglia to other neural structures, it is involved in a wide range of cognitive acts. Prefrontal cortical areas, through basal ganglia circuits and direct cortical-to-cortical circuits, connect to information-storing regions of the brain, pulling memory traces of images, words, and probably other stored information, into short-term working memory (e.g., Postle, 2006; Badre and Wagner, 2006; Miller and Wallis, 2009). Although some studies comparing chimpanzee and human brains using MRIs claim that humans have a disproportionately larger prefrontal cortex, Semenderferi et al. (2002) point out that MRIs inherently cannot show that humans have a disproportionately larger prefrontal cortex. It is impossible to differentiate prefrontal cortical areas from motor regions of the frontal cortex on an MRI, and the human frontal cortex, which includes prefrontal as well as posterior areas involved in motor control, is not disproportionately larger than an ape's.

Many proposals have been made for why hominin brains became larger over time. One reoccurring theory hinges on abrupt climate changes, alternating periods of glacial cold and heat, taking place in Africa where early hominins evolved. However, there is no evidence for glacial cold in Africa. Another proposal suggests alternating periods of drought and heavy rainfall that resulted in alternating desert like and lush rain forests in the Great Rift Valley of Africa as the causal element driving hominin brain size enlargement. However, archaic hominins most likely would have moved away when the climate became inhospitable, as was the case for other species. Another theory stresses group size in primates and communication within the group. However, the data point generally overlooked is the brain size of solitary orangutans. Given the interlocked neural structures implicated in motor control, cognition, and language (as well as emotional regulation), it is improbable that any single factor was responsible for increasing hominin brain size.

Transcriptional Genes

Genetic studies have opened up a new avenue of inquiry. Transcriptional factors essentially are "master" genes that affect the way that other genes are activated. Members of the extended KE family in London, who had severe speech production, sentence comprehension, and cognitive deficits, had only one copy of the FOXP2^{human} transcriptional factor instead of the normal two (Fisher et al., 1998). The information in the double helix of DNA that constitutes the genetic code has to be transcribed into single-stranded mRNA that is later translated into proteins and the structures of living organisms. Transcription factors are genes that govern this process. The FOXP2 gene is one of many transcription factors existing in all mammals, birds, and other creatures. The mouse form Foxp2 controls the embryonic development of the lungs, the intestinal system, heart and other muscles, and spinal column of mice (Shu et al., 2001). The areas of expression of FOXP2^{human} and Foxp2 in both the human and mouse brain are similar and include the cortical–striatal–cortical circuits involved in motor control and cognition—the thalamus, caudate nucleus, and putamen as well as other subcortical structures (Lai et al., 2003). The cortical plate (layer 6), the input level of the cortex, is also affected by the FOXP2 mutation.

The subsequent focus on the role of FOXP2^{human} follows from it being one of the few genes that has been shown to differ from its chimpanzee version. Since the last common ancestor with chimpanzees, FOXP2^{human} underwent two substitutions in its DNA sequences, causing two amino acid changes in FOXP2 protein. An additional change occurred during the period when anatomically modern human beings appeared; a selective sweep occurred as early as 260,000 years ago, resulting in FOXP2^{human} spreading throughout the human population (Enard et al., 2002). In most instances it is unclear what genes do when we find a difference between the chimpanzee and human versions. However, in this instance the FOXP2^{human} anomaly in the KE family showed that it plays a role in the attributes—speech, language, and cognition—that distinguish humans from other living species. When FOXP2^{human} was inserted into mouse pups, the significant neural difference included increased synaptic plasticity in basal ganglia neurons as well as increased dendritic lengths in the basal ganglia, thalamus, and layer VI of the cortex (Enard et al., 2009; Reimers-Kipping et al., 2011). The process by which we learn anything involves modifying synaptic "weights"—the degree to which a synapse transmits information to a neuron (Hebb, 1949).

Comparisons of the Neanderthal genome with humans and a second extinct hominin species, the Denisovans, show that the nature of FOXP2^{human} and the genes that it acts on are not fully understood. Denisovans, who lived in Siberia, diverged from Neanderthals ~200,000 years ago (Reich et al., 2010). The two amino acid substitutions that were thought to distinguish FOXP2^{human} from chimpanzee Foxp2 have been found in a Denisovan fossil as well as in Neanderthal fossils (Krause et al., 2007; Meyer et al., 2012). If FOXP2^{human} only involved these two amino acid substitutions, then the selective sweep on this gene would date to the last common ancestor of Neanderthals and humans, which is not consistent with the date of the FOXP2^{human} selective sweep 260,000 years ago. Other "highly accelerated regions" of the human genome appear to be implicated in neural development (Konopka et al., 2009). These genetic studies suggest that neural circuits humans share with other primates were in effect "supercharged" through the action of transcriptional genes.

When Did Fully Human Language Appear?

The genetic evidence that we have reviewed, the archaeological record, and fossil evidence provide a time line for the evolution of fully human language. Artifacts preserved in the archaeological record inherently cannot serve as absolute markers of their maker's cognitive or linguistic abilities. The computer I am now using to write represents a technology far more complex than Jane Austen's pen, yet it does not constitute evidence for a leap in human cognitive ability. However, there are periods extending over millions of years in which the Oldowan tools attributed to early *Homo* are virtually identical (see the chapter by Toth and Schick). Stone tools attributed to *Homo erectus* are more complex, and it is difficult to see how the technique necessary to make them could have been transmitted without some form of language, but they conform to the same pattern over a long period. Neanderthal stone tools are more complex, and Neanderthals survived in a cold and difficult environment (see the chapter by Ahern). Their brains were as large as those of humans, and they undoubtedly talked. But there is no evidence of the patterns of innovation and imitation that mark human culture. In contrast, in Africa where modern humans first appeared, a paint "factory" that employed a complex process was operating 100,000 years ago (Henshilwood et al., 2011). Stone tools with inscribed decorative lines dating back 75,000 years were found in the same cave.

"Hard" fossil evidence for the evolution of the unique human tongue provides supporting evidence for an African origin for fully human cognitive and linguistic capabilities. As Darwin noted, human anatomy is not optimal for the avoidance of choking to death when we eat. A set of acrobatic maneuvers must be initiated to avoid having food block the larynx. The larynx and hyoid bone that supports it must be pulled forward and upward. At the same time, the epiglottis must be flipped down to cover the larynx. Long necks turn out to have a purpose. The length of neck must accommodate the portion of the human tongue in the neck (SVTv) that is necessary to produce the quantal vowel [i], plus the larynx. If neck length were shorter than what is needed to accommodate the SVTv and larynx, the larynx would be positioned in the chest and blocked by the collarbone, making it impossible to swallow.

Robert McCarthy (unpublished data) measured the cervical vertebrae of the necks of 73 specimens of modern humans from populations distributed around the globe, as well as a sample of Neanderthals and early anatomically modern humans. The evidence that precludes Neanderthals and earlier extinct hominin species from having anatomically modern human tongues necessary to produce quantal vowels is their long mouth lengths, determined from boney landmarks on their skulls. If the oral cavity is long, the part of the tongue in the neck must match its length to produce quantal vowels. Neck lengths can be determined from surviving cervical vertebrae. The vocal tract anatomy necessary to produce quantal vowels becomes evident in Upper Paleolithic fossil hominins unearthed in Europe who lived about 40,000 years ago. We can be certain that they possessed neural circuits that enabled them to use their peculiar human tongues to full advantage; otherwise there would have been only negative consequences (choking to death) from their development of human SVTs.

Thus, through the lens of anatomy we can discern the presence of the brain mechanisms that regulate the voluntary, rapid, complex, internally guided motor acts that underlie human speech. Neural tissue and morphology coevolved to make human speech possible. But we can infer more than that, since the cortical–basal ganglia circuits that allow humans to learn and sequence the complex motor acts involved in speech also act in the realm of cognition. The burst of technological change and the appearance of art in the European Upper Paleolithic has often been interpreted as evidence for a "cultural revolution" (Klein, 1999), signifying an abrupt increase in hominin cognitive capabilities at that time. However, that cannot be the case. The neural bases for human speech, motor control, cognition, and language must have been present in Africa probably 260,000 years ago at the time of the $FOXP2^{human}$ selective sweep. Every human on earth has African ancestors, though some left Africa earlier than others. The ancestors of the "non-African" European humans whose fossil remains McCarthy examined had emigrated from Africa to Europe, displacing the indigenous Neanderthal population, and contemporary Africans have normal human tongues and SVTs (c.f. McCarthy and Strait, 2005; Lieberman and McCarthy, 2007; Lieberman, 2013). And any child of recent African descent can master any language on earth, and has normal cognitive capacities, barring insult to the brain that can strike anyone.

The evolutions of human language, cognition, and motor control are commingled. Some aspects of language have a long evolutionary history and are shared with other living species. Other features have recent bases and are defining characteristics of present-day humans. But the evolutionary mechanisms proposed by Charles Darwin appear to account for the evolution of all aspects of human language and cognition.

REFERENCES

Alexander, G.E., DeLong, M.R., Strick, P.L., 1986. Parallel organization of segregated circuits linking basal ganglia and cortex. Annual Review of Neuroscience 9, 357–381.

Alexander, M.P., Naeser, M.A., Palumbo, C.L., 1987. Correlations of subcortical CT lesion sites and aphasia profiles. Brain 110, 961–991.

Armstrong, L.E., Ward, I.C., 1926. Handbook of English Intonation. Leipzig and Berlin: Teubner.

Badre, D., Wagner, A.D., 2006. Computational and neurobiological mechanisms underlying cognitive flexibility. PNAS 103, 7186–7190.

Bouhuys, A., 1974. Breathing. Grune and Stratton, New York.

Brainard, M.S., Doupe, A.J., 2000. Interruption of a basal-ganglia-forebrain circuit prevents plasticity of learned vocalizations. Nature 404, 762–766.

Broca, P., 1861. Nouvelle observation d'aphémie produite par une lésion de la moitié postérieure des deuxième et troisième circonvolutions frontales. Bulletin Societé Anatomie, 2nd Ser 6, 398–407.

Carre, R., Lindblom, B., MacNeilage, P., 1995. Acoustic factors in the evolution of the human vocal tract. C. R. Academie des Sciences Paris, t320(Serie IIb) t320, 471–476.

Cheney, D.L., Seyfarth, R.M., 1990. How Monkeys See the World: Inside the Mind of Another Species. University of Chicago Press, Chicago.

Chomsky, N., 1972. Language and Mind. Harcourt Brace Jovanovich, San Diego, Ca.

Chomsky, 1976. On the nature of language. In: Steklis, H.B., Harnad, S.R., Lancaster, J. (Eds.), Origins and Evolution of Language and Speech. New York Academy of Sciences, New York, pp. 46–57.

Chomsky, N., 2012. The Science of Language (Interviews with James McGilvray). Cambridge University Press, Cambridge.

Crelin, E.S., 1969. Anatomy of the Newborn: An Atlas. Lea and Febiger, Philadelphia.

Cummings, J.L., 1993. Frontal-subcortical circuits and human behavior. Archives of Neurology 50, 873–880.

Darwin, C., 1859/1964. On the Origin of Species. Harvard University Press, Cambridge, MA.

De Boer, B., 2010. Modeling vocal anatomy's significant effect on speech. Journal of Evolutionary Psychology 8, 351–366.

Dronkers, N.F., Plaisant, O., Iba-Zizain, M.T., Cananis, E.A., 2007. Paul Broca's historic cases: high resolution MR imaging of the brains of Leborgne and Lelong. Brain 130, 143–144.

Duncan, J., Owen, A.M., 2000. Common regions of the human frontal lobe recruited by diverse cognitive demands. TINS 10, 475–483.

Enard, W., Przeworski, M., Fisher, S.E., Lai, C.S.L., Wiebe, V., Kitano, T., Monaco, A.P., Paabo, S., 2002. Molecular evolution of FOXP2, a gene involved in speech and language. Nature 41, 869–872.

Enard, W., Gehre, S., Hammerschmidt, K., Hölter, S.M., Blass, T., Somel, M., Brückner, M.K., Schreiweis, C., Winter, C., Sohr, R., et al., 2009. A humanized version of Foxp2 affects cortico-basal ganglia circuits in mice. Cell 137, 961–971.

Fant, G., 1960. Acoustic Theory of Speech Production. Mouton, The Hague.

Fernald, A., Taeschner, T., Dunn, J., Papousek, M., de Boysson-Bardies, B., Fukui, I., 1989. A cross-language study of prosodic modifications in mothers' and fathers' speech to preverbal infants. Journal of Child Language 16, 477–501.

Finlay, B., Darlington, R., 1995. Linked regularities in the development and evolution of mammalian brains. Science 268, 1578–1584.

Fisher, S.E., Vargha-Khadem, F., Watkins, K.E., Monaco, A.P., Pembrey, M.E., 1998. Localization of a gene implicated in a severe speech and language disorder. Nature Genetics 18, 168–170.

Fitch, W.T., 2000. Skull dimensions in relation to body size in nonhuman mammals: the causal bases for acoustic allometry. Zoology 103, 40–58.

Fitch, W.T., 2010. The Evolution of Language. Cambridge University Press, New York.

Fitch III, W.T., Reby, D., 2001. The descended larynx is not uniquely human. Proceedings of the Royal Society of London B 268, 1669–1675.

Flowers, K.A., Robertson, C., 1985. The effects of Parkinson's disease on the ability to maintain a mental set. Journal of Neurology, Neurosurgery, and Psychiatry 48, 517–529.

Gardner, R.A., Gardner, B.T., 1969. Teaching sign language to a chimpanzee. Science 165, 664–672.

Graybiel, A.M., 1995. Building action repertoires: memory and learning functions of the basal ganglia. Current Opinion in Neurobiology 5, 733–741.

Greenberg, J., 1963. Universals of Language. MIT Press, Cambridge, MA.

Harrington, D.L., Haaland, K.Y., 1991. Sequencing in Parkinson's disease: abnormalities programming and controlling movement. Brain 114, 99–115.

Hebb, D.O., 1949. The organization of behavior: A neuropsychological theory. Wiley, New York.

Henshilwood, C.S., d'Errico, F., van Niekerk, K.L., Coquinot, Y., Jacobs, Z., Lauritzen, S.E., Menu, M., García-Moreno, R., et al., 2011. A 100,000 year old ochre processing workshop at Blombos Cave, South Africa. Science 334, 219–221.

Herculano-Houzel, S., 2009. The human brain in numbers: a linearly scaled-up primate brain. Frontiers in Human Neuroscience 3, 1–11.

Hewes, G.W., 1973. Primate communication and the gestural origin of language. Current Anthropology 14, 5–24.

Hillenbrand, J.L., Getty, A., Clark, M.J., Wheeler, K., 1995. Acoustic characteristics of American English vowels. Journal of the Acoustical Society of America 97, 3099–3111.

Jin, X., Costa, R.M., 2010. Start/stop signals emerge in nigrostriatal circuits during sequence learning. Nature 466, 457–462.

Jellinger, K., 1990. New developments in the pathology of Parkinson's disease. In: Streifler, M.B., Korezyn, A.D., Melamed, J., Youdim, M.B.H. (Eds.), Advances in Neurology. Parkinson's Disease: Anatomy, Pathology and Therapy, vol. 53. Raven, New York, pp. 1–15.

Kaminski, J., Call, J., Fisher, J., 2004. Word learning in a domestic dog: Evidence for "fast mapping." Science 304, 1682–1683.

Klein, R.G., 1999. The Human Career, second ed. Chicago University, Chicago.

Konopka, G., Bomarl, J., Winden, K., Coppola, G., Jonsson, Z.O., Gao, F., Peng, S., Preuss, T.M., 2009. CNS development genes by FOXP2. Nature 462, 213–217.

Krause, J., Lalueza-Fox, C., Orlando, L., Enard, W., Green, R.E., Burbano, H.A., Hublin, J.J., Hänni, C., Fortea, J., de la Rasilla, M., Bertranpetit, J., Rosas, A., Pääbo, S., 2007. The derived FOXP2 variant of modern humans was shared with Neandertals. Current Biology 17, 1908–1912.

Lai, C.S., Gerrelli, D., Monaco, A.P., Fisher, S.E., Copp, A.J., 2003. FOXP2 expression during brain development coincides with adult sites of pathology in a severe speech and language disorder. Brain 126, 2455–2462.

Lange, K.W., Robbins, T.W., Marsden, C.D., James, M., Owen, A., Paul, G.M., 1992. L-Dopa withdrawal in Parkinson's disease selectively impairs cognitive performance in tests sensitive to frontal lobe dysfunction. Psychopharmacology 107, 394–404.

Lehericy, S., et al., 2004. Diffusion tensor fiber tracking shows distinct corticostriatal circuits in humans. Annals of Neurology 55, 522–527.

Liberman, A.M., Cooper, F.S., Shankweiler, D.P., Studdert-Kennedy, M., 1967. Perception of the speech code. Psychological Review 74, 431–461.

Lieberman, D.E., 2011. The Evolution of the Human Head. Harvard University Press, Cambridge, MA.

Lieberman, D.E., McCarthy, R.C., 1999. The ontogeny of cranial base angulation in humans and chimpanzees and its implications for reconstructing pharyngeal dimensions. Journal of Human Evolution 36, 487–517.

Lieberman, D.E., McCarthy, R.C., Hiiemae, K.M., Palmer, J.B., 2001. Ontogeny of postnatal hyoid and laryngeal descent: implications for deglutition and vocalization. Archives of Oral Biology 46, 117–128.

Lieberman, M.R., Lieberman, P., 1973. Olson's "projective verse" and the use of breath control as a structural element. Language and Style 5, 287–298.

Lieberman, P., 1967. Intonation, Perception and Language. MIT Press, Cambridge, MA.

Lieberman, P., 2012. Vocal tract anatomy and the neural bases of talking. Journal of Phonetics 40, 608–622.

Lieberman, P., 2013. The Unpredictable Species: What Makes Humans Unique. Princeton University Press, Princeton NJ.

Lieberman, P., Klatt, D.H., Wilson, W.H., 1969. Vocal tract limitations on the vowel repertoires of rhesus monkey and other nonhuman primates. Science 164, 1185–1187.

Lieberman, P., Crelin, E.S., Klatt, D.H., 1972. Phonetic ability and related anatomy of the newborn, adult human, Neanderthal man, and the chimpanzee. American Anthropologist 74, 287–307.

Lieberman, P., Kako, E.T., Friedman, J., Tajchman, G., Feldman, L.S., Jiminez, E.B., 1992. Speech production, syntax comprehension, and cognitive deficits in Parkinson's disease. Brain and Language 43, 169–189.

Lieberman, P., Crelin, E.S., 1971. On the speech of Neanderthal man. Linguistic Inquiry 2, 203–222.

Lieberman, P., McCarthy, R.M., 2007. Tracking the evolution of language and speech. Expedition 49, 15–20.

MacLean, P.D., Newman, J.D., 1988. Role of midline frontolimbic cortex in the production of the isolation call of squirrel monkeys. Brain Research 450, 111–123.

McCarthy, R.C., Strait, D.S., 2005. Morphological constraints on hominin speech production. Paleoanthropology 3, AO2.

Meyer, et al., 2012. A high coverage genome sequence from an archaic Denisovan individual. Science 338. http://dx.doi.org/10.1126/science.1224344.

Miller, E.K., Wallis, J.D., 2009. Executive function and higher order cognition: definition and neural substrates. In: Squire, L. (Ed.), Encyclopedia of Neuroscience, vol. 4, pp. 99–104.

Mirenowicz, J., Schultz, W., 1996. Preferential activation of midbrain dopamine neurons by appetitive rather than aversive stimuli. Nature 379, 449–451.

Monchi, O., Petrides, M., Petre, K., Worsley, K.J., Dagher, A., 2001. Wisconsin card sorting revisited; distinct neural circuits participating in different stages of the test as evidenced by event-related functional magnetic resonance imaging. The Journal of Neuroscience 21, 7739–7741.

Monchi, O., Petrides, M., Strafella, A.P., Worsely, K.J., Doyon, A., 2006. Functional role of the basal ganglia in the planning and execution of actions. Annals of Neurology 59, 257–264.

Nearey, T., 1978. Phonetic Features for Vowels. Indiana University Linguistics Club, Bloomington.

Negus, V.E., 1949. The Comparative Anatomy and Physiology of the Larynx. Hafner, New York.

Nisimura, T., 2005. Developmental changes in the shape of the supralaryngeal vocal tract in chimpanzees. American Journal of Physical Anthropology 126, 193–204.

Peterson, G.E., Barney, H.L., 1952. Control methods used in a study of the vowels. Journal of the Acoustical Society of America 24, 175–184.

Petrides, M., 2005. Lateral prefrontal cortex: architectonic and functional organization. Philosophical Transactions of the Royal Society B 360, 781–795.

Pickett, E.R., Kuniholm, E., Protopapas, A., Friedman, J., Lieberman, P., 1998. Selective speech motor, syntax and cognitive deficits associated with bilateral damage to the putamen and the head of the caudate nucleus: a case study. Neuropsychologia 36, 173–188.

Pike, K.E., 1945. The Intonation of American English. University of Michigan Press, Ann Arbor.

Postle, B.R., 2006. Working memory as an emergent property of the mind and brain. Neuroscience 139, 23–38.

Reich, dD., Green, R.E., Kircher, M., Krause, J., Patterson, N., Y.Durand, E., et al., 2010. Genetic history of an archaic hominen group from Denisova cave in Siberia. Nature 468, 1053—1060.

Reimers-Kipping, S., Hevers, S., Paabo, S., Enard, W., 2011. Humanized Foxp2 specifically affects cortico-basal ganglia circuits. Neuroscience 175, 75–84.

Savage-Rumbaugh, S., Rumbaugh, D., McDonald, K., 1985. Language learning in two species of apes. Neuroscience and Biobehavioral Reviews 9, 653–665.

Semendeferi, K., Damasio, H., Frank, R., Van Hoesen, G.W., 1997. The evolution of the frontal lobes: a volumetric analysis based on three-dimensional reconstructions of magnetic resonance scans of human and ape brains. Journal of Human Evolution 32, 375–378.

Semendeferi, K., Lu, A., Schenker, N., Damasio, H., 2002. Humans and apes share a large frontal cortex. Nature Neuroscience 5, 272–276.

Shu, W., Yang, H., Zhang, L., Lu, M.M., Morrisey, E.E., 2001. Characterization of a new subfamily of winged-helix/forkhead (Fox) genes that are expressed in the lungs and act as transcriptional repressors. Journal of Biological Chemistry 276, 27488–27497.

Simard, F., Joanette, Y., Petrides, M., Jubault, T., Madjar, C., Monchi, O., 2011. Fronto-striatal contributions to lexical set-shifting. Cerebral Cortex 21, 1084–1093.

Stevens, K.N., 1972. Quantal nature of speech. In: David Jr., E.E., Denes, P.B. (Eds.), Human Communication: A Unified View. McGraw Hill, New York, pp. 51–66.

Stuss, D.T., Benson, D.F., 1986. The Frontal Lobes. Raven, New York.

Tishkoff, S.A., Reed, F.A., Ranciarol, A., Voight, B.F., Babbitt, C.C., Silverman, J.S., et al., 2007. Convergent adaptation of human lactose persistence in Africa and Europe. Nature Genetics 39, 31–39.

Truby, H.L., F Bosma, J., Lind, J., 1965. Newborn Infant Cry. Almquist and Wiksell, Upsalla.

Tseng, C.Y., 1981. An Acoustic Study of Tones in Mandarin (Ph.D. diss.). Brown University.

Chapter 37

Evolution of Moral Systems

Douglas Allchin
Minnesota Center for the Philosophy of Science University of Minnesota, St. Paul, MN, USA

SYNOPSIS

Biologically, morality is a form of behavior. How did it evolve? A complete explanation involves genetic, organismal, and social levels. Some cooperative systems can originate through kin selection or mutualism. However, once primates formed societies, reciprocal cooperation can persist through selective relationships, punishment, rewards, coordination, or reputation. Such social environments can then select for cooperative dispositions or moral sentiments.

INTRODUCTION

When Charles Darwin finally turned to the question of human evolution in *The Descent of Man and Selection in Relation to Sex* (1871), he treated anatomy only briefly, then quickly advanced to the more engaging topic of explaining mental powers. The moral sense or conscience, in particular, he described as the most important difference between humans and other animals. It was also an evolutionary puzzle. Natural selection seems to amplify only "selfish" behavior (see the chapter by Futuyma in this volume). How then could cooperation or altruism ever evolve? Darwin's early scientific sketch proved exceptionally insightful. Our modern biological understanding of human morality builds on and extends the basic concepts that he introduced in 1871.

First, Darwin adopted a biological perspective. He sought to explain morality, not to decide what was right or wrong or to justify particular moral principles. His orientation helped to initiate an objective understanding of morality as a form of behavior (first section below). Second, Darwin noted the basis of morality in social organization. What conditions lead to mutually beneficial interactions, what he called social instincts? Today, we can characterize these precisely in mathematical terms. Stable moral systems have particular properties (second section). Finally, Darwin considered the conditions for moral motivations, or intent, that are so central to many philosophical views of morality. Researchers are now able to explore neural activity in the brain as well as developmental aspects of moral behavior, which helps clarify the respective roles of instinct and learning. This understanding is coupled with findings from anthropology in developing accounts of the evolutionary origin of our moral sentiments (third section).

MORAL BEHAVIOR

Philosophers and theologians have considered morality for centuries. They focus on ethical principles and their justification. Darwin introduced a different perspective. In his new context of natural history, morality is simply a form of behavior. One can thus ask questions about it as one does with such behaviors as aggression, parenting (see chapter by Flinn), tool use (see the chapter by Toth and Schick), or hunting and cooking. What do we observe? Why does it occur? The biological study of morality sets aside questions of ultimate values, or right or wrong, associated with our everyday conceptions of morality. The scientist aims, rather, to describe and explain.

Biologists benefit from studying the behavior of nonhumans. Apparent kindness or helping has been observed among other animals. Some cases are widely known in popular culture. In two cases, gorillas have rescued human children who have fallen into the gorilla enclosures at zoos. A troop of snow monkeys once accommodated a handicapped member who, born without hands or feet, was unable to climb. Dolphins and whales have helped injured peers escape from human threats, sometimes buoying them to the surface to breathe. Chimps have plunged into water to save another who was drowning, only to drown themselves. Caution is appropriate when interpreting such stories. Humans can easily project their own emotions or values onto animals. But enough striking examples are sufficiently documented to warrant serious consideration (de Waal, 2009). Together, they belie the stereotypical image of nature red in tooth and class, governed only by competition, violence, and selfishness. Humans may not be the only species to exhibit moral or proto-moral behavior.

In some cases, a formal study of animals can provide simple models for interpreting the more complex human behavior. Behavior among primates can be especially informative by offering clues to our shared evolutionary heritage (see chapter by Watts). Together, the comparison of behaviors among humans and other organisms can contribute to an understanding of how modern human behavior emerged gradually through successive stages or layers of complexity. Other relevant information comes from lab studies of human behavior, mathematical analysis and modeling, developmental observations, neurophysiological diagnostics, and cultural and historical anthropology.

From an evolutionary perspective, cooperation and the helping of others at a cost to oneself, as forms of moral behavior, are puzzling. Traits preserved in successive generations are just those that enhance the relative survival of the individual. It is hard to imagine how a trait that benefits other organisms, especially at one's own expense, can persist by comparison. However, biologists recognize that the interpretation of cost depends on the genetic context. In some cases, an individual helps the survival of its own offspring. Of course, this readily promotes the ultimate persistence of those traits. In other cases, the organism receiving help is more distantly related, but still shares a proportion of the genetic makeup. The trait is thereby enhanced indirectly in the lineage as a whole. Biologists call this kin selection. In multigenerational contexts especially, what matters is not fitness per se, but inclusive fitness. Certainly, humans also frequently help members of their family and extended family. But it is not yet clear to what degree this behavior is due to instinct and kin selection, or to other mechanisms at the social level, as described below.

In some cases, cooperation and helping occur conspicuously among non-kin (Clutton-Brock, 2009). In these cases, one may find that long-term benefits outweigh the apparent short-term costs. Benefits may be returned. Mutualisms, for example, are found widely in nature: in pollination ecology, seed dispersal, and other well-known partnerships, such as those between ants and acacia trees, *Rhizobium* bacteria and legumes, and cleaner wrasse and larger fish with parasites.

FIGURE 1 Musk oxen benefit from shared defense, an example of mutualistic cooperation. *From US govt.*

FIGURE 2 Macaques groom one another, a gesture that is returned later; an example of direct reciprocity. (See color plate section). *From US govt.*

In the same way, organisms within a species may cooperate to their mutual benefit. For example, threatened musk oxen gather together in a circle to defend themselves more effectively against predators (Figure 1). Some predators, such as lions, pelicans, and whales, on the other hand, band together to improve their chances at capturing prey. Cliff swallows, osprey, and others will share information about where to find food. All benefit. In many cases, cooperation can enhance fitness rather than diminish it.

In other cases the benefits for the helper are not immediate. One may still talk in terms of reciprocity, however (Trivers, 1971). Chimpanzees, for example, help groom each other and share food (Figure 2). In the long run, primatologists observe, such exchanges even out (de Waal, 1989). Implicit debts are paid back. Benefits and costs balance. Help, costly at first, is reciprocated. Another fascinating case occurs among vampire bats. The bats need sustenance daily to survive, but do not find food every night. Thus, after an unsuccessful night, some bats will nuzzle others in a gesture of begging. The second bat generally regurgitates and shares a portion of blood. Later the situation may be reversed, and the return exchange is made. By cooperating, the bats individually and collectively survive more effectively (Wilkinson, 1984). Among humans, one can see the same sense of trade-offs in the familiar expression, "you scratch my back, I'll scratch yours." In old-fashioned community barn raisings or quilting bees, one finds reciprocity in action. Everybody pitches in to help one person on the mutual understanding that the recipient will help everyone else out when the need shifts to another person. The biological concept of reciprocity resonates with the view of some moral philosophers that ethics are based on an implicit social contract.

MORAL SYSTEMS

A relationship of reciprocity is vulnerable, however. It is based on trust. Trust may be broken. It may be misplaced. Organisms can cheat. They may not meet their end of the bargain. One can easily envision how cheaters might very well prosper and proliferate at the expense of generous individuals—until nothing is left but cheaters. Any system of cooperation may seem doomed to ultimate failure. Nevertheless, several conditions stabilize cooperation, as described below: network reciprocity, strong reciprocity, and indirect reciprocity (Sigmund, 2010).

Network Reciprocity

If one shifts focus from the individual to the social level, the prospects change dramatically. Social interactions can shape how individuals behave. Consider again the case of the vampire bats (Figure 3). Some bats do try to cheat. When the request for payback comes, they refuse to share in return. But bats can recognize each other. They can identify free riders. Cheaters do not get repeated handouts. Cooperators ultimately interact only with each other. They exclude and thus insulate themselves from the defectors. Socially, they form a network of reciprocity. Social organization, in a sense, trumps individual tendencies. Capuchin monkeys, too, exhibit awareness of the failure to reciprocate. In one study, two capuchin monkeys learned to cooperate to get food. Then, only one was given the reward. Generally, that privileged capuchin shared the reward. On one occasion, however, he did not. On the next round, the partner went "on strike," refusing to cooperate further (de Waal and Suchak, 2010). Violating the implicit rules of fair cooperation had consequences. Humans, too, in laboratory settings, act to join cooperative groups. If their group becomes infected by defectors, they migrate to another group. Stable cooperation may thus emerge through "network reciprocity," an

FIGURE 3 Vampire bats both share and exclude non-sharers, an example of network reciprocity. (See color plate section). *Image courtesy of the Ontario Specialized Species Centre.*

emergent property of the social system. It contradicts what one might expect if one attended only to the properties of the individuals.

The realm of cooperation has been explored in particular through computer models. Once computational power became available in the 1980s, biologists could simulate the effects of repeated encounters among individuals with certain dispositions or strategies (Axelrod, 1984). Modelers have asked about what types of interactions lead to sustained cooperation and which ones disrupt it. The earliest model results indicated that the most successful simple approach is to cooperate when a partner cooperates, but stop when they stop—a strategy known as "tit for tat." Tit for tat behavior is perhaps the simplest form of interaction that establishes network reciprocity.

Strong Reciprocity

Another social behavior that can limit cheaters is sanctions. Organisms may actively punish others who do not heed the expectations of cooperation. Thus, there can be a social cost to not cooperating. Such punitive interactions have been observed in the cooperative foraging of a semicaptive macaque troop, in the cooperative breeding of fairy wrens, and in the shared nesting of paper wasps (Clutton-Brock and Parker, 1995). Humans, too, are willing to punish noncooperators in lab-based scenarios (Fehr and Gächter, 2002). Punishments critically help keep free riders in check. Sanctions enforce cooperation. Biologists call this "strong reciprocity."

Of course, punishment itself bears an additional cost. Does that cost outweigh the ultimate benefit? Anthropological studies have shown that individuals from different cultures around the world accept such costs. Pacific Islanders, African pastoralists, Siberian hunters, US college students, and others—all will bear some added burden in order to sanction noncooperators. Indeed, the degree of enforced punishment seems correlated with the culture's norms of cooperation. The greater the willingness to punish, the higher the overall level of generosity in that culture (Henrich et al., 2010; Heinrich et al., 2006).

FIGURE 4 Rewards for service to the community are incentives for moral behavior; an example of strong reciprocity. *From Wikimedia.*

Rewards also shape behavior (Figure 4). In the one laboratory study, subjects could both reward and punish others based on their degree of sharing in the immediate past. With time and increasing levels of trust, rewards became more common. The amount of sharing increased. A reward system outperformed an exclusively punitive system

FIGURE 5 The cleaner wrasse mutualism. When observed by other fish, this is example of reputation, which is key to indirect reciprocity. (See color plate section). *From Wikimedia.*

(Rand et al., 2009). Incentives thus seem to function to shape moral behavior just as much as the threat of negative consequences. Just as Darwin noted in 1871, then, a human seems to respond to "the wishes, approbation, and blame of his fellow-men," as well as the "fear of punishment" (pp. 86, 92).

Socially coordinated actions can also be important. First, collaboration allows the cost of punishment to be distributed and thus reduced for each individual. According to mathematical models, once multiple individuals can act together, they can leverage a system of cooperation into a population where none had existed previously (Boyd et al., 2010). Coordinated action can also keep free riders in check. Indeed, this factor may have been critical in human evolution. Chimpanzees have strongly hierarchical societies, as did, presumably, early hominins. According to one anthropologist, communication and coordination—facilitated by the emergence of language (see chapter by Lieberman)—enabled early humans to level such hierarchies (Boehm, 1999). That would have initiated a distinctly human social structure such as one finds among the more egalitarian Paleolithic hunter-gatherers.

Indirect Reciprocity

In the cases noted above, reciprocity and trust are regulated through direct interactions. Cheating may also be governed indirectly. Organisms may rely on information about another organism's trustworthiness rather than on direct experience. For example, in the cleaner wrasse mutualism noted above, the small cleaner fish typically eats just the parasites on the larger fish (Figure 5). But sometimes, they take a small bite of flesh. That is an extra cost to the large fish. But the larger fish keep their eyes on the cleaners as they help other fish. They avoid cleaners who cheat. The cleaners, in response, cheat less when they are aware of observers. Here, information substitutes for a direct encounter (Bshary and Gutter, 2006). The reciprocity is maintained indirectly.

In humans, too, reputation matters. Neighborhood gossip, some biologists contend, fulfills this important social function. Psychologists have certainly documented that humans behave differently when they know they are being watched (Milinski and Rockenbach, 2007). Laboratory studies confirm that even without direct encounters, cooperation can persist in a group if information about others' histories of cooperative (or selfish) behavior is available.

Morality and Social Organization

The various relationships described above have been neatly summarized mathematically (Nowak, 2006). The persistence of cooperation basically reduces to an economic question of fair exchange or reciprocity. It can be sustained when the benefit-to-cost ratio of the behaviors exceeds some critical variable. The key parameter varies in each instance. It may be kin relatedness. It may be the probability of a repeat encounter. It may be proximity in a network (or the number of close neighbors). Or it may be reputation. All enable and set limits on sustainable cooperation. Numerical analysis can sometimes help us understand the dynamics in particular cases.

Still, the apparent rigor of mathematical expressions should not obscure the assumptions and the nature of the variables on which they are based. In all these cases, the social organization significantly shapes individual behavior (Richerson and Boyd, 2005). Individual decision-making and choice are context dependent, and interactions among individuals contribute to that context. In this sense, it is more appropriate to speak of moral systems at the social level than of moral choice at an individual level.

The biology of moral systems cannot dictate specific moral principles. Values cannot be derived from facts alone. Nevertheless, an evolutionary perspective can fruitfully inform an understanding of human morality. For example, consider the apparently universal golden rule: do unto others as you would have them do unto you. This precept's details are indeterminate, perhaps cryptic. It refers instead to the structure of morality as a context, one essentially based on reciprocal interaction. Consider also the numerous cases when humans try to share a socially shared resource. Economists call these public goods, because no single person owns them. If individuals act independently, each can take more than a proportional share. When the resource is the quality of the air or water, or the atmosphere as a reservoir of ozone or greenhouse gases, the collective effect of separate actions can lead to environmental degradation. Philosopher Garret Hardin (1968) called this "the tragedy of the commons." Hardin offered several possible solutions. One was "mutual coercion, mutually agreed upon." That is, groups could adopt and enforce consensual regulatory rules. Hardin's moral perspective was essentially an articulation of the dilemma of defection and the effect of strong reciprocity. He described how morality might function at a social level. Again, the science, in describing the case, does not thereby tell us specifically what we ought to do. But understanding the framework of moral systems can help inform our social interactions and ethical discourse.

MORAL SENTIMENTS

The biological analysis of morality in the previous sections has focused on behavior only. It has not addressed what an organism may feel, think, or intend. For some moral philosophers (known as consequentialists or utilitarians), actions and their consequences are what matter. They may be fully satisfied by accounts that disregard the mental dimension. However, for other moral philosophers (known as deontologists or virtue ethicists), motivation or intent is paramount. For them, accounts of behavior alone, without an understanding of intentions, are markedly insufficient. Since biologists cannot preemptively stipulate the concept of morality, they must be able to accommodate or address multiple conceptions of morality such as those emerging from divergent philosophical traditions. Indeed, Darwin and his fellow Victorians centered their chief concerns on moral feelings. They wanted to understand such emotions as empathy, a felt sense of duty to help others, shame, and remorse (Figure 6). For Darwin, the challenge was not merely to describe the origin of social instincts, but ultimately to explain human moral sentiments.

Scientific investigation of moral emotions and moral cognition unfolds primarily on two fronts. One approach is to study the brain (see the chapter by Holloway). What is involved physiologically in moral perceptions and in generating moral behavior? A second strategy is to study psychological development. How do moral feelings and responses unfold with age? To what degree might they be inborn? Each perspective is addressed below.

Morality and Neurophysiology

To interpret how humans think morally, philosophers have tended to rely heavily on introspection. Yet our self-awareness is limited. The brain functions extensively beyond the reach of its consciousness. In addition, interpreting our own thoughts or those of others is easily susceptible to ideological filtering. Cognitively, we tend to find confirmation of imagined patterns of thought that do not accurately reflect the relevant mental processes. For these reasons, scientists seek more objective means to study mental processes. They examine the anatomy and physiology of the brain and nervous system, here related to moral behaviors.

FIGURE 6 William-Adolphe Bourguereau's *The Remorse of Orestes* (1862), an illustration of a moral sentiment that Darwin considered central to understanding morality. (See color plate section). *From Wikimedia.*

Most notably, neurological imaging techniques (such as fMRI) have been used in recent years to monitor what areas of the brain are active as subjects engage in moral processing (Greene et al., 2001). For example, viewing images that evoke moral impressions but do not require actual moral judgment activates the ventromedial prefontal cortex. This region seems to be part of a network that processes feelings associated with social interactions. When this area is damaged, subjects are less able to integrate their emotions with analyses of costs and benefits, and their resulting moral judgments are accordingly biased.

On the other hand, when subjects try to interpret other people's intentions as part of a moral assessment, the right temporoparietal junction is activated. When they reflect on moral problems that are more personal, the medial frontal gyrus becomes more involved. When they address more abstract or hypothetical problems, the dorsolateral prefontal cortex (and other areas) shows activity. When they integrate emotions, impressions, and memory, parts of constructing coherent narratives, the posterior cingulate seems involved. When different forms of thinking conflict, as one often finds in mediating moral dilemmas, activity rises in the anterior cingulate cortex. In short, moral mental processing is distributed throughout the brain. No single region of the brain is devoted exclusively to moral thinking. Indeed, every area that shows activity seems to have a primary role in some other mental process not directly related to morality (Greene and Haidt, 2002). That is, while we may tend to regard moral thinking as a special category, brain imaging techniques indicate that it is a melange of other mental processes.

The neuroimaging studies foster two important conclusions. First, moral thinking seems to borrow from and build on other mental processes. Thus, scientists do not need to explain the evolutionary emergence of some new moral organ in the brain. Morality could have evolved, with small modifications, based on preexisting structures and functions. Consequently, the challenge of deciphering the evolutionary transition to human moral behavior seems far less formidable.

Second, the neuroimaging studies indicate that moral reasoning involves not just emotion or logic, but both. Philosophers have long debated which is (or should be) primary in moral judgment. Biologically, as least, both seem relevant. This scientific finding may challenge philosophers to revisit their disagreements and reconsider instead how one might resolve and integrate the two philosophical systems of moral assessment. Can humans benefit from the merits of each, perhaps using the alternative frameworks synergistically?

Morality and Development

A major evolutionary question for any form of behavior is whether it is innate or learned. That is, does it arise somehow from the species' inborn genetic heritage, or is it a product of enculturation? The neurophysiological findings indicate a deep evolutionary legacy of capacities. At the same time, they leave open the possibility that these capacities may have been enlisted in a moral context only recently in human evolution, through cultural practices. One standard strategy for investigating this type of question is to examine behavior among young, naive organisms, before any opportunity for learning can occur. Do innate behaviors exhibit themselves?

The results of such studies on human infants' helping behavior are striking. When infants 12–18 months of age encounter adults having problems completing simple tasks in a lab setting, they spontaneously offer help. They may open a cabinet door, fetch a relevant object hidden from the adult's view, or point to a tool that had been moved during the adult's brief absence. The researchers ensured that the adults did not ask for help, which could have prompted a specific response, or reward the behavior, which alternatively might have motivated an infant's act. A series of these experiments demonstrate that an unschooled impetus to help seems rooted somehow in human genetics. This conclusion is strengthened by findings that young chimps exhibit the same behavior, although at a later developmental stage (3 to 4½ years of age). Ironically, the researchers also noted, children apply such behaviors more selectively as they age. They learn the social contexts where such behaviors yield benefits and where they do not (Warneken and Tomasello, 2006; Tomasello, 2009). Helping and sharing seem to have a strongly innate component in humans, later shaped through learning.

The Evolution of Moral Psychology

Given that some basic moral impulses seem innate, how did they evolve? On a reductionistic view, the process begins with the parts: individuals and the random genes or cooperative instincts that might bring them together into a stable, functional society. As the discussion above indicates, this process is immensely difficult without an existing social context, unless one appeals to kin selection as the initial first step. Decades ago, most biologists accepted that reductionistic view as inevitable. However, Darwin had already proposed an alternative.

In Darwin's perspective, now widely endorsed, the social organization originated first. Societies can emerge as adaptive ensembles without moral sentiments being completely in place at the outset. The social context can then shape subsequent evolution. While Darwin relied on views of inheritance and learning that are now no longer accepted, his vision provides an informative guide. For example, Darwin assigned important roles to memory and language. Organisms did not need to act exclusively on instinct. Memory critically allowed individual organisms to learn. The flexibility of behavior was itself adaptive. In addition, communication allowed organisms to convey their needs more effectively. Under these circumstances, organisms would respond to the demands of the social environment. For Darwin, an individual's social learning could be inherited and alter the future of the species. Today, we recognize that no such mechanisms of heredity exist. However, the social context can have a long-term effect. It contributes via natural selection. Organisms that fare well in a social environment—for example, those rewarded for being inclined to help others—are more likely to reproduce. Individuals with selfish dispositions are less likely to reproduce. Over time, the other members of the society select for prosocial emotions and neurological organizations that are inscribed at the genetic level. Cooperative, or moral, sentiments seem to be the result of social organization, not the prerequisite for it.

Genetic selection at the social level can thus explain how humans exhibit spontaneous other-regarding behavior even without genetic relatedness or the prospect of direct reciprocity. Our genes are a legacy of ancestral humans who valued cooperative dispositions from other members of their species. It is the same process humans used to domesticate wild animals—dogs, pigs, goats, and horses—and to tame their undesirable behaviors.

Any satisfactory explanation of moral sentiments in humans must also include an understanding of occasions where individuals do not express the idealized (or possibly romanticized) moral sentiments. Not everyone in human society embodies the society's moral ideals. Wartime atrocities and bullying in schools are also part of the spectrum of observed human behaviors. Spite and runaway punishment may, ironically, also be periodic outcomes of the normal dynamics of the human moral system (Jensen, 2010). One fascinating case is the multiple effects of the hormone oxytocin. Oxytocin enhances parental bonding with offspring and seems to also facilitate trust among unrelated individuals (non-kin). At the same time, oxytocin increases antagonism toward non-group members (De Dreu et al., 2010). In-group loyalty seems coupled with out-group aggression. That seems to ensure continuing problematic social relationships. Biologists have yet to fully explore all of these contradictory tendencies.

MORALITY AND LEVELS OF SELECTION

In 1976 Richard Dawkins prominently introduced the notion of "the selfish gene." In that view, an organism was just how one set of genes made another set of genes. All natural selection was reduced to the genetic level. That included, notably, forms of behavior. E.O. Wilson soon published his now landmark volume, *Sociobiology*, that profiles the promise of explaining the spectrum of all animal behaviors as gene-based adaptations. That view was epitomized in explanations for the distinctive social structure known as eusociality, where a single reproductive female was supported by a whole colony of nonreproducing or sterile individuals—as found in honeybees and naked mole rats (Figure 7), among others. The apparent exception to the principle that natural selection acted only for the good of the individual was dramatically explained by kin selection acting on the species' unusual shared genetic structures. In this intellectual climate, buoyed by Cold War politics, cooperation as exhibited in the cooperative breeding of these societies and in human moral behavior was generally ascribed to the ultimate selfishness of genes.

More recently, however, Wilson has embraced a different explanation for eusociality (Nowak et al., 2010). Three decades of research have shown that many cooperative breeding societies (such as termites) do not exhibit the required haplodiploid genetic structure. Moreover, many species that do (including sawflies and horntails) are not social. The documented cases and the explanation do not align. Rather, the societies—from ants and honeybees to beetles and shrimp, along with the naked mole rats—all seem to have nests with restricted access, guarded by just a few individuals. The social cooperation seems just an ordinary adaptation to certain environmental conditions. Wilson has renounced kin selection as the explanation. The striking reproductive and genetic structure, he now contends, is an evolutionary consequence—not a cause—of the social organization.

Wilson's turnabout nicely reflects an overall shift in evolutionary perspectives. As exemplified in the discussion above, biologists are now more attuned to the role of multiple levels of selection (Richerson and Boyd, 2005). They acknowledge that higher levels of organization can govern genetics, just as humans could domesticate wild animals through selective breeding. In a sense, the once selfish gene has now been domesticated. The broader perspectives have opened new understanding of human morality. In case after case, reciprocal interactions—immediate or deferred, direct or indirect, and mediated by rewards, sanctions, or reputation—have emerged as significant. Moral systems at the social level, and moral sentiments and learning at the psychological level, are just as important as moral behavior and its genetic context.

FIGURE 7 Naked mole rats live in communities supporting only one reproductive female. Once considered to be the result of kin selection, the social structure is now considered an outcome of ecology with the unusual genetic structure being a consequence of the social structure. (See color plate section). *Image courtesy of Chris Faulkes.*

As noted earlier, science can describe and even interpret moral behavior, but it cannot thereby justify or dictate moral principles or values. Similarly, scientists can adopt consequentialist or deontological perspectives from moral philosophy and comment on their implications, but they cannot resolve the long-standing tensions between them. Still, a biological understanding of moral behavior, moral systems, and moral sentiments seems to offer an informative context for how humans might continue and conduct their ongoing moral discourse.

REFERENCES

Axelrod, R., 1984. The Evolution of Cooperation. Basic Books, New York, NY.

Boehm, C., 1999. Hierarchy in the Forest: The Evolution of Egalitarian Behavior. Harvard University Press, Cambridge, MA.

Boyd, R., Gintis, H., Bowles, S., 2010. Coordinated punishment of defectors sustains cooperation and can proliferate when rare. Science 328, 617–620.

Bshary, R., Gutter, A.S., 2006. Image scoring and cooperation in a cleaner fish mutualism. Nature 441, 975–978.

Clutton-Brock, T., 2009. Cooperation between non-kin in animal societies. Nature 462, 51–57.

Clutton-Brock, T.H., Parker, G.A., 1995. Punishment in animal societies. Nature 373, 209–216.

Darwin, C., 1871. The Descent of Man. John Murray, London.

De Dreu, C.K.W., Greer, L.L., Handgraaf, M.J.J., Shalvi, S., Van Kleef, G.A., Baas, M., et al., 2010. The neuropeptide oxytocin regulates parochial altruism in intergroup conflict among humans. Science 328, 1408–1411.

de Waal, F., 1989. Food sharing and reciprocal obligations among chimpanzees. Journal of Human Evolution 18, 433–459.

de Waal, F., 2009. The Age of Empathy: Nature's Lessons for a Kinder Society. Harmony Books, New York.

de Waal, F.B.M., Suchak, M., 2010. Prosocial primates: selfish and unselfish motivations. Philosophical Transactions of the Royal Society B 365, 2711–2722.

Fehr, E., Gächter, S., 2002. Altruistic punishment in humans. Nature 415, 137–140.

Greene, J.D., Sommerville, R.B., Nystrom, L.E., Darley, J.M., Cohen, J.D., 2001. An fMRI investigation of emotional engagement in moral judgment. Science 293, 2105–2108.

Greene, J.D., Haidt, J., 2002. How (and where) does moral judgment work? Trends in Cognition 6, 517–523.

Hardin, G., 1968. The tragedy of the commons. Science 162, 1243–1248.
Henrich, J., Ensminger, J., McElreath, R., Barr, A., Barrett, C., Bolyanatz, A., et al., 2010. Markets, religion, community size, and the evolution of fairness and punishment. Science 327, 1480–1484.
Heinrich, J., McElrath, R., Barr, A., Ensminger, J., Barrett, C., Bolyanatz, A., et al., 2006. Costly punishment across human societies. Science 312, 1767–1770.
Jensen, K., 2010. Punishment and spite, the dark side of cooperation. Philosophical Transactions of the Royal Society B 365, 2593–2598.
Milinski, M., Rockenbach, B., 2007. Spying on others evolves. Science 317, 464–465.
Nowak, M.A., 2006. Five rules for the evolution of cooperation. Science 314, 1560–1563.
Nowak, M.A., Tarnita, C.E., Wilson, E.O., 2010. The evolution of eusociality. Nature 466, 1057–1062.
Rand, D.G., Dreber, A., Ellingsen, T., Fudenberg, D., Nowak, M.A., 2009. Positive interactions promote public cooperation. Science 325, 1272–1275.
Richerson, P.J., Boyd, R., 2005. Not by Genes Alone: How Culture Transformed Human Evolution. University of Chicago Press, Chicago, IL.
Sigmund, K., 2010. The Calculus of Selfishness. Princeton University Press, Princeton, NJ.
Tomasello, M., 2009. Why We Cooperate. MIT Press, Cambridge, MA.
Trivers, R.L., 1971. The evolution of reciprocal altruism. Quarterly Review of Biology 46, 35–57.
Warneken, F., Tomasello, M., 2006. Altruistic helping in human infants and young chimpanzees. Science 311, 1301–1303.
Wilkinson, G.S., 1984. Reciprocal food sharing in the vampire bat. Nature 308, 181–184.

Chapter 38

Race and Ethnicity

Catherine Bliss
Department of Social and Behavioral Sciences, University of California, San Francisco, CA, USA

SYNOPSIS

Race and ethnicity are historically determined systems of human classification that originated in the premodern era to distinguish natives from foreigners. For centuries, scientists characterized race and ethnicity in ways that supported systems of hierarchy and discrimination such as slavery. After eugenic Nazism was exposed in the twentieth century, scientists attempted to replace "race" with "population," and to introduce the notion of clinal variation. Though experts now agree that race and ethnicity are social constructs, a genetic debate over their innate dimensions has emerged. Some offer race-based medicine and commercial products as a solution to the persisting problem of racial health disparities.

INTRODUCTION

Race and ethnicity are historically determined systems of classification by which humans are categorized in science and society. Race typically connotes discrete physiological or genetic groupings along a continental gradient; however, it has historically also referred to culture, nationality, and geographical origin. These groupings are often conceived as biologically essential, each possessing its own physical and behavioral traits. Ethnicity connotes a more dynamic typology of cultural identity that may or may not be based on national origin. In the postcolonial era, race and ethnicity are frequently used interchangeably, and thereby have common social implications.

Race and ethnicity have long histories and etymologies; however, their modern use can be traced to the late seventeenth century, when a number of natural historians and philosophers began mapping human variation. The first human taxonomies posited people of European descent as superior and all others as inferior. As the disciplines of biology and anthropology and the theory of genetic evolution emerged in the nineteenth century, scientists reconceived race in genetic terms. Colonial governments adopted early biological and anthropological classifications in order to fashion censuses. National governments also used scientific definitions of race and ethnicity in public administration such as taxation and immigration policy.

In the wake of World War II, as Nazi crimes against Jews were exposed, hierarchical notions of race and ethnicity fell out of favor in the biological and anthropological sciences of Europe, the United Kingdom, and the United States. This led to a series of debates in which natural scientists questioned the utility of biologically grouping humans by race, and social scientists interrogated folk concepts of race and ethnicity. By the mid-twentieth century, many biologists and anthropologists had concluded that "population" was a better classification system for humans. Social scientists began examining the social constructions of race and ethnicity, or their production as a system of classification in specific sociohistorical contexts.

Race and ethnicity are now understood as products of political and scientific frameworks. Scholars of race and ethnicity study how an elite's need to maintain social hierarchies determines which physical and behavioral traits experts herald as distinct markers of difference. Those who study the persistence of race and ethnicity over time also analyze social patterns in perception, ascription, and identification that make race and ethnicity salient for everyone. Focusing on the dynamics of seeing and enacting race, they distinguish between racism, or discrimination by race according to assumptions of superiority, and racialism, or a basic, possibly nonhierarchical, belief in race.

Despite these advances in understanding, the presence of stark racial health disparities confounds experts, making knowledge about the relationship between race, ethnicity, and health a top priority for national governments and bioethical associations. Public health departments around the world have devised racial classification systems to collect data on disease treatment, progression, and outcomes in different racial groups. However, lingering misassumptions about race-specific genetics cause researchers to hunt for biological causal factors at the expense of social explanations.

Since the revolution in DNA science of the late twentieth century, biologists and anthropologists have reopened the debate over the biology of race. Some have argued that according to zoological standards, the human species does not have races. Others have countered that as a taxonomy of fuzzy-set groupings, race can be useful in biomedicine and health care. As the US Department of Health and Human Services ramped up efforts to extend federal race standards to all biomedicine in the 1990s, race became a mandated framework for human genomic research. Biologists from all sides of the genetic debate over race have thus been given incentives to structure their studies with federal race standards, and to focus their efforts on finding ways to put an end to health care disparities.

In 2005, the US Food and Drug Administration approved the first race-based medicine, BiDil, to treat cardiovascular disease in people of African descent. Since then, post-genomic research, or studies that attempt to annotate the data provided by the genome sequencing projects of the early part of the twenty-first century, has also been given incentives to adopt a racial framework. Scientists working in emerging fields of drug development and gene–environmental research target specific racial groups in research and marketing, furthering the collective notion that race and ethnicity are products of our genomes.

Commercial genetics has also compounded the marriage of race, ethnicity, and genetics in science and society by offering racial- and ethnic-specific products. Genetic genealogy companies sell ancestry kits to provide genetic proof of particular racial lineages. Forensic technologies utilize similar race-based software platforms to determine a suspect's racial identity. This results in a public sphere wherein race is assumed to be essentially genetic, and identity and politics are refashioned in biological terms.

A HISTORY OF RACIAL SCIENCE

Race and ethnicity predate the scientific revolution and all modern biological concepts of variation and diversity (Bernasconi and Lott, 2000). Race and ethnicity were originally used to systematize a distinction between native and foreign populations in early modern and mercantile societies. Racial concepts were particularly important to Spanish and English royals in their genocidal campaigns on Irish, Moorish, and Jewish inhabitants. The modern concept of race emerged amid debates by colonial authorities over the humanity of foreign populations such as indigenous Americans and African slaves (Smedley, 1993).

Francois Bernier was the first expert to talk about race in terms of a taxonomy of human kinds. As secretary to the French ambassador of Poland and Germany, Bernier obtained a license to practice medicine outside France. His travels to the Middle East and Northern Africa inspired him to publish the "New division of Earth by the different species or races which inhabit it" in 1684. In it Bernier specified physiological traits for each of his races, paying special attention to differences in skin pigmentation.

The Swedish naturalist and father of modern taxonomy, Carl Linnaeus, devised an alternative classification system while developing his broader biological taxonomy in the early eighteenth century. In successive editions of *Systema Naturae*, Linnaeus elaborated a human taxonomy based on the European, African, Asian, and American continents. Linnaeus also had a category for the "feral" and "monstrous" humans for which sightings were reported in the European periodicals of his day. Linnaeus schematized his races by physiology, behavior, and dress, attributing positive qualities to Europeans and negative qualities to all other of his races.

Subsequent naturalists such as Georges Louis-LeClerc and Johannes Blumenbach maintained Linnaeus's continental stratification, yet they argued that individuals fell on a racial spectrum. Essential to their theory was the belief that humans were one stock adapted to different environments, and that non-European races were degenerate forms of the European race. In their taxonomies, philosophers such as David Hume and Immanuel Kant also emphasized the hereditary descent of the human species. Yet their taxonomies more greatly detail the relationship between physiology, behavioral traits, and social status. For example, Kant attributed European superiority to fluids in white skin that produced a unique capacity to reason. Similarly, Hume maintained that Africans were destined for slavery due to a lack of intelligence and morality.

Throughout the nineteenth century, European and American political elites debated the nature of white supremacy (Shipman, 2002; Stepan, 1982). Scientists from a range of backgrounds contributed biological theories of race to these debates. For example, Swiss geologist Louis Aggasiz popularized the theory of polygenism, or the notion that human races were separate species created for different stations in society. When Charles Darwin introduced the genetic concept of race in *On the Origin of the Species by Means of Natural Selection, or the Preservation of Favoured Races in the Struggle for Life*, polygenism prevailed across the sciences. Darwin was in strong disfavor of slavery and any innate hierarchy of humans. However, in *The Descent of Man, and Selection in Relation to Sex*, Darwin also portrayed Africans as evolutionarily situated between apes and Europeans.

As Darwinian evolution gained ground, Darwin's cousin Francis Galton created the statistical system that would lead to the field of racial eugenics. Galton initially measured physical traits, but later trained his efforts on behavior. Racial eugenics became the mainstay of early twentieth-century genetics and social policy. Leading genetics institutes such as Cold Spring Harbor provided the biological rationale for racist government policies, including anti-immigration and race-based sterilization, into the first half of the twentieth century.

POSTWAR DEBATES

Eugenics found political favor with political leaders around the world and became the basis of the Rassenhygiene program of the Nazi movement in Germany (Tucker, 1994). When the atrocities of Nazi science and experimentation were uncovered at the close of World War II, European and American sentiments changed course. Leading scientists of all ilk allied with a new generation of anthropologists who promoted a cultural model of racial and ethnic difference (Reardon, 2005).

Drawing on studies of morphological changes in immigrant ethnic groups, anthropologist Franz Boas led efforts in his discipline to challenge the notion that racial and ethnic characteristics were genetically determined. With Ashley Montagu, the author of *Man's Most Dangerous Myth: The Fallacy of Race*, Boas produced an official definition of race for global governance. *The Race Question*, the first of four statements created for the United Nations Educational, Scientific, and Cultural Organization, stated that there were no superior or inferior races. Though the series outlined cultural reasons for the many ostensible differences in social status and attainment between racial and ethnic groups, it did not challenge the basic assumption that biological races existed.

Debates among evolutionary biologists, who specialized in population genetics, intimated the first scientific doubts about the biological veracity of race in humans. While Darwin conceptualized races as varieties of a species, mid-century scientists had a more stringent definition of race as subspecies. Approaching race as a zoological constant, these biologists argued that humans could only be divided into races if they were endogamous populations. Since levels of isolation and morphological divergence in the human species were still under study, many biologists supported replacing race with the less politically charged terminology of "population."

Anthropologists Frank Livingstone and Loring Brace also introduced an alternative concept of gradational human variation called "clinal" variation. Livingstone and Brace argued that the physical traits by which races had been mapped were distributed continuously across the globe. Traits such as skin pigmentation and hair texture were not the property of one or another population, but rather varied over geographical distances irrespective of continent (see the chapter by Elias and Williams in this volume). Since the 1970s, population geneticists have used this alternative system to map the distribution and covariance of traits.

Sociologists also reconceptualized race and ethnicity in the postwar period (Steinberg, 2007). One influential theory was Robert Park's *race relations cycle*, an elaboration of four stages of group relations as immigrant ethnic groups integrate into a host country. Park argued that groups unknowingly compete until they develop a strong ethnic identity. Oppositional identities lead to open conflict. When conflict becomes too costly to endure, the host group finds ways to accommodate the foreign group. Park believed that assimilation to the host culture was the natural and just endpoint for immigrant groups.

Gunnar Myrdal's *An American Dilemma* supported this social conflict model of race. However, Myrdal focused on structures of discrimination that prevented the assimilation of black Americans, such as racism in education and employment. While many of Myrdal's most prominent successors claimed that nonWhites were also responsible for their own lower social status by way of cultural habits and attitudes, the notion that race and ethnicity were dynamic, socially structured entities took hold in the sciences.

THE SOCIAL CONSTRUCTION OF RACE

Since the turn of the twenty-first century, natural and social scientists alike have approached race and ethnicity as social constructs, or fluid concepts that structure social interaction. Scholarship on race and ethnicity shows that social constructs about human difference shape the world, even when they are not based on objective biological foundations. Race and ethnicity persist as structuring logics despite genetic evidence of clinal variation, and demographic evidence of unremitting exogamy throughout history. Scientific debates over innate intellectual capacities continue to rage (Gould, 2008). Experts thus agree that race and ethnicity are likely to continue to inform social conduct for some time to come (PBS, 2003).

Sociologists Michael Omi and Howard Winant (1994) have suggested that people view race and ethnicity in terms of historical formations, or changing sets of ideas and politics. Race and ethnicity are concepts born of political struggles in which dominant groups define the terms of difference. For example, governments create census taxonomies that condition how wealth is taxed and distributed. Social institutions such as medicine and education reinforce official categories. In the postwar United States, the government's taxonomy produced a system in which citizens who were classified as "white" received a disproportionate level of government assistance, such as the GI Bill (Lipsitz, 2006). During the 1970s, the US Office of Management and Budget devised Directive No. 15, a system of racial classifications that monitors the participation of racial groups in public institutions, in an effort to counter such institutionalized discrimination (Yanow, 2003).

Authoritative entities like governments are not the only social entities to construct definitions of race and ethnicity. The concepts are internalized and embodied as members of society build their identities and interpret others (Omi and Winant, 1994). Race and ethnicity are common senses that structure perceptions, choices, and actions. Some interactions around race and ethnicity are subconscious and routine. Others involve overt political action, such as projects to redistribute resources along racial or ethnic lines in ways that service or disservice the status quo. The balance of these projects amount to broad historical formations, such as the racial dictatorship of nineteenth-century America or the campaign for racial democracy in twentieth-century Brazil.

Research has also shown that race and ethnicity are equally reliant on the naturalization of categories in a given social context and a belief in those categories regardless of whether they connote the supremacy or inferiority of a given group (Appiah, 1990; Frederickson, 2002). Racism in the form of prejudice and discrimination assumes a hierarchy between groups, but racialism can exist even in the absence of a hierarchical bias. Members of society need only collectively believe in the existence of groups that are mutually exclusive and in possession of certain immutable characteristics.

Research into racial ideologies has shown that individuals develop a sense of groupness from the belief that membership is obligatory (Appiah, 2005). In the wider society, the idea that membership is obligatory is usually based on an assumption of an essential biological truth to race and ethnicity as witnessed in physical and behavioral variation. De facto social policies such as the one-drop rule, or the idea that one drop of African ancestry classifies an individual as black, perpetuate the assumption of obligatory membership even as de jure segregation recedes (Gilroy, 2000). Individuals also develop a sense of groupness due to the way they experience particular classifications (Nobles, 2000). As individuals make sense of themselves with the available classifications, they reinforce the notion that races and ethnic groups are immutably real.

Race and ethnicity are belief systems, and thus their measurement requires attention to how people know and interpret their surroundings (Daynes and Lee, 2008). Race and ethnicity are constructs that condition the social institutions in which people interact, but they do so via the ritualistic enactment of beliefs. As individuals perceive the world through the lenses of race and ethnicity, they form memories of the world as racially and ethnically real. Institutions ascribe members of society racial and ethnic identities, while individuals continually calibrate their senses of self to their memories of racial and ethnic experiences, and to social norms about the group to which they are ascribed. For example, hospitals that require that patients register a racial identity, and treat patients based on their racial affiliation, legitimize the belief that race is real while generating experiences that confirm that belief. Likewise, patients who identify as a particular race or ethnicity conform to the classification system, thus experiencing the collective norms of groupness as real.

RACE AND HEALTH

Despite the emergence of a constructionist framework in the analyses of race and ethnicity, experts have had difficulty explaining the relationship of either system to health (Epstein, 2007). Research into racial health outcomes consistently shows deep and lasting disparities in disease and life expectancies among various racial groups in different countries (Krieger, 2005).

Addressing the problem of disparities in health outcomes has been a leading priority for public health systems around the world since the late twentieth century (Carter-Pokras and Baquet, 2002). Governments have funded research to study the relationships among genetics, ancestry, health-related behaviors, and social practices. Though many governments have devised official racial taxonomies to facilitate research into causal factors and promote equal treatment across health institutions, biomedical researchers and practitioners still do not have a standardized operational definition of race. Operational definitions fall along the lines of essentialism and antiessentialism, or conceptions that use biological reasoning to prove or disprove the existence of discrete racial groups. Single studies often deploy contradictory operational definitions at different points in the research process.

One reason for confusion about the relationships among race, ethnicity, and health is the common misassumption that racial groups possess unique genetic disorders (Montoya, 2011). This misunderstanding is the result of a trend among scientists to portray high-profile monogenetic diseases as race-specific (Hubbard and Wald, 1999; Morning, 2011). For example, in the twentieth century scientists depicted cystic fibrosis as a "white" disease, while they depicted sickle cell anemia as "black." Today, geneticists attribute cystic fibrosis to the farthest northwestern region of Eurasia, and they have mapped hemoglobin disorders like sickle cell disease to malarial regions around the globe. And though some diseases are more frequent in certain ethnic groups that have stricter traditions of endogamy, successful campaigns to prevent carriers from partnering have reduced and redistributed disease prevalence among groups. For example, once primarily associated with Ashkenazi Jews, Tay–Sachs disease is now on the rise in non-Ashkenazi and non-European populations such as Irish Americans and Han Chinese.

Common cardiovascular diseases, cancers, and autoimmune disorders have also been depicted as racial diseases (Hunt and Megyesi, 2008). For example, people of African descent have been marked as a high-risk population for hypertension, prostate cancer, and lupus, while people of European descent have been marked as being at low risk for these diseases. While researchers acknowledge that environmental factors such as living conditions, health care delivery, cultural behaviors, discrimination, and self-identity all play roles in the prevalence, prevention, and progression of disease, a greater amount of funding is available for research that investigates genetic components of common disease than for research focused on social or environmental factors. In the United States, the administrative seat of most genome projects and genetic research conducted worldwide, the National Institutes of Health sets aside a larger amount of funding for those genetic studies that promise to address diseases that exhibit a higher prevalence in African Americans than it does for any other kind of biomedical research. This is despite the fact that most African Americans possess recent European ancestry, and that Americans who identify as black are 10% less likely to have health insurance than those who identify as white.

The large body of social research into race, ethnicity, and health that has accumulated in recent decades associates racial and ethnic minorities with lower socioeconomic status and educational attainment, and poorer diets and access to health care. Racial and ethnic minorities are more likely to live in the dense urban areas associated with toxic working and living conditions as well as a lack of healthy food and exercise options (see the chapter by Dressler). An emerging yet underfunded research area examines the ways that these institutionalized forms of racism, and everyday experiences of discrimination, compound the stress that contributes to poorer health outcomes and lower life expectancies in racial and ethnic minorities.

NEW GENETIC SCIENCES

In the final decades of the twentieth century, with the rise of recombinant DNA technology and science, the biological debate over race reopened (Braun, 2002). Though many of the scientists leading the DNA revolution declared that study of the genome proved that humans were 99.9% the same, and that there was no genetic justification for race, other influential geneticists began discussing the possibility of race-based medicine (Bliss, 2012). A central driver of this debate was the simultaneous rise of health disparities and genomic research as integrated priorities of public health in the United States (Bliss, 2014).

When the National Institutes of Health, US Department of Energy, and world's leading human genome research institutes launched the Human Genome Project in 1986, the project took no notice of minority health or population-based inequalities. The goal of the HGP was to assemble the sequence of a single set of human chromosomes that would represent half the DNA that one human possesses. The project utilized preexisting immortalized cell lines from populations living in Utah, France, and Venezuela. Project planners did not address the racial identity of whom they would sample from, and there were no attempts to include or exclude people of non-European descent. None of the project's early communiqués discussed ethical issues associated with race, ethnicity, or population representation.

However, in 1990 the US Department of Health and Human Services released the report *Healthy People 2000*, a set of national health goals wherein race and ethnicity were of primary interest. *Healthy People 2000* endeavored to increase life expectancies, secure access to preventative health care, and reduce health disparities for all Americans. Just as the second global genome initiative, the Human Genome Diversity Project, began planning its launch, the US federal government issued a series of public health policies designed to increase the use of Directive No. 15 racial classifications in public health and biomedicine.

Throughout the 1990s, US public health agencies ramped up efforts to racialize all of their programs worldwide. For example, the Food and Drug Administration directed all new drug applicants to comparatively test efficacy and dosage differentials for the different federal racial groups, even for research conducted internationally. The National Institutes of Health, the largest sponsor of the Human Genome Project, required all publicly funded researchers to publish how their research affected minorities and to conduct outreach with minority communities. They also created an in-house Polymorphism Discovery Resource based on Directive No. 15 classifications.

By the end of the decade, the Directive No. 15 taxonomy was a requirement for all publicly funded biomedical entities, and the foundation of international genome projects. At the same time, the US government began seeking more aggressive measures for solving the issue of health disparities, including providing new funding mechanisms to biomedical institutes that would design research models for understanding the genomic component of racial health disparities. The Human Genome Diversity Project, which planned to use an array of alternative ethnolinguistic groupings to sample DNA, was denied federal support and brought to a close in 1997. Meanwhile, the newly minted International HapMap Project, the latest endeavor by the leaders of the Human Genome Project, was rewarded for its use of a racial sampling framework, and officially launched in 2002.

These measures solidified a genetic conception of race in science and society across the world, and made race-based medicine a reality (Kahn, 2012). In 2001, the US Food and Drug Administration condoned the race-based clinical trial of the heart disease drug BiDil, a combination vasodilator designated for use in black Americans. In 2005, the agency approved the drug despite its prior use in a non-combined generic form in people of all racial affiliations, and despite the fact that its makers had no causal rationale for a race-based efficacy. Proponents of BiDil, such as the NAACP and the Association of Black Cardiologists, argue that race-targeted treatment is a superior form of health justice. However, critics warn that with no evidence that BiDil does not work for patients who are not black, BiDil's sale is tantamount to an unwarranted biologization of race.

POSTGENOMIC DEVELOPMENTS

Following BiDil's success, pharmacogenomics, the science of drug response markers, has been a pivotal focus of racialization efforts in biomedicine (Lee, 2007; Whitmarsh, 2008). Though BiDil's pharmacogenomic mechanisms were never assessed, it has been heralded as providing a resource for the pharmacogenomics industry to extend health rights to racial minorities across the world. Members of the National Institutes of Health-sponsored Pharmacogenomics Research Network have made a series of statements instructing all biomedical researchers to use US federal classifications in all drug studies, and to use disparities as starting points for further research.

The scientists who spearheaded the recombinant DNA revolution in biomedicine and public health have now come to laud racial health justice as a central goal for the postgenomic sciences of functional DNA analysis, epigenetics, and gene–environment research (Bliss, 2013a). Since the publication of the draft map of the human genome in 2000, all of the major genomic projects, even those internationally led, have appropriated a racial health disparities optic. The 1000 Genomes Project, which set out to sequence 1000 whole genomes but has now sequenced a far greater number, used the Directive No. 15-inspired sampling protocol established by the International HapMap Project in Phase I of its research. During project planning, a number of scientists made statements supportive of using common racial classifications as a last resort in a proactive health justice science. Leaders of the epigenetic ENCODE and Envirome projects have also argued that geneticists have a responsibility to do whatever it takes to foreground the study of racial health disparities.

Today postgenomic sciences hold the largest budgets for health disparities research and are using those resources to create racial and ethnic genome projects. Founded in 2010, the Human Heredity and Health in Africa project is the first international large-scale sequencing undertaking that is dedicated to a singular racial group. Human Heredity and Health in Africa models how to use next-generation genome sequencing and gene–environment frameworks in the fight against disparities. However, the project focuses on genetics, only studying the environment insofar as it relates to a limited range of health behaviors and urban versus rural habitation. The project does not address the social construction of race, racism, or racialism. Critical issues like transgenerational health effects, institutionalized forms of discrimination, and social environments thus continue to be poorly understood.

COMMERCIAL ANCESTRY ESTIMATION

The growing market in consumer applications, such as genetic ancestry tests and forensic kits, also compounds collective preoccupations with race, ethnicity, and genetics to the detriment of concerted constructionist analyses (Bolnick et al., 2007; Nelson, 2008). Both kinds of products rest on the development of software platforms that link specific genetic markers with racial and ethnic groups.

The original concept of racial- and ethnic-specific alleles was devised by scientists who were studying disease in racial minorities while determining forensic markers for the Federal Bureau of Investigation (Bliss, 2008). These researchers were charged with the task of choosing which autosomal alleles could serve as indexing variants for a national database that would faithfully represent the nation's criminal populations. As they conducted disease research in African American and Mexican American populations alongside their forensic genomic technological development, they generated a panel of ancestry-informative markers that could not only represent and serve the US government in forensics and public health, but also be repurposed for personal genealogy searches.

In addition to autosomal tests, consumers can buy traditional population genetics technologies such as mitochondrial DNA and Y chromosome tests. These tests assess the nonrecombinant portions of the genome to map personal sequences onto ancestral lineages. Many companies sell ancestry estimation tests that focus on a specific racial lineage. Though mitochondrial DNA and Y chromosome tests only report less than 2% of a client's ancestry, passed on to them from their maternal or paternal family line, they are popular among racial groups for having limited historical records. Individuals seeking confirmation of sub-Saharan African or indigenous American ancestry use tests to envision precolonial and preslavery ancestry.

A shortcoming of autosomal testing is that ancestry-informative markers are actually distributed in different frequencies across a range of populations. Indeed, the markers used in autosomal tests have been chosen because they hold a 30% greater frequency in one racial group. As such, they can neither predict with any certainty whether the marker's ancestry is in fact indicative of a certain genetic heritage, nor explain a person's complete genomic ancestry. A limitation of mitochondrial DNA and Y chromosome tests is that for African Americans, Latin Americans, and Native Americans who typically comprise up to 30% recent European ancestry, and for white Americans who often have more than 10% non-European recent ancestry, they reveal an insignificant amount of data about a person's ancestry.

Nevertheless, docudramas and television series portraying personal genealogy quests have abounded in recent years. Though much footage has been devoted to interracial family reunions, as a result of news of yet unknown multiracial heritage, many shows have convened individuals who are pursuing genealogy quests along racial lines. For example, the American Public Broadcasting System miniseries *African American Lives* told the stories of famous black American celebrities and of their struggles to use DNA data to fill in the gaps of their history due to slavery. Similarly, various prime-time television channels in the United Kingdom have traced the Scandinavian legacies of the British Isles. While racialized genealogy shows do not reinforce negative stereotypes about various racial groups, and while many create positive stereotypes in their respectful depictions of the survival of ancestral populations, they nonetheless support the notion that racial groups have specific genetic histories worth discretely mapping.

Since 2010, companies and research groups have been selling whole-genome sequencing on the open market. Whole-genome sequencing determines the entire DNA code of a person's autosomal and mitochondrial genome. Proponents of whole-genome sequencing claim that it makes race, ethnicity, and ancestry-informative based methods unnecessary. However, companies that require clients to affiliate with a race or ethnicity continue to interpret whole genomes vis-à-vis the racialized reference data provided by international genome projects such as the International HapMap Project (Bliss, 2012). The first sequenced whole genomes of James Watson, Craig Venter, and anonymous individuals in Nigeria and China, which are now circulated as reference whole genomes for public use, were also introduced in inaugural publications with racial labels such as "Caucasian genome," "Asian diploid genome," and "African genome." Thus, commercial testing follows the public health and market trends of racializing its data for the public.

The role genealogy kits play in presenting an inaccurate depiction of human population variation and promoting biologically essentialized conceptions of race and ethnicity continues to be debated. Meanwhile, a wholesale commercialization of genetic races has ensued in the realm of forensic services, such as services that use DNA to inform law enforcement agencies about a suspect's potential racial identity or those that provide craniofacial portraits of suspects. In 2005, the genetic genealogy and forensic genomics company DNAPrint Genomics used its DNAWitness software to dispute an eyewitness report claiming that a long-sought murderer, the "Louisiana Serial Killer," was white. According to DNAPrint's ancestry-informative marker test, the suspect was almost entirely of African descent.

Louisiana State Police used these test results to successfully apprehend Derrick Todd Lee. DNAPrint Genomics was soon after hired by the London Metropolitan Police to construct an ancestry-informative marker panel that would aid in the apprehension of Afro-Caribbean suspects.

In the United States and the United Kingdom, where the criminal justice system disproportionately arrests and incarcerates racial minorities, and where racial profiling is condemned as unethical and unconstitutional, a number of prominent lawyers and bioethicists have argued that race-based forensic services lead to racial misconduct and institutionalized racism (Duster, 2004; Ossorio, 2006). They have been most critical of cases in which race-based forensic services have been used for race-based dragnets, or the stop-and-sequence measures that target members of a specific race when officials lack an "appropriate" suspect. (Krimsky and Simoncelli, 2010). Critics also take issue with the use of forensic databases for behavioral genetic research and criminology, arguing that the racialized nature of these databases portends disproportionate negative findings for racial minorities. Criminologists have indeed begun publishing studies that find a genetic predisposition to criminal behavior in Latin American males and people of African descent. Furthermore, in courtroom forensics, population genetics have been advanced for predicting.

By racial frameworks companies' profit have serious implications for the future of race and ethnicity. Similar to the political mainstream of the late twentieth century, markets in genetic ancestry estimation are now important sites for the ascription and maintenance of racial identities (Bliss, 2013b). Companies appropriate the racial frameworks of public health and justice governance, lending credence to a genetically deterministic way of interpreting racial and ethnic differences. Clients then use these services to ascribe, enact, and embody classifications. Test results are now being used to justify personal affiliations in college admissions and tribal enrollments, and online communities are being formed around specific groupings. In the future, racial and ethnic groups may dispense with political organizing to instead formulate around social spheres arising from knowledge about one's DNA.

REFERENCES

Appiah, Kwame Anthony, 1990. Racisms. In: Goldberg, David Theo (Ed.), Anatomy of Racism. University of Minnesota Press, Minneapolis.

Appiah, Kwame Anthony, 2005. The Ethics of Identity. Princeton University Press, Princeton, NJ.

Bernasconi, Robert, Lott, Tommy, 2000. The Idea of Race. Hackett Readings in Philosophy. Hackett Publishing, Indianapolis.

Bliss, Catherine, 2008. Mapping race through admixture. International Journal of Technology, Knowledge, and Society 4 (4), 79–83.

Bliss, Catherine, 2012. Race Decoded: The Genomic Fight for Social Justice. Stanford University Press.

Bliss, Catherine, 2013a. Translating racial genomics: passages in and beyond the lab. Qualitative Sociology 1–21.

Bliss, Catherine, 2013b. The marketization of identity politics. Sociology 47 (5), 1011–1025.

Bliss, Catherine, 2014. Defining health justice in the postgenomic era. In: Richardson, Sarah S., Stevens, Hallam (Eds.), Postgenomics. Duke University Press, Durham, NC.

Bolnick, Deborah A., Fullwiley, Duana, Dusler, Troy, Cooper, Richard S., Fujimura, Joan H., Kahn, Jonathan, Kaufman, Jay S., et al., 2007. The science and business of genetic ancestry testing. Science 318 (5849), 2.

Braun, Lundy, 2002. Race, ethnicity, and health: can genetics explain disparities? Perspectives in Biology and Medicine 45 (2), 159–174.

Carter-Pokras, Olivia, Baquet, Claudia, 2002. What is a 'Health Disparity'? Public Health Reports 117 (5), 426–434.

Daynes, Sarah, Lee, Orville, 2008. Desire for Race. Cambridge University Press, Cambridge, UK; New York.

Duster, T., 2004. Selective arrests, an ever-expanding DNA forensic database, and the specter of an early-twenty-first-century equivalent of phrenology. In: Lazer, David (Ed.), DNA and the Criminal Justice System: The Technology of Justice. MIT Press, Cambridge.

Epstein, Steven, 2007. Inclusion: The Politics of Difference in Medical Research. Chicago Studies in Practices of Meaning. University of Chicago Press, Chicago.

Fredrickson, George M., 2002. Racism: A Short History. Princeton University Press, Princeton.

Gilroy, Paul, 2000. Against Race: Imagining Political Culture beyond the Color Line. Belknap Press of Harvard University Press, Cambridge, Mass.

Gould, Stephen Jay, 2008. The Mismeasure of Man Revised and Expanded, with a New Introduction. W.W. Norton, New York.

Hubbard, R., Wald, E., 1999. Exploding the Gene Myth: How Genetic Information Is Produced and Manipulated by Scientists, Physicians, Employers, Insurance Companies, Educators, and Law Enforcers. Beacon Press, Boston, MA.

Hunt, L.M., Megyesi, M.S., 2008. The ambiguous meanings of the racial/ethnic categories routinely used in human genetics research. Social Science and Medicine 66 (2), 349–361.

Kahn, Jonathan, 2012. Race in a Bottle: The Story of BiDil and Racialized Medicine in a Post-Genomic Age. Columbia University Press.

Krieger, Nancy, 2005. Stormy weather: race, gene expression, and the science of health disparities. American Journal of Public Health 95 (12), 2155–2160. http://dx.doi.org/10.2105/AJPH.2005.067108 (December).

Krimsky, Sheldon, Simoncelli, Tania, 2010. Genetic Justice: DNA Data Banks, Criminal Investigations, and Civil Liberties. Columbia University Press, New York.

Lee, S.S., 2007. The ethical implications of stratifying by race in pharmacogenomics. Clinical Pharmacological Therapeutics 81 (1), 122–125.

Lipsitz, G., 2006. The Possessive Investment in WhiteNess: How White People Profit from Identity Politics. Temple University Press.

Montoya, M.J., 2011. Making the Mexican Diabetic: Race, Science, and the Genetics of Inequality. University of California Press.

Morning, A.J., 2011. The Nature of Race: How Scientists Think and Teach about Human Difference. University of California Press, Berkeley, CA.

Nelson, Alondra, 2008. Bio science: genetic genealogy testing and the pursuit of African ancestry. Social Studies of Science (Sage) 38 (5), 759–783.

Nobles, Melissa, 2000. Shades of Citizenship: Race and the Census in Modern Politics/Melissa Nobles. Stanford University Press, Stanford, CA.

Omi, Michael, Winant, Howard, 1994. Racial Formation in the United States: From the 1960s to the 1990s. Routledge, NY.

Ossorio, P.N., 2006. About face: forensic genetic testing for race and visible traits. The Journal of Law, Medicine and Ethics 34 (2), 277–292.

PBS, 2003. Race: The Power of an Illusion. DVD. PBS.

Reardon, Jenny, 2005. Race to the Finish Identity and Governance in an Age of Genomics. In-formation Series. Princeton University Press, Princeton.

Shipman, P., 2002. The Evolution of Racism: Human Differences and the Use and Abuse of Science. Harvard University Press.

Smedley, Audrey, 1993. Race in North America: Origin and Evolution of a Worldview. Westview Press, Boulder, CO.

Steinberg, S., 2007. Race Relations: A Critique. Stanford Social Sciences.

Stepan, Nancy, 1982. The Idea of Race in Science: Great Britain, 1800–1960. Archon Books.

Tucker, William H., 1994. The Science and Politics of Racial Research. University of Illinois Press, Urbana.

Whitmarsh, Ian, 2008. Biomedical Ambiguity: Race, Asthma, and the Contested Meaning of Genetic Research in the Caribbean. Cornell University Press, Ithaca.

Yanow, Dvora, 2003. Constructing "Race" and "Ethnicity" in America: Category-making in Public Policy and Administration. M.E. Sharpe, Armonk, NY.

Chapter 39

Evolution of Culture

Robert S. Walker

Department of Anthropology, University of Missouri, Columbia, MO, USA

SYNOPSIS

Studies of cultural adaptation and evolution are important for advancing our knowledge of what it means to be human and helping to move us toward a richer explanation of human variation. Considerable progress has been made in understanding cultural evolution using linguistic phylogenies to track the evolution of inheritance, residence, kinship, marriage, and other cultural systems. Two patterns have emerged from these studies. First, much cultural change at the macrolevel among cultures appears to occur relatively slowly, averaging only a few changes per 10,000 years. Second, cultural traits that are hierarchical in nature tend to change through time in a stairstepping fashion, passing through intermediate stages of intensity or complexity.

HUMAN EXPANSIONS

The nature and success of human expansions around the world (see the chapter by Lao and Kayser in the current volume) were enabled by our unparalleled abilities for cultural adaptation (Richerson and Boyd, 2005; Boyd et al., 2011), first as hunter-gatherers (see the chapter by Little and Blumler) and more recently as pastoralists (see the chapter by Little) and agriculturalists (see the chapter by Blumler). Humans are extremely cooperative within groups, and with sophisticated cultural adaptations have been highly successful in achieving ecological dominance (Alexander, 1990; Flinn et al., 2005) in a wide range of environments, generating an impressive degree of cross-cultural variation. Two general themes underlie explanations for global human demographics. The first includes ecological explanations emphasizing similarities between humans and other species, while the second emphasizes human uniqueness, namely the invention of agriculture, as the primary driving force behind the current distribution of human populations. In general, the least complex human systems are hunter-gatherer societies with population sizes averaging a few thousand (Binford, 2001), while the most complex are large-scale states with population sizes that are often several orders of magnitude larger (Carneiro, 1967; Currie and Mace, 2009; Turchin and Gavrilets, 2009). One comprehensive model is that agricultural dispersals were the engine behind much contemporary human linguistic, cultural, and genetic diversity (Renfrew, 1987, 1989; Diamond and Bellwood, 2003). However, humans are still subject to environmental constraints and, at least coarsely, appear to follow ecogeographic "rules" similar to those of other life forms. For example, strong latitudinal gradients in the diversity of indigenous cultures and languages (Mace and Pagel, 1995; Nettle, 1998; Cashdan, 2001; Collard and Foley, 2002; Moore et al., 2002) mirror the diversity in other species (Hawkins et al., 2003), and are highest in topographically diverse tropical regions and lowest in polar and desert regions.

COMPARATIVE CULTURAL DATA

Comparative cultural data combined with linguistic phylogenies are useful for reconstructing ancestral cultures, modeling transition rates of cultural traits over time, and examining correlated evolution between cultural traits (Mace and Pagel, 1994; Jordan, 2007; Nunn, 2011; Mace and Jordan, 2011). It is an exciting time for comparative studies of human culture, as more and larger cultural databases become matched with linguistic phylogenies of deeper time depths. Rigorous phylogenetic methods are now being applied in a more systematic fashion to larger samples of the world's language families, and with a broad array of cultural variables. We are beginning to more fully understand how and why cultures change, with increasingly extensive analyses of the sequences and trajectories of human cultural change around the world. Cultures are like "clouds" of traits, interdependent information bits, each one subject to change by individuals, with potential impact on how the interdependent cloud subsequently changes. This is an extraordinarily complex process to examine on the microscale of individual thought processes (see the chapter by Vonk and Adadhye), but more tractable to study at the macrolevel of cross-cultural comparisons.

Anthropologists have a long tradition of cross-cultural comparative studies, but such work has been hampered by limited sampling of the world's enormous cultural diversity. The commonly used Ethnographic Atlas (EA, Murdock, 1967) only covers about 12% of the world's 7000 or so languages, and the coverage of large language families varies from nearly complete coverage of cultures (e.g., Uto-Aztecan and Algic) to only a few percent (e.g., Oto-Manguean and Pama-Nyungan). The Standard Cross-Cultural Sample (SCCS) is a further subset of only 186 cultures, albeit with some 2000 coded traits. Comparative methods that include a narrow sample of cultures suffer from the detrimental effects of loss of statistical power and incomplete coverage of ethnolinguistic diversity. Following Mace and Pagel (1994), a number of studies have now accumulated that trace the histories of cultural features over linguistic phylogenies (reviewed in Mace and Jordan, 2011), where effort is made to include as many cultures as possible in a language family instead of arbitrarily choosing "representative" cultures as in the EA and A number of cross-cultural resources now exist (Table 1), although none are fully comprehensive.

PHYLOGENETIC COMPARATIVE METHODS

As the genomic revolution proceeds to unravel complex phylogenetic relationships in the tree or network of life (Dunn et al., 2008), analogous comparative methods are available to interpret nested patterns of relatedness among the world's nearly 7000 languages (Lewis, 2009) and cultures. Phylogenetic methods using cognate codings of basic vocabulary words can help us infer historical relationships among human languages, such as the internal classifications of recent agricultural expansions and their phylogeography. Character-coded language phylogenies are perhaps the "gold standard" for internal classification of language families and are readily available for eight agricultural language families around the world (Figure 1), with an increasing number expected in the near future.

The processes by which cultural similarities and differences emerge among human societies over time and space have long been a central focus of anthropological inquiry (Boas, 1896; Kroeber, 1948), and linguistic phylogenies can help reconstruct ancestral histories of cultural traits through the use of evolutionary models of culture change over

TABLE 1 Important Resources Covering Global Ethnolinguistic Variation

Comparative Resource	Abbreviation	Data	N Societies	N Variables	Strengths	Website/ Citation
Corrected ethnographic atlas	EA	Cultural	1267	99	Wide coverage of societies	eclectic.ss.uci.edu/~drwhite
Standard cross-cultural sample	SCCS	Cultural	186	2000	Large number of detailed cultural traits	eclectic.ss.uci.edu/~drwhite
Electronic human relation area files	eHRAF	Cultural	258+		Search/view original ethnographic sources	ehrafworldcultures.yale.edu
Atlas of world cultures		Cultural	3500		Encyclopedic entries	everyculture.com
Database for indigenous cultural evolution	DICE	Cultural	550+	100	Provides ethnographic questionnaires	dice.missouri.edu
Binford's hunter-gatherer dataset		Cultural	339	40	Comprehensive for hunter-gatherers	Binford (2001).
Apostolou's hunter-gatherer dataset		Cultural	190	40	Focus on marriage arrangement/ decisions	Apostolou (2007).
Dziebel's kinship studies		Cultural	2500	4	Provides raw sibling terms (kinship)	kinshipstudies.org
Automated similarity judgment program	ASJP	Linguistic	6013+	40	Automatic distance-based classification	lingweb.eva.mpg.de/asjp
World atlas of language structures	WALS	Linguistic	2678+	192	Coded language structure features	wals.info
Ethnologue		Linguistic	6909		Comprehensive, gives N speakers	ethnologue.com
World language mapping system	WLMS	Geographic	6909		Gives area polygons for ethnologue entries	worldgeodatasets.com

linguistic phylogenies (Mace et al., 2005). Following Mace and Pagel (1994), a large number of studies have accumulated that trace the histories of cultural features by mapping their evolution onto linguistic trees, mainly in order to test for correlated evolution between cultural features (Table 2). In particular, the relationships among systems of kinship, marriage, inheritance, and residence have drawn much attention (e.g., Cowlishaw and Mace, 1996; Borgerhoff Mulder et al., 2001; Holden and Mace, 2003; Fortunato et al., 2006; Jordan, 2007; Mace and Jordan, 2011).

The phylogenetic comparative method applied to cultural evolution is a two-step process, first requiring as input some phylogenetic hypothesis about the historical relationships among cultures, usually using cognate sets in basic vocabulary word lists. With linguistic phylogeny in hand, the second step is to reconstruct the evolution of a cultural trait over the phylogeny to infer ancestral state and transition rate parameters using a model of trait evolution. Conventional wisdom suggests that cultural change is often fast and innovative, and it has been claimed that rapid rates of cultural adaptation are the most distinctive of all human characteristics (Harris, 1989; Diamond, 2001). Phylogenetic analyses are useful for quantifying rates of cultural change in order to make valid cross-cultural comparisons of cultural dynamics over relatively deep periods. Phylogenetic methods have the advantage of directly estimating the rates of change or transition rates of cultural traits between different states (e.g., warlike to peaceful). Studies of cultural evolution tend to examine fundamental (core) cultural traits, and therefore comparisons of transition rates

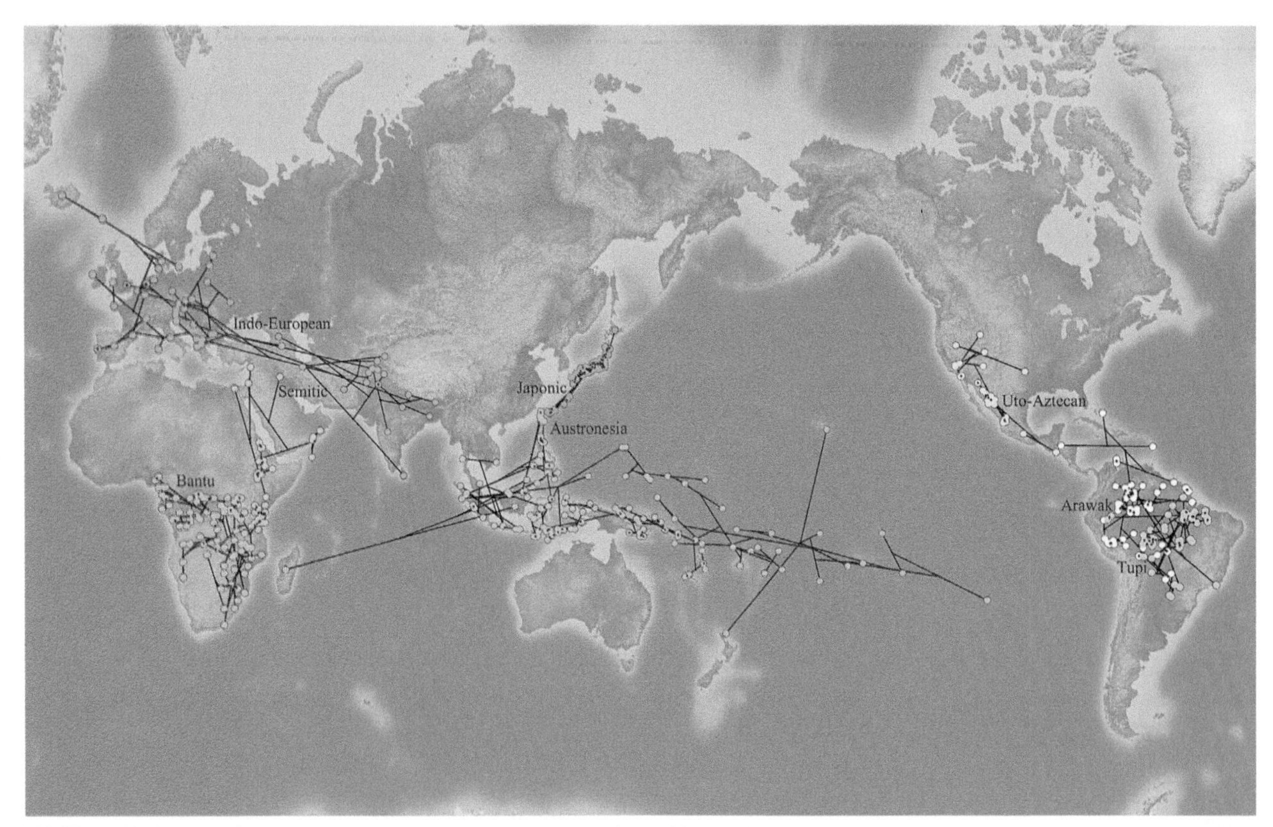

FIGURE 1 Phylogeography of eight agricultural language family expansions with published character-coded language phylogenies. The cultures of each language family are connected by a consensus tree, with internal nodes interpolated as the average spatial location of descendant taxa. Language family names are placed approximately near purported homelands. (See color plate section).

investigate on how key defining characteristics of individual cultures change through time. Phylogenetic comparative methods are ideal for studying the mode and tempo of human cultural evolution. These methods help answer the following research questions concerning traditional cultural traits: (1) ancestral trait reconstruction (what were ancient cultures like?), (2) tempo of trait evolution (do traits persevere over long periods or change rapidly?), (3) evolutionary sequence (what is the sequence of change in cultural traits?), and (4) correlated traits (do different traits change together?). Quantitative answers to these research questions help determine patterns of cultural trait evolution, leading to a more holistic understanding of cultural dynamics over relatively deep periods.

The Problem of Borrowing

Healthy caution and criticism of phylogenetic methods applied to human language and culture stem from many concrete examples and the apparent ease of diffusion and borrowing of ethnolinguistic traits (Terrell, 1988; Borgerhoff Mulder et al., 2006; Temkin and Eldridge, 2007). In particular, concern has been raised that the speed of cultural change and the effects of borrowing may pose problems for cultural phylogenies (Boyd et al., 1997; Nunn et al., 2006, 2010). Quantitative comparisons between cultural and biological (i.e., genetic and morphological) sequence data have concluded that both are similarly treelike using parsimony-based consistency and retention indices (Collard et al., 2006). Sophisticated Bayesian phylogenetic methods are robust against some levels of horizontal transmission (Gray et al., 2007; Greenhill et al., 2009). Borrowing does not necessarily invalidate phylogenetic methods, because transitions in likelihood models include change originating from borrowing. The general conclusion of these papers is that phylogenetic inference is often validated even in the face of considerable borrowing.

CULTURAL TRANSITION RATES

As mentioned above, phylogenetic comparative methods are ideal for studying the tempo of trait evolution, answering whether traits persevere over long periods or tend to change more rapidly. Traits that change at slower rates are more easily reconstructed farther back in prehistory and further attest to the importance of using phylogenetic methods in the first place. A number of cultural phylogenetic studies have now emerged, primarily from Austronesian, Bantu, Indo-European, and Tupi language families, to allow for a comparison of transition rates. The traits studied include

TABLE 2 Summary of Human Studies Showing Correlated Evolution between Cultural Traits

Phylogeny	Trait 1	Trait 2	Data type	*n*	Direction of Relationship	Source
Global nations	Sex ratio in adults (SR)	Marriage payments	Continuous	74	Higher SR with bridewealth	Mace and Jordan (2005).
Global nations	Sex ratio at birth (SRB)	Fertility and mortality	Continuous	74	Lower SRB with higher mortality/ fertility	Mace and Jordan (2005).
Global cultures	Total fertility rate	Dependence on agriculture	Continuous	69	Positive correlation	Sellen and Mace (1997).
Global cultures	Lactose tolerance	Dairying	Discrete	62	Dairying leads to lactose tolerance	Holden and Mace (1997).
Lowland South Am	Matrilocality	Partible paternity	Discrete	31	Positive correlation	Walker et al. (2010)
Austronesia	Political complexity	Class stratification	Discrete	86	Positive correlation	Currie and Mace, 2011
Austronesia	Patrilineal descent	Patrilocal residence	Discrete	67	Positive correlation	Jordan (2007).
Bantu	Inheritance system	Cattle herding	Discrete	68	Cattle herding leads to loss of matriliny	Holden and Mace (2003).
Indo-European	Marriage system	Marriage payments	Discrete	51	Monogamy with dowry; bridewealth with polygyny	Pagel and Meade (2005).

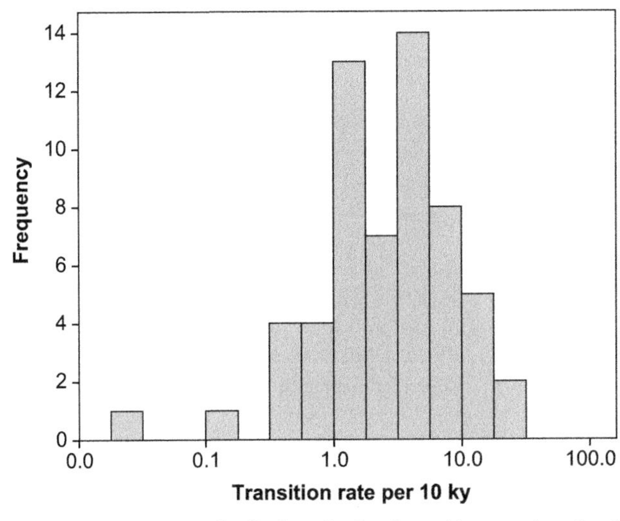

FIGURE 2 Frequency distribution of cultural transition rates in units of per 10,000 years from studies with four different linguistic phylogenies (Austronesia, Bantu, Indo-European, and Tupi, $n = 59$). The traits examined include patterns of residence, inheritance, warfare, religion, body modification, and political complexity (Walker et al., 2012). Note the log scale on the x-axis.

patterns of residence, inheritance, warfare, religion, body modification, and political complexity (Walker et al., 2012). A frequency distribution of cultural transition rates from these studies collectively shows a right-skewed distribution with a median transition rate across a diverse set of cultural traits of only three changes per 10,000 years ($n = 48$, Figure 2).

While most binary or multistate cultural traits appear to fluctuate relatively slowly between possible states, another interesting evolutionary pattern of incremental change emerges for three phylogenetic studies that have looked at transition rates for multilevel hierarchical data. For example, warfare intensity (520. War) in Austronesian societies was coded from the ethnographic record into three hierarchical levels of (1) mostly peaceful, (2) small-scale raiding, and (3) intense large-scale warfare (Cisco et al., n.d.). Multiple ascensions or "double jumps" from mostly peaceful to intense warfare and vice versa were rare and not significantly different from zero. Analogously, in both Austronesia and Bantu language families, political complexity often tends to increase in a stairstepping or incremental process across four hierarchical levels from acephalous to state-level political organization (Currie et al., 2010; Currie and Mace, 2011; Walker and Hamilton, 2011). Therefore, studies that have examined hierarchical characters have so far all shown that cultural change occurs in an incremental process across phylogenies, but further testing of this idea requires the type of comprehensive data-collection scheme proposed here.

Certainly, some peripheral cultural traits can change quickly, such as the rapid diffusion of technological innovations like the bow and arrow, but it appears that at least some cultural traditions often change at a much slower pace. Estimated transition rates of the cultural traits examined here, at least at the macroscale between populations, do *not* support a rapid, dynamic, and innovative nature to human cultural change. These cultural traits actually appear to persevere over fairly long periods and adhere to a pattern of long-term cultural tradition. This is not to argue that cultural phylogenies necessarily represent high-fidelity transmission of arbitrary cultural practices in a process of blind copying. Instead, preserved cultural practices are likely adaptations to common social circumstances of human populations that persist over long periods and form the roots of human cultural institutions. The slow cultural transition rates reported here highlight the importance of using comparative phylogenetic methods in the first place to study human variation, for the simple reason that many cross-cultural similarities might often arise from shared common ancestry even over considerable periods.

FUTURE WORK

Comparative cultural data combined with linguistic phylogenies based on basic vocabulary words are useful for reconstructing ancestral cultures and modeling transition rates of cultural traits over time, thereby greatly expanding the possibilities of comparative anthropological studies. While considerable progress has been made in understanding cultural evolution using linguistic phylogenies, most studies are confined to only 4 of the some 30 major language families of the world. It seems likely that rigorous phylogenetic methods will soon be applied in a systematic fashion to all of the world's large language families using an increasingly broader array of interesting cultural variables.

The availability of lexical phylogenies is extremely useful for evaluating phylogeographic hypotheses and the processes of cultural evolution that must have existed in the past given patterns seen in the present. The Automated Similarity Judgment Program (Table 1), with data and methods for the construction of phylogenies for nearly all of the world's language families, now allows comparative phylogenetic modeling of cultural traditions around the world. We need to explain why some cultural traits change faster than others, and under what socioenvironmental contexts, and we need to improve our ability to distinguish phylogenetic signal from borrowing and diffusion with phylogenetic network methods. A primary objective for the future is to expand the scope of cultural evolution studies to include all of the world's large language families, not only providing more statistical power and geographic scale, but also opening up inquiry into a wider array of cultural phenomena.

REFERENCES

Alexander, R.D., 1990. How Did Humans Evolve? Reflections on the Uniquely Unique Species. University of Michigan, Ann Arbor.

Apostolou, M., 2007. Sexual selection under parental choice: the role of parents in the evolution of human mating. Evolution and Human Behavior 28, 403–409.

Binford, L.R., 2001. Constructing Frames of Reference: An Analytical Method for Archaeological Theory Building Using Ethnographic and Environmental Data Sets. University of California Press, Berkeley.

Boas, F., 1896. The growth of Indian mythologies. Journal of American Folklore 9, 1–11.

Borgerhoff Mulder, M., George-Cramer, M., Eshleman, J., Ortolani, A., 2001. A study of East African kinship and marriage using a phylogenetically based comparative method. American Anthropologist 103, 1059–1082.

Borgerhoff Mulder, M., Nunn, C.L., Towner, M.C., 2006. Cultural macroevolution and the transmission of traits. Evolutionary Anthropology 15, 52–64.

Boyd, R., Borgerhoff Mulder, M., Durham, W.H., Richerson, P.J., 1997. Are cultural phylogenies possible? In: Weingart, P., Mitchell, S.D., Richerson, P.J., Maasen, S. (Eds.), Human by Nature: Between Biology and the Social Sciences. Lawrence Erlbaum Associates, Mahwah, NJ, pp. 355–386.

Boyd, R., Richerson, P.J., Henrich, J., 2011. The cultural niche: why social learning is essential for human adaptation. Proceedings of the National Academy of Sciences of the USA 108, 10918–10925.

Carneiro, R.L., 1967. On the relationship between size of population and complexity of social organization. Southwestern Journal of Anthropology 23, 234–243.

Cashdan, E., 2001. Ethnic diversity and its environmental determinants: effects of, climate, pathogens, and habitat diversity. American Anthropologist 103, 968–991.

Cisco, J., Dombroski, A.M., Roush, C.C., Pierce, D.E., Kelly, D.E., Boulanger, M.T., Walker, R.S., n.d. Cultural Phylogenetics of Warfare and Tattooing across the Austronesian Expansion. University of Missouri.

Collard, I.F., Foley, R.A., 2002. Latitudinal patterns and environmental determinants of recent human cultural diversity: do humans follow biogeographical rules? Evolutionary Ecology Research 4, 371–383.

Collard, M., Shennan, S.J., Tehrani, J.J., 2006. Branching, blending, and the evolution of cultural similarities and differences among human populations. Evolution and Human Behavior 27, 169–184.

Cowlishaw, G., Mace, R., 1996. Cross-cultural patterns of marriage and inheritance: a phylogenetic approach. Ethnology and Sociobiology 17, 87–97.

Currie, T.E., Mace, R., 2009. Political complexity predicts the spread of ethnolinguistic groups. Proceedings of the National Academy of Sciences of the USA 106, 7339–7344.

Currie, T.E., Greenhill, S.J., Gray, R.D., Hasegawa, T., Mace, R., 2010. Rise and fall of political complexity in island South-East Asia and the Pacific. Nature 467, 801–804.

Currie, T.E., Mace, R., 2011. Mode and tempo in the evolution of socio-political organization: reconciling "Darwinian" and "Spencerian" evolutionary approaches in anthropology. Philosophical Transactions of the Royal Society B 366, 1108–1117.

Diamond, J., 2001. The great leap forward. In: Whitten, P. (Ed.), Anthropology: Contemporary Perspectives, eighth ed. Allyn and Bacon, Boston, pp. 24–33.

Diamond, J., Bellwood, P., 2003. Farmers and their languages: the first expansions. Science 300, 597–603.
Dunn, C.W., Hejnol, A., Matus, D.Q., et al., 2008. Broad phylogenomic sampling improves resolution of the animal tree of life. Nature 452, 745–749.
Flinn, M.V., Geary, D.C., Ward, C.V., 2005. Ecological dominance, social competition, and coalitionary arms races: why humans evolved extraordinary intelligence. Evolution and Human Behavior 26, 10–46.
Fortunato, L., Holden, C., Mace, R., 2006. From bridewealth to dowry? A Bayesian estimation of ancestral states of marriage transfers in Indo-European groups. Human Nature 17, 355–376.
Gray, R.D., Greenhill, S.J., Ross, R.M., 2007. The pleasures and perils of Darwinizing culture (with phylogenies). Biological Theory 2, 360–375.
Greenhill, S.J., Currie, T.E., Gray, R.D., 2009. Does horizontal transmission invalidate cultural phylogenies? Proceedings of the Royal Society of London Series B 276, 2299–2306.
Harris, M., 1989. Our Kind. Harper Collins, London.
Hawkins, B.A., Field, R., Cornell, H.V., Currie, D.J., Guegan, J.F., Kaufman, D.M., Kerr, J.T., Mittelbach, G.G., Oberdorff, T., O'Brien, E.M., Porter, E.E., Turner, J.R.G., 2003. Energy, water, and broad-scale geographic patterns of species richness. Ecology 84, 3105–3117.
Holden, C., Mace, R., 1997. Phylogenetic analysis of the evolution of lactose digestion in adults. Human Biology 69, 605–628.
Holden, C.J., Mace, R., 2003. Spread of cattle led to the loss of matrilineal descent in Africa: a coevolutionary analysis. Proceedings of the Royal Society of London Series B 270, 2425–2433.
Jordan, F.M., 2007. A Comparative Phylogenetic Approach to Austronesian Cultural Evolution (Ph.D. thesis). University College London.
Kroeber, A.L., 1948. Anthropology. Harcourt and Brace, New York.
Lewis, M.P. (Ed.), 2009. Ethnologue: Languages of the World, 16th ed. SIL International, Dallas, TX.
Mace, R., Holden, C.J., Shennan, S., 2005. The Evolution of Cultural Diversity: A Phylogenetic Approach. Left Coast Press, Walnut Creek, California.
Mace, R., Jordan, F., 2005. The evolution of human sex ratio at birth: a bio-cultural analysis. In: Mace, R., Holden, C., Shennan, S. (Eds.), The Evolution of Cultural Diversity: A Phylogenetic Approach. Left Coast Press, Walnut Creek, California, pp. 207–216.
Mace, R., Jordan, F.M., 2011. Macro-evolutionary studies of cultural diversity: a review of empirical studies of cultural transmission and cultural adaptation. Philosophical Transactions of the Royal Society B 366, 402–411.
Mace, R., Pagel, M., 1994. The comparative method in anthropology. Current Anthropology 35, 549–564.
Mace, R., Pagel, M., 1995. A latitudinal gradient in the density of human languages in North America. Proceedings of the Royal Society B 261, 117–121.
Moore, J.L., Manne, L., Brooks, T., Burgess, N.D., Davies, R., Rahbek, C., Williams, P., Balmford, A., 2002. The distribution of cultural and biological diversity in Africa. Proceedings of the Royal Society B 269, 1645–1653.
Murdock, G.P., 1967. Ethnographic Atlas: A Summary. University of Pittsburgh Press, Pittsburgh.
Nettle, D., 1998. Explaining global patterns of language diversity. Journal of, Anthropological Archaeology 17, 354–374.
Nunn, C.L., Borgerhoff Mulder, M., Langley, S., 2006. Comparative methods for studying cultural trait evolution: a simulation study. Cross-Cultural Research 40, 177–209.
Nunn, C.L., Arnold, C., Matthews, L., Borgerhoff Mulder, M., 2010. Simulating trait evolution for cross-cultural comparison. Philosophical Transactions of the Royal Society B 365, 3807–3819.
Nunn, C.L., 2011. The Comparative Approach in Evolutionary Anthropology and Biology. University of Chicago Press, Chicago.
Pagel, M., Meade, A., 2005. Bayesian estimation of correlated evolution across cultures: a case study of marriage systems and wealth transfer at marriage. In: Mace, R., Holden, C.J., Shennan, S. (Eds.), The Evolution of Cultural Diversity: A Phylogenetic Approach. Left Coast Press, Walnut Creek, California, pp. 235–256.
Richerson, P.J., Boyd, R., 2005. Not by Genes Alone: How Culture Transformed Human Evolution. University of Chicago Press, Chicago.
Renfrew, C., 1987. Archaeology and Language: The Puzzle of Indo-European Origin. Jonathan Cape, London.
Renfrew, C., 1989. The origins of Indo-European languages. Scientific American 261, 106–114.
Sellen, D.W., Mace, R., 1997. Fertility and mode of subsistence: a phylogenetic analysis. Current Anthropology 38, 878–889.
Temkin, I., Eldridge, N., 2007. Phylogenetics and material culture evolution. Current Anthropology 48, 146–153.
Terrell, J.E., 1988. History as a family tree, history as an entangled bank: constructing images and interpretations of prehistory in the South Pacific. Antiquity 62, 642–657.
Turchin, P., Gavrilets, S., 2009. Evolution of complex hierarchical societies. Social Evolution and History 8, 168–198.
Walker, R.S., Mark, V.F., Kim, R.H., 2010. Evolutionary history of partible paternity in lowland South America. Proceedings of the National Academy of Sciences 107 (45), 19195–19200.
Walker, R.S., Wichmann, S., Mailund, T., Atkisson, C.J., 2012. Cultural phylogenetics of the Tupi language family in lowland South America. PLoS One 7, e35025.
Walker, R.S., Hamilton, M.J., 2011. Social complexity and linguistic diversity in the Austronesian and Bantu population expansions. Proceedings of the Royal Society Series B 278, 1399–1404.

Glossary

ABO blood group A genetic system that codes for the human blood types A, B, AB, and O.

Acclimation Adaptive responses to a single stressor, often in an experimental/laboratory context.

Acclimatization Adaptive responses to multiple stressors in a "real world" environmental context.

Aché Eastern Paraguay forest dwellers, whose subsistence includes hunting mammals and collecting honey, palm starch, insect larvae, and fruits. Their fertility is high, and life expectancy is moderate (see Hill and Hurtado, 1996).

Acheulean The handaxe/cleaver industries of the later Lower Palaeolithic (Early Stone Age) seen primarily in Africa, the Near East, Western Europe, and the Indian subcontinent, dating from approximately 1.75 million years ago until 250,000 years ago. Associated primarily with *Homo erectus* and *Homo heidelbergensis*. In some parts of the world, notably Eastern Europe and East Asia, contemporaneous industries are often found with no Acheulean tool, but rather simpler, more Oldowan-like technologies. The earliest wooden spears are known from this time period, ca. 400,000 years ago.

Adaptation A process of genetic change in a population whereby, as a result of natural selection, the average state of a character becomes improved with reference to a specific function, or whereby a population has become better suited to some aspect of its environment. Also, an adaptation: a feature that has become prevalent because of a selective advantage conveyed by that feature in the improvement of some function. Variations that prove advantageous will tend to spread throughout the population.

Adaptive radiation An evolutionary pattern produced when a lineage of organisms diversifies to fill increasingly specialized ecological niches.

Admixture The interbreeding or mixing of individuals from separate populations.

Adolescence A stage in human life cycle covering the years after the onset of puberty until the onset of adulthood (approximately ages 9–19 years). The adolescent phase is characterized by a growth spurt in height and weight, the development of secondary sexual characteristics, sociosexual maturation, and intensification of interest and practice adult of social, economic, and sexual activities.

Adrenarche An increase in the production of androgens by the adrenal cortex typically occurring at around 6–8 years of age in boys and girls. It elicits the development of androgen-dependent signs of puberty including pubic hair, axillary hair, acne, and adult body odor. Adrenarche is a process distinct from hypothalamic–pituitary–gonadal maturation and function.

Affiliative neuropeptides Endogenous chemicals involved with the formation of social and emotional bonds with others; primarily the nanopeptides oxytocin (OT) and arginine vasopressin (AVP).

Aggression Hostile behavior directed at another individual or group; an evolved component of reproductive competition.

Agro-pastoralism A mixed subsistence of raising livestock and cultivating plants.

Agta and Aeta Two "Negrito" societies from the tropical forested areas of the Island of Luzon and other Philippine islands. They hunt and fish and occasionally cultivate manioc, sweet potato, and other foods, also trading with their neighbors (see Griffin and Griffin, 1999; Migliano et al., 2013).

Aka (Western Pygmies) They are tropical forest dwellers who live in the southwest region of the Central African Republic and northern Congo. Large mammals and monkeys are hunted, and insects, fruits, honey, and other foods are gathered. Trade is with Bantu agriculturalists for food and material items (see Bahuchet, 1988).

Alien species A species introduced to a new region/habitat usually due to human transport, either intentional or otherwise, or environmental change.

Allele One of several forms of the same gene that differ in DNA sequence.

Allen's rule The "ecological rule" outlining the relationship between body proportions (e.g., limb lengths) and environmental temperature in mammalian species with wide geographic distribution. In these species there is positive association between limb lengths and temperature, such that animals in warmer climates have longer limbs and those in colder climates have shorter limb lengths.

Alliance Complex, often long-term coalitionary relationship. Key aspect of human sociality.

Allocare Provision of offspring care by individuals other than mothers. Also referred to as "alloparental care."

Allopolyploidy A doubling of the number of sets of chromosomes that happens to occur when two species hybridize; see "Polyploid."

Alpha-melanocyte-stimulating hormone (α-MSH) A naturally occurring endogenous peptide hormone of the melanocortin family, that stimulates melanogenesis, a process which in mammals (including humans) is responsible for pigmentation primarily of the hair and skin.

Altricial offspring Young that are hatched or born in a very immature and helpless condition so as to require care and feeding from parents or others for some time after birth.

Alveolus (alveoli, pl) Small, balloon-like sac in the lungs where oxygen is exchanged into the bloodstream.

Amenorrhea The absence or suppression of menstruation.

Ancestry-informative markers A set of genetic polymorphisms that exhibits different frequencies between populations from different geographical regions.

Ancient DNA (aDNA) Genetic material recovered from ancient remains.

Andamanese Sometimes referred to as "Negritos" on the Andaman Islands in the Bay of Bengal. They fish and gather shellfish, honey, turtle eggs, and fruits, and hunt turtles and manatees at sea. The Ongees are one of 13 Andamanese groups (see Pandya, 1999).

Androgen Compound, usually a steroid hormone, that stimulates or controls the development and maintenance of male characteristics in mammals by binding to androgen receptors. Testosterone (T) and dehydroepiandrosterone (DHEA) are important androgens in humans.

Antagonistic pleiotropy A trait having advantageous, fitness-promoting effects in one aspect of life (e.g., early) but which may incur costs in a different aspect of life (e.g., late). For example, a trait may confer an advantage to reproduction but a cost to survivorship.

Anthropoids Members of the primate infraorder Anthropoidea, also often designated as Simiiformes, that includes the Old World monkeys, New World monkeys, and hominoids (i.e., all of the haplorhines except tarsiers).

Antibiotic resistance A situation where a bacterial species develops genetic adaptations that reduce or eliminate its sensitivity to antibiotics.

Antibody Proteins from the immune system that attach to foreign cells so that they can be recognized and destroyed by the body.

Antiessentialism Concepts of race that use biological reasoning to disprove the existence of discreet racial groups.

Antigen A substance that provokes an immune system response.

Antigenic drift A small change in the surface antigens of an influenza virus due to mutation.

Antigenic shift A large change in the surface antigens of an influenza virus due to genetic combination of material from multiple viruses.

Archaeobotany The study of botanical remains found at archaeological sites.

Arterial oxygen saturation (SaO_2) Percentage of hemoglobin in the arterial blood that is bound with oxygen molecules.

Artifact An object that has been modified by a human or proto-human, whether it is used for a tool or not. Handaxes, waste flakes from stone flaking, and bones with cut-marks from butchery with stone tools are examples of artifacts.

Artificial selection Selective breeding (the preservation of favorable characteristics as chosen by humans) of domestic animals or cultivated plants.

Ascription The social process by which individuals are attributed racial affiliations by others.

Atresia The disappearance of an anatomical part by degeneration.

Aurochs The wild ancestor of domestic cattle.

Australian aborigines All of Australia and Tasmania and Torres Strait natives were hunter–gatherers and fishers with considerable cultural and linguistic diversity. Modern *Homo sapiens* hunter–gatherers have inhabited these areas for at least 50,000 years (see Peterson, 1999).

Australopithecine Colloquial term for species attributed to the genus *Australopithecus*, and sometimes *Parathropus* and *Kenyanthropus*.

Autonoetic consciousness Ability to mentally place ourselves in the past, in the future, or in counterfactual situations, and to analyze our own thoughts; the type of consciousness involved with reliving past moments.

Balancing selection Selection in favor of intermediate types and against extreme types; contributes to the maintenance of variation in a population.

Basal ganglia Structures deep within the brain that form part of neural circuits regulating motor control, cognition, attention, and emotion.

Basal metabolic rate (BMR) The quantity of energy required to maintain an organism (e.g., respiration, circulation, maintenance of body temperature) when at complete rest. It is measured in humans by the heat given off per unit time, and expressed as the calories released per kilogram of body weight or per square meter of body surface per hour.

Batak Another "Negrito" population from the interior forested Philippine Palawan Island. They hunt wild pigs, gliding squirrels, jungle fowl, gather a variety of plant foods, and gather shellfish along rivers and the coast. They also engage in trade and practice some cultivation (see Eder, 1999).

Before the Common Era (BCE) A modern term substituted for "BC."

Bergmann's rule The "ecological rule" outlining the relationship between body mass (weight) and environmental temperature in mammalian species with wide geographic distribution. In these species there is a negative (inverse) relationship between body mass and temperatures, such that animals living in colder climates are heavier, while those living in warmer climates are lighter.

Bicondylar angle Angle between long axis of femoral shaft and horizontal axis of the knee joint when viewed from the front.

Bioarchaeology The study of human remains with regard to their archaeological context.

Biocultural anthropology An approach to research in anthropology that examines the interaction between biology and culture.

Biocultural reproduction The human type of system for feeding and care of pregnant women and their offspring. Many members of the social group pool resources, such as food and labor, to increase the reproductive success of most adults.

Biohistory The study of stature and other health indicators in the context of economic or social history.

Biomarker A naturally occurring molecule, gene, or characteristic that can be objectively measured and evaluated as an indicator of normal biological processes, pathogenic processes, or pharmacologic responses to a therapeutic intervention.

Biomechanics The study of structure and function of animals and plants.

Biostratigraphy Using animal remains found at an archeological site to estimate the age of a specific layer or specimen.

Black Death A major epidemic of bubonic plague that occurred throughout Europe in the middle of the fourteenth century.

Body mass index (BMI) A measure of height and weight relationship and body composition, commonly used to assess physical nutritional status. It is calculated by dividing weight in kilograms by height in squared meters.

Brachiation Hand-over-hand locomotion wherein the body is suspended beneath an overhead support.

Breeding system A system specifying not only sexual unions but also reproductive output and offspring care.

Brown adipose tissue (BAT) A distinct form of fat that plays a central role in metabolic heat production (non-shivering thermogenesis) in response to cold stress.

Calvarium A partial skull that includes only the braincase, including the cranial base. It is distinguished from the cranium (the entire skull minus the mandible) and the calotte (a braincase without the cranial base).

Carbon dating Dating of organic materials by calculating the decay of the isotope carbon 14.

Cardiovascular disease A general diagnostic category consisting of several separate diseases of the heart and circulatory system.

Carotid body A small cluster of cells in the carotid artery in the neck that detects chemical changes in the blood stream (e.g., changes in oxygen pressure) to stimulate responses to low oxygen availability.

Carrying capacity The maximum number of individuals of a species that the environment can support.

Catarrhines Members of the primate parvorder Cattarhini that includes the Old World Monkeys and hominoids.

CCR5-Δ32 A variant of the chemokine receptor number 5 (CCR5) that contains a 32 base pair deletion that prevents some strains of HIV from entering the white blood cell.

Cercopithecoids Members of the primate superfamily Cercopithecoidea that includes the Old World Monkeys, which are native to Africa and Asia today and include a single family, Cercopithecidae, which in turn includes the subfamilies Cercopithecinae and Colobinae.

Chemoreceptor Specialized cells that detect chemical changes in the body and send a signal to the nervous system to initiate a physiological response.

Childhood A stage in the human life cycle that occurs between the end of infancy and the start of the juvenile growth period (about the ages 3.0–6.9 years). Children are weaned from all breast feeding (or bottle feeding) but must be provided specially prepared foods due to immaturity of their dentition and digestive systems. Children require intensive care by older individuals due to the child's motor, neurological, and cognitive immaturity.

Clade A group of organisms that includes all of the species derived from a common ancestor.

Cladistics Method of determining evolutionary relationships that relies on evolutionarily derived characteristics.

Classical markers Type of genetic variants that, because they can be inferred without using DNA techniques, were initially used in the analysis of the genetic diversity of the human species. These include blood groups and HLA polymorphisms, among others.

Cline A gradation in one or more characteristics within a species in which the frequency of characteristic varies gradually over space.

Closed canopy forest Dense forests not dependent on river- or lake-watering. In these forests, trees with interlocking crowns (closed canopy) cover >95% of the ground surface, though openings due to tree falls, rocky areas, or animal activity may be seen. Trees are ≥5 m in stature, often exceeding 50 m. Undergrowth is often sparse; forests with shorter and therefore less light-blocking canopies may have a significant herb layer. Canopy is never completely without foliage, though individual trees may drop leaves. More seasonal forests may undergo clear reductions in foliage during dry season. Forests that are most closed, tallest, and densest have few vines and few palm trees.

Coalescence theory A model that tries to trace back in time the history of how a genetic trait arose.

Coalition Support given from one individual to a second in an aggressive interaction between the second individual and a third party; pairs of individuals who repeatedly form coalitions are allies. Human coalitions are unusually complex, nested, and extensive.

Coevolution The joint evolution of two species that interact (e.g., livestock and humans) or of two attributes of a species (e.g., culture and biology).

Cognitive niche Reasoning about the causal structure of the world, cooperating with other individuals, and sharing that knowledge and negotiating those agreements via language.

Common Era (CE) A modern term substituted for "AD."

Complexity The organization of interrelated parts into a whole and the observable outcomes of their interactions (see Dembski, 1998).

Composite tool A tool made up of more than one component, for example, a spear composed of a wooden shaft, a stone point, and sinew or mastic to hold the parts together.

Conception The formation of a viable zygote by the union of the male sperm and female ovum; fertilization.

Conceptus The whole product of conception at any stage of development, from fertilization of the ovum to birth, including extraembryonic membranes as well as the embryo or fetus.

Conduction The transfer of heat due to direct physical contact.

Consensus tree Convenient way to summarize into a single tree (phylogeny) the agreement among a large sample of alternative trees, or potential scenarios of evolutionary relationships among taxa (e.g., populations, species, languages, cultures, or artifacts).

Convection The transfer to heat through the movement of air molecules.

Convergence Evolution in distinct lineages of similar morphology in response to similar function.

Cooperation Behavior by one individual that provides benefits to another and that may or may not be phenotypically costly.

Corpus luteum A yellow, progesterone-secreting, glandular body formed in the mammalian ovary from a Graafian follicle after extrusion of an ovum.

Correlated evolution When a change in one trait reconstructed over a phylogeny is correlated with a change in another trait; can also be used to evaluate causal relationships between traits.

Cortex (neocortex) The outermost layer of the mammalian brain.

Cranial buttressing Reinforced, robust structures of the cranium, especially present among premodern humans. Examples include structures sometimes called tori and keels.

Creationism A diverse set of beliefs that the universe and all life were created by a supernatural power (e.g., God of the Bible) with the features of living organisms pretty much in the same form as we see today.

Cultivation The intentional planting and growing of plants.

Cultural consensus The degree to which knowledge about some particular cultural domain is shared.

Cultural consonance The degree to which individuals approximate, in their own beliefs and behaviors, prototypes for belief and behavior encoded in shared cultural models.

Cultural phylogenetics Using methods developed in evolutionary biology (e.g., biological characters evolving on gene trees) in an anthropological context (e.g., cultural characters evolving on language trees).

7-Dehydrocholesterol (7DHC) A zoosterol that functions in the serum as a cholesterol precursor, and is converted to vitamin D_3 in the skin, therefore functioning as provitamin-D_3. 7DHC in human skin enables humans to manufacture vitamin D_3 from ultraviolet rays in sunlight.

Descent with modification Darwin's basic evolutionary model based on the observation that offspring differed from their parents in various features, and that these features, if providing an advantage in survival and/or reproduction, would tend to be preserved disproportionately in future generations, resulting in a change in the form of descendants compared to that of their ancestors.

Development A progression of changes, either quantitative or qualitative, that lead from an undifferentiated or immature state to a highly organized, specialized, and mature state.

Developmental acclimatization Adaptive responses that can only be acquired with exposure to the stressors during the growth period.

Developmental plasticity Adaptive responses to environmental cues that enable an organism to adjust its phenotypic development to match the environment. Molecular mechanisms include epigenetic changes that regulate gene expression during development through to maturity.

Dichromatic Having or exhibiting two colors (e.g., red and blue).

Divergence The evolution of greater difference between lineages in one or more characters or in DNA sequence.

Dry fallow Repeatedly plowing a field to prevent weed growth, so that rain will be stored in the soil for a crop that will be planted subsequently.

Duffy antigen A protein located on the surface of red blood cells. Loss of this protein, which is observed in many Africans, results in protection from *Plasmodium vivax* malaria.

Early fetal (pregnancy) loss Loss of conceptus within the first 12 completed weeks of pregnancy.

Efe (Western Pygmies) A group of the Bambuti who live in the Ituri Forest of the Democratic Republic of the Congo. They are bow hunters who live in symbiotic relations with neighboring Lese Bantu farmers (see Bailey and Peacock, 1988).

Effective population size Number of haploid organisms (understanding that diploid organisms such as humans can be considered as two haploid organisms) that reproduce and pass copies of their genetic material to the next generation.

Electron spinning resonance (ESR) Dating technique that uses the accumulation of electrons trapped in defects in dental enamel to estimate the time elapsed since death.

Emergence New patterns and properties that appear at a higher level of organization due to the organization and interaction of lower-level parts.

Emerging infectious disease Infectious diseases that have increased in incidence over the past decades and could increase in the near future. This includes newly identified pathogens, but also reemerging infections.

Environment of Evolutionary Adaptedness (EEA) The environment to which organisms are adapted. For humans this is often considered the Paleolithic, when *Homo sapiens* evolved into its modern form.

Epigenetics Functional changes in gene activity that are heritable across cell divisions or between generations *without* changes in the cell's nuclear DNA sequence. Epigenetic mechanisms include alterations of gene expression by structural or blocker molecules that prevent or retard reading of genes. Use of the term has expanded to include modifications reproducibly acquired during sensitive periods of development.

Episodic memory The memory of autobiographical events (times, places, associated emotions, and other contextual knowledge) that is the collection of past personal experiences that occurred at a particular time and place.

Erectine A nontaxonomic term that is roughly synonymous with *Homo erectus*. The term refers to the early populations of the genus *Homo*, which may or may not represent multiple species such as *Homo antecessor*, *Homo erectus*, and *Homo ergaster*.

Erythropoiesis The process of red blood cell production in the body.

Erythropoietin A hormone secreted by the kidneys that stimulates the production of red blood cells in the bone marrow.

Essentialism The notion that there is a set of biological attributes essential to every racial group.

Estradiol A steroid hormone that is produced by the ovaries and, in lesser amounts, by the adrenal cortex, placenta, and male testes. It helps control and guide female sexual development, including the physical changes associated with puberty. It also influences the course of ovulation in the monthly menstrual cycle, lactation after pregnancy, aspects of mood, and the aging process.

Ethics The justifying of principles of human moral conduct.

Eugenics The racist process of trying to "improve" a racial group's genetic composition by way of controlled breeding.

Evaporation The loss of heat due to conversion of water into vapor.

Evolutionary medicine (also Darwinian medicine) The application of evolutionary principles to understanding health and disease.

Evolutionary mismatch When a species or genotype becomes maladapted in some way to its environment as a result of environmental change (including the evolution of other species).

Evolutionary science Contemporary multidisciplinary research focused on the phylogenetic interrelationships among living organisms, their current biologic variation, and their historic descent from common ancestors.

Exome The overall amount of expressed DNA covered in exons (those portions of DNA that code for amino acids).

Extra-pair copulation (EPC) Mating events which occur outside of the structured unit of a mating system.

Extra-pair paternity (EPP) Siring of an offspring by an extra-pair male, resulting from female extra-pair copulation.

F_{ST} A relative measure of genetic differentiation that is the proportion of total variation due to variation among groups. The quantity $(1-F_{ST})$ is the proportion of total variation due to variation within groups.

Fat-free mass (FFM) The lean component (muscle, bone, water) of the total body weight (mass). Calculated by subtracting fat mass from body mass.

Fecal–oral transmission Spread of a pathogen as a consequence of inadvertent ingestion of the fecal material of an infected individual, usually through contaminated food or water.

Fecundity The ability to produce offspring.

Feeding competition Feeding-related behavior by one or more individuals that reduces the net energy gain rate of one or more other individuals.

Fertility The number of offspring produced in a given period of time.

Filaggrin (FLG) A filament-associated protein that binds to keratin fibers in epithelial cells.

Fitness The contribution of an allele or genotype to the next generation or succeeding generations, as measured by its rate of increase, which is based on both survival and reproduction.

Fixation Attainment of a frequency of 1 (i.e., 100%) by an allele in a population.

Follicle-stimulating hormone (FSH) A gonadotropic hormone of the anterior pituitary gland that stimulates the growth of follicles in the ovary and induces the formation of sperm in the testis.

Follicles Spherical groups of cells found in the ovary that contain a single oocyte (immature egg). Follicles grow and develop, produce hormones, and, in humans, produce a single egg for ovulation each month.

Follicular phase The phase of the menstrual cycle during which follicles in the ovary mature. It ends with ovulation.

Fomite An inanimate object that can be contaminated with pathogens (e.g., doorknobs, computer keyboards, etc.).

Foramen magnum Opening in the bottom of the skull for passage of the spinal cord and associated structures.

Founder effect A type of genetic drift resulting from a small group forming a new population. Because of sampling effects, the founding population will not likely have the exact same allele frequencies as the parental population.

Frequency As usually used in evolutionary biology, a proportion (e.g., the frequency of an allele is the proportion of gene copies with that allelic state).

G6PD deficiency Glucose-6-phosphate dehydrogenous (G6PD) deficiency is a common human enzyme deficiency coded by a gene on the X chromosome. Individuals with the deficiency can be resistant against *Plasmodium falciparum* malaria.

Gene–environment interactions The phenotypic effects of genes interacting with the environment.

Gene flow A change in allele frequency due to the movement of genes from one population to another. Gene flow acts to make populations more similar over time.

Gene tree A diagram representing the history by which copies of a gene have been derived from gene copies in previous generations.

Genetic correlation Correlated differences among genotypes in two or more phenotypic characters, due to pleiotropy or linkage disequilibrium.

Genetic drift Random changes in the frequencies of two or more alleles or genotypes within a population; usually restricted to random changes that arise through random sampling (i.e., the genes carried in one generation are a sample of those present in the population in the previous generation).

Genic selection A form of natural selection in which the single gene is the unit of selection (see "selfish gene"), such that the outcome is determined by fitness values assigned to different alleles. May be contrasted with *individual selection*, consisting of nonrandom differences among organisms with genetically different phenotypes within a population in their contribution to subsequent generations, and with *group selection*, the differential rate of origination or extinction of whole groups or populations of organisms, on the basis of average differences among groups in the characteristics of component individuals.

Glacial refugia Geographic regions that during glacial epochs were used as refuge by different species, including *Homo sapiens* and archaic *Homo* species.

Glycoprotein A type of protein that has a sugar attached to it; examples include neuraminidase and hemagglutinin, the major influenza surface antigens.

Gonadotropin-releasing hormone (GnRH) A trophic peptide hormone produced by the hypothalamus that signals the anterior pituitary gland to secrete luteinizing hormone (LH) and follicle-stimulating hormone (FSH).

Grade Organisms that are fairly closely related and share a similar suite of morphological and/or behavioral adaptations.

Groupness The belief among a collection of people that they are a group.

Growth A quantitative increase in size or mass. Measurements of height in centimeters or weight in kilograms indicate how much growth has taken place in a child.

Gut microbiota The collection of microorganisms that live in an organism's gut. Identification of the role played by the human gut microbiota is a growing and active area of research.

Hadza This group inhabits the dry savanna of northern Tanzania. They speak a click language and had been thought to be related to the southern African Ju'/hoansi, but that has not been verified. They hunt with bow and poison arrows for a variety of game, and collect honey and a variety of plant foods (Kaare and Woodburn, 1999).

Haplogroup A set of haplotypes that share the same alleles in all loci (and so a common ancestor).

Haplorhines Members of the primate suborder Haplorhini ("dry-nosed" primates) that includes the tarsiers and the anthropoids.

Haplotype A combination of adjacent alleles in a genomic region that are inherited as a block. Usually used for mtDNA and nonrecombining Y-chromosome DNA analyses.

Health disparities Gaps in the quality of health and health care across populations.

Hemoglobin (Hb) A component of red blood cells that serves to carry oxygen to the body's tissues. Normal hemoglobin is designated as hemoglobin A (Hb A); numerous human variants of this molecule are known, many of which have evolved in response to malaria (e.g., Hb S, Hb C, Hb E).

Herbivore An animal that feeds on plants. Most nonhuman primates are largely herbivores.

Heritability The proportion of the total phenotypic variance in a population that is attributable to genetic causes.

Heterozygous Refers to a person whose DNA for a particular gene consists of two different variants (alleles)

Hiwi (Cuiva) These nomadic people live in the river forests of the llanos (Orinoco River basin plains that flood during the rainy season) of Colombia and Venezuela. Hunting, fishing, and plant food gathering contribute to a favorable diet that is provided by a relatively simple technology (canoes, bows, and arrows) (see Arcand, 1999).

Holocene The most recent geologic epoch that began at the end of the Pleistocene Epoch about 10,000 years ago, i.e., since the glaciers of the last Glacial period melted off.

Homeland Inferred geographic area from which a preexpansion population, or the node at the base of a language phylogeny, first arose before diverging out to other areas.

Homeostasis The maintenance of biological function in normal ranges (e.g., body temperature, blood pH, tissue oxygenation) in the face of environmental stressors. The sympathetic nervous system plays a central role in the regulation of homeostasis.

Hominin Members of the tribe Hominini (of the subfamily Homininae, family Hominidae, superfamily Hominoidea) that includes modern humans and our extinct relatives up until the last common ancestor with chimpanzees.

Hominoid Members of the primate superfamily Hominoidea that comprises the families Hylobatidae ("lesser" ages) and Hominidae ("great" apes plus humans). The Hominidae includes the Asian Ponginae (orangutans) and African Homininae (bonobos, chimpanzees, gorillas, and humans). Due to the old idea of a scala naturae leading to humans, the term "apes" was and often continues to be used (the group "apes" includes nonhuman hominoids and therefore is nonmonophyletic because chimpanzees are more closely related to humans than they are to other nonhuman hominoids such as gorillas, orangutans, and hylobatids).

Homo The taxonomic genus that includes all humans, fossil and living.

Homologous Similarity in structure and sometimes function that indicates a common origin.

Homozygous Refers to a person whose DNA for a particular gene consists of two copies of the same variant (allele).

Horticulturalist A form of subsistence where crops are grown in small gardens, often in forest areas. Many hunting–gathering groups practice some horticulture to supplement hunting and foraging for plants.

Hot flashes A sudden brief sensation of heat, often over the entire body or part of the body, caused by a transient dilation of the blood vessels of the skin and experienced by some women during menopause.

Human chorionic gonadotropin (hCG) A hormone produced by the placenta that maintains the corpus luteum during pregnancy.

Human leukocyte antigen (HLA) A group of immune genes that are the human version of the major histocompatibility complex (MHC). These immune genes encode for proteins on the surface of cells presenting antigens to immune cells. The HLA system helps at differentiating the self cells to the nonself cells (often microbes).

Hypobaric Referring to low or reduced barometric pressure.

Hypoplasia Defects in teeth caused by the disruption of dental enamel deposition during growth, as a result of stress factors such as nutritional deficiency or illness.

Hypothalamus–pituitary–ovarian (HPO) axis Refers to the effects of the hypothalamus, pituitary gland, and ovaries as if these individual endocrine glands were a single entity as a whole. The axis is functionally involved in reproduction through the development of primary and secondary sexual characteristics and the control of oogenesis. It also controls development and aging in animals.

Hypothalamus–pituitary–testicular (HPT) axis Refers to the effects of the hypothalamus, pituitary gland, and testicles as if these individual endocrine glands were a single entity as a whole. The axis is functionally involved in reproduction through the development of primary and secondary sexual characteristics and the control of spermatogenesis. It also controls development and aging in animals.

Hypoxia A condition in which there is oxygen deficiency in a habitat or a body part. At high altitude, reduced oxygen availability is due to low atmospheric/barometric pressure (hypobaric hypoxia).

Hypoxia inducible factors (HIF) A set of proteins that regulate a variety of genes associated with oxygen homeostasis.

Identical by state distance (IBS) A genetic distance computed between two individuals by counting for a given locus the number of nonshared alleles.

Identification The process by which an individual identifies with a racial category.

Ilium (pl: ilia) The upper blade-like portions of the pelvis.

Immunocompetence The ability of the body to resist, clear, or tolerate infectious organisms.

Implantation The attachment and embedding of the fertilized ovum in the endometrium.

Inclusive fitness The measure of the reproductive success of an individual in terms of the survival and reproductive success of its kin (e.g., children, grandchildren, nieces, and nephews). The value of each relative is the probability of shared genetic information (e.g., children share 50% of their genetic information with their mother, grandchildren share 25% of their genetic information with their grandmother).

Indehiscence The retention of seeds on the plant.

Indirect reciprocity Reciprocity maintained through a diffuse network of interactions in a society, rather than through direct one-on-one interaction.

Industry The technology represented at an archaeological site at a given time horizon. The term is sometimes used to denote a broader cultural–historical entity, such as the Acheulean (handaxe/cleaver) Industry or Acheulean Industrial Complex.

Infant mortality The number of infant deaths during the first year of life per 1000 births.

Information theory A research framework that examines the ways in which messages are encoded, transmitted, and decoded, and the relationship between the structure of the message and the likelihood that an error will occur in this process; note that the content of the message (meaning) is not a factor in information theory.

Intelligent design An antievolutionary proposal that some complex features of living things, particularly molecular machines and biochemical pathways, are impossible to construct following existing models of evolutionary change and thus provide evidence for the intervention into the history of life by an intelligent agent.

Intermembral Index (IMI) Upper limb length as a percentage of lower limb length. A value of 100 or more indicates that arms are longer than legs, and values less than 100 mean lower limbs are longer than upper limbs.

Introgressive hybridization Hybridization that leads to the spread of genes from one species into another.

Inuit and Inupiac (Eskimo) These peoples inhabit and exploit Arctic sea mammals and land mammals (especially caribou) from the coastal and inland areas of Alaska, across Canada, and throughout coastal Greenland. The hunter–gatherer traditional tools were complex (bone, ivory, wood, slate), and have been replaced with more effective modern tools (rifle, steel knives, and harpoons) (see Burch and Csonka, 1999; Worl, 1999).

Kin selection Natural selection acting through the survival and reproduction of organisms in the same lineage, who share a proportion of the same genes. It is a form of selection whereby alleles differ in their rate of propagation by influencing the impact of their bearers on the reproductive success of individuals (kin) who carry the same alleles by common descent. It is an evolutionary process by which phenotypically costly forms of cooperation can emerge and become common because by performing them, individuals confer compensatory benefits on close relatives.

Lactase The small intestine enzyme responsible for cleaving the milk sugar lactose into its component sugars, glucose and galactose.

Lactase impersistence When lactase production ceases before adulthood. The most common phenotype for humans.

Lactase persistence When lactase production continues throughout life. Common only in some human populations who have a long history of keeping dairy animals.

Lactation Refers to the ability of mammals to nourish their young with milk produced by the mammary glands. A complete period of milk secretion from about the time of parturition to that of weaning.

Lactational amenorrhea Absence of menstruation associated with breastfeeding.

Lewis hunting phenomenon The alternation of vasodilation and vasoconstriction in response to acute peripheral cold stress. Allows for maintenance of skin temperature that is sufficiently high to prevent frostbite (tissue death).

Life expectancy (e_x) Life expectancy is the average number of years remaining at any given age, x years. Life expectancy at birth (e_0) is the average number of years remaining at birth.

Life history The sequence and timing of events related to survival and reproduction that occur from birth through death of an organism. Childhood and menopause are unusual aspects of human life history.

Life history theory That part of evolutionary theory that aims to explain the set of characteristics that make up different species life histories, including pattern and rate of development, adult size, timing and pace of reproduction, investments in offspring, and life expectancy. It posits that the schedule and duration of key events in an organism's lifetime are shaped by natural selection to produce the largest possible number of surviving offspring.

Life history traits Characteristics that affect the life table of an organism, and can be imagined as various investments in growth, reproduction, and survivorship. Typically, size at birth; growth pattern; age and size at maturity; number, size, and sex of offspring; age-, stage-, or size-specific reproductive effort; age-, stage-, or size-specific rates of survival; and lifespan.

Ligament Fibrous, sheath-like tissue linking bones.

Linkage equilibrium The random association of alleles at one locus with those at another locus, such that each combination occurs at the frequency predicted by their individual frequencies; in linkage disequilibrium, alleles at different loci are associated more or less frequently (nonrandom association) than predicted by their individual frequencies.

Lordosis Posterior concavity of the spine with a curvature throughout the vertebral column.

Lower Palaeolithic The technological stage including the Oldowan core and flake industries and the subsequent Acheulean handaxe and cleaver industries. In sub-Saharan Africa this stage is usually called the Early Stone Age. Dating from approximately 2.6 million years ago to 300,000 years ago. Contemporary with such hominins as *Australopithecus garhi*, *Homo rudolfensis*, *Homo habilis*, *Homo erectus*, and *Homo heidelbergensis*.

Luteal phase The portion of the menstrual cycle that begins with the formation of the corpus luteum after ovulation and ends, in the absence of fertilization, with the start of the menstrual flow.

Luteinizing hormone (LH) A hormone produced by the anterior lobe of the pituitary gland that stimulates ovulation and the development of the corpus luteum in the female and the production of testosterone by the interstitial cells of the testis in the male.

Macroevolution Large-scale evolutionary changes such as the origin of new species and other higher-level taxonomic groups.

Maltase The small intestine enzyme that cleaves maltose into two glucose units.

Maltose A double sugar containing two glucose molecules.

Marine Isotope Stage (MIS) The Marine Isotope Stages form a chronology that is based on fluctuations in oxygen–isotope ratios measured from fossil foraminifera. Odd numbered stages represent warmer periods, while even numbered ones reflect colder, glacial periods. Sometimes referred to as OIS for "Oxygen Isotope Stage."

Mastoid process The bony protuberance that protrudes inferiorly behind the ear.

Maternal constraint Fetal growth-limiting maternal and uteroplacental factors in mammals that operate to minimize risk of dystocia (obstructed labor).

Mating System Composite patterns of mating relationships in a group; includes general categories of polygyny and monogamy, and more specific biases such as cross-cousin marriage.

Maturation A measure of functional capacity, for example, the maturation of bipedal walking results from changes with age in skeletal, muscular, and motor skills of the infant and child.

Maximal oxygen consumption (VO_2max) Peak level of energy expenditure for an individual when he/she is working as hard as possible. Typically measured in liters of O_2 per minute, or milliliters of O_2 per kilogram mass per minute while the subject is performing a standardized exercise test.

Megadontia Absolutely and relatively large size of post canine dentition.

Megafauna Wild animals weighing at least 50 kg. About 3/4 to 4/5 of such species suffered extinction in North and South America about the time of the arrival of humans.

Menarche The onset of menstruation defined by the appearance of the first menstrual flow.

Mendelian disease A genetically inherited disease produced by single (usually rare) mutations in a single gene.

Menopause The irreversible cessation of regular menstruation in some female mammal species, marking the end of a female's fecundability.

Mesolithic The postglacial technological stage succeeding the Upper Palaeolithic in the Old World, characterized by geometric microlithic stone tools and bow-and-arrow technologies. Outside of Europe the term "Epipalaeolithic" is sometimes used. The postglacial Holocene epoch began ca. 12,000 years ago, and many Mesolithic cultures were replaced by Neolithic farmers in many parts of the Old World by 6000 years ago.

Metabolomics The study of the unique chemical fingerprints that specific cellular processes leave behind, i.e., metabolite profiles. A metabolic profile gives information about the physiology of a cell or organ system.

Middle Palaeolithic The technological stage characterized by a range of flake tools (side scrapers, points, denticulates, backed knives) often made on flake blanks struck from special core forms, such as disc cores and Levallois prepared cores. In sub-Saharan Africa, the term "Middle Stone Age" is usually used. It is likely that some stone points were hafted onto spear shafts during this time. Dating from approximately 300,000 years ago until 30,000 years ago. Contemporary with *Homo neandertalensis*, *Homo helmei*, and early *Homo sapiens*.

Miocene A period of time that lasted between 23.0 and 5.3 million years ago.

Mitochondrial DNA (mtDNA) Genetic material that lies in the mitochondria organelles of cells. It is inherited maternally and tends to preserve better than nuclear DNA in the fossil record.

Mode of transmission The mechanism by which a pathogen transfers from one host to another.

Monogamous The practice or state of having a sexual relationship with only one partner.

Monophyletic A clade of organisms all derived from a common ancestor.

Moral sentiment Emotions related to moral behavior, such as conscience, empathy, or remorse.

Moral system Morality viewed at the social level, such as systems of rewards or sanctions for maintaining valued behaviors.

Morality The normative differential valuing of behavior.

Multidimensional scaling analysis (MDS) Also known as principal coordinates analysis, is a mathematical technique to reduce the number of dimensions needed to represent the data. MDS produces a set of coordinates out of a distance matrix by means of matrix transformations.

Multimale polyandrous In multimale groups, no single male has full reproductive access to all females. Rather, the top ranking males in the groups have an approximately equal share of sharing all offspring born to females within the group.

Mutation An error in the replication of a nucleotide sequence, or any other alteration of the genome other than reciprocal recombination.

Mutualism A form of cooperation in which all participants gain immediate net benefits.

Natural fertility population A population that makes no conscious effort to limit fertility.

Natural selection One of the mechanisms of evolution. Natural selection is a process by which an individual possesses traits that are better suited or adapted to a specific environmental pressure, which includes changes in climate, or competition for food or mates, etc. Those individuals that are better adapted will tend to survive and reproduce in greater numbers than others of their kind, thus ensuring the perpetuation of those favorable traits in succeeding generations. This is the central tenant of the thesis of Charles Darwin's (1859) *On the Origin of Species by Means of Natural Selection*. It represents consistent (nonrandom, or biased) differences in the rate of survival or reproduction among classes of entities that differ in inherited characteristics.

Neo-Darwinism The prevailing evolutionary model in the early and mid-twentieth century based on the successful merging of natural history and population genetics to provide a plausible mechanism for studying biologic change among and within populations; also a term used by antievolutionists to refer to evolutionary science.

Neolithic The "New Stone Age; " technological stage associated with early food-producing societies in the Old World, dating from about 12,000 years ago to 4000 years ago. Technologies often are characterized by blade and bladelet technologies, ground stone axes, grinding stones for processing cereals (mortars and pestles), sickles, and pottery for storage and cooking. The Neolithic saw larger populations, more permanent villages, and more substantial architecture, including megaliths.

Network analysis Application of the network theory to visualize relationships in social, ecological, or epidemiological contexts. Several metrics have been developed to analyze the architecture of network such as centrality.

Network reciprocity Reciprocity maintained by limiting interactions to reciprocators.

Neutral alleles Alleles that do not differ in their effect on fitness.

New World North and South Americas (see Old World).

Nonshivering thermogenesis Increase in metabolic heat production due to hormonal, biochemical pathways rather than stimulation/movement of skeletal muscles.

Norepinephrine (NE) A chemical released from the sympathetic nervous system in response to stress. Norepinephrine is classified as both a neurotransmitter and a hormone.

Northern Algonquians Often referred to as Arctic Native Americans, these Boreal hunters and trappers inhabit much of the inland forested areas of Canada. Hunting large mammals such as the moose and trapping small mammals as well as foraging for berries and other seasonal fruits contributed to the diet (see Steegmann, 1983).

Northwest Pacific Coast There were numerous societies in the Pacific Northwest that subsisted on river salmon and sea mammals and plants from the temperate rainforests. Because of the rich food resources these societies had complex social structure, art, and architecture that set them apart from many other hunter–gatherer populations (see Boas, 1921).

Nuchal crest Ridge of bone running transversely across the back of the skull for attachment of muscles.

Nuchal plane The portion of the occipital bone that faces inferiorly and is the area of attachment for the nuchal (neck) muscles.

Nuchal torus A bar of bone running mediolaterally across the back of the skull. It is one of the cranial buttresses.

Nulliparity The status of a woman who has never borne a child.

Occipital bun A posterior projection of the back of the skull that is characteristic of European Neandertals. The bun involves a flattening of the posterior parietals, a vertical occipital plane, and a horizontal nuchal plane.

Occipital plane The more or less vertical portion of the posterior occipital. It lies above the nuchal plane and forms part of the back of the cranium.

Occipitomastoid crest A crest that lies between the mastoid process of the temporal bone and occipital condyle on either side of the skull base. A large occipitomastoid crest is common among Neandertals.

Oldowan The earliest known stage of human technology, beginning about 2.6 million years ago, named after the famous site of Olduvai Gorge in Tanzania. Characterized by battered percussors (hammersones, spheroids), simple cores on cobbles or chunks (choppers, discoids, polyhedrons, heavy-duty scrapers) and simple flake tools (light-duty scrapers, awls). Contemporary with hominins such as *Australopithecus garhi*, *Homo rudolfensis*, *Homo habilis*, and *Homo erectus*.

Old World Africa, Europe, Asia, and Australia (see New World).

Omnivore An animal that feeds on a variety of foods, both plant and animal. Humans are omnivores.

Omnivore An organism with a mixed diet of animal and vegetable foods.

Onge A "Negrito" population from Little Andaman Island, the southernmost and relatively isolated island in the Andaman Islands.

Ontogeny Origin and development of an organism from fertilization to maturity.

Oocyte A germ cell of an animal ovary that undergoes meiosis to form an ovum.

Opistocranion The most posterior point on the midline of the back of the cranium.

Orthognathic Refers to the tucked position of the face in anatomically modern humans under the brain case, as opposed to the projecting face of nonhuman primates.

Osteobiography The study of human remains for the purpose of describing the life experience of an ancient person.

Osteoporosis A bone disease characterized by decrease in bone mass and density, resulting in a predisposition to fractures and bone deformities. It occurs most commonly in women after menopause as a result of estrogen deficiency.

Outcrossers Species that cannot self-fertilize.

Ovarian cycle The normal sequence of physiological changes in the ovary that includes development of an ovarian follicle, rupture of the follicle, discharge of the ovum, and formation and regression of a corpus luteum.

Ovarian reserve The number and quality of oocytes in the ovaries of a woman of childbearing age.

Ovary The gonad of a female animal; vertebrates have a pair of ovaries. The ovary contains oocytes from which ova are produced, and it also secretes various steroid hormones. Its activity is cyclical and is controlled by gonadotropins.

Ovulation A phase that occurs in a female's menstrual cycle, in which a mature ovum (ovule) is released from the ovary into the oviduct.

Pair-bonded The temporary or permanent association formed between a female and a male animal during courtship, mating, and parenting. Human pair-bonds are unusually variable and complex.

Paleoanthropology The study of human evolution through the exploration of the fossil and archaeological record.

Palaeolithic (Paleolithic) The "Old Stone Age," usually divided into the Lower Palaeolithic (Early Stone Age), Middle Palaeolithic (Middle Stone Age), and Upper Palaeolithic (Later Stone Age). It was succeeded in the Old World by the Mesolithic and Neolithic. During the Palaeolithic, hominin brain size tripled, and humans spread from Africa to most of the inhabitable world.

Paleoparasitology The study of ancient parasitism in animals, including humans.

Paleopathology The study of ancient health and disease.

Parent–offspring conflict A term used to signify the evolutionary conflict arising from differences in optimal fitness of parents and their offspring. Whereas parents tend to maximize the number of offspring, the offspring can increase their fitness by getting a greater share of parental investment often by competing with their siblings.

Parental care (maternal or paternal) Parental investment (e.g., calories, protection, hygienic behavior, etc.) provided to offspring. Parental care is posited to increase growth rates, quality, and/or survival of young, and hence ultimately increase the inclusive fitness of parents.

Parkinson disease A neurodegenerative disease that primarily affects basal ganglia, disrupting neural circuits that regulate motor control, cognitive flexibility, and other aspects of cognition, attention, and behavior.

Partial pressure of oxygen (PO_2) The contribution of the oxygen component to air pressure. Calculated as: (O_2 fraction = 20.9%) × (atmospheric pressure).

Pastoralism The practice of keeping domestic animals that graze in grasslands.

Perimenopause Means "around menopause" and refers to the time period during which a woman's body makes its natural transition toward permanent infertility (menopause). Also called the menopausal transition.

Phenotype The morphological, physiological, biochemical, behavioral, and other properties of an organism, or any subset of such properties.

Phenotypic accommodation Adjustment of the phenotype following disruption to the organism's developmental program by environmental factors.

Phylogeny The history of descent of a group of populations or taxa (such as species) from their common ancestors, including the order of branching and sometimes the absolute times at which they branched, as well as fusion of branches in some cases; also applied to the genealogy of genes derived from a common ancestral gene, or generally to objects that have a history of descent from common ancestors (e.g., languages, in part).

Phylogeography Using phylogenetic information to reconstruct the geographic expansion from a homeland to the current or recent locations of all descendant taxa.

Platyrrhines Members of the primate parvorder Platyrrhini that includes the New World monkeys, which comprise five extant families found in Central and South America: Callitrichidae, Cebidae, Aotidae, Pitheciidae, and Atelidae.

Pleistocene The geological epoch (time period) preceding the Holocene, characterized by a long series of glaciations, alternating with much shorter periods of climate similar to that of today, between 2.59 and 11.7 million years before present. Early Pleistocene: 2.59 million to 781 thousand years before present. Middle Pleistocene: 781 to 126 thousand years before present. Late Pleistocene: 126 to 11.7 thousand years before present.

Pliocene A period of time that lasted between 5.33 and 2.58 million years ago.

Polar T_3 syndrome Marked reductions in circulating thyroid hormone (especially free T_3) levels during severe winter conditions in arctic and Antarctic environments. These declines are the result of tissue-level uptake of thyroid hormones exceeding production. This syndrome results in increased metabolic heat production and alterations in mood and behavior.

Political economy The study of the relationship between economy and political power, and especially the effects of that relationship.

Polyandry Individual females have exclusive sexual access to two or more males.

Polycythemia Elevated red blood cell content in the bloodstream, characterized by high levels of hemoglobin.

Polygamy Sexual exclusivity occurs, but is not monogamous.

Polygenic character A feature whose variation is based wholly or in part on allelic variation at more than a few loci.

Polygynandry Two or more males have exclusive sexual access to two or more females, and vice versa.

Polygyny Individual males have exclusive sexual access to two or more females.

Polyploid Having more than two sets of chromosomes; humans are diploid; i.e., we get one set from our mother and another set from our father, but many plant and other species have more sets of chromosomes. A tetraploid has four sets, a hexaploid has six, and so on.

Population bottleneck Sharp contraction of the effective population size of a population.

Positron emission tomography (PET) A medical imaging technique that uses radioactive tracers to measure variation in blood flow and metabolic activity of tissues.

Postcranial Skeletal elements below the skull.

Postgenomic sciences Sciences that study the expression of genes.

Prehensile An appendage or limb adapted for grasping.

Preterm delivery A birth that takes place before 37 weeks of pregnancy.

Primates Order of mammals that originated at least 65 million years ago, and includes a very diverse range of taxa, from the lemurs of Madagascar to our own geographically very broad species.

Principal component analysis A mathematical technique to reduce the number of required dimensions to explain the variation present in a matrix (i.e., of individuals by genetic variants) by means of matrix transformations.

Progesterone A steroid hormone produced in the corpus luteum and placenta. It has an important physiological role in the luteal phase of the menstrual cycle and in the maintenance of pregnancy. In addition, progesterone produced in the testis and adrenals has a key role as an intermediate in the biosynthesis of androgens, estrogens, and the corticoids (adrenal cortex steroids).

Prognathism The anterior projection (or jutting) of the jaws (maxilla and mandible) relative to the skull base.

Pro-opiomelanocortin (POMC) A precursor polypeptide with 241 amino acid residues, which can be cleaved into alpha-melanocyte-stimulating hormone (α-MSH).

Prosimians An old designation that refers to a group of primates formed by the strepsirrhines and the tarsiers. Although tarsiers are now considered to be haplorhines, authors often continue to use the term "prosimians" (a nonmonophyletic, or not natural, group) as a convenient way of referring to nonanthropoid primates.

Proteomics The application of molecular biology, biochemistry, and genetics to the analysis of the structure, function, and interactions of proteins produced by the genes of a particular cell, tissue, or organism.

Pseudogene A DNA sequence that is derived from a functional sequence, but is thought not to be functional; often, it is not transcribed or translated.

Punctuated equilibrium A model of macroevolution developed by Stephen J. Gould and Niles Eldridge in the early 1970s. This model contends that most major evolutionary changes happen during speciation ("punctuated"), while little or no significant directional evolution takes place between speciation events ("equilibrium").

Punishment Actions that tend to negatively reinforce behaviors in others.

Quantitative trait A phenotypic characteristic that varies more or less continuously rather than as discretely different character states.

Race-based medicine Medicine produced for use in specific racial groups.

Racialism A belief in race, despite disagreement with hierarchical or discriminatory concepts.

Radiation Transmission of heat from an object's surface in the form of electromagnetic waves.

Rassenhygiene The Nazi eugenics program to "improve" the Aryan race.

Reaction norm The manifestation of a single genotype as a range of phenotypes depending on variation in the environment during development.

Reciprocity Mutual exchange of help or benefits, sometimes displaced in time.

Red Queen hypothesis An hypothesis about how host–pathogen interactions lead to coevolutionary changes that may stimulate the evolution and maintenance of sexual reproduction.

Regulatory gene A gene whose product governs the transcription of other genes, and hence their expression in different cells or at different times or in different environments.

Relative fitness The average number of surviving offspring of individuals with a particular genotype (combination of alleles) compared with average number of surviving offspring from individuals with different genotypes after a single generation.

Relative sitting height (RSH) The ratio of seated height to total stature (RSH = 100 × (sitting height)/(total stature)). This index is a good measure of differences in body proportions among humans.

Reproductive skew The degree to which total reproductive output of a population is distributed unevenly among its members.

Reproductive success The relative production of fecund offspring by a genotype.

Residual volume The amount of air remaining in the lungs after forced exhalation.

Respiratory transmission Transmission of a pathogen through the air, usually as a consequence of coughing or sneezing.

Rewards Actions that tend to positively reinforce behaviors in others.

Sagittal crest Ridge of bone running along the top of the skull for attachment of muscles.

Sahel A belt of land directly south of the Sahara Desert that is semiarid to very arid.

San (Ju'/hoansi; Bushmen) Live in the area of southern Africa that includes Namibia, Botswana, and the Republic of South Africa. In the past they inhabited most of the land south of the Zambezi River but now are limited to a number of enclaves including the Kalahari Desert (see Lee and DeVore, 1976).

Savanna Grasses dominate ground surface area; woody plants cover less than 30% of the surface; trees are variably deciduous. Tree crowns rarely or never touch. Trees vary in height but rarely exceed 15 m. Bare soil is uncommon; palms are variably common. Tree leaves and stems may trend toward succulence in drier or more seasonal habitats; tree thorns are common.

Scramble competition polygyny One individual mates with as many individuals as possible within mate guarding.

Secondary products revolution A term that refers to livestock products that were developed (such as milk) sometime after the domestication process took place.

Secondary sexual characteristics The sex-associated phenotypes of somatic tissues in sexually dimorphic animals that usually appear at puberty.

Sectorial premolar A lower first premolar with a distinctive elongated shape featuring one very prominent cusp, the protocone, and a long surface anteriorly, giving the tooth a somewhat right-triangle shape as viewed from the side. The long edge rubs against the back of the upper canine, sharpening it. Living apes and most primates possess this canine/premolar honing complex.

Selection coefficient A measure of the strength of natural selection operating on a genetic variant.

Selective sweep When an advantageous gene rapidly spreads throughout the population that constitutes a species.

Selfish gene Reductionistic view of evolution, as defended by Richard Dawkins, that regards genes as the primary units of selection, due to their causally determining all relevant higher-level properties.

Semang These Malayan "Negritos" are a small population who combine hunting, fishing, gathering vegetable foods, and engaging in some trade of forest products. They hunt primates and small mammals with blowguns and poison darts (see Endicott, 1999).

Senescence Age-related changes, both structural and functional, that decrease the ability of an individual to withstand stress, reduce the probability of reproduction, and increase the likelihood that an individual will die.

Serial founding A process by which new populations diverge from previous populations, causing repeated examples of the founder effect over space.

Sexual conflict Conflicts of interest between males and females with regard to mate choice, investment in offspring, provision of parental care, and other aspects of reproductive strategies; can take the form of behavioral conflicts, including infanticide.

Sexual dimorphism Different shapes in the two sexes. In most sexually dimorphic mammals including primates, males are larger. In many primates, males also have larger canines.

Sexual transmission The spread of a pathogen from one person to another as a consequence of sexual activities.

Shivering thermogenesis Increase in metabolic heat production that involves stimulation/movement of skeletal muscles.

Short tandem repeat (STR) A type of genetic variation produced by differences in the number of particular short nucleotide motifs repeated in a tandem-like fashion (e.g., tetranucleotide STRs consist on repetitions of a sequence of four nucleotides). Also known as microsatellite.

Shoshone No longer practicing a foraging economy, the Shoshone and Piute lived in the arid Great Basin area of California and Nevada. They hunted bighorn sheep and small animals and they collected mesquite pods, pine nuts, annual and perennial seeds, and other plant foods. Timbisha Shoshone were seasonally transhumant and maintained temporary camps at certain times of the year (see Fowler, 1999).

Sickle cell trait A change in the hemoglobin beta gene that confers some protection against malaria infection.

Single-male multifemale polygyny One male has an exclusive reproductive access with two or more females.

Single nucleotide polymorphism (SNP) A common type of genetic variation widely present in the human genome, consisting of a change of one nucleotide base (A, T, C, G) for another at a single sequence position.

Social construction The process by which concepts are given meaning through the interactions of groups of people living in a particular sociohistorical context.

Social monogamy Rearing of offspring by one breeding male and one breeding female, even in the absence of sexual monogamy.

Social organization How members of a population or species are dispersed in space and time: whether they are clustered in space or solitary; whether stable groups form and, if so, whether these are cohesive; etc.

Social relationships Descriptions of the patterning and quality of social interactions between pairs of individuals; can be seen as the outcomes of fitness-maximizing strategies.

Social selection Selection (differential reproduction) occurring as a consequence of intraspecific social competition. Key aspect of within-species evolutionary arms races; proposed as primary selective pressure for evolution of human intelligence.

Social structure Patterning of social relationships within a group or population.

Soma The parts of the body (organs, tissues, and cells) that carry out all biological functions except reproduction.

Sperm The male gamete or reproductive cell involved in sexual reproduction.

Sperm competition The occurrence of sperm of multiple males within a female's reproductive tract, and the associated competition for fertilization of the ovum.

Spermatogenesis The production of male sex gametes; the maturational changes that transform spermatids into spermatozoa.

Squamous cell carcinoma (SCC) A cancer of a kind of epithelial cell, the squamous cell, which composes the main part of the epidermis of the skin, and this cancer is one of the major forms of skin cancer.

SRY (sex-determining region Y) gene A single gene on the Y chromosome that codes for testis-determining factor (or the SRY protein) and is responsible for determining the sex in placental mammals and marsupials. In humans, its presence will cause the development of embryos into male.

Status incongruity A state in which an individual occupies different ranks on two or more dimensions of social status.

Stone Age A term generally used to denote human and proto-human societies that had a strong reliance on stone as a principal raw material for tools. In the Old World, the Stone Age is usually divided into the Lower Palaeolithic, Middle Palaeolithic, Upper Palaeolithic, Mesolithic, and Neolithic.

Strepsirrhines Members of the primate suborder Strepsirrhini ("wet-nosed" primates) that includes the phylogenetically most basal extant primates (in the sense that they are the sister-group of all other extant primates). The infraorder Lemuriformes includes all of the five families found on the island of Madagascar (Cheirogaleidae, Daubentoniidae, Lemuridae, Lepilemuridae, and Indriidae) while the infraorder Lorisiformes includes two families (the Asian Lorisidae and African Galagidae).

Stress reactivity The cognitive–emotional and associated neuroendocrine sensitivity to stressors. High reactivity is linked to greater endocrine and autonomic stress responsiveness that increase strength and duration of hormonal (cortisol) and physical (heart rate) reactions to stress. As such, reactivity reflects the psychological and physiological impact of stressors on an individual.

Stress response A generalized suite of neuroendocrine and autonomic reactions that are triggered by perceptions of threat or challenge. Increased activity of the sympathetic nervous system and release of cortisol during the response organize a characteristic set of physiological, cognitive, and behavioral shifts that allocate the body's resources toward meeting immediate needs rather than long-term goals.

Stressor An internal or external factor that disrupts homeostasis. Stressors include physical trauma, disease, social events and situations, and the demands of exercise and competition.

Strong reciprocity Reciprocity maintained through punishment.

Subsistence How a people procure food, through hunting and gathering, or fishing, or cultivating plants, or raising domestic animals, or any combination of these.

Support phase The part of the walking phase during which the hindlimb is bearing weight and is rotating backward.

Supraorbital torus The usual browridge form for all adult primates except most modern humans. It is a continuous bar of bone above the orbits (eye holes).

Swing phase The part of the walking phase during which the hindlimb is not bearing weight but is swinging forward.

Sympathetic nervous system Part of the autonomic (involuntary) nervous system that is responsible for regulating key aspects of physiological function (e.g., heart rate, blood pressure, constriction/dilation of blood vessels) in the face of environmental stressors. (see also homeostasis).

Sympatry Two species or populations live in sympatry when they are encountered in the same geographic area or in the same habitat.

Synaptic plasticity Malleability of synapses that transmit information between neurons. It plays a key role in associative learning.

Synchrony Two species or populations living at the same time.

Synonymous mutation A DNA base pair change that does not alter the amino acid in the protein product of a gene; nonsynonymous mutations do alter amino acids.

Taphonomy The science of studying the processes through which fossils and archaeological sites form and are preserved.

Tarsiiformes Members of the primate infraorder Tarsiiformes that includes taxa that once ranged across Europe, northern Africa, Asia, and North America, but today represented by a single extant genus, *Tarsius* (the tarsier).

Taxon (pl: taxa) The named taxonomic unit (e.g., *Homo sapiens*, or Hominidae) to which individuals, or sets of species, are assigned. Higher taxa are those above the species level.

Technology The repertoire of tool-making and tool-using behavior of a human or proto-human group, maintained by cultural norms.

Tendon Fibrous, cable, or sheath-like tissue that links muscle to bone.

Terrestrial Living on the ground as opposed to living in the trees (arboreal). Some nonhuman primates are partly terrestrial, while humans are wholly terrestrial.

Testis (pl. testes) The male gonad; either of the paired, egg-shaped glands normally situated in the scrotum; called also testicle. The testes produce the spermatozoa, the male reproductive cells, and the male sex hormone, testosterone, which is responsible for the secondary sex characters of the male.

Testosterone Steroid hormone and androgen produced by the Leydig cells of the testes in response to luteinizing hormone (LH) secreted by the pituitary. Other sources include the adrenal gland and conversion of precursors in peripheral tissues. Its functions include regulation of development of male reproductive organs, growth of bone and muscle, secondary sex characters, neural development, and behavior.

Thalassemia Types of inherited blood disorders that cause anemia; occur when a person carries alleles that interfere with the production of normal hemoglobin. Thalassemias are found in areas with a high prevalence of malaria, most commonly in the Mediterranean region, and provide some genetic protection against the disease.

Thelarche The first stage of secondary (postnatal) breast development, usually occurring at the beginning of puberty in girls.

Theory of mind The ability to represent and reason about the mental states of other beings, including thoughts, feelings, beliefs, intentions.

Thermoneutral Temperature range in which the minimum amount of metabolic energy is needed to maintain core body temperature. For humans (unclothed), the thermoneutral range is 25–27 °C.

Thyroid stimulating hormone (thyrotropin; TSH) Hormone secreted by the anterior pituitary that stimulates the thyroid gland to produce thyroxine (T_4) and triiodothryronine (T_3).

Thyrotropin releasing hormone (TRH) Hormone secreted by the hypothalamus that promotes the release of thyroid stimulating hormone (TSH) from the anterior pituitary.

Thyroxine (T_4) An iodine-containing hormone released by the thyroid gland that promotes oxidative metabolism at the cellular level.

Tool An extrasomatic object (outside of one's own body) that has been used toward some functional end, whether it is modified or not. Examples would include a thrown unmodified cobble as a hunting weapon, a sharp stone flake used as a butchery knife, and a piece of deer antler used as a soft hammer to make a handaxe.

Torus (pl. tori) A bony bar that protrudes from the surface of a bone. Tori are biomechanical reinforcements whose size and development are not due to the direct attachment of muscles.

Trade-off The fitness cost incurred when a beneficial change in one trait is accompanied by a disadvantageous change in another. This involves "choosing" to allocate resources based on cost–benefit ratios between competing demands or goals, such as whether to respond to a current opportunity or demand, or to ignore it, save time/energy, but perhaps suffer some consequences. Evolutionary theory predicts that such trade-offs are subject to selection pressures determined by their influence on fitness of the individual.

Transhumance Seasonal movement of livestock to seek available pastures.

Transition rate Instantaneous rate of change at which a trait reconstructed over a phylogeny switches from one particular state to another state (e.g., from polygyny to monogamy or from monogamy to polygyny).

Transmissible spongiform encephalopathies (TSE) A class of prion-caused diseases in animals and humans that affect the nervous system and are uniformly fatal.

Trichromatic Having or exhibiting three colors (e.g., red, green, and blue).

Triiodothyronine (T_3) An iodine-containing hormone released by the thyroid gland that promotes oxidative metabolism at the cellular level.

Twa (Pygmies) This Pygmy population is found in the tropical forest east of the Congo River and close to Lake Tumba in the Democratic Republic of the Congo. They trade with and live among the Oto Bantu, and because of substantial admixture and acculturation they are often referred to as pygmoid (see Pagezy, 1988).

Ultraviolet-B (UV-B) Electromagnetic radiation with wavelengths between 290 and 320 nm. It causes sunburn and a number of damaging photochemical changes within cells, including damage to DNA, leading to premature aging of the skin, premalignant and malignant changes, and various photosensitivity reactions. It is also used therapeutically in the treatment of skin disorders.

Uncoupling of oxidative phosphorylation A reduction in the cellular efficiency of energy transfer in the cell's mitochondrion that results in greater loss of energy as heat.

Upper Palaeolithic The technological stage characterized by systematic blade production, blade tools such as end scrapers, awls, backed knives, and points. In sub-Saharan Africa contemporary hunter–gatherer groups are usually called Later Stone Age. This stage dates from approximately 50,000 to the end of the Ice Age approximately 12,000 years ago. The first representational art comes from this stage, best seen in painting, engraving, and sculpture. Personal ornamentation such as beads and pendants becomes common for the first time. Many artifacts of bone, antler, and ivory (and presumably wood) are seen in the form of points, harpoons, spear-throwers, and eyed needles. Associated primarily with anatomically modern humans, *Homo sapiens*, but late Neandertals in Europe may also have adopted some Upper Palaeolithic elements in their technologies, such as the "Chatelperronian" of France.

Uranium-series (U-series) Radiometric dating technique that relies on the gradual and constant decay of a number of uranium isotopes.

Vasoconstriction Reduction in the diameter of a blood vessel due to smooth muscle contraction.

Vasodilation Expansion of the diameter of a blood vessel due to smooth muscle relaxation.

Vector-borne Pathogen transmission that occurs by means of another living organism such as a mosquito, tick, flea, or other arthropod.

Vertical (or congenital) transmission Transmission of a pathogen from mother to offspring either in utero or during the birth process.

Virulence The severity of a disease, sometimes measured as the amount of damage done to the host.

Vital capacity The volume of air that can be expelled from the lungs on forced exhalation, after a maximal inspiration.

Vitamin D_3 (VD_3) A colorless, crystalline steroid hormone that the body synthesizes in the skin when its precursor, a derivative of cholesterol, is irradiated by sunlight. VD_3 is also found in fish-liver oils, irradiated milk, and all irradiated animal foodstuffs.

Wallace's Line A biogeographical line recognized by Alfred R. Wallace that runs between the Asian and Wallacean faunal zones. During peaks of Ice Age glaciation lower sea levels meant that Borneo, Java, and Sumatra were all connected to the Asian mainland. Deep water east of these regions meant that the islands of Wallacea as well as Papau New Guinea and Australia were never connected to the Asian mainland during the course of human evolution.

WEIRD societies Western, Educated, Industrialized, Rich, and Democratic nations of the world. These societies provide the majority of human participants in research, but represent only a minority of the world's population. Results of research with WEIRD societies are likely not representative of human nature.

Whole genome sequencing Sequencing of the entire DNA code of an individual's genome.

Woodland Relatively dry or distinctly seasonal habitats with short trees, predominantly ≥5–20 m in height. The canopy is typically "open;" i.e., crowns are not dense and may touch without interlocking, allowing abundant sunlight to reach the ground; woody plants cover as little as 30% of the ground surface area but less than 95% of the surface; grass covers the ground between trees.

Yanomamo A horticultural tropical forest people from Venezuela and Brazil. They cultivate and hunt (see Chagnon, 1968).

Zoonotic disease Diseases caused by various kinds of pathogens that are transmitted between animals and humans.

Index

Note: Page numbers followed by "f" and "t" indicate figures and tables respectively.

A

A. (P.) aethiopicus, 194–195
ABO blood groups, 418–419, 419t
Abstraction, 484
ACC. *See* Anterior cingulate cortex (ACC)
Acclimation, 252
  to cold stress, 256–258
  to heat stress, 261–262
  to hypoxic stress, 264–265
Acclimatization, 252
Acheulean
  and contemporaneous industries, 201–202
  technology, 201
  tool industry, 201
Adaptation. *See also* Human mating systems
  discriminative care, 473
    male testosterone and care, 474
    oxytocin, 473–474
  female estrus and extended sexuality, 474–475
  for mating, 473
    male adaptations for pair-bonding and parenting, 473
    mutual mate choice, 473
  modeling, 9
Adaptive factor, 49
Adaptive radiations, 11–12
Adenosine triphosphate (ATP), 257–258
Admixture, 213, 215–216
aDNA. *See* ancient DNA (aDNA)
Adolescence, 289
  contributions to reproductive success of adults, 292
  evolution, 289–290
  risks, 292
  separate paths through, 290–291
Adult female reproductive function, variation in, 300
Adult male reproductive function, variation in, 302–303
Adult milk digestion, variation in, 397–398
Affiliation, 455
Africa, 86, 178, 181, 185
  Archaic *Homo*, 169–171
  climatic patterns, 91–92
  continental plate, 86
  hominin, 86
  South African fossil sites, 88
  Tethys Sea, 87
  vegetation map, 91f
African genome, 521
African Hunter-gatherer, 325–326
"African Replacement" hypothesis, 220–221
Aging, 311
  eggs, 317
  mothers, 317
Agriculturalism
  agricultural origins, 351
  Columbian exchange, 356
  competing hypotheses, 351–354
  industrial revolution, 357
  Paleo diet hypothesis evaluation, 360–361
  scientific breeding, 357
  spread of agriculture, 354–356
Agriculture/agricultural, 350, 396, 399–400
  origins, 351, 352f
  transition, 396
    adult milk digestion, variation in, 397–398
    starch digestion, 396–397
Allen's rule, 253–254
Alliances, 62–64, 455, 458–459
Allogrooming, 61
Allometric equation, 238–240, 243
Alloparental care, 457–459, 461
Allostasis, 442
Allostatic load, 442
Alveolar, 116
  air pressure, 494
AMH. *See* Anatomically modern humans (AMH)
AMHS. *See* Anatomically modern *Homo sapiens* (AMHS)
Amylase (AMY1), 397
Anatomically modern *Homo sapiens* (AMHS), 179t–180t, 196
  climatic conditions for spread, 180–181
  earliest fossil evidence of modern humans, 178–180
  history of discovery, 178
  origins, 181
  paleobiology, 181–189
  relationships with archaic *homo*, 181
Anatomically modern humans (AMH), 164, 181–182, 220
  Out of Africa migration, 220–222
ancient DNA (aDNA), 173
  techniques, 430–431
Animal(s), 93–94
  domestication, 409–410
  tools in animal world, 196
Antagonistic pleiotropy, 315
Anterior cingulate cortex (ACC), 494
Anthropoid(s), 45
  skeleton, 45
Anthropologists, 526
Antinutritional compounds, 361
Ape foot, 107
Aquatic ape hypothesis, 110
"Arable", 350
Archaic *Homo*, 164. *See also* Genus *Homo*
  chronological variation, 164–167
  chronology and classification of Pleistocene *Homo*, 165t
  evolutionary relationships and taxonomy, 173
  fossil sites, 166f
  geographical distribution, 164
  growth, development, and energetics, 172
  language, behavior, and culture, 172–173
  regional variation, 168–172
Archaic humans, 212
  Out of Africa migration, 220
  genetic admixture between archaic and modern humans, 222–225
Ardipithecus kadabba (Ar. kadabba), 114, 122
Ardipithecus ramidus (Ar. ramidus), 124–125
Arginine vasopressin (AVP), 458–459
Arm-hanging, 111
Artifact, 196–197
Artificial selection model, 339
ASD. *See* Autism spectrum disorders (ASD)
Asian diploid genome, 521
Association cortex, 236–237, 245
Astronomical parameters, 87
Atherosclerosis, 432
ATP. *See* Adenosine triphosphate (ATP)
Attraction, 474–475
Auditory system, 51
Australasia, 178
  Archaic *Homo*, 171–172
Australopithecines, 130, 194
  A. (K.) platyops, 136–137
  A. afarensis, 133–135
  A. anamensis, 135–136
  A. bahrelghazali, 135
  A. garhi, 136
  A. sediba, 137
  adaptations, 137–138
  Australopithecus genus, 130–133
  early, 95
  timeline of species, 130f

Australopithecus afarensis (*A. afarensis*), 75–76, 76f, 108, 130–131, 131f, 133–135
cranium A. L. 444–2, 134f
rib cage, 107f
Australopithecus anamensis (*A. anamensis*), 95, 135–136
Australopithecus bahrelghazali (*A. bahrelghazali*), 135
Australopithecus garhi (*A. garhi*), 136
Australopithecus genus, 130–133
Australopithecus sediba (*A. sediba*), 137
Australopithecus transvaalensis, 130–131
Autism spectrum disorders (ASD), 485
AVP. *See* Arginine vasopressin (AVP)

B

Backbone of History Project, 432
Balanced Treatment, 19
Balancing selection, 7
Barbary "ape", 45–46
Barrier requirements, 276
Basal ganglia, 497–498
Basal metabolic rate (BMR), 172, 252–253, 253f, 300, 394–395
BAT. *See* Brown adipose tissue (BAT)
Bergmann's rule, 253–254
BiDil, 520
Bioarchaeology, 432–433
Biocultural feedback system, 194
Biocultural reproduction, 288–290
Biodiversity loss, 411
Biogeochemistry, 93
Biohistory, 432
Biological species concept (BSC), 10–11
Biomarkers, 304
Biparental care, 470
Bipedalism, 76–77, 104, 120–122, 125, 394
anatomical features, 105–107
evolutionary stages, 107–108
evolving, 108–112
humans walk, 104–105
lever arm for hip flexors, 106f
Bipeds, habitats of earliest, 94–95
Blend view, 472
Blood pressure, 384
BMR. *See* Basal metabolic rate (BMR)
Body Size, 329–331
Bones, 44–46
Bonobo mating, 471
Bonobos (*Pan paniscus*), 58–59
Borrowing problem, 528
Bovine spongiform encephalopathy (BSE), 345
Brain. *See* Human brain
Breeding systems, 468
Broca's areas, 247–248, 497
Brother–sister relationships, 458
Brown adipose tissue (BAT), 258
Brown lemurs (*Eulemur* species), 33
BSC. *See* Biological species concept (BSC)
BSE. *See* Bovine spongiform encephalopathy (BSE)
Buffering effect, 387

C

Caloric restriction (CR), 393
Care (C), 9
Catarrhine skeleton, 45
Caucasian genome, 521
Causal reasoning, 485–486
CCR5-Δ32 allele, 422–423
Center for Science and Culture (CRSC), 23
Chacma baboons (*Papio ursinus*), 64–65
Character-coded language phylogenies, 526, 528f
Child Growth, 329–331
Childbirth, 184
Childhood, 456–457
mortality, 361
risks, 292
Chimpanzees (*Pan troglodytes*), 59–60, 66
growth *vs.* human growth, 286–289
mating, 470–471
as model for early hominins, 196–197
morphology, 116–118
Chondrodysplasia calcificans punctate, 185
Chronic disease
developmental plasticity, 371–372
DOHaD, 370–371
epigenetics, 372–374
evolutionary and developmental mismatch, 374–377
excessive nutrition, 376–377
GDM, 376–377
human disease, evolutionary perspective on, 370
maternal constraint, 375–376
PAR, 371–372
transgenerational inheritance, 377
Chronological variation, Archaic *Homo*, 164–167
CJD. *See* Creutzfeldt–Jakob disease (CJD)
Clade, 12
Classification, 516–517
Cleaner wrasse mutualism, 509, 509f
Climate, 182–183
change proxy evidence, 88–89
"Clinal" variation, 517–518
Coalitions, 455, 459, 462
"Coastal express" hypothesis, 222
Coevolution, 359, 415
coevolutionary process, 345
examples of pathogens, 418
ABO blood groups and infectious diseases, 418–419, 419t
CCR5-Δ32 allele, 422–423
interacting species and influenza viruses evolution, 421–422
malaria, genetic adaptations to, 419–421
host–pathogen mechanisms, 417–418
importance, 416
modes of transmission, 416–417, 417t
Cognition evolution, 480
abstraction, 484
architecture of human brain, 480
causal reasoning, 485–486
consciousness, 482–483
cooperation, 487
economic mind, 486–487
human intelligence, 480–481
language, 483–484
multitude of proposals, 482
primates, 480
productive approach, 480
theory of mind, 485
timeline of phylogenetic distance within primate order, 481f
Cold stress. *See also* Heat stress
acclimation, 256–258
physiological adaptations to, 256
responses among indigenous arctic populations, 258–261
Colugos, 40
Columbian exchange, 356
Commensal pathway, 338–339
Commercial ancestry estimation, 521–522
Community average blood pressures, 384–385, 385f
Comparative cultural data, 526, 527t
Comparative studies, evidence from, 494
human speech, 494–495
speech capacities of species, 497
speech encoding, 495–496
SVT, 495–496
Complexity, 21
Computed tomography (CT), 236
Consciousness, 482–483
Consensus, 351
Consequentialists, 510
Contemporary human genetic variation
genetic distance, 214–215
genetic diversity, 213
genetic variation level, 213
higher genetic variation, 213
human biology, 212
human evolution, 212–213
human genome, 215–216
low levels of genetic differentiation, 213–214
natural selection, 216–217
neutral variation, 212
Continental drift, 86
Cooperation, 65–66, 487
Cooperative breeding, 288
Corpus luteum, 298
Cortisol, 442, 446–447
CR. *See* Caloric restriction (CR)
"Cradle of Humankind", 88
Cranial morphology, 181–182
Creation science. *See* Scientific creationism
Creationism, 18. *See also* Scientific creationism
Creutzfeldt–Jakob disease (CJD), 345
Crop
domestication, 350–351, 359
geneticists, 350–351
Crop evolution, 357. *See also* Agriculturalism
coevolution, 359
crop choice, 357–358
crop mimics, 358
loss of competitive, protective, and dispersal functions, 358–359
palatability, 360
speciation, 359
unconscious *vs.* conscious evolution, 358
Cross-cultural variation, 526

CRSC. *See* Center for Science and Culture (CRSC)
CT. *See* Computed tomography (CT)
Cuisine, 340–342
Cultural
change, 526–528
consensus, 388–389
consonance, 388–389
models, 386–388
Culture evolution. *See also* Language evolution; Moral systems evolution
comparative cultural data, 526
cultural transition rates, 528–530, 529t
human expansions, 526
phylogenetic comparative methods, 526–528

D

Dairying, 341
Dark skin, 276
Darwin, Charles, 506. *See also* Evolutionary theory; Natural selection
Darwin's theory of evolution, 4, 23
sexual selection, 8
Darwin-Bateman effect, 469–470
De-localization process, 386
"Deep skull of Niah", 178
Defect (D), 9
7-dehydrocholesterol (7DHC), 274
"Delayed expansion" hypothesis, 222
Demographic characteristics, 327–328
Denisovans, 224–225
population, 216
Dentition, 75–76
Deontologists, 510
Descent with modification, 16, 20–21
Detection–assessment, 444–445
Developmental acclimatization, 252
Developmental mismatch, 374–377
Developmental Origins of Health and Disease (DOHaD), 370–371
Developmental plasticity, 371–372
Developmental processes, 448–449
7DHC. *See* 7-dehydrocholesterol (7DHC)
Diabetes, 370, 374
Diet, 60, 75–76, 186, 328–329, 340–342, 393
agricultural transition, 396–398
in hominin evolution, 394
bipedalism, 394
brain and body size, 394–396
human biology, industrialization of diet and consequences for, 398–399
paleolithic prescriptions, 399–402
Dietary intake of milk, 341–342
Differential mortality, 323, 327–328
Diffuse idiopathic skeletal hyperostosis (DISH), 428–429
Digestive system, 51–52
Digestive tract, 395
Direct benefits, 469
Direct evidence, 236–237
Directed pathway, 338–339
Directional selection, 7
Directive No.15 taxonomy, 520
Discriminative care, 473
male testosterone and care, 474
oxytocin, 473–474
Disease, 331–332
of civilization, 431–432
DISH. *See* Diffuse idiopathic skeletal hyperostosis (DISH)
Dispersal, 64
Disposable Soma theory, 313–314
Divergence, 5, 11–12
Diversity, 4, 13
DNA methylation, 372–373
DNAPrint Genomics, 521–522
DOHaD. *See* Developmental Origins of Health and Disease (DOHaD)
Domestication process, 338
Dominance, 61–62

E

EA. *See* Ethnographic Atlas (EA)
Early experience, 449–450
Early hominins. *See also* Hominin
adaptations, 124–125
African vegetation map, 91f
chimpanzee morphology, 116–118
chimpanzee tool as model for, 196–197
early hominin record, 86
features of critical species, 119t
geological evidence, 89–94
hominin adaptive radiations, 88f
human evolution patterns, 86
inferring hominin habitats and adaptations, 94–95
Late Pliocene adaptive radiations, 96–98
limits and potential of paleoecological knowledge, 98
macro paleoenvironmental context, 86–89
Miocene apes
late, 120
middle, 118–120
place in nature, 125–126
purported early hominin species, 114
morphology, 120–124
reconstructing terrestrial habitats, 89
and related species, 116f
taxonomy, 114, 115t
Early Pleistocene, 144–148, 151, 155–156
Early Possible Hominins (EPHs), 114
Early Upper Paleolithic (EUP), 182
East Asia, Archaic *Homo*, 171
EBV. *See* Epstein–Barr virus (EBV)
Eccrine gland destruction, 274
Ecological entrepreneurs, 96–98
Ecological imperialism, 356
Economic mind, 486–487
Edge species, 77
Education, 16–17
EEA. *See* Environment of evolutionary adaptedness (EEA)
Emerging infectious diseases, 406, 411
Encephalization coefficient (E.Q.), 240–241
Encephalization quotient (EQ), 480–481
ENCODE project. *See* Encyclopedia of DNA Elements project (ENCODE project)
Encoding, 494
Encyclopedia of DNA Elements project (ENCODE project), 375
Endocasts, 236
Energy conservation, 318
Enhancer effect, 343
Environment of evolutionary adaptedness (EEA), 399–400
Environmental determinism, 328
Environmental stressors, physiological adaptations to. *See also* Cold stress; Heat stress; High altitude hypoxia
body morphology, 253–256
climate morphology, 253–256
long-term studies, 252
thermoregulation principles, 252–253
EPC. *See* Extra-pair copulation (EPC)
EPHs. *See* Early Possible Hominins (EPHs)
Epigenetic mechanisms, 449
Epigenetics, 372–373
human evidence, 373–374
nonhuman animal evidence, 373
Episodic memory, 482–483
"Episodic-*like*" memory, 482–483
EPP. *See* Extra-pair paternity (EPP)
Epstein–Barr virus (EBV), 388
EQ. *See* Encephalization quotient (EQ)
Equatorial climate, 87
Erythemogenic UV-B, 277
Estrus, 474–475, 507
Ethnicity, 516
social constructions, 516
Ethnographic Atlas (EA), 526
Euarchonta, 40
Euarchontoglires, 40
Eugenics, 517
EUP. *See* Early Upper Paleolithic (EUP)
Euprimates, 37
Europe, 178, 181, 183, 185
Archaic *Homo*, 169
neolithic sites and autosomal genetic variation in, 226f
European farmers, 225
Eusociality, 511–512
Evolution of tool usage
Acheulean and contemporaneous industries, 201–202
chimpanzee tool, 196–197
earliest known stone tools, 197
hominin forms and technologies, 194–196, 195t
humans, 194
later developments, 205–206
Later Stone Age, 204–205
Middle Palaeolithic, 202–204
Middle Stone Age, 202–204
Oldowan tools, 197–201
tools in animal world, 196
Upper Paleolithic, 204–205
Evolutionary mismatch, 350, 354–355, 374–377
Evolutionary models in anthropology, 353–354
Evolutionary science, 17–18, 20, 23–24

Evolutionary theory. *See also* Darwin, Charles
- Evolutionary Synthesis, 4
- genetic drift, 5–6
- levels of selection, 9–10
- from microevolution to macroevolution, 12–13
- natural selection, 6–9
- origin of genetic variation, 4
- speciation, 10–12
- today, 13
- variation within populations, 4–5

Evolutionary tree, 33, 38–40
Evolutionary trend, 12
Excessive nutrition, 376–377
Expensive tissue hypothesis, 200, 395
Extant primates, 32–33, 46
Extended kinship, 459–460
Extended sexuality, 474–475
External features, 49–53
Extra-pair copulation (EPC), 468
Extra-pair mating system, 468–470
Extra-pair paternity (EPP), 472–473

F

Fair Treatment, 19
Fat-free mass (FFM), 253
Fathers of human family, 458–459
Feeding competition, 58–60
Female
- estrus, 474–475
- primates, 52
- reproductive ecology, 299–302
- reproductive senescence, 316–318
  - aging eggs, 317
  - aging mothers, 317
  - energy conservation, 318
  - grandmother hypothesis, 317–318
  - menopause, 318
  - reproductive cancers prevention, 318

Femur, 107–108, 134–135
Fermentation, 51–52, 361
Fertile crescent, 351, 353
FFM. *See* Fat-free mass (FFM)
Fiber, 396–397
Filaggrin (FLG), 279
Finnish adults, 449
First globalizations, 410–411
Fitness, 66–67, 370–372
Fitness components, 8–9
FLG. *See* Filaggrin (FLG)
fMRI. *See* functional magnetic resonance imaging (fMRI)
Folate degradation, 274–276
Follicle-stimulating hormone (FSH), 298
Food, 340–342
- distribution, 60
- food-carriage hypothesis, 109–110
- resource hypothesis, 343

Foraging/ecological hypothesis, 480
Foregut fermentation, 52
Forelimb muscles of primates, 48–49
Formant frequencies, 495
Fossa (*Cryptoprocta ferox*), 60–61
Fossil(s), 91–94
- and living primates, 79–80
- primates, 37–38

Founder effect, 213
FOXP2 gene, 500
Frequency of phonation, 494–495
Fresh milk, 341
Frozen ancestors, 34
FSH. *See* Follicle-stimulating hormone (FSH)
functional magnetic resonance imaging (fMRI), 461, 498
Funerary behavior, 187–189

G

G6PD. *See* Glucose-6-phosphate-dehydrogenase (G6PD)
GDM. *See* Gestational diabetes mellitus (GDM)
Gender differences, 450
Gene–culture coevolution, 194
Genetic admixture between archaic and modern humans, 222–225
Genetic distance, 214–215
Genetic diversity, 213, 223f
Genetic drift, 5–6, 11, 212–213, 323, 327–328
Genetic sciences, 519–520
Genetic variation, 6, 221, 224–225, 227
Genetics, 516, 519, 521
Genic selection, 10
Genomics, 521–522
Genotoxic hypothesis, 274
Gentle lemurs (*Hapalemur* species), 33
Genus *Homo*, 144, 220. *See also* Archaic *Homo*
- association with stone tools, 155–156
- context of origins and existence, 154–155
- early *Homo*
  - discovery of fossil, 144–145
  - musculoskeletal adaptations in, 155
  - phyletic origins and dating, 153–154
- taxonomic and phylogenetic issues, 151–153
- taxonomic diversity in, 145–151

Geochemistry, 90
Geographical distribution, 32–33
- Archaic *Homo*, 164
- extant nonhuman primates, 32f

Geographical parameters, 87–88
Geological evidence, 89–94
Gestational diabetes mellitus (GDM), 376–377
Gestational weight gain (GWG), 376
Glucose-6-phosphate-dehydrogenase (G6PD), 420
Gluten, 401
"Good genes" hypothesis, 8–9, 469
Gorilla (*Gorilla*), 49–50
Grandmother hypothesis, 317–318
Grandparents, 459
Gravettian, 186
Great apes, 74, 78–79
Great toe, 107–108
Green Revolution, 357
Green Sahara, 87–88
Group living, 58–61
Growth, 286
- and evolution, 286
- human *vs*. chimpanzee growth, 286–289

Gustatory system, 51
GWG. *See* Gestational weight gain (GWG)

H

H-G. *See* Hunter-gatherer (H-G)
H1N1/2009 pandemic strain, 422
HA. *See* Hemagglutinin (HA)
Hand morphology, 183
Hansen's disease. *See* Leprosy
Haplorhines, 33–36, 38
Hardy–Weinberg theorem (H–W theorem), 4, 7
Hb. *See* Hemoglobin (Hb)
hCG. *See* human chorionic gonadotropin, (hCG)
Head muscles of primates, 48
Health
- care disparities, 516
- race and, 519
- stress and, 450

Heat stress. *See also* Cold stress
- acclimation, 261–262
- physiological adaptations, 261
- responses among indigenous populations of tropics, 262–263

Heel strike, 104, 104f
Hemagglutinin (HA), 421
Hemoglobin (Hb), 264
- Hb E, 420
- Hb S, 419–420

Hemoglobin variants, 418, 420
Hepatitis G virus (HGV), 406–408
Herding, 338–339, 343
hGH. *See* human growth hormone (hGH)
HGV. *See* Hepatitis G virus (HGV)
HIFs. *See* Hypoxia inducible factors (HIFs)
High altitude hypoxia
- acclimation to hypoxic stress, 264–265
- physiological adaptations to, 263
- responses among indigenous high altitude populations, 265–269

HIV, 418
HLAs. *See* Human leukocyte antigens (HLAs)
Homelands, 528f
Homin brain endocasts, 247
Hominids, 114
Hominin, 212. *See also* Early hominins
- forms and technologies, 194–196, 195t
- habitats and adaptations, inferring, 94–95

Hominoid skeleton, 45–46
Homo erectus (*H. erectus*), 75, 78, 148–150, 181, 195–196
Homo ergaster (*H. ergaster*), 150–151
Homo floresiensis (*H. floresiensis*), 429
Homo habilis (*H. habilis*), 144–148, 151–153
Homo neanderthalensis (*H. neanderthalensis*), 220
Homo rudolfensis (*H. rudolfensis*), 145, 148, 151–153
Host–pathogen mechanisms, 417–418
HPA. *See* Hypothalamic–pituitary–adrenal system (HPA)

HPG axis. *See* Hypothalamic–pituitary–gonadal axis (HPG axis)
HPO axis. *See* Hypothalamus–pituitary–ovarian axis (HPO axis)
HPT axis. *See* Hypothalamus–pituitary–testicular axis (HPT axis)
HRE. *See* Human reproductive ecology (HRE)
Human adaptability, 384, 389
Human biology, industrialization of diet and consequences for, 398–399
Human brain, 456–457, 480
  asymmetry, 246–247
  characteristics
    brain organization and reorganization, 241–246
    brain size, absolute and relative, 237–238
    encephalization, 238–241
    fossil hominid brain volumes, 242t
    neandertal cerebral asymmetries, 246f
    neural structures for *Homo sapiens*, 244t
  evolution, 236
    brain size changes in hominid evolution, 248t
    lines of evidence, 236–237
    synthesis, 247–249
  volume, 243
Human cerebral cortex, 243
human chorionic gonadotropin, (hCG), 298
Human cognition, 480–481, 486
Human communities, 459
Human disease, evolutionary perspective on, 370
Human evidence, 373–374
Human evolution, 212–213, 394–395, 480
  patterns, 86
  primate models for
    dentition, 75–76
    diet, 75–76
    fossils and living primates, 79–80
    habitat of earliest ancestors, 77–78
    locomotion, 76–77
    macaque model, 78–79
    man the hunted, 80
    models, 74–75
    predation by leopards, 74f
Human expansions, 526
Human family, 456–458
  extended kinship and control of mating relationships, 459–460
  fathers, 458–459
  grandparents, 459
Human genome, 215–216
Human growth
  adolescence, 289
    contributions to reproductive success of adults, 292
    evolution, 289–290
    risks, 292
    separate paths through, 290–291
  chimpanzee growth *vs.*, 286–289
  and development, 286
    growth and evolution, 286
    human life cycle and life history, 287t
    human *vs.* chimpanzee growth, 286–289
human growth hormone (hGH), 330–331
Human history, 16
Human infectious diseases, 405
  diversity, 406
  epidemiological transition
    out of Africa, 406–409
    animal domestication, 409–410
    emergences and homogenization of infectious diseases, 411
    first globalizations, 410–411
  in nonhuman primates, 406
Human intelligence, 480–481
Human language, 483–484
  neural circuits and evolution of, 497
    Broca's area, 497
    circuits linking cortex and basal ganglia, 497–498
    neuroimaging, 498
Human leukocyte antigens (HLAs), 408–409
Human life history, 296–297
Human marital systems, 468, 470–471
  Bonobo mating, 471
  breeding system, 468
  chimpanzee mating, 470–471
  EPP, 472–473
  human case, 471
    blend view, 472
    hunting-as-mating-effort views, 472
    hunting-as-paternal-effort views, 471–472
  human marital systems, 471
  mating in hominoids and ancestral state, 470
  mixed systems, 468
  selection pressures, 469
  sexual selection, 468–469
  traditional view, 468
  variants with stable and unstable unions, 468
Human migration, 227–229
Human niche construction, 345
Human origins, 215
Human population movements, 220
  Out of Africa migration
    anatomically modern humans, 220–222
    archaic humans, 220
  genetic admixture, 222–225
  genetic diversity in worldwide human populations, 223f
  Polynesians, 227–229
  Romani, 227
  technological advances and migration, 225
Human reproductive ecology (HRE), 296
  challenges and future directions, 304
  female reproductive ecology, 299–302
  human life history, 296–297
  human reproductive physiology, 297–299
  life history theory studies, 297f
  testicular function, 302–303
Human reproductive physiology, 297–299
Human speech, 494–495
Human variation, 212
Human–animal interactions, 422
Humans (*Homo sapiens*), 144, 194, 394–395
Hunter-gatherer (H-G), 323
  and evolution, 327–332
  historical era, 325–326
  migration, biogeography, and contemporary populations, 326–327
  subsistence, 324–325
Hunting-as-mating-effort views, 472
Hunting-as-paternal-effort views, 471–472
H–W theorem. *See* Hardy–Weinberg theorem (H–W theorem)
Hydration hypothesis, 343
Hypertrophy, 241
Hypobaric hypoxia, 263
Hypolactasia. *See* Lactose intolerance
Hypothalamic–pituitary–adrenal system (HPA), 461
Hypothalamic–pituitary–gonadal axis (HPG axis), 289, 443, 445–446
Hypothalamus–pituitary–ovarian axis (HPO axis), 297
Hypothalamus–pituitary–testicular axis (HPT axis), 297
Hypoxia inducible factors (HIFs), 268–269
Hypoxic stress, acclimation to, 264–265

I

IAVs. *See* Influenza A viruses (IAVs)
ICAZ. *See* International Council for Archaeozoology (ICAZ)
Ice cores, 89
ID. *See* Intelligent design (ID)
IGF-I. *See* Insulin-like growth factor I (IGF-I)
IMI. *See* Intermembral index (IMI)
Immunocompetence, 388, 390
Incipient tillers, 326–327
Indigenous arctic populations, responses among, 258–261
Indigenous high altitude populations, responses among, 265–269
Indigenous populations of tropics, responses among, 262–263
Indirect evidence, 237
Indirect reciprocity, 509. *See also* Network reciprocity; Strong reciprocity
Industrial revolution, 357
Industrialization/globalization of diet, 393, 398–399
Infancy, 286–288
Infant-carriage hypothesis, 109
Infectious diseases, 418–419, 419t
Influenza A viruses (IAVs), 421, 421t
Influenza viruses evolution, 421–422
Information theory, 21–22. *See also* Evolutionary theory
"Inheritance" of maternal ranks, 62
Initial Upper Paleolithic (IUP), 178–180
Insulin-like growth factor I (IGF-I), 330–331
Intelligent design (ID), 16–17, 20
  pillar 1, 20–23
  pillar 2, 23–24
  pillar 3, 24
  Smoking Gun, 22f
  sputters along, 24–25
Interacting species, 421–422
Intermembral index (IMI), 117
Internal organs, 49–53
International Council for Archaeozoology (ICAZ), 434
International HapMap Project, 521
Intersexual mutualism, 64–65

Isolation by distance, 215
IUP. *See* Initial Upper Paleolithic (IUP)

J

Jacobson's organ, 35–36
Juvenility, 286

K

Kenyanthropus (*Australopithecus*) *platyops* (*A.* (*K.*) *platyops*), 136–137
Kenyanthropus platyops (*K. platyops*), 154
Killer-ape hypothesis, 109
Kin selection, 10, 65, 506, 511–512
Kinship, 455, 457–459
Knuckle walking, 45–46
Koala lemurs (*Megaladipis*), 44
kya. *See* thousand years ago (kya)

L

Lactase enzyme (LCT enzyme), 397–398
Lactase persistence, 342
Lactase-phlorizin hydrolase (LPH), 342
Lactation, 301
Lactose, 397
  intolerance, 342
  lactose tolerance, milk and evolutionary basis for, 342–343
"Landscape archaeology" studies, 90–91
Language, 189, 483–484
  families, 526, 529
Language evolution. *See also* Culture evolution; Moral systems evolution
  comparative studies, evidence from, 494–497
  fully human linguistic and cognitive capability, 500–501
    brain size, 499
    transcriptional genes, 500
  human language, neural circuits and evolution of, 497–498
  universal grammar, implausibility of, 498–499
Last common ancestor (LCA), 114
Late Pliocene adaptive radiations, 96
  early homo, 96–98
  later "robust" hominins, 96
Late Stone Age (LSA), 181–182
Late Upper Paleolithic (LUP), 182
Late-acting genes, 314–315
Later Stone Age, 204–205
LCA. *See* Last common ancestor (LCA)
LCT enzyme. *See* Lactase enzyme (LCT enzyme)
Leprosy, 411, 430
Levels of selection, 9–10
Leverage, 61–62
Lewis Hunting Phenomenon, 256–257
LH. *See* Luteinizing hormone (LH)
Life expectancy, 310
Life history theory, 442–443
Lifestyle incongruity, 387
Livestock, 337, 345
Locomotion, 34, 107–110
Locomotor efficiency hypothesis, 110
Long-tailed macaques (*Macaca fascicularis*), 78–79, 79f
Longevity, 314, 318
Lordosis, 105
Louisiana Serial Killer, 521–522
LPH. *See* Lactase-phlorizin hydrolase (LPH)
LSA. *See* Late Stone Age (LSA)
Lunate sulcus, 237, 245
LUP. *See* Late Upper Paleolithic (LUP)
Luteinizing hormone (LH), 298

M

Macaque model, 78–79
Macaques, 75, 78
Macro paleoenvironmental context, 86–89
Macrobotanical fossils, 92
Macroevolution, 12–13
  equilibrium theory, 164
Magnetic resonance imaging (MRI), 245–246, 497
Major histocompatibility complex (MHC), 473
Malaria, genetic adaptations to, 419–421
Male adaptations for pair-bonding and parenting, 473
Male primates, 36
Male reproductive ecology, 302
Male reproductive endocrinology, 297
Male reproductive senescence, 303, 316
Maltase–glucoamylase gene (MGAM gene), 397
"Man the hunted" model, 80
"Man the Hunter" model, 75
Marriage, 459
Matai, 385
Mate guarding, 64
Maternal constraint, 375–376
Maternal nutrition, 370–371, 373–374
Maternal undernutrition rat model, 372
MC1R. *See* Melanocortin 1 receptor (MC1R)
McCarthy, Robert, 501
Measles virus (MeV), 344
Melanocortin 1 receptor (MC1R), 274
Melanocytes, 274, 276
Menarche, 299–300
Menopause, 301–302, 316, 318
Mental time travel, 482–483
Mesolithic stage, 205
Messinian Salinity Crisis, 86
Metabolic disease, 373–375
Metabolic energy conservation, 279–280
Metabolism, 394, 399–400
Metacognition, 482
Methylation, 372–373
MeV. *See* Measles virus (MeV)
MGAM gene. *See* Maltase–glucoamylase gene (MGAM gene)
MHC. *See* Major histocompatibility complex (MHC)
Microbotanical fossils, 92
Microevolution, 12–13
Middle Palaeolithic, 202–204
Middle Stone Age, 202–204
Midlatitude climate, 87–88
Migration, 384–386, 390–391
  Polynesians, 227–229
  Romani, 227
Milk, 342–343, 394, 398
million years ago (mya), 86, 169, 220, 236, 394
Miocene apes
  late, 120
  middle, 118–120
mitochondrial DNA (mtDNA), 173, 181, 213, 221, 261
Mixed systems, 468
Mobility, 186
Mode and tempo, 527–528
Modern humans, genetic admixture between archaic and modern humans, 222–225
Modern Hunter-gatherer, 325
Modernization, 383–384, 387–388
  critiques of studies of, 386–387
  and disease, 384–386
  new approaches, 387–390
Molecular evidence, 33–34, 40
Molecular evolution neutral theory, 6
Monogamy, 468–469
Moral behavior, 506–507
Moral psychology evolution, 511
Moral sentiments, 510
  moral psychology evolution, 511
  morality and
    development, 511
    neurophysiology, 510–511
Moral systems evolution, 507. *See also* Culture evolution; Language evolution
  indirect reciprocity, 509
  moral behavior, 506–507
  moral sentiments, 510–511
  morality
    and levels of selection, 511–512
    and social organization, 509
  network reciprocity, 507–508
  strong reciprocity, 508–509
Morality, 509
  and development, 511
  and neurophysiology, 510–511
Morbidity, 331–332
Morphogenetic factor, 49
Mortality rates, 327
Mother–offspring relationship, 460
Mounted hunters, 326–327
MRI. *See* Magnetic resonance imaging (MRI)
MRM. *See* Multiregional models (MRM)
mtDNA. *See* mitochondrial DNA (mtDNA)
Multiregional hypothesis, 220–221
Multiregional models (MRM), 181
Muscles
  head and neck muscles, 48
  pectoral and forelimb muscles, 48–49
Mutation, 4–6, 8, 12–13, 212
  accumulation, 314–315
Mutual mate choice, 473
Mutualisms, 65–66, 506–507
mya. *See* million years ago (mya)
Mycobacterioses, 429–430
Mycobacterium species, 430
Myology, 46. *See also* Osteology
  head, neck, pectoral, and forelimb muscles, 47t
  head and neck muscles of primates, 48
  pectoral and forelimb muscles of primates, 48–49
  primate muscles evolution, 46–48
  variation and evolutionary history of primate muscles, 49

N

NA. *See* Neuraminidase (NA)
Natural selection, 6–7, 216–217, 342, 419, 498–499. *See also* Darwin, Charles; Evolutionary theory
  fitness components, 8–9
  modeling adaptation, 9
  modes of selection, 7–8
NE. *See* Norepinephrine (NE)
Neandertals, 215–216, 245
Neanderthals, 164, 169, 172–173
Neck muscles of primates, 48
Negative feedback, 257
Neo-Darwinism, 24
neo-Europes, 356
Neolithic Revolution, 225
Neolithic stage, 205
Nervous system, 50–51
Network reciprocity, 507–508. *See also* Indirect reciprocity; Strong reciprocity
Neural circuits, 494–495, 497–498
Neuraminidase (NA), 421
Neuroendocrine systems, 460
Neurogenomics, 236
Neurological and physiological mechanisms, 460
  hormonal basis for attachment and family love, 460–462
Neurological imaging techniques, 510
Neurophysiology, 510–511
Neutral variation, 212
Next generation sequencing (NGS), 222–224
NGS. *See* Next generation sequencing (NGS)
Niche construction, 355
Niche partitioning, 448
Nomadic pastoralism, 338
Non-pigment-based mechanisms, 279
Nonhuman animal evidence, 373
Nonhuman primates, 483
Nonprimate species, 196
Nonrecombining part of Y chromosome (NRY), 221
Nonshivering thermogenesis (NST), 257
Norepinephrine (NE), 257
NRY. *See* Nonrecombining part of Y chromosome (NRY)
NST. *See* Nonshivering thermogenesis (NST)
Null hypothesis, 6
Nutrition, 328–329

O

Obesity, 369, 373–377
OIS. *See* Oxygen Isotope Stages (OIS)
Old Stone Age. *See* Paleolithic
Oldowan tools, 197–201
Olfactory system, 35–36, 50–51
OOPART. *See* Out-of-place artifact (OOPART)
Open system, 186
Optimization, 9
Orangutan (*Pongo*), 49–50
Orrorin humerus, 122
Orrorin tugenensis (*O. tugenensis*), 114
Osteobiography, 433–434
Osteology, 44–46. *See also* Myology
OT. *See* Oxytocin (OT)
Ouranopithecus, 114, 116, 120–121, 123, 126
"Out of Africa I" hypothesis, 220
Out of Africa migration
  anatomically modern humans, 220–222
  archaic humans, 220
Out of Africa model. *See* Recent African Origin model (RAO model)
Out-of-place artifact (OOPART), 17
Ovarian cycle, 298
Overnutrition, 376
Ovulation, 298
Oxygen Isotope Stages (OIS), 180–181
Oxytocin (OT), 458–459, 473–474, 511

P

Pair-bonding, male adaptations for, 473
PAL. *See* Physical activity levels (PAL)
Palaeolithic. *See* Paleolithic
Palaeopathology. *See* Paleopathology
Palatability, 360
Paleo diet, 350, 399–400
  hypothesis evaluation, 360–361
Paleobiology of anatomically modern *Homo sapiens*
  childbirth, 184
  cranial morphology, 181–182
  diet, 186
  funerary behavior, 187–189
  hand morphology, 183
  health and diseases, 184–185
  language, 189
  life history, 183–184
  mobility, 186
  robusticity, 182–183
  stature, 183
  symbolic behavior, 186–187
  technology, 185–186
Paleoclimate, 87–89
Paleogeography, 86–87
Paleolandscapes, 90–91
Paleolithic, 196–197
Paleolithic prescriptions, 399–402
Paleoneurology, 236
Paleontology, 428
Paleoparasitology, 431
Paleopathology, 427–428
  bioarchaeology, 432–433
  diseases in past, 429–432
  fossil humans and disease, 428–429
  interdisciplinarity and difficulties in communication, 434
  osteobiography, 433–434
Paleosols, 90
Pan troglodytes (*P. troglodytes*), 324
Panins, 117–118
PAR. *See* Predictive adaptive response (PAR)
Paranthropus genus, 131–132
Parenting, 456–457
  male adaptations for, 473
Parkinson disease (PD), 494–495
Pastoralism, 337
  biogeography of, 339–340
  coevolution of livestock and human hosts, 345
  food, diet, and cuisine, 340–342
  health, disease, and, 343–345
  lactose tolerance, milk and evolutionary basis for, 342–343
  prehistory, 338–339
  as subsistence, 338
Paternal care, 461
Paternity, 468
PCA. *See* Principal component analysis (PCA)
PD. *See* Parkinson disease (PD)
Peak height velocity (PHV), 290
Pectoral muscles of primates, 48–49
Pelvis, 105, 107
Penn State Samoa Project, 385
Peripatric speciation, 11
Permissive effect, 303
PET. *See* Positron emission tomography (PET)
PHV. *See* Peak height velocity (PHV)
Phylogenetic comparative methods, 526
  advantage, 527–528
  borrowing problem, 528
  cultural similarities and differences, 526–527
Phylogenetic constraint, 357–358
Phylogenetic factor, 49
Phylogenetic issues, *Homo*, 151–153
Phylogenetic species concept (PSC), 10–11
Phylogenetically ancient limbic system, 444–445
Phylogeny, 12–13
Phylogeography, 526, 528f
Physical activity levels (PAL), 329
Phytoliths, 92
Pigment dilution in modern humans, 277
  cultural theories for, 277–278
  non-pigment-based mechanisms, 279
  for vitamin D, 278–279
Placentation, 36
Plague hypothesis, 423
Plants, 91–93
Platyrrhine skeleton, 45
Plesianthropus transvaalensis (*P. transvaalensis*), 130–131
Polar T3 syndrome, 258
Political economy and health, 389–390
Polyandry, 468
  revolution within behavioral biology, 469–470
Polygamy, 468–469
Polygynandry, 468
Polygyny, 468
Polymorphism, 4
Polynesians, 227–229
Populations, 516–517, 519–520
  genetic techniques, 274
Positional behavior, 104, 109–110
Positron emission tomography (PET), 258, 258f
Postgenomic developments, 520–521
Postorbital wall, 45
Postural feeding hypothesis, 111
Posture, 107–109, 111–112
Postwar debates, 517–518
Power, 61–62
PPARα, 373
Pre-australopithecines, 194
pre-VD3. *See* pre-vitamin D3 (pre-VD3)
pre-vitamin D3 (pre-VD3), 274
Predation, 58–61, 74f, 79–80

Predictive adaptive response (PAR), 371–372
Pregnancy, 300–301
Premaxillary, 116
Prey pathway, 338–339
Primary visual striate cortex (PVC), 236
Primate(s), 480
  behavior
    cooperation, 65–66
    dominance, leverage and power, 61–62
    fitness, 66–67
    group living, 58–61
    intersexual mutualism, 64–65
    order primates, 58
    predation, 58–61
    sexual conflict, 64–65
    social organization, 58–61
    social relationships, 61
    sociality, 66–67
    socioecological model, 62–64, 63f
  comparative anatomy
    external features and internal organs, 49–53
    myology, 46–49
    osteology, 44–46
    phylogenetic relationships, 44f
  evolution, 32
    on ascending scale, 34
    evolutionary relationships, 38–40, 39f
    extant groups, 32–33
    extant nonhuman primates geographical distribution, 32f
    features, 34–37
    fossil, 37–38
    higher-level relationships, 33–34
  fossils and living, 79–80
  muscles evolution, 46–48
  social evolution, 65
Principal component analysis (PCA), 225
Productive approach, 480
Promiscuity, 468
Prosimians, 33–34
PSC. *See* Phylogenetic species concept (PSC)
Psychosocial stress, 384, 388, 390
Puberty, 302
Pulmonary diffusion, 264
Punishment, 508
PVC. *See* Primary visual striate cortex (PVC)

Q
Quantal, 497

R
Race, 516
  and health, 519
  social construction, 518–519
Race-based medicine, 516, 519
Racial profiling, 522
Racial science, 516–517
Radiation, 253
Rank acquisition, 62
RAO model. *See* Recent African Origin model (RAO model)
Reaction norms, 371
Recent African Origin model (RAO model), 181, 220, 406–409
Receptors, 446–447
Reciprocal altruism, 65–66
Reciprocity, 65–66, 507
Recognition, 444
Red Queen hypotheses, 416
"Refuging" species, 79
"Regional Continuity" hypothesis. *See* Multiregional hypothesis
Regional variation, Archaic *Homo*, 168–172
Reinterpretation hypothesis, 485
Relative sitting heights (RSH), 254
Reproductive cancers prevention, 318
Reproductive isolating barriers (RIBs), 10–11
Reproductive isolation (RI), 10–11
Reproductive senescence
  female, 316–318
  male, 316
Reproductive skew, 471
Reproductive system, 52–53
Resource allocation, 443
Response to demand, 442
Rewards, 508–509, 508f, 512
RI. *See* Reproductive isolation (RI)
RIBs. *See* Reproductive isolating barriers (RIBs)
"Ricochet" brachiation, 45–46
Rinderpest virus (RPV), 344
RNA editing, 372–373
"Robust" Australopithecines, 131–133
Robusticity, 120, 182–183
Roma. *See* Romani
Romani, 227
RPV. *See* Rinderpest virus (RPV)
RSH. *See* Relative sitting heights (RSH)
Rubella virus. *See* Measles virus (MeV)

S
SA. *See* Surface area (SA)
Sahelanthropus tchadensis (*S. tchadensis*), 94, 114, 120, 126
Santorum Amendment, 17
Savannas, 94–95
Scanning frequency. *See* Vigilance
Scapula, 111
SCCS. *See* Standard Cross-Cultural Sample (SCCS)
Scheuerman's disease, 428–429
Scientific breeding, 357
Scientific creationism, 16, 19, 23
  atheism, 18–19
  creation science winds down, 19–20
  creation–evolution continuum, 18f
  fairness, 19
  pillars, 17f
  theory in crisis, 17–18
Secondary domestication, 358
Sedentary fisherman, 326–327
Sediment cores, 88–89
Seed eating hypothesis, 110
Segregation distortion, 9–10
Selection pressures, 469
"Selfish gene", 511–512
Senescence, 309, 311–313
  evolution, 313–315
  reproductive, 316–318
Sensory system, 50–51
Serial founder model, 213, 215
Sex differences in optimal social dispersion, diets, and feeding competition, 58–61
Sex-hormone binding globulin (SHBG), 316
Sexual conflict, 64–65, 469
Sexual selection, 8–9, 468–469
  hypothesis, 277–278
  new models of, 470
  theory, 277–278
SHBG. *See* Sex-hormone binding globulin (SHBG)
Shivering thermogenesis, 258
Short tandem repeat polymorphisms (STRs), 221–222
Simian shelf, 116
Simians, 33, 37–38
Single-nucleotide polymorphisms (SNPs), 213, 222, 339–340, 422
Site taphonomy, 89–90
Skin cancer, 274
Skin color evolution
  epidermal pigmentation evolving, 274
    barrier requirements, 276
    eccrine gland destruction, protection against, 274
    erythemogenic UV-B, 277
    folate degradation, protection against, 274–276
    skin cancer, protection against, 274
    sweating, 276–277
    vitamin D toxicity, protection against, 274–276
  metabolic energy conservation, 279–280
  pigment dilution in modern humans, 277–279
Small-object feeding hypothesis, 110–111
Smallpox, 423
SNPs. *See* Single-nucleotide polymorphisms (SNPs)
SNS. *See* Sympathetic nervous system (SNS)
Social bonds, 58, 65–67
Social construction of race, 518–519
Social dispersion, 58–60
Social display hypothesis, 109
Social intelligence hypothesis, 480
Social organization, 58–61, 509
  nonhuman primate, 59t
Social regulation, 447, 449
Social relationships, 61
Social structure, 58
Social support, 388
Social threat, 444
Sociality, 66–67
Sociobiology, 511–512
Socioecological model, 62–64, 63f
South American Hunter-gatherer, 325–326, 331
South Asia, Archaic *Homo*, 171
Southeast Asian Hunter-gatherer, 329–330
Specialization, 480

Speciation, 10–12, 359
Species fitness, 288
Species life history, 296
Species selection, 10
Specified complexity, 21
Speech capacities of species, 497
Speech encoding, 495–496
Spermatogenesis, 302
Spread of Western diet, 398–399
Squirrel monkey (*Saimiri*), 49–50
Standard Cross-Cultural Sample (SCCS), 471, 526
Starch digestion, 396–397
Stature, 183
Status incongruity, 390
Sterkfontein, 130–131
Sticky rice, 361
Stone tools, earliest known, 197
Strepsirrhines, 33–36, 44–45, 50–51
  skeleton, 44–45
Stress, 387–388, 442–443
  adaptations, 442
  and body talks back, 447
  brain talks to body, 445–447
  buffers, 447–448
  detection–assessment, 444–445
  developmental and intergenerational processes, 448–449
    gender differences, 450
    transgenerational effects, 449–450
  endocrine architecture of life history, 443
  and health, 450
  life history, 442–443
  and life history, 450–451
  moderators, 447–448
  response, 445, 445f, 446t
  stressors, 443–444
Strong reciprocity, 508–509. *See also* Indirect reciprocity; Network reciprocity
STRs. *See* Short tandem repeat polymorphisms (STRs)
Structural violence, 433
Sub-Saharan African populations, higher genetic variation in, 213
“Support the high-born” rule, 62
Supralaryngeal vocal tract (SVT), 495–496
  unique human tongue and, 496–497
Supralaryngeal vocal tract horizontal (SVTh), 496
Supralaryngeal vocal tract vertical (SVTv), 496
Supraorbital torus, 116
Surface area (SA), 253
Suspensory behavior, 107–109
Suspensory hypothesis, 109
SVT. *See* Supralaryngeal vocal tract (SVT)
SVTh. *See* Supralaryngeal vocal tract horizontal (SVTh)
SVTv. *See* Supralaryngeal vocal tract vertical (SVTv)
Sweating, 276–277
Sweet rice. *See* Sticky rice
Symbolic behavior, 186–187
Sympathetic nervous system (SNS), 445–446
Syphilis, 430
Systolic blood pressure, 387, 388f

T

t-UCA. *See trans*-urocanic acid (t-UCA)
“Tap foraging” method, 44
Taphonomy, 89–90
Tarsiiform skeleton, 45
Taung Child, 90
Taung specimen, 130–131
“Taxon-free” approach, 93
Taxonomy/taxonomic, 516–517
  diversity in *Homo*, 145–151
  issues in *Homo*, 151–153
Techno-organic evolution, 194
Temperate climate, 88
Tephrochronology, 90
Terrestrial bipedalism, 76
Terrestrial habitats, reconstructing, 89
Testicular function, 302–303
Testosterone, 460–461
  male, 474
  production, 302
Thalassemias, 420
Theory of mind (ToM), 456, 485
Thermal radiation avoidance hypothesis, 110
Thermoregulation principles, 252–253
1000 Genomes Project, 520
thousand years ago (kya), 277
“Thrifty gene” hypothesis, 399–400
Thrifty phenotype hypothesis, 370–371
Thyroid hormones, 257
Thyroid-stimulating hormone (TSH), 257
Thyroxine (T_4), 257
Time of most recent common ancestor (TMRCA), 221
“Tit for tat” strategy, 508
TMRCA. *See* Time of most recent common ancestor (TMRCA)
Tokelau Island Migrant Study, 385–386
ToM. *See* Theory of mind (ToM)
Tool-use hypothesis, 108–109
Tools in animal world, 196
Tooth decay, 431
Trade-offs, 442–443
trans-urocanic acid (t-UCA), 279
Transcriptional genes, 500
Transgenerational effects, 449–450
Transgenerational inheritance, 377
Transmissible spongiform encephalopathies (TSEs), 345
Tree shrews, 34–35, 40
Treponematoses, 429–430
TRH. *See* Tyrotropin-releasing hormone (TRH)
Tri-iodo thyronine (T_3), 257
“Troubled middle”, 18
Tyrotropin-releasing hormone (TRH), 257
TSEs. *See* Transmissible spongiform encephalopathies (TSEs)
TSH. *See* Thyroid-stimulating hormone (TSH)
Tuberculosis, 430, 430f

U

UG. *See* Universal grammar (UG)
Undernutrition, 370–371
Universal grammar (UG), 498
  implausibility of, 498–499
UP. *See* Upper Paleolithic (UP)
Upper Palaeolithic. *See* Upper Paleolithic (UP)
Upper Paleolithic (UP), 182, 204–205
Utilitarians. *See* Consequentialists

V

Vasti, 104
Vegetarian diet, 401
Vegetarianism, 401
Vegetation, 91, 264
Venus figurines, 186–187
Vertical climbing, 110
Vigilance, 444, 446
Virtue ethicists. *See* Deontologists
Vision, 35
Vitamin D
  non-pigment-based mechanisms, 279
  pigment dilution, 278–279
  toxicity, 274–276
  VD3, 274
Voice “cracking”, 289
Volcanic sediments, 90

W

WCST. *See* Wisconsin card sorting task (WCST)
WEIRD societies. *See* Western, Educated, Industrialized, Rich, and Democratic societies (WEIRD societies)
West Asia, Archaic *Homo*, 171
Western, Educated, Industrialized, Rich, and Democratic societies (WEIRD societies), 288
Western Samoa, 385
“What-where-when” memory, 482–483
WHO. *See* World Health Organization (WHO)
Whole-genome sequencing, 521
Wisconsin card sorting task (WCST), 498
Woodlands, 94–95
Working memory, 499
World Health Organization (WHO), 344
World systems theory, 386

Y

Yersinia pestis (*Y. pestis*), 430–431

Z

Zoonotic diseases, 337, 344, 410–411

Color Plates

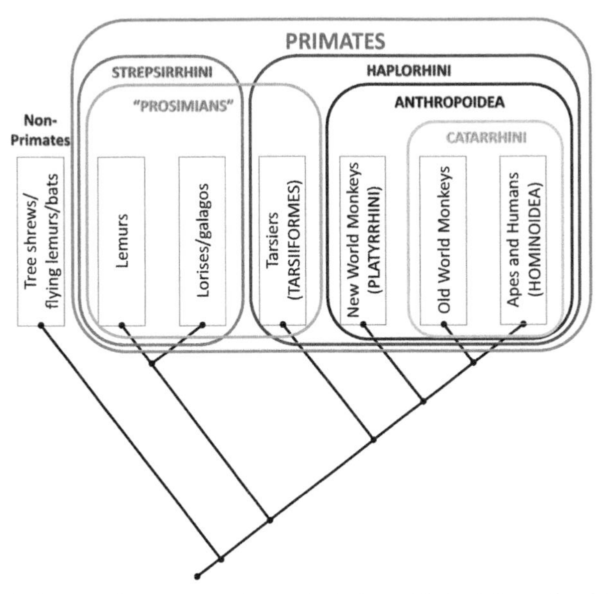

FIGURE 4.1 Scheme showing the main primate clades and their phylogenetic relationships.

FIGURE 6.2 A reconstruction of two *Australopithecus afarensis* (used by permission of American Museum of Natural History AMNH Library 4936[2]).

FIGURE 6.3 Long-tailed macaques (*Macaca fascicularis*).

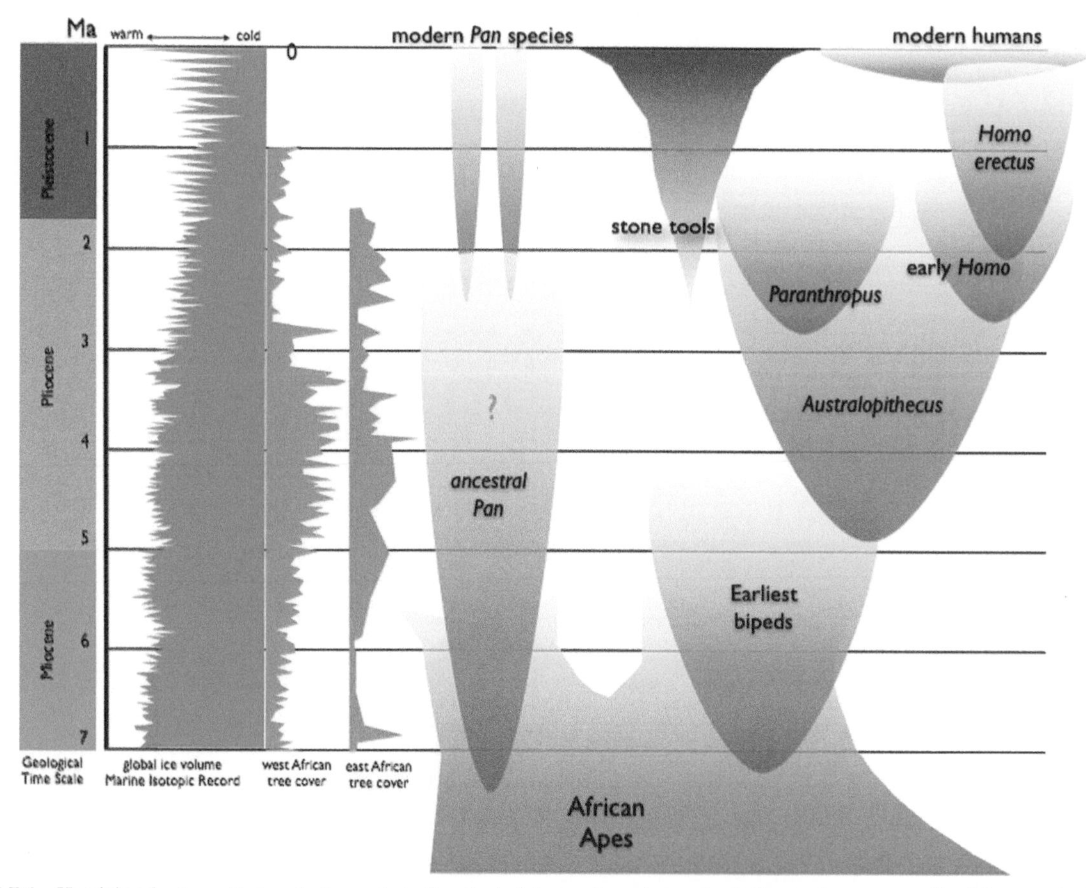

FIGURE 7.1 Hominin adaptive radiations in the context of geological timescale, and records of climate change generalized from marine isotope and paleobotanical evidence, after Bonnefille (2010), Kingston (2007), and Potts (2013).

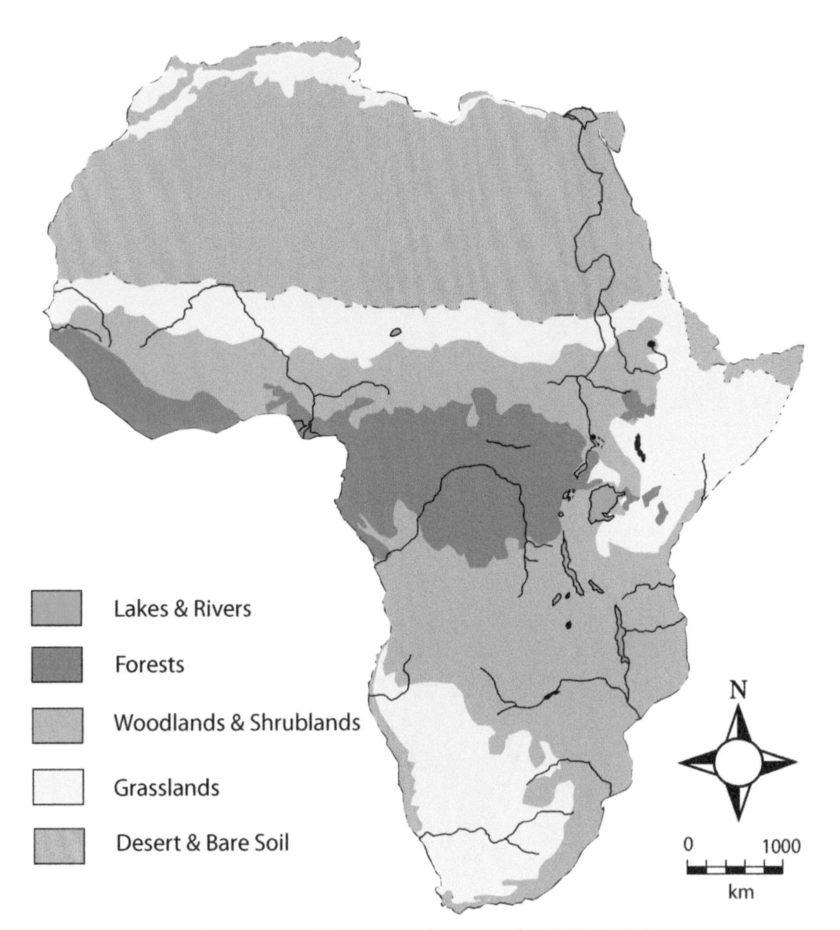

FIGURE 7.2 African vegetation map, after White (1983).

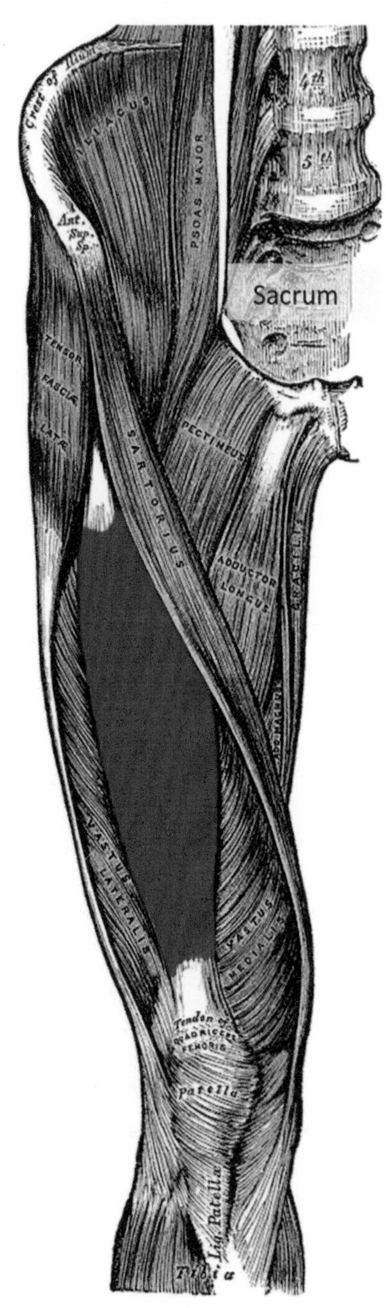

FIGURE 8.2 Muscles of the thigh. *From Gray and Carter (1858).*

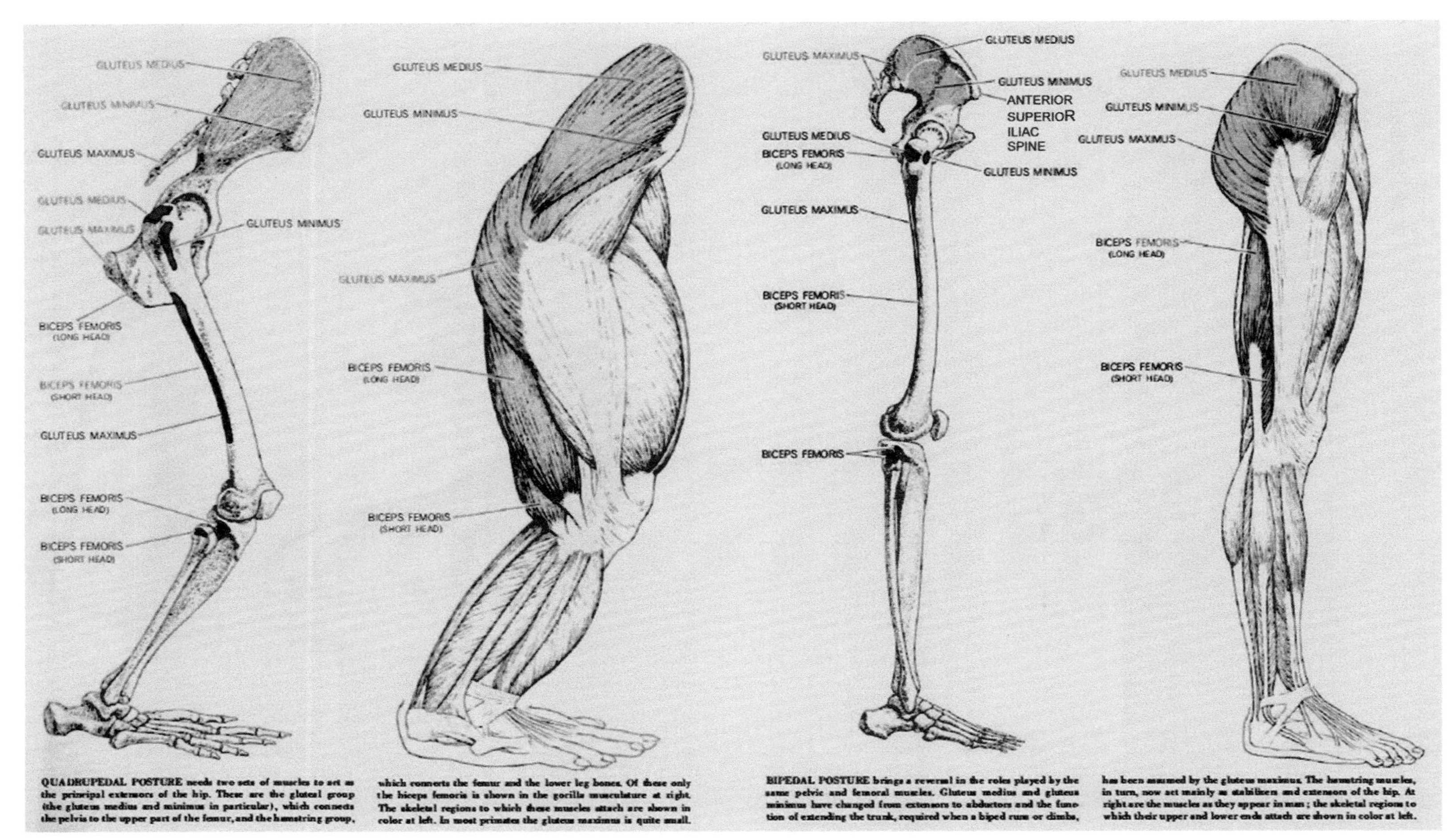

FIGURE 8.3 To maintain a lever arm for hip flexors, compared with the ape condition (left), the human anterior superior iliac spine is pulled forward, helping to reshape the pelvis from blade-shaped to curved and bowl-shaped. *From Napier (1967).*

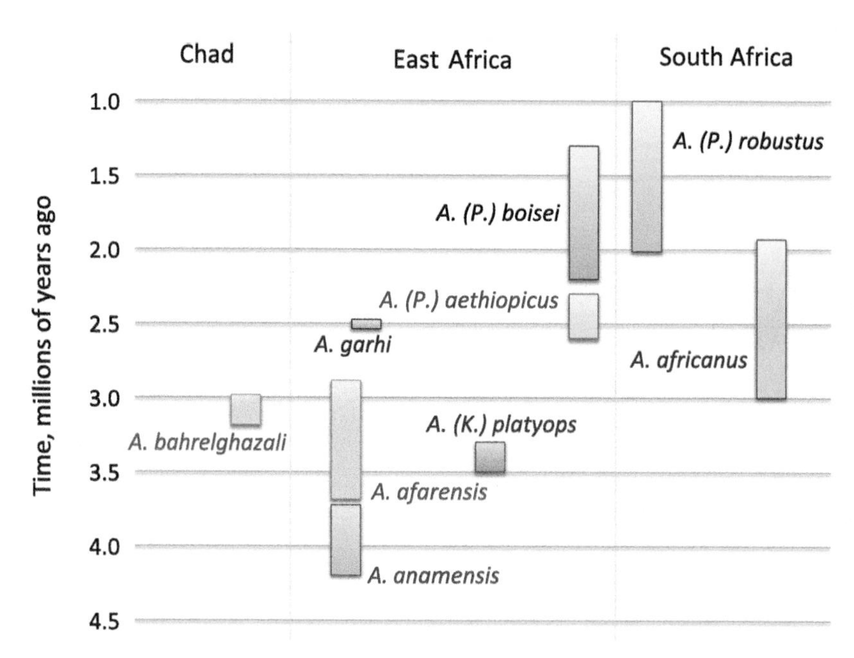

FIGURE 10.1 Timeline of australopithecine species. Bars represent the range of dates for which each species is known. *Dates summarized from Brown et al. (2013) and Herries et al. (2013).*

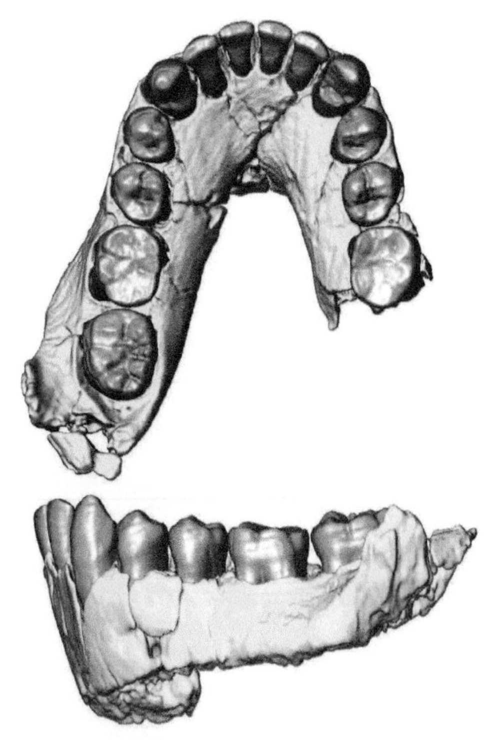

FIGURE 11.1 Computer-based reconstruction of the *Homo habilis* type specimen OH 7 including occlusal (top) and left lateral (bottom) views. *Image used by permission of F. Spoor.*

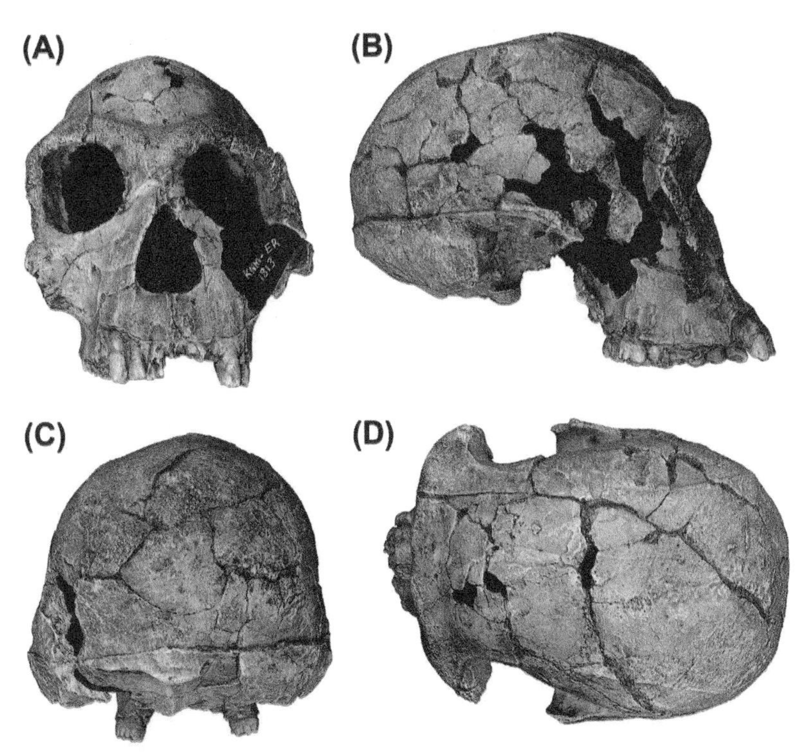

FIGURE 11.2 KNM-ER 1813 cranium (cast) attributed to *Homo habilis*, Turkana Basin, Kenya, 1.86 mya. (A) Facial view, (B) right lateral view, (C) posterior view, (D) superior view.

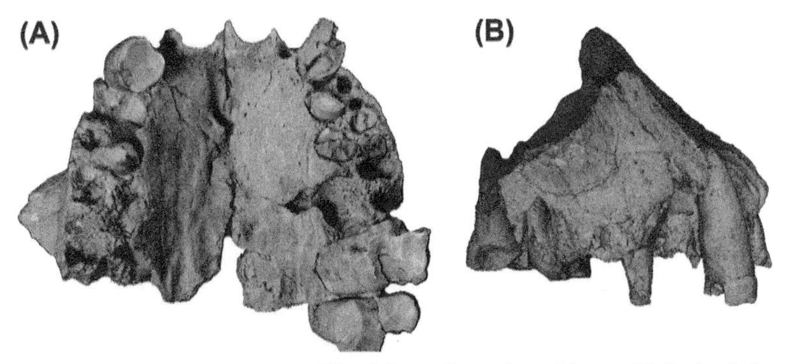

FIGURE 11.3 OH 62 maxilla (cast) attributed to *Homo habilis*, Olduvai Gorge, Tanzania, c. 1.8 mya. (A) Occlusal view, (B) right lateral view.

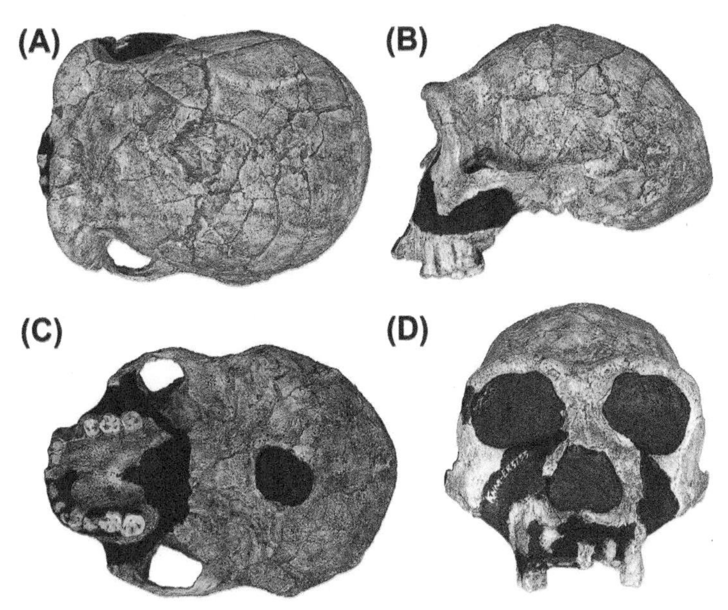

FIGURE 11.4 KNM-ER 3733 cranium attributed to *Homo erectus*, Turkana Basin, Kenya, c. 1.7 mya. (A) Superior view, (B) left lateral view, (C) basal view, (D) facial view.

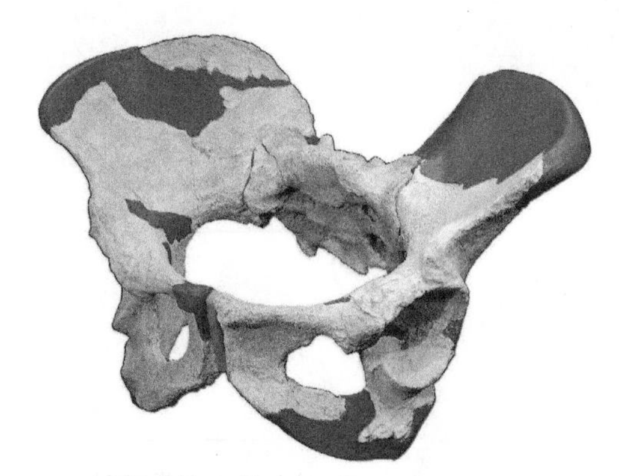

FIGURE 11.5 BSN49/P27 reconstructed pelvis, attributed to *Homo erectus*, Gona, Ethiopia, 0.9–1.4 mya. Oblique view.

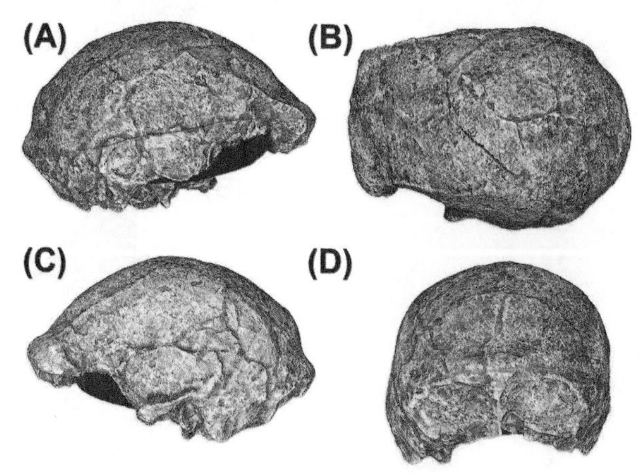

FIGURE 11.6 Ngandong XX cranium attributed to *Homo erectus*, Java, Indonesia. (A) Right lateral view, (B) superior view, (C) left lateral view, (D) posterior view.

FIGURE 11.7 OH 65 maxilla attributed to *Homo habilis*, Olduvai Gorge, Tanzania, 1.79–184 mya. Occlusal view. *Image provided by R. Blumenschine.*

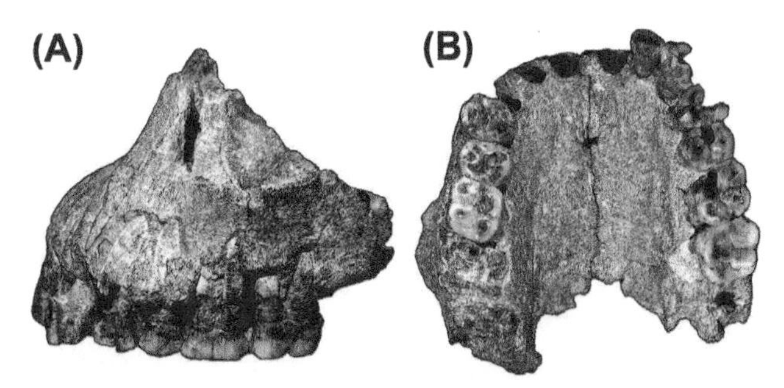

FIGURE 11.8 A.L. 666-1 maxilla attributed to *Homo* cf. *habilis*, Hadar, Ethiopia, 2.33 mya. (A) Left lateral view, (B) occlusal view.

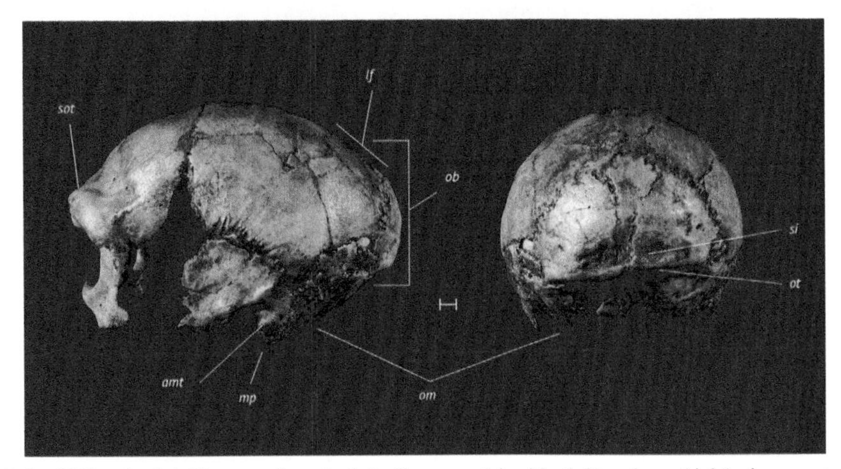

FIGURE 12.8 The La Quina V Neanderthal (European Late Archaic *Homo,* cast) in side (left) and rear (right) views. *amt:* anterior mastoid tubercle, *lf*: lambdoidal flattening, *mp*: mastoid process, *ob*: occipital bun, *om*: occipitomastoid crest, *ot*: occipital torus, *si*: suprainiac fossa, and *sot*: supraorbital torus. The anterior mastoid tubercle, a large occipitomastoid crest, an oblong suprainiac fossa, and an occipital bun with associated lambdoidal flattening are all Neanderthal characteristics.

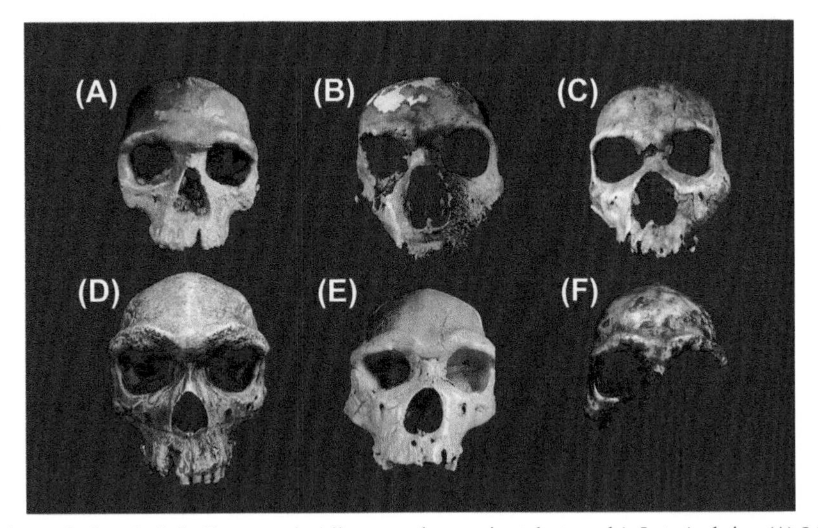

FIGURE 12.9 Anterior views of select Archaic *Homo* crania (all casts and approximately to scale). Late Archaics: (A) Jebel Irhoud (Morocco), (B) Guattari 1 (Italy), and (C) Shanidar 5 (Iraq). Early Archaics: (D) Kabwe (Zambia), (E) Arago 21 (France), and (F) Zuttiyeh (Israel).

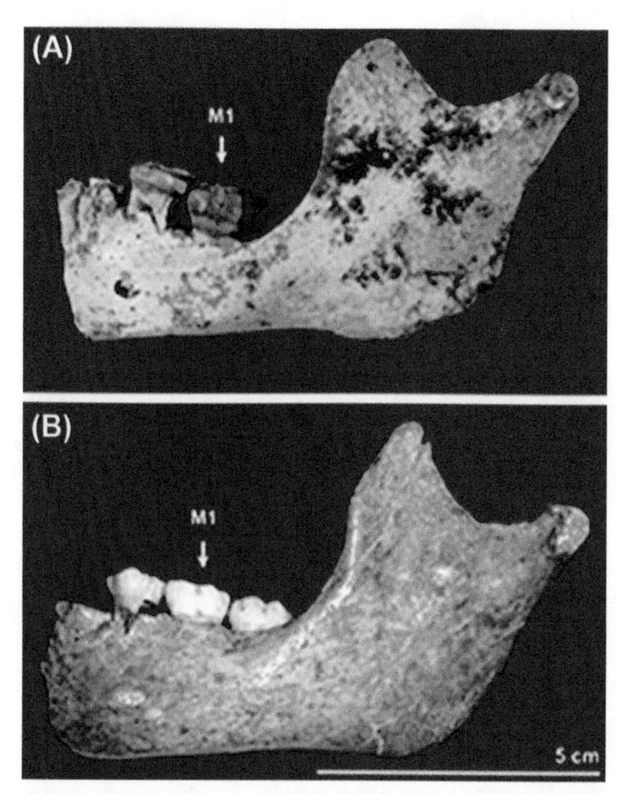

FIGURE 13.4 Comparison of degree of development in the anatomically modern human juvenile from Jebel Irhoud (A) and the Neandertal juvenile from Scladina (B) Both are 8 years old, but the Scladina child already has a second molar. *From Smith et al. (2007); courtesy of the National Academy of Sciences (Figure 3)—Copyright (2013) National Academy of Sciences, USA.*

FIGURE 13.5 A rare example of Paleolithic dental enamel hypoplasia (Grotte des Enfants 6). *Courtesy of the Musée d'Anthropologie Préhistorique de Monaco.*

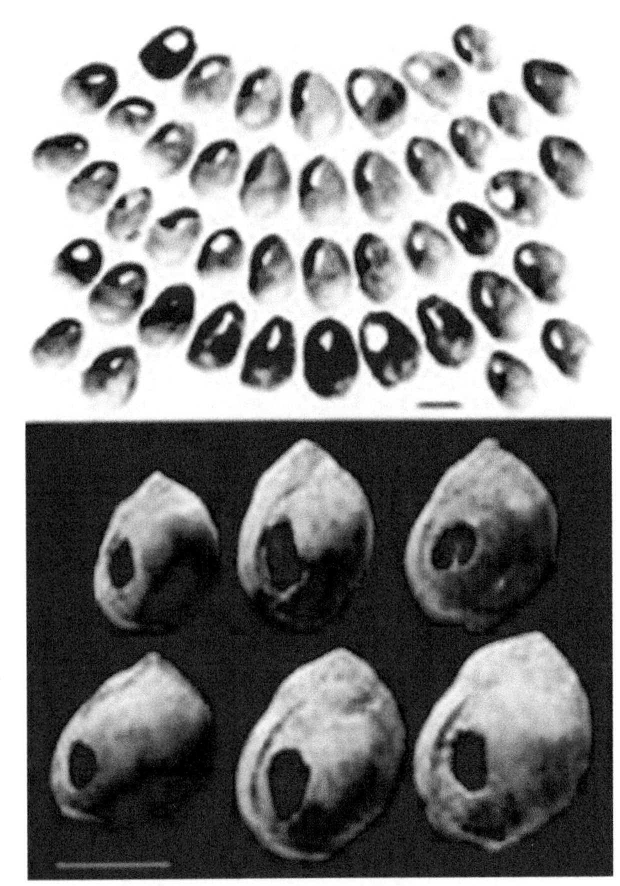

FIGURE 13.9 Perforated marine shell beads from the Middle Stone Age of Blombos Cave (South Africa). Scale bar: 5 mm. *From Henshilwood et al. (2004); Courtesy of the American Association for the Advancement of Science.*

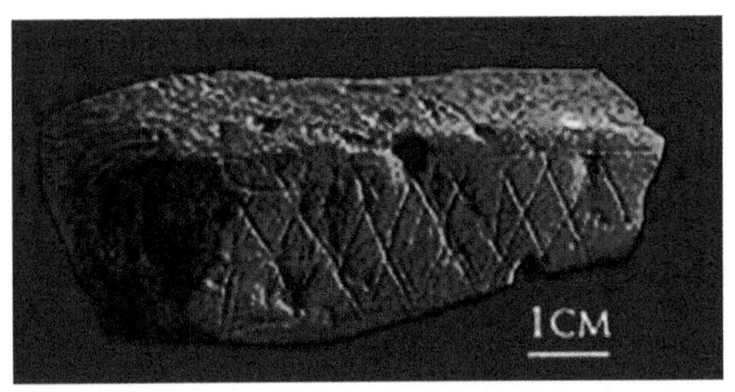

FIGURE 13.10 Engraved red ochre fragment from the Middle Stone Age of Blombos Cave (South Africa).

FIGURE 13.11 Example of Upper Paleolithic (Gravettian) figurine from Willendorf (Austria). These so-called "Venus figures" are found throughout Europe between 28,000 and 20,000 years ago and often exhibit exaggerated sexual characteristics, such as enlarged buttocks, abdomen, and breasts.

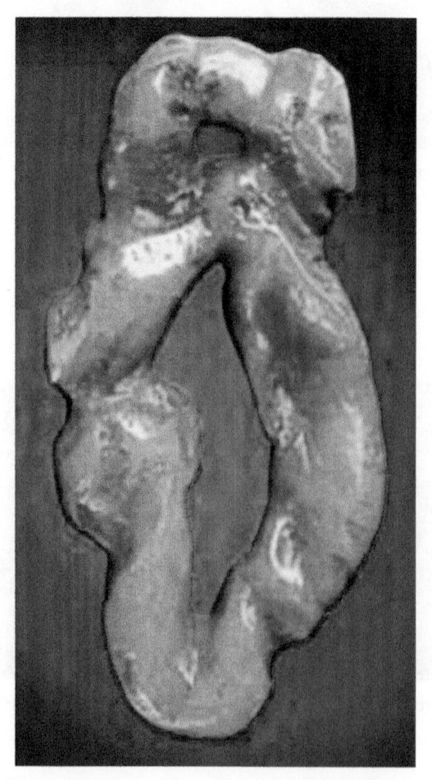

FIGURE 13.12 The "Doublet" figurine from the Upper Paleolithic (Gravettian) site of Balzi Rossi (Italy). Carved from steatite, this figure represents a woman (left) joined at the head, shoulder, and lower limb to a nonhuman being (right), possibly symbolizing a ritualistic transformation. *Alexandre Louis Jullien Collection, copyright Pierre Bolduc.*

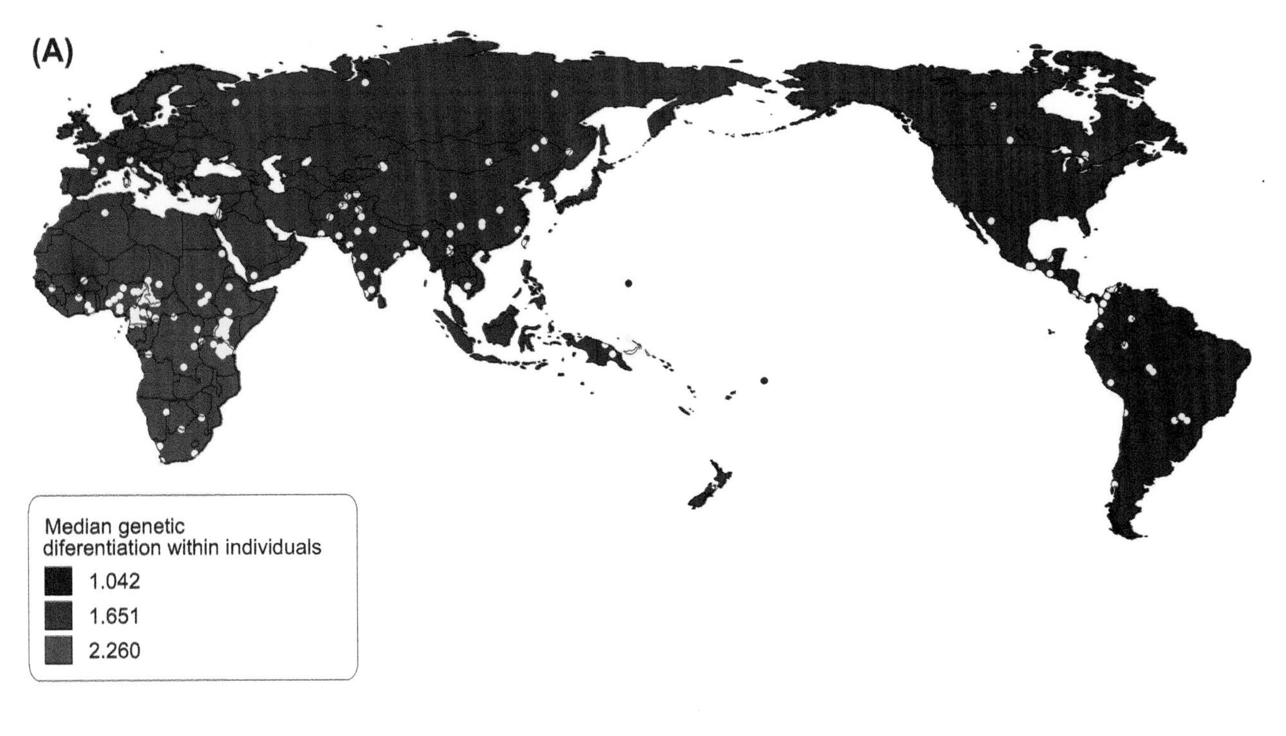

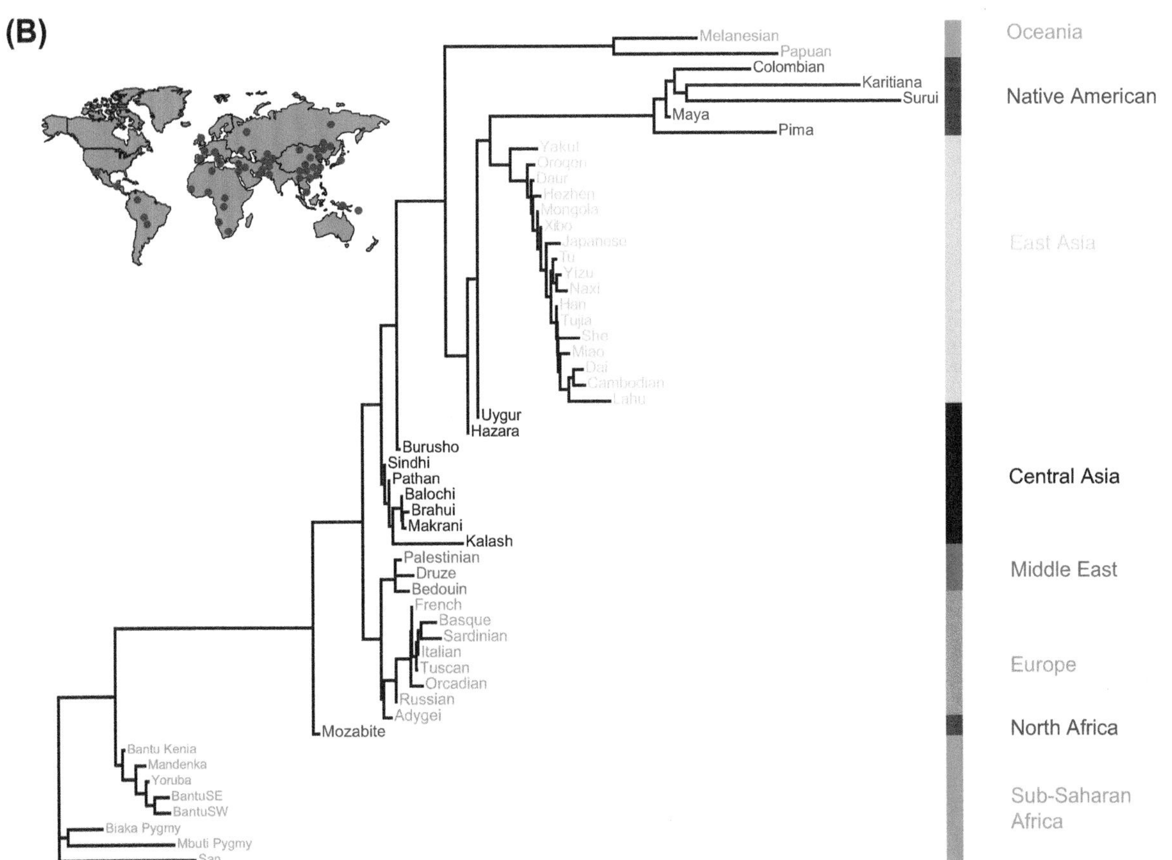

FIGURE 16.2 Genetic diversity in worldwide human populations. (A) Density map showing the genetic differentiation (estimated as Euclidean distance) within individuals from each population using 658 autosomal STRs in 5879 individuals from 250 populations (Pemberton et al., 2013). Sub-Saharan African chromosomes show the greatest genetic differentiation within individuals, and this level decays with geographic distance from the African continent, with the lowest levels occurring in native South American populations. (B) Phylogenetic neighbor-joining tree (Saitou and Nei, 1987) showing the genetic affinities of 53 populations from the HGDP-CEPH panel (Cann et al., 2002) estimated by means of Fst distance (Weir and Cockerham, 1984) on ~650,000 autosomal SNPs in ~1000 individuals from ~50 worldwide populations (Li et al., 2008). Sub-Saharan African populations are separated from the other worldwide populations. Outside the African continent, populations are clustered by sampling continent.

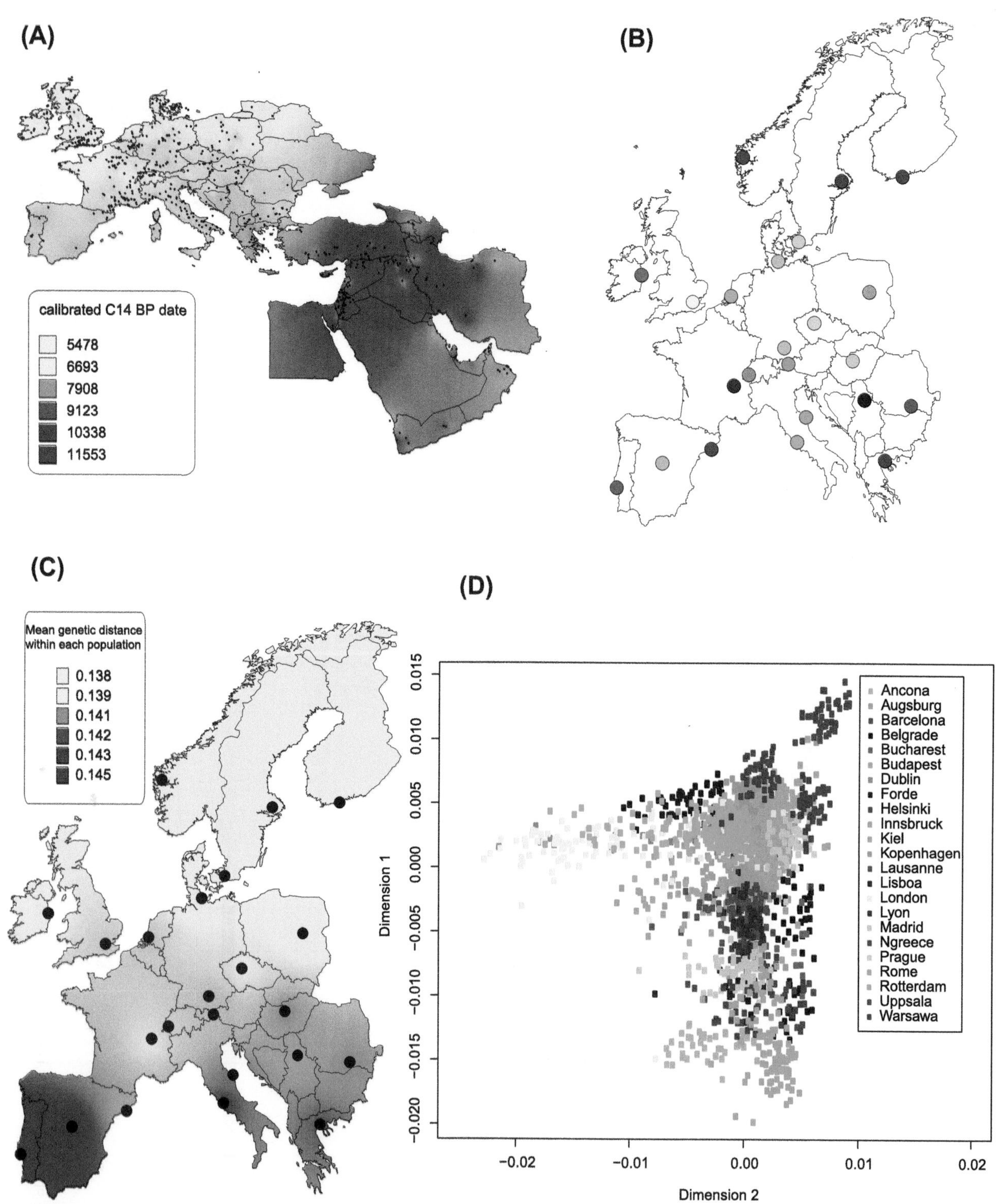

FIGURE 16.4 Neolithic sites and autosomal genetic variation in Europe. (A) Estimated C14 dates of Neolithic sites in Europe. *(Data taken from Pinhasi et al., 2005.)* The oldest Neolithic sites are found in the Fertile Crescent, while younger Neolithic sites are found as one moves away from the Middle East. (B) Geographic location of 23 European sampling sites/subpopulations containing 2457 individuals analyzed by Lao et al. (2008) with ~133,000 autosomal SNPs after linkage disequilibrium pruning. Each dot represents a sampling site or subpopulation. (C) Density map of the mean genetic diversity among individuals from the same European sampling site/subpopulation; within-population genetic differentiation decreases as latitude increases. (D) Plot of the first two dimensions computed by means of a classical multidimensional scaling analysis using a genetic distance (identical by state) matrix between pairs of European individuals; each dot is one individual. The relative position of each individual in the two dimensions resembles the sampling site location.

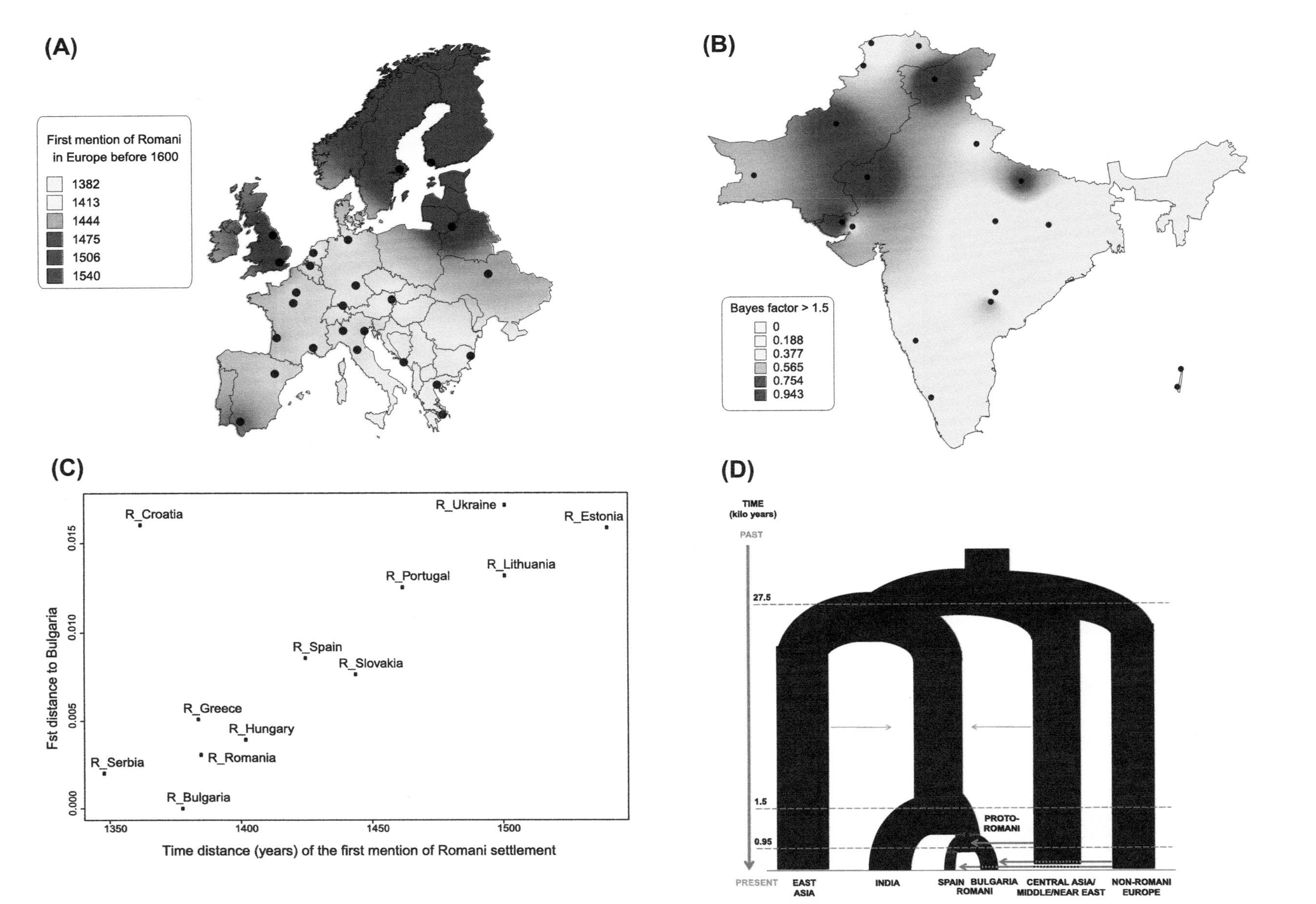

FIGURE 16.5 Analysis of the genetic variation of Romani populations in Europe using SNP data (Mendizabal et al., 2012). (A) Density map based on the first written mention of Romani in Europe before 1600 (Tcherenkov and Laederich, 2004). (B) Density map of Pakistan and India showing the number of times each population (a dot) is supported by statistical modeling as the homeland of the European Romani, compared with the other populations. The Northwestern India region (including Meghawal and Kashmiri Pandit populations) has the highest probability of representing the homeland of the European Romani. (C) Plot between the oldest historical records of the Romani settlements in each European country and the genetic distances (estimated by means of Fst Weir and Cockerham, 1984) between each Romani population and the Balkans (i.e., Bulgaria). (D) Inferred demographic history of the European Romani assuming the Out of Africa model and a generation time of 25 years. Branch width is proportional to the estimated effective population size; the presence of bottlenecks in the European Romani branch are indicated by red lines. Arrow width indicates migration rates, measured as the number of migrant chromosomes per generation from the donor population.

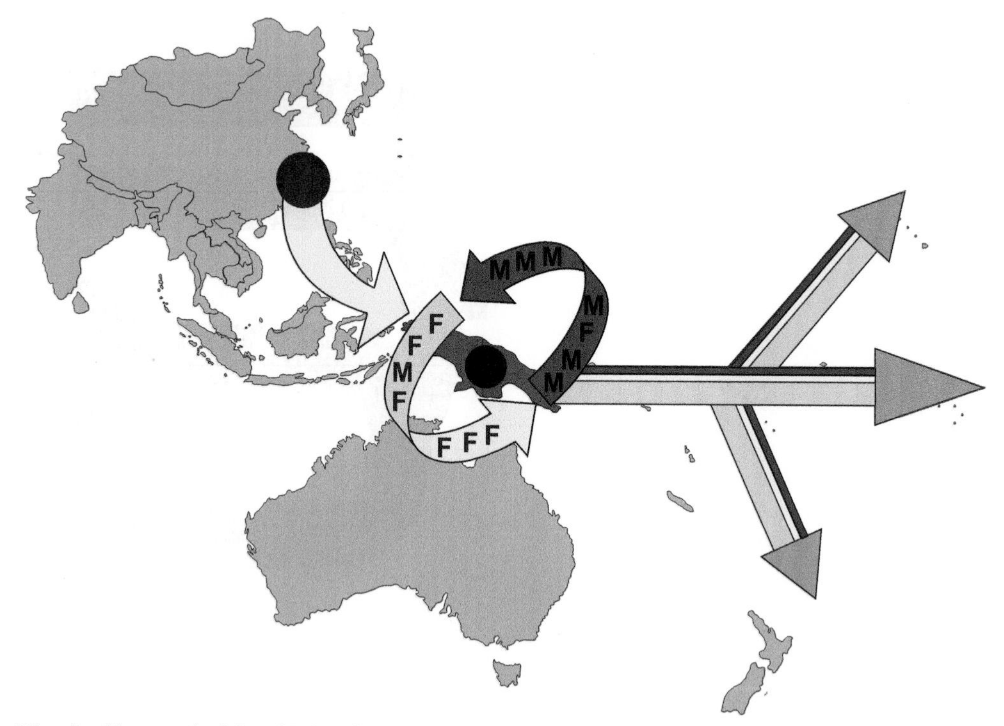

FIGURE 16.6 Migration history and origins of Polynesians proposed by the "Slow Boat" from Asia model (Kayser et al., 2000) according to which Polynesian ancestors originated in East Asia, perhaps Taiwan, and migrated via Southeast Asia toward New Guinea (Kayser et al., 2006, Kayser et al., 2000), likely the Bismarck Archipelago (Kayser et al., 2008a), where they mixed in a sex-biased way that included more Asian women and New Guinean men (Kayser et al., 2006, Kayser et al., 2000, Kayser et al., 2008a, Kayser et al., 2008b; Wollstein et al., 2010), before the admixed population migrated further eastwards to Polynesia.

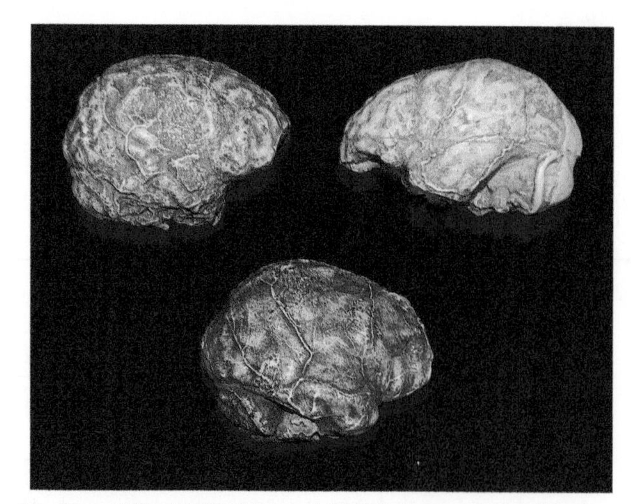

FIGURE 17.1 Casts of the Taung (left), STS60 (right), and SK1585 (bottom) "natural" endocasts of Australopithecines.

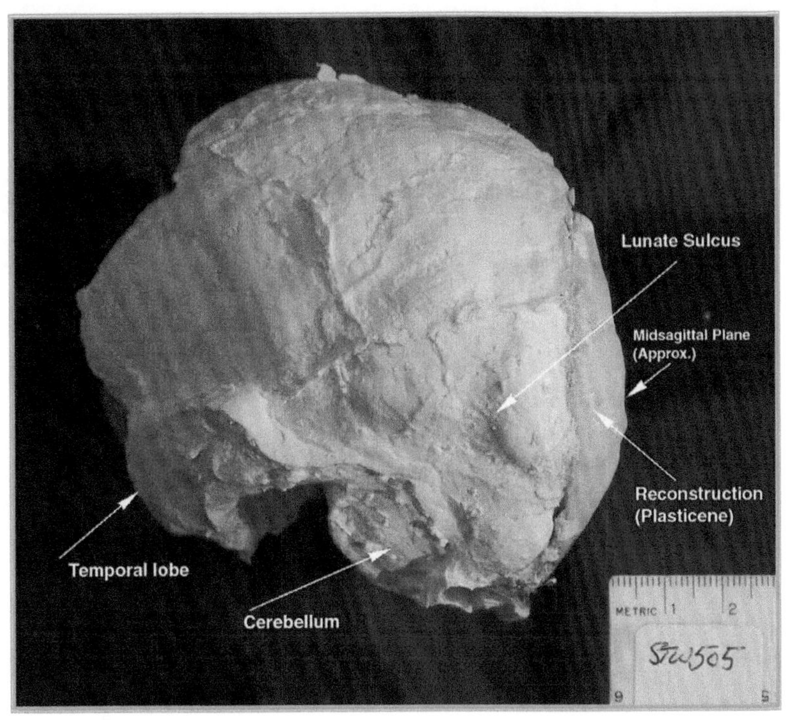

FIGURE 17.4 (B) Oblique view of the Stw 505 *A. africanus* specimen, showing a prominent lunate sulcus in a posterior position. This specimen makes it clear that at least some australopithecines had reduced primary visual cortices and expanded posterior parietal lobes, evidence showing that reorganization probably preceded brain size increases.

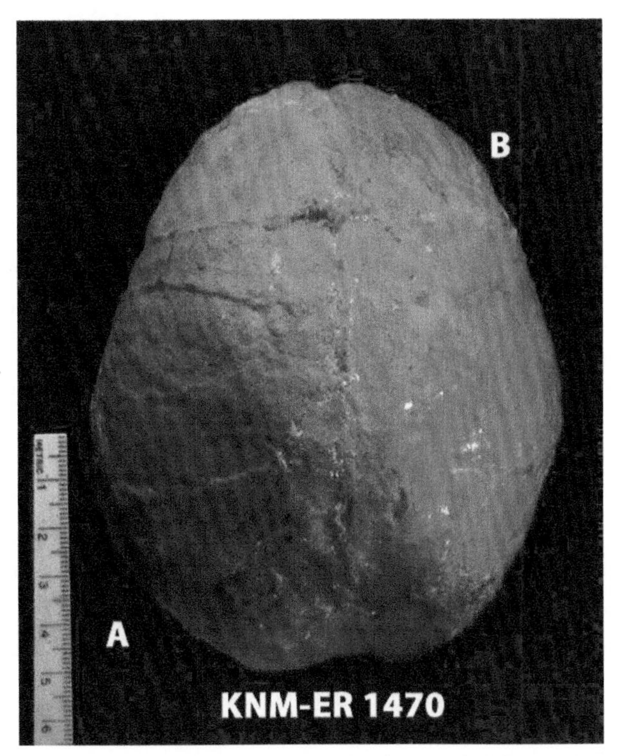

FIGURE 17.8 Dorsal view of KNM-ER 1470, *Homo rudolfensis* (1.8 mya), showing a typical *Homo* pattern of petalias, the left occipital projecting more posterior and being wider than the right side (A) and the right frontal being wider than the left (B).

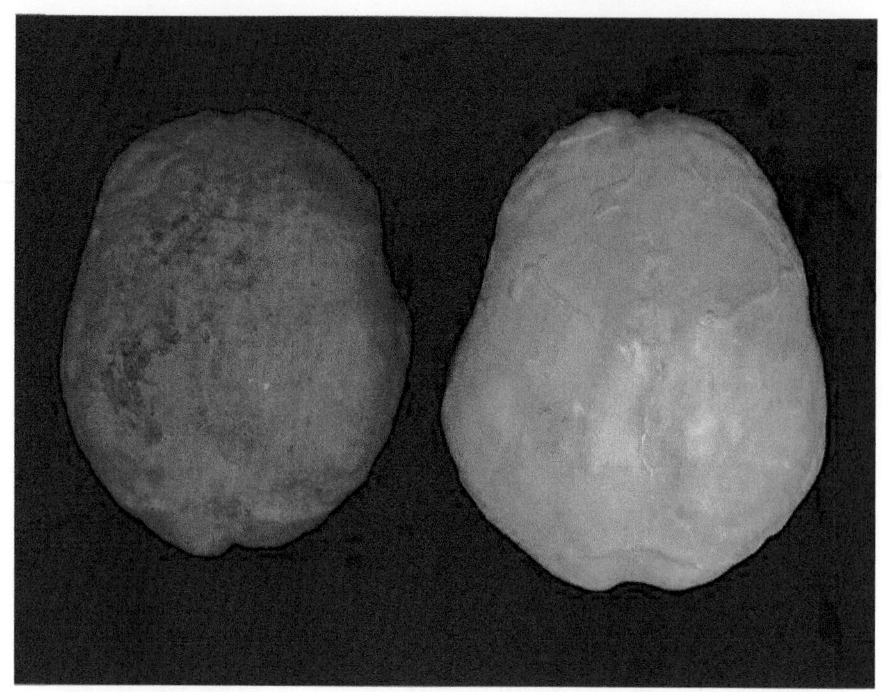

FIGURE 17.9 Neandertal cerebral asymmetries. Left is Monte Circeo and right is LaFerrassie. Both show a larger width of the right frontal lobe and a larger left occipital region (as in Figure 8).

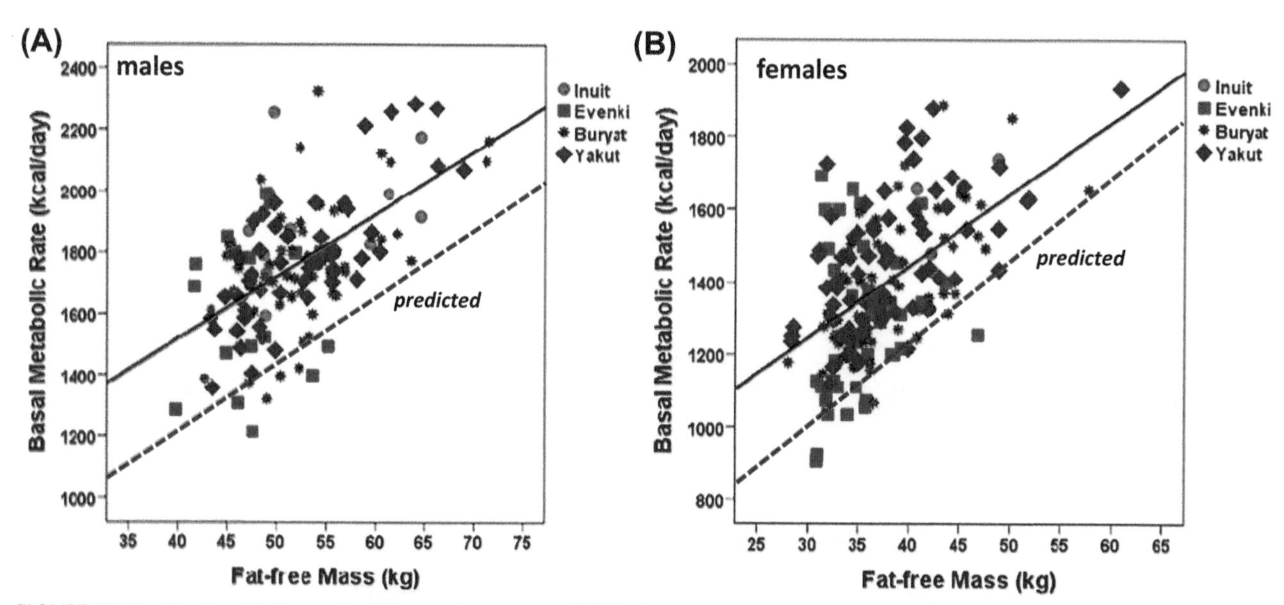

FIGURE 18.11 Basal metabolic rate (kcal/day) vs. fat-free mass (FFM, kg) among (A) men and (B) women from indigenous arctic populations (North American, Inuit; Siberian, Evenki, Buryat, Yakut). The solid lines denote the best fit regression lines for BMR versus FFM, while the dotted lines denote the predicted BMR values based on the equation of Cunningham (1991). Measured BMR values are significantly greater than those predicted based on FFM in both men and women (+18.2 and +18.3%, respectively; $P < 0.001$), indicating greater rates of metabolic heat production in these cold-adapted populations.

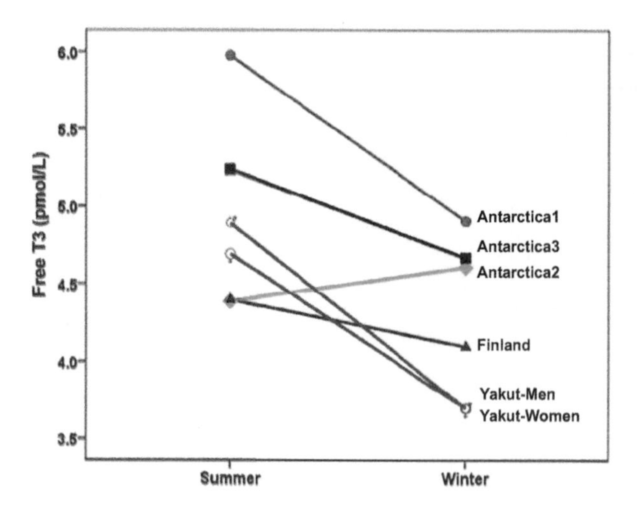

FIGURE 18.12 Seasonal change in free T_s (fT_3; pmol/L) levels in selected circumpolar groups. These groups are characterized by a pattern known as "Polar T_3 Syndrome"—marked wintertime declines in fT_3 levels associated with tissue update of thyroid hormones exceeding production. The seasonal changes are most pronounced in the indigenous, Yakut population, who show declines of 21 and 24% in men and women, respectively, as compared to the average declines of 8% in the three studies of Antarctic sojourners (all men) and declines of 7% in men of northern Finland. *Adapted from Levy et al. (2013).*

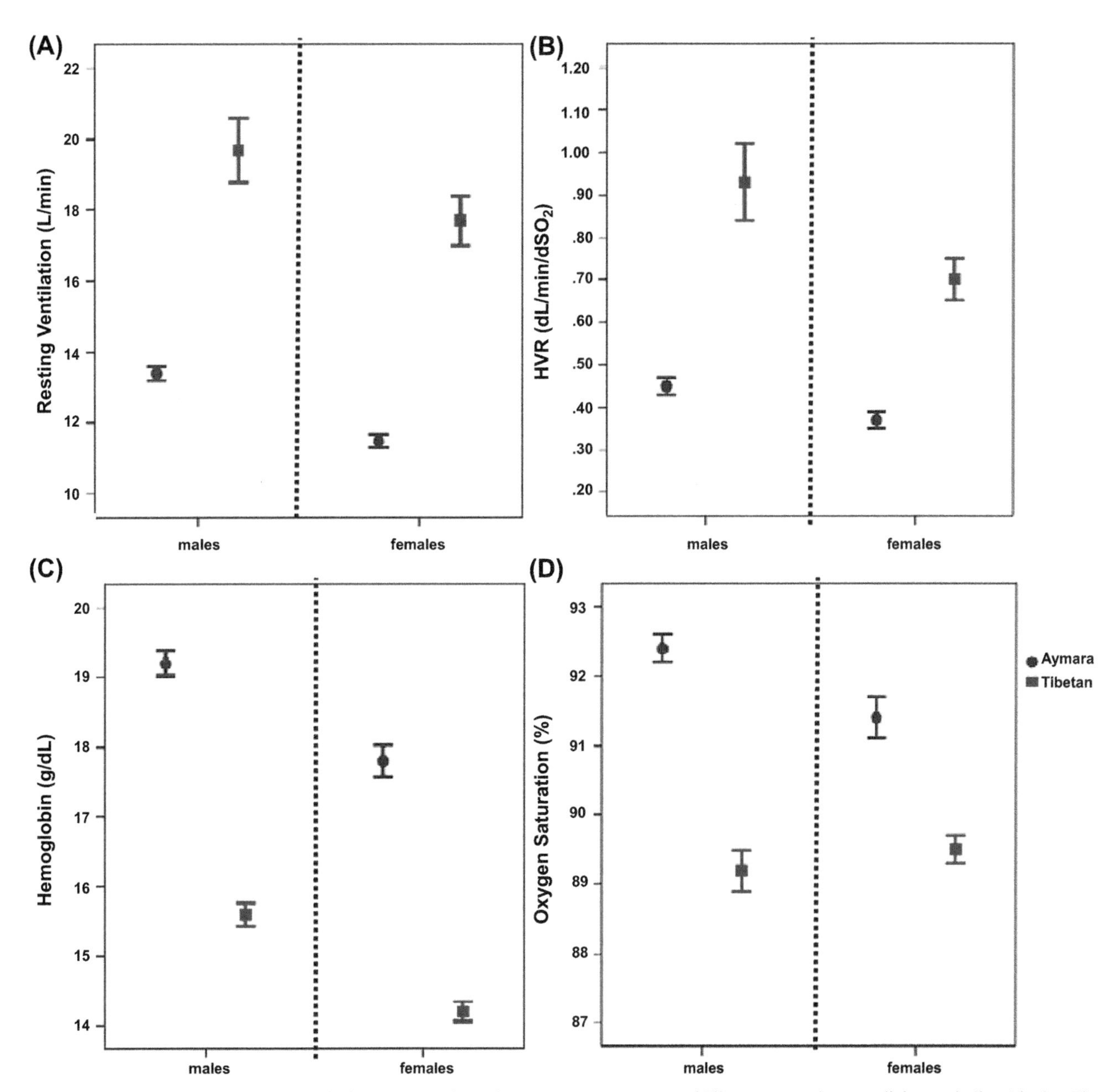

FIGURE 18.19 Comparison of physiological responses to hypoxic stress among Aymara and Tibetan men and women living at similar altitudes. The measures include (A) resting ventilation (L/min), (B) hypoxic ventilatory response (HVR; $\Delta L/min/\Delta SaO_2$), (C) hemoglobin (Hb) levels (g/dL), and (D) percentage oxygen saturation (SaO_2; %). Tibetan men and women have significantly higher resting ventilatory rates and HVRs than their Aymara counterparts ($P < 0.05$ for all sex-specific comparisons). In contrast, the Aymara have significantly higher Hb and SaO_2 levels ($P < 0.05$ for all sex-specific comparisons). Consequently, the Tibetan adaptive pattern emphasizes greater ventilatory responses and more moderate increases in pulmonary diffusion and red blood cell production. The Aymara, in turn, rely more heavily on enhanced pulmonary diffusion and show greater increases in red cell production. *Data from Beall et al. (1997a,b, 1998, 1999).*

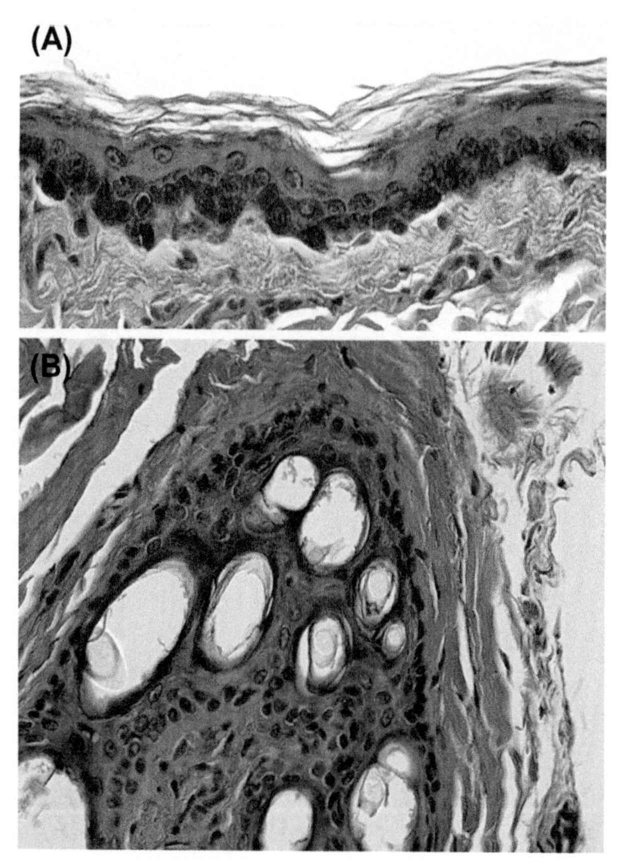

FIGURE 19.1 Polar bear skin: note dense carpet of melanocytes containing abundant melanin in interfollicular epidermis (A, arrows). (B) In contrast, the follicular epithelium lacks evidence of melanocytes, corresponding to the white fur of this species. *Illustration courtesy of Elizabeth Maudlin, DVM, University of Pennsylvania.*

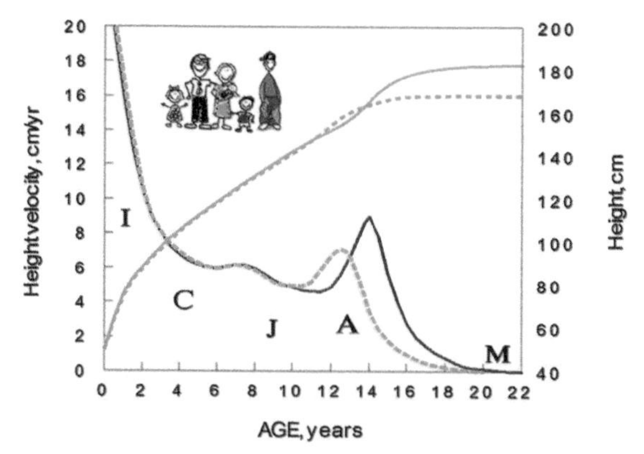

FIGURE 20.1 Distance and velocity curves of growth for healthy, well-nourished human beings. Boys, solid line; girls, dashed line. These are modal curves based on height data for Western European and North American populations. The stages of postnatal growth are abbreviated as follows: I, infancy; C, childhood; J, juvenility A, adolescence; M, mature adult. Weaning takes place at a mean age of 30–36 months—at the transition from infancy to childhood. *(Modified from Bogin, 1999.)* The distance curve (right *y*-axis) indicates the amount of height achieved at a given age. The velocity curve (left *y*-axis) indicates the rate of growth at a given age. Growth velocity during infancy is rapid with a steep deceleration. Childhood growth is relatively constant at about 6 cm per year. Growth rate slows during the juvenility stage and then accelerates during the first phase of adolescence—the adolescent growth spurt. Growth rates decline during the second phase until all growth in height stops at the onset of the adult stage. The image of the "family" is not meant to promote any particular type of family as desirable or normal; rather, the cartoon figures illustrate the stages of human life history between birth and adulthood from left to right—juvenile, adult male, adult female with infant, child, and adolescent.

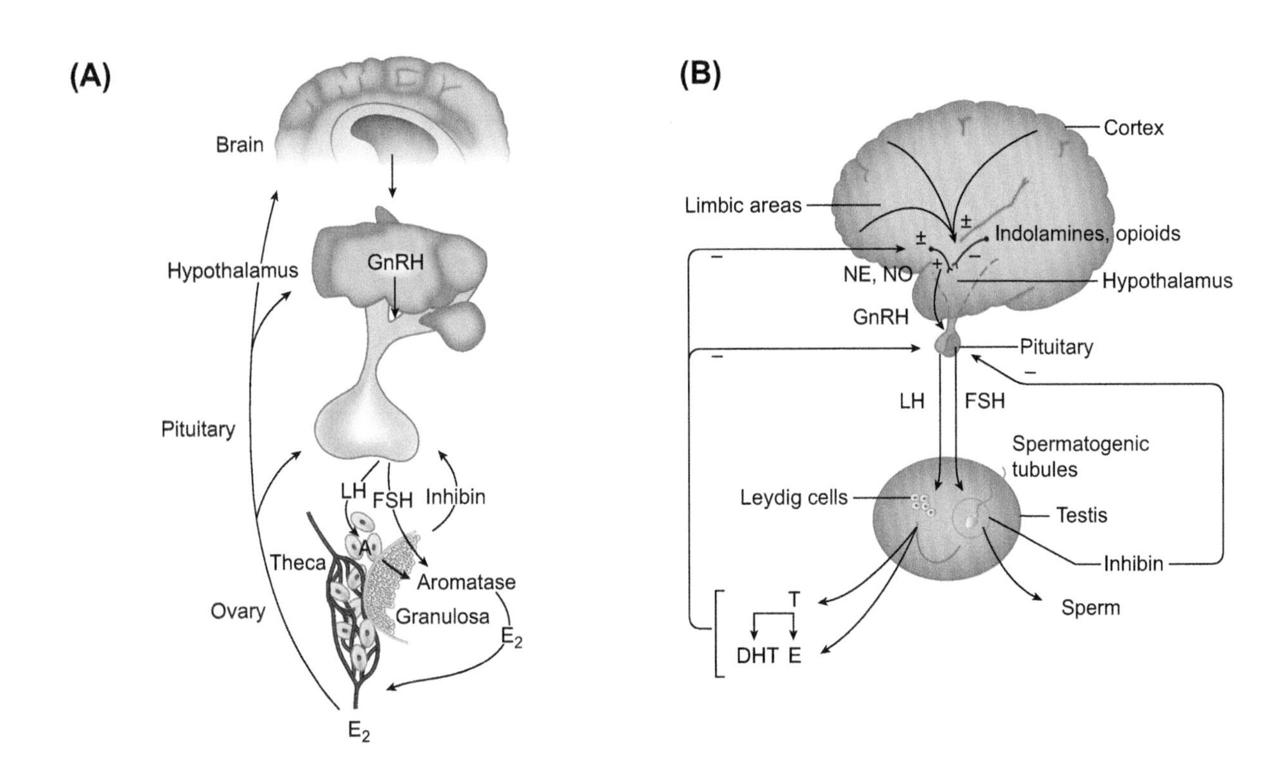

FIGURE 21.3 (A) The hypothalamic–pituitary–ovarian axis. GnRH = gonadotropin-releasing hormone; LH = luteinizing hormone; FSH = follicle-stimulating hormone; E_2 = estradiol; (B) The hypothalamic–pituitary–testicular axis. DHT = dihydrotestosterone; E = estrogen; FSH = follicle-stimulating hormone; GnRH = gonadotropin-releasing hormone; LH = luteinizing hormone; NE = norepinephrine; NO = nitric oxide; T = testosterone.

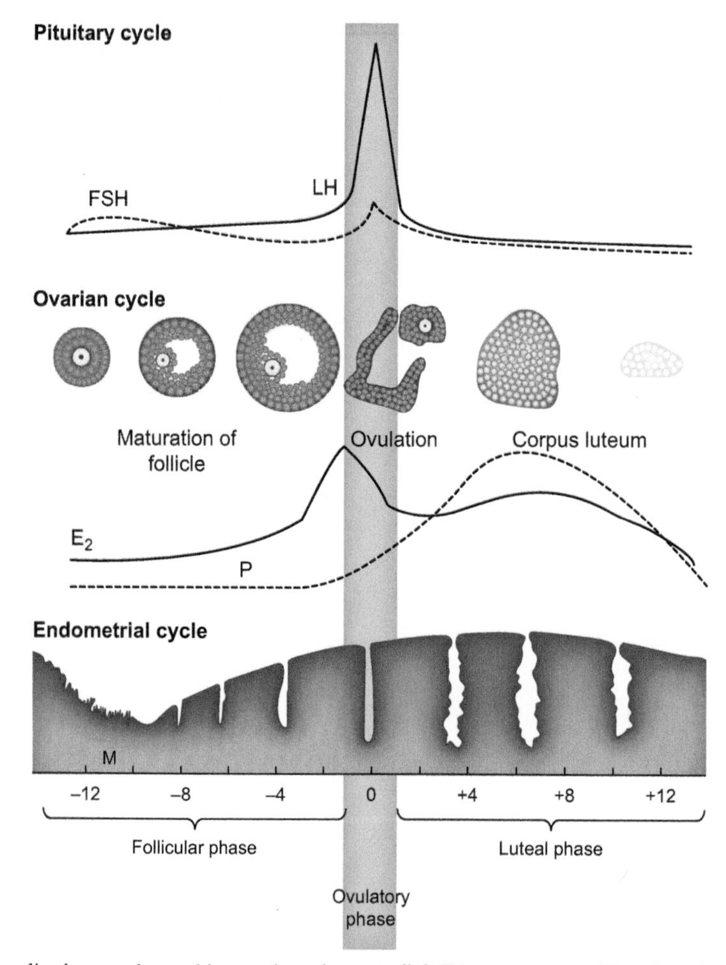

FIGURE 21.4 The idealized cyclic changes observed in gonadotropins, estradiol (E_2), progesterone (P), and uterine endometrium during the normal menstrual cycle. The data are centered on the day of the luteinizing hormone (LH) surge (day 0). Days of menstrual bleeding are indicated by M. FSH=follicle-stimulating hormone. *Reprinted with permission from Endocrine and Metabolism Continuing Education Quality Control Program, 1982. Copyright American Association for Clinical Chemistry, Inc.*

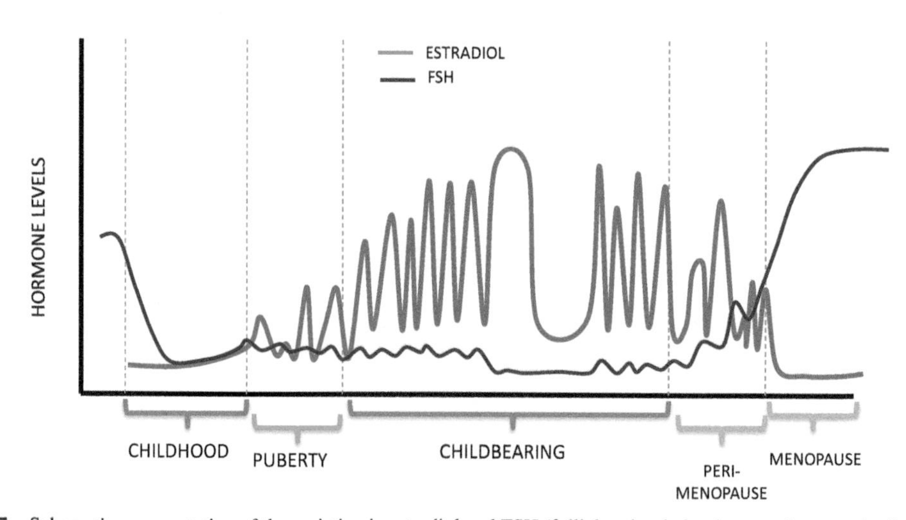

FIGURE 21.5 Schematic representation of the variation in estradiol and FSH (follicle-stimulating hormone) across the female life span.

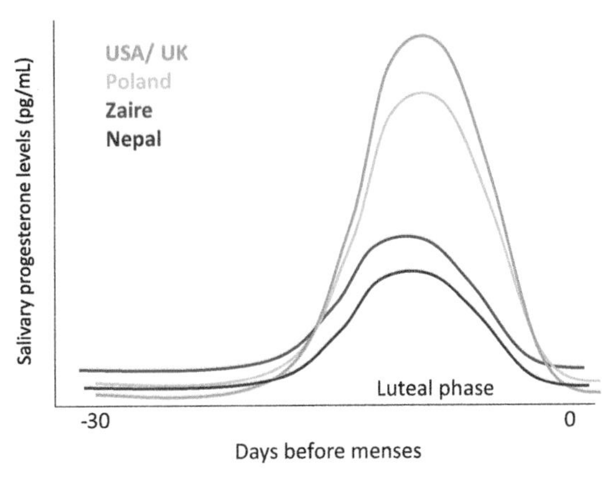

FIGURE 21.6 Interpopulation variation in ovarian function. *Adapted from Ellison et al. (1993a).*

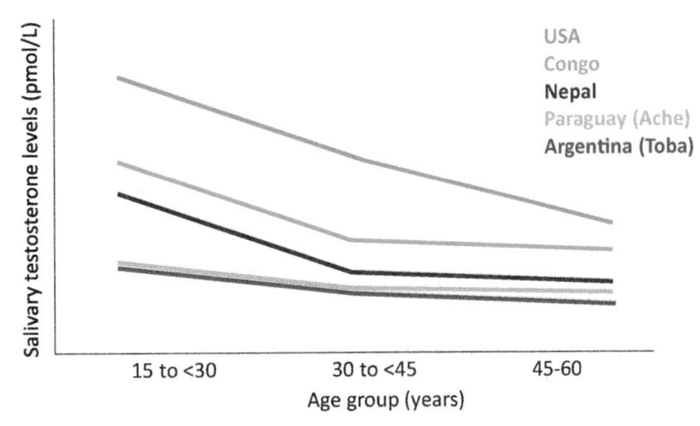

FIGURE 21.7 Interpopulation variation in testicular function across age groups. *Adapted from Ellison et al. (2002) and Valeggia et al. (2009).*

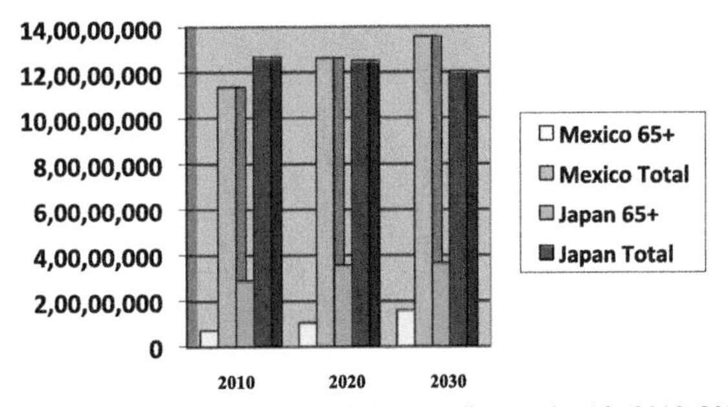

FIGURE 22.1 The number of people aged 65 or older and the total population (medium variant) in 2010, 2020, and 2030 in Mexico and Japan (United Nations, Population Division, 2011).

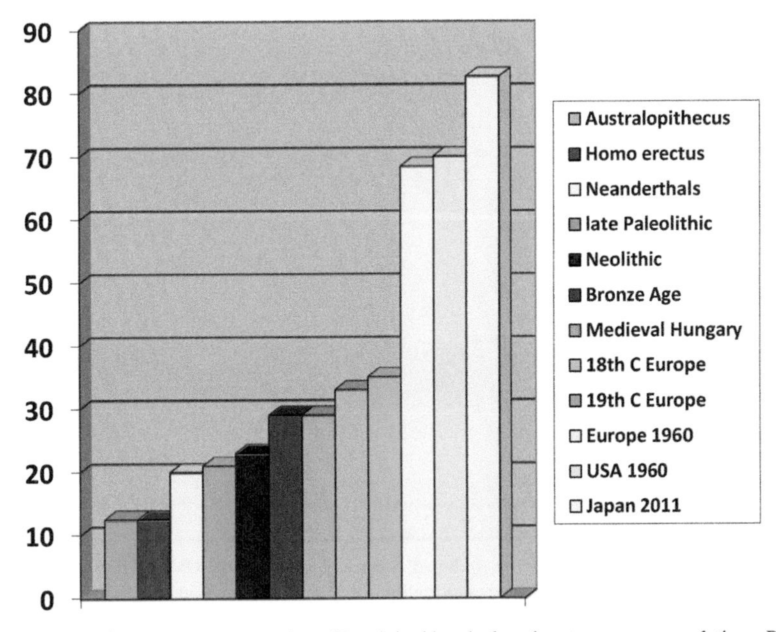

FIGURE 22.2 Life expectancy at birth, in years, across a number of hominin, historical, and contemporary populations. *Data from Kennedy (2003) and World Bank.*

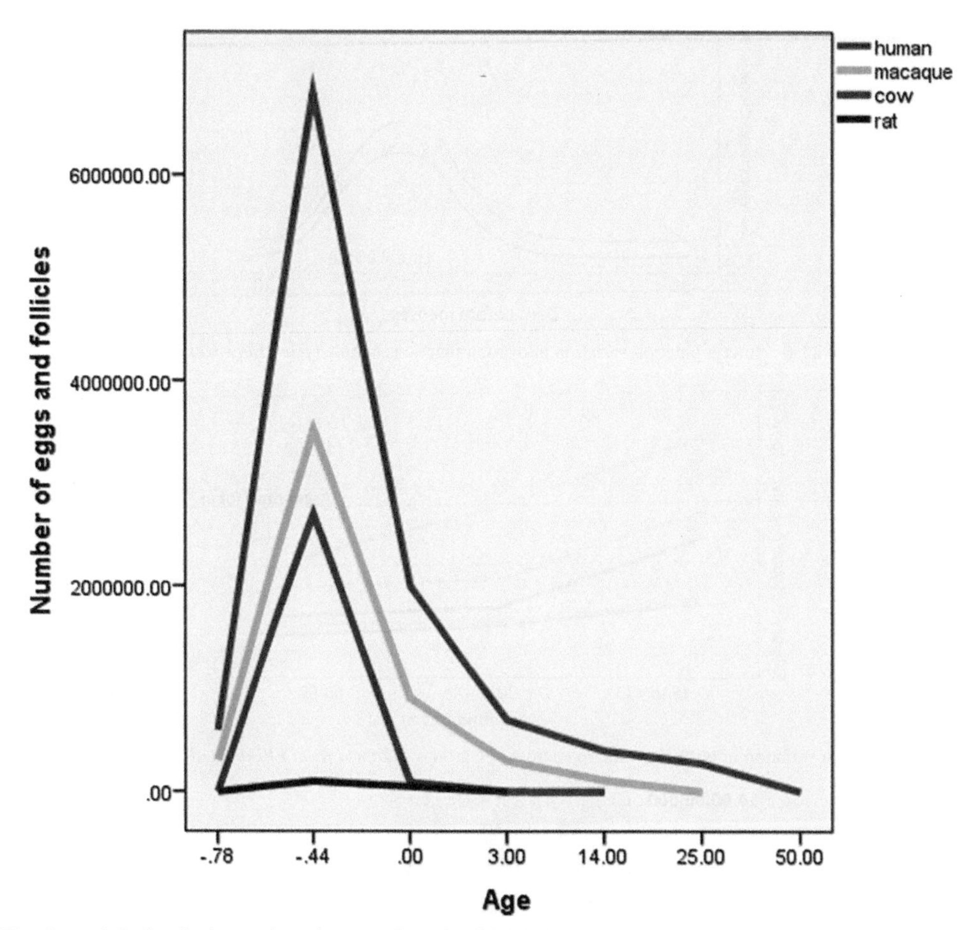

FIGURE 22.6 The rise and decline in the number of eggs and ovarian follicles in human, rhesus macaque, cow, and rat ovaries during the prenatal period, at birth, and at ages 3, 14, 25, and 50 (Baker, 1986; Erickson, 1966a,b; Hansen et al., 2008; Ryan, 1981).

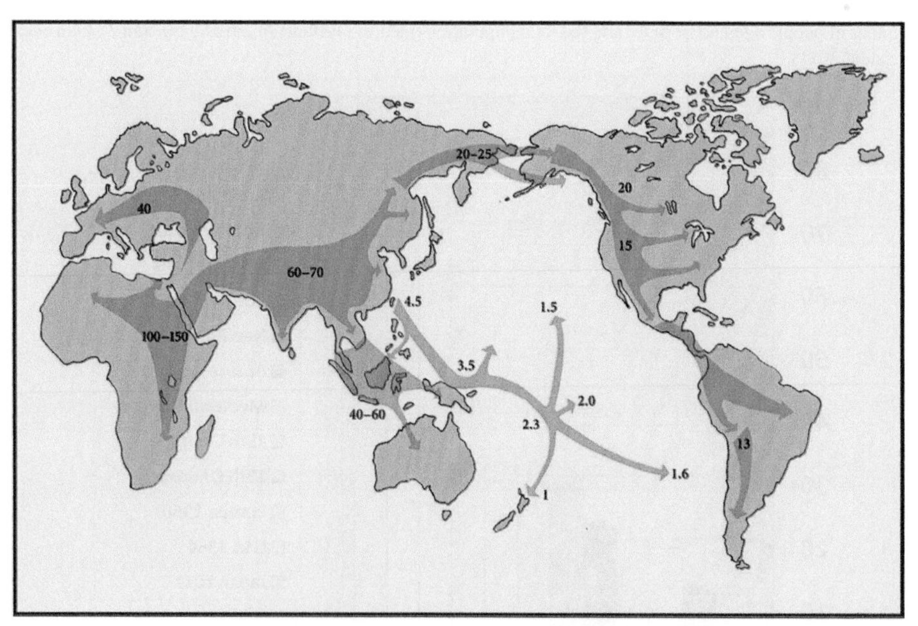

FIGURE 23.1 Approximations of some major migratory routes of hunter-gatherers following the Out of Africa exodus of anatomically modern *Homo sapiens* in the Middle Paleolithic. Values are in thousands of years ago. Arctic migrations in Eurasia and North America are not shown.

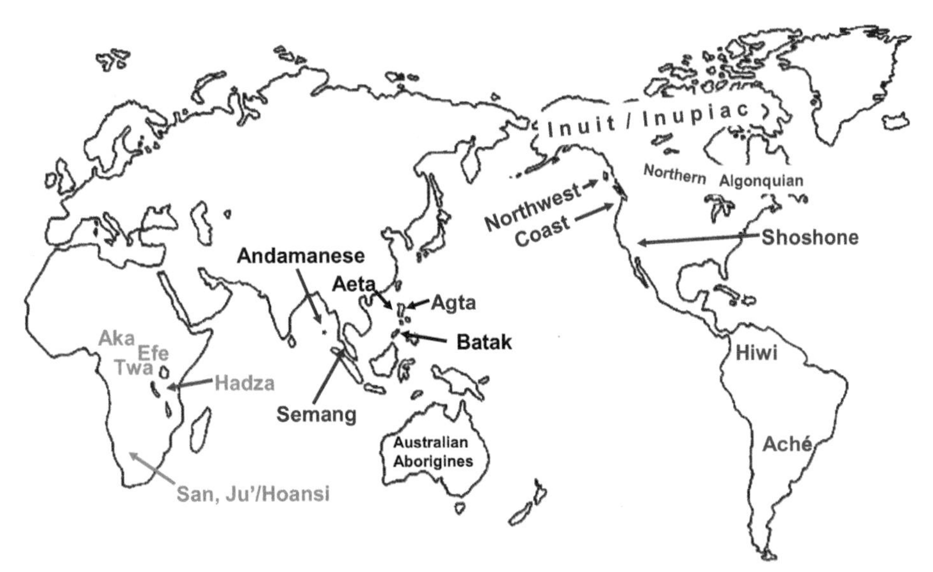

FIGURE 23.2 Map showing the location of major hunting-gathering groups discussed here.

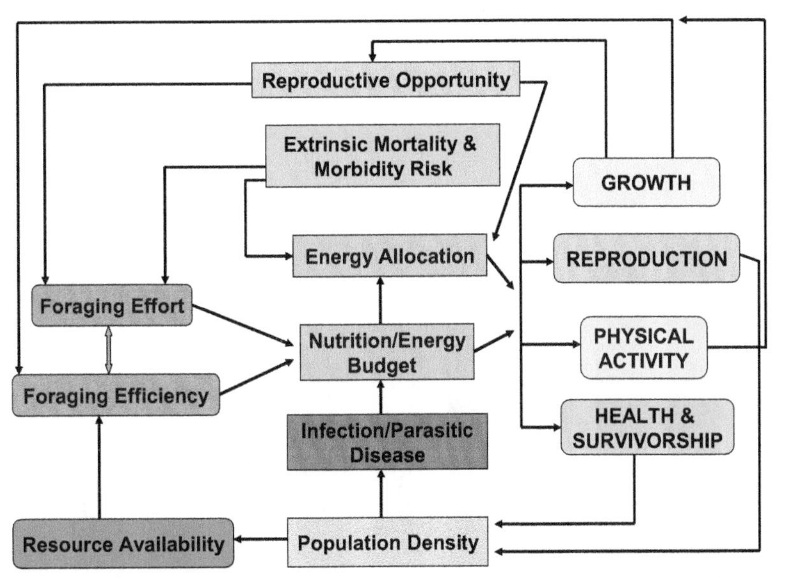

FIGURE 23.3 A diagram representing relationships among the environment, nutrition, energy intake and expenditure, and other variables associated with foraging in hunter-gatherers. *Redrawn and modified from Jenike (2001).*

FIGURE 24.1 Painting of an Aurochs (*Bos primigenius*) by Gilles Tosello. *With permission, ©Gilles Tosello.*

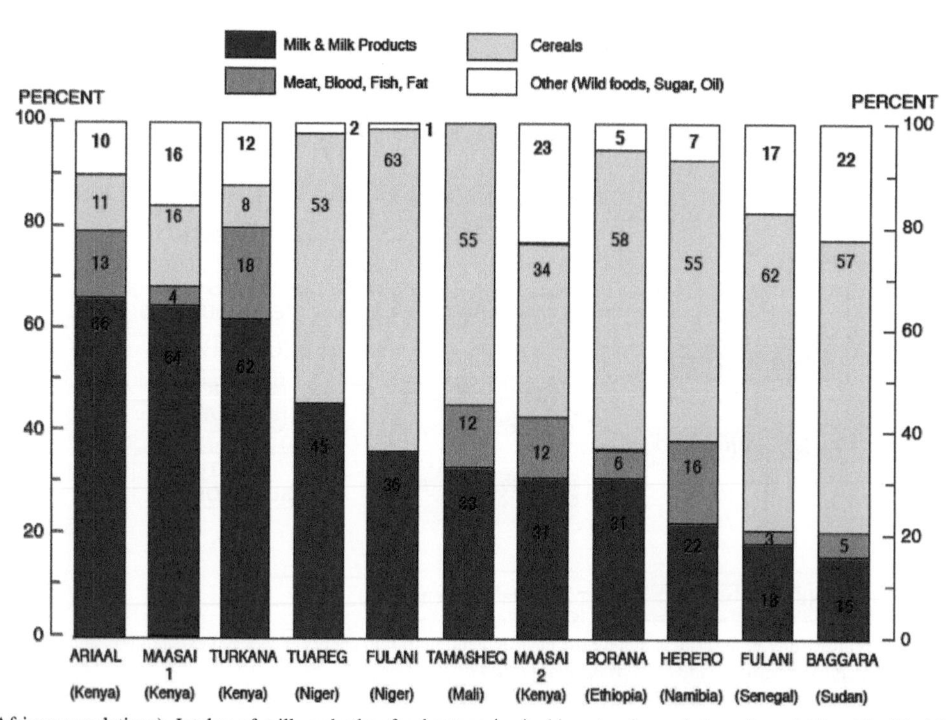

FIGURE 24.2 (African populations). Intakes of milk and other food categories in 11 pastoral populations from Africa. *Modified from Galvin (1992). Data are drawn from Ariaal (Fratkin, 1991), Baggara (Holter, 1988), Borana (Galvin et al., 1994), Herero (O'Keefe et al., 1988), Maasai 1 and 2 (Nestel, 1985, 1986), Senegal Fulani (Bénéfice et al., 1984), Tamasheq (Wagenaar-Brouwer, 1985), Tuareg and Niger Fulani (Bernus, 1988), and Turkana (Galvin, 1985).*

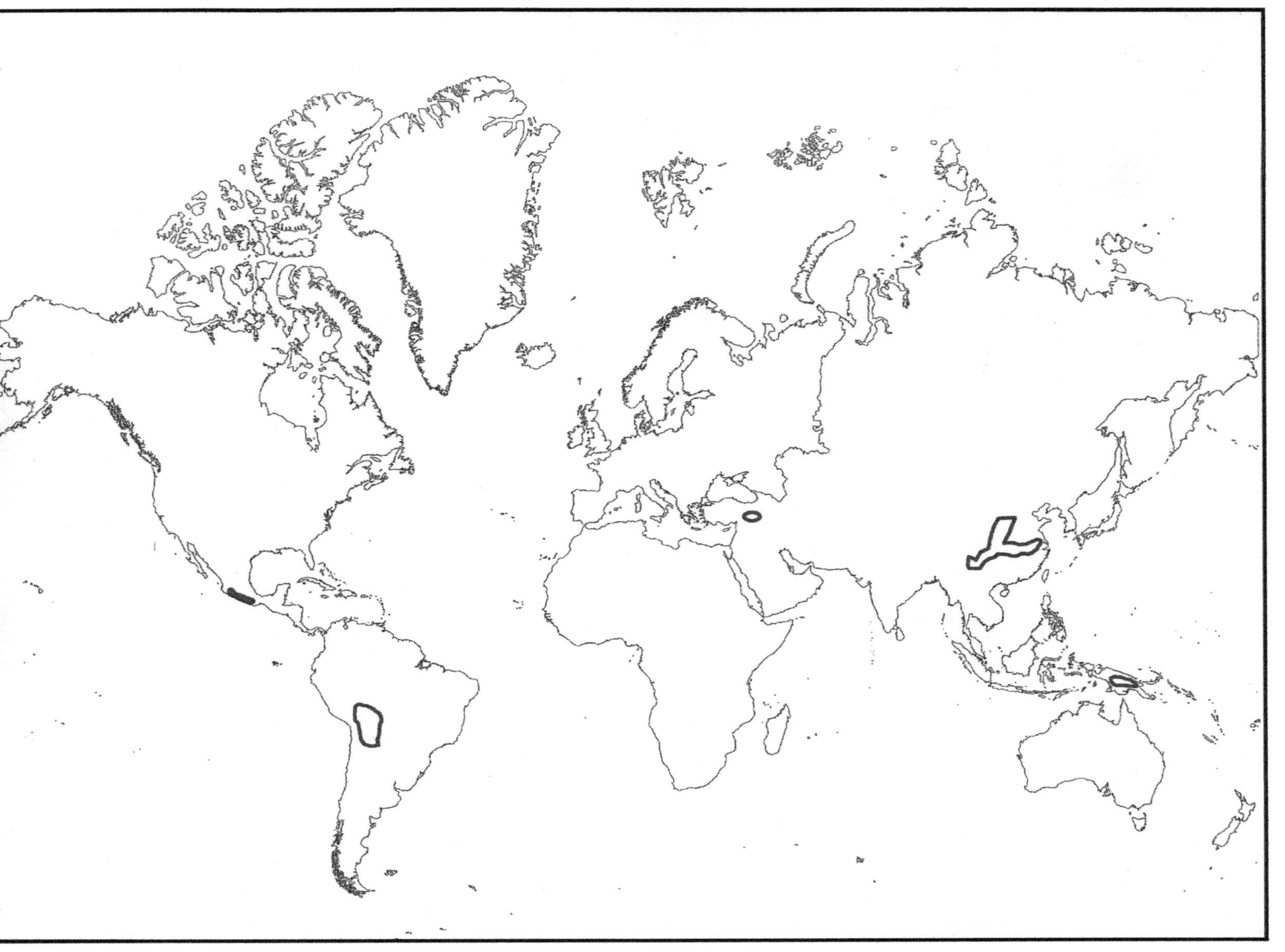

FIGURE 25.1 Centers of agricultural origin, based on current evidence.

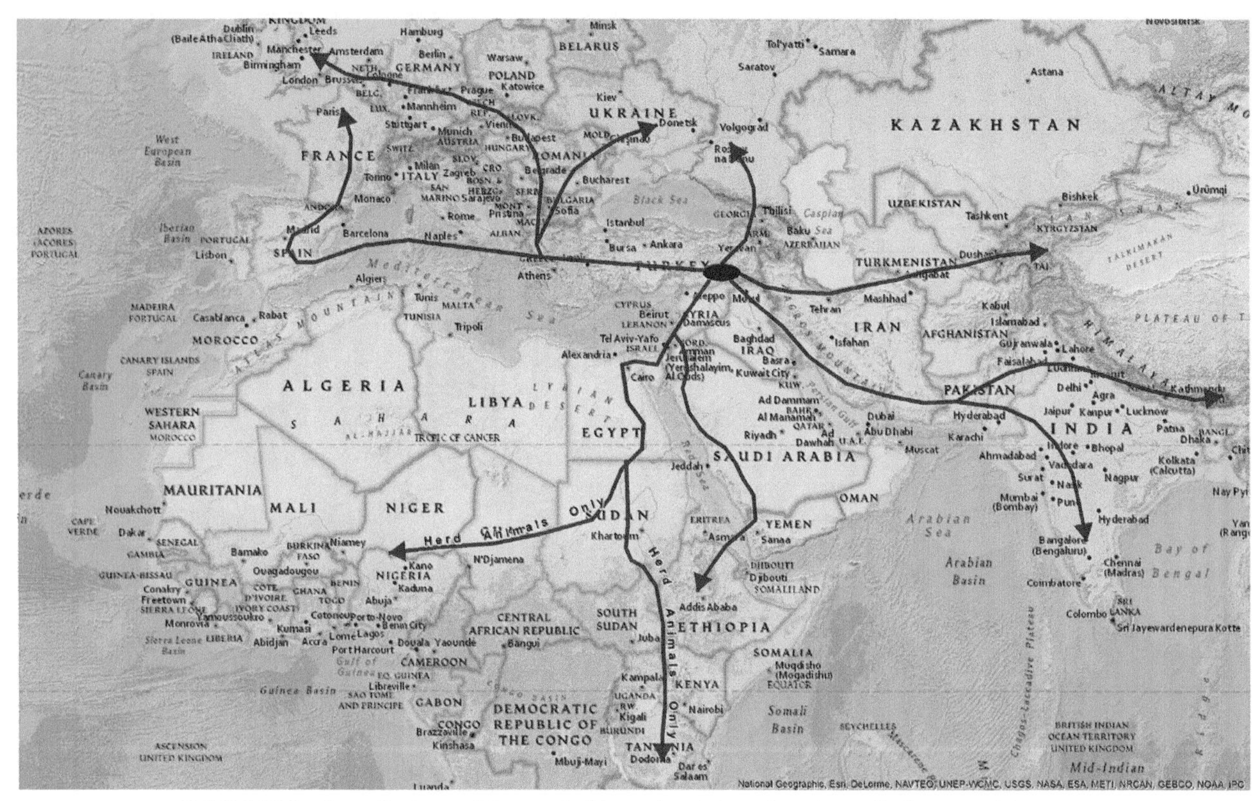

FIGURE 25.2 Generalized map of the spread of domesticated plants from the Fertile Crescent origin center.

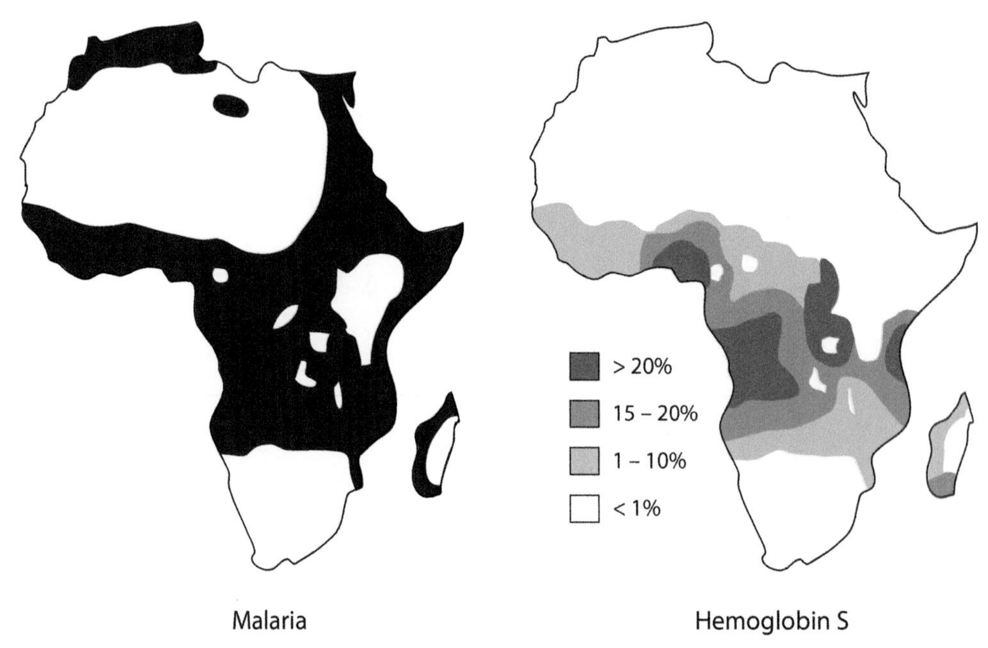

FIGURE 30.2 The distribution of malaria (left) and the hemoglobin S (Hb S) allele (right) in Africa. Note the absence of Hb S in regions that do not experience malaria. *Adapted from the original figure of Anthony Allison, dedicated by him to the public domain and made available at http://commons.wikimedia.org/wiki/File:Malaria_versus_sickle-cell_trait_distributions.png.*

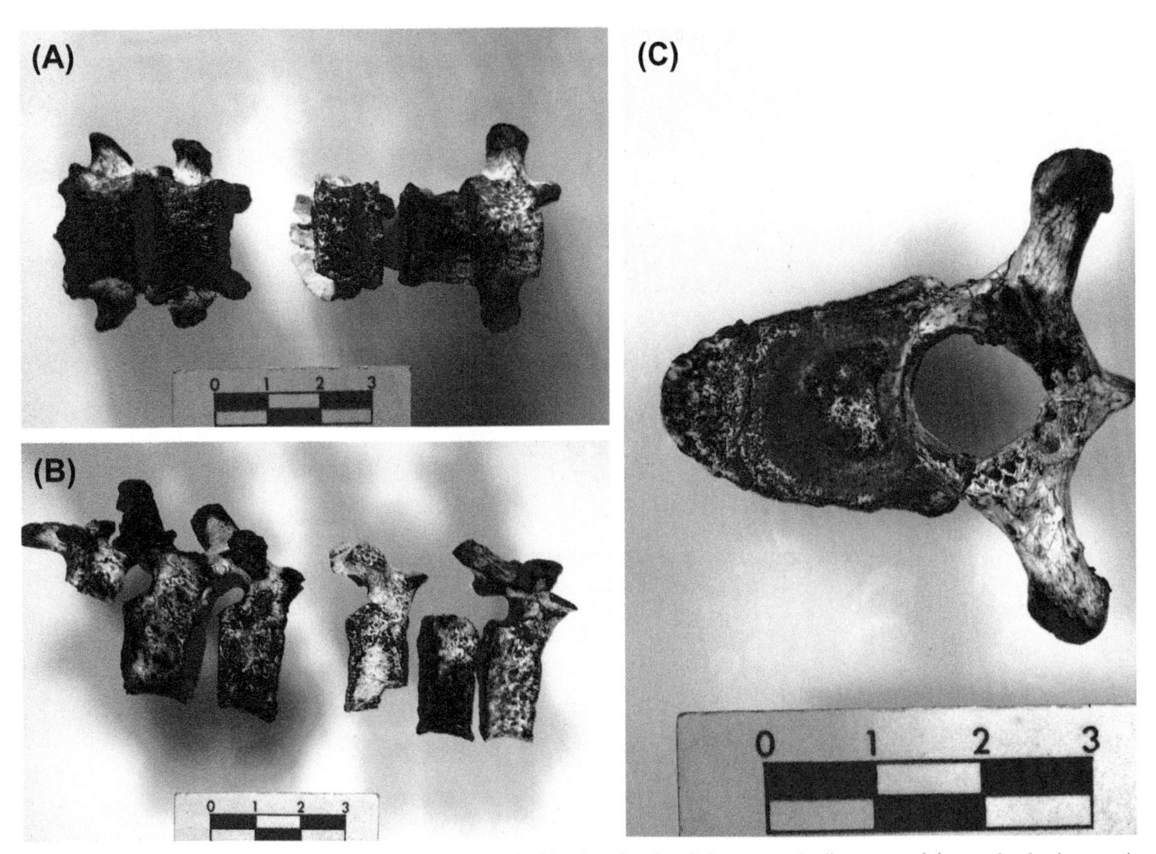

FIGURE 31.1 Vertebrae of the 3.5 million-year-old "Lucy" australopithecine showing Scheuermann's disease or adolescent kyphosis, associated with lifting heavy weights in modern people. (A) Anterior, (B) right side, (C) upper surface of TV6. The light-colored anterior third of the vertebral body is pathological new bone.

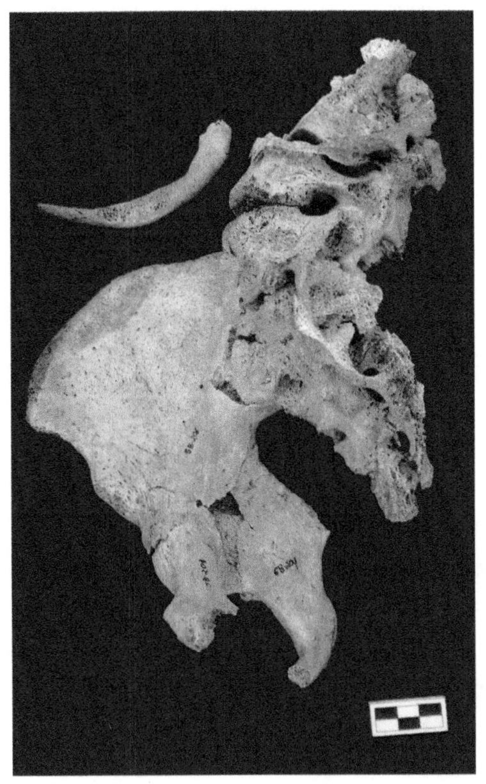

FIGURE 31.2 Tuberculosis in the spine and hip bone of a young Mississippian woman from Illinois, 1000 AD.

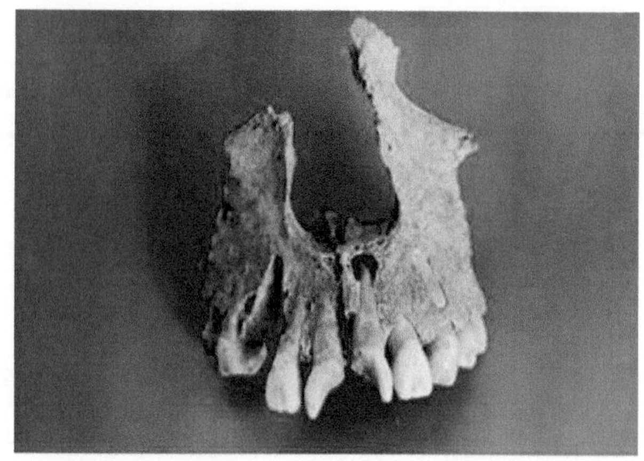

FIGURE 32.3 Decorative notching of the central incisors that has resulted in an abscessed tooth in an enslaved young woman from Cemeterio Pretos Novos, Rio de Janeiro, Brazil.

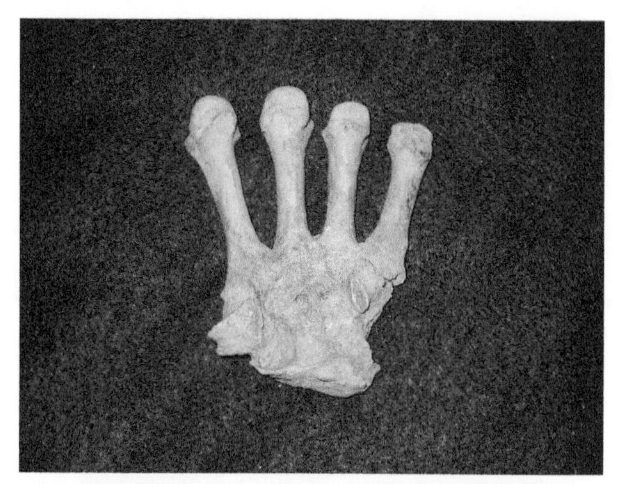

FIGURE 33.4 Man with a fused wrist, perhaps a complication of immobilization after an injury that would have posed a substantial disability, Milwaukee Poorhouse. *Photo courtesy of Sean Dougherty.*

FIGURE 37.2 Macaques groom one another, a gesture that is returned later; an example of direct reciprocity. *From US govt.*

FIGURE 37.3 Vampire bats both share and exclude nonsharers, an example of network reciprocity. *Image courtesy of the Ontario Specialized Species Centre.*

FIGURE 37.5 The cleaner wrasse mutualism. When observed by other fish, this is example of reputation, which is key to indirect reciprocity. *From Wikimedia.*

FIGURE 37.6 William-Adolphe Bourguereau's *The Remorse of Orestes* (1862), an illustration of a moral sentiment that Darwin considered central to understanding morality. *From Wikimedia.*

FIGURE 37.7 Naked mole rats live in communities supporting only one reproductive female. Once considered to be the result of kin selection, the social structure is now considered an outcome of ecology with the unusual genetic structure being a consequence of the social structure. *Image courtesy of Chris Faulkes.*

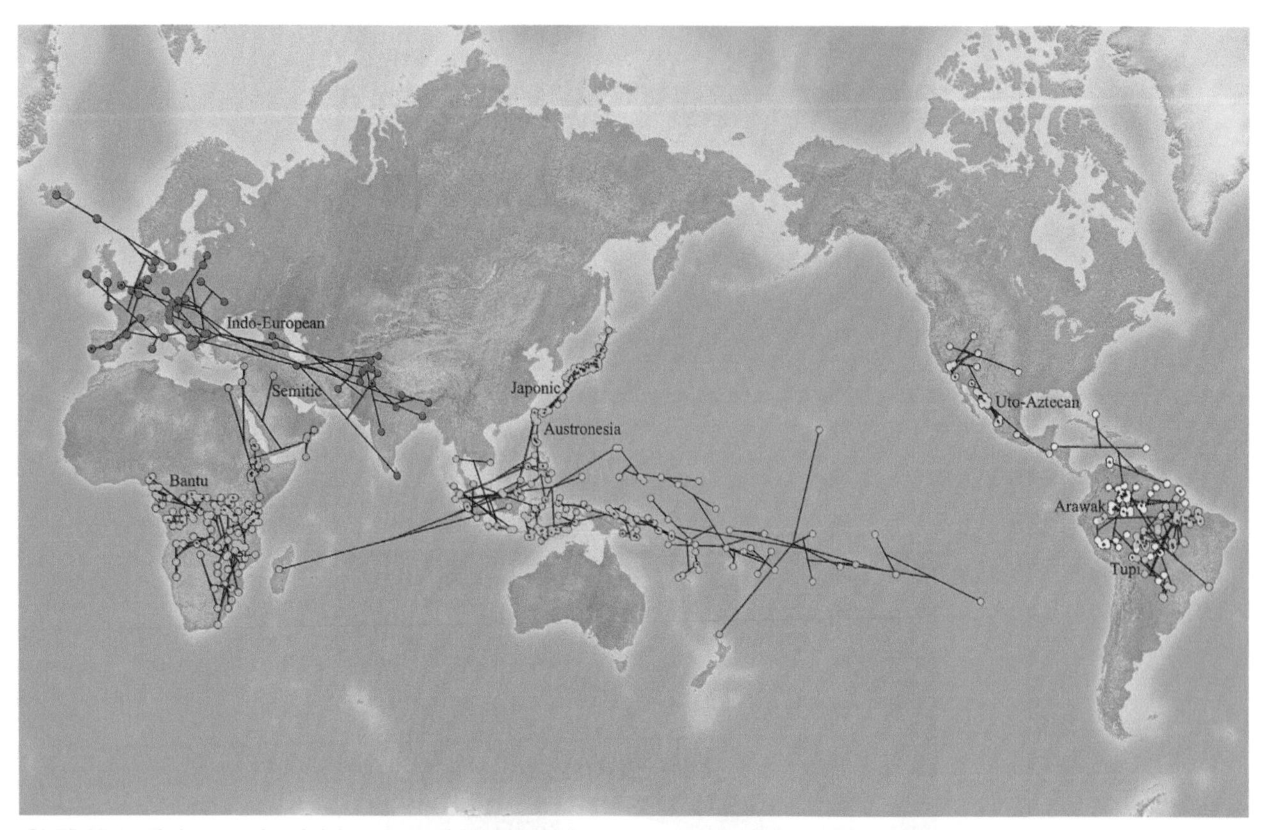

FIGURE 39.1 Phylogeography of eight agricultural language family expansions with published character-coded language phylogenies. The cultures of each language family are connected by a consensus tree, with internal nodes interpolated as the average spatial location of descendant taxa. Language family names are placed approximately near purported homelands.

Printed and bound by CPI Group (UK) Ltd, Croydon, CR0 4YY
08/05/2025
01865031-0002